TABLES FOR ACTIVE FILTER DESIGN

TABLES FOR ACTIVE FILTER DESIGN

Based on Cauer MCPER Functions

Mario Biey, Dott.Ing.

Amedeo Premoli, Dott.Ing.

Foreword by George S. Moschytz

International Standard Book Number: 1-58053-126-1
Library of Congress Catalog Card Number: 84-073279

To the Memory of our Parents

Foreword

G.S. Moschytz

Swiss Federal Institute of Technology

Zurich, Switzerland

Over more than a decade inductor-less filters have increasingly replaced conventional LC filters because of their compatibility with modern integrated circuit technology. Active RC filters — and their close relatives, switched-capacitor filters — currently dominate the field of inductor-less filters and are probably produced in larger quantities than any other filter type — at least for frequency ranges below approximately 100 kHz. The earliest active filters consisted of second- or third-order filter sections. When connected in cascade, such sections, or building blocks, readily provide a prescribed nth-order filter function. However, when pole Q factors become too high, difficulties arise because of the high sensitivity of filter characteristics to component tolerances.

As a result, a vast array of alternative active filter structures have been developed. Foremost among them are those that simulate conventional LC ladder structures because it can be shown that the superior sensitivity properties in the passband of LC ladder filters are also maintained in the corresponding active RC filter structures. This superior behavior is a result of the inherent coupling between resonators within an LC ladder structure. By contrast, in a cascade of second-order sections, each section is decoupled and, therefore, nominally independent of any other section in the cascade. Interestingly enough, however, it is precisely this decoupling that continues to make the cascade approach to active RC (and switched-capacitor) filter design so attractive; manufacture, tuning, and optimization of individual second-order sections is a relatively simple matter — provided, that is, that the pole Q factors remain relatively low. Thus, because low cost and efficient manufacture are very real and tangible factors, whereas the sensitivity to component tolerances often appears to be of more academic than practical interest, active RC and switched-capacitor filter design using second-order filter sections in cascade continues to be the preferred method of design in a majority of industrial applications.

With active RC filters, the problem of high pole Q factors remains, however. (With switched-capacitor filters, the pole frequency stability is determined by a crystal clock; thus, the problem of high-Q filter stability is less severe). In fact, until recently, it was regarded as an insurmountable obstacle to the design of high-selectivity filters using the method of cascaded second-order building blocks. It is to the credit of the authors of this book that they have made a significant contribution towards the solution of this problem. They have accomplished this by taking a fresh look at the approximation problem in filter design with a view to extending the design of active RC filters beyond the limits imposed by the Q factors of the dominant poles. By increasing the order of a given network function only slightly, such that only one or two additional second-order low-Q filter sections are required for its realization, they have shown that the Q factor of the dominant pole pair of the resulting higher-order filter network can be reduced considerably. Thus, if the dominant pole Q of an initial seventh-order Cauer (elliptic) low-pass function is close to 50, a modified (MCPER) function of ninth-order with a dominant *triple* pole of Q_{max} less than about 15 may be derived, which satisfies the originally specified low-pass characteristics. Whereas the first version with a Q_{max} of 50 would be unsuitable for cascade design, the second version with a Q_{max} of 15 would present no problems.

The authors of this book have persevered over several years, and have successfully extended their early work on all-pole filter design to encompass every possible filter type. In this book they make the fruits of their considerable efforts available to the engineering community at large.

This book provides clear and well-documented filter tables that will undoubtedly become a major factor in advancing the capabilities of cascaded active RC filter design well beyond the range of currently established applications. One obvious reason for the impact which this book is expected to have on active RC filter design is that it readily permits the utmost utilization of the already vast quantity of accumulated *know-how* in cascade filter design. Thus, by applying the newly available pole-zero values given in this book, improved filters can be obtained *without requiring any technological improvement in the actual filter sections used.* From this point of view, the book must be regarded as a valuable and important addition to the existing literature on active filter design. It provides design material and presents a design approach that, besides being very useful, is quite original and new; nothing comparable is currently available in any other book.

Preface

The advent of large-scale integration (LSI) has made it possible to integrate all the major components of communication systems, including electrical filters, now typically used in large numbers. In this situation, the possibility of producing these filters (e.g., RC active, digital, switched-capacitor, *et cetera*) by cascading low (second, third) order sections is highly appreciated because it permits building-block standardization, ease of tuning, and design simplification.

In the cascade structure, the construction of high-Q (i.e., highly selective) blocks results in serious sensitivity and technology problems which may be avoided by using more sophisticated and expensive structures. Unfortunately, the classical transfer functions, developed to solve the approximation problem for LC ladder filters, do not allow the designer to reduce the Q factors of the most critical pole-pairs.

Only recently, new numerical methods have been studied for solving the low-Q approximation problem for cascaded filters. In particular, a new class of low-pass rational transfer functions (MCPER) has been introduced which turns out to be a generalization of the well-known Cauer (elliptic) transfer functions. These functions provide a low-Q approximation by replacing the most critical pole-pair with a less critical multiple pole-pair.

The aim of this book is to supply electrical engineers and research workers with all the data pertaining to Cauer and MCPER functions.

The book is divided into two parts. Part I, comprising sections 1 through 12, contains the essential theory on Cauer and MCPER functions, the frequency transformations for designing different types of filters, and typical examples illustrating the practical use of the material supplied. Part II, which represents the major portion of the book, presents poles and zeros of Cauer and MCPER low-pass functions of orders 2 to 11, 4 to 12, and 6 to 13, depending on the respective values 1, 2, and 3 of the multiplicity of the critical pole-pair. The pass-band attenuation ripple values and the stop-band edges have been chosen specifically to meet practical requirements.

We would like to express our thanks to Professor G.S. Moschytz of the Swiss Federal Institute of Technology for his encouragement during the development of this work and for reviewing the manuscript.

We would also like to express our appreciation to Dr. R. Gadenz and Dr. J. Tow of Bell Laboratories, Holmdel, N.J. (USA) for some stimulating discussions. We also wish to thank Professor C. Beccari (Politecnico di Torino) for his valuable comments and suggestions on the manuscript.

In particular, A. Premoli is grateful to his colleagues of the Gruppo Sistemi e Circuiti of the Istituto Elettrotecnico Nazionale "Galileo Ferraris" (IENGF) for their much appreciated encouragement.

We wish to thank Professor D. Biey and the staff of the Computer Center of the Politecnico di Torino for their assistance in preparing the program for editing the original version of these tables. Thanks are particularly due to Mrs. Piera Bonola and Mrs. Caterina Susenna who typed the manuscript, and to Mr. G. Viarengo who carefully prepared the figures in the book.

We also wish to acknowledge financial aid from the Consiglio Nazionale delle Ricerche, Roma, Italy, supporting research on MCPER functions.

M. Biey, A. Premoli

Contents

Part I: Fundamentals

1 Introduction

The cascade connection of second-order blocks is considered a very attractive solution in the production of electrical active filters because it is characterized by ease of design and tuning, and it allows for a high degree of standardization which, in turn, implies a remarkable reduction in cost. Using this approach, a rational transfer function, which approximates the required electrical characteristics, is broken down into a product of sub-functions, each of which is constituted by a second-order pole-zero pair (negative real poles may be included as well): each elementary sub-function is then implemented by a single block manufactured by suitable electronic technology, and the cascade of these blocks produces the required overall transfer function [1-6].

At the present time, the technology most widely used is perhaps that related to the production of RC active linear networks, based on the development of hybrid integrated biquadratic building blocks [1, 3, 5, 6].

However, today, due to advances in MOS integrated-circuit technology, high-order filters can be realized as micro-miniature silicon integrated-circuit chips, using switched-capacitor [5, 7] and digital filter techniques [2, 4].

Also, the cascade structure may turn out to be profitable for these technologies.

For these reasons, and since we are concerned only with the approximation problem, in the following we will refer to a general cascade connection of second- (or third-) order blocks, without focusing our attention on a particular type of technology.

Unfortunately, the cascade structure, with respect to other more sophisticated topologies [3, 5, 8], has a higher sensitivity which may result in a strict limitation in the production of very selective filters.

This undesirable property depends on the sections producing the pole-pairs with the highest Q factor (highest selectivity). On the other hand, even if sophisticated design procedures are used, the attenuation variations of these sections cannot be reduced below a fixed irreducible term which depends on the chosen technology and increases with the pole Q.

Consequently, if a further reduction in overall sensitivity is desired — yet still preserving the advantages of a cascade connection — some authors [9-12] have suggested attacking the sensitivity problem at the approximation stage, i.e., to approximate the assigned specification by a transfer function of higher order, but with lower Q factors in the critical pole-pairs.

This approach leads to a cascade configuration employing a non-minimum number of less critical sections. In large scale production this may be the only way to achieve acceptable tolerances and may, in fact, turn out to be cheaper than using precision components in a minimum-order realization.

A general solution has been given by Khazanov [13] which substitutes each critical pole-pair by an approximating function possessing an nth-order pole-pair with lower Q factor and n-1 complex zero-pairs. Somewhat similar results were obtained recently by Levi and Zeheb [14, 15] and by Filanovsky and Stromsmoe [16] using different algorithms. Any filtering function with an arbitrary frequency behavior can be handled by this technique. However, the zeros introduced in this low-Q approximation lie off the $j\omega$-axis and are partially lost from an attenuation point of view. Furthermore, a known approximation error is introduced, which may turn out to be unacceptable. Moreover, the complex conjugate zeros require still more complicated networks and tuning procedures.

To overcome these difficulties, it seems more advantageous to maintain all the transmission zeros on the $j\omega$-axis. This may be achieved by replacing the critical pole-pair with a less critical multiple pole-pair and by readjusting the poles and the zeros in order to meet the assigned filtering specifications exactly while still maintaining all the transmission zeros on the $j\omega$-axis.

This philosophy has led to the generation of the new classes of MUCROMAF [17] and MUCROER [18,19] polynomials and of MCPER rational functions [20, 21]. The MUCROMAF and MUCROER polynomials are essentially based on the fact that the multiplicity of their most critical root is greater than one; they allow realizations of low-pass all-pole transfer functions which are maximally flat at the origin and equiripple in the pass band, respectively. The MCPER functions are low-pass, stable rational functions with equiripple behavior, both in the pass band and in the stop band, with a multiple critical pole-pair and an arbitrary number (but not greater than that of the poles) of transmission zeros. These functions have a higher order, but a lower critical Q factor than classical equivalent approximations [22 -25], i.e., satisfying the same low-pass specification.

Independently, the same results have been obtained by Massad and Yarlagadda [26] and by Fritzsche and Jentschel [27, 28] following different computer procedures. The interested reader may find more details on low pole-Q approximation methods in the review paper [29].

These functions may be computed through iterative procedures which require complex computer programs. This disadvantage may be overcome through the availability of tables containing all the parameters needed for the identification of an adequate number of functions. Tables and graphs providing MUCROMAF and MUCROER polynomials have already been published [30, 31].

The aim of this book is to cover the more important and wider class of those MCPER functions having the maximum number of transmission zeros compatible with a given order.

Furthermore, since Cauer (elliptic) functions are a special case of MCPER functions and, moreover, they are suitable for the design of less selective filters, these functions will also be reported in this book.

2 Definition of Cauer Transfer Functions

In this section we intend to review briefly the definition and the main properties of Cauer transfer functions. The corresponding filters are often called "elliptic" because these functions are generally derived using Chebyschev rational functions in conjunction with Jacobian elliptic functions [23, 24, 32]. In the following, in view of future developments, we prefer to give a different, but equivalent, definition, based on attenuation properties.

Let us consider the following rational transfer functions of the complex frequency $s = \Sigma + j\Omega$;

$$F_n(s) = \frac{K \prod_{i=1}^{w} (s^2 + \Omega_{zi}^2)}{(s + \Omega_r)^u \prod_{i=1}^{w} (s^2 + s\Omega_{pi}/Q_i + \Omega_{pi}^2)} \tag{1}$$

where the introduced parameters are defined as follows:

$n = 2w + u$ = order of the function, equal to the number of poles;

u = index equal to 1 or 0 for n odd or even, respectively; obviously, for n even (i.e., $u = 0$), the real pole vanishes;

K = positive multiplicative constant;

w = number of imaginary zero-pairs and complex pole-pairs;

Ω_{zi} ($i = 1, 2, \ldots, w$) = frequencies of finite imaginary zero-pairs, located on the $j\Omega$-axis ($\Omega_{zi} > 0$);

Ω_r = real-pole frequency ($\Omega_r > 0$);

Ω_{pi}, Q_i ($i = 1, 2, \ldots, w$) = pole frequencies and Q factors, respectively, of the complex pole-pairs ($\Omega_{pi} > 0$, $Q_i > 0.5$).

The zero-frequency Ω_{zi} obviously coincides with the absolute value of the imaginary part of the imaginary zero-pair, whereas the pole frequency Ω_{pi} and pole Q factor Q_i are related to real and imaginary parts of the complex pole-pair $a_i \pm jb_i$, through the relation

$$\Omega_{pi} = \sqrt{a_i^2 + b_i^2}; \qquad Q_i = -\sqrt{a_i^2 + b_i^2}/(2a_i) \tag{2}$$

as illustrated in Fig. 1.

The inverse relations are

$$a_i = -\Omega_{pi}/(2Q_i); \qquad b_i = \Omega_{pi}\sqrt{1 - 1/(4Q_i^2)} \tag{3}$$

Hereafter, frequencies (Ω_{pi}) and Q factors (Q_i) will be adopted to characterize poles, instead of real and imaginary parts, which are less significant from a physical point of view.

Each $F_n(s)$ may be associated by a one-to-one correspondence to the attenuation function $\alpha_F(\Omega)$:

$$\alpha_F(\Omega) = -10 \log |F_n(j\Omega)|^2 = -10 \log [F_n(j\Omega) F_n(-j\Omega)] \tag{4}$$

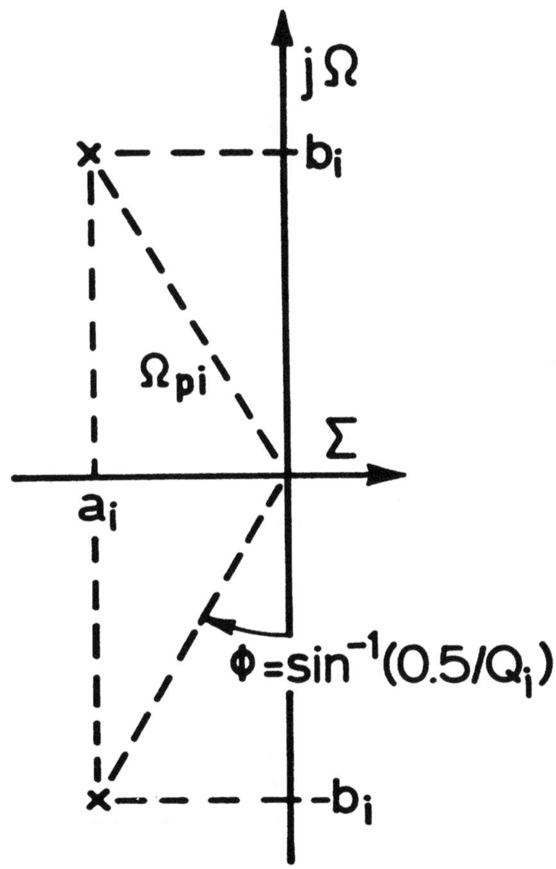

Fig. 1 Geometrical definition of pole frequency and pole-Q.

The one-to-one correspondence is assured by K being positive and by confining the poles inside the left half of the s-plane and the zeros on the $j\Omega$-axis.

A normalized low-pass Cauer transfer function of order n is a function $F_n(s)$, whose corresponding attenuation $\alpha_F(\Omega)$ satisfies the following properties (Fig. 2):

(i) $\alpha_F(\Omega)$ oscillates between the values 0 and α_M ($\alpha_M > 0$) with n half-cycles* in the interval $0 \leq \Omega \leq 1$ (pass band);

(ii) $\alpha_F(\Omega)$ oscillates between α_m ($\alpha_m > \alpha_M$) and infinity for $\Omega \geq \Omega_s > 1$ (stop band), with n half-cycles. Obviously, this property implies that all the zero-frequencies Ω_{zi} are greater than Ω_s because they correspond to attenuation poles.

It turns out [23, 24, 32] that only two parameters among α_M, α_m, and Ω_s, which characterize the filtering behavior of the function considered, are independent. According to common practice, the maximum pass-band attenuation α_M and the (normalized) stop-band edge Ω_s are chosen as independent design parameters.

Furthermore, instead of the maximum pass-band attenuation α_M, it is preferable to characterize the frequency response in the pass band by the maximum admissible modulus ϱ_M of the reflection coefficient ϱ, defined as

$$\varrho_M = \sqrt{1 - 10^{0.1\alpha_M}} \tag{5}$$

Moreover, it is convenient to assume the modular angle

$$\Theta = \frac{180°}{\pi} \sin^{-1}(1/\Omega_s) \tag{6}$$

*The half-cycles of a curve, equiripplé in a given interval, are the segments where the curve is monotonic.

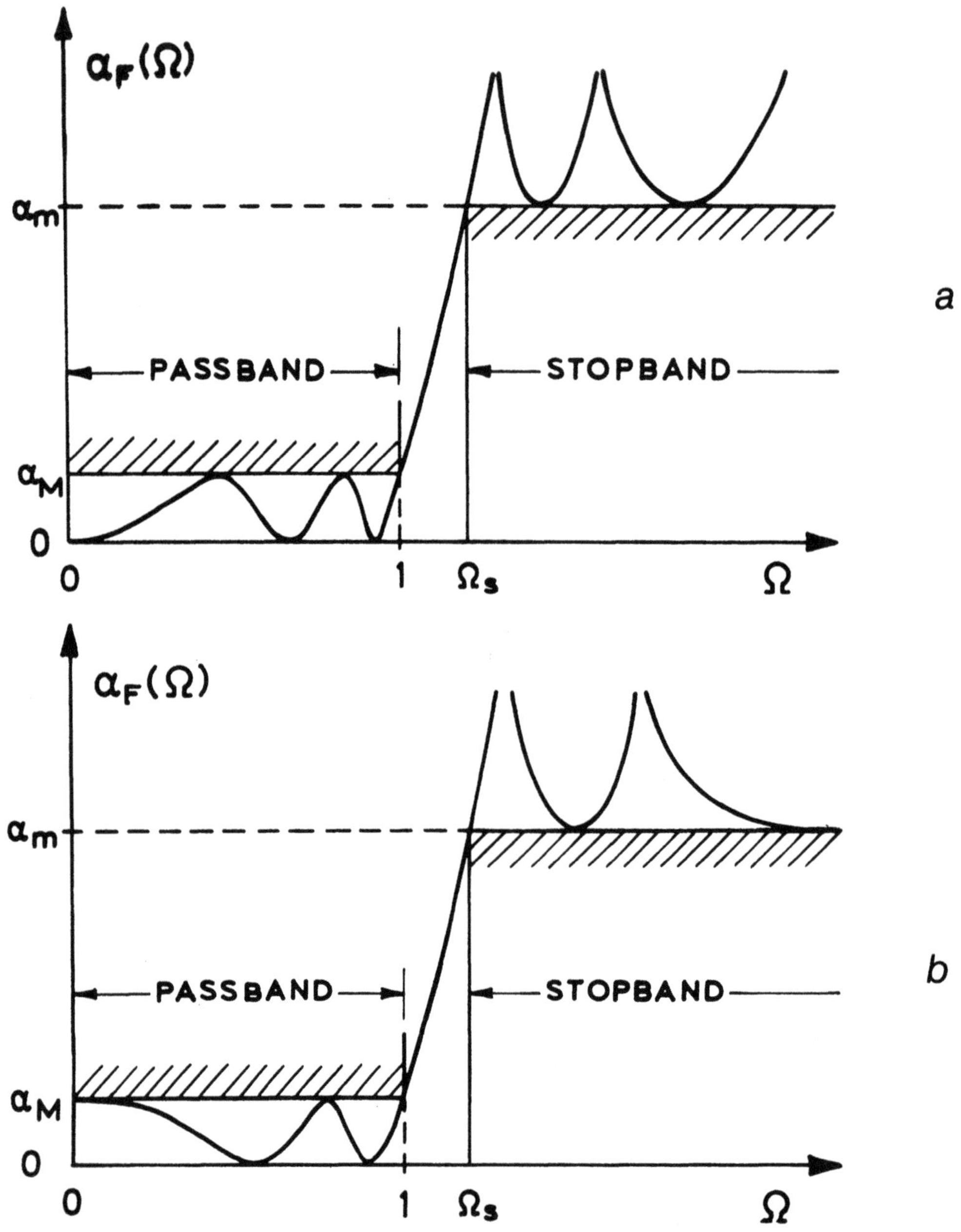

Fig. 2 Examples of attenuation of Cauer functions:
(a) $n = 5$, $w = 2$; (b) $n = 4$, $w = 2$.

as a measure of the normalized stop-band edge Ω_s; they have been found to be more practical as input parameters for the tables, in accordance with the experience of previous authors [22, 23, 25, 33].

The main property of filters based on Cauer transfer functions, which is appreciated by filter designers, is that they maximize (for given n, α_M, and Ω_s) the minimum stop-band attenuation α_m, in the sense that any other rational function of the same order n, with the same stop-band edge Ω_s and with attenuation not higher than α_M in the pass band, offers a minimum of attenuation $\alpha_F(\Omega)$ lower than α_m in the stop band.

It is worth noting that in the case of n odd (Fig. 2a), the attenuation $\alpha_F(\Omega)$ equals zero at the origin and becomes infinite at infinity; in the case of n even (Fig. 2b) it reaches a value α_M at the origin and a value α_m at infinity. As a consequence, functions with n even cannot be used to design equally terminated LC ladder circuits without coupled coils. Therefore, other (non-optimal) equiripple functions of even order derived from the previous ones through suitable frequency transformations, have been tabulated [22, 23, 25] with transmission zeros at infinity or zero value at the origin.

Because these tables are devoted to the case of active filters, the modifications required in the case where n is even will not be considered here.

3 From Cauer to MCPER Transfer Functions

The optimum property of Cauer transfer functions (in the sense specified at the end of section 2) is generally achieved at the expense of a high value for the Q factor of the complex pole-pair closest to the cut-off frequency. As mentioned in section 1, this fact may present serious practical problems.

A first non-optimal, but straightforward, method for overcoming the previous difficulty may be to extend the idea of cascading identical low-Q resonators [34]. In fact, a high-Q resonator, which is described by a transfer function of the form $s/(s^2+s\Omega_0/Q+\Omega_0^2)$ with $Q\gg 1$, may be advantageously replaced by the cascade of z low-Q identical resonators described by the function $[s/(s^2+s\Omega_0/\overline{Q}+\Omega_0^2)]^z$. For integer values of $z>1$, $\overline{Q}$ turns out to be substantially lower than Q, if the same selectivity is required.

Generalizing this idea, a nth-order Cauer transfer function $F_n(s)$ (with a high-Q pole pair) satisfying a low-pass specification fixed by α'_M, α'_m, and Ω'_s, may be substituted by a new function $F(s)=[F_n(s)]^s$ where $F_n(s)$ is evaluated to meet the modified attenuation specification $\overline{\alpha}_M=\alpha'_M/z$, $\overline{\alpha}_m=\alpha'_m/z$, $\overline{\Omega}_s=\Omega'_s$.

Obviously $F_{\overline{n}}(s)$, which can be supplied by classical tables [22, 23, 25], has lower pole Q factors and lower order than $F_n(s)$ (i.e., $\overline{n}<n$), since it has to meet less severe requirements. Unfortunately, the order $z\overline{n}$ of the global approximating function $\overline{F}(s)=[F_{\overline{n}}(s)]^z$ is rather high compared to that of $F_n(s)$.

The previous idea is illustrated by an example which considers a normalized low-pass filter with a maximum pass-band attenuation $\alpha'_M=0.570$ dB, pass-band edge $\Omega'_s=1.110$, minimum stop-band attenuation $\alpha'_m=47.0$ dB. These specifications are met by the Cauer transfer function $F_7(s)$ with $\alpha_M=0.568$ dB (i.e., $\rho_M=35\%$) and $\Omega_s=1.103$ (i.e., $\Theta=65°$); its stop-band attenuation is $\alpha_m=47.6$ dB and the maximum pole Q is $Q_c=23.3$. The same specification is also met by a function $[F_5(s)]^2$, where $F_5(s)$ is a Cauer function having $\alpha_M=0.280$ dB (i.e., $\rho_M=25\%$), $\alpha_m=24.9$ dB, $Q_c=11.1$, and the same value of Ω_s; as required by this approach, the attenuation levels α_M and α_m of $F_5(s)$ are nearly equal to $\overline{\alpha}_M=0.570/2$ dB and $\alpha_m=47.0/2$ dB, respectively.

The main disadvantage of this technique is the loss of some degrees of freedom due to the equal multiplicity assigned to all poles and zeros of $F_n(s)$, causing an order increase, from 7 to 10 in the example considered, which is higher than that obtainable with other more sophisticated approaches described in the following.

To overcome this last difficulty it may be more advantageous to approximate the assigned specification by a new transfer function of non-minimum order, according to the theory described in section 1. This may be accomplished by increasing the multiplicity of the most critical pole-pair and by readjusting poles and zeros to meet the assigned attenuation specification exactly, with an equiripple behavior in the pass band and in the stop band, and with all the zeros on the $j\omega$-axis [21]. Details on this class of functions will be the subject of the next section.

4 Definition of MCPER Transfer Functions

According to the considerations of the previous section, let us introduce the following low-pass transfer function:

$$P^{(w)}_{m+\delta}(s)=\frac{K\prod_{i=1}^{w}(s^2+\Omega_{zi}^2)}{(s+\Omega_r)^u\,(s^2+s\Omega_{pc}/Q_c+\Omega_{pc}^2)^{\mu_c}\prod_{i=1}^{(m-u-2)/2}(s^2+s\Omega_{pi}/Q_i+\Omega_{pi}^2)} \qquad (7)$$

Some of the symbols are the same as those of equation (1); those symbols introduced for the first time are defined as follows:

Ω_{pc}, Q_c, μ_c = frequency, Q factor, and multiplicity of the critical pole-pair ($\Omega_{pc}>0$, $Q_c>0.5$, $\mu_c\geq 1$);

m = number of distinct poles (the multiple pole-pair is considered only once);

δ = order increment (≥ 0), equal to $2(\mu_c-1)$, due to the imposed multiplicity μ_c;

in this case the number w of imaginary zero-pairs satisfies the constraint $w\leq(m+\delta-u)/2$ for the sake of stability; and, hence, w may be lower than the number of complex pole-pairs.

The order of the functions under consideration is equal to $m+\delta$ and, if necessary, will be denoted by n, as in Eq. (1). However, in practical situations, the order will be denoted as the sum of two digits, the first specifying m and the second specifying δ, to recognize immediately the number of distinct poles and the order increment due to the multiplicity of the critical pole-pair. In fact, functions of the same order, but with different multiplicity of the critical pole-pair, are completely different (for example, $P_{3+4}^{(3)} \neq P_{5+2}^{(3)}$ because $\mu_c = 3$ in the first case, and $\mu_c = 2$ in the second).

At this point, we introduce the definition of MCPER transfer functions in a similar fashion to that used for Cauer functions in the previous section.

Let us introduce the attenuation function $\alpha_P(\Omega)$:

$$\alpha_P(\Omega) = -10 \log | P_{m+\delta}^{(w)} (j\Omega) |^2 = -10 \log [P_{m+\delta}^{(w)} (j\Omega) P_{m+\delta}^{(w)} (-j\Omega)] \qquad (8)$$

Because in this case $P_{m+\delta}^{(w)}$ and $\alpha_P(\Omega)$ are again related by a one-to-one correspondence, any definition may be developed by using $\alpha_P(\Omega)$.

A MCPER transfer function [21] of order $n = m + \delta$ (i.e., with m distinct poles and multiplicity $\mu_c = \delta/2 + 1$) and with w imaginary zero-pairs, is a function of type $P_{m+\delta}^{(w)}(s)$ whose corresponding attenuation $\alpha_P(\Omega)$ satisfies the following properties (Fig. 3):

(i) $\alpha_P(\Omega)$ oscillates between the values 0 and $\alpha_M (\alpha_M > 0)$ with m half-cycles in the interval $0 \leq \Omega \leq 1$ (pass band);

(ii) $\alpha_P(\Omega)$ oscillates between $\alpha_m (\alpha_m > \alpha_M)$ and infinity for $\Omega \geq \Omega_s > 1$ (stop band), with $2w$ (if $n = 2w$) or $2w + 1$ (in all other cases) half-cycles.

As in the case of Cauer functions (section 2), α_m again depends on α_M and Ω_s, which are assumed as independent design parameters. Moreover, these functions may also be conveniently tabulated with respect to ρ_M and Θ, defined by Eqs. (5) and (6), respectively.

Although the MCPER functions have been defined [21] for any integer $w \leq (m + \delta - u)/2$, hereafter we will focus our attention on the case $w = (m + \delta - u)/2$, since, as in the Cauer case, the best attenuation performance is obtained if the number of finite zeros is the maximum allowed. As a consequence of this assumption, the number of half-cycles equals n in the stop band and, if $\delta = 0$ (i.e., $n = m$), MCPER functions coincide with Cauer functions as previously defined.

Accordingly, a Cauer function $F_n(s)$ will hereafter be denoted with the symbol $P_{m+0}^{(w)}(s)$, where $w = m/2$ (m even) or $w = (m - 1)/2$ (m odd), where m coincides with n.

5 Comparison between Cauer and MCPER Transfer Functions

This section is devoted to the comparison of attenuation performances of Cauer and MCPER functions. For the sake of clarity, let us indicate with $n_1 = m_1$, $n_2 = m_2 + 2$, and $n_3 = m_3 + 4$ the order n of functions with single (i.e., Cauer), double, and triple (i.e., MCPER) critical pole-pairs, respectively.

Cauer functions of order n_1 obviously provide more stop-band attenuation α_m than MCPER transfer functions of the same total order (i.e., $n_2 = n_1$, or $n_3 = n_1$), pass-band attenuation α_M and stop-band edge Ω_s. On the other hand, Cauer functions of order n_1 offer lower stop-band attenuation (again with equal values of α_M and Ω_s) if they are compared with MCPER functions having the same number of distinct poles (i.e., $n_2 = n_1 + 2$, or $n_3 = n_1 + 4$).

However, from the point of view of a practical design, it is more significant to compare different solutions satisfying the same attenuation specifications (i.e., with equal values of α_M and Ω_s, and a nearly identical value of α_m). In this circumstance, according to the previous considerations, the required orders n_2 or n_3 (for MCPER solutions), in the majority of the cases, turn out to be

$$n_2 = n_1 + 1 \qquad (9)$$

$$n_3 = n_1 + 2 \qquad (10)$$

As expected, these values of n_2 and n_3 are intermediate between those shown in the two previous types of comparison. In other words, the degree increment δ due to introduction of multiplicity μ_c for the critical pole-pair is partially compensated by a decrease in the required number of distinct poles. In passing, it is worth noting that the number of zero-pairs generally increases, in agreement with the total order.

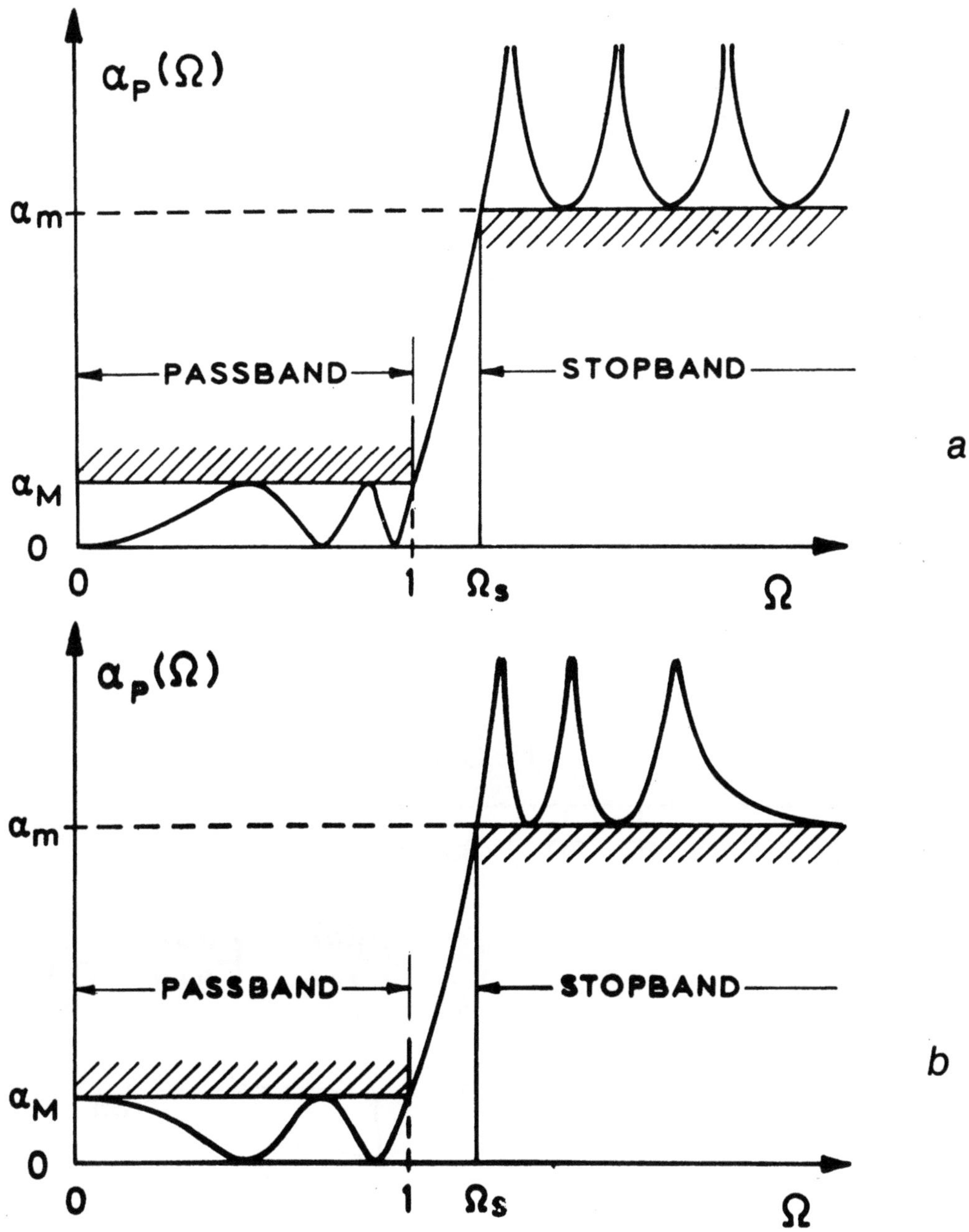

Fig. 3 Examples of attenuation of MCPER functions:
(a) $P^{(3)}_{5+2}$; (b) $P^{(3)}_{4+2}$.

To evaluate the limits of Eqs. (9) and (10) and to gain more insight into the behavior of MCPER functions, a comparison has been made for several values of ρ_M and Θ, between stop-band attenuation α_m and critical pole Q factor Q_c of Cauer and MCPER function, chosen according to Eqs. (9) and (10), which are comparable from the point of view of filtering specifications. An example of such a comparison is shown in Fig. 4 where the sets of functions $P_{9+0}{}^{(4)}$, $P_{8+2}{}^{(5)}$ and $P_{7+4}{}^{(5)}$ have been considered. The following interesting conclusions may be drawn:

(i) For any Θ, for high ρ_M the stop-band attenuation α_m of MCPER solutions is greater than that offered by Cauer solutions, whereas the opposite is true for low values of ρ_M (0.02 - 0.05); in this case MCPER solutions of higher order, with respect to that given by Eqs. (9) and (10), may be required.

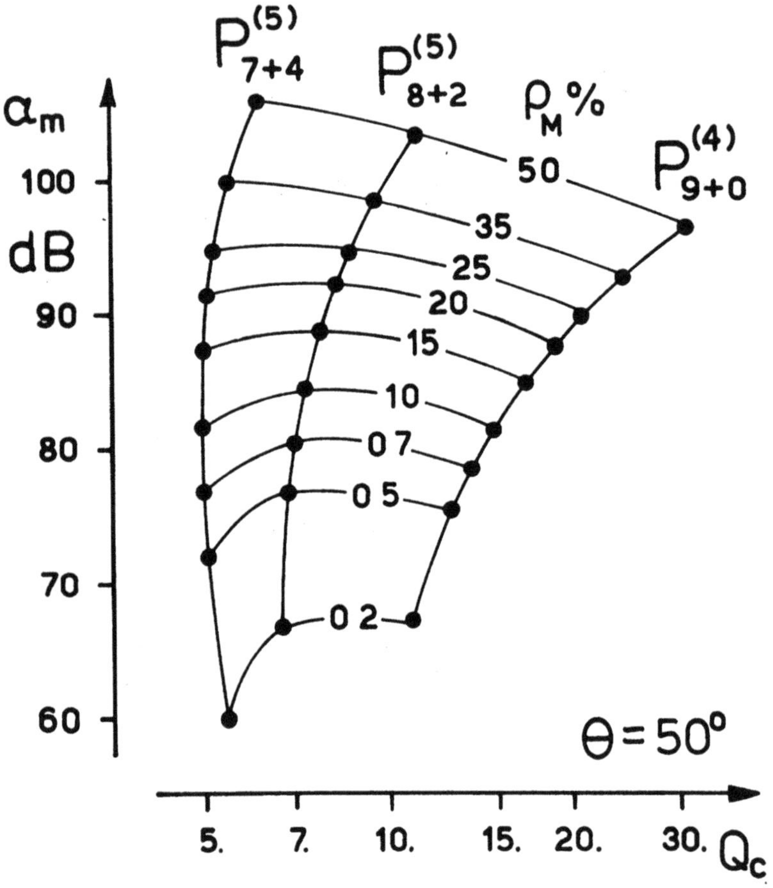
$P^{(5)}_{7+4}$
$P^{(5)}_{8+2}$
ρ_M%
$P^{(4)}_{9+0}$
α_m
dB
100
90
80
70
60
50
35
25
20
15
10
07
05
02
$\Theta=50°$
5.
7.
10.
15.
20.
30.
Q_c

a

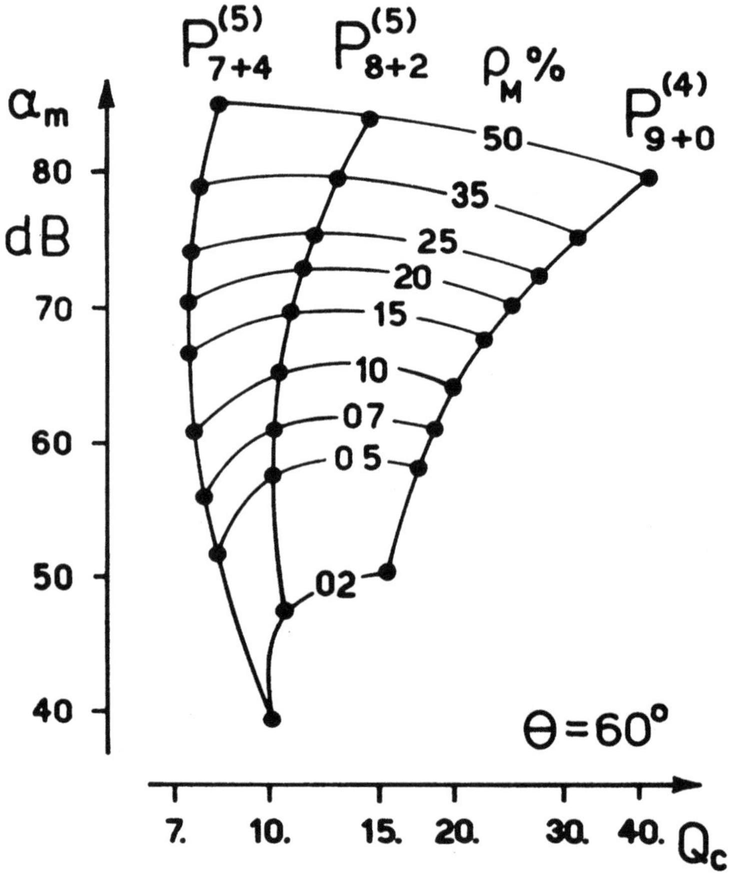
$P^{(5)}_{7+4}$
$P^{(5)}_{8+2}$
ρ_M%
$P^{(4)}_{9+0}$
α_m
dB
80
70
60
50
40
50
35
25
20
15
10
07
05
02
$\Theta=60°$
7.
10.
15.
20.
30.
40.
Q_c

b

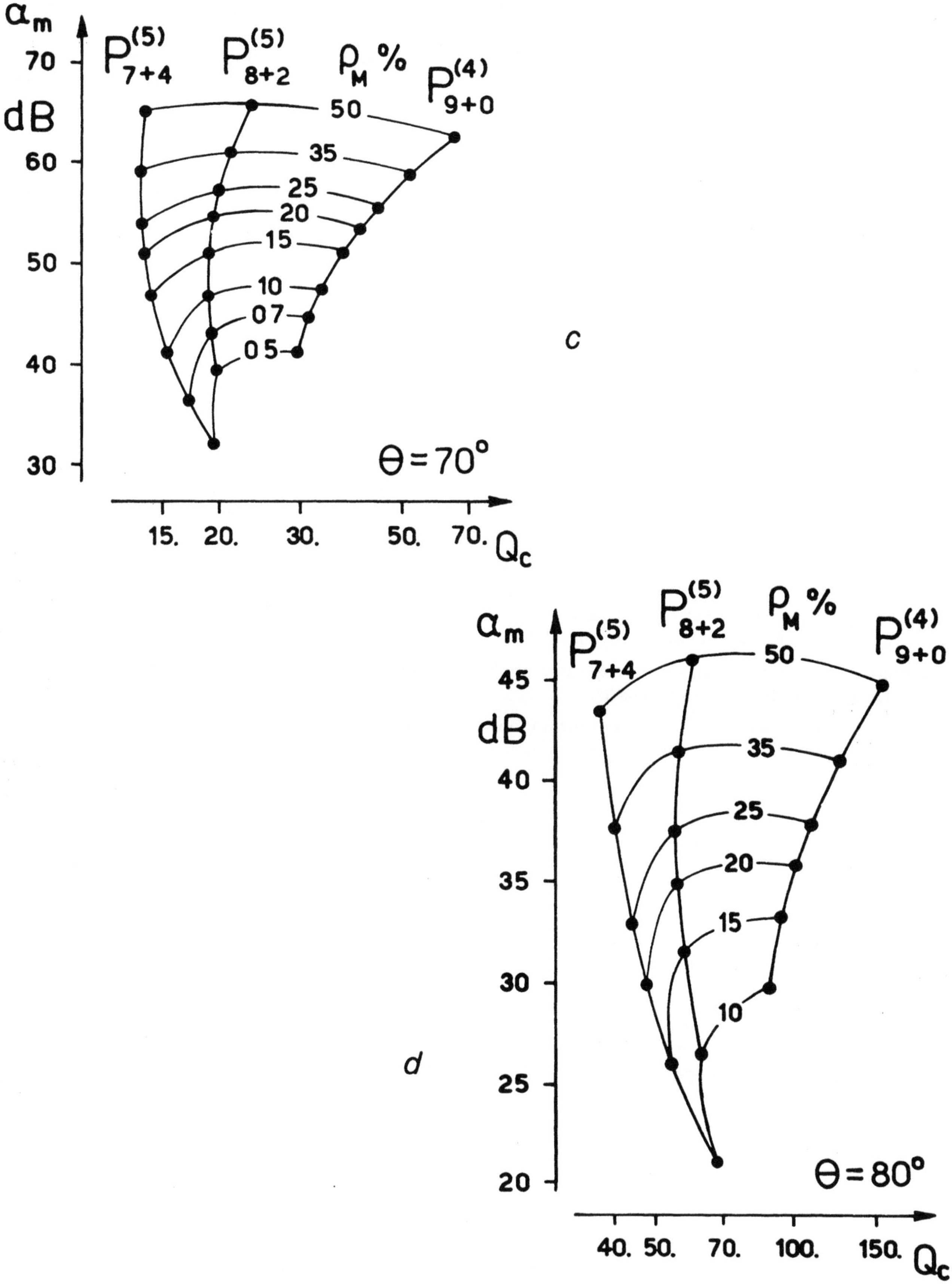

Fig. 4 Comparative diagrams of stop-band attenuation α_m *versus* critical pole-Q factor Q_c of the three sets $P^{(4)}_{9+0}$, $P^{(5)}_{8+2}$, $P^{(5)}_{7+4}$ for several values of ρ_M and the values of Θ: (a) 50°, (b) 60°, (c) 70°, and (d) 80°.

(ii) The typical relative reduction in Q_c (with respect to a corresponding Cauer solution, i.e., $\mu_c = 1$) varies within the range 45% - 55% and 65% - 75% for $\mu_c = 2,3$, respectively. In particular, for high (low) values of Q_M, a higher (lower) reduction in Q_c may occur. This fact, as mentioned in section 1, reduces sensitivity problems occurring in the production of the corresponding second-order active filter sections, allowing the use of cheaper and less sophisticated building blocks to be connected in cascade.

Other interesting features, deduced by a direct analysis of MCPER functions, are the following:

(iii) The value of Ω_{pc} increases with μ_c: that is, the multiple pole-pair is moved towards the transition band, where higher values of the overall sensitivity of the filter may be tolerated. Moreover, this fact yields a stronger Q reduction in the case of band-pass filters obtained through classical frequency transformation, as will be shown later [see section 7, Eq. (12a)].

(iv) The group delay has a smoother behavior in the pass band, according to the reduction in Q_c.

Solutions with μ_c greater than three show an order increment that is generally larger than that given by Eq. (9) and a less significant decrease in Q_c. Hence, they have not been considered in this book.

6 Generation of MCPER Transfer Functions

Whereas Cauer functions may be generated analytically through the use of Jacobian elliptic functions [32, 23, 24], MCPER functions cannot be generated in the same way and it is necessary to resort to an iterative procedure. The definition given previously is particularly suitable for the direct implementation of the iterative algorithm.

The algorithm proposed in the original work [21] assumes the coefficients of the squared magnitude of the transfer functions as unknowns and is directly concerned with the generation of a single function with assigned values of m, δ, w, Q_M, and Θ. When generating tables where groups of functions are produced with the same values of m, δ, w, Q_M, and with Θ gradually increasing from a starting value, it was necessary to implement a proper algorithm in order to decrease computing time and increase the number of significant digits.

As unknowns, this algorithm uses pole frequencies, pole Q factors, zero-frequencies, and the multiplicative constant shown in Eq. (7). It is divided into two parts:

(i) Given a suitable set of initial values of the unknowns, this part iteratively finds the exact values which solve the approximation problem for given values of m, δ, w, Q_M, Θ. This part is based on the linear approximation of the attenuation $\alpha_P(\Omega)$ with respect to the unknown parameters and on the use of linear programming techniques [35, 36].

(ii) This part gives accurate initial values of the unknowns to be supplied to part (i) when a new function (i.e., with a new value of Θ) is to be computed. These initial values are computed by extrapolating the final values obtained for functions corresponding to lower values of Θ (and with the same values of m, δ, w, Q_M). The extrapolation uses a gradually increasing number (from 1 to 8) of values of the independent variable Θ.

When the number of interpolating values of Θ reaches the value 8, only 3 or 4 iterations in part (i) are necessary to obtain the unknown parameters with an accuracy of 10^{-11} to 10^{-12}.

7 Frequency Transformations

The data supplied by the present tables correspond to normalized low-pass (NLP) filters; i.e., filters, whose pass-band edge occurs at $\Omega = 1$. From the tabulated data for an NLP/filter, data for the unnormalized low-pass, high-pass, geometrically-symmetrical band-pass and band-stop filters may be obtained by using frequency transformations [22-25, 37-39].

Although other types of transformations have been reported in the literature [22-25, 37-41], only the abovementioned most popular transformations have been summarized in Table I, where the new symbols introduced have the followin-

f_i = pass-band and stop-band edges of the four different types of specifications reported at the top of the table; i is equal to 1, 2 for low-pass and high-pass specifications and equal to 1, 2, 3, 4 for band-pass and band-stop specifications;

α'_M = maximum pass-band attenuation assigned by the designer;

$\rho'_M = \sqrt{1 - 10^{-0.1}\ \alpha_M}$ = maximum admissible modulus of the reflection coefficient, assigned by the designer;

α'_m = minimum stop-band attenuation assigned by the designer;

Ω'_s = nominal stop-band edge of NLP, obtained from the specifications assigned by the designer;

$p = \sigma + j\omega$ = complex frequency of transformed filters;

$f = \omega/(2\pi)$ = frequency of transformed filters;

n_t = order of the transformed function;

w_t = number of finite, non-zero, imaginary zero-pairs;

ν_0 = number of transmission zeros at origin;

ν_∞ = number of transmission zeros at infinity; obviously $n_t = \nu_0 + \nu_\infty + 2w_t$;

k = multiplicative constant defined such that the attenuation of the transformed filters equals zero in its pass-band minima; k is computed by assuming the transformed transfer functions written as a ratio of polynomials with the terms of highest degree having unity coefficients;

f_r = real-pole frequency of unnormalized low-pass and high-pass functions, corresponding to the real pole of the NLP model;

f_{pi}, q_i = pole frequencies and Q factors of complex pole-pairs of unnormalized low-pass and high-pass functions, corresponding to the (non-critical) complex pole-pairs of the NLP model;

f_{pc}, q_c = pole frequency and Q factor of the (multiple) critical complex pole-pair of unnormalized low-pass and high-pass functions, corresponding to the (multiple) critical complex pole-pair of the NLP model;

f_{zi} = frequencies of finite imaginary zero-pairs of unnormalized low-pass and high-pass functions, corresponding to the finite imaginary zero-pairs of the NLP model;

f_{pr}, q_r = pole frequency and Q factor of the complex pole-pair of band-pass and band-stop functions, corresponding to the real pole of the NLP model. In some cases this complex pole-pair degenerates into a pair of real poles (i.e., $q_r \leq 0.5$) whose frequencies will be denoted by f_{r+} and f_{r-}, ($f_{r+} \geq f_{r-}$);

f_{pi+}, q_{i+}, f_{pi-}, q_{i-} = pole frequencies ($f_{pi+} > f_{pi-}$) and Q factors ($q_{i+} = q_{i-}$) of the two complex pole-pairs of band-pass and band-stop functions, corresponding to each (non-critical) complex pole-pair of the NLP model;

f_{pc+}, q_{c+}, f_{pc-}, q_{c-} = pole frequencies ($f_{pc+} > f_{pc-}$) and Q factors ($q_{c+} = q_{c-}$) of the two (multiple) critical complex pole-pairs of band-pass and band-stop functions, corresponding to the (multiple) critical complex pole-pair of the NLP model;

f_{zi+}, f_{zi-} = frequencies ($f_{zi+} > f_{zi-}$) of finite imaginary zero-pairs of band-pass and band-stop functions, corresponding to each finite imaginary zero-pair of the NLP model.

The other symbols pertinent to NLP filters have already been defined in sections 2 and 4.

Because α_M (i.e., ρ_M), α_m, and Ω_s (i.e., Θ) are not independent (see sections 2 and 4), and in any case only a finite discrete set of their values may be tabulated, in general the actual NLP/model cannot meet exactly the design requirements and must be chosen in such a way that $\alpha_M \leq \alpha'_M$, $\alpha_m \geq \alpha'_m$, $\Omega_s \leq \Omega'_s$. Of course, between the many possible solutions, the designer will choose the solution of minimum order, offering the highest margins in both the pass band and in the stop band [42]. These considerations may be repeated for any assigned multiplicity of the critical pole-pair.

The final expressions of the transformed transfer functions $\mathscr{P}^{(w)}_{m+\delta}(p)$ are shown in the last row of Table I.

Table I
Formulas for the Computation of the Transformed Transfer Functions, Starting from the Attenuation Specification

	Frequency Transformation	LOW PASS	HIGH PASS	BAND PASS	BAND STOP
	Specification Diagram	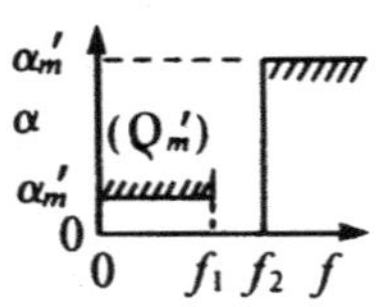	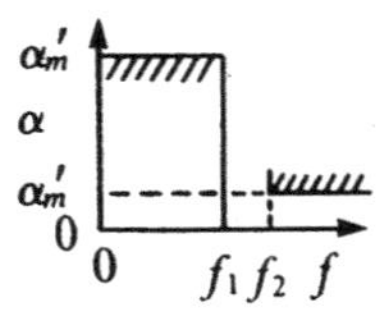	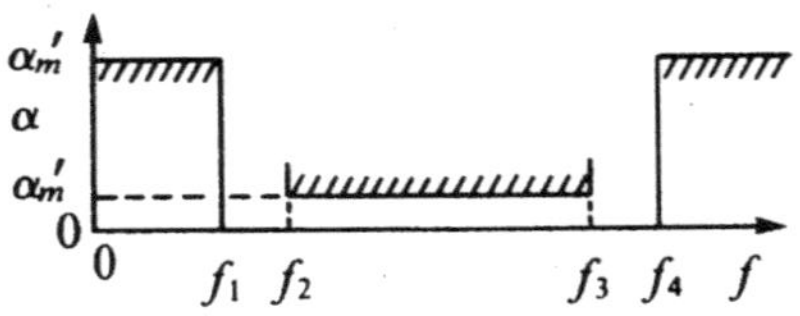	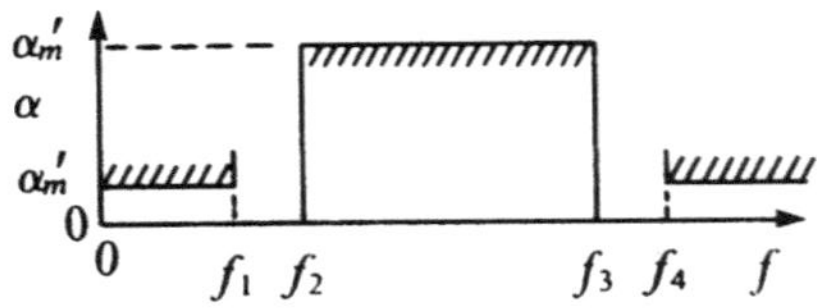
		$p = \sigma + j2\pi f$		$p = \sigma + j2\pi f,\ f_0 = \sqrt{f_1 f_4} = \sqrt{f_2 f_3},\ B = (f_3 - f_2)/f_0$	$p = \sigma + j2\pi f,\ f_0 = \sqrt{f_1 f_4} = \sqrt{f_2 f_3},\ B = (f_4 - f_1)/f_0$
NORMALIZED LOW PASS (NLP)	complex frequency s	$p/(2\pi f_1)$	$2\pi f_2/p$	$[p/(2\pi f_0) + 2\pi f_0/P]/B$	$B/[p/(2\pi f_0) + 2\pi f_0/p]$
	real frequency Ω	f/f_1	f_2/f	$(f/f_0 - f_0/f)/B$	$B/(f/f_0 - f_0/f)$
	nominal stopband edge Ω_s'	f_2/f_1		$(f_4 - f_1)/(f_3 - f_2)$	
	choice of ρ_m and Θ	choose	ρ_m and Θ such that $\rho_m \le \rho_m'$ (i.e., $\alpha_m \le \alpha_m'$), $\Omega_s \le \Omega_s'$		
	choice of the order n	choose	$n = m + \delta$ such that $\alpha_m \ge \alpha_m'$, with $\delta = 0$ or 2 or 4		
	NLP transfer function $P_{m+\delta}^{w}(s)$		$\dfrac{K \prod_{i=1}^{w} (S^2 + \Omega_{Zi}^2)}{(s+\Omega_r)^u (s^2 + s\Omega_{pc}/Q_c + \Omega_{pc}^2)^{\mu_c} \prod_{i=1}^{(m-u-2)/2} (s^2 + s\Omega_{pi}/Q_i + \Omega_{pi}^2)}$		$u = 0\ (1)$ for n even (odd) $w = (n-u)/2$ $\mu_c = \delta/2 + 1$
FUNCTION	number of poles n_t	n		$2n$	
	number of finite zero pairs w_t	w		$2w$	
	number of transmission zeros at origin ν_0	0	u	u	0
	number of transmission zeros at infinity ν_∞	u	0	u	0
	multiplicative factor k	$K(2\pi f_1)^u$	$(1-\rho_M^2)^{(1-u)/2}$	$K[B(2\pi f_0)]^u$	$(1-\rho_m^2)^{(1-u)/2}$

TRANSFORMED	Poles corresponding to a real pole Ω_r of NLP model (only for $u = 1$)	$f_r = f_1 \Omega_r$ real pole	$f_r = f_1 \Omega_r$ real pole	$V_r = B\Omega_r$ $V_r = B\Omega_r$ $V_r < 2: f_{pr} = f_0,\ q_r = 1/V_r$, complex pole-pair $V_r \geq 2: f_{r^+} = f_0 [\ V_r/2 + \sqrt{(V_r/2)^2 - 1}\ ]; f_{r-} = f_0^2/f_{r^+}$, two real poles
	Poles corresponding to a (non-critical) complex pole-pair Ω_{pi}, Q_i of NLP model. $i = 1, 2, \ldots, (m-u-2/2)$	$f_{pi} = f_1 \Omega_{pi},\ q_i = Q_i$	$f_{pi} = f_2 / \Omega_{pi},\ q_i = Q_i$	$V_{pi} = B\Omega_{pi}$ $V_{pi} = B\Omega_{pi}$ $H_{pi} = V_{pi}^2 (4 - 1/Q_i^2)/[\ 2 - V_{pi}^2/2 - 2)^2 + \sqrt{V_{pi}^2 (4 - 1/Q_i^2)}\ ]$ $f_{pi^+} = f_0 [\ \sqrt{H_{pi}}/2 + \sqrt{H_{pi}/4 + 1}\ ],\ q_{i^+} = 2Q_i \sqrt{H_{pi}/4 + 1}/V_{pi}; f_{pi-} = f_0^2/f_{pi^+},\ q_{i-} = q_{i^+}$
	Poles corresponding to a critical complex pole-pair Ω_{pc}, Q_c, with multiplicity μ_c, of NLP model	$f_{pc} = f_1 \Omega_{pc},\ q_c = Q_c$	$f_{pc} = f_2 / \Omega_{pc},\ q_c = Q_c$	$V_{pc} = B\ \Omega_{pc}$ $V_{pc} = B\ \Omega_{pc}$ $H_{pc} = V_{pc}^2 (4 - 1/Q_i^2/\ [\ 2 - V_{pc}^2/2 + \sqrt{(V_{pc}^2/2 - 2)^2 + V_{pc}^2 (4 - 1/Q_i^2)}\ ]$ $f_{pc^+} = f_0 [\ \sqrt{H_{pc}}/2 + \sqrt{H_{pc}/4 + 1}\ ],\ q_+ = 2Q_c \sqrt{H_{pc}/4 + 1}/V_{pc}; f_{pc-} = f_0^2/f_{pc^+},\ q_{c-} = q_{c^+}$
	Zeros corresponding to a finite imaginary conjugate zero-pair $\pm j\ \Omega_{Zi}$	$f_{zi} = f_1 \Omega_{zi}$	$f_{zi} = f_2 \Omega_{zi}$	$V_{zi} = B\Omega_{pc}$ $V_{zi} = B\Omega_{pc}$ $f_{zi^+} = f_0 [\ V_{zi}/2 + \sqrt{(V_{zi}/2)^2 + 1}\ ]; f_{zi-} = f_0^2/f_{zi^+}$
	Transformed transfer function $Q_m{}^{(w)}\delta(p)$	$\dfrac{kp^{v_0} \prod_{i+1}^{w_t} [p^2 + (2\pi f_{zi})_2]}{(p + 2\pi f_r)^u [p^2 + p 2\pi f_{pc}/q_c + (2\pi f_{pc})^2]^{\mu_c} \prod_{i=1}^{(m-u-2)/2} [p^2 + p 2\pi f_{pi}/q_i + (2\pi_{pi})^2]}$		$\dfrac{kp^{v_0} \prod_{i=1}^{w_t/2} [p^2 + (2\pi f_{zi^+})^2][p^2 + (2\pi f_{zi-})^2]}{[p^2 + p 2\pi f_{pr}/q_r + (2\pi f_{pr})^2]\ [p^2 + p 2\pi f_{pc}/q_{c^+} + (2\pi f_{pc})^2]\ [p^2 + p 2\pi f_{pc-}/q_{c-} + (2\pi f_{pc-})^2] \prod_{i=1}^{(m-u-2/2)} [p^2 + p 2\pi f_{pi^+}/q_{i^+} + (2\pi f_{pi^+})^2]\ [p^2 + p 2\pi f_{pi--}/q_{i-} + (2\pi f_{pi-})^2]}$

In the case where a fast and rough evaluation of pole and zero frequencies and pole Q factors of band-pass or band-stop filters is required, the diagram of Fig. 5 [43] may be used instead of the exact formulas listed in Table I. Assigning band-pass or band-stop specifications, it is possible to compute the auxiliary parameter A defined as

$$A = B\,\Omega_0 \; (band\text{-}pass) \tag{11a}$$

$$A = B\,/\,\Omega_0 \; (band\text{-}stop) \tag{11b}$$

where B is the relative bandwidth and Ω_0 is the pole or zero frequency of the NLP model.

The entries for the diagram are the auxiliary parameter A and the Q factor (Q) of the considered pole or zero of the NLP model; in the case of real poles (or zeros) and imaginary zero-pairs, the values 0.5 and ∞, respectively, must be chosen for the entry Q.

As output, the designer may read the normalized frequencies (with respect to the center frequency f_0) f_{0+}/f_0 and f_{0-}/f_0 and Q factor (q) of the transformed pairs of poles (or zeros). Obviously, the entry $Q = \infty$ produces $q = \infty$.

In the case of band-pass or band-stop filters, if $B \ll 1$, it is always possible to give an approximate expression for the Q factor (q) of the transformed pair of poles,

$$q \simeq \frac{2\,Q}{B\,\Omega_p} \; (band\text{-}pass) \tag{12a}$$

$$q \simeq \frac{2\,Q\Omega_p}{B} \; (band\text{-}stop) \tag{12b}$$

where Ω_p and Q are the corresponding pole frequency and Q factor of the NLP. Equation (12a), applied to the case of the critical pole-pair Q_c, Ω_{pc}, shows that the reduction in q_{c+} and q_{c-} due to the reduction in Q_c is enhanced by the increment in Ω_{pc}, when the multiplicity of the critical pole-pair increases, as observed under point (*iv*) of section 5. The opposite is true in the case of band-stop filters [Eq. (12b)].

The practical use of the abovementioned table and diagram will be considered in section 9.

8 Description of the Tables

In the tables of Part II, poles and zeros are reported for Cauer (odd and even type A) and MCPER low-pass transfer functions of orders

n = 2 to 11 for μ_c = 1 (i.e., 2 + 0 to 11 + 0),

n = 4 to 12 for μ_c = 2 (i.e., 2 + 2 to 10 + 2),

n = 6 to 13 for μ_c = 3 (i.e., 2 + 4 to 9 + 4),

with nine values of the reflection coefficient: i.e., ρ_M = 2, 5, 7, 10, 15, 20, 25, 35, and 50 percent.

The range of Θ is within the lower and upper bounds Θ_L and Θ_U, respectively; it is chosen such that α_m reaches about 90 dB for $\rho_M = 2\%$ and $\Theta = \Theta_L$, and 25-33 dB for $\rho_M = 50\%$ and $\Theta = \Theta_U$. In ány case, Θ_U never exceeds 86°, corresponding to Ω_s = 1.0024 and the incremental step of Θ is equal to 1. Moreover, functions with α_m lower than 5-7.5 dB (depending on the order) have not been compiled.

Ten significant figures have been provided for all the characteristic parameters, with the exception of α_M and K, extremely accurate knowledge of which is rarely required. However, if required, they may be derived from the other data with the same accuracy.

Because the original print-out of the tables has been supplied directly by the computer, the symbols adopted in the tables differ from conventional symbols used in the literature and in Part I of this book. Table II provides the correspondence between the symbols used in the tables and those defined in the previous sections. However, for completeness, at the beginning of each new set of functions characterized by given values of m, δ, (and w), self-explanatory diagrams are

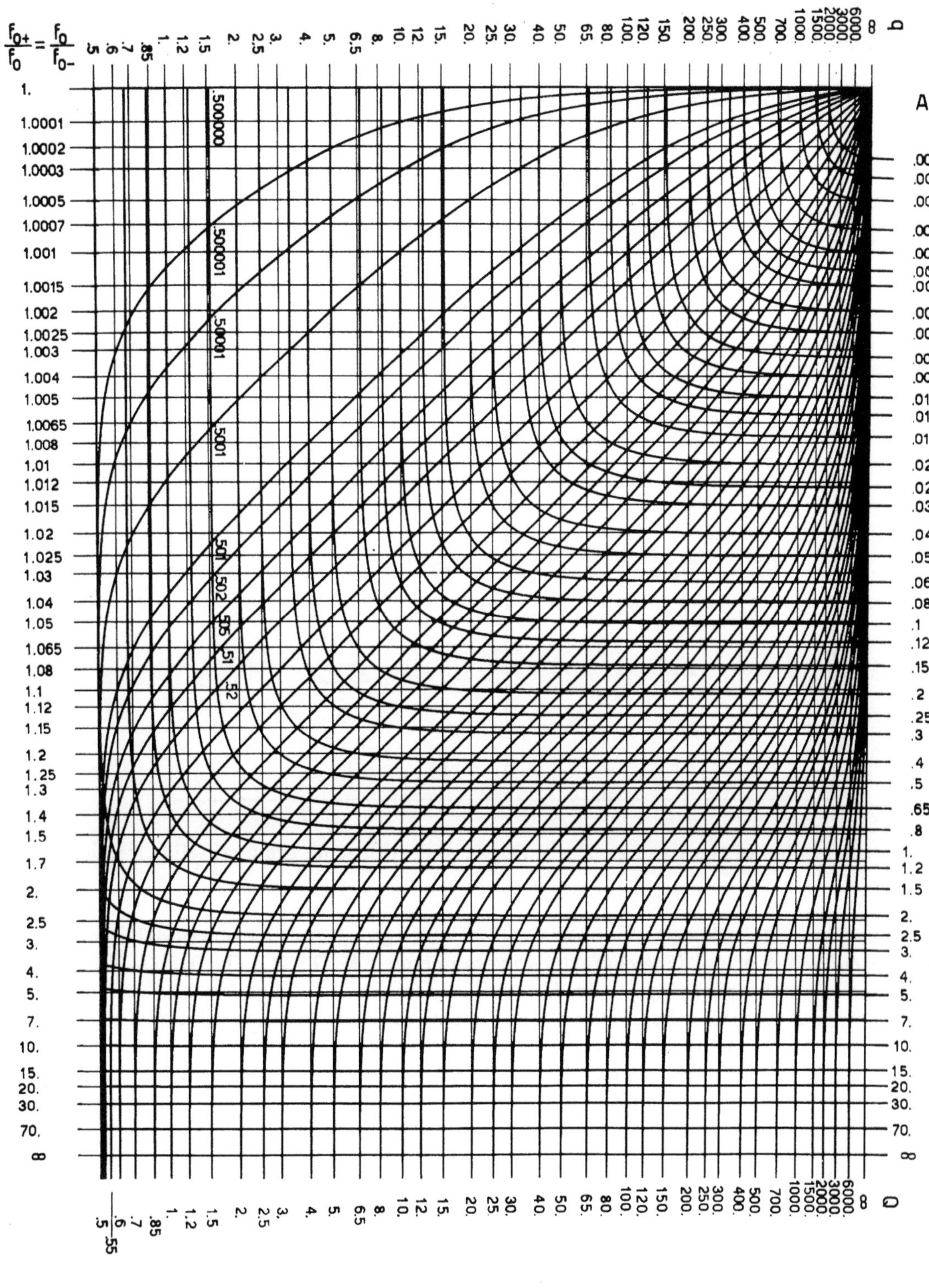

Fig. 5 Frequency transformation diagram:

Input parameters:

A: auxiliary parameter equal to $B\Omega_0$ (band-pass) or to B/Ω_0 (band-stop), where B is the relative bandwidth and Ω_0 is the pole or zero frequency of the NLP model;

Q:Q factor of the considered pole or zero of the NLP model. ($Q = \infty$ for imaginary zero-pairs.)

Output parameters:

$f_{0+}/f_0, f_{0-}/f_0$: pole or zero frequencies, normalized with respect to the center frequency f_0 of the band-pass or band-stop filter;

q:Q factor of the transformed pairs of poles or zeros. ($q = \infty$ for imaginary zero-pairs.)

Table II

Correspondence between Symbols Used in the Tables of Part II (1st column) and those Used in the Introduction (2nd column)

(*) μ is the multiplicity of each pole and zero, equal to μ_c for the multiple pole-pair and equal to 1 in all the other cases

TH	Θ
SBE	Ω_s
RC	$100\ \varrho_M$
AMAX	α_M
AMIN	α_m
MU	μ (*)
K	K
PFR	Ω_r
PF1, PF2, ...	$\Omega_{p1}, \Omega_{p2}, \ldots$
PQ1, PQ2, ...	$Q_1, Q_2, \ldots$
PFC	Ω_{pc}
PQC	Q_c
ZF1, ZF2, ...	$\Omega_{z1}, \Omega_{z2}, \ldots$

appended, defining the symbol of the set (e.g., $P_{6+0}^{(3)}$, $P_{7+2}^{(4)}$), and evidencing the qualitative frequency response and pole-zero locations.

Each of the abovementioned sets is divided into subsets, each of which groups the nine functions having the same value of Θ and the nine different values of ϱ_M. The characteristic parameters (K, Ω_r, Ω_{pi}, Q_i, Ω_{pc}, Q_c, Ω_{zi}) of each subset of functions are compiled columnwise in a single table. The symbol of the set, including the considered subset, is repeated in the upper left-hand corner of the table, whereas the value of Θ (with the corresponding value of Ω_s) defining the subset is shown in the first row. In the second, third, and fourth rows of the table are respectively reported the values of ϱ_M, the corresponding α_M, and α_m pertinent to each single function. Moreover, on the left of the symbols indicating the poles and zeros, the multiplicity is reported, which is equal to $\mu_c \neq 1$ only for critical pole-pairs of MCPER functions, and equal to 1 in all other cases.

9 Approximation Examples

This section illustrates the practical use of the tables presented (Part II) and of frequency transformations (section 7) through two examples.

As a first example, let us consider the case of a low-pass transfer function satisfying the following specifications:

$$\varrho_M' = 0.153 \text{ (i.e., } \alpha_M' = 0.10 \text{ dB)}, f_1 = 2000 \text{ Hz}, f_2 = 2150 \text{ Hz}, \alpha_m' = 59 \text{ dB}$$

The nominal stop-band edge Ω_s' of the NLP filter is computed according to the fourth row of Table I as

$$\Omega_s' = \frac{2150}{2000} = 1.075000 \tag{13}$$

Because this value does not appear exactly in the tables of Part II, we choose for Ω_s the greatest value from those available lower than 1.075000. As a consequence, we obtain $\Omega_s = 1.071145$, corresponding to $\Theta = 69°$. Similar considerations may be repeated for the choice of Q_M, that may be assumed equal to 0.15. No attempt has been made to optimize the choice of Q_M and Θ in order to maximize the margins in the pass band and stop band; this subject has been investigated elsewhere [42].

According to the above considerations, we restrict ourselves to considering solutions with $\Theta = 69°$, $Q_M = 0.15$, and $\alpha_m \geq \alpha_m'$. Three possibilities are left to the designer, corresponding to the three different values of the multiplicity μ_c (i.e., δ) of the critical pole-pair. Imposing the restriction of using functions of the lowest possible order, the following three NLP transfer functions may be obtained, using the parameters reported on pages II-381, II-429, II-468, respectively.

$$P_{10+0}^{(5)} = \frac{0.0008239\,[s^2 + (1.07540)^2]}{[s^2 + (0.48833/0.61045)\,s + (0.48833)^2]} \times$$

$$\frac{[s^2 + (1.11640)^2]\,[s^2 + (1.24841)^2]}{[s^2 + (0.74103/1.48079)\,s + (0.74103)^2]\,[s^2 + (0.90681/3.85736)\,s + (0.90681)^2]} \times$$

$$\frac{[s^2 + (1.68970)^2]\,[s^2 + (4.43440)^2]}{[s^2 + (0.98089/10.5077)\,s + (0.98089)^2]\,[s^2 + (1.00722/41.7745)\,s + (1.00722)^2]}$$

$$P_{9+2}^{(5)} = \frac{0.0050749\,[s^2 + (1.07401)^2]}{(s + 0.42662)\,[s^2 + (0.61701/0.92354)\,s + (0.61701)^2]} \times$$

$$\frac{[s^2 + (1.10081)^2]\,[s^2 + (1.18241)^2]}{[s^2 + (0.84355/2.34788)\,s + (0.84355)^2]\,[s^2 + (0.96183/6.08751)\,s + (0.96183)^2]} \times$$

$$\frac{[s^2 + (1.42661)^2]\,[s^2 + (2.39325)^2]}{[s^2 + (1.01974/20.4563)\,s + (1.01974)^2]^2}$$

$$P_{8+4}^{(6)} = \frac{0.0010400\,[s^2 + (1.07313)^2]\,[s^2 + (1.09060)^2]}{[s^2 + (0.51574/0.61177)\,s + (0.51574)^2]\,[s^2 + (0.77701/1.50471)\,s + (0.77701)^2]} \times$$

$$\frac{[s^2 + (1.13878)^2]\,[s^2 + (1.27194)^2]\,[s^2 + (1.71246)^2]\,[s^2 + (4.47117)^2]}{[s^2 + (0.94462/3.99631)\,s + (0.94462)^2]\,[s^2 + (1.03606/14.4931)\,s + (1.03606)^2]^3}$$

The expressions of the unnormalized transfer functions computed with the aid of Table I and satisfying the original specifications, are listed below.

$$\mathscr{P}_{10+0}^{(5)} = \frac{0.0008239\,[p^2 + (2\pi 2150.80)^2]}{[p^2 + (2\pi 976.65/0.61045)\,p + (2\pi 976.65)^2]} \times$$

$$\frac{[p^2 + (2\pi 2232.80)^2]\,[p^2 + (2\pi 2496.82)^2]}{[p^2 + (2\pi 1482.05/1.48079)\,p + (2\pi 1482.05)^2]\,[p^2 + (2\pi 1813.62/3.85736)\,p + (2\pi 1813.62)^2]} \times$$

$$\frac{[p^2 + (2\pi 3379.40)^2]\,[p^2 + (2\pi 8868.81)^2]}{[p^2 + (2\pi 1961.78/10.5077)\,p + (2\pi 1961.78)^2]\,[p^2 + (2\pi 2014.44/41.7745)\,p + (2\pi 2014.44)^2]}$$

$$\mathscr{P}_{9+2}^{(5)} = \frac{63.773\,[p^2 + (2\pi 2148.02)^2]}{(p + 853.24)\,[p^2 + (2\pi 1234.01/0.92354)\,p + (2\pi 1234.01)^2]} \times$$

$$\frac{[p^2 + (2\pi 2201.61)^2]\,[p^2 + (2\pi 2364.82)^2]}{[p^2 + (2\pi 1687.09/2.34788)\,p + (2\pi 1687.09)^2]} \times$$

$$\frac{[p^2 + (2\pi 2853.22)^2]}{[p^2 + (2\pi 1923.65/6.08751)\,p + (2\pi 1923.65)^2]} \times$$

$$\frac{[p^2 + (2\pi 4786.50)^2]}{[p^2 + (2\pi 2039.48/20.4563)p + (2\pi 2039.48)^2]^2}$$

$$\mathscr{P}^{(6)}_{8+4} = \frac{0.0010400\,[p^2 + (2\pi 2146.27)^2]}{[p^2 + (2\pi 1031.48/0.61177)p + (2\pi 1031.48)^2]} \times$$

$$\frac{[p^2 + (2\pi 2181.20)^2][p^2 + (2\pi 2277.57)^2]}{[p^2 + (2\pi 1554.02/1.50471)p + (2\pi 1554.02)^2]} \times$$

$$\frac{[p^2 + (2\pi 2543.88)^2]}{[p^2 + (2\pi 1889.24/3.99631)p + (2\pi 1889.24)^2]} \times$$

$$\frac{[p^2 + (2\pi 3424.92)^2][p^2 + (2\pi 8942.35)^2]}{[p^2 + (2\pi 2072.11/14.4931)p + (2\pi 2072.11)^2]^3}$$

Each of the above functions satisfies the assigned specification, with different values of the critical Q factor. In particular, the second solution reduces the highest Q factor by about 50% (i.e., from 41.7745 to 20.4563), without increasing the total number of second- or third-order blocks to be cascaded. A highest reduction (i.e., from 41.7745 to 14.4931) in the value of the critical Q factor is offered by the third solution, which, on the other hand, requires one more second-order block, as compared to the first (i.e., Cauer) solution. The choice from among these different possibilities is left to the designer on the basis of technological or economic considerations.

The second example concerns the design of a band-pass transfer function with $f_1 = 2215$ Hz, $f_2 = 2280$ Hz, $f_3 = 2520$ Hz, and $f_4 = 2585$ Hz. The other specifications are $\rho'_M = 0.267$ (i.e., $\alpha'_M = 0.3$ dB) and $\alpha'_m = 60$ dB. Since $f_1 f_4 \neq f_2 f_3$, the assigned specifications are not geometrically symmetrical (in fact, they are arithmetically symmetrical). By slightly increasing f_1, it is possible to reduce the specifications to a geometrically symmetrical case without violating the assigned specification. Hence, we assume as the lower stop-band edge $f_1 = f_2 f_3 / f_4 = 2222.67$ Hz. Thus, the formulas of Table I yield:

$$f_0 = 2397.00 \text{ Hz}, \qquad B = 0.100125, \qquad \Omega'_s = 1.509715 \tag{14}$$

Since the values of Ω'_s and ρ'_M do not appear exactly in the tables of Part II, we will look for functions with $\alpha_m \geq \alpha'_m$ and $\Omega_s = 1.494477$ (i.e., $\Theta = 42°$) and $\rho_M = 0.25$ (i.e., $\alpha_M = 0.280$). Also in this case no attempt will be made to optimize the choice of ρ_M and Θ in order to maximize the margins in the pass band and stop band.

As in the previous example, three possibilities are available to the designer, corresponding to the three different values of the multiplicity μ_c (i.e., δ) of the critical pole-pair. Imposing the restriction of using functions of minimum order, the following NLP transfer functions are obtained through the parameters reported on pages II-115, II-154, II-193, respectively.

$$P^{(3)}_{6+0} = \frac{0.0007873\,[s^2 + (1.53346)^2]}{[s^2 + (0.50573/0.65901)s + (0.50573)^2][s^2 + (0.84419/1.89757)s + (0.84419)^2]} \times$$

$$\frac{[s^2 + (1.97313)^2][s^2 + (5.05691)^2]}{[s^2 + (1.02194/7.82126)s + (1.02194)^2]}$$

$$P^{(3)}_{5+2} = \frac{0.004758\,[s^2 + (1.52024)^2]}{(s + 0.43067)[s^2 + (0.72113/1.07638)s + (0.72113)^2]} \times$$

$$\frac{[s^2 + (1.78456)^2][s^2 + (2.95565)^2]}{[s^2 + (1.08107/3.54737)s + (1.08107)^2]^2}$$

$$P^{(4)}_{4+4} = \frac{0.0008999\,[s^2 + (1.51296)^2][s^2 + (1.68625)^2]}{[s^2 + (0.63392/0.65813)s + (0.63392)^2]} \times$$

$$\frac{[s^2 + (2.28284)^2][s^2 + (5.95767)^2]}{[s^2 + (1.18610/2.30197)s + (1.18610)^2]^3}$$

The values of the parameters of the transformed functions (whose general expression is given in Table I) are shown in Table III. From this table it may be seen that the reduction in critical Q factor obtained through the use of MCPER functions is equal to 57% and 75% for the cases $\mu_c = 2$ and $\mu_c = 3$, respectively. The three solutions require 6, 7, and 8 second-order blocks, respectively. However, the increment in the number of blocks due to the use of MCPER solutions is abundantly counterbalanced by a less critical realization. For example, in the case of RC active filters, the third solution may be implemented using single-amplifier biquadratic blocks, whereas in the first case it is very likely that more complex blocks with multiple amplifiers would be required.

An approximate evaluation of pole and zero locations may be carried out using the diagram of Fig. 5. As an example, in the case of the real pole of the function $P_{5+2}^{(3)}$, we obtain

$$A = \Omega_r\ B = 0.43067 \times 0.100125 \cong 0.043 \tag{15}$$

Intersecting the curve $Q = 0.5$ with the curve $A = 0.043$ (interpolated), we obtain $q_r \cong 23$ and $f_{pr}/f_0 = 1$ (i.e., $f_{pr} = f_0$). In the case of the critical pole-pair of $P_{6+0}^{(3)}$, we obtain

$$A = \Omega_{pc}\ B = 1.02194 \times 0.100125 \cong 0.102 \tag{16}$$

Intersecting the curve $Q = 7.82$ with the curve $A = 0.102$ (interpolated), we obtain $q_{c+} = q_{c-} \cong 150$ and $f_{pc+}/f_0 = f_0/f_{pc-} \cong 1.055$ (i.e., $f_{pc+} \cong 2529$ Hz, f_{pc-} 2272 Hz). These results are sufficiently close to the accurate values reported in Table III.

Since $B \ll 1$, it is possible to compute the value of $q_{c+} = q_{c-}$ through the approximate expression given in Eq. (12a). For example, in the case of $P_{4+4}^{(4)}$, we obtain

$$q_{c+} \simeq \frac{2\,Q_c}{B\Omega_{pc}} + 38.77 \tag{17}$$

This agrees very closely with the value given in Table III.

10 Design Example

In the previous section, the use of the tables of Part II and of frequency transformations has been described without focusing on a particular technological implementation of the filter.

However, because in many applications the functions obtained in the approximation stage are realized by RC active circuits, this section will be devoted to a complete design example, from specification to circuit implementation, of an RC active filter in the cascade configuration.

To this end, let us consider the design of a geometrically symmetrical band-pass filter, satisfying the following specifications:

(a) center frequency: $f_0 = 1000$ Hz;

(b) pass-band edges: $f_2 = 951.2492$ and $f_3 = 1051.2492$ Hz, corresponding to a relative bandwidth $B = 10\%$, with $\sqrt{f_2 f_3} = f_0$;

(c) stop-band edges: $f_1 = 907.6210$ and $f_4 = 1101.7814$ Hz, with $\sqrt{f_1 f_4} = f_0$;

(d) pass-band attenuation lower than $\alpha'_M = 0.280$ dB;

(e) stop-band attenuation greater than $\alpha'_m = 40$ dB;

(f) pole Q factors lower than 20-30, in consideration of an implementation with single-amplifier blocks.

According to the cascade approach, the design considered consists of the three main steps illustrated in Fig. 6 [1, 3]. They will be considered separately in what follows, with reference to the particular specifications (a) through (f), previously reported.

Table III

Values of Parameters of the Three Transformed Transfer Functions, Solving the Second Example. Frequencies in Hz

$\varrho_M = 25\%$ $\Theta = 42°$	$\mathscr{P}^{(3)}_{6+0}$	$\mathscr{P}^{(3)}_{5+2}$	$\mathscr{P}^{(4)}_{4+4}$
u	0	1	0
μ_c	1	2	3
n_t	12	14	16
w_t	6	6	8
ν_0	0	1	0
k	0.7872732E-3	7.174878	0.8998515E-3
f_{pr} q_r	-	2396.998 23.19054	-
f_{p1+} q_{1+} f_{p1-} q_{1-}	2436.865 26.03283 2357.784 26.03283	2474.867 29.83052 2321.579 29.83052	2446.987 20.74207 2348.031 20.74207
f_{p2+} q_{2+} f_{p2-} q_{2-}	2496.718 44.93698 2301.261 44.93698	-	-
f_{pc+} q_{c+} f_{pc-} q_{c-}	2522.503 153.0740 2277.737 153.0740	2528.874 65.63889 2271.999 65.63889	2539.967 38.83241 2262.077 38.83241
f_{z1+} f_{z1-}	2588.066 2220.036	2586.359 2221.501	2585.420 2222.309
f_{z2+} f_{z2-}	2645.440 2171.888	2620.693 2192.397	2607.874 2203.174
f_{z3+} f_{z3-}	3079.448 1865.789	2777.775 2068.418	2686.542 2138.660
f_{z4+} f_{z4-}	-	-	3216.262 1786.421

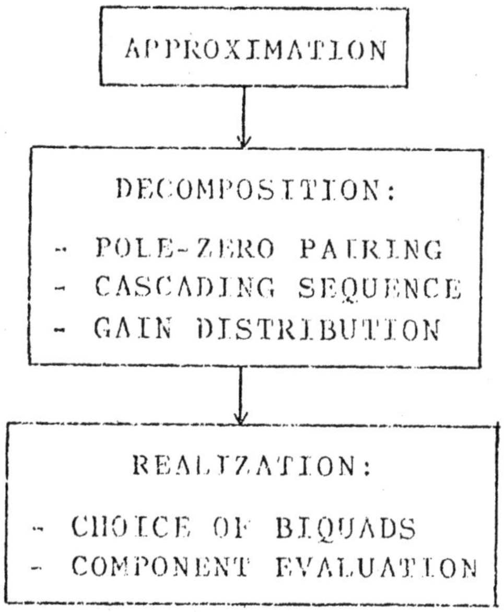

Fig. 6 The three main steps of the cascade design.

Step 1: Approximation

This step consists of finding a suitable rational function which satisfies the given constraints (a) through (f). This may be accomplished following the guidelines shown in the previous section and repeated here for the sake of clarity.

The parameters of the NLP function corresponding to (a) through (e) are

$$\Omega_s' = 1.941\,604, \; \alpha_M' = 0.280 \text{ dB}, \; \alpha_m' = 40 \text{ dB} \tag{18}$$

This specification is satisfied by functions belonging to sets $P_{4+0}^{(2)}$ and $P_{3+2}^{(2)}$.

From the frequency transformation chart of Fig. 5, it follows that the functions with simple poles (i.e., $P_{4+0}^{(2)}$) lead to $q_{c+} = q_{c-}$ as high as about 60 ($A \cong 0.1$ and $Q \cong 3.0$), whereas functions with a double pole-pair (i.e., $P_{3+2}^{(2)}$) lead to $q_{c+} = q_{c-}$ equal to about 25 ($A \cong 0.12$, $Q \cong 1.5$), which satisfies condition (f). Solutions with a triple pole-pair offer a further reduction in the Q factor at the expense of a further order increase.

Deciding that a solution of set $P_{3+2}^{(2)}$ is adequate to solve the present problem, specification (18) is met by a function with $\rho_M = 25\%$ and $\Theta = 31°$. Obviously, other solutions, such as $P_{3+2}^{(2)}$ with $\rho_M = 20\%$, $\Theta = 31°$ or $\rho_M = 20\%$, $\Theta = 32°$, are acceptable as well, and a selection between them may be done only in a specific real case on the basis of the knowledge of component tolerances [42]. However, this particular aspect of the problem is beyond the scope of this book and will not be considered further.

The values of the parameters of the transformed band-pass function $\mathcal{P}_{3+2}^{(2)}$, computed through the aid of Table I, are shown in Table IV.

Step 2: Decomposition

In this step, the transfer function previously found is decomposed into the product of second-order (biquadratic) functions. This problem may be divided into three parts (Fig. 6).

(i) *Pole-zero pairing*. To form biquadratic functions, each pole-pair of the transfer function must be assigned to one of the zero-pairs. If n_t (even) is the total order of the function, there are $(n_t/2)!$ different combinations of pole-zero pairs.

(ii) *Cascading sequence*. Having found the elementary biquadratic functions to within a multiplicative constant, the designer still has a degree of freedom in choosing the sequence in which the biquadratic blocks are to be cascaded. Again, there are $(n_t/2)!$ different sequences.

(iii) *Gain distribution*. The multiplicative constants k_j of the biquadratic functions are subject to the constraint

$$\prod_{j=1}^{n_t/2} k_j = k \tag{19}$$

Table IV
Values of Parameters Specifying the Approximating Function of the Design Example. Frequencies in Hz

$\mathscr{P}^{(2)}_{3+2}$ $\varrho_M = 25\%$ $\Theta = 31°$	
$u = 1$, $\mu_c = 2$, $n_t = 10$ $w_t = 4$, $\nu_0 = 1$ $k = 30.38913$	
f_{pr}	1000.000
q_r	13.69004
f_{pc+}	1061.350
q_{c+}	24.09684
f_{pc-}	942.1962
q_{c-}	24.09684
f_{z1+}	1106.203
f_{z1-}	903.9930
f_{z2+}	1164.613
f_{z2-}	858.6540

where k is the multiplicative constant of the overall transfer function. Hence, there are an infinite number of ways of distributing the single k_j among the different sections in the cascade.

There are many criteria according to which the decomposition problem may be solved. However, because, among all the filter performance factors, dynamic range is the most affected by the decomposition of the overall transfer function, it is generally chosen as the primary objective of this step [1].

A detailed discussion of the problem can be found elsewhere [1, 3, 6]. Here it is sufficient to say that pole-zero pairing, cascading sequence, and gain distribution should be chosen such that the magnitude of each intermediate transfer function $T_j(p)$ (Fig. 7) is as flat as possible and with peak value equal to that of the overall transfer function over the filter pass band.

The exact solution of this problem may be obtained by dynamic programming [44]. However, in many cases nearly optimum solutions can be found using the simple rules of thumb given by Lueder [44] and reported below:

First rule — Pole-zero pairing: the pole-pair with the highest Q is assigned to the zero-pair nearest to it. Then the pole-pair with the second highest Q is assigned to the zero pair as close to it as possible, and so on, until all poles and zeros are paired.

Second rule — Cascading sequence: the sections are ordered from input to output with decreasing "flatness" of their amplitude response. That is, the first section is that with the flattest amplitude response, followed by sections with progressively lesser flatness. The flatness is evaluated as the ratio between maximum (over the interval 0 — ∞) and minimum (over the pass band) of the amplitude response.

Third rule — Gain distribution: referring to Fig. 7, the multiplicative constants of the single biquadratic functions are given by

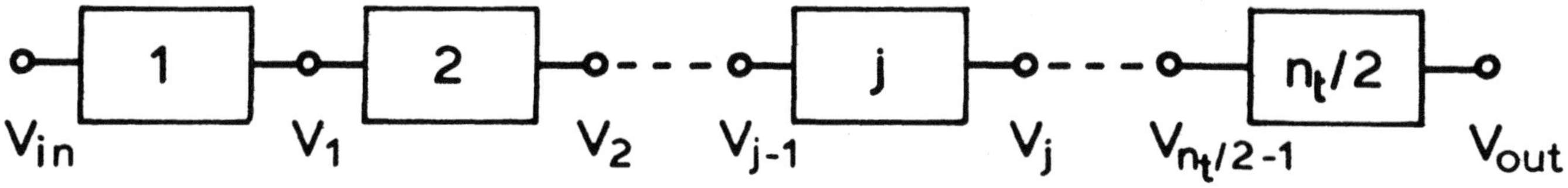

Fig. 7 Definition of intermediate transfer functions: $T_j = V_j / V_{in}, j = 1, 2, \ldots, n_t/2$.

$$k_j = \frac{Max|T_{j-1}(j\omega)|}{Max|T_j(j\omega)|}, \qquad j = 2, \ldots, n_t/2 \tag{20a}$$

$$k_1 = k / \prod_{j=2}^{n_t/2} k_j \tag{20b}$$

being the maxima evaluated over the whole frequency range 0 — ∞ and the function $T_j(p)$ considered as the ratio of monic polynomials (i.e., with the leading coefficient equal to unity).

According to the first rule, poles and zeros of transfer function $\mathcal{T}_{3+2}^{(2)}(p)$ are paired as shown in Fig. 8 in a qualitative way. Then the expressions of the five biquadratic functions are:

$$t_a(p) = \frac{p^2 + (2\pi\, 858.6540)^2}{p^2 + (2\pi\, 942.1962/24.09684)p + (2\pi\, 942.1962)^2} \tag{21a}$$

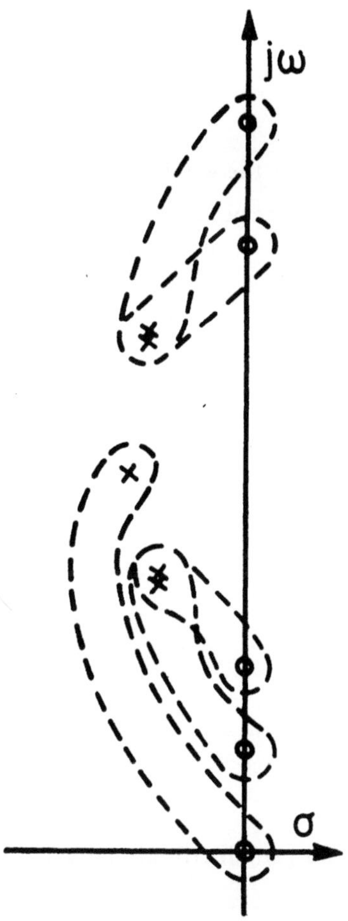

Fig. 8 Pole-zero pairing for the function $\mathcal{P}_{3+2}^{(2)}$ considered in the example.

$$t_b(p) = \frac{p^2 + (2\pi\, 903.9930)^2}{p^2 + (2\pi\, 942.1962/24.09684)p + (2\pi\, 942.1962)^2} \quad (21b)$$

$$t_c(p) = \frac{p}{p^2 + (2\pi\, 1000.000/13.69004)p + (2\pi\, 1000.000)^2} \quad (21c)$$

$$t_d(p) = \frac{p^2 + (2\pi \cdot 1106.203)^2}{p^2 + (2\pi\, 1061.350/24.09684)p + (2\pi\, 1061.350)^2} \quad (21d)$$

$$t_e(p) = \frac{p^2 + (2\pi\, 1164.613)^2}{p^2 + (2\pi\, 1061.350/24.09684)p + (2\pi\, 1061.350)^2} \quad (21e)$$

According to the second rule, the cascading sequence is found by computing the flatness of the five functions t_a, t_b, t_c, t_d, and t_e. Computing the amplitude of the considered functions, the following values are obtained:

$$\frac{Max|t_a(j\omega)|}{Min|t_a(j\omega)|} = 2.52 \quad (22a)$$

$$\frac{Max|t_b(j\omega)|}{Min|t_b(j\omega)|} = 1.65 \quad (22b)$$

$$\frac{Max|t_c(j\omega)|}{Min|t_c(j\omega)|} = 1.70 \quad (22c)$$

$$\frac{Max|t_d(j\omega)|}{Min|t_d(j\omega)|} = 1.65 \quad (22d)$$

$$\frac{Max|t_e(j\omega)|}{Min|t_e(j\omega)|} = 2.52 \quad (22e)$$

with maxima and minima computed, respectively, over the whole frequency range $0 — \infty$ and in the pass band.

In this case there are many equivalent sequences from the point of view of maximum dynamic range, insofar as there are two couples of functions with equal flatness, because of the geometrical symmetry of the frequency response. The following sequence is chosen:

$$t_d(p)\, t_b(p)\, t_c(p)\, t_a(p)\, t_e(p) \quad (23)$$

The multiplicative constants of the functions considered may now be computed according to the third rule:

$$k_2 = \frac{Max|T_1(j\omega)|}{Max|T_2(j\omega)|} = \frac{Max|t_d|}{Max|t_d t_b|} = 0.765 \quad (24a)$$

$$k_3 = \frac{Max|T_2(j\omega)|}{Max|T_3(j\omega)} = \frac{Max|t_d t_b|}{Max|t_d t_b t_c|} = 543 \quad (24b)$$

$$k_4 = \frac{Max|T_3(j\omega)|}{Max|T_4(j\omega)|} = \frac{Max|t_d t_b t_c|}{Max|t_d t_b t_c t_a|} = 0.347 \quad (24c)$$

$$k_5 = \frac{Max|T_4(j\omega)|}{Max|T_5(j\omega)|} = \frac{Max|t_d t_b t_c t_a|}{Max|t_d t_b t_c t_a t_e|} = 0.490 \quad (24d)$$

$$k_1 = \frac{k}{k_2 k_3 k_4 k_5} = 0.430 \quad (24e)$$

The final pole-zero pairing, cascading sequence, and gain distribution are summarized in Table V.

Step 3: Realization

After having solved the approximation problem and decomposed the overall transfer function, the designer must choose

Table V

Pole-Zero Pairing, Cascading Sequence, and Gain Distribution for the Function $\mathscr{P}^{(2)}_{3+2}$ Considered in the Example. Frequencies in Hz

BIQUADRATIC - FUNCTION PARAMETERS				
sequence	f_{zj}	f_{pj}	q_j	k_j
1	1106.203	1061.350	24.09684	0.430
2	903.9930	942.1962	24.09684	0.765
3	band-pass	1000.000	13.69004	543.0
4	858.6540	942.1962	24.09684	0.347
5	1164.613	1061.350	24.09684	0.490

suitable biquadratic blocks (biquads) for realizing each of the elementary biquadratic functions previously determined and, successively, compute the component values.

This problem has been treated extensively elsewhere [1, 3, 5, 6, 45] and, for the example considered, the design procedure suggested in [6] will be adopted.

The block chosen to implement the biquadratic functions of Table V is the popular STAR biquadratic block (STAR biquad) [46, 47]. Figures 9 and 10 show the band-pass, low-pass notch, and high-pass notch versions of STAR biquad, respectively.

The design formulas proposed in [6] and used for computing component values of the circuits of Figs. 9 and 10 are listed in Tables VI and VII, respectively. Note that in the band-pass case, the peak amplitude H is used as the design parameter; it is related to the multiplicative constant given in Table V by the relation

$$H = k_3 \frac{q_3}{2\pi f_{p3}} \tag{25}$$

In designing these circuits, there are many degrees of freedom which can be used by the designer for different purposes. In the design formulas of Tables VI and VII, the degrees of freedom have been used to allow the designer to fix preferred nominal values for the capacitors, and to minimize the active sensitivities by minimizing the gain-sensitivity product Γ [6].

Note that Γ depends on the parameter φ, which may be freely assigned by the designer. In the case of the band-pass section (Table VI), the value of φ which minimizes Γ can be determined explicitly, whereas in the case of the notch sections (Table VII) the value of φ for minimum Γ can be found only by numerical iterations. Furthermore, the choice of the capacitor

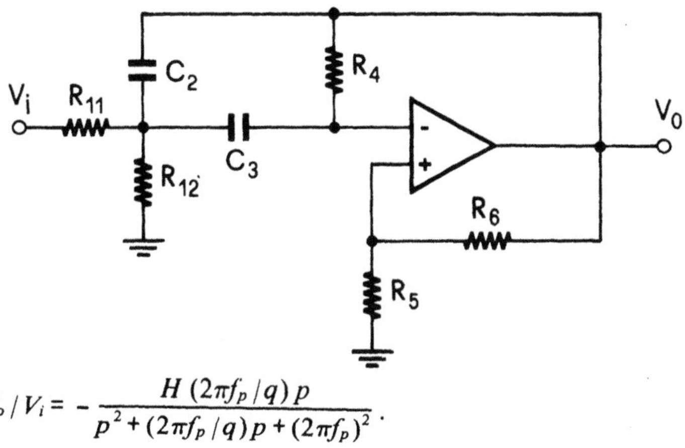

Fig. 9 Band-pass circuit $V_o/V_i = -\dfrac{H\,(2\pi f_p/q)\,p}{p^2 + (2\pi f_p/q)p + (2\pi f_p)^2}$.

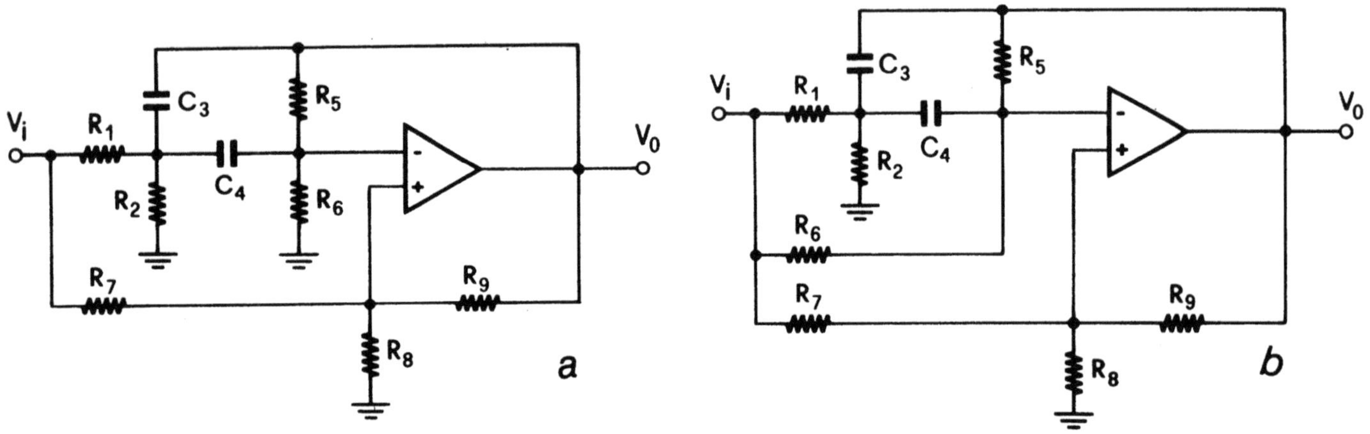

Fig. 10 (a) Low-pass notch circuit: $f_z > f_p$
(b) High-pass notch circuit: $f_z < f_p$

$$\frac{V_o}{V_i} = K \frac{p^2 + (2\pi f_z)^2}{p^2 + (2\pi f_p / q)p + (2\pi f_p)^2}$$

Table VI
Design Formulas for the Band-Pass Section of Fig. 9. *H* is the Peak Amplitude at the Pole Frequency

Design parameters: f_p, q, H ($H < H_M$)

Free parameters: C_2, C_3, R_6, φ

$$R_1 = (2\pi f_p \sqrt{\varphi C_2 C_3})^{-1}$$

$$R_4 = \varphi R_1$$

$$R_5 = R_6 \left[\frac{1}{\varphi} \left(1 + \frac{C_2}{C_3}\right) - \sqrt{\frac{C_2}{\varphi C_3}} \frac{1}{q} \right]$$

$$H_M = q\left(1 + \frac{R_5}{R_6}\right)\sqrt{\frac{\varphi C_3}{C_2}}$$

$$R_{11} = \frac{H_M}{H} R_1$$

$$R_{12} = \frac{H_M}{H_M - H} R_1 \qquad (H < H_M)$$

Gain-Sensitivity Product $\Gamma = (1 + R_5/R_6)\, H_M$

For minimum Γ: $\varphi = \dfrac{C_2/C_3}{4q^2}\left[\sqrt{1 + 12\, q^2\left(1 + \dfrac{C_3}{C_2}\right)} - 1\right]^2$

Table VII

Design Formulas for the Notch Sections of Fig. 10

Design parameters: f_z, f_p, q, K $\quad (K < K_M,\ K < 1)$

Free parameters: C_3, C_4, R_9, φ

$$\gamma = \frac{C_3}{\varphi}\frac{\pi f_p}{q}\left[\sqrt{1 + 4q^2\varphi\left(1 + \frac{C_4}{C_3}\right)} - 1\right]$$

$$K_M = \frac{1 + \varphi}{1 + 4\pi^2\left(1 + \frac{C_4}{C_3}\right) f_z^2 C_3^2/\gamma^2}$$

$$R_1 = \frac{K_M}{\gamma K} \qquad R_2 = \frac{1}{\gamma(1 - K/K_M)} \qquad (K < K_M)$$

Low-pass notch: $$R_6 = \frac{\gamma(1 + \varphi)}{4\pi^2 C_3 C_4 (f_z^2 - f_p^2)}$$

High-pass notch: $$R_6 = \frac{\gamma(1 + \varphi)(1 - 1/K)}{4\pi^2 C_3 C_4 (f_z^2 - f_p^2)}$$

$$R_5 = \left[4\pi^2 C_3 C_4 f_p^2/\gamma + \frac{\varphi}{R_6}\right]^{-1}; \quad R_7 = \frac{\varphi R_9}{K}; \quad R_8 = \frac{\varphi R_9}{1 - K}$$

Gain-Sensitivity Product:

$$\Gamma = \frac{1 + \varphi}{2}\left\{\left|1 - \frac{q}{2\pi f_p}\left[\frac{\gamma}{C_3} + \frac{(C_3 + C_4)(R_5 + R_6)}{C_3 C_4 R_5 R_6}\right]\right| + q\left|1 - \frac{\gamma(R_5 + R_6)}{4\pi^2 f_p^2 C_3 C_4 R_5 R_6}\right|\right\}$$

values also influences Γ and, hence, some attempts are necessary before finding the final optimum solution. In this situation, the availability of a computer program for a repeated evaluation of component values and of the gain-sensitivity product [6] may shorten design time considerably.

In the present case the capacitor values have been chosen over the following set:

1.0, 1.5, 2.2, 3.3, 4.7, 6.8, 10.0, 15.0, 22.0, *nF*

and only the circuits with minimum resistor spread, with resistors lower than 300 $k\Omega$, and with minimum Γ, have been accepted for the final solution shown in Fig. 11. The resulting value of the design parameter φ is equal to 0.30, 0.27, 0.30, 0.28 for the notch blocks 1, 2, 4, and 5 described by the biquadratic functions $t_a(p)$, $t_b(p)$, $t_d(p)$, and $t_e(p)$, respectively.

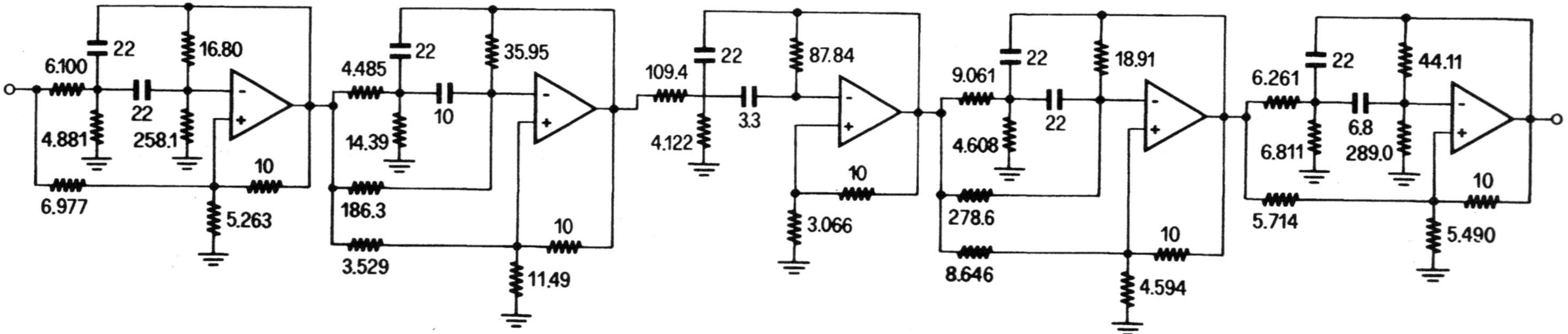

Fig. 11 Final realization of the band-pass RC active filter, with transfer function $\mathscr{P}_{3+2}^{(2)}$. Resistances in $k\Omega$, capacitances in nF.

11 Extension to Coupled-Section Structures

Several methods for the design of selective active filters attempt to provide internal interactions between the different biquads of the filter by coupling them through feedback paths [3, 5]. These methods retain in part the modularity of the cascade structure while achieving the desirable property of lower sensitivity. Thus, they are used to reduce the sensitivity of rather selective filters, in particular the band-pass ones, to a level allowable by the chosen technology.

Among the previous methods, partial coupling of the two most critical blocks has been proposed by Tow [48]. Although this method is applicable for each type of filter [49-51], hereafter our attention will be focused on the very frequent case of geometrically-symmetrical ones. In this case, all the biquads (except, possibly, one for an odd-order NLP prototype) realizing the cascade structure are two-by-two geometrically symmetrical, in the sense that each biquad corresponds to another biquad with symmetrical values of zero and pole frequencies, and coincident values of pole-Q, as may be observed in the examples of sections 9 and 10.

Tow has proposed the partial coupling of critical pairs of symmetrical biquads, originating a fourth-order section, hereafter referred to as the biquartic section (Fig. 12). The overall structure so derived, referred to as cascade biquartic (CBQ) structure, is characterized by the cascade of biquads (with non-critical values of pole-Q) and of biquartic sections.

The CBQ structure is a good compromise between the two conflicting requirements of low-sensitivity, characteristic of all coupled-section structures, and of modularity, characteristic of the cascade structure.

The initial motivation of the CBQ structure is similar to that of the MCPER functions. Both try to reduce the effects of the high sensitivity related to the critical pole-pairs. The first acts at the topological stage, the second at the approximation stage. Then, if the two techniques are simultaneously adopted, both advantages can be obtained [52, 53].

On the other hand, when a multiple-critical-pole transfer function is adopted, the CBQ structure compares favorably with the other coupled-section configuration, except for the leapfrog structure, which offers superior sensitivity performances [53]. However, the latter configuration requires infinite-Q blocks, which are less convenient and, for RC active technology, biquads with two or three amplifiers.

Since the two biquads embedded in a biquartic section are geometrically symmetrical, each biquartic section can be derived from a low-pass second-order prototype through the low-pass-to-band-pass frequency transformation. Let the low-pass biquadratic transfer function $t_j(s)$ of the prototype be

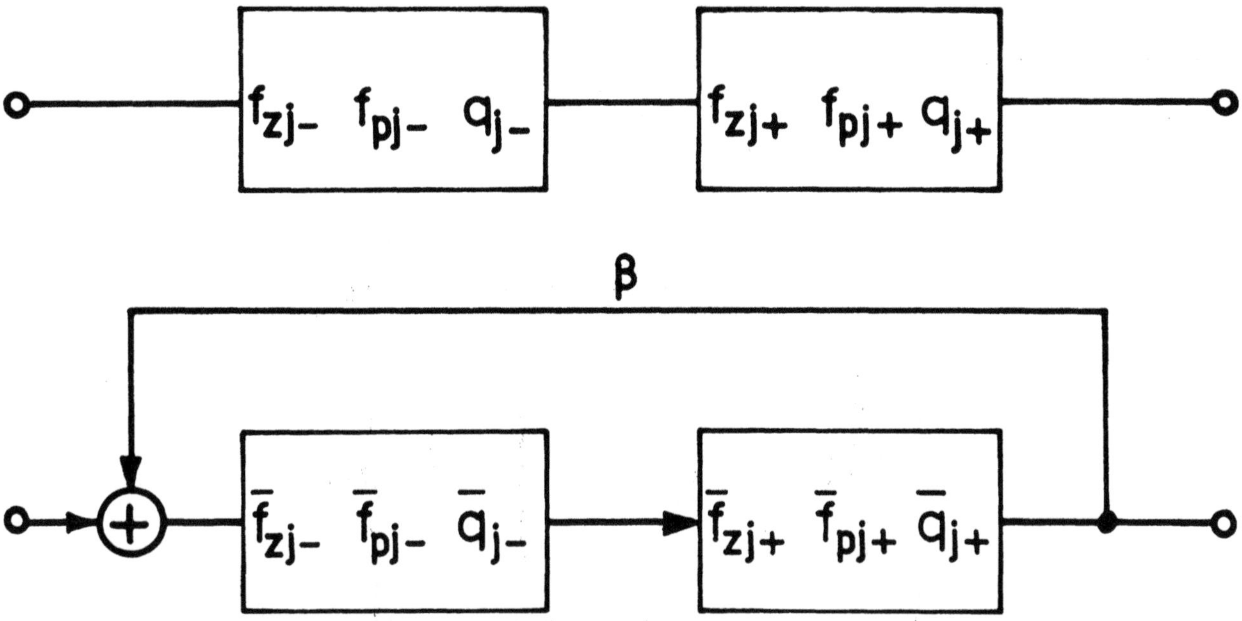

Fig. 12 (a) Two geometrically symmetrical biquads connected in cascade $(f_{zj-} f_{zj+} = f_0^2, f_{pj-} f_{pj+} = f_0^2, q_{j-} = q_{j+})$.
(b) Equivalent biquartic section $(\overline{f_{zj-}}\, \overline{f_{zj+}} = f_0^2, \overline{f_{pj-}}\, \overline{f_{pj+}} = f_0^2, \overline{q_{j-}} = \overline{q_{j+}})$.

$$t_j(s) = k_j \frac{s^2 + \Omega_{zj}^2}{s^2 + s\,\Omega_{pj}/Qj + \Omega_{pj}^2} \tag{26}$$

Even if this transfer function may be directly implemented by a biquad, in consideration of a CBQ configuration, it is realized by inserting a feedback path around another biquad (Fig. 13), which is described by a different transfer function:

$$\overline{t}_j(s) = \overline{k}_j \frac{s^2 + \Omega_{zj}^2}{s^2 + s\,\overline{\Omega}_{pj}/\overline{Q}_j + \overline{\Omega}_{pj}} \tag{27}$$

Obviously, $\overline{t}_j(s)$ is related to the original transfer function $t_j(s)$ so that the two realizations are equivalent.

By inspection of Fig. 13, the following relation is established:

$$\overline{t}_j(s) = t_j(s) / [1 + \beta_j\, t_j(s)] \tag{28}$$

Thus, the formulas

$$\overline{k}_j = k_j / (1 + \beta_j k_j) \tag{29a}$$

$$\overline{\Omega}_{zj} = \Omega_{zj} \tag{29b}$$

$$\overline{\Omega}_{pj} = \Omega_{pj} \sqrt{[1 + \beta_j k_j (\Omega_{zj}/\Omega_{pj})^2] / [1 + \beta_j k_j]} \tag{29c}$$

$$\overline{Q}_j = Q_j \sqrt{[1 + \beta_j k_j (\Omega_{zj}/\Omega_{pj})^2] [1 + \beta_j k_j]} \tag{29d}$$

are derived for the parameters of $\overline{t}_j(s)$.

Now, the biquartic section is derived from the low-pass prototype of Fig. 13, by applying the low-pass-to-band-pass frequency transformation to the internal biquadratic block with transfer function $\overline{t}_j(s)$. Thus, this block is transformed in the cascade of two geometrically symmetrical biquads. The feedback factor β_j is a free parameter to be evaluated so that the sensitivity of the biquartic section is minimized.

It has been shown in Reference [48] that the best solution, from the sensitivity viewpoint, corresponds to value $\hat{\beta}_j$ of β_j such that $\overline{Q}_j = 0.5$. In other words, the minimum sensitivity is reached when the internal block (i.e., transfer function $t_j(s)$) exhibits two coincident real poles. From Eq. (29d) we have

$$\hat{\beta}_j = \Lambda_1 [(\sqrt{\Lambda_2} + \Lambda_3)\, k_j] \tag{30a}$$

where

$$\Lambda_1 = (0.5/Q_j - 1)(0.5/Q_j + 1)(\Omega_{pj}/\Omega_{zj})^2 \tag{30b}$$

$$\Lambda_2 = [0.5(\Omega_{pj}/\Omega_{zj} - 1)(\Omega_{pj}/\Omega_{zj} + 1)]^2 + [0.5\,\Omega_{pj}/(\Omega_{zj}\,Q_j)]^2 \tag{30c}$$

$$\Lambda_3 = 0.5(\Omega_{pj}/\Omega_{zj})^2 + 0.5 \tag{30d}$$

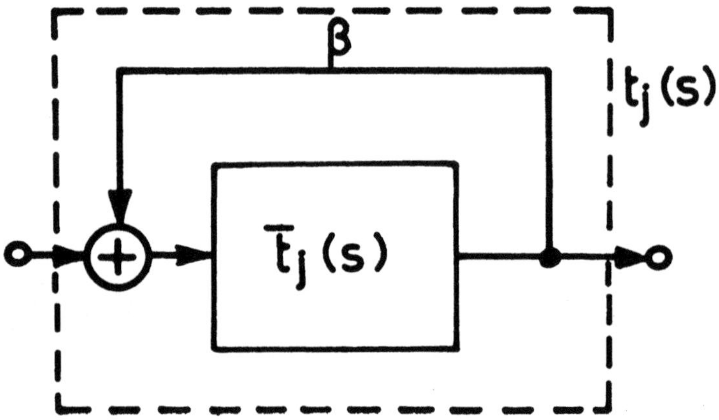

Fig. 13 A biquadratic section of the low-pass prototype with the insertion of a feedback path.

Since the assumption $\beta_j = \hat{\beta}_j$ forces the poles of $\overline{t_j}(s)$ to be real and coincident, the applied frequency transformation produces two pole-pairs with frequencies coincident with the center frequency of the filter and equal Q factors.

For example, two biquartic sections are introduced in the symmetrical band-pass filter designed in the previous section. In the cascade structure it exhibits two pairs of symmetrical biquads, all with pole-Q equal to 24.09684 besides one band-pass biquad with pole-Q = 13.69004 and pole frequency centered at 1000 Hz [see Eqs. (21a-e)].

Then the CBQ structure can be implemented by transforming the four highest pole-Q blocks into two biquartic sections and leaving unaltered the biquad with lowest Q. Consequently, only the design of the two biquartic sections will be considered in the sequel.

Table VIII shows the parameters of the two biquadratic functions of the low-pass prototype producing the two biquartic sections in the band-pass filter. They are derived directly from Part II of this book (see $P_{3+2}^{(2)}$, $\rho_M = 25\%$, $\Theta = 31°$). The same parameters, modified by the insertion of the feedback path with $\beta_j = \hat{\beta}_j$, are reported too. Finally, the parameters of the internal biquads of the two biquartic sections are computed with the aid of Table I and are shown in the last rows of Table VIII.

The multiplicative constants of the different sections are not reported in Table VIII because their design depends on the optimization of dynamic range. However, they can be evaluated as indicated in Reference [54].

It is worth noting that the pole-Q factors of the internal biquads of the biquartic sections are not much reduced with respect to those of the cascade structure (compare Table IV of Section 10 and Table VIII). Nevertheless, sensitivities of biquartic sections are generally reduced to a greater extent, since the action of the inserted feedback path is also involved in determining the overall sensitivity of these sections.

Table VIII
Parameters of the Two Biquadratic Sections of the NLP Prototype and of the Two Corresponding Biquartic Sections of the Band-Pass Filter

LOW-PASS PROTOTYPE			
2ND - ORDER SECTION 1		2ND ORDER SECTION 2	
WITHOUT FEEDBACK	WITH FEEDBACK	WITHOUT FEEDBACK	WITH FEEDBACK
$\Omega_{z1} = 2.022102$ $\Omega_{p1} = 1.261738$ $Q_1 = 1.517505$	$\bar{\Omega}_{z1} = 2.022102$ $\bar{\Omega}_{p1} = 0.617335$ $\bar{Q}_1 = 0.500000$ $K_1\hat{\beta}_1 = -0.326577$	$\Omega_{z2} = 3.059594$ $\Omega_{p2} = 1.261738$ $Q_2 = 1.517505$	$\bar{\Omega}_{z2} = 3.059594$ $\bar{\Omega}_{p2} = 0.488164$ $\bar{Q}_2 = 0.500000$ $K_2\hat{\beta}_2 = -0.148384$

BAND-PASS FILTER			
BIQUARTIC SECTION 1		BIQUARTIC SECTION 2	
FEEDBACK FACTOR $K_1\hat{\beta}_1 = -0.326577$		FEEDBACK FACTOR $K_2\hat{\beta}_2 = -0.148384$	
TWO INTERNAL BIQUADS		TWO INTERNAL BIQUADS	
$\bar{f}_{z1-} = 903.9930$ $\bar{f}_{p1-} = 1000.000$ $\bar{q}_{1-} = 16.19865$	$\bar{f}_{z1+} = 1106.203$ $\bar{f}_{p1+} = 1000.000$ $\bar{q}_{1+} = 16.19865$	$\bar{f}_{z2-} = 858.6540$ $\bar{f}_{p2-} = 1000.000$ $\bar{q}_{2-} = 20.48493$	$\bar{f}_{z2+} = 1164.613$ $\bar{f}_{p2+} = 1000.000$ $\bar{q}_{2+} = 20.48493$

12 References

1. G.S. Moschytz, *Linear Integrated Networks: Fundamentals* and *Linear Integrated Networks: Design*, New York, Van Nostrand Reinhold, 1974 and 1975.
2. A. Peled and B. Liu, *Digital Signal Processing: Theory, Design, and Implementation*, New York, Wiley, 1976.
3. A.S. Sedra and P.O. Brackett, *Filter Theory and Design: Active and Passive*, Portland, OR, Matrix Publishers, 1978.
4. A. Antoniou, *Digital Filter Analysis and Design*, New York, McGraw-Hill, 1979.
5. M.S. Ghausi and K.R. Laker, *Modern Filter Design: Active RC and Switched Capacitor*, Englewood Cliffs, NJ, Prentice-Hall, 1981.
6. G.S. Moschytz and P. Horn, *Active Filter Design Handbook*, New York, Wiley, 1981.
7. P. Allen and E. Sinencio, *Switched Capacitor Circuits*, New York, Van Nostrand Reinhold, 1983.
8. L.T. Bruton, *RC-Active Circuits: Theory and Design*, Englewood Cliffs, NJ, Prentice-Hall, 1980.
9. S.C. Lee, "Sensitivity minimization in active RC integrated circuit design," in *Proc. IV Allerton Conf. on Circuit and System Theory,* Monticello, IL, Oct. 5-7 1966, pp. 269-281.
10. J. Gorsky-Popiel and A.J. Drew, "Reduction of network sensitivity through the use of higher-order approximating functions," *Electron. Lett.*, vol. 3, 1967, pp. 365-366 and p.439 (errata).
11. Z. Kominek, "Low-pass approximation convenient for circuits with low Q," in *Proc. 4th Colloq. Microwave Communication*, vol. 2, Apr. 1970.
12. A.Ye. Znamenskiy, "Polynomial approximation of the characteristics of filters with localization of the transmission poles," *Telecommun.* vol. 25, no. 6, June 1971, pp. 46-49.
13. G.L. Khazanov, "A method for synthesizing active RC networks with transfer function poles having a given Q factor," *Telecommun.*, vol. 23. no. 12, Dec. 1969, pp. 20-25.
14. A. Levy and E. Zeheb, "Stable high-Q active networks," *Int. J. of Circuit Theory and Applications*, vol. 4, Apr. 1976, pp. 137-149.
15. J. Tow and R.N. Gadenz, "A note on high-Q active networks," *Int. J. of Circuit Theory and Applications*, vol. 6, Apr. 1978, pp. 147-152.
16. I.M. Filanovsky and K.A. Stromsmoe, "On the realization of highly selective networks by low Q-factor stages," *IEEE Trans. on Circuits and Systems*, vol. CAS-25, Aug. 1978, pp. 580-585.
17. A. Premoli, "The MUCROMAF polynomials: an approach to the maximally flat approximation of RC-active-filters with low sensitivity," *IEEE Trans. on Circuit Theory*, vol. CT-20, Jan. 1973, pp. 77-80.
18. A. Premoli and F. Germano, "New approach to the approximation of RC-active-filter transmission functions with low Q factors," *Electron. Lett.*, vol. 7, 1971, pp. 456-458.
19. A. Premoli, "A new class of equal-ripple filtering functions with low Q-factors: the MUCROER polynomials," *IEEE Trans. on Circuits and Systems*, vol. CAS-21, Sept. 1974. pp. 609-613.
20. M. Biey and A. Premoli, "Equal-ripple rational functions with multiple critical pole-pair," *Electron. Lett.*, vol. 12, May 1976, pp. 263-264.
21. M. Biey and A. Premoli, "A new class of multiple-critical-pole equal-ripple rational (MCPER) functions for the design of cascaded RC-active filters with low Q-factors," *IEEE Trans. on Circuits and Systems*, vol. CAS-25, Jan. 1978. pp. 19-27.
22. E. Christian and E. Eisenmann, *Filter Design Tables and Graphs*, New York, Wiley, 1966.
23. A.I. Zverev, *Handbook of Filter Synthesis*, New York, Wiley, 1967.

24. R.W. Daniels, *Approximation Methods for Electronic Filter Design,* New York, McGraw-Hill, 1974.

25. R. Saal, *Handbook of Filter Design*, Berlin, AEG-Telefunken, 1979.

26. K. Massad and R. Yarlagadda, "Modified Butterworth functions with low Q factor," *Proc. IEE,* vol. 122, Feb. 1975, pp. 135-136.

27. G. Fritzsche and H.J. Jentschel, "*Ein neuer Vorschlag furdie gleichmassige Approximation bei Selektivschaltungen mit minimaler Polgute,*" *Nachrichtentechnik Elektronik,* vol. 24, July 1974, pp. 202-205.

28. G. Fritzsche and H.J. Jentschel, "*PN-Daten und Koeffizienten von Ubertragungsfaktoren mit minimaler Polgute* (C_M-Approximation)," *Nachrichtentechnik Elektronik*, vol. 27, May 1977, pp. 208-210.

29. M. Biey and A. Premoli, "A review of low pole-Q approximation methods for active filter design," *Circuits, Systems, and Signal Processing*, vol. 3, no. 1, 1984, pp. 79-104.

30. A. Premoli, "Design of maximally-flat RC-active filters with low Q-factors: tables and graphs of the MUCROM-AF polynomials," *Alta Frequenza,* vol. 44, Nov. 1975, pp. 684-689.

31. A. Premoli, "Tables and graphs of the MUCROER polynomials for the realization of equal-ripple RC-active filters with low sensitivity," Note, *Recensioni, Notizie*, vol. 24, suppl. of no. 5-6, Sept. -Dec. 1975, pp. 399-518.

32. W. Cauer, *Synthesis of Linear Communication Networks*, New York, McGraw-Hill, 1958.

33. E. Glowatzki, "*Katalogisierte Filter,*" *Nach. Tech. Z.*, vol. 9, Nov. 1956, pp. 508-513.

34. S.K. Mitra and D. Huey, "Q of cascaded identical resonators," *Proc. IEEE,* vol. 60, Nov. 1972, pp. 1463-1464.

35. H.M. Wagner, *Principles of Operations Research*, Englewood Cliffs, NJ, Prentice-Hall, 1969, Ch. 4.

36. A. Premoli, "Multiple-pole-and-zero approximation of general attenuation specifications," *IEEE Trans. on Circuits and Systems*, vol. CAS-27, Dec. 1980. pp. 1218-1224.

37. J.K. Skwirzynski, *Design Theory and Data for Electrical Filters*, London, Van Nostrand, 1965, Ch. 5.

38. D.S. Humpherys, *The Analysis, Design and Synthesis of Electrical Filters*, Englewood Cliffs, N.J, Prentice-Hall, 1970.

39. G.C. Temes and J.W. La Patra, *Circuit Synthesis and Design*, New York, McGraw-Hill, 1977, Ch. 12.

40. I. Zdunek, "Generation of filter functions from a given model," *Proc. IEE,* vol. 110, Feb. 1963, pp. 282-294.

41. G.C. Temes and M. Gyi, "Design of filters with arbitrary pass-band and Chebyschev stopband attenuation," *IEEE Internat. Conv. Rec.*, vol. 15, Pt. 5, 1967, pp. 2-12.

42. J. Neirynck and Ph. Van Bastelaer, "The minimization of LC filters sensitivity," *Proc. of IEEE Int. Symp. on Circuits and Systems* San Francisco, CA, Apr. 1974, pp. 590-594.

43. A. Premoli, "A frequency transformation chart for RC-active band-pass filters," *Alta Frequenza*, vol. 41, June 1972, pp. 468-469.

44. E. Lueder, "Optimization of the dynamic range and the noise distance of RC-active filters by dynamic programming," *Int. J. of Circuit Theory and Applications*, vol. 3, Dec. 1975, pp. 365-370.

45. A.B. Williams, *Active Filter Design*, Dedham, MA, Artech House, 1975.

46. T. Deliyannis, "High-Q factor circuit with reduced sensitivity," *Electron. Lett.*, vol. 4, no. 26, Dec. 1968, pp. 577-579.

47. J.J. Friend, C.A. Harris, and D. Hilberman, "STAR: an active biquadratic filter section," *IEEE Trans. on Circuits and Systems*, vol. CAS-22, Feb. 1975, pp. 115-121.

48. J. Tow, "Some results on two-section generalized FLF active filters," *IEEE Trans. on Circuits and Systems*, vol. CAS-25, Apr. 1978, pp. 181-184.

49. A.N. Gonuleren, "Multiloop feedback unsymmetrical active filters using quads," *Int. J. of Circuit Theory and Applications*, vol. 10, Jan. 1982, pp. 1-8.

50. M. Biey, "Design of lowpass two-section generalized FLF active filters," *Electron. Lett.*, vol. 19, Aug. 1983, pp. 639-640.

51. S. Fotopoulos and T. Deliyannis, "Active RC realization of high-order bandpass filter functions by cascading biquartic sections," *Int. J. of Circuit Theory and Applications*, vol. 12, July 1984, pp. 223-238.

52. M. Biey and A. Premoli, "The use of MCPER functions in the design of the two-section generalized FLF active filters," *Proc. of IEEE Int. Symposium on Circuits and Systems*, Houston, TX, Apr. 28-30, 1980, pp. 91-94.

53. N. Mijat and G.S. Moschytz, "Multiple-critical-pole coupled active filters," *Int. J. of Circuit Theory and Applications*, vol. 12, July 1984, pp. 249-268.

54. D.J. Perry, "Scaling transformation of multiple-feedback filters," *IEE Int. J. of Electronic Circuits and Systems*, vol. 128, Part G, 1981, pp. 176-179.

Part II: Tables

$P_{2+0}^{(1)}$

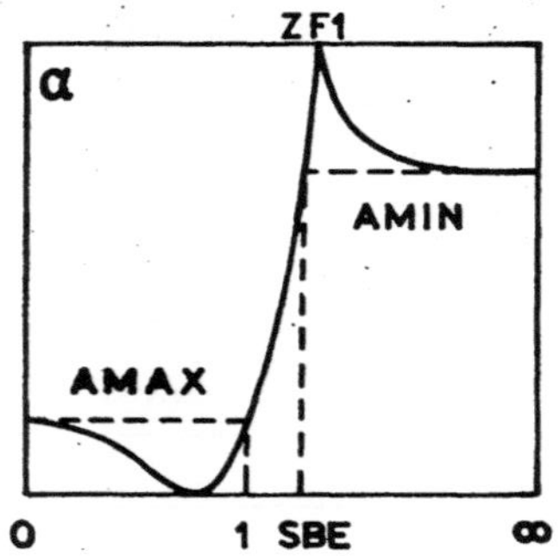

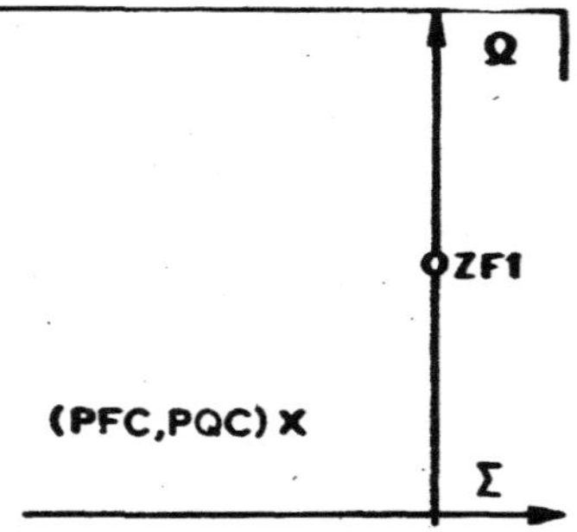

$P_{2+0}^{(1)}$	TH = 1	SBE = 57.29868850						CAUER FUNCTIONS	
	RC = 02	RC = 05	RC = 07	RC = 10	RC = 15	RC = 20	RC = 25	RC = 35	RC = 50
AMAX DB	.0017375255	.0108709564	.0213327376	.0436480540	.0988323393	.1772876696	.2802872360	.5675287486	1.249387366
AMIN DB	48.38806452	56.35594523	59.28896279	62.40931468	65.98632275	68.56355232	70.60475192	73.81455395	77.59445166
MU K	.3807122D-2	.1521258D-2	.1085305D-2	.7577645D-3	.5019771D-3	.3730975D-3	.2949595D-3	.2038320D-3	.1319099D-3
1 PFC	5.000172273	3.162396245	2.672713401	2.236152802	1.825811265	1.581198982	1.414267383	1.195274109	1.000038074
1 PQC	.7156768727	.7260570594	.7336627234	.7456683842	.7671889860	.7907500952	.8166520767	.8771868621	1.000114253
1 ZF1	81.02949691	81.02949691	81.02949691	81.02949691	81.02949691	81.02949691	81.02949691	81.02949691	81.02949691

$P_{2+0}^{(1)}$	TH = 2	SBE = 28.65370835						CAUER FUNCTIONS	
	RC = 02	RC = 05	RC = 07	RC = 10	RC = 15	RC = 20	RC = 25	RC = 35	RC = 50
AMAX DB	.0017375255	.0108709564	.0213327376	.0436480540	.0988323393	.1772376696	.2802872360	.5675237486	1.249387366
AMIN DB	36.34648606	44.31357315	47.24651667	50.36682922	53.94381629	56.52103851	58.56223471	61.77203378	65.55192992
MU K	.1522915D-1	.6085851D-2	.4341844D-2	.3031507D-2	.2008210D-2	.1492616D-2	.1180017D-2	.8154518D-3	.5277199D-3
1 PFC	5.000471662	3.162730078	2.673006935	2.236403455	1.826018129	1.581378801	1.414428494	1.195410477	1.000152258
1 PQC	.7198992287	.7278079414	.7349487114	.7466079638	.7678621895	.7912929065	.8171191210	.8775737508	1.000457263
1 ZF1	40.51629119	40.51629119	40.51629119	40.51629119	40.51629119	40.51629119	40.51629119	40.51629119	40.51629119

$P_{2+0}^{(1)}$	TH = 3	SBE = 19.10732261						CAUER FUNCTIONS	
	RC = 02	RC = 05	RC = 07	RC = 10	RC = 15	RC = 20	RC = 25	RC = 35	RC = 50
AMAX DB	.0017375255	.0108709564	.0213327376	.0436480540	.0988323393	.1772876696	.2802872360	.5675287486	1.249387366
AMIN DB	29.30472289	37.26837121	40.20099370	43.32113570	46.89803172	49.47522208	51.51640353	54.72618975	58.50607907
MU K	.3425815D-1	.1369561D-1	.9771254D-2	.6822495D-2	.4519583D-2	.3359223D-2	.2655705D-2	.1835230D-2	.1187671D-2
1 PFC	5.000245788	3.163213317	2.673464753	2.236808455	1.826358382	1.581676371	1.414695850	1.195637318	1.000342440
1 PQC	.7271004757	.7307543553	.7371072112	.7481819627	.7689882426	.7922001665	.8178993874	.8782197684	1.001029789
1 ZF1	27.01257507	27.01257507	27.01257507	27.01257507	27.01257507	27.01257507	27.01257507	27.01257507	27.01257507

(1) P 2+0	TH = 4 SBE = 14.33558703							CAUER FUNCTIONS	
	RC = 02	RC = 05	RC = 07	RC = 10	RC = 15	RC = 20	RC = 25	RC = 35	RC = 50
AMAX DB	.0017375255	.0108709564	.0213327376	.0436480540	.0988323393	.1772876696	.2802872360	.5675287486	1.249387366
AMIN DB	24.31509467	32.26949495	35.20125271	38.32093518	41.89758589	44.47469038	46.51583208	49.72558370	53.50545463
MU I K	.6084785D-1	.2435147D-1	.1737550D-1	.1213258D-1	.8037495D-2	.5974004D-2	.4722896D-2	.3263780D-2	.2112162D-2
1 I PFC	4.998410380	3.163736015	2.674039628	2.237348618	1.826825224	1.582088488	1.415067697	1.195953969	1.000608428
1 I PQC	.7375299641	.7349390472	.7401611493	.7504024953	.7705732438	.7934757522	.8189956751	.8791267259	1.001833099
1 I ZF1	20.26123148	20.26123148	20.26123148	20.26123148	20.26123148	20.26123148	20.26123148	20.26123148	20.26123148

(1) P 2+0	TH = 5 SBE = 11.47371325							CAUER FUNCTIONS	
	RC = 02	RC = 05	RC = 07	RC = 10	RC = 15	RC = 20	RC = 25	RC = 35	RC = 50
AMAX DB	.0017375255	.0108709564	.0213327376	.0436480540	.0988323393	.1772876696	.2802872360	.5675287486	1.249387366
AMIN DB	20.45787376	28.39283900	31.32277293	34.44143575	38.01761963	40.59454191	42.63559971	45.84527826	49.62511038
MU I K	.9486507D-1	.3805030D-1	.2715572D-1	.1396382D-1	.1256374D-1	.9338409D-2	.7382781D-2	.5101949D-2	.3301752D-2
1 I PFC	4.993465283	3.164143996	2.674665253	2.237996990	1.827409095	1.582610652	1.415541573	1.196359500	1.000949957
1 I PQC	.7515423875	.7404224967	.7441429096	.7532866492	.7726257847	.7951251225	.8204119236	.8802971705	1.002868975
1 I ZF1	16.21083706	16.21083706	16.21083706	16.21083706	16.21083706	16.21083706	16.21083706	16.21083706	16.21083706

(1) P 2+0	TH = 6 SBE = 9.566772234							CAUER FUNCTIONS	
	RC = 02	RC = 05	RC = 07	RC = 10	RC = 15	RC = 20	RC = 25	RC = 35	RC = 50
AMAX DB	.0017375255	.0108709564	.0213327376	.0436480540	.0988323393	.1772876696	.2802872360	.5675287486	1.249387366
AMIN DB	17.32761666	25.22749306	28.15411146	31.27105996	34.84625031	37.42284350	39.46374851	42.67329402	46.45305544
MU I K	.1360251D 0	.5478042D-1	.3911060D-1	.2731788D-1	.1810037D-1	.1345420D-1	.1063684D-1	.7350812D-2	.4757154D-2
1 I PFC	4.983538987	3.164239050	2.675256145	2.238718771	1.828097645	1.583237044	1.416114294	1.196852708	1.001366681
1 I PQC	.7696006469	.7472834896	.7490946057	.7568566110	.7751570048	.7971553490	.8221532338	.8817343980	1.004139724
1 I ZF1	13.51091739	13.51091739	13.51091739	13.51091739	13.51091739	13.51091739	13.51091739	13.51091739	13.51091739

(1) P 2+0	TH = 7 SBE = 8.205509048							CAUER FUNCTIONS	
	RC = 02	RC = 05	RC = 07	RC = 10	RC = 15	RC = 20	RC = 25	RC = 35	RC = 50
AMAX DB	.0017375255	.0108709564	.0213327376	.0436480540	.0988323393	.1772876696	.2802872360	.5675287486	1.249387366
AMIN DB	14.71217683	22.55500975	25.47617515	28.59021739	32.16385380	34.73990267	36.78055567	39.98938169	43.76952650
MU I K	.1838193D 0	.7451600D-1	.5323426D-1	.3719539D-1	.2464945D-1	.1832335D-1	.1448679D-1	.1001166D-1	.6479234D-2
1 I PFC	4.966480571	3.163779835	2.675707668	2.239471243	1.828875689	1.583960504	1.416781941	1.197432110	1.001858172
1 I PQC	.7922755133	.7556197596	.7550684091	.7611398220	.7781806597	.7995751561	.8242258947	.8834424688	1.005648181
1 I ZF1	11.58269774	11.58269774	11.58269774	11.58269774	11.58269774	11.58269774	11.58269774	11.58269774	11.58269774

(1) P 2+0	TH = 8 SBE = 7.185296534								CAUER FUNCTIONS
	RC = 02	RC = 05	RC = 07	RC = 10	RC = 15	RC = 20	RC = 25	RC = 35	RC = 50
AMAX DB	.0017375255	.0108709564	.0213327376	.0436480540	.0988323393	.1772876696	.2802872360	.5675287486	1.249387366
AMIN DB	12.48862438	20.24575716	23.15858109	26.26816679	29.83941725	32.41462985	34.45489557	37.66388425	41.44334981
MU I K	.2374481D 0	.9721027D-1	.6951379D-1	.4859501D-1	.3221285D-1	.2394796D-1	.1893456D-1	.1308597D-1	.8469007D-2
1 I PFC	4.940015507	3.162483940	2.675396274	2.240203726	1.829725164	1.584772510	1.417539846	1.198095939	1.002423923
1 I PQC	.8202392063	.7655486259	.7621269133	.7661691628	.7817132047	.8023949697	.8266374163	.8854262272	1.007397732
1 I ZF1	10.13679080	10.13679080	10.13679080	10.13679080	10.13679080	10.13679080	10.13679080	10.13679080	10.13679080

(1) P 2+0	TH = 9 SBE = 6.392453221								CAUER FUNCTIONS
	RC = 02	RC = 05	RC = 07	RC = 10	RC = 15	RC = 20	RC = 25	RC = 35	RC = 50
AMAX DB	.0017375255	.0108709564	.0213327376	.0436480540	.0988323393	.1772876696	.2802872360	.5675287486	1.249387366
AMIN DB	10.58057939	18.21688173	21.11763334	24.22074591	27.78852406	30.36251852	32.40221991	35.61071695	39.38992123
MU I K	.2957815D 0	.1227880D 0	.8792621D-1	.6151240D-1	.4079189D-1	.3033012D-1	.2398220D-1	.1657541D-1	.1072763D-1
1 I PFC	4.901970092	3.160031665	2.675680101	2.240857591	1.830625092	1.585663146	1.418382574	1.198842136	1.003063338
1 I PQC	.8542500827	.7772075265	.7703435166	.7719831601	.7857738929	.8056269750	.8293965692	.8876913246	1.009392318
1 I ZF1	9.012425833	9.012425833	9.012425833	9.012425833	9.012425833	9.012425833	9.012425833	9.012425833	9.012425833

(1) P 2+0	TH = 10 SBE = 5.758770483								CAUER FUNCTIONS
	RC = 02	RC = 05	RC = 07	RC = 10	RC = 15	RC = 20	RC = 25	RC = 35	RC = 50
AMAX DB	.0017375255	.0108709564	.0213327376	.0436480540	.0988323393	.1772876696	.2802872360	.5675287486	1.249387366
AMIN DB	8.937669181	16.41255427	19.29659007	22.39069390	25.95362634	28.52591865	30.56483109	33.77264061	37.55147942
MU I K	.3573687D 0	.1511375D 0	.1084353D 0	.7593908D-1	.5038702D-1	.3747176D-1	.2963183D-1	.2048179D-1	.1325641D-1
1 I PFC	4.850546836	3.156072063	2.674900060	2.241366351	1.831551538	1.586621078	1.419303904	1.199668339	1.003775734
1 I PQC	.8951267447	.7907543325	.7798027804	.7786262113	.7903848853	.8092851832	.8325134306	.8902442472	1.011636460
1 I ZF1	8.113140441	8.113140441	8.113140441	8.113140441	8.113140441	8.113140441	8.113140441	8.113140441	8.113140441

(1) P 2+0	TH = 11 SBE = 5.240843064								CAUER FUNCTIONS
	RC = 02	RC = 05	RC = 07	RC = 10	RC = 15	RC = 20	RC = 25	RC = 35	RC = 50
AMAX DB	.0017375255	.0108709564	.0213327376	.0436480540	.0988323393	.1772876696	.2802872360	.5675287486	1.249387366
AMIN DB	7.524379462	14.79365601	17.65537990	20.73737551	24.29377064	26.86376250	28.90160799	32.10848732	35.88683154
MU I K	.4205145D 0	.1821030D 0	.1309878D 0	.9186101D-1	.6099742D-1	.4537450D-1	.3588555D-1	.2480708D-1	.1605678D-1
1 I PFC	4.784604917	3.150231750	2.673381581	2.241655881	1.832477594	1.587633521	1.420296811	1.200571878	1.004560335
1 I PQC	.9437122115	.8063673111	.7906007487	.7861488177	.7955713749	.8133855075	.8359994361	.8930923468	1.014135278
1 I ZF1	7.377549475	7.377549475	7.377549475	7.377549475	7.377549475	7.377549475	7.377549475	7.377549475	7.377549475

(1) P 2+0	TH = 12	SBE = 4.809734345						CAUER FUNCTIONS	
	RC = 02	RC = 05	RC = 07	RC = 10	RC = 15	RC = 20	RC = 25	RC = 35	RC = 50
AMAX DB	.0017375255	.0108709564	.0213327376	.0436480540	.0988323393	.1772876696	.2802872360	.5675287486	1.249387366
AMIN DB	6.313510688	13.33194014	16.16480342	19.23099368	22.77881302	25.34578001	27.38222128	30.58787575	34.36556846
MU K	.4834198D 0	.2154780D 0	.1555105D 0	.1092573D 0	.7262052D-1	.5403946D-1	.4274536D-1	.2955332D-1	.1913029D-1
1 PFC	4.703878630	3.142126800	2.670937154	2.241644792	1.833373372	1.588686218	1.421353442	1.201549761	1.005416271
1 PQC	1.000831951	.8242446004	.8028451698	.7946078196	.8013617214	.8179458478	.8398674392	.8962438769	1.016894510
1 ZF1	6.764729520	6.764729520	6.764729520	6.764729520	6.764729520	6.764729520	6.764729520	6.764729520	6.764729520

(1) P 2+0	TH = 13	SBE = 4.445411483						CAUER FUNCTIONS	
	RC = 02	RC = 05	RC = 07	RC = 10	RC = 15	RC = 20	RC = 25	RC = 35	RC = 50
AMAX DB	.0017375255	.0108709564	.0213327376	.0436480540	.0988323393	.1772876696	.2802872360	.5675287486	1.249387366
AMIN DB	5.282260872	12.00650232	14.80305329	17.84912448	21.38595751	23.94903755	25.98367230	29.18774990	32.96460343
MU K	.5443609D 0	.2510007D 0	.1819061D 0	.1280984D 0	.8525152D-1	.6346702D-1	.5021302D-1	.3472262D-1	.2247863D-1
1 PFC	4.609068494	3.131377753	2.667369814	2.241245007	1.834206028	1.589763416	1.422465098	1.202598664	1.006342569
1 PQC	1.067252367	.8446030719	.8166555863	.8040666211	.8077875962	.8229861838	.8441317772	.8997080330	1.019920542
1 ZF1	6.246349001	6.246349001	6.246349001	6.246349001	6.246349001	6.246349001	6.246349001	6.246349001	6.246349001

(1) P 2+0	TH = 14	SBE = 4.133565494						CAUER FUNCTIONS	
	RC = 02	RC = 05	RC = 07	RC = 10	RC = 15	RC = 20	RC = 25	RC = 35	RC = 50
AMAX DB		.0108709564	.0213327376	.0436480540	.0988323393	.1772876696	.2802872360	.5675287486	1.249387366
AMIN DB		10.80152437	13.55351531	16.57453686	20.09758055	22.65576305	24.68811821	27.89020426	31.66599727
MU K		.2883525D 0	.2100507D 0	.1483451D 0	.9888285D-1	.7365663D-1	.5829000D-1	.4031715D-1	.2610358D-1
1 PFC		3.117627299	2.662477623	2.240362582	1.834939821	1.590847854	1.423622216	1.203714919	1.007338153
1 PQC		.8676764860	.8321632482	.8145953943	.8148841356	.8285286778	.8438083445	.9034949971	1.023220438
1 ZF1		5.802171091	5.802171091	5.802171091	5.802171091	5.802171091	5.802171091	5.802171091	5.802171091

(1) P 2+0	TH = 15	SBE = 3.863703305						CAUER FUNCTIONS	
	RC = 02	RC = 05	RC = 07	RC = 10	RC = 15	RC = 20	RC = 25	RC = 35	RC = 50
AMAX DB		.0108709564	.0213327376	.0436480540	.0988323393	.1772876696	.2802872360	.5675287486	1.249387366
AMIN DB		9.704760356	12.40331618	15.39376286	18.89980740	21.45192323	23.48144940	26.69106119	30.45553560
MU K		.3271613D 0	.2397917D 0	.1699464D 0	.1135036D 0	.8460652D-1	.6697728D-1	.4633903D-1	.3000704D-1
1 PFC		3.100559704	2.656059181	2.238898799	1.835536213	1.591920753	1.424814346	1.204894501	1.008401836
1 PQC		.8937129065	.8495108058	.8262712486	.8226901010	.8345977855	.8539146727	.9076159880	1.026801968
1 ZF1		5.417355468	5.417355468	5.417355468	5.417355468	5.417355468	5.417355468	5.417355468	5.417355468

(1) P 2+0	TH = 16	SBE = 3.627955279						CAUER FUNCTIONS	
	RC = 02	RC = 05	RC = 07	RC = 10	RC = 15	RC = 20	RC = 25	RC = 35	RC = 50
AMAX DB		.0108709564	.0213327376	.0436480540	.0988323393	.1772876696	.2802872360	.5675287486	1.249387366
AMIN DB		8.706478776	11.34232596	14.29612601	17.78154816	20.32626084	22.35232693	25.54890840	29.32176629
MU I K		.3670084D 0	.2709466D 0	.1928385D 0	.1290989D 0	.9631345D-1	.7627525D-1	.5279035D-1	.3419099D-1
1 I PFC		3.079920473	2.647920048	2.236751562	1.835954029	1.592961828	1.426030143	1.206133019	1.009532315
1 I PQC		.9229714178	.8688517512	.8391783540	.8312480429	.8412203753	.8594700188	.9120333166	1.030673649
1 I ZF1		5.080771904	5.080771904	5.080771904	5.080771904	5.080771904	5.080771904	5.080771904	5.080771904

(1) P 2+0	TH = 17	SBE = 3.420303620						CAUER FUNCTIONS	
	RC = 02	RC = 05	RC = 07	RC = 10	RC = 15	RC = 20	RC = 25	RC = 35	RC = 50
AMAX DB		.0108709564	.0213327376	.0436480540	.0988323393	.1772876696	.2802872360	.5675287486	1.249387366
AMIN DB		7.798700819	10.36244875	13.27306009	16.73382575	19.26962409	21.29151231	24.48442876	28.25532924
MU I K		.4074412D 0	.3033036D 0	.2169437D 0	.1456494D 0	.1087724D 0	.8618355D-1	.5967309D-1	.3865748D-1
1 I PFC		3.055534339	2.637879878	2.233817117	1.836149675	1.593949306	1.427257351	1.207425699	1.010728167
1 I PQC		.9557182852	.8903495886	.8534080034	.8406044670	.8484258552	.8654954609	.9169104466	1.034844781
1 I ZF1		4.783909061	4.783909061	4.783909061	4.783909061	4.783909061	4.783909061	4.783909061	4.783909061

(1) P 2+0	TH = 18	SBE = 3.236067977						CAUER FUNCTIONS	
	RC = 02	RC = 05	RC = 07	RC = 10	RC = 15	RC = 20	RC = 25	RC = 35	RC = 50
AMAX DB		.0108709564	.0213327376	.0436480540	.0988323393	.1772876696	.2802872360	.5675287486	1.249387366
AMIN DB		6.974643027	9.457102367	12.31761766	15.74929515	18.27448784	20.29138907	23.47992204	27.24847805
MU I K		.4479895D 0	.3366238D 0	.2421693D 0	.1631305D 0	.1219763D 0	.9670091D-1	.6698906D-1	.4340863D-1
1 I PFC		3.027319505	2.625779903	2.229992080	1.836077424	1.594859966	1.428482798	1.208767372	1.011987839
1 I PQC		.9922227946	.9141767363	.8690586027	.8508099994	.8562463087	.8720140017	.9221120620	1.039325496
1 I ZF1		4.520147021	4.520147021	4.520147021	4.520147021	4.520147021	4.520147021	4.520147021	4.520147021

(1) P 2+0	TH = 19	SBE = 3.071553487						CAUER FUNCTIONS	
	RC = 02	RC = 05	RC = 07	RC = 10	RC = 15	RC = 20	RC = 25	RC = 35	RC = 50
AMAX DB		.0108709564	.0213327376	.0436480540	.0988323393	.1772876696	.2802872360	.5675287486	1.249387366
AMIN DB		6.228310542	8.620826981	11.42410622	14.82189160	17.33460405	19.34561383	22.52895609	26.29473121
MU I K		.4881850D 0	.3706454D 0	.2684075D 0	.1815120D 0	.1359158D 0	.1078250D 0	.7473985D-1	.4844662D-1
1 I PFC		2.995296255	2.611490301	2.225175753	1.835689787	1.595669206	1.429692400	1.210152464	1.013309646
1 I PQC		1.032753083	.9405131840	.8862355803	.8619195475	.8647166365	.8790506803	.9277041398	1.044126805
1 I ZF1		4.284259547	4.284259547	4.284259547	4.284259547	4.284259547	4.284259547	4.284259547	4.284259547

$P_{2+2}^{(2)}$

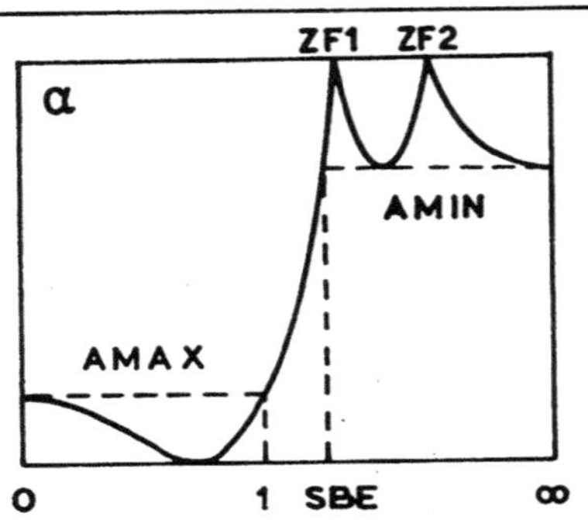

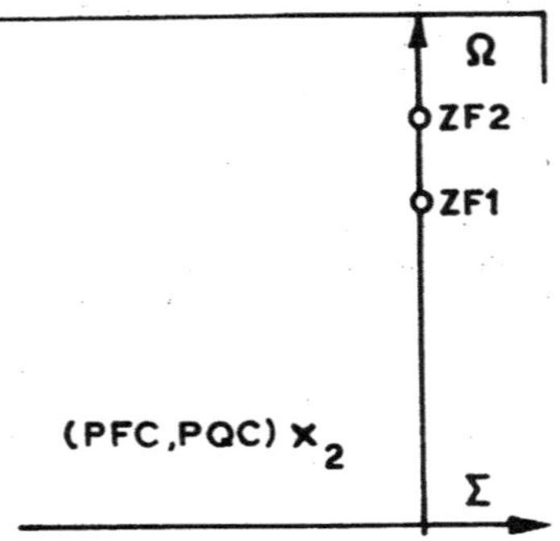

(2) P 2+2	TH = 1	SBE = 57.29868850						MCPER FUNCTIONS	
	RC = 02	RC = 05	RC = 07	RC = 10	RC = 15	RC = 20	RC = 25	RC = 35	RC = 50
AMAX DB	.0017375255	.0108709564	.0213327376	.0436480540	.0988323393	.1772876696	.2802872360	.5675287486	1.249387366
AMIN DB	96.77513432	112.7064332	118.5672424	124.7967729	131.9230865	137.0380073	141.0682289	147.3407087	154.5417647
MU I K	.1449584D-4	.2315679D-5	.1179337D-5	.5756538D-6	.2534228D-6	.1406370D-6	.8842775D-7	.4295014D-7	.1874614D-7
2 I PFC	5.946091765	3.760155992	3.177434760	2.657581047	2.168190494	1.875606076	1.675140970	1.410054460	1.168815394
2 I PQC	.7141131636	.7207563305	.7258819715	.7339737444	.7483031075	.7636775039	.7802081799	.8174670952	.8882000607
1 I ZF1	62.01905494	62.01897672	62.01896943	62.01896555	62.01896348	62.01896276	62.01896242	62.01896213	62.01896197
1 I ZF2	149.7176461	149.7187467	149.7188494	149.7189039	149.7189331	149.7189432	149.7189480	149.7189521	149.7189543

(2) P 2+2	TH = 2	SBE = 28.65370835						MCPER FUNCTIONS	
	RC = 02	RC = 05	RC = 07	RC = 10	RC = 15	RC = 20	RC = 25	RC = 35	RC = 50
AMAX DB	.0017375255	.0108709564	.0213327376	.0436480540	.0988323393	.1772876696	.2802872360	.5675287486	1.249387366
AMIN DB	72.69008865	88.62138703	94.48219627	100.7117268	107.8380404	112.9529611	116.9831828	123.2556626	130.4567185
MU I K	.2320041D-3	.3706215D-4	.1887514D-4	.9213267D-5	.4056000D-5	.2250878D-5	.1415275D-5	.6874116D-6	.3000296D-6
2 I PFC	5.946447679	3.760552905	3.177783708	2.657878924	2.168436138	1.875819366	1.675331791	1.410215325	1.168948843
2 I PQC	.7200673024	.7231841291	.7276482598	.7352465362	.7491944803	.7643798419	.7807983962	.8179323338	.8885790542
1 I ZF1	31.01391174	31.01328594	31.01322756	31.01319654	31.01317999	31.01317419	31.01317151	31.01316918	31.01316793
1 I ZF2	74.84580012	74.85460521	74.85542679	74.85586326	74.85609624	74.85617779	74.85621554	74.85624842	74.85626592

(2) P 2+2	TH = 3	SBE = 19.10732261						MCPER FUNCTIONS	
	RC = 02	RC = 05	RC = 07	RC = 10	RC = 15	RC = 20	RC = 25	RC = 35	RC = 50
AMAX DB	.0017375255	.0108709564	.0213327376	.0436480540	.0988323393	.1772876696	.2802872360	.5675287486	1.249387366
AMIN DB	58.59838700	74.52967415	80.39048317	86.62001362	93.74632724	98.86124795	102.8914696	109.1639494	116.3650054
MU I K	.1175116D-2	.1877225D-3	.9560395D-4	.4666586D-4	.2054394D-4	.1140086D-4	.7168471D-5	.3481790D-5	.1519672D-5
2 I PFC	5.946176963	3.761127365	3.178327899	2.658360187	2.168840145	1.876172296	1.675648423	1.410482395	1.169171096
2 I PQC	.7303216049	.7272848096	.7306205953	.7373824718	.7506871093	.7655546547	.7817850185	.8187094583	.8892117568
1 I ZF1	20.68205615	20.67994301	20.67974595	20.67964127	20.67958540	20.67956584	20.67955679	20.67954890	20.67954471
1 I ZF2	49.86527108	49.89497658	49.89774991	49.89922337	49.90000991	49.90023522	49.90041266	49.90052368	49.90058276

(2) P 2+2 — TH = 4 — SBE = 14.33558703 — MCPER FUNCTIONS

			RC = 02	RC = 05	RC = 07	RC = 10	RC = 15	RC = 20	RC = 25	RC = 35	RC = 50
		AMAX DB	.0017375255	.0108709564	.0213327376	.0436480540	.0988323393	.1772876696	.2802872360	.5675287486	1.249387366
		AMIN DB	48.59721522	64.52839712	70.38920410	76.61873400	83.74504748	88.85996817	92.89018982	99.16266963	106.3637256
MU		K	.3716544D-2	.5937179D-3	.3023708D-3	.1475922D-3	.6497520D-4	.3605800D-4	.2267204D-4	.1101201D-4	.4806332D-5
2		PFC	5.943975560	3.761748327	3.179011068	2.659001968	2.169394390	1.876661029	1.676088756	1.410856358	1.169481907
2		PQC	.7454020762	.7331419075	.7348425321	.7404037621	.7527916029	.7672083630	.7831724666	.8198010717	.8900997445
1		ZF1	15.51992694	15.51491171	15.51444446	15.51419628	15.51406381	15.51401745	15.51399599	15.51397730	15.51396735
1		ZF2	37.33802137	37.40832982	37.41490397	37.41839752	37.42026258	37.42091543	37.42121765	37.42148090	37.42162101

(2) P 2+2 — TH = 5 — SBE = 11.47371325 — MCPER FUNCTIONS

			RC = 02	RC = 05	RC = 07	RC = 10	RC = 15	RC = 20	RC = 25	RC = 35	RC = 50
		AMAX DB	.0017375255	.0108709564	.0213327376	.0436480540	.0988323393	.1772876696	.2802872360	.5675287486	1.249387366
		AMIN DB	40.83705794	56.76765859	62.62845428	68.85798117	75.98429390	81.09921445	85.12943607	91.40191587	98.60297179
MU		K	.9081281D-2	.1450832D-2	.7388857D-3	.3606625D-3	.1587762D-3	.8811286D-4	.5540235D-4	.2690941D-4	.1174496D-4
2		PFC	5.937993275	3.762231274	3.179754050	2.659772105	2.170087449	1.877280163	1.676649820	1.411334577	1.169880930
2		PQC	.7661088811	.7408765172	.7403766237	.7443420854	.7555230101	.7693500502	.7849669829	.8212108432	.8912452357
1		ZF1	12.42727757	12.41745717	12.41654400	12.41605909	12.41580032	12.41570976	12.41566784	12.41563132	12.41561189
1		ZF2	29.77051960	29.90735449	29.92019123	29.92701566	29.93065975	29.93193547	29.93252605	29.93304050	29.93331431

(2) P 2+2 — TH = 6 — SBE = 9.566772234 — MCPER FUNCTIONS

			RC = 02	RC = 05	RC = 07	RC = 10	RC = 15	RC = 20	RC = 25	RC = 35	RC = 50
		AMAX DB	.0017375255	.0108709564	.0213327376	.0436480540	.0988323393	.1772876696	.2802872360	.5675287486	1.249387366
		AMIN DB	34.49519916	50.42349044	56.28424129	62.51375620	69.64006590	74.75498596	78.78520743	85.05768717	92.25874307
MU		K	.1884691D-1	.3011795D-2	.1533868D-2	.7487075D-3	.3296072D-3	.1829156D-3	.1150111D-3	.5586189D-4	.2438164D-4
2		PFC	5.925792398	3.762336843	3.180454315	2.660628978	2.170904544	1.878022717	1.677327777	1.411916075	1.170367719
2		PQC	.7935886916	.7506517189	.7473060825	.7492391934	.7589010220	.7719915589	.7871766884	.8229435342	.8926511039
1		ZF1	10.37110732	10.35406223	10.35248259	10.35164417	10.35119685	10.35104032	10.35096787	10.35090476	10.35087118
1		ZF2	24.66063626	24.89545495	24.91761827	24.92941013	24.93570932	24.93791495	24.93893610	24.93982565	24.94029912

(2) P 2+2 — TH = 7 — SBE = 8.205509048 — MCPER FUNCTIONS

			RC = 02	RC = 05	RC = 07	RC = 10	RC = 15	RC = 20	RC = 25	RC = 35	RC = 50
		AMAX DB	.0017375255	.0108709564	.0213327376	.0436480540	.0988323393	.1772876696	.2802872360	.5675287486	1.249387366
		AMIN DB	29.13546641	45.05642905	50.91703733	57.14651417	64.27281426	69.38773270	73.41795375	79.69043326	86.89148911
MU		K	.3493226D-1	.5586998D-2	.2845431D-2	.1388911D-2	.6114477D-3	.3393231D-3	.2133547D-3	.1036283D-3	.4522989D-4
2		PFC	5.904279349	3.761768611	3.180985265	2.661521279	2.171827472	1.878880091	1.678117897	1.412599036	1.170941721
2		PQC	.8294368987	.7626786702	.7557370294	.7551477221	.7629502399	.7751476178	.7898116556	.8250050319	.8943208932
1		ZF1	8.909097181	8.881844692	8.879332055	8.877999436	8.877288755	8.877040112	8.876925031	8.876824798	8.876771456
1		ZF2	20.93357100	21.30197811	21.33711176	21.35582854	21.36593381	21.36933822	21.37096086	21.37237450	21.37312695

MCPER FUNCTIONS

(2) P 2+2	TH = 8 SBE = 7.185296534								
	RC = 02	RC = 05	RC = 07	RC = 10	RC = 15	RC = 20	RC = 25	RC = 35	RC = 50
AMAX DB	.0017375255	.0108709564	.0213327376	.0436480540	.0988323393	.1772876696	.2802872360	.5675287486	1.249387366
AMIN DB	24.50306183	40.40429518	46.26451782	52.49389169	59.62016576	64.73507986	68.76529973	75.03777865	82.23883435
MU K	.5954522D-1	.9545205D-2	.4861543D-2	.2373042D-2	.1044700D-2	.5797570D-3	.3645315D-3	.1770562D-3	.7727843D-4
2 PFC	5.869624703	3.760169753	3.181195218	2.662387697	2.172834511	1.879842028	1.679014542	1.413381294	1.171602278
2 PQC	.8758263524	.7772246402	.7658014215	.7621322327	.7677005154	.7788360018	.7928840001	.8274023910	.8962588384
1 ZF1	7.820476236	7.779412480	7.775652468	7.773660553	7.772598933	7.772227618	7.772055782	7.771906125	7.771826484
1 ZF2	18.05180256	18.59104104	18.64332744	18.67123795	18.68617354	18.69140741	18.69383128	18.69594318	18.69706739

MCPER FUNCTIONS

(2) P 2+2	TH = 9 SBE = 6.392453221								
	RC = 02	RC = 05	RC = 07	RC = 10	RC = 15	RC = 20	RC = 25	RC = 35	RC = 50
AMAX DB	.0017375255	.0108709564	.0213327376	.0436480540	.0988323393	.1772876696	.2802872360	.5675287486	1.249387366
AMIN DB	20.44370248	36.29824598	42.15754528	48.38667253	55.51288428	60.62778797	64.65800503	70.93048253	78.13153786
MU K	.9501997D-1	.1531397D-1	.7800505D-2	.3807732D-2	.1676316D-2	.9302734D-3	.5849244D-3	.2841030D-3	.1240004D-3
2 PFC	5.817243544	3.757118335	3.180906008	2.663156496	2.173900302	1.880896559	1.680011134	1.414260320	1.172348620
2 PQC	.9356351398	.7946232715	.7776607578	.7702705133	.7731873688	.7830777281	.7964079932	.8301438846	.8984698887
1 ZF1	6.982950224	6.923816256	6.918443895	6.915602348	6.914089258	6.913560263	6.913315496	6.913102343	6.912988921
1 ZF2	15.71971508	16.46450163	16.53859313	16.57826107	16.59952124	16.60697678	16.61043048	16.61344015	16.61504244

MCPER FUNCTIONS

(2) P 2+2	TH = 10 SBE = 5.758770483								
	RC = 02	RC = 05	RC = 07	RC = 10	RC = 15	RC = 20	RC = 25	RC = 35	RC = 50
AMAX DB	.0017375255	.0108709564	.0213327376	.0436480540	.0988323393	.1772876696	.2802872360	.5675287486	1.249387366
AMIN DB	16.86796457	32.62346266	38.48075261	44.70934280	51.83541884	56.95029987	60.98051079	67.25298520	74.45403974
MU K	.1434174D 0	.2337905D-1	.1191139D-1	.5814774D-2	.2559936D-2	.1420643D-2	.8932530D-3	.4338611D-3	.1893643D-3
2 PFC	5.742020430	3.752121211	3.179911116	2.663744976	2.174995700	1.882029945	1.681100124	1.415233213	1.173179863
2 PQC	1.012490738	.8152872773	.7915106719	.7796551757	.7794524975	.7878972914	.8004001948	.8332390643	.9009597366
1 ZF1	6.323369411	6.241331351	6.233927392	6.230019511	6.227941140	6.227214948	6.226879013	6.226586510	6.226430879
1 ZF2	13.76691366	14.74318354	14.84409478	14.89835263	14.92749617	14.93772661	14.94246756	14.94659991	14.94880023

MCPER FUNCTIONS

(2) P 2+2	TH = 11 SBE = 5.240843064								
	RC = 02	RC = 05	RC = 07	RC = 10	RC = 15	RC = 20	RC = 25	RC = 35	RC = 50
AMAX DB	.0017375255	.0108709564	.0213327376	.0436480540	.0988323393	.1772876696	.2802872360	.5675287486	1.249387366
AMIN DB	13.73307913	29.29880670	35.15204847	41.37955524	48.50535740	53.62019269	57.65039124	63.92285941	71.12391235
MU K	.2057529D 0	.3428149D-1	.1747421D-1	.8531438D-2	.3756057D-2	.2084445D-2	.1310631D-2	.6365859D-3	.2778461D-3
2 PFC	5.639097075	3.744606942	3.177973308	2.664058779	2.176087591	1.883226599	1.682272952	1.416296683	1.174095002
2 PQC	1.110564575	.8397235216	.8075865071	.7903955873	.7865443823	.7933229434	.8048796116	.8366988311	.9037348500
1 ZF1	5.795084088	5.685086434	5.675173048	5.669954259	5.667183086	5.666215591	5.665768168	5.665378662	5.665171447
1 ZF2	12.09603235	13.31293458	13.44588136	13.51779107	13.55653276	13.57015131	13.57646562	13.58197104	13.58490312

(2) P 2+2 — TH = 12 — SBE = 4.809734345 — MCPER FUNCTIONS

(2) P 2+2		RC = 02	RC = 05	RC = 07	RC = 10	RC = 15	RC = 20	RC = 25	RC = 35	RC = 50
	AMAX DB	.0017375255	.0108709564	.0213327376	.0436480540	.0988323393	.1772876696	.2802872360	.5675287486	1.249387366
	AMIN DB	11.02789989	26.26553158	32.11112359	38.33657780	45.46186076	50.57660930	54.60678439	60.87924073	68.08029064
MU	K	.2809345D 0	.4860975D-1	.2479955D-1	.1211075D-1	.5332207D-2	.2959167D-2	.1860633D-2	.9037285D-3	.3944441D-3
2	PFC	5.505470187	3.733919076	3.174821827	2.663991036	2.177138669	1.884468995	1.683520001	1.417447035	1.175092906
2	PQC	1.233928170	.8685498576	.8261699228	.8026201776	.7945190072	.7993870209	.8098678811	.8405355167	.9068025109
1	ZF1	5.366324322	5.223877603	5.210914234	5.204110269	5.200504499	5.199246903	5.198665556	5.198159583	5.197890454
1	ZF2	10.65530715	12.09771435	12.26788118	12.36067740	12.41087545	12.42855367	12.43675590	12.44391028	12.44772164

(2) P 2+2 — TH = 13 — SBE = 4.445411483 — MCPER FUNCTIONS

(2) P 2+2		RC = 02	RC = 05	RC = 07	RC = 10	RC = 15	RC = 20	RC = 25	RC = 35	RC = 50
	AMAX DB	.0017375255	.0108709564	.0213327376	.0436480540	.0988323393	.1772876696	.2802872360	.5675287486	1.249387366
	AMIN DB	8.752852496	23.48072275	29.31263066	35.53439604	42.65874470	47.77333713	51.80346998	58.07590502	65.27694948
MU	K	.3650542D 0	.6698289D-1	.3422697D-1	.1672169D-1	.7363135D-2	.4086327D-2	.2569369D-2	.1247972D-2	.5446939D-3
2	PFC	5.341970966	3.719312743	3.170149333	2.663421352	2.178107175	1.885737563	1.684830545	1.418680149	1.176172312
2	PQC	1.385528411	.9025119735	.8475964805	.8164791562	.8034407030	.8061263284	.8153894851	.8447629781	.9101708581
1	ZF1	5.013801277	4.836143468	4.819524618	4.810829498	4.806232366	4.804631043	4.803891179	4.803247440	4.802905110
1	ZF2	9.419195351	11.04517281	11.25738933	11.37439978	11.43803804	11.46050349	11.47093616	11.48004077	11.48489289

(2) P 2+2 — TH = 14 — SBE = 4.133565494 — MCPER FUNCTIONS

(2) P 2+2		RC = 02	RC = 05	RC = 07	RC = 10	RC = 15	RC = 20	RC = 25	RC = 35	RC = 50
	AMAX DB	.0017375255	.0108709564	.0213327376	.0436480540	.0988323393	.1772876696	.2802872360	.5675287486	1.249387366
	AMIN DB	6.896960034	20.91337686	26.72196771	32.93739644	40.06013672	45.17446024	49.20452034	55.47691869	62.67795376
MU	K	.4520141D 0	.9001837D-1	.4612131D-1	.2254915D-1	.9931004D-2	.5511591D-2	.3465564D-2	.1683271D-2	.7346869D-3
2	PFC	5.154164521	3.699959948	3.163609058	2.662214630	2.178946574	1.887010561	1.686192685	1.419991457	1.177331818
2	PQC	1.566358771	.9424965955	.8722639483	.8321476650	.8133831322	.8135825818	.8214719959	.8493967049	.9138489370
1	ZF1	4.719372506	4.506518180	4.485576285	4.474650442	4.468889386	4.466885715	4.465960539	4.465155875	4.464728088
1	ZF2	8.371684614	10.11848672	10.37679526	10.52133205	10.60049556	10.62852834	10.64156123	10.65294261	10.65901093

(2) P 2+2 — TH = 15 — SBE = 3.863703305 — MCPER FUNCTIONS

(2) P 2+2		RC = 02	RC = 05	RC = 07	RC = 10	RC = 15	RC = 20	RC = 25	RC = 35	RC = 50
	AMAX DB	.0017375255	.0108709564	.0213327376	.0436480540	.0988323393	.1772876696	.2802872360	.5675287486	1.249387366
	AMIN DB	5.424450537	18.54199499	24.31259802	30.51756405	37.63763900	42.75151652	46.78145592	53.05379342	60.25481290
MU	K	.5355222D 0	.1182770D 0	.6086535D-1	.2979352D-1	.1312557D-1	.7284910D-2	.4580651D-2	.2224899D-2	.9710897D-3
2	PFC	4.950958890	3.674972170	3.154813087	2.660219791	2.179605198	1.888263938	1.687593277	1.421375920	1.178569874
2	PQC	1.775520153	.9895345949	.9006407146	.8498293615	.8244304288	.8218029206	.8281463594	.8544539420	.9178467548
1	ZF1	4.468712300	4.223755563	4.197773989	4.184243120	4.177128876	4.174659040	4.173519490	4.172528848	4.172002371
1	ZF2	7.494869415	9.291534278	9.598643173	9.773858102	9.870700203	9.905128372	9.921157505	9.935167102	9.942641112

MCPER FUNCTIONS

(2) P 2+2	TH = 16	SBE = 3.627955279							
	RC = 02	RC = 05	RC = 07	RC = 10	RC = 15	RC = 20	RC = 25	RC = 35	RC = 50
AMAX DB		.0108709564	.0213327376	.0436480540	.0988323393	.1772876696	.2802872360	.5675287486	1.249387366
AMIN DB		16.35297999	22.06431815	28.25261147	35.36841411	40.48157573	44.51132137	50.78356114	57.98455560
MU I K		.1521777D 0	.7884680D-1	.3866958D-1	.1704430D-1	.9460655D-2	.5948863D-2	.2889495D-2	.1261166D-2
2 I PFC		3.643451652	3.143333230	2.657268502	2.180025818	1.889471156	1.689017850	1.422827996	1.179884774
2 I PQC		1.044784385	.9332731372	.8697603813	.8366785075	.8308404959	.8354472183	.8599538288	.9221753431
1 I ZF1		3.979419786	3.947662749	3.931118412	3.922444107	3.919438854	3.918053533	3.916849924	3.916210536
1 I ZF2		8.545923339	8.902570865	9.111268373	9.227967458	9.269660771	9.289107027	9.306120717	9.315203951

MCPER FUNCTIONS

(2) P 2+2	TH = 17	SBE = 3.420303620							
	RC = 02	RC = 05	RC = 07	RC = 10	RC = 15	RC = 20	RC = 25	RC = 35	RC = 50
AMAX DB		.0108709564	.0213327376	.0436480540	.0988323393	.1772876696	.2802872360	.5675287486	1.249387366
AMIN DB		14.33928872	19.96212323	26.12470482	33.23385834	38.34590334	42.37534662	48.64743388	55.84838929
MU I K		.1918826D 0	.1004370D 0	.4940430D-1	.2179250D-1	.1209776D-1	.7607337D-2	.3695118D-2	.1612800D-2
2 I PFC		3.604582214	3.128706820	2.653174088	2.180145184	1.890603009	1.690450514	1.424341609	1.181274643
2 I PQC		1.109484347	.9707898562	.8922135450	.8502365524	.8407551434	.8434132817	.8659175571	.9268468288
1 I ZF1		3.767022702	3.728789321	3.708792786	3.698333181	3.694717562	3.693052647	3.691607102	3.690839580
1 I ZF2		7.869045883	8.273354912	8.517758920	8.656462418	8.706324223	8.729631459	8.750048863	8.760958879

MCPER FUNCTIONS

(2) P 2+2	TH = 18	SBE = 3.236067977							
	RC = 02	RC = 05	RC = 07	RC = 10	RC = 15	RC = 20	RC = 25	RC = 35	RC = 50
AMAX DB		.0108709564	.0213327376	.0436480540	.0988323393	.1772876696	.2802872360	.5675287486	1.249387366
AMIN DB		12.49886828	17.99543563	24.11958555	31.21866204	36.32900904	40.35799226	46.62984742	53.83074339
MU I K		.2371683D 0	.1259587D 0	.6223300D-1	.2748317D-1	.1525982D-1	.9596224D-2	.4661306D-2	.2034524D-2
2 I PFC		3.557761547	3.110450492	2.647730956	2.179893520	1.891627401	1.691873850	1.425910113	1.182737428
2 I PQC		1.184866696	1.013900128	.9175025495	.8652286924	.8516141489	.8520877447	.8723685493	.9318745126
1 I ZF1		3.581433438	3.536139734	3.512231161	3.499742296	3.495435381	3.493454530	3.491736041	3.490824143
1 I ZF2		7.252650051	7.699627975	7.931105503	8.143858839	8.202815609	8.230448212	8.254691735	8.267660033

MCPER FUNCTIONS

(2) P 2+2	TH = 19	SBE = 3.071553487							
	RC = 02	RC = 05	RC = 07	RC = 10	RC = 15	RC = 20	RC = 25	RC = 35	RC = 50
AMAX DB		.0108709564	.0213327376	.0436480540	.0988323393	.1772876696	.2802872360	.5675287486	1.249387366
AMIN DB		10.83255565	16.15751388	22.22596124	29.31013290	34.41795574	38.44625442	44.71776413	51.91857162
MU I K		.2873242D 0	.1556411D 0	.7739305D-1	.3423681D-1	.1901526D-1	.1195879D-1	.5809139D-2	.2535546D-2
2 I PFC		3.502755998	3.088085463	2.640714994	2.179194003	1.892509113	1.693268794	1.427526250	1.184270884
2 I PQC		1.272038194	1.063382336	.9459856970	.8817958553	.8634931156	.8615187652	.8793326597	.9372729580
1 I ZF1		3.418466579	3.365749779	3.337465220	3.322684807	3.317599407	3.315263647	3.313239104	3.312165554
1 I ZF2		6.691608897	7.173008152	7.491764959	7.680423333	7.749411374	7.781851118	7.810364672	7.825636554

(2) P 2+2	TH = 20	SBE = 2.923804400						MCPER FUNCTIONS	
	RC = 02	RC = 05	RC = 07	RC = 10	RC = 15	RC = 20	RC = 25	RC = 35	RC = 50
AMAX DB		.0108709564	.0213327376	.0436480540	.0988323393	.1772876696	.2802872360	.5675287486	1.249387366
AMIN DB		9.341461481	14.44487900	20.43507986	27.49770248	32.60184912	36.62914948	42.90015517	50.10083355
MU I K		.3411355D 0	.1895641D 0	.9511434D-1	.4218081D-1	.2343730D-1	.1474153D-1	.7161306D-2	.3125779D-2
2 I PFC		3.439831297	3.061177021	2.631885617	2.177962268	1.893209537	1.694614502	1.429182103	1.185872561
2 I PQC		1.371854273	1.120059167	.9780684738	.9000977810	.8764769388	.8717600027	.8868384017	.9430580899
1 I ZF1		3.274602022	3.214426937	3.181325446	3.163974544	3.158016967	3.155284578	3.152918696	3.151665159
1 I ZF2		6.182733462	6.687472159	7.042254522	7.258376234	7.338325044	7.376067097	7.409314452	7.427148791

(2) P 2+2	TH = 21	SBE = 2.790428110						MCPER FUNCTIONS	
	RC = 02	RC = 05	RC = 07	RC = 10	RC = 15	RC = 20	RC = 25	RC = 35	RC = 50
AMAX DB		.0108709564	.0213327376	.0436480540	.0988323393	.1772876696	.2802872360	.5675287486	1.249387366
AMIN DB		8.024332755	12.85661225	18.74042430	25.77256391	30.87145562	34.89732614	41.16760949	48.36810278
MU I K		.3969935D 0	.2275985D 0	.1156056D 0	.5144839D-1	.2860403D-1	.1799425D-1	.8742176D-2	.3815881D-2
2 I PFC		3.369798696	3.029387913	2.620990336	2.176105993	1.893686404	1.695888200	1.430869046	1.187539786
2 I PQC		1.484830080	1.184756220	1.014203999	.9203151433	.8906608938	.8828712255	.8949172036	.9492473056
1 I ZF1		3.146812047	3.079546769	3.041248637	3.021035263	3.014105154	3.010931550	3.008186789	3.006733848
1 I ZF2		5.723657747	6.238355981	6.626715382	6.871437246	6.963248932	7.006796866	7.045259532	7.065928461

(2) P 2+2	TH = 22	SBE = 2.669467163						MCPER FUNCTIONS	
	RC = 02	RC = 05	RC = 07	RC = 10	RC = 15	RC = 20	RC = 25	RC = 35	RC = 50
AMAX DB		.0108709564	.0213327376	.0436480540	.0988323393	.1772876696	.2802872360	.5675287486	1.249387366
AMIN DB		6.875686789	11.39342310	17.13747393	24.12740490	29.21891354	33.24276958	39.51203433	46.71226653
MU I K		.4531225D 0	.2693574D 0	.1390357D 0	.6217700D-1	.3459827D-1	.2177015D-1	.1057787D-1	.4617285D-2
2 I PFC		3.293938660	2.992539630	2.607772816	2.173524651	1.893893492	1.697065028	1.432577687	1.189269648
2 I PQC		1.611126447	1.258247935	1.054890100	.9426516978	.9061518371	.8949189936	.9036036964	.9558595991
1 I ZF1		3.032473330	2.958902264	2.915136481	2.891761966	2.883752304	2.880089921	2.876926399	2.875253527
1 I ZF2		5.311955408	5.824403579	6.240598786	6.514497365	6.619021555	6.668881173	6.713055745	6.736844214

(2) P 2+2	TH = 23	SBE = 2.559304665						MCPER FUNCTIONS	
	RC = 02	RC = 05	RC = 07	RC = 10	RC = 15	RC = 20	RC = 25	RC = 35	RC = 50
AMAX DB		.0108709564	.0213327376	.0436480540	.0988323393	.1772876696	.2802872360	.5675287486	1.249387366
AMIN DB		5.885308763	10.05648829	15.62348371	22.55621125	27.63751310	31.65857347	37.92642284	45.12629204
MU I K		.5078490D 0	.3141779D 0	.1655106D 0	.7450569D-1	.4150729D-1	.2612590D-1	.1269635D-1	.5542241D-2
2 I PFC		3.213819439	2.950668368	2.591985352	2.170109550	1.893780352	1.698117861	1.434297804	1.191058976
2 I PQC		1.750616662	1.341197723	1.100661630	.9673363195	.9230695151	.9079774196	.9129360377	.9629156993
1 I ZF1		2.929333269	2.850594537	2.801249345	2.774418246	2.765215592	2.761013794	2.757389194	2.755474724
1 I ZF2		4.944644894	5.442322108	5.880431684	6.183379865	6.301382637	6.358054402	6.408449255	6.435654458

TH = 24 SBE = 2.458593336 — MCPER FUNCTIONS

(2) P 2+2	RC = 02	RC = 05	RC = 07	RC = 10	RC = 15	RC = 20	RC = 25	RC = 35	RC = 50
AMAX DB		.0108709564	.0213327376	.0436480540	.0988323393	.1772876696	.2802872360	.5675287486	1.249387366
AMIN DB		5.039090797	8.846203172	14.19723282	21.05412204	26.12152853	30.13876276	36.40467174	43.60404292
MU K		.5598162D 0	.3611518D 0	.1950466D 0	.8857148D-1	.4942237D-1	.3112160D-1	.1512747D-1	.6603860D-2
2 PFC		3.131076633	2.904057761	2.573406157	2.165744331	1.893292056	1.699017132	1.436018272	1.192904319
2 PQC		1.903003073	1.434107879	1.152076898	.9946247302	.9415479647	.9221290158	.9229562752	.9704382250
1 ZF1		2.835496348	2.752958646	2.698124875	2.667559079	2.657044204	2.652249228	2.648118713	2.645939805
1 ZF2		4.618104946	5.091345546	5.543630108	5.874657721	6.006790251	6.070760125	6.127891349	6.158821983

TH = 25 SBE = 2.366201583 — MCPER FUNCTIONS

(2) P 2+2	RC = 02	RC = 05	RC = 07	RC = 10	RC = 15	RC = 20	RC = 25	RC = 35	RC = 50
AMAX DB			.0213327376	.0436480540	.0988323393	.1772876696	.2802872360	.5675287486	1.249387366
AMIN DB			7.761105790	12.85870094	19.61732141	24.66608983	28.67815536	34.94143693	42.14013311
MU K			.4092086D 0	.2275438D 0	.1045042D 0	.5843802D-1	.3682072D-1	.1790310D-1	.7816158D-2
2 PFC			2.853233199	2.551860914	2.160306122	1.892368997	1.699730640	1.437726993	1.194801921
2 PQC			1.537294375	1.209697833	1.024800640	.9617369783	.9374656266	.9337107543	.9784518589
1 ZF1			2.664519389	2.604514735	2.569971855	2.558020913	2.552575874	2.547891999	2.545424602
1 ZF2			4.770347083	5.228336163	5.585530816	5.732282852	5.804011306	5.868397760	5.903373085

TH = 26 SBE = 2.281172033 — MCPER FUNCTIONS

(2) P 2+2	RC = 02	RC = 05	RC = 07	RC = 10	RC = 15	RC = 20	RC = 25	RC = 35	RC = 50
AMAX DB			.0213327376	.0436480540	.0988323393	.1772876696	.2802872360	.5675287486	1.249387366
AMIN DB			6.797241891	11.60864677	18.24295374	23.26708531	27.27225366	33.53201818	40.72980960
MU K			.4572334D 0	.2627651D 0	.1224200D 0	.6865082D-1	.4328997D-1	.2105713D-1	.9194106D-2
2 PFC			2.798914216	2.527246385	2.153667603	1.890946800	1.700223372	1.439410807	1.196747697
2 PQC			1.650895198	1.274065029	1.058175944	.9838035834	.9540894462	.9452505756	.9869835412
1 ZF1			2.583970051	2.519335391	2.480630646	2.467117070	2.460962067	2.455674715	2.452893530
1 ZF2			4.478055060	4.933261629	5.313715284	5.475374183	5.555283213	5.627440543	5.666789220

TH = 27 SBE = 2.202689265 — MCPER FUNCTIONS

(2) P 2+2	RC = 02	RC = 05	RC = 07	RC = 10	RC = 15	RC = 20	RC = 25	RC = 35	RC = 50
AMAX DB			.0213327376	.0436480540	.0988323393	.1772876696	.2802872360	.5675287486	1.249387366
AMIN DB			5.948105783	10.44809468	16.92904915	21.92108781	25.91715975	32.17226648	39.36885755
MU K			.5041906D 0	.3003276D 0	.1424123D 0	.8015777D-1	.5059901D-1	.2462559D-1	.1075368D-1
2 PFC			2.741936641	2.499552202	2.145700216	1.888956383	1.700457326	1.441055402	1.198737200
2 PQC			1.774909070	1.345670767	1.095089569	1.007933469	.9721141163	.9576321051	.9960626873
1 ZF1			2.510164692	2.441630644	2.398660191	2.383457529	2.376529888	2.370586221	2.367464664
1 ZF2			4.212907658	4.657529572	5.057370634	5.233972613	5.322430604	5.402864089	5.446922740

(2) P 2+2	TH = 28	SBE = 2.130054468						MCPER FUNCTIONS	
	RC = 02	RC = 05	RC = 07	RC = 10	RC = 15	RC = 20	RC = 25	RC = 35	RC = 50
AMAX DB			.0213327376	.0436480540	.0988323393	.1772876696	.2802872360	.5675287486	1.249387366
AMIN DB			5.205090802	9.377775946	15.67444631	20.62529881	24.60950925	30.85850918	38.05352282
MU I K			.5492189D 0	.3397122D 0	.1645423D 0	.9305400D-1	.5881993D-1	.2864670D-1	.1251192D-1
2 I PFC			2.683166011	2.468875760	2.136278741	1.886324250	1.700391380	1.442645220	1.200765594
2 I PQC			1.909251423	1.424934838	1.135904511	1.034332253	.9916658907	.9709175443	1.005721431
1 I ZF1			2.442114529	2.370544606	2.323307119	2.306292970	2.298527679	2.291872147	2.288382304
1 I ZF2			3.973040682	4.400515314	4.815036275	5.006318770	5.103623186	5.192819256	5.241930711

(2) P 2+2	TH = 29	SBE = 2.062665340						MCPER FUNCTIONS	
	RC = 02	RC = 05	RC = 07	RC = 10	RC = 15	RC = 20	RC = 25	RC = 35	RC = 50
AMAX DB				.0436480540	.0988323393	.1772876696	.2802872360	.5675287486	1.249387366
AMIN DB				8.397607816	14.47869934	19.37750543	23.34642012	29.58748945	36.78044830
MU I K				.3802941D 0	.1888274D 0	.1074298D 0	.6802664D-1	.3316084D-1	.1448697D-1
2 I PFC				2.435425543	2.125287325	1.882973096	1.699981194	1.444163356	1.202827612
2 I PQC				1.512188944	1.181002737	1.063226463	1.012884844	.9851755640	1.015994898
1 I ZF1				2.305304031	2.253916627	2.234977825	2.226308230	2.218882629	2.214995223
1 I ZF2				4.161697385	4.585578495	4.790936949	4.897294987	4.995711264	5.050222448

(2) P 2+2	TH = 30	SBE = 2.000000000						MCPER FUNCTIONS	
	RC = 02	RC = 05	RC = 07	RC = 10	RC = 15	RC = 20	RC = 25	RC = 35	RC = 50
AMAX DB				.0436480540	.0988323393	.1772876696	.2802872360	.5675287486	1.249387366
AMIN DB				7.506305862	13.34195696	18.17604518	22.12545342	28.35631719	35.54662140
MU I K				.4213905D 0	.2152297D 0	.1233666D 0	.7829379D-1	.3821062D-1	.1669817D-1
2 I PFC				2.399510537	2.112626832	1.878822833	1.699179200	1.445591447	1.204917526
2 I PQC				1.607672685	1.230777777	1.094864060	1.035926093	1.000482007	1.026921510
1 I ZF1				2.245208446	2.189913446	2.168952446	2.159311272	2.151054896	2.146739254
1 I ZF2				3.940534197	4.368142003	4.586596827	4.702104418	4.810158167	4.870417556

(2) P 2+2	TH = 31	SBE = 1.941604026						MCPER FUNCTIONS	
	RC = 02	RC = 05	RC = 07	RC = 10	RC = 15	RC = 20	RC = 25	RC = 35	RC = 50
AMAX DB				.0436480540	.0988323393	.1772876696	.2802872360	.5675287486	1.249387366
AMIN DB				6.701201054	12.26480575	17.01977355	20.94458335	27.16242934	34.34933036
MU I K				.4623171D 0	.2436462D 0	.1409326D 0	.8969554D-1	.4384081D-1	.1916609D-1
2 I PFC				2.361517050	2.098223046	1.873792121	1.697934704	1.446909570	1.207029098
2 I PQC				1.711543871	1.285625241	1.129514303	1.060960978	1.016920661	1.038543330
1 I ZF1				2.189625888	2.130786275	2.107728543	2.097049316	2.087899230	2.083123260
1 I ZF2				3.736380327	4.162102300	4.392282722	4.516902631	4.634957532	4.701312134

(2) P 2+2	TH = 32	SBE = 1.887079915						MCPER FUNCTIONS	
	RC = 02	RC = 05	RC = 07	RC = 10	RC = 15	RC = 20	RC = 25	RC = 35	RC = 50
AMAX DB				.0436480540	.0988323393	.1772876696	.2802872360	.5675287486	1.249387366
AMIN DB				5.978280499	11.24807615	15.90802888	19.80217431	26.00355806	33.18612785
MU K				.5024420D 0	.2739026D 0	.1601764D 0	.1023037D 0	.5009820D-1	.2191258D-1
2 PFC				2.321877617	2.082034906	1.867800510	1.696194133	1.448096138	1.209155536
2 PQC				1.823898880	1.345931971	1.167466765	1.088178143	1.034584113	1.050906445
1 ZF1				2.137991596	2.076075241	2.050877141	2.039096072	2.028987542	2.023717720
1 ZF2				3.548448177	3.967016611	4.207168103	4.340708300	4.469059562	4.541851340

(2) P 2+2	TH = 33	SBE = 1.836078459						MCPER FUNCTIONS	
	RC = 02	RC = 05	RC = 07	RC = 10	RC = 15	RC = 20	RC = 25	RC = 35	RC = 50
AMAX DB				.0436480540	.0988323393	.1772876696	.2802872360	.5675287486	1.249387366
AMIN DB				5.332413441	10.29261974	14.84058906	18.69696255	24.87770527	32.05480064
MU K				.5412268D 0	.3057518D 0	.1811217D 0	.1161855D 0	.5703149D-1	.2496088D-1
2 PFC				2.281038308	2.064061598	1.860771228	1.693901495	1.449127797	1.211289446
2 PQC				1.944799207	1.412065100	1.209029306	1.117784425	1.053574674	1.064061400
1 ZF1				2.089807281	2.025362121	1.998018526	1.985076904	1.973944026	1.968145403
1 ZF2				3.375810462	3.782572922	4.030593392	4.172687389	4.311545306	4.391106985

(2) P 2+2	TH = 34	SBE = 1.788291650						MCPER FUNCTIONS	
	RC = 02	RC = 05	RC = 07	RC = 10	RC = 15	RC = 20	RC = 25	RC = 35	RC = 50
AMAX DB					.0988323393	.1772876696	.2802872360	.5675287486	1.249387366
AMIN DB					9.399076834	13.81761466	17.62803997	23.78312277	30.95334416
MU K					.3388802D 0	.2037602D 0	.1314008D 0	.6469100D-1	.2833562D-1
2 PFC					2.044347204	1.852634597	1.690999091	1.449979349	1.213422776
2 PQC					1.484362640	1.254524896	1.150005456	1.074005376	1.078063679
1 ZF1					1.978263142	1.948813790	1.934660863	1.922437444	1.916073681
1 ZF2					3.608538690	3.862045363	4.012136688	4.161608343	4.248259128

(2) P 2+2	TH = 35	SBE = 1.743446796						MCPER FUNCTIONS	
	RC = 02	RC = 05	RC = 07	RC = 10	RC = 15	RC = 20	RC = 25	RC = 35	RC = 50
AMAX DB					.0988323393	.1772876696	.2802872360	.5675287486	1.249387366
AMIN DB					8.567661631	12.83957438	16.59483734	22.71829683	29.87994131
MU K					.3729211D 0	.2280454D 0	.1479988D 0	.7312825D-1	.3206291D-1
2 PFC					2.022981809	1.843331948	1.687428523	1.450623687	1.215546758
2 PQC					1.563127307	1.304287306	1.185085858	1.096001052	1.092974252
1 ZF1					1.934424163	1.902957713	1.887553991	1.874174720	1.867208160
1 ZF2					3.444712486	3.701136741	3.858470122	4.018543331	4.112580366

(2) P 2+2	TH = 36	SBE = 1.701301617						MCPER FUNCTIONS	
	RC = 02	RC = 05	RC = 07	RC = 10	RC = 15	RC = 20	RC = 25	RC = 35	RC = 50
AMAX DB					.0988323393	.1772876696	.2802872360	.5675287486	1.249387366
AMIN DB					7.797993987	11.90715063	15.59710414	21.68193664	28.83294474
MU I K					.4074744D 0	.2538882D 0	.1660140D 0	.8239544D-1	.3617035D-1
2 I PFC					2.000098507	1.832819788	1.683132065	1.451031765	1.217651849
2 I PQC					1.648624841	1.358655862	1.223288910	1.119699458	1.108860180
1 I ZF1					1.893517899	1.860172811	1.843493630	1.828895590	1.821287368
1 I ZF2					3.290882661	3.547585014	3.711206916	3.881728213	3.983425727

(2) P 2+2	TH = 37	SBE = 1.661640141						MCPER FUNCTIONS	
	RC = 02	RC = 05	RC = 07	RC = 10	RC = 15	RC = 20	RC = 25	RC = 35	RC = 50
AMAX DB					.0988323393	.1772876696	.2802872360	.5675287486	1.249387366
AMIN DB					7.089000880	11.02112659	14.63488195	20.67296583	27.81086233
MU I K					.4421300D 0	.2811536D 0	.1854624D 0	.9254473D-1	.4068711D-1
2 I PFC					1.975866510	1.821073811	1.678054385	1.451172624	1.219727663
2 I PQC					1.741087333	1.417969716	1.264895611	1.145252448	1.125795304
1 I ZF1					1.855242698	1.820204476	1.802243554	1.786368085	1.778078312
1 I ZF2					3.146796678	3.401189995	3.569960830	3.750620557	3.860217159

(2) P 2+2	TH = 38	SBE = 1.624269245						MCPER FUNCTIONS	
	RC = 02	RC = 05	RC = 07	RC = 10	RC = 15	RC = 20	RC = 25	RC = 35	RC = 50
AMAX DB					.0988323393	.1772876696	.2802872360	.5675287486	1.249387366
AMIN DB					6.438897721	10.18226035	13.70846864	19.69051629	26.81234515
MU I K					.4764915D 0	.3096613D 0	.2063367D 0	.1036273D 0	.4564390D-1
2 I PFC					1.950481366	1.808092286	1.672144623	1.451013470	1.221762908
2 I PQC					1.840721091	1.482562253	1.310203041	1.172827172	1.143861006
1 I ZF1					1.819322258	1.782817145	1.763589778	1.746384711	1.737372721
1 I ZF2					3.012143186	3.261810317	3.434429736	3.624746642	3.742439759

(2) P 2+2	TH = 39	SBE = 1.589015729						MCPER FUNCTIONS	
	RC = 02	RC = 05	RC = 07	RC = 10	RC = 15	RC = 20	RC = 25	RC = 35	RC = 50
AMAX DB					.0988323393	.1772876696	.2802872360	.5675287486	1.249387366
AMIN DB					5.845244402	9.391155584	12.81837108	18.73392353	25.83617784
MU I K					.5101969D 0	.3391894D 0	.2286027D 0	.1156921D 0	.5107297D-1
2 I PFC					1.924153728	1.793898284	1.665358751	1.450519845	1.223745311
2 I PQC					1.947717827	1.552756447	1.359522080	1.202607253	1.163147066
1 I ZF1					1.785505716	1.747791461	1.727336968	1.708759175	1.698983877
1 I ZF2					2.886546711	3.129339726	3.304384947	3.503695537	3.629631922

$P_{2+4}^{(3)}$

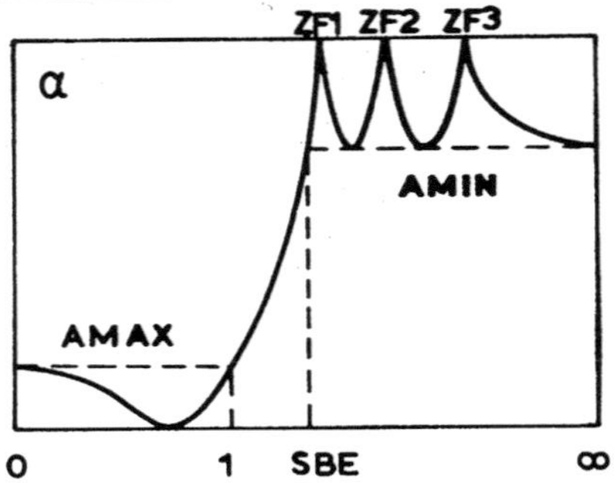

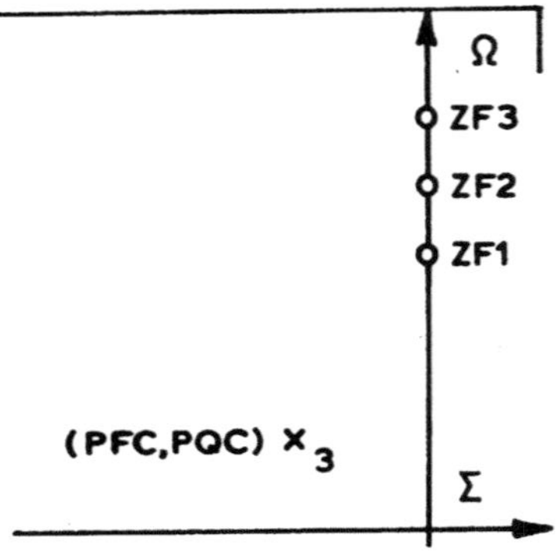

$P_{2+4}^{(3)}$	TH = 3	SBE = 19.10732261						MCPER FUNCTIONS	
	RC = 02	RC = 05	RC = 07	RC = 10	RC = 15	RC = 20	RC = 25	RC = 35	RC = 50
AMAX DB	.0017375255	.0108709564	.0213327376	.0436480540	.0988323393	.1772876696	.2802872360	.5675287486	1.249387366
AMIN DB	85.62470045	109.5193550	118.3079519	127.6466630	138.3223059	145.9749865	151.9943807	161.3303166	171.9564849
MU I K	.5233172D-4	.3342199D-5	.1215073D-5	.4146359D-6	.1213067D-6	.5026326D-7	.2513512D-7	.8579938D-8	.2524502D-8
3 I PFC	6.579506194	4.162056927	3.516996134	2.941326456	2.399074095	2.074562896	1.851927611	1.556749808	1.286322581
3 I PQC	.7335656319	.7264663514	.7283040375	.7330946555	.7431935376	.7546665958	.7671839145	.7953734362	.8478530175
1 I ZF1	19.78247066	19.78077079	19.78061238	19.78052824	19.78048333	19.78046761	19.78046034	19.78045400	19.78045062
1 I ZF2	27.01257507	27.01257507	27.01257507	27.01257507	27.01257507	27.01257507	27.01257507	27.01257507	27.01257507
1 I ZF3	73.67284756	73.76102646	73.76926090	73.77363600	73.77597148	73.77678897	73.77716740	73.77749706	73.77767257

$P_{2+4}^{(3)}$	TH = 4	SBE = 14.33558703						MCPER FUNCTIONS	
	RC = 02	RC = 05	RC = 07	RC = 10	RC = 15	RC = 20	RC = 25	RC = 35	RC = 50
AMAX DB	.0017375255	.0108709564	.0213327376	.0436480540	.0988323393	.1772876696	.2802872360	.5675287486	1.249387366
AMIN DB	70.62287262	94.51743764	103.3060328	112.6447434	123.3203863	130.9730669	136.9924610	146.3283970	156.9545652
MU I K	.2943448D-3	.1879871D-4	.6834368D-5	.2332184D-5	.6823084D-6	.2827136D-6	.1413764D-6	.4825920D-7	.1419946D-7
3 I PFC	6.574991148	4.162532863	3.517661105	2.941999349	2.399673774	2.075096846	1.852410619	1.557160485	1.286663967
3 I PQC	.7523881512	.7336351384	.7334379235	.7367398641	.7457037913	.7566179158	.7688035661	.7966199688	.8488307487
1 I ZF1	14.84488861	14.84084646	14.84047072	14.84027120	14.84016473	14.84012747	14.84011022	14.84009519	14.84008719
1 I ZF2	20.26123148	20.26123148	20.26123148	20.26123148	20.26123148	20.26123148	20.26123148	20.26123148	20.26123148
1 I ZF3	55.07710612	55.28568345	55.30520129	55.31557424	55.32111224	55.32305083	55.32394826	55.32473006	55.32514629

(3) P 2+4	TH = 5	SBE = 11.47371325						MCPER FUNCTIONS	
	RC = 02	RC = 05	RC = 07	RC = 10	RC = 15	RC = 20	RC = 25	RC = 35	RC = 50
AMAX DB	.0017375255	.0108709564	.0213327376	.0436480540	.0988323393	.1772876696	.2802872360	.5675287486	1.249387366
AMIN DB	58.98225551	82.87631988	91.66490549	101.0036136	111.6792558	119.3319363	125.3513304	134.6872663	145.3134345
MU K	.1124313D-2	.7180985D-4	.2610687D-4	.8908802D-5	.2606377D-5	.1079949D-5	.5400494D-6	.1843473D-6	.5424107D-7
3 PFC	6.564024817	4.162621889	3.518291219	2.942772916	2.400412101	2.075767804	1.853023003	1.557685097	1.287101773
3 PQC	.7786166773	.7431572748	.7401954331	.7415052969	.7489681140	.7591486648	.7709008000	.7982310917	.8500926725
1 ZF1	11.88639219	11.87844475	11.87770983	11.87731986	11.87711183	11.87703904	11.87700535	11.87697600	11.87696037
1 ZF2	16.21083706	16.21083706	16.21083706	16.21083706	16.21083706	16.21083706	16.21083706	16.21083706	16.21083706
1 ZF3	43.76914837	44.17454568	44.21264736	44.23290840	44.24372879	44.24751703	44.24927082	44.25079868	44.25161213

(3) P 2+4	TH = 6	SBE = 9.566772234						MCPER FUNCTIONS	
	RC = 02	RC = 05	RC = 07	RC = 10	RC = 15	RC = 20	RC = 25	RC = 35	RC = 50
AMAX DB	.0017375255	.0108709564	.0213327376	.0436480540	.0988323393	.1772876696	.2802872360	.5675287486	1.249387366
AMIN DB	49.46799983	73.36002827	82.14857573	91.48727366	102.1629133	109.8155934	115.8349874	125.1709232	135.7970915
MU K	.3362018D-2	.2147823D-3	.7808565D-4	.2664626D-4	.7795686D-5	.3230132D-5	.1615290D-5	.5513836D-6	.1622353D-6
3 PFC	6.542754736	4.161929620	3.518716720	2.943577976	2.401264375	2.076564016	1.853758241	1.558321068	1.287635152
3 PQC	.8140810230	.7552799191	.7487009688	.7474524562	.7530149451	.7622756467	.7734869447	.8002132259	.8516424603
1 ZF1	9.918923277	9.905024871	9.903751676	9.903076917	9.902717209	9.902591380	9.902533146	9.902482423	9.902455421
1 ZF2	13.51091739	13.51091739	13.51091739	13.51091739	13.51091739	13.51091739	13.51091739	13.51091739	13.51091739
1 ZF3	36.04035910	36.73438043	36.80013464	36.83513655	36.85383959	36.36038925	36.86342177	36.86606376	36.86747046

(3) P 2+4	TH = 7	SBE = 8.205509048						MCPER FUNCTIONS	
	RC = 02	RC = 05	RC = 07	RC = 10	RC = 15	RC = 20	RC = 25	RC = 35	RC = 50
AMAX DB	.0017375255	.0108709564	.0213327376	.0436480540	.0988323393	.1772876696	.2802872360	.5675287486	1.249387366
AMIN DB	41.42399856	65.30931196	74.09773725	83.43640281	94.11203429	101.7647130	107.7841067	117.1200423	127.7462105
MU K	.8487896D-2	.5426682D-3	.1972937D-3	.6732554D-4	.1969692D-4	.8161394D-5	.4081264D-5	.1393150D-5	.4099109D-6
3 PFC	6.506528422	4.159972195	3.518729010	2.944329451	2.402200203	2.077471018	1.854608309	1.559065231	1.288263062
3 PQC	.8614242509	.7703271551	.7591145868	.7546593510	.7573799591	.7660198238	.7765760850	.8025743237	.8534846458
1 ZF1	8.519474487	8.496972733	8.494942457	8.493868666	8.493296864	8.493096946	8.493004441	8.492923978	8.492880994
1 ZF2	11.58269774	11.58269774	11.58269774	11.58269774	11.58269774	11.58269774	11.58269774	11.58269774	11.58269774
1 ZF3	30.29576961	31.38039172	31.48454436	31.54008163	31.56978463	31.58019085	31.58500974	31.58920946	31.59144418

(3) P 2+4	TH = 8	SBE = 7.185296534						MCPER FUNCTIONS	
	RC = 02	RC = 05	RC = 07	RC = 10	RC = 15	RC = 20	RC = 25	RC = 35	RC = 50
AMAX DB	.0017375255	.0108709564	.0213327376	.0436480540	.0988323393	.1772876696	.2802872360	.5675287486	1.249387366
AMIN DB	34.46462955	58.33077954	67.11887056	76.45744803	87.13305742	94.78573246	100.8051251	110.1410603	120.7672283
MU I K	.1891335D-1	.1211884D-2	.4406122D-3	.1503584D-3	.4398931D-4	.1822692D-4	.9114729D-5	.3111337D-5	.9154583D-6
3 I PFC	6.449974218	4.156176042	3.518080180	2.944926081	2.403183384	2.078471578	1.855563642	1.559913810	1.288984256
3 I PQC	.9244355061	.7887156442	.7716374611	.7632223706	.7636066526	.7704065843	.7801852110	.8053239334	.8556246537
1 I ZF1	7.477097480	7.442526156	7.439476128	7.437868032	7.437013143	7.436714488	7.436576336	7.436456040	7.436392015
1 I ZF2	10.13679080	10.13679080	10.13679080	10.13679080	10.13679080	10.13679080	10.13679080	10.13679080	10.13679080
1 I ZF3	25.74024702	27.31861421	27.47342514	27.55619454	27.60052431	27.61606512	27.62326351	27.62953642	27.63287695

(3) P 2+4	TH = 9	SBE = 6.392453221						MCPER FUNCTIONS	
	RC = 02	RC = 05	RC = 07	RC = 10	RC = 15	RC = 20	RC = 25	RC = 35	RC = 50
AMAX DB	.0017375255	.0108709564	.0213327376	.0436480540	.0988323393	.1772876696	.2802872360	.5675287486	1.249387366
AMIN DB	28.35396929	52.17092735	60.95820440	70.29656910	80.97212534	88.62479156	94.64418186	103.9801158	114.6062835
MU I K	.3822096D-1	.2462939D-2	.8955499D-3	.3056128D-3	.8941157D-4	.3704763D-4	.1852639D-4	.6324034D-5	.1860741D-5
3 I PFC	6.367156391	4.149878337	3.516482605	2.945250134	2.404171777	2.079545622	1.856613095	1.560862401	1.289797284
3 I PQC	1.008490753	.8109755587	.7865189916	.7732586648	.7702470765	.7754660734	.7843344017	.8084732785	.8580688332
1 I ZF1	6.675305694	6.624085777	6.619702870	6.617402401	6.616182406	6.615756691	6.615559851	6.615388496	6.615297313
1 I ZF2	9.012425833	9.012425833	9.012425833	9.012425833	9.012425833	9.012425833	9.012425833	9.012425833	9.012425833
1 I ZF3	21.94469165	24.10681595	24.32579328	24.44332774	24.50640661	24.52854158	24.53879808	24.54773784	24.55249928

(3) P 2+4	TH = 10	SBE = 5.758770483						MCPER FUNCTIONS	
	RC = 02	RC = 05	RC = 07	RC = 10	RC = 15	RC = 20	RC = 25	RC = 35	RC = 50
AMAX DB	.0017375255	.0108709564	.0213327376	.0436480540	.0988323393	.1772876696	.2802872360	.5675287486	1.249387366
AMIN DB	22.95596431	46.65710676	55.44257027	64.78046596	75.45590578	83.10855273	89.12793783	98.46386916	109.0900362
MU I K	.7115440D-1	.4646700D-2	.1689941D-2	.5767355D-3	.1687348D-3	.6991532D-4	.3496255D-4	.1193456D-4	.3511545D-5
3 I PFC	6.251937066	4.140328483	3.513608710	2.945167100	2.405117149	2.080670165	1.857743903	1.561905958	1.290700477
3 I PQC	1.120975415	.8377781800	.8040659320	.7849091241	.7778627311	.7812335933	.7890470450	.8120353510	.8608245000
1 I ZF1	6.045030261	5.971142450	5.965053280	5.961876942	5.960198128	5.959613249	5.959342979	5.959107789	5.958982668
1 I ZF2	8.113140441	8.113140441	8.113140441	8.113140441	8.113140441	8.113140441	8.113140441	8.113140441	8.113140441
1 I ZF3	18.67720946	21.47899041	21.77648610	21.93705214	22.02347523	22.05384281	22.06792115	22.08019580	22.08673485

(3) P 2+4	TH = 11 SBE = 5.240843064								MCPER FUNCTIONS
	RC = 02	RC = 05	RC = 07	RC = 10	RC = 15	RC = 20	RC = 25	RC = 35	RC = 50
AMAX DB	.0017375255	.0108709564	.0213327376	.0436480540	.0988323393	.1772876696	.2802872360	.5675287486	1.249387366
AMIN DB	18.21614685	41.66694135	50.44863377	59.78556761	70.46077061	78.11337851	84.13275310	93.46867914	104.0948449
MU I K	.1227984D 0	.8253781D-2	.3003090D-2	.1024995D-2	.2998896D-3	.1242599D-3	.6213873D-4	.2121124D-4	.6241051D-5
3 I PFC	6.098947535	4.126690976	3.509090976	2.944525412	2.405965018	2.081819213	1.858941621	1.563038768	1.291691948
3 I PQC	1.271191741	.8699716792	.8246540035	.7983420785	.7865256530	.7877500819	.7943500992	.8160250226	.8638999830
1 I ZF1	5.542982750	5.438884119	5.430641231	5.426376352	5.424132357	5.423352263	5.422992084	5.422678807	5.422512203
1 I ZF2	7.377549475	7.377549475	7.377549475	7.377549475	7.377549475	7.377549475	7.377549475	7.377549475	7.377549475
1 I ZF3	15.82957547	19.26573881	19.65636310	19.86881048	19.98361127	20.02402429	20.04277260	20.05912558	20.06783981

(3) P 2+4	TH = 12 SBE = 4.809734345								MCPER FUNCTIONS
	RC = 02	RC = 05	RC = 07	RC = 10	RC = 15	RC = 20	RC = 25	RC = 35	RC = 50
AMAX DB	.0017375255	.0108709564	.0213327376	.0436480540	.0988323393	.1772876696	.2802872360	.5675287486	1.249387366
AMIN DB	14.14908933	37.11135468	45.88562344	55.22069528	65.89544452	73.54797802	79.56733262	88.90324862	99.52941178
MU I K	.1961307D 0	.1394544D-1	.5078306D-2	.1733665D-2	.5072567D-3	.2101847D-3	.1051074D-3	.3587877D-4	.1055673D-4
3 I PFC	5.905863999	4.108050268	3.502522278	2.943156195	2.406654483	2.082963678	1.860190074	1.564254424	1.292769574
3 I PQC	1.468690508	.9086262897	.8487425295	.8137573590	.7963197216	.7950626811	.8002744000	.8204591724	.8673046796
1 I ZF1	5.140093569	4.997502994	4.986566397	4.980965986	4.978036494	4.977020972	4.976552601	4.976145484	4.975929075
1 I ZF2	6.764729520	6.764729520	6.764729520	6.764729520	6.764729520	6.764729520	6.764729520	6.764729520	6.764729520
1 I ZF3	13.37558110	17.35464926	17.85245488	18.12600522	18.27460609	18.32704483	18.35139432	18.37264430	18.38397240

(3) P 2+4	TH = 13 SBE = 4.445411483								MCPER FUNCTIONS
	RC = 02	RC = 05	RC = 07	RC = 10	RC = 15	RC = 20	RC = 25	RC = 35	RC = 50
AMAX DB	.0017375255	.0108709564	.0213327376	.0436480540	.0988323393	.1772876696	.2802872360	.5675287486	1.249387366
AMIN DB	10.80044470	32.92478000	41.68507197	51.01670394	61.69062548	69.34302419	75.36234267	84.69824056	95.32439911
MU I K	.2883884D 0	.2258193D-1	.8236570D-2	.2812968D-2	.8231306D-3	.3410741D-3	.1705622D-3	.5822211D-4	.1713089D-4
3 I PFC	5.676792523	4.083419018	3.493456679	2.940873074	2.407118049	2.084071268	1.861471290	1.565545805	1.293930994
3 I PQC	1.719444624	.9550886055	.8768926365	.8313943927	.8073422252	.8032254078	.8068550206	.8253568361	.8710491185
1 I ZF1	4.814310439	4.626441174	4.612155276	4.604931246	4.601180234	4.599884613	4.599287889	4.598769635	4.598494313
1 I ZF2	6.246349001	6.246349001	6.246349001	6.246349001	6.246349001	6.246349001	6.246349001	6.246349001	6.246349001
1 I ZF3	11.32703931	15.66918028	16.28656141	16.63052058	16.81866442	16.88526683	16.91622967	16.94326996	16.95769192

(3) P 2+4	TH = 14	SBE = 4.133565494						MCPER FUNCTIONS	
	RC = 02	RC = 05	RC = 07	RC = 10	RC = 15	RC = 20	RC = 25	RC = 35	RC = 50
AMAX DB	.0017375255	.0108709564	.0213327376	.0436480540	.0988323393	.1772876696	.2802872360	.5675287486	1.249387366
AMIN DB	8.177117691	29.05952266	37.79447705	47.11999415	57.79246612	65.44463076	71.46388674	80.79975334	91.42590391
MU I K	.3900714D 0	.3523902D-1	.1289069D-1	.4405552D-2	.1289367D-2	.5342794D-3	.2671811D-3	.9120367D-4	.2683520D-4
3 I PFC	5.423539433	4.051753188	3.481410884	2.937472065	2.407281450	2.085106380	1.862765435	1.566905036	1.295173590
3 I PQC	2.022501793	1.011040650	.9097894242	.8515340235	.8197057313	.8122999432	.8141316898	.8307393765	.8751450317
1 I ZF1	4.546688922	4.311103827	4.292671588	4.283486136	4.278759601	4.277134373	4.276387148	4.275738865	4.275394723
1 I ZF2	5.802171091	5.802171091	5.802171091	5.802171091	5.802171091	5.802171091	5.802171091	5.802171091	5.802171091
1 I ZF3	9.681600057	14.15683684	14.90308784	15.32646478	15.56012332	15.64317137	15.68183762	15.71563522	15.73367251

(3) P 2+4	TH = 15	SBE = 3.863703305						MCPER FUNCTIONS	
	RC = 02	RC = 05	RC = 07	RC = 10	RC = 15	RC = 20	RC = 25	RC = 35	RC = 50
AMAX DB	.0017375255	.0108709564	.0213327376	.0436480540	.0988323393	.1772876696	.2802872360	.5675287486	1.249387366
AMIN DB	6.204097061	25.48273039	34.17330365	43.48829562	54.15831400	61.81008644	67.82923811	77.16505262	87.79118995
MU I K	.4895478D 0	.5319410D-1	.1955847D-1	.6692451D-2	.1959225D-2	.8118872D-3	.4060113D-3	.1385949D-3	.4077937D-4
3 I PFC	5.161528406	4.011981954	3.465866797	2.932731572	2.407063471	2.086029983	1.864050733	1.568323465	1.296494481
3 I PQC	2.371355718	1.078550197	.9482680156	.8745117604	.8335403119	.8223565587	.8221492790	.8366306797	.8796054352
1 I ZF1	4.321040661	4.040883579	4.017343591	4.005798776	3.999921850	3.997912174	3.996990188	3.996191320	3.995767635
1 I ZF2	5.417355468	5.417355468	5.417355468	5.417355468	5.417355468	5.417355468	5.417355468	5.417355468	5.417355468
1 I ZF3	8.394952966	12.78232120	13.66182018	14.17283301	14.45808683	14.55998829	14.60752172	14.64911579	14.67133129

(3) P 2+4	TH = 16	SBE = 3.627955279						MCPER FUNCTIONS	
	RC = 02	RC = 05	RC = 07	RC = 10	RC = 15	RC = 20	RC = 25	RC = 35	RC = 50
AMAX DB		.0108709564	.0213327376	.0436480540	.0988323393	.1772876696	.2802872360	.5675287486	1.249387366
AMIN DB		22.17502373	30.79047972	40.08785200	50.75383523	58.40497022	64.42395318	73.75968368	84.38579971
MU I K		.7784824D-1	.2887194D-1	.9899366D-2	.2899401D-2	.1201577D-2	.6009002D-3	.2051237D-3	.6035455D-4
3 I PFC		3.963068317	3.446276129	2.926412550	2.406375783	2.086799500	1.865303396	1.569791617	1.297890507
3 I PQC		1.160081721	.9933422188	.9007251323	.8489961874	.8334752009	.8309583653	.8430573791	.8844447218
1 I ZF1		3.807912181	3.778131715	3.763757729	3.756531109	3.754076202	3.752952910	3.751981156	3.751466367
1 I ZF2		5.080771904	5.080771904	5.080771904	5.080771904	5.080771904	5.080771904	5.080771904	5.080771904
1 I ZF3		11.52343282	12.53349227	13.13894189	13.48182636	13.60509108	13.66272346	13.71322315	13.74022109

(3) P 2+4	TH = 17	SBE = 3.420303620						MCPER FUNCTIONS	
	RC = 02	RC = 05	RC = 07	RC = 10	RC = 15	RC = 20	RC = 25	RC = 35	RC = 50
AMAX DB		.0108709564	.0213327376	.0436480540	.0988323393	.1772876696	.2802872360	.5675287486	1.249387366
AMIN DB		19.12986794	27.62290639	36.89151090	47.55101914	55.20114481	61.21986217	70.55546087	81.18154357
MU I K		.1105367D 0	.4157715D-1	.1430291D-1	.4192268D-2	.1737572D-2	.8689742D-3	.2966381D-3	.8728162D-4
3 I PFC		3.904125679	3.422069051	2.918258909	2.405122778	2.087368676	1.866497533	1.571299162	1.299358210
3 I PQC		1.258413151	1.046232113	.9306457337	.8662468541	.8457467607	.8406158847	.8500491116	.8896787643
1 I ZF1		3.606222417	3.568930498	3.551174108	3.542369776	3.539402151	3.538048519	3.536879728	3.536261411
1 I ZF2		4.783909061	4.783909061	4.783909061	4.783909061	4.783909061	4.783909061	4.783909061	4.783909061
1 I ZF3		10.36837915	11.49700291	12.20149792	12.60780883	12.75501837	12.82404209	12.88462349	12.91704916

(3) P 2+4	TH = 18	SBE = 3.236067977						MCPER FUNCTIONS	
	RC = 02	RC = 05	RC = 07	RC = 10	RC = 15	RC = 20	RC = 25	RC = 35	RC = 50
AMAX DB		.0108709564	.0213327376	.0436480540	.0988323393	.1772876696	.2802872360	.5675287486	1.249387366
AMIN DB		16.35235977	24.65468006	33.87742664	44.52676388	52.17532701	58.19363592	67.52903277	78.15506456
MU I K		.1521886D 0	.5851484D-1	.2023619D-1	.5938296D-2	.2461692D-2	.1231171D-2	.4202893D-3	.1236650D-3
3 I PFC		3.834615418	3.392670634	2.907998363	2.403201435	2.087687457	1.867605065	1.572834369	1.300893825
3 I PQC		1.376388221	1.108382507	.9648323350	.8854927681	.8592745548	.8511858875	.8576388093	.8953250327
1 I ZF1		3.431139090	3.385033566	3.363249933	3.352606215	3.349050759	3.347435010	3.346043053	3.345307888
1 I ZF2		4.520147021	4.520147021	4.520147021	4.520147021	4.520147021	4.520147021	4.520147021	4.520147021
1 I ZF3		9.313475537	10.53763719	11.34266837	11.81772028	11.99148969	12.07324828	12.14515051	12.18368973

(3) P 2+4	TH = 19	SBE = 3.071553487						MCPER FUNCTIONS	
	RC = 02	RC = 05	RC = 07	RC = 10	RC = 15	RC = 20	RC = 25	RC = 35	RC = 50
AMAX DB		.0108709564	.0213327376	.0436480540	.0988323393	.1772876696	.2802872360	.5675287486	1.249387366
AMIN DB		13.85567105	21.87676873	31.02819810	41.66185744	49.30804344	55.32574195	64.66083608	75.28679183
MU I K		.2028694D 0	.8056781D-1	.2809248D-1	.8258613D-2	.3424503D-2	.1712825D-2	.5847338D-3	.1720523D-3
3 I PFC		3.754616131	3.357531262	2.895344036	2.400501203	2.087701854	1.868595630	1.574386563	1.302493254
3 I PQC		1.516470904	1.181457548	1.003944988	.9069656566	.8741760544	.8627404106	.8658630327	.9014027262
1 I ZF1		3.278804589	3.222760655	3.196214453	3.183430648	3.179203520	3.177290829	3.175647447	3.174781179
1 I ZF2		4.284259547	4.284259547	4.284259547	4.284259547	4.284259547	4.284259547	4.284259547	4.284259547
1 I ZF3		8.360366570	9.645889197	10.54878872	11.09711905	11.30004951	11.39592578	11.48044577	11.52582415

(3) P 2+4	TH = 20	SBE = 2.923804400						MCPER FUNCTIONS	
	RC = 02	RC = 05	RC = 07	RC = 10	RC = 15	RC = 20	RC = 25	RC = 35	RC = 50
AMAX DB		.0108709564	.0213327376	.0436480540	.0988323393	.1772876696	.2802872360	.5675287486	1.249387366
AMIN DB		11.65401861	19.28631383	28.33032946	38.94023636	46.58288720	52.59966994	61.93431846	72.56016275
MU I K		.2613961D 0	.1085574D 0	.3832512D-1	.1129765D-1	.4636576D-2	.2344318D-2	.8003576D-3	.2355005D-3
3 I PFC		3.665070271	3.316179308	2.879997482	2.396903937	2.087353820	1.869436487	1.575941072	1.304152055
3 I PQC		1.680202645	1.267289143	1.048753770	.9309335090	.8905848962	.8753604874	.8747623503	.9079329209
1 I ZF1		3.145820774	3.079184039	3.047069298	3.031800382	3.026807724	3.024559857	3.022634475	3.021621869
1 I ZF2		4.072065660	4.072065660	4.072065660	4.072065660	4.072065660	4.072065660	4.072065660	4.072065660
1 I ZF3		7.512368115	8.816589563	9.809486089	10.43449923	10.66911989	10.78052116	10.87900715	10.93198903

(3) P 2+4	TH = 21	SBE = 2.790428110						MCPER FUNCTIONS	
	RC = 02	RC = 05	RC = 07	RC = 10	RC = 15	RC = 20	RC = 25	RC = 35	RC = 50
AMAX DB		.0108709564	.0213327376	.0436480540	.0988323393	.1772876696	.2802872360	.5675287486	1.249387366
AMIN DB		9.753635622	16.88856803	25.77393488	36.34844577	43.98589232	50.00134824	59.33535150	69.96103519
MU I K		.3253256D 0	.1430776D 0	.5144027D-1	.1522572D-1	.6319830D-2	.3161787D-2	.1079524D-2	.3176495D-3
3 I PFC		3.567833835	3.268302874	2.861654160	2.392283917	2.086581114	1.870092414	1.577484180	1.305865412
3 I PQC		1.867834967	1.367754133	1.100174503	.9577062720	.9086532137	.8891373123	.8843817716	.9149387375
1 I ZF1		3.029045033	2.951914923	2.913405497	2.895257915	2.889394383	2.886769262	2.884528772	2.883353556
1 I ZF2		3.880180768	3.880180768	3.880180768	3.880180768	3.880180768	3.880180768	3.880180768	3.880180768
1 I ZF3		6.770570851	8.048074256	9.117078974	9.820628469	10.08932588	10.21766594	10.33150934	10.39289633

(3) P 2+4	TH = 22	SBE = 2.669467163						MCPER FUNCTIONS	
	RC = 02	RC = 05	RC = 07	RC = 10	RC = 15	RC = 20	RC = 25	RC = 35	RC = 50
AMAX DB		.0108709564	.0213327376	.0436480540	.0988323393	.1772876696	.2802872360	.5675287486	1.249387366
AMIN DB		8.146136168	14.69035008	23.35261566	33.87525056	41.50514870	47.51869840	56.85178084	67.47723664
MU I K		.3914652D 0	.1842818D 0	.6797813D-1	.2024126D-1	.8408965D-2	.4207897D-2	.1436848D-2	.4228031D-3
3 I PFC		3.465411869	3.213856999	2.840013050	2.386507996	2.085317189	1.870525605	1.579000567	1.307628121
3 I PQC		2.078403170	1.484569112	1.159221861	.9876421961	.9285543295	.9041735844	.8947712406	.9224455283
1 I ZF1		2.925565377	2.838928167	2.793268444	2.771798705	2.764945847	2.761897058	2.759305559	2.757950361
1 I ZF2		3.705835670	3.705835670	3.705835670	3.705835670	3.705835670	3.705835670	3.705835670	3.705835670
1 I ZF3		6.131267171	7.341157116	8.466157915	9.248073914	9.553007257	9.699684466	9.830309859	9.900938979

(3) P 2+4	TH = 23 SBE = 2.559304665								MCPER FUNCTIONS
	RC = 02	RC = 05	RC = 07	RC = 10	RC = 15	RC = 20	RC = 25	RC = 35	RC = 50
AMAX DB		.0108709564	.0213327376	.0436480540	.0988323393	.1772876696	.2802872360	.5675287486	1.249387366
AMIN DB		6.808020339	12.70207249	21.06341979	31.51136380	39.13044682	45.14129160	54.47307693	65.09821371
MU I K		.4566663D 0	.2316842D 0	.8847672D-1	.2657246D-1	.1105294D-1	.5532678D-2	.1889497D-2	.5560186D-3
3 I PFC		3.360475833	3.153170253	2.814792659	2.379435993	2.083491084	1.870695561	1.580473751	1.309434557
3 I PQC		2.310218482	1.619033437	1.227047970	1.021154657	.9504858401	.9205850556	.9059861996	.9304810894
1 I ZF1		2.832796105	2.738421434	2.685054692	2.659773443	2.651798005	2.648274225	2.645292725	2.643738995
1 I ZF2		3.546742776	3.546742776	3.546742776	3.546742776	3.546742776	3.546742776	3.546742776	3.546742776
1 I ZF3		5.585808655	6.697778455	7.853272971	8.710859830	9.053860215	9.220231160	9.369084081	9.449825381

(3) P 2+4	TH = 24 SBE = 2.458593336								MCPER FUNCTIONS
	RC = 02	RC = 05	RC = 07	RC = 10	RC = 15	RC = 20	RC = 25	RC = 35	RC = 50
AMAX DB		.0108709564	.0213327376	.0436480540	.0988323393	.1772876696	.2802872360	.5675287486	1.249387366
AMIN DB		5.705685228	10.93112860	18.90675214	29.24926944	36.85303275	42.86008154	52.19006074	62.81475643
MU I K		.5184606D 0	.2840819D 0	.1134129D 0	.3447756D-1	.1436641D-1	.7194422D-2	.2457519D-2	.7232063D-3
3 I PFC		3.255410722	3.087000292	2.785755817	2.370921474	2.081027348	1.870558984	1.581886026	1.311278656
3 I PQC		2.561470217	1.771808453	1.304882758	1.058719064	.9746731152	.9385023069	.9180882320	.9390758985
1 I ZF1		2.748579030	2.648719868	2.587429491	2.557814684	2.548566956	2.544511319	2.541097388	2.539325309
1 I ZF2		3.400995771	3.400995771	3.400995771	3.400995771	3.400995771	3.400995771	3.400995771	3.400995771
1 I ZF3		5.122399007	6.119455588	7.276660697	8.204219474	8.586671530	8.774019609	8.942551360	9.034305438

(3) P 2+4	TH = 25 SBE = 2.366201583								MCPER FUNCTIONS
	RC = 02	RC = 05	RC = 07	RC = 10	RC = 15	RC = 20	RC = 25	RC = 35	RC = 50
AMAX DB			.0213327376	.0436480540	.0988323393	.1772876696	.2802872360	.5675287486	1.249387366
AMIN DB			9.378454078	16.88604954	27.08312118	34.66542143	40.66719616	49.99468641	60.61877891
MU I K			.3396857D 0	.1431191D 0	.4424294D-1	.1848115D-1	.9260623D-2	.3164213D-2	.9312388D-3
3 I PFC			3.016489014	2.752744527	2.360813185	2.077846011	1.870069673	1.583218394	1.313153877
3 I PQC			1.942843952	1.393973096	1.100879137	1.001373210	.9580727827	.9311457967	.9482633842
1 I ZF1			2.568239504	2.499257726	2.464780800	2.454093315	2.449442716	2.445550093	2.443538482
1 I ZF2			3.266993358	3.266993358	3.266993358	3.266993358	3.266993358	3.266993358	3.266993358
1 I ZF3			5.605953668	6.735943963	7.724413843	8.147119692	8.356617923	8.546269974	8.649962311

(3) P 2+4	TH = 26	SBE = 2.281172033						MCPER FUNCTIONS	
	RC = 02	RC = 05	RC = 07	RC = 10	RC = 15	RC = 20	RC = 25	RC = 35	RC = 50
AMAX DB			.0213327376	.0436480540	.0988323393	.1772876696	.2802872360	.5675287486	1.249387366
AMIN DB			8.036464289	15.00701292	25.00870125	32.56126111	38.55577519	47.87986741	58.50314323
MU I K			.3964394D 0	.1776844D 0	.5617783D-1	.2354707D-1	.1180895D-1	.4036516D-2	.1188072D-2
3 I PFC			2.943012683	2.715722528	2.348957519	2.073862637	1.869178419	1.584450494	1.315053182
3 I PQC			2.131503275	1.495484208	1.148251367	1.030879145	.9794631045	.9452350664	.9580802293
1 I ZF1			2.495505344	2.419544563	2.379712493	2.367399369	2.362083742	2.357661889	2.355388075
1 I ZF2			3.143380605	3.143380605	3.143380605	3.143380605	3.143380605	3.143380605	3.143380605
1 I ZF3			5.154650409	6.231745543	7.268599683	7.731625358	7.964292726	8.176472813	8.293052305

(3) P 2+4	TH = 27	SBE = 2.202689265						MCPER FUNCTIONS	
	RC = 02	RC = 05	RC = 07	RC = 10	RC = 15	RC = 20	RC = 25	RC = 35	RC = 50
AMAX DB			.0213327376	.0436480540	.0988323393	.1772876696	.2802872360	.5675287486	1.249387366
AMIN DB			6.889664340	13.27626387	23.02341889	30.53524062	36.51984470	45.83933676	56.46151678
MU I K			.4523939D 0	.2168637D 0	.7060396D-1	.2973295D-1	.1492821D-1	.5105440D-2	.1502880D-2
3 I PFC			2.867982352	2.674817221	2.335202505	2.068988516	1.867832926	1.585560530	1.316968991
3 I PQC			2.336827836	1.610380780	1.201525874	1.063524429	1.002861683	.9604408850	.9685667137
1 I ZF1			2.429196688	2.347386020	2.301798157	2.287655733	2.281597343	2.276590957	2.274030647
1 I ZF2			3.029003316	3.029003316	3.029003316	3.029003316	3.029003316	3.029003316	3.029003316
1 I ZF3			4.760749001	5.765187879	6.834729332	7.337238674	7.593889376	7.829948818	7.960380394

(3) P 2+4	TH = 28	SBE = 2.130054468						MCPER FUNCTIONS	
	RC = 02	RC = 05	RC = 07	RC = 10	RC = 15	RC = 20	RC = 25	RC = 35	RC = 50
AMAX DB			.0213327376	.0436480540	.0988323393	.1772876696	.2802872360	.5675287486	1.249387366
AMIN DB			5.917245029	11.69951761	21.12631987	28.58303284	34.55422182	43.86753405	54.48825570
MU I K			.5059851D 0	.2600304D 0	.8783832D-1	.3722617D-1	.1871927D-1	.6406536D-2	.1886198D-2
3 I PFC			2.792671441	2.630346377	2.319403980	2.063131065	1.865977742	1.586525188	1.318893158
3 I PQC			2.557817896	1.739317869	1.261461221	1.099687619	1.028481637	.9768578593	.9797671024
1 I ZF1			2.368185680	2.281933044	2.230345563	2.214155089	2.207267903	2.201616384	2.198743512
1 I ZF2			2.922872192	2.922872192	2.922872192	2.922872192	2.922872192	2.922872192	2.922872192
1 I ZF3			4.418061555	5.337315100	6.421469152	6.961554410	7.242739550	7.533943304	7.649202487

$P^{(3)}_{2+4}$	TH = 29	SBE = 2.062665340						MCPER FUNCTIONS	
	RC = 02	RC = 05	RC = 07	RC = 10	RC = 15	RC = 20	RC = 25	RC = 35	RC = 50
AMAX DB			.0213327376	.0436480540	.0988323393	.1772876696	.2802872360	.5675287486	1.249387366
AMIN DB			5.096223433	10.27967643	19.31806518	26.70126877	32.65444509	41.95951409	52.57830910
MU K			.5561460D 0	.3062078D 0	.1081675D 0	.4623135D-1	.2329581D-1	.7980393D-2	.2350090D-2
3 PFC			2.718116570	2.582814236	2.301434523	2.056194582	1.863554241	1.587319562	1.320816922
3 PQC			2.793634415	1.882581112	1.328870327	1.139796504	1.056563993	.9945916039	.9917300851
1 ZF1			2.311549723	2.222373421	2.164757864	2.146291216	2.138480489	2.132117382	2.128903937
1 ZF2			2.824134395	2.824134395	2.824134395	2.824134395	2.824134395	2.824134395	2.824134395
1 ZF3			4.119911638	4.948548484	6.028121614	6.602648086	6.908589811	7.196084732	7.357147984

$P^{(3)}_{2+4}$	TH = 30	SBE = 2.000000000						MCPER FUNCTIONS	
	RC = 02	RC = 05	RC = 07	RC = 10	RC = 15	RC = 20	RC = 25	RC = 35	RC = 50
AMAX DB				.0436480540	.0988323393	.1772876696	.2802872360	.5675287486	1.249387366
AMIN DB				9.015439145	17.60082246	24.88753578	30.81672685	40.11087298	50.72713975
MU K				.3541833D 0	.1318132D 0	.5696698D-1	.2878483D-1	.9873164D-2	.2908326D-2
3 PFC				2.532870467	2.281195486	2.048081482	1.860500673	1.587917067	1.322730876
3 PQC				2.040104111	1.404594855	1.184331333	1.087381135	1.013760157	1.004509274
1 ZF1				2.167932412	2.104512577	2.083541997	2.074704224	2.067556662	2.063972494
1 ZF2				2.732050808	2.732050808	2.732050808	2.732050808	2.732050808	2.732050808
1 ZF3				4.598320178	5.654535913	6.259027578	6.589546419	6.904319229	7.082157920

$P^{(3)}_{2+4}$	TH = 31	SBE = 1.941604026						MCPER FUNCTIONS	
	RC = 02	RC = 05	RC = 07	RC = 10	RC = 15	RC = 20	RC = 25	RC = 35	RC = 50
AMAX DB				.0436480540	.0988323393	.1772876696	.2802872360	.5675287486	1.249387366
AMIN DB				7.900887542	15.97800626	23.14039177	29.03792511	38.31768395	48.93065827
MU K				.4026759D 0	.1588911D 0	.6965951D-1	.3532675D-1	.1213712D-1	.3576573D-2
3 PFC				2.481240953	2.258631722	2.038694277	1.856752316	1.588289353	1.324624914
3 PQC				2.211559546	1.489467012	1.233826214	1.121240374	1.034495588	1.018163770
1 ZF1				2.117886975	2.049142803	2.025455370	2.015478850	2.007467031	2.003479642
1 ZF2				2.645977691	2.645977691	2.645977691	2.645977691	2.645977691	2.645977691
1 ZF3				4.284980517	5.300992807	5.929595357	6.284032850	6.626367381	6.822435174

(3) P 2+4	TH = 32 SBE = 1.887079915							MCPER FUNCTIONS	
	RC = 02	RC = 05	RC = 07	RC = 10	RC = 15	RC = 20	RC = 25	RC = 35	RC = 50
AMAX DB				.0436480540	.0988323393	.1772876696	.2802872360	.5675287486	1.249387366
AMIN DB				6.926095146	14.45381588	21.45938257	27.31553231	36.57647552	47.18516807
MU I K				.4505005D 0	.1893691D 0	.8453389D-1	.4307481D-1	.1483120D-1	.4372619D-2
3 I PFC				2.428651282	2.233747328	2.027938513	1.852241802	1.588406228	1.326488138
3 I PQC				2.396490058	1.584261034	1.288867528	1.158487476	1.056945810	1.032758805
1 I ZF1				2.071584141	1.998220692	1.971637425	1.960403746	1.951440512	1.947014825
1 I ZF2				2.565351779	2.565351779	2.565351779	2.565351779	2.565351779	2.565351779
1 I ZF3				4.005965049	4.968054453	5.613616520	5.990757224	6.362193180	6.576404127

(3) P 2+4	TH = 33 SBE = 1.836078459							MCPER FUNCTIONS	
	RC = 02	RC = 05	RC = 07	RC = 10	RC = 15	RC = 20	RC = 25	RC = 35	RC = 50
AMAX DB				.0436480540	.0988323393	.1772876696	.2802872360	.5675287486	1.249387366
AMIN DB				6.078399120	13.03256417	19.84504473	25.64767841	34.88414533	45.48731941
MU I K				.4966839D 0	.2230344D 0	.1018000D 0	.5219346D-1	.1802157D-1	.5316601D-2
3 I PFC				2.375763824	2.206619302	2.015726953	1.846899677	1.588235578	1.328309057
3 I PQC				2.594435743	1.689641380	1.350038008	1.199509852	1.081276622	1.048366473
1 I ZF1				2.028454792	1.951343875	1.921742036	1.909128839	1.899119438	1.894217550
1 I ZF2				2.489678091	2.489678091	2.489678091	2.489678091	2.489678091	2.489678091
1 I ZF3				3.758114786	4.656381198	5.310687517	5.708687362	6.108923674	6.342677787

(3) P 2+4	TH = 34 SBE = 1.788291650							MCPER FUNCTIONS	
	RC = 02	RC = 05	RC = 07	RC = 10	RC = 15	RC = 20	RC = 25	RC = 35	RC = 50
AMAX DB				.0436480540	.0988323393	.1772876696	.2802872360	.5675287486	1.249387366
AMIN DB				5.343835501	11.71786709	18.29886929	24.03314461	33.23798371	43.83407089
MU I K				.5405156D 0	.2594816D 0	.1216344D 0	.6285543D-1	.2178215D-1	.6431266D-2
3 I PFC				2.323139537	2.177404968	2.001985159	1.840655308	1.587743301	1.330075034
3 I PQC				2.805023831	1.806119063	1.418155135	1.244738983	1.107673982	1.065066561
1 I ZF1				1.988018473	1.908126036	1.875461558	1.861346963	1.850189124	1.844770049
1 I ZF2				2.418519883	2.418519883	2.418519883	2.418519883	2.418519883	2.418519883
1 I ZF3				3.538021734	4.366533314	5.020700007	5.437031382	5.865924315	6.120030858

$P^{(3)}_{2+4}$	TH = 35	SBE = 1.743446796						MCPER FUNCTIONS	
	RC = 02	RC = 05	RC = 07	RC = 10	RC = 15	RC = 20	RC = 25	RC = 35	RC = 50
AMAX DB					.0988323393	.1772876696	.2802872360	.5675287486	1.249387366
AMIN DB					10.51184408	16.82319808	22.47138224	31.63563074	42.22265735
MU I K					.2981315D 0	.1441584D 0	.7523690D-1	.2619500D-1	.7742249D-2
3 I PFC					2.146339512	1.986658429	1.833438270	1.586893266	1.331772723
3 I PQC					1.934028661	1.493752992	1.294651499	1.136346514	1.082947499
1 I ZF1					1.863192616	1.832518337	1.816787309	1.804371785	1.798391192
1 I ZF2					2.351490335	2.351490335	2.351490335	2.351490335	2.351490335
1 I ZF3					4.098790357	4.743794268	5.175221787	5.632179347	5.907377594

$P^{(3)}_{2+4}$	TH = 36	SBE = 1.701301617						MCPER FUNCTIONS	
	RC = 02	RC = 05	RC = 07	RC = 10	RC = 15	RC = 20	RC = 25	RC = 35	RC = 50
AMAX DB					.0988323393	.1772876696	.2802872360	.5675287486	1.249387366
AMIN DB					9.414516875	15.42102518	20.96252888	30.07507149	40.65056317
MU I K					.3382783D 0	.1694138D 0	.8951041D-1	.3135064D-1	.9278373D-2
3 I PFC					2.113722295	1.969719667	1.825180330	1.585647293	1.333387763
3 I PQC					2.073534302	1.577557777	1.349768193	1.167528242	1.102107439
1 I ZF1					1.831181711	1.792656973	1.775209676	1.761421460	1.754831424
1 I ZF2					2.288245611	2.288245611	2.288245611	2.288245611	2.288245611
1 I ZF3					3.853024919	4.480297665	4.922900817	5.406826316	5.703753525

$P^{(3)}_{2+4}$	TH = 37	SBE = 1.661640141						MCPER FUNCTIONS	
	RC = 02	RC = 05	RC = 07	RC = 10	RC = 15	RC = 20	RC = 25	RC = 35	RC = 50
AMAX DB					.0988323393	.1772876696	.2802872360	.5675287486	1.249387366
AMIN DB					8.423553368	14.09568838	19.50741042	28.55463360	39.11550038
MU I K					.3791598D 0	.1973402D 0	.1058350D 0	.3734808D-1	.1107197D-1
3 I PFC					2.079894110	1.951177209	1.815818175	1.583965199	1.334904762
3 I PQC					2.224664226	1.670209026	1.410650129	1.201481490	1.122655482
1 I ZF1					1.796748917	1.755637532	1.736399294	1.721119742	1.713868518
1 I ZF2					2.228479056	2.228479056	2.228479056	2.228479056	2.228479056
1 I ZF3					3.628655612	4.230646590	4.679904586	5.189133989	5.508300360

(3) P 2+4	TH = 38	SBE = 1.624269245						MCPER FUNCTIONS	
	RC = 02	RC = 05	RC = 07	RC = 10	RC = 15	RC = 20	RC = 25	RC = 35	RC = 50
AMAX DB					.0988323393	.1772876696	.2802872360	.5675287486	1.249387366
AMIN DB					7.534395377	12.85045998	18.10751458	27.07299180	37.61539092
MU I K					.4200299D 0	.2277598D 0	.1243438D 0	.4429456D-1	.1315923D-1
3 I PFC					2.045210348	1.931081159	1.805296947	1.581804900	1.336307225
3 I PQC					2.387364721	1.772280658	1.477891083	1.238499903	1.144713085
1 I ZF1					1.764573900	1.721230108	1.700162056	1.683272150	1.675304006
1 I ZF2					2.171916299	2.171916299	2.171916299	2.171916299	2.171916299
1 I ZF3					3.424680686	3.995295153	4.446243291	4.978493027	5.320253510

(3) P 2+4	TH = 39	SBE = 1.589015729						MCPER FUNCTIONS	
	RC = 02	RC = 05	RC = 07	RC = 10	RC = 15	RC = 20	RC = 25	RC = 35	RC = 50
AMAX DB					.0988323393	.1772876696	.2802872360	.5675287486	1.249387366
AMIN DB					6.740691919	11.68808062	16.76491953	25.62917334	36.14835292
MU I K					.4602199D 0	.2603730D 0	.1451289D 0	.5230474D-1	.1558053D-1
3 I PFC					2.010015807	1.909526537	1.793574537	1.579122636	1.337577491
3 I PQC					2.561559855	1.884257604	1.552105836	1.278911450	1.168415649
1 I ZF1					1.734366376	1.689211227	1.666320079	1.647705011	1.638960154
1 I ZF2					2.118311118	2.118311118	2.118311118	2.118311118	2.118311118
1 I ZF3					3.239771026	3.774619800	4.222074801	4.774408876	5.138931791

(3) P 2+4	TH = 40	SBE = 1.555723827						MCPER FUNCTIONS	
	RC = 02	RC = 05	RC = 07	RC = 10	RC = 15	RC = 20	RC = 25	RC = 35	RC = 50
AMAX DB					.0988323393	.1772876696	.2802872360	.5675287486	1.249387366
AMIN DB					6.034890967	10.61031055	15.48216245	24.22259818	34.71269041
MU I K					.4991780D 0	.2947708D 0	.1682255D 0	.61499290-1	.1838085D-1
3 I PFC					1.974625368	1.886651857	1.780626400	1.575873338	1.338696656
3 I PQC					2.747204948	2.006523886	1.633914491	1.323081224	1.193914335
1 I ZF1					1.705869890	1.659362412	1.634707607	1.614262756	1.604677393
1 I ZF2					2.067441918	2.067441918	2.067441918	2.067441918	2.067441918
1 I ZF3					3.072390267	3.568833681	4.007669536	4.576496292	4.963728969

(3) P 2+4	TH = 41	SBE = 1.524253087						MCPER FUNCTIONS	
	RC = 02	RC = 05	RC = 07	RC = 10	RC = 15	RC = 20	RC = 25	RC = 35	RC = 50
AMAX DB					.0988323393	.1772876696	.2802872360	.5675287486	1.249387366
AMIN DB					5.408842289	9.617584858	14.26203957	22.85304678	33.30688658
MU I K					.5364854D 0	.3304614D 0	.1935967D 0	.7200251D-1	.2161004D-1
3 I PFC					1.939312248	1.862632675	1.766450386	1.572011190	1.339644507
3 I PQC					2.944326931	2.139365945	1.723923947	1.371413820	1.221378110
1 I ZF1					1.678862867	1.631470913	1.605167363	1.582805513	1.572312131
1 I ZF2					2.019108731	2.019108731	2.019108731	2.019108731	2.019108731
1 I ZF3					2.920912449	3.377925285	3.803365937	4.384474912	4.794106875

(3) P 2+4	TH = 42	SBE = 1.494476550						MCPER FUNCTIONS	
	RC = 02	RC = 05	RC = 07	RC = 10	RC = 15	RC = 20	RC = 25	RC = 35	RC = 50
AMAX DB						.1772876696	.2802872360	.5675287486	1.249387366
AMIN DB						8.708840646	13.10734227	21.52072741	31.92960033
MU I K						.3669087D 0	.2211225D 0	.8393897D-1	.2532328D-1
3 I PFC						1.837670924	1.751070850	1.567490460	1.340399453
3 I PQC						2.282991389	1.822708867	1.424354924	1.250996069
1 I ZF1						1.605332110	1.577547580	1.553206948	1.541734861
1 I ZF2						1.973130639	1.973130639	1.973130639	1.973130639
1 I ZF3						3.201631345	3.609518102	4.198165166	4.629589853

(3) P 2+4	TH = 43	SBE = 1.466279186						MCPER FUNCTIONS	
	RC = 02	RC = 05	RC = 07	RC = 10	RC = 15	RC = 20	RC = 25	RC = 35	RC = 50
AMAX DB						.1772876696	.2802872360	.5675287486	1.249387366
AMIN DB						7.881541147	12.02055148	20.22626354	30.57966617
MU I K						.4035738D 0	.2505950D 0	.9742868D-1	.2958126D-1
3 I PFC						1.811981917	1.734541163	1.562266636	1.340938470
3 I PQC						2.437559449	1.930795414	1.482391753	1.282980036
1 I ZF1						1.580752751	1.551699939	1.525352260	1.512828543
1 I ZF2						1.929343559	1.929343559	1.929343559	1.929343559
1 I ZF3						3.039445770	3.426439952	4.017483711	4.469760383

(3) P 2+4	TH = 44	SBE = 1.439556540						MCPER FUNCTIONS	
	RC = 02	RC = 05	RC = 07	RC = 10	RC = 15	RC = 20	RC = 25	RC = 35	RC = 50
AMAX DB						.1772876696	.2802872360	.5675287486	1.249387366
AMIN DB						7.131871525	11.00352995	18.97070035	29.25609735
MU I K						.4399531D 0	.2817238D 0	.1125810D 0	.3445047D-1
3 I PFC						1.785781321	1.716943834	1.556297933	1.341237055
3 I PQC						2.603217148	2.048651327	1.546051921	1.317567472
1 I ZF1						1.557554141	1.527478603	1.499136299	1.485487194
1 I ZF2						1.887598333	1.887598333	1.887598333	1.887598333
1 I ZF3						2.890657835	3.254352652	3.842437452	4.314255709

(3) P 2+4	TH = 45	SBE = 1.414213562						MCPER FUNCTIONS	
	RC = 02	RC = 05	RC = 07	RC = 10	RC = 15	RC = 20	RC = 25	RC = 35	RC = 50
AMAX DB						.1772876696	.2802872360	.5675287486	1.249387366
AMIN DB						6.455045861	10.05726158	17.75548812	27.95809238
MU I K						.4756064D 0	.3141499D 0	.1294868D 0	.4000326D-1
3 I PFC						1.759274072	1.698387856	1.549547182	1.341269207
3 I PQC						2.780135052	2.176685183	1.615900358	1.355024694
1 I ZF1						1.535574532	1.504740419	1.474461774	1.459614651
1 I ZF2						1.847759065	1.847759065	1.847759065	1.847759065
1 I ZF3						2.754407670	3.093342851	3.673115108	4.162765358

(3) P 2+4	TH = 46	SBE = 1.390163591						MCPER FUNCTIONS	
	RC = 02	RC = 05	RC = 07	RC = 10	RC = 15	RC = 20	RC = 25	RC = 35	RC = 50
AMAX DB						.1772876696	.2802872360	.5675287486	1.249387366
AMIN DB						5.845654626	9.181682484	16.58243970	26.68504461
MU I K						.5101728D 0	.3474688D 0	.1482102D 0	.4631778D-1
3 I PFC						1.732646382	1.679003408	1.541984075	1.341007439
3 I PQC						2.968538262	2.315255894	1.692534063	1.395650404
1 I ZF1						1.514670378	1.483346089	1.451237540	1.435123486
1 I ZF2						1.809701684	1.809701684	1.809701684	1.809701684
1 I ZF3						2.629746340	2.943337971	3.509675344	4.015029400

$P^{(1)}_{3+0}$

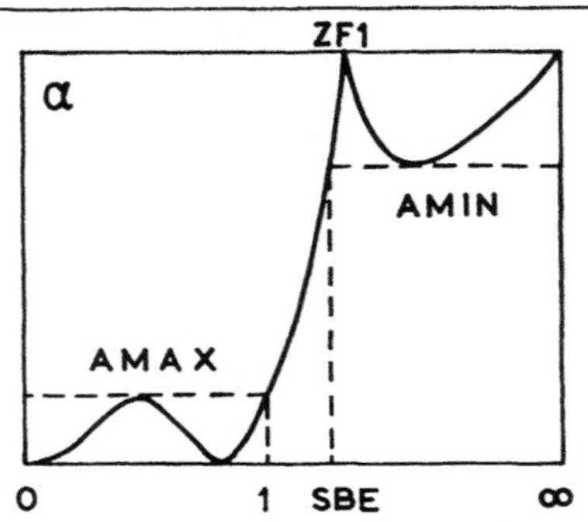

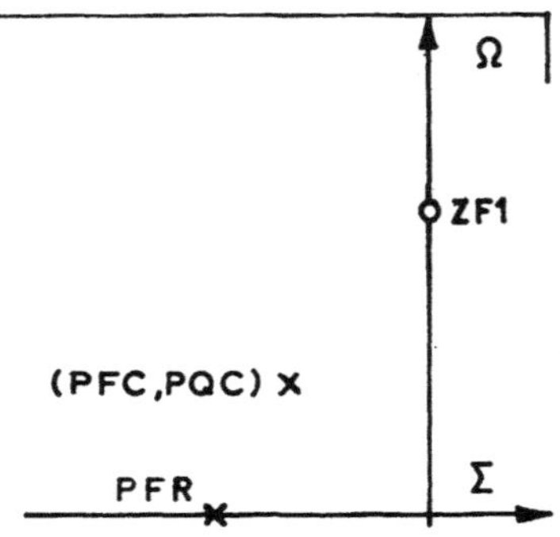

$P^{(1)}_{3+0}$ TH = 1 SBE = 57.29868850		CAUER FUNCTIONS								
		RC = 02	RC = 05	RC = 07	RC = 10	RC = 15	RC = 20	RC = 25	RC = 35	RC = 50
AMAX DB		.0017375255	.0108709564	.0213327376	.0436480540	.0988323393	.1772876696	.2802872360	.5675287486	1.249387366
AMIN DB		95.59143354	103.5593671	106.4923896	109.6127442	113.1897536	115.7669837	117.8081835	121.0179857	124.7978836
MU	K	.2855362D-2	.1140944D-2	.8139793D-3	.5683236D-3	.3764828D-3	.2798232D-3	.2212196D-3	.1528740D-3	.9893243D-4
1	PFR	2.213986032	1.563798192	1.364605476	1.171930067	.9722779023	.8408207428	.7434300089	.6017573636	.4532634534
1	PFC	2.376013674	1.787095168	1.615879297	1.456977215	1.301916599	1.206968433	1.141294123	1.054537098	.9774617508
1	PQC	1.074569543	1.143625814	1.184843359	1.243831995	1.339556228	1.435942504	1.535630552	1.752874381	2.156968839
1	ZF1	66.16156682	66.16156682	66.16156632	66.16156682	66.16156682	66.16156682	66.16156682	66.16156682	66.16156682

$P^{(1)}_{3+0}$ TH = 2 SBE = 28.65370835		CAUER FUNCTIONS								
		RC = 02	RC = 05	RC = 07	RC = 10	RC = 15	RC = 20	RC = 25	RC = 35	RC = 50
AMAX DB		.0017375255	.0108709564	.0213327376	.0436480540	.0988323393	.1772876696	.2802872360	.5675287486	1.249387366
AMIN DB		77.52764901	85.49558255	88.42860504	91.54895955	95.12596902	97.70319908	99.74439890	102.9542911	106.7340989
MU	K	.1142319D-1	.4564473D-2	.3256413D-2	.2273641D-2	.1506161D-2	.1119463D-2	.8350134D-3	.6115891D-3	.3957900D-3
1	PFR	2.216974256	1.565029949	1.365498281	1.172565840	.9727101966	.8411487724	.7436938214	.6019446873	.4533883203
1	PFC	2.374818768	1.786697917	1.615627795	1.456831801	1.301850341	1.206939876	1.141287239	1.054553700	.9774946363
1	PQC	1.076746217	1.144977537	1.186006562	1.244844495	1.340449888	1.436783053	1.536448178	1.753693091	2.157860386
1	ZF1	33.08393284	33.08393284	33.08393284	33.08393284	33.08393284	33.08393284	33.08393284	33.08393284	33.08393284

$P^{(1)}_{3+0}$ TH = 3 SBE = 19.10732261		CAUER FUNCTIONS								
		RC = 02	RC = 05	RC = 07	RC = 10	RC = 15	RC = 20	RC = 25	RC = 35	RC = 50
AMAX DB		.0017375255	.0108709564	.0213327376	.0436480540	.0988323393	.1772876696	.2802872360	.5675287486	1.249387366
AMIN DB		66.95886493	74.92679780	77.85982022	80.98017470	84.55718415	87.13441421	89.17561403	92.38541625	96.16531407
MU	K	.2570870D-1	.1027267D-1	.7328791D-2	.5116991D-2	.3389722D-2	.2519432D-2	.1991786D-2	.1376425D-2	.8907537D-3
1	PFR	2.221972313	1.567087086	1.366988813	1.173626941	.9734315051	.8416960297	.7441339029	.6022571368	.4535965741
1	PFC	2.372824058	1.786034948	1.615208070	1.456589108	1.301739722	1.206892159	1.141275682	1.054581324	.9775494266
1	PQC	1.080391282	1.147236904	1.187949962	1.246535477	1.341941910	1.438186148	1.537812847	1.755059371	2.159348042
1	ZF1	22.05945544	22.05945544	22.05945544	22.05945544	22.05945544	22.05945544	22.05945544	22.05945544	22.05945544

(1) P 3+0	TH = 4	SBE = 14.33558703							CAUER FUNCTIONS
	RC = 02	RC = 05	RC = 07	RC = 10	RC = 15	RC = 20	RC = 25	RC = 35	RC = 50
AMAX DB	.0017375255	.0108709564	.0213327376	.0436480540	.0988323393	.1772876696	.2802872360	.5675287486	1.249387366
AMIN DB	59.45790913	67.42583860	70.35886071	73.47921502	77.05622438	79.63345440	81.67465421	84.88445642	88.66435423
MU K	.4572062D-1	.1826903D-1	.1303360D-1	.9100108D-2	.6028317D-2	.4480583D-2	.3542212D-2	.2447848D-2	.1584125D-2
1 PFR	2.229006916	1.569975948	1.369080867	1.175115596	.9744430584	.8424633286	.7447508456	.6026950732	.4538884289
1 PFC	2.370024757	1.785104936	1.614619293	1.456248631	1.301584462	1.206825100	1.141259324	1.054619903	.9776260932
1 PQC	1.085530907	1.150413723	1.190680685	1.248910195	1.344036192	1.440155071	1.539727511	1.756975876	2.161434428
1 ZF1	16.54826814	16.54826814	16.54826814	16.54826814	16.54826814	16.54826814	16.54826814	16.54826814	16.54826814

(1) P 3+0	TH = 5	SBE = 11.47371325							CAUER FUNCTIONS
	RC = 02	RC = 05	RC = 07	RC = 10	RC = 15	RC = 20	RC = 25	RC = 35	RC = 50
AMAX DB	.0017375255	.0108709564	.0213327376	.0436480540	.0988323393	.1772876696	.2802872360	.5675287486	1.249387366
AMIN DB	53.63735768	61.60527549	64.53829651	67.65865025	71.23565929	73.81288921	75.85408897	79.06389114	82.84378892
MU K	.7147116D-1	.2855842D-1	.2037431D-1	.1422543D-1	.9423556D-2	.7004116D-2	.5537240D-2	.3826513D-2	.2476329D-2
1 PFR	2.238115832	1.573705469	1.371779788	1.177034939	.9757465858	.8434518124	.7455454813	.6032590250	.4542641849
1 PFC	2.366414185	1.783906026	1.613860297	1.455809661	1.301384163	1.206738443	1.141237988	1.054669340	.9777245960
1 PQC	1.092202248	1.154521845	1.194208791	1.251976050	1.346738226	1.442694441	1.542196321	1.759446341	2.164123234
1 ZF1	13.24239795	13.24239795	13.24239795	13.24239795	13.24239795	13.24239795	13.24239795	13.24239795	13.24239795

(1) P 3+0	TH = 6	SBE = 9.566772234							CAUER FUNCTIONS
	RC = 02	RC = 05	RC = 07	RC = 10	RC = 15	RC = 20	RC = 25	RC = 35	RC = 50
AMAX DB	.0017375255	.0108709564	.0213327376	.0436480540	.0988323393	.1772876696	.2802872360	.5675287486	1.249387366
AMIN DB	48.87922357	56.84710993	59.78012802	62.90048019	66.47748841	69.05471803	71.09591766	74.30571971	78.08561743
MU K	.1029761D 0	.4114714D-1	.2935543D-1	.2049608D-1	.1357752D-1	.1009157D-1	.7978088D-2	.5513263D-2	.3567909D-2
1 PFR	2.249348265	1.578287233	1.375092496	1.179389020	.9773443220	.8446629583	.7465188840	.6039496404	.4547242296
1 PFC	2.361983801	1.782435832	1.612929583	1.455271286	1.301138318	1.206631859	1.141211443	1.054729511	.9778448834
1 PQC	1.100453969	1.159579297	1.198547346	1.255742644	1.350055127	1.445810242	1.545224649	1.762475596	2.167419228
1 ZF1	11.03918745	11.03918745	11.03918745	11.03918745	11.03918745	11.03918745	11.03918745	11.03918745	11.03918745

(1) P 3+0	TH = 7	SBE = 8.205509048							CAUER FUNCTIONS
	RC = 02	RC = 05	RC = 07	RC = 10	RC = 15	RC = 20	RC = 25	RC = 35	RC = 50
AMAX DB	.0017375255	.0108709564	.0213327376	.0436480540	.0988323393	.1772876696	.2802872360	.5675287486	1.249387366
AMIN DB	44.85386892	52.82168317	55.75469453	58.87504313	62.45204943	65.02927839	67.07047771	70.28027949	74.06017707
MU K	.1402547D 0	.5604290D-1	.3998246D-1	.2791591D-1	.1849274D-1	.1374484D-1	.1086625D-1	.7509131D-2	.4859536D-2
1 PFR	2.262765334	1.583735545	1.379027522	1.182182828	.9792390163	.8460985817	.7476723737	.6047677386	.4552690388
1 PFC	2.356723238	1.780691438	1.611825318	1.454632390	1.300846300	1.206504943	1.141179408	1.054800264	.9779868919
1 PQC	1.110346946	1.165608441	1.203712528	1.260221848	1.353995675	1.449509851	1.548819120	1.766069590	2.171328283
1 ZF1	9.466069419	9.466069419	9.466069419	9.466069419	9.466069419	9.466069419	9.466069419	9.466069419	9.466069419

(1) P 3+0	TH = 8	SBE = 7.185296534							CAUER FUNCTIONS
	RC = 02	RC = 05	RC = 07	RC = 10	RC = 15	RC = 20	RC = 25	RC = 35	RC = 50
AMAX DB	.0017375255	.0108709564	.0213327376	.0436480540	.0988323393	.1772876696	.2802872360	.5675287486	1.249387366
AMIN DB	41.36455303	49.33222007	52.26521770	55.38555900	58.96256141	61.53978901	63.58098769	66.79078892	70.57068621
MU I K	.1833298D 0	.7325486D-1	.5226192D-1	.3648948D-1	.2417225D-1	.1796618D-1	.1420351D-1	.9815344D-2	.6352002D-2
1 I PFR	2.278440678	1.590067517	1.383595059	1.185422314	.9814339427	.8477608434	.7490075200	.6057142877	.4558991775
1 I PFC	2.350620360	1.778669399	1.610545329	1.453891649	1.300507369	1.206357215	1.141141548	1.054881418	.9781505458
1 I PQC	1.121955130	1.172636183	1.209723757	1.265427882	1.358570372	1.453802081	1.552987641	1.770235423	2.175857395
1 I ZF1	8.286760418	8.286760418	8.286760418	8.286760418	8.286760418	8.286760418	8.286760418	8.286760418	8.286760418

(1) P 3+0	TH = 9	SBE = 6.392453221							CAUER FUNCTIONS
	RC = 02	RC = 05	RC = 07	RC = 10	RC = 15	RC = 20	RC = 25	RC = 35	RC = 50
AMAX DB	.0017375255	.0108709564	.0213327376	.0436480540	.0988323393	.1772876696	.2802872360	.5675287486	1.249387366
AMIN DB	38.28440818	46.25180000	49.18477195	52.30509961	55.88209474	58.45931979	60.50051729	63.71031749	67.49021424
MU I K	.2322281D 0	.9279362D-1	.6620137D-1	.4622205D-1	.3061955D-1	.2275817D-1	.1799191D-1	.1243332D-1	.8046227D-2
1 I PFR	2.296461171	1.597303179	1.388807007	1.189114413	.9839329130	.8496522568	.7505261481	.6067904121	.4566153019
1 I PFC	2.343661331	1.776365745	1.609087110	1.453047531	1.300120666	1.206188120	1.141097476	1.054972765	.9783357573
1 I PQC	1.135366625	1.180694219	1.216603849	1.271377413	1.363791502	1.458697231	1.557739444	1.774981373	2.181014724
1 I ZF1	7.369992055	7.369992055	7.369992055	7.369992055	7.369992055	7.369992055	7.369992055	7.369992055	7.369992055

(1) P 3+0	TH = 10	SBE = 5.758770483							CAUER FUNCTIONS
	RC = 02	RC = 05	RC = 07	RC = 10	RC = 15	RC = 20	RC = 25	RC = 35	RC = 50
AMAX DB	.0017375255	.0108709564	.0213327376	.0436480540	.0988323393	.1772876696	.2802872360	.5675287486	1.249387366
AMIN DB	35.52685645	43.49376770	46.42669480	49.54699864	53.12398105	55.70120165	57.74239709	60.95219550	64.73209129
MU I K	.2869797D 0	.1146712D 0	.8180942D-1	.5711966D-1	.3783861D-1	.2812378D-1	.2223380D-1	.1536468D-1	.9943256D-2
1 I PFR	2.316927750	1.605465602	1.394677048	1.193267085	.9867402921	.8517756968	.7522303441	.6079973962	.4574181606
1 I PFC	2.335830722	1.773775979	1.607447814	1.452098296	1.299685215	1.205997023	1.141046750	1.055074064	.9785424260
1 I PQC	1.150684960	1.189819339	1.224379202	1.278089675	1.369673207	1.464207140	1.563085133	1.780316941	2.186809621
1 I ZF1	6.637003450	6.637003450	6.637003450	6.637003450	6.637003450	6.637003450	6.637003450	6.637003450	6.637003450

(1) P 3+0	TH = 11	SBE = 5.240843064							CAUER FUNCTIONS
	RC = 02	RC = 05	RC = 07	RC = 10	RC = 15	RC = 20	RC = 25	RC = 35	RC = 50
AMAX DB	.0017375255	.0108709564	.0213327376	.0436480540	.0988323393	.1772876696	.2802872360	.5675287486	1.249387366
AMIN DB	33.03020623	40.99632303	43.92917599	47.04944044	50.62640183	53.20361506	55.24480710	58.45460255	62.23449676
MU I K	.3476185D 0	.1389013D 0	.9909575D-1	.6918904D-1	.4583390D-1	.3406633D-1	.2693180D-1	.1861124D-1	.1204427D-1
1 I PFR	2.339956368	1.614581040	1.401220710	1.197889344	.9898610146	.8541344103	.7541224622	.6093366878	.4583085973
1 I PFC	2.327111630	1.770895083	1.605624255	1.451041990	1.299199918	1.205783214	1.140988872	1.055185046	.9787704384
1 I PQC	1.168030601	1.200053765	1.233080010	1.285586605	1.376231574	1.470345257	1.569036744	1.786252893	2.193252676
1 I ZF1	6.037674190	6.037674190	6.037674190	6.037674190	6.037674190	6.037674190	6.037674190	6.037674190	6.037674190

(1) P 3+0 — TH = 12 SBE = 4.809734345 CAUER FUNCTIONS

	RC = 02	RC = 05	RC = 07	RC = 10	RC = 15	RC = 20	RC = 25	RC = 35	RC = 50
AMAX DB	.0017375255	.0108709564	.0213327376	.0436480540	.0988323393	.1772876696	.2802872360	.5675287486	1.249387366
AMIN DB	30.74898410	38.71384520	41.64658094	44.76678311	48.34371125	50.92091285	52.96209950	56.17189026	59.95178198
MU K	.4141822D 0	.1654988D 0	.1180711D 0	.8243771D-1	.5461041D-1	.4058952D-1	.3208884D-1	.2217501D-1	.1435056D-1
1 PFR	2.365679057	1.624679095	1.408455457	1.202991303	.9933006050	.8567320276	.7562051326	.6108099028	.4592875525
1 PFC	2.317485855	1.767717528	1.603612910	1.449876449	1.298663554	1.205545898	1.140923288	1.055305411	.9790196678
1 PQC	1.187542716	1.211445561	1.242740516	1.293892999	1.383484735	1.477126716	1.575607803	1.792801319	2.200355762
1 ZF1	5.538590797	5.538590797	5.538590797	5.538590797	5.538590797	5.538590797	5.538590797	5.538590797	5.538590797

(1) P 3+0 — TH = 13 SBE = 4.445411483 CAUER FUNCTIONS

	RC = 02	RC = 05	RC = 07	RC = 10	RC = 15	RC = 20	RC = 25	RC = 35	RC = 50
AMAX DB	.0017375255	.0108709564	.0213327376	.0436480540	.0988323393	.1772876696	.2802872360	.5675287486	1.249387366
AMIN DB	28.64875315	36.61170238	39.54425956	42.66436687	46.24124437	48.81842825	50.85960669	54.06939030	57.84927823
MU K	.4867125D 0	.1944804D 0	.1387473D 0	.9687394D-1	.6417361D-1	.4769743D-1	.3770813D-1	.2605823D-1	.1686359D-1
1 PFR	2.394245095	1.635792899	1.416400777	1.208584221	.9970651992	.8595725755	.7584812700	.6124188302	.4603560661
1 PFC	2.306934109	1.764237277	1.601409914	1.448599291	1.298074781	1.205284203	1.140349388	1.055434827	.9792899738
1 PQC	1.209381220	1.224049097	1.253399298	1.303036694	1.391452982	1.484568422	1.582813405	1.799975686	2.208132094
1 ZF1	5.116620990	5.116620990	5.116620990	5.116620990	5.116620990	5.116620990	5.116620990	5.116620990	5.116620990

(1) P 3+0 — TH = 14 SBE = 4.133565494 CAUER FUNCTIONS

	RC = 02	RC = 05	RC = 07	RC = 10	RC = 15	RC = 20	RC = 25	RC = 35	RC = 50
AMAX DB	.0017375255	.0108709564	.0213327376	.0436480540	.0988323393	.1772876696	.2802872360	.5675287486	1.249387366
AMIN DB	26.70286200	34.66299117	37.59528484	40.71525211	44.29205487	46.86921258	48.91037892	52.12015199	55.90003431
MU K	.5652548D 0	.2258643D 0	.1611374D 0	.1125068D 0	.7452951D-1	.5539451D-1	.4379321D-1	.3026333D-1	.1958492D-1
1 PFR	2.425822284	1.647959316	1.425078295	1.214680555	1.001161570	.8626604921	.7609540839	.6141654369	.4615152809
1 PFC	2.295436297	1.760447801	1.599011069	1.447207919	1.297432127	1.204997171	1.140766502	1.055572930	.9795812015
1 PQC	1.233729122	1.237925574	1.265099595	1.313048777	1.400158898	1.492689153	1.590670294	1.807790910	2.216596286
1 ZF1	4.755241832	4.755241832	4.755241832	4.755241832	4.755241832	4.755241832	4.755241832	4.755241832	4.755241832

(1) P 3+0 — TH = 15 SBE = 3.863703305 CAUER FUNCTIONS

	RC = 02	RC = 05	RC = 07	RC = 10	RC = 15	RC = 20	RC = 25	RC = 35	RC = 50
AMAX DB	.0017375255	.0108709564	.0213327376	.0436480540	.0988323393	.1772876696	.2802872360	.5675287486	1.249387366
AMIN DB	24.89032314	32.84640458	35.77831968	38.89808578	42.47478114	45.05190126	47.09305019	50.30280811	54.08268238
MU K	.6498587D 0	.2596703D 0	.1852555D 0	.1293461D 0	.8568463D-1	.6368562D-1	.5034791D-1	.3479295D-1	.2251627D-1
1 PFR	2.460598291	1.661219170	1.434511878	1.221294021	1.005597153	.8660006428	.7636270889	.6160518744	.4627664449
1 PFC	2.282971856	1.756342091	1.596411838	1.445699516	1.296733996	1.204683760	1.140673899	1.055719322	.9798931817
1 PQC	1.260795207	1.253143639	1.277889680	1.323963811	1.409627499	1.501509668	1.599196957	1.816263430	2.225764423
1 ZF1	4.442336609	4.442336609	4.442336609	4.442336609	4.442336609	4.442336609	4.442336609	4.442336609	4.442336609

(1) P 3+0	TH = 16	SBE = 3.627955279						CAUER FUNCTIONS	
	RC = 02	RC = 05	RC = 07	RC = 10	RC = 15	RC = 20	RC = 25	RC = 35	RC = 50
AMAX DB	.0017375255	.0108709564	.0213327376	.0436480540	.0988323393	.1772876696	.2802872360	.5675287486	1.249387366
AMIN DB	23.19438246	31.14479038	34.07617421	37.19565794	40.77220253	43.34926988	45.39039439	48.60013103	52.37999400
MU K	.7405777D 0	.2959197D 0	.2111168D 0	.1474026D 0	.9764604D-1	.7257602D-1	.5737638D-1	.3964998D-1	.2565950D-1
1 PFR	2.498782056	1.675617490	1.444727775	1.228439653	1.010380080	.8695983385	.7665041173	.6180804850	.4641109155
1 PFC	2.269520189	1.751912685	1.593607357	1.444071048	1.295978659	1.204342841	1.140570788	1.055873570	.9802257294
1 PQC	1.290817066	1.269780064	1.291823280	1.335820103	1.419886407	1.511052836	1.608413725	1.825411296	2.235654138
1 ZF1	4.168817895	4.168817895	4.168817895	4.168817895	4.168817895	4.168817895	4.168817895	4.168817895	4.168817895

(1) P 3+0	TH = 17	SBE = 3.420303620						CAUER FUNCTIONS	
	RC = 02	RC = 05	RC = 07	RC = 10	RC = 15	RC = 20	RC = 25	RC = 35	RC = 50
AMAX DB	.0017375255	.0108709564	.0213327376	.0436480540	.0988323393	.1772876696	.2802872360	.5675287486	1.249387366
AMIN DB	21.60152798	29.54414823	32.47480159	35.59389698	39.17023421	41.74722896	43.78831987	46.99802726	50.77787469
MU K	.8374698D 0	.3346358D 0	.2387379D 0	.1666877D 0	.1104214D 0	.8207136D-1	.6488311D-1	.4483751D-1	.2901661D-1
1 PFR	2.540605205	1.691203777	1.455754747	1.236133880	1.015519206	.8734593552	.7695893321	.6202538095	.4655501630
1 PFC	2.255061190	1.747151689	1.590592434	1.442319258	1.295164254	1.203973196	1.140456313	1.056035207	.9805786441
1 PQC	1.324064510	1.287920527	1.306960057	1.348659999	1.430966027	1.521343770	1.618342896	1.835254260	2.246284694
1 ZF1	3.927736584	3.927736584	3.927736584	3.927736584	3.927736584	3.927736584	3.927736584	3.927736584	3.927736584

(1) P 3+0	TH = 18	SBE = 3.236067977						CAUER FUNCTIONS	
	RC = 02	RC = 05	RC = 07	RC = 10	RC = 15	RC = 20	RC = 25	RC = 35	RC = 50
AMAX DB	.0017375255	.0108709564	.0213327376	.0436480540	.0988323393	.1772876696	.2802872360	.5675287486	1.249387366
AMIN DB	20.10078666	28.03291494	30.96258189	34.08115272	37.65720981	40.23410649	42.27515200	45.48481985	49.26464628
MU K	.9405970D 0	.3758434D 0	.2681364D 0	.1872139D 0	.1240188D 0	.9217775D-1	.7287291D-1	.5035887D-1	.3258975D-1
1 PFR	2.586323423	1.708032300	1.467624228	1.244394601	1.021024155	.8775899558	.7728872417	.6225745950	.4670857748
1 PFC	2.239575883	1.742050815	1.587361563	1.440440668	1.294288786	1.203573517	1.140329554	1.056203729	.9809517085
1 PQC	1.360843366	1.307660493	1.323366136	1.362530216	1.442899759	1.532409990	1.629008860	1.845813884	2.257677085
1 ZF1	3.713687911	3.713687911	3.713687911	3.713687911	3.713687911	3.713687911	3.713687911	3.713687911	3.713687911

(1) P 3+0	TH = 19	SBE = 3.071553487						CAUER FUNCTIONS	
	RC = 02	RC = 05	RC = 07	RC = 10	RC = 15	RC = 20	RC = 25	RC = 35	RC = 50
AMAX DB	.0017375255	.0108709564	.0213327376	.0436480540	.0988323393	.1772876696	.2802872360	.5675287486	1.249387366
AMIN DB	18.68321542	26.60144419	29.52980035	32.64767384	36.22335843	38.80012469	40.84110983	44.05072510	47.83052360
MU K	.1050026D 1	.4195690D 0	.2993314D 0	.2089943D 0	.1384472D 0	.1029017D 0	.8135093D-1	.5621762D-1	.3638124D-1
1 PFR	2.636217715	1.726162399	1.480370484	1.253241269	1.026905351	.8819969134	.7764027156	.6250458035	.4687194601
1 PFC	2.223047188	1.736601423	1.583908930	1.438431581	1.293350121	1.203142401	1.140189524	1.056378593	.9813446877
1 PQC	1.401499671	1.329106201	1.341114717	1.377482217	1.455724230	1.544281593	1.640438246	1.857113657	2.269854136
1 ZF1	3.522405053	3.522405053	3.522405053	3.522405053	3.522405053	3.522405053	3.522405053	3.522405053	3.522405053

(1) P 3+0 — TH = 20 SBE = 2.923804400 — CAUER FUNCTIONS

	RC = 02	RC = 05	RC = 07	RC = 10	RC = 15	RC = 20	RC = 25	RC = 35	RC = 50
AMAX DB	.0017375255	.0108709564	.0213327376	.0436480540	.0988323393	.1772876696	.2802872360	.5675287486	1.249387366
AMIN DB	17.34152530	25.24162102	28.16826024	31.28521982	34.86041609	37.43701135	39.47791733	42.68746366	46.46722554
MU I K	.1165828D 1	.4658410D 0	.3323431D 0	.2320432D 0	.1537158D 0	.1142502D 0	.9032270D-1	.6241756D-1	.4039354D-1
1 I PFR	2.690595458	1.745658826	1.494030800	1.262694985	1.033174068	.8866875380	.7801410020	.6276706221	.4704530552
1 I PFC	2.205460819	1.730794573	1.580228431	1.436288079	1.292345987	1.202678350	1.140035166	1.056559217	.9817573292
1 I PQC	1.446424240	1.352375778	1.360286750	1.393572629	1.469479552	1.556991449	1.652660083	1.869179127	2.282840623
1 I ZF1	3.350474640	3.350474640	3.350474640	3.350474640	3.350474640	3.350474640	3.350474640	3.350474640	3.350474640

(1) P 3+0 — TH = 21 SBE = 2.790428110 — CAUER FUNCTIONS

	RC = 02	RC = 05	RC = 07	RC = 10	RC = 15	RC = 20	RC = 25	RC = 35	RC = 50
AMAX DB	.0017375255	.0108709564	.0213327376	.0436480540	.0988323393	.1772876696	.2802872360	.5675287486	1.249387366
AMIN DB	16.06979855	23.94657147	26.87099078	29.98676791	33.56133213	36.13770603	38.17850952	41.38796661	45.16768106
MU I K	.1288078D 1	.5146899D 0	.3671931D 0	.2563757D 0	.1698347D 0	.1262306D 0	.9979409D-1	.6896277D-1	.4462927D-1
1 I PFR	2.749791123	1.766592091	1.508645665	1.272778597	1.039842473	.8916697039	.7841077466	.6304524728	.4722885283
1 I PFC	2.186806326	1.724621092	1.576313689	1.434006029	1.291273969	1.202179769	1.139865352	1.056744979	.9821893609
1 I PQC	1.496057584	1.377600492	1.380971698	1.410863714	1.484209614	1.570575421	1.665705980	1.882038043	2.296663405
1 I ZF1	3.195133559	3.195133559	3.195133559	3.195133559	3.195133559	3.195133559	3.195133559	3.195133559	3.195133559

(1) P 3+0 — TH = 22 SBE = 2.669467163 — CAUER FUNCTIONS

	RC = 02	RC = 05	RC = 07	RC = 10	RC = 15	RC = 20	RC = 25	RC = 35	RC = 50
AMAX DB	.0017375255	.0108709564	.0213327376	.0436480540	.0988323393	.1772876696	.2802872360	.5675287486	1.249387366
AMIN DB	14.86327054	22.71044098	25.63202379	28.74628869	32.32004421	34.89613481	36.93680713	40.14614998	43.92580373
MU I K	.1416858D 1	.5661478D 0	.4039045D 0	.2820077D 0	.1868144D 0	.1388510D 0	.1097713D 0	.7585756D-1	.4909124D-1
1 I PFR	2.814166542	1.789038829	1.524258986	1.283516809	1.046923681	.8969518808	.7883090133	.6333950241	.4742279861
1 I PFC	2.167078276	1.718071655	1.572158079	1.431581081	1.290131510	1.201644963	1.139678892	1.056935214	.9826404912
1 I PQC	1.550895109	1.404926151	1.403268394	1.429423396	1.499962402	1.585072602	1.679610326	1.895720518	2.311351561
1 I ZF1	3.054121166	3.054121166	3.054121166	3.054121166	3.054121166	3.054121166	3.054121166	3.054121166	3.054121166

(1) P 3+0 — TH = 23 SBE = 2.559304665 — CAUER FUNCTIONS

	RC = 02	RC = 05	RC = 07	RC = 10	RC = 15	RC = 20	RC = 25	RC = 35	RC = 50
AMAX DB	.0017375255	.0108709564	.0213327376	.0436480540	.0988323393	.1772876696	.2802872360	.5675287486	1.249387366
AMIN DB	13.71815645	21.52822314	24.44622060	27.55857188	31.13130342	33.70703521	35.74754138	38.95673951	42.73631637
MU I K	.1552254D 1	.6202489D 0	.4425016D 0	.3089564D 0	.2046665D 0	.1521196D 0	.1202611D 0	.8310651D-1	.5378240D-1
1 I PFR	2.884110540	1.813082179	1.540918300	1.294936298	1.054431808	.9025431673	.7927513067	.6365022035	.4762736800
1 I PFC	2.146277556	1.711136874	1.567754753	1.429008682	1.288915908	1.201072131	1.139474478	1.057129213	.9831104073
1 I PQC	1.611492517	1.434514670	1.427285997	1.449328349	1.516790368	1.600525588	1.694410512	1.910259203	2.326936554
1 I ZF1	2.925570058	2.925570058	2.925570058	2.925570058	2.925570058	2.925570058	2.925570058	2.925570058	2.925570058

(1) P 3+0	TH = 24	SBE = 2.458593336						CAUER FUNCTIONS	
	RC = 02	RC = 05	RC = 07	RC = 10	RC = 15	RC = 20	RC = 25	RC = 35	RC = 50
AMAX DB	.0017375255	.0108709564	.0213327376	.0436480540	.0988323393	.1772876696	.2802872360	.5675287486	1.249387366
AMIN DB	12.63150795	20.39562592	23.30913654	26.41908960	29.99053687	32.56581849	34.60611616	37.81513269	41.59461305
MU K	.1694355D 1	.6770296D 0	.4830104D 0	.3372398D 0	.2234026D 0	.1660454D 0	.1312704D 0	.9071450D-1	.5870591D-1
1 PFR	2.960037770	1.838812168	1.558675010	1.307065837	1.062382036	.9084533266	.7974415962	.6397782105	.4784280128
1 PFC	2.124412770	1.703807419	1.563096681	1.426284073	1.287624313	1.200459371	1.139250788	1.057326220	.9835987748
1 PQC	1.678471265	1.466545821	1.453145072	1.470659659	1.534750826	1.616980777	1.710147179	1.925689493	2.343452406
1 ZF1	2.807924150	2.807924150	2.807924150	2.807924150	2.807924150	2.807924150	2.807924150	2.807924150	2.807924150

(1) P 3+0	TH = 25	SBE = 2.366201583						CAUER FUNCTIONS	
	RC = 02	RC = 05	RC = 07	RC = 10	RC = 15	RC = 20	RC = 25	RC = 35	RC = 50
AMAX DB	.0017375255	.0108709564	.0213327376	.0436480540	.0988323393	.1772876696	.2802872360	.5675287486	1.249387366
AMIN DB	11.60108840	19.30896628	22.21691366	25.32388789	28.89373847	31.46846013	33.50849843	36.71728901	40.49664929
MU K	.1843258D 1	.7365283D 0	.5254584D 0	.3668771D 0	.2430357D 0	.1806378D 0	.1428067D 0	.9868667D-1	.6386510D-1
1 PFR	3.042386577	1.866326096	1.577584628	1.319936430	1.070790673	.9146928263	.8023873425	.6432275316	.4806935470
1 PFC	2.101501675	1.696074147	1.558176691	1.423402308	1.286253729	1.199804672	1.139006375	1.057525434	.9841052356
1 PQC	1.752523942	1.501219181	1.480978791	1.493508568	1.553906409	1.634488705	1.726864491	1.942049736	2.360935888
1 ZF1	2.699876408	2.699876408	2.699876408	2.699876408	2.699876408	2.699876408	2.699876408	2.699876408	2.699876408

(1) P 3+0	TH = 26	SBE = 2.281172033						CAUER FUNCTIONS	
	RC = 02	RC = 05	RC = 07	RC = 10	RC = 15	RC = 20	RC = 25	RC = 35	RC = 50
AMAX DB	.0017375255	.0108709564	.0213327376	.0436480540	.0988323393	.1772876696	.2802872360	.5675287486	1.249387366
AMIN DB	10.62525810	18.26508618	21.16619531	24.26949998	27.83738129	30.41141196	32.45113013	35.65964179	39.43885384
MU K	.1999065D 1	.7987857D 0	.5698744D 0	.3978885D 0	.2635791D 0	.1959068D 0	.1548779D 0	.1070285D 0	.6926351D-1
1 PFR	3.131615743	1.895728910	1.597707024	1.333581458	1.079675227	.9212728803	.8075965263	.6468549557	.4830730123
1 PFC	2.077572591	1.687928259	1.552987521	1.420358260	1.284801013	1.199105911	1.138739723	1.057726003	.9846294078
1 PQC	1.834419369	1.538756296	1.510934282	1.517974796	1.574325575	1.653104418	1.744610445	1.959381486	2.379426743
1 ZF1	2.600320966	2.600320966	2.600320966	2.600320966	2.600320966	2.600320966	2.600320966	2.600320966	2.600320966

(1) P 3+0	TH = 27	SBE = 2.202689265						CAUER FUNCTIONS	
	RC = 02	RC = 05	RC = 07	RC = 10	RC = 15	RC = 20	RC = 25	RC = 35	RC = 50
AMAX DB	.0017375255	.0108709564	.0213327376	.0436480540	.0988323393	.1772876696	.2802872360	.5675287486	1.249387366
AMIN DB	9.702862763	17.26128520	20.15405740	23.25287625	26.81834678	29.39153096	31.43085685	34.63902668	38.41805704
MU K	.2161884D 1	.8638448D 0	.6162893D 0	.4302956D 0	.2850470D 0	.2118629D 0	.1674923D 0	.1157457D 0	.7490485D-1
1 PFR	3.228199993	1.927133559	1.619106691	1.348036828	1.089054484	.9282054953	.8130776791	.6506655915	.4855693150
1 PFC	2.052665671	1.679361483	1.547521884	1.417146640	1.283262873	1.198360855	1.138449227	1.057927021	.9851708836
1 PQC	1.925007221	1.579403075	1.543174132	1.544167978	1.596083169	1.672887893	1.763437208	1.977729767	2.398967916
1 ZF1	2.508315866	2.508315866	2.508315866	2.508315866	2.508315866	2.508315866	2.508315866	2.508315866	2.508315866

(1) P 3+0 — TH = 28 SBE = 2.130054468 — CAUER FUNCTIONS

	RC = 02	RC = 05	RC = 07	RC = 10	RC = 15	RC = 20	RC = 25	RC = 35	RC = 50
AMAX DB	.0017375255	.0108709564	.0213327376	.0436480540	.0988323393	.1772876696	.2802872360	.5675287486	1.249387366
AMIN DB	8.833121068	16.29526579	19.17795279	22.27132738	25.83386690	28.40602106	30.44486946	33.65262315	37.43143229
MU K	.2331828D 1	.9317511D 0	.6647353D 0	.4641208D 0	.3074543D 0	.2285173D 0	.1806588D 0	.1248444D 0	.8079306D-1
1 PFR	3.332624265	1.960661308	1.641853003	1.363341133	1.098948583	.9355035202	.8188399170	.6546648858	.4881855473
1 PFC	2.026833937	1.670366280	1.541772536	1.413762016	1.281635873	1.197567157	1.138133198	1.058127533	.9857292282
1 PQC	2.025221990	1.623432440	1.577878078	1.572208706	1.619261059	1.693904502	1.783401505	1.997143377	2.419605825
1 ZF1	2.423053796	2.423053796	2.423053796	2.423053796	2.423053796	2.423053796	2.423053796	2.423053796	2.423053796

(1) P 3+0 — TH = 29 SBE = 2.062665340 — CAUER FUNCTIONS

	RC = 02	RC = 05	RC = 07	RC = 10	RC = 15	RC = 20	RC = 25	RC = 35	RC = 50
AMAX DB	.0017375255	.0108709564	.0213327376	.0436480540	.0988323393	.1772876696	.2802872360	.5675287486	1.249387366
AMIN DB	8.015509017	15.36508833	18.23566603	21.32247790	24.88147674	27.45238539	29.49065619	32.69790636	36.47644778
MU K	.2509018D 1	.1002552D 1	.7152468D 0	.4993881D 0	.3308169D 0	.2458817D 0	.1943866D 0	.1343310D 0	.8693231D-1
1 PFR	3.445376813	1.996442016	1.666020484	1.379535824	1.109379114	.9431806997	.8248929771	.6588586434	.4909249982
1 PFC	2.000143916	1.660936076	1.535732369	1.410198836	1.279916431	1.196722352	1.137789853	1.058326523	.9863039783
1 PQC	2.136086159	1.671147233	1.615244877	1.602229694	1.643948848	1.716225534	1.804565037	2.017675225	2.441390654
1 ZF1	2.343838876	2.343838876	2.343838876	2.343838876	2.343838876	2.343838876	2.343838876	2.343838876	2.343838876

(1) P 3+0 — TH = 30 SBE = 2.000000000 — CAUER FUNCTIONS

	RC = 02	RC = 05	RC = 07	RC = 10	RC = 15	RC = 20	RC = 25	RC = 35	RC = 50
AMAX DB	.0017375255	.0108709564	.0213327376	.0436480540	.0988323393	.1772876696	.2802872360	.5675287486	1.249387366
AMIN DB	7.249641497	14.46913338	17.32527620	20.40422801	23.95897536	26.52838665	28.56596273	31.77260711	35.55082640
MU K	.2693578D 1	.1076299D 1	.7678596D 0	.5361226D 0	.3551515D 0	.2639685D 0	.2086854D 0	.1442122D 0	.9332696D-1
1 PFR	3.566941386	2.034614324	1.691689068	1.396665388	1.120369202	.9512517326	.8312472564	.6632530496	.4937911645
1 PFC	1.972675777	1.651065525	1.529394501	1.406451457	1.278100825	1.195823859	1.137417318	1.058522918	.9868946405
1 PQC	2.258712515	1.722883417	1.655494410	1.634377098	1.670244668	1.739928783	1.826994962	2.039382703	2.464376677
1 ZF1	2.270068086	2.270068086	2.270068086	2.270068086	2.270068086	2.270068086	2.270068086	2.270068086	2.270068086

(1) P 3+0 — TH = 31 SBE = 1.941604026 — CAUER FUNCTIONS

	RC = 02	RC = 05	RC = 07	RC = 10	RC = 15	RC = 20	RC = 25	RC = 35	RC = 50
AMAX DB	.0017375255	.0108709564	.0213327376	.0436480540	.0988323393	.1772876696	.2802872360	.5675287486	1.249387366
AMIN DB	6.535153893	13.60606951	16.44512628	19.51472209	23.06439331	25.63201419	27.66875913	30.87467854	34.65251220
MU K	.2885644D 1	.1153045D 1	.8226119D 0	.5743508D 0	.3804756D 0	.2827908D 0	.2235657D 0	.1544953D 0	.9998164D-1
1 PFR	3.697788868	2.075325770	1.718944348	1.414777529	1.131943618	.9597323339	.8379138550	.6678546928	.4967877633
1 PFC	1.944522835	1.640750795	1.522752398	1.402514180	1.276185198	1.194868978	1.137013622	1.058715587	.9875006896
1 PQC	2.394305696	1.779013556	1.698870010	1.668811980	1.698256077	1.765099199	1.850764429	2.062328102	2.488622623
1 ZF1	2.201216296	2.201216296	2.201216296	2.201216296	2.201216296	2.201216296	2.201216296	2.201216296	2.201216296

(1) P 3+0	TH = 32	SBE = 1.887079915						CAUER FUNCTIONS	
	RC = 02	RC = 05	RC = 07	RC = 10	RC = 15	RC = 20	RC = 25	RC = 35	RC = 50
AMAX DB	.0017375255	.0108709564	.0213327376	.0436480540	.0988323393	.1772876696	.2802872360	.5675287486	1.249387366
AMIN DB	5.871588883	12.77482477	15.59379758	18.65232259	22.19596548	24.76145641	26.79721194	30.00226798	33.77964228
MU K	.3085355D 1	.1232845D 1	.8795435D 0	.6141007D 0	.4068077D 0	.3023623D 0	.2390384D 0	.1651876D 0	.1069012D 0
1 PFR	3.838368902	2.118732774	1.747877808	1.433923357	1.144128878	.9686393023	.8449046215	.6726705911	.4999187451
1 PFC	1.915790359	1.629989876	1.515800000	1.398381290	1.274165565	1.193854890	1.136576697	1.058903330	.9881215669
1 PQC	2.544163196	1.839950618	1.745641064	1.705711969	1.728101063	1.791829631	1.875953170	2.086579085	2.514192084
1 ZF1	2.136824093	2.136824093	2.136824093	2.136824093	2.136824093	2.136824093	2.136824093	2.136824093	2.136824093

(1) P 3+0	TH = 33	SBE = 1.836078459						CAUER FUNCTIONS	
	RC = 02	RC = 05	RC = 07	RC = 10	RC = 15	RC = 20	RC = 25	RC = 35	RC = 50
AMAX DB	.0017375255	.0108709564	.0213327376	.0436480540	.0988323393	.1772876696	.2802872360	.5675287486	1.249387366
AMIN DB	5.258295154	11.97456062	14.77008804	17.81558834	21.35210847	23.91507761	25.94966078	29.15369334	32.93052290
MU K	.3292859D 1	.1315759D 1	.9386968D 0	.6554017D 0	.4341673D 0	.3226975D 0	.2551148D 0	.1762972D 0	.1140908D 0
1 PFR	3.989102114	2.165000498	1.778587027	1.454157590	1.156953362	.9779905927	.8522322032	.6777082193	.5031883087
1 PFC	1.886593676	1.618782910	1.508531866	1.394047104	1.272037817	1.192778655	1.136104373	1.059084885	.9887566782
1 PQC	2.709676242	1.906152077	1.796105895	1.745273113	1.759909185	1.820221650	1.902648178	2.112209201	2.541153962
1 ZF1	2.076487829	2.076487829	2.076487829	2.076487829	2.076487829	2.076487829	2.076487829	2.076487829	2.076487829

(1) P 3+0	TH = 34	SBE = 1.788291650						CAUER FUNCTIONS	
	RC = 02	RC = 05	RC = 07	RC = 10	RC = 15	RC = 20	RC = 25	RC = 35	RC = 50
AMAX DB		.0108709564	.0213327376	.0436480540	.0988323393	.1772876696	.2802872360	.5675287486	1.249387366
AMIN DB		11.20464695	13.97299369	17.00325642	20.53140157	23.09139839	25.12459848	28.32742293	32.10360924
MU K		.1401850D 1	.1000116D 1	.6982849D 0	.4625750D 0	.3438117D 0	.2718070D 0	.1878324D 0	.1215558D 0
1 PFR		2.214302537	1.811175854	1.475538740	1.170447439	.9878053943	.8599100998	.6829755390	.5066009166
1 PFC		1.607132523	1.500943334	1.389506028	1.269797737	1.191637216	1.135594380	1.059258918	.9894053919
1 PQC		1.978124358	1.850594966	1.787711965	1.793822841	1.850386488	1.930944454	2.139298480	2.569582981
1 ZF1		2.019851423	2.019851423	2.019851423	2.019851423	2.019851423	2.019851423	2.019851423	2.019851423

(1) P 3+0	TH = 35	SBE = 1.743446796						CAUER FUNCTIONS	
	RC = 02	RC = 05	RC = 07	RC = 10	RC = 15	RC = 20	RC = 25	RC = 35	RC = 50
AMAX DB		.0108709564	.0213327376	.0436480540	.0988323393	.1772876696	.2802872360	.5675287486	1.249387366
AMIN DB		10.46463734	13.20169219	16.21422694	19.73257075	22.28907917	24.32065426	27.52205836	31.29748805
MU K		.1491182D 1	.1063848D 1	.7427828D 0	.4920524D 0	.3657209D 0	.2891278D 0	.1998020D 0	.1293019D 0
1 PFR		2.266820421	1.845754525	1.498129318	1.184643589	.9981042140	.8679527216	.6884810320	.5101613121
1 PFC		1.595044169	1.493030703	1.384752619	1.267441012	1.190427399	1.135044342	1.059424025	.9900670365
1 PQC		2.056427605	1.909474426	1.833267926	1.829998715	1.882446082	1.960945865	2.167934085	2.599560251
1 ZF1		1.966599567	1.966599567	1.966599567	1.966599567	1.966599567	1.966599567	1.966599567	1.966599567

TH = 36   SBE = 1.701301617   CAUER FUNCTIONS

(1) P 3+0	RC = 02	RC = 05	RC = 07	RC = 10	RC = 15	RC = 20	RC = 25	RC = 35	RC = 50
AMAX DB		.0108709564	.0213327376	.0436480540	.0988323393	.1772876696	.2802872360	.5675287486	1.249387366
AMIN DB		9.754243538	12.45552785	15.44755012	18.95447523	21.50690610	23.53657933	26.73631981	30.51086283
MU I K		.1583825D 1	.1129942D 1	.7889297D 0	.5226222D 0	.3884422D 0	.3070905D 0	.2122151D 0	.1373350D 0
1 I PFR		2.322742928	1.882439729	1.521996021	1.199576543	1.008908966	.8763754519	.6942337347	.5138745380
1 I PFC		1.582526453	1.484791420	1.379781659	1.264963246	1.189145912	1.134451778	1.059578726	.9907408990
1 I PQC		2.141680811	1.973150043	1.882205881	1.868609406	1.916534272	1.992766095	2.198211055	2.631173902
1 I ZF1		1.916452067	1.916452067	1.916452067	1.916452067	1.916452067	1.916452067	1.916452067	1.916452067

TH = 37   SBE = 1.661640141   CAUER FUNCTIONS

(1) P 3+0	RC = 02	RC = 05	RC = 07	RC = 10	RC = 15	RC = 20	RC = 25	RC = 35	RC = 50
AMAX DB		.0108709564	.0213327376	.0436480540	.0988323393	.1772876696	.2802872360	.5675287486	1.249387366
AMIN DB		9.073308648	11.73399754	14.70241535	18.19609607	20.74377920	22.77123466	25.96903347	29.74254095
MU I K		.1679851D 1	.1198449D 1	.8367620D 0	.5543084D 0	.4119931D 0	.3257092D 0	.2250815D 0	.1456616D 0
1 I PFR		2.382265177	1.921354598	1.547209918	1.215283426	1.020243071	.8851947156	.7002432775	.5177459568
1 I PFC		1.569591428	1.476224283	1.374588239	1.262359986	1.187789356	1.133814101	1.059721463	.9914262217
1 I PQC		2.234567305	2.042071540	1.934819157	1.909845267	1.952798141	2.026529736	2.230233142	2.664519799
1 I ZF1		1.869159100	1.869159100	1.869159100	1.869159100	1.869159100	1.869159100	1.869159100	1.869159100

TH = 38   SBE = 1.624269245   CAUER FUNCTIONS

(1) P 3+0	RC = 02	RC = 05	RC = 07	RC = 10	RC = 15	RC = 20	RC = 25	PC = 35	RC = 50
AMAX DB		.0108709564	.0213327376	.0436480540	.0988323393	.1772876696	.2802872360	.5675287486	1.249387366
AMIN DB		8.421778513	11.03673674	13.97814161	17.45652669	19.99870219	22.02358048	25.21912062	28.99142261
MU I K		.1779337D 1	.1269425D 1	.8863176D 0	.5871363D 0	.4363926D 0	.3449987D 0	.2384116D 0	.1542881D 0
1 I PFR		2.445587521	1.962628620	1.573846622	1.231803908	1.032131553	.8944280516	.7065199259	.5217812726
1 I PFC		1.556254847	1.467329658	1.369167852	1.259626739	1.186354224	1.133128615	1.059850599	.9921222003
1 I PQC		2.335840651	2.116737396	1.991432852	1.953916497	1.991399540	2.062373510	2.264113753	2.699702348
1 I ZF1		1.824497221	1.824497221	1.824497221	1.824497221	1.824497221	1.824497221	1.824497221	1.824497221

TH = 39   SBE = 1.589015729   CAUER FUNCTIONS

(1) P 3+0	RC = 02	RC = 05	RC = 07	RC = 10	RC = 15	RC = 20	RC = 25	RC = 35	RC = 50
AMAX DB		.0108709564	.0213327376	.0436480540	.0988323393	.1772876696	.2802872360	.5675287486	1.249387366
AMIN DB		7.799671311	10.36350536	13.27416898	16.73496471	19.27077397	21.29266731	24.48558827	28.25649114
MU I K		.1882363D 1	.1342927D 1	.9376367D 0	.6211323D 0	.4616604D 0	.3649746D 0	.2522159D 0	.1632216D 0
1 I PFR		2.512914238	2.006397443	1.601986450	1.249180355	1.044601157	.9040941925	.7130746266	.5259865549
1 I PFC		1.542536349	1.458109686	1.363516493	1.256759004	1.184836911	1.132392518	1.059964411	.9928279808
1 I PQC		2.446330992	2.197700143	2.052407586	2.001055503	2.032516821	2.100447662	2.299977017	2.736835404
1 I ZF1		1.782265986	1.782265986	1.782265986	1.782265986	1.782265986	1.782265986	1.782265986	1.782265986

(1) P 3+0	TH = 40 SBE = 1.555723827 RC = 02	RC = 05	RC = 07	RC = 10	RC = 15	RC = 20	RC = 25	RC = 35	RC = 50 (CAUER FUNCTIONS)
AMAX DB		.0108709564	.0213327376	.0436480540	.0988323393	.1772876696	.2802872360	.5675287486	1.249387366
AMIN DB		7.207045499	9.714172770	12.59005087	16.03070522	18.55918126	20.57762830	23.76752102	27.53680467
MU I K		.1989015D 1	.1419014D 1	.9907615D 0	.6563245D 0	.4878172D 0	.3856534D 0	.2665060D 0	.1724694D 0
1 I PFR		2.584452062	2.052802591	1.631714552	1.267457994	1.057680460	.9142131494	.7199190562	.5303682651
1 I PFC		1.528459560	1.448568501	1.357630776	1.253752303	1.183233723	1.131602901	1.060061089	.9935426569
1 I PQC		2.566951925	2.285572226	2.118143715	2.051519598	2.076346807	2.140917548	2.337958998	2.776043294
1 I ZF1		1.742285078	1.742285078	1.742285078	1.742285078	1.742285078	1.742285078	1.742285078	1.742285078

(1) P 3+0	TH = 41 SBE = 1.524253087 RC = 02	RC = 05	RC = 07	RC = 10	RC = 15	RC = 20	RC = 25	RC = 35	RC = 50 (CAUER FUNCTIONS)
AMAX DB		.0108709564	.0213327376	.0436480540	.0988323393	.1772876696	.2802872360	.5675287486	1.249387366
AMIN DB		6.643966731	9.088701796	11.92544673	15.34313497	17.86319248	19.87767260	23.06407404	26.83148872
MU I K		.2099380D 1	.1497752D 1	.1045736D 1	.6927424D 0	.5148850D 0	.4070523D 0	.2812938D 0	.1820393D 0
1 I PFR		2.660408605	2.101991052	1.663121015	1.286685076	1.071400002	.9248063039	.7270656755	.5349332849
1 I PFC		1.514052091	1.438712429	1.351508055	1.250602220	1.181540885	1.130756751	1.060138736	.9942652668
1 I PQC		2.698708017	2.381032474	2.189086084	2.105594057	2.123107034	2.183965435	2.378209074	2.817461983
1 I ZF1		1.704391854	1.704391854	1.704391854	1.704391854	1.704391854	1.704391854	1.704391854	1.704391854

(1) P 3+0	TH = 42 SBE = 1.494476550 RC = 02	RC = 05	RC = 07	RC = 10	RC = 15	RC = 20	RC = 25	RC = 35	RC = 50 (CAUER FUNCTIONS)
AMAX DB		.0108709564	.0213327376	.0436480540	.0988323393	.1772876696	.2802872360	.5675287486	1.249387366
AMIN DB		6.110474664	8.487131320	11.28011471	14.67172748	17.18215256	19.19207986	22.37446696	26.13972980
MU I K		.2213554D 1	.1579207D 1	.1102608D 1	.7304169D 0	.5428869D 0	.4291898D 0	.2965919D 0	.1919395D 0
1 I PFR		2.740990716	2.154114766	1.696300938	1.306913048	1.085792412	.9358965071	.7345277876	.5396889473
1 I PFC		1.499345417	1.428550171	1.345146551	1.247304445	1.179754559	1.129850954	1.060195360	.9949947900
1 I PQC		2.842703112	2.484833284	2.265729387	2.163595607	2.173038319	2.229792558	2.420891502	2.861240396
1 I ZF1		1.668439251	1.668439251	1.668439251	1.668439251	1.668439251	1.668439251	1.668439251	1.668439251

(1) P 3+0	TH = 43 SBE = 1.466279186 RC = 02	RC = 05	RC = 07	RC = 10	RC = 15	RC = 20	RC = 25	RC = 35	RC = 50 (CAUER FUNCTIONS)
AMAX DB		.0108709564	.0213327376	.0436480540	.0988323393	.1772876696	.2802872360	.5675287486	1.249387366
AMIN DB		5.606550880	7.909557439	10.65390423	14.01603882	16.51547859	18.52019541	21.69797857	25.46076975
MU I K		.2331636D 1	.1663450D 1	.1161427D 1	.7693810D 0	.5718471D 0	.4520848D 0	.3124135D 0	.2021785D 0
1 I PFR		2.826402881	2.209329998	1.731354452	1.328196716	1.100892555	.9475081859	.7423196021	.5446430710
1 I PFC		1.484374630	1.418092956	1.338545497	1.243854821	1.177870855	1.128882296	1.060228374	.9957301444
1 I PQC		3.000149649	2.597808605	2.348624216	2.225876410	2.226407693	2.278621485	2.466187217	2.907541922
1 I ZF1		1.634293972	1.634293972	1.634293972	1.634293972	1.634293972	1.634293972	1.634293972	1.634293972

$P^{(2)}_{3+2}$

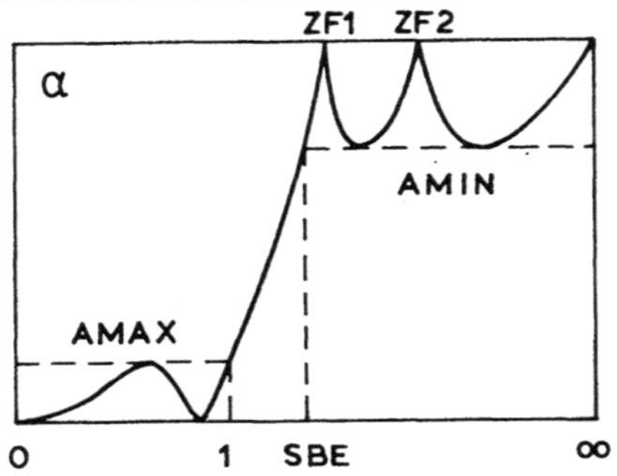

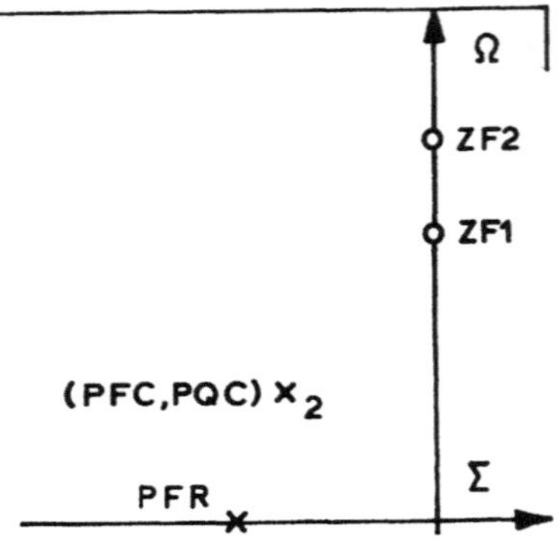

(2) P 3+2		TH = 5	SBE = 11.47371325						MCPER FUNCTIONS	
		RC = 02	RC = 05	RC = 07	RC = 10	RC = 15	RC = 20	RC = 25	RC = 35	RC = 50
	AMAX DB	.0017375255	.0108709564	.0213327376	.0436480540	.0988323393	.1772876696	.2802872360	.5675287486	1.249387366
	AMIN DB	88.29253003	101.5716419	106.4584351	111.6548319	117.6042983	121.8804499	125.2555899	130.5253849	136.6199710
MU	K	.2190754D-2	.4762824D-3	.2715271D-3	.1493604D-3	.7533138D-4	.4605636D-4	.3123294D-4	.1703045D-4	.8444782D-5
1	PFR	2.049731507	1.431872226	1.243215798	1.060818592	.8718514171	.7474951557	.6554811837	.5220823512	.3836278991
2	PFC	2.766633938	2.068502315	1.862461921	1.669197662	1.477648747	1.358009798	1.273577713	1.158569889	1.050156642
2	PQC	.9852127226	1.017176957	1.038771552	1.070175981	1.120785369	1.170585638	1.220679849	1.325014896	1.503804538
1	ZF1	12.05550148	12.05841905	12.05913197	12.05973863	12.06028140	12.06059181	12.06079838	12.06106469	12.06130292
1	ZF2	19.42092116	19.45546137	19.46370115	19.47065372	19.47682880	19.48034150	19.48267159	19.48566680	19.48833793

(2) P 3+2		TH = 6	SBE = 9.566772234						MCPER FUNCTIONS	
		RC = 02	RC = 05	RC = 07	RC = 10	RC = 15	RC = 20	RC = 25	RC = 35	RC = 50
	AMAX DB	.0017375255	.0108709564	.0213327376	.0436480540	.0988323393	.1772876696	.2802872360	.5675287486	1.249387366
	AMIN DB	80.36244419	93.64141448	98.52818638	103.7245689	109.6740255	113.9501727	117.3253103	122.5951028	128.6896872
MU	K	.4535054D-2	.9872777D-3	.5630177D-3	.3097816D-3	.1562765D-3	.9555695D-4	.6480697D-4	.3534119D-4	.1752607D-4
1	PFR	2.058676854	1.435538802	1.245872916	1.062711770	.8731401021	.7484736638	.6562681972	.5226403342	.3839975064
2	PFC	2.760446909	2.066425091	1.861130140	1.668407833	1.477263455	1.357817576	1.273496789	1.158610125	1.050281767
2	PQC	.9948420796	1.022791771	1.043468887	1.074122281	1.124097645	1.173564816	1.223459274	1.327585452	1.506288594
1	ZF1	10.04881277	10.05222740	10.05306966	10.05378872	10.05443382	10.05480351	10.05504982	10.05536771	10.05565241
1	ZF2	16.15648622	16.19812782	16.20804464	16.21640715	16.22383056	16.22805172	16.23085109	16.23444877	16.23765645

(2) P 3+2	TH = 7	SBE = 8.205509048						MCPER FUNCTIONS	
	RC = 02	RC = 05	RC = 07	RC = 10	RC = 15	RC = 20	RC = 25	RC = 35	RC = 50
AMAX DB	.0017375255	.0108709564	.0213327376	.0436480540	.0988323393	.1772876696	.2802872360	.5675287486	1.249387366
AMIN DB	73.65370764	86.93244358	91.81918021	97.01553913	102.9649796	107.2411196	110.6162532	115.8860415	121.9806231
MU K	.8384012D-2	.1828246D-2	.1042992D-2	.5740499D-3	.2896713D-3	.1771496D-3	.1201553D-3	.6553265D-4	.3250199D-4
1 PFR	2.069363976	1.439899068	1.249029310	1.064958645	.8746683418	.7496335608	.6572008392	.5233013378	.3844352313
2 PFC	2.753045500	2.063948392	1.859543231	1.667466915	1.476804144	1.357587912	1.273399437	1.158656747	1.050429218
2 PQC	1.006465730	1.029509668	1.049077225	1.078825589	1.128039148	1.177106834	1.226761903	1.330637685	1.509236290
1 ZF1	8.615950393	8.619815102	8.620779490	8.621606074	8.622350129	8.622777554	8.623062750	8.623431306	8.623761828
1 ZF2	13.82025565	13.86910496	13.88071482	13.89049792	13.89917695	13.90410980	13.90738025	13.91158227	13.91532779

(2) P 3+2	TH = 8	SBE = 7.185296534						MCPER FUNCTIONS	
	RC = 02	RC = 05	RC = 07	RC = 10	RC = 15	RC = 20	RC = 25	RC = 35	RC = 50
AMAX DB	.0017375255	.0108709564	.0213327376	.0436480540	.0988323393	.1772876696	.2802872360	.5675287486	1.249387366
AMIN DB	67.83839997	81.11677369	86.00345602	91.19977867	97.14919423	101.4253232	104.8004507	110.0702326	116.1648099
MU K	.1426585D-1	.3117178D-2	.1779114D-2	.9795650D-3	.4944568D-3	.3024410D-3	.2051608D-3	.1119113D-3	.5551153D-4
1 PFR	2.081853706	1.444966967	1.252693273	1.067564057	.8764388061	.7509766123	.6582803948	.5240661559	.3849415370
2 PFC	2.744387099	2.061061494	1.857694761	1.666371179	1.476268830	1.357319554	1.273284796	1.158709287	1.050598781
2 PQC	1.020213979	1.037373491	1.055626053	1.084306252	1.132623720	1.181222518	1.230596850	1.334178882	1.512653725
1 ZF1	7.541767600	7.546027690	7.547105806	7.548034239	7.548873294	7.549356674	7.549679753	7.550097903	7.550473500
1 ZF2	12.06406983	12.12025147	12.13357306	12.14478924	12.15473243	12.16038078	12.16412439	12.16893293	12.17321778

(2) P 3+2	TH = 9	SBE = 6.392453221						MCPER FUNCTIONS	
	RC = 02	RC = 05	RC = 07	RC = 10	RC = 15	RC = 20	RC = 25	RC = 35	RC = 50
AMAX DB	.0017375255	.0108709564	.0213327376	.0436480540	.0988323393	.1772876696	.2802872360	.5675287486	1.249387366
AMIN DB	62.70502883	75.98287001	80.86947293	86.06574261	92.01512188	96.29123471	99.66635326	104.9361259	111.0306968
MU K	.2278013D-1	.4989696D-2	.2849359D-2	.1569512D-2	.7925413D-3	.4848688D-3	.3289552D-3	.1794696D-3	.8903602D-4
1 PFR	2.096218825	1.450758926	1.256874537	1.070533675	.8784546113	.7525048765	.6595083605	.5249357121	.3855169614
2 PFC	2.734424201	2.057752161	1.855577353	1.665116334	1.475655222	1.357011056	1.273151870	1.158767200	1.050790208
2 PQC	1.036244188	1.046434064	1.063150204	1.090588218	1.137867599	1.185924547	1.234974777	1.338217551	1.516548011
1 ZF1	6.706743537	6.711336657	6.712518914	6.713542716	6.714472261	6.715009554	6.715369372	6.715835893	6.716255706
1 ZF2	10.69449505	10.75815196	10.77320678	10.78587034	10.79708749	10.80345567	10.80767484	10.81309242	10.81791831

(2) P 3+2	TH = 10 SBE = 5.758770483								MCPER FUNCTIONS
	RC = 02	RC = 05	RC = 07	RC = 10	RC = 15	RC = 20	RC = 25	RC = 35	RC = 50
AMAX DB	.0017375255	.0108709564	.0213327376	.0436480540	.0988323393	.1772876696	.2802872360	.5675287486	1.249387366
AMIN DB	58.10921490	71.38630153	76.27279288	81.46898827	87.41831668	91.69440690	95.06951298	100.3392725	106.4338346
MU K	.3459214D-1	.7598722D-2	.4341928D-2	.2392865D-2	.1208821D-2	.7397197D-3	.5019344D-3	.2738961D-3	.1359049D-3
1 PFR	2.112545343	1.457294012	1.261584355	1.073874026	.8807193368	.7542207130	.6608864503	.5259110627	.3861621196
2 PFC	2.723105137	2.054006709	1.853182712	1.663697532	1.474960722	1.356660775	1.272999523	1.158829870	1.051003215
2 PQC	1.054744233	1.056750844	1.071690231	1.097699254	1.143789546	1.191227540	1.239907959	1.342763476	1.520927312
1 ZF1	6.039170279	6.044026465	6.045302093	6.046413983	6.047428945	6.048017855	6.048413126	6.048926641	6.049389705
1 ZF2	9.595460701	9.666754022	9.683566320	9.697693442	9.710195607	9.717288531	9.721985955	9.728015420	9.733384267

(2) P 3+2	TH = 11 SBE = 5.240843064								MCPER FUNCTIONS
	RC = 02	RC = 05	RC = 07	RC = 10	RC = 15	RC = 20	RC = 25	RC = 35	RC = 50
AMAX DB	.0017375255	.0108709564	.0213327376	.0436480540	.0988323393	.1772876696	.2802872360	.5675287486	1.249387366
AMIN DB	53.94800898	67.22405537	72.11039478	77.30648923	83.25574864	87.53180821	90.90689739	96.17663923	102.2711893
MU K	.5042584D-1	.1111404D-1	.6355133D-2	.3504373D-2	.1771197D-2	.1084149D-2	.7357744D-3	.4015866D-3	.1993020D-3
1 PFR	2.130934059	1.464594123	1.266835585	1.077592543	.8832370435	.7561267939	.6624166026	.5269934007	.3868777048
2 PFC	2.710374918	2.049810077	1.850501646	1.662109375	1.474182425	1.356266869	1.272826482	1.158896599	1.051237483
2 PQC	1.075936758	1.068392700	1.081292848	1.105671199	1.150410987	1.197148154	1.245410362	1.347827764	1.525800886
1 ZF1	5.493419670	5.498461374	5.499818430	5.501010325	5.502105074	5.502743050	5.503172344	5.503731322	5.504236562
1 ZF2	8.693094492	8.772203193	8.790800030	8.806408751	8.820208267	8.828031406	8.833210101	8.839854599	8.845768550

(2) P 3+2	TH = 12 SBE = 4.809734345								MCPER FUNCTIONS
	RC = 02	RC = 05	RC = 07	RC = 10	RC = 15	RC = 20	RC = 25	RC = 35	RC = 50
AMAX DB	.0017375255	.0108709564	.0213327376	.0436480540	.0988323393	.1772876696	.2802872360	.5675287486	1.249387366
AMIN DB	50.14543472	63.42007503	68.30621253	73.50217330	79.45134151	83.72736062	87.10242752	92.37214599	98.46668026
MU K	.7105513D-1	.1572174D-1	.8997236D-2	.4964548D-2	.2510596D-2	.1537201D-2	.1043450D-2	.5696566D-3	.2827734D-3
1 PFR	2.151502474	1.472684207	1.272642801	1.081697611	.8860122954	.7582261160	.6641009885	.5281840599	.3876644913
2 PFC	2.696176170	2.045145911	1.847524099	1.660345920	1.473317119	1.355827295	1.272631335	1.158966615	1.051492654
2 PQC	1.100084379	1.081438833	1.092011453	1.114540263	1.157756177	1.203705196	1.251497731	1.353422912	1.531179127
1 ZF1	5.039076759	5.044218861	5.045644228	5.046907252	5.048075596	5.048759834	5.049221578	5.049824335	5.050370569
1 ZF2	7.938140339	8.025260496	8.045671719	8.062781953	8.077892442	8.086451835	8.092115128	8.099378139	8.105839559

(2) P 3+2	TH = 13 SBE = 4.445411483							MCPER FUNCTIONS	
	RC = 02	RC = 05	RC = 07	RC = 10	RC = 15	RC = 20	RC = 25	RC = 35	RC = 50
AMAX DB	.0017375255	.0108709564	.0213327376	.0436480540	.0988323393	.1772876696	.2802872360	.5675287486	1.249387366
AMIN DB	46.64384230	59.91660752	64.80248186	69.99826918	75.94731941	80.22328628	83.59832439	88.86801275	94.96252664
MU I K	.9729349D-1	.2162349D-1	.1238623D-1	.6839622D-2	.3460988D-2	.2119832D-2	.1439253D-2	.7859547D-3	.3902346D-3
1 I PFR	2.174387083	1.481592521	1.279022412	1.086198625	.8890501846	.7605220151	.6659420212	.5294845200	.3885233368
2 I PFC	2.680450187	2.039996663	1.844239184	1.658400693	1.472361283	1.355339812	1.272412524	1.159039061	1.051768334
2 I PQC	1.127496009	1.095979863	1.103906737	1.124347368	1.165852394	1.210919761	1.258187690	1.359562875	1.537073619
1 I ZF1	4.655088590	4.660238381	4.661717775	4.663042255	4.664277449	4.665004889	4.665497369	4.666142066	4.666728004
1 I ZF2	7.296491404	7.391834958	7.414093178	7.432726717	7.449163091	7.458465353	7.464616883	7.472502219	7.479513694

(2) P 3+2	TH = 14 SBE = 4.133565494							MCPER FUNCTIONS	
	RC = 02	RC = 05	RC = 07	RC = 10	RC = 15	RC = 20	RC = 25	RC = 35	RC = 50
AMAX DB	.0017375255	.0108709564	.0213327376	.0436480540	.0988323393	.1772876696	.2802872360	.5675287486	1.249387366
AMIN DB	43.39848145	56.66876667	61.55430320	66.74986921	72.69876949	76.97467008	80.34967174	85.61932195	91.71381005
MU I K	.1299819D 0	.2903565D-1	.1664956D-1	.9201499D-2	.4659397D-2	.2854928D-2	.1938815D-2	.1059078D-2	.5259803D-3
1 I PFR	2.199746136	1.491350939	1.285992805	1.091106057	.8923563598	.7630181817	.6679423656	.5308964119	.3894551848
2 I PFC	2.663138066	2.034343697	1.840635221	1.656266695	1.471311090	1.354801971	1.272168349	1.159113003	1.052064089
2 I PQC	1.158534547	1.112119104	1.117047390	1.135138548	1.174730156	1.218815375	1.265499858	1.366263148	1.543497197
1 I ZF1	4.326420683	4.331477808	4.332995778	4.334371250	4.335665990	4.336433321	4.336954677	4.337639324	4.338263564
1 I ZF2	6.743780821	6.847573650	6.871714192	6.891894696	6.909673164	6.919725483	6.926369208	6.934381010	6.942445350

(2) P 3+2	TH = 15 SBE = 3.863703305							MCPER FUNCTIONS	
	RC = 02	RC = 05	RC = 07	RC = 10	RC = 15	RC = 20	RC = 25	RC = 35	RC = 50
AMAX DB	.0017375255	.0108709564	.0213327376	.0436480540	.0988323393	.1772876696	.2802872360	.5675287486	1.249387366
AMIN DB	40.37395941	53.64097720	58.52608517	63.72137258	69.67008502	73.94590283	77.32085900	82.59046165	88.68491763
MU I K	.1699752D 0	.3818820D-1	.2192378D-1	.1212767D-1	.6145912D-2	.3767330D-2	.2559124D-2	.1398338D-2	.6946935D-3
1 I PFR	2.227762932	1.501995298	1.293574509	1.096431527	.8959370576	.7657186792	.6701049508	.5324215240	.3904610685
2 I PFC	2.644181954	2.028167408	1.836699786	1.653936416	1.470162408	1.354211120	1.271896964	1.159137422	1.052379446
2 I PQC	1.193626176	1.129974058	1.131510921	1.146965404	1.184423471	1.227418168	1.273455976	1.373540858	1.550464015
1 I ZF1	4.042050948	4.046907293	4.048447225	4.049862435	4.051208863	4.052012519	4.052560749	4.053283200	4.053944234
1 I ZF2	6.262136672	6.374615751	6.400676558	6.422429531	6.441567596	6.452377742	6.459517939	6.468660682	6.476780919

(2) P 3+2	TH = 16 SBE = 3.627955279							MCPER FUNCTIONS	
	RC = 02	RC = 05	RC = 07	RC = 10	RC = 15	RC = 20	RC = 25	RC = 35	RC = 50
AMAX DB	.0017375255	.0108709564	.0213327376	.0436480540	.0988323393	.1772876696	.2802872360	.5675287486	1.249387366
AMIN DB	37.54185497	50.80456943	55.68913838	60.88407884	66.83255883	71.10827455	74.48317472	79.75271889	85.84713540
MU K	.2181272D 0	.4932349D-1	.2835412D-1	.1570114D-1	.7963693D-2	.4883854D-2	.3318551D-2	.1814026D-2	.9014561D-3
1 PFR	2.258649694	1.513565814	1.301790382	1.102187893	.8997991392	.7686279640	.6724329828	.5340618089	.3915421132
2 PFC	2.623526390	2.021447358	1.832419756	1.651401850	1.468910794	1.353564401	1.271596372	1.159261211	1.052713888
2 PQC	1.233271567	1.149678160	1.147384613	1.159885627	1.194970121	1.236757068	1.282080058	1.381414364	1.557989617
1 ZF1	3.793715914	3.798255437	3.799799551	3.801242454	3.802632158	3.803468320	3.804041280	3.804799235	3.805495443
1 ZF2	5.838154512	5.959563865	5.987585364	6.010938112	6.031454554	6.043030886	6.050672151	6.060450644	6.069130039

(2) P 3+2	TH = 17 SBE = 3.420303620							MCPER FUNCTIONS	
	RC = 02	RC = 05	RC = 07	RC = 10	RC = 15	RC = 20	RC = 25	RC = 35	RC = 50
AMAX DB	.0017375255	.0108709564	.0213327376	.0436480540	.0988323393	.1772876696	.2802872360	.5675287486	1.249387366
AMIN DB	34.87907122	48.13610534	53.02000044	58.21451240	64.16270760	68.43829880	71.81313079	77.08260388	83.17697246
MU K	.2752743D 0	.6269468D-1	.3609395D-1	.2001023D-1	.1015897D-1	.6233309D-2	.4236869D-2	.2316933D-2	.1151761D-2
1 PFR	2.292652116	1.526107546	1.310665825	1.108389341	.9039501306	.7717509078	.6749299589	.5358193914	.3926995410
2 PFC	2.601119764	2.014162425	1.827781366	1.648654510	1.467551506	1.352858746	1.271264425	1.159333178	1.053066856
2 PQC	1.278059324	1.171382828	1.164766621	1.173963600	1.206411987	1.246864017	1.291398553	1.389905870	1.566091022
1 ZF1	3.575099501	3.579197990	3.580727332	3.582185093	3.583609107	3.584473704	3.585069107	3.585860110	3.586589766
1 ZF2	5.461586301	5.592171407	5.622196300	5.647177867	5.669092766	5.681444192	5.689591443	5.700010831	5.709252871

(2) P 3+2	TH = 18 SBE = 3.236067977							MCPER FUNCTIONS	
	RC = 02	RC = 05	RC = 07	RC = 10	RC = 15	RC = 20	RC = 25	RC = 35	RC = 50
AMAX DB	.0017375255	.0108709564	.0213327376	.0436480540	.0988323393	.1772876696	.2802872360	.5675287486	1.249387366
AMIN DB	32.36667772	45.61618493	50.49924070	55.69322681	61.64107591	65.91651661	69.29126642	74.56065396	80.65496495
MU K	.3422191D 0	.7856413D-1	.4530422D-1	.2514848D-1	.1278101D-1	.7846518D-2	.5335278D-2	.2918868D-2	.1451522D-2
1 PFR	2.330054582	1.539670949	1.320229029	1.115051503	.9083982687	.7750928236	.6775996842	.5376965763	.3939346747
2 PFC	2.576915924	2.006290961	1.822770271	1.645685447	1.466079495	1.352090874	1.270898823	1.159402039	1.053437747
2 PQC	1.328681973	1.195259840	1.183767243	1.189271087	1.218795411	1.257774216	1.301440530	1.399036559	1.574786822
1 ZF1	3.381292225	3.384816982	3.386311417	3.387770410	3.389219217	3.390107924	3.390723338	3.391544782	3.392306046
1 ZF2	5.124467359	5.264467575	5.296540557	5.323181631	5.346516280	5.359652313	5.368310789	5.379376555	5.389184952

(2) P 3+2			TH = 19 SBE = 3.071553487								MCPER FUNCTIONS
			RC = 02	RC = 05	RC = 07	RC = 10	RC = 15	RC = 20	RC = 25	RC = 35	RC = 50
AMAX DB			.0017375255	.0108709564	.0213327376	.0436480540	.0988323393	.1772876696	.2802872360	.5675287486	1.249387366
AMIN DB			29.98908851	43.22857715	48.11058924	53.30393286	59.25136408	63.52662430	66.90127585	72.17056132	78.26480372
MU		K	.4197130D 0	.9720137D-1	.5615267D-1	.3121442D-1	.1588215D-1	.9756327D-2	.6636425D-2	.3632429D-2	.1807089D-2
1		PFR	2.371186063	1.554312494	1.330511254	1.122191571	.9131525520	.7786594932	.6804462893	.5396958582	.3952489422
2		PFC	2.550875973	1.997810972	1.317371618	1.642485269	1.464489412	1.351257296	1.270497106	1.159466418	1.053825911
2		PQC	1.385954793	1.221504133	1.204510394	1.205888019	1.232171618	1.269526406	1.312237891	1.408831730	1.584097278
1		ZF1	3.208421190	3.211230666	3.212668865	3.214114667	3.215578194	3.216486433	3.217119287	3.217968408	3.218759330
1		ZF2	4.820520218	4.970158726	5.004326098	5.032658377	5.057435782	5.071366493	5.080541751	5.092259715	5.102638412

(2) P 3+2			TH = 20 SBE = 2.923804400								MCPER FUNCTIONS
			RC = 02	RC = 05	RC = 07	RC = 10	RC = 15	RC = 20	RC = 25	RC = 35	RC = 50
AMAX DB			.0017375255	.0108709564	.0213327376	.0436480540	.0988323393	.1772876696	.2802872360	.5675287486	1.249387366
AMIN DB			27.73347856	40.95957578	45.84029007	51.03285086	56.97978009	61.25482526	64.62936015	69.89852476	75.99268611
MU		K	.5084397D 0	.1188810D 0	.6881307D-1	.3831129D-1	.1951771D-1	.1199762D-1	.8164424D-2	.4471070D-2	.2225263D-2
1		PFR	2.416426572	1.570095386	1.341547152	1.129828445	.9182227984	.7824571990	.6834742508	.5418199319	.3966438820
2		PFC	2.522970334	1.988700312	1.811570116	1.639044168	1.462775610	1.350354304	1.270056657	1.159524844	1.054230650
2		PQC	1.450837639	1.250337046	1.227135307	1.223903394	1.246597180	1.282163175	1.323825599	1.419318460	1.594044445
1		ZF1	3.053391069	3.055334480	3.056693893	3.058111283	3.059578902	3.060501844	3.061149421	3.062023300	3.062841822
1		ZF2	4.544738721	4.704209423	4.740518578	4.770576592	4.796819404	4.811555421	4.821253337	4.833629654	4.844582821

(2) P 3+2			TH = 21 SBE = 2.790428110								MCPER FUNCTIONS
			RC = 02	RC = 05	RC = 07	RC = 10	RC = 15	RC = 20	RC = 25	RC = 35	RC = 50
AMAX DB			.0017375255	.0108709564	.0213327376	.0436480540	.0988323393	.1772876696	.2802872360	.5675287486	1.249387366
AMIN DB			25.58937453	38.79751490	43.67661315	48.86822139	54.81454979	59.08934004	62.46373728	67.73275972	73.82682597
MU		K	.6089985D 0	.1438805D 0	.8346428D-1	.4654682D-1	.2374597D-1	.1460734D-1	.9944882D-2	.5449127D-2	.2713307D-2
1		PFR	2.466213920	1.587090387	1.353375126	1.137982888	.9236197094	.7864927589	.6866884129	.5440717033	.3981211484
2		PFC	2.493181187	1.978936894	1.805350123	1.635351944	1.460932146	1.349377977	1.269574695	1.159575747	1.054651214
2		PQC	1.524459630	1.282010128	1.251798502	1.243416315	1.262134555	1.295731317	1.336241949	1.430526287	1.604652291
1		ZF1	2.913699035	2.914616020	2.915872853	2.917245797	2.918706325	2.919638887	2.920298328	2.921193892	2.922037844
1		ZF2	4.293092531	4.462543280	4.501042040	4.532859914	4.560593350	4.576145844	4.586372603	4.599413764	4.610945800

TH = 22 SBE = 2.669467163 MCPER FUNCTIONS

MU	(2) P 3+2	RC = 02	RC = 05	RC = 07	RC = 10	RC = 15	RC = 20	RC = 25	RC = 35	RC = 50
	AMAX DB	.0017375255	.0108709564	.0213327376	.0436480540	.0988323393	.1772876696	.2802872360	.5675287486	1.249387366
	AMIN DB	23.54837630	36.73239983	41.60948098	46.79993020	52.74554128	57.02003063	60.39426625	65.66312237	71.75707771
	K	.7218888D 0	.1724772D 0	.1002892D 0	.5603282D-1	.2862811D-1	.1762447D-1	.1200491D-1	.6581835D-2	.3278964D-2
1	PFR	2.521050290	1.605376741	1.366037746	1.146677706	.9293549423	.7907735653	.6900940124	.5464543023	.3996825181
2	PFC	2.461505413	1.968498924	1.798695737	1.631398035	1.458952787	1.348324175	1.269048276	1.159617456	1.055086803
2	PQC	1.608146158	1.316809552	1.278676065	1.264537174	1.278652695	1.310282222	1.349528854	1.442487403	1.615946847
1	ZF1	2.787300026	2.787020483	2.788149662	2.789461304	2.790902999	2.791839845	2.792508148	2.793422169	2.794289269
1	ZF2	4.062313930	4.241825527	4.282561278	4.316174426	4.345424169	4.361804828	4.372566923	4.386279747	4.398395278

TH = 23 SBE = 2.559304665 MCPER FUNCTIONS

MU	(2) P 3+2	RC = 02	RC = 05	RC = 07	RC = 10	RC = 15	RC = 20	RC = 25	RC = 35	RC = 50
	AMAX DB	.0017375255	.0108709564	.0213327376	.0436480540	.0988323393	.1772876696	.2802872360	.5675287486	1.249387366
	AMIN DB	21.60397642	34.75562365	39.63017952	44.81921706	50.76397293	55.03810825	58.41215488	63.68081739	69.77464408
	K	.8474954D 0	.2049464D 0	.1194737D 0	.6638483D-1	.3422816D-1	.2109003D-1	.1437316D-1	.7885358D-2	.3930474D-2
1	PFR	2.581507821	1.625043229	1.379582218	1.155937946	.9354411906	.7953076284	.6936967058	.5489710966	.4013298965
2	PFC	2.427958163	1.957365159	1.791590897	1.627171550	1.456831014	1.347188540	1.268474283	1.159643194	1.055536561
2	PQC	1.703446870	1.355061266	1.307966282	1.287389008	1.296827730	1.325872332	1.363732186	1.455236880	1.627956364
1	ZF1	2.672506944	2.670851214	2.671826343	2.673058992	2.674469547	2.675405087	2.676079107	2.677008200	2.677896056
1	ZF2	3.849741897	4.039302425	4.082321031	4.117765266	4.148557824	4.165778826	4.177083044	4.191474678	4.204178554

TH = 24 SBE = 2.458593336 MCPER FUNCTIONS

MU	(2) P 3+2	RC = 02	RC = 05	RC = 07	RC = 10	RC = 15	RC = 20	RC = 25	RC = 35	RC = 50
	AMAX DB	.0017375255	.0108709564	.0213327376	.0436480540	.0988323393	.1772876696	.2802872360	.5675287486	1.249387366
	AMIN DB	19.75144948	32.85974810	37.73113210	42.91844708	48.86218377	53.13590351	56.50972990	61.77816792	67.87184608
	K	.9860750D 0	.2415576D 0	.1412052D 0	.7922170D-1	.4061287D-1	.2504710D-1	.1707980D-1	.9376803D-2	.4676589D-2
1	PFR	2.648231953	1.646189346	1.394060908	1.165791129	.9418922743	.8001036247	.6975025998	.5516257069	.4030653255
2	PFC	2.392577181	1.945515193	1.784019498	1.622661306	1.454560031	1.345966493	1.267849434	1.159666080	1.055999576
2	PQC	1.812162207	1.397136967	1.339892696	1.312109065	1.316143758	1.342563644	1.378902144	1.468812921	1.640711488
1	ZF1	2.567915441	2.564695120	2.565488431	2.566623540	2.567990078	2.568918469	2.569594917	2.570535545	2.571441652
1	ZF2	3.653206562	3.852681004	3.898025431	3.935336467	3.967699007	3.985772982	3.997626394	4.012704299	4.026001595

(2) P 3+2 — TH = 25 SBE = 2.366201583 — MCPER FUNCTIONS

MU	3+2	RC = 02	RC = 05	RC = 07	RC = 10	RC = 15	RC = 20	RC = 25	RC = 35	RC = 50
	AMAX DB	.0017375255	.0108709564	.0213327376	.0436480540	.0988323393	.1772876696	.2802872360	.5675287486	1.249387366
	AMIN DB	17.98778199	31.03833409	35.90572133	41.09092929	47.03345089	51.30668352	54.68025389	59.94843251	66.04193988
MU	K	.1137744D 1	.2825725D 0	.1656716D 0	.9316509D-1	.4785163D-1	.2954079D-1	.2015659D-1	.1107425D-1	.5526594D-2
1	PFR	2.721940806	1.668926603	1.409531940	1.176267500	.9487232411	.8051709505	.7015182843	.5544220235	.4048909910
2	PFC	2.355427863	1.932929792	1.775965511	1.617855869	1.452132765	1.344653236	1.267170266	1.159669116	1.056474873
2	PQC	1.936364668	1.443460995	1.374707627	1.338850595	1.336893733	1.360424282	1.395093675	1.483257139	1.654245459
1	ZF1	2.472346246	2.467365983	2.467948287	2.468966431	2.470275500	2.471190641	2.471866085	2.472814553	2.473736295
1	ZF2	3.470942096	3.680037846	3.727746460	3.766959296	3.800919426	3.819859404	3.832269346	3.848041292	3.861937301

(2) P 3+2 — TH = 26 SBE = 2.281172033 — MCPER FUNCTIONS

MU	3+2	RC = 02	RC = 05	RC = 07	RC = 10	RC = 15	RC = 20	RC = 25	RC = 35	RC = 50
	AMAX DB	.0017375255	.0108709564	.0213327376	.0436480540	.0988323393	.1772876696	.2802872360	.5675287486	1.249387366
	AMIN DB	16.31160828	29.28581126	34.14814812	39.33077161	45.27184291	49.54450499	52.91777830	58.18565799	64.27896967
MU	K	.1302468D 1	.3282413D 0	.1930598D 0	.1088389D 0	.5601632D-1	.3461819D-1	.2363682D-1	.1299677D-1	.6490326D-2
1	PFR	2.803418422	1.693379961	1.426059857	1.187400318	.9559504788	.8105197821	.7057508701	.5573642252	.4068092327
2	PFC	2.316608861	1.919591261	1.767413130	1.612743601	1.449541886	1.343243748	1.266433143	1.159655194	1.056961420
2	PQC	2.078409336	1.494518270	1.412696258	1.367784916	1.359180482	1.379529150	1.412366950	1.498614863	1.668594328
1	ZF1	2.384800137	2.377860854	2.378201490	2.379082338	2.380319901	2.381215434	2.381886300	2.382838758	2.383773406
1	ZF2	3.301519178	3.519749141	3.569853607	3.611001756	3.646587255	3.666406605	3.679380659	3.695854708	3.710354938

(2) P 3+2 — TH = 27 SBE = 2.202689265 — MCPER FUNCTIONS

MU	3+2	RC = 02	RC = 05	RC = 07	RC = 10	RC = 15	RC = 20	RC = 25	RC = 35	RC = 50
	AMAX DB	.0017375255	.0108709564	.0213327376	.0436480540	.0988323393	.1772876696	.2802872360	.5675287486	1.249387366
	AMIN DB	14.72310985	27.59737885	32.45331984	37.63276406	43.57210130	47.84409513	51.21702419	56.48456014	62.57764828
MU	K	.1480055D 1	.3788000D 0	.2235541D 0	.1263687D 0	.6518120D-1	.4032339D-1	.2755538D-1	.1516446D-1	.7578191D-2
1	PFR	2.893499544	1.719689360	1.443716360	1.199226177	.9635918421	.8161611417	.7102080298	.5604567993	.4088225528
2	PFC	2.276257658	1.905483890	1.758346926	1.607312714	1.446779805	1.341732788	1.265634247	1.159622085	1.057458114
2	PQC	2.240927070	1.550863351	1.454181336	1.399103783	1.383117868	1.399960664	1.430787897	1.514935500	1.683797201
1	ZF1	2.304422109	2.295326131	2.295392924	2.296115213	2.297266634	2.298135944	2.298798510	2.299750952	2.300695667
1	ZF2	3.143790099	3.370436528	3.422959250	3.466073798	3.503312281	3.524024634	3.537570600	3.554755091	3.569865258

(2) P 3+2	TH = 28	SBE = 2.130054468						MCPER FUNCTIONS	
	RC = 02	RC = 05	RC = 07	RC = 10	RC = 15	RC = 20	RC = 25	RC = 35	RC = 50
AMAX DB	.0017375255	.0108709564	.0213327376	.0436480540	.0988323393	.1772876696	.2802872360	.5675287486	1.249387366
AMIN DB	13.22382989	25.96893199	30.81676192	35.99228423	41.92954368	46.20075438	49.57328472	54.84042604	60.93325944
MU I K	.1670153D 1	.4344676D 0	.2573346D 0	.1458813D 0	.7542270D-1	.4672237D-1	.3194873D-1	.1759846D-1	.8801190D-2
1 I PFR	2.993044110	1.748011328	1.462581127	1.211785366	.9716667931	.8221069725	.7148980436	.5637045647	.4109336276
2 I PFC	2.234555147	1.890594468	1.748752035	1.601551327	1.443838695	1.340114895	1.264769576	1.159567435	1.057963787
2 I PQC	2.426793785	1.613130660	1.499528588	1.433022126	1.408832109	1.421809589	1.450428802	1.532272915	1.699896513
1 I ZF1	2.230472425	2.219030891	2.218790128	2.219331633	2.220381668	2.221217876	2.221868277	2.222816542	2.223768370
1 I ZF2	2.996840851	3.230924805	3.285875846	3.330984300	3.369902813	3.391521964	3.405647827	3.423551353	3.439277372

(2) P 3+2	TH = 29	SBE = 2.062665340						MCPER FUNCTIONS	
	RC = 02	RC = 05	RC = 07	RC = 10	RC = 15	RC = 20	RC = 25	RC = 35	RC = 50
AMAX DB	.0017375255	.0108709564	.0213327376	.0436480540	.0988323393	.1772876696	.2802872360	.5675287486	1.249387366
AMIN DB	11.81635767	24.39700939	29.23454844	34.40522060	40.33998433	44.61027626	47.98234482	53.24903355	59.34157729
MU I K	.1872258D 1	.4954423D 0	.2945759D 0	.1675036D 0	.8681924D-1	.5385307D-1	.3685492D-1	.2032096D-1	.1017094D-1
1 I PFR	3.102901132	1.778520593	1.482742693	1.225122266	.9801965593	.8283702203	.7198298501	.5671126961	.4131453187
2 I PFC	2.191727726	1.874912895	1.738614366	1.595447533	1.440710503	1.338384388	1.263834943	1.159438767	1.058477196
2 I PQC	2.639071828	1.682045902	1.549152935	1.469781191	1.436463302	1.445175986	1.471368998	1.550685383	1.716938331
1 I ZF1	2.162303015	2.148345716	2.147762161	2.148099667	2.149032450	2.149828412	2.150462638	2.151402407	2.152358284
1 I ZF2	2.859945852	3.100208678	3.157582100	3.204706998	3.245331514	3.267871307	3.282585196	3.301216578	3.317564553

(2) P 3+2	TH = 30	SBE = 2.000000000						MCPER FUNCTIONS	
	RC = 02	RC = 05	RC = 07	RC = 10	RC = 15	RC = 20	RC = 25	RC = 35	RC = 50
AMAX DB	.0017375255	.0108709564	.0213327376	.0436480540	.0988323393	.1772876696	.2802872360	.5675287486	1.249387366
AMIN DB	10.50385743	22.87875839	27.70324860	32.86791039	38.79966871	43.06888106	46.44041455	51.70653451	57.79879941
MU I K	.2085730D 1	.5618993D 0	.3354452D 0	.1913622D 0	.9945105D-1	.6177518D-1	.4231355D-1	.2335525D-1	.1169968D-1
1 I PFR	3.223864324	1.811411630	1.504299414	1.239285801	.9892043094	.8349649262	.7250131023	.5706367519	.4154606860
2 I PFC	2.148045272	1.858432919	1.727920852	1.588989483	1.437386963	1.336535369	1.262825971	1.159383467	1.058997026
2 I PQC	2.880925328	1.758438555	1.603525589	1.509652160	1.466167150	1.470170297	1.493695633	1.570238581	1.734972694
1 I ZF1	2.099338422	2.082725729	2.081762692	2.081871969	2.082670994	2.083419298	2.084033192	2.084959991	2.085916739
1 I ZF2	2.732522628	2.977426439	3.037196028	3.086353269	3.128708090	3.152182271	3.167492409	3.186960661	3.203236974

(2) P 3+2			TH = 31	SBE = 1.941604026							MCPER FUNCTIONS
			RC = 02	RC = 05	RC = 07	RC = 10	RC = 15	RC = 20	RC = 25	RC = 35	RC = 50
AMAX DB			.0017375255	.0108709564	.0213327376	.0436480540	.0988323393	.1772876696	.2802872360	.5675287486	1.249387366
AMIN DB			9.289457358	21.41191437	26.21988655	31.37708923	37.30521912	41.57316054	44.94407345	50.20964886	56.30149086
MU		K	.2309831D 1	.6339871D 0	.3801008D 0	.2175829D 0	.1133999D 0	.7054520D-1	.4836580D-1	.2672575D-1	.1340035D-1
1		PFR	3.356625161	1.846899984	1.527360465	1.254329923	.9987153492	.8419063287	.7304582309	.5744327047	.4178830023
2		PFC	2.103813646	1.841153005	1.716659744	1.582165469	1.433859620	1.334561725	1.261738093	1.159248790	1.059521880
2		PQC	3.155520109	1.843255233	1.663182109	1.552940318	1.498116962	1.496914575	1.517504552	1.591001150	1.754053991
1		ZF1	2.041060946	2.021696853	2.020316347	2.020172144	2.020820250	2.021513213	2.022102468	2.023011666	2.023965995
1		ZF2	2.614086060	2.861839013	2.923953383	2.975150242	3.019257270	3.043679320	3.059593963	3.079708253	3.097319145

(2) P 3+2			TH = 32	SBE = 1.887079915							MCPER FUNCTIONS
			RC = 02	RC = 05	RC = 07	RC = 10	RC = 15	RC = 20	RC = 25	RC = 35	RC = 50
AMAX DB			.0017375255	.0108709564	.0213327376	.0436480540	.0988323393	.1772876696	.2802872360	.5675287486	1.249387366
AMIN DB			8.175568896	19.99479083	24.78191266	29.92985097	35.85358965	40.12003167	43.49022387	48.75511765	54.84653716
MU		K	.2543771D 1	.7118251D 0	.4286903D 0	.2462893D 0	.1287488D 0	.8022127D-1	.5505439D-1	.3045797D-1	.1528653D-1
1		PFR	3.501731818	1.885223213	1.552046884	1.270314149	1.008757340	.8492109777	.7361765132	.5783569744	.4204157682
2		PFC	2.059361441	1.823077355	1.704820956	1.574964035	1.430119850	1.332457123	1.260566547	1.159081846	1.060050278
2		PQC	3.465927535	1.937573499	1.728731468	1.599989831	1.532505930	1.525543891	1.542901296	1.613050328	1.774241395
1		ZF1	1.986999888	1.964844594	1.963007615	1.962583665	1.963063021	1.963692683	1.964252846	1.965139654	1.966088161
1		ZF2	2.504204379	2.752813146	2.817190274	2.870423071	2.916300945	2.941683867	2.958211231	2.979080832	2.997332985

(2) P 3+2			TH = 33	SBE = 1.836078459							MCPER FUNCTIONS
			RC = 02	RC = 05	RC = 07	RC = 10	RC = 15	RC = 20	RC = 25	RC = 35	RC = 50
AMAX DB			.0017375255	.0108709564	.0213327376	.0436480540	.0988323393	.1772876696	.2802872360	.5675287486	1.249387366
AMIN DB			7.163254065	18.62627600	23.38718462	28.52361595	34.44202874	38.70669771	42.07605156	47.34016330	53.43110435
MU		K	.2786763D 1	.7955000D 0	.4813486D 0	.2776022D 0	.1455819D 0	.9086313D-1	.6242357D-1	.3457863D-1	.1737254D-1
1		PFR	3.659562896	1.926641185	1.578492584	1.287304149	1.019360540	.8568968618	.7421801507	.5824664658	.4230627299
2		PFC	2.015022252	1.804217090	1.692396485	1.567374100	1.426158891	1.330215065	1.259306379	1.158879602	1.060580653
2		PQC	3.815055941	2.042615468	1.800866154	1.651189238	1.569549765	1.556207951	1.570002240	1.636470167	1.795599337
1		ZF1	1.936724704	1.911804767	1.909471746	1.909740809	1.909032902	1.909591019	1.910117486	1.910976958	1.911916130
1		ZF2	2.402461047	2.649807770	2.716329108	2.771580504	2.819243564	2.845599639	2.862747811	2.884382041	2.903282157

(2) P 3+2			TH = 34 SBE = 1.788291650								MCPER FUNCTIONS
			RC = 02	RC = 05	RC = 07	RC = 10	RC = 15	RC = 20	RC = 25	RC = 35	RC = 50
AMAX DB			.0017375255	.0108709564	.0213327376	.0436480540	.0988323393	.1772876696	.2802872360	.5675287486	1.249387366
AMIN DB			6.251770794	17.30583079	22.03395690	27.15610623	33.06804818	37.33061555	40.69899232	45.96220588	52.05260512
MU		K	.3038088D 1	.8850636D 0	.5381969D 0	.3116390D 0	.1639840D 0	.1025320D 0	.7051910D-1	.3911558D-1	.1967345D-1
1		PFR	3.830322031	1.971435464	1.606845311	1.305372364	1.030558076	.8649835493	.7484823552	.5867686100	.4258278981
2		PFC	1.971115293	1.784591559	1.679380901	1.559385108	1.421967872	1.327828794	1.257952435	1.158638872	1.061111345
2		PQC	4.205629345	2.159760299	1.880373272	1.706977713	1.609489710	1.589072935	1.598935890	1.661352844	1.818198050
1		ZF1	1.889841516	1.862255759	1.859387240	1.858321184	1.858406812	1.858884856	1.859372871	1.860199902	1.861126111
1		ZF2	2.308427010	2.552362744	2.620867168	2.678103066	2.727560144	2.754900629	2.772677464	2.795085618	2.814640489

(2) P 3+2			TH = 35 SBE = 1.743446796								MCPER FUNCTIONS
			RC = 02	RC = 05	RC = 07	RC = 10	RC = 15	RC = 20	RC = 25	RC = 35	RC = 50
AMAX DB			.0017375255	.0108709564	.0213327376	.0436480540	.0988323393	.1772876696	.2802872360	.5675287486	1.249387366
AMIN DB			5.438389527	16.03348142	20.72087698	25.82532871	31.72939752	35.98946803	39.35670365	44.61888422	50.70866975
MU		K	.3297137D 1	.9805296D 0	.5993399D 0	.3485124D 0	.1840401D 0	.1152904D 0	.7938819D-1	.4409791D-1	.2220507D-1
1		PFR	4.014054133	2.019907441	1.637267442	1.324598675	1.042386236	.8734923455	.7550974446	.5912714098	.4287155686
2		PFC	1.927927842	1.764229739	1.665771925	1.550987202	1.417537851	1.325291405	1.256499367	1.158356316	1.061640597
2		PQC	4.640221278	2.290554350	1.968146508	1.767852220	1.652596026	1.624323626	1.629844374	1.687799585	1.842114178
1		ZF1	1.845991985	1.815911992	1.812469586	1.811039538	1.810898811	1.811287961	1.811732614	1.812521942	1.813431450
1		ZF2	2.221646102	2.460089264	2.530367278	2.589533358	2.640786349	2.669121133	2.687534128	2.710725393	2.730941868

(2) P 3+2			TH = 36 SBE = 1.701301617								MCPER FUNCTIONS
			RC = 02	RC = 05	RC = 07	RC = 10	RC = 15	RC = 20	RC = 25	RC = 35	RC = 50
AMAX DB				.0108709564	.0213327376	.0436480540	.0988323393	.1772876696	.2802872360	.5675287486	1.249387366
AMIN DB				14.80979874	19.44698637	24.52956210	30.42404331	34.68114073	38.04704062	43.30803120	49.39712119
MU		K		.1081872D 1	.6648647D 0	.3883298D 0	.2058355D 0	.1292021D 0	.8907949D-1	.4955589D-1	.2498403D-1
1		PFR		2.072374885	1.669936528	1.345071079	1.054884798	.8824464673	.7620409489	.5959834903	.4317303458
2		PFC		1.743171621	1.651571104	1.542171434	1.412859868	1.322595805	1.254941630	1.158028431	1.062166551
2		PQC		2.436717612	2.065198750	1.834375618	1.699171991	1.662165840	1.662885136	1.715921712	1.867431465
1		ZF1		1.772518347	1.768466006	1.766642588	1.766254943	1.766546084	1.766942310	1.767688512	1.768577471
1		ZF2		2.372661322	2.444450134	2.505468053	2.558510269	2.587847453	2.606903601	2.630886953	2.651771899

(2) P 3+2	TH = 37	SBE = 1.661640141						MCPER FUNCTIONS	
	RC = 02	RC = 05	RC = 07	RC = 10	RC = 15	RC = 20	RC = 25	RC = 35	RC = 50
AMAX DB		.0108709564	.0213327376	.0436480540	.0988323393	.1772876696	.2802872360	.5675287486	1.249387366
AMIN DB		13.63585507	18.21172434	23.26735055	29.15015242	33.40370245	36.76803529	42.02765248	48.11595349
MU I K		.1189024D 1	.7348386D 0	.4311919D 0	.2294551D 0	.1443321D 0	.9964302D-1	.5552105D-1	.2802780D-1
1 I PFR		2.129166683	1.705045447	1.366886377	1.068097387	.8918712368	.7693297293	.6009141562	.4348771683
2 I PFC		1.721469429	1.636784581	1.532930018	1.407924994	1.319734737	1.253273485	1.157651550	1.062687239
2 I PQC		2.600145010	2.172674967	1.907185789	1.749558518	1.702829239	1.698232899	1.745841844	1.894241538
1 I ZF1		1.731845348	1.727150999	1.724904680	1.724248917	1.724432635	1.724775215	1.725472709	1.726337157
1 I ZF2		2.289807666	2.362787912	2.425551280	2.480365550	2.510710967	2.530416578	2.555200670	2.576760924

(2) P 3+2	TH = 38	SBE = 1.624269245						MCPER FUNCTIONS	
	RC = 02	RC = 05	RC = 07	RC = 10	RC = 15	RC = 20	RC = 25	RC = 35	RC = 50
AMAX DB		.0108709564	.0213327376	.0436480540	.0988323393	.1772876696	.2802872360	.5675287486	1.249387366
AMIN DB		12.51314994	17.01493091	22.03750122	27.90607915	32.15538890	35.51787919	40.77590810	46.86331313
MU I K		.1301878D 1	.8093075D 0	.4771921D 0	.2549828D 0	.1607460D 0	.1111301D 0	.6202616D-1	.3135468D-1
1 I PFR		2.190615666	1.742802002	1.390150834	1.082071864	.9017942959	.7769821093	.6060734535	.4381613372
2 I PFC		1.699188479	1.621423937	1.523256623	1.402724404	1.316700799	1.251489000	1.157221832	1.063200583
2 I PQC		2.782901481	2.291864835	1.987005860	1.804139463	1.746570564	1.736081904	1.77769[illegible]264	1.922644789
1 I ZF1		1.693685016	1.688322534	1.685624122	1.684678466	1.684745052	1.685028606	1.685671652	1.686507517
1 I ZF2		2.211303782	2.285098339	2.349469145	2.406025664	2.437382304	2.457742804	2.483335823	2.505578141

(2) P 3+2	TH = 39	SBE = 1.589015729						MCPER FUNCTIONS	
	RC = 02	RC = 05	RC = 07	RC = 10	RC = 15	RC = 20	RC = 25	RC = 35	RC = 50
AMAX DB		.0108709564	.0213327376	.0436480540	.0988323393	.1772876696	.2802872360	.5675287486	1.249387366
AMIN DB		11.44349794	15.85684553	20.83908582	26.69035571	30.93458938	34.29490844	39.55109662	45.63748276
MU I K		.1420287D 1	.8882940D 0	.5264152D 0	.2825018D 0	.1785105D 0	.1235934D 0	.6910525D-1	.3498390D-1
1 I PFR		2.257049723	1.783427794	1.414980778	1.096860741	.9122458430	.7850180215	.6114722400	.4415885475
2 I PFC		1.676407427	1.605507090	1.513146731	1.397249455	1.313486469	1.249582055	1.156735262	1.063704382
2 I PQC		2.987210500	2.424214397	2.074655504	1.863347742	1.793677377	1.776648505	1.811631486	1.952751390
1 I ZF1		1.657847299	1.651798750	1.648620062	1.647362269	1.647301719	1.647520710	1.648103407	1.648906510
1 I ZF2		2.136963575	2.211142422	2.276945167	2.335199088	2.367566443	2.388586133	2.414995638	2.437926626

(2) P 3+2	TH = 40	SBE = 1.555723827						MCPER FUNCTIONS	
	RC = 02	RC = 05	RC = 07	RC = 10	RC = 15	RC = 20	RC = 25	RC = 35	RC = 50
AMAX DB		.0108709564	.0213327376	.0436480540	.0988323393	.1772876696	.2802872360	.5675287486	1.249387366
AMIN DB		10.42887689	14.73809648	19.67144511	25.50168586	29.73983589	33.09759116	38.35164137	44.43686695
MU I K		.1544071D 1	.9717963D 0	.5789362D 0	.3120933D 0	.1976925D 0	.1370866D 0	.7679364D-1	.3893561D-1
1 I PFR		2.328781751	1.827156201	1.441503069	1.112521628	.9232588939	.7934591703	.6171222629	.4451649231
2 I PFC		1.653217676	1.589059190	1.502598035	1.391491787	1.310084128	1.247546350	1.156187641	1.064196313
2 I PQC		3.215437022	2.571335965	2.171063310	1.927672354	1.844472382	1.820174158	1.847816061	1.984682437
1 I ZF1		1.624157067	1.617415088	1.613729817	1.612137325	1.611939351	1.612088085	1.612604372	1.613370423
1 I ZF2		2.066630616	2.140715069	2.207736467	2.267625263	2.300998564	2.322680336	2.349913068	2.373539107

(2) P 3+2	TH = 41	SBE = 1.524253087						MCPER FUNCTIONS	
	RC = 02	RC = 05	RC = 07	RC = 10	RC = 15	RC = 20	RC = 25	RC = 35	RC = 50
AMAX DB		.0108709564	.0213327376	.0436480540	.0988323393	.1772876696	.2802872360	.5675287486	1.249387366
AMIN DB		9.471241668	13.65967582	18.53419529	24.33894148	28.56979469	31.92451701	37.17607347	43.25997979
MU I K		.1673021D 1	.1059787D 1	.6348195D 0	.3438365D 0	.2183595D 0	.1516646D 0	.8512793D-1	.4323091D-1
1 I PFR		2.406099378	1.874229321	1.469855402	1.129117730	.9348695673	.8023292119	.6230362452	.4488970561
2 I PFC		1.629721761	1.572113428	1.491610920	1.385443441	1.306486102	1.245375414	1.155574586	1.064673918
2 I PQC		3.470067367	2.735015434	2.277280106	1.997666446	1.899318428	1.866928883	1.886432648	2.018571266
1 I ZF1		1.592451685	1.585021762	1.580806561	1.578856712	1.578510759	1.578583386	1.579027043	1.579751646
1 I ZF2		2.000169096	2.073645834	2.141630548	2.203071168	2.237440532	2.259785546	2.287847199	2.312174358

(2) P 3+2	TH = 42	SBE = 1.494476550						MCPER FUNCTIONS	
	RC = 02	RC = 05	RC = 07	RC = 10	RC = 15	RC = 20	RC = 25	RC = 35	RC = 50
AMAX DB		.0108709564	.0213327376	.0436480540	.0988323393	.1772876696	.2802872360	.5675287486	1.249387366
AMIN DB		8.572317991	12.62289452	17.42723467	23.20116190	27.42326005	30.77438848	36.02304655	42.10543385
MU I K		.1806915D 1	.1152213D 1	.6941173D 0	.3778079D 0	.2405791D 0	.1673834D 0	.9414602D-1	.4789194D-1
1 I PFR		2.489255680	1.924893837	1.500186327	1.146718180	.9471173970	.8116539543	.6292279329	.4527920497
2 I PFC		1.606030680	1.554711658	1.480188975	1.379097005	1.302684698	1.243062613	1.154891521	1.065134600
2 I PQC		3.753690225	2.917216325	2.394493045	2.073956546	1.958624305	1.917215275	1.927685406	2.054564966
1 I ZF1		1.562579165	1.554481546	1.549717315	1.547387659	1.546882936	1.546873452	1.547233104	1.547916756
1 I ZF2		1.937454873	2.009792079	2.078442529	2.141328554	2.176677911	2.199635215	2.228580181	2.253614111

(2) P 3+2	TH = 43 RC = 02	SBE = 1.466279186 RC = 05	RC = 07	RC = 10	RC = 15	RC = 20	RC = 25	MCPER FUNCTIONS RC = 35	RC = 50
AMAX DB		.0108709564	.0213327376	.0436480540	.0988323393	.1772876696	.2802872360	.5675287486	1.249387366
AMIN DB		7.733397930	11.62931351	16.35074857	22.08755577	26.29915007	29.64601404	34.89127786	40.97193062
MU I K		.1945521D 1	.1248996D 1	.7568686D 0	.4140811D 0	.2644185D 0	.1842996D 0	.1038871D 0	.5294186D-1
1 I PFR		2.578462184	1.979395869	1.532654923	1.165398816	.9600456696	.8214615783	.6357124520	.4568575665
2 I PFC		1.582260282	1.536904701	1.468339575	1.372445780	1.298672267	1.240601173	1.154133676	1.065575616
2 I PQC		4.068984038	3.120080307	2.524040194	2.157253099	2.022851460	1.971373163	1.971801770	2.092826102
1 I ZF1		1.534396912	1.525667830	1.520341181	1.517609882	1.516935400	1.516837658	1.517116774	1.517744870
1 I ZF2		1.878367237	1.949034879	2.018012701	2.082211324	2.118517426	2.142183520	2.171914581	2.197660394

(2) P 3+2	TH = 44 RC = 02	SBE = 1.439556540 RC = 05	RC = 07	RC = 10	RC = 15	RC = 20	RC = 25	MCPER FUNCTIONS RC = 35	RC = 50
AMAX DB		.0108709564	.0213327376	.0436480540	.0988323393	.1772876696	.2802872360	.5675287486	1.249387366
AMIN DB		6.955163087	10.68064869	15.30521009	20.99750525	25.19650452	28.53830281	33.77959061	39.85825189
MU I K		.2088622D 1	.1350037D 1	.8230983D 0	.4527257D 0	.2899449D 0	.2024710D 0	.1143917D 0	.5840493D-1
1 I PFR		2.673885243	2.037975071	1.567430014	1.185242354	.9737017869	.8317828804	.6425059295	.4611018830
2 I PFC		1.558527068	1.518752195	1.456074473	1.365483990	1.294441264	1.237984199	1.153296085	1.065994066
2 I PQC		4.418716091	3.345924594	2.667425252	2.248362448	2.092521784	2.029785030	2.019035662	2.133534719
1 I ZF1		1.507771018	1.498462955	1.492567779	1.489414133	1.488558772	1.488366491	1.488553391	1.489126223
1 I ZF2		1.822782084	1.891274116	1.960204281	2.025553696	2.062784805	2.087103142	2.117671118	2.144133246

(2) P 3+2	TH = 45 RC = 02	SBE = 1.414213562 RC = 05	RC = 07	RC = 10	RC = 15	RC = 20	RC = 25	MCPER FUNCTIONS PC = 35	RC = 50
AMAX DB		.0108709564	.0213327376	.0436480540	.0988323393	.1772876696	.2802872360	.5675287486	1.249387366
AMIN DB		6.237559425	9.778651604	14.29137349	19.93057218	24.11448461	27.45026082	32.69638254	38.76325227
MU I K		.2236018D 1	.1455214D 1	.8928161D 0	.4938068D 0	.3172246D 0	.2219557D 0	.1257016D 0	.6430652D-1
1 I PFR		2.775646318	2.100858410	1.604688844	1.206338956	.9881376515	.8426515396	.6496261271	.4655339495
2 I PFC		1.534943840	1.500321898	1.443410390	1.358207018	1.289984330	1.235204701	1.152373582	1.066386891
2 I PQC		4.805757543	3.597237565	2.826331984	2.348200394	2.168226652	2.092882332	2.069671229	2.176890648
1 I ZF1		1.482576008	1.472756830	1.466295847	1.462700941	1.461653535	1.461360314	1.461448176	1.461960938
1 I ZF2		1.770567088	1.836423289	1.904901227	1.971208379	2.009322939	2.034283356	2.065686702	2.092868732

(2) P 3+2	TH = 46	SBE = 1.390163591						MCPER FUNCTIONS	
	RC = 02	RC = 05	RC = 07	RC = 10	RC = 15	RC = 20	RC = 25	RC = 35	RC = 50
AMAX DB		.0108709564	.0213327376	.0436480540	.0988323393	.1772376696	.2802872360	.5675287486	1.249387366
AMIN DB		5.579739725	8.924972292	13.31025699	18.88650571	23.05237470	26.33098375	31.61212549	37.68585249
MU I K		.2387546D 1	.1564396D 1	.9660157D 0	.5373846D 0	.3463227D 0	.2428124D 0	.1378600D 0	.7067322D-1
1 I PFR		2.883826095	2.168254225	1.644615132	1.228786560	1.003410071	.8541044100	.6570923409	.4701634578
2 I PFC		1.511615721	1.481688361	1.430369552	1.350611684	1.285294396	1.232255630	1.151360799	1.066750856
2 I PQC		5.233116547	3.876674988	3.002637994	2.457807452	2.250637385	2.161152876	2.124027208	2.223116178
1 I ZF1		1.458694894	1.448445850	1.441431997	1.437379493	1.436128946	1.435728293	1.435710160	1.436157952
1 I ZF2		1.721579169	1.784404236	1.852006018	1.919045122	1.957990324	1.983578406	2.015812738	2.043717219

(2) P 3+2	TH = 47	SBE = 1.367327461						MCPER FUNCTIONS	
	RC = 02	RC = 05	RC = 07	RC = 10	RC = 15	RC = 20	RC = 25	RC = 35	RC = 50
AMAX DB			.0213327376	.0436480540	.0988323393	.1772876696	.2802872360	.5675287486	1.249387366
AMIN DB			8.121015422	12.36311170	17.86525978	22.00958584	25.32968122	30.55436105	36.62503353
MU I K			.1677442D 1	.1042675D 1	.5835133D 0	.3773032D 0	.2651004D 0	.1509112D 0	.7753279D-1
1 I PFR			2.240347282	1.687396513	1.252691119	1.019581171	.8661818367	.6649256188	.4750009173
2 I PFC			1.462931026	1.416980105	1.342696567	1.280364796	1.229129927	1.150252172	1.067082549
2 I PQC			4.187060137	3.198427547	2.578365914	2.340517389	2.235149436	2.182462029	2.272459175
1 I ZF1			1.425432115	1.417889593	1.413366648	1.411902096	1.411387455	1.411256253	1.411634085
1 I ZF2			1.735142054	1.801437287	1.868949445	1.908659740	1.934856100	1.967913659	1.996541886

(2) P 3+2	TH = 48	SBE = 1.345632730						MCPER FUNCTIONS	
	RC = 02	RC = 05	RC = 07	RC = 10	RC = 15	RC = 20	RC = 25	RC = 35	RC = 50
AMAX DB			.0213327376	.0436480540	.0988323393	.1772876696	.2802872360	.5675287486	1.249387366
AMIN DB			7.367804492	11.45137403	16.86695635	20.98566103	24.29562758	29.51269711	35.57983138
MU I K			.1794215D 1	.1122757D 1	.6322405D 0	.4102283D 0	.2888788D 0	.1649010D 0	.8491430D-1
1 I PFR			2.317295453	1.733221387	1.278166687	1.036718816	.8789279975	.6731489460	.4800577403
2 I PFC			1.444131843	1.403276341	1.334462355	1.275189419	1.225820572	1.149041939	1.067378368
2 I PQC			4.531391607	3.416003480	2.711218793	2.438736196	2.315499853	2.245379802	2.325196706
1 I ZF1			1.403622959	1.395587764	1.390586056	1.388897064	1.388261867	1.388010419	1.388313218
1 I ZF2			1.688560590	1.753127230	1.820821515	1.861217128	1.887996619	1.921865661	1.951217412

(2) P 3+2	TH = 49 SBE = 1.325012993							MCPER FUNCTIONS	
	RC = 02	RC = 05	RC = 07	RC = 10	RC = 15	RC = 20	RC = 25	RC = 35	RC = 50
AMAX DB			.0213327376	.0436480540	.0988323393	.1772876696	.2802872360	.5675287486	1.249387366
AMIN DB			6.665869587	10.57660033	15.89198228	19.98028219	23.27821418	28.48630544	34.54933252
MU I K			.1914587D 1	.1206209D 1	.6836066D 0	.4451577D 0	.3142069D 0	.1798762D 0	.9284813D-1
1 I PFR			2.399228507	1.782275339	1.305335294	1.054897006	.8923912676	.6817874524	.4853463371
2 I PFC			1.425372631	1.389298664	1.325912242	1.269762873	1.222320660	1.147724151	1.067634512
2 I PQC			4.912862577	3.657898688	2.857890707	2.546285690	2.402918861	2.313237361	2.381639299
1 I ZF1			1.382930743	1.374450547	1.368967364	1.367044184	1.366281906	1.365902962	1.366125579
1 I ZF2			1.644578870	1.707018792	1.774575116	1.815560637	1.842891499	1.877555596	1.907628839

(2) P 3+2	TH = 50 SBE = 1.305407289							MCPER FUNCTIONS	
	RC = 02	RC = 05	RC = 07	RC = 10	RC = 15	RC = 20	RC = 25	RC = 35	RC = 50
AMAX DB			.0213327376	.0436480540	.0988323393	.1772876696	.2802872360	.5675287486	1.249387366
AMIN DB			6.015171929	9.740384607	14.94090305	18.99327835	22.27692825	27.47442023	33.53266996
MU I K			.2038446D 1	.1292967D 1	.7376437D 0	.4821489D 0	.3411437D 0	.1958852D 0	.1013661D 0
1 I PFR			2.486249182	1.834737359	1.334326556	1.074196254	.9066246031	.6908686426	.4908802250
2 I PFC			1.406732432	1.375093228	1.317052330	1.264080694	1.218623483	1.146292681	1.067846973
2 I PQC			5.334894528	3.926883311	3.020110744	2.664298841	2.498221969	2.386552576	2.442135933
1 I ZF1			1.363272858	1.354406172	1.348445493	1.346279380	1.345383618	1.344869887	1.345007110
1 I ZF2			1.603108727	1.663062634	1.730136646	1.771599822	1.799442763	1.834880025	1.865670575

(2) P 3+2	TH = 51 SBE = 1.286759566							MCPER FUNCTIONS	
	RC = 02	RC = 05	RC = 07	RC = 10	RC = 15	RC = 20	RC = 25	RC = 35	RC = 50
AMAX DB			.0213327376	.0436480540	.0988323393	.1772876696	.2802872360	.5675287486	1.249387366
AMIN DB			5.415073137	8.944263079	14.01450647	18.02463490	21.29136326	26.47633756	32.52901980
MU I K			.2165707D 1	.1382959D 1	.7943754D 0	.5212558D 0	.3697475D 0	.2129774D 0	.1105013D 0
1 I PFR			2.578436388	1.890776198	1.365276971	1.094703894	.9216859392	.7004226511	.4966741489
2 I PFC			1.388285134	1.360711206	1.307892047	1.258139576	1.214722636	1.144741240	1.068011527
2 I PQC			5.801187340	4.226004528	3.199937361	2.794071271	2.602341757	2.465914196	2.507079931
1 I ZF1			1.344571878	1.335386500	1.328959983	1.326543569	1.325508151	1.324852346	1.324898908
1 I ZF2			1.564053749	1.621214010	1.687444098	1.729254940	1.757562195	1.793744396	1.825245521

(2) P 3+2	TH = 52 SBE = 1.269018215							MCPER FUNCTIONS	
	RC = 02	RC = 05	RC = 07	RC = 10	RC = 15	RC = 20	RC = 25	RC = 35	RC = 50
AMAX DB				.0436480540	.0988323393	.1772376696	.2802872360	.5675287486	1.249387366
AMIN DB				8.189611850	13.11378514	17.07450315	20.32122579	25.49141602	31.53759841
MU I K				.1476108D 1	.8538157D 0	.5625286D 0	.4000760D 0	.2312032D 0	.1202885D 0
1 I PFR				1.950547243	1.398328851	1.116514314	.9376385945	.7104825240	.5027442195
2 I PFC				1.346207703	1.298444527	1.251937637	1.210612144	1.143063394	1.068123721
2 I PQC				4.558556605	3.399285443	2.937086034	2.716347028	2.551993529	2.576915939
1 I ZF1				1.317326618	1.310454393	1.307782131	1.306601236	1.305796137	1.305746736
1 I ZF2				1.581429696	1.646445952	1.688456337	1.717170722	1.754062336	1.786264309

(2) P 3+2	TH = 53 SBE = 1.252135658							MCPER FUNCTIONS	
	RC = 02	RC = 05	RC = 07	RC = 10	RC = 15	RC = 20	RC = 25	RC = 35	RC = 50
AMAX DB				.0436480540	.0988323393	.1772876696	.2802872360	.5675287486	1.249387366
AMIN DB				7.477546638	12.23991845	16.14320952	19.36634384	24.51907834	30.55766009
MU I K				.1572336D 1	.9159695D 0	.6060134D 0	.4321856D 0	.2506145D 0	.1307639D 0
1 I PFR				2.014190286	1.433628891	1.139729055	.9545516691	.7210845295	.5091080676
2 I PFC				1.331640357	1.288726944	1.245474725	1.206286615	1.141252600	1.068178872
2 I PQC				4.928159473	3.620955798	3.095042046	2.841465324	2.645558350	2.652148213
1 I ZF1				1.300164602	1.292875752	1.289944411	1.288612738	1.287651264	1.287500581
1 I ZF2				1.543665194	1.607099948	1.649143886	1.678197894	1.715755052	1.748644631

(2) P 3+2	TH = 54 SBE = 1.236067977							MCPER FUNCTIONS	
	RC = 02	RC = 05	RC = 07	RC = 10	RC = 15	RC = 20	RC = 25	RC = 35	RC = 50
AMAX DB				.0436480540	.0988323393	.1772876696	.2802872360	.5675287486	1.249387366
AMIN DB				6.808834316	11.39424309	15.23126309	18.42667632	23.55881428	29.58849532
MU I K				.1671573D 1	.9808322D 0	.6517511D 0	.4661312D 0	.2712640D 0	.1419655D 0
1 I PFR				2.081828477	1.471326379	1.164456755	.9725004212	.7322684983	.5157850201
2 I PFC				1.317067736	1.278760735	1.238752735	1.201741418	1.139302240	1.068172052
2 I PQC				5.338774490	3.867667633	3.269886640	2.979109407	2.747489527	2.733350516
1 I ZF1				1.283841424	1.276174076	1.272983276	1.271496231	1.270371543	1.270114269
1 I ZF2				1.507872406	1.569371692	1.611266432	1.640581448	1.678750316	1.712310661

(2) P 3+2 — TH = 55 SBE = 1.220774589 — MCPER FUNCTIONS

(2) P 3+2	RC = 02	RC = 05	RC = 07	RC = 10	RC = 15	RC = 20	RC = 25	RC = 35	RC = 50
AMAX DB				.0436480540	.0988323393	.1772876696	.2802872360	.5675287486	1.249387366
AMIN DB				6.183825454	10.57821085	14.33936027	17.50232340	22.61018472	28.62942957
MU I K				.1773760D 1	.1048391D 1	.6997772D 0	.5019657D 0	.2932054D 0	.1539328D 0
1 I PFR				2.153568655	1.511571127	1.190812892	.9915666025	.7440781933	.5227962983
2 I PFC				1.302547649	1.268571684	1.231775954	1.196972893	1.137205671	1.068098085
2 I PQC				5.794765969	4.142595009	3.463852841	3.130908437	2.858300927	2.821177976
1 I ZF1				1.268300988	1.260301932	1.256854691	1.255208626	1.253914245	1.253545113
1 I ZF2				1.473997962	1.533233030	1.574781222	1.604266928	1.642984537	1.677192543

(2) P 3+2 — TH = 56 SBE = 1.206217949 — MCPER FUNCTIONS

(2) P 3+2	RC = 02	RC = 05	RC = 07	RC = 10	RC = 15	RC = 20	RC = 25	RC = 35	RC = 50
AMAX DB				.0436480540	.0988323393	.1772876696	.2802872360	.5675287486	1.249387366
AMIN DB				5.602414573	9.793333713	13.46838479	16.59353730	21.67232718	27.67982274
MU I K				.1878852D 1	.1118624D 1	.7501215D 0	.5397397D 0	.3164934D 0	.1667072D 0
1 I PFR				2.229503007	1.554511235	1.218919278	1.011838725	.7565617097	.5301652420
2 I PFC				1.288135547	1.258189816	1.224551405	1.191978594	1.134956291	1.067951543
2 I PQC				6.300977109	4.449308888	3.679502091	3.298744678	2.980663361	2.916381374
1 I ZF1				1.253490244	1.245214077	1.241517341	1.239709820	1.238239780	1.237753608
1 I ZF2				1.441982300	1.498660406	1.539653276	1.569207349	1.608397410	1.643225961

(2) P 3+2 — TH = 57 SBE = 1.192363293 — MCPER FUNCTIONS

(2) P 3+2	RC = 02	RC = 05	RC = 07	RC = 10	RC = 15	RC = 20	RC = 25	RC = 35	RC = 50
AMAX DB				.0436480540	.0988323393	.1772876696	.2802872360	.5675287486	1.249387366
AMIN DB				5.064031148	9.041117516	12.61940124	15.70073258	20.74646278	26.73906930
MU I K				.1986828D 1	.1191505D 1	.8028069D 0	.5795008D 0	.3411833D 0	.1803319D 0
1 I PFR				2.309711926	1.600290878	1.248903301	1.033412232	.7697719021	.5379175644
2 I PFC				1.273883127	1.247649079	1.217089172	1.186757555	1.132547618	1.067726742
2 I PQC				6.862828388	4.791827201	3.919773404	3.484796785	3.114433462	3.019824415
1 I ZF1				1.239359364	1.230867175	1.226932270	1.224962370	1.223311410	1.222703148
1 I ZF2				1.411759506	1.465632901	1.505854652	1.535362872	1.574936627	1.610351759

$P_{3+4}^{(3)}$

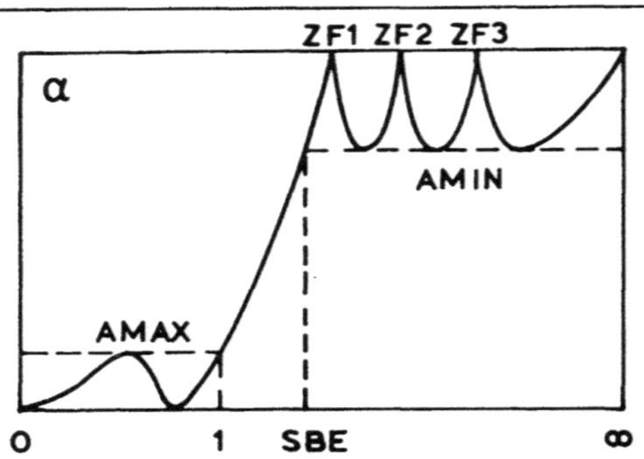

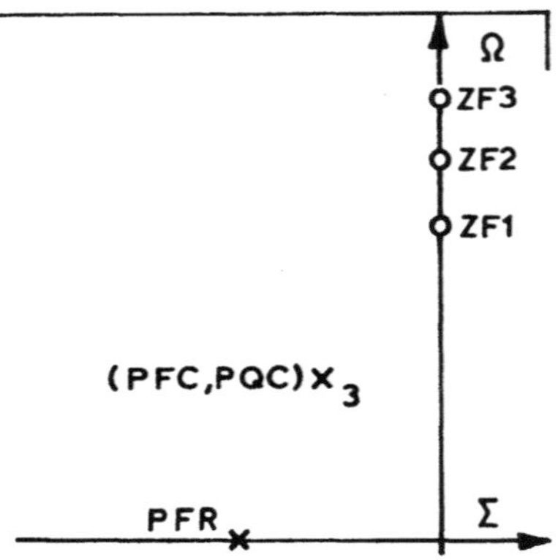

(3) P 3+4	TH = 8	SBE = 7.185296534						MCPER FUNCTIONS	
	RC = 02	RC = 05	RC = 07	RC = 10	RC = 15	RC = 20	RC = 25	RC = 35	RC = 50
AMAX DB	.0017375255	.0108709564	.0213327376	.0436480540	.0988323393	.1772876696	.2802872360	.5675287486	1.249387366
AMIN DB	91.60823048	110.1983138	117.0401307	124.3150193	132.6418904	138.6226984	143.3382057	150.6329729	159.1215402
MU I K	.1284378D-2	.1528272D-3	.6970130D-4	.3022950D-4	.1161187D-4	.5838734D-5	.3395032D-5	.1458777D-5	.5525858D-6
1 I PFR	2.008055569	1.390097506	1.203055587	1.022735389	.8363585761	.7139363662	.6234925247	.4926812597	.3576169570
3 I PFC	3.016253529	2.258593144	2.031901119	1.817817419	1.603901829	1.469099548	1.373189798	1.241043123	1.113912305
3 I PQC	.9754758660	.9783340918	.9893402100	1.008087259	1.040710256	1.073724811	1.107105382	1.176170025	1.291700106
1 I ZF1	7.361674731	7.364468716	7.365199827	7.365836540	7.366417453	7.366754453	7.366980645	7.367274528	7.367539587
1 I ZF2	9.101294834	9.133377917	9.141258431	9.147978570	9.154002775	9.157453587	9.159752456	9.162719276	9.165376410
1 I ZF3	16.19440913	16.33666606	16.36965730	16.39720132	16.42143827	16.43513047	16.44417540	16.45575879	16.46604862

(3) P 3+4	TH = 9	SBE = 6.392453221						MCPER FUNCTIONS	
	RC = 02	RC = 05	RC = 07	RC = 10	RC = 15	RC = 20	RC = 25	RC = 35	RC = 50
AMAX DB	.0017375255	.0108709564	.0213327376	.0436480540	.0988323393	.1772876696	.2802872360	.5675287486	1.249387366
AMIN DB	84.42256629	103.0111502	109.8527451	117.1274859	125.4542559	131.4350189	136.1505014	143.4952426	151.9337923
MU I K	.2591248D-2	.3095203D-3	.1412806D-3	.6131354D-4	.2356507D-4	.1185272D-4	.6893333D-5	.2962667D-5	.1122505D-5
1 I PFR	2.021187745	1.395401775	1.206888069	1.025459835	.8382096095	.7153401342	.6246202878	.4934787450	.3581423108
3 I PFC	3.004035400	2.254567878	2.029321690	1.816279045	1.603134254	1.468697739	1.372998154	1.241079595	1.114107938
3 I PQC	.9936115598	.9882565598	.9974687648	1.014761969	1.046155252	1.078510051	1.111478979	1.180066905	1.295270517
1 I ZF1	6.547691914	6.550625336	6.551415255	6.552109487	6.552747639	6.553119813	6.553370393	6.553696868	6.553992171
1 I ZF2	8.077025640	8.112459892	8.121220396	8.128708725	8.135435482	8.139294660	8.141867967	8.145191786	8.148171337
1 I ZF3	14.31608603	14.47970520	14.51739009	14.54876499	14.57630232	14.59182837	14.60207218	14.61517602	14.62680231

(3) P 3+4	TH = 10	SBE = 5.758770483						MCPER FUNCTIONS	
	RC = 02	RC = 05	RC = 07	RC = 10	RC = 15	RC = 20	RC = 25	RC = 35	RC = 50
AMAX DB	.0017375255	.0108709564	.0213327376	.0436480540	.0988323393	.1772876696	.2802872360	.5675287436	1.249387366
AMIN DB	77.98993003	96.57637873	103.4176608	110.6921940	119.0188220	124.9995221	129.7149698	137.0596746	145.4981998
MU I K	.4847219D-2	.5816904D-3	.2657685D-3	.1154279D-3	.4439193D-4	.2233605D-4	.1299322D-4	.5585920D-5	.2116938D-5
1 I PFR	2.036115947	1.401387066	1.211205274	1.028524552	.8402892978	.7169162502	.6258859675	.4943732852	.3587313496
3 I PFC	2.990025699	2.249982864	2.026387820	1.814530582	1.602261123	1.468239123	1.372777419	1.241116783	1.114324891
3 I PQC	1.014699561	.9995999090	1.006723931	1.022336093	1.052315531	1.083914805	1.116413345	1.184457295	1.299288235
1 I ZF1	5.897065448	5.900067066	5.900904591	5.901648715	5.902338756	5.902743692	5.903017302	5.903374918	5.903699451
1 I ZF2	7.256639744	7.295213353	7.304818680	7.313050341	7.320462847	7.324722443	7.327565700	7.331241685	7.334540242
1 I ZF3	12.80289474	12.98900273	13.03157090	13.06690784	13.09783831	13.11524047	13.12670686	13.14135645	13.15433680

(3) P 3+4	TH = 11	SBE = 5.240843064						MCPER FUNCTIONS	
	RC = 02	RC = 05	RC = 07	RC = 10	RC = 15	RC = 20	RC = 25	RC = 35	RC = 50
AMAX DB	.0017375255	.0108709564	.0213327376	.0436480540	.0988323393	.1772876696	.2802872360	.5675287486	1.249387366
AMIN DB	72.16632464	90.74981307	97.59066794	104.8649183	113.1913536	119.1719681	123.8873688	131.2320243	139.6705161
MU I K	.8526215D-2	.1028890D-2	.4706215D-3	.2045819D-3	.7873800D-4	.3963350D-4	.2306141D-4	.9917558D-5	.3759587D-5
1 I PFR	2.052934276	1.408073654	1.216019101	1.031936420	.8426014009	.7186671895	.6272913595	.4953659848	.3593847130
3 I PFC	2.974117139	2.244810170	2.023082764	1.812562369	1.601277330	1.467720493	1.372525385	1.241153475	1.114562589
3 I PQC	1.039060895	1.012457343	1.017167348	1.030850402	1.059217459	1.089958900	1.121924621	1.189353365	1.303762674
1 I ZF1	5.365283702	5.368275582	5.369148416	5.369934040	5.370670104	5.371105107	5.371400246	5.371787396	5.372140037
1 I ZF2	6.584640834	6.626129654	6.636541783	6.645491265	6.653569571	6.658220850	6.661329109	6.665351920	6.668965708
1 I ZF3	11.55470389	11.76446937	11.81212469	11.85156599	11.88599095	11.90531567	11.91803075	11.93425402	11.94860792

(3) P 3+4	TH = 12	SBE = 4.809734345						MCPER FUNCTIONS	
	RC = 02	RC = 05	RC = 07	RC = 10	RC = 15	RC = 20	RC = 25	RC = 35	RC = 50
AMAX DB	.0017375255	.0108709564	.0213327376	.0436480540	.0988323393	.1772876696	.2802872360	.5675287486	1.249387366
AMIN DB	66.84554624	85.42501133	92.26529628	99.53917110	107.8653511	113.8458526	118.5611911	125.9057816	134.3442294
MU I K	.1424914D-1	.1730855D-2	.7927495D-3	.3449688D-3	.1328827D-3	.6691855D-4	.3894914D-4	.1675621D-4	.6354010D-5
1 I PFR	2.071753135	1.415484686	1.221343054	1.035703203	.8451501467	.7205957286	.6288384755	.4964580788	.3601031156
3 I PFC	2.956197329	2.239019386	2.019388162	1.810363758	1.600177220	1.467138290	1.372239599	1.241188322	1.114820386
3 I PQC	1.067079406	1.026936858	1.028869894	1.040351481	1.066890992	1.096664790	1.128031043	1.194768830	1.308704425
1 I ZF1	4.922675556	4.925573211	4.926467991	4.927285981	4.928061638	4.928523793	4.928838818	4.929253744	4.929633259
1 I ZF2	6.024034831	6.068205587	6.079383966	6.089022270	6.097746379	6.102779843	6.106147712	6.110511513	6.114436395
1 I ZF3	10.50476443	10.73936733	10.79232537	10.83602329	10.87405215	10.89534983	10.90934205	10.92716951	10.94291829

(3) P 3+4 — TH = 13 — SBE = 4.445411483 — MCPER FUNCTIONS

MU		RC = 02	RC = 05	RC = 07	RC = 10	RC = 15	RC = 20	RC = 25	RC = 35	RC = 50
	AMAX DB	.0017375255	.0108709564	.0213327376	.0436480540	.0988323393	.1772876696	.2802872360	.5675287486	1.249387366
	AMIN DB	61.94708719	80.52116276	87.36070172	94.63408767	102.9599367	108.9402921	113.6555503	121.0000568	129.4384479
MU	K	.2280208D-1	.2791307D-2	.1280400D-2	.5578337D-3	.2150883D-3	.1083729D-3	.6309810D-4	.2715642D-4	.1030142D-4
1	PFR	2.092701883	1.423646442	1.227192359	1.039833611	.8479402551	.7227049599	.6305295523	.4976509375	.3608873488
3	PFC	2.936150999	2.232577797	2.015284093	1.807923120	1.598954589	1.466488597	1.371917355	1.241219834	1.115097572
3	PQC	1.099213333	1.043163039	1.041912660	1.050892249	1.075369946	1.104057734	1.134753081	1.200719040	1.314125306
1	ZF1	4.548702549	4.551414958	4.552317238	4.553157787	4.553966094	4.554452234	4.554785363	4.555226155	4.555631204
1	ZF2	5.549212000	5.595825607	5.607727501	5.618024327	5.627372287	5.632777710	5.636399364	5.641097836	5.645329318
1	ZF3	9.606889566	9.867435385	9.925970520	9.974085832	10.01583530	10.03916010	10.05446014	10.07392480	10.09109162

(3) P 3+4 — TH = 14 — SBE = 4.133565494 — MCPER FUNCTIONS

MU		RC = 02	RC = 05	RC = 07	RC = 10	RC = 15	RC = 20	RC = 25	RC = 35	RC = 50
	AMAX DB	.0017375255	.0108709564	.0213327376	.0436480540	.0988323393	.1772876696	.2802872360	.5675287486	1.249387366
	AMIN DB	57.40854733	75.97547763	82.81405464	90.08681454	98.41224172	104.3924115	109.1075678	116.4519678	124.8902871
MU	K	.3514969D-1	.4341858D-2	.1995154D-2	.3704118D-3	.3359826D-3	.1693853D-3	.9865803D-4	.4248029D-4	.1612058D-4
1	PFR	2.115932148	1.432588662	1.233584116	1.044337366	.8509769652	.7249983059	.6323670612	.4989460720	.3617382839
3	PFC	2.913862626	2.225450580	2.010749164	1.805227871	1.597602682	1.465767138	1.371555695	1.241246377	1.115393362
3	PQC	1.136009602	1.061279199	1.056388084	1.062532571	1.084692317	1.112166001	1.142113538	1.207221074	1.320038431
1	ZF1	4.228697344	4.231126823	4.232021299	4.232873744	4.233707248	4.234213967	4.234563282	4.235027880	4.235457016
1	ZF2	5.141879588	5.190694516	5.203275371	5.214198821	5.224147329	5.229913795	5.233782993	5.238809342	5.243342579
1	ZF3	8.828131674	9.115810737	9.180052803	9.232753091	9.278346409	9.303755997	9.320396660	9.341533947	9.360143760

(3) P 3+4 — TH = 15 — SBE = 3.863703305 — MCPER FUNCTIONS

MU		RC = 02	RC = 05	RC = 07	RC = 10	RC = 15	RC = 20	RC = 25	RC = 35	RC = 50
	AMAX DB	.0017375255	.0108709564	.0213327376	.0436480540	.0988323393	.1772876696	.2802872360	.5675287486	1.249387366
	AMIN DB	53.18068519	71.73821043	78.57556145	85.84753009	94.17242735	100.1523647	104.8673936	112.2116608	120.6498906
MU	K	.5244141D-1	.6545920D-2	.3014014D-2	.1316932D-2	.5089763D-3	.2567686D-3	.1496163D-3	.6445472D-4	.2447003D-4
1	PFR	2.141621982	1.442344920	1.240537462	1.049225269	.8542660674	.7274795376	.6343537195	.5003451395	.3626568746
3	PFC	2.889219498	2.217601048	2.005760541	1.802264483	1.596114193	1.464969273	1.371151402	1.241266171	1.115706899
3	PQC	1.178121434	1.081449908	1.072401282	1.075339960	1.094900640	1.121021108	1.150137967	1.214293853	1.326458287
1	ZF1	3.951906903	3.953948792	3.954819013	3.955672154	3.956522906	3.957046564	3.957410013	3.957896210	3.958347882
1	ZF2	4.788621481	4.839397050	4.852610886	4.864127673	4.874652195	4.880768130	4.884878232	4.890225207	4.895055007
1	ZF3	8.144390956	8.460132550	8.530362985	8.587820790	8.637386804	8.664942047	8.682958146	8.705805815	8.725885312

(3) P 3+4	TH = 16	SBE = 3.627955279							MCPER FUNCTIONS
	RC = 02	RC = 05	RC = 07	RC = 10	RC = 15	RC = 20	RC = 25	RC = 35	RC = 50
AMAX DB	.0017375255	.0108709564	.0213327376	.0436480540	.0988323393	.1772876696	.2802872360	.5675287486	1.249387366
AMIN DB	49.22408856	67.76929413	74.60509683	81.87607599	90.20031528	96.17996530	100.8948371	108.2389407	116.6770604
MU K	.7600789D-1	.9602462D-2	.4431518D-2	.1939677D-2	.7507117D-3	.3789977D-3	.2209399D-3	.9523450D-4	.3617257D-4
1 PFR	2.169981001	1.452953074	1.248073772	1.054509296	.8578139384	.7301527935	.6364925031	.5018499503	.3636441605
3 PFC	2.862115302	2.208990922	2.000294102	1.799018512	1.594481263	1.464089993	1.370700995	1.241277281	1.116037252
3 PQC	1.226329959	1.103863991	1.090071608	1.089390393	1.106042404	1.130658085	1.158854374	1.221958267	1.333400813
1 ZF1	3.710269596	3.711811717	3.712640295	3.713482205	3.714341766	3.714878496	3.715253895	3.715759340	3.716231891
1 ZF2	4.479373106	4.531872624	4.545672430	4.557748106	4.568822989	4.575276209	4.579620194	4.585280099	4.590400930
1 ZF3	7.537673040	7.882294553	7.958742059	8.021132713	8.074804820	8.104569469	8.123997681	8.148595698	8.170173271

(3) P 3+4	TH = 17	SBE = 3.420303620							MCPER FUNCTIONS
	RC = 02	RC = 05	RC = 07	RC = 10	RC = 15	RC = 20	RC = 25	RC = 35	RC = 50
AMAX DB	.0017375255	.0108709564	.0213327376	.0436480540	.0988323393	.1772876696	.2802872360	.5675287486	1.249387366
AMIN DB	45.50687955	64.03599864	70.86985954	78.13961180	86.46304209	92.44234070	97.15702085	104.5009253	112.9389112
MU K	.1073453D 0	.1374945D-1	.6361864D-2	.2790083D-2	.1081543D-2	.5464656D-3	.3187298D-3	.1374714D-3	.5224236D-4
1 PFP	2.201256736	1.464455809	1.256216891	1.060202693	.8616275822	.7330226023	.6387866607	.5034624750	.3647012707
3 PFC	2.832454264	2.199580665	1.994324519	1.795474630	1.592695485	1.463123916	1.370200723	1.241277620	1.116383408
3 PQC	1.281570627	1.128738070	1.109534475	1.104769254	1.118170526	1.141115777	1.168293931	1.230237317	1.340883497
1 ZF1	3.497624021	3.498545894	3.499314442	3.500132527	3.500991983	3.501537691	3.501922728	3.502444928	3.502936597
1 ZF2	4.206434785	4.260428384	4.274766532	4.287365715	4.298964325	4.305742075	4.310312560	4.316277271	4.321683267
1 ZF3	6.994316363	7.368417730	7.451303390	7.518802589	7.576717551	7.608757860	7.629636566	7.656026990	7.679132674

(3) P 3+4	TH = 18	SBE = 3.236067977							MCPER FUNCTIONS
	RC = 02	RC = 05	RC = 07	RC = 10	RC = 15	RC = 20	RC = 25	RC = 35	RC = 50
AMAX DB	.0017375255	.0108709564	.0213327376	.0436480540	.0988323393	.1772876696	.2802872360	.5675287486	1.249387366
AMIN DB	42.00310500	60.51126198	67.34270031	74.61094149	82.93338506	88.91225770	93.62670625	100.9703705	109.4081955
MU K	.1480851D 0	.1926681D-1	.8940949D-2	.3929869D-2	.1526047D-2	.7717587D-3	.4503883D-3	.1943899D-3	.7391463D-4
1 PFR	2.235742337	1.476901269	1.264993406	1.066320093	.8657146762	.7360939076	.6412397292	.5051848526	.3658294277
3 PFC	2.800155936	2.189329858	1.987825401	1.791616663	1.590747903	1.462065282	1.369646563	1.241264937	1.116744275
3 PQC	1.344965250	1.156320765	1.130943484	1.121572424	1.131343895	1.152437192	1.178490976	1.239156271	1.348925481
1 ZF1	3.309181683	3.309353464	3.310042574	3.310823570	3.311673537	3.312223912	3.312616145	3.313152469	3.313661393
1 ZF2	3.963814231	4.019081248	4.033909900	4.046996520	4.059091369	4.066180375	4.070969636	4.077230615	4.082915586
1 ZF3	6.503811854	6.907715851	6.997247687	7.070028763	7.132325371	7.166709676	7.189078786	7.217305624	7.241971037

(3) P 3+4 — TH = 19 SBE = 3.071553487 — MCPER FUNCTIONS

MU	K	RC = 02	RC = 05	RC = 07	RC = 10	RC = 15	RC = 20	RC = 25	RC = 35	RC = 50
	AMAX DB	.0017375255	.0108709564	.0213327376	.0436480540	.0988323393	.1772876696	.2802872360	.5675287486	1.249387366
	AMIN DB	38.69159711	57.17247610	64.00090355	71.26729343	79.58854141	85.56690158	90.28107251	97.62444971	106.0620826
	K	.1999476D 0	.2647859D-1	.1232828D-1	.5432172D-2	.2113553D-2	.1069951D-2	.6247991D-3	.2698689D-3	.1026783D-3
1	PFR	2.273785764	1.490343813	1.274432961	1.072877646	.8700836241	.7393720962	.6438555524	.5070193992	.3670299519
3	PFC	2.765160656	2.178197652	1.980769457	1.787427632	1.588629017	1.460907954	1.369034213	1.241236818	1.117118676
3	PQC	1.417860435	1.186897652	1.154472914	1.139907540	1.145627988	1.164669889	1.189483348	1.248742843	1.357547674
1	ZF1	3.141166277	3.140447432	3.141036628	3.141766600	3.142597229	3.143147742	3.143544608	3.144092283	3.144616501
1	ZF2	3.746776995	3.803107728	3.818379241	3.831916784	3.844479673	3.851866187	3.856866182	3.863414499	3.869371936
1	ZF3	6.057996233	6.491685449	6.588052405	6.666283287	6.733100863	6.769899105	6.793799853	6.823908907	6.850167177

(3) P 3+4 — TH = 20 SBE = 2.923804400 — MCPER FUNCTIONS

MU	K	RC = 02	RC = 05	RC = 07	RC = 10	RC = 15	RC = 20	RC = 25	RC = 35	RC = 50
	AMAX DB	.0017375255	.0108709564	.0213327376	.0436480540	.0988323393	.1772876696	.2802872360	.5675287486	1.249387366
	AMIN DB	35.55516806	54.00058783	60.82528335	68.08941416	76.40922118	82.38696872	87.10080915	94.44384565	102.8812513
	K	.2646790D 0	.3575424D-1	.1670863D-1	.7382668D-2	.2878700D-2	.1458915D-2	.8525204D-3	.3685312D-3	.1403115D-3
1	PFR	2.315800435	1.504844916	1.284568634	1.079893177	.8747436146	.7428630300	.6466382996	.5089686188	.3683042663
3	PFC	2.727435738	2.166143274	1.973128674	1.782889810	1.586328794	1.459645409	1.368359087	1.241190679	1.117505346
3	PQC	1.501872895	1.220797140	1.180320634	1.159895448	1.161095585	1.177866427	1.201312703	1.259027399	1.366772890
1	ZF1	2.990561256	2.988798998	2.989266674	2.989930997	2.990731974	2.991277884	2.991676696	2.992232817	2.992770268
1	ZF2	3.551532201	3.608728811	3.624396129	3.638347882	3.651350045	3.659019899	3.664222293	3.671048641	3.677271726
1	ZF3	5.650487487	6.113539189	6.216904532	6.300744383	6.372221106	6.411504297	6.436979015	6.469017702	6.496903392

(3) P 3+4 — TH = 21 SBE = 2.790428110 — MCPER FUNCTIONS

MU	K	RC = 02	RC = 05	RC = 07	RC = 10	RC = 15	RC = 20	RC = 25	RC = 35	RC = 50
	AMAX DB	.0017375255	.0108709564	.0213327376	.0436480540	.0988323393	.1772876696	.2802872360	.5675287486	1.249387366
	AMIN DB	32.58004951	50.97942342	57.79950178	65.06088419	73.37896192	79.35598096	84.06943023	91.41206496	99.84920355
	K	.3439722D 0	.4750851D-1	.2229340D-1	.9880645D-2	.3862008D-2	.1959652D-2	.1145991D-2	.4958395D-3	.1889205D-3
1	PFR	2.362277017	1.520474227	1.295437365	1.087386359	.8797046900	.7465730816	.6495924890	.5110352135	.3696539022
3	PFC	2.686982372	2.153126632	1.964874541	1.777984788	1.583836669	1.458270739	1.367616309	1.241123760	1.117902926
3	PQC	1.598941446	1.258397415	1.208711515	1.181671894	1.177827568	1.192084863	1.214024870	1.270043172	1.376625988
1	ZF1	2.854929504	2.851957095	2.852280419	2.852863756	2.853624302	2.854160658	2.854558608	2.855120136	2.855668659
1	ZF2	3.375008554	3.432884891	3.448901914	3.463231210	3.476643464	3.484582123	3.489978309	3.497073029	3.503554642
1	ZF3	5.276280954	5.767801500	5.878295652	5.967891159	6.044162213	6.086001843	6.113093706	6.147110356	6.176659871

(3) P 3+4	TH = 22	SBE = 2.669467163						MCPER FUNCTIONS	
	RC = 02	RC = 05	RC = 07	RC = 10	RC = 15	RC = 20	RC = 25	RC = 35	RC = 50
AMAX DB	.0017375255	.0108709564	.0213327376	.0436480540	.0988323393	.1772876696	.2802872360	.5675287486	1.249387366
AMIN DB	29.75551631	48.09517503	54.90954792	62.16759378	70.48360311	76.45975964	81.17274827	88.51491182	96.95173831
MU I K	.4393746D 0	.6219999D-1	.2932156D-1	.1304000D-1	.5110453D-2	.2596632D-2	.1519747D-2	.6581995D-3	.2509815D-3
1 I PFR	2.413795589	1.537310818	1.307080471	1.095378918	.8849778229	.7505091742	.6527230115	.5132220967	.3710805048
3 I PFC	2.643843176	2.139108998	1.955978303	1.772693551	1.581141566	1.456776651	1.366800709	1.241033120	1.118309964
3 I PQC	1.711384236	1.300134643	1.239901421	1.205389471	1.195913858	1.207389330	1.227670259	1.281826525	1.387134034
1 I ZF1	2.732282058	2.727917157	2.728071955	2.728558217	2.729267081	2.729788722	2.730182885	2.730746649	2.731303987
1 I ZF2	3.214692026	3.273072110	3.289394020	3.304064480	3.317857372	3.326049992	3.331631120	3.338984217	3.345716949
1 I ZF3	4.931453688	5.450014837	5.567727647	5.663208969	5.744404491	5.788871865	5.817624609	5.853670244	5.884919733

(3) P 3+4	TH = 23	SBE = 2.559304665						MCPER FUNCTIONS	
	RC = 02	RC = 05	RC = 07	RC = 10	RC = 15	RC = 20	RC = 25	RC = 35	RC = 50
AMAX DB	.0017375255	.0108709564	.0213327376	.0436480540	.0988323393	.1772876696	.2802872360	.5675287486	1.249387366
AMIN DB	27.07364822	45.33600697	52.14333486	59.39733623	67.71087787	73.68601686	78.39846530	85.74007884	94.17654256
MU I K	.5521878D 0	.8032765D-1	.3805995D-1	.1699013D-1	.6678050D-2	.3398152D-2	.1990656D-2	.8630714D-3	.3293869D-3
1 I PFR	2.471036653	1.555444648	1.319544232	1.103894869	.8905750034	.7546788274	.6560351594	.5155324060	.3725858399
3 I PFC	2.598110220	2.124053800	1.946411249	1.766996569	1.578231908	1.455155460	1.365906818	1.240915632	1.118724907
3 I PQC	1.841957463	1.346512643	1.274181884	1.231219881	1.215454470	1.223850685	1.242304322	1.294417226	1.398326489
1 I ZF1	2.620980774	2.615024139	2.614984748	2.615357044	2.616002492	2.616504047	2.616891378	2.617454079	2.618017882
1 I ZF2	3.068506505	3.127221413	3.143804962	3.158780720	3.172924679	3.181356162	3.187113164	3.194714335	3.201690496
1 I ZF3	4.612942482	5.156522881	5.281495379	5.382971740	5.469214564	5.516380068	5.546837672	5.584962757	5.617950997

(3) P 3+4	TH = 24	SBE = 2.458593336						MCPER FUNCTIONS	
	RC = 02	RC = 05	RC = 07	RC = 10	RC = 15	RC = 20	RC = 25	RC = 35	RC = 50
AMAX DB	.0017375255	.0108709564	.0213327376	.0436480540	.0988323393	.1772876696	.2802872360	.5675287486	1.249387366
AMIN DB	24.52918663	42.69175232	49.49038526	56.73948869	65.05009187	71.02403387	75.73585102	83.07682512	91.51286908
MU I K	.6833671D 0	.1024255D 0	.4880302D-1	.2187664D-1	.8626420D-2	.4396700D-2	.2578154D-2	.1119089D-2	.4274946D-3
1 I PFR	2.534788437	1.574978270	1.332880591	1.112960788	.8965093394	.7590902080	.6595346566	.5179695189	.3741718009
3 I PFC	2.549933114	2.107927523	1.936145058	1.760873908	1.575095638	1.453399098	1.364928864	1.240767976	1.119146096
3 I PQC	1.993908424	1.398114271	1.311885586	1.259356566	1.236560736	1.241547254	1.257988066	1.307858772	1.410235404
1 I ZF1	2.519664739	2.511899795	2.511638889	2.511879462	2.512449256	2.512925145	2.513302483	2.513860693	2.514428516
1 I ZF2	2.934724897	2.993607887	3.010411612	3.025657526	3.040123013	3.048778069	3.054701688	3.062540343	3.069751971
1 I ZF3	4.318371704	4.884308203	5.016523895	5.124078825	5.215482012	5.265414307	5.297620645	5.337376306	5.372643078

(3) P 3+4 — TH = 25 — SBE = 2.366201583 — MCPER FUNCTIONS

MU	K	RC = 02	RC = 05	RC = 07	RC = 10	RC = 15	RC = 20	RC = 25	RC = 35	RC = 50
	AMAX DB	.0017375255	.0108709564	.0213327376	.0436480540	.0988323393	.1772876696	.2802872360	.5675287486	1.249387366
	AMIN DB	22.11943608	40.15367919	46.94158399	54.18475939	62.49186825	68.46440529	73.17548681	80.51572006	88.95128025
MU	K	.8334309D 0	.1290548D 0	.6187164D-1	.2786186D-1	.1102534D-1	.5629331D-2	.3304500D-2	.1436184D-2	.5491814D-3
1	PFR	2.605946695	1.596028787	1.347147947	1.122606125	.9027951700	.7637521876	.6632276943	.5205370694	.3758404167
3	PFC	2.499526482	2.090700742	1.925152190	1.754305353	1.571720249	1.451499112	1.363860766	1.240586634	1.119571766
3	PQC	2.171010987	1.455614736	1.353392768	1.290017749	1.259356696	1.260565670	1.274788646	1.322198743	1.422895652
1	ZF1	2.427193447	2.417387428	2.416875812	2.416965958	2.417447333	2.417891761	2.418255830	2.418805995	2.419375300
1	ZF2	2.811902033	2.870781893	2.887766252	2.903248096	2.918005748	2.926868953	2.932949774	2.941015060	2.948453936
1	ZF3	4.045913763	4.630869111	4.770244889	4.883931125	4.980595272	5.033360398	5.067358853	5.109798070	5.146382515

(3) P 3+4 — TH = 26 — SBE = 2.281172033 — MCPER FUNCTIONS

MU	K	RC = 02	RC = 05	RC = 07	RC = 10	RC = 15	RC = 20	RC = 25	RC = 35	RC = 50
	AMAX DB	.0017375255	.0108709564	.0213327376	.0436480540	.0988323393	.1772876696	.2802872360	.5675287486	1.249387366
	AMIN DB	19.84414280	37.71431173	44.48898296	51.72498573	60.02794298	65.99883381	70.70906008	78.04843742	86.48344200
MU	K	.1002391D 1	.1607939D 0	.7761108D-1	.3512510D-1	.1395325D-1	.7138035D-2	.4195038D-2	.1825726D-2	.6989012D-3
1	PFR	2.685502140	1.618730085	1.362412091	1.132863565	.9094481948	.7686744080	.6671209696	.5232389669	.3775938606
3	PFC	2.447175583	2.072349286	1.913406341	1.747270560	1.568092810	1.449446669	1.362696135	1.240367882	1.120000036
3	PQC	2.377566465	1.519797110	1.399138735	1.323449985	1.283980713	1.281001843	1.292780032	1.337489211	1.436345179
1	ZF1	2.342601791	2.330509387	2.329715776	2.329635749	2.330015381	2.330422337	2.330769739	2.331308180	2.331876338
1	ZF2	2.698823183	2.757516158	2.774643560	2.790328182	2.805348955	2.814404812	2.820633287	2.828914115	2.836571774
1	ZF3	3.794169238	4.394125062	4.540501841	4.660335903	4.762346227	4.818006628	4.853839658	4.898513922	4.936957385

(3) P 3+4 — TH = 27 — SBE = 2.202689265 — MCPER FUNCTIONS

MU	K	RC = 02	RC = 05	RC = 07	RC = 10	RC = 15	RC = 20	RC = 25	RC = 35	RC = 50
	AMAX DB	.0017375255	.0108709564	.0213327376	.0436480540	.0988323393	.1772876696	.2802872360	.5675287486	1.249387366
	AMIN DB	17.70526166	35.36729524	42.12564686	49.35297207	57.65099939	63.61996375	68.32919753	75.66758842	84.10195677
MU	K	.1189719D 1	.1982257D 0	.9638786D-1	.4386257D-1	.1749772D-1	.8970115D-2	.5278461D-2	.2300657D-2	.8817473D-3
1	PFR	2.774510422	1.643235306	1.378747262	1.143769445	.9164856220	.7738673544	.6712217294	.5260794167	.3794344598
3	PFC	2.393238303	2.052855551	1.900882955	1.739749229	1.564200007	1.447232564	1.361428265	1.240107785	1.120428906
3	PQC	2.618349522	1.591570206	1.449622612	1.359932279	1.310587322	1.302962059	1.312043763	1.353787189	1.450625285
1	ZF1	2.265063469	2.250434027	2.249324805	2.249053703	2.249317671	2.249680914	2.250008137	2.250531050	2.251095341
1	ZF2	2.594463556	2.652764681	2.669999395	2.685854835	2.701110134	2.710343132	2.716709613	2.725194684	2.733062428
1	ZF3	3.562056927	4.172343146	4.325476395	4.451432861	4.558856045	4.617469499	4.655178177	4.702139107	4.742482959

(3) P 3+4		TH = 28	SBE = 2.130054468							MCPER FUNCTIONS
		RC = 02	RC = 05	RC = 07	RC = 10	RC = 15	RC = 20	RC = 25	RC = 35	RC = 50
AMAX DB		.0017375255	.0108709564	.0213327376	.0436480540	.0988323393	.1772876696	.2802872360	.5675287486	1.249387366
AMIN DB		15.70651031	33.10729746	39.84553178	47.06235876	55.35453340	61.32124525	66.02932889	73.36658504	81.80022672
MU	K	.1394362D 1	.2419227D 0	.1185854D 0	.5428695D-1	.2175585D-1	.1117855D-1	.6587091D-2	.2875649D-2	.1103519D-2
1	PFR	2.874041426	1.669719537	1.396237367	1.155364235	.9239263369	.7793424390	.6755378200	.5290629437	.3813647059
3	PFC	2.338141141	2.032209955	1.887559818	1.731721301	1.560028190	1.444847229	1.360050138	1.239802193	1.120856249
3	PQC	2.898481336	1.671988938	1.505417546	1.399780895	1.339349363	1.326564239	1.332669818	1.371155148	1.465780946
1	ZF1	2.193860594	2.176449721	2.174988714	2.174504354	2.174638088	2.174951146	2.175254554	2.175758010	2.176315626
1	ZF2	2.497955194	2.555630529	2.572938431	2.588933997	2.604395793	2.613790467	2.620285238	2.628963071	2.637031980
1	ZF3	3.348707451	3.964080281	4.123630642	4.255636163	4.368516980	4.430135435	4.469758921	4.519057350	4.561343292

(3) P 3+4		TH = 29	SBE = 2.062665340							MCPER FUNCTIONS
		RC = 02	RC = 05	RC = 07	RC = 10	RC = 15	RC = 20	RC = 25	RC = 35	RC = 50
AMAX DB		.0017375255	.0108709564	.0213327376	.0436480540	.0988323393	.1772876696	.2802872360	.5675287486	1.249387366
AMIN DB		13.85262967	30.92994004	37.64339062	44.84751626	53.13274299	59.09682252	63.80357467	71.13952755	79.57234105
MU	K	.1614813D 1	.2924295D 0	.1445985D 0	.6662648D-1	.2683455D-1	.1382233D-1	.8157148D-2	.3567258D-2	.1370795D-2
1	PFR	2.985109430	1.698382597	1.414977340	1.167693082	.9317910940	.7851120953	.6800777426	.5321944177	.3833872664
3	PFC	2.282366852	2.010412552	1.873417725	1.723167198	1.555563426	1.442280747	1.358554420	1.239446728	1.121279812
3	PQC	3.223225582	1.762277098	1.567182522	1.443354889	1.370460450	1.351939384	1.354757593	1.389661596	1.481861167
1	ZF1	2.128358409	2.107944247	2.106092515	2.105371296	2.105359522	2.105615674	2.105891510	2.106371453	2.106919496
1	ZF2	2.408558581	2.465340379	2.482688527	2.498794803	2.514435741	2.523976729	2.530590033	2.539448993	2.547709942
1	ZF3	3.153358785	3.763137158	3.933661435	4.071588554	4.189946256	4.254614563	4.296189550	4.347375056	4.392144903

(3) P 3+4		TH = 30	SBE = 2.000000000							MCPER FUNCTIONS
		RC = 02	RC = 05	RC = 07	RC = 10	RC = 15	RC = 20	RC = 25	RC = 35	RC = 50
AMAX DB		.0017375255	.0108709564	.0213327376	.0436480540	.0988323393	.1772876696	.2802872360	.5675287486	1.249387366
AMIN DB		12.14834349	28.83175498	35.51470104	42.70345965	50.98043727	56.94144141	61.64665321	68.98111103	77.41298235
MU	K	.1849239D 1	.3502442D 0	.1748266D 0	.8112358D-1	.3285077D-1	.1696680D-1	.1002903D-1	.4394089D-2	.1691008D-2
1	PFR	3.108593229	1.729451767	1.435074654	1.180806435	.9401027372	.7911898850	.6848507157	.5354790825	.3855049981
3	PFC	2.226432420	1.987474778	1.858441240	1.714068086	1.550791572	1.439522868	1.356933464	1.239036785	1.121697199
3	PQC	3.597728599	1.863852191	1.635675991	1.491062698	1.404137819	1.379233239	1.378417025	1.409381738	1.498919383
1	ZF1	2.067984988	2.044388234	2.042103944	2.041120702	2.040947375	2.041139636	2.041384014	2.041836264	2.042371749
1	ZF2	2.325637268	2.381224245	2.398580249	2.414769036	2.430562514	2.440234611	2.446956683	2.455985010	2.464428681
1	ZF3	2.975258190	3.583521001	3.754463910	3.898124750	4.021949238	4.089703789	4.133263876	4.187384282	4.233679717

(3) P 3+4			TH = 31	SBE = 1.941604026							MCPER FUNCTIONS
			RC = 02	RC = 05	RC = 07	RC = 10	RC = 15	RC = 20	RC = 25	RC = 35	RC = 50
		AMAX DB	.0017375255	.0108709564	.0213327376	.0436480540	.0988323393	.1772876696	.2802872360	.5675287486	1.249387366
		AMIN DB	10.59713516	26.81016110	33.45561227	40.62577987	48.89296132	54.85037255	59.55380307	66.88654731	75.31734833
MU		K	.2095654D 1	.4157983D 0	.2096652D 0	.9803285D-1	.3993150D-1	.2068393D-1	.1224759D-1	.5376970D-2	.2072531D-2
	1	PFR	3.245162812	1.763184185	1.456650960	1.194760756	.9488864492	.7975906193	.6898667456	.5389225881	.3877209611
	3	PFC	2.170858563	1.963421302	1.842619534	1.704406188	1.545698346	1.436563035	1.355179313	1.238567521	1.122105877
	3	PQC	4.026756318	1.978351540	1.711771434	1.543369586	1.440625633	1.408608193	1.403769881	1.430398219	1.517013911
	1	ZF1	2.012216427	1.985321751	1.982560161	1.981288031	1.980936261	1.981057368	1.981266268	1.981686515	1.982206370
	1	ZF2	2.248634995	2.302699234	2.320030398	2.336274566	2.352194795	2.361983005	2.368804102	2.377989936	2.386606835
	1	ZF3	2.813580954	3.409414866	3.585102068	3.734242006	3.863489795	3.934357044	3.979932072	4.036532881	4.084895192

(3) P 3+4			TH = 32	SBE = 1.887079915							MCPER FUNCTIONS
			RC = 02	RC = 05	RC = 07	RC = 10	RC = 15	RC = 20	RC = 25	RC = 35	RC = 50
		AMAX DB	.0017375255	.0108709564	.0213327376	.0436480540	.0988323393	.1772876696	.2802872360	.5675287486	1.249387366
		AMIN DB	9.200086842	24.86345456	31.46290841	38.61058917	46.86613397	52.81934704	57.52071804	64.85149940	73.28108600
MU		K	.2352110D 1	.4894369D 0	.2494972D 0	.1176186D 0	.4821370D-1	.2505257D-1	.1486237D-1	.6539128D-2	.2524761D-2
	1	PFR	3.395231956	1.799868518	1.479843808	1.209619315	.9581700373	.8043304969	.6951367065	.5425310266	.3900384346
	3	PFC	2.116135180	1.938291931	1.825947326	1.694165130	1.540269429	1.433390407	1.353283703	1.238033851	1.122503156
	3	PQC	4.514500268	2.107658260	1.796474948	1.600806440	1.480198804	1.440245485	1.430951227	1.452801976	1.536208457
	1	ZF1	1.960567956	1.930343297	1.927056948	1.925467235	1.924919207	1.924961594	1.925130859	1.925514661	1.926015726
	1	ZF2	2.177055911	2.229256428	2.246528673	2.262801908	2.278823936	2.288713517	2.295623946	2.304955362	2.313735834
	1	ZF3	2.667378673	3.245151699	3.424784807	3.579076274	3.713666274	3.787661155	3.835276473	3.894400260	3.944870041

(3) P 3+4			TH = 33	SBE = 1.836078459							MCPER FUNCTIONS
			RC = 02	RC = 05	RC = 07	RC = 10	RC = 15	RC = 20	RC = 25	RC = 35	RC = 50
		AMAX DB	.0017375255	.0108709564	.0213327376	.0436480540	.0988323393	.1772876696	.2802872360	.5675287486	1.249387366
		AMIN DB	7.955074442	22.99080555	29.53398588	36.65447866	44.89619641	50.84450272	55.54349237	62.87202600	71.30023591
MU		K	.2616867D 1	.5713997D 0	.2946821D 0	.1401514D 0	.5784400D-1	.3015863D-1	.1792789D-1	.7906374D-2	.3058217D-2
	1	PFR	3.558950399	1.839825373	1.504808383	1.225453075	.9679842567	.8114272597	.7006724301	.5463109728	.3924609349
	3	PFC	2.062689534	1.912143470	1.808425909	1.683330351	1.534490569	1.429993894	1.351238079	1.237430444	1.122886195
	3	PQC	5.064515487	2.253924867	1.890944796	1.663979878	1.523167432	1.474347741	1.460111122	1.476693200	1.556572690
	1	ZF1	1.912590633	1.879100642	1.875239836	1.873301927	1.872538813	1.872494590	1.872619914	1.872962694	1.873441723
	1	ZF2	2.110449022	2.160450227	2.177626799	2.193903244	2.210002945	2.219979449	2.226969593	2.236434628	2.245368871
	1	ZF3	2.535565707	3.090191642	3.272846160	3.431882763	3.571691898	3.648816135	3.698491813	3.760177557	3.812794386

(3) P 3+4		TH = 34	SBE = 1.788291650							MCPER FUNCTIONS
		RC = 02	RC = 05	RC = 07	RC = 10	RC = 15	RC = 20	RC = 25	RC = 35	RC = 50
	AMAX DB	.0017375255	.0108709564	.0213327376	.0436480540	.0988323393	.1772876696	.2802872360	.5675287486	1.249387366
	AMIN DB	6.856525674	21.19225045	27.66684265	34.75448636	42.97976979	48.92233887	53.61857447	60.94453453	69.37118478
MU	K	.2888516D 1	.6618046D 0	.3455449D 0	.1659045D 0	.6897814D-1	.3609522D-1	.2150384D-1	.9507287D-2	.3684640D-2
1	PFR	3.736236632	1.883405808	1.531719127	1.242341666	.9783631780	.8189003709	.7064868076	.5502695291	.3949922343
3	PFC	2.010863802	1.885051386	1.790064264	1.671889547	1.528347711	1.426362201	1.349033597	1.236751715	1.123251985
3	PQC	5.679812665	2.419591408	1.996512681	1.733583905	1.569881962	1.511141907	1.491416560	1.502182443	1.578182890
1	ZF1	1.867872122	1.831283078	1.826796768	1.824478102	1.823479977	1.823340901	1.823417822	1.823714863	1.824168523
1	ZF2	2.048397737	2.095889600	2.112929639	2.129183394	2.145337410	2.155386712	2.162447060	2.172033739	2.181111822
1	ZF3	2.416942753	2.944101361	3.128728739	3.292019963	3.436878693	3.517118999	3.568868972	3.633151330	3.687953403

(3) P 3+4		TH = 35	SBE = 1.743446796							MCPER FUNCTIONS
		RC = 02	RC = 05	RC = 07	RC = 10	RC = 15	RC = 20	RC = 25	RC = 35	RC = 50
	AMAX DB	.0017375255	.0108709564	.0213327376	.0436480540	.0988323393	.1772876696	.2802872360	.5675287486	1.249387366
	AMIN DB	5.895763862	19.46866502	25.86007681	32.90807470	41.11382065	47.04967837	51.74272776	59.06574090	67.49062493
MU	K	.3166036D 1	.7606358D 0	.4023647D 0	.1951490D 0	.8178026D-1	.4296264D-1	.2565532D-1	.1137341D-1	.4417096D-2
1	PFR	3.926839312	1.930987204	1.560771097	1.260374434	.9893446014	.8267712155	.7125939046	.5544143769	.3976363836
3	PFC	1.960905639	1.857111054	1.770880221	1.659833189	1.521827148	1.422483879	1.346661147	1.235991821	1.123597345
3	PQC	6.363075704	2.607394118	2.114706207	1.810413257	1.620739168	1.550882646	1.525053717	1.529391874	1.601122691
1	ZF1	1.826039323	1.786614763	1.781451974	1.778718110	1.777463868	1.777221312	1.777245200	1.777491641	1.777916509
1	ZF2	1.990514297	2.035230794	2.052087867	2.068292362	2.084477991	2.094586309	2.101707477	2.111403844	2.120615722
1	ZF3	2.310246252	2.806534295	2.991969577	3.158936437	3.308624218	3.391950388	3.445781542	3.512690050	3.569713791

(3) P 3+4		TH = 36	SBE = 1.701301617							MCPER FUNCTIONS
		RC = 02	RC = 05	RC = 07	RC = 10	RC = 15	RC = 20	RC = 25	RC = 35	RC = 50
	AMAX DB	.0017375255	.0108709564	.0213327376	.0436480540	.0988323393	.1772876696	.2802872360	.5675287486	1.249387366
	AMIN DB	5.061778614	17.82170089	24.11289094	31.11311638	39.29563304	45.22363584	49.91299732	57.23263521	65.65551953
MU	K	.3448795D 1	.8677378D 0	.4653631D 0	.2281488D 0	.9642176D-1	.5086833D-1	.3045293D-1	.1353943D-1	.5270092D-2
1	PFR	4.130408936	1.982965869	1.592180792	1.279651541	1.000970523	.9350633285	.7190090922	.5587538332	.4003977358
3	PFC	1.912971292	1.828438300	1.750901667	1.647155085	1.514915693	1.418347381	1.344111371	1.235144663	1.123918915
3	PQC	7.116942472	2.820360205	2.247271602	1.895378575	1.676189108	1.593856286	1.561230537	1.558456730	1.625483911
1	ZF1	1.786760663	1.744848898	1.738960786	1.735775615	1.734242900	1.733887821	1.733853867	1.734044694	1.734437261
1	ZF2	1.936437901	1.978171123	1.994791908	2.010919145	2.027114164	2.037268070	2.044440819	2.054234961	2.063570496
1	ZF3	2.214207094	2.677211009	2.862187686	3.032159776	3.186400565	3.272763470	3.328674654	3.398232874	3.457512497

I (3) P 3+4 I	TH = 37	SBE = 1.661640141						MCPER FUNCTIONS	
	RC = 02	RC = 05	RC = 07	RC = 10	RC = 15	RC = 20	RC = 25	RC = 35	RC = 50
AMAX DB		.0108709564	.0213327376	.0436480540	.0988323393	.1772876696	.2802872360	.5675287486	1.249387366
AMIN DB		16.25366655	22.42509733	29.36788778	37.52278659	43.44159118	48.12668168	55.44245224	63.86307249
MU I K		.9828172D 0	.5346937D 0	.2651545D 0	.1130800D 0	.5992670D-1	.3597300D-1	.1604338D-1	.6259691D-2
1 I PFR		2.039745981	1.626186188	1.300285084	1.013287656	.8438026532	.7257491939	.5632969158	.4032809737
3 I PFC		1.799168882	1.730167702	1.633853012	1.507600878	1.413941137	1.341374687	1.234203877	1.124213148
3 I PQC		3.061784882	2.396195273	1.989523475	1.736743190	1.640385394	1.600179736	1.589526970	1.651367509
1 I ZF1		1.705762598	1.699105213	1.695431369	1.693596525	1.693119431	1.693022634	1.693152673	1.693509338
1 I ZF2		1.924443526	1.940766896	1.956786571	1.972968994	1.983155403	1.990370656	2.000250730	2.009699702
1 I ZF3		2.555899094	2.739072721	2.911287291	3.069745190	3.159074683	3.217055658	3.289280240	3.350847270

I (3) P 3+4 I	TH = 38	SBE = 1.624269245						MCPER FUNCTIONS	
	RC = 02	RC = 05	RC = 07	RC = 10	RC = 15	RC = 20	RC = 25	RC = 35	RC = 50
AMAX DB		.0108709564	.0213327376	.0436480540	.0988323393	.1772876696	.2802872360	.5675287486	1.249387366
AMIN DB		14.76733652	20.79711761	27.67106885	35.79314005	41.70116773	46.38130891	53.69264603	62.11070252
MU I K		.1105454D 1	.6104327D 0	.3063976D 0	.1319367D 0	.7025878D-1	.4229753D-1	.1892682D-1	.7403643D-2
1 I PFR		2.101725083	1.663045578	1.322400161	1.026348003	.8530178316	.7328326534	.5680534152	.4062911401
3 I PFC		1.769456546	1.708729689	1.619929397	1.499871185	1.409253639	1.338441324	1.233162836	1.124476300
3 I PQC		3.335198767	2.563722252	2.094043495	1.802983543	1.690834102	1.642162276	1.622769192	1.678884675
1 I ZF1		1.669152368	1.661690087	1.657489627	1.655327690	1.654718608	1.654553759	1.654617665	1.654934735
1 I ZF2		1.873811611	1.889768422	1.905646983	1.921794731	1.932000876	1.939249730	1.949203985	1.958756112
1 I ZF3		2.442392597	2.622374235	2.795978029	2.958253243	3.050455980	3.110486302	3.185385972	3.249268724

I (3) P 3+4 I	TH = 39	SBE = 1.589015729						MCPER FUNCTIONS	
	RC = 02	RC = 05	RC = 07	RC = 10	RC = 15	RC = 20	RC = 25	RC = 35	RC = 50
AMAX DB		.0108709564	.0213327376	.0436480540	.0988323393	.1772876696	.2802872360	.5675287486	1.249387366
AMIN DB		13.36568277	19.22996773	26.02174850	34.10481966	40.00021448	44.67461644	51.98086801	60.39602046
MU I K		.1235121D 1	.6925720D 0	.3520835D 0	.1531758D 0	.8199174D-1	.4951430D-1	.2223501D-1	.8721510D-2
1 I PFR		2.169277129	1.703034879	1.346135816	1.040209491	.8627405328	.7402797235	.5730339774	.4094336714
3 I PFC		1.739469413	1.686652053	1.605392050	1.491716303	1.404273542	1.335301359	1.232014652	1.124704422
3 I PQC		3.644257579	2.752369054	2.210306762	1.875573849	1.745614308	1.687471407	1.658368838	1.708158074
1 I ZF1		1.634830232	1.626539672	1.621775087	1.619259837	1.618508276	1.618269941	1.618262193	1.618535981
1 I ZF2		1.826065026	1.841578912	1.857273606	1.873369105	1.883582490	1.890856216	1.900873014	1.910517966
1 I ZF3		2.336491545	2.511891033	2.685945813	2.851571107	2.946528316	3.008576144	3.086150620	3.152373624

(3) P 3+4 — TH = 40 — SBE = 1.555723827 — MCPER FUNCTIONS

(3) P 3+4	RC = 02	RC = 05	RC = 07	RC = 10	RC = 15	RC = 20	RC = 25	RC = 35	RC = 50
AMAX DB		.0108709564	.0213327376	.0436480540	.0988323393	.1772876696	.2802872360	.5675287486	1.249387366
AMIN DB		12.05153835	17.72521808	24.41943347	32.45621227	38.33679208	43.00453419	50.30494313	58.71680959
MU I K		.1371218D 1	.7810160D 0	.4023850D 0	.1769808D 0	.9525822D-1	.5771672D-1	.2601709D-1	.1023481D-1
1 I PFR		2.242734990	1.746443055	1.371645732	1.054936653	.8730058200	.7481126800	.5782501962	.4127144367
3 I PFC		1.709384631	1.664012675	1.590254928	1.483127430	1.398989779	1.331944765	1.230752179	1.124893353
3 I PQC		3.992771278	2.964928276	2.339876004	1.955271862	1.805192900	1.736437371	1.696532743	1.739323278
1 I ZF1		1.602620566	1.593494639	1.588130256	1.585234345	1.584329273	1.584011781	1.583926664	1.584153087
1 I ZF2		1.781015068	1.796004451	1.811482491	1.827492179	1.837700505	1.844990548	1.855058374	1.864785793
1 I ZF3		2.237982860	2.407460283	2.580952987	2.749390929	2.846956151	2.910976999	2.991215817	3.059799198

(3) P 3+4 — TH = 41 — SBE = 1.524253087 — MCPER FUNCTIONS

(3) P 3+4	RC = 02	RC = 05	RC = 07	RC = 10	RC = 15	RC = 20	RC = 25	RC = 35	RC = 50
AMAX DB		.0108709564	.0213327376	.0436480540	.0988323393	.1772876696	.2802872360	.5675287486	1.249387366
AMIN DB		10.82722491	16.28491728	22.86405856	30.84596273	36.70916223	41.36917075	48.66287857	57.07100838
MU I K		.1513105D 1	.8755815D 0	.4574356D 0	.2035328D 0	.1101955D 0	.6700370D-1	.3032621D-1	.1196715D-1
1 I PFR		2.322375031	1.793565476	1.399098523	1.070601342	.8838525599	.7563560631	.5837147183	.4161397798
3 I PFC		1.679381589	1.640902698	1.574538912	1.474097606	1.393391701	1.328361469	1.229368018	1.125038711
3 I PQC		4.384535057	3.204462505	2.484531274	2.042943801	1.870100178	1.789432890	1.737492104	1.772530406
1 I ZF1		1.572357783	1.562409340	1.556413150	1.553108350	1.552038174	1.551635597	1.551467194	1.551642367
1 I ZF2		1.738490564	1.752871943	1.768079920	1.783983678	1.794174740	1.801472703	1.811580185	1.821379695
1 I ZF3		2.146624447	2.308946140	2.480304635	2.651445962	2.751442801	2.817378224	2.900259479	2.971218235

(3) P 3+4 — TH = 42 — SBE = 1.494476550 — MCPER FUNCTIONS

(3) P 3+4	RC = 02	RC = 05	RC = 07	RC = 10	RC = 15	RC = 20	RC = 25	RC = 35	RC = 50
AMAX DB		.0108709564	.0213327376	.0436480540	.0988323393	.1772876696	.2802872360	.5675287486	1.249387366
AMIN DB		9.694193688	14.91147024	21.35599455	29.27297542	35.11578038	39.76680211	47.05279978	55.45669537
MU I K		.1660144D 1	.9760042D 0	.5173240D 0	.2330066D 0	.1269444D 0	.7747934D-1	.3521969D-1	.1394439D-1
1 I PFR		2.408406462	1.844695315	1.428677453	1.087283483	.8953238760	.7650369494	.5894413620	.4197165682
3 I PFC		1.649634355	1.617425532	1.558272541	1.464622081	1.387469232	1.324541423	1.227854528	1.125135884
3 I PQC		4.823327841	3.474286205	2.646292445	2.139580833	1.940939678	1.846879578	1.781505917	1.807946027
1 I ZF1		1.543884898	1.533149377	1.526495262	1.522752863	1.521505423	1.521011562	1.520753740	1.520873576
1 I ZF2		1.698334180	1.712026464	1.726910183	1.742680711	1.752842267	1.760139890	1.770275797	1.780137020
1 I ZF3		2.062134359	2.216227910	2.385342992	2.557506550	2.659726483	2.727502718	2.812991724	2.886335037

(3) P 3+4	TH = 43	SBE = 1.466279186						MCPER FUNCTIONS	
	RC = 02	RC = 05	RC = 07	RC = 10	RC = 15	RC = 20	RC = 25	RC = 35	RC = 50
AMAX DB		.0108709564	.0213327376	.0436480540	.0988323393	.1772876696	.2802872360	.5675287486	1.249387366
AMIN DB		8.652737612	13.60746571	19.89604925	27.73641957	33.55529144	38.19586309	45.47298860	53.87207591
MU I K		.1811740D 1	.1081949D 1	.5820883D 0	.2655678D 0	.1456482D 0	.8925257D-1	.4075913D-1	.1619477D-1
1 I PFR		2.500967477	1.900113478	1.460579359	1.105071804	.9074676459	.7741852569	.5954452521	.4234522475
3 I PFC		1.620304278	1.593694929	1.541492670	1.454698739	1.381213057	1.320474684	1.226203833	1.125180024
3 I PQC		5.312883211	3.777936299	2.827440174	2.246317807	2.018399617	1.909255436	1.828864993	1.845755347
1 I ZF1		1.517052990	1.505589471	1.498259749	1.494051142	1.492613727	1.492022100	1.491668496	1.491728801
1 I ZF2		1.660399322	1.673328745	1.687828663	1.703435830	1.713555428	1.720844546	1.730997790	1.740910360
1 I ZF3		1.984186396	2.129188063	2.294441848	2.467376624	2.571576960	2.641103520	2.729151376	2.804881755

(3) P 3+4	TH = 44	SBE = 1.439556540						MCPER FUNCTIONS	
	RC = 02	RC = 05	RC = 07	RC = 10	RC = 15	RC = 20	RC = 25	RC = 35	RC = 50
AMAX DB		.0108709564	.0213327376	.0436480540	.0988323393	.1772876696	.2802872360	.5675287486	1.249387366
AMIN DB		7.701822818	12.37545743	18.48545591	26.23573781	32.02652861	36.65494103	43.92184811	52.31547047
MU I K		.1967372D 1	.1193029D 1	.6517133D 0	.3013691D 0	.1664505D 0	.1024366D 0	.4701047D-1	.1874911D-1
1 I PFR		2.600128851	1.960077984	1.495012607	1.124064537	.9203370405	.7838340862	.6017429717	.4273549032
3 I PFC		1.591533746	1.569832055	1.524244946	1.444328539	1.374614827	1.316151510	1.224407848	1.125166036
3 I PQC		5.856912492	4.119135125	3.030533828	2.364454382	2.103266225	1.977103651	1.879896655	1.886164757
1 I ZF1		1.491721379	1.479611695	1.471599784	1.466897271	1.465256666	1.464560498	1.464104507	1.464100974
1 I ZF2		1.624547812	1.636652747	1.650705140	1.666115360	1.676180133	1.683452628	1.693612253	1.703565821
1 I ZF3		1.912412454	2.047700742	2.208000731	2.380890592	2.486792654	2.557960906	2.648502977	2.726615386

(3) P 3+4	TH = 45	SBE = 1.414213562						MCPER FUNCTIONS	
	RC = 02	RC = 05	RC = 07	RC = 10	RC = 15	RC = 20	RC = 25	RC = 35	RC = 50
AMAX DB		.0108709564	.0213327376	.0436480540	.0988323393	.1772876696	.2802872360	.5675287486	1.249387366
AMIN DB		6.839064046	11.21771427	17.12584214	24.77065747	30.52851512	35.14277187	42.39789930	50.78530444
MU I K		.2126615D 1	.1308821D 1	.7261281D 0	.3405464D 0	.1894943D 0	.1171483D 0	.5404403D-1	.2164093D-1
1 I PFR		2.705904067	2.024814133	1.532193898	1.144370004	.9339910986	.7940201004	.6083527356	.4314333304
3 I PFC		1.563441876	1.545961655	1.506584036	1.433516008	1.367667400	1.311562477	1.222458296	1.125088575
3 I PQC		6.459171656	4.501752680	3.258424708	2.495478599	2.196439253	2.051042898	1.934970239	1.929404796
1 I ZF1		1.467758271	1.455104114	1.446417052	1.441194914	1.439337482	1.438529704	1.437964467	1.437892673
1 I ZF2		1.590648379	1.601883331	1.615422262	1.630597973	1.640594392	1.647842124	1.657997282	1.667981529
1 I ZF3		1.846410689	1.971621701	2.125938722	2.297910488	2.405198170	2.477879906	2.570834223	2.651315116

(3) P 3+4	TH = 46	SBE = 1.390163591						MCPER FUNCTIONS	
	RC = 02	RC = 05	RC = 07	RC = 10	RC = 15	RC = 20	RC = 25	RC = 35	RC = 50
AMAX DB		.0108709564	.0213327376	.0436480540	.0988323393	.1772876696	.2802872360	.5675287486	1.249387366
AMIN DB		6.060837552	10.13596568	15.81917252	23.34120318	29.06046895	33.65823854	40.89977424	49.28009909
MU I K		.2289155D 1	.1428897D 1	.8052073D 0	.3832149D 0	.2149195D 0	.1335073D 0	.6193450D-1	.2490661D-1
1 I PFR		2.818263821	2.094506899	1.572343889	1.166107044	.9484953291	.8047839439	.6152945851	.4356971135
3 I PFC		1.536122488	1.522207615	1.488573470	1.422269748	1.360365111	1.306698612	1.220346738	1.124942035
3 I PQC		7.123565721	4.929777650	3.514263270	2.641092834	2.298949960	2.131779453	1.994503557	1.975733599
1 I ZF1		1.445041563	1.431959901	1.422620377	1.416856202	1.414768024	1.413841279	1.413159673	1.413015073
1 I ZF2		1.558575907	1.568914111	1.581874125	1.596773448	1.606687050	1.613901765	1.624041688	1.634046326
1 I ZF3		1.785757549	1.900780714	2.048187850	2.218323291	2.326642134	2.400688186	2.495953769	2.578780048

(3) P 3+4	TH = 47	SBE = 1.367327461						MCPER FUNCTIONS	
	RC = 02	RC = 05	RC = 07	RC = 10	RC = 15	RC = 20	RC = 25	RC = 35	RC = 50
AMAX DB		.0108709564	.0213327376	.0436480540	.0988323393	.1772876696	.2802872360	.5675287486	1.249387366
AMIN DB		5.362499106	9.131176064	14.56765958	21.94770947	27.62181028	32.20037162	39.42621081	47.79846376
MU I K		.2454790D 1	.1552843D 1	.8887750D 0	.4294657D 0	.2428606D 0	.1516348D 0	.7076093D-1	.2858559D-1
1 I PFR		2.937152277	2.169296799	1.615681717	1.189405168	.9639223218	.8161707016	.6225906094	.4401567157
3 I PFC		1.509644282	1.498688348	1.470285011	1.410602951	1.352704079	1.301551546	1.218064610	1.124720547
3 I PQC		7.854279369	5.407307578	3.801499735	2.803241925	2.411981922	2.220121418	2.058970521	2.025440949
1 I ZF1		1.423459537	1.410076921	1.400124477	1.393800735	1.391467819	1.390414483	1.389609116	1.389387041
1 I ZF2		1.528211266	1.537645578	1.549964902	1.564541608	1.574356687	1.581529911	1.591643862	1.601658636
1 I ZF3		1.730021334	1.834977333	1.974686125	2.142038281	2.250995299	2.326234223	2.423689339	2.508827227

(3) P 3+4	TH = 48	SBE = 1.345632730						MCPER FUNCTIONS	
	RC = 02	RC = 05	RC = 07	RC = 10	RC = 15	RC = 20	RC = 25	RC = 35	RC = 50
AMAX DB			.0213327376	.0436480540	.0988323393	.1772876696	.2802872360	.5675287486	1.249387366
AMIN DB			8.203380526	13.37364074	20.59083075	26.21217160	30.76835243	37.97604885	46.33908900
MU I K			.1680281D 1	.9766109D 0	.4793618D 0	.2734444D 0	.1716530D 0	.8060650D-1	.3272050D-1
1 I PFR			2.249279950	1.662418753	1.214404350	.9803523436	.8282303957	.6302651955	.4448235823
3 I PFC			1.475512535	1.451797468	1.398533902	1.344682537	1.296113697	1.215603268	1.124417978
3 I PQC			5.938566952	4.123878927	2.984143081	2.536894997	2.316995430	2.128910147	2.078853029
1 I ZF1			1.389357755	1.378848878	1.371954677	1.369363249	1.368175474	1.367238682	1.366934332
1 I ZF2			1.507983587	1.519607514	1.533811380	1.543510670	1.550633580	1.560710788	1.570725472
1 I ZF3			1.773980455	1.905370439	2.068984345	2.178148831	2.254385740	2.353886111	2.441289937

TH = 49 SBE = 1.325012993 MCPER FUNCTIONS

(3) P 3+4	MU	RC = 02	RC = 05	RC = 07	RC = 10	RC = 15	RC = 20	RC = 25	RC = 35	RC = 50
AMAX DB				.0213327376	.0436480540	.0988323393	.1772876696	.2802872360	.5675287486	1.249387366
AMIN DB				7.351605198	12.23942428	19.27154578	24.83141001	29.36151844	36.54822791	44.90074067
K				.1810891D 1	.1068461D 1	.5329357D 0	.3067875D 0	.1936828D 0	.9155841D-1	.3735730D-1
PFR	1			2.334512237	1.712752114	1.241254332	.9978738854	.8410185146	.6383453091	.4497102566
PFC	3			1.452775720	1.433194913	1.386086432	1.336301208	1.290378471	1.212954046	1.124027920
PQC	3			6.527958313	4.485432060	3.186317016	2.675252789	2.423466302	2.204937222	2.136338025
ZF1	1			1.369710035	1.358717005	1.351249918	1.348386819	1.347056592	1.345980451	1.345588892
ZF2	1			1.479838287	1.490722310	1.504499957	1.514064332	1.521127603	1.531157181	1.541161566
ZF3	1			1.717531540	1.840169718	1.999107102	2.108012729	2.185028340	2.286405334	2.376016221

TH = 50 SBE = 1.305407289 MCPER FUNCTIONS

(3) P 3+4	MU	RC = 02	RC = 05	RC = 07	RC = 10	RC = 15	RC = 20	RC = 25	RC = 35	RC = 50
AMAX DB				.0213327376	.0436480540	.0988323393	.1772876696	.2802872360	.5675287486	1.249387366
AMIN DB				6.573879856	11.16711403	17.99115261	23.47962114	27.97937114	35.14178640	43.48225504
K				.1944421D 1	.1164051D 1	.5901866D 0	.3429934D 0	.2178431D 0	.1037076D 0	.4254546D-1
PFR	1			2.425016795	1.766858663	1.270113338	1.016584113	.8545965659	.6468608112	.4548305137
PFC	3			1.430558081	1.414564373	1.373290266	1.327563691	1.284340501	1.210108322	1.123543699
PQC	3			7.180146884	4.890470362	3.412619587	2.828853922	2.540760073	2.287754979	2.198312754
ZF1	1			1.351046968	1.339655502	1.331623300	1.328476494	1.326995739	1.325772081	1.325288245
ZF2	1			1.453123489	1.463235793	1.476532048	1.485940250	1.492933886	1.502904732	1.512888607
ZF3	1			1.665350748	1.778993876	1.932365782	2.040514302	2.118064311	2.221123143	2.312867607

TH = 51 SBE = 1.286759566 MCPER FUNCTIONS

(3) P 3+4	MU	RC = 02	RC = 05	RC = 07	RC = 10	RC = 15	RC = 20	RC = 25	RC = 35	RC = 50
AMAX DB				.0213327376	.0436480540	.0988323393	.1772876696	.2802872360	.5675287486	1.249387366
AMIN DB				5.867333715	10.15842864	16.75124967	22.15715353	26.62158630	33.75586239	42.08253459
K				.2080692D 1	.1263097D 1	.6510796D 0	.3821497D 0	.2442483D 0	.1171481D 0	.4833809D-1
PFR	1			2.520793541	1.824890302	1.301146116	1.036589164	.8690326376	.6558448117	.4601995127
PFC	3			1.408923498	1.395993123	1.360181232	1.318476887	1.277995912	1.207057599	1.122958374
PQC	3			7.900175207	5.343587319	3.666273262	2.999767385	2.670291027	2.378170192	2.265250542
ZF1	1			1.333287865	1.321593798	1.313015889	1.309575095	1.307935811	1.306556262	1.305974946
ZF2	1			1.427756414	1.437079422	1.449839163	1.459067620	1.465980769	1.475331436	1.485834567
ZF3	1			1.617144845	1.721754137	1.868729767	1.975596649	2.053411555	2.157929552	2.251718008

(3) P 3+4		TH = 52	SBE = 1.269018215						MCPER FUNCTIONS
		RC = 02	RC = 05	RC = 07	RC = 10	RC = 15	RC = 20	RC = 25	RC = 35	RC = 50
AMAX DB				.0213327376	.0436480540	.0988323393	.1772876696	.2802872360	.5675287486	1.249387366
AMIN DB				5.228352649	9.214536073	15.55369842	20.86462219	25.28802638	32.38969617	40.70054481
MU	K			.2219599D 1	.1365326D 1	.7155461D 0	.4243257D 0	.2730068D 0	.1319771D 0	.5479203D-1
1	PFR			2.621829356	1.886971244	1.334521271	1.058004216	.8844019475	.6653340632	.4658339701
3	PFC			1.387919805	1.377565835	1.346801245	1.309051408	1.271342611	1.203793608	1.122264746
3	PQC			8.693602167	5.849677287	3.950897907	3.190372156	2.813693320	2.477111193	2.337690608
1	ZF1			1.316358581	1.304463879	1.295372299	1.291629749	1.289824193	1.288280231	1.287596101
1	ZF2			1.403657734	1.412188550	1.424358930	1.433381704	1.440202466	1.450021017	1.459933105
1	ZF3			1.572615657	1.668310242	1.808174795	1.913217050	1.991002597	2.096727598	2.192452770

(3) P 3+4		TH = 53	SBE = 1.252135658						MCPER FUNCTIONS
		RC = 02	RC = 05	RC = 07	RC = 10	RC = 15	RC = 20	RC = 25	RC = 35	RC = 50
AMAX DB					.0436480540	.0988323393	.1772876696	.2802872360	.5675287486	1.249387366
AMIN DB					8.335923733	14.40056391	19.60291892	23.97875499	31.04263447	39.33531182
MU	K				.1470486D 1	.7834852D 0	.4695695D 0	.3042190D 0	.1482937D 0	.6196796D-1
1	PFR				1.953197702	1.370407962	1.080953253	.9007873479	.6753693960	.4717523590
3	PFC				1.359361854	1.333198014	1.299301990	1.264380622	1.200308419	1.121455365
3	PQC				6.413978030	4.270541026	3.403401232	2.972857887	2.585549444	2.416249303
1	ZF1				1.288200444	1.278640082	1.274591362	1.272612304	1.270895343	1.270102941
1	ZF2				1.388501554	1.400034421	1.408823327	1.415538573	1.425262407	1.435123045
1	ZF3				1.618519836	1.750678861	1.853345225	1.930783644	2.037432618	2.134967869

(3) P 3+4		TH = 54	SBE = 1.236067977						MCPER FUNCTIONS
		RC = 02	RC = 05	RC = 07	RC = 10	RC = 15	RC = 20	RC = 25	RC = 35	RC = 50
AMAX DB					.0436480540	.0988323393	.1772876696	.2802872360	.5675287486	1.249387366
AMIN DB					7.522320781	13.29403119	18.37321690	22.69405251	29.71413681	37.98592089
MU	K				.1578361D 1	.8547676D 0	.5179063D 0	.3379747D 0	.1661985D 0	.6993052D-1
1	PFR				2.023640063	1.408972110	1.105568434	.9182797477	.6859961954	.4779751365
3	PFC				1.341452873	1.319424420	1.289247850	1.257112430	1.196594586	1.120522553
3	PQC				7.042143146	4.629708529	3.641990168	3.149975401	2.705025405	2.501633609
1	ZF1				1.272741037	1.262769175	1.258414137	1.256255171	1.254356683	1.253450442
1	ZF2				1.365959193	1.376813466	1.385338432	1.391933639	1.401549302	1.41134790[illegible]
1	ZF3				1.572215035	1.696217951	1.795961365	1.872713621	1.979971633	2.079169217

(3) P 3+4	TH = 55	SBE = 1.220774589						MCPER FUNCTIONS	
	RC = 02	RC = 05	RC = 07	RC = 10	RC = 15	RC = 20	RC = 25	RC = 35	RC = 50
AMAX DB				.0436480540	.0988323393	.1772876696	.2802872360	.5675287486	1.249387366
AMIN DB				6.772680530	12.23629887	17.17696606	21.43443219	28.40378337	36.65151595
MU I K				.1688777D 1	.9292412D 0	.5693368D 0	.3743511D 0	.1857926D 0	.7874825D-1
1 I PFR				2.098347196	1.450372431	1.131989004	.9369783991	.6972649174	.4845250043
3 I PFC				1.323901246	1.305537538	1.278912973	1.249543355	1.192645306	1.119458422
3 I PQC				7.740348155	5.033398627	3.909730248	3.347586107	2.836679633	2.594657443
1 I ZF1				1.258026454	1.247711449	1.243055108	1.240711040	1.238622720	1.237596991
1 I ZF2				1.344504224	1.354647964	1.362877642	1.369336787	1.378829759	1.388555506
1 I ZF3				1.529210844	1.644761825	1.741053896	1.816763164	1.924282826	2.024972094

(3) P 3+4	TH = 56	SBE = 1.206217949						MCPER FUNCTIONS	
	RC = 02	RC = 05	RC = 07	RC = 10	RC = 15	RC = 20	RC = 25	RC = 35	RC = 50
AMAX DB				.0436480540	.0988323393	.1772876696	.2802872360	.5675287486	1.249387366
AMIN DB				6.085221376	11.22945458	16.01587543	20.20065514	27.11128815	35.33130023
MU I K				.1801610D 1	.1006738D 1	.6238363D 0	.4134105D 0	.2071766D 0	.8849370D-1
1 I PFR				2.177352272	1.494756667	1.160359687	.9569909875	.7092316389	.4914272057
3 I PFC				1.306759012	1.291597314	1.268326267	1.241681936	1.138454611	1.118254911
3 I PQC				8.515431847	5.487143224	4.210726640	3.568637422	2.982290238	2.696261413
1 I ZF1				1.244001111	1.233420367	1.228473703	1.225941007	1.223654992	1.222504073
1 I ZF2				1.324081232	1.333493191	1.341395849	1.347701373	1.357055854	1.366697569
1 I ZF3				1.489309869	1.596270158	1.688616934	1.762913486	1.870315093	1.972300667

(3) P 3+4	TH = 57	SBE = 1.192363293						MCPER FUNCTIONS	
	RC = 02	RC = 05	RC = 07	RC = 10	RC = 15	RC = 20	RC = 25	RC = 35	RC = 50
AMAX DB				.0436480540	.0988323393	.1772876696	.2802872360	.5675287486	1.249387366
AMIN DB				5.457516017	10.27534181	14.89187893	18.99374229	25.83650696	34.02453798
MU I K				.1916789D 1	.1087085D 1	.6813553D 0	.4551987D 0	.2304493D 0	.9924328D-1
1 I PFR				2.260679354	1.542258494	1.190828566	.9784334577	.7219586333	.4987098657
3 I PFC				1.290067667	1.277664964	1.257521547	1.233540293	1.184017588	1.116903824
3 I PQC				9.375073272	5.997063118	4.549662345	3.816550346	3.143818085	2.807536881
1 I ZF1				1.230613384	1.219850762	1.214631335	1.211908665	1.209417822	1.208136011
1 I ZF2				1.304636649	1.313307145	1.320851794	1.326984676	1.336183372	1.345729430
1 I ZF3				1.452307654	1.550689373	1.638647435	1.711155077	1.818027640	1.921087607

(3) P 3+4 — TH = 58 SBE = 1.179178403 — MCPER FUNCTIONS

MU		RC = 02	RC = 05	RC = 07	RC = 10	RC = 15	RC = 20	RC = 25	RC = 35	RC = 50
	AMAX DB					.0988323393	.1772876696	.2802872360	.5675287486	1.249387366
	AMIN DB					9.375430395	13.80708158	17.81498059	24.57945783	32.73055768
	K					.1170108D 1	.7418212D 0	.4997425D 0	.2557064D 0	.1110772D 0
1	PFR					1.592995512	1.223544510	1.001429504	.7355149537	.5064043783
3	PFC					1.263801212	1.246537303	1.225134467	1.179330627	1.115396888
3	PQC					6.569945047	4.931869653	4.095295875	3.323561423	2.929755406
1	ZF1					1.206958769	1.201491030	1.198579773	1.195378048	1.194459712
1	ZF2					1.294049935	1.301207631	1.307147605	1.316171545	1.325609732
1	ZF3					1.507950494	1.591142093	1.661486172	1.767389603	1.871273794

(3) P 3+4 — TH = 59 SBE = 1.166633397 — MCPER FUNCTIONS

MU		RC = 02	RC = 05	RC = 07	RC = 10	RC = 15	RC = 20	RC = 25	RC = 35	RC = 50
	AMAX DB					.0988323393	.1772876696	.2802872360	.5675287486	1.249387366
	AMIN DB					8.530704640	12.76368491	16.66592047	23.34033606	31.44875678
	K					.1255649D 1	.8051409D 0	.5470493D 0	.2830389D 0	.1240793D 0
1	PFR					1.647068655	1.258654301	1.026109664	.7499770045	.5145458467
3	PFC					1.250064511	1.235416207	1.216484687	1.174391711	1.113725820
3	PQC					7.213349846	5.363412201	4.409483007	3.524222041	3.064404939
1	ZF1					1.194701870	1.189017103	1.185921931	1.183004781	1.181444452
1	ZF2					1.275693275	1.282428484	1.288154419	1.296982816	1.306300181
1	ZF3					1.467968246	1.546094109	1.613910946	1.718379670	1.822808096

(3) P 3+4 — TH = 60 SBE = 1.154700538 — MCPER FUNCTIONS

MU		RC = 02	RC = 05	RC = 07	RC = 10	RC = 15	RC = 20	RC = 25	RC = 35	RC = 50
	AMAX DB					.0988323393	.1772876696	.2802872360	.5675287486	1.249387366
	AMIN DB					7.741581503	11.76389281	15.54836081	22.11953417	30.17860825
	K					.1343566D 1	.8712054D 0	.5971058D 0	.3125315D 0	.1383367D 0
1	PFR					1.704563151	1.296299709	1.052609985	.7654290674	.5231735820
3	PFC					1.236509413	1.224204341	1.207615531	1.169200725	1.111882414
3	PQC					7.935761693	5.851182700	4.764460493	3.748985493	3.213234552
1	ZF1					1.183039064	1.177174903	1.173904274	1.170769173	1.169061668
1	ZF2					1.258170089	1.264481984	1.269972442	1.278582632	1.287765321
1	ZF3					1.430641498	1.503490015	1.568437412	1.670985653	1.775647215

(3) P 3+4		TH = 61	SBE = 1.143354068						MCPER FUNCTIONS
		RC = 02	RC = 05	RC = 07	RC = 10	RC = 15	RC = 20	RC = 25	RC = 35	RC = 50
AMAX DB						.0988323393	.1772876696	.2802872360	.5675287486	1.249387366
AMIN DB						7.007867565	10.90980244	14.46431802	20.91766247	28.91966925
MU	K					.1433750D 1	.9398957D 0	.6498787D 0	.3442609D 0	.1539394D 0
1	PFR					1.765550948	1.336614808	1.081070255	.7819637422	.5323316653
3	PFC					1.223185239	1.212950178	1.198555942	1.163759803	1.109858654
3	PQC					8.746788763	6.403023928	5.166435576	4.001618582	3.378309988
1	ZF1					1.171931108	1.165930636	1.162497185	1.159144193	1.157284783
1	ZF2					1.241474261	1.247337804	1.252571759	1.260939266	1.269972337
1	ZF3					1.395854881	1.463306790	1.525075017	1.625203988	1.729755609

(3) P 3+4		TH = 62	SBE = 1.132570051						MCPER FUNCTIONS
		RC = 02	RC = 05	RC = 07	RC = 10	RC = 15	RC = 20	RC = 25	RC = 35	RC = 50
AMAX DB						.0988323393	.1772876696	.2802872360	.5675287486	1.249387366
AMIN DB						6.328757802	9.903288048	13.41597644	19.73556954	27.67159226
MU	K					.1526122D 1	.1011090D 1	.7053161D 0	.3782939D 0	.1709801D 0
1	PFR					1.830094356	1.379723896	1.111631875	.7996822511	.5420695757
3	PFC					1.210135130	1.201703339	1.189339052	1.158073693	1.107646843
3	PQC					9.657426199	7.027883606	5.622614580	4.286588077	3.562083031
1	ZF1					1.161340788	1.155251275	1.151672036	1.148104413	1.146089026
1	ZF2					1.225560513	1.230967212	1.235924906	1.244023649	1.252890888
1	ZF3					1.363481550	1.425509535	1.483832008	1.581039094	1.685105456

(3) P 3+4		TH = 63	SBE = 1.122326238						MCPER FUNCTIONS
		RC = 02	RC = 05	RC = 07	RC = 10	RC = 15	RC = 20	RC = 25	RC = 35	RC = 50
AMAX DB						.0988323393	.1772876696	.2802872360	.5675287486	1.249387366
AMIN DB						5.702873314	9.045887990	12.40561877	18.57436085	26.43413906
MU	K					.1620641D 1	.1084669D 1	.7633497D 0	.4146859D 0	.1895531D 0
1	PFR					1.898250506	1.425740336	1.144435510	.8186945483	.5524428886
3	PFC					1.197395545	1.190513236	1.180001782	1.152150123	1.105239770
3	PQC					10.68039381	7.736017104	6.141372715	4.609205741	3.767478618
1	ZF1					1.151233193	1.145104573	1.141400971	1.137625800	1.135451284
1	ZF2					1.210394411	1.215342674	1.220006533	1.227809220	1.236492960
1	ZF3					1.333386383	1.390049945	1.444712662	1.538502567	1.641676676

$P^{(2)}_{4+0}$

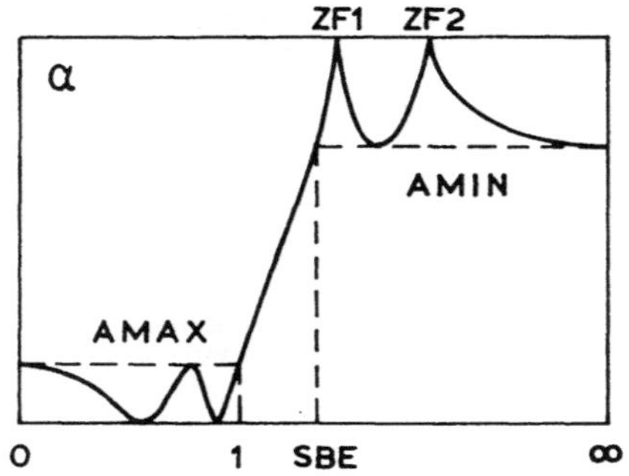

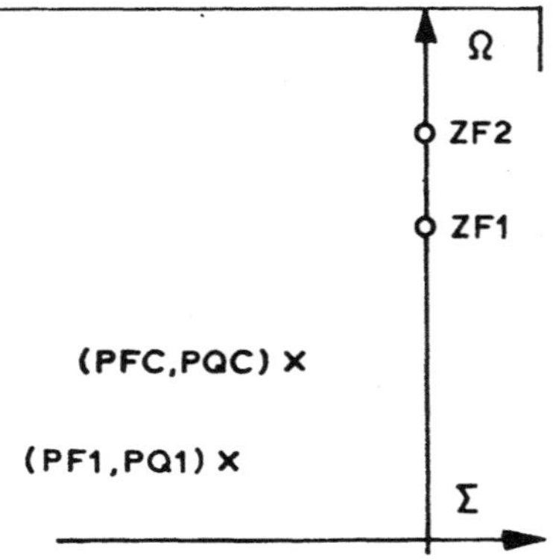

(2) P 4+0	TH = 5	SBE = 11.47371325						CAUER FUNCTIONS	
	RC = 02	RC = 05	RC = 07	RC = 10	RC = 15	RC = 20	RC = 25	RC = 35	RC = 50
AMAX DB	.0017375255	.0108709564	.0213327376	.0436480540	.0988323393	.1772876696	.2802872360	.5675287486	1.249387366
AMIN DB	86.85607265	94.82400624	97.75702874	100.8773833	104.4543927	107.0316228	109.0728226	112.2826248	116.0625227
MU I K	.4541469D-4	.1814678D-4	.1294639D-4	.9039218D-5	.5987980D-5	.4450603D-5	.3518511D-5	.2431469D-5	.1573526D-5
1 I PF1	1.479971004	1.128981557	1.017123496	.9073087935	.7926153549	.7173902291	.6624069960	.5848940129	.5096477104
1 I PQ1	.5613791863	.5761707044	.5850044571	.5977739269	.6189380894	.6408640517	.6641891487	.7171214540	.8218505209
1 I PFC	1.692443180	1.403170521	1.316314286	1.234602584	1.153915988	1.104108737	1.069516248	1.023696229	.9829404827
1 I PQC	1.573224919	1.746344777	1.843924904	1.978865082	2.189841092	2.395584908	2.603555448	3.045607610	3.844548021
1 I ZF1	12.41559567	12.41559567	12.41559567	12.41559567	12.41559567	12.41559567	12.41559567	12.41559567	12.41559567
1 I ZF2	29.93354278	29.93354278	29.93354278	29.93354278	29.93354278	29.93354278	29.93354278	29.93354278	29.93354278

(2) P 4+0	TH = 6	SBE = 9.566772234						CAUER FUNCTIONS	
	RC = 02	RC = 05	RC = 07	RC = 10	RC = 15	RC = 20	RC = 25	RC = 35	RC = 50
AMAX DB	.0017375255	.0108709564	.0213327376	.0436480540	.0988323393	.1772876696	.2802872360	.5675287486	1.249387366
AMIN DB	80.51184395	88.47977752	91.41280002	94.53315453	98.11016400	100.6873941	102.7285939	105.9383961	109.7182939
MU I K	.9427744D-4	.3767134D-4	.2687571D-4	.1876473D-4	.1243059D-4	.9239113D-5	.7304160D-5	.5047545D-5	.3266520D-5
1 I PF1	1.482812104	1.130541725	1.018385293	.9083213197	.7934086930	.7180601504	.6629958813	.5853805647	.5100468472
1 I PQ1	.5618045595	.5764744831	.5852777591	.5980216649	.6191645448	.6410804292	.6644008853	.7173328590	.8220760180
1 I PFC	1.690617975	1.402410018	1.315786616	1.234261253	1.153729558	1.104004338	1.069463012	1.023703721	.9829955095
1 I PQC	1.580093122	1.751678488	1.848922249	1.983628625	2.194500281	2.400291580	2.608387411	3.050837583	3.850695680
1 I ZF1	10.35084315	10.35084315	10.35084315	10.35084315	10.35084315	10.35084315	10.35084315	10.35084315	10.35084315
1 I ZF2	24.94069420	24.94069420	24.94069420	24.94069420	24.94069420	24.94069420	24.94069420	24.94069420	24.94069420

(2) P 4+0	TH = 7 SBE = 8.205509048							CAUER FUNCTIONS	
	RC = 02	RC = 05	RC = 07	RC = 10	RC = 15	RC = 20	RC = 25	RC = 35	RC = 50
AMAX DB	.0017375255	.0108709564	.0213327376	.0436480540	.0988323393	.1772876696	.2802872360	.5675287486	1.249387366
AMIN DB	75.14459007	83.11252356	86.04554605	89.16590056	92.74291002	95.32014008	97.36133991	100.5711421	104.3510400
MU I K	.1748922D-3	.6988337D-4	.4985660D-4	.3481008D-4	.2305974D-4	.1713930D-4	.1354980D-4	.9363601D-5	.6059656D-5
1 I PF1	1.486179123	1.132389610	1.019879545	.9095202079	.7943479246	.7188532060	.6636929687	.5859564758	.5105192605
1 I PQ1	.5623117519	.5768355742	.5856023596	.5983156925	.6194331356	.6413369681	.6646518573	.7175833557	.8223431380
1 I PFC	1.688461704	1.401510946	1.315162662	1.233857541	1.153508974	1.103880752	1.069399930	1.023712466	.9830604833
1 I PQC	1.588271429	1.758017455	1.854858469	1.989284615	2.200030105	2.405876430	2.614120016	3.057041126	3.857986479
1 I ZF1	8.876726949	8.876726949	8.876726949	8.876726949	8.876726949	8.876726949	8.876726949	8.876726949	8.876726949
1 I ZF2	21.37375485	21.37375485	21.37375485	21.37375485	21.37375485	21.37375485	21.37375485	21.37375485	21.37375485

(2) P 4+0	TH = 8 SBE = 7.185296534							CAUER FUNCTIONS	
	RC = 02	RC = 05	RC = 07	RC = 10	RC = 15	RC = 20	RC = 25	RC = 35	RC = 50
AMAX DB	.0017375255	.0108709564	.0213327376	.0436480540	.0988323393	.1772876696	.2802872360	.5675287486	1.249387366
AMIN DB	70.49193552	78.45986880	81.39289127	84.51324577	88.09025522	90.66748528	92.70868511	95.91848733	99.69838515
MU I K	.2988156D-3	.1194006D-3	.8518349D-4	.5947545D-4	.3939919D-4	.2928369D-4	.2315079D-4	.1599837D-4	.1035335D-4
1 I PF1	1.490076746	1.134527246	1.021607779	.9109065987	.7954338801	.7197700658	.6644988283	.5866221972	.5110653060
1 I PQ1	.5629030771	.5772550417	.5859790790	.5986566446	.6197443505	.6416340847	.6649424390	.7178732758	.8226521964
1 I PFC	1.685974804	1.400473163	1.314442258	1.233391287	1.153254107	1.103737875	1.069326922	1.023722410	.9831353749
1 I PQC	1.597791384	1.765379812	1.861749006	1.995846471	2.206442523	2.412350853	2.620764493	3.064229747	3.866433297
1 I ZF1	7.771760036	7.771760036	7.771760036	7.771760036	7.771760036	7.771760036	7.771760036	7.771760036	7.771760036
1 I ZF2	18.69800556	18.69800556	18.69800556	18.69800556	18.69800556	18.69800556	18.69800556	18.69800556	18.69800556

(2) P 4+0	TH = 9 SBE = 6.392453221							CAUER FUNCTIONS	
	RC = 02	RC = 05	RC = 07	RC = 10	RC = 15	RC = 20	RC = 25	RC = 35	RC = 50
AMAX DB	.0017375255	.0108709564	.0213327376	.0436480540	.0988323393	.1772876696	.2802872360	.5675287486	1.249387366
AMIN DB	66.38463955	74.35257232	77.28559474	80.40594921	83.98295865	86.56018871	88.60138853	91.81119076	95.59108857
MU I K	.4794773D-3	.1915894D-3	.1366848D-3	.9543387D-4	.6321965D-4	.4698839D-4	.3714758D-4	.2567086D-4	.1661290D-4
1 I PF1	1.494510382	1.136956990	1.023571765	.9124818135	.7966675223	.7208115062	.6654141208	.5873782516	.5116853959
1 I PQ1	.5635812636	.5777341327	.5864088769	.5990452623	.6200987597	.6419722640	.6652730661	.7182030051	.8230035602
1 I PFC	1.683157807	1.399296515	1.313625213	1.232862311	1.152964809	1.103575591	1.069243890	1.023733489	.9832201506
1 I PQC	1.608690079	1.773786787	1.869611912	2.003329870	2.213751493	2.419728141	2.628333935	3.072416858	3.876051140
1 I ZF1	6.912894291	6.912894291	6.912894291	6.912894291	6.912894291	6.912894291	6.912894291	6.912894291	6.912894291
1 I ZF2	16.61637966	16.61637966	16.61637966	16.61637966	16.61637966	16.61637966	16.61637966	16.61637966	16.61637966

(2) P 4+0 — TH = 10 — SBE = 5.758770483 — CAUER FUNCTIONS

(2) P 4+0	RC = 02	RC = 05	RC = 07	RC = 10	RC = 15	RC = 20	RC = 25	RC = 35	RC = 50
AMAX DB	.0017375255	.0108709564	.0213327376	.0436480540	.0988323393	.1772876696	.2802872360	.5675287486	1.249387366
AMIN DB	62.70714256	70.67507421	73.60809652	76.72845094	80.30546035	82.88269040	84.92389022	88.13369243	91.91359025
MU K	.7322222D-3	.2925811D-3	.2087350D-3	.1457396D-3	.9654438D-4	.7175720D-4	.5672905D-4	.3920265D-4	.2537000D-4
1 PF1	1.499486160	1.139681521	1.025773516	.9142473573	.7980499479	.7219784118	.6664395992	.5882252346	.5123800000
1 PQ1	.5643494823	.5782742856	.5868928580	.5994823977	.6204970170	.6423520620	.6656442379	.7185729853	.8233976492
1 PFC	1.680011356	1.397980833	1.312711314	1.232270407	1.152640913	1.103393764	1.069150728	1.023745630	.9833147721
1 PQC	1.621010465	1.783262843	1.878467966	2.011752835	2.221973043	2.428023551	2.636843365	3.081617839	3.886857208
1 ZF1	6.226301041	6.226301041	6.226301041	6.226301041	6.226301041	6.226301041	6.226301041	6.226301041	6.226301041
1 ZF2	14.95063676	14.95063676	14.95063676	14.95063676	14.95063676	14.95063676	14.95063676	14.95063676	14.95063676

(2) P 4+0 — TH = 11 — SBE = 5.240843064 — CAUER FUNCTIONS

(2) P 4+0	RC = 02	RC = 05	RC = 07	RC = 10	RC = 15	RC = 20	RC = 25	RC = 35	RC = 50
AMAX DB	.0017375255	.0108709564	.0213327376	.0436480540	.0988323393	.1772876696	.2802872360	.5675287486	1.249387366
AMIN DB	59.37701743	67.34494683	70.27796893	73.39832323	76.97533259	79.55256261	81.59376242	84.80356463	88.58346244
MU K	.1074358D-2	.4292919D-3	.3062680D-3	.2138375D-3	.1416555D-3	.1052863D-3	.8323613D-4	.5752038D-4	.3722433D-4
1 PF1	1.505010933	1.142703847	1.028215294	.9162049210	.7995823898	.7232717772	.6675761099	.5891638160	.5131496465
1 PQ1	.5652113782	.5788771401	.5874322789	.5999690180	.6209398637	.6427741081	.6660565192	.7189837157	.8238349375
1 PFC	1.676536217	1.396525937	1.311700331	1.231615349	1.152282233	1.103192244	1.069047313	1.023758755	.9834191962
1 PQC	1.634801732	1.793835834	1.888340790	2.021135828	2.231125352	2.437254373	2.646309803	3.091850104	3.898870969
1 ZF1	5.664998589	5.664998589	5.664998589	5.664998589	5.664998589	5.664998589	5.664998589	5.664998589	5.664998589
1 ZF2	13.58735074	13.58735074	13.58735074	13.58735074	13.58735074	13.58735074	13.58735074	13.58735074	13.58735074

(2) P 4+0 — TH = 12 — SBE = 4.809734345 — CAUER FUNCTIONS

(2) P 4+0	RC = 02	RC = 05	RC = 07	RC = 10	RC = 15	RC = 20	RC = 25	RC = 35	RC = 50
AMAX DB	.0017375255	.0108709564	.0213327376	.0436480540	.0988323393	.1772876696	.2802872360	.5675287486	1.249387366
AMIN DB	56.33340002	64.30132514	67.23434684	70.35470094	73.93171018	76.50894016	78.55013995	81.75994215	85.53983995
MU K	.1525211D-2	.6094439D-3	.4347931D-3	.3035743D-3	.2011011D-3	.1494696D-3	.1181661D-3	.8165878D-4	.5284550D-4
1 PF1	1.511092265	1.146027308	1.030899613	.9133563848	.8012662191	.7246927089	.6688245949	.5901947409	.5139949233
1 PQ1	.5661711082	.5795445494	.5880285563	.6005062112	.6214281319	.6432391082	.6665105428	.7194357549	.8243159551
1 PFC	1.672733292	1.394931640	1.310592012	1.230896889	1.151888563	1.102970867	1.068933512	1.023772773	.9835333749
1 PQC	1.650119741	1.805537194	1.899257003	2.031501876	2.241228847	2.447440024	2.656752348	3.103133186	3.912114244
1 ZF1	5.197665974	5.197665974	5.197665974	5.197665974	5.197665974	5.197665974	5.197665974	5.197665974	5.197665974
1 ZF2	12.45090385	12.45090385	12.45090385	12.45090385	12.45090385	12.45090385	12.45090385	12.45090385	12.45090385

(2) P 4+0	TH = 13	SBE = 4.445411483						CAUER FUNCTIONS	
	RC = 02	RC = 05	RC = 07	RC = 10	RC = 15	RC = 20	RC = 25	RC = 35	RC = 50
AMAX DB	.0017375255	.0108709564	.0213327376	.0436480540	.0988323393	.1772876696	.2802872360	.5675287486	1.249387366
AMIN DB	53.53006660	61.49798402	64.43100500	67.55135872	71.12836776	73.70559767	75.74679743	78.95659959	82.73649738
MU K	.2106186D-2	.8415905D-3	.6004125D-3	.4192104D-3	.2777037D-3	.2064050D-3	.1631774D-3	.1127629D-3	.7297517D-4
1 PF1	1.517738425	1.149655575	1.033829240	.9207038211	.8031029482	.7262424274	.6701860932	.5913188316	.5149164790
1 PQ1	.5672333850	.5802785930	.5886832757	.6010951922	.6219627484	.6437478475	.6670070121	.7199297233	.8248412901
1 PFC	1.668603631	1.393197750	1.309386089	1.230114757	1.151459680	1.102729451	1.068809177	1.023787587	.9936572553
1 PQC	1.667027531	1.818402150	1.911246378	2.042876692	2.252306311	2.458602138	2.668192273	3.115488823	3.926611307
1 ZF1	4.802619615	4.802619615	4.802619615	4.802619615	4.802619615	4.802619615	4.802619615	4.802619615	4.802619615
1 ZF2	11.48894500	11.48894500	11.48894500	11.48894500	11.48894500	11.48894500	11.48894500	11.48894500	11.48894500

(2) P 4+0	TH = 14	SBE = 4.133565494						CAUER FUNCTIONS	
	RC = 02	RC = 05	RC = 07	RC = 10	RC = 15	RC = 20	RC = 25	RC = 35	RC = 50
AMAX DB	.0017375255	.0108709564	.0213327376	.0436480540	.0988323393	.1772876696	.2802872360	.5675287486	1.249387366
AMIN DB	50.93108422	58.89898837	61.83200812	64.95236117	68.52936986	71.10659965	73.14779935	76.35760147	80.13749922
MU K	.2840834D-2	.1135143D-2	.8098407D-3	.5654340D-3	.3745687D-3	.2784005D-3	.2200949D-3	.1520967D-3	.9842945D-4
1 PF1	1.524958372	1.153592660	1.037007203	.9232494980	.8050942332	.7279222698	.6716617434	.5925369892	.5159150243
1 PQ1	.5684035275	.5810815921	.5893982016	.6017373093	.6225447398	.6443011947	.6675467041	.7204663055	.8254115906
1 PFC	1.664148446	1.391324076	1.308082280	1.229268666	1.150995342	1.102467801	1.068674150	1.023803089	.9837907796
1 PQC	1.685595904	1.832469962	1.924342035	2.055288833	2.264383003	2.470764682	2.680653128	3.128941062	3.942388990
1 ZF1	4.464371388	4.464371388	4.464371388	4.464371388	4.464371388	4.464371388	4.464371388	4.464371388	4.464371388
1 ZF2	10.66408023	10.66408023	10.66408023	10.66408023	10.66408023	10.66408023	10.66408023	10.66408023	10.66408023

(2) P 4+0	TH = 15	SBE = 3.863703305						CAUER FUNCTIONS	
	RC = 02	RC = 05	RC = 07	RC = 10	RC = 15	RC = 20	RC = 25	RC = 35	RC = 50
AMAX DB	.0017375255	.0108709564	.0213327376	.0436480540	.0988323393	.1772876696	.2802872360	.5675287486	1.249387366
AMIN DB	48.50796548	56.47584763	59.40886532	62.52921729	66.10622539	68.68345498	70.72465458	73.93445662	77.71435433
MU K	.3754929D-2	.1500402D-2	.1070426D-2	.7473760D-3	.4950952D-3	.3679826D-3	.2909158D-3	.2010375D-3	.1301015D-3
1 PF1	1.532761734	1.157842916	1.040436790	.9259958833	.8072418777	.7297336926	.6732527857	.5938501952	.5169913338
1 PQ1	.5696875199	.5819561275	.5901752896	.6024349530	.6231752374	.6449001039	.6681304732	.7210462530	.8260275673
1 PFC	1.659369129	1.389310427	1.306680288	1.228358309	1.150495288	1.102185703	1.068528258	1.023819163	.9839338849
1 PQC	1.705904092	1.847784203	1.938580649	2.068769860	2.277486795	2.483954075	2.694160857	3.143516374	3.959476812
1 ZF1	4.171563479	4.171563479	4.171563479	4.171563479	4.171563479	4.171563479	4.171563479	4.171563479	4.171563479
1 ZF2	9.948886992	9.948886992	9.948886992	9.948886992	9.948886992	9.948886992	9.948886992	9.948886992	9.948886992

(2) P 4+0	TH = 16	SBE = 3.627955279							CAUER FUNCTIONS
	RC = 02	RC = 05	RC = 07	RC = 10	RC = 15	RC = 20	RC = 25	RC = 35	RC = 50
AMAX DB	.0017375255	.0108709564	.0213327376	.0436480540	.0988323393	.1772876696	.2802872360	.5675287486	1.249387366
AMIN DB	46.23774369	54.20559051	57.13860491	60.25895512	63.83596229	66.41319155	68.45439101	71.66419291	75.44409055
MU I K	.4876552D-2	.1948590D-2	.1390176D-2	.9706267D-3	.6429865D-3	.4779037D-3	.3778161D-3	.2610901D-3	.1689645D-3
1 I PF1	1.541158776	1.162411038	1.044121557	.9289456483	.8095478354	.7316782742	.6749605645	.5952595142	.5181462473
1 I PQ1	.5710920786	.5829050585	.5910166990	.6031870636	.6238554832	.6455456239	.6687592544	.7216703871	.8266899962
1 I PFC	1.654267259	1.387156618	1.305179806	1.227383359	1.149959241	1.101882931	1.068371314	1.023835685	.9840865029
1 I PQC	1.728040533	1.864393067	1.954002696	2.083354537	2.291648326	2.498199330	2.708743931	3.159243780	3.977907112
1 I ZF1	3.915677664	3.915677664	3.915677664	3.915677664	3.915677664	3.915677664	3.915677664	3.915677664	3.915677664
1 I ZF2	9.322798088	9.322798088	9.322798088	9.322798088	9.322798088	9.322798088	9.322798088	9.322798088	9.322798088

(2) P 4+0	TH = 17	SBE = 3.420303620							CAUER FUNCTIONS
	RC = 02	RC = 05	RC = 07	RC = 10	RC = 15	RC = 20	RC = 25	RC = 35	RC = 50
AMAX DB	.0017375255	.0108709564	.0213327376	.0436480540	.0988323393	.1772876696	.2802872360	.5675287486	1.249387366
AMIN DB	44.10163282	52.06942449	55.00243374	58.12278122	61.69978693	64.27701568	66.31821490	69.52301660	73.30791413
MU I K	.6236176D-2	.2491889D-2	.1777781D-2	.1241255D-2	.8222628D-3	.6111520D-3	.4831581D-3	.3338867D-3	.2160749D-3
1 I PF1	1.550160364	1.167302069	1.048065331	.9321016723	.8120142143	.7337577183	.6767865316	.5967660956	.5193806717
1 I PQ1	.5726247307	.5839315458	.5919248080	.6039981422	.6245868364	.6462388981	.6694340681	.7223396028	.8273997213
1 I PFC	1.648844616	1.384862475	1.303580516	1.226343475	1.149386904	1.101559241	1.068203120	1.023852520	.9842485601
1 I PQC	1.752103751	1.882349727	1.970652718	2.099081035	2.306901177	2.513532206	2.724433494	3.176154999	3.997715203
1 I ZF1	3.690200132	3.690200132	3.690200132	3.690200132	3.690200132	3.690200132	3.690200132	3.690200132	3.690200132
1 I ZF2	8.770085370	8.770085370	8.770085370	8.770085370	8.770085370	8.770085370	8.770085370	8.770085370	8.770085370

(2) P 4+0	TH = 18	SBE = 3.236067977							CAUER FUNCTIONS
	RC = 02	RC = 05	RC = 07	RC = 10	RC = 15	RC = 20	RC = 25	RC = 35	RC = 50
AMAX DB	.0017375255	.0108709564	.0213327376	.0436480540	.0988323393	.1772876696	.2802872360	.5675287486	1.249387366
AMIN DB	42.08407129	50.05177903	52.98478045	56.10512377	59.68212726	62.25935524	64.30055409	67.51035548	71.29025285
MU I K	.7866770D-2	.3143483D-2	.2242647D-2	.1565827D-2	.1037274D-2	.7709607D-3	.6094980D-3	.4211940D-3	.2725758D-3
1 I PF1	1.559777911	1.172521399	1.052272215	.9354670463	.8146432805	.7359738575	.6787322484	.5983711761	.5206955832
1 I PQ1	.5742939031	.5850390767	.5929022301	.6048692609	.6253707807	.6469811729	.6701560246	.7230548724	.8281576535
1 I PFC	1.643103193	1.382427836	1.301882093	1.225238297	1.148777965	1.101214375	1.068023463	1.023869523	.9844199774
1 I PQC	1.778203368	1.901712728	1.988579628	2.115991175	2.323282058	2.529987385	2.741263524	3.194284598	4.018939542
1 I ZF1	3.490064714	3.490064714	3.490064714	3.490064714	3.490064714	3.490064714	3.490064714	3.490064714	3.490064714
1 I ZF2	8.278515560	8.278515560	8.278515560	8.278515560	8.278515560	8.278515560	8.278515560	8.278515560	8.278515560

(2) P 4+0	TH = 19	SBE = 3.071553487							CAUER FUNCTIONS
	RC = 02	RC = 05	RC = 07	RC = 10	RC = 15	RC = 20	RC = 25	RC = 35	RC = 50
AMAX DB	.0017375255	.0108709564	.0213327376	.0436480540	.0988323393	.1772876696	.2802872360	.5675287486	1.249387366
AMIN DB	40.17202511	48.13960792	51.07259768	54.19293481	57.76993500	60.34716181	62.38836014	65.59816105	69.37805817
MU I K	.9803897D-2	.3917596D-2	.2794925D-2	.1951431D-2	.1292716D-2	.9608197D-3	.7595948D-3	.5249186D-3	.3397012D-3
1 I PF1	1.570023303	1.178074766	1.056746587	.9390450784	.8174374622	.7383286566	.6807993899	.6000760827	.5220920291
1 I PQ1	.5761090258	.5862314933	.5939518328	.6058025757	.6262089324	.6477738035	.6709263293	.7238172493	.8289647985
1 I PFC	1.637045205	1.379852557	1.300084207	1.224067452	1.148132093	1.100848060	1.067832119	1.023886543	.9846006703
1 I PQC	1.806461268	1.922546441	2.007837051	2.134130692	2.340831028	2.547602657	2.759271018	3.213670178	4.041621917
1 I ZF1	3.311271920	3.311271920	3.311271920	3.311271920	3.311271920	3.311271920	3.311271920	3.311271920	3.311271920
1 I ZF2	7.838430534	7.838430534	7.838430534	7.838430534	7.838430534	7.838430534	7.838430534	7.838430534	7.838430534

(2) P 4+0	TH = 20	SBE = 2.923804400							CAUER FUNCTIONS
	RC = 02	RC = 05	RC = 07	RC = 10	RC = 15	RC = 20	RC = 25	RC = 35	RC = 50
AMAX DB	.0017375255	.0108709564	.0213327376	.0436480540	.0988323393	.1772876696	.2802872360	.5675287486	1.249387366
AMIN DB	38.35447030	46.32187080	49.25484355	52.37517165	55.95216701	58.52939213	60.57058967	63.78038991	67.56028667
MU I K	.1208583D-1	.4829548D-2	.3445544D-2	.2405700D-2	.1593645D-2	.1184487D-2	.9364196D-3	.6471136D-3	.4187797D-3
1 I PF1	1.580908814	1.183968253	1.061493108	.9428392974	.8203993543	.7408242165	.6829897478	.6018822355	.5235711305
1 I PQ1	.5780806502	.5875130249	.5950767590	.6068004402	.6271030496	.6486182605	.6717462885	.7246278732	.8298222115
1 I PFC	1.630673096	1.377136512	1.298186525	1.222830552	1.147448941	1.100460007	1.067628849	1.023903415	.9847905482
1 I PQC	1.837012917	1.944921569	2.028483709	2.153549536	2.359591731	2.566419141	2.778496191	3.234352562	4.065807656
1 I ZF1	3.150622274	3.150622274	3.150622274	3.150622274	3.150622274	3.150622274	3.150622274	3.150622274	3.150622274
1 I ZF2	7.442103525	7.442103525	7.442103525	7.442103525	7.442103525	7.442103525	7.442103525	7.442103525	7.442103525

(2) P 4+0	TH = 21	SBE = 2.790428110							CAUER FUNCTIONS
	RC = 02	RC = 05	RC = 07	RC = 10	RC = 15	RC = 20	RC = 25	RC = 35	RC = 50
AMAX DB	.0017375255	.0108709564	.0213327376	.0436480540	.0988323393	.1772876696	.2802872360	.5675287486	1.249387366
AMIN DB	36.62200227	44.58914140	47.52208977	50.64240491	54.21939335	56.79661606	58.83781248	62.04761174	65.82750798
MU I K	.1475366D-1	.5895803D-2	.4206254D-2	.2936836D-2	.1945496D-2	.1446003D-2	.1143166D-2	.7899860D-3	.5112397D-3
1 I PF1	1.592446984	1.190208286	1.066516725	.9468534582	.8235317229	.7434627788	.6853052343	.6037911506	.5251340846
1 I PQ1	.5802205862	.5888883245	.5962804499	.6078654211	.6280550419	.6495161377	.6726173158	.7254379749	.8307310513
1 I PFC	1.623989547	1.374279601	1.296188710	1.221527194	1.146728146	1.100049911	1.067413400	1.023919968	.9849895148
1 I PQC	1.870008891	1.968915715	2.050583848	2.174302203	2.379611659	2.586481518	2.798982703	3.256376014	4.091545859
1 I ZF1	3.005525835	3.005525835	3.005525835	3.005525835	3.005525835	3.005525835	3.005525835	3.005525835	3.005525335
1 I ZF2	7.083279273	7.083279273	7.083279273	7.083279273	7.083279273	7.083279273	7.083279273	7.083279273	7.083279273

(2) P 4+0 — TH = 22 — SBE = 2.669467163 — CAUER FUNCTIONS

(2) P 4+0			RC = 02	RC = 05	RC = 07	RC = 10	RC = 15	RC = 20	RC = 25	RC = 35	RC = 50
AMAX DB			.0017375255	.0108709564	.0213327376	.0436480540	.0988323393	.1772876696	.2802872360	.5675287486	1.249387366
AMIN DB			34.96653664	42.93330708	45.86622104	48.98651790	52.56349659	55.14071588	57.18191072	60.39170861	64.17160412
MU		K	.1785144D-1	.7134025D-2	.5089662D-2	.3553646D-2	.2354101D-2	.1749702D-2	.1383262D-2	.9559046D-3	.6186141D-3
	1	PF1	1.604650479	1.196801627	1.071822667	.9510915455	.8268375099	.7462467298	.6877478861	.6058044435	.5267821675
	1	PQ1	.5825420590	.5903625094	.5975666721	.6090003150	.6290669819	.6504691610	.6735409391	.7263988822	.8316925599
	1	PFC	1.616997478	1.371281751	1.294090430	1.220156964	1.145969328	1.099617453	1.067185508	1.023936017	.9851974674
	1	PQC	1.905616621	1.994614033	2.074207720	2.196448106	2.400942457	2.607838295	2.820777903	3.279788478	4.118889648
	1	ZF1	2.873863673	2.873863673	2.873863673	2.873863673	2.873863673	2.873863673	2.873863673	2.873863673	2.873863673
	1	ZF2	6.756839575	6.756839575	6.756839575	6.756839575	6.756839575	6.756839575	6.756839575	6.756839575	6.756839575

(2) P 4+0 — TH = 23 — SBE = 2.559304665 — CAUER FUNCTIONS

(2) P 4+0			RC = 02	RC = 05	RC = 07	RC = 10	RC = 15	RC = 20	RC = 25	RC = 35	RC = 50
AMAX DB			.0017375255	.0108709564	.0213327376	.0436480540	.0988323393	.1772876696	.2802872360	.5675287486	1.249387366
AMIN DB			33.38107745	41.34733528	44.28020141	47.40047285	50.97743797	53.55465251	55.59584515	58.80564113	62.58553562
MU		K	.2142625D-1	.8563144D-2	.6109279D-2	.4265563D-2	.2825713D-2	.2100232D-2	.1660381D-2	.1147408D-2	.7425458D-3
	1	PF1	1.617531899	1.203755364	1.077416452	.9555577790	.8303198385	.7491786052	.6903198689	.6079238329	.5235167371
	1	PQ1	.5850598908	.5919412079	.5989395470	.6102081685	.6301411174	.6514791971	.6745188088	.7273620255	.8327080736
	1	PFC	1.609700053	1.368142918	1.291891352	1.218719439	1.145172091	1.099162297	1.066944893	1.023951370	.9854142970
	1	PQC	1.944022394	2.022109948	2.099432129	2.220051997	2.423640244	2.630542101	2.843933106	3.304641847	4.147896455
	1	ZF1	2.753885470	2.753885470	2.753885470	2.753885470	2.753885470	2.753885470	2.753885470	2.753885470	2.753885470
	1	ZF2	6.458556093	6.458556093	6.458556093	6.458556093	6.458556093	6.458556093	6.458556093	6.458556093	6.458556093

(2) P 4+0 — TH = 24 — SBE = 2.458593336 — CAUER FUNCTIONS

(2) P 4+0			RC = 02	RC = 05	RC = 07	RC = 10	RC = 15	RC = 20	RC = 25	RC = 35	RC = 50
AMAX DB			.0017375255	.0108709564	.0213327376	.0436480540	.0988323393	.1772876696	.2802872360	.5675287486	1.249387366
AMIN DB			31.85953531	39.82508987	42.75789034	45.87812691	49.45507341	52.03228144	54.07347107	57.28326441	61.06315751
MU		K	.2552838D-1	.1020341D-1	.7279566D-2	.5082690D-2	.3367025D-2	.2502568D-2	.1978456D-2	.1367215D-2	.8847939D-3
	1	PF1	1.631103553	1.211076899	1.083303882	.9602566166	.8339820180	.7522610955	.6930234818	.6101511447	.5303392364
	1	PQ1	.5877907103	.5936306119	.6004035843	.6114922991	.6312798853	.6525482646	.6755527056	.7283789454	.8337790280
	1	PFC	1.602100681	1.364863094	1.289591149	1.217214182	1.144336024	1.098684093	1.066691262	1.023965823	.9856398881
	1	PQC	1.985433657	2.051505984	2.126341036	2.245184427	2.447765985	2.654650009	2.868503901	3.330992248	4.178629326
	1	ZF1	2.644132736	2.644132736	2.644132736	2.644132736	2.644132736	2.644132736	2.644132736	2.644132736	2.644132736
	1	ZF2	6.184904957	6.184904957	6.184904957	6.184904957	6.184904957	6.184904957	6.184904957	6.184904957	6.184904957

(2) P 4+0 — TH = 25 — SBE = 2.366201583 — CAUER FUNCTIONS

	RC = 02	RC = 05	RC = 07	RC = 10	RC = 15	RC = 20	RC = 25	RC = 35	RC = 50
AMAX DB	.0017375255	.0108709564	.0213327376	.0436480540	.0988323393	.1772876696	.2802872360	.5675287486	1.249387366
AMIN DB	30.39658377	38.36118506	41.29389654	44.41408582	47.99100709	50.56820628	52.60939182	55.81918160	59.59907281
MU K	.3021140D-1	.1207649D-1	.8615990D-2	.6015832D-2	.3985196D-2	.2962032D-2	.2341695D-2	.1618233D-2	.1047240D-2
1 PF1	1.645377166	1.218773924	1.089491042	.9651927583	.8378275493	.7554970507	.6958611618	.6124883155	.5322511967
1 PQ1	.5907531942	.5954375359	.6019637196	.6128563192	.6324859268	.6536785452	.6766445511	.7294512997	.8349069650
1 PFC	1.594203014	1.361442305	1.287189502	1.215640748	1.143460700	1.098182476	1.066424308	1.023979162	.9858741193
1 PQC	2.030081664	2.082914693	2.155026245	2.271922273	2.473385902	2.680223902	2.894550487	3.358900380	4.211152272
1 ZF1	2.543380440	2.543380440	2.543380440	2.543380440	2.543380440	2.543380440	2.543380440	2.543380440	2.543380440
1 ZF2	5.932925857	5.932925857	5.932925857	5.932925857	5.932925857	5.932925857	5.932925857	5.932925857	5.932925857

(2) P 4+0 — TH = 26 — SBE = 2.281172033 — CAUER FUNCTIONS

	RC = 02	RC = 05	RC = 07	RC = 10	RC = 15	RC = 20	RC = 25	RC = 35	RC = 50
AMAX DB	.0017375255	.0108709564	.0213327376	.0436480540	.0988323393	.1772876696	.2802872360	.5675287486	1.249387366
AMIN DB	28.98754481	36.95086828	39.88346043	43.00358630	46.58047372	49.15766107	51.19884112	54.40862613	58.18851480
MU K	.3553225D-1	.1420550D-1	.1013508D-1	.7076535D-2	.4687873D-2	.3484311D-2	.2754596D-2	.1903569D-2	.1231897D-2
1 PF1	1.660363533	1.226854403	1.095984296	.9703711493	.8418601303	.7588894860	.6988354898	.6149373978	.5342542415
1 PQ1	.5939683464	.5973694844	.6036253569	.6143041638	.6337621045	.6548723964	.6777964175	.7305808720	.8360935398
1 PFC	1.586010943	1.357880617	1.284686097	1.213998684	1.142545676	1.097657065	1.066143709	1.023991160	.9861168580
1 PQC	2.078224527	2.116459703	2.185588177	2.300349318	2.500571930	2.707330874	2.922138053	3.388431864	4.245540647
1 ZF1	2.450592118	2.450592118	2.450592118	2.450592118	2.450592118	2.450592118	2.450592118	2.450592118	2.450592118
1 ZF2	5.700113662	5.700113662	5.700113662	5.700113662	5.700113662	5.700113662	5.700113662	5.700113662	5.700113662

(2) P 4+0 — TH = 27 — SBE = 2.202689265 — CAUER FUNCTIONS

	RC = 02	RC = 05	RC = 07	RC = 10	RC = 15	RC = 20	RC = 25	RC = 35	RC = 50
AMAX DB	.0017375255	.0108709564	.0213327376	.0436480540	.0988323393	.1772876696	.2802872360	.5675287486	1.249387366
AMIN DB	27.62829711	35.58992517	38.52235892	41.64240064	45.21924313	47.79641475	49.83758753	53.04736620	56.82725151
MU K	.4155135D-1	.1661513D-1	.1185447D-1	.8277134D-2	.5483247D-2	.4075485D-2	.3221964D-2	.2226546D-2	.1440912D-2
1 PF1	1.676072079	1.235326533	1.102790277	.9757969823	.8460836612	.7624415876	.7019491950	.6175005643	.5363500906
1 PQ1	.5974598218	.5994347284	.6053944159	.6158401199	.6351115214	.6561323663	.6790105397	.7317695807	.8373405287
1 PFC	1.577528602	1.354178138	1.282080634	1.212287530	1.141590497	1.097107462	1.065849131	1.024001583	.9863679703
1 PQC	2.130150728	2.152276915	2.218136738	2.330556914	2.529402228	2.736043682	2.951337194	3.419657650	4.281871573
1 ZF1	2.364884957	2.364884957	2.364884957	2.364884957	2.364884957	2.364884957	2.364884957	2.364884957	2.364884957
1 ZF2	5.484334108	5.484334108	5.484334108	5.484334108	5.484334108	5.484334108	5.484334108	5.484334108	5.484334108

(2) P 4+0	TH = 28	SBE = 2.130054468						CAUER FUNCTIONS	
	RC = 02	RC = 05	RC = 07	RC = 10	RC = 15	RC = 20	RC = 25	RC = 35	RC = 50
AMAX DB	.0017375255	.0108709564	.0213327376	.0436480540	.0988323393	.1772876696	.2802872360	.5675287486	1.249387366
AMIN DB	26.31520242	34.27460220	37.20682768	40.32675872	43.90354213	46.48069308	48.52185628	51.73162662	55.51150749
MU K	.4833257D-1	.1933169D-1	.1379300D-1	.9630793D-2	.6380033D-2	.4742041D-2	.3748929D-2	.2590709D-2	.1676581D-2
1 PF1	1.692510332	1.244198706	1.109915875	.9814756980	.8505022502	.7661567183	.7052051617	.6201801131	.5385405640
1 PQ1	.6012542991	.6016423918	.6072773863	.6174683606	.6365375419	.6574612088	.6802893281	.7330194891	.8386498375
1 PFC	1.568760357	1.350335019	1.279372821	1.210506817	1.140594689	1.096533256	1.065540223	1.024010180	.9866273108
1 PQC	2.186183179	2.190515857	2.252792304	2.362644715	2.559961752	2.766441243	2.992224383	3.452654460	4.320229402
1 ZF1	2.285502360	2.285502360	2.285502360	2.285502360	2.285502360	2.285502360	2.285502360	2.285502360	2.285502360
1 ZF2	5.283757569	5.283757569	5.283757569	5.283757569	5.283757569	5.283757569	5.283757569	5.283757569	5.283757569

(2) P 4+0	TH = 29	SBE = 2.062665340						CAUER FUNCTIONS	
	RC = 02	RC = 05	RC = 07	RC = 10	RC = 15	RC = 20	RC = 25	RC = 35	RC = 50
AMAX DB	.0017375255	.0108709564	.0213327376	.0436480540	.0988323393	.1772876696	.2802872360	.5675287486	1.249387366
AMIN DB	25.04504625	33.00154305	35.93349702	39.05328377	42.62999016	45.20711413	47.24826486	50.45802433	54.23789943
MU K	.5594325D-1	.2238323D-1	.1597074D-1	.1115156D-1	.7387551D-2	.5490910D-2	.4340970D-2	.2999845D-2	.1941355D-2
1 PF1	1.709683285	1.253479454	1.117368224	.9874129855	.8551202184	.7700384240	.7086064344	.6229784734	.5408275866
1 PQ1	.6053819139	.6040025500	.6092813883	.6191954833	.6380438156	.6588619012	.6816353829	.7343328169	.8400235113
1 PFC	1.559710810	1.346351455	1.276562382	1.208656073	1.139557767	1.095934018	1.065216620	1.024016693	.9868947267
1 PQC	2.246683906	2.231341224	2.289686832	2.396721509	2.592342895	2.798609199	3.014882486	3.487505285	4.360705244
1 ZF1	2.211792183	2.211792183	2.211792183	2.211792183	2.211792183	2.211792183	2.211792183	2.211792183	2.211792183
1 ZF2	5.096806514	5.096806514	5.096806514	5.096806514	5.096806514	5.096806514	5.096806514	5.096806514	5.096806514

(2) P 4+0	TH = 30	SBE = 2.000000000						CAUER FUNCTIONS	
	RC = 02	RC = 05	RC = 07	RC = 10	RC = 15	RC = 20	RC = 25	RC = 35	RC = 50
AMAX DB	.0017375255	.0108709564	.0213327376	.0436480540	.0988323393	.1772876696	.2802872360	.5675287486	1.249387366
AMIN DB	23.81499037	31.76773608	34.69933891	37.81893905	41.39554579	43.97263490	46.01376948	49.22351489	53.00338252
MU K	.6445409D-1	.2579961D-1	.1840912D-1	.1285444D-1	.8515746D-2	.6329483D-2	.5003933D-2	.3457994D-2	.2237849D-2
1 PF1	1.727592622	1.263177380	1.125154679	.9936147800	.8599421054	.7740904400	.7121562250	.6258982107	.5432131933
1 PQ1	.6098767588	.6065263421	.6114142416	.6210255524	.6396343037	.6603376642	.6830515104	.7357119519	.8414637446
1 PFC	1.550384795	1.342227690	1.273649054	1.206734820	1.138479230	1.095309306	1.064877943	1.024020848	.9871700573
1 PQC	2.312059468	2.274934635	2.328965119	2.432906153	2.626646204	2.832640539	3.049401348	3.524299933	4.403397535
1 ZF1	2.143189335	2.143189335	2.143189335	2.143189335	2.143189335	2.143189335	2.143189335	2.143189335	2.143189335
1 ZF2	4.922113486	4.922113486	4.922113486	4.922113486	4.922113486	4.922113486	4.922113486	4.922113486	4.922113486

(2) P 4+0	TH = 31	SBE = 1.941604026							CAUER FUNCTIONS
	RC = 02	RC = 05	RC = 07	RC = 10	RC = 15	RC = 20	RC = 25	RC = 35	RC = 50
AMAX DB	.0017375255	.0108709564	.0213327376	.0436480540	.0988323393	.1772876696	.2802872360	.5675287486	1.249387366
AMIN DB	22.62253493	30.57047068	33.50162279	36.62098333	40.19746215	42.77450647	44.81562032	48.02534768	51.80520573
MU K	.7393895D-1	.2961259D-1	.2113094D-1	.1475539D-1	.9775228D-2	.7265653D-2	.5744060D-2	.3969471D-2	.2568856D-2
1 PF1	1.746235793	1.273301080	1.133282788	1.000087259	.8649726737	.7783166973	.7158579187	.6289420337	.5456995342
1 PQ1	.6147774622	.6092260988	.6136845431	.6229651471	.6413133081	.6618919832	.6845407401	.7371594644	.8429728929
1 PFC	1.540787382	1.337964016	1.270632590	1.204742574	1.137358563	1.094658659	1.064523795	1.024022361	.9874531328
1 PQC	2.382767224	2.321496633	2.370786232	2.471328634	2.662981189	2.868636302	3.085878436	3.563135642	4.448412687
1 ZF1	2.079201735	2.079201735	2.079201735	2.079201735	2.079201735	2.079201735	2.079201735	2.079201735	2.079201735
1 ZF2	4.758487199	4.758487199	4.758487199	4.758487199	4.758487199	4.758487199	4.758487199	4.758487199	4.758487199

(2) P 4+0	TH = 32	SBE = 1.887079915							CAUER FUNCTIONS
	RC = 02	RC = 05	RC = 07	RC = 10	RC = 15	RC = 20	RC = 25	RC = 35	RC = 50
AMAX DB	.0017375255	.0108709564	.0213327376	.0436480540	.0988323393	.1772876696	.2802872360	.5675287486	1.249387366
AMIN DB	21.46548865	29.40730101	32.33787850	35.45693355	39.03324925	41.61023646	43.65132388	46.86102822	50.64087404
MU K	.8447449D-1	.3385595D-1	.2416051D-1	.1687149D-1	.1117732D-1	.8307841D-2	.6568010D-2	.4538879D-2	.2937354D-2
1 PF1	1.765604920	1.283859036	1.141760257	1.006836834	.8702169132	.7827213291	.7197150817	.6321128003	.5482888804
1 PQ1	.6201278578	.6121154882	.6161017555	.6250209160	.6430855050	.6635286328	.6861063448	.7386781219	.8445534858
1 PFC	1.530923882	1.333560775	1.267512761	1.202678850	1.136195235	1.093981601	1.064153767	1.024020933	.9877437744
1 PQC	2.459322596	2.371248973	2.415325132	2.512131255	2.701467226	2.906706361	3.124419572	3.604117772	4.495865797
1 ZF1	2.019398909	2.019398909	2.019398909	2.019398909	2.019398909	2.019398909	2.019398909	2.019398909	2.019398909
1 ZF2	4.604885001	4.604885001	4.604885001	4.604885001	4.604885001	4.604885001	4.604885001	4.604885001	4.604885001

(2) P 4+0	TH = 33	SBE = 1.836078459							CAUER FUNCTIONS
	RC = 02	RC = 05	RC = 07	RC = 10	RC = 15	RC = 20	RC = 25	RC = 35	RC = 50
AMAX DB	.0017375255	.0108709564	.0213327376	.0436480540	.0988323393	.1772876696	.2802872360	.5675287486	1.249387366
AMIN DB	20.34194572	28.27601559	31.20586528	34.32453330	37.90064229	40.47755715	42.51861107	45.72828625	49.50811657
MU K	.9613969D-1	.3856552D-1	.2752369D-1	.1922088D-1	.1273409D-1	.9465033D-2	.7482891D-2	.5171133D-2	.3346526D-2
1 PF1	1.785685501	1.294859491	1.150594905	1.013870142	.8756800447	.7873086778	.7237314681	.6354135250	.5509836299
1 PQ1	.6259777561	.6152096828	.6186763076	.6272001380	.6449559814	.6652517038	.6877518671	.7402709062	.8462082404
1 PFC	1.520799861	1.329018361	1.264289355	1.200543157	1.134988705	1.093277640	1.063767430	1.024016252	.9880417939
1 PQC	2.542307469	2.424437252	2.462774529	2.555469997	2.742234584	2.946970309	3.165139741	3.647360563	4.545881446
1 ZF1	1.963402658	1.963402658	1.963402658	1.963402658	1.963402658	1.963402658	1.963402658	1.963402658	1.963402658
1 ZF2	4.460390342	4.460390342	4.460390342	4.460390342	4.460390342	4.460390342	4.460390342	4.460390342	4.460390342

(2) P 4+0	TH = 34	SBE = 1.788291650						CAUER FUNCTIONS	
	RC = 02	RC = 05	RC = 07	RC = 10	RC = 15	RC = 20	RC = 25	RC = 35	RC = 50
AMAX DB	.0017375255	.0108709564	.0213327376	.0436480540	.0988323393	.1772876696	.2802872360	.5675287486	1.249387366
AMIN DB	19.25026858	27.17461205	30.10354543	33.22172603	36.79757469	39.37439839	41.41541012	44.62504855	48.40485935
MU K	.1090151D 0	.4377936D-1	.3124804D-1	.2182296D-1	.1445843D-1	.1074682D-1	.8496293D-2	.5871480D-2	.3799768D-2
1 PF1	1.806454906	1.306310299	1.159794609	1.021194033	.8813675227	.7920833013	.7279110272	.6388473860	.5537863143
1 PQ1	.6323838348	.6185255503	.6214197088	.6295107917	.6469302773	.6670656335	.6894811196	.7419410317	.8479400774
1 PFC	1.510421164	1.324337221	1.260962180	1.198335005	1.133738413	1.092546268	1.063364342	1.024007993	.9883469928
1 PQC	2.632379933	2.481333907	2.513346988	2.601516051	2.785425578	2.989558450	3.208164010	3.692988000	4.598594591
1 ZF1	1.910879374	1.910879374	1.910879374	1.910879374	1.910879374	1.910879374	1.910879374	1.910879374	1.910879374
1 ZF2	4.324194210	4.324194210	4.324194210	4.324194210	4.324194210	4.324194210	4.324194210	4.324194210	4.324194210

(2) P 4+0	TH = 35	SBE = 1.743446796						CAUER FUNCTIONS	
	RC = 02	RC = 05	RC = 07	RC = 10	RC = 15	RC = 20	RC = 25	RC = 35	RC = 50
AMAX DB	.0017375255	.0108709564	.0213327376	.0436480540	.0988323393	.1772876696	.2802872360	.5675287486	1.249387366
AMIN DB	18.18907551	26.10127598	29.02906222	32.14663241	35.72215495	38.29886447	40.33982334	43.54941574	47.32920209
MU K	.1231817D 0	.4953774D-1	.3536282D-1	.2469837D-1	.1636410D-1	.1216345D-1	.9616318D-2	.6645524D-2	.4300707D-2
1 PF1	1.827880637	1.318218736	1.169367228	1.028815545	.8872850370	.7970499794	.7322579119	.6424177335	.5566996061
1 PQ1	.6394106615	.6220818725	.6243446791	.6319616340	.6490144318	.6689752401	.6912982612	.7436919669	.8497521380
1 PFC	1.499793944	1.319517856	1.257531064	1.196053899	1.132443790	1.091786959	1.062944042	1.023995815	.9886591622
1 PQC	2.730285555	2.542241683	2.567277354	2.650457568	2.831195872	3.034612920	3.253628555	3.741134768	4.654151560
1 ZF1	1.861533672	1.861533672	1.861533672	1.861533672	1.861533672	1.861533672	1.861533672	1.861533672	1.861533672
1 ZF2	4.195579755	4.195579755	4.195579755	4.195579755	4.195579755	4.195579755	4.195579755	4.195579755	4.195579755

(2) P 4+0	TH = 36	SBE = 1.701301617						CAUER FUNCTIONS	
	RC = 02	RC = 05	RC = 07	RC = 10	RC = 15	RC = 20	RC = 25	RC = 35	RC = 50
AMAX DB	.0017375255	.0108709564	.0213327376	.0436480540	.0988323393	.1772876696	.2802872360	.5675287486	1.249387366
AMIN DB	17.15723242	25.05436339	27.98072114	31.09753084	34.67264702	37.24921423	39.29010722	42.49964225	46.27939811
MU K	.1387198D 0	.5588327D-1	.3989918D-1	.2786913D-1	.1846578D-1	.1372585D-1	.1085162D-1	.7499251D-2	.4853221D-2
1 PF1	1.849918348	1.330591278	1.179320526	1.036741884	.8934385125	.8022137202	.7367764862	.6461280979	.5597263260
1 PQ1	.6471318641	.6258995965	.6274652981	.6345622889	.6512150368	.6709857602	.6932077787	.7455274566	.8516478034
1 PFC	1.488924709	1.314560826	1.253995857	1.193699345	1.131104247	1.090999169	1.062506052	1.023979364	.9889780821
1 PQC	2.836870425	2.607497615	2.624825513	2.702501649	2.879715965	3.082288955	3.301681819	3.791947344	4.712711170
1 ZF1	1.815103086	1.815103086	1.815103086	1.815103086	1.815103086	1.815103086	1.815103086	1.815103086	1.815103086
1 ZF2	4.073909472	4.073909472	4.073909472	4.073909472	4.073909472	4.073909472	4.073909472	4.073909472	4.073909472

(2) P 4+0	TH = 37	SBE = 1.661640141							CAUER FUNCTIONS
	RC = 02	RC = 05	RC = 07	RC = 10	RC = 15	RC = 20	RC = 25	RC = 35	RC = 50
AMAX DB	.0017375255	.0108709564	.0213327376	.0436480540	.0988323393	.1772876696	.2802872360	.5675287486	1.249387366
AMIN DB	16.15384793	24.03238624	26.95697403	30.07284094	33.64745317	36.22384388	38.26465515	41.47411902	45.25383707
MU I K	.1557068D 0	.6286091D-1	.4489017D-1	.3135869D-1	.2077913D-1	.1544571D-1	.1221145D-1	.8439060D-2	.5461452D-2
1 I PF1	1.872509643	1.343433329	1.189662058	1.044980390	.8998341079	.8075797659	.7414713336	.6499821988	.5628694513
1 I PQ1	.6556314665	.6300021225	.6307971746	.6373233485	.6535392959	.6731028924	.6952145452	.7474515489	.8536307159
1 I PFC	1.477820386	1.309466751	1.250356433	1.191270846	1.129719186	1.090182338	1.062049877	1.023958267	.9893035209
1 I PQC	2.953096242	2.677477629	2.686279580	2.757876625	2.931172869	3.132756322	3.352485825	3.845585205	4.774445984
1 I ZF1	1.771353621	1.771353621	1.771353621	1.771353621	1.771353621	1.771353621	1.771353621	1.771353621	1.771353621
1 I ZF2	3.958614460	3.958614460	3.958614460	3.958614460	3.958614460	3.958614460	3.958614460	3.958614460	3.958614460

(2) P 4+0	TH = 38	SBE = 1.624269245							CAUER FUNCTIONS
	RC = 02	RC = 05	RC = 07	RC = 10	RC = 15	RC = 20	RC = 25	RC = 35	RC = 50
AMAX DB	.0017375255	.0108709564	.0213327376	.0436480540	.0988323393	.1772876696	.2802872360	.5675287486	1.249387366
AMIN DB	15.17827113	23.03400073	25.95640585	29.07110930	32.64509940	35.22127219	37.26198256	40.47135855	44.25102991
MU I K	.1742154D 0	.7051800D-1	.5037090D-1	.3519205D-1	.2332089D-1	.1733550D-1	.1370569D-1	.9471790D-2	.6129831D-2
1 I PF1	1.895579659	1.356748903	1.200399045	1.053538495	.9064782110	.8131535977	.7463472651	.6539839541	.5661321247
1 I PQ1	.6650054017	.6344156360	.6343576416	.6402564886	.6559950922	.6753328451	.6973238542	.7494686234	.8557048030
1 I PFC	1.466488401	1.304236316	1.246612689	1.188767908	1.128287992	1.089335887	1.061575002	1.023932138	.9896352345
1 I PQC	3.080057724	2.752601866	2.751959562	2.816834662	2.985772026	3.186200946	3.406217657	3.902222211	4.839543721
1 I ZF1	1.730076012	1.730076012	1.730076012	1.730076012	1.730076012	1.730076012	1.730076012	1.730076012	1.730076012
1 I ZF2	3.849185379	3.849185379	3.849185379	3.849185379	3.849185379	3.849185379	3.849185379	3.849185379	3.849185379

(2) P 4+0	TH = 39	SBE = 1.589015729							CAUER FUNCTIONS
	RC = 02	RC = 05	RC = 07	RC = 10	RC = 15	RC = 20	RC = 25	RC = 35	RC = 50
AMAX DB	.0017375255	.0108709564	.0213327376	.0436480540	.0988323393	.1772876696	.2802872360	.5675287486	1.249387366
AMIN DB	14.23009095	22.05799788	24.97772344	28.09099737	31.66422271	34.24012754	36.28071384	39.48998177	43.26959572
MU I K	.1943101D 0	.7890420D-1	.5637854D-1	.3939582D-1	.2610892D-1	.1940857D-1	.1534491D-1	.1060476D-1	.6863098D-2
1 I PF1	1.919034503	1.370540239	1.211538216	1.062423672	.9133774318	.8189409408	.7514093276	.6581374907	.5695176638
1 I PQ1	.6753632151	.6391694889	.6381659793	.6433745998	.6585910650	.6776823916	.6995414627	.7515834238	.8578743034
1 I PFC	1.454936783	1.298870276	1.242764551	1.186190034	1.126810036	1.088459220	1.061080896	1.023900572	.9899729661
1 I PQC	3.219002673	2.833340847	2.822221578	2.879654752	3.043739495	3.242826761	3.463071140	3.962043154	4.908208850
1 I ZF1	1.691082559	1.691082559	1.691082559	1.691082559	1.691082559	1.691082559	1.691082559	1.691082559	1.691082559
1 I ZF2	3.745164792	3.745164792	3.745164792	3.745164792	3.745164792	3.745164792	3.745164792	3.745164792	3.745164792

(2) P 4+0	TH = 40	SBE = 1.555723827						CAUER FUNCTIONS	
	RC = 02	RC = 05	RC = 07	RC = 10	RC = 15	RC = 20	RC = 25	RC = 35	RC = 50
AMAX DB	.0017375255	.0108709564	.0213327376	.0436480540	.0988323393	.1772876696	.2302872360	.5675237486	1.249387366
AMIN DB	13.30913618	21.10329633	24.01974643	27.13127107	30.70356006	33.27913675	35.31957108	38.52870665	42.30825026
MU I K	.2160445D 0	.8807146D-1	.6295246D-1	.4399836D-1	.2916232D-1	.2167920D-1	.1714042D-1	.1184581D-1	.7666330D-2
1 I PF1	1.942758630	1.384807350	1.223085623	1.071643370	.9205385917	.8249477672	.7566628124	.6624471542	.5730295713
1 I PQ1	.6868299652	.6442966388	.6422436714	.6466919379	.6613366963	.6801589315	.7018736400	.7538010948	.8601437972
1 I PFC	1.443174294	1.293369464	1.238811973	1.183536731	1.125284675	1.087551719	1.060567004	1.023363147	.9903164450
1 I PQC	3.371355035	2.920222643	2.897462757	2.946646161	3.105324452	3.302857817	3.523258757	4.025270524	4.930664395
1 I ZF1	1.654204428	1.654204428	1.654204428	1.654204428	1.654204428	1.654204428	1.654204428	1.654204428	1.654204428
1 I ZF2	3.646140658	3.646140658	3.646140658	3.646140658	3.646140658	3.646140658	3.646140658	3.646140658	3.646140658

(2) P 4+0	TH = 41	SBE = 1.524253087						CAUER FUNCTIONS	
	RC = 02	RC = 05	RC = 07	RC = 10	RC = 15	RC = 20	RC = 25	RC = 35	RC = 50
AMAX DB	.0017375255	.0108709564	.0213327376	.0436480540	.0988323393	.1772876696	.2802872360	.5675287486	1.249387366
AMIN DB	12.41547478	20.16893703	23.08139965	26.19079201	29.76193887	32.33711515	34.37736404	37.58633807	41.36579585
MU I K	.2394563D 0	.9807383D-1	.7013423D-1	.4902983D-1	.3250147D-1	.2416263D-1	.1910433D-1	.1320332D-1	.8544963D-2
1 I PF1	1.966612318	1.399547487	1.235046411	1.081204936	.9279687078	.8311802978	.7621132623	.6669175207	.5766715461
1 I PQ1	.6995483207	.6498341552	.6466147000	.6502242958	.6642424100	.6827705607	.7043272229	.7561272224	.8625182391
1 I PFC	1.431210578	1.287734804	1.234754943	1.180807507	1.123711251	1.086612750	1.060032754	1.023819422	.9906653860
1 I PQC	3.538741325	3.013841213	2.978126906	3.018152402	3.170802062	3.366540691	3.587013825	4.092115510	5.057153970
1 I ZF1	1.619289364	1.619289364	1.619289364	1.619289364	1.619289364	1.619289364	1.619289364	1.619289364	1.619289364
1 I ZF2	3.551740773	3.551740773	3.551740773	3.551740773	3.551740773	3.551740773	3.551740773	3.551740773	3.551740773

(2) P 4+0	TH = 42	SBE = 1.494476550						CAUER FUNCTIONS	
	RC = 02	RC = 05	RC = 07	RC = 10	RC = 15	RC = 20	RC = 25	RC = 35	RC = 50
AMAX DB	.0017375255	.0108709564	.0213327376	.0436480540	.0988323393	.1772876696	.2802872360	.5675287486	1.249387366
AMIN DB	11.54941105	19.25407974	22.16170726	25.26851005	28.83826868	31.41295809	33.45298145	36.66175902	40.44111238
MU I K	.2645632D 0	.1089673D 0	.7796768D-1	.5452234D-1	.3614819D-1	.2687522D-1	.2124961D-1	.1468629D-1	.9504831D-2
1 I PF1	1.990429424	1.414754515	1.247424550	1.091115512	.9356749724	.8376450025	.7677664798	.6715534078	.5804474952
1 I PQ1	.7136808458	.6558238031	.6513058854	.6539892009	.6673196346	.6855261512	.7069096734	.7585678807	.8650029953
1 I PFC	1.419056347	1.281967319	1.230593486	1.178001875	1.122089091	1.085641658	1.059477552	1.023768936	.9910194885
1 I PQC	3.723020826	3.114866124	3.064711096	3.094555826	3.240476778	3.434147251	3.654592993	4.162835282	5.137944101
1 I ZF1	1.586199712	1.586199712	1.586199712	1.586199712	1.586199712	1.586199712	1.586199712	1.586199712	1.586199712
1 I ZF2	3.461628002	3.461628002	3.461628002	3.461628002	3.461628002	3.461628002	3.461628002	3.461628002	3.461628002

(2) P 4+0	TH = 43 SBE = 1.466279186								CAUER FUNCTIONS
	RC = 02	RC = 05	RC = 07	RC = 10	RC = 15	RC = 20	RC = 25	RC = 35	RC = 50
AMAX DB	.0017375255	.0108709564	.0213327376	.0436480540	.0988323393	.1772876696	.2802872360	.5675287486	1.249387366
AMIN DB	10.71147871	18.35800117	21.25978825	24.36345712	27.93153400	30.50563331	32.54538328	35.75392265	39.53314943
MU K	.2913574D 0	.1208092D 0	.8649890D-1	.6051000D-1	.4012576D-1	.2983447D-1	.2359016D-1	.1630436D-1	.1055219D-1
1 PF1	2.014015712	1.430418190	1.260222511	1.101381922	.9436647251	.8443485979	.7736285339	.6763598873	.5843615463
1 PQ1	.7294124506	.6623127148	.6563472772	.6580061419	.6705811824	.6884354411	.7096291745	.7611296840	.8676038864
1 PFC	1.406723574	1.276068152	1.226327672	1.175119353	1.120417508	1.084637765	1.058900781	1.023711209	.9913784359
1 PQC	3.926319962	3.224053918	3.157773351	3.176282964	3.314686148	3.505977823	3.726279090	4.237700593	5.223326863
1 ZF1	1.554810727	1.554810727	1.554810727	1.554810727	1.554810727	1.554810727	1.554810727	1.554810727	1.554810727
1 ZF2	3.375496180	3.375496180	3.375496180	3.375496180	3.375496180	3.375496180	3.375496180	3.375496180	3.375496180

(2) P 4+0	TH = 44 SBE = 1.439556540								CAUER FUNCTIONS
	RC = 02	RC = 05	RC = 07	RC = 10	RC = 15	RC = 20	RC = 25	RC = 35	RC = 50
AMAX DB	.0017375255	.0108709564	.0213327376	.0436480540	.0988323393	.1772876696	.2802872360	.5675287486	1.249387366
AMIN DB	9.902428000	17.48009465	20.37485327	23.47474215	27.04078808	29.61417436	31.65359391	34.86184537	38.64091912
MU K	.3198001D 0	.1336581D 0	.9577614D-1	.6702902D-1	.4445909D-1	.3305912D-1	.2614089D-1	.1806790D-1	.1169376D-1
1 PF1	2.037148071	1.446523323	1.273440889	1.112010523	.9519454173	.8512980420	.7797057654	.6813422977	.5884180622
1 PQ1	.7469529705	.6693541647	.6617726069	.6622968304	.6740408970	.6915091386	.7124946609	.7638198457	.8703272341
1 PFC	1.394225719	1.270038587	1.221957623	1.172159466	1.118695800	1.083600376	1.058301803	1.023645737	.9917418939
1 PQC	4.151071289	3.342261413	3.257941615	3.263810721	3.393805221	3.582364856	3.802384415	4.317013764	5.313622890
1 ZF1	1.525009115	1.525009115	1.525009115	1.525009115	1.525009115	1.525009115	1.525009115	1.525009115	1.525009115
1 ZF2	3.293066563	3.293066563	3.293066563	3.293066563	3.293066563	3.293066563	3.293066563	3.293066563	3.293066563

(2) P 4+0	TH = 45 SBE = 1.414213562								CAUER FUNCTIONS
	RC = 02	RC = 05	RC = 07	RC = 10	RC = 15	RC = 20	RC = 25	RC = 35	RC = 50
AMAX DB	.0017375255	.0108709564	.0213327376	.0436480540	.0988323393	.1772876696	.2802872360	.5675287486	1.249387366
AMIN DB	9.123204697	16.61987114	19.50620277	22.60154707	26.16514767	28.73767483	30.77669616	33.98460057	37.76348984
MU K	.3498161D 0	.1475728D 0	.1058498D 0	.7411782D-1	.4917480D-1	.3656927D-1	.2891780D-1	.1998803D-1	.1293676D-1
1 PF1	2.059575027	1.463048820	1.287077948	1.123007031	.9605245656	.8585005250	.7860047926	.6865062573	.5926216550
1 PQ1	.7665398213	.6770084617	.6676198102	.6663855019	.6777143235	.6947590395	.7155159610	.7666462462	.8731799150
1 PFC	1.381577955	1.263880079	1.217483519	1.169121751	1.116923251	1.082528770	1.057679956	1.023571997	.9921095101
1 PQC	4.400057594	3.470461303	3.365924272	3.357673620	3.478251653	3.663677142	3.883254516	4.401107110	5.409184830
1 ZF1	1.496691757	1.496691757	1.496691757	1.496691757	1.496691757	1.496691757	1.496691757	1.496691757	1.496691757
1 ZF2	3.214084753	3.214084753	3.214084753	3.214084753	3.214084753	3.214084753	3.214084753	3.214084753	3.214084753

(2) P 4+0	TH = 46	SBE = 1.390163591						CAUER FUNCTIONS	
	RC = 02	RC = 05	RC = 07	RC = 10	RC = 15	RC = 20	RC = 25	RC = 35	RC = 50
AMAX DB	.0017375255	.0108709564	.0213327376	.0436480540	.0988323393	.1772876696	.2802872360	.5675287486	1.249387366
AMIN DB	8.374918986	15.77696156	18.65322630	21.74312382	25.30378842	27.87528322	29.91382587	33.12131304	36.89998048
MU K	.3812888D 0	.1626117D 0	.1167720D 0	.8181705D-1	.5430134D-1	.4038646D-1	.3193807D-1	.2207671D-1	.1428897D-1
1 PF1	2.081018939	1.479966612	1.301129102	1.134376310	.9694096944	.8659634557	.7925325132	.6918576782	.5969772028
1 PQ1	.7884406619	.6853439745	.6739316320	.6717992649	.6816186536	.6981981629	.7187038772	.7696175084	.8761694218
1 PFC	1.368797390	1.257594284	1.212905619	1.166005759	1.115099131	1.081422208	1.057034553	1.023489438	.9924809124
1 PQC	4.676461621	3.609760496	3.482522491	3.458472271	3.568491661	3.750324720	3.969272562	4.490347882	5.510401297
1 ZF1	1.469764615	1.469764615	1.469764615	1.469764615	1.469764615	1.469764615	1.469764615	1.469764615	1.469764615
1 ZF2	3.138318020	3.138318020	3.138318020	3.138318020	3.138318020	3.138318020	3.138318020	3.138318020	3.138318020

(2) P 4+0	TH = 47	SBE = 1.367327461						CAUER FUNCTIONS	
	RC = 02	RC = 05	RC = 07	RC = 10	RC = 15	RC = 20	RC = 25	RC = 35	RC = 50
AMAX DB	.0017375255	.0108709564	.0213327376	.0436480540	.0988323393	.1772876696	.2802872360	.5675287486	1.249387366
AMIN DB	7.658802712	14.95112013	17.81540302	20.89879222	24.45594120	27.02619857	29.06416725	32.27115400	36.04955535
MU K	.4140567D 0	.1788315D 0	.1285967D 0	.9016965D-1	.5986913D-1	.4453383D-1	.3522019D-1	.2434682D-1	.1575877D-1
1 PF1	2.101180275	1.497240445	1.315586292	1.146122117	.9786082630	.8736944421	.7992961058	.6974027796	.6014898667
1 PQ1	.8129559866	.6944383106	.6807563262	.6770685047	.6857730012	.7018409074	.7220703115	.7727430858	.8793039318
1 PFC	1.355903266	1.251183107	1.208224269	1.162811061	1.113222695	1.080279926	1.056364880	1.023397488	.9928557080
1 PQC	4.983922103	3.761421695	3.608644772	3.566883311	3.665046958	3.842764572	4.060864420	4.585142829	5.617701434
1 ZF1	1.444141765	1.444141765	1.444141765	1.444141765	1.444141765	1.444141765	1.444141765	1.444141765	1.444141765
1 ZF2	3.065552963	3.065552963	3.065552963	3.065552963	3.065552963	3.065552963	3.065552963	3.065552963	3.065552963

(2) P 4+0	TH = 48	SBE = 1.345632730						CAUER FUNCTIONS	
	RC = 02	RC = 05	RC = 07	RC = 10	RC = 15	RC = 20	RC = 25	RC = 35	RC = 50
AMAX DB	.0017375255	.0108709564	.0213327376	.0436480540	.0988323393	.1772876696	.2802872360	.5675287486	1.249387366
AMIN DB	6.976154282	14.14222856	16.99230333	20.06793891	23.62088896	26.18966664	28.22694864	31.43333655	35.21141941
MU K	.4479116D 0	.1962857D 0	.1413790D 0	.9922088D-1	.6591064D-1	.4903618D-1	.3878400D-1	.2681224D-1	.1735518D-1
1 PF1	2.119744224	1.514824567	1.330437285	1.158246803	.9881275753	.8817012641	.8063030277	.7031481029	.6061651098
1 PQ1	.8404215657	.7043796671	.6881484667	.6827273517	.6901986639	.7057032300	.7256284042	.7760333620	.8825923862
1 PFC	1.342917113	1.244648751	1.203439925	1.159537253	1.111293190	1.079101137	1.055670199	1.023295545	.9932334819
1 PQC	5.326596920	3.926888854	3.745324119	3.683671122	3.768502898	3.941507279	4.158504553	4.685943480	5.731560180
1 ZF1	1.419744560	1.419744560	1.419744560	1.419744560	1.419744560	1.419744560	1.419744560	1.419744560	1.419744560
1 ZF2	2.995593463	2.995593463	2.995593463	2.995593463	2.995593463	2.995593463	2.995593463	2.995593463	2.995593463

(2) P 4+0	TH = 49	SBE = 1.325012993						CAUER FUNCTIONS	
	RC = 02	RC = 05	RC = 07	RC = 10	RC = 15	RC = 20	RC = 25	RC = 35	RC = 50
AMAX DB	.0017375255	.0108709564	.0213327376	.0436480540	.0988323393	.1772876696	.2802872360	.5675287486	1.249387366
AMIN DB	6.328271738	13.35030077	16.18359149	19.25001702	22.79796437	25.36497657	27.40143884	30.60711163	34.38481409
MU I K	.4825990D 0	.2150230D 0	.1551745D 0	.1090182D 0	.7246058D-1	.5392016D-1	.4265089D-1	.2948794D-1	.1908795D-1
1 I PF1	2.136389738	1.532662314	1.345664854	1.170750953	.9979746685	.8899918386	.8135610082	.7091005249	.6110087173
1 I PQ1	.8712106663	.7152683756	.6961698878	.6888142273	.6949194247	.7098028546	.7293926938	.7794997656	.8860445798
1 I PFC	1.329862833	1.237993780	1.198553180	1.156183967	1.109309848	1.077885030	1.054949744	1.023182982	.9936137952
1 I PQC	5.709234532	4.107817245	3.893738384	3.809701661	3.879518036	4.047124803	4.262722917	4.793252290	5.852504374
1 I ZF1	1.396500895	1.396500895	1.396500895	1.396500895	1.396500895	1.396500895	1.396500895	1.396500895	1.396500895
1 I ZF2	2.928258878	2.928258878	2.928258878	2.928258878	2.928258878	2.928258878	2.928258878	2.928258878	2.928258878

(2) P 4+0	TH = 50	SBE = 1.305407289						CAUER FUNCTIONS	
	RC = 02	RC = 05	RC = 07	RC = 10	RC = 15	RC = 20	RC = 25	RC = 35	RC = 50
AMAX DB	.0017375255	.0108709564	.0213327376	.0436480540	.0988323393	.1772876696	.2802872360	.5675287486	1.249387366
AMIN DB	5.716376209	12.57548766	15.38902935	18.44454690	21.98654806	24.55145824	26.58694386	29.79176433	33.56901326
MU I K	.5178228D 0	.2350854D 0	.1700390D 0	.1196114D 0	.7955594D-1	.5921437D-1	.4684387D-1	.3239006D-1	.2096763D-1
1 I PF1	2.150800785	1.550684635	1.361245852	1.183632962	1.008156176	.8985741721	.8210780391	.7152672730	.6160268187
1 I PQ1	.9057360154	.7272186633	.7048907729	.6953724779	.6999619054	.7141595121	.7333793013	.7831549020	.8896712641
1 I PFC	1.316766686	1.231221194	1.193564792	1.152750379	1.107271894	1.076630772	1.054202719	1.023059141	.9939961840
1 I PQC	6.137255294	4.306109041	4.055234420	3.945958875	3.999835417	4.160259653	4.374113038	4.907629815	5.981119854
1 I ZF1	1.374344551	1.374344551	1.374344551	1.374344551	1.374344551	1.374344551	1.374344551	1.374344551	1.374344551
1 I ZF2	2.863382452	2.863382452	2.863382452	2.863382452	2.863382452	2.863382452	2.863382452	2.863382452	2.863382452

(2) P 4+0	TH = 51	SBE = 1.286759566						CAUER FUNCTIONS	
	RC = 02	RC = 05	RC = 07	RC = 10	RC = 15	RC = 20	RC = 25	RC = 35	RC = 50
AMAX DB		.0108709564	.0213327376	.0436480540	.0988323393	.1772876696	.2802872360	.5675287486	1.249387366
AMIN DB		11.81808153	14.60848086	17.65111771	21.18606748	23.74847989	25.78280410	28.98661052	32.76331970
MU I K		.2565051D 0	.1860270D 0	.1310521D 0	.8723618D-1	.6494953D-1	.5138777D-1	.3553608D-1	.2300562D-1
1 I PF1		1.568808605	1.377150175	1.196883531	1.018678158	.9074563012	.8288623570	.7216559377	.6212259109
1 I PQ1		.7403606535	.7143909175	.7024511163	.7053559761	.7187952208	.7376061456	.7870127057	.8934842661
1 I PFC		1.224334513	1.188475722	1.149237721	1.105178547	1.075337505	1.053428299	1.022923334	.9943801572
1 I PQC		4.523955506	4.231356853	4.093564218	4.127295953	4.281635682	4.493341516	5.029703119	6.118059751
1 I ZF1		1.353214630	1.353214630	1.353214630	1.353214630	1.353214630	1.353214630	1.353214630	1.353214630
1 I ZF2		2.800809912	2.800809912	2.800809912	2.800809912	2.800809912	2.800809912	2.800809912	2.800809912

(2) P 4+0	RC = 02	RC = 05	RC = 07	RC = 10	RC = 15	RC = 20	RC = 25	RC = 35	RC = 50
TH = 52	SBE = 1.269018215							CAUER FUNCTIONS	
AMAX DB		.0108709564	.0213327376	.0436480540	.0988323393	.1772876696	.2802872360	.5675287486	1.249387366
AMIN DB		11.07851932	13.84191744	16.86938990	20.39599649	22.95544643	24.98839189	28.19099385	31.96706166
MU K		.2793020D 0	.2031908D 0	.1433938D 0	.9554329D-1	.7115865D-1	.5630934D-1	.3894487D-1	.2521430D-1
1 PF1		1.586936006	1.393339606	1.210510089	1.029545903	.9166462158	.8369224196	.7282744864	.6266128839
1 PQ1		.7548426286	.7247611914	.7101056871	.7111352361	.7237346132	.7420931917	.7910886175	.8974966250
1 PFC		1.217337879	1.183287180	1.145644302	1.103029025	1.074004349	1.052625631	1.022774840	.9947651945
1 PQC		4.763887092	4.423882443	4.253799976	4.265854323	4.412070896	4.621159254	5.160175663	6.264054199
1 ZF1		1.333055049	1.333055049	1.333055049	1.333055049	1.333055049	1.333055049	1.333055049	1.333055049
1 ZF2		2.740398209	2.740398209	2.740398209	2.740398209	2.740398209	2.740398209	2.740398209	2.740398209

(2) P 4+0	RC = 02	RC = 05	RC = 07	RC = 10	RC = 15	RC = 20	RC = 25	RC = 35	RC = 50
TH = 53	SBE = 1.252135658							CAUER FUNCTIONS	
AMAX DB		.0108709564	.0213327376	.0436480540	.0988323393	.1772876696	.2802872360	.5675287486	1.249387366
AMIN DB		10.35738393	13.08942378	16.09909868	19.61585554	22.17179813	24.20310947	27.40428293	31.17958973
MU K		.3034805D 0	.2215791D 0	.1566914D 0	.1045219D 0	.7787716D-1	.6163743D-1	.4263692D-1	.2760708D-1
1 PF1		1.604952083	1.409766563	1.224486106	1.040763673	.9261517629	.8452668714	.7351312738	.6321950483
1 PQ1		.7708335777	.7361052312	.7183992809	.7173375768	.7290053180	.7468627429	.7953097891	.9017227507
1 PFC		1.210236164	1.178000676	1.141970524	1.100822548	1.072630402	1.051793827	1.022612903	.9951507449
1 PQC		5.028833068	4.634861263	4.428137211	4.415597971	4.552492681	4.758414781	5.299839984	6.419921770
1 ZF1		1.313814087	1.313814087	1.313814087	1.313814087	1.313814087	1.313814087	1.313814087	1.313814087
1 ZF2		2.682014413	2.682014413	2.682014413	2.682014413	2.682014413	2.682014413	2.682014413	2.682014413

(2) P 4+0	RC = 02	RC = 05	RC = 07	RC = 10	RC = 15	RC = 20	RC = 25	RC = 35	RC = 50
TH = 54	SBE = 1.236067977							CAUER FUNCTIONS	
AMAX DB		.0108709564	.0213327376	.0436480540	.0988323393	.1772876696	.2802872360	.5675287486	1.249387366
AMIN DB		9.655402556	12.35120391	15.34005842	18.84521263	21.39700982	23.42638725	26.62536886	30.40027386
MU K		.3290257D 0	.2412347D 0	.1710004D 0	.1142193D 0	.8514311D-1	.6740322D-1	.4663442D-1	.3019857D-1
1 PF1		1.622724600	1.426372762	1.238800310	1.052334411	.9359805254	.8539044984	.7422350524	.6379801649
1 PQ1		.7885260515	.7485413981	.7274037225	.7240058443	.7346384095	.7519397814	.7999653236	.9061785079
1 PFC		1.203035093	1.172618090	1.138216407	1.098558344	1.071214739	1.050931971	1.022436730	.9955362238
1 PQC		5.322192636	4.866666176	4.618269408	4.577769889	4.703956011	4.906070115	5.449586554	6.586582976
1 ZF1		1.295443995	1.295443995	1.295443995	1.295443995	1.295443995	1.295443995	1.295443995	1.295443995
1 ZF2		2.625534707	2.625534707	2.625534707	2.625534707	2.625534707	2.625534707	2.625534707	2.625534707

$P^{(3)}_{4+2}$

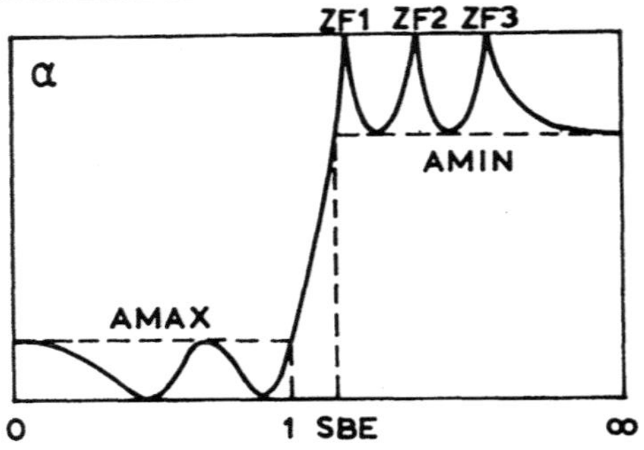

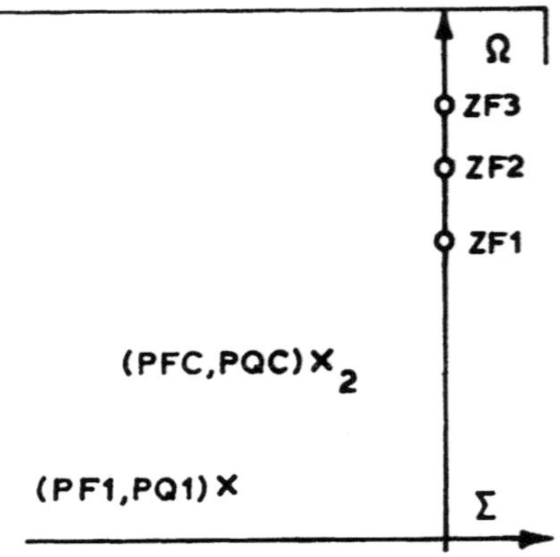

(3) P 4+2			TH = 10	SBE = 5.758770483						MCPER FUNCTIONS
			RC = 02	RC = 05	RC = 07	RC = 10	RC = 15	RC = 20	RC = 25	RC = 35	RC = 50
AMAX DB			.0017375255	.0108709564	.0213327376	.0436480540	.0988323393	.1772876696	.2802872360	.5675287486	1.249387366
AMIN DB			92.44867799	104.4097499	108.8146345	113.5016609	118.8736881	122.7404207	125.7971537	130.5818419	136.1434491
MU		K	.2385427D-4	.6018836D-5	.3624668D-5	.2113085D-5	.1138454D-5	.7294222D-6	.5130295D-6	.2957385D-6	.1558933D-6
	1	PF1	1.381224019	1.038444128	.9294228089	.8223850588	.7104957973	.6369931050	.5831636820	.5069990185	.4323666445
	1	PQ1	.5502809578	.5630327038	.5709207464	.5824928176	.6019414639	.6223124615	.6441460419	.6940841886	.7937380135
	2	PFC	1.924585136	1.579305914	1.472469133	1.369897462	1.265748865	1.199300298	1.151680386	1.085710346	1.022101708
	2	PQC	1.343682020	1.419624022	1.465741863	1.529801919	1.628398371	1.721733259	1.813041380	1.997472538	2.302204472
	1	ZF1	5.953935551	5.955741388	5.956234310	5.956680017	5.957107048	5.957367690	5.957549865	5.957797926	5.958036286
	1	ZF2	8.061400721	8.080585521	8.085762099	8.090421826	8.094868097	8.097573374	8.099460452	8.102025142	8.104484360
	1	ZF3	21.82213554	21.92345357	21.95055661	21.97486513	21.99798082	22.01200699	22.02177376	22.03502487	22.04770664

(3) P 4+2			TH = 11	SBE = 5.240843064						MCPER FUNCTIONS
			RC = 02	RC = 05	RC = 07	RC = 10	RC = 15	RC = 20	RC = 25	RC = 35	RC = 50
AMAX DB			.0017375255	.0108709564	.0213327376	.0436480540	.0988323393	.1772876696	.2802872360	.5675287486	1.249387366
AMIN DB			87.45399910	99.41475971	103.8195851	108.5065670	113.8785595	117.7452747	120.8019973	125.5866735	131.1482715
MU		K	.4239358D-4	.1069700D-4	.6442000D-5	.3755534D-5	.2023355D-5	.1296392D-5	.9118011D-6	.5256133D-6	.2770680D-6
	1	PF1	1.386245656	1.041170678	.9316195253	.8241405486	.7118642373	.6381438263	.5841713977	.5078257783	.4330377936
	1	PQ1	.5508634850	.5634426249	.5712881795	.5828246204	.6022431787	.6225992422	.6444251037	.6943593422	.7940249981
	2	PFC	1.919407086	1.577091558	1.470905096	1.368855444	1.265142972	1.198927058	1.151453462	1.085659569	1.022193303
	2	PQC	1.358178944	1.429928868	1.475028857	1.538247335	1.636151086	1.729156061	1.820303231	2.004670944	2.309664730
	1	ZF1	5.416979739	5.418941124	5.419478435	5.419964960	5.420431688	5.420716838	5.420916266	5.421187981	5.421449238
	1	ZF2	7.320758706	7.341792756	7.347472855	7.352587570	7.357469562	7.360440718	7.362513609	7.365331302	7.368033628
	1	ZF3	19.77732824	19.88915847	19.91905419	19.94585906	19.97134065	19.98679838	19.99756012	20.01215866	20.02612727

(3) P 4+2 — TH = 12 SBE = 4.809734345 — MCPER FUNCTIONS

MU	K	RC = 02	RC = 05	RC = 07	RC = 10	RC = 15	RC = 20	RC = 25	RC = 35	RC = 50
	AMAX DB	.0017375255	.0108709564	.0213327376	.0436480540	.0988323393	.1772876696	.2802872360	.5675287436	1.249387366
	AMIN DB	82.88924144	94.84959159	99.25433882	103.9412621	109.3132090	113.1799013	116.2366101	121.0212705	126.5828564
MU	K	.7170310D-4	.1809341D-4	.1089640D-4	.6352386D-5	.3422469D-5	.2192830D-5	.1542302D-5	.8890711D-6	.4686592D-6
1	PF1	1.391785676	1.044172917	.9340371230	.8260716814	.7133689629	.6394088388	.5852790181	.5087343108	.4337751902
1	PQ1	.5515120780	.5638963761	.5716942830	.5831908564	.6025757981	.6229151742	.6447323875	.6946621415	.7943406562
2	PFC	1.913720763	1.574658503	1.469186199	1.367709380	1.264476454	1.198516108	1.151203246	1.085602867	1.022293017
2	PQC	1.374370289	1.441372069	1.485326046	1.547598720	1.644724569	1.737358507	1.828323634	2.012615920	2.317893623
1	ZF1	4.969917004	4.972026635	4.972606913	4.973133166	4.973638722	4.973947930	4.974164332	4.974459367	4.974743252
1	ZF2	6.702920659	6.725785897	6.731965693	6.737532385	6.742847629	6.746083360	6.748341251	6.751410964	6.754355596
1	ZF3	18.06774210	18.19016846	18.22287710	18.25219499	18.28005669	18.29695372	18.30871539	18.32466749	18.33992812

(3) P 4+2 — TH = 13 SBE = 4.445411483 — MCPER FUNCTIONS

MU	K	RC = 02	RC = 05	RC = 07	RC = 10	RC = 15	RC = 20	RC = 25	RC = 35	RC = 50
	AMAX DB	.0017375255	.0108709564	.0213327376	.0436480540	.0988323393	.1772876696	.2802872360	.5675287486	1.249387366
	AMIN DB	78.68509928	90.64492007	95.04956658	99.73641429	105.1083023	108.9749652	112.0316563	116.8162962	122.3778665
MU	K	.1163443D-3	.2935986D-4	.1768160D-4	.1030812D-4	.5553732D-5	.3558375D-5	.2502748D-5	.1442730D-5	.7605131D-6
1	PF1	1.397855469	1.047455360	.9366788875	.8281808326	.7150116454	.6407894591	.5864876441	.5097254642	.4345794851
1	PQ1	.5522299403	.5643953616	.5721401258	.5835923431	.6029399441	.6232607825	.6450683617	.6949929951	.7946853673
2	PFC	1.907522808	1.572004743	1.467310887	1.366459619	1.263748519	1.198066857	1.150929278	1.085539940	1.022400682
2	PQC	1.392350859	1.453999813	1.496670172	1.557885595	1.654142616	1.746361516	1.837121968	2.021324923	2.326907735
1	ZF1	4.592008986	4.594258834	4.594880514	4.595445302	4.595988731	4.596321504	4.596554575	4.596872564	4.597178780
1	ZF2	6.179557042	6.204234523	6.210909940	6.216925381	6.222671215	6.226170114	6.228612119	6.231932785	6.235118848
1	ZF3	16.61597071	16.74907359	16.78461574	16.81646389	16.84672064	16.86506519	16.87783206	16.89514427	16.91170249

(3) P 4+2 — TH = 14 SBE = 4.133565494 — MCPER FUNCTIONS

MU	K	RC = 02	RC = 05	RC = 07	RC = 10	RC = 15	RC = 20	RC = 25	RC = 35	RC = 50
	AMAX DB	.0017375255	.0108709564	.0213327376	.0436480540	.0988323393	.1772876696	.2802872360	.5675287486	1.249387366
	AMIN DB	74.78770514	86.74685587	91.15137493	95.83812707	101.2099407	105.0765663	108.1332349	112.9178490	118.4793995
MU	K	.1822278D-3	.4598934D-4	.2769691D-4	.1614707D-4	.8699674D-5	.5574061D-5	.3920471D-5	.2259995D-5	.1191324D-5
1	PF1	1.404467617	1.051022986	.9395484357	.8304706157	.7167941219	.6422871338	.5877984849	.5108001702	.4354513924
1	PQ1	.5530207050	.5649411535	.5726269004	.5840299897	.6033363064	.6236366483	.6454335439	.6953523542	.7950595501
2	PFC	1.900809901	1.569128150	1.465277497	1.365103415	1.262958305	1.197578661	1.150631054	1.035470458	1.022516115
2	PQC	1.412228086	1.467863879	1.509102307	1.569140874	1.664431694	1.756188324	1.846719421	2.030817296	2.336725422
1	ZF1	4.268443121	4.270824412	4.271485790	4.272087811	4.272668074	4.273023874	4.273273281	4.273613827	4.273942050
1	ZF2	5.730432840	5.756902913	5.764069649	5.770530409	5.776703987	5.780464542	5.783089710	5.786660176	5.790086723
1	ZF3	15.36672664	15.51058014	15.54897634	15.58337233	15.61603959	15.63584025	15.64961791	15.66829718	15.68615896

(3) P 4+2		TH = 15	SBE = 3.863703305							MCPER FUNCTIONS
		RC = 02	RC = 05	RC = 07	RC = 10	RC = 15	RC = 20	RC = 25	RC = 35	RC = 50
AMAX DB		.0017375255	.0108709564	.0213327376	.0436480540	.0988323393	.1772876696	.2802872360	.5675287486	1.249387366
AMIN DB		71.15436164	83.11267751	87.51703791	92.20367107	97.57539209	101.4419712	104.4986119	109.2831938	114.8447199
MU	K	.2768738D-3	.6988213D-4	.4208701D-4	.2453672D-4	.1321997D-4	.8470352D-5	.5957573D-5	.3434316D-5	.1810356D-5
1	PF1	1.411635935	1.054881248	.9426497282	.8329438899	.7187184011	.6439034438	.5892128615	.5119594461	.4363916917
1	PQ1	.5538884814	.5655355044	.5731559313	.5845048030	.6037656463	.6240434121	.6458285042	.6957407150	.7954636638
2	PFC	1.893578846	1.566026487	1.463084259	1.363639936	1.262104887	1.197050825	1.150308031	1.085394065	1.022639114
2	PQC	1.434123354	1.483022091	1.522668172	1.581401097	1.675621112	1.766864621	1.857139610	2.041114373	2.347366906
1	ZF1	3.988357389	3.990860591	3.991559817	3.992197661	3.992813631	3.993191877	3.993457258	3.993819931	3.994169811
1	ZF2	5.340709297	5.368951796	5.376605346	5.383507808	5.390106112	5.394126710	5.396934024	5.400753054	5.404419067
1	ZF3	14.27946999	14.43413891	14.47540922	14.51237068	14.54746430	14.56873000	14.58352428	14.60357793	14.62274959

(3) P 4+2		TH = 16	SBE = 3.627955279							MCPER FUNCTIONS
		RC = 02	RC = 05	RC = 07	RC = 10	RC = 15	RC = 20	RC = 25	RC = 35	RC = 50
AMAX DB		.0017375255	.0108709564	.0213327376	.0436480540	.0988323393	.1772876696	.2802872360	.5675287486	1.249387366
AMIN DB		67.75065376	79.70794348	84.11210905	88.79859613	94.17020346	98.03672565	101.0933320	105.8778745	111.4393704
MU	K	.4097013D-3	.1034196D-3	.6228659D-4	.3631367D-4	.1956545D-4	.1253614D-4	.8817255D-5	.5082838D-5	.2679363D-5
1	PF1	1.419375533	1.059036097	.9459870825	.8356037700	.7207866693	.6456401088	.5907322104	.5132043974	.4374012299
1	PQ1	.5548379093	.5661803630	.5737286854	.5850178937	.6042288011	.6244817767	.6462538672	.6961586200	.7958982103
2	PFC	1.885826661	1.562697430	1.460729309	1.362067766	1.261187271	1.196482600	1.149959620	1.085310374	1.022769462
2	PQC	1.458173549	1.499538838	1.537418511	1.594706699	1.687743207	1.778418706	1.868408465	2.052239596	2.358854372
1	ZF1	3.743605779	3.746220584	3.746955662	3.747627805	3.748278268	3.748678332	3.748959301	3.749343635	3.749714796
1	ZF2	4.999257502	5.029251938	5.037387622	5.044728000	5.051747846	5.056026778	5.059015155	5.063081431	5.066985818
1	ZF3	13.32380129	13.48933793	13.53350151	13.57304584	13.61058184	13.63332176	13.64913870	13.67057440	13.69106257

(3) P 4+2		TH = 17	SBE = 3.420303620							MCPER FUNCTIONS
		RC = 02	RC = 05	RC = 07	RC = 10	RC = 15	RC = 20	RC = 25	RC = 35	RC = 50
AMAX DB		.0017375255	.0108709564	.0213327376	.0436480540	.0988323393	.1772876696	.2802872360	.5675287486	1.249387366
AMIN DB		64.54843818	76.50448134	80.90841049	85.59472038	90.96618983	94.83264295	97.88920774	102.6737024	108.2351618
MU	K	.5923496D-3	.1495464D-3	.9006986D-4	.5251266D-4	.2829375D-4	.1812875D-4	.1275086D-4	.7350466D-5	.3874734D-5
1	PF1	1.427702865	1.063494002	.9495651879	.8384536360	.7230012969	.6474989926	.5923580877	.5145362213	.4384809233
1	PQ1	.5558742222	.5668778910	.5743467826	.5855704839	.6047266887	.6249525110	.6467103155	.6966066612	.7963637362
2	PFC	1.877550667	1.559138579	1.458210694	1.360385404	1.260204398	1.195873187	1.149585193	1.085218971	1.022906925
2	PQC	1.484532844	1.517485666	1.553409508	1.609102305	1.700833566	1.790881664	1.880554435	2.064218636	2.371212082
1	ZF1	3.527959473	3.530674781	3.531443565	3.532148372	3.532832023	3.533253231	3.533549370	3.533954869	3.534346905
1	ZF2	4.697567181	4.729292941	4.737905929	4.745680282	4.753118334	4.757653797	4.760822092	4.765134213	4.769275814
1	ZF3	12.47648047	12.65292136	12.69999594	12.74214002	12.78213442	12.80635789	12.82320371	12.84602940	12.86784105

(3) P 4+2		TH = 18 SBE = 3.236067977							MCPER FUNCTIONS
		RC = 02	RC = 05	RC = 07	RC = 10	RC = 15	RC = 20	RC = 25	RC = 35	RC = 50
AMAX DB		.0017375255	.0108709564	.0213327376	.0436480540	.0988323393	.1772876696	.2802872360	.5675287486	1.249387366
AMIN DB		61.52440834	73.47895249	77.88259776	82.56869496	87.93999896	91.80636920	94.86288411	99.64732137	105.2087369
MU	K	.8390340D-3	.2118617D-3	.1276057D-3	.7439868D-4	.4008668D-4	.2568512D-4	.1806574D-4	.1041439D-4	.5489884D-5
1	PF1	1.436635797	1.068261971	.9533891216	.8414971444	.7253648460	.6494821083	.5940921737	.5159562096	.4396317607
1	PQ1	.5570033218	.5676304826	.5750120089	.5861639153	.6052603130	.6254564545	.6471985927	.6970854823	.7968608350
2	PFC	1.868748595	1.555347479	1.455526382	1.358591272	1.259155145	1.195221732	1.149184077	1.085119411	1.023051250
2	PQC	1.513374780	1.536941958	1.570703259	1.624637067	1.714931259	1.804287559	1.893608654	2.077079545	2.384466505
1	ZF1	3.336574300	3.339378204	3.340178394	3.340914114	3.341629558	3.342071190	3.342382053	3.342808185	3.343220663
1	ZF2	4.429019226	4.462455764	4.471541110	4.479745361	4.487598137	4.492388241	4.495735245	4.500291732	4.504669311
1	ZF3	11.71943964	11.90680206	11.95680347	12.00156330	12.04403186	12.06974831	12.08762935	12.11185321	12.13499559

(3) P 4+2		TH = 19 SBE = 3.071553487							MCPER FUNCTIONS
		RC = 02	RC = 05	RC = 07	RC = 10	RC = 15	RC = 20	RC = 25	RC = 35	RC = 50
AMAX DB		.0017375255	.0108709564	.0213327376	.0436480540	.0988323393	.1772876696	.2802872360	.5675287486	1.249387366
AMIN DB		58.65904821	70.61180557	75.01511310	79.70095736	85.07206467	88.93833640	91.99479203	96.77916109	102.3405246
MU	K	.1166937D-2	.2947201D-3	.1775188D-3	.1035028D-3	.5576950D-4	.3573413D-4	.2513393D-4	.1448912D-4	.7637897D-5
1	PF1	1.446193660	1.073347581	.9574643657	.8447382400	.7278800782	.6515916245	.5959362776	.5174657527	.4408548052
1	PQ1	.5582318649	.5684407873	.5757263307	.5867996589	.6058307701	.6259945211	.6477195076	.6975957817	.7973901494
2	PFC	1.859418687	1.551321639	1.452674273	1.356683718	1.258038326	1.194527332	1.148755557	1.085011217	1.023202165
2	PQC	1.544894686	1.557995694	1.589368308	1.641365047	1.730079112	1.818673654	1.907605132	2.090352907	2.398646451
1	ZF1	3.165626371	3.168506136	3.169335277	3.170100042	3.170845791	3.171307079	3.171632192	3.172078387	3.172510846
1	ZF2	4.188387951	4.223514984	4.233067652	4.241697607	4.249961492	4.255004259	4.258528707	4.263328000	4.267940253
1	ZF3	11.03842334	11.23670181	11.28964341	11.33703378	11.38199182	11.40921064	11.42813331	11.45376371	11.47824430

(3) P 4+2		TH = 20 SBE = 2.923804400							MCPER FUNCTIONS
		RC = 02	RC = 05	RC = 07	RC = 10	RC = 15	RC = 20	RC = 25	RC = 35	RC = 50
AMAX DB		.0017375255	.0108709564	.0213327376	.0436480540	.0988323393	.1772876696	.2802872360	.5675287486	1.249387366
AMIN DB		55.93585485	67.88649862	72.28940755	76.97495348	82.34582888	86.21198450	89.26837026	94.05265895	99.61396114
MU	K	.1596641D-2	.4033435D-3	.2429571D-3	.1416617D-3	.7633234D-4	.4891035D-4	.3440183D-4	.1983202D-4	.1045447D-4
1	PF1	1.456397315	1.078758997	.9617968272	.8481811695	.7305499635	.6538298720	.5978923431	.5190663431	.4421511978
1	PQ1	.5595673648	.5693117349	.5764919109	.5874793247	.6064392549	.6265677046	.6482739381	.6981383161	.7979523743
2	PFC	1.849559806	1.547058555	1.449652204	1.354661018	1.256852691	1.193739029	1.148298874	1.084893885	1.023359381
2	PQC	1.579312481	1.580744334	1.609480255	1.659345643	1.746324009	1.834030654	1.922580969	2.105572018	2.413783232
1	ZF1	3.012057012	3.014999061	3.015854540	3.016646363	3.017420833	3.017900961	3.018239817	3.018705474	3.019157423
1	ZF2	3.971492689	4.008290376	4.018305267	4.027356633	4.036027892	4.041321265	4.045021832	4.050062294	4.054907843
1	ZF3	10.42203687	10.63119952	10.68709164	10.73712574	10.78458785	10.81331829	10.83328905	10.86033450	10.88616101

(3) P 4+2	TH = 21	SBE = 2.790428110						MCPER FUNCTIONS	
	RC = 02	RC = 05	RC = 07	RC = 10	RC = 15	RC = 20	RC = 25	RC = 35	RC = 50
AMAX DB	.0017375255	.0108709564	.0213327376	.0436480540	.0988323393	.1772876696	.2802872360	.5675287486	1.249387366
AMIN DB	53.34075122	65.28891122	69.69135312	74.37654972	79.74715371	83.61317348	86.66947751	91.45367221	97.01490267
MU I K	.2152596D-2	.5439443D-3	.3276667D-3	.1910612D-3	.1029538D-3	.6596922D-4	.4640087D-4	.2674954D-4	.1410116D-4
1 I PF1	1.467269210	1.084505010	.9663928576	.8518304952	.7333776894	.6561993510	.5999624541	.5207595802	.4435221603
1 I PQ1	.5610183113	.5702465648	.5773111276	.5882046744	.6070870680	.6271770844	.6488628356	.6987139038	.7985482598
2 I PFC	1.839171550	1.542555732	1.446457970	1.352521385	1.255596930	1.193005813	1.147813223	1.084766875	1.023522588
2 I PQC	1.616875955	1.605295811	1.631122443	1.678644070	1.763717229	1.850552981	1.938576587	2.121273086	2.429910829
1 I ZF1	2.873390585	2.876380479	2.877259521	2.878076290	2.878877803	2.879375905	2.879727970	2.880212448	2.880683367
1 I ZF2	3.774948927	3.813398038	3.823870024	3.833338427	3.842413216	3.847955059	3.851830365	3.857110281	3.862187679
1 I ZF3	9.861066388	10.08105131	10.13990067	10.19258970	10.24256950	10.27282061	10.29384584	10.32231493	10.34949526

(3) P 4+2	TH = 22	SBE = 2.669467163						MCPER FUNCTIONS	
	RC = 02	RC = 05	RC = 07	RC = 10	RC = 15	RC = 20	RC = 25	RC = 35	RC = 50
AMAX DB	.0017375255	.0108709564	.0213327376	.0436480540	.0988323393	.1772876696	.2802872360	.5675287486	1.249387366
AMIN DB	50.86163602	62.80689358	67.20879172	71.89358196	77.26387040	81.12973229	84.18594134	88.97002682	94.53117396
MU I K	.2863639D-2	.7238612D-3	.4360742D-3	.2542851D-3	.1370271D-3	.8780381D-4	.6175938D-4	.3560399D-4	.1876901D-4
1 I PF1	1.478833432	1.090595063	.9712592757	.8556911109	.7363666721	.6587027387	.6021488411	.5225471746	.4449689985
1 I PQ1	.5625943116	.5712488588	.5781865943	.5889776339	.6077756254	.6278238315	.6494872308	.6993234286	.7991786144
2 I PFC	1.828254369	1.537810707	1.443089326	1.350262971	1.254269670	1.192176624	1.147297758	1.084629617	1.023691456
2 I PQC	1.657864582	1.631769670	1.654386737	1.699331903	1.782314830	1.868139079	1.955635995	2.137995443	2.447066087
1 I ZF1	2.747601986	2.750624405	2.751524070	2.752363553	2.753190331	2.753705491	2.754070198	2.754572821	2.755062160
1 I ZF2	3.595987332	3.636069398	3.646993347	3.656874353	3.666348729	3.672136834	3.676185445	3.681703025	3.687010752
1 I ZF3	9.347984582	9.578696168	9.640505247	9.695858009	9.748367887	9.780148332	9.802234326	9.832135656	9.860677845

(3) P 4+2	TH = 23	SBE = 2.559304665						MCPER FUNCTIONS	
	RC = 02	RC = 05	RC = 07	RC = 10	RC = 15	RC = 20	RC = 25	RC = 35	RC = 50
AMAX DB	.0017375255	.0108709564	.0213327376	.0436480540	.0988323393	.1772876696	.2802872360	.5675287486	1.249387366
AMIN DB	48.48803411	60.42991597	64.83118459	69.51550482	74.88542857	78.75110806	81.80720742	86.59116678	92.15221771
MU I K	.3763555D-2	.9517091D-3	.5733777D-3	.3343680D-3	.1801891D-3	.1154635D-3	.8121563D-4	.4682111D-4	.2468250D-4
1 I PF1	1.491115744	1.097039282	.9764033917	.8597682589	.7395205668	.6613428981	.6044538881	.5244309534	.4464931067
1 I PQ1	.5643062563	.5723225798	.5791211843	.5898003085	.6085064663	.6285092154	.6501482396	.6999678446	.7998443087
2 I PFC	1.816809679	1.532821080	1.439544008	1.347883872	1.252869484	1.191300346	1.146751585	1.084481509	1.023865636
2 I PQC	1.702593986	1.660298364	1.679374408	1.721487692	1.802178068	1.886891751	1.973807083	2.155781793	2.465288924
1 I ZF1	2.633018711	2.636057439	2.636974619	2.637834456	2.638684623	2.639215874	2.639592625	2.640112678	2.640619857
1 I ZF2	3.432320000	3.474017432	3.485388240	3.495677372	3.505547304	3.511579396	3.515799824	3.521553205	3.527089671
1 I ZF3	8.876585161	9.117890463	9.182656875	9.240679346	9.295730135	9.329048105	9.352200992	9.383543122	9.413455308

(3) P 4+2	TH = 24	SBE = 2.453593336						MCPER FUNCTIONS	
	RC = 02	RC = 05	RC = 07	RC = 10	RC = 15	RC = 20	RC = 25	RC = 35	RC = 50
AMAX DB	.0017375255	.0108709564	.0213327376	.0436480540	.0988323393	.1772876696	.2802872360	.5675287486	1.249387366
AMIN DB	46.21082206	58.14879287	62.54933625	67.23311572	72.60262021	76.46809012	79.52406347	84.30787799	89.86881844
MU I K	.4891690D-2	.1237543D-2	.7456468D-3	.4348547D-3	.2343522D-3	.1501743D-3	.1056323D-3	.6089843D-4	.3210399D-4
1 I PF1	1.504143613	1.103848517	.9818330323	.8640675476	.7428432807	.6641228874	.6068801403	.5264128661	.4480959710
1 I PQ1	.5661665159	.5734721145	.5801180568	.5906749996	.6092812644	.6292346122	.6508470649	.7006481806	.8005462797
2 I PFC	1.804839987	1.527584539	1.435819745	1.345382138	1.251394885	1.190375814	1.146173766	1.094321913	1.024044755
2 I PQC	1.751421164	1.691028741	1.706197134	1.745197650	1.823373879	1.906868547	1.993141948	2.174678483	2.484622569
1 I ZF1	2.528247413	2.531285313	2.532216730	2.533094432	2.533966011	2.534512334	2.534900498	2.535437229	2.535961634
1 I ZF2	3.282040154	3.325336336	3.337148949	3.347841704	3.358103088	3.364376828	3.368767535	3.374754781	3.380518323
1 I ZF3	8.441708755	8.693433929	8.761149768	8.821844647	8.879445334	8.914308404	8.938534073	8.971325437	9.002615821

(3) P 4+2	TH = 25	SBE = 2.366201583						MCPER FUNCTIONS	
	RC = 02	RC = 05	RC = 07	RC = 10	RC = 15	RC = 20	RC = 25	RC = 35	RC = 50
AMAX DB	.0017375255	.0108709564	.0213327376	.0436480540	.0988323393	.1772876696	.2802872360	.5675287486	1.249387366
AMIN DB	44.02201044	55.95546351	60.35517505	65.03833518	70.40735999	74.27259027	77.32841959	82.11206858	87.67288281
MU I K	.6293605D-2	.1593041D-2	.9599337D-3	.5598649D-3	.3017394D-3	.1933617D-3	.1360126D-3	.7841453D-4	.4133861D-4
1 I PF1	1.517946205	1.111034373	.9875565685	.8685949717	.7463389858	.6670459696	.6094303121	.5284949899	.4497791740
1 I PQ1	.5681891741	.5747023236	.5811806879	.5916042237	.6101018393	.6300015129	.6515850134	.7013655457	.8012855346
2 I PFC	1.792349006	1.522098891	1.431914273	1.342755778	1.249844332	1.189401812	1.145563318	1.084150157	1.024228418
2 I PQC	1.804750624	1.724123744	1.734978144	1.770556435	1.845975418	1.928132184	2.013697262	2.194735800	2.505113820
1 I ZF1	2.432118076	2.435137072	2.436079274	2.436972223	2.437863132	2.438423455	2.438822369	2.439374987	2.439915974
1 I ZF2	3.143545942	3.188425269	3.200674708	3.211766579	3.222415248	3.228928234	3.233487636	3.239706737	3.245695621
1 I ZF3	8.039034544	8.300961386	8.371612557	8.434978799	8.495136202	8.531551192	8.556855213	8.591104051	8.623780845

(3) P 4+2	TH = 26	SBE = 2.281172033						MCPER FUNCTIONS	
	RC = 02	RC = 05	RC = 07	RC = 10	RC = 15	RC = 20	RC = 25	RC = 35	RC = 50
AMAX DB	.0017375255	.0108709564	.0213327376	.0436480540	.0988323393	.1772876696	.2802872360	.5675287486	1.249387366
AMIN DB	41.91456988	53.84281561	58.24157662	62.92403035	68.29250871	72.15746620	75.21313161	79.99659230	85.55726296
MU I K	.8021794D-2	.2031698D-2	.1224394D-2	.7141649D-3	.3849236D-3	.2466759D-3	.1735176D-3	.1000392D-3	.5273960D-4
1 I PF1	1.532554342	1.118609248	.9935829447	.8733569328	.7500121335	.6701156240	.6121072961	.5306795367	.4515444001
1 I PQ1	.5703903040	.5760185995	.5823129051	.5925907335	.6109701695	.6308115329	.6523634935	.7021211355	.8020631565
2 I PFC	1.779341773	1.516362092	1.427825354	1.340002767	1.248216232	1.188377070	1.144919212	1.083965534	1.024416207
2 I PQC	1.863041589	1.759764359	1.765853518	1.797668037	1.870062656	1.950751029	2.035534676	2.216008310	2.526813334
1 I ZF1	2.343641079	2.346622132	2.347571491	2.348476934	2.349384987	2.349958185	2.350367153	2.350934825	2.351491717
1 I ZF2	3.015481838	3.061929704	3.074611089	3.086097576	3.097129309	3.103879088	3.108605552	3.115054429	3.121266848
1 I ZF3	7.664919922	7.936782541	8.010348085	8.076380425	8.139098852	8.177071678	8.203459227	8.239173500	8.273244878

(3) P 4+2 — TH = 27 SBE = 2.202689265 — MCPER FUNCTIONS

(3) P 4+2			RC = 02	RC = 05	RC = 07	RC = 10	RC = 15	RC = 20	RC = 25	RC = 35	RC = 50
AMAX DB			.0017375255	.0108709564	.0213327376	.0436480540	.0988323393	.1772876696	.2802872360	.5675287486	1.249387366
AMIN DB			39.88229116	51.80454248	56.20222085	60.98387178	66.25173004	70.11637824	73.17185789	77.95510523	83.51561329
MU		K	.1013644D-1	.2569052D-2	.1548421D-2	.9032468D-3	.4868705D-3	.3120190D-3	.2194861D-3	.1265449D-3	.6671436D-4
1		PF1	1.548000405	1.126586373	.9999217093	.8783602625	.7538674699	.6733355575	.6149141726	.5329688605	.4533934397
1		PQ1	.5727882980	.5774269316	.5835189266	.5936375421	.6118884070	.6316664230	.6531840284	.7029162380	.8028803090
2		PFC	1.765824767	1.510372282	1.423550795	1.337121055	1.246508941	1.187300271	1.144240370	1.083767301	1.024607682
2		PQC	1.926816488	1.798151859	1.798973666	1.826646772	1.895723064	1.974799633	2.058721230	2.238555228	2.549775949
1		ZF1	2.261973798	2.264896884	2.265849590	2.266764641	2.267687547	2.268272438	2.268690730	2.269272585	2.269844672
1		ZF2	2.896693093	2.944695815	2.957804377	2.969681004	2.981091540	2.988075605	2.992967454	2.999643959	3.006078034
1		ZF3	7.316275730	7.597757404	7.674208812	7.742897268	7.808178170	7.847713685	7.875189457	7.912376763	7.947850804

(3) P 4+2 — TH = 28 SBE = 2.130054468 — MCPER FUNCTIONS

(3) P 4+2			RC = 02	RC = 05	RC = 07	RC = 10	RC = 15	RC = 20	RC = 25	RC = 35	RC = 50
AMAX DB			.0017375255	.0108709564	.0213327376	.0436480540	.0988323393	.1772876696	.2802872360	.5675287486	1.249387366
AMIN DB			37.91967247	49.83502622	54.23147483	58.91221631	64.27937332	68.14367211	71.19894202	75.98194854	81.54227313
MU		K	.1270622D-1	.3222914D-2	.1942792D-2	.1133416D-2	.6109861D-3	.3915763D-3	.2754564D-3	.1588190D-3	.8373101D-4
1		PF1	1.564318162	1.134979847	1.006583047	.8836122465	.7579100527	.6767097180	.6178542203	.5353654645	.4553281961
1		PQ1	.5754042604	.5789339833	.5848034069	.5947479503	.6128588936	.6325680809	.6540482656	.7037522415	.8037382428
2		PFC	1.751806021	1.504127822	1.419088462	1.334108574	1.244720768	1.186179046	1.143525671	1.083554678	1.024802374
2		PQC	1.996670973	1.839510386	1.834505029	1.857619429	1.923052380	2.000359336	2.083330115	2.262440840	2.574061042
1		ZF1	2.186394375	2.189238452	2.190190515	2.191112149	2.192047507	2.192642855	2.193069708	2.193664831	2.194251369
1		ZF2	2.786189983	2.835734635	2.849265731	2.861528059	2.873313105	2.880528902	2.885584415	2.892486328	2.899140109
1		ZF3	6.990468174	7.281198403	7.360498928	7.431823297	7.499669899	7.540771729	7.569339849	7.608007331	7.644891962

(3) P 4+2 — TH = 29 SBE = 2.062665340 — MCPER FUNCTIONS

(3) P 4+2			RC = 02	RC = 05	RC = 07	RC = 10	RC = 15	RC = 20	RC = 25	RC = 35	RC = 50
AMAX DB			.0017375255	.0108709564	.0213327376	.0436480540	.0988323393	.1772876696	.2802872360	.5675287486	1.249387366
AMIN DB			36.02182838	47.92924168	52.32429649	57.00401051	62.37037696	66.23428231	69.28931617	74.07205181	79.63217009
MU		K	.1580915D-1	.4013635D-2	.2419832D-2	.1411885D-2	.7611691D-3	.4878495D-3	.3431897D-3	.1978780D-3	.1043257D-3
1		PF1	1.581542493	1.143804676	1.013577813	.8891206506	.7621452697	.6802423078	.6209309273	.5378720100	.4573506906
1		PQ1	.5782624772	.5805471799	.5861714885	.5959255775	.6138841796	.6335185648	.6549579866	.7046306422	.8046383017
2		PFC	1.737295232	1.497627325	1.414436306	1.330963251	1.242849979	1.184984976	1.142773946	1.083326846	1.024999792
2		PQC	2.073285754	1.884089925	1.872632012	1.890721553	1.952155470	2.027518941	2.109440746	2.287734965	2.599732928
1		ZF1	2.116280899	2.119023886	2.119971126	2.120896181	2.121841480	2.122445992	2.122880605	2.123488041	2.124088251
1		ZF2	2.683119493	2.734193643	2.748142759	2.760786394	2.772941632	2.780386564	2.785603979	2.792729015	2.799600475
1		ZF3	6.685240954	6.984792758	7.066896725	7.140846064	7.211242988	7.253913361	7.283577286	7.323731535	7.362034471

$P^{(3)}_{4+2}$			TH = 30	SBE = 2.000000000						MCPER FUNCTIONS
			RC = 02	RC = 05	RC = 07	RC = 10	RC = 15	RC = 20	RC = 25	RC = 35	RC = 50
		AMAX DB	.0017375255	.0108709564	.0213327376	.0436480540	.0988323393	.1772876696	.2802872360	.5675287486	1.249387366
		AMIN DB	34.18441689	46.08267683	50.47615457	55.15471052	60.52018814	64.38365177	67.43842081	72.22085269	77.78073974
MU		K	.1953346D-1	.4964393D-2	.2993590D-2	.1746886D-2	.9418692D-3	.6036948D-3	.4246968D-3	.2448823D-3	.1291109D-3
1		PF1	1.599708962	1.153076805	1.020917565	.8948937480	.7665788577	.6839377993	.6241480040	.5404913253	.4594630699
1		PQ1	.5813909774	.5822748108	.5876288611	.5971743969	.6149670439	.6345201085	.6559151193	.7055530535	.8055819303
2		PFC	1.722303864	1.490869700	1.409592382	1.327683016	1.240894798	1.183743595	1.141983979	1.083082948	1.025199415
2		PQC	2.157440585	1.932169759	1.913559219	1.926108924	1.983147312	2.056375479	2.137139905	2.314513477	2.626861301
1		ZF1	2.051094771	2.053713504	2.054651555	2.055576725	2.056529343	2.057141670	2.057583208	2.058201959	2.058815028
1		ZF2	2.586742705	2.639334028	2.653696780	2.666717382	2.679238478	2.686909905	2.692287419	2.699633222	2.706720263
1		ZF3	6.398652921	6.706540379	6.791392505	6.867934609	6.940877477	6.985117042	7.015879464	7.057526413	7.097255086

$P^{(3)}_{4+2}$			TH = 31	SBE = 1.941604026						MCPER FUNCTIONS
			RC = 02	RC = 05	RC = 07	RC = 10	RC = 15	RC = 20	RC = 25	RC = 35	RC = 50
		AMAX DB	.0017375255	.0108709564	.0213327376	.0436480540	.0988323393	.1772876696	.2802872360	.5675287486	1.249387366
		AMIN DB	32.40358168	44.29126655	48.68296189	53.36021504	58.72469573	62.58766476	65.64213751	70.42422974	75.98385835
MU		K	.2397844D-1	.6101501D-2	.3680035D-2	.2147777D-2	.1158151D-2	.7423638D-3	.5222676D-3	.3011539D-3	.1587841D-3
1		PF1	1.618853215	1.162813147	1.028614602	.9009403492	.7712169243	.6878009506	.6275093958	.5432264157	.4616676131
1		PQ1	.5848222056	.5841261488	.5891818303	.5984987754	.6161105180	.6355751373	.6569217512	.7065212155	.8065706815
2		PFC	1.706845254	1.483854189	1.404554871	1.324265814	1.238853413	1.182444391	1.141154506	1.082822087	1.025400694
2		PQC	2.250030828	1.984062454	1.957514011	1.963949235	2.016154100	2.087035071	2.166522221	2.342858893	2.655521734
1		ZF1	1.990367272	1.992837475	1.993761776	1.994683613	1.995640816	1.996259550	1.996707141	1.997336166	1.997961243
1		ZF2	2.496416625	2.550512371	2.565284493	2.578677775	2.591560381	2.599455627	2.604991397	2.612555546	2.619855990
1		ZF3	6.129027720	6.444703788	6.532238507	6.611339377	6.686814411	6.732622043	6.764484783	6.807629601	6.848791091

$P^{(3)}_{4+2}$			TH = 32	SBE = 1.887079915						MCPER FUNCTIONS
			RC = 02	RC = 05	RC = 07	RC = 10	RC = 15	RC = 20	RC = 25	RC = 35	RC = 50
		AMAX DB	.0017375255	.0108709564	.0213327376	.0436480540	.0988323393	.1772876696	.2802872360	.5675287486	1.249387366
		AMIN DB	30.67590743	42.55133714	46.94101932	51.61680905	56.98017385	60.84259037	63.89673243	68.67844585	74.23778641
MU		K	.2925530D-1	.7454751D-2	.4497271D-2	.2625183D-2	.1415765D-2	.9075498D-3	.6385036D-3	.3681948D-3	.1941381D-3
1		PF1	1.639010111	1.173031604	1.036681998	.9072698322	.7760659701	.6918368242	.6310192983	.5460804742	.4639667397
1		PQ1	.5885938276	.5861115881	.5908373957	.5999035193	.6173179119	.6366862876	.6579801442	.7075370064	.8076062258
2		PFC	1.690934706	1.476580410	1.399322108	1.320709618	1.236723978	1.181085803	1.140284217	1.082543325	1.025603053
2		PQC	2.352087043	2.040118506	2.004749459	2.004429018	2.051314511	2.119613900	2.197691033	2.372861022	2.685796230
1		ZF1	1.933688664	1.935984907	1.936890701	1.937805611	1.938764549	1.939388221	1.939840959	1.940479171	1.941115370
1		ZF2	2.411579468	2.467165762	2.482343077	2.496104804	2.509344567	2.517460917	2.523153061	2.530933067	2.538444663
1		ZF3	5.874912830	6.197767435	6.287908233	6.369526538	6.447515142	6.494887751	6.527851646	6.572498618	6.615099576

(3) P 4+2		TH = 33	SBE = 1.836078459						MCPER FUNCTIONS	
		RC = 02	RC = 05	RC. = 07	RC = 10	RC = 15	RC = 20	RC = 25	RC = 35	RC = 50
AMAX DB		.0017375255	.0108709564	.0213327376	.0436480540	.0988323393	.1772876696	.2802872360	.5675287486	1.249387366
AMIN DB		28.99838699	40.85955997	45.24696849	49.92111615	55.28323407	59.14503466	62.19880850	66.98010043	72.53912069
MU	K	.3548793D-1	.9057785D-2	.5465773D-2	.3191128D-2	.1721228D-2	.1103439D-2	.7763536D-3	.4477081D-3	.2360717D-3
1	PF1	1.660212530	1.183751077	1.045133630	.9138921751	.7811329135	.6960508051	.6346821726	.5490568934	.4663630181
1	PQ1	.5927496941	.5882428061	.5926033424	.6013939265	.6185928441	.6378564275	.6590927510	.7086024545	.8086903608
2	PFC	1.674589592	1.469048400	1.393892605	1.317012441	1.234504621	1.179666229	1.139371755	1.082245683	1.025805884
2	PQC	2.464798152	2.100731745	2.055547757	2.047754848	2.088781131	2.154239316	2.230759312	2.404617710	2.717773846
1	ZF1	1.880699264	1.882794917	1.883677249	1.884581490	1.885539197	1.886166281	1.886623220	1.887269489	1.887915886
1	ZF2	2.331738698	2.388799627	2.404378019	2.418504000	2.432096555	2.440431259	2.446277857	2.454271161	2.461991578
1	ZF3	5.635045984	5.964404395	6.057063173	6.141149715	6.221628044	6.270560364	6.304625152	6.350777560	6.394824127

(3) P 4+2		TH = 34	SBE = 1.788291650						MCPER FUNCTIONS	
		RC = 02	RC = 05	RC = 07	RC = 10	RC = 15	RC = 20	RC = 25	RC = 35	RC = 50
AMAX DB		.0017375255	.0108709564	.0213327376	.0436480540	.0988323393	.1772876696	.2802872360	.5675287486	1.249387366
AMIN DB		27.36839935	39.21291242	43.59775183	48.27005818	53.63078481	57.49190002	60.54526463	65.32608858	70.88475350
MU	K	.4281343D-1	.1094849D-1	.6608645D-2	.3859198D-2	.2081904D-2	.1334766D-2	.9391539D-3	.5416211D-3	.2856027D-3
1	PF1	1.682489752	1.194991461	1.053984219	.9208179894	.7864251156	.7004486225	.6385027625	.5521592782	.4688591752
1	PQ1	.5973409929	.5905329524	.5944883462	.6029758464	.6199392754	.6390886808	.6602622341	.7097197521	.8098250222
2	PFC	1.657829454	1.461258657	1.388265078	1.313172350	1.232193447	1.178184023	1.138415717	1.081928137	1.026008545
2	PQC	2.589538777	2.166345639	2.110224174	2.094155875	2.128722096	2.191051087	2.265850712	2.438235665	2.751551387
1	ZF1	1.831082090	1.832949291	1.833802998	1.834692679	1.835646072	1.836274981	1.836735138	1.837388288	1.838043922
1	ZF2	2.256461254	2.314977725	2.330953081	2.345439149	2.359380120	2.367930391	2.373929486	2.382133460	2.390060286
1	ZF3	5.408327474	5.743448957	5.838525401	5.925022576	6.007961100	6.058445481	6.093609675	6.141269665	6.186767387

(3) P 4+2		TH = 35	SBE = 1.743446796						MCPER FUNCTIONS	
		RC = 02	RC = 05	RC = 07	RC = 10	RC = 15	RC = 20	RC = 25	RC = 35	RC = 50
AMAX DB		.0017375255	.0108709564	.0213327376	.0436480540	.0988323393	.1772876696	.2802872360	.5675287486	1.249387366
AMIN DB		25.78369780	37.60864522	41.99057864	46.66082066	52.01999653	55.88035027	58.93326090	63.71356622	69.27183773
MU	K	.5138249D-1	.1316945D-1	.7951903D-2	.4644714D-2	.2506110D-2	.1606876D-2	.1130673D-2	.6521112D-3	.3438809D-3
1	PF1	1.705865292	1.206773613	1.063249346	.9280585542	.7919504137	.7050363716	.6424861133	.5553914595	.4714581065
1	PQ1	.6024276228	.5929968706	.5965020954	.6046557484	.6213615475	.6403864537	.6614914865	.7108912706	.8110122958
2	PFC	1.640676102	1.453212190	1.382438479	1.309187481	1.229788544	1.176637498	1.137414651	1.081539622	1.026210364
2	PQC	2.727901405	2.237460676	2.169131639	2.143886749	2.171322958	2.230202821	2.303100746	2.473831398	2.787234191
1	ZF1	1.784556777	1.786166384	1.786986097	1.787857174	1.788303052	1.789432137	1.789894491	1.790553300	1.791217171
1	ZF2	2.185365542	2.245313858	2.261681996	2.276523987	2.290808978	2.299571994	2.305721585	2.314133533	2.322264273
1	ZF3	5.193797146	5.533873896	5.631254880	5.720096145	5.805459201	5.857485378	5.893746143	5.942914588	5.989868320

(3) P 4+2 — TH = 36 — SBE = 1.701301617 — MCPER FUNCTIONS

MU	(3) P 4+2	RC = 02	RC = 05	RC = 07	RC = 10	RC = 15	RC = 20	RC = 25	RC = 35	RC = 50
	AMAX DB	.0017375255	.0108709564	.0213327376	.0436480540	.0988323393	.1772876696	.2802872360	.5675287486	1.249387366
	AMIN DB	24.24240801	36.04425520	40.42289618	45.09082333	50.44827195	54.30778076	57.36018852	62.13992010	67.69775681
MU	K	.6135919D-1	.1576839D-1	.9524785D-2	.5564919D-2	.3003215D-2	.1925799D-2	.1355160D-2	.7816350D-3	.4122040D-3
1	PF1	1.730354051	1.219119299	1.072945480	.9356258511	.7977171407	.7098205382	.6466375920	.5587575100	.4741628873
1	PQ1	.6080798228	.5956513586	.5986554329	.6064408008	.6228644264	.6417534648	.6627836553	.7121195777	.8122544312
2	PFC	1.623153738	1.444910566	1.376412023	1.305056059	1.227287990	1.175024931	1.136367061	1.081229027	1.026410633
2	PQC	2.881734054	2.314643018	2.232666083	2.197230985	2.216788826	2.271863592	2.342658136	2.511532278	2.824937011
1	ZF1	1.740874482	1.742196052	1.742976188	1.743824465	1.744759506	1.745387057	1.745850549	1.746513747	1.747184816
1	ZF2	2.118114857	2.179464985	2.196221559	2.211415273	2.226039858	2.235012754	2.241310802	2.249927952	2.258260027
1	ZF3	4.990615173	5.334771533	5.434330548	5.525439885	5.613185224	5.666740093	5.704093107	5.754769460	5.803183271

(3) P 4+2 — TH = 37 — SBE = 1.661640141 — MCPER FUNCTIONS

MU	(3) P 4+2	RC = 02	RC = 05	RC = 07	RC = 10	RC = 15	RC = 20	RC = 25	RC = 35	RC = 50
	AMAX DB	.0017375255	.0108709564	.0213327376	.0436480540	.0988323393	.1772876696	.2802872360	.5675287486	1.249387366
	AMIN DB	22.74303570	34.51746274	38.89236521	43.55769471	48.91322029	52.77179251	55.82364398	60.60274179	66.16009879
MU	K	.7292026D-1	.1879866D-1	.1136009D-1	.6639193D-2	.3583761D-2	.2298319D-2	.1617401D-2	.9329598D-3	.4920339D-3
1	PF1	1.755958639	1.232051089	1.083089982	.9435325984	.8037341672	.7148080248	.6509629090	.5622617598	.4769767851
1	PQ1	.6143800939	.5985154748	.6009605224	.6083389623	.6244531528	.6431937803	.6641421683	.7134074567	.8135538565
2	PFC	1.605289078	1.436355957	1.370185222	1.300776411	1.224689859	1.173344560	1.135271406	1.080845193	1.026608608
2	PQC	3.053184088	2.398534665	2.301272669	2.254504873	2.265346826	2.316219793	2.384686339	2.551477742	2.864785008
1	ZF1	1.699813619	1.700815398	1.701550158	1.702371279	1.703292042	1.703916285	1.704379818	1.705046089	1.705723277
1	ZF2	2.054411977	2.117125457	2.134265837	2.149806993	2.164766694	2.173946560	2.180390983	2.189210488	2.197741228
1	ZF3	4.798045868	5.145337868	5.246934471	5.340225869	5.430304188	5.485371564	5.523810878	5.575993023	5.625870087

(3) P 4+2 — TH = 38 — SBE = 1.624269245 — MCPER FUNCTIONS

MU	(3) P 4+2	RC = 02	RC = 05	RC = 07	RC = 10	RC = 15	RC = 20	RC = 25	RC = 35	RC = 50
	AMAX DB	.0017375255	.0108709564	.0213327376	.0436480540	.0988323393	.1772876696	.2802872360	.5675287486	1.249387366
	AMIN DB	21.28448328	33.02619355	37.39683925	42.05925033	47.41263496	51.27016979	54.32140654	59.09930516	64.65663369
MU	K	.8625332D-1	.2231980D-1	.1349454D-1	.7889282D-2	.4259594D-2	.2732068D-2	.1922780D-2	.1109200D-2	.5850168D-3
1	PF1	1.782664711	1.245592205	1.093701105	.9517922838	.8100109300	.7200061789	.6554681422	.5659088152	.4799032725
1	PQ1	.6214254450	.6016108996	.6034310409	.6103590875	.6261334990	.6447118524	.6555707643	.7147579279	.8149131953
2	PFC	1.587111505	1.427551193	1.363757917	1.296346988	1.221992231	1.171594594	1.134126099	1.080436919	1.026803506
2	PQC	3.244748724	2.489865427	2.375453080	2.316062005	2.317248934	2.363477261	2.429365301	2.593820649	2.906914876
1	ZF1	1.661176242	1.661825193	1.662508553	1.663298006	1.664200926	1.664820027	1.665282463	1.665950443	1.666632631
1	ZF2	1.993994704	2.058022192	2.075541310	2.091425478	2.106715745	2.116099610	2.122688286	2.131707221	2.140433865
1	ZF3	4.615443931	4.964859205	5.068338507	5.163715442	5.256069919	5.312630289	5.352148180	5.405832265	5.457174753

(3) P 4+2 — TH = 39 SBE = 1.589015729 — MCPER FUNCTIONS

MU		RC = 02	RC = 05	RC = 07	RC = 10	RC = 15	RC = 20	RC = 25	RC = 35	RC = 50
	AMAX DB	.0017375255	.0108709564	.0213327376	.0436480540	.0988323393	.1772876696	.2802872360	.5675287486	1.249387366
	AMIN DB	19.86607481	31.56856423	35.93434708	40.59347395	45.94447423	49.80086057	52.85141862	57.62904677	63.18529386
MU	K	.1015538D 0	.2639805D-1	.1596918D-1	.9339558D-2	.5044014D-2	.3235616D-2	.2277346D-2	.1313856D-2	.6930033D-3
1	PF1	1.810435178	1.259766293	1.104797966	.9604191949	.8165574686	.7254228233	.6601597627	.5697035783	.4829460424
1	PQ1	.6293299839	.6049623622	.6060824062	.6125110489	.6279118345	.6463125640	.6670735268	.7161742733	.8163352854
2	PFC	1.568653245	1.418499820	1.357130317	1.291766384	1.219193198	1.169773213	1.132929511	1.080002951	1.026994506
2	PQC	3.459332542	2.589467043	2.455774071	2.382298558	2.372775238	2.413863717	2.476893452	2.638728344	2.951476122
1	ZF1	1.624784966	1.625046845	1.625672553	1.626425665	1.627307054	1.627919116	1.628379279	1.629047556	1.629733584
1	ZF2	1.936632199	2.001910601	2.019802758	2.036025284	2.051641457	2.061226316	2.067957047	2.077172401	2.086092089
1	ZF3	4.442242635	4.792700844	4.897893018	4.995247945	5.089813762	5.147844032	5.188430850	5.243611118	5.296420076

(3) P 4+2 — TH = 40 SBE = 1.555723827 — MCPER FUNCTIONS

MU		RC = 02	RC = 05	RC = 07	RC = 10	RC = 15	RC = 20	RC = 25	RC = 35	RC = 50
	AMAX DB	.0017375255	.0108709564	.0213327376	.0436480540	.0988323393	.1772876696	.2802872360	.5675287486	1.249387366
	AMIN DB	18.48758731	30.14287162	34.50307830	39.15850137	44.50684438	48.36195949	51.41176866	56.18854856	61.74415669
MU	K	.1190202D 0	.3110688D-1	.1882982D-1	.1101729D-1	.5951930D-2	.3818581D-2	.2687890D-2	.1550860D-2	.8180732D-3
1	PF1	1.839203233	1.274597105	1.116400501	.9694284447	.8233844621	.7310662888	.6650446629	.5736512682	.4861090235
1	PQ1	.6382278578	.6085981441	.6089320422	.6148058791	.6297952015	.6480012798	.6686549233	.7176600633	.8178231998
2	PFC	1.549949554	1.409206151	1.350303033	1.287033357	1.216290377	1.167878570	1.131679970	1.079541989	1.027180745
2	PQC	3.700311903	2.698289866	2.542877517	2.453659444	2.432237734	2.467631580	2.527490001	2.686336924	2.998632509
1	ZF1	1.590480312	1.590319826	1.590881397	1.591593330	1.592449379	1.593052439	1.593509112	1.594176226	1.594864895
1	ZF2	1.882121918	1.948571151	1.966829784	1.983385688	1.999322959	2.009105705	2.015976262	2.025334935	2.034494704
1	ZF3	4.277943494	4.628297455	4.735017282	4.834231139	4.930935003	4.990408232	5.032052239	5.088720849	5.142996072

(3) P 4+2 — TH = 41 SBE = 1.524253087 — MCPER FUNCTIONS

MU		RC = 02	RC = 05	RC = 07	RC = 10	RC = 15	RC = 20	RC = 25	RC = 35	RC = 50
	AMAX DB	.0017375255	.0108709564	.0213327376	.0436480540	.0988323393	.1772876696	.2802872360	.5675287486	1.249387366
	AMIN DB	17.14928550	28.74758560	33.10137152	37.75260666	43.09798499	46.95169287	50.00067604	54.77652273	60.33142939
MU	K	.1388468D 0	.3652756D-1	.2212745D-1	.1295298D-1	.7000044D-2	.4491748D-2	.3162032D-2	.1824626D-2	.9625616D-3
1	PF1	1.868864231	1.290108056	1.128529373	.9788359902	.8305032673	.7369454494	.6701301871	.5777574449	.4893963982
1	PQ1	.6482765137	.6125506765	.6119996921	.6172559369	.6317914018	.6497839039	.6703198488	.7192191885	.8193802708
2	PFC	1.531038938	1.399675330	1.343277123	1.282146856	1.213283416	1.165908802	1.130375763	1.079052684	1.027361316
2	PQC	3.971605625	2.817422602	2.637492261	2.530645531	2.495984708	2.525061204	2.581397579	2.736998269	3.048563706
1	ZF1	1.558118396	1.557499470	1.557990182	1.558655935	1.559482707	1.560074741	1.560526669	1.561191111	1.561881177
1	ZF2	1.830287041	1.897806455	1.916423858	1.933307709	1.949561065	1.959538510	1.966546591	1.976145383	1.985442163
1	ZF3	4.122107025	4.471144877	4.579191312	4.680133032	4.778892689	4.839777821	4.882465022	4.940611858	4.996351753

(3) P 4+2	TH = 42	SBE = 1.494476550						MCPER FUNCTIONS	
	RC = 02	RC = 05	RC = 07	RC = 10	RC = 15	RC = 20	RC = 25	RC = 35	RC = 50
AMAX DB	.0017375255	.0108709564	.0213327376	.0436480540	.0988323393	.1772876696	.2802872360	.5675287486	1.249387366
AMIN DB	15.85195504	27.38134557	31.72770511	36.37419038	41.71625609	45.56840557	48.61647783	53.39129839	58.94543558
MU K	.1612138D 0	.4274967D-1	.2591879D-1	.1518065D-1	.8207052D-2	.5267199D-2	.3708311D-2	.2140105D-2	.1129089D-2
1 PF1	1.899266680	1.306321627	1.141205841	.9886586416	.8379259578	.7430697593	.6754241643	.5820280351	.4928126218
1 PQ1	.6596602032	.6168572459	.6153077861	.6198751002	.6339090975	.6516669474	.6720736768	.7208558945	.8210101161
2 PFC	1.511963358	1.389913400	1.336054130	1.277106044	1.210169009	1.163862027	1.129015139	1.078533634	1.027535269
2 PQC	4.277750418	2.948115691	2.740448119	2.613822099	2.564405866	2.586464643	2.638885273	2.790787373	3.101467177
1 ZF1	1.527568890	1.526455091	1.526867981	1.527482382	1.528275818	1.528854739	1.529300627	1.529960839	1.530651018
1 ZF2	1.780974228	1.849438807	1.868405799	1.885611553	1.902175713	1.912344557	1.919487787	1.929273392	1.938753994
1 ZF3	3.974344225	4.320793070	4.429948842	4.532474874	4.633198622	4.695460206	4.739174178	4.798786648	4.855988090

(3) P 4+2	TH = 43	SBE = 1.466279186						MCPER FUNCTIONS	
	RC = 02	RC = 05	RC = 07	RC = 10	RC = 15	RC = 20	RC = 25	RC = 35	RC = 50
AMAX DB	.0017375255	.0108709564	.0213327376	.0436480540	.0988323393	.1772876696	.2802872360	.5675287486	1.249387366
AMIN DB	14.59692708	26.04296043	30.38069045	35.02176971	40.36012696	44.21054940	47.25761697	52.03130959	57.58460332
MU K	.1862746D 0	.4987145D-1	.3026673D-1	.1773828D-1	.9593866D-2	.6158466D-2	.4336298D-2	.2502848D-2	.1320596D-2
1 PF1	1.930202928	1.323258569	1.154451568	.9989140557	.8456653625	.7494492932	.6809349443	.5864693601	.4963624438
1 PQ1	.6725935918	.6215608291	.6188818770	.6226789931	.6361579270	.6536576045	.6739223167	.7225748220	.8227166696
2 PFC	1.492768408	1.379927369	1.328636131	1.271910325	1.206945907	1.161736356	1.127596312	1.077983388	1.027701606
2 PQC	4.623978686	3.091809019	2.852692451	2.703828819	2.637938301	2.652190015	2.700252162	2.848002519	3.157560329
1 ZF1	1.498713201	1.497068355	1.497396220	1.497953931	1.498709845	1.499273501	1.499712014	1.500366391	1.501055356
1 ZF2	1.734051555	1.803308093	1.822613570	1.840134430	1.857003760	1.867360568	1.874636486	1.884605480	1.894266592
1 ZF3	3.834308405	4.176840066	4.286871300	4.390825146	4.493411343	4.557009244	4.601730951	4.662793780	4.721451957

(3) P 4+2	TH = 44	SBE = 1.439556540						MCPER FUNCTIONS	
	RC = 02	RC = 05	RC = 07	RC = 10	RC = 15	RC = 20	RC = 25	RC = 35	RC = 50
AMAX DB	.0017375255	.0108709564	.0213327376	.0436480540	.0988323393	.1772876696	.2802872360	.5675287486	1.249387366
AMIN DB	13.38608411	24.73141243	29.05906759	33.69397031	39.02816635	42.87667286	45.92263190	50.69508481	56.24745442
MU K	.2141390D 0	.5800018D-1	.3524087D-1	.2066814D-1	.1118386D-1	.7180693D-2	.5056714D-2	.2919078D-2	.1540378D-2
1 PF1	1.961400538	1.340936863	1.168288338	1.009620711	.8537351045	.7560947881	.6866714360	.5910881675	.5000509314
1 PQ1	.6873252554	.6267110813	.6227511536	.6256852520	.6385486407	.6557638416	.6758722798	.7243810522	.8245042150
2 PFC	1.473503420	1.369725294	1.321025786	1.266559375	1.203612432	1.159529897	1.126117461	1.077400442	1.027859278
2 PQC	5.016295348	3.250164753	2.975309854	2.801391524	2.717073505	2.722626587	2.765831395	2.908918875	3.217082993
1 ZF1	1.471442820	1.469231878	1.469467280	1.469962790	1.470676871	1.471223048	1.471652812	1.472299699	1.472986082
1 ZF2	1.689406478	1.759270015	1.778900543	1.796728674	1.813897087	1.824438257	1.831844304	1.841993130	1.851831303
1 ZF3	3.701687101	4.038926731	4.149582613	4.254794363	4.359130931	4.424020038	4.469727643	4.532222646	4.592330897

(3) P 4+2	TH = 45	SBE = 1.414213562						MCPER FUNCTIONS	
	RC = 02	RC = 05	RC = 07	RC = 10	RC = 15	RC = 20	RC = 25	RC = 35	RC = 50
AMAX DB	.0017375255	.0108709564	.0213327376	.0436480540	.0988323393	.1772876696	.2802872360	.5675287486	1.249387366
AMIN DB	12.22183528	23.44586490	27.76170345	32.38951994	37.71903404	41.56541213	44.61014726	49.38123744	54.93259493
MU I K	.2448546D 0	.6725224D-1	.4091804D-1	.2401729D-1	.1300314D-1	.8350825D-2	.5881561D-2	.3395769D-2	.1792133D-2
1 I PF1	1.992515852	1.359370370	1.182737681	1.020797856	.8621496362	.7630176874	.6926431491	.5958916643	.5038834949
1 I PQ1	.7041407682	.6323655001	.6269490488	.6289138385	.6410932590	.6579944998	.6779307563	.7262801592	.8263774250
2 I PFC	1.454221408	1.359316367	1.313226396	1.261053174	1.200166995	1.157240764	1.124576735	1.076783239	1.028007188
2 I PQC	5.461549781	3.425106258	3.109545555	2.907336160	2.802365604	2.798210726	2.835994951	2.973842070	3.280300272
1 I ZF1	1.445657809	1.442848005	1.442983277	1.443410907	1.444078718	1.444605141	1.445024744	1.445662437	1.446344829
1 I ZF2	1.646943666	1.717194593	1.737133808	1.755260118	1.772720957	1.783442682	1.790976181	1.801301135	1.811312776
1 I ZF3	3.576193800	3.906732223	4.017744701	4.124030588	4.229994510	4.296124430	4.342793098	4.406698949	4.468248587

(3) P 4+2	TH = 46	SBE = 1.390163591						MCPER FUNCTIONS	
	RC = 02	RC = 05	RC = 07	RC = 10	RC = 15	RC = 20	RC = 25	RC = 35	RC = 50
AMAX DB	.0017375255	.0108709564	.0213327376	.0436480540	.0988323393	.1772876696	.2802872360	.5675287486	1.249387366
AMIN DB	11.10704816	22.18567403	26.48759256	31.10724363	36.43147357	40.27548305	43.31886560	48.08845733	53.63870647
MU I K	.2783861D 0	.7775285D-1	.4738276D-1	.2783799D-1	.1508087D-1	.9687815D-2	.6824278D-2	.3940734D-2	.2080006D-2
1 I PF1	2.023131705	1.378567109	1.197820354	1.032465418	.8709242718	.7702301873	.6988602392	.6008875547	.5078659170
1 I PQ1	.7233650286	.6385907958	.6315139597	.6323874091	.6438052561	.6603594142	.6801057023	.7282782695	.8283414053
2 I PFC	1.434978800	1.348711015	1.305241957	1.255392043	1.196608108	1.154867085	1.122972256	1.076130171	1.028144181
2 I PQC	5.967499280	3.618864120	3.256833308	3.022605361	2.894441073	2.879432872	2.911159203	3.043112352	3.347505832
1 I ZF1	1.421265428	1.417827740	1.417855010	1.418208912	1.418825893	1.419330224	1.419738214	1.420364964	1.421041912
1 I ZF2	1.606582563	1.676964902	1.697192885	1.715606704	1.733352593	1.744250816	1.751908955	1.762406164	1.772587518
1 I ZF3	3.457559408	3.779970027	3.891053575	3.998215529	4.105672345	4.172987091	4.220588782	4.285880768	4.348860891

(3) P 4+2	TH = 47	SBE = 1.367327461						MCPER FUNCTIONS	
	RC = 02	RC = 05	RC = 07	RC = 10	RC = 15	RC = 20	RC = 25	RC = 35	RC = 50
AMAX DB	.0017375255	.0108709564	.0213327376	.0436480540	.0988323393	.1772876696	.2802872360	.5675287486	1.249387366
AMIN DB	10.04492595	20.95040494	25.23586075	29.84606060	35.16430603	39.00567408	42.04755998	46.81550305	52.36453848
MU I K	.3145964D 0	.8963544D-1	.5472767D-1	.3218822D-1	.1744957D-1	.1121286D-1	.7899907D-2	.4562731D-2	.2408647D-2
1 I PF1	2.052761483	1.398527094	1.213555661	1.044643869	.8800752117	.7777452838	.7053335556	.6060840811	.5120043838
1 I PQ1	.7453634619	.6454644955	.6364901005	.6361317515	.6466997771	.6628695524	.6824059427	.7303821308	.8304017449
2 I PFC	1.415834852	1.337921009	1.297077231	1.249576678	1.192934410	1.152407008	1.121302125	1.075439574	1.028269050
2 I PQC	6.542862171	3.834030506	3.418828665	3.148278210	2.994010239	2.966845755	2.991791426	3.117109431	3.419025713
1 I ZF1	1.398178898	1.394089812	1.394001035	1.394275199	1.394836666	1.395316507	1.395711395	1.396325404	1.396995415
1 I ZF2	1.568254577	1.638476001	1.658968554	1.677657268	1.695679943	1.706750303	1.714530107	1.725195508	1.735542647
1 I ZF3	3.345523550	3.658384452	3.769235930	3.877061151	3.985864441	4.054302107	4.102805365	4.169455123	4.233852417

(3) P 4+2 — TH = 48 SBE = 1.345632730 — MCPER FUNCTIONS

MU	I	4+2	RC = 02	RC = 05	RC = 07	RC = 10	RC = 15	RC = 20	RC = 25	RC = 35	RC = 50
		AMAX DB	.0017375255	.0108709564	.0213327376	.0436480540	.0988323393	.1772876696	.2802872360	.5675287486	1.249387366
		AMIN DB	9.038824975	19.73985173	24.00577166	28.60498290	33.91642504	37.75484015	40.79506738	45.56119499	51.10890106
MU	I	K	.3532310D 0	.1030404D 0	.6305382D-1	.3713221D-1	.2014553D-1	.1294965D-1	.9125289D-2	.5271573D-2	.2783267D-2
1	I	PF1	2.080861525	1.419239652	1.229960555	1.057354029	.8896195587	.7855768209	.7120746931	.6114900686	.5163055200
1	I	PQ1	.7705418246	.6530768107	.6419285119	.6401763023	.6497938915	.6655371764	.6848412890	.7325991911	.8325645741
2	I	PFC	1.396850696	1.326959589	1.288737821	1.243608201	1.189144679	1.149858718	1.119564428	1.074709733	1.028380527
2	I	PQC	7.197362313	4.073623133	3.597448736	3.285593850	3.101880947	3.061074096	3.078417475	3.196258141	3.495222767
1	I	ZF1	1.376316336	1.371559845	1.371346853	1.371535115	1.372036266	1.372489158	1.372869419	1.373468845	1.374130383
1	I	ZF2	1.531899892	1.601634034	1.622361872	1.641310505	1.659600607	1.670838371	1.678736677	1.689565991	1.700074789
1	I	ZF3	3.239826050	3.541747526	3.652046173	3.760306713	3.870297577	3.939790010	3.989159735	4.057134978	4.122933503

(3) P 4+2 — TH = 49 SBE = 1.325012993 — MCPER FUNCTIONS

MU	I	4+2	RC = 02	RC = 05	RC = 07	RC = 10	RC = 15	RC = 20	RC = 25	RC = 35	RC = 50
		AMAX DB	.0017375255	.0108709564	.0213327376	.0436480540	.0988323393	.1772876696	.2802872360	.5675287486	1.249387366
		AMIN DB	8.092017479	18.55406131	22.79673657	27.38311591	32.68679263	36.52189727	39.56028277	44.32440904	49.87065848
MU	I	K	.3939119D 0	.1181128D 0	.7247082D-1	.4274095D-1	.2320921D-1	.1492468D-1	.1051928D-1	.6078264D-2	.3209719D-2
1	I	PF1	2.106852894	1.440680172	1.247048483	1.070616777	.8995753184	.7937395375	.7190960457	.6171149744	.5207764279
1	I	PQ1	.7993445308	.6615327933	.6478882559	.6445547605	.6531068935	.6683760329	.6874226775	.7349376900	.8348366302
2	I	PFC	1.378088002	1.315841604	1.280230251	1.237488199	1.185237865	1.147220446	1.117757236	1.073938880	1.028477283
2	I	PQC	7.941772216	4.341160267	3.794919718	3.435979796	3.218974858	3.162826120	3.171630829	3.281035091	3.576501827
1	I	ZF1	1.355599903	1.350169616	1.349824197	1.349920257	1.350356171	1.350779601	1.351143673	1.351726628	1.352378118
1	I	ZF2	1.497463997	1.566355471	1.587283346	1.606474067	1.625020904	1.636420893	1.644434307	1.655423006	1.666089126
1	I	ZF3	3.140199185	3.429856155	3.539263813	3.647716184	3.758722713	3.829195163	3.879392379	3.948656610	4.015837568

(3) P 4+2 — TH = 50 SBE = 1.305407289 — MCPER FUNCTIONS

MU	I	4+2	RC = 02	RC = 05	RC = 07	RC = 10	RC = 15	RC = 20	RC = 25	RC = 35	RC = 50
		AMAX DB	.0017375255	.0108709564	.0213327376	.0436480540	.0988323393	.1772876696	.2802872360	.5675287486	1.249387366
		AMIN DB	7.207418010	17.39336009	21.60832771	26.17966087	31.47443633	35.30581789	38.34215385	43.10407077	48.64872317
MU	I	K	.4361432D 0	.1349994D 0	.8309667D-1	.4909270D-1	.2668567D-1	.1716758D-1	.1210298D-1	.6995141D-2	.3694570D-2
1	I	PF1	2.130151973	1.462806250	1.264827903	1.084452674	.9099613802	.8022491131	.7264108645	.6229689420	.5254247306
1	I	PQ1	.8322517170	.6709547987	.6544378254	.6493058175	.6566606567	.6714015758	.6901623312	.7374067655	.8372253347
2	I	PFC	1.359607312	1.304583658	1.271562063	1.231218785	1.181213110	1.144490481	1.115878621	1.073125193	1.028557925
2	I	PQC	8.787968468	4.640748176	4.013833741	3.601085976	3.346346950	3.272907259	3.272103329	3.371976490	3.663315760
1	I	ZF1	1.335955208	1.329856390	1.329370401	1.329367845	1.329733493	1.330124889	1.330471177	1.331035730	1.331675558
1	I	ZF2	1.464894163	1.532566447	1.553652220	1.573063789	1.591855052	1.603411548	1.611536413	1.622679680	1.633498544
1	I	ZF3	3.046361500	3.322529506	3.430691152	3.539075979	3.650912708	3.722283475	3.773265086	3.843777285	3.912318780

(3) P 4+2 — TH = 51 — SBE = 1.286759566 — MCPER FUNCTIONS

MU			RC = 02	RC = 05	RC = 07	RC = 10	RC = 15	RC = 20	RC = 25	RC = 35	RC = 50
		AMAX DB	.0017375255	.0108709564	.0213327376	.0436480540	.0988323393	.1772876696	.2802872360	.5675287486	1.249387366
		AMIN DB	6.387304563	16.25838195	20.44029512	24.99391966	30.27844728	34.10562709	37.13967637	41.89915018	47.44205012
MU		K	.4793302D 0	.1538441D 0	.9505725D-1	.5627351D-1	.3062511D-1	.1971145D-1	.1390004D-1	.8036047D-2	.4245194D-2
1		PF1	2.150207158	1.485553248	1.283300427	1.098881436	.9207974706	.8111222081	.7340333172	.6290628601	.5302586201
1		PQ1	.8697756127	.6814852624	.6616568037	.6544740250	.6604800549	.6746312302	.6930739485	.7400165771	.8397388807
2		PFC	1.341466180	1.293204270	1.262741923	1.224802651	1.177069777	1.141667190	1.113926658	1.072266801	1.028620994
2		PQC	9.749018992	4.977182475	4.257216845	3.782825761	3.485208963	3.392236537	3.380597956	3.469687415	3.756172592
1		ZF1	1.317311000	1.310562311	1.309927829	1.309820170	1.310110420	1.310467160	1.310794037	1.311338216	1.311964729
1		ZF2	1.434136199	1.500202181	1.521395886	1.541003032	1.560024468	1.571731103	1.579963445	1.591256133	1.602222897
1		ZF3	2.958013957	3.219606466	3.326151222	3.434192968	3.546660332	3.618840388	3.670558921	3.742273224	3.812149997

(3) P 4+2 — TH = 52 — SBE = 1.269018215 — MCPER FUNCTIONS

MU			RC = 02	RC = 05	RC = 07	RC = 10	RC = 15	RC = 20	RC = 25	RC = 35	RC = 50
		AMAX DB	.0017375255	.0108709564	.0213327376	.0436480540	.0988323393	.1772876696	.2802872360	.5675287486	1.249387366
		AMIN DB	5.633073638	15.15009510	19.29258713	23.82530208	29.09797957	32.92039944	35.95189000	40.70865681	46.24963168
MU		K	.5228129D 0	.1747814D 0	.1084852D 0	.6437762D-1	.3508335D-1	.2259332D-1	.1593696D-1	.9216524D-2	.4869882D-2
1		PF1	2.166536999	1.508829382	1.302458515	1.113921240	.9321040680	.8203764976	.7419785494	.6354084278	.5352869113
1		PQ1	.9124570715	.6932897761	.6696378082	.6601108294	.6645934627	.6780847031	.6961729272	.7427784511	.8423863359
2		PFC	1.323717326	1.281724023	1.253779736	1.218243141	1.172807489	1.138749032	1.111899436	1.071361782	1.028664960
2		PQC	10.83932486	5.356064734	4.528610193	3.983425560	3.636957678	3.521866247	3.497984110	3.574852811	3.855643905
1		ZF1	1.299599148	1.292233844	1.291443379	1.291224105	1.291433739	1.291753151	1.292058957	1.292580752	1.293192263
1		ZF2	1.405131845	1.469206424	1.490449381	1.510222114	1.529457143	1.541306778	1.549642247	1.561078826	1.572189344
1		ZF3	2.874838980	3.120943129	3.225485924	3.332892726	3.445776500	3.518669108	3.571072439	3.643937793	3.715120937

(3) P 4+2 — TH = 53 — SBE = 1.252135658 — MCPER FUNCTIONS

MU			RC = 02	RC = 05	RC = 07	RC = 10	RC = 15	RC = 20	RC = 25	RC = 35	RC = 50
		AMAX DB		.0108709564	.0213327376	.0436480540	.0988323393	.1772876696	.2802872360	.5675287486	1.249387366
		AMIN DB		14.06982407	18.16537434	22.67333593	27.93225096	31.74925665	34.77787481	39.53163525	45.07049265
MU		K		.1979287D 0	.1235183D 0	.7350776D-1	.4012245D-1	.2585456D-1	.1824342D-1	.1055403D-1	.5577960D-2
1		PF1		1.532510569	1.322282668	1.129587822	.9439022672	.8300306929	.7502627464	.6420182254	.5405191017
1		PQ1		.7065604240	.6784887563	.6662758041	.6690333535	.6817843543	.6994766278	.7457050502	.8451777617
2		PFC		1.270165707	1.244686782	1.211544327	1.168426161	1.135734582	1.109795069	1.070408164	1.028688223
2		PQC		5.783935641	4.832166973	4.205484907	3.803209168	3.663005677	3.625255961	3.688250626	3.962374800
1		ZF1		1.274821248	1.273868012	1.273530666	1.273654398	1.273933766	1.274216815	1.274714180	1.275308964
1		ZF2		1.439530898	1.460754967	1.480657831	1.500087116	1.512071688	1.520505492	1.532079988	1.543326778
1		ZF3		3.026410173	3.128554297	3.235017991	3.348088731	3.421589034	3.474620108	3.548579902	3.621036566

(3) P 4+2	TH = 54	SBE = 1.236067977						MCPER FUNCTIONS	
	RC = 02	RC = 05	RC = 07	RC = 10	RC = 15	RC = 20	RC = 25	RC = 35	RC = 50
AMAX DB		.0108709564	.0213327376	.0436480540	.0988323393	.1772876696	.2802872360	.5675287486	1.249387366
AMIN DB		13.01926196	17.05907620	21.53768013	26.78054506	30.59136611	33.61674820	38.36716096	43.90368552
MU I K		.2233762D 0	.1402963D 0	.8377530D-1	.4581131D-1	.2954144D-1	.2085271D-1	.1206818D-1	.6379927D-2
1 I PF1		1.556435440	1.342738084	1.145893297	.9562135762	.8401045466	.7589031933	.6489057918	.5459654389
1 I PQ1		.7215193066	.6883354833	.6730381174	.6738370141	.6857556373	.7030046871	.7488105730	.8481243525
2 I PFC		1.258554420	1.235475856	1.204711093	1.163926046	1.132622551	1.107611713	1.069403934	1.028689106
2 I PQC		6.268426026	5.172767792	4.452049417	3.985840473	3.817049863	3.763554584	3.810767543	4.077095753
1 I ZF1		1.258278085	1.257156343	1.256694623	1.256727122	1.256963695	1.257222275	1.257693131	1.258269433
1 I ZF2		1.411134674	1.432261758	1.452253062	1.471854010	1.483964360	1.492491187	1.504197109	1.515575307
1 I ZF3		2.935890057	3.035230876	3.140427284	3.253439787	3.327434387	3.381030911	3.456022586	3.529715641

(3) P 4+2	TH = 55	SBE = 1.220774589						MCPER FUNCTIONS	
	RC = 02	RC = 05	RC = 07	RC = 10	RC = 15	RC = 20	RC = 25	RC = 35	RC = 50
AMAX DB		.0108709564	.0213327376	.0436480540	.0988323393	.1772876696	.2802872360	.5675287486	1.249387366
AMIN DB		12.00046678	15.97438930	20.41814127	25.64221521	29.44594034	32.46766254	37.21433640	42.74828605
MU I K		.2511751D 0	.1539573D 0	.9530001D-1	.5222630D-1	.3370567D-1	.2380219D-1	.1378108D-1	.7287620D-2
1 I PF1		1.580401117	1.363770778	1.162844667	.9690596236	.8506188322	.7679183316	.6560857082	.5516369957
1 I PQ1		.7384221470	.6993247407	.6804782791	.6790474038	.6900276275	.7067793914	.7521109916	.8512385993
2 I PFC		1.246917604	1.226161415	1.197749237	1.159307780	1.129411816	1.105347574	1.068347035	1.028665853
2 I PQC		6.818427273	5.556157759	4.726699546	4.187040489	3.985614582	3.914194914	3.943417946	4.200636804
1 I ZF1		1.242560754	1.241266245	1.240674145	1.240610077	1.240801070	1.241033450	1.241475690	1.242031725
1 I ZF2		1.383983445	1.404925388	1.424956431	1.444702638	1.456928309	1.465542234	1.477372494	1.488875800
1 I ZF3		2.849273958	2.945404050	3.048993681	3.161686473	3.236052981	3.290147110	3.366101739	3.440989430

(3) P 4+2	TH = 56	SBE = 1.206217949						MCPER FUNCTIONS	
	RC = 02	RC = 05	RC = 07	RC = 10	RC = 15	RC = 20	RC = 25	RC = 35	RC = 50
AMAX DB		.0108709564	.0213327376	.0436480540	.0988323393	.1772876696	.2802872360	.5675287486	1.249387366
AMIN DB		11.01583494	14.91231508	19.31469370	24.51669042	28.31223746	31.32980330	36.07228749	41.60338884
MU I K		.2813250D 0	.1796322D 0	.1082095D 0	.5945186D-1	.3840503D-1	.2713375D-1	.1571758D-1	.8314393D-2
1 I PF1		1.604160590	1.385303246	1.180441931	.9824617497	.8615952900	.7773278080	.6635736896	.5575457549
1 I PQ1		.7575618456	.7116275872	.6886902157	.6847141861	.6946336563	.7108261232	.7556243318	.8545344830
2 I PFC		1.235284992	1.216759731	1.190665581	1.154572444	1.126101451	1.103000932	1.067235372	1.028616629
2 I PQC		7.444283491	5.989108917	5.033658759	4.409372413	4.170579176	4.078697961	4.087366870	4.333944617
1 I ZF1		1.227628053	1.226158485	1.225430477	1.225264551	1.225407166	1.225611597	1.226023090	1.226557048
1 I ZF2		1.358048636	1.378707681	1.398722029	1.418582670	1.430911678	1.439606052	1.451552864	1.463174489
1 I ZF3		2.766458404	2.858974381	2.960603688	3.072698583	3.147305153	3.201823143	3.278664983	3.354700545

(3) P 4+2	TH = 57 SBE = 1.192363293								MCPER FUNCTIONS
	RC = 02	RC = 05	RC = 07	RC = 10	RC = 15	RC = 20	RC = 25	RC = 35	RC = 50
AMAX DB		.0108709564	.0213327376	.0436480540	.0988323393	.1772876696	.2802872360	.5675287486	1.249387366
AMIN DB		10.06804530	13.87418407	18.22750319	23.40348362	27.18956293	30.20238798	34.94016029	40.46810305
MU I K		.3137601D 0	.2024374D 0	.1226379D 0	.6758119D-1	.4370407D-1	.3089446D-1	.1790573D-1	.9475341D-2
1 I PF1		1.627422633	1.407229822	1.198675755	.9964404475	.8730565238	.7871525107	.6713866825	.5637047044
1 I PQ1		.7792718416	.7254431707	.6977837281	.6908949741	.6996120759	.7151738978	.7593710093	.8580277027
2 I PFC		1.223688418	1.207239029	1.183468094	1.149721622	1.122690759	1.100570159	1.066066821	1.028539513
2 I PQC		8.158009461	6.479612218	5.377925551	4.655850662	4.374139198	4.258831910	4.243957953	4.478103095
1 I ZF1		1.213440776	1.211796380	1.210927651	1.210654689	1.210746120	1.210920844	1.211299443	1.211809491
1 I ZF2		1.333306333	1.353576298	1.373509190	1.393448351	1.405866920	1.414634246	1.426689015	1.438421608
1 I ZF3		2.687341644	2.775852783	2.875156189	2.986357962	3.061062810	3.115924656	3.193570667	3.270701910

(3) P 4+2	TH = 58 SBE = 1.179178403								MCPER FUNCTIONS
	RC = 02	RC = 05	RC = 07	RC = 10	RC = 15	RC = 20	RC = 25	RC = 35	RC = 50
AMAX DB		.0108709564	.0213327376	.0436480540	.0988323393	.1772876696	.2802872360	.5675287486	1.249387366
AMIN DB		9.159968825	12.86167239	17.15695406	22.30220257	26.07727277	29.08466584	33.81711797	39.34154819
MU I K		.3483386D 0	.2274659D 0	.1387242D 0	.7671669D-1	.4967483D-1	.3513716D-1	.2037718D-1	.1078754D-1
1 I PF1		1.649855286	1.429412054	1.217524626	1.011014609	.8850258333	.7974145857	.6795429690	.5701279440
1 I PQ1		.8039291678	.7410028734	.7078873923	.6976568356	.7050071848	.7198560117	.7633742319	.8617359474
2 I PFC		1.212161466	1.197769599	1.176166033	1.144757481	1.119179321	1.098053745	1.064839232	1.028432499
2 I PQC		8.973541553	7.037104084	5.765434774	4.930036066	4.598871506	4.456660527	4.414747646	4.634358441
1 I ZF1		1.199961358	1.198145468	1.197132205	1.196747234	1.196784691	1.196927947	1.197271492	1.197755781
1 I ZF2		1.309736013	1.329504316	1.349282291	1.369258275	1.381750533	1.390582321	1.402735510	1.414571078
1 I ZF3		2.611819843	2.695958498	2.792561429	2.902557695	2.977208589	3.032327637	3.110686958	3.188855814

(3) P 4+2	TH = 59 SBE = 1.166633397								MCPER FUNCTIONS
	RC = 02	RC = 05	RC = 07	RC = 10	RC = 15	RC = 20	RC = 25	RC = 35	RC = 50
AMAX DB		.0108709564	.0213327376	.0436480540	.0988323393	.1772876696	.2802872360	.5675287486	1.249387366
AMIN DB		8.294542727	11.87680512	16.10367887	21.21256395	24.97477847	27.97591854	32.70233813	38.22284981
MU I K		.3848335D 0	.2547767D 0	.1566088D 0	.8697047D-1	.5639766D-1	.3992124D-1	.2316771D-1	.1227037D-1
1 I PF1		1.671093692	1.451674564	1.236951474	1.026200527	.8975269571	.8081374247	.6880622770	.5768308061
1 I PQ1		.8319571619	.7585747737	.7191519632	.7050781140	.7108703519	.7249108281	.7676604847	.8656792218
2 I PFC		1.200738950	1.188223858	1.168770085	1.139682853	1.115567038	1.095450325	1.063550442	1.028293492
2 I PQC		9.907033648	7.672733801	6.203255017	5.236154179	4.847815172	4.674604903	4.601547361	4.804149798
1 I ZF1		1.187153589	1.185173185	1.184012937	1.183511304	1.183492035	1.183602069	1.183908400	1.184365070
1 I ZF2		1.287319100	1.306469713	1.326010582	1.345975196	1.358522837	1.367409437	1.379650413	1.391580229
1 I ZF3		2.539783332	2.619216814	2.712739991	2.821201361	2.895635121	2.950917652	3.029891035	3.109033068

(3) P 4+2	TH = 60	SBE = 1.154700538						MCPER FUNCTIONS	
	RC = 02	RC = 05	RC = 07	RC = 10	RC = 15	RC = 20	RC = 25	RC = 35	RC = 50
AMAX DB		.0108709564	.0213327376	.0436480540	.0988323393	.1772876696	.2802872360	.5675287486	1.249387366
AMIN DB		7.474613579	10.92194029	15.06858960	20.13441099	23.88155398	26.87546195	31.59501037	37.11113524
MU I K		.4229308D 0	.2843826D 0	.1764292D 0	.9846445D-1	.6396204D-1	.4531343D-1	.2631779D-1	.1394579D-1
1 I PF1		1.690752619	1.473802048	1.256899754	1.042010589	.9105836978	.8193456095	.6969658943	.5838299917
1 I PQ1		.8638279357	.7784683564	.7317543467	.7132506369	.7172613873	.7303827341	.7722601179	.8698802367
2 I PFC		1.189456208	1.178676340	1.161292527	1.134501332	1.111854193	1.092758715	1.062198281	1.028120305
2 I PQC		10.97721705	8.399680086	6.699830373	5.579242989	5.124572643	4.915519733	4.806475712	4.989146949
1 I ZF1		1.174982406	1.172848571	1.171540651	1.170918181	1.170839506	1.170914582	1.171181548	1.171608733
1 I ZF2		1.266037439	1.284454715	1.303668002	1.323565884	1.336147793	1.345078203	1.357395057	1.369409554
1 I ZF3		2.471113176	2.545556472	2.635621696	2.742202391	2.816244383	2.871589169	2.951068366	3.031112229

(3) P 4+2	TH = 61	SBE = 1.143354068						MCPER FUNCTIONS	
	RC = 02	RC = 05	RC = 07	RC = 10	RC = 15	RC = 20	RC = 25	RC = 35	RC = 50
AMAX DB		.0108709564	.0213327376	.0436480540	.0988323393	.1772876696	.2802872360	.5675287486	1.249387366
AMIN DB		6.702760472	9.999727174	14.05290787	19.06773515	22.79714528	25.78264943	30.49433435	36.00552917
MU I K		.4622341D 0	.3162377D 0	.1983146D 0	.1113303D 0	.7246741D-1	.5138869D-1	.2987331D-1	.1583885D-1
1 I PF1		1.708443276	1.495538237	1.277289075	1.058451591	.9242193901	.8310647973	.7062767839	.5911437239
1 I PQ1		.9000648888	.8010394033	.7459021968	.7222823961	.7242502202	.7363233125	.7772080626	.8743648823
2 I PFC		1.178348265	1.169153580	1.153747377	1.129217393	1.108041517	1.089977952	1.060780593	1.027910657
2 I PQC		12.20585219	9.233528563	7.265277028	5.965338164	5.433436954	5.182789204	5.032023709	5.191297001
1 I ZF1		1.163413791	1.161141989	1.159687931	1.158941122	1.158800473	1.158838889	1.159064358	1.159460198
1 I ZF2		1.245871784	1.263444977	1.282232988	1.302001019	1.314592861	1.323554505	1.335933850	1.348022491
1 I ZF3		2.405678446	2.474906783	2.561144386	2.665483452	2.738947128	2.794244960	2.874112055	2.954978894

(3) P 4+2	TH = 62	SBE = 1.132570051						MCPER FUNCTIONS	
	RC = 02	RC = 05	RC = 07	RC = 10	RC = 15	RC = 20	RC = 25	RC = 35	RC = 50
AMAX DB		.0108709564	.0213327376	.0436480540	.0988323393	.1772876696	.2802872360	.5675287486	1.249387366
AMIN DB		5.981115595	9.113033443	13.05819114	18.01270203	21.72118292	24.69687682	29.39951827	34.90514916
MU I K		.5022781D 0	.3502260D 0	.2223773D 0	.1257086D 0	.8202398D-1	.5823126D-1	.3388629D-1	.1797805D-1
1 I PF1		1.723793121	1.516587688	1.298010551	1.075522596	.9384561626	.8433215214	.7160196969	.5987919212
1 I PQ1		.9412457709	.8266950279	.7618391927	.7323008004	.7319189598	.7427927832	.7825447085	.8791628034
2 I PFC		1.167448915	1.159683867	1.146150529	1.123836512	1.104130265	1.087107344	1.059295247	1.027662177
2 I PQC		13.61831038	10.19272733	7.911747908	6.401706447	5.779552721	5.480448176	5.281136704	5.412882625
1 I ZF1		1.152414776	1.150024882	1.148428911	1.147555174	1.147350152	1.147350258	1.147532136	1.147894781
1 I ZF2		1.226800453	1.243428600	1.261688216	1.281255109	1.293828892	1.302807377	1.315234106	1.327385240
1 I ZF3		2.343334493	2.407194517	2.489252515	2.590975887	2.663662388	2.718795580	2.798922260	2.880525053

TH = 63 SBE = 1.122326238 MCPER FUNCTIONS

(3) P 4+2	RC = 02	RC = 05	RC = 07	RC = 10	RC = 15	RC = 20	RC = 25	RC = 35	RC = 50
AMAX DB			.0213327376	.0436480540	.0988323393	.1772876696	.2802872360	.5675287486	1.249387366
AMIN DB			8.264838864	12.08635008	16.96968191	20.65339823	23.61758955	28.30977800	33.80910088
MU I K			.3861518D 0	.2487038D 0	.1417477D 0	.9275345D-1	.6593569D-1	.3841591D-1	.2039600D-1
1 I PF1			1.536621221	1.318922193	1.093212243	.9533139319	.8561428760	.7262212758	.6067963924
1 I PQ1			.8558988941	.7798510449	.7434566257	.7403644364	.7498617850	.7883169854	.8843081036
2 I PFC			1.150296853	1.138519872	1.118365318	1.100122311	1.084146523	1.057740164	1.027372396
2 I PQC			11.29914612	8.653882520	6.897141111	6.169121165	5.813336442	5.557318166	5.656595238
1 I ZF1			1.139469573	1.137739053	1.136737009	1.136465452	1.136425675	1.136561923	1.136889550
1 I ZF2			1.224394988	1.242020272	1.261306429	1.273830052	1.282808895	1.295265906	1.307466602
1 I ZF3			2.342340711	2.419895476	2.518619135	2.590317030	2.645158909	2.725405662	2.807648495

TH = 64 SBE = 1.112601940 MCPER FUNCTIONS

(3) P 4+2	RC = 02	RC = 05	RC = 07	RC = 10	RC = 15	RC = 20	RC = 25	RC = 35	RC = 50
AMAX DB			.0213327376	.0436480540	.0988323393	.1772876696	.2802872360	.5675287486	1.249387366
AMIN DB			7.458097420	11.13965153	15.93928445	19.59364371	22.54429257	27.22433700	32.71647319
MU I K			.4237358D 0	.2773431D 0	.1596011D 0	.1047895D 0	.7460800D-1	.4352945D-1	.2313004D-1
1 I PF1			1.555285440	1.339844839	1.111495452	.9688090525	.8695560429	.7369101398	.6151810574
1 I PQ1			.8891768020	.8002722766	.7559288069	.7497013439	.7576135902	.7945797049	.8898402129
2 I PFC			1.141023005	1.130875342	1.112811748	1.096020255	1.081095519	1.056113339	1.027038756
2 I PQC			12.57877681	9.509364803	7.462337843	6.609662984	6.187296573	5.864762175	5.925627733
1 I ZF1			1.129449128	1.127594936	1.126464759	1.126124832	1.126043709	1.126132366	1.126423188
1 I ZF2			1.206333620	1.223219201	1.242136969	1.254573786	1.263534107	1.276001995	1.288237839
1 I ZF3			2.280257636	2.353025628	2.448360131	2.518845362	2.573259749	2.653474998	2.736252246

TH = 65 SBE = 1.103377919 MCPER FUNCTIONS

(3) P 4+2	RC = 02	RC = 05	RC = 07	RC = 10	RC = 15	RC = 20	RC = 25	RC = 35	RC = 50
AMAX DB			.0213327376	.0436480540	.0988323393	.1772876696	.2802872360	.5675287486	1.249387366
AMIN DB			6.695575842	10.22070024	14.92239690	18.54191852	21.47656354	26.14242735	31.62633302
MU I K			.4626166D 0	.3082939D 0	.1794238D 0	.1182780D 0	.8436685D-1	.4930360D-1	.2622306D-1
1 I PF1			1.572216295	1.360559288	1.130329456	.9849525292	.8835876032	.7481169366	.6239721940
1 I PQ1			.9271230251	.8234938444	.7699302469	.7600661405	.7661468712	.8013072369	.8958049656
2 I PFC			1.131892934	1.123238909	1.107185218	1.091827548	1.077954938	1.054412376	1.026658606
2 I PQC			14.06262980	10.49962021	8.110375511	7.110357750	6.609429691	6.208524018	6.223792958
1 I ZF1			1.119937299	1.117974055	1.116717859	1.116308168	1.116184383	1.116223597	1.116475880
1 I ZF2			1.189232806	1.205277917	1.223732377	1.236040797	1.244961003	1.257417703	1.269672570
1 I ZF3			2.220846176	2.288595974	2.380152614	2.449188764	2.503029479	2.583043636	2.666244052

$P^{(4)}_{4+4}$

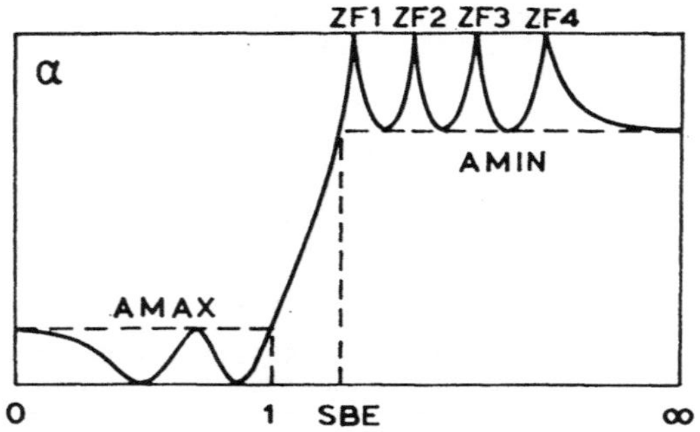

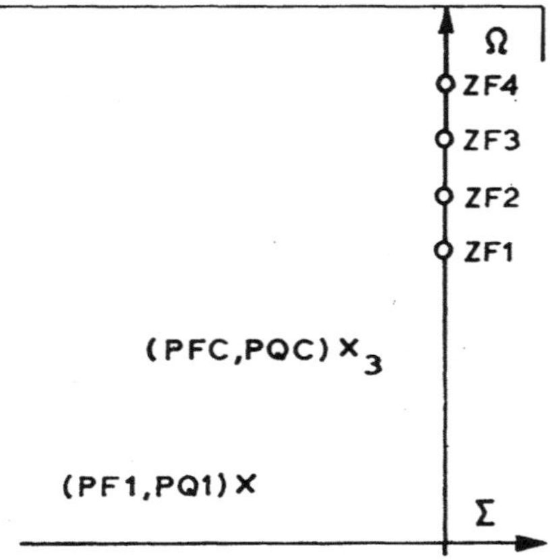

(4) P 4+4	TH = 15	SBE = 3.863703305							MCPER FUNCTIONS
	RC = 02	RC = 05	RC = 07	RC = 10	RC = 15	RC = 20	RC = 25	RC = 35	RC = 50
AMAX DB	.0017375255	.0108709564	.0213327376	.0436480540	.0988323393	.1772876696	.2802872360	.5675287486	1.249387366
AMIN DB	90.82389828	106.7785305	112.6602818	118.9229178	126.1062143	131.2794554	135.3692523	141.7655661	149.1703560
MU I K	.2876107D-4	.4582194D-5	.2328016D-5	.1132020D-5	.4950959D-6	.2729149D-6	.1704266D-6	.8160593D-7	.3479222D-7
1 I PF1	1.358590856	1.009655980	.8996188347	.7919006066	.6795496693	.6058307920	.5518560330	.4754336517	.4002947912
1 I PQ1	.5477049609	.5589433575	.5662509502	.5771419946	.5956738914	.6152471514	.6363331107	.6848028616	.7820291297
3 I PFC	2.061018754	1.695625661	1.579340234	1.466082880	1.349169159	1.273310911	1.218167710	1.140346240	1.063030112
3 I PQC	1.317753762	1.330772190	1.352408630	1.387130810	1.444876599	1.501298923	1.556930503	1.668842036	1.849758443
1 I ZF1	3.931013299	3.932994092	3.933564624	3.934091302	3.934605408	3.934923694	3.935148140	3.935456297	3.935754997
1 I ZF2	4.562775225	4.583983325	4.589877652	4.595247268	4.600427632	4.603606600	4.605836028	4.608881146	4.611816238
1 I ZF3	6.655674041	6.736511711	6.758217078	6.777713060	6.796276718	6.807551503	6.815406425	6.826067015	6.836269685
1 I ZF4	18.57645779	18.92140843	19.01252629	19.09376121	19.17054801	19.21690593	19.24907447	19.29256205	19.33399573

(4) P 4+4	TH = 16	SBE = 3.627955279							MCPER FUNCTIONS
	RC = 02	RC = 05	RC = 07	RC = 10	RC = 15	RC = 20	RC = 25	RC = 35	RC = 50
AMAX DB	.0017375255	.0108709564	.0213327376	.0436480540	.0988323393	.1772876696	.2802872360	.5675287486	1.249387366
AMIN DB	86.28892251	102.2401672	108.1212795	114.3834374	121.5663623	126.7394175	130.8291024	137.2252876	144.6299794
MU I K	.4847902D-4	.7726657D-5	.3925871D-5	.1909098D-5	.8349912D-6	.4602874D-6	.2874385D-6	.1376371D-6	.5868148D-7
1 I PF1	1.366007721	1.013613543	.9027912046	.7944231719	.6815051896	.6074684055	.5532851082	.4765988298	.4012322613
1 I PQ1	.5484955185	.5594808710	.5667284365	.5775695791	.5960592105	.6156109213	.6366849879	.6851459527	.7823808252
3 I PFC	2.051167844	1.691394637	1.576335807	1.464060792	1.347966192	1.272544411	1.217675621	1.140184625	1.063136083
3 I PQC	1.343783832	1.347811714	1.367326840	1.400276424	1.456489916	1.512089203	1.567219153	1.678595995	1.859276677
1 I ZF1	3.690214948	3.692252290	3.692845888	3.693396202	3.693935409	3.694270188	3.694506685	3.694831932	3.695147777
1 I ZF2	4.274281166	4.296519420	4.302728474	4.308395464	4.313872306	4.317237843	4.319600203	4.322829655	4.325945410
1 I ZF3	6.211237357	6.297478457	6.320637833	6.341440021	6.361247239	6.373277101	6.381657948	6.393032142	6.403917610
1 I ZF4	17.27999055	17.65081738	17.74868497	17.83588320	17.91824859	17.96794262	18.00241061	18.04898508	18.09333567

(4) P 4+4	TH = 17	SBE = 3.420303620						MCPER FUNCTIONS	
	RC = 02	RC = 05	RC = 07	RC = 10	RC = 15	RC = 20	RC = 25	RC = 35	RC = 50
AMAX DB	.0017375255	.0108709564	.0213327376	.0436480540	.0988323393	.1772876696	.2802872360	.5675287486	1.249387366
AMIN DB	82.02332715	97.97044236	103.8507779	110.1123549	117.2948288	122.4676584	126.5572076	132.9532368	140.3578096
MU K	.7921978D-4	.1263217D-4	.6418907D-5	.3121636D-5	.1365396D-5	.7526916D-6	.4700452D-6	.2250806D-6	.9596426D-7
1 PF1	1.374001326	1.017863920	.9061951602	.7971276930	.6836001928	.6092220578	.5548149962	.4778457356	.4022351633
1 PQ1	.5493588828	.5600624055	.5672437985	.5780301272	.5964734535	.6160015733	.6370625964	.6855137969	.7827576023
3 PFC	2.040605982	1.686855560	1.573111904	1.461890138	1.346673683	1.271719753	1.217145072	1.140008227	1.063246402
3 PQC	1.372494945	1.366398742	1.383552101	1.414536406	1.469056753	1.523748268	1.578324988	1.689110561	1.869524425
1 ZF1	3.478087248	3.480167138	3.480780986	3.481352762	3.481915324	3.482265691	3.482513679	3.482855351	3.483187806
1 ZF2	4.019692769	4.042894530	4.049404148	4.055357370	4.061121575	4.064668897	4.067161209	4.070571424	4.073864949
1 ZF3	5.817354444	5.908989021	5.933602108	5.955710710	5.976761960	5.989547265	5.998454310	6.010542497	6.022111151
1 ZF4	16.12669975	16.52368852	16.62840055	16.72164581	16.80966480	16.86273689	16.89953145	16.94922648	16.99652237

(4) P 4+4	TH = 18	SBE = 3.236067977						MCPER FUNCTIONS	
	RC = 02	RC = 05	RC = 07	RC = 10	RC = 15	RC = 20	RC = 25	RC = 35	RC = 50
AMAX DB	.0017375255	.0108709564	.0213327376	.0436480540	.0988323393	.1772876696	.2802872360	.5675287486	1.249387366
AMIN DB	77.99614860	93.93827792	99.81767868	106.0785575	113.2604896	118.4330482	122.5224346	128.9182765	136.3227066
MU K	.1259484D-3	.2009491D-4	.1021212D-4	.4966748D-5	.2172579D-5	.1197699D-5	.7479598D-6	.3581675D-6	.1527090D-6
1 PF1	1.382592154	1.022414736	.9098361407	.8000180280	.6858373388	.6110938158	.5564474076	.4791756669	.4033044746
1 PQ1	.5503001571	.5606900011	.5677985523	.5785247734	.5969174656	.6164198107	.6374665569	.6859069269	.7831599444
3 PFC	2.029323475	1.682002521	1.569664011	1.459567624	1.345289407	1.270835290	1.216574781	1.139816205	1.063360583
3 PQC	1.404124177	1.386633002	1.401159809	1.429968136	1.482620500	1.536312473	1.590280116	1.700412996	1.880525472
1 ZF1	3.289859407	3.291967154	3.292598284	3.293189229	3.293773299	3.294138296	3.294397182	3.294754573	3.295103069
1 ZF2	3.793385212	3.817481963	3.824277400	3.830505199	3.836547189	3.840271257	3.842890370	3.846477568	3.849945793
1 ZF3	5.465605794	5.562618623	5.588684514	5.612099382	5.634394926	5.647935959	5.657369439	5.670171973	5.682424182
1 ZF4	15.09265388	15.51602352	15.62766910	15.72704316	15.82079096	15.87728385	15.91643290	15.96928336	16.01955420

(4) P 4+4	TH = 19	SBE = 3.071553487						MCPER FUNCTIONS	
	RC = 02	RC = 05	RC = 07	RC = 10	RC = 15	RC = 20	RC = 25	RC = 35	RC = 50
AMAX DB	.0017375255	.0108709564	.0213327376	.0436480540	.0988323393	.1772876696	.2802872360	.5675287486	1.249387366
AMIN DB	74.18149047	90.11764954	95.99593533	102.2559823	109.4372695	114.6095058	118.6986984	125.0943177	132.4985781
MU K	.1954004D-3	.3119734D-4	.1585635D-4	.7712601D-5	.3373934D-5	.1860050D-5	.1161623D-5	.5562681D-6	.2371762D-6
1 PF1	1.391802611	1.027274277	.9137200427	.8030983524	.6882195014	.6130859106	.5581841872	.4805900220	.4044412475
1 PQ1	.5513251303	.5613659221	.5683943711	.5790547639	.5973921706	.6168664007	.6378975452	.6863259209	.7835883753
3 PFC	2.017311462	1.676829474	1.565987443	1.457089792	1.343811006	1.269889271	1.215963381	1.139607659	1.063478102
3 PQC	1.438940927	1.408625735	1.420233641	1.446635008	1.497228889	1.549821703	1.603119702	1.712533092	1.892305751
1 ZF1	3.121765392	3.123885639	3.124530937	3.125138639	3.125742271	3.126120889	3.126390046	3.126762409	3.127126343
1 ZF2	3.590918410	3.615840140	3.622906122	3.629396362	3.635706121	3.639601640	3.642344241	3.646104435	3.649744116
1 ZF3	5.149347191	5.251716639	5.279233681	5.303954227	5.327494061	5.341791004	5.351751104	5.365268291	5.378204401
1 ZF4	14.15899294	14.60888067	14.72754119	14.83312251	14.93267361	14.99263047	15.03416251	15.09020432	15.14348093

(4) P 4+4		TH = 20	SBE = 2.923804400						MCPER FUNCTIONS	
MU		RC = 02	RC = 05	RC = 07	RC = 10	RC = 15	RC = 20	RC = 25	RC = 35	RC = 50
	AMAX DB	.0017375255	.0108709564	.0213327376	.0436480540	.0988323393	.1772876696	.2802872360	.5675287486	1.249387366
	AMIN DB	70.55748874	86.48654968	92.36351581	98.62257954	105.8031051	110.9749611	115.0639251	121.4592819	128.8633423
	K	.2965689D-3	.4738845D-4	.2408930D-4	.1171847D-4	.5126781D-5	.2826519D-5	.1765240D-5	.8453487D-6	.3604399D-6
1	PF1	1.401657168	1.032451526	.9178532475	.8063731757	.6907497791	.6152007447	.5600273203	.4820903042	.4056466124
1	PQ1	.5524403780	.5620926809	.5690331002	.5796214659	.5978985773	.6173421785	.6383562957	.6867714060	.7840434617
3	PFC	2.004562182	1.671330283	1.562077359	1.454453033	1.342235995	1.268879840	1.215309421	1.139381627	1.063598400
3	PQC	1.477251541	1.432500992	1.440866412	1.464607006	1.512934375	1.564319671	1.616882212	1.725503354	1.904893491
1	ZF1	2.970792732	2.972909480	2.973565689	2.974187624	2.974808776	2.975199952	2.975478721	2.975865270	2.976244007
1	ZF2	3.408740696	3.434416284	3.441737064	3.448477164	3.455044255	3.459105691	3.461968305	3.465897305	3.469705020
1	ZF3	4.863266445	4.970963194	4.999928827	5.025953923	5.050737719	5.065790626	5.076277464	5.090509550	5.104129870
1	ZF4	13.31066638	13.78711244	13.91285924	14.02472147	14.13014872	14.19361279	14.23755669	14.29682663	14.35314088

(4) P 4+4		TH = 21	SBE = 2.790428110						MCPER FUNCTIONS	
MU		RC = 02	RC = 05	RC = 07	RC = 10	RC = 15	RC = 20	RC = 25	RC = 35	RC = 50
	AMAX DB	.0017375255	.0108709564	.0213327376	.0436480540	.0988323393	.1772876696	.2802872360	.5675287486	1.249387366
	AMIN DB	67.10553071	83.02620434	88.90161910	95.15952867	102.3391614	107.5105718	111.5992683	117.9943178	125.3981441
	K	.4412894D-3	.7058132D-4	.3588550D-4	.1745917D-4	.7639095D-5	.4211836D-5	.2630490D-5	.1259749D-5	.5371466D-6
1	PF1	1.412182512	1.037956213	.9222426500	.8098473607	.6934315062	.6174409009	.5619789393	.4836781266	.4069217811
1	PQ1	.5536533819	.5628730656	.5697167744	.5802263787	.5984377862	.6178480525	.6388436055	.6872440610	.7845258159
3	PFC	1.991069271	1.665498775	1.557928796	1.451653589	1.340561764	1.267805038	1.214611363	1.139137090	1.063720878
3	PQC	1.519404740	1.458397122	1.463161057	1.483961352	1.529794560	1.579854243	1.631609686	1.739359209	1.918319381
1	ZF1	2.834503093	2.836599724	2.837263452	2.837896983	2.838533519	2.838936139	2.839223830	2.839623739	2.840016609
1	ZF2	3.243977170	3.270334766	3.277894183	3.284871154	3.291684743	3.295906324	3.298885319	3.302978735	3.306950889
1	ZF3	4.603066852	4.716052452	4.746463034	4.773790905	4.799817940	4.815626705	4.826640321	4.841587472	4.855892266
1	ZF4	12.53553175	13.03846382	13.17135575	13.28956619	13.40093975	13.46795388	13.51433869	13.57687418	13.63625885

(4) P 4+4		TH = 22	SBE = 2.669467163						MCPER FUNCTIONS	
MU		RC = 02	RC = 05	RC = 07	RC = 10	RC = 15	RC = 20	RC = 25	RC = 35	RC = 50
	AMAX DB	.0017375255	.0108709564	.0213327376	.0436480540	.0988323393	.1772876696	.2802872360	.5675287486	1.249387366
	AMIN DB	63.80965689	79.72047237	85.59407417	91.85063736	99.02922998	104.2001218	108.2885070	114.6831992	122.0867535
	K	.6449368D-3	.1032705D-3	.5251656D-4	.2555454D-4	.1118249D-4	.6165864D-5	.3851010D-5	.1844336D-5	.7864341D-6
1	PF1	1.423407712	1.043798862	.9268956912	.8135261442	.6962682664	.6198091515	.5640413316	.4853552177	.4082680507
1	PQ1	.5549726726	.5637101718	.5704476375	.5808711453	.5990109968	.6183850096	.6393603389	.6877446202	.7850360985
3	PFC	1.976828080	1.659328795	1.553536688	1.448687569	1.338785577	1.266662802	1.213867581	1.138872962	1.063844899
3	PQC	1.565797999	1.486468485	1.487231753	1.504783248	1.547872682	1.596477806	1.647348032	1.754139239	1.932616759
1	ZF1	2.710901796	2.712961092	2.713628812	2.714271193	2.714920885	2.715333787	2.715629675	2.716042077	2.716448380
1	ZF2	3.094275764	3.121243178	3.129024713	3.136225194	3.143274075	3.147649804	3.150741397	3.154994640	3.159127467
1	ZF3	4.365237047	4.483462303	4.515312913	4.543941015	4.571210091	4.587774420	4.599314754	4.614977043	4.629966517
1	ZF4	11.82369913	12.35291662	12.49299751	12.61761540	12.73500169	12.80560792	12.85446260	12.92030149	12.98279016

(4) P 4+4	TH = 23 SBE = 2.559304665								MCPER FUNCTIONS
	RC = 02	RC = 05	RC = 07	RC = 10	RC = 15	RC = 20	RC = 25	RC = 35	RC = 50
AMAX DB	.0017375255	.0108709564	.0213327376	.0436480540	.0988323393	.1772876696	.2802872360	.5675287486	1.249387366
AMIN DB	60.65609721	76.55537859	82.42687283	88.68187395	95.85926174	101.0295536	105.1175788	111.5118580	118.9150978
MU I K	.9272464D-3	.1486726D-3	.7562343D-4	.3680496D-4	.1610783D-4	.8882236D-5	.5547803D-5	.2657095D-5	.1133040D-5
1 I PF1	1.435364382	1.049990847	.9318203941	.8174151601	.6992639065	.6223084689	.5662169469	.4871234265	.4096868076
1 I PQ1	.5564079988	.5646074393	.5712281645	.5815575669	.5996195165	.6189541222	.6399074327	.6882738769	.7855750215
3 I PFC	1.961836010	1.652814270	1.548895898	1.445550961	1.336904581	1.265450964	1.213076362	1.138588095	1.063969784
3 I PQC	1.616885047	1.516887419	1.513205188	1.527166709	1.567238150	1.614247677	1.664147374	1.769885443	1.947821812
1 I ZF1	2.598341364	2.600345518	2.601013578	2.601661958	2.602322489	2.602744459	2.603047790	2.603471782	2.603890782
1 I ZF2	2.957693472	2.985198551	2.993185393	3.000595694	3.007868310	3.012391975	3.015592232	3.020000523	3.024290083
1 I ZF3	4.146880970	4.270284153	4.303568362	4.333493245	4.362002610	4.379321982	4.391388858	4.407766241	4.423440527
1 I ZF4	11.16704647	11.72220495	11.86950097	12.00057577	12.12403628	12.19827532	12.24962853	12.31880882	12.38443571

(4) P 4+4	TH = 24 SBE = 2.458593336								MCPER FUNCTIONS
	RC = 02	RC = 05	RC = 07	RC = 10	RC = 15	RC = 20	RC = 25	RC = 35	RC = 50
AMAX DB	.0017375255	.0108709564	.0213327376	.0436480540	.0988323393	.1772876696	.2802872360	.5675287486	1.249387366
AMIN DB	57.63290730	73.51874671	79.38780214	85.64099972	92.81699892	97.98660002	102.0742111	108.4680155	115.8708941
MU I K	.1313272D-2	.2108932D-3	.1073025D-3	.5223361D-4	.2286389D-4	.1260869D-4	.7875705D-5	.3772239D-5	.1608627D-5
1 I PF1	1.448086849	1.056544452	.9370254026	.8215204639	.7024225525	.6249420363	.5685084069	.4889847293	.4111795319
1 I PQ1	.5579705314	.5655686939	.5720610873	.5822876183	.6002647699	.6195565541	.6404859016	.6888326878	.7861433518
3 I PFC	1.946092868	1.645949277	1.544001257	1.442239645	1.334915804	1.264167254	1.212235904	1.138281275	1.064094812
3 I PQC	1.673184711	1.549846501	1.541222028	1.551215521	1.587967165	1.633226571	1.682062427	1.786643525	1.963973807
1 I ZF1	2.495449199	2.497379824	2.498044445	2.498695874	2.499364837	2.499794615	2.500104602	2.500539239	2.500970169
1 I ZF2	2.832611034	2.860582030	2.868757134	2.876363272	2.883847746	2.888512930	2.891817776	2.896376143	2.900818327
1 I ZF3	3.945590409	4.074095283	4.108804907	4.140022065	4.169769310	4.187842938	4.200436040	4.217528337	4.233887485
1 I ZF4	10.55885588	11.13945191	11.29396880	11.43153831	11.56112838	11.63903911	11.69291890	11.76547852	11.83427834

(4) P 4+4	TH = 25 SBE = 2.366201583								MCPER FUNCTIONS
	RC = 02	RC = 05	RC = 07	RC = 10	RC = 15	RC = 20	RC = 25	RC = 35	RC = 50
AMAX DB	.0017375255	.0108709564	.0213327376	.0436480540	.0988323393	.1772876696	.2802872360	.5675287486	1.249387366
AMIN DB	54.72968061	70.59990748	76.46615216	82.71727633	89.89168234	95.06049203	99.14762899	105.5408900	112.9433556
MU I K	.1834493D-2	.2951241D-3	.1502078D-3	.7313684D-4	.3201960D-4	.1765938D-4	.1103109D-4	.5283911D-5	.2253369D-5
1 I PF1	1.461612321	1.063472936	.9425200244	.8258485613	.7057486261	.6277132607	.5709185142	.4909412363	.4127478024
1 I PQ1	.5596731061	.5665981969	.5729494240	.5830634655	.6009483093	.6201935693	.6410968444	.6894219777	.7867419152
3 I PFC	1.929601241	1.638728119	1.538847594	1.438749410	1.332816164	1.262809301	1.211344315	1.137951221	1.064219218
3 I PQC	1.735291342	1.585561170	1.571438589	1.577044317	1.610143405	1.653483118	1.701152921	1.804463219	1.981115344
1 I ZF1	2.401072608	2.402910735	2.403568019	2.404219446	2.404894351	2.405330629	2.405646456	2.406090756	2.406532816
1 I ZF2	2.717668094	2.746034015	2.754380168	2.762167914	2.769852073	2.774652161	2.778057381	2.782760670	2.787351199
1 I ZF3	3.759348194	3.892861860	3.928986685	3.961490381	3.992472338	4.011299123	4.024417976	4.042224845	4.059268799
1 I ZF4	9.993536908	10.59889353	10.76061379	10.90470224	11.04046973	11.12209863	11.17852219	11.25449868	11.32650645

(4) P 4+4		TH = 26	SBE = 2.281172033						MCPER FUNCTIONS	
		RC = 02	RC = 05	RC = 07	RC = 10	RC = 15	RC = 20	RC = 25	RC = 35	RC = 50
	AMAX DB	.0017375255	.0108709564	.0213327376	.0436480540	.0988323393	.1772876696	.2802872360	.5675287486	1.249387366
	AMIN DB	51.93731864	67.78946449	73.65248119	79.90123071	87.07381613	92.24172285	96.32831929	102.7209614	110.1229566
MU	K	.2530079D-2	.4078747D-3	.2076710D-3	.1011436D-3	.4429036D-4	.2442946D-4	.1526104D-4	.7310582D-5	.3117828D-5
1	PF1	1.475981025	1.070790604	.9483142781	.8304064374	.7092468638	.6306257859	.5734502637	.4929951993	.4143933016
1	PQ1	.5615305158	.5677007016	.5738965122	.5838874864	.6016718273	.6208665402	.6417414512	.6900427450	.7873716001
3	PFC	1.912366885	1.631145408	1.533429782	1.435075969	1.330602471	1.261374629	1.210399613	1.137596582	1.064342192
3	PQC	1.803887131	1.624272738	1.604028752	1.604779813	1.633858816	1.675092453	1.721484083	1.823398649	1.999292639
1	ZF1	2.314236525	2.315962593	2.316608527	2.317256806	2.317935079	2.318376503	2.318697323	2.319150268	2.319602625
1	ZF2	2.611713334	2.640404276	2.648904163	2.656859081	2.664730483	2.669658678	2.673159925	2.678002799	2.682737229
1	ZF3	3.586453792	3.724864355	3.762391829	3.796174912	3.828387532	3.847966017	3.861609956	3.880130867	3.897859451
1	ZF4	9.466413195	10.09566644	10.26454643	10.41516246	10.55714640	10.64250700	10.70152030	10.78095050	10.85620137

(4) P 4+4		TH = 27	SBE = 2.202689265						MCPER FUNCTIONS	
		RC = 02	RC = 05	RC = 07	RC = 10	RC = 15	RC = 20	RC = 25	RC = 35	RC = 50
	AMAX DB	.0017375255	.0108709564	.0213327376	.0436480540	.0988323393	.1772876696	.2802872360	.5675287486	1.249387366
	AMIN DB	49.24784645	65.07910454	70.93842554	77.18446451	84.35497692	89.52185724	93.60783986	99.99977957	107.4012416
MU	K	.3448321D-2	.5572432D-3	.2838433D-3	.1382855D-3	.6056910D-4	.3341236D-4	.2087411D-4	.1000025D-4	.4265185D-5
1	PF1	1.491236313	1.078512885	.9544189448	.8352015909	.7129223372	.6336835077	.5761068536	.4951490198	.4161178215
1	PQ1	.5635598619	.5688815191	.5749060480	.5847622930	.6024371698	.6215769573	.6424210109	.6906960670	.7880333627
3	PFC	1.894399128	1.623196153	1.527742783	1.431214985	1.328271433	1.259860664	1.209399726	1.137215937	1.064462874
3	PQC	1.879756670	1.666251892	1.639186169	1.634562216	1.659214493	1.698136873	1.743127172	1.843508738	2.018555836
1	ZF1	2.234110633	2.235704468	2.236334924	2.236976818	2.237655305	2.238100976	2.238425912	2.238886446	2.239348235
1	ZF2	2.513765710	2.542713155	2.551349416	2.559456910	2.567502375	2.572552208	2.576145007	2.581121954	2.585995673
1	ZF3	3.425465483	3.568639559	3.607554453	3.642608145	3.676046367	3.696374681	3.710542822	3.729777029	3.748189928
1	ZF4	8.973556066	9.625642806	9.801609853	9.958744439	10.10697358	10.19610577	10.25772324	10.34064300	10.41917205

(4) P 4+4		TH = 28	SBE = 2.130054468						MCPER FUNCTIONS	
		RC = 02	RC = 05	RC = 07	RC = 10	RC = 15	RC = 20	RC = 25	RC = 35	RC = 50
	AMAX DB	.0017375255	.0108709564	.0213327376	.0436480540	.0988323393	.1772876696	.2302872360	.5675287486	1.249387366
	AMIN DB	46.65426365	62.46144299	68.31654415	74.55949826	81.72765780	86.89337534	90.97866332	97.36980888	104.7706684
MU	K	.4648222D-2	.7532304D-3	.3838599D-3	.1870790D-3	.8196286D-4	.4522007D-4	.2825315D-4	.1353660D-4	.5773864D-5
1	PF1	1.507424711	1.086656413	.9608456233	.8402420700	.7167804758	.6368905899	.5788916985	.4974052577	.4179232703
1	PQ1	.5657809771	.5701465951	.5759821302	.5856907583	.6032463517	.6223264404	.6431369203	.6913831068	.7887282317
3	PFC	1.875711278	1.614875859	1.521781696	1.427162085	1.325819666	1.258264733	1.208342489	1.136807794	1.064580357
3	PQC	1.963804178	1.711802743	1.677126804	1.666546824	1.686321690	1.722706590	1.766160086	1.864857663	2.038959365
1	ZF1	2.159983531	2.161424313	2.162035052	2.162667234	2.163344204	2.163791678	2.164119825	2.164586855	2.165057180
1	ZF2	2.422983994	2.452121076	2.460876376	2.469121731	2.477329371	2.482492710	2.486172465	2.491277801	2.496286034
1	ZF3	3.275154968	3.422935038	3.463219078	3.499532745	3.534190352	3.555266147	3.569957358	3.589903862	3.609000599
1	ZF4	8.511653071	9.185300640	9.368250102	9.531874411	9.686365691	9.779295098	9.843539297	9.929983094	10.01182510

(4) P 4+4	TH = 29	SBE = 2.062665340						MCPER FUNCTIONS	
	RC = 02	RC = 05	RC = 07	RC = 10	RC = 15	RC = 20	RC = 25	RC = 35	RC = 50
AMAX DB	.0017375255	.0108709564	.0213327376	.0436480540	.0988323393	.1772876696	.2802872360	.5675287486	1.249387366
AMIN DB	44.15042398	59.92989664	65.78019072	72.01964308	79.18513967	84.34954394	88.43404827	94.82429879	102.2244800
MU I K	.6201244D-2	.1008104D-2	.5140324D-3	.2506212D-3	.1098356D-3	.6060700D-4	.3787020D-4	.1814617D-4	.7740624D-5
1 I PF1	1.524595863	1.095239114	.9676067908	.8455365119	.7208270914	.6402514820	.5818084431	.4997666413	.4198116784
1 I PQ1	.5682169346	.5715025997	.5771293113	.5866760458	.6041015741	.6231167510	.6438906933	.6921051202	.7894573142
3 I PFC	1.856321030	1.606180633	1.515541814	1.422912893	1.323243699	1.256584068	1.207225646	1.136370585	1.064693683
3 I PQC	2.057073894	1.761267538	1.718091849	1.700905874	1.715302964	1.748900573	1.790668037	1.887515357	2.060562328
1 I ZF1	2.091242253	2.092508469	2.093095142	2.093714198	2.094386343	2.094834636	2.095165061	2.095637458	2.096115391
1 I ZF2	2.338642648	2.367904371	2.376761457	2.385129886	2.393486133	2.398756199	2.402518200	2.407746074	2.412883886
1 I ZF3	3.134471384	3.286673046	3.328304503	3.365865403	3.401734857	3.423555242	3.438768108	3.459425624	3.479205535
1 I ZF4	8.077902981	8.771620939	8.961413319	9.131476535	9.292233516	9.388980781	9.455871999	9.545872555	9.631061783

(4) P 4+4	TH = 30	SBE = 2.000000000						MCPER FUNCTIONS	
	RC = 02	RC = 05	RC = 07	RC = 10	RC = 15	RC = 20	RC = 25	RC = 35	RC = 50
AMAX DB	.0017375255	.0108709564	.0213327376	.0436480540	.0988323393	.1772876696	.2802872360	.5675287486	1.249387366
AMIN DB	41.73093788	57.47857880	63.32340788	69.55889424	76.72138450	81.88430952	85.96793232	92.35717707	99.75659684
MU I K	.8193192D-2	.1336814D-2	.6820710D-3	.3327019D-3	.1458582D-3	.8049789D-4	.5030410D-4	.2410689D-4	.1028419D-4
1 I PF1	1.542802335	1.104280295	.9747158683	.8510941868	.7250684051	.6437709391	.5848609775	.5022360781	.4217852069
1 I PQ1	.5708946631	.5729570315	.5783526576	.5877216438	.6050052437	.6239498058	.6446839722	.6928634638	.7902218019
3 I PFC	1.836250862	1.597107298	1.509018687	1.418463053	1.320539985	1.254815806	1.206046854	1.135902668	1.064801840
3 I PQC	2.160774223	1.815032137	1.762351115	1.737830646	1.746293473	1.776827503	1.816744331	1.911558092	2.083428939
1 I ZF1	2.027355887	2.028425273	2.028983422	2.029585852	2.030250291	2.030697876	2.031029619	2.031506218	2.031990801
1 I ZF2	2.260112529	2.289435937	2.298377698	2.306854384	2.315346025	2.320715405	2.324554831	2.329899233	2.335161527
1 I ZF3	3.002512522	3.158921666	3.201874920	3.240667921	3.277740189	3.300301658	3.316034436	3.337401354	3.357863560
1 I ZF4	7.669930910	8.382005393	8.578463515	8.754890618	8.921901838	9.022481884	9.092037737	9.185625625	9.274195565

(4) P 4+4	TH = 31	SBE = 1.941604026						MCPER FUNCTIONS	
	RC = 02	RC = 05	RC = 07	RC = 10	RC = 15	RC = 20	RC = 25	RC = 35	RC = 50
AMAX DB	.0017375255	.0108709564	.0213327376	.0436480540	.0988323393	.1772876696	.2802872360	.5675287486	1.249387366
AMIN DB	39.39109414	55.10221205	60.94083893	67.17184235	74.33094615	79.49220903	83.57484267	89.96296020	97.36152756
MU I K	.1072619D-1	.1757476D-2	.8973421D-3	.4379332D-3	.1920670D-3	.1060204D-3	.6626098D-4	.3175792D-4	.1354951D-4
1 I PF1	1.562099204	1.113800740	.9821872905	.8569250447	.7295110773	.6474540426	.5880534538	.5048166661	.4238461547
1 I PQ1	.5738456876	.5745183408	.5796578168	.5888314046	.6059599953	.6248276922	.6455185396	.6936596037	.7910229779
3 I PFC	1.815528419	1.587653514	1.502208181	1.413808262	1.317704911	1.252956996	1.204803674	1.135402322	1.064903760
3 I PQC	2.276306306	1.873532407	1.810206937	1.777533890	1.779442467	1.806606864	1.844491226	1.937069116	2.107629022
1 I ZF1	1.967862368	1.968711831	1.969236886	1.969819108	1.970472891	1.970918200	1.971250274	1.971729377	1.972220120
1 I ZF2	2.186845363	2.216169643	2.225179166	2.233749308	2.242363012	2.247824171	2.251736100	2.257190865	2.262572387
1 I ZF3	2.878501597	3.038871570	3.083116633	3.123123913	3.161388269	3.184686616	3.200937197	3.223011534	3.244154915
1 I ZF4	7.285718898	8.014209978	8.217116271	8.399805816	8.573043092	8.677464338	8.749699346	8.846902555	8.938885650

(4) P 4+4	TH = 32	SBE = 1.887079915						MCPER FUNCTIONS	
	RC = 02	RC = 05	RC = 07	RC = 10	RC = 15	RC = 20	RC = 25	RC = 35	RC = 50
AMAX DB	.0017375255	.0108709564	.0213327376	.0436480540	.0988323393	.1772876696	.2802872360	.5675287486	1.249387366
AMIN DB	37.12679763	52.79605553	58.62765403	64.85359885	72.00889541	77.16829462	81.24982090	87.63667813	95.03429363
MU I K	.1392067D-1	.2291908D-2	.1171163D-2	.5718999D-3	.2509315D-3	.1385432D-3	.8659822D-4	.4151128D-4	.1771272D-4
1 I PF1	1.582543347	1.123822802	.9900365798	.8630397670	.7341622394	.6513062243	.5913903046	.5075117071	.4259969682
1 I PQ1	.5771070234	.5761960730	.5810510987	.5900095894	.6069687169	.6257526866	.6463963323	.6944951259	.7918622249
3 I PFC	1.794186845	1.577817914	1.495106561	1.408944305	1.314734808	1.251004601	1.203493581	1.134867749	1.064998319
3 I PQC	2.405297795	1.937261685	1.861998720	1.820252618	1.814914979	1.838370174	1.874020929	1.964139386	2.133238555
1 I ZF1	1.912357749	1.912963283	1.913450561	1.914008908	1.914649015	1.915090444	1.915421835	1.915903211	1.916398093
1 I ZF2	2.118361139	2.147627657	2.156688281	2.165337136	2.174059487	2.179604783	2.183584197	2.189143011	2.194638354
1 I ZF3	2.761768324	2.925817151	2.971319159	3.012519871	3.051963683	3.075993914	3.092759773	3.115539127	3.137362288
1 I ZF4	6.923548463	7.666290948	7.875384885	8.064206784	8.243623465	8.351887012	8.426812148	8.527655620	8.623082973

(4) P 4+4	TH = 33	SBE = 1.836078459						MCPER FUNCTIONS	
	RC = 02	RC = 05	RC = 07	RC = 10	RC = 15	RC = 20	RC = 25	RC = 35	RC = 50
AMAX DB	.0017375255	.0108709564	.0213327376	.0436480540	.0988323393	.1772876696	.2802872360	.5675287486	1.249387366
AMIN DB	34.93452101	50.55584418	56.37948810	62.59973309	69.75075664	74.90807023	78.98835932	85.37331048	92.77036552
MU I K	.1791736D-1	.2966250D-2	.1517140D-2	.7413330D-3	.3254334D-3	.1797200D-3	.1123523D-3	.5386535D-4	.2298697D-4
1 I PF1	1.604192311	1.134370494	.9982804251	.8694498223	.7390295304	.6553332917	.5948762630	.5103247202	.4282402507
1 I PQ1	.5807222521	.5780010378	.5825395672	.5912609203	.6080345786	.6267272734	.6473194564	.6953717469	.7927410331
3 I PFC	1.772265083	1.567600241	1.487710560	1.403867091	1.311625972	1.248955505	1.202113961	1.134297070	1.065084331
3 I PQC	2.549642691	2.006779527	1.918108215	1.866251330	1.852893773	1.872262395	1.905456709	1.992868383	2.160340295
1 I ZF1	1.860487429	1.860824011	1.861268709	1.861799435	1.862422779	1.862858683	1.863188353	1.863670237	1.864168709
1 I ZF2	2.054237829	2.083390086	2.092485450	2.101198372	2.110015904	2.115637601	2.119679394	2.125335800	2.130939402
1 I ZF3	2.651733374	2.819141113	2.865859782	2.908229701	2.948838194	2.973594428	2.990872570	3.014354067	3.036855301
1 I ZF4	6.581952502	7.336560562	7.551536316	7.746329645	7.931858829	8.043957604	8.121579814	8.226084952	8.324986019

(4) P 4+4	TH = 34	SBE = 1.788291650						MCPER FUNCTIONS	
	RC = 02	RC = 05	RC = 07	RC = 10	RC = 15	RC = 20	RC = 25	RC = 35	RC = 50
AMAX DB	.0017375255	.0108709564	.0213327376	.0436480540	.0988323393	.1772876696	.2802872360	.5675287486	1.249387366
AMIN DB	32.81126887	48.37773780	54.19238846	60.40621914	67.55245399	72.70743748	76.78634683	83.17023233	90.56560833
MU I K	.2287897D-1	.3811651D-2	.1951554D-2	.9543091D-3	.4191576D-3	.2315411D-3	.1447714D-3	.6942060D-4	.2962918D-4
1 I PF1	1.627102614	1.145469566	1.006936763	.8761675262	.7441211344	.6595414561	.5985163846	.5132594575	.4305787732
1 I PQ1	.5847428137	.5799455089	.5841311489	.5925906398	.6091610649	.6277541681	.6482902052	.6962913260	.7936610103
3 I PFC	1.749808089	1.557001500	1.480017472	1.398572700	1.308374669	1.246806515	1.200662113	1.133688325	1.065160551
3 I PQC	2.711548179	2.082721957	1.978965671	1.915825739	1.893581572	1.908443541	1.938934172	2.023365045	2.189024477
1 I ZF1	1.811938923	1.811980402	1.812377596	1.812876869	1.813480297	1.813908999	1.814235885	1.814716979	1.815217962
1 I ZF2	1.994102946	2.023086445	2.032200525	2.040963009	2.049862233	2.055552515	2.059651502	2.065398908	2.071105059
1 I ZF3	2.547895455	2.718301793	2.766190853	2.809701985	2.851457987	2.876933345	2.894720250	2.918900480	2.942077729
1 I ZF4	6.259674520	7.023550520	7.244054878	7.444625755	7.636178470	7.752096337	7.832418041	7.940602187	8.043004433

(4) P 4+4	TH = 35	SBE = 1.743446796						MCPER FUNCTIONS	
	RC = 02	RC = 05	RC = 07	RC = 10	RC = 15	RC = 20	RC = 25	RC = 35	RC = 50
AMAX DB	.0017375255	.0108709564	.0213327376	.0436480540	.0988323393	.1772876696	.2802872360	.5675287486	1.249387366
AMIN DB	30.75455354	46.25827841	52.06277053	58.26939049	65.41026549	70.56264955	74.64002267	81.02216781	88.41623536
MU I K	.2899161D-1	.4865036D-2	.2493799D-2	.1220479D-2	.5363975D-3	.2963927D-3	.1853527D-3	.8889792D-4	.3794794D-4
1 I PF1	1.651327288	1.157147577	1.016024859	.8832061058	.7494458244	.6639373639	.6023160715	.5163199209	.4330154858
1 I PQ1	.5892295533	.5820434588	.5858347593	.5940045802	.6103520127	.6288363425	.6493110782	.6972558801	.7946238923
3 I PFC	1.726866963	1.546024111	1.472025241	1.393057426	1.304977164	1.244554374	1.199135252	1.133039469	1.065225665
3 I PQC	2.893589424	2.165813518	2.045057022	1.969307097	1.937203625	1.947090514	1.974602714	2.055748820	2.219389601
1 I ZF1	1.766435871	1.766154858	1.766499489	1.766963395	1.767543692	1.767963475	1.768286492	1.768765470	1.769267856
1 I ZF2	1.937626575	1.966388587	1.975505738	1.984303452	1.993270890	1.999021876	2.003172798	2.009004483	2.014807327
1 I ZF3	2.449820493	2.622822677	2.671829275	2.716449448	2.759333110	2.785519597	2.803811154	2.828686110	2.852536923
1 I ZF4	5.955633718	6.725981525	6.951612145	7.157731691	7.355195060	7.474905899	7.557924458	7.669800337	7.775728881

(4) P 4+4	TH = 36	SBE = 1.701301617						MCPER FUNCTIONS	
	RC = 02	RC = 05	RC = 07	RC = 10	RC = 15	RC = 20	RC = 25	RC = 35	RC = 50
AMAX DB	.0017375255	.0108709564	.0213327376	.0436480540	.0988323393	.1772876696	.2802872360	.5675287486	1.249387366
AMIN DB	28.76238173	44.19435471	49.98738032	56.18590137	63.32078374	68.47027160	72.54593663	78.92615018	86.31876812
MU I K	.3646539D-1	.6169959D-2	.3166875D-2	.1551333D-2	.6822771D-3	.3771260D-3	.2358865D-3	.1131599D-3	.4831273D-4
1 I PF1	1.676912440	1.169433933	1.025565393	.8905797689	.7550130076	.6685281300	.6062810994	.5195103807	.4355535311
1 I PQ1	.5942545592	.5843108375	.5876604506	.5955092429	.6116116539	.6299770540	.6503848036	.6982675984	.7956315548
3 I PFC	1.703498970	1.534672079	1.463732563	1.387317830	1.301429734	1.242195765	1.197530512	1.132348373	1.065278295
3 I PQC	3.098773223	2.256881457	2.116932289	2.027067220	1.984010666	1.988399218	2.012627176	2.090150865	2.251543323
1 I ZF1	1.723733038	1.723100798	1.723387667	1.723812204	1.724366088	1.724775206	1.725093244	1.725568748	1.726071401
1 I ZF2	1.884515587	1.913004827	1.922109802	1.930928623	1.939950838	1.945754599	1.949952132	1.955861253	1.961754788
1 I ZF3	2.357132440	2.532283679	2.582347765	2.628040187	2.672028683	2.698917058	2.717708493	2.743273498	2.767794982
1 I ZF4	5.668894747	6.442737770	6.673041832	6.884444260	7.087679660	7.211146409	7.296853573	7.412428706	7.521905929

(4) P 4+4	TH = 37	SBE = 1.661640141						MCPER FUNCTIONS	
	RC = 02	RC = 05	RC = 07	RC = 10	RC = 15	RC = 20	RC = 25	RC = 35	RC = 50
AMAX DB	.0017375255	.0108709564	.0213327376	.0436480540	.0988323393	.1772876696	.2802872360	.5675287486	1.249387366
AMIN DB	26.83325184	42.18317250	47.96326271	54.15269381	61.28088207	66.42714658	70.50091476	76.87898739	84.27000173
MU I K	.4553417D-1	.7777524D-2	.3997945D-2	.1960493D-2	.8628909D-3	.4771365D-3	.2985068D-3	.1432355D-3	.6116457D-4
1 I PF1	1.703892569	1.182359888	1.035580534	.8983037760	.7608327762	.6733213756	.6104176464	.5228353959	.4381962577
1 I PQ1	.5999033325	.5867659015	.5896195852	.5971118923	.6129446646	.6311798787	.6515143636	.6993288605	.7966860269
3 I PFC	1.679767431	1.522951154	1.455138992	1.381350797	1.297728695	1.239727324	1.195844950	1.131612825	1.065316993
3 I PQC	3.330611230	2.356872453	2.195215450	2.089524343	2.034282331	2.032586980	2.053189749	2.126715407	2.285603466
1 I ZF1	1.683612110	1.682598476	1.682822228	1.683203303	1.683727433	1.684124104	1.684436034	1.684906676	1.685408435
1 I ZF2	1.834508825	1.862675011	1.871752982	1.880579025	1.889642649	1.895491218	1.899729981	1.905709582	1.911687667
1 I ZF3	2.269505389	2.446313840	2.497367545	2.544090344	2.589157539	2.616737171	2.636022975	2.662272603	2.687461367
1 I ZF4	5.398641286	6.172845357	6.407318690	6.623699558	6.832540792	6.959714467	7.048095797	7.167371872	7.280417005

(4) P 4+4	TH = 38	SBE = 1.624269245						MCPER FUNCTIONS	
	RC = 02	RC = 05	RC = 07	RC = 10	RC = 15	RC = 20	RC = 25	RC = 35	RC = 50
AMAX DB	.0017375255	.0108709564	.0213327376	.0436480540	.0988323393	.1772876696	.2802872360	.5675287486	1.249387366
AMIN DB	24.96616110	40.22223053	45.98773465	52.16696933	59.28768542	64.43036579	68.50202971	74.87773220	82.26697493
MU I K	.5645364D-1	.9747393D-2	.5018955D-2	.2464061D-2	.1085465D-2	.6004567D-3	.3757496D-3	.1803489D-3	.7702847D-4
1 I PF1	1.732284364	1.195958496	1.046094013	.9063945168	.7669159626	.6783252686	.6147323253	.5262998368	.4409472364
1 I PQ1	.6062773153	.5894296048	.5917250401	.5988206630	.6143562221	.6324487490	.6527030225	.7004422562	.7977895057
3 I PFC	1.655741458	1.510869009	1.446245057	1.375153602	1.293870424	1.237145647	1.194075548	1.130830524	1.065340235
3 I PQC	3.593203071	2.466872364	2.280616031	2.157149961	2.088331130	2.079895351	2.096492152	2.165601303	2.321699166
1 I ZF1	1.645878129	1.644451456	1.644606569	1.644939993	1.645430964	1.645813375	1.646118045	1.646582413	1.647082087
1 I ZF2	1.787373059	1.815166379	1.824202936	1.833022582	1.842114344	1.847999735	1.852274298	1.858317316	1.864373673
1 I ZF3	2.186656701	2.364585187	2.416552200	2.464257933	2.510374041	2.538632753	2.558406592	2.585334580	2.611186682
1 I ZF4	5.144152764	5.915453921	6.153540669	6.374555338	6.588806839	6.719625528	6.810659789	6.933632006	7.050260685

(4) P 4+4	TH = 39	SBE = 1.589015729						MCPER FUNCTIONS	
	RC = 02	RC = 05	RC = 07	RC = 10	RC = 15	RC = 20	RC = 25	RC = 35	RC = 50
AMAX DB	.0017375255	.0108709564	.0213327376	.0436480540	.0988323393	.1772876696	.2802872360	.5675287486	1.249387366
AMIN DB	23.16062152	38.30930101	44.05836264	50.22616476	57.33854513	62.47724320	66.54657487	72.91965612	80.30694396
MU I K	.6949746D-1	.1214884D-1	.6267320D-2	.3081000D-2	.1358541D-2	.7518615D-3	.4706210D-3	.2259525D-3	.9652789D-4
1 I PF1	1.762078732	1.210264473	1.057131177	.9148695880	.7732741984	.6835485685	.6192322192	.5299089098	.4438102765
1 I PQ1	.6134967915	.5923260624	.5939914466	.6006446862	.6158520698	.6337879973	.6539543598	.7016106082	.7989443735
3 I PFC	1.631495538	1.498435412	1.437052377	1.368723972	1.289851393	1.234447309	1.192219225	1.129999085	1.065346426
3 I PQC	3.891328983	2.588129583	2.373942772	2.230476836	2.146507045	2.130593333	2.142758145	2.206983820	2.359972178
1 I ZF1	1.610356447	1.608483636	1.608564402	1.608845887	1.609300228	1.609666536	1.609962775	1.610419428	1.610915802
1 I ZF2	1.742899576	1.770270063	1.779251213	1.788051121	1.797157875	1.803072094	1.807376987	1.813476259	1.819604478
1 I ZF3	2.108340928	2.286807533	2.339602495	2.388237647	2.435368852	2.464292754	2.484547381	2.512146534	2.538657413
1 I ZF4	4.904783707	5.669820817	5.910913741	6.136176091	6.355611173	6.489999010	6.583657541	6.710313916	6.830537705

(4) P 4+4	TH = 40	SBE = 1.555723827						MCPER FUNCTIONS	
	RC = 02	RC = 05	RC = 07	RC = 10	RC = 15	RC = 20	RC = 25	RC = 35	RC = 50
AMAX DB	.0017375255	.0108709564	.0213327376	.0436480540	.0988323393	.1772876696	.2802872360	.5675287486	1.249387366
AMIN DB	21.41668204	36.44241458	42.17294404	48.32793158	55.43101713	60.56529322	64.63204184	71.00222670	78.38735957
MU I K	.8495049D-1	.1506188D-1	.7786688D-2	.3833570D-2	.1692190D-2	.9369908D-3	.5866754D-3	.2817661D-3	.1204015D-3
1 I PF1	1.793230911	1.225313959	1.068719019	.9237478727	.7799199786	.6890006755	.6239249199	.5336681846	.4467894456
1 I PQ1	.6217041430	.5954831013	.5964354742	.6025942371	.6174385933	.6352024051	.6552723069	.7028369973	.8001532162
3 I PFC	1.607108930	1.485662391	1.427563797	1.362060166	1.285668198	1.231628875	1.190272837	1.129116034	1.065333890
3 I PQC	4.230550575	2.722082674	2.476119754	2.310108401	2.209202901	2.184981129	2.192236425	2.251056684	2.400578361
1 I ZF1	1.576890083	1.574536713	1.574537222	1.574762369	1.575176548	1.575524878	1.575811495	1.576258967	1.576750803
1 I ZF2	1.700901278	1.727798122	1.736710261	1.745477386	1.754586132	1.760521194	1.764850913	1.770999187	1.777192728
1 I ZF3	2.034344344	2.212724057	2.266251978	2.315756448	2.363864502	2.393437809	2.414164959	2.442427045	2.469591451
1 I ZF4	4.679945299	5.435297423	5.678738888	5.907820332	6.132179528	6.270045652	6.366291716	6.496612348	6.620438234

(4) P 4+4	TH = 41	SBE = 1.524253087						MCPER FUNCTIONS	
	RC = 02	RC = 05	RC = 07	RC = 10	RC = 15	RC = 20	RC = 25	RC = 35	RC = 50
AMAX DB	.0017375255	.0108709564	.0213327376	.0436480540	.0988323393	.1772876696	.2802872360	.5675287486	1.249387366
AMIN DB	19.73495254	34.61984935	40.32949173	46.47011834	53.56284300	58.69221125	62.75610074	69.12308748	76.50584695
MU I K	.1030985D 0	.1857837D-1	.9627764D-2	.4747818D-2	.2098253D-2	.1162491D-2	.7281066D-3	.3498208D-3	.1495229D-3
1 I PF1	1.825648688	1.241144121	1.080886182	.9330496168	.7868667312	.6946916839	.6288185716	.5375836246	.4498890896
1 I PQ1	.6310673940	.5989329148	.5990761658	.6046809077	.6191229075	.6366972609	.6566611900	.7041247914	.8014188449
3 I PFC	1.582664885	1.472564402	1.417783511	1.355161053	1.281317598	1.228686920	1.188233189	1.128178814	1.065300871
3 I PQC	4.617316853	2.870393110	2.588205497	2.396729832	2.276860639	2.243394493	2.245203981	2.298034431	2.443689387
1 I ZF1	1.545337409	1.542468022	1.542382145	1.542546437	1.542916852	1.543245301	1.543521089	1.543957889	1.544443924
1 I ZF2	1.661210190	1.687581018	1.696410892	1.705132489	1.714230394	1.720178334	1.724527351	1.730717292	1.736969489
1 I ZF3	1.964479921	2.142107506	2.196263237	2.246569811	2.295611620	2.325816450	2.347006734	2.375922369	2.403734285
1 I ZF4	4.469088820	5.211317133	5.456400876	5.688829737	5.917819234	6.059056744	6.157844856	6.291801155	6.419231002

(4) P 4+4	TH = 42	SBE = 1.494476550						MCPER FUNCTIONS	
	RC = 02	RC = 05	RC = 07	RC = 10	RC = 15	RC = 20	RC = 25	RC = 35	RC = 50
AMAX DB	.0017375255	.0108709564	.0213327376	.0436480540	.0988323393	.1772876696	.2802872360	.5675287486	1.249387366
AMIN DB	18.11662247	32.84012399	38.52622195	44.65075578	51.73193363	56.85585666	60.91658291	67.28004046	74.66018789
MU I K	.1242135D 0	.2280310D-1	.1184920D-1	.5854128D-2	.2590618D-2	.1436174D-2	.8998515D-3	.4325118D-3	.1849229D-3
1 I PF1	1.859179161	1.257792561	1.093662899	.9427965010	.7941288898	.7006324411	.6339219175	.5416616201	.4531138565
1 I PQ1	.6417839121	.6027128401	.6019353346	.6069178092	.6209129588	.6382784256	.6581257787	.7054776788	.8027443199
3 I PFC	1.558249663	1.459158471	1.407717204	1.348026198	1.276796554	1.225618047	1.186097044	1.127184780	1.065245527
3 I PQC	5.059070893	3.034984037	2.711415580	2.491121115	2.349978665	2.306209802	2.301969989	2.348155124	2.489494688
1 I ZF1	1.515570082	1.512148681	1.511970050	1.512068845	1.512391825	1.512698462	1.512962194	1.513386810	1.513865758
1 I ZF2	1.623675263	1.649465456	1.658200108	1.666863725	1.675938138	1.681891025	1.686253797	1.692477996	1.698782063
1 I ZF3	1.898582598	2.074756924	2.129424690	2.180458527	2.230385703	2.261201869	2.282844653	2.312403174	2.340855730
1 I ZF4	4.271690688	4.997384755	5.243358489	5.478619788	5.711910001	5.856394913	5.957670145	6.095224024	6.226253997

(4) P 4+4	TH = 43	SBE = 1.466279186						MCPER FUNCTIONS	
	RC = 02	RC = 05	RC = 07	RC = 10	RC = 15	RC = 20	RC = 25	RC = 35	RC = 50
AMAX DB	.0017375255	.0108709564	.0213327376	.0436480540	.0988323393	.1772876696	.2802872360	.5675287486	1.249387366
AMIN DB	16.56346349	31.10199484	36.76154504	42.86804442	49.93635503	55.05423800	59.11146573	65.47103050	72.84830500
MU I K	.1485343D 0	.2785481D-1	.1451853D-1	.7187830D-2	.3185534D-2	.1767210D-2	.1107712D-2	.5326580D-3	.2278163D-3
1 I PF1	1.893594947	1.275296452	1.107080877	.9530117031	.8017219724	.7068346111	.6392443518	.5459090256	.4564687215
1 I PQ1	.6540840434	.6068662810	.6050380343	.6093198115	.6228176433	.6399524094	.6596713418	.7068997055	.8041329780
3 I PFC	1.533951357	1.445464337	1.397372185	1.340655960	1.272102276	1.222418912	1.183861134	1.126131204	1.065165933
3 I PQC	5.564350682	3.218086186	2.847149526	2.594172510	2.429120487	2.373849991	2.362880342	2.401683491	2.538203706
1 I ZF1	1.487471167	1.483461996	1.483183990	1.483212507	1.483484308	1.483767173	1.484017606	1.484428504	1.484899059
1 I ZF2	1.588160427	1.613312537	1.621939228	1.630532696	1.639571163	1.645521110	1.649892085	1.656143068	1.662492101
1 I ZF3	1.836504728	2.010494800	2.065547831	2.117225953	2.167984357	2.199389136	2.221472411	2.251661740	2.280747118
1 I ZF4	4.087238887	4.793067042	5.039135964	5.276671677	5.513896000	5.661486210	5.765183485	5.906286515	6.040906460

(4) P 4+4	TH = 44	SBE = 1.439556540						MCPER FUNCTIONS	
	RC = 02	RC = 05	RC = 07	RC = 10	RC = 15	RC = 20	RC = 25	RC = 35	RC = 50
AMAX DB	.0017375255	.0108709564	.0213327376	.0436480540	.0988323393	.1772876696	.2802872360	.5675287486	1.249387366
AMIN DB	15.07780215	29.40445718	35.03405907	41.12034441	48.17431615	53.28549997	57.33885917	63.69413154	71.06824773
MU I K	.1762422D 0	.3386703D-1	.1771320D-1	.8789877D-2	.3901972D-2	.2166332D-2	.1358492D-2	.6535720D-3	.2796325D-3
1 I PF1	1.928581535	1.293691340	1.121173072	.9637199429	.8096626626	.7133107438	.6447959774	.5503332005	.4599590156
1 I PQ1	.6682343672	.6114438003	.6084131179	.6119038239	.6248469459	.6417264593	.6613037129	.7083953182	.8055884634
3 I PFC	1.509858500	1.431504549	1.386757530	1.333051584	1.267232273	1.219086249	1.181522175	1.125015276	1.065060070
3 I PQC	6.142876375	3.422292161	2.997022771	2.706902907	2.514924904	2.446791504	2.428322945	2.458914580	2.590048483
1 I ZF1	1.460933423	1.456302077	1.455917811	1.455871122	1.456087925	1.456345028	1.456580902	1.456976528	1.457437364
1 I ZF2	1.554542785	1.578996172	1.587502289	1.596013690	1.605003960	1.610943139	1.615316764	1.621586997	1.627973980
1 I ZF3	1.778111586	1.949164564	2.004464840	2.056695656	2.108224921	2.140192814	2.162703054	2.193509546	2.223218876
1 I ZF4	3.915220702	4.597984160	4.843315411	5.082525263	5.323279020	5.473813298	5.579856663	5.724449192	5.862641980

(4) P 4+4	TH = 45	SBE = 1.414213562						MCPER FUNCTIONS	
	RC = 02	RC = 05	RC = 07	RC = 10	RC = 15	RC = 20	RC = 25	RC = 35	RC = 50
AMAX DB	.0017375255	.0108709564	.0213327376	.0436480540	.0988323393	.1772876696	.2802872360	.5675287486	1.249387366
AMIN DB	13.66244669	27.74675076	33.34254630	39.40616745	46.44415839	51.54791209	55.59699402	61.94753435	69.31817975
MU I K	.2074329D 0	.4098854D-1	.2152151D-1	.1070759D-1	.4762029D-2	.2646089D-2	.1660161D-2	.7991408D-3	.3420511D-3
1 I PF1	1.963728217	1.313009492	1.135973342	.9749475024	.8179688936	.7200743508	.6505876681	.5549420548	.4635904575
1 I PQ1	.6845401727	.6165044104	.6120939005	.6146891292	.6270121019	.6436086631	.6630293641	.7099694144	.8071147632
3 I PFC	1.486058472	1.417304555	1.375884210	1.325215312	1.262184408	1.215616900	1.179076880	1.123834110	1.064925829
3 I PQC	6.805616280	3.650620528	3.162904753	2.830481663	2.608118054	2.525572483	2.498733920	2.520177996	2.645286658
1 I ZF1	1.435857710	1.430572641	1.430074959	1.429947978	1.430105880	1.430335203	1.430555247	1.430934028	1.431383800
1 I ZF2	1.522710916	1.546401709	1.554774651	1.563192286	1.572122313	1.578042959	1.582413686	1.588695589	1.595113399
1 I ZF3	1.723276899	1.890628360	1.946026505	1.998709368	2.050942420	2.083444889	2.106366903	2.137775180	2.168098427
1 I ZF4	3.755111802	4.411801924	4.655530048	4.895772888	5.139612542	5.292909556	5.401211455	5.549221687	5.690962514

(4) P 4+4	TH = 46	SBE = 1.390163591						MCPER FUNCTIONS	
	RC = 02	RC = 05	RC = 07	RC = 10	RC = 15	RC = 20	RC = 25	RC = 35	RC = 50
AMAX DB	.0017375255	.0108709564	.0213327376	.0436480540	.0988323393	.1772876696	.2802872360	.5675287486	1.249387366
AMIN DB	12.32055298	26.12836974	31.68597263	37.72417078	44.74434682	49.83985877	53.88421147	60.22953559	67.59636773
MU I K	.2420875D 0	.4938346D-1	.2604362D-1	.1299545D-1	.5791388D-2	.3221121D-2	.2022039D-2	.9739199D-3	.4170437D-3
1 I PF1	1.998525734	1.333277710	1.151515919	.9867222087	.8266599333	.7271399856	.6566311379	.5597440992	.4673691881
1 I PQ1	.7033467319	.6221170902	.6161189468	.6176977772	.6293257882	.6456080684	.6648554936	.7116273985	.8087162474
3 I PFC	1.462635713	1.402892735	1.364765224	1.317150492	1.256956959	1.212007848	1.176521980	1.122584743	1.064761007
3 I PQC	7.564826627	3.906591228	3.346964329	2.966254665	2.709527711	2.610802426	2.574604892	2.585842869	2.704204947
1 I ZF1	1.412151545	1.406185953	1.405567442	1.405354914	1.405449929	1.405649426	1.405852349	1.406212696	1.406650041
1 I ZF2	1.492563242	1.515424755	1.523651801	1.531964135	1.540822072	1.546716493	1.551078787	1.557364739	1.563806148
1 I ZF3	1.671878397	1.834765041	1.890100395	1.943125225	1.995987793	2.028992993	2.052309758	2.084302533	2.115228367
1 I ZF4	3.606366979	4.234224669	4.475458095	4.716053926	4.962496582	5.118353971	5.228814512	5.380157551	5.525413193

(4) P 4+4 — TH = 47, SBE = 1.367327461 — MCPER FUNCTIONS

MU	K	RC = 02	RC = 05	RC = 07	RC = 10	RC = 15	RC = 20	RC = 25	RC = 35	RC = 50
	AMAX DB	.0017375255	.0108709564	.0213327376	.0436480540	.0988323393	.1772876696	.2802872360	.5675287486	1.249387366
	AMIN DB	11.05542071	24.54907713	30.06349010	36.07315313	43.07346276	48.15983081	52.19895393	58.53852802	65.90117119
	K	.2800457D 0	.5923060D-1	.3139247D-1	.1571601D-1	.7019834D-2	.3908485D-2	.2455005D-2	.1183242D-2	.5069224D-3
1	PF1	2.032373896	1.354514472	1.167834656	.9990733625	.8357564673	.7345233307	.6629390166	.5647485019	.4713018105
1	PQ1	.7250390109	.6283625569	.6205330065	.6209550539	.6318023474	.6477348236	.6667901265	.7133752478	.8103977157
3	PFC	1.439669790	1.388300399	1.353415711	1.308861689	1.251548688	1.208256255	1.173854244	1.121264144	1.064563301
3	PQC	8.434066421	4.194313956	3.551723960	3.115775490	2.820100307	2.703173635	2.656491596	2.656323668	2.767123193
1	ZF1	1.389727819	1.383061907	1.382314917	1.382011416	1.382039464	1.382207060	1.382391560	1.382731868	1.383155410
1	ZF2	1.464006441	1.485970144	1.494038305	1.502233908	1.511008092	1.516868673	1.521217019	1.527499363	1.533957045
1	ZF3	1.623793539	1.781468343	1.836569253	1.889816230	1.943226365	1.976698857	2.000391346	2.032949226	2.064464887
1	ZF4	3.468413076	4.064988642	4.302817222	4.543049923	4.791573176	4.949766695	5.062272911	5.216849778	5.365577803

(4) P 4+4 — TH = 48, SBE = 1.345632730 — MCPER FUNCTIONS

MU	K	RC = 02	RC = 05	RC = 07	RC = 10	RC = 15	RC = 20	RC = 25	RC = 35	RC = 50
	AMAX DB	.0017375255	.0108709564	.0213327376	.0436480540	.0988323393	.1772876696	.2802872360	.5675287486	1.249387366
	AMIN DB	9.870222812	23.00892335	28.47444269	34.45205277	41.43019786	46.50641795	50.53975690	56.87299174	64.23103328
	K	.3209880D 0	.7072189D-1	.3769449D-1	.1894076D-1	.8481841D-2	.4728018D-2	.2971749D-2	.1433344D-2	.6143959D-3
1	PF1	2.064601710	1.376726287	1.184961976	1.012031594	.8452806770	.7422412894	.6695249319	.5699651503	.4753954334
1	PQ1	.7500396612	.6353353196	.6253881228	.6244900374	.6344580531	.6500003422	.6688422323	.7152195888	.8121644500
3	PFC	1.417233420	1.373561730	1.341853059	1.300354805	1.245958913	1.204359505	1.171070503	1.119869220	1.064330310
3	PQC	9.428198542	4.518591135	3.780124380	3.280842790	2.940921269	2.803474827	2.745024062	2.732087040	2.834399102
1	ZF1	1.368503749	1.361127198	1.360243897	1.359843824	1.359800731	1.359934319	1.360099082	1.360417731	1.360826080
1	ZF2	1.436953937	1.457951045	1.465846909	1.473914368	1.482593301	1.488412509	1.492741414	1.499012469	1.505479005
1	ZF3	1.578895636	1.730645180	1.785329542	1.838668904	1.892536520	1.926436970	1.950483963	1.983585240	2.015676383
1	ZF4	3.340644851	3.903855824	4.137359435	4.376480240	4.626522413	4.786805162	4.901230290	5.058926904	5.211074832

(4) P 4+4 — TH = 49, SBE = 1.325012993 — MCPER FUNCTIONS

MU	K	RC = 02	RC = 05	RC = 07	RC = 10	RC = 15	RC = 20	RC = 25	RC = 35	RC = 50
	AMAX DB	.0017375255	.0108709564	.0213327376	.0436480540	.0988323393	.1772876696	.2802872360	.5675287486	1.249387366
	AMIN DB	8.767687876	21.50826841	26.91837570	32.85994767	39.81334952	44.87830266	48.90524195	55.23148631	62.58447246
	K	.3644312D 0	.8405946D-1	.4509010D-1	.2275111D-1	.1021721D-1	.5702757D-2	.3587054D-2	.1731513D-2	.7426366D-3
1	PF1	2.094500640	1.399903176	1.202927445	1.025628616	.8552563064	.7503120822	.6764035998	.5754047206	.4796577208
1	PQ1	.7788054621	.6431460302	.6307449439	.6283362617	.6373114237	.6524174966	.6710218633	.7171677850	.8140222760
3	PFC	1.395390591	1.358713665	1.330096979	1.291637194	1.240187589	1.200315250	1.168167677	1.118396823	1.064059530
3	PQC	10.56340024	4.885037034	4.035601723	3.463545219	3.073239405	2.912607373	2.840918717	2.813659891	2.906433796
1	ZF1	1.348400108	1.340314584	1.339287046	1.338784633	1.338666125	1.338763570	1.338907267	1.339202627	1.339594380
1	ZF2	1.411324515	1.431288170	1.438997742	1.446925573	1.455497890	1.461268270	1.465572269	1.471824335	1.478292214
1	ZF3	1.537050745	1.682214036	1.736290143	1.789582108	1.843808537	1.878093415	1.902471304	1.936091720	1.968742250
1	ZF4	3.222424547	3.750608101	3.978866322	4.216098096	4.467058928	4.629160689	4.745363463	4.906049601	5.061554016

(4) P 4+4 — TH = 50, SBE = 1.305407289, MCPER FUNCTIONS

MU	K	RC = 02	RC = 05	RC = 07	RC = 10	RC = 15	RC = 20	RC = 25	RC = 35	RC = 50
	AMAX DB	.0017375255	.0108709564	.0213327376	.0436480540	.0988323393	.1772876696	.2802872360	.5675287486	1.249387366
	AMIN DB	7.749772620	20.04780600	25.39504879	31.29605797	38.22181761	43.27425490	47.29411055	53.61264370	60.96007480
MU	K	.4097428D 0	.9945112D-1	.5373380D-1	.2723937D-1	.1227182D-1	.6859418D-2	.4318118D-2	.2086257D-2	.8953571D-3
1	PF1	2.121368753	1.424013203	1.221755878	1.039896842	.8657087115	.7587553463	.6835909219	.5810787551	.4840969465
1	PQ1	.8118227893	.6519241394	.6366742710	.6325325057	.6403835956	.6550008467	.6733403162	.7192280401	.8159776335
3	PFC	1.374194966	1.343795708	1.318169570	1.282717784	1.234235390	1.196121467	1.165142810	1.116843763	1.063748355
3	PQC	11.85721684	5.300214333	4.322179413	3.666315561	3.218496239	3.031604762	2.944992839	2.901638967	2.983678351
1	ZF1	1.329340790	1.320562213	1.319382551	1.318771883	1.318573584	1.318632720	1.318754010	1.319024441	1.319398187
1	ZF2	1.387041143	1.405909075	1.413417628	1.421194172	1.429648597	1.435362770	1.439636428	1.445861790	1.452323420
1	ZF3	1.498115695	1.636103400	1.689371148	1.742465990	1.796943593	1.831564855	1.856247432	1.890359938	1.923551824
1	ZF4	3.113085696	3.605041711	3.827144599	4.061686955	4.312928785	4.476555489	4.594379459	4.757907690	4.916693304

(4) P 4+4 — TH = 51, SBE = 1.286759566, MCPER FUNCTIONS

MU	K	RC = 02	RC = 05	RC = 07	RC = 10	RC = 15	RC = 20	RC = 25	RC = 35	RC = 50
	AMAX DB	.0017375255	.0108709564	.0213327376	.0436480540	.0988323393	.1772876696	.2802872360	.5675287486	1.249387366
	AMIN DB	6.817372896	18.62858716	23.90445291	29.75975103	36.65460272	41.69312799	45.70513872	52.01516191	59.35648704
MU	K	.4561749D 0	.1171037D 0	.6379364D-1	.3250966D-1	.1469839D-1	.8228934D-2	.5184932D-2	.2507506D-2	.1076901D-2
1	PF1	2.144560413	1.448996120	1.241464874	1.054868822	.8766648824	.7675922362	.6911040899	.5869997479	.4887220569
1	PQ1	.8496030351	.6618208374	.6432588730	.6371237365	.6436987673	.6577669111	.6758103233	.7214095185	.8180376590
3	PFC	1.353688777	1.328849670	1.306095342	1.273607184	1.228103808	1.191776510	1.161993102	1.115206815	1.063394074
3	PQC	13.32869792	5.771789042	4.644577426	3.891996067	3.378361431	3.161656005	3.058181858	2.996702252	3.066641520
1	ZF1	1.311252733	1.301813005	1.300473566	1.299748612	1.299466048	1.299484677	1.299582209	1.299826061	1.300180380
1	ZF2	1.364030075	1.381747527	1.389039481	1.396652782	1.404978078	1.410628734	1.414866648	1.421057584	1.427505288
1	ZF3	1.461937543	1.592250214	1.644502734	1.697241045	1.751852875	1.786757654	1.811715888	1.846290378	1.880003453
1	ZF4	3.011941158	3.466961896	3.682021871	3.913057168	4.163906682	4.328740037	4.448012926	4.614217530	4.776196190

(4) P 4+4 — TH = 52, SBE = 1.269018215, MCPER FUNCTIONS

MU	K	RC = 02	RC = 05	RC = 07	RC = 10	RC = 15	RC = 20	RC = 25	RC = 35	RC = 50
	AMAX DB	.0017375255	.0108709564	.0213327376	.0436480540	.0988323393	.1772876696	.2802872360	.5675287486	1.249387366
	AMIN DB	5.970121388	17.25203948	22.44683090	28.25054920	35.11080574	40.13385542	44.13717261	50.43779924	57.77241003
MU	K	.5029142D 0	.1372139D 0	.7544986D-1	.3867876D-1	.1755738D-1	.9847075D-2	.6210712D-2	.3006838D-2	.1292348D-2
1	PF1	2.163533980	1.474756299	1.262061685	1.070576445	.8881534242	.7768455218	.6989616983	.5931812401	.4935427401
1	PQ1	.8926790883	.6730122359	.6505956051	.6421622327	.6472847295	.6607344892	.6784462779	.7237224860	.8202102810
3	PFC	1.333902355	1.313919313	1.293901205	1.264317799	1.221795242	1.187279182	1.158715955	1.113482742	1.062993871
3	PQC	14.99865240	6.308704203	5.008341844	4.143917471	3.554775723	3.304133911	3.181560191	3.099622608	3.155898912
1	ZF1	1.294066147	1.284014070	1.282507706	1.281662377	1.281290978	1.281266872	1.281339285	1.281554902	1.281888372
1	ZF2	1.342220286	1.358742922	1.365801764	1.373239454	1.381424359	1.387004243	1.391201039	1.397349822	1.403775848
1	ZF3	1.428354542	1.550598307	1.601624090	1.653837271	1.708456813	1.743587104	1.768788919	1.803791940	1.838003680
1	ZF4	2.918294658	3.336177728	3.543342579	3.770042844	4.019793427	4.185490726	4.306023839	4.474719723	4.639789364

(4) P 4+4	TH = 53 SBE = 1.252135658								MCPER FUNCTIONS
	RC = 02	RC = 05	RC = 07	RC = 10	RC = 15	RC = 20	RC = 25	RC = 35	RC = 50
AMAX DB		.0108709564	.0213327376	.0436480540	.0988323393	.1772876696	.2802872360	.5675287486	1.249387366
AMIN DB		15.91997637	21.02270115	26.76814081	33.58962917	38.59544873	42.58912478	48.87936914	56.20659270
MU I K		.1599562D 0	.8889246D-1	.4587678D-1	.2091792D-1	.1175513D-1	.7422390D-2	.3597755D-2	.1547641D-2
1 I PF1		1.501155360	1.283539314	1.087049835	.9002044788	.7865396795	.7071838626	.5996379265	.4985695054
1 I PQ1		.6857027048	.6587978558	.6477089243	.6511735007	.6639250448	.6812645031	.7261784766	.8225043316
3 I PFC		1.299049906	1.281616413	1.254863921	1.215313112	1.182628805	1.155309016	1.111668304	1.062544827
3 I PQC		6.921372296	5.419997893	4.425994675	3.750003157	3.460629385	3.316366411	3.211284141	3.252103924
1 I ZF1		1.267116158	1.265436586	1.264464821	1.263999935	1.263930837	1.263976760	1.264162483	1.264473677
1 I ZF2		1.336839727	1.343648019	1.350897191	1.358930344	1.364432239	1.368582571	1.374681471	1.381077995
1 I ZF3		1.511096759	1.560682358	1.612193379	1.666684401	1.701976762	1.727386795	1.762781245	1.797466528
1 I ZF4		3.212497051	3.410964060	3.632498881	3.880413636	4.046607792	4.168195489	4.339177097	4.507220643

(4) P 4+4	TH = 54 SBE ≐ 1.236067977								MCPER FUNCTIONS
	RC = 02	RC = 05	RC = 07	RC = 10	RC = 15	RC = 20	RC = 25	RC = 35	RC = 50
AMAX DB		.0108709564	.0213327376	.0436480540	.0988323393	.1772876696	.2802872360	.5675287486	1.249387366
AMIN DB		14.63458919	19.63288330	25.31239443	32.09038018	37.07699657	41.05997133	47.33873563	54.65782649
MU I K		.1854687D 0	.1043172D 0	.5424757D-1	.2485885D-1	.1400071D-1	.8851185D-2	.4295990D-2	.1849731D-2
1 I PF1		1.528005189	1.305871810	1.104315869	.9128495650	.7967009696	.7157923425	.6063857727	.5038137718
1 I PQ1		.7001282417	.6679983459	.6538349882	.6554020893	.6673631661	.6842835705	.7287904891	.8249296774
3 I PFC		1.284287674	1.269272456	1.245261817	1.208661957	1.177825296	1.151770232	1.109760286	1.062043920
3 I PQC		7.621886187	5.887229910	4.742842744	3.966694863	3.632993217	3.464033827	3.332701960	3.356000843
1 I ZF1		1.251073141	1.249215391	1.248111286	1.247548200	1.247431825	1.247449880	1.247604048	1.247891542
1 I ZF2		1.315986947	1.322526429	1.329573536	1.337443387	1.342860094	1.346958637	1.352999924	1.359359048
1 I ZF3		1.473698197	1.521631586	1.572256072	1.626472596	1.661857864	1.687437226	1.723182031	1.758312873
1 I ZF4		3.095721576	3.284752692	3.500298113	3.745613616	3.911913437	4.034332690	4.207372927	4.378257137

(4) P 4+4	TH = 55 SBE = 1.220774589								MCPER FUNCTIONS
	RC = 02	RC = 05	RC = 07	RC = 10	RC = 15	RC = 20	RC = 25	RC = 35	RC = 50
AMAX DB		.0108709564	.0213327376	.0436480540	.0988323393	.1772876696	.2802872360	.5675287486	1.249387366
AMIN DB		13.39841410	18.27852354	23.88337692	30.61247573	35.57766488	39.54874986	45.81430918	53.12493995
MU I K		.2138352D 0	.1219197D 0	.6394862D-1	.2946973D-1	.1663860D-1	.1053325D-1	.5119877D-2	.2206749D-2
1 I PF1		1.555062347	1.329008754	1.122396216	.9261213052	.8073574902	.7248106678	.6134421457	.5092879692
1 I PQ1		.7165597122	.6783522847	.6606237420	.6600134105	.6710771175	.6875246827	.7315732207	.8274973744
3 I PFC		1.269679136	1.256902894	1.235529789	1.201847554	1.172869259	1.148097909	1.107755519	1.061488026
3 I PQC		8.424249068	6.419091887	5.099917630	4.207967216	3.823387236	3.626227778	3.465046135	3.468440604
1 I ZF1		1.235841509	1.233802481	1.232560461	1.231894437	1.231728479	1.231717280	1.231838235	1.232100609
1 I ZF2		1.296137593	1.302389425	1.309220190	1.316914916	1.322239220	1.326280664	1.332256603	1.338570362
1 I ZF3		1.438357003	1.484431639	1.533979338	1.587765783	1.623168822	1.648874852	1.684924623	1.720469891
1 I ZF4		2.985642172	3.164580058	3.373328553	3.615259390	3.781250151	3.904260195	4.079109377	4.252683632

(4) P 4+4	TH = 56	SBE = 1.206217949						MCPER FUNCTIONS	
	RC = 02	RC = 05	RC = 07	RC = 10	RC = 15	RC = 20	RC = 25	PC = 35	RC = 50
AMAX DB		.0108709564	.0213327376	.0436480540	.0988323393	.1772876696	.2802872360	.5675287486	1.249387366
AMIN DB		12.21426548	16.96111654	22.48137507	29.15545000	34.09669851	38.05455832	44.30654312	51.60679375
MU I K		.2450681D 0	.1418875D 0	.7515039D-1	.3485198D-1	.1973173D-1	.1251043D-1	.6090779D-2	.2628212D-2
1 I PF1		1.582025166	1.352869086	1.141304807	.9400529984	.8185391943	.7342642618	.6208259580	.5150056528
1 I PQ1		.7353057864	.6900408657	.6681728860	.6650573925	.6750995045	.6910121323	.7345433459	.8302198496
3 I PFC		1.255270334	1.244543115	1.225688224	1.194877030	1.167762073	1.144290777	1.105650911	1.060873926
3 I PQC		9.344626022	7.026252459	5.503686923	4.477497999	4.034347219	3.804891392	3.609670895	3.590399873
1 I ZF1		1.221379911	1.219159007	1.217774062	1.217000391	1.216782527	1.216740686	1.216826774	1.217062617
1 I ZF2		1.277248136	1.283193313	1.289792692	1.297300090	1.302524732	1.306503768	1.312406617	1.318666981
1 I ZF3		1.405027471	1.449047066	1.497323770	1.550515289	1.585854777	1.611640805	1.647945477	1.683870576
1 I ZF4		2.882034536	3.050319114	3.251490670	3.489234832	3.654479166	3.777821282	3.954206120	4.130301162

(4) P 4+4	TH = 57	SBE = 1.192363293						MCPER FUNCTIONS	
	RC = 02	RC = 05	RC = 07	RC = 10	RC = 15	RC = 20	RC = 25	RC = 35	RC = 50
AMAX DB		.0108709564	.0213327376	.0436480540	.0988323393	.1772876696	.2802872360	.5675287486	1.249387366
AMIN DB		11.08512991	15.68251896	21.10692091	27.71896463	32.63342430	36.57655487	42.81293055	50.10227584
MU I K		.2790895D 0	.1643895D 0	.8803471D-1	.4111987D-1	.2335225D-1	.1483106D-1	.7233583D-2	.3125260D-2
1 I PF1		1.608534982	1.377334563	1.161044627	.9546779891	.8302778532	.7441805584	.6285578254	.5209816326
1 I PQ1		.7567154204	.7032750629	.6765971426	.6705923111	.6794680787	.6947738554	.7377198502	.8331111191
3 I PFC		1.241105966	1.232230019	1.215759596	1.187758975	1.162505996	1.140348069	1.103443480	1.060198311
3 I PQC		10.40162419	7.721278660	5.961837010	4.779645137	4.268860627	4.002301982	3.768150423	3.723004238
1 I ZF1		1.207648738	1.205248546	1.203716532	1.202830611	1.202558514	1.202484644	1.202534222	1.202742133
1 I ZF2		1.259277968	1.264897909	1.271250115	1.278557510	1.283675143	1.287586450	1.293408453	1.299607320
1 I ZF3		1.373661951	1.415445880	1.462255880	1.514678945	1.549867215	1.575682328	1.612186787	1.648453321
1 I ZF4		2.784655459	2.941840322	3.134694675	3.367439884	3.531479045	3.654876479	3.832499125	4.010925739

(4) P 4+4	TH = 58	SBE = 1.179178403						MCPER FUNCTIONS	
	RC = 02	RC = 05	RC = 07	RC = 10	RC = 15	RC = 20	RC = 25	RC = 35	RC = 50
AMAX DB		.0108709564	.0213327376	.0436480540	.0988323393	.1772876696	.2802872360	.5675287486	1.249387366
AMIN DB		10.01401908	14.44494844	19.76081970	26.30282198	31.18725612	35.11395879	41.33300175	48.61029683
MU I K		.3157178D 0	.1895626D 0	.1027919D 0	.4840151D-1	.2758273D-1	.1755101D-1	.8577286D-2	.3710950D-2
1 I PF1		1.634182826	1.402243393	1.181603766	.9700287676	.8426069397	.7545891014	.6366602403	.5272321219
1 I PQ1		.7811798077	.7182996505	.6860313453	.6766864046	.6842267132	.6988421022	.7411244321	.8361870461
3 I PFC		1.227228452	1.220001614	1.205768445	1.180503556	1.157104275	1.136269597	1.101130395	1.059457790
3 I PQC		11.61661571	8.518963656	6.483524272	5.119593835	4.530463159	4.221140967	3.942322969	3.867556589
1 I ZF1		1.194609772	1.192036744	1.190354767	1.189352206	1.189023563	1.188916282	1.188927719	1.189106316
1 I ZF2		1.242188860	1.247466183	1.253554796	1.260648952	1.265652095	1.269490320	1.275223696	1.281352900
1 I ZF3		1.344209113	1.383598253	1.428747381	1.480220670	1.515163627	1.540952452	1.577596152	1.614161554
1 I ZF4		2.693240042	2.839007774	3.022857778	3.249788816	3.412144363	3.535302417	3.713839588	3.894387232

(4) P 4+4	TH = 59	SBE = 1.166633397						MCPER FUNCTIONS	
	RC = 02	RC = 05	RC = 07	RC = 10	RC = 15	RC = 20	RC = 25	RC = 35	RC = 50
AMAX DB		.0108709564	.0213327376	.0436480540	.0988323393	.1772876696	.2802872360	.5675287486	1.249387366
AMIN DB		9.003786923	13.25096085	18.44417944	24.90698211	29.75770194	33.66605282	39.86582210	47.12978549
MU I K		.3546587D 0	.2174963D 0	.1196165D 0	.5683958D-1	.3251733D-1	.2073468D-1	.1015568D-1	.4400588D-2
1 I PF1		1.658522378	1.427384836	1.202950660	.9861357213	.8555614005	.7655216169	.6451577594	.5337749052
1 I PQ1		.8091338692	.7353973512	.6966340217	.6834198274	.6894265922	.7032542531	.7447819892	.8394656526
3 I PFC		1.213676989	1.207896505	1.195741307	1.173122624	1.151561255	1.132055850	1.098709025	1.058648897
3 I PQC		13.01412773	9.436705640	7.079679491	5.503539576	4.823359282	4.464581042	4.134345568	4.025572051
1 I ZF1		1.182225935	1.179490980	1.177657849	1.176534618	1.176147150	1.176005089	1.175976780	1.176124702
1 I ZF2		1.225944451	1.230863887	1.236672089	1.243539138	1.248420125	1.252179863	1.257816792	1.263868091
1 I ZF3		1.316612449	1.353475113	1.396774431	1.447110074	1.481707208	1.507409721	1.544126289	1.580943428
1 I ZF4		2.607500238	2.741675348	2.915901366	3.136208519	3.296384453	3.418990787	3.598092984	3.780528385

(4) P 4+4	TH = 60	SBE = 1.154700538						MCPER FUNCTIONS	
	RC = 02	RC = 05	RC = 07	RC = 10	RC = 15	RC = 20	RC = 25	RC = 35	RC = 50
AMAX DB		.0108709564	.0213327376	.0436480540	.0988323393	.1772876696	.2802872360	.5675287486	1.249387366
AMIN DB		8.056924904	12.10339810	17.15843923	23.53158368	28.34437358	32.23218694	38.41049073	45.65968434
MU I K		.3955066D 0	.2482162D 0	.1387005D 0	.6659181D-1	.3826320D-1	.2445629D-1	.1200813D-1	.5212137D-2
1 I PF1		1.681089082	1.452495823	1.225028590	1.003025444	.8691772732	.7770120380	.6540772052	.5406295305
1 I PQ1		.8410575497	.7548930011	.7085915117	.6908870156	.6951276627	.7080538152	.7487212100	.8429674945
3 I PFC		1.200486660	1.195953301	1.185706584	1.165629813	1.145882505	1.127708091	1.096176988	1.057768112
3 I PQC		14.62233628	10.49494881	7.763376529	5.938916509	5.152573482	4.736395512	4.346762322	4.198821235
1 I ZF1		1.170461146	1.167580047	1.165596787	1.164349412	1.163900912	1.163722726	1.163653091	1.163769006
1 I ZF2		1.210509792	1.215059178	1.220570118	1.227195523	1.231946451	1.235622216	1.241154822	1.247119899
1 I ZF3		1.290809189	1.325046665	1.366316807	1.415322066	1.449466586	1.475017953	1.511734799	1.548751550
1 I ZF4		2.527125064	2.649683025	2.813748080	3.026636777	3.184122218	3.305847368	3.485138238	3.669203931

(4) P 4+4	TH = 61	SBE = 1.143354068						MCPER FUNCTIONS	
	RC = 02	RC = 05	RC = 07	RC = 10	RC = 15	RC = 20	RC = 25	RC = 35	RC = 50
AMAX DB		.0108709564	.0213327376	.0436480540	.0988323393	.1772876696	.2802872360	.5675287486	1.249387366
AMIN DB		7.175356505	11.00530053	15.90539290	22.17696934	26.94699945	30.81178411	36.96613973	44.19894538
MU I K		.4377561D 0	.2816664D 0	.1602250D 0	.7783081D-1	.4494175D-1	.2880121D-1	.1418055D-1	.6166699D-2
1 I PF1		1.701424093	1.477260828	1.247749562	1.020718494	.8834910909	.7890964602	.6634478782	.5478175286
1 I PQ1		.8774774091	.7771576491	.7221226435	.6991995484	.7014004148	.7132916461	.7529752964	.8467161168
3 I PFC		1.187687683	1.184209915	1.175694358	1.158040626	1.140074942	1.123228465	1.093532219	1.056811872
3 I PQC		16.47371465	11.71770397	8.550278905	6.434683485	5.524141115	5.041096266	4.582590207	4.389384192
1 I ZF1		1.159280311	1.156273861	1.154144273	1.152770084	1.152258460	1.152042842	1.151930342	1.152012955
1 I ZF2		1.195850974	1.200022225	1.205219544	1.211588116	1.216200791	1.219786986	1.225207312	1.231077765
1 I ZF3		1.266729776	1.298280894	1.337356990	1.384836452	1.418415574	1.443746043	1.480383976	1.517542770
1 I ZF4		2.451782687	2.562853564	2.716318748	2.921020503	3.075292946	3.195791132	3.374866983	3.560279833

(4) P 4+4 — TH = 62 SBE = 1.132570051 — MCPER FUNCTIONS

(4) P 4+4	RC = 02	RC = 05	RC = 07	RC = 10	RC = 15	RC = 20	RC = 25	RC = 35	RC = 50
AMAX DB		.0108709564	.0213327376	.0436480540	.0988323393	.1772876696	.2802872360	.5675287486	1.249387366
AMIN DB		6.360256767	9.959780910	14.68720261	20.84371548	25.56544106	29.40434828	35.53193436	42.74652575
MU I K		.4808251D 0	.3176954D 0	.1843486D 0	.9074323D-1	.5268997D-1	.3386746D-1	.1672643D-1	.7289097D-2
1 I PF1		1.719100527	1.501316168	1.270987939	1.039226479	.8985390029	.8018129923	.6733017772	.5553626633
1 I PQ1		.9189691956	.8026125793	.7374839793	.7084896073	.7083280692	.7190274632	.7575828530	.8507386096
3 I PFC		1.175304876	1.172702810	1.165736125	1.150372502	1.134146956	1.118620126	1.090773039	1.055776596
3 I PQC		18.60590028	13.13317664	9.459181536	7.001683501	5.945350552	5.384109048	4.845427656	4.599718169
1 I ZF1		1.148649433	1.145543228	1.143274433	1.141771875	1.141195218	1.140940916	1.140784066	1.140832124
1 I ZF2		1.181934881	1.185724819	1.190593318	1.196689307	1.201155192	1.204646080	1.209946060	1.215713391
1 I ZF3		1.244297989	1.273142124	1.309879145	1.355637502	1.388532933	1.413567792	1.450040662	1.487277997
1 I ZF4		2.381124375	2.480989821	2.623529173	2.819313895	2.969843141	3.088753381	3.267132908	3.453632601

(4) P 4+4 — TH = 63 SBE = 1.122326238 — MCPER FUNCTIONS

(4) P 4+4	RC = 02	RC = 05	RC = 07	RC = 10	RC = 15	RC = 20	RC = 25	RC = 35	RC = 50
AMAX DB		.0108709564	.0213327376	.0436480540	.0988323393	.1772876696	.2802872360	.5675287486	1.249387366
AMIN DB		5.611921471	8.969863053	13.50639581	19.53266639	24.19971381	28.00947526	34.10707423	41.30138357
MU I K		.5240877D 0	.3560466D 0	.2111933D 0	.1055278D 0	.6166153D-1	.3976731D-1	.1970817D-1	.8608566D-2
1 I PF1		1.733748722	1.524259605	1.294574360	1.058548358	.9143555165	.8152014568	.6836738182	.5632912174
1 I PQ1		.9661622194	.8317333791	.7549756266	.7189141472	.7160092738	.7253317129	.7625889867	.8550662911
3 I PFC		1.163357402	1.161466203	1.155864438	1.142644857	1.128108540	1.113887358	1.087898239	1.054658714
3 I PQC		21.06285770	14.77454613	10.51267032	7.653097013	6.425052153	5.771997607	5.139593011	4.832743377
1 I ZF1		1.138535834	1.135359673	1.132962590	1.131331598	1.130688263	1.130394107	1.130191491	1.130203797
1 I ZF2		1.168729070	1.172139991	1.176666435	1.182473714	1.186783891	1.190173555	1.195344985	1.201000587
1 I ZF3		1.223431729	1.249589762	1.283867996	1.327713480	1.359802137	1.384461762	1.420676130	1.457922066
1 I ZF4		2.314790008	2.403873018	2.535286794	2.721476482	2.867729327	2.984676916	3.162001172	3.349148700

(4) P 4+4 — TH = 64 SBE = 1.112601940 — MCPER FUNCTIONS

(4) P 4+4	RC = 02	RC = 05	RC = 07	RC = 10	RC = 15	RC = 20	RC = 25	RC = 35	RC = 50
AMAX DB			.0213327376	.0436480540	.0988323393	.1772876696	.2802872360	.5675287486	1.249387366
AMIN DB			8.038295043	12.36583787	18.24497157	22.85001273	26.62686700	32.69079584	39.86247375
MU I K			.3963558D 0	.2408286D 0	.1223915D 0	.7202767D-1	.4662906D-1	.2319852D-1	.1015959D-1
1 I PF1			1.545665431	1.318290794	1.078665854	.9309717481	.8293028789	.6946020417	.5716323189
1 I PQ1			.8650543792	.7749476124	.7306599147	.7245614346	.7322878969	.7680466754	.8597355524
3 I PFC			1.150531298	1.146112471	1.134879095	1.121971418	1.109035720	1.084907176	1.053454698
3 I PQC			16.68095963	11.73793083	8.405015752	6.974054882	6.212752408	5.470302472	5.091952437
1 I ZF1			1.125695352	1.123185034	1.121427465	1.120716183	1.120381120	1.120131417	1.120106841
1 I ZF2			1.159241673	1.163415661	1.168918031	1.173063184	1.176345493	1.181379991	1.186915130
1 I ZF3			1.227577349	1.259307589	1.301056100	1.332211128	1.356411141	1.392266011	1.429443639
1 I ZF4			2.331262291	2.451487283	2.627470997	2.768916786	2.883515208	3.059247884	3.246724035

(4) P 4+4	TH = 65	SBE = 1.103377919						MCPER FUNCTIONS	
	RC = 02	RC = 05	RC = 07	RC = 10	RC = 15	RC = 20	RC = 25	RC = 35	RC = 50
AMAX DB			.0213327376	.0436480540	.0988323393	.1772876696	.2802872360	.5675287486	1.249387366
AMIN DB			7.167354325	11.26867306	16.98212452	21.51674400	25.25635028	31.28237680	38.42874382
MU I K			.4381596D 0	.2732539D 0	.1415448D 0	.8397747D-1	.5459872D-1	.2728231D-1	.1198292D-1
1 I PF1			1.565104251	1.341867775	1.099537938	.9484130441	.8441586817	.7061277896	.5804183121
1 I PQ1			.9031740321	.7978068473	.7439494625	.7341248378	.7399954817	.7740184831	.8647889101
3 I PFC			1.139925612	1.136513471	1.127098564	1.115749160	1.104072184	1.081799879	1.052161108
3 I PQC			18.89883218	13.16775062	9.277172132	7.605638503	6.716165271	5.843900758	5.381551276
1 I ZF1			1.116523071	1.113918817	1.112038919	1.111258940	1.110882079	1.110584088	1.110521581
1 I ZF2			1.147004421	1.150819248	1.156000888	1.159971306	1.163139886	1.168028850	1.173434644
1 I ZF3			1.207052068	1.236180000	1.275659901	1.305752042	1.329403623	1.364790257	1.401815153
1 I ZF4			2.262895753	2.372011230	2.537261054	2.673378212	2.785231545	2.958859620	3.146263500

(4) P 4+4	TH = 66	SBE = 1.094636279						MCPER FUNCTIONS	
	RC = 02	RC = 05	RC = 07	RC = 10	RC = 15	RC = 20	RC = 25	RC = 35	RC = 50
AMAX DB			.0213327376	.0436480540	.0988323393	.1772876696	.2802872360	.5675287486	1.249387366
AMIN DB			6.358667021	10.21822794	15.74599959	20.20056241	23.89790042	29.88114227	36.99912990
MU I K			.4809131D 0	.3083817D 0	.1631924D 0	.9771739D-1	.6384178D-1	.3205848D-1	.1412679D-1
1 I PF1			1.582165644	1.364985007	1.121094441	.9666958175	.8598094783	.7182958233	.5896851792
1 I PQ1			.9467620528	.8240247837	.7590483339	.7448677625	.7485735514	.7805787232	.8702763275
3 I PFC			1.129672454	1.127100143	1.119328471	1.109457288	1.099005297	1.078577176	1.050774639
3 I PQC			21.48357683	14.84178080	10.29387107	8.336219347	7.294319327	6.268162797	5.706642217
1 I ZF1			1.107816388	1.105141569	1.103146467	1.102297736	1.101878408	1.101531088	1.101429700
1 I ZF2			1.135403224	1.138856641	1.143702697	1.147488330	1.150536536	1.155271102	1.160538494
1 I ZF3			1.187954767	1.214464029	1.251521511	1.280420898	1.303431279	1.338233122	1.375012800
1 I ZF4			2.198493119	2.296721114	2.450808564	2.581092231	2.689798144	2.860782981	3.047680589

(4) P 4+4	TH = 67	SBE = 1.086360377						MCPER FUNCTIONS	
	RC = 02	RC = 05	RC = 07	RC = 10	RC = 15	RC = 20	RC = 25	RC = 35	RC = 50
AMAX DB			.0213327376	.0436480540	.0988323393	.1772876696	.2802872360	.5675287486	1.249387366
AMIN DB			5.613064657	9.217875631	14.53888196	18.90241499	22.55167140	28.48647420	35.57255287
MU I K			.5240187D 0	.3460240D 0	.1875236D 0	.1134695D 0	.7454464D-1	.3764231D-1	.1664839D-1
1 I PF1			1.596481064	1.387276437	1.143228994	.9858234328	.8762933242	.7311543445	.5994730167
1 I PQ1			.9965693873	.8541460305	.7762736155	.7569928262	.7581654178	.7878162072	.8762568877
3 I PFC			1.119790625	1.117903967	1.111595716	1.103113363	1.093845329	1.075240829	1.049292191
3 I PQC			24.50193103	16.80815377	11.48518776	9.186221614	7.962235090	6.752692132	6.073464298
1 I ZF1			1.099549819	1.096831375	1.094731508	1.093814881	1.093352716	1.092955236	1.092814139
1 I ZF2			1.124413422	1.127508178	1.132005491	1.135596063	1.138516968	1.143087966	1.148207695
1 I ZF3			1.170220560	1.194133986	1.228638779	1.256217214	1.278490412	1.312583185	1.349016560
1 I ZF4			2.137759747	2.225458839	2.368070858	2.492041732	2.597195167	2.764974168	2.950897064

(4) P 4+4	TH = 68 SBE = 1.078534743							MCPER FUNCTIONS	
	RC = 02	RC = 05	RC = 07	RC = 10	RC = 15	RC = 20	RC = 25	RC = 35	RC = 50
AMAX DB				.0436480540	.0988323393	.1772876696	.2802872360	.5675287486	1.249387366
AMIN DB				3.270865407	13.36348378	17.62358995	21.21803307	27.09782435	34.14791483
MU I K				.3858840D 0	.2146969D 0	.1314681D 0	.8691572D-1	.4416811D-1	.1961572D-1
1 I PF1				1.408340504	1.165791751	1.005780984	.8936432561	.7447548567	.6098265688
1 I PQ1				.3887984252	.7960041012	.7707448743	.7689444617	.7958377634	.8828009291
3 I PFC				1.108954495	1.103928646	1.096737037	1.088604429	1.071793696	1.047710940
3 I PQC				19.12560164	12.88851960	10.18122618	8.738729882	7.309452471	6.489712150
1 I ZF1				1.088966700	1.086776171	1.085793664	1.085288690	1.084840494	1.084659009
1 I ZF2				1.116754816	1.120892752	1.124277948	1.127064354	1.131462269	1.136424834
1 I ZF3				1.175158697	1.207009769	1.233143541	1.254581382	1.287833383	1.323810262
1 I ZF4				2.158044160	2.288997510	2.406211950	2.507409629	2.671398556	2.855842666

(4) P 4+4	TH = 69 SBE = 1.071144994							MCPER FUNCTIONS	
	RC = 02	RC = 05	RC = 07	RC = 10	RC = 15	RC = 20	RC = 25	RC = 35	RC = 50
AMAX DB				.0436480540	.0988323393	.1772876696	.2802872360	.5675287486	1.249387366
AMIN DB				7.380129377	12.22293783	16.36576909	19.89761679	25.71473211	32.72409608
MU I K				.4275565D 0	.2448235D 0	.1519538D 0	.1011857D 0	.5179208D-1	.2310975D-1
1 I PF1				1.427755534	1.188582607	1.026528859	.9118839073	.7591517839	.6207958190
1 I PQ1				.9287054202	.8186923871	.7864207993	.7811215707	.8047727768	.8899927989
3 I PFC				1.100278687	1.096356719	1.090350068	1.083296780	1.068239905	1.046028435
3 I PQC				21.86628793	14.55061760	11.35349786	9.647571808	7.953484719	6.964964084
1 I ZF1				1.081526402	1.079263150	1.078218221	1.077670988	1.077171874	1.076949506
1 I ZF2				1.106577895	1.110349213	1.113518964	1.116163444	1.120378386	1.125174007
1 I ZF3				1.157500836	1.186631617	1.211204875	1.231708379	1.263981066	1.299381710
1 I ZF4				2.094274269	2.213526886	2.323588239	2.420434118	2.580030251	2.762454869

(4) P 4+4	TH = 70 SBE = 1.064177772							MCPER FUNCTIONS	
	RC = 02	RC = 05	RC = 07	RC = 10	RC = 15	RC = 20	RC = 25	RC = 35	RC = 50
AMAX DB				.0436480540	.0988323393	.1772876696	.2802872360	.5675287486	1.249387366
AMIN DB				6.548085268	11.12075906	15.13108002	18.59136976	24.33684864	31.29995303
MU I K				.4705391D 0	.2779470D 0	.1751645D 0	.1176066D 0	.6069565D-1	.2722716D-1
1 I PF1				1.445099321	1.211345950	1.047995076	.9310269616	.7744017227	.6324366340
1 I PQ1				.9747020861	.8448793537	.8043817874	.7949546514	.8147790968	.8979344387
3 I PFC				1.091900342	1.088910052	1.083976274	1.077938730	1.064585049	1.044242713
3 I PQC				25.11966445	16.53027996	12.74403502	10.71904500	8.703885720	7.511264117
1 I ZF1				1.074489824	1.072175558	1.071073402	1.070485128	1.069935355	1.069671838
1 I ZF2				1.096958964	1.100360639	1.103305509	1.105800495	1.109822187	1.114440768
1 I ZF3				1.141116716	1.167499279	1.190407928	1.209879106	1.241028061	1.275722840
1 I ZF4				2.033924736	2.141582505	2.244153445	2.336265281	2.490851592	2.670678661

(4) P 4+4 — TH = 71 SBE = 1.057620681 — MCPER FUNCTIONS

(4) P 4+4	RC = 02	RC = 05	RC = 07	RC = 10	RC = 15	RC = 20	RC = 25	RC = 35	RC = 50
AMAX DB				.0436480540	.0988323393	.1772876696	.2802872360	.5675287486	1.249387366
AMIN DB				5.776458332	10.06076508	13.92214209	17.30061815	22.96396909	29.87431746
MU K				.5142533D 0	.3140232D 0	.2013228D 0	.1364486D 0	.7108886D-1	.3208368D-1
1 PF1				1.459970917	1.233768252	1.070066585	.9510651933	.7905621591	.6448114458
1 PQ1				1.027756589	.8752117426	.8250686423	.8107608574	.8260507911	.9067501038
3 PFC				1.083839690	1.081618888	1.077641411	1.072548903	1.060836402	1.042352431
3 PQC				28.99781011	18.90198071	14.40535034	11.99209635	9.585161400	8.143924255
1 ZF1				1.067836961	1.065496807	1.064344634	1.063717383	1.063117804	1.062813153
1 ZF2				1.087879699	1.090913596	1.093625275	1.095963207	1.099781007	1.104212091
1 ZF3				1.125956583	1.149604227	1.170760224	1.189104367	1.218980724	1.252829939
1 ZF4				1.976751820	2.073069386	2.167884827	2.254901967	2.403852571	2.580466342

(4) P 4+4 — TH = 72 SBE = 1.051462224 — MCPER FUNCTIONS

(4) P 4+4	RC = 02	RC = 05	RC = 07	RC = 10	RC = 15	RC = 20	RC = 25	RC = 35	RC = 50
AMAX DB					.0988323393	.1772876696	.2802872360	.5675287486	1.249387366
AMIN DB					9.046949530	12.74209754	16.02713737	21.59607517	28.44599802
MU K					.3529007D 0	.2306190D 0	.1579949D 0	.8321397D-1	.3781813D-1
1 PF1					1.255479856	1.092579965	.9719648673	.8076894107	.6579899420
1 PQ1					.9104639550	.8490210704	.8289323981	.8388284188	.9165926507
3 PFC					1.074512965	1.071372963	1.067148274	1.057003158	1.040357037
3 PQC					21.76084972	16.40529303	13.51731624	10.62912364	8.882647097
1 ZF1					1.059210527	1.058017786	1.057354661	1.056706889	1.056361472
1 ZF2					1.081995210	1.084467098	1.086640645	1.090243613	1.094476351
1 ZF3					1.132933198	1.152269014	1.169397494	1.197849974	1.230703919
1 ZF4					2.007870640	2.094750456	2.176342944	2.319030107	2.491777330

(4) P 4+4 — TH = 73 SBE = 1.045691756 — MCPER FUNCTIONS

(4) P 4+4	RC = 02	RC = 05	RC = 07	RC = 10	RC = 15	RC = 20	RC = 25	RC = 35	RC = 50
AMAX DB					.0988323393	.1772876696	.2802872360	.5675287486	1.249387366
AMIN DB					8.083308647	11.59461630	14.77322583	20.23339072	27.01378502
MU K					.3943071D 0	.2631899D 0	.1825319D 0	.9734877D-1	.4459752D-1
1 PF1					1.276062131	1.115312377	.9936563640	.8258354764	.6720497085
1 PQ1					.9515658835	.8769021801	.8499571194	.8534127820	.9276520202
3 PFC					1.067620841	1.065199816	1.061760182	1.053096676	1.038256971
3 PQC					25.22965030	18.83238367	15.36113525	11.87753992	9.753122703
1 ZF1					1.053300540	1.052079043	1.051384398	1.050691001	1.050305627
1 ZF2					1.073592940	1.075820781	1.077823152	1.081200190	1.085223316
1 ZF3					1.117467110	1.134940015	1.150773594	1.177651276	1.209350640
1 ZF4					1.945844603	2.024705106	2.100584044	2.236387098	2.404577939

$P_{5+0}^{(2)}$

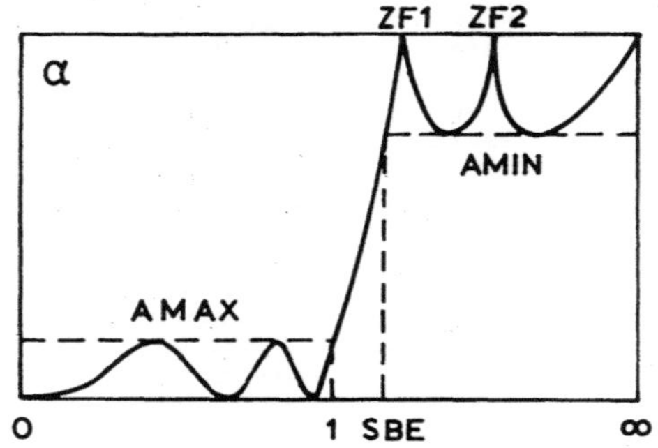

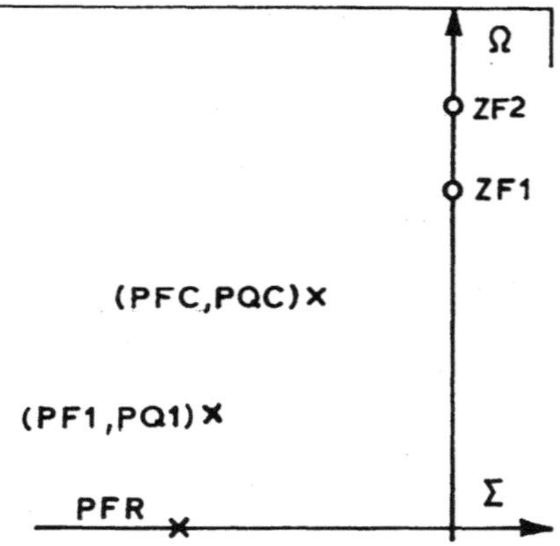

(2) P 5+0	TH = 10	SBE = 5.758770483							CAUER FUNCTIONS
	RC = 02	RC = 05	RC = 07	RC = 10	RC = 15	RC = 20	RC = 25	RC = 35	RC = 50
AMAX DB	.0017375255	.0108709564	.0213327376	.0436480540	.0988323393	.1772876696	.2802872360	.5675287486	1.249387366
AMIN DB	89.88864089	97.85657449	100.7895970	103.9099515	107.4869610	110.0641910	112.1053909	115.3151931	119.0950909
MU I K	.9152713D-3	.3657237D-3	.2609168D-3	.1821731D-3	.1206796D-3	.8969584D-4	.7091079D-4	.4900295D-4	.3171227D-4
1 I PFR	1.074167164	.8166475664	.7300038273	.6419028655	.5456133790	.4791878763	.4283492581	.3518866180	.2686372890
1 I PF1	1.218510431	1.004387043	.9364547373	.8704673680	.8028910251	.7597274324	.7289831393	.6871450487	.6486954430
1 I PQ1	.7152558428	.7715541745	.8036243726	.8484475908	.9195870317	.9900634131	1.062233771	1.218075252	1.505452907
1 I PFC	1.413057374	1.242819920	1.190835359	1.141576621	1.092616796	1.062246994	1.041093515	1.013004516	.9879608564
1 I PQC	2.241706683	2.559932098	2.734119446	2.971058234	3.335146347	3.685106090	4.035318297	4.771885018	6.087565300
1 I ZF1	6.050729094	6.050729094	6.050729094	6.050729094	6.050729094	6.050729094	6.050729094	6.050729094	6.050729094
1 I ZF2	9.748602539	9.748602539	9.748602539	9.748602539	9.748602539	9.748602539	9.748602539	9.748602539	9.748602539

(2) P 5+0	TH = 11	SBE = 5.240843064							CAUER FUNCTIONS
	RC = 02	RC = 05	RC = 07	RC = 10	RC = 15	RC = 20	RC = 25	RC = 35	RC = 50
AMAX DB	.0017375255	.0108709564	.0213327376	.0436480540	.0988323393	.1772876696	.2802872360	.5675287486	1.249387366
AMIN DB	85.72598113	93.69391473	96.62693722	99.74729174	103.3243012	105.9015313	107.9427311	111.1525333	114.9324311
MU I K	.1342934D-2	.5366091D-3	.3828308D-3	.2672940D-3	.1770674D-3	.1316065D-3	.1040440D-3	.7189971D-4	.4652991D-4
1 I PFR	1.077871200	.8188331642	.7318018895	.6433603009	.5467526414	.4801363352	.4291655026	.3525235823	.2691013506
1 I PF1	1.220437686	1.005756630	.9376741569	.8715521579	.8038487224	.7606093228	.7298134565	.6879082732	.6494000040
1 I PQ1	.7169998515	.7730053006	.8050130081	.8497969348	.9209299965	.9914327916	1.063647377	1.219613879	1.507266454
1 I PFC	1.411232008	1.241962565	1.190213438	1.141153810	1.092369810	1.062098456	1.041008902	1.012999047	.9880205684
1 I PQC	2.255745431	2.572401802	2.746383990	2.983350240	3.347842445	3.698407987	4.049344562	4.787645238	6.106757077
1 I ZF1	5.505703409	5.505703409	5.505703409	5.505703409	5.505703409	5.505703409	5.505703409	5.505703409	5.505703409
1 I ZF2	8.862521196	8.862521196	8.862521196	8.862521196	8.862521196	8.862521196	8.862521196	8.862521196	8.862521196

(2) P 5+0	TH = 12 SBE = 4.809734345							CAUER FUNCTIONS	
	RC = 02	RC = 05	RC = 07	RC = 10	RC = 15	RC = 20	RC = 25	RC = 35	RC = 50
AMAX DB	.0017375255	.0108709564	.0213327376	.0436480540	.0988323393	.1772876696	.2802872360	.5675287486	1.249387366
AMIN DB	81.92145302	89.88938661	92.82240910	95.94276361	99.51977308	102.0970031	104.1382030	107.3480052	111.1279030
MU I K	.1906487D-2	.7617934D-3	.5434832D-3	.3794621D-3	.2513726D-3	.1868342D-3	.1477054D-3	.1020719D-3	.6605587D-4
1 I PFR	1.081956434	.8212397681	.7337808799	.6449637377	.5480055492	.4811791670	.4300628247	.3532236754	.2696113158
1 I PF1	1.222546867	1.007257539	.9390109373	.8727416765	.8048991276	.7615767088	.7307243455	.6887456438	.6501730749
1 I PQ1	.7189303333	.7746073507	.8065449454	.8512845389	.9224096361	.9929409663	1.065203865	1.221307465	1.509262076
1 I PFC	1.409235509	1.241023854	1.189532283	1.140690572	1.092099089	1.061935567	1.040916047	1.012992918	.9880858794
1 I PQC	2.271282056	2.586177640	2.759926562	2.996917143	3.361849486	3.713079717	4.064812634	4.805021695	6.127912982
1 I ZF1	5.051963012	5.051963012	5.051963012	5.051963012	5.051963012	5.051963012	5.051963012	5.051963012	5.051963012
1 I ZF2	8.124130175	8.124130175	8.124130175	8.124130175	8.124130175	8.124130175	8.124130175	8.124130175	8.124130175

(2) P 5+0	TH = 13 SBE = 4.445411483							CAUER FUNCTIONS	
	RC = 02	RC = 05	RC = 07	RC = 10	RC = 15	RC = 20	RC = 25	RC = 35	RC = 50
AMAX DB	.0017375255	.0108709564	.0213327376	.0436480540	.0988323393	.1772876696	.2802872360	.5675287486	1.249387366
AMIN DB	78.41727483	86.38520838	89.31823087	92.43858538	96.01559485	98.59282491	100.6340247	103.8438270	107.6237248
MU I K	.2632683D-2	.1051966D-2	.7505000D-3	.5240020D-3	.3471223D-3	.2580008D-3	.2039675D-3	.1409519D-3	.9121704D-4
1 I PFR	1.086431029	.8238710276	.7359435395	.6467152071	.5493735500	.4823175075	.4310421620	.3539875881	.2701676625
1 I PF1	1.224837408	1.008889980	.9404653942	.8740363005	.8060426469	.7626300016	.7317162156	.6896575614	.6510150445
1 I PQ1	.7210531999	.7763639715	.8082233417	.8529131806	.9240284420	.9945903124	1.066905571	1.223158390	1.511442414
1 I PFC	1.407068640	1.240003874	1.188791895	1.140186866	1.091804577	1.061758271	1.040814901	1.012986090	.9881567651
1 I PQC	2.288363250	2.601293703	2.774778710	3.011788774	3.377196543	3.729150654	4.081752716	4.824047201	6.151071791
1 I ZF1	4.668444006	4.668444006	4.668444006	4.668444006	4.668444006	4.668444006	4.668444006	4.668444006	4.668444006
1 I ZF2	7.499346206	7.499346206	7.499346206	7.499346206	7.499346206	7.499346206	7.499346206	7.499346206	7.499346206

(2) P 5+0	TH = 14 SBE = 4.133565494							CAUER FUNCTIONS	
	RC = 02	RC = 05	RC = 07	RC = 10	RC = 15	RC = 20	RC = 25	RC = 35	RC = 50
AMAX DB	.0017375255	.0108709564	.0213327376	.0436480540	.0988323393	.1772876696	.2802872360	.5675287486	1.249387366
AMIN DB	75.16852718	83.13646067	86.06948316	89.18983767	92.76684713	95.34407719	97.38527702	100.5950792	104.3749771
MU I K	.3550955D-2	.1418889D-2	.1012272D-2	.7067726D-3	.4681976D-3	.3479908D-3	.2751108D-3	.1901155D-3	.1230333D-3
1 I PFR	1.091304044	.8267309744	.7382928927	.6486169482	.5508582379	.4835526069	.4321045451	.3548160800	.2707709155
1 I PF1	1.227308636	1.010654164	.9420378622	.8754364357	.8072797206	.7637696481	.7327895124	.6906444637	.6519263370
1 I PQ1	.7233750399	.7782792073	.8100516937	.8546859319	.9257891667	.9963834523	1.068755075	1.225169281	1.513810380
1 I PFC	1.404732241	1.238902726	1.187992280	1.139642648	1.091486211	1.061566510	1.040705408	1.012978520	.9882331988
1 I PQC	2.307041068	2.617787851	2.790975431	3.027998198	3.393915806	3.746653311	4.100198223	4.844758046	6.176276372
1 I ZF1	4.340100916	4.340100916	4.340100916	4.340100916	4.340100916	4.340100916	4.340100916	4.340100916	4.340100916
1 I ZF2	6.963824019	6.963824019	6.963824019	6.963824019	6.963824019	6.963824019	6.963824019	6.963824019	6.963824019

(2) P 5+0	TH = 15	SBE = 3.863703305						CAUER FUNCTIONS	
	RC = 02	RC = 05	RC = 07	RC = 10	RC = 15	RC = 20	RC = 25	RC = 35	RC = 50
AMAX DB	.0017375255	.0108709564	.0213327376	.0436480540	.0988323393	.1772876696	.2802872360	.5675287486	1.249387366
AMIN DB	72.13959615	80.10752953	83.04055201	86.16090651	89.73791597	92.31514603	94.35634586	97.56614808	101.3460459
MU K	.4693518D-2	.1875434D-2	.1337983D-2	.9341852D-3	.6188459D-3	.4599610D-3	.3636311D-3	.2512875D-3	.1626208D-3
1 1 PFR	1.096585486	.8298240406	.7408322591	.6506714154	.5524613582	.4848858327	.4332511017	.3557099809	.2714216483
1 1 PF1	1.229959757	1.012550303	.9437286922	.8769425147	.8086108228	.7649961309	.7339447182	.6917068246	.6529074131
1 1 PQ1	.7259031645	.7803575225	.8120338552	.8566061744	.9276948363	.9983232666	1.070755205	1.227343019	1.516369164
1 1 PFC	1.402227230	1.237720519	1.187133446	1.139057873	1.091143927	1.061360218	1.040587510	1.012970164	.9883151517
1 1 PQC	2.327373293	2.635701935	2.808555370	3.045581886	3.412042756	3.765623498	4.120185946	4.867194167	6.203573882
1 1 ZF1	4.055900547	4.055900547	4.055900547	4.055900547	4.055900547	4.055900547	4.055900547	4.055900547	4.055900547
1 1 ZF2	6.499710337	6.499710337	6.499710337	6.499710337	6.499710337	6.499710337	6.499710337	6.499710337	6.499710337

(2) P 5+0	TH = 16	SBE = 3.627955279						CAUER FUNCTIONS	
	RC = 02	RC = 05	RC = 07	RC = 10	RC = 15	RC = 20	RC = 25	RC = 35	RC = 50
AMAX DB	.0017375255	.0108709564	.0213327376	.0436480540	.0988323393	.1772876696	.2802872360	.5675287486	1.249387366
AMIN DB	69.30176660	77.26969978	80.20272223	83.32307673	86.90008618	89.47731624	91.51851606	94.72831829	98.50821610
MU K	.6095465D-2	.2435623D-2	.1737637D-2	.1213225D-2	.8036941D-3	.5973506D-3	.4722471D-3	.3263467D-3	.2111953D-3
1 1 PFR	1.102286371	.8331550784	.7435652681	.6528812884	.5541848138	.4863186749	.4344830592	.3566701940	.2721204842
1 1 PF1	1.232789838	1.014578602	.9455382492	.8785549966	.8100364606	.7663099687	.7351823519	.6928451557	.6539587701
1 1 PQ1	.7286456596	.7826038265	.8141740579	.8586776154	.9297487636	1.000412906	1.072909059	1.229682750	1.519122253
1 1 PFC	1.399554605	1.236457379	1.186215405	1.138432493	1.090777654	1.061139326	1.040461143	1.012960973	.9884025923
1 1 PQC	2.349423855	2.655082051	2.827561040	3.064579916	3.431616337	3.786100502	4.141756232	4.891399335	6.233015985
1 1 ZF1	3.807568154	3.807568154	3.807568154	3.807568154	3.807568154	3.807568154	3.807568154	3.807568154	3.807568154
1 1 ZF2	6.093615131	6.093615131	6.093615131	6.093615131	6.093615131	6.093615131	6.093615131	6.093615131	6.093615131

(2) P 5+0	TH = 17	SBE = 3.420303620						CAUER FUNCTIONS	
	RC = 02	RC = 05	RC = 07	RC = 10	RC = 15	RC = 20	RC = 25	RC = 35	RC = 50
AMAX DB	.0017375255	.0108709564	.0213327376	.0436480540	.0988323393	.1772876696	.2802872360	.5675287486	1.249387366
AMIN DB	66.63154639	74.59947921	77.53250163	80.65285610	84.22986555	86.80709560	88.84829543	92.05809765	95.83799547
MU K	.7794885D-2	.3114677D-2	.2222091D-2	.1551473D-2	.1027765D-2	.7638925D-3	.6039100D-3	.4173324D-3	.2700768D-3
1 1 PFR	1.108418785	.8367293822	.7464958733	.6552494818	.5560306715	.4878527506	.4358017493	.3576976979	.2728680985
1 1 PF1	1.235797788	1.016739254	.9474669089	.8802743647	.8115571743	.7677117157	.7365029695	.6940600057	.6550809428
1 1 PQ1	.7316114447	.7850235016	.8164769329	.8609043061	.9319545636	1.002655807	1.075220006	1.232191902	1.522073436
1 1 PFC	1.396715448	1.235113441	1.185238172	1.137766457	1.090387317	1.060903759	1.040326240	1.012950894	.9884954870
1 1 PQC	2.373263303	2.675978821	2.848039065	3.085036191	3.452679167	3.808127288	4.164953185	4.917421374	6.264659095
1 1 ZF1	3.588776130	3.588776130	3.588776130	3.588776130	3.588776130	3.588776130	3.588776130	3.588776130	3.588776130
1 1 ZF2	5.735298898	5.735298898	5.735298898	5.735298898	5.735298898	5.735298898	5.735298898	5.735298898	5.735298898

(2) P 5+0	TH = 18	SBE = 3.236067977						CAUER FUNCTIONS	
	RC = 02	RC = 05	RC = 07	RC = 10	RC = 15	RC = 20	RC = 25	RC = 35	RC = 50
AMAX DB	.0017375255	.0108709564	.0213327376	.0436480540	.0988323393	.1772876696	.2802872360	.5675287486	1.249387366
AMIN DB	64.10947035	72.07740254	75.01042490	78.13077935	81.70778878	84.28501883	86.32621865	89.53602087	93.31591868
MU I K	.9832993D-2	.3929063D-2	.2803096D-2	.1957132D-2	.1296492D-2	.9636253D-3	.7618127D-3	.5264512D-3	.3406930D-3
1 I PFR	1.114995964	.8405527135	.7496283693	.6577791573	.5580011701	.4894898094	.4372086116	.3587935502	.2736652205
1 I PF1	1.238982338	1.019032439	.9495150545	.8821011252	.8131735360	.7692019615	.7379071638	.6953519615	.6562745039
1 I PQ1	.7348103390	.7876224347	.8189475360	.8632906619	.9343161697	1.005055703	1.077691711	1.234874193	1.525226826
1 I PFC	1.393710928	1.233688857	1.184201766	1.137059715	1.089972837	1.060653439	1.040182730	1.012939871	.9885937993
1 I PQC	2.398969341	2.698447712	2.870040460	3.106998690	3.475277764	3.831750717	4.189824889	4.945312389	6.298564653
1 I ZF1	3.394603138	3.394603138	3.394603138	3.394603138	3.394603138	3.394603138	3.394603138	3.394603138	3.394603138
1 I ZF2	5.416797510	5.416797510	5.416797510	5.416797510	5.416797510	5.416797510	5.416797510	5.416797510	5.416797510

(2) P 5+0	TH = 19	SBE = 3.071553487						CAUER FUNCTIONS	
	RC = 02	RC = 05	RC = 07	RC = 10	RC = 15	RC = 20	RC = 25	RC = 35	RC = 50
AMAX DB	.0017375255	.0108709564	.0213327376	.0436480540	.0988323393	.1772876696	.2802872360	.5675287486	1.249387366
AMIN DB	61.71922799	69.68715914	72.62018140	75.74053580	79.31754520	81.89477524	83.93597506	87.14577727	90.92567509
MU I K	.1225426D-1	.4896552D-2	.3493327D-2	.2439055D-2	.1615738D-2	.1200907D-2	.9494007D-3	.6560840D-3	.4245849D-3
1 I PFR	1.122032375	.8446313279	.7529674110	.6604737353	.5600987275	.4912317396	.4387051991	.3599588905	.2745126351
1 I PF1	1.242342015	1.021458310	.9516830728	.8840358052	.8148861494	.7707813316	.7393955652	.6967216485	.6575400647
1 I PQ1	.7382531356	.7904070515	.8215913755	.8658414849	.9368378526	1.007616644	1.080328145	1.237733649	1.528586873
1 I PFC	1.390542301	1.232183795	1.183106211	1.136312215	1.089534133	1.060388284	1.040030535	1.012927844	.9886974906
1 I PQC	2.426627430	2.722549386	2.893620937	3.130519738	3.499462796	3.857021797	4.216423654	4.975129030	6.334799418
1 I ZF1	3.221164045	3.221164045	3.221164045	3.221164045	3.221164045	3.221164045	3.221164045	3.221164045	3.221164045
1 I ZF2	5.131823431	5.131823431	5.131823431	5.131823431	5.131823431	5.131823431	5.131823431	5.131823431	5.131823431

(2) P 5+0	TH = 20	SBE = 2.923804400						CAUER FUNCTIONS	
	RC = 02	RC = 05	RC = 07	RC = 10	RC = 15	RC = 20	RC = 25	RC = 35	RC = 50
AMAX DB	.0017375255	.0108709564	.0213327376	.0436480540	.0988323393	.1772876696	.2802872360	.5675287486	1.249387366
AMIN DB	59.44701524	67.41494470	70.34796681	73.46832112	77.04533048	79.62256050	81.66376031	84.87356252	88.65346033
MU I K	.1510657D-1	.6036275D-2	.4306435D-2	.3006770D-2	.1991818D-2	.1480431D-2	.1170384D-2	.8087944D-3	.5234115D-3
1 I PFR	1.129543806	.8489720057	.7565180336	.6633369094	.5623259504	.4930805745	.4402931827	.3611949438	.2754111856
1 I PF1	1.245875115	1.024016992	.9539713503	.8860789505	.8166956491	.7724504866	.7409688412	.6981697311	.6588782758
1 I PQ1	.7419516860	.7933843563	.8244144430	.8685619883	.9395242404	1.010343016	1.083133601	1.240774625	1.532158380
1 I PFC	1.387210913	1.230598437	1.181951538	1.135523904	1.089071117	1.060108205	1.039869576	1.012914749	.9888065194
1 I PQC	2.456331466	2.748350101	2.918841249	3.155656311	3.525289365	3.883995954	4.244806293	5.006932785	6.373435808
1 I ZF1	3.065350869	3.065350869	3.065350869	3.065350869	3.065350869	3.065350869	3.065350869	3.065350869	3.065350869
1 I ZF2	4.875346566	4.875346566	4.875346566	4.875346566	4.875346566	4.875346566	4.875346566	4.875346566	4.875346566

(2) P 5+0	TH = 21 SBE = 2.790428110								CAUER FUNCTIONS
	RC = 02	RC = 05	RC = 07	RC = 10	RC = 15	RC = 20	RC = 25	RC = 35	RC = 50
AMAX DB	.0017375255	.0108709564	.0213327376	.0436480540	.0988323393	.1772876696	.2802872360	.5675287486	1.249387366
AMIN DB	57.28104452	65.24897130	68.18199316	71.30234734	74.87935662	77.45658662	79.49778642	82.70758862	86.48748643
MU I K	.1844137D-1	.7368794D-2	.5257088D-2	.3670520D-2	.2431516D-2	.1807239D-2	.1428748D-2	.9873372D-3	.6389555D-3
1 I PFR	1.137547473	.8535820855	.7602856754	.6663726614	.5646856436	.4950384996	.4419743572	.3625030248	.2763617763
1 I PF1	1.249579679	1.026708571	.9563802682	.8882311229	.8186026994	.7742101222	.7426276971	.6996969132	.6602898281
1 I PQ1	.7459189939	.7965619745	.8274232474	.8714578245	.9423803417	1.013239558	1.086112718	1.244001816	1.535946528
1 I PFC	1.383718207	1.228932986	1.180737784	1.134694728	1.088583700	1.059813112	1.039699766	1.012900521	.9889208414
1 I PQC	2.488184550	2.775922152	2.945767570	3.182470382	3.552817319	3.912733338	4.275034427	5.040790293	6.414552264
1 I ZF1	2.924647748	2.924647748	2.924647748	2.924647748	2.924647748	2.924647748	2.924647748	2.924647748	2.924647748
1 I ZF2	4.643294870	4.643294870	4.643294870	4.643294870	4.643294870	4.643294870	4.643294870	4.643294870	4.643294870

(2) P 5+0	TH = 22 SBE = 2.669467163								CAUER FUNCTIONS
	RC = 02	RC = 05	RC = 07	RC = 10	RC = 15	RC = 20	RC = 25	RC = 35	RC = 50
AMAX DB	.0017375255	.0108709564	.0213327376	.0436480540	.0988323393	.1772876696	.2802872360	.5675287486	1.249387366
AMIN DB	55.21116889	63.17909150	66.11211297	69.23246694	72.80947611	75.38670607	77.42790585	80.63770804	84.41760583
MU I K	.2231387D-1	.8916167D-2	.6361024D-2	.4441292D-2	.2942110D-2	.2186741D-2	.1728771D-2	.1194668D-2	.7731297D-3
1 I PFR	1.146062139	.8584695010	.7642762030	.6695852778	.5671808211	.4971078608	.4437506472	.3638845417	.2773653749
1 I PF1	1.253453457	1.029533085	.9589101974	.8904928980	.8206079933	.7760609690	.7443728758	.7013039389	.6617754534
1 I PQ1	.7501693222	.7999482016	.8306248529	.8745351155	.9454115699	1.016311387	1.089270502	1.247420281	1.539956892
1 I PFC	1.380065719	1.227187661	1.179464992	1.133824637	1.088071787	1.059502907	1.039521018	1.012885087	.9890404098
1 I PQC	2.522299853	2.805344365	2.974471926	3.211029293	3.582111592	3.943299157	4.307174818	5.076773704	6.458233654
1 I ZF1	2.796996367	2.796996367	2.796996367	2.796996367	2.796996367	2.796996367	2.796996367	2.796996367	2.796996367
1 I ZF2	4.432336604	4.432336604	4.432336604	4.432336604	4.432336604	4.432336604	4.432336604	4.432336604	4.432336604

(2) P 5+0	TH = 23 SBE = 2.559304665								CAUER FUNCTIONS
	RC = 02	RC = 05	RC = 07	RC = 10	RC = 15	RC = 20	RC = 25	RC = 35	RC = 50
AMAX DB	.0017375255	.0108709564	.0213327376	.0436480540	.0988323393	.1772876696	.2802872360	.5675287486	1.249387366
AMIN DB	53.22858977	61.19650602	64.12952689	67.24988055	70.82688956	73.40411946	75.44531921	78.65512137	82.43501916
MU I K	.2678322D-1	.1070203D-1	.7635104D-2	.5330860D-2	.3531399D-2	.2624734D-2	.2075034D-2	.1433954D-2	.9279836D-3
1 I PFR	1.155108237	.8636428229	.7684959390	.6729793689	.5698147175	.4992911724	.4456241143	.3653410014	.2784230157
1 I PF1	1.257493872	1.032490518	.9615614927	.8928648617	.8227122515	.7780037920	.7462051581	.7029915926	.6633359252
1 I PQ1	.7547183125	.8035520565	.8340269216	.8778004865	.9486237716	1.019564022	1.092612344	1.251035470	1.544195467
1 I PFC	1.376255082	1.225362703	1.178133213	1.132913576	1.087535283	1.059177490	1.039333236	1.012868372	.9891651748
1 I PQC	2.558801598	2.836702652	3.005032670	3.241406177	3.613242599	3.975764054	4.341299750	5.114961068	6.504571728
1 I ZF1	2.680696497	2.680696497	2.680696497	2.680696497	2.680696497	2.680696497	2.680696497	2.680696497	2.680696497
1 I ZF2	4.239719401	4.239719401	4.239719401	4.239719401	4.239719401	4.239719401	4.239719401	4.239719401	4.239719401

(2) P 5+0 — TH = 24 SBE = 2.458593336 — CAUER FUNCTIONS

	RC = 02	RC = 05	RC = 07	RC = 10	RC = 15	RC = 20	RC = 25	RC = 35	RC = 50
AMAX DB	.0017375255	.0108709564	.0213327376	.0436480540	.0988323393	.1772876696	.2802872360	.5675287486	1.249387366
AMIN DB	51.32562702	59.29353373	62.22655371	65.34690690	68.92391565	71.50114547	73.54234518	76.75214730	80.53204507
MU K	.3191273D-1	.1275168D-1	.9097375D-2	.6351823D-2	.4207730D-2	.3127421D-2	.2472444D-2	.1708584D-2	.1105710D-2
1 PFR	1.164708024	.8691113038	.7729516925	.6765598881	.5725908011	.5015911271	.4475969641	.3668740139	.2795358032
1 PF1	1.261697984	1.035580783	.9643344863	.8953476070	.8249162211	.7800393904	.7481253626	.7047607004	.6649720599
1 PQ1	.7595831210	.8073833419	.8376377601	.8812611033	.9520232561	1.023003411	1.096144052	1.254853239	1.548668697
1 PFC	1.372288034	1.223458372	1.176742510	1.131961495	1.086974086	1.058836756	1.039136323	1.012850298	.9892950837
1 PQC	2.597826165	2.870090633	3.037535011	3.273680430	3.646286651	4.010204518	4.377487435	5.155436771	6.553665612
1 ZF1	2.574331394	2.574331394	2.574331394	2.574331394	2.574331394	2.574331394	2.574331394	2.574331394	2.574331394
1 ZF2	4.063149555	4.063149555	4.063149555	4.063149555	4.063149555	4.063149555	4.063149555	4.063149555	4.063149555

(2) P 5+0 — TH = 25 SBE = 2.366201583 — CAUER FUNCTIONS

	RC = 02	RC = 05	RC = 07	RC = 10	RC = 15	RC = 20	RC = 25	RC = 35	RC = 50
AMAX DB	.0017375255	.0108709564	.0213327376	.0436480540	.0988323393	.1772876696	.2802872360	.5675287486	1.249387366
AMIN DB	49.49553593	57.46342855	60.39644722	63.51679970	67.09380808	69.67103777	71.71223742	74.92203949	78.70193723
MU K	.3777008D-1	.1509215D-1	.1076713D-1	.7517654D-2	.4980029D-2	.3701437D-2	.2926243D-2	.2022182D-2	.1308656D-2
1 PFR	1.174885743	.8748849290	.7776507928	.6803321552	.5755127881	.5040106061	.4496715542	.3684852987	.2807049150
1 PF1	1.266062446	1.038803715	.9672294807	.8979417301	.8272206736	.7821685968	.7501343460	.7066121299	.6666847175
1 PQ1	.7647825711	.8114527110	.8414663716	.8849247141	.9556168300	1.026635962	1.099871874	1.258379886	1.553383499
1 PFC	1.368166412	1.221474954	1.175292950	1.130968344	1.086388094	1.058480597	1.038930176	1.012830783	.9894300806
1 PQC	2.639523356	2.905610326	3.072071610	3.307938227	3.681326438	4.046703342	4.415822478	5.198292014	6.605622358
1 ZF1	2.476711104	2.476711104	2.476711104	2.476711104	2.476711104	2.476711104	2.476711104	2.476711104	2.476711104
1 ZF2	3.900700293	3.900700293	3.900700293	3.900700293	3.900700293	3.900700293	3.900700293	3.900700293	3.900700293

(2) P 5+0 — TH = 26 SBE = 2.281172033 — CAUER FUNCTIONS

	RC = 02	RC = 05	RC = 07	RC = 10	RC = 15	RC = 20	RC = 25	RC = 35	RC = 50
AMAX DB	.0017375255	.0108709564	.0213327376	.0436480540	.0988323393	.1772876696	.2802872360	.5675287486	1.249387366
AMIN DB	47.73236019	55.70023228	58.63324903	61.75360050	65.33060834	67.90783783	69.94903740	73.15883939	76.93873709
MU K	.4442759D-1	.1775236D-1	.1266499D-1	.8842747D-2	.5857829D-2	.4353868D-2	.3442035D-2	.2378620D-2	.1539325D-2
1 PFR	1.185667803	.8809744727	.7826011268	.6843018802	.5785846579	.5065526902	.4518504037	.3701766900	.2819316066
1 PF1	1.270583458	1.042159055	.9702467408	.9006478261	.8296264040	.7843922767	.7522330027	.7085467914	.6684748029
1 PQ1	.7703373264	.8157717421	.8455225141	.8887996953	.9594118337	1.030468575	1.103802534	1.263122178	1.558347297
1 PFC	1.363892159	1.219412757	1.173784616	1.129934075	1.085777202	1.058108900	1.038714688	1.012809738	.9895701065
1 PQC	2.684057810	2.943372928	3.108743241	3.344273108	3.718451551	4.085350138	4.456396382	5.243625346	6.660557551
1 ZF1	2.386828852	2.386828852	2.386828852	2.386828852	2.386828852	2.386828852	2.386828852	2.386828852	2.386828852
1 ZF2	3.750741203	3.750741203	3.750741203	3.750741203	3.750741203	3.750741203	3.750741203	3.750741203	3.750741203

(2) P 5+0	TH = 27	SBE = 2.202689265							CAUER FUNCTIONS
	RC = 02	RC = 05	RC = 07	RC = 10	RC = 15	RC = 20	RC = 25	RC = 35	RC = 50
AMAX DB	.0017375255	.0108709564	.0213327376	.0436480540	.0988323393	.1772876696	.2802872360	.5675287486	1.249387366
AMIN DB	46.03081265	53.99865523	56.93166924	60.05201924	63.62902630	66.20625552	68.24745495	71.45725684	75.23715448
MU K	.5196248D-1	.2076315D-1	.1481297D-1	.1034247D-1	.6851313D-2	.5092281D-2	.4025801D-2	.2782033D-2	.1800393D-2
1 PFR	1.197082993	.8873915597	.7878111798	.6884751905	.5818106704	.5092206725	.4541362019	.3719501439	.2832172150
1 PF1	1.275256713	1.045646438	.9733864849	.9034664842	.8321342277	.7867113280	.7544222652	.7105656382	.6703432667
1 PQ1	.7762700871	.8203530227	.8498167649	.8928951026	.9634161829	1.034508681	1.107943264	1.267587383	1.563568056
1 PFC	1.359467323	1.217272113	1.172217600	1.128858644	1.085141300	1.057721547	1.038489748	1.012787073	.9897150993
1 PQC	2.731610637	2.983499691	3.147659535	3.382786625	3.757759070	4.126241898	4.499308117	5.291543254	6.718595986
1 ZF1	2.303827126	2.303827126	2.303827126	2.303827126	2.303827126	2.303827126	2.303827126	2.303827126	2.303827126
1 ZF2	3.611883354	3.611883354	3.611883354	3.611883354	3.611883354	3.611883354	3.611883354	3.611883354	3.611883354

(2) P 5+0	TH = 28	SBE = 2.130054468							CAUER FUNCTIONS
	RC = 02	RC = 05	RC = 07	RC = 10	RC = 15	RC = 20	RC = 25	RC = 35	RC = 50
AMAX DB	.0017375255	.0108709564	.0213327376	.0436480540	.0988323393	.1772876696	.2802872360	.5675287486	1.249387366
AMIN DB	44.38617786	52.35397853	55.28698863	58.40733656	61.98434250	64.56157134	66.60277059	69.81257232	73.59246988
MU K	.6045715D-1	.2415745D-1	.1723455D-1	.1203323D-1	.7971346D-2	.5924752D-2	.4683928D-2	.3236831D-2	.2094716D-2
1 PFR	1.209162708	.8941487351	.7932900814	.6928586611	.5851953845	.5120180716	.4565318196	.3738077453	.2845631636
1 PF1	1.280077345	1.049265371	.9766488753	.9063982824	.8347449794	.7891266797	.7567031036	.7126696674	.6722911066
1 PQ1	.7826058121	.8252102426	.8543605929	.8972207275	.9676384138	1.038764279	1.112301842	1.272283306	1.569054320
1 PFC	1.354894060	1.215053383	1.170592005	1.127742008	1.084480278	1.057318417	1.038255240	1.012762693	.9898649933
1 PQC	2.782381254	3.026122902	3.188939812	3.423589066	3.799354218	4.169483629	4.544664742	5.342160814	6.779872410
1 ZF1	2.226971021	2.226971021	2.226971021	2.226971021	2.226971021	2.226971021	2.226971021	2.226971021	2.226971021
1 ZF2	3.482936168	3.482936168	3.482936168	3.482936168	3.482936168	3.482936168	3.482936168	3.482936168	3.482936168

(2) P 5+0	TH = 29	SBE = 2.062665340							CAUER FUNCTIONS
	RC = 02	RC = 05	RC = 07	RC = 10	RC = 15	RC = 20	RC = 25	RC = 35	RC = 50
AMAX DB	.0017375255	.0108709564	.0213327376	.0436480540	.0988323393	.1772876696	.2802872360	.5675287486	1.249387366
AMIN DB	42.79423177	50.76197359	53.69497819	56.81532320	60.39232759	62.96955588	65.01075488	68.22055639	72.00045383
MU K	.6999953D-1	.2797039D-1	.1995480D-1	.1393252D-1	.9229520D-2	.6859898D-2	.5423225D-2	.3747723D-2	.2425340D-2
1 PFR	1.221941219	.9012595399	.7990476559	.6974593469	.5887436788	.5149486468	.4590403204	.3757517160	.2859709671
1 PF1	1.285039861	1.053015224	.9800340071	.9094437816	.8374595097	.7916392911	.7590765255	.7148599204	.6743193683
1 PQ1	.7893719725	.8303582987	.8591664384	.9017871608	.9720877340	1.043243984	1.116886633	1.277218332	1.574815255
1 PFC	1.350174635	1.212756955	1.168907949	1.126584129	1.083794024	1.056899387	1.038011044	1.012736498	.9900197192
1 PQC	2.836589504	3.071386990	3.232714011	3.466800266	3.843351088	4.215189050	4.592582107	5.395602425	6.844532352
1 ZF1	2.155627103	2.155627103	2.155627103	2.155627103	2.155627103	2.155627103	2.155627103	2.155627103	2.155627103
1 ZF2	3.362873219	3.362873219	3.362873219	3.362873219	3.362873219	3.362873219	3.362873219	3.362873219	3.362873219

(2) P 5+0	TH = 30	SBE = 2.000000000						CAUER FUNCTIONS	
	RC = 02	RC = 05	RC = 07	RC = 10	RC = 15	RC = 20	RC = 25	RC = 35	RC = 50
AMAX DB	.0017375255	.0108709564	.0213327376	.0436480540	.0988323393	.1772876696	.2802872360	.5675287486	1.249387366
AMIN DB	41.25117518	49.21883517	52.15183214	55.27217310	58.84917532	61.42640285	63.46760151	66.67740271	70.45729999
MU I K	.8068339D-1	.3223944D-1	.2300045D-1	.1605900D-1	.1063820D-1	.7906907D-2	.6250959D-2	.4319729D-2	.2795514D-2
1 I PFR	1.235455962	.9087385972	.8050944777	.7022848192	.5924607739	.5180164142	.4616649732	.3777844230	.2874422375
1 I PF1	1.290138077	1.056895200	.9835418960	.9126035188	.8402786824	.7942501504	.7615435759	.7171374839	.6764291472
1 I PQ1	.7965988393	.8358134129	.8642478035	.9066058622	.9767740783	1.047957074	1.121706639	1.282401466	1.580860692
1 I PFC	1.345311421	1.210383247	1.167165565	1.125384974	1.083082424	1.056464326	1.037757034	1.012708384	.9901792043
1 I PQC	2.894478058	3.119449775	3.279123741	3.512550514	3.889873456	4.263481375	4.643185631	5.452002609	6.912733043
1 I ZF1	2.089246502	2.089246502	2.089246502	2.089246502	2.089246502	2.089246502	2.089246502	2.089246502	2.089246502
1 I ZF2	3.250804875	3.250804875	3.250804875	3.250804875	3.250804875	3.250804875	3.250804875	3.250804875	3.250804875

(2) P 5+0	TH = 31	SBE = 1.941604026						CAUER FUNCTIONS	
	RC = 02	RC = 05	RC = 07	RC = 10	RC = 15	RC = 20	RC = 25	RC = 35	RC = 50
AMAX DB	.0017375255	.0108709564	.0213327376	.0436480540	.0988323393	.1772876696	.2802872360	.5675287486	1.249387366
AMIN DB	39.75357821	47.72112554	50.65411200	53.77444737	57.35144661	59.92867310	61.96987127	65.17967205	68.95956910
MU I K	.9260872D-1	.3700456D-1	.2640001D-1	.1843259D-1	.1221057D-1	.9075580D-2	.7174876D-2	.4958203D-2	.3208702D-2
1 I PFR	1.249747877	.9166017071	.8114419335	.7073432055	.5963522577	.5212256644	.4644092665	.3799083884	.2889786894
1 I PF1	1.295365042	1.060904322	.9871724654	.9158779995	.8432033713	.7969602736	.7641053364	.7195034900	.6786215894
1 I PQ1	.8043198125	.8415932641	.8696193525	.9116892392	.9817081706	1.052913548	1.126771545	1.287842385	1.587201182
1 I PFC	1.340306902	1.207932706	1.165365000	1.124144513	1.082345361	1.056013104	1.037493083	1.012678242	.9903433718
1 I PQC	2.956315188	3.170483879	3.328323458	3.560981573	3.939055694	4.314494189	4.696611164	5.511506916	6.984644439
1 I ZF1	2.027351277	2.027351277	2.027351277	2.027351277	2.027351277	2.027351277	2.027351277	2.027351277	2.027351277
1 I ZF2	3.145956226	3.145956226	3.145956226	3.145956226	3.145956226	3.145956226	3.145956226	3.145956226	3.145956226

(2) P 5+0	TH = 32	SBE = 1.887079915						CAUER FUNCTIONS	
	RC = 02	RC = 05	RC = 07	RC = 10	RC = 15	RC = 20	RC = 25	RC = 35	RC = 50
AMAX DB	.0017375255	.0108709564	.0213327376	.0436480540	.0988323393	.1772876696	.2802872360	.5675287486	1.249387366
AMIN DB	38.29833378	46.26572733	49.19869944	52.31902719	55.89602237	58.47324743	60.51444494	63.72424515	67.50414190
MU I K	.1058822D 0	.4230835D-1	.3018388D-1	.2107450D-1	.1396069D-1	.1037637D-1	.8203238D-2	.5668853D-2	.3668600D-2
1 I PFR	1.264861787	.9248659521	.8181022911	.7126432343	.6004241123	.5245809825	.4672769230	.3821262999	.2905821475
1 I PF1	1.300712956	1.065041401	.9909255308	.9192676897	.8462344564	.7997707024	.7667629249	.7219591173	.6808978931
1 I PQ1	.8125717977	.8477171371	.8752970244	.9170507339	.9869015935	1.058124184	1.132091782	1.293551492	1.593848048
1 I PFC	1.335163669	1.205405815	1.163506419	1.122862722	1.081582718	1.055545583	1.037219056	1.012645959	.9905121411
1 I PQC	3.022397939	3.224678323	3.380481800	3.612247823	3.991043784	4.368372419	4.753005957	5.574272927	7.060450356
1 I ZF1	1.969523329	1.969523329	1.969523329	1.969523329	1.969523329	1.969523329	1.969523329	1.969523329	1.969523329
1 I ZF2	3.047649159	3.047649159	3.047649159	3.047649159	3.047649159	3.047649159	3.047649159	3.047649159	3.047649159

(2) P 5+0		TH = 33	SBE = 1.836078459							CAUER FUNCTIONS
		RC = 02	RC = 05	RC = 07	RC = 10	RC = 15	RC = 20	RC = 25	RC = 35	RC = 50
AMAX DB		.0017375255	.0108709564	.0213327376	.0436480540	.0988323393	.1772876696	.2802872360	.5675287486	1.249387366
AMIN DB		36.88261840	44.84980381	47.78275650	50.90307393	54.48006360	57.05728673	59.09848335	62.30828279	66.08817912
MU	K	.1206174D 0	.4819627D-1	.3438446D-1	.2400737D-1	.1590355D-1	.1182041D-1	.9344855D-2	.6457769D-2	.4179146D-2
1	PFR	1.280846823	.9335498163	.8250887759	.7181942833	.6046827444	.5280872694	.4702719169	.3844410182	.2922545527
1	PF1	1.306173079	1.069305014	.9948007837	.9227730062	.8493728196	.8026825024	.7695174949	.7245055912	.6832593101
1	PQ1	.8213956369	.8542060909	.8812981600	.9227049211	.9923668648	1.063600609	1.137678585	1.299539973	1.600913450
1	PFC	1.329884425	1.202803085	1.161590001	1.121539585	1.080794378	1.055061625	1.036934815	1.012611418	.9906954273
1	PQC	3.093055791	3.282240335	3.435783087	3.666517547	4.045996467	4.425273432	4.812529747	5.640471369	7.140349742
1	ZF1	1.915395346	1.915395346	1.915395346	1.915395346	1.915395346	1.915395346	1.915395346	1.915395346	1.915395346
1	ZF2	2.955287688	2.955287688	2.955287688	2.955287688	2.955287688	2.955287688	2.955287688	2.955287688	2.955287688

(2) P 5+0		TH = 34	SBE = 1.788291650							CAUER FUNCTIONS
		RC = 02	RC = 05	RC = 07	RC = 10	RC = 15	RC = 20	RC = 25	RC = 35	RC = 50
AMAX DB		.0017375255	.0108709564	.0213327376	.0436480540	.0988323393	.1772876696	.2802872360	.5675287486	1.249387366
AMIN DB		35.50385916	43.47076498	46.40369157	49.52399514	53.10097742	55.67819796	57.71939338	60.92919177	64.70908755
MU	K	.1369358D 0	.5471675D-1	.3903635D-1	.2725533D-1	.1805514D-1	.1341960D-1	.1060912D-1	.7331441D-2	.4744544D-2
1	PFR	1.297756919	.9426733171	.8324156558	.7240064348	.6091350190	.5317497659	.4733984918	.3868555982	.2939979712
1	PF1	1.311735635	1.073693471	.9987977730	.9263943063	.8526193401	.8056967609	.7723702343	.7271441845	.6857071476
1	PQ1	.8308366045	.8610831483	.8876416439	.9286676178	.9981175240	1.069355372	1.143544064	1.305819864	1.608110453
1	PFC	1.324471977	1.200125068	1.159615948	1.120175092	1.079980221	1.054561087	1.036640219	1.012574495	.9908631412
1	PQC	3.168654884	3.343397413	3.494429021	3.723974386	4.104086522	4.485368255	4.875355964	5.710287376	7.224558095
1	ZF1	1.864643330	1.864643330	1.864643330	1.864643330	1.864643330	1.864643330	1.864643330	1.864643330	1.864643330
1	ZF2	2.868345871	2.868345871	2.868345871	2.868345871	2.868345871	2.868345871	2.868345871	2.868345871	2.868345871

(2) P 5+0		TH = 35	SBE = 1.743446796							CAUER FUNCTIONS
		RC = 02	RC = 05	RC = 07	RC = 10	RC = 15	RC = 20	RC = 25	RC = 35	RC = 50
AMAX DB		.0017375255	.0108709564	.0213327376	.0436480540	.0988323393	.1772876696	.2802872360	.5675287486	1.249387366
AMIN DB		34.15970573	42.12623866	45.05913046	48.17941554	51.75638794	54.33360503	56.37479886	59.58459585	63.36449089
MU	K	.1549666D 0	.6192149D-1	.4417639D-1	.3084413D-1	.2043253D-1	.1518660D-1	.1200606D-1	.8296797D-2	.5369274D-2
1	PFR	1.315651356	.9522581545	.8400983367	.7300905352	.6137882955	.5355740785	.4766611814	.3893732935	.2958146022
1	PF1	1.317389702	1.078204784	1.002915884	.9301318760	.8559748883	.8088145840	.7753223649	.7298762182	.6882427693
1	PQ1	.8409449779	.8683735114	.8943480653	.9349560056	1.004168228	1.075402031	1.149701284	1.312404127	1.615753101
1	PFC	1.318929240	1.197372349	1.157584478	1.118769239	1.079140128	1.054043823	1.036335119	1.012535062	.9910451889
1	PQC	3.249602918	3.408399672	3.556640621	3.784818980	4.165502218	4.548842959	4.941673095	5.783921884	7.313309048
1	ZF1	1.816980417	1.816980417	1.816980417	1.816980417	1.816980417	1.816980417	1.816980417	1.816980417	1.816980417
1	ZF2	2.786357802	2.786357802	2.786357802	2.786357802	2.786357802	2.786357802	2.786357802	2.786357802	2.786357802

(2) P 5+0	TH = 36 SBE = 1.701301617							CAUER FUNCTIONS	
	RC = 02	RC = 05	RC = 07	RC = 10	RC = 15	RC = 20	RC = 25	RC = 35	RC = 50
AMAX DB	.0017375255	.0108709564	.0213327376	.0436480540	.0988323393	.1772876696	.2802872360	.5675287486	1.249387366
AMIN DB	32.84800673	40.81404567	43.74689137	46.86715195	50.44411128	53.02132379	55.06251550	58.27231065	62.05220471
MU I K	.1748479D 0	.6986564D-1	.4984396D-1	.3480125D-1	.2305390D-1	.1713495D-1	.1354637D-1	.9361226D-2	.6058119D-2
1 I PFR	1.334595401	.9623278771	.8481534680	.7364582628	.6186504691	.5395662088	.4800648314	.3919975757	.2977067877
1 I PF1	1.323123097	1.082836627	1.007154317	.9339859171	.8594403204	.8120370939	.7783751403	.7327030617	.6908675974
1 I PQ1	.8517766965	.8761048052	.9014398991	.9415887694	1.010534858	1.081755246	1.156164350	1.319306728	1.623756503
1 I PFC	1.313259233	1.194545550	1.155495833	1.117322032	1.078273981	1.053509684	1.036019366	1.012492987	.9912314721
1 I PQC	3.336354832	3.477522519	3.622660415	3.849270827	4.230448942	4.615900202	5.011686212	5.861593211	7.406856145
1 I ZF1	1.772151721	1.772151721	1.772151721	1.772151721	1.772151721	1.772151721	1.772151721	1.772151721	1.772151721
1 I ZF2	2.708909268	2.708909268	2.708909268	2.708909268	2.708909268	2.708909268	2.708909268	2.708909268	2.708909268

(2) P 5+0	TH = 37 SBE = 1.661640141							CAUER FUNCTIONS	
	RC = 02	RC = 05	RC = 07	RC = 10	RC = 15	RC = 20	RC = 25	RC = 35	RC = 50
AMAX DB	.0017375255	.0108709564	.0213327376	.0436480540	.0988323393	.1772876696	.2802872360	.5675287486	1.249387366
AMIN DB	31.56678981	39.53217843	42.46496342	45.58519175	49.16213386	51.73934035	53.78052927	56.99032199	60.77021476
MU I K	.1967269D 0	.7860806D-1	.5608103D-1	.3915599D-1	.2593867D-1	.1927908D-1	.1524145D-1	.1053261D-1	.6816183D-2
1 I PFR	1.354661025	.9729080697	.8565990622	.7431222021	.6237300162	.5437325856	.4836146241	.3947321503	.2996770232
1 I PF1	1.328922247	1.087586298	1.011512058	.9379565323	.8630164710	.8153654250	.7815298450	.7356261330	.6935831142
1 I PQ1	.8633941239	.8843073560	.9089417116	.9485862524	1.017234640	1.088430880	1.162948505	1.326542731	1.632136926
1 I PFC	1.307465076	1.191645334	1.153350273	1.115833487	1.077381660	1.052958516	1.035692802	1.012448131	.9914218870
1 I PQC	3.429419430	3.551069729	3.692754939	3.917570391	4.299151050	4.686760988	5.085618696	5.943538830	7.505474842
1 I ZF1	1.729930015	1.729930015	1.729930015	1.729930015	1.729930015	1.729930015	1.729930015	1.729930015	1.729930015
1 I ZF2	2.635630759	2.635630759	2.635630759	2.635630759	2.635630759	2.635630759	2.635630759	2.635630759	2.635630759

(2) P 5+0	TH = 38 SBE = 1.624269245							CAUER FUNCTIONS	
	RC = 02	RC = 05	RC = 07	RC = 10	RC = 15	RC = 20	RC = 25	RC = 35	RC = 50
AMAX DB	.0017375255	.0108709564	.0213327376	.0436480540	.0988323393	.1772876696	.2802872360	.5675287486	1.249387366
AMIN DB	30.31424497	38.27878246	41.21148798	44.33167409	47.90859367	50.48579227	52.52697753	55.73676708	59.51665816
MU I K	.2207610D 0	.8821160D-1	.6293244D-1	.4393967D-1	.2910760D-1	.2163440D-1	.1710349D-1	.1181938D-1	.7648915D-2
1 I PFR	1.375927722	.9840265644	.8654546274	.7500959280	.6290360442	.5480801000	.4873161058	.3975309748	.3017279687
1 I PF1	1.334772052	1.092450673	1.015987857	.9420437088	.8667041449	.8188007200	.7847877921	.7386468995	.6963908638
1 I PQ1	.8758669329	.8930145080	.9168803931	.9559706322	1.024286284	1.095446122	1.170070238	1.334128396	1.640911899
1 I PFC	1.301549988	1.188672402	1.151148086	1.114303628	1.076463048	1.052390166	1.035355267	1.012400349	.9916163249
1 I PQC	3.529367109	3.629376963	3.767217602	3.989981502	4.371853951	4.761666638	5.163714182	6.030017358	7.609464747
1 I ZF1	1.690112096	1.690112096	1.690112096	1.690112096	1.690112096	1.690112096	1.690112096	1.690112096	1.690112096
1 I ZF2	2.566191583	2.566191583	2.566191583	2.566191583	2.566191583	2.566191583	2.566191583	2.566191583	2.566191583

(2) P 5+0	TH = 39	SBE = 1.589015729						CAUER FUNCTIONS	
	RC = 02	RC = 05	RC = 07	RC = 10	RC = 15	RC = 20	RC = 25	RC = 35	RC = 50
AMAX DB	.0017375255	.0108709564	.0213327376	.0436480540	.0988323393	.1772876696	.2802872360	.5675287486	1.249387366
AMIN DB	29.08871067	37.05214035	39.98474241	43.10487356	46.68176379	49.25895212	51.30013263	54.50991804	58.28980692
MU K	.2471181D 0	.9874337D-1	.7044608D-1	.4918573D-1	.3258282D-1	.2421738D-1	.1914552D-1	.1323052D-1	.8562136D-2
1 PFR	1.398483455	.9957136788	.8747413173	.7573940986	.6345783477	.5526161450	.4911752170	.4005482784	.3038624615
1 PF1	1.340655730	1.097426153	1.020580188	.9462472998	.8705041086	.8223441252	.7881503208	.7417668778	.6992924540
1 PQ1	.8892731337	.9022629852	.9252854224	.9637661195	1.031710129	1.102819618	1.177547405	1.342081295	1.650100331
1 PFC	1.295517284	1.185627500	1.148889579	1.112732491	1.075518027	1.051804475	1.035006596	1.012349493	.9918146714
1 PQC	3.636838924	3.712815826	3.846371952	4.066794095	4.448826491	4.840881035	5.246238760	6.121310819	7.719152154
1 ZF1	1.652515709	1.652515709	1.652515709	1.652515709	1.652515709	1.652515709	1.652515709	1.652515709	1.652515709
1 ZF2	2.500294881	2.500294881	2.500294881	2.500294881	2.500294881	2.500294881	2.500294881	2.500294881	2.500294881

(2) P 5+0	TH = 40	SBE = 1.555723827						CAUER FUNCTIONS	
	RC = 02	RC = 05	RC = 07	RC = 10	RC = 15	RC = 20	RC = 25	RC = 35	RC = 50
AMAX DB	.0017375255	.0108709564	.0213327376	.0436480540	.0988323393	.1772876696	.2802872360	.5675287486	1.249387366
AMIN DB	27.88866266	35.85065789	38.78312595	41.90318590	45.48003812	48.05721315	50.09838750	53.30316755	57.08805358
MU K	.2759779D 0	.1102751D 0	.7867313D-1	.5492989D-1	.3638801D-1	.2704561D-1	.2138143D-1	.1477565D-1	.9562066D-2
1 PFR	1.422425730	1.008002485	.8844820993	.7650325599	.6403674711	.5573486594	.4951983254	.4036385845	.3060835305
1 PF1	1.346554661	1.102508612	1.025287222	.9505670040	.8744170799	.8259967849	.7916187942	.7449876340	.7022895586
1 PQ1	.9037002740	.9120933069	.9341891692	.9719991830	1.039528322	1.110571617	1.185399366	1.350420440	1.659722639
1 PFC	1.289370369	1.182511412	1.146575087	1.111120122	1.074546483	1.051201282	1.034646620	1.012295407	.9920168062
1 PQC	3.752557232	3.801798548	3.930575439	4.148327338	4.530363653	4.924693170	5.333483472	6.217727186	7.834892904
1 ZF1	1.616976932	1.616976932	1.616976932	1.616976932	1.616976932	1.616976932	1.616976932	1.616976932	1.616976932
1 ZF2	2.437673400	2.437673400	2.437673400	2.437673400	2.437673400	2.437673400	2.437673400	2.437673400	2.437673400

(2) P 5+0	TH = 41	SBE = 1.524253087						CAUER FUNCTIONS	
	RC = 02	RC = 05	RC = 07	RC = 10	RC = 15	RC = 20	RC = 25	RC = 35	RC = 50
AMAX DB	.0017375255	.0108709564	.0213327376	.0436480540	.0988323393	.1772876696	.2802872360	.5675287486	1.249387366
AMIN DB	26.71270509	34.67285195	37.60514727	40.72511542	44.30191865	46.87907652	48.92024294	52.13001607	55.90989843
MU K	.3075322D 0	.1228836D 0	.8766833D-1	.6121037D-1	.4054848D-1	.3013790D-1	.2382611D-1	.1646505D-1	.1065536D-1
1 PFR	1.447862835	1.020929117	.8947019444	.7730284634	.6464147779	.5622861766	.4993922634	.4068567348	.3083944108
1 PF1	1.352448209	1.107693332	1.030106781	.9550023422	.8784437161	.8297598356	.7951945954	.7483107832	.7053839185
1 PQ1	.9192468404	.9225502640	.9436272385	.9806988062	1.047765009	1.118724140	1.193647137	1.359166421	1.669800893
1 PFC	1.283112739	1.179324968	1.144204973	1.109466581	1.073548302	1.050580424	1.034275164	1.012237930	.9922226029
1 PQC	3.877338241	3.896783406	4.020223747	4.234933219	4.616789645	5.013420039	5.425767149	6.319603285	7.957075625
1 ZF1	1.583347943	1.583347943	1.583347943	1.583347943	1.583347943	1.583347943	1.583347943	1.583347943	1.583347943
1 ZF2	2.378085870	2.378085870	2.378085870	2.378085870	2.378085870	2.378085870	2.378085870	2.378085870	2.378085870

(2) P 5+0 — TH = 42 SBE = 1.494476550 — CAUER FUNCTIONS

(2) P 5+0	RC = 02	RC = 05	RC = 07	RC = 10	RC = 15	RC = 20	RC = 25	RC = 35	RC = 50
AMAX DB	.0017375255	.0108709564	.0213327376	.0436480540	.0988323393	.1772876696	.2802872360	.5675287486	1.249387366
AMIN DB	25.55956394	33.51734000	36.44941364	39.56926399	43.14600433	45.72314019	47.76429642	50.97406068	54.75393832
MU K	.3419865D 0	.1366508D 0	.9749023D-1	.6806806D-1	.4509132D-1	.3351440D-1	.2649546D-1	.1830970D-1	.1184913D-1
1 PFR	1.474915260	1.034533112	.9054280421	.7814003979	.6527325291	.5674378786	.5037643691	.4102079173	.3107985615
1 PF1	1.358313537	1.112974936	1.035036297	.9595526312	.8825846012	.8336343982	.7988791242	.7517379896	.7085773445
1 PQ1	.9360239019	.9336834687	.9536388656	.9898967790	1.056446557	1.127301169	1.202313568	1.368341572	1.680358983
1 PFC	1.276747975	1.176069044	1.141779623	1.107771941	1.072523372	1.049941735	1.033892049	1.012176893	.9924319288
1 PQC	4.012106858	3.998281020	4.115755791	4.327000670	4.708461434	5.107409959	5.523439643	6.427308074	8.086125404
1 ZF1	1.551495107	1.551495107	1.551495107	1.551495107	1.551495107	1.551495107	1.551495107	1.551495107	1.551495107
1 ZF2	2.321313910	2.321313910	2.321313910	2.321313910	2.321313910	2.321313910	2.321313910	2.321313910	2.321313910

(2) P 5+0 — TH = 43 SBE = 1.466279186 — CAUER FUNCTIONS

(2) P 5+0	RC = 02	RC = 05	RC = 07	RC = 10	RC = 15	RC = 20	RC = 25	RC = 35	RC = 50
AMAX DB	.0017375255	.0108709564	.0213327376	.0436480540	.0988323393	.1772876696	.2802872360	.5675287486	1.249387366
AMIN DB	24.42808271	32.38283107	35.31462142	38.43432121	42.01098116	44.58803889	46.62923209	49.83898501	53.61885663
MU K	.3795607D 0	.1516647D 0	.1082015D 0	.7554674D-1	.5004552D-1	.3719664D-1	.2940653D-1	.2032140D-1	.1315100D-1
1 PFR	1.503717332	1.048857811	.9166900424	.7901685382	.6593339702	.5728136572	.5083225328	.4136976957	.3132996842
1 PF1	1.364125409	1.118347308	1.040072762	.9642169546	.8868402310	.8376215704	.8026737924	.7552709657	.7118717190
1 PQ1	.9541570393	.9455479891	.9642673703	.9996280316	1.065601804	1.136328864	1.211423533	1.377970146	1.691422797
1 PFC	1.270279739	1.172744560	1.139299456	1.106036289	1.071471584	1.049285047	1.033497092	1.012112123	.9926446439
1 PQC	4.157914300	4.106861686	4.217659518	4.424960320	4.805772802	5.207046348	5.626885525	6.541246409	8.222507979
1 ZF1	1.521297331	1.521297331	1.521297331	1.521297331	1.521297331	1.521297331	1.521297331	1.521297331	1.521297331
1 ZF2	2.267159358	2.267159358	2.267159358	2.267159358	2.267159358	2.267159358	2.267159358	2.267159358	2.267159358

(2) P 5+0 — TH = 44 SBE = 1.439556540 — CAUER FUNCTIONS

(2) P 5+0	RC = 02	RC = 05	RC = 07	RC = 10	RC = 15	RC = 20	RC = 25	RC = 35	RC = 50
AMAX DB	.0017375255	.0108709564	.0213327376	.0436480540	.0988323393	.1772376696	.2802872360	.5675287486	1.249387366
AMIN DB	23.31722029	31.26811792	34.19954764	37.31905576	40.89561338	43.47268528	45.51381190	48.72355038	52.50341433
MU K	.4204906D 0	.1680195D 0	.1198694D 0	.8369331D-1	.5544218D-1	.4120774D-1	.3257758D-1	.2251276D-1	.1456914D-1
1 PFR	1.534419097	1.063950805	.9285203322	.7993548134	.6662334300	.5784241819	.5130752480	.4173320440	.3159017441
1 PF1	1.369855985	1.123803513	1.045212674	.9689941294	.8912109958	.8417224168	.8065800185	.7589114716	.7152689984
1 PQ1	.9737886184	.9582050848	.9755606794	1.009931016	1.075262346	1.145835801	1.221004154	1.388078523	1.703020433
1 PFC	1.263711776	1.169352483	1.136764919	1.104259726	1.070392834	1.048610191	1.033090104	1.012043437	.9928606014
1 PQC	4.315959052	4.223163951	4.326478641	4.529289998	4.909159012	5.312752081	5.736528318	6.661863320	8.366734512
1 ZF1	1.492644646	1.492644646	1.492644646	1.492644646	1.492644646	1.492644646	1.492644646	1.492644646	1.492644646
1 ZF2	2.215441968	2.215441968	2.215441968	2.215441968	2.215441968	2.215441968	2.215441968	2.215441968	2.215441968

(2) P 5+0	TH = 45 SBE = 1.414213562							CAUER FUNCTIONS	
	RC = 02	RC = 05	RC = 07	RC = 10	RC = 15	RC = 20	RC = 25	RC = 35	RC = 50
AMAX DB	.0017375255	.0108709564	.0213327376	.0436480540	.0988323393	.1772876696	.2802872360	.5675287486	1.249387366
AMIN DB	22.22605108	30.17207039	33.10304275	36.22230774	39.79873554	42.37576200	44.41686759	47.62658775	51.40644197
MU I K	.4650288D 0	.1858160D 0	.1325660D 0	.9255809D-1	.6131460D-1	.4557245D-1	.3602819D-1	.2489730D-1	.1611229D-1
1 I PFR	1.567188479	1.079864446	.9409543483	.8089830965	.6734464307	.5842809764	.5180316692	.4211173836	.3186089933
1 I PF1	1.375474607	1.129335702	1.050451976	.9738826695	.8956971617	.8459379585	.8105992216	.7626613135	.7187712146
1 I PQ1	.9950804714	.9717230613	.9875719313	1.020848147	1.085462863	1.155853258	1.231085055	1.398695442	1.715182431
1 I PFC	1.257047910	1.165893829	1.134176491	1.102442370	1.069287019	1.047916993	1.032670892	1.011970648	.9930796464
1 I PQC	4.487611908	4.347904675	4.442820494	4.640521107	5.019102200	5.424994491	5.852835361	6.789648941	8.519367050
1 I ZF1	1.465436967	1.465436967	1.465436967	1.465436967	1.465436967	1.465436967	1.465436967	1.465436967	1.465436967
1 I ZF2	2.165997415	2.165997415	2.165997415	2.165997415	2.165997415	2.165997415	2.165997415	2.165997415	2.165997415

(2) P 5+0	TH = 46 SBE = 1.390163591							CAUER FUNCTIONS	
	RC = 02	RC = 05	RC = 07	RC = 10	RC = 15	RC = 20	RC = 25	RC = 35	RC = 50
AMAX DB	.0017375255	.0108709564	.0213327376	.0436480540	.0988323393	.1772876696	.2802872360	.5675287486	1.249387366
AMIN DB	21.15376760	29.09362984	32.02402416	35.14298181	38.71924549	41.29621449	43.33729348	46.54699049	50.32683240
MU I K	.5134468D 0	.2051629D 0	.1463685D 0	.1021951D 0	.6769856D-1	.5031737D-1	.3977937D-1	.2748956D-1	.1778988D-1
1 I PFR	1.602213778	1.096656442	.9540309350	.8190794204	.6809898131	.5903965048	.5232016772	.4250606256	.3214259971
1 I PF1	1.380947578	1.134935001	1.055785983	.9788807435	.9002988486	.8502691599	.8147328140	.7665223423	.7223804778
1 I PQ1	1.018217070	.9861782667	1.000360175	1.032426306	1.096241493	1.166415523	1.241698647	1.409852259	1.727942036
1 I PFC	1.250292049	1.162369661	1.131534683	1.100584356	1.068154043	1.047205281	1.032239258	1.011393559	.9933016161
1 I PQC	4.674445962	4.481890864	4.567365232	4.759246049	5.136137612	5.544291156	5.976323415	6.925144155	8.681024798
1 I ZF1	1.439583032	1.439583032	1.439583032	1.439583032	1.439583032	1.439583032	1.439583032	1.439583032	1.439583032
1 I ZF2	2.118675554	2.118675554	2.118675554	2.118675554	2.118675554	2.118675554	2.118675554	2.118675554	2.118675554

(2) P 5+0	TH = 47 SBE = 1.367327461							CAUER FUNCTIONS	
	RC = 02	RC = 05	RC = 07	RC = 10	RC = 15	RC = 20	RC = 25	RC = 35	RC = 50
AMAX DB	.0017375255	.0108709564	.0213327376	.0436480540	.0988323393	.1772876696	.2802872360	.5675287486	1.249387366
AMIN DB	20.09968548	28.03180463	30.96147073	34.08004110	37.65609794	40.23299454	42.27404001	45.48370783	49.26353424
MU I K	.5660359D 0	.2261764D 0	.1613601D 0	.1126622D 0	.7463249D-1	.5547106D-1	.4385372D-1	.3030514D-1	.1961198D-1
1 I PFR	1.639706535	1.114390524	.9677927520	.8296722235	.6888818780	.5967842682	.5285959511	.4291692185	.3243576632
1 I PF1	1.386237955	1.140591407	1.061209312	.9839861272	.9050160052	.8547169153	.8189821931	.7704954515	.7260989786
1 I PQ1	1.043409284	1.001656257	1.013991187	1.044717427	1.107640262	1.177560263	1.252880457	1.421583252	1.741335502
1 I PFC	1.243448187	1.158781097	1.128840040	1.098685838	1.066993812	1.046474878	1.031794998	1.011811968	.9935263388
1 I PQC	4.878272665	4.626033669	4.700876625	4.886126903	5.260860860	5.671216591	6.107565151	7.068947123	8.852391334
1 I ZF1	1.414999464	1.414999464	1.414999464	1.414999464	1.414999464	1.414999464	1.414999464	1.414999464	1.414999464
1 I ZF2	2.073338907	2.073338907	2.073338907	2.073338907	2.073338907	2.073338907	2.073338907	2.073338907	2.073338907

ρ (2) 5+0	TH = 48 SBE = 1.345632730							CAUER FUNCTIONS	
	RC = 02	RC = 05	RC = 07	RC = 10	RC = 15	RC = 20	RC = 25	RC = 35	RC = 50
AMAX DB	.0017375255	.0108709564	.0213327376	.0436480540	.0988323393	.1772876696	.2802872360	.5675287486	1.249387366
AMIN DB	19.06325126	26.98566655	29.91441801	33.03250183	36.60829878	39.18510438	41.22610773	44.43573886	48.21554579
MU K	.6231095D 0	.2489818D 0	.1776301D 0	.1240220D 0	.8215772D-1	.6106423D-1	.4827551D-1	.3336032D-1	.2158946D-1
1 PFR	1.679904829	1.133137227	.9822867428	.8407926290	.6971425468	.6034589150	.5342260511	.4334512009	.3274092742
1 PF1	1.391305337	1.146293650	1.066715789	.9891961510	.9098483808	.8592820314	.8233487315	.7745355746	.7299289897
1 PQ1	1.070898848	1.018253163	1.028538418	1.057779177	1.119705582	1.189328941	1.264669505	1.433925959	1.755402433
1 PFC	1.236520414	1.155129302	1.126093142	1.096746987	1.065806239	1.045725608	1.031337905	1.011725662	.9937536337
1 PQC	5.101185302	4.781364972	4.844214794	5.021905587	5.393936380	5.806410038	6.247196668	7.221720825	9.034222942
1 ZF1	1.391609952	1.391609952	1.391609952	1.391609952	1.391609952	1.391609952	1.391609952	1.391609952	1.391609952
1 ZF2	2.029861330	2.029861330	2.029861330	2.029861330	2.029861330	2.029861330	2.029861330	2.029861330	2.029861330

ρ (2) 5+0	TH = 49 SBE = 1.325012993							CAUER FUNCTIONS	
	RC = 02	RC = 05	RC = 07	RC = 10	RC = 15	RC = 20	RC = 25	RC = 35	RC = 50
AMAX DB	.0017375255	.0108709564	.0213327376	.0436480540	.0988323393	.1772876696	.2802872360	.5675287486	1.249387366
AMIN DB	18.04405297	25.95434833	28.88195427	31.99942850	35.57489977	38.15159134	40.19254190	43.40212706	47.18190956
MU K	.6850052D 0	.2737141D 0	.1952747D 0	.1363416D 0	.9031873D-1	.6712995D-1	.5307089D-1	.3667467D-1	.2373402D-1
1 PFR	1.723077066	1.152974782	.9975646711	.8524747652	.7057935432	.6104363648	.5401045113	.4379152620	.3305865246
1 PF1	1.396105687	1.152029056	1.072298359	.9945076395	.9147954933	.8639652093	.8278337652	.7787916812	.7338728683
1 PQ1	1.100963670	1.036077300	1.044084104	1.071675742	1.132488822	1.201767291	1.277108734	1.446921573	1.770186176
1 PFC	1.229512931	1.151415498	1.123294607	1.094767996	1.064591241	1.044957292	1.030867766	1.011634422	.9939833100
1 PQC	5.345611587	4.949057158	4.998351278	5.167415829	5.536107324	5.950584553	6.395926252	7.384201828	9.227358263
1 ZF1	1.369344542	1.369344542	1.369344542	1.369344542	1.369344542	1.369344542	1.369344542	1.369344542	1.369344542
1 ZF2	1.988126856	1.988126856	1.988126856	1.988126856	1.988126856	1.988126856	1.988126856	1.988126856	1.988126856

ρ (2) 5+0	TH = 50 SBE = 1.305407289							CAUER FUNCTIONS	
	RC = 02	RC = 05	RC = 07	RC = 10	RC = 15	RC = 20	RC = 25	RC = 35	RC = 50
AMAX DB	.0017375255	.0108709564	.0213327376	.0436480540	.0988323393	.1772876696	.2802872360	.5675287486	1.249387366
AMIN DB	17.04183377	24.93704214	27.86321725	30.97992970	34.55499389	37.13154290	39.17242746	42.38195515	46.16170711
MU K	.7520865D 0	.3005184D 0	.2143976D 0	.1496933D 0	.9916349D-1	.7370387D-1	.5826803D-1	.4026616D-1	.2605825D-1
1 PFR	1.769526309	1.173990151	1.013683740	.8647561317	.7148586022	.6177339500	.5462449452	.4425708089	.3338955622
1 PF1	1.400591182	1.157783393	1.077948973	.9999168434	.9198565921	.8687670218	.8324385809	.7831167735	.7379330579
1 PQ1	1.133924151	1.055251072	1.060720578	1.086478754	1.146046977	1.214925882	1.290245513	1.460615387	1.785734268
1 PFC	1.222430070	1.147640963	1.120445091	1.092749080	1.063348745	1.044169752	1.030384362	1.011538018	.9942151663
1 PQC	5.614377466	5.130446750	5.164386952	5.323597296	5.688207198	6.104537634	6.554544604	7.557210487	9.432729515
1 ZF1	1.348138999	1.348138999	1.348138999	1.348138999	1.348138999	1.348138999	1.348138999	1.348138999	1.348138999
1 ZF2	1.948028660	1.948028660	1.948028660	1.948028660	1.948028660	1.948028660	1.948028660	1.948028660	1.948028660

(2) P 5+0	TH = 51 SBE = 1.286759566								CAUER FUNCTIONS
	RC = 02	RC = 05	RC = 07	RC = 10	RC = 15	RC = 20	RC = 25	RC = 35	RC = 50
AMAX DB	.0017375255	.0108709564	.0213327376	.0436480540	.0988323393	.1772876696	.2802872360	.5675287486	1.249387366
AMIN DB	16.05650870	23.93299928	26.85739163	29.97315439	33.54771094	36.12408215	38.16488439	41.37434040	45.15405427
MU I K	.8247461D 0	.3295517D 0	.2351108D 0	.1641552D 0	.1087437D 0	.8082446D-1	.6389734D-1	.4415630D-1	.2857576D-1
1 I PFR	1.819595229	1.196280222	1.030707304	.8776780215	.7243637076	.6253705769	.5526621654	.4474280440	.3373430349
1 I PF1	1.404710118	1.163540698	1.083658472	1.005419363	.9250306163	.8736878887	.8371643999	.7875628806	.7421120906
1 I PQ1	1.170150723	1.075913222	1.078551803	1.102268368	1.160443443	1.228860757	1.304132209	1.475057321	1.802098954
1 I PFC	1.215276325	1.143807033	1.117545291	1.090690476	1.062078683	1.043362807	1.029887470	1.011436211	.9944489903
1 I PQC	5.910784768	5.327062798	5.343573396	5.491512371	5.851173566	6.269163731	6.723936839	7.741662892	9.651375589
1 I ZF1	1.327934257	1.327934257	1.327934257	1.327934257	1.327934257	1.327934257	1.327934257	1.327934257	1.327934257
1 I ZF2	1.909468157	1.909468157	1.909468157	1.909468157	1.909468157	1.909468157	1.909468157	1.909468157	1.909468157

(2) P 5+0	TH = 52 SBE = 1.269018215								CAUER FUNCTIONS
	RC = 02	RC = 05	RC = 07	RC = 10	RC = 15	RC = 20	RC = 25	RC = 35	RC = 50
AMAX DB	.0017375255	.0108709564	.0213327376	.0436480540	.0988323393	.1772876696	.2802872360	.5675287486	1.249387366
AMIN DB	15.08818447	22.94153099	25.86370734	28.97828880	32.55221364	35.12836357	37.16906336	40.37843013	44.15809659
MU I K	.9034081D 0	.3609834D 0	.2575349D 0	.1798119D 0	.1191154D 0	.8853326D-1	.6999169D-1	.4836780D-1	.3130123D-1
1 I PFR	1.873671718	1.219953195	1.048705699	.8912860063	.7343373650	.6333669105	.5593723204	.4524980519	.3409361442
1 I PF1	1.408406902	1.169283092	1.089416449	1.011010061	.9303161463	.8787280476	.8420123601	.7921320522	.7464125882
1 I PQ1	1.210072833	1.098221518	1.097695198	1.119134538	1.175748923	1.243634189	1.318826864	1.490302515	1.819337782
1 I PFC	1.208056394	1.139915107	1.114595946	1.088592447	1.060780993	1.042536277	1.029376861	1.011328754	.9946845573
1 I PQC	6.238705938	5.540661134	5.537338539	5.672366139	6.026064306	6.445469004	6.905096618	7.938584879	9.884457375
1 I ZF1	1.308675924	1.308675924	1.308675924	1.308675924	1.308675924	1.308675924	1.308675924	1.308675924	1.308675924
1 I ZF2	1.872354197	1.872354197	1.872354197	1.872354197	1.872354197	1.872354197	1.872354197	1.872354197	1.872354197

(2) P 5+0	TH = 53 SBE = 1.252135658								CAUER FUNCTIONS
	RC = 02	RC = 05	RC = 07	RC = 10	RC = 15	RC = 20	RC = 25	RC = 35	RC = 50
AMAX DB	.0017375255	.0108709564	.0213327376	.0436480540	.0988323393	.1772876696	.2802872360	.5675287486	1.249387366
AMIN DB	14.13718208	21.96201071	24.88143865	27.99455371	31.56769403	34.14356908	36.18414159	39.39339750	43.17300506
MU I K	.9885312D 0	.3949969D 0	.2818010D 0	.1967546D 0	.1303390D 0	.9687526D-1	.7658662D-1	.5292523D-1	.3425058D-1
1 I PFR	1.932195232	1.245130194	1.067757195	.9056304992	.7448109168	.6417455852	.5663930515	.4577928993	.3446827050
1 I PF1	1.411622148	1.174990573	1.095211100	1.016682965	.9357113494	.8838875204	.8469834949	.7968263509	.7508372644
1 I PQ1	1.254189654	1.122355958	1.118283804	1.137178513	1.192042495	1.259315555	1.334393970	1.506412035	1.837514290
1 I PFC	1.200775232	1.135966655	1.111597843	1.086455282	1.059455627	1.041689982	1.028852300	1.011215387	.9949216298
1 I PQC	6.602699926	5.773265897	5.747317527	5.867530305	6.214076933	6.634588812	7.099142853	8.149128561	10.13327579
1 I ZF1	1.290313851	1.290313851	1.290313851	1.290313851	1.290313851	1.290313851	1.290313851	1.290313851	1.290313851
1 I ZF2	1.836602352	1.836602352	1.836602352	1.836602352	1.836602352	1.836602352	1.836602352	1.836602352	1.836602352

(2) P 5+0		TH = 54 SBE = 1.236067977							CAUER FUNCTIONS	
		RC = 02	RC = 05	RC = 07	RC = 10	RC = 15	RC = 20	RC = 25	RC = 35	RC = 50
AMAX DB		.0017375255	.0108709564	.0213327376	.0436480540	.0988323393	.1772876696	.2802872360	.5675287486	1.249387366
AMIN DB		13.20406166	20.99387775	23.90990418	27.02120241	30.59337017	33.16890436	35.20931902	38.41843744	42.19797195
MU	K	.1080613D 1	.4317909D 0	.3080508D 0	.2150823D 0	.1424801D 0	.1058992D 0	.8372066D-1	.5785522D-1	.3744102D-1
1	PFR	1.995663889	1.271947146	1.087949118	.9207674058	.7558189043	.6505314488	.5737436719	.4633257492	.3485912147
1	PF1	1.414292921	1.180640801	1.101029060	1.022431155	.9412139158	.8891660740	.8520787084	.8016478430	.7553889250
1	PQ1	1.303082858	1.148522610	1.140468868	1.156514630	1.209412866	1.275982371	1.350905390	1.523453680	1.856698811
1	PFC	1.193438126	1.131963216	1.108551818	1.084279299	1.058102543	1.040823739	1.028313549	1.011095839	.9951599565
1	PQC	7.008154280	6.027220107	5.975390109	6.078571977	6.416571707	6.837808552	7.307339557	8.374591896	10.39929309
1	ZF1	1.272801742	1.272801742	1.272801742	1.272801742	1.272801742	1.272801742	1.272801742	1.272801742	1.272801742
1	ZF2	1.802134280	1.802134280	1.802134280	1.802134280	1.802134280	1.802134280	1.802134280	1.802134280	1.802134280

(2) P 5+0		TH = 55 SBE = 1.220774589							CAUER FUNCTIONS	
		RC = 02	RC = 05	RC = 07	RC = 10	RC = 15	RC = 20	RC = 25	RC = 35	RC = 50
AMAX DB		.0017375255	.0108709564	.0213327376	.0436480540	.0988323393	.1772876696	.2802872360	.5675287486	1.249387366
AMIN DB		12.28964834	20.03664282	22.94846795	26.05751915	29.62848319	32.20359536	34.24381456	37.45276272	41.23220675
MU	K	.1180193D 1	.4715810D 0	.3364381D 0	.2349024D 0	.1556098D 0	.1156579D 0	.9143563D-1	.6318665D-1	.4089126D-1
1	PFR	2.064642334	1.300556970	1.109379146	.9367588830	.7673994888	.6597518430	.5814453737	.4691109912	.3526709323
1	PF1	1.416353184	1.186208859	1.106855211	1.028246636	.9468209873	.8945631758	.8572987465	.8065985865	.7600704697
1	PQ1	1.357431813	1.176958242	1.164422939	1.177272428	1.227959857	1.293721505	1.368441430	1.541502941	1.876969406
1	PFC	1.186050782	1.127906417	1.105458761	1.082064847	1.056721711	1.039937370	1.027760361	1.010969831	.9953992710
1	PQC	7.461459755	6.305247581	6.223726112	6.307288450	6.635099378	7.056588577	7.531119528	8.616441958	10.68415810
1	ZF1	1.256096814	1.256096814	1.256096814	1.256096814	1.256096814	1.256096814	1.256096814	1.256096814	1.256096814
1	ZF2	1.768877156	1.768877156	1.768877156	1.768877156	1.768877156	1.768877156	1.768877156	1.768877156	1.768877156

(2) P 5+0		TH = 56 SBE = 1.206217949							CAUER FUNCTIONS	
		RC = 02	RC = 05	RC = 07	RC = 10	RC = 15	RC = 20	RC = 25	RC = 35	RC = 50
AMAX DB		.0017375255	.0108709564	.0213327376	.0436480540	.0988323393	.1772876696	.2802872360	.5675287486	1.249387366
AMIN DB		11.39505751	19.08989544	21.99654155	25.10281819	28.67229459	31.24688499	33.28686249	36.49560010	40.27493223
MU	K	.1287858D 1	.5146017D 0	.3671302D 0	.2563317D 0	.1698056D 0	.1262090D 0	.9977699D-1	.6895095D-1	.4462163D-1
1	PFR	2.139770347	1.331132130	1.132156832	.9536742243	.7795949400	.6694369292	.5895214677	.4751643929	.3569319684
1	PF1	1.417734476	1.191667002	1.112672480	1.034120200	.9525290729	.9000779415	.8626441631	.8116806179	.7648848916
1	PQ1	1.418031659	1.207935901	1.190343596	1.199599199	1.247796176	1.312630621	1.387092111	1.560644119	1.898412958
1	PFC	1.178619442	1.123797978	1.102319623	1.079812310	1.055313115	1.039030696	1.027192485	1.010837066	.9956392911
1	PQC	7.970225304	6.610529122	6.494841076	6.555749494	6.871434695	7.292594180	7.772112753	8.876342763	10.98973633
1	ZF1	1.240159485	1.240159485	1.240159485	1.240159485	1.240159485	1.240159485	1.240159485	1.240159485	1.240159485
1	ZF2	1.736763166	1.736763166	1.736763166	1.736763166	1.736763166	1.736763166	1.736763166	1.736763166	1.736763166

(2) P 5+0	TH = 57	SBE = 1.192363293							CAUER FUNCTIONS
	RC = 02	RC = 05	RC = 07	RC = 10	RC = 15	RC = 20	RC = 25	RC = 35	RC = 50
AMAX DB	.0017375255	.0108709564	.0213327376	.0436480540	.0988323393	.1772876696	.2802872360	.5675287486	1.249387366
AMIN DB	10.52171661	18.15331373	21.05358771	24.15644361	27.72408392	30.29802999	32.33770896	35.54618647	39.32538037
MU I K	.1404247D 1	.5611084D 0	.4003093D 0	.2794975D 0	.1851516D 0	.1376151D 0	.1087943D 0	.7518234D-1	.4865427D-1
1 I PFR	2.221772107	1.363867631	1.156405392	.9715908970	.7924522069	.6796200677	.5979976603	.4815032736	.3613853895
1 I PF1	1.418366899	1.196984400	1.118461608	1.040041256	.9583339536	.9057090740	.8681152799	.8168959354	.7698352768
1 I PQ1	1.485814737	1.241771704	1.218457956	1.223663042	1.269049527	1.332819888	1.406958675	1.580971651	1.921126466
1 I PFC	1.171151013	1.119639726	1.099135420	1.077522109	1.053876751	1.038103539	1.026609666	1.010697238	.9958797171
1 I PQC	8.543543289	6.946796784	6.791664697	6.826349024	7.127617079	7.547731821	8.032180659	9.156188688	11.31814604
1 I ZF1	1.224953108	1.224953108	1.224953108	1.224953108	1.224953108	1.224953108	1.224953108	1.224953108	1.224953108
1 I ZF2	1.705729048	1.705729048	1.705729048	1.705729048	1.705729048	1.705729048	1.705729048	1.705729048	1.705729048

(2) P 5+0	TH = 58	SBE = 1.179178403							CAUER FUNCTIONS
	RC = 02	RC = 05	RC = 07	RC = 10	RC = 15	RC = 20	RC = 25	RC = 35	RC = 50
AMAX DB	.0017375255	.0108709564	.0213327376	.0436480540	.0988323393	.1772876696	.2802872360	.5675287486	1.249387366
AMIN DB	9.671379876	17.22667678	20.11912544	23.21776994	26.78314669	29.35629793	31.39560855	34.60376508	38.38278836
MU I K	.1530058D 1	.6113798D 0	.4361742D 0	.3045385D 0	.2017399D 0	.1499444D 0	.1185415D 0	.8191316D-1	.5301335D-1
1 I PFR	2.311465977	1.398984512	1.182263800	.9905957609	.8060235858	.6903382595	.6069023771	.4881467072	.3660433390
1 I PF1	1.418180447	1.202126865	1.124200904	1.045997654	.9642305719	.9114547931	.8737121404	.8222464785	.7749248035
1 I PQ1	1.561875963	1.278833091	1.249028152	1.249656562	1.291865149	1.354414032	1.428155382	1.602591682	1.945218565
1 I PFC	1.163653234	1.115433620	1.095907245	1.075194708	1.052412633	1.037155724	1.026011640	1.010559026	.9961202307
1 I PQC	9.192317256	7.318451150	7.117625512	7.121868644	7.406000280	7.824193190	8.313457619	9.458144836	11.67180163
1 I ZF1	1.210443720	1.210443720	1.210443720	1.210443720	1.210443720	1.210443720	1.210443720	1.210443720	1.210443720
1 I ZF2	1.675715680	1.675715680	1.675715680	1.675715680	1.675715680	1.675715680	1.675715680	1.675715680	1.675715680

(2) P 5+0	TH = 59	SBE = 1.166633397							CAUER FUNCTIONS
	RC = 02	RC = 05	RC = 07	RC = 10	RC = 15	RC = 20	RC = 25	RC = 35	RC = 50
AMAX DB	.0017375255	.0108709564	.0213327376	.0436480540	.0988323393	.1772876696	.2802872360	.5675287486	1.249387366
AMIN DB	8.846131070	16.30988003	19.19273710	22.28620367	25.84879274	28.42096433	30.45982081	33.66758155	37.44639443
MU I K	.1666053D 1	.6657206D 0	.4749424D 0	.3316066D 0	.2196710D 0	.1632718D 0	.1290777D 0	.8919923D-1	.5772530D-1
1 I PFR	2.409774586	1.436733953	1.209889265	1.010786506	.8203675061	.7016326654	.6162671416	.4951157578	.3709191771
1 I PF1	1.417106715	1.207056591	1.129865966	1.051975473	.9702129040	.9173127529	.8794344547	.8277341042	.7801567390
1 I PQ1	1.647502770	1.319548931	1.282358017	1.277801365	1.316408882	1.377554801	1.450811663	1.625623939	1.970811344
1 I PFC	1.156134847	1.111181770	1.092636279	1.072830619	1.050920790	1.036187079	1.025398140	1.010395094	.9953604932
1 I PQC	9.929667890	7.730708092	7.476756345	7.445556236	7.709313383	8.124509083	8.618401583	9.784696079	12.05346624
1 I ZF1	1.196599825	1.196599825	1.196599825	1.196599825	1.196599825	1.196599825	1.196599825	1.196599825	1.196599825
1 I ZF2	1.646667711	1.646667711	1.646667711	1.646667711	1.646667711	1.646667711	1.646667711	1.646667711	1.646667711

(2) P 5+0	TH = 60 SBE = 1.154700538								CAUER FUNCTIONS
	RC = 02	RC = 05	RC = 07	RC = 10	RC = 15	RC = 20	RC = 25	RC = 35	RC = 50
AMAX DB	.0017375255	.0108709564	.0213327376	.0436480540	.0988323393	.1772876696	.2802872360	.5675287486	1.249387366
AMIN DB	8.048368472	15.40295395	18.27407762	21.36118599	24.92034496	27.49130998	29.52960692	32.73637988	36.51543343
MU I K	.1813068D 1	.7244648D 0	.5168520D 0	.3608681D 0	.2390551D 0	.1776792D 0	.1404677D 0	.9707030D-1	.6281907D-1
1 I PFR	2.517734937	1.477402064	1.239460152	1.032273355	.8355494591	.7135492171	.6261270197	.5024337564	.3760276444
1 I PF1	1.415081015	1.211731900	1.135429388	1.057958786	.9762738124	.9232799460	.8852815346	.8333605578	.7855344361
1 I PQ1	1.744210402	1.364421928	1.318801291	1.308353549	1.342870888	1.402403939	1.475074712	1.650203985	1.998042504
1 I PFC	1.148605797	1.106886471	1.089323803	1.070430407	1.049401276	1.035197437	1.024768892	1.010232089	.9966001439
1 I PQC	10.77143699	8.189783532	7.873826450	7.801223838	8.040736226	8.451615720	8.949856186	10.13970699	12.46631579
1 I ZF1	1.183392198	1.183392198	1.183392198	1.183392198	1.183392198	1.183392198	1.183392198	1.183392198	1.183392198
1 I ZF2	1.618533215	1.618533215	1.618533215	1.618533215	1.618533215	1.618533215	1.618533215	1.618533215	1.618533215

(2) P 5+0	TH = 61 SBE = 1.143354068								CAUER FUNCTIONS
	RC = 02	RC = 05	RC = 07	RC = 10	RC = 15	RC = 20	RC = 25	RC = 35	RC = 50
AMAX DB	.0017375255	.0108709564	.0213327376	.0436480540	.0988323393	.1772876696	.2802872360	.5675287486	1.249387366
AMIN DB	7.280765716	14.50608647	17.36288642	20.44219693	23.99713855	26.56661831	28.60422616	31.81089825	35.58913227
MU I K	.1972021D 1	.7879792D 0	.5621648D 0	.3925056D 0	.2600132D 0	.1932564D 0	.1527826D 0	.1055805D 0	.6832646D-1
1 I PFR	2.636508215	1.521315514	1.271179446	1.055181085	.8516430988	.7261393404	.6365211443	.5101266255	.3813850538
1 I PF1	1.412044864	1.216107017	1.140860445	1.063929398	.9824048754	.9293525901	.8912522173	.8391274373	.7910613273
1 I PQ1	1.853783457	1.414043933	1.358771749	1.341610457	1.371470202	1.429146811	1.501112631	1.676485933	2.027067969
1 I PFC	1.141077421	1.102550246	1.085971219	1.067994702	1.047854167	1.034186636	1.024123618	1.010060643	.9968387985
1 I PQC	11.73681517	8.703127500	8.314508232	8.193370373	8.403993290	8.808936915	9.311127468	10.52349554	12.91401746
1 I ZF1	1.170793708	1.170793708	1.170793708	1.170793708	1.170793708	1.170793708	1.170793708	1.170793708	1.170793708
1 I ZF2	1.591263392	1.591263392	1.591263392	1.591263392	1.591263392	1.591263392	1.591263392	1.591263392	1.591263392

(2) P 5+0	TH = 62 SBE = 1.132570051								CAUER FUNCTIONS
	RC = 02	RC = 05	RC = 07	RC = 10	RC = 15	RC = 20	RC = 25	RC = 35	RC = 50
AMAX DB	.0017375255	.0108709564	.0213327376	.0436480540	.0988323393	.1772876696	.2802872360	.5675287486	1.249387366
AMIN DB	6.546202647	13.61964915	16.45900247	19.52876115	23.07852088	25.64617300	27.68293244	30.88886451	34.66670490
MU I K	.2143923D 1	.8566678D 0	.6111690D 0	.4267206D 0	.2826787D 0	.2101027D 0	.1661008D 0	.1147840D 0	.7428252D-1
1 I PFR	2.767388991	1.568848088	1.305278858	1.079651437	.8687315545	.7394608150	.6474933389	.5182232641	.3870095165
1 I PF1	1.407948761	1.220131906	1.146124755	1.069866545	.9885961893	.9355259962	.8973447749	.8450361511	.7967409172
1 I PQ1	1.978324867	1.469114894	1.402755758	1.377919019	1.402460317	1.457996838	1.529118265	1.704645741	2.058065028
1 I PFC	1.133562635	1.098175894	1.082580075	1.065524207	1.046279568	1.033154521	1.023462034	1.009880370	.9970760464
1 I PQC	12.84912697	9.279722569	8.805589117	8.627336671	8.803471426	9.200486620	9.706079295	10.94292444	13.40082658
1 I ZF1	1.158779161	1.158779161	1.158779161	1.158779161	1.158779161	1.158779161	1.158779161	1.158779161	1.158779161
1 I ZF2	1.564812280	1.564812280	1.564812280	1.564812280	1.564812280	1.564812280	1.564812280	1.564812280	1.564812280

(2) P 5+0	TH = 63	SBE = 1.122326238						CAUER FUNCTIONS	
	RC = 02	RC = 05	RC = 07	RC = 10	RC = 15	RC = 20	RC = 25	RC = 35	RC = 50
AMAX DB	.0017375255	.0108709564	.0213327376	.0436480540	.0988323393	.1772876696	.2802872360	.5675287486	1.249387366
AMIN DB	5.847662183	12.74422728	15.56238298	18.62045589	22.16385210	24.72925579	26.76497077	29.96999143	33.74734690
MU I K	.2329890D 1	.9309766D 0	.6641828D 0	.4637350D 0	.3071988D 0	.2283274D 0	.1805086D 0	.1247406D 0	.8072591D-1
1 I PFR	2.911813620	1.620428361	1.342023711	1.105846005	.8869090020	.7535788020	.6590928600	.5267560041	.3929212101
1 I PF1	1.402755106	1.223752178	1.151183942	1.075746562	.9948361392	.9417944131	.9035568068	.8510878661	.8025767710
1 I PQ1	2.120313878	1.530466441	1.451327950	1.417686154	1.436136109	1.489200954	1.559313914	1.734385241	2.091236159
1 I PFC	1.126076078	1.093766561	1.079152092	1.063019713	1.044677619	1.032100949	1.022783852	1.009590363	.9973114491
1 I PQC	14.13681854	9.930466969	9.355242813	9.109502842	9.244368639	9.630997881	10.13925295	11.40151505	13.93170705
1 I ZF1	1.147325153	1.147325153	1.147325153	1.147325153	1.147325153	1.147325153	1.147325153	1.147325153	1.147325153
1 I ZF2	1.539136496	1.539136496	1.539136496	1.539136496	1.539136496	1.539136496	1.539136496	1.539136496	1.539136496

(2) P 5+0	TH = 64	SBE = 1.112601940						CAUER FUNCTIONS	
	RC = 02	RC = 05	RC = 07	RC = 10	RC = 15	RC = 20	RC = 25	RC = 35	RC = 50
AMAX DB		.0108709564	.0213327376	.0436480540	.0988323393	.1772876696	.2802872360	.5675287486	1.249387366
AMIN DB		11.88065336	14.67312621	17.71692151	21.25250679	23.81514453	25.84957356	29.05347152	32.83022943
MU I K		.1011400D 1	.7215588D 0	.5037952D 0	.3337364D 0	.2480516D 0	.1961020D 0	.1355164D 0	.8769948D-1
1 I PFR		1.676548606	1.381718753	1.133949703	.9062825556	.7685670792	.6713752885	.5357611574	.3991426971
1 I PF1		1.226909123	1.155995292	1.081542514	1.001111132	.9481508433	.9098851112	.8572834452	.8085725005
1 I PQ1		1.599091349	1.505170887	1.461391787	1.472842463	1.523046389	1.591957147	1.767437097	2.126813723
1 I PFC		1.089325814	1.075689208	1.060482116	1.043048499	1.031025786	1.022088782	1.009491697	.9975445367
1 I PQC		10.66866988	9.973379397	9.647541618	9.732883665	10.10608633	10.61601823	11.90459063	14.51248218
1 I ZF1		1.136409950	1.136409950	1.136409950	1.136409950	1.136409950	1.136409950	1.136409950	1.136409950
1 I ZF2		1.514194999	1.514194999	1.514194999	1.514194999	1.514194999	1.514194999	1.514194999	1.514194999

(2) P 5+0	TH = 65	SBE = 1.103377919						CAUER FUNCTIONS	
	RC = 02	RC = 05	RC = 07	RC = 10	RC = 15	RC = 20	RC = 25	RC = 35	RC = 50
AMAX DB		.0108709564	.0213327376	.0436480540	.0988323393	.1772876696	.2802872360	.5675287486	1.249387366
AMIN DB		11.03004308	13.79149913	16.81787519	20.34387681	22.90311165	24.93595694	28.13847136	31.91449255
MU I K		.1098486D 1	.7836885D 0	.5471744D 0	.3624727D 0	.2694100D 0	.2129873D 0	.1471850D 0	.9525084D-1
1 I PFR		1.737775140	1.424715085	1.164174950	.9269745554	.7845095328	.6844036049	.5452796724	.4056993082
1 I PF1		1.229539912	1.160511431	1.087223779	1.007405290	.9545868268	.9163255327	.8636233707	.8147317450
1 I PQ1		1.676180500	1.565099901	1.509605292	1.512985107	1.559869146	1.627348068	1.802570949	2.165065744
1 I PFC		1.084857748	1.072193622	1.057912438	1.041392439	1.029928919	1.021376531	1.009282425	.9977748058
1 I PQC		11.51069656	10.67210092	10.25074646	10.27645969	10.63245927	11.14276594	12.45845801	15.15002520
1 I ZF1		1.126013362	1.126013362	1.126013362	1.126013362	1.126013362	1.126013362	1.126013362	1.126013362
1 I ZF2		1.489948860	1.489948860	1.489948860	1.489948860	1.489948860	1.489948860	1.489948860	1.489948860

$P^{(3)}_{5+2}$

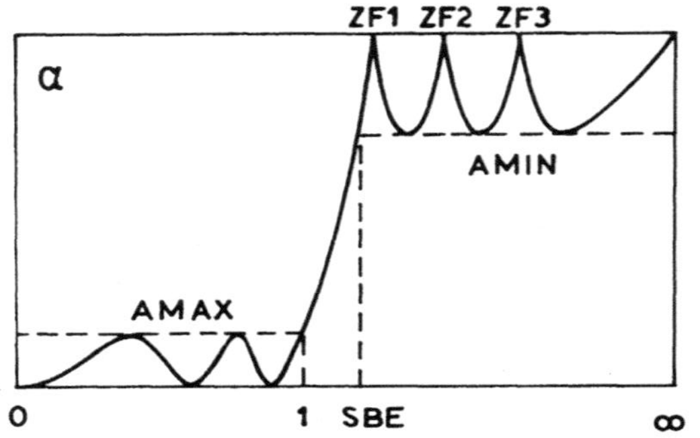

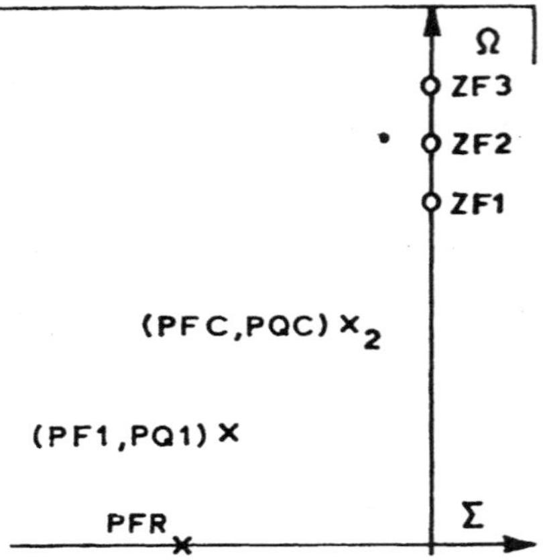

(3) P 5+2	TH = 17	SBE = 3.420303620						MCPER FUNCTIONS	
	RC = 02	RC = 05	RC = 07	RC = 10	RC = 15	RC = 20	RC = 25	RC = 35	RC = 50
AMAX DB	.0017375255	.0108709564	.0213327376	.0436480540	.0988323393	.1772876696	.2802872360	.5675287486	1.249387366
AMIN DB	91.45471554	102.6314504	106.7504438	111.1356724	116.1657807	119.7899384	122.6577218	127.1535436	132.3953324
MU I K	.6117961D-3	.1702200D-3	.1061682D-3	.6421026D-4	.3605495D-4	.2378508D-4	.1711231D-4	.1021103D-4	.5591549D-5
1 I PFR	1.000979640	.7471822748	.6627841142	.5775389553	.4851617058	.4220406352	.3741513195	.3029886495	.2270379565
1 I PF1	1.158519225	.9406516396	.8712828534	.8035807509	.7336566018	.6884428212	.6557998183	.6103785914	.5666478014
1 I PQ1	.6835456941	.7308102492	.7585882828	.7978787664	.8608867327	.9237853762	.9885196258	1.129047728	1.389828704
2 I PFC	1.562329796	1.360869017	1.296511201	1.233729598	1.168903380	1.126918484	1.096499034	1.053844683	1.012056438
2 I PQC	1.844418310	1.971826222	2.047129017	2.150128652	2.305991024	2.451338517	2.591978825	2.872643465	3.329808661
1 I ZF1	3.500705724	3.501866283	3.502206033	3.502525324	3.502844805	3.503047940	3.503194349	3.503400648	3.503607880
1 I ZF2	4.299968318	4.311690441	4.315098490	4.318292594	4.321480659	4.323503805	4.324960181	4.327009344	4.329066024
1 I ZF3	7.598661369	7.644028217	7.657166007	7.669456430	7.681701121	7.689459662	7.695038872	7.702832551	7.710741364

(3) P 5+2	TH = 18	SBE = 3.236067977						MCPER FUNCTIONS	
	RC = 02	RC = 05	RC = 07	RC = 10	RC = 15	RC = 20	RC = 25	RC = 35	RC = 50
AMAX DB	.0017375255	.0108709564	.0213327376	.0436480540	.0988323393	.1772876696	.2802872360	.5675287486	1.249387366
AMIN DB	87.92488404	99.10104390	103.2199118	107.6050394	112.6350630	116.2591752	119.1269296	123.6227164	128.8644763
MU I K	.8641966D-3	.2406832D-3	.1501591D-3	.9083940D-4	.5102067D-4	.3366324D-4	.2422200D-4	.1445576D-4	.7917231D-5
1 I PFR	1.006405358	.7503468417	.6653794513	.5796361437	.4867949639	.4233960667	.3753143198	.3038903376	.2276891133
1 I PF1	1.162115077	.9431301487	.8734650354	.8054994449	.7353286732	.6899687930	.6572267739	.6116769882	.5678334266
1 I PQ1	.6858473897	.7326684934	.7603483199	.7995699795	.8625466343	.9254587855	.9902297010	1.130875159	1.391927478
2 I PFC	1.557647277	1.358574226	1.294805002	1.232525444	1.168147967	1.126418276	1.096167230	1.053726695	1.012123808
2 I PQC	1.869329394	1.991670127	2.065773443	2.167837038	2.323046477	2.468212319	2.608896469	2.890015309	3.348455388
1 I ZF1	3.311305872	3.312523013	3.312880090	3.313215954	3.313552291	3.313766279	3.313920575	3.314138076	3.314356663
1 I ZF2	4.059989742	4.072342025	4.075936491	4.079306671	4.082671316	4.084808039	4.086346157	4.088511349	4.090684004
1 I ZF3	7.157203127	7.205155423	7.219050151	7.232051798	7.245007937	7.253218699	7.259123774	7.267426504	7.275746328

(3) P 5+2 — TH = 19 SBE = 3.071553487 — MCPER FUNCTIONS

MU	K	RC = 02	RC = 05	RC = 07	RC = 10	RC = 15	RC = 20	RC = 25	RC = 35	RC = 50
	AMAX DB	.0017375255	.0108709564	.0213327376	.0436480540	.0988323393	.1772876696	.2802872360	.5675287486	1.249387366
	AMIN DB	84.57982136	95.75529805	99.87401684	104.2590246	109.2889475	112.9130056	115.7807257	120.2764707	125.5181964
MU	K	.1198450D-2	.3341273D-3	.2085200D-3	.1261801D-3	.7088948D-4	.4678064D-4	.3366461D-4	.2009459D-4	.1100742D-4
1	PFR	1.012212771	.7537238621	.6681467684	.5818706686	.4885339826	.4248386813	.3765517851	.3048504496	.2283815005
1	PF1	1.165927140	.9457598558	.8757806920	.8075357138	.7371033322	.6915884321	.6587413363	.6130551074	.5690918472
1	PQ1	.6883273925	.7346600649	.7622320898	.8013778767	.8643190014	.9272443328	.9920535041	1.132822908	1.394163189
2	PFC	1.552699169	1.356145294	1.292998113	1.231249490	1.167346872	1.125887374	1.095814676	1.053600654	1.012194197
2	PQC	1.896275310	2.013030661	2.085815348	2.186848832	2.341334969	2.486292479	2.627014113	2.908606188	3.368397293
1	ZF1	3.142137753	3.143409310	3.143783225	3.144135262	3.144488104	3.144712749	3.144874801	3.145103336	3.145333127
1	ZF2	3.845245997	3.858218832	3.861997374	3.865541638	3.869082081	3.871330361	3.872949541	3.875229386	3.877517730
1	ZF3	6.761182563	6.811702736	6.826351092	6.840061519	6.853727263	6.862389372	6.868619815	6.877381084	6.886161617

(3) P 5+2 — TH = 20 SBE = 2.923804400 — MCPER FUNCTIONS

MU	K	RC = 02	RC = 05	RC = 07	RC = 10	RC = 15	RC = 20	RC = 25	RC = 35	RC = 50
	AMAX DB	.0017375255	.0108709564	.0213327376	.0436480540	.0988323393	.1772876696	.2802872360	.5675287486	1.249387366
	AMIN DB	81.40022676	92.57489859	96.69344171	101.0783082	106.1081126	109.7321068	112.5997865	117.0954823	122.3371678
MU	K	.1634764D-2	.4562849D-3	.2848462D-3	.1724178D-3	.9689451D-4	.6395329D-4	.4602853D-4	.2747969D-4	.1505557D-4
1	PFR	1.018415587	.7573192004	.6710904182	.5842457219	.4903810129	.4263702326	.3778651542	.3058685362	.2291158353
1	PF1	1.169956804	.9485423068	.8782312804	.8096908991	.7389817846	.6933028534	.6603445545	.6145139086	.5704239417
1	PQ1	.6909932305	.7367901347	.7642439615	.8033062051	.8662070978	.9291450621	.9939939751	1.134893879	1.396538933
2	PFC	1.547486287	1.353581837	1.291090066	1.229901266	1.166499684	1.125325425	1.095441072	1.053466337	1.012267461
2	PQC	1.925378544	2.035982085	2.107318585	2.207219619	2.360905381	2.505624566	2.646375506	2.928458349	3.389677041
1	ZF1	2.990169875	2.991493515	2.991883747	2.992251525	2.992620496	2.992855586	2.993025256	2.993264646	2.993505482
1	ZF2	3.651956924	3.665540331	3.669500497	3.673216753	3.676930616	3.679289876	3.680989398	3.683382965	3.685786166
1	ZF3	6.403787828	6.456856376	6.472254692	6.486671193	6.501044502	6.510156986	6.516712246	6.525931481	6.535172376

(3) P 5+2 — TH = 21 SBE = 2.790428110 — MCPER FUNCTIONS

MU	K	RC = 02	RC = 05	RC = 07	RC = 10	RC = 15	RC = 20	RC = 25	RC = 35	RC = 50
	AMAX DB	.0017375255	.0108709564	.0213327376	.0436480540	.0988323393	.1772876696	.2802872360	.5675287486	1.249387366
	AMIN DB	78.36962429	89.54335500	93.66169272	98.04639411	103.0760598	106.6999795	109.5676119	114.0632502	119.3048885
MU	K	.2196965D-2	.6139373D-3	.3833943D-3	.2321418D-3	.1304983D-3	.8614949D-4	.6201220D-4	.3702933D-4	.2029152D-4
1	PFR	1.025028799	.7611392160	.6742151099	.5867647490	.4923384795	.4279926077	.3792559744	.3069462256	.2298928871
1	PF1	1.174205405	.9514791135	.8808183329	.8119664190	.7409653102	.6951132420	.6620375449	.6160544144	.5718306463
1	PQ1	.6938553570	.7390643499	.7663886929	.8053590352	.8682144614	.9311642681	.9960542933	1.137091211	1.399058047
2	PFC	1.542009622	1.350883485	1.289080387	1.228480289	1.165605974	1.124732060	1.095046101	1.053323507	1.012343447
2	PQC	1.956774686	2.060605888	2.130353059	2.229010112	2.381810986	2.526258175	2.667028212	2.949617667	3.412340910
1	ZF1	2.852948377	2.854321595	2.854727584	2.855110644	2.855495344	2.855740654	2.855917792	2.856167847	2.856419556
1	ZF2	3.477062614	3.491246270	3.495385500	3.499271557	3.503156866	3.505625971	3.507405077	3.509911383	3.512428557
1	ZF3	6.079505213	6.135100528	6.151244750	6.166364334	6.181442953	6.191004735	6.197884199	6.207560758	6.217261625

(3) P 5+2		TH = 22	SBE = 2.669467163						MCPER FUNCTIONS	
		RC = 02	RC = 05	RC = 07	RC = 10	RC = 15	RC = 20	RC = 25	RC = 35	RC = 50
AMAX DB		.0017375255	.0108709564	.0213327376	.0436480540	.0988323393	.1772876696	.2802872360	.5675287486	1.249387366
AMIN DB		75.47383664	86.64647434	90.76457351	95.14908312	100.1785877	103.8024208	106.6699983	111.1655699	116.4071535
MU	K	.2912938D-2	.8150482D-3	.5091678D-3	.3083990D-3	.1734229D-3	.1145101D-3	.8243881D-4	.4923674D-4	.2698648D-4
1	PFR	1.032068798	.7651907978	.6775259319	.5894314649	.4944089913	.4297078342	.3807259064	.3080347278	.2307134804
1	PF1	1.178674195	.9545719483	.8835434545	.8143637686	.7430552638	.6970208549	.6638214930	.6176777115	.5733129575
1	PQ1	.6969247436	.7414888731	.7686714619	.8075407856	.8703449228	.9333055130	.9982378921	1.139418285	1.401724127
2	PFC	1.536270358	1.348049885	1.286968598	1.226986058	1.164665293	1.124106892	1.094629427	1.053171912	1.012421992
2	PQC	1.990613848	2.086991459	2.154995246	2.252286575	2.404109796	2.548247236	2.689023908	2.972133914	3.436439040
1	ZF1	2.728465759	2.729885870	2.730307020	2.730704870	2.731104872	2.731360162	2.731544609	2.731805125	2.732067527
1	ZF2	3.318059704	3.332832972	3.337148603	3.341202173	3.345256860	3.347834616	3.349692506	3.352310515	3.354940728
1	ZF3	5.783824187	5.841922375	5.858808027	5.874627395	5.890408833	5.900418723	5.907621713	5.917754882	5.927915276

(3) P 5+2		TH = 23	SBE = 2.559304665						MCPER FUNCTIONS	
		RC = 02	RC = 05	RC = 07	RC = 10	RC = 15	RC = 20	RC = 25	RC = 35	RC = 50
AMAX DB		.0017375255	.0108709564	.0213327376	.0436480540	.0988323393	.1772876696	.2802872360	.5675287486	1.249387366
AMIN DB		72.70057603	83.87195203	87.98977585	92.37406411	97.40338282	101.0271159	103.8946301	108.3901246	113.6316450
MU	K	.3815091D-2	.1068908D-2	.6680101D-3	.4047508D-3	.2276835D-3	.1503706D-3	.1082725D-3	.6467992D-4	.3545843D-4
1	PFR	1.039553496	.7694814033	.6810283781	.5922498708	.4965953517	.4315180881	.3822767310	.3092853387	.2315784970
1	PF1	1.183364302	.9578225391	.8864083217	.8168845197	.7452530762	.6990270227	.6656976550	.6193849532	.5748719332
1	PQ1	.7002134863	.7440704269	.7710979005	.8098562492	.8726026263	.9355726445	1.000548477	1.141878747	1.404541042
2	PFC	1.530269893	1.345080707	1.284754220	1.225418060	1.163677175	1.123449515	1.094190700	1.053011288	1.012502923
2	PQC	2.027062279	2.115236838	2.181328799	2.277121308	2.427864952	2.571650363	2.712418693	2.996061047	3.462025718
1	ZF1	2.615063849	2.616527982	2.616963656	2.617375774	2.617790623	2.618055638	2.618247227	2.618517989	2.618790890
1	ZF2	3.172880389	3.188232345	3.192721615	3.196940314	3.201162217	3.203847372	3.205783207	3.208511826	3.211254094
1	ZF3	5.513019396	5.573594081	5.591216234	5.607731750	5.624213260	5.634669944	5.642195707	5.652784692	5.663404113

(3) P 5+2		TH = 24	SBE = 2.458593336						MCPER FUNCTIONS	
		RC = 02	RC = 05	RC = 07	RC = 10	RC = 15	RC = 20	RC = 25	RC = 35	RC = 50
AMAX DB		.0017375255	.0108709564	.0213327376	.0436480540	.0988323393	.1772876696	.2802872360	.5675287486	1.249387366
AMIN DB		70.03912224	81.20904985	85.32655761	89.71059177	94.73969710	98.36331547	101.2307569	105.7261629	110.9676108
MU	K	.4940864D-2	.1386290D-2	.8667050D-3	.5253361D-3	.2956239D-3	.1952856D-3	.1406360D-3	.8403237D-4	.4607812D-4
1	PFR	1.047502476	.7740191018	.6847283763	.5952242732	.4989005710	.4334257025	.3839103556	.3105494444	.2324888794
1	PF1	1.188276696	.9612326624	.8894146794	.8195303211	.7475602559	.7011331510	.6676673599	.6211773615	.5765086957
1	PQ1	.7037349263	.7468163443	.7736741324	.8123106228	.8749920533	.9379698163	1.002990042	1.144476521	1.407512952
2	PFC	1.524009861	1.341975654	1.282436776	1.223775770	1.162641134	1.122759504	1.093729551	1.052841356	1.012586056
2	PQC	2.066304209	2.145449567	2.209445215	2.303593183	2.453145160	2.596531233	2.737273449	3.021457535	3.489159688
1	ZF1	2.511361036	2.512866125	2.513315645	2.513741474	2.514170688	2.514445157	2.514643709	2.514924489	2.515207686
1	ZF2	3.039801669	3.055721137	3.060381188	3.064762538	3.069149402	3.071940646	3.073953545	3.076791629	3.079644914
1	ZF3	5.263987165	5.327009283	5.345362510	5.362570174	5.379748736	5.390650763	5.398498469	5.409542385	5.420620269

(3) P 5+2 — TH = 25 SBE = 2.366201583 — MCPER FUNCTIONS

MU		RC = 02	RC = 05	RC = 07	RC = 10	RC = 15	RC = 20	RC = 25	RC = 35	RC = 50
	AMAX DB	.0017375255	.0108709564	.0213327376	.0436480540	.0988323393	.1772876696	.2802872360	.5675287486	1.249387366
	AMIN DB	67.48006640	78.64833965	82.76548642	87.14923039	92.17809203	95.80157935	98.66893769	103.1642427	108.4056078
	K	.6333268D-2	.1779624D-2	.1113087D-2	.6749398D-3	.3799570D-3	.2510555D-3	.1808300D-3	.1080749D-3	.5927561D-4
1	PFR	1.055937150	.7788126225	.6886323205	.5983593047	.5013278792	.4354331772	.3856288219	.3118785268	.2334456338
1	PF1	1.193412140	.9648041355	.8925643380	.8223028977	.7499783902	.7033407229	.6697320115	.6230562287	.5782244331
1	PQ1	.7075037868	.7497346240	.7764068157	.8149095396	.8775180476	.9405015106	1.005566893	1.147215832	1.410644329
2	PFC	1.517492150	1.338734463	1.280015798	1.222058655	1.161556669	1.122036418	1.093245595	1.052661822	1.012671198
2	PQC	2.108543968	2.177747647	2.239444588	2.331788248	2.480025176	2.622959015	2.763654236	3.048386721	3.517904492
1	ZF1	2.416196960	2.417739738	2.418202383	2.418641333	2.419084400	2.419368036	2.419573363	2.419863917	2.420157194
1	ZF2	2.917376384	2.933851969	2.938679850	2.943221283	2.947770758	2.950666724	2.952755765	2.955702110	2.958665321
1	ZF3	5.034121244	5.099558821	5.118637160	5.136532579	5.154404869	5.165750643	5.173919372	5.185417241	5.196952955

(3) P 5+2 — TH = 26 SBE = 2.281172033 — MCPER FUNCTIONS

MU		RC = 02	RC = 05	RC = 07	RC = 10	RC = 15	RC = 20	RC = 25	RC = 35	RC = 50
	AMAX DB	.0017375255	.0108709564	.0213327376	.0436480540	.0988323393	.1772876696	.2802872360	.5675287486	1.249387366
	AMIN DB	65.01510512	76.18149734	80.29823374	84.68164780	89.71023241	93.33357075	96.20083466	100.6960248	105.9372957
	K	.8041463D-2	.2263177D-2	.1416162D-2	.8590676D-3	.4838075D-3	.3197554D-3	.2303548D-3	.1377085D-3	.7554755D-4
1	PFR	1.064880943	.7838714084	.6927471059	.6016599471	.5038807406	.4375431887	.3874343138	.3132741686	.2344498335
1	PF1	1.198771139	.9685388075	.8958591696	.8252040507	.7525091464	.7056513002	.6718930908	.6250229208	.5800204025
1	PQ1	.7115363298	.7528339937	.7793031906	.8176591064	.8801858444	.9431725624	1.008283669	1.150101222	1.413939973
2	PFC	1.510718931	1.335356916	1.277490829	1.220266171	1.160423264	1.121279796	1.092738428	1.052472379	1.012758144
2	PQC	2.154008410	2.212260622	2.271436458	2.361800398	2.508586347	2.651008851	2.791632725	3.076917218	3.548328860
1	ZF1	2.328589972	2.330166960	2.330641965	2.331093411	2.331549788	2.331842289	2.332054190	2.332354264	2.332657392
1	ZF2	2.804380163	2.821400280	2.826392955	2.831091815	2.835801461	2.838800722	2.840964942	2.844018287	2.847090279
1	ZF3	4.821217236	4.889035156	4.908832065	4.927410415	4.945972784	4.957760548	4.966249290	4.978200026	4.990192861

(3) P 5+2 — TH = 27 SBE = 2.202689265 — MCPER FUNCTIONS

MU		RC = 02	RC = 05	RC = 07	RC = 10	RC = 15	RC = 20	RC = 25	RC = 35	RC = 50
	AMAX DB	.0017375255	.0108709564	.0213327376	.0436480540	.0988323393	.1772876696	.2802872360	.5675287486	1.249387366
	AMIN DB	62.63687348	73.80113581	77.91740762	82.30044822	87.32871923	90.95188893	93.81904594	98.31410605	103.5552704
	K	.1012138D-1	.2853253D-2	.1786232D-2	.1084027D-2	.6107594D-3	.4037673D-3	.2909329D-3	.1739687D-3	.9546506D-4
1	PFR	1.074359506	.7892056758	.6970801680	.6051315569	.5065628696	.4397586025	.3893291661	.3147380597	.2355026231
1	PF1	1.204353878	.9724385487	.8993011038	.8282356558	.7551542730	.7080665258	.6741521580	.6270788791	.5818979321
1	PQ1	.7158505347	.7561239805	.7823711318	.8205659436	.8830011013	.9459881874	1.011145368	1.153137576	1.417405040
2	PFC	1.503692676	1.331842846	1.274861426	1.218397773	1.159240383	1.120489162	1.092207630	1.052272703	1.012846678
2	PQC	2.202949705	2.249130810	2.305540771	2.393732139	2.538917219	2.680762385	2.821286697	3.107123359	3.580507127
1	ZF1	2.247704045	2.249311546	2.249798101	2.250261378	2.250730493	2.251031537	2.251249802	2.251559124	2.251871863
1	ZF2	2.699770160	2.717323075	2.722477423	2.727330966	2.732198252	2.735299323	2.737537716	2.740696743	2.743876313
1	ZF3	4.623398270	4.693558024	4.714066333	4.733322329	4.752570773	4.764798598	4.773606237	4.786008641	4.798457803

(3) P 5+2	TH = 28 SBE = 2.130054468								MCPER FUNCTIONS
	RC = 02	RC = 05	RC = 07	RC = 10	RC = 15	RC = 20	RC = 25	RC = 35	RC = 50
AMAX DB	.0017375255	.0108709564	.0213327376	.0436480540	.0988323393	.1772876696	.2802872360	.5675287486	1.249387366
AMIN DB	60.33880843	71.50066817	75.61641606	79.99903552	85.02695289	88.64993243	91.51696889	96.01188240	101.2529266
MU I K	.1263638D-1	.3568413D-2	.2235043D-2	.1357018D-2	.7649079D-3	.5058147D-3	.3645353D-3	.2180411D-3	.1196825D-3
1 I PFR	1.084400948	.7948264813	.7016395261	.6087798933	.5093782482	.4420824848	.3913158748	.3162720034	.2366052226
1 I PF1	1.210160155	.9765052383	.9028921219	.8313996629	.7579156010	.7105881258	.6765108554	.6292256236	.5838584244
1 I PQ1	.7204663028	.7596149903	.7856192087	.8236372307	.8859699341	.9489540123	1.014157373	1.156330147	1.421045064
2 I PFC	1.496416181	1.328192146	1.272127170	1.216452912	1.158007481	1.119664021	1.091652765	1.052062456	1.012936572
2 I PQC	2.255648557	2.288514692	2.341888957	2.427695437	2.571114219	2.712308361	2.852700583	3.139085691	3.614519708
1 I ZF1	2.172822772	2.174456864	2.174954109	2.175428518	2.175909764	2.176219013	2.176443418	2.176761704	2.177083799
1 I ZF2	2.602652638	2.620726494	2.626039319	2.631044719	2.636067022	2.639268358	2.641579875	2.644843207	2.648129097
1 I ZF3	4.439056598	4.511516026	4.532727887	4.552655742	4.572585876	4.585251643	4.594376955	4.607229700	4.620134306

(3) P 5+2	TH = 29 SBE = 2.062665340								MCPER FUNCTIONS
	RC = 02	RC = 05	RC = 07	RC = 10	RC = 15	RC = 20	RC = 25	RC = 35	RC = 50
AMAX DB	.0017375255	.0108709564	.0213327376	.0436480540	.0988323393	.1772876696	.2802872360	.5675287486	1.249387366
AMIN DB	58.11503621	69.27419504	73.38935418	77.77150040	82.79902041	86.42178628	89.28868729	93.78343607	99.02434525
MU I K	.1565798D-1	.4429705D-2	.2775934D-2	.1686224D-2	.9509159D-3	.6290012D-3	.4534090D-3	.2712782D-3	.1489471D-3
1 I PFR	1.095036109	.8007457964	.7064338321	.6126111493	.5123311453	.4445181168	.3933971073	.3178779243	.2377589317
1 I PF1	1.216189299	.9807407499	.9066342506	.8346980937	.7607950449	.7132179111	.6789709092	.6314647554	.5859033538
1 I PQ1	.7254056919	.7633183976	.7890567517	.8268807573	.8890989553	.9520761083	1.017325436	1.159684586	1.424865989
2 I PFC	1.488892594	1.324404777	1.269287663	1.214431038	1.156723996	1.118803863	1.091073378	1.051841284	1.013027586
2 I PQC	2.312417914	2.330584496	2.380625154	2.463812685	2.605282417	2.745743288	2.885966079	3.172891529	3.650453624
1 I ZF1	2.103328752	2.104985278	2.105492305	2.105977105	2.106469843	2.106786940	2.107017251	2.107344200	2.107675383
1 I ZF2	2.512257250	2.530840111	2.536308142	2.541462490	2.546637099	2.549937096	2.552320647	2.555686845	2.559077737
1 I ZF3	4.266807294	4.341520310	4.363427142	4.384020528	4.404627551	4.417728938	4.427170574	4.440472200	4.453831263

(3) P 5+2	TH = 30 SBE = 2.000000000								MCPER FUNCTIONS
	RC = 02	RC = 05	RC = 07	RC = 10	RC = 15	RC = 20	RC = 25	RC = 35	RC = 50
AMAX DB	.0017375255	.0108709564	.0213327376	.0436480540	.0988323393	.1772876696	.2802872360	.5675287486	1.249387366
AMIN DB	55.96027892	67.11641089	71.23091054	75.61252674	80.63960171	84.26212828	87.12887760	91.62344190	96.86419987
MU I K	.1926658D-1	.5460922D-2	.3424003D-2	.2080926D-2	.1174075D-2	.7768523D-3	.5601082D-3	.3352189D-3	.1841094D-3
1 I PFR	1.106298866	.8069765913	.7114724238	.6166319864	.5154261378	.4470690099	.3955757146	.3195578755	.2389651351
1 I PF1	1.222440083	.9851469346	.9105295547	.8381330399	.7637946036	.7159577800	.6815341328	.6337979602	.5880342951
1 I PQ1	.7306931860	.7672466462	.7926939271	.8303049795	.8923953177	.9553610284	1.020655959	1.163206971	1.428874194
2 I PFC	1.481125432	1.320480777	1.266342541	1.212331605	1.155389355	1.117903159	1.090468997	1.051608817	1.013119469
2 I PQC	2.373607274	2.375529997	2.421907593	2.502217782	2.641536381	2.781172188	2.921182827	3.208635572	3.688403082
1 I ZF1	2.038687095	2.040361652	2.040877503	2.041371912	2.041875468	2.042200036	2.042436006	2.042771303	2.043111292
1 I ZF2	2.427916478	2.446996357	2.452616256	2.457916565	2.463240678	2.466637673	2.469092123	2.472559688	2.476054203
1 I ZF3	4.105451209	4.182367511	4.204959944	4.226211946	4.247490610	4.261025073	4.270781555	4.284530452	4.298342877

(3) P 5+2	TH = 31	SBE = 1.941604026						MCPER FUNCTIONS	
	RC = 02	RC = 05	RC = 07	RC = 10	RC = 15	RC = 20	RC = 25	RC = 35	RC = 50
AMAX DB	.0017375255	.0108709564	.0213327376	.0436480540	.0988323393	.1772876696	.2802872360	.5675287486	1.249387366
AMIN DB	53.86977649	65.02252575	69.13628890	73.51731323	78.54389127	82.16615065	85.03273060	89.52708890	94.76767804
MU K	.2355231D-1	.6688871D-2	.4196289D-2	.2551609D-2	.1440376D-2	.9533611D-3	.6875274D-3	.4116087D-3	.2261353D-3
1 PFR	1.118226486	.8135329285	.7167653854	.6208495729	.5186681345	.4497389222	.3978547438	.3213140476	.2402253082
1 PF1	1.228910622	.9897256021	.9145801275	.8417066599	.7669163612	.7188097202	.6842024294	.6362270111	.5902528769
1 PQ1	.7363560062	.7714133641	.7965418215	.8339190833	.8958667629	.9588158494	1.024155533	1.166903840	1.433076533
2 PFC	1.473118610	1.316420269	1.263291478	1.210154073	1.154002975	1.116976368	1.089839133	1.051364669	1.013211955
2 PQC	2.439607682	2.423560576	2.465910172	2.543057363	2.680001145	2.818709434	2.958459179	3.246420594	3.728470131
1 ZF1	1.978432113	1.980120051	1.980643714	1.981146908	1.981660572	1.981992216	1.982233585	1.982576895	1.982925393
1 ZF2	2.349049046	2.368613935	2.374382302	2.379825508	2.385296237	2.388788506	2.391312676	2.394880046	2.398476743
1 ZF3	3.953945100	4.033009871	4.056277685	4.078180759	4.100125329	4.114090085	4.124159792	4.138354187	4.152618758

(3) P 5+2	TH = 32	SBE = 1.887079915						MCPER FUNCTIONS	
	RC = 02	RC = 05	RC = 07	RC = 10	RC = 15	RC = 20	RC = 25	RC = 35	RC = 50
AMAX DB	.0017375255	.0108709564	.0213327376	.0436480540	.0988323393	.1772876696	.2802872360	.5675287486	1.249387366
AMIN DB	51.83922108	62.98819936	67.10114220	71.48150742	76.50753214	80.12949400	82.99588537	87.49001432	92.73041548
MU K	.2861589D-1	.8143677D-2	.5111970D-2	.3110091D-2	.1756577D-2	.1163039D-2	.8389390D-3	.5024223D-3	.2761182D-3
1 PFR	1.130860028	.8204300680	.7223236146	.6252716265	.5220624014	.4525318772	.4002374524	.3231487784	.2415410225
1 PF1	1.235598250	.9944784976	.9187880808	.8454211757	.7701624878	.7217758108	.6869777945	.6387537726	.5925608353
1 PQ1	.7424244689	.7758334929	.8006125369	.8377330559	.8995216753	.9624482177	1.027831481	1.170782234	1.437480369
2 PFC	1.464876462	1.312223473	1.260134187	1.207897913	1.152564263	1.116007929	1.089183281	1.051108436	1.013304766
2 PQC	2.510857541	2.474907570	2.512824246	2.586492186	2.720813299	2.858479694	2.997913062	3.286358214	3.770765384
1 ZF1	1.922156521	1.923852929	1.924383337	1.924894448	1.925417474	1.925755777	1.926002270	1.926353246	1.926709942
1 ZF2	2.275146438	2.295184345	2.301097720	2.306680687	2.312295056	2.315880816	2.318473481	2.322139030	2.325836405
1 ZF3	3.811377377	3.892530966	3.916463027	3.939008947	3.961613164	3.976005170	3.986386327	4.001024272	4.015739643

(3) P 5+2	TH = 33	SBE = 1.836078459						MCPER FUNCTIONS	
	RC = 02	RC = 05	RC = 07	RC = 10	RC = 15	RC = 20	RC = 25	RC = 35	RC = 50
AMAX DB	.0017375255	.0108709564	.0213327376	.0436480540	.0988323393	.1772876696	.2802872360	.5675287486	1.249387366
AMIN DB	49.86470176	61.00948537	65.12151685	69.50114995	74.52656010	78.14819155	81.01437348	85.50824772	90.74844012
MU K	.3456958D-1	.9859102D-2	.6192574D-2	.3769663D-2	.2130292D-2	.1410970D-2	.1018034D-2	.6098892D-3	.3352938D-3
1 PFR	1.144244802	.8276845861	.7281588976	.6299064615	.5256145912	.4554521840	.4027273241	.3250645630	.2429139527
1 PF1	1.242499390	.9994072761	.9231555318	.8492788685	.7735352396	.7248582246	.6898623194	.6413802036	.5949599928
1 PQ1	.7489324019	.7805234348	.8049192978	.8417577655	.9033691426	.9662664009	1.031691652	1.174849735	1.442093617
2 PFC	1.456403764	1.307890715	1.256870435	1.205562608	1.151072615	1.115002270	1.088500918	1.050839701	1.013397609
2 PQC	2.587849409	2.529826962	2.562860668	2.632698707	2.764122210	2.900618991	3.039672937	3.328569765	3.815408835
1 ZF1	1.869502587	1.871202286	1.871738317	1.872256434	1.872788035	1.873132560	1.873383891	1.873742164	1.874106730
1 ZF2	2.205761884	2.226260856	2.232315725	2.238035245	2.243790191	2.247467597	2.250127489	2.253889525	2.257686008
1 ZF3	3.676948247	3.760125848	3.784710036	3.807889845	3.831146885	3.845962818	3.856653481	3.871732837	3.886897516

(3) P 5+2	TH = 34	SBE = 1.788291650						MCPER FUNCTIONS	
	RC = 02	RC = 05	RC = 07	RC = 10	RC = 15	RC = 20	RC = 25	RC = 35	RC = 50
AMAX DB	.0017375255	.0108709564	.0213327376	.0436480540	.0988323393	.1772876696	.2802872360	.5675287486	1.249387366
AMIN DB	47.94265754	59.08278395	63.19380519	67.57262693	72.59735613	76.21862148	79.08457137	83.57816342	88.81812453
MU I K	.4153813D-1	.1187290D-1	.7462214D-2	.4545230D-2	.2570073D-2	.1702874D-2	.1228965D-2	.7365209D-3	.4050555D-3
1 I PFR	1.158430905	.8353145090	.7342839937	.6347630419	.5293307744	.4585044602	.4053280862	.3270640654	.2443458837
1 I PF1	1.249609400	1.004513472	.9276845886	.8532820727	.7770369591	.7280592301	.6928581941	.6441083621	.5974522675
1 I PQ1	.7559176263	.7855012207	.8094765728	.8460050513	.9074190233	.9702793456	1.035744527	1.179114515	1.446924790
2 I PFC	1.447705760	1.303422439	1.253500042	1.203147657	1.149527426	1.113958802	1.087791503	1.050558025	1.013490176
2 I PQC	2.671137950	2.588602481	2.616252117	2.681870875	2.810091422	2.945275904	3.083878892	3.373187261	3.862530766
1 I ZF1	1.820154855	1.821852390	1.822392863	1.822917027	1.823456379	1.823806667	1.824062534	1.824427718	1.824799811
1 I ZF2	2.140501278	2.161449425	2.167642227	2.173495022	2.179387397	2.183154545	2.185880347	2.189737112	2.193631064
1 I ZF3	3.549953381	3.635084689	3.660307833	3.684111781	3.708014210	3.723250441	3.734248485	3.749766906	3.765379247

(3) P 5+2	TH = 35	SBE = 1.743446796						MCPER FUNCTIONS	
	RC = 02	RC = 05	RC = 07	RC = 10	RC = 15	RC = 20	RC = 25	RC = 35	RC = 50
AMAX DB	.0017375255	.0108709564	.0213327376	.0436480540	.0988323393	.1772876696	.2802872360	.5675287486	1.249387366
AMIN DB	46.06983751	57.20480121	61.31470487	65.69262937	70.71660566	74.33746629	77.20315965	81.69643974	86.93614517
MU I K	.4965994D-1	.1422718D-1	.8947841D-2	.5453483D-2	.3085510D-2	.2045168D-2	.1476398D-2	.8851412D-3	.4869714D-3
1 I PFR	1.173473826	.8433394636	.7407127290	.6398510400	.5332174753	.4616936561	.4080437281	.3291501313	.2458387185
1 I PF1	1.256922399	1.009798464	.9323773331	.8574331697	.7806700746	.7313811939	.6959677102	.6469404084	.6000396777
1 I PQ1	.7634225185	.7907867009	.8143002123	.8504878249	.9116820226	.9744967414	1.039999273	1.183585387	1.451983050
2 I PFC	1.438788180	1.298819218	1.250022895	1.200652584	1.147928083	1.112876926	1.087054481	1.050262954	1.013582146
2 I PQC	2.761349277	2.651549166	2.673255774	2.734222167	2.858900226	2.992612920	3.130683866	3.420354496	3.912272774
1 I ZF1	1.773834108	1.775523741	1.776067416	1.776596620	1.777142858	1.777498427	1.777758513	1.778130205	1.778509465
1 I ZF2	2.079015666	2.100401175	2.106728305	2.112711033	2.118737603	2.122592523	2.125382873	2.129332540	2.133322253
1 I ZF3	3.429770384	3.516779241	3.542627035	3.567044527	3.591584251	3.607236825	3.618539928	3.634494842	3.650553028

(3) P 5+2	TH = 36	SBE = 1.701301617						MCPER FUNCTIONS	
	RC = 02	RC = 05	RC = 07	RC = 10	RC = 15	RC = 20	RC = 25	RC = 35	RC = 50
AMAX DB	.0017375255	.0108709564	.0213327376	.0436480540	.0988323393	.1772876696	.2802872360	.5675287486	1.249387366
AMIN DB	44.24326677	55.37251430	59.48118391	63.85811810	68.88126360	72.50167769	75.36708804	79.86002397	85.09944734
MU I K	.5908814D-1	.1696887D-1	.1067952D-1	.6513070D-2	.3687340D-2	.2445046D-2	.1765566D-2	.1058920D-2	.5828034D-3
1 I PFR	1.189435158	.8517808485	.7474601039	.6451809029	.5372817110	.4650250832	.4108785231	.3313258021	.2473944872
1 I PF1	1.264431069	1.015263437	.9372358006	.8617345797	.7844370998	.7348265820	.6991932646	.6498786103	.6027243470
1 I PQ1	.7714946648	.7964017636	.8194076048	.8552201852	.9161697781	.9789290932	1.044465813	1.188271867	1.457278263
2 I PFC	1.429657262	1.294081764	1.246438950	1.198076936	1.146273970	1.111756026	1.086289279	1.049954017	1.013673181
2 I PQC	2.859191970	2.719017481	2.734156387	2.789987929	2.910745456	3.042807969	3.180255036	3.470228270	3.964788915
1 I ZF1	1.730292341	1.731968042	1.732513618	1.733046805	1.733599022	1.733959367	1.734223341	1.734601118	1.734987166
1 I ZF2	2.020994980	2.042806132	2.049263940	2.055373194	2.061530639	2.065471303	2.068324790	2.072365460	2.076449154
1 I ZF3	3.315847513	3.404651549	3.431108476	3.456128003	3.481296219	3.497360826	3.508966453	3.525355045	3.541857074

(3) P 5+2	TH = 37	SBE = 1.661640141						MCPER FUNCTIONS	
	RC = 02	RC = 05	RC = 07	RC = 10	RC = 15	RC = 20	RC = 25	RC = 35	RC = 50
AMAX DB	.0017375255	.0108709564	.0213327376	.0436480540	.0988323393	.1772876696	.2802872360	.5675287486	1.249387366
AMIN DB	42.46021749	53.58314137	57.69045053	62.06629360	67.08852405	70.70844636	73.57354507	78.06610202	83.30521481
MU I K	.6999187D-1	.2015008D-1	.1269074D-1	.7744799D-2	.4387559D-2	.2910554D-2	.2102325D-2	.1261410D-2	.6945288D-3
1 I PFR	1.206383408	.8606620279	.7545424120	.6507639259	.5415310355	.4685044442	.4138370510	.3335943300	.2490153561
1 I PF1	1.272126432	1.020909330	.9422619565	.8661887522	.7883406317	.7383979620	.7025373624	.6529253471	.6055085089
1 I PQ1	.7801876299	.8023705857	.8248178551	.8602175480	.9208949554	.9835878019	1.049154891	1.193184234	1.462821069
2 I PFC	1.420319777	1.289210945	1.242748248	1.195420293	1.144564472	1.110595477	1.085495307	1.049630723	1.013762926
2 I PQC	2.965470085	2.791398089	2.799269801	2.849428036	2.965843530	3.096056170	3.232775392	3.522979790	4.020247001
1 I ZF1	1.689308551	1.690963989	1.691510102	1.692046166	1.692603411	1.692968003	1.693235519	1.693618936	1.694011378
1 I ZF2	1.966162803	1.988387964	1.994972767	2.001205076	2.007489993	2.011514307	2.014429470	2.018559174	2.022734992
1 I ZF3	3.207694246	3.298204501	3.325253746	3.350862820	3.376649962	3.393121909	3.405027294	3.421846483	3.438790149

(3) P 5+2	TH = 38	SBE = 1.624269245						MCPER FUNCTIONS	
	RC = 02	RC = 05	RC = 07	RC = 10	RC = 15	RC = 20	RC = 25	RC = 35	RC = 50
AMAX DB	.0017375255	.0108709564	.0213327376	.0436480540	.0988323393	.1772876696	.2802872360	.5675287486	1.249387366
AMIN DB	40.71818427	51.83411551	55.93992696	60.31456973	65.33579397	68.95517565	71.81993176	76.31207213	81.55084352
MU I K	.8255763D-1	.2382869D-1	.1501872D-1	.9171844D-2	.5199554D-2	.3450681D-2	.2493226D-2	.1496589D-2	.8243628D-3
1 I PFR	1.224394949	.8700085517	.7619773761	.6566123359	.5459735884	.4721378676	.4169242246	.3359591953	.2507036389
1 I PF1	1.279997593	1.026736789	.9474576685	.8707981540	.7923833494	.7420980049	.7060026203	.6560831146	.6083945131
1 I PQ1	.7895618569	.8087199200	.8305519880	.8654967933	.9258713566	.9884852558	1.054078161	1.198333605	1.468622943
2 I PFC	1.410783043	1.284207793	1.238950914	1.192682270	1.142798978	1.109394644	1.084671960	1.049292562	1.013851010
2 I PQC	3.081098607	2.869127383	2.868947034	2.912829970	3.024432781	3.152571813	3.288445520	3.578796240	4.078830061
1 I ZF1	1.650685185	1.652313723	1.652858943	1.653396726	1.653958005	1.654326291	1.654596984	1.654985578	1.655383999
1 I ZF2	1.914271955	1.936899572	1.943607647	1.949959482	1.956368379	1.960474186	1.963449516	1.967666208	1.971932218
1 I ZF3	3.104873336	3.196993871	3.224617233	3.250802317	3.277197995	3.294072174	3.306274303	3.323520722	3.340903600

(3) P 5+2	TH = 39	SBE = 1.589015729						MCPER FUNCTIONS	
	RC = 02	RC = 05	RC = 07	RC = 10	RC = 15	RC = 20	RC = 25	RC = 35	RC = 50
AMAX DB	.0017375255	.0108709564	.0213327376	.0436480540	.0988323393	.1772876696	.2802872360	.5675287486	1.249387366
AMIN DB	39.01486336	50.12306206	54.22722665	58.60055085	63.62067032	67.23945856	70.10383864	74.59552182	79.83391856
MU I K	.9699066D-1	.2806881D-1	.1770481D-1	.1081999D-1	.6138246D-2	.4075458D-2	.2945582D-2	.1768902D-2	.9747849D-3
1 I PFR	1.243555116	.8798484068	.7697843016	.6627393846	.5506181499	.4759319453	.4201453184	.3384241253	.2524618079
1 I PF1	1.288031455	1.032746100	.9528246745	.8755652556	.7965680106	.7459294857	.7095917702	.6593545305	.6113848310
1 I PQ1	.7996857290	.8154794273	.8366331809	.8710764331	.9311140432	.9936349324	1.059248276	1.203732019	1.474696281
2 I PFC	1.401054949	1.279073520	1.235047175	1.189862524	1.140976877	1.108152881	1.083818618	1.048939007	1.013937044
2 I PQC	3.207121825	2.952693947	2.943579001	2.980512339	3.086776128	3.212590624	3.347485633	3.637882574	4.140737993
1 I ZF1	1.614245144	1.615839827	1.616382660	1.616920950	1.617485223	1.617856624	1.618130116	1.618523400	1.618927369
1 I ZF2	1.865100773	1.888119342	1.894946934	1.901414701	1.907944005	1.912129080	1.915163014	1.919464574	1.923818762
1 I ZF3	3.006994091	3.100621593	3.128799389	3.155545820	3.182538759	3.199809617	3.212305207	3.229975175	3.247794603

(3) P 5+2	TH = 40	SBE = 1.555723827						MCPER FUNCTIONS	
	RC = 02	RC = 05	RC = 07	RC = 10	RC = 15	RC = 20	RC = 25	RC = 35	RC = 50
AMAX DB	.0017375255	.0108709564	.0213327376	.0436480540	.0988323393	.1772876696	.2802872360	.5675287486	1.249387366
AMIN DB	37.34813535	48.44777887	52.55013433	56.92201177	61.94091989	65.55905765	68.42302564	72.91420779	78.15219397
MU K	.1135165D 0	.3294140D-1	.2079485D-1	.1271787D-1	.7220244D-2	.4796062D-2	.3467550D-2	.2083314D-2	.1148567D-2
1 PFR	1.263959484	.8902123038	.7779842499	.6691594538	.5554742021	.4798937754	.4235060003	.3409931151	.2542925073
1 PF1	1.296212396	1.038937121	.9583645458	.8804925147	.8008974478	.7498952850	.7133076622	.6627423396	.6144820622
1 PQ1	.8106368226	.8226820590	.8430870311	.8769768026	.9366394749	.9990515147	1.064678990	1.209392525	1.481054487
2 PFC	1.391143966	1.273809536	1.231037368	1.186960762	1.139097571	1.106869534	1.082934645	1.048569510	1.014020621
2 PQC	3.344735283	3.042646094	3.023601991	3.052828958	3.153164137	3.276372353	3.410137895	3.700463549	4.206189439
1 ZF1	1.579829231	1.581382778	1.581921662	1.582459194	1.583025373	1.583399285	1.583675179	1.584072648	1.584481709
1 ZF2	1.818449948	1.841847972	1.848791289	1.855371336	1.862017383	1.866279437	1.869370362	1.873754588	1.878194858
1 ZF3	2.913706676	3.008730042	3.037441013	3.064732920	3.092310895	3.109972399	3.122757872	3.140847373	3.159100427

(3) P 5+2	TH = 41	SBE = 1.524253087						MCPER FUNCTIONS	
	RC = 02	RC = 05	RC = 07	RC = 10	RC = 15	RC = 20	RC = 25	RC = 35	RC = 50
AMAX DB	.0017375255	.0108709564	.0213327376	.0436480540	.0988323393	.1772876696	.2802872360	.5675287486	1.249387366
AMIN DB	35.71605106	46.80621907	50.90658851	55.27688016	60.29446168	63.91188736	66.77540429	71.26603812	76.50357503
MU K	.1323828D 0	.3852491D-1	.2433962D-1	.1489729D-1	.8464018D-2	.5624941D-2	.4068223D-2	.2445365D-2	.1348807D-2
1 PFR	1.285715359	.9011340048	.7866002369	.6758881747	.5605519977	.4840310096	.4270123678	.3436704512	.2561985672
1 PF1	1.304521901	1.045309197	.9640786430	.8855823566	.8053745634	.7539983895	.7171532682	.6662494191	.6176889417
1 PQ1	.8225033916	.8303645012	.8499418636	.8832202787	.9424656675	1.004751022	1.070385275	1.215329287	1.487712069
2 PFC	1.381059171	1.268417456	1.226921947	1.183976744	1.137160468	1.105543944	1.082019392	1.048183504	1.014101315
2 PQC	3.495312085	3.139600707	3.109504049	3.130173563	3.223918551	3.344203745	3.476669067	3.766786044	4.275423914
1 ZF1	1.547293969	1.548798768	1.549332072	1.549867524	1.550434476	1.550810269	1.551088150	1.551489272	1.551902955
1 ZF2	1.774139845	1.797905761	1.804960975	1.811649587	1.818408628	1.822745301	1.825891548	1.830356156	1.834880323
1 ZF3	2.824697247	2.920997163	2.950218367	2.978038590	3.006188352	3.024233949	3.037305415	3.055910067	3.074493540

(3) P 5+2	TH = 42	SBE = 1.494476550						MCPER FUNCTIONS	
	RC = 02	RC = 05	RC = 07	RC = 10	RC = 15	RC = 20	RC = 25	RC = 35	RC = 50
AMAX DB	.0017375255	.0108709564	.0213327376	.0436480540	.0988323393	.1772876696	.2802872360	.5675287486	1.249387366
AMIN DB	34.11682064	45.19647592	49.29466610	53.66322101	58.67935120	62.29599827	65.15902210	69.64905658	74.88610246
MU K	.1538607D 0	.4490596D-1	.2839531D-1	.1739349D-1	.9890087D-2	.6575941D-2	.4757723D-2	.2861232D-2	.1578962D-2
1 PFR	1.308943515	.9126507002	.7956574579	.6829425622	.5658626378	.4883519064	.4306709878	.3464607369	.2581830198
1 PF1	1.312938158	1.051861065	.9699680659	.8908371509	.8100023234	.7582418916	.7211316846	.6698787852	.6210083468
1 PQ1	.8353861315	.8385676903	.8572290854	.8898315307	.9486123730	1.010750959	1.076383454	1.221557699	1.494684756
2 PFC	1.370810258	1.262899123	1.222701498	1.180910294	1.135164995	1.104175446	1.081072195	1.047780403	1.014178677
2 PQC	3.660434515	3.244253646	3.201832442	3.212985290	3.299396359	3.416401954	3.547373555	3.837121717	4.348704236
1 ZF1	1.516509742	1.517957836	1.518483860	1.519015355	1.519582398	1.519959413	1.520238849	1.520643073	1.521060981
1 ZF2	1.732008201	1.756130287	1.763293529	1.770086927	1.776955120	1.781363984	1.784563827	1.789106449	1.793712240
1 ZF3	2.739683795	2.837132295	2.866838998	2.895168998	2.923876202	2.942298778	2.955652007	2.974567031	2.993677409

(3) P 5+2 — TH = 43 — SBE = 1.466279186 — MCPER FUNCTIONS

MU	K	RC = 02	RC = 05	RC = 07	RC = 10	RC = 15	RC = 20	RC = 25	RC = 35	RC = 50
	AMAX DB	.0017375255	.0108709564	.0213327376	.0436480540	.0988323393	.1772876696	.2802872360	.5675287486	1.249387366
	AMIN DB	32.54880562	43.61676966	47.71256884	52.07922284	57.09376669	60.70956325	63.57204861	68.06142868	73.29793849
MU	K	.1782471D 0	.5218009D-1	.3302405D-1	.2024554D-1	.1152124D-1	.7664455D-2	.5547314D-2	.3337797D-2	.1842389D-2
1	PFR	1.333780222	.9248034415	.8051835458	.6903411685	.5714181592	.4928653909	.4344889415	.3493689220	.2602491173
1	PF1	1.321435607	1.058590747	.9760335943	.8962591839	.8147837483	.7626289889	.7252461357	.6736335991	.6244433058
1	PQ1	.8494002799	.8473374148	.8649835970	.8968378084	.9551012849	1.017070485	1.082691352	1.228094519	1.501989615
2	PFC	1.360407553	1.257256619	1.218376751	1.177761305	1.133110594	1.102763373	1.080092378	1.047359600	1.014252241
2	PQC	3.841932190	3.357392031	3.301202401	3.301755059	3.379994511	3.493318486	3.622576907	3.911770040	4.426319297
1	ZF1	1.487359186	1.488742267	1.489259238	1.489786341	1.490351243	1.490728795	1.491009335	1.491416084	1.491837503
1	ZF2	1.691908173	1.716374416	1.723641765	1.730536103	1.737509513	1.741988066	1.745239722	1.749357902	1.754542946
1	ZF3	2.658412572	2.756872581	2.787038147	2.815857910	2.845107030	2.863898871	2.877529265	2.896349452	2.916382887

(3) P 5+2 — TH = 44 — SBE = 1.439556540 — MCPER FUNCTIONS

MU	K	RC = 02	RC = 05	RC = 07	RC = 10	RC = 15	RC = 20	RC = 25	RC = 35	RC = 50
	AMAX DB	.0017375255	.0108709564	.0213327376	.0436480540	.0988323393	.1772876696	.2802872360	.5675287486	1.249387366
	AMIN DB	31.01051439	42.06543594	46.15861130	50.52318551	55.53599667	59.15086504	62.01276292	66.50142920	71.73735437
MU	K	.2058664D 0	.6045267D-1	.3829450D-1	.2349667D-1	.1338274D-1	.8907585D-2	.6449525D-2	.3882724D-2	.2144888D-2
1	PFR	1.360379618	.9376376405	.8152088660	.6981042567	.5772316335	.4975811224	.4384738745	.3524003348	.2624003511
1	PF1	1.329984431	1.065495422	.9822756201	.9018506268	.8197219040	.7671629827	.7294999761	.6775171733	.6279970056
1	PQ1	.8646781260	.8567250178	.8732442688	.9042692743	.9619562737	1.023730605	1.089328463	1.234958013	1.509645203
2	PFC	1.349862030	1.251492287	1.213948592	1.174529747	1.130996731	1.101307055	1.079079253	1.046920467	1.014321516
2	PQC	4.041928224	3.479908795	3.408307416	3.397033031	3.466155388	3.575343753	3.702639858	3.991061798	4.508587239
1	ZF1	1.459735803	1.461045199	1.461551275	1.462071993	1.462633972	1.463011345	1.463292522	1.463701194	1.464125686
1	ZF2	1.653706652	1.678504573	1.685872053	1.692863412	1.699938006	1.704483670	1.707785296	1.712476483	1.717238312
1	ZF3	2.580655019	2.679979878	2.710575648	2.739863586	2.769637832	2.788790572	2.802693134	2.822412813	2.842365094

(3) P 5+2 — TH = 45 — SBE = 1.414213562 — MCPER FUNCTIONS

MU	K	RC = 02	RC = 05	RC = 07	RC = 10	RC = 15	RC = 20	RC = 25	RC = 35	RC = 50
	AMAX DB	.0017375255	.0108709564	.0213327376	.0436480540	.0988323393	.1772876696	.2802872360	.5675287486	1.249387366
	AMIN DB	29.50060101	40.54091586	44.63121028	48.99350929	54.00442896	57.61828510	60.47954257	64.96743096	70.20271912
MU	K	.2370727D 0	.6983973D-1	.4428244D-1	.2719474D-1	.1550264D-1	.1032432D-1	.7478279D-2	.4504544D-2	.2489753D-2
1	PFR	1.388916481	.9512036463	.8257668553	.7062540003	.5833172792	.5025095700	.4426340539	.3555607185	.2646404743
1	PF1	1.338550014	1.072571280	.9886940678	.9076134967	.8248198884	.7718472751	.7338966928	.6815329786	.6316728017
1	PQ1	.8813720158	.8667882229	.8820544967	.9121593880	.9692036573	1.030754394	1.096316144	1.242168130	1.517671719
2	PFC	1.339185328	1.245608743	1.209418074	1.171215681	1.128822899	1.099805827	1.078032120	1.046462355	1.014385990
2	PQC	4.262895263	3.612819997	3.523931388	3.499437362	3.558373187	3.662912365	3.787962998	4.075363116	4.595859109
1	ZF1	1.433542758	1.434769426	1.435262689	1.435775471	1.436333194	1.436709648	1.436990971	1.437400941	1.437827945
1	ZF2	1.617282832	1.642399264	1.649862822	1.656947202	1.664118846	1.668728963	1.672078653	1.676840202	1.681676245
1	ZF3	2.506205111	2.606238075	2.637233244	2.666966088	2.697247313	2.716751889	2.730921195	2.751034190	2.771400709

(3) P 5+2 — TH = 46 SBE = 1.390163591 — MCPER FUNCTIONS

MU		RC = 02	RC = 05	RC = 07	RC = 10	RC = 15	RC = 20	RC = 25	RC = 35	RC = 50
	AMAX DB	.0017375255	.0108709564	.0213327376	.0436480540	.0988323393	.1772876696	.2802872360	.5675287486	1.249387366
	AMIN DB	28.01786809	39.04174733	43.12887540	47.48868512	52.49754069	56.11029362	58.97085350	63.45789480	68.69248945
	K	.2722521D 0	.8046903D-1	.5107153D-1	.3139267D-1	.1791201D-1	.1193575D-1	.8649052D-2	.5212749D-2	.2882829D-2
1	PFR	1.419589466	.9655574135	.8368944124	.7148147099	.5896905888	.5076620992	.4469784316	.3588562721	.2669735267
1	PF1	1.347092341	1.079813358	.9952883020	.9135496124	.8300808159	.7766853653	.7384399078	.6856846513	.6354742277
1	PQ1	.8996579563	.8775921053	.8914628514	.9205453519	.9768725122	1.038167248	1.103677842	1.249746691	1.526091188
2	PFC	1.328389763	1.239608900	1.204786438	1.167819261	1.126588623	1.098259026	1.076950271	1.045984597	1.014445128
2	PQC	4.507723708	3.757285495	3.648963037	3.609664501	3.657201378	3.756509290	3.878992205	4.165080124	4.688523058
1	ZF1	1.408691839	1.409826352	1.410304812	1.410808046	1.411360129	1.411734891	1.412015854	1.412426472	1.412855403
1	ZF2	1.582526975	1.607947781	1.615503270	1.622676581	1.629941029	1.634612861	1.638008641	1.642837805	1.647745382
1	ZF3	2.434877066	2.535450777	2.566812261	2.596964951	2.627733554	2.647580149	2.662010310	2.682509899	2.703285620

(3) P 5+2 — TH = 47 SBE = 1.367327461 — MCPER FUNCTIONS

MU		RC = 02	RC = 05	RC = 07	RC = 10	RC = 15	RC = 20	RC = 25	RC = 35	RC = 50
	AMAX DB	.0017375255	.0108709564	.0213327376	.0436480540	.0988323393	.1772876696	.2802872360	.5675287486	1.249387366
	AMIN DB	26.56127396	37.56655784	41.65020089	46.00728586	51.01388934	54.62544047	57.48524090	61.97136037	67.20520054
	K	.3118246D 0	.9248114D-1	.5875404D-1	.3614899D-1	.2064528D-1	.1376527D-1	.9979035D-2	.6017903D-2	.3330066D-2
1	PFR	1.452624874	.9807612784	.8486323488	.7238130944	.5963684743	.5130510689	.4515167172	.3622936965	.2694038619
1	PF1	1.355565363	1.087215345	1.002057020	.9196605407	.8355077986	.7816808441	.7431333784	.6899760002	.6394050063
1	PQ1	.9197399485	.8892102402	.9015238397	.9294686304	.9849950338	1.045997173	1.111439343	1.257717608	1.534927673
2	PFC	1.317488353	1.233495983	1.200055122	1.164340751	1.124293469	1.096665998	1.075832989	1.045486502	1.014498366
2	PQC	4.779804947	3.914633730	3.784413051	3.728501366	3.763261454	3.856677052	3.976224967	4.260664360	4.787009209
1	ZF1	1.385102540	1.386135079	1.386596670	1.387088686	1.387633693	1.388005963	1.388286038	1.388696625	1.389126877
1	ZF2	1.549339365	1.575049079	1.582692235	1.589950283	1.597303170	1.602033890	1.605473719	1.610367646	1.615343962
1	ZF3	2.366503354	2.467439278	2.499131579	2.529677145	2.560911966	2.581089951	2.595774580	2.616653450	2.637832868

(3) P 5+2 — TH = 48 SBE = 1.345632730 — MCPER FUNCTIONS

MU		RC = 02	RC = 05	RC = 07	RC = 10	RC = 15	RC = 20	RC = 25	RC = 35	RC = 50
	AMAX DB	.0017375255	.0108709564	.0213327376	.0436480540	.0988323393	.1772876696	.2802872360	.5675287486	1.249387366
	AMIN DB	25.12994513	36.11405848	40.19385832	44.54795849	49.55210461	53.16234693	56.02132094	60.50643779	65.73945766
	K	.3562464D 0	.1060307D 0	.6743174D-1	.4152838D-1	.2374062D-1	.1583884D-1	.1148733D-1	.6931759D-2	.3838100D-2
1	PFR	1.488281048	.9968848627	.8610259099	.7332785614	.6033694333	.5186899431	.4562594596	.3658802463	.2719361791
1	PF1	1.363916333	1.094769366	1.008998130	.9259475343	.8411039231	.7868373862	.7479809979	.6944110145	.6434690601
1	PQ1	.9418552005	.9017260642	.9122988003	.9389755559	.9936069538	1.054275126	1.119629069	1.266107140	1.544207503
2	PFC	1.306494835	1.227273551	1.195225781	1.160780534	1.121937043	1.095026099	1.074679552	1.044967360	1.014545119
2	PQC	5.083133162	4.086391563	3.931434569	3.856839787	3.877253247	3.964024170	4.080217776	4.362619074	4.891795297
1	ZF1	1.362701264	1.363621608	1.364064191	1.364543258	1.365079702	1.365448649	1.365727289	1.366137143	1.366568081
1	ZF2	1.517629415	1.543610805	1.551337210	1.558675680	1.566112514	1.570899202	1.574380964	1.579336690	1.584378828
1	ZF3	2.300932972	2.402040805	2.434025864	2.464935304	2.496613510	2.517111385	2.532043552	2.553293751	2.574870854

(3) P 5+2			TH = 49	SBE = 1.325012993						MCPER FUNCTIONS
			RC = 02	RC = 05	RC = 07	RC = 10	RC = 15	RC = 20	RC = 25	RC = 35	RC = 50
		AMAX DB	.0017375255	.0108709564	.0213327376	.0436480540	.0988323393	.1772876696	.2802872360	.5675287486	1.249387366
		AMIN DB	23.72319462	34.68303928	38.75859032	43.10941703	48.11088100	51.71969817	54.57777316	59.06180000	64.29392846
MU		K	.4060123D 0	.1212877D 0	.7721682D-1	.4760235D-1	.2724030D-1	.1818528D-1	.1319514D-1	.7967391D-2	.4414321D-2
1		PFR	1.526853451	1.014006128	.8741253812	.7432435629	.6107137384	.5245934174	.4612181403	.3696237891	.2745755582
1		PF1	1.372085135	1.102465726	1.016108601	.9324114585	.8468722222	.7921587404	.7529867952	.6989938711	.6476705246
1		PQ1	.9662804030	.9152344952	.9238569623	.9491180404	1.002748025	1.063035403	1.128278414	1.274944181	1.553959552
2		PFC	1.295423700	1.220945522	1.190300302	1.157139124	1.119519006	1.093338700	1.073489233	1.044426438	1.014584771
2		PQC	5.422429997	4.274320350	4.091347775	3.995693704	3.999967120	4.079235099	4.191594802	4.471506578	5.003413226
1		ZF1	1.341420619	1.342218143	1.342639503	1.343103831	1.343630172	1.343994935	1.344271572	1.344679963	1.345110932
1		ZF2	1.487314919	1.513548442	1.521353484	1.528767918	1.536284064	1.541123701	1.544645204	1.549659647	1.554764565
1		ZF3	2.238029946	2.339106975	2.371344024	2.402586166	2.434683138	2.455488460	2.470660653	2.492273535	2.514241755

(3) P 5+2			TH = 50	SBE = 1.305407289						MCPER FUNCTIONS
			RC = 02	RC = 05	RC = 07	RC = 10	RC = 15	RC = 20	RC = 25	RC = 35	RC = 50
		AMAX DB	.0017375255	.0108709564	.0213327376	.0436480540	.0988323393	.1772876696	.2802872360	.5675287486	1.249387366
		AMIN DB	22.34054738	33.27236605	37.34320528	41.69043627	46.68897103	50.29623629	53.15333345	57.63617564	62.86733582
MU		K	.4616575D 0	.1384391D 0	.8823300D-1	.5444994D-1	.3119118D-1	.2083657D-1	.1512606D-1	.9139352D-2	.5066970D-2
1		PFR	1.568680522	1.032212608	.8879867934	.7537439965	.6184236566	.5307775637	.4664052799	.3735328719	.2773274998
1		PF1	1.380003636	1.110292620	1.023384302	.9390527048	.8528156415	.7976487170	.7581549319	.7037289428	.6520137613
1		PQ1	.9933392808	.9298438682	.9362767006	.9599544113	1.012462591	1.072316087	1.137422137	1.284260593	1.564215545
2		PFC	1.284290226	1.214516188	1.185280828	1.153417181	1.117039076	1.091603191	1.072261302	1.043862983	1.014616679
2		PQC	5.803297403	4.480459747	4.265669530	4.146219759	4.132298442	4.203081972	4.311058119	4.587956886	5.122456743
1		ZF1	1.321198799	1.321862474	1.322260318	1.322708058	1.323222705	1.323582394	1.323856441	1.324262612	1.324692928
1		ZF2	1.458321419	1.484784571	1.492663394	1.500149164	1.507739827	1.512629286	1.516188255	1.521258206	1.526422728
1		ZF3	2.177672019	2.278502451	2.310947847	2.342489213	2.374978415	2.396077732	2.411481806	2.433447971	2.455800141

(3) P 5+2			TH = 51	SBE = 1.286759566						MCPER FUNCTIONS
			RC = 02	RC = 05	RC = 07	RC = 10	RC = 15	RC = 20	RC = 25	RC = 35	RC = 50
		AMAX DB	.0017375255	.0108709564	.0213327376	.0436480540	.0988323393	.1772876696	.2802872360	.5675287486	1.249387366
		AMIN DB	20.98177342	31.88097863	35.94657293	40.28984609	45.28517903	48.89075389	51.74678753	56.22834243	61.45845116
MU		K	.5237591D 0	.1576903D 0	.1006167D 0	.6215850D-1	.3564518D-1	.2382821D-1	.1730628D-1	.1046384D-1	.5805232D-2
1		PFR	1.614150355	1.051602855	.9026727446	.7648196704	.6265237004	.5372599971	.4718345599	.3776167978	.2801979705
1		PF1	1.387595108	1.118235809	1.030819801	.9458710890	.8589369979	.8033111714	.7634896994	.7086208057	.6565033720
1		PQ1	1.023411667	.9456782567	.9496470299	.9715504005	1.022800247	1.082159584	1.147098816	1.294091594	1.575010416
2		PFC	1.273110526	1.207990248	1.180169770	1.149615524	1.114497041	1.089818988	1.070995033	1.043276223	1.014640171
2		PQC	6.232405000	4.707181147	4.456149267	4.309742026	4.275264859	4.336438543	4.439399791	4.712677870	5.249590421
1		ZF1	1.301979024	1.302497431	1.302869393	1.303298636	1.303799946	1.304153642	1.304424491	1.304827659	1.305256613
1		ZF2	1.430581694	1.457248224	1.465195657	1.472747932	1.480408141	1.485344178	1.488938247	1.494060367	1.499281171
1		ZF3	2.119749501	2.220103762	2.252710817	2.284515473	2.317368317	2.338747085	2.354374216	2.376683450	2.399411743

(3) P 5+2 — TH = 52 SBE = 1.269018215 — MCPER FUNCTIONS

MU	K	RC = 02	RC = 05	RC = 07	RC = 10	RC = 15	RC = 20	RC = 25	RC = 35	RC = 50
	AMAX DB	.0017375255	.0108709564	.0213327376	.0436480540	.0988323393	.1772876696	.2802872360	.5675287486	1.249387366
	AMIN DB	19.64692992	30.50789097	34.56762098	38.90652653	43.89835537	47.50208809	50.35696481	54.83712095	60.06608814
MU	K	.5929375D 0	.1792668D 0	.1145181D 0	.7082462D-1	.4065986D-1	.2719960D-1	.1976492D-1	.1195889D-1	.6639355D-2
1	PFR	1.663708239	1.072288133	.9182533642	.7765148447	.6350409198	.5440600673	.4775209620	.3818857129	.2831934544
1	PF1	1.394773800	1.126278238	1.038408146	.9528657318	.8652389304	.8091499838	.7689955123	.7136742472	.6611442141
1	PQ1	1.056944375	.9628802668	.9640693914	.9839803167	1.033816624	1.092613233	1.157351382	1.304476206	1.586382724
2	PFC	1.261901609	1.201372828	1.174969837	1.145735148	1.111892765	1.087985533	1.069689699	1.042665364	1.014654544
2	PQC	6.717719571	4.957253162	4.664812650	4.487781883	4.430026977	4.480296800	4.577516239	4.846467279	5.385560245
1	ZF1	1.283709052	1.284070406	1.284414044	1.284822823	1.285309104	1.285655858	1.285922881	1.286322235	1.286749094
1	ZF2	1.404035333	1.430874318	1.438884801	1.446498505	1.454223083	1.459202327	1.462829037	1.467999841	1.473273455
1	ZF3	2.064164239	2.163798270	2.196517065	2.228546463	2.261732162	2.283374666	2.299215291	2.321856494	2.344952370

(3) P 5+2 — TH = 53 SBE = 1.252135658 — MCPER FUNCTIONS

MU	K	RC = 02	RC = 05	RC = 07	RC = 10	RC = 15	RC = 20	RC = 25	RC = 35	RC = 50
	AMAX DB	.0017375255	.0108709564	.0213327376	.0436480540	.0988323393	.1772876696	.2802872360	.5675287486	1.249387366
	AMIN DB	18.33641235	29.15219326	33.20533280	37.53940337	42.52739113	46.12911491	48.98273264	53.46136876	58.68909668
MU	K	.6698570D 0	.2034167D 0	.1301032D 0	.8055514D-1	.4629904D-1	.3099455D-1	.2253438D-1	.1364461D-1	.7580778D-2
1	PFR	1.717865016	1.094394419	.9348074432	.7888788641	.6440052404	.5511990804	.4834809295	.3863507059	.2863210123
1	PF1	1.401444735	1.134399611	1.046140597	.9600349184	.8717238411	.8151690328	.7746768999	.7188942737	.6659414176
1	PQ1	1.094464169	.9816144137	.9796597979	.9973284432	1.045574317	1.103730038	1.168227739	1.315457778	1.598375135
2	PFC	1.250681447	1.194669516	1.169684055	1.141777240	1.109226197	1.086102308	1.068344585	1.042029593	1.014659060
2	PQC	7.268785529	5.233922240	4.894014973	4.682094254	4.597913264	4.635786872	4.726425379	4.990227010	5.531206121
1	ZF1	1.266340717	1.266532927	1.266845724	1.267232015	1.267701525	1.268040359	1.268302909	1.268697611	1.269121613
1	ZF2	1.378628401	1.405603172	1.413670655	1.421340414	1.429123951	1.434142890	1.437799677	1.443015531	1.448338319
1	ZF3	2.010828681	2.109483262	2.142260453	2.174473258	2.207958672	2.229847933	2.245891696	2.263352806	2.292306944

(3) P 5+2 — TH = 54 SBE = 1.236067977 — MCPER FUNCTIONS

MU	K	RC = 02	RC = 05	RC = 07	RC = 10	RC = 15	RC = 20	RC = 25	RC = 35	RC = 50
	AMAX DB	.0017375255	.0108709564	.0213327376	.0436480540	.0988323393	.1772876696	.2802872360	.5675287486	1.249387366
	AMIN DB	17.05101458	27.81305646	31.85874636	36.18744448	41.17121321	44.77074391	47.62299082	52.09997468	57.32635718
MU	K	.7552250D 0	.2304121D 0	.1475548D 0	.9146825D-1	.5263357D-1	.3526171D-1	.2565072D-1	.1554340D-1	.8642274D-2
1	PFR	1.777206166	1.118064743	.9524237643	.8019668992	.6534498569	.5587005550	.4897325527	.3910239236	.2895883486
1	PF1	1.407503881	1.142575914	1.054006329	.9673759326	.8783938233	.8213721640	.7805384945	.7242361171	.6709004026
1	PQ1	1.136593139	1.002071218	.9965514199	1.011690710	1.058143976	1.115569521	1.179781498	1.327084603	1.611034976
2	PFC	1.239469064	1.187886386	1.164315793	1.137743201	1.106497390	1.084168834	1.066958986	1.041368078	1.014652951
2	PQC	7.897066534	5.541012392	5.146506782	4.894711820	4.780450168	4.804200971	4.887287172	5.144980135	5.687476727
1	ZF1	1.249829522	1.249840277	1.250119643	1.250481366	1.250932316	1.251262225	1.251519635	1.251908821	1.252329179
1	ZF2	1.354313163	1.381380080	1.389497906	1.397217985	1.405054805	1.410109770	1.413793957	1.419051059	1.424419212
1	ZF3	1.959664987	2.057065157	2.089843763	2.122195671	2.155945140	2.178062319	2.194298499	2.217566414	2.241368641

(3) P 5+2 — TH = 55 — SBE = 1.220774589 — MCPER FUNCTIONS

MU	K	RC = 02	RC = 05	RC = 07	RC = 10	RC = 15	RC = 20	RC = 25	RC = 35	RC = 50
	AMAX DB	.0017375255	.0108709564	.0213327376	.0436480540	.0988323393	.1772876696	.2802872360	.5675287486	1.249387366
	AMIN DB	15.79199599	26.48973984	30.52695462	34.84965679	39.82877974	43.42591312	46.27666625	50.75185331	55.97677488
MU	K	.8497908D 0	.2605519D 0	.1670744D 0	.1036948D 0	.5974208D-1	.4005525D-1	.2915409D-1	.1768027D-1	.9838126D-2
1	PFR	1.842401355	1.143461965	.9712026713	.8158408177	.6634116940	.5665905217	.4962957852	.3959187031	.2930038896
1	PF1	1.412838842	1.150778869	1.061992080	.9748848620	.8852505758	.8277631497	.7865850147	.7298552415	.6760268992
1	PQ1	1.184066772	1.024472191	1.014897718	1.027176703	1.071605618	1.128198746	1.192072838	1.339410631	1.624414898
2	PFC	1.228284630	1.181030041	1.158868794	1.133634665	1.103706505	1.082184686	1.065532211	1.040679968	1.014635412
2	PQC	8.616359531	5.883049132	5.425514945	5.127999222	4.979398746	4.987022367	5.061428391	5.311891305	5.855447237
1	ZF1	1.234134242	1.233951155	1.234194431	1.234529453	1.234960007	1.235279960	1.235531544	1.235914323	1.236330224
1	ZF2	1.331047849	1.358154953	1.366315713	1.374079944	1.381964055	1.387051204	1.390759989	1.396054362	1.401463884
1	ZF3	1.910604122	2.006458816	2.039177986	2.071621531	2.105596692	2.127922983	2.144338430	2.167898912	2.192038126

(3) P 5+2 — TH = 56 — SBE = 1.206217949 — MCPER FUNCTIONS

MU	K	RC = 02	RC = 05	RC = 07	RC = 10	RC = 15	RC = 20	RC = 25	RC = 35	RC = 50
	AMAX DB	.0017375255	.0108709564	.0213327376	.0436480540	.0988323393	.1772876696	.2802872360	.5675287486	1.249387366
	AMIN DB	14.56115167	25.18160213	29.20910759	33.52508407	38.49907582	42.09358420	44.94270780	49.41593963	54.63927434
MU	K	.9543422D 0	.2941641D 0	.1888847D 0	.1173796D 0	.6771194D-1	.4543542D-1	.3308927D-1	.2008314D-1	.1118431D-1
1	PFR	1.914214058	1.170772032	.9912579250	.8305702125	.6739319459	.5748978717	.5031926942	.4010497259	.2965768727
1	PF1	1.417330272	1.158975340	1.070081750	.9825563691	.8922952994	.8343456396	.7928212445	.7356073484	.6813269677
1	PQ1	1.237754967	1.049075930	1.034876256	1.043912097	1.086050194	1.141693524	1.205169539	1.352496321	1.638573636
2	PFC	1.217149551	1.174107650	1.153347202	1.129453523	1.100853832	1.080149497	1.064063595	1.039964398	1.014605601
2	PQC	9.443293451	6.265414260	5.734843389	5.384719913	5.196799456	5.185960611	5.250372615	5.492291346	6.036349560
1	ZF1	1.219216550	1.218827375	1.219031836	1.219337972	1.219746253	1.220055190	1.220300245	1.220675700	1.221086306
1	ZF2	1.308796432	1.335882013	1.344077374	1.351879060	1.359804108	1.364919401	1.368649847	1.373977317	1.379424011
1	ZF3	1.863584902	1.957586933	1.990181704	2.022666042	2.056825637	2.079339152	2.095921207	2.119758786	2.144222868

(3) P 5+2 — TH = 57 — SBE = 1.192363293 — MCPER FUNCTIONS

MU	K	RC = 02	RC = 05	RC = 07	RC = 10	RC = 15	RC = 20	RC = 25	RC = 35	RC = 50
	AMAX DB	.0017375255	.0108709564	.0213327376	.0436480540	.0988323393	.1772876696	.2802872360	.5675287486	1.249387366
	AMIN DB	13.36087832	23.88811711	27.90441665	32.21280567	37.18110963	40.77273768	43.62008127	48.09118360	53.31279383
MU	K	.1069702D 1	.3316086D 0	.2132312D 0	.1326833D 0	.7664033D-1	.5146939D-1	.3750621D-1	.2278324D-1	.1269874D-1
1	PFR	1.993510646	1.200207822	1.012718903	.8462336182	.6850567126	.5836547659	.5104477546	.4064331964	.3003174503
1	PF1	1.420854207	1.167126676	1.078255936	.9903834214	.8995285729	.8411230999	.7992520049	.7415483807	.6868070211
1	PQ1	1.298686141	1.076185588	1.056693357	1.062041603	1.101581477	1.156139868	1.219148200	1.366409654	1.653576927
2	PFC	1.206086559	1.167126990	1.147755602	1.125201946	1.097939804	1.078062971	1.062552497	1.039220487	1.014562638
2	PQC	10.39792551	6.694540076	6.078998928	5.668119082	5.435027244	5.402994621	5.455876768	5.687707049	6.231552949
1	ZF1	1.205040653	1.204433591	1.204596452	1.204871473	1.205255561	1.205552402	1.205790205	1.206157397	1.206561842
1	ZF2	1.287528375	1.314519538	1.322740038	1.330571845	1.338531045	1.343670220	1.347419239	1.352775422	1.358254873
1	ZF3	1.818552879	1.910379512	1.942780540	1.975251221	2.009550882	2.032228531	2.048962941	2.073060806	2.097836526

(3) P 5+2 — TH = 58 SBE = 1.179178403 — MCPER FUNCTIONS

MU		RC = 02	RC = 05	RC = 07	RC = 10	RC = 15	RC = 20	RC = 25	RC = 35	RC = 50
	AMAX DB	.0017375255	.0108709564	.0213327376	.0436480540	.0988323393	.1772876696	.2802872360	.5675287436	1.249387366
	AMIN DB	12.19422444	22.60889467	26.61216141	30.91193628	35.87390892	39.46236842	42.30776427	46.77654475	51.99627969
	K	.1196720D 1	.3732797D 0	.2403849D 0	.1497839D 0	.8663542D-1	.5823208D-1	.4246070D-1	.2581552D-1	.1440147D-1
1	PFR	2.081268126	1.232013669	1.035733208	.8629199560	.6968377520	.5928971176	.5180881932	.4120870514	.3042368105
1	PF1	1.423285518	1.175188003	1.086491408	.9983569745	.9069502034	.8480987380	.8058821183	.7476345241	.6924738493
1	PQ1	1.368074480	1.106158063	1.080589828	1.081732570	1.118318348	1.171635738	1.234095707	1.381227324	1.669498578
2	PFC	1.195119762	1.160096508	1.142099055	1.120882419	1.094965015	1.075924894	1.060998314	1.038447342	1.014505604
2	PQC	11.50444880	7.178154190	6.463349411	5.982027159	5.696859721	5.640425693	5.679975890	5.899897461	6.442685014
1	ZF1	1.191572919	1.190737055	1.190855482	1.191097113	1.191455055	1.191738696	1.191968511	1.192326475	1.192723868
1	ZF2	1.267218298	1.294029651	1.302264458	1.310118281	1.318104347	1.323262889	1.327027222	1.332407503	1.337915061
1	ZF3	1.775459025	1.864773408	1.896906687	1.929305407	1.963697429	1.986514279	2.003385605	2.027725485	2.052798395

(3) P 5+2 — TH = 59 SBE = 1.166633397 — MCPER FUNCTIONS

MU		RC = 02	RC = 05	RC = 07	RC = 10	RC = 15	RC = 20	RC = 25	RC = 35	RC = 50
	AMAX DB	.0017375255	.0108709564	.0213327376	.0436480540	.0988323393	.1772876696	.2802872360	.5675287486	1.249387366
	AMIN DB	11.06490853	21.34370849	25.33169993	29.62162712	34.57651782	38.16148111	41.00474127	45.47098661	50.68868054
	K	.1336269D 1	.4196093D 0	.2706451D 0	.1688792D 0	.9781774D-1	.6580715D-1	.4801518D-1	.2921914D-1	.1631507D-1
1	PFR	2.178579604	1.266470664	1.060469779	.8807302553	.7093333724	.6026651618	.5261443952	.4180312053	.3083473182
1	PF1	1.424502596	1.183107484	1.094760506	1.006465599	.9145590465	.8552754100	.8127163618	.7540222062	.6983346446
1	PQ1	1.447350254	1.139415333	1.106848023	1.103179408	1.136397575	1.188293157	1.250111004	1.397036188	1.686421754
2	PFC	1.184274623	1.153025371	1.136383142	1.116497769	1.091930239	1.073735148	1.059400485	1.037644057	1.014433539
2	PQC	12.79202655	7.725589516	6.894323612	6.330989837	5.985562070	5.900943135	5.925038339	6.130898373	6.671579540
1	ZF1	1.178781515	1.177707387	1.177778512	1.177984446	1.178314255	1.178583573	1.178804648	1.179152398	1.179541827
1	ZF2	1.247845503	1.274378161	1.282614788	1.290481594	1.298486646	1.303659749	1.307435949	1.312835455	1.318366216
1	ZF3	1.734258116	1.820711927	1.852498498	1.884762824	1.919195914	1.942125043	1.959116564	1.983678595	2.009032909

(3) P 5+2 — TH = 60 SBE = 1.154700538 — MCPER FUNCTIONS

MU		RC = 02	RC = 05	RC = 07	RC = 10	RC = 15	RC = 20	RC = 25	RC = 35	RC = 50
	AMAX DB	.0017375255	.0108709564	.0213327376	.0436480540	.0988323393	.1772876696	.2802872360	.5675287486	1.249387366
	AMIN DB	9.977285348	20.09253164	24.06248319	28.34106892	33.28799428	36.86908592	39.70999839	44.17347098	49.38894113
	K	.1489239D 1	.4710696D 0	.3043423D 0	.1901886D 0	.1103218D 0	.7428819D-1	.5423955D-1	.3303810D-1	.1846489D-1
1	PFR	2.286656507	1.303902851	1.087122576	.8997797112	.7226094976	.6130041330	.5346503849	.4242878395	.3126626809
1	PF1	1.424393243	1.190825565	1.103030463	1.014695045	.9223527892	.8626555069	.8197594086	.7605680918	.7043970290
1	PQ1	1.538193103	1.176458470	1.135800589	1.126609040	1.155977226	1.206240783	1.267307235	1.413934994	1.704440511
2	PFC	1.173577850	1.145923545	1.130614019	1.112051203	1.088836457	1.071493723	1.057758506	1.036809724	1.014345440
2	PQC	14.29577307	8.348178519	7.379665475	6.720432516	6.304993535	6.187706093	6.193834293	6.383077192	6.920367857
1	ZF1	1.166636037	1.165316373	1.165337319	1.165505220	1.165804885	1.166058740	1.166270311	1.166606841	1.166987370
1	ZF2	1.229393294	1.255534433	1.263758416	1.271628053	1.279643510	1.284826036	1.288610444	1.294024021	1.299572798
1	ZF3	1.694906784	1.778144481	1.809500130	1.841563209	1.875982214	1.898994549	1.916088145	1.940850724	1.966469194

(3) P 5+2		TH = 61	SBE = 1.143354068							MCPER FUNCTIONS
		RC = 02	RC = 05	RC = 07	RC = 10	RC = 15	RC = 20	RC = 25	RC = 35	RC = 50
	AMAX DB	.0017375255	.0108709564	.0213327376	.0436480540	.0988323393	.1772876696	.2802872360	.5675287486	1.249387366
	AMIN DB	8.936238546	18.85558142	22.80407475	27.06949707	32.00740796	35.58419417	38.42251817	42.88295182	48.09599586
MU	K	.1656529D 1	.5281762D 0	.3418413D 0	.2139563D 0	.1242978D 0	.8378001D-1	.6121227D-1	.3732183D-1	.2087954D-1
1	PFR	2.406826938	1.344684413	1.115914988	.9202001523	.7367409415	.6239650735	.5436443988	.4308817449	.3171981438
1	PF1	1.422861520	1.198274264	1.111262656	1.023027728	.9303276892	.8702408137	.8270157533	.7673290728	.7106690832
1	PQ1	1.642568344	1.217884994	1.167841378	1.152287683	1.177240861	1.225627061	1.285814357	1.432036481	1.723661655
2	PFC	1.163057130	1.138801871	1.124798471	1.107546350	1.085684883	1.069200740	1.056071934	1.035343429	1.014240260
2	PQC	16.05791153	9.059756646	7.928759480	7.156869765	6.659741932	6.504446272	6.489621379	6.659201101	7.191527090
1	ZF1	1.155107170	1.153537777	1.153505687	1.153633206	1.153900697	1.154137938	1.154339230	1.154663517	1.155034187
1	ZF2	1.211848055	1.237471302	1.245665828	1.253526812	1.261543263	1.266729689	1.270518403	1.275940582	1.281501883
1	ZF3	1.657361235	1.737026265	1.767861247	1.799651491	1.833997095	1.857061231	1.874237265	1.899176890	1.925040652

(3) P 5+2		TH = 62	SBE = 1.132570051							MCPER FUNCTIONS
		RC = 02	RC = 05	RC = 07	RC = 10	RC = 15	RC = 20	RC = 25	RC = 35	RC = 50
	AMAX DB	.0017375255	.0108709564	.0213327376	.0436480540	.0988323393	.1772876696	.2802872360	.5675287486	1.249387366
	AMIN DB	7.946982964	17.63337467	21.55617704	25.80619968	30.73383905	34.30531423	37.14127418	41.59336889	46.80876182
MU	K	.1839054D 1	.5914908D 0	.3835447D 0	.2404540D 0	.1399139D 0	.9440020D-1	.6902151D-1	.4212605D-1	.2359128D-1
1	PFR	2.540530151	1.389247951	1.147105056	.9421430079	.7518129422	.6356058036	.5531695706	.4378407308	.3219707221
1	PF1	1.419834963	1.205376554	1.119411775	1.031442136	.9384782593	.8780323373	.8344896173	.7743122517	.7171593776
1	PQ1	1.762766903	1.264410411	1.203439081	1.180529309	1.200402744	1.246624115	1.305782347	1.451469951	1.744206982
2	PFC	1.152740696	1.131672158	1.118943978	1.102987304	1.082476991	1.066856466	1.054340407	1.035044261	1.014116907
2	PQC	18.12915934	9.877307680	8.553049657	7.648174073	7.055294865	6.855597896	6.816252486	6.962522324	7.487951338
1	ZF1	1.144166375	1.142347153	1.142259242	1.142344035	1.142577314	1.142796780	1.142987011	1.143298017	1.143657852
1	ZF2	1.195198062	1.220165012	1.228310517	1.236149769	1.244156821	1.249341180	1.253130025	1.258554984	1.264122981
1	ZF3	1.621574748	1.697317962	1.727536764	1.758977509	1.793185907	1.816267915	1.833505092	1.858596183	1.884684598

(3) P 5+2		TH = 63	SBE = 1.122326238							MCPER FUNCTIONS
		RC = 02	RC = 05	RC = 07	RC = 10	RC = 15	RC = 20	RC = 25	RC = 35	RC = 50
	AMAX DB	.0017375255	.0108709564	.0213327376	.0436480540	.0988323393	.1772876696	.2802872360	.5675287486	1.249387366
	AMIN DB	7.014771295	16.42679414	20.31866578	24.55052950	29.46637813	33.03294750	35.86522539	40.31864068	45.52613108
MU	K	.2037748D 1	.6616236D 0	.4298970D 0	.2699841D 0	.1573585D 0	.1062809D 0	.7776655D-1	.4751365D-1	.2663666D-1
1	PFR	2.689308234	1.438093876	1.180991692	.9657828862	.7679230157	.6479920914	.5632747551	.4451961136	.3269994766
1	PF1	1.415271314	1.212045951	1.127424948	1.039912131	.9467968878	.8860300925	.8421848286	.7815249159	.7238770043
1	PQ1	1.901450459	1.316894923	1.243154379	1.211706286	1.225714348	1.269432571	1.327385175	1.472384411	1.766216025
2	PFC	1.142656704	1.124547290	1.113058796	1.098378686	1.079214552	1.064461341	1.052563658	1.034111319	1.013974243
2	PQC	20.57042726	10.82179397	9.266582663	8.203923383	7.498259439	7.246463429	7.178312615	7.296885625	7.813040824
1	ZF1	1.133785666	1.131721680	1.131575304	1.131615050	1.131812081	1.132012610	1.132190993	1.132487669	1.132835676
1	ZF2	1.179432073	1.203595175	1.211668910	1.219471465	1.227457566	1.232633371	1.236417859	1.241839382	1.247407874
1	ZF3	1.587495136	1.658985432	1.688486627	1.719495789	1.753498323	1.776561528	1.793836752	1.819051449	1.845341918

(3) P 5+2 — TH = 64 — SBE = 1.112601940 — MCPER FUNCTIONS

MU	P 5+2	RC = 02	RC = 05	RC = 07	RC = 10	RC = 15	RC = 20	RC = 25	RC = 35	RC = 50
	AMAX DB	.0017375255	.0108709564	.0213327376	.0436480540	.0988323393	.1772876696	.2802872360	.5675287486	1.249387366
	AMIN DB	6.144520342	15.23716564	19.09163431	23.30192063	28.20412744	31.76458473	34.59331032	39.04265691	44.24696221
	K	.2253584D 1	.7392354D 0	.4813881D 0	.3028839D 0	.1768430D 0	.1195711D 0	.8755945D-1	.5355580D-1	.3005711D-1
1	PFR	2.854797314	1.491800928	1.217922028	.9913218986	.7851832073	.6611990712	.5740155258	.4529333062	.3323058460
1	PF1	1.409163680	1.218186429	1.135240824	1.048406154	.9552733795	.8942328395	.8501046710	.7889745009	.7308316101
1	PQ1	2.061705014	1.376376588	1.287661573	1.246262835	1.253472536	1.294287598	1.350825733	1.494952469	1.789849422
2	PFC	1.132832450	1.117441331	1.107152044	1.093725698	1.075899676	1.062016005	1.050741529	1.033143716	1.013811084
2	PQC	23.45497158	11.91922810	10.08671620	8.835854110	7.996646647	7.683426761	7.581294061	7.666364975	8.170814480
1	ZF1	1.123937485	1.121639988	1.121432737	1.121425178	1.121583944	1.121764375	1.121930122	1.122211415	1.122546587
1	ZF2	1.164537822	1.187744734	1.195720334	1.203469006	1.211421231	1.216581395	1.220356677	1.225768105	1.231330477
1	ZF3	1.555062512	1.621999359	1.650675631	1.681165355	1.714888119	1.737892860	1.755181062	1.780489000	1.806956759

(3) P 5+2 — TH = 65 — SBE = 1.103377919 — MCPER FUNCTIONS

MU	P 5+2	RC = 02	RC = 05	RC = 07	RC = 10	RC = 15	RC = 20	RC = 25	RC = 35	RC = 50
	AMAX DB		.0108709564	.0213327376	.0436480540	.0988323393	.1772876696	.2802872360	.5675287486	1.249387366
	AMIN DB		14.06634353	17.87544973	22.05991173	26.94620423	30.49970276	33.32444094	37.76927008	42.97007072
	K		.8250390D 0	.5385576D 0	.3395295D 0	.1986053D 0	.1344386D 0	.9852695D-1	.6033322D-1	.3389976D-1
1	PFR		1.551037693	1.258300075	1.018994896	.8037228383	.6753129729	.5854553875	.4612425321	.3379140462
1	PF1		1.223692847	1.142788681	1.056886307	.9638944034	.9026377586	.8582516926	.7966685368	.7380334309
1	PQ1		1.444112435	1.337776000	1.284732179	1.284029941	1.321465520	1.376342040	1.519375192	1.815293091
2	PFC		1.110369645	1.101233805	1.089034200	1.072534855	1.059521326	1.048874002	1.032140592	1.013626195
2	PQC		13.20206069	11.03504911	9.558456675	8.560242846	8.174229957	8.031822713	8.077938836	8.566053332
1	ZF1		1.112081988	1.111811818	1.111754801	1.111873321	1.112032507	1.112184835	1.112449689	1.112771010
1	ZF2		1.172599914	1.180447002	1.188122016	1.196025823	1.201162555	1.204923372	1.210317539	1.215866717
1	ZF3		1.586334816	1.614073241	1.643949579	1.677312990	1.700216348	1.717490300	1.742858362	1.769476245

(3) P 5+2 — TH = 66 — SBE = 1.094636279 — MCPER FUNCTIONS

MU	P 5+2	RC = 02	RC = 05	RC = 07	RC = 10	RC = 15	RC = 20	RC = 25	RC = 35	RC = 50
	AMAX DB		.0108709564	.0213327376	.0436480540	.0988323393	.1772876696	.2802872360	.5675287486	1.249387366
	AMIN DB		12.91679928	16.67082235	20.32417742	25.69174685	29.23726195	32.05749638	36.49728612	41.69421816
	K		.9198003D 0	.6019990D 0	.3803408D 0	.2229134D 0	.1510736D 0	.1108128D 0	.6793775D-1	.3821836D-1
1	PFR		1.616574877	1.302596857	1.049075813	.8236918738	.6904332438	.5976672630	.4700196972	.3438515573
1	PF1		1.228452088	1.149987627	1.065307323	.9726428248	.9112400481	.8666274637	.8046145757	.7454933246
1	PQ1		1.521629415	1.394488895	1.327758516	1.317808227	1.351298503	1.404215082	1.545338204	1.842763438
2	PFC		1.103348996	1.095315242	1.084310797	1.069123020	1.056978443	1.046961215	1.031101122	1.013418298
2	PQC		14.71098404	12.13865061	10.38976529	9.201099738	8.728336074	8.537953258	8.536717144	9.004484824
1	ZF1		1.103028695	1.102694102	1.102585644	1.102662002	1.102798816	1.102936951	1.103184316	1.103490765
1	ZF2		1.158150123	1.165834016	1.173412615	1.181251564	1.186356250	1.190096864	1.195466034	1.200994430
1	ZF3		1.551970681	1.578653422	1.607816073	1.640734403	1.663489905	1.680720009	1.706112039	1.732850214

(3) P 5+2	TH = 67 RC = 02	SBE = 1.086360377 RC = 05	RC = 07	RC = 10	RC = 15	RC = 20	RC = 25	MCPER FUNCTIONS RC = 35	RC = 50
AMAX DB		.0108709564	.0213327376	.0436480540	.0988323393	.1772876696	.2802872360	.5675287486	1.249387366
AMIN DB		11.79170329	15.47888977	19.59457018	24.43992458	27.97620469	30.79131640	35.22545335	40.41809958
MU I K		.1024339D 1	.6723635D 0	.4257867D 0	.2500696D 0	.1696917D 0	.1245805D 0	.7647416D-1	.4307443D-1
1 I PFR		1.689297978	1.351362134	1.081885365	.8452650718	.7066751672	.6107353246	.4793674611	.3501497178
1 I PF1		1.232345163	1.156746023	1.073615419	.9814969042	.9200324261	.8752322706	.8128200908	.7532228035
1 I PQ1		1.610787509	1.459011872	1.376125404	1.355315195	1.384176971	1.434778834	1.574769429	1.872513923
2 I PFC		1.096397631	1.089408724	1.079562932	1.065667604	1.054388806	1.045003499	1.030024534	1.013186065
2 I PQC		16.49728585	13.43169573	11.35241503	9.934187399	9.357409885	9.109558337	9.051238271	9.493022149
1 I ZF1		1.094462045	1.094062292	1.093900666	1.093933042	1.094046391	1.094169575	1.094338414	1.094688972
1 I ZF2		1.144387781	1.151869376	1.159325426	1.167080853	1.172143920	1.175858041	1.181193822	1.186693272
1 I ZF3		1.518888843	1.544394422	1.572736602	1.605117493	1.627674772	1.644828827	1.670205308	1.697030975

(3) P 5+2	TH = 68 RC = 02	SBE = 1.078534743 RC = 05	RC = 07	RC = 10	RC = 15	RC = 20	RC = 25	MCPER FUNCTIONS RC = 35	RC = 50
AMAX DB		.0108709564	.0213327376	.0436480540	.0988323393	.1772876696	.2802872360	.5675287486	1.249387366
AMIN DB		10.69498458	14.30131478	18.37117555	23.18995283	26.71545549	29.52469492	33.95245308	39.14032887
MU I K		.1139528D 1	.7503648D 0	.4763910D 0	.2804161D 0	.1905381D 0	.1400166D 0	.8606241D-1	.4853862D-1
1 I PFR		1.770219836	1.405237823	1.117800368	.8686471196	.7241731142	.6247572699	.4893465639	.3568444596
1 I PF1		1.235250489	1.162961269	1.081747064	.9904293338	.9290045127	.8840647265	.8212923401	.7612340628
1 I PQ1		1.713857959	1.532833886	1.430792654	1.397167037	1.420575705	1.468433180	1.606348996	1.904843396
2 I PFC		1.089535310	1.083527961	1.074799001	1.062172620	1.051754229	1.043001412	1.028910120	1.012928124
2 I PQC		18.62593814	14.95765640	12.47507039	10.77827430	10.07596204	9.758847393	9.631361887	10.04007338
1 I ZF1		1.086364708	1.085900105	1.085683956	1.085670674	1.085759514	1.085867014	1.086076312	1.086349966
1 I ZF2		1.131308015	1.138543974	1.145847599	1.153498262	1.158509014	1.162189713	1.167482965	1.172944653
1 I ZF3		1.487073124	1.511278490	1.538687028	1.570430991	1.592735416	1.609778345	1.635096037	1.661973078

(3) P 5+2	TH = 69 RC = 02	SBE = 1.071144994 RC = 05	RC = 07	RC = 10	RC = 15	RC = 20	RC = 25	MCPER FUNCTIONS RC = 35	RC = 50
AMAX DB		.0108709564	.0213327376	.0436480540	.0988323393	.1772876696	.2802872360	.5675287486	1.249387366
AMIN DB		9.631346669	13.14039430	17.15438362	21.94111543	25.45392366	28.25637370	32.67638091	37.85942158
MU I K		.1266297D 1	.8367822D 0	.5327390D 0	.3143410D 0	.2138923D 0	.1573345D 0	.9684034D-1	.5469236D-1
1 I PFR		1.860492595	1.464973088	1.157265010	.8940790210	.7430846122	.6398471681	.5000274820	.3639772202
1 I PF1		1.237048483	1.168520201	1.089627706	.9994060882	.9381420633	.8931212772	.8300381816	.7695400024
1 I PQ1		1.833620642	1.617794410	1.492944654	1.444117578	1.461069905	1.505660752	1.641022064	1.940106815
2 I PFC		1.082783252	1.077688139	1.070028491	1.058642746	1.049076955	1.040955787	1.027757266	1.012643061
2 I PQC		21.17968328	16.77225887	13.79437536	11.75712734	10.90222287	10.50106694	10.28929499	10.65598365
1 I ZF1		1.078719896	1.078192132	1.077920634	1.077860221	1.077923574	1.078014693	1.078203468	1.078459222
1 I ZF2		1.118908185	1.125851557	1.132968862	1.140490554	1.145436983	1.149076588	1.154317311	1.159731685
1 I ZF3		1.456507860	1.479291463	1.505647252	1.536647191	1.558639452	1.575533004	1.600744521	1.627633103

(3) P 5+2 — TH = 70 SBE = 1.064177772 — MCPER FUNCTIONS

MU	K	RC = 02	RC = 05	RC = 07	RC = 10	RC = 15	RC = 20	RC = 25	RC = 35	RC = 50
	AMAX DB		.0108709564	.0213327376	.0436480540	.0988323393	.1772876696	.2802872360	.5675287486	1.249387366
	AMIN DB		8.606213630	11.99917156	15.94497986	20.69279685	24.19050953	26.98503660	31.39723620	36.57377432
MU	K		.1405638D 1	.9324652D 0	.5954843D 0	.3522849D 0	.2400746D 0	.1767794D 0	.1089670D 0	.6162996D-1
1	PFR		1.961418770	1.531441027	1.200804429	.9218460761	.7635954656	.6561390503	.5114925147	.3715960894
1	PF1		1.237627466	1.173300359	1.097170532	1.008385061	.9474260161	.9023955713	.8390638247	.7781542385
1	PQ1		1.973486263	1.716177846	1.564054135	1.497097092	1.506364075	1.547049127	1.679265596	1.978729151
2	PFC		1.076163972	1.071906032	1.065262121	1.055083434	1.046359727	1.038867779	1.026565470	1.012329424
2	PQC		24.26449973	18.94750874	15.35764411	12.90116911	11.85934344	11.35545830	11.04030551	11.35354625
1	ZF1		1.071511178	1.070923690	1.070596744	1.070488012	1.070524995	1.070599080	1.070766397	1.071003280
1	ZF2		1.107187206	1.113788614	1.120681576	1.128046725	1.132915303	1.136505276	1.141682475	1.147039145
1	ZF3		1.427176107	1.448422151	1.473601143	1.503741960	1.525357613	1.542060014	1.567113346	1.593969456

(3) P 5+2 — TH = 71 SBE = 1.057620681 — MCPER FUNCTIONS

MU	K	RC = 02	RC = 05	RC = 07	RC = 10	RC = 15	RC = 20	RC = 25	RC = 35	RC = 50
	AMAX DB		.0108709564	.0213327376	.0436480540	.0988323393	.1772876696	.2802872360	.5675287486	1.249387366
	AMIN DB		7.625579496	10.88153779	14.74425790	19.44452785	22.92411628	25.70930522	30.11190147	35.28164009
MU	K		.1558612D 1	.1038338D 1	.6653568D 0	.3947498D 0	.2694522D 0	.1986342D 0	.1226266D 0	.6946114D-1
1	PFR		2.074461746	1.605656747	1.249040989	.9522878897	.7859262529	.6737914754	.5238384360	.3797572617
1	PF1		1.236890551	1.177172420	1.104275421	1.017314462	.9568313108	.9118776564	.8483744945	.7870910975
1	PQ1		2.137652002	1.830836008	1.645966968	1.557264405	1.557329372	1.593320456	1.721660594	2.021223674
2	PFC		1.069700964	1.066200094	1.060512011	1.051501036	1.043605883	1.036738939	1.025334382	1.011985729
2	PQC		28.01702607	21.57722212	17.22661589	14.24979806	12.97707016	12.34658659	11.90370443	12.14881805
1	ZF1		1.064722331	1.064080665	1.063699153	1.063541307	1.063551159	1.063607622	1.063752609	1.063969639
1	ZF2		1.096144634	1.102354143	1.108980783	1.116158079	1.120933516	1.124464314	1.129565850	1.134853471
1	ZF3		1.399057519	1.418661436	1.442536421	1.471694771	1.492863747	1.509329315	1.534167275	1.560942175

(3) P 5+2 — TH = 72 SBE = 1.051462224 — MCPER FUNCTIONS

MU	K	RC = 02	RC = 05	RC = 07	RC = 10	RC = 15	RC = 20	RC = 25	RC = 35	RC = 50
	AMAX DB		.0108709564	.0213327376	.0436480540	.0988323393	.1772876696	.2802872360	.5675287486	1.249387366
	AMIN DB		6.695742626	9.792301447	13.55415486	18.19604906	21.65366971	24.42773681	28.81912227	33.98109824
MU	K		.1726374D 1	.1155407D 1	.7431715D 0	.4423085D 0	.3024488D 0	.2232264D 0	.1380340D 0	.7831425D-1
1	PFR		2.201257066	1.688796635	1.302713657	.9858109733	.8103406259	.6929933904	.5371799010	.3885268979
1	PF1		1.234762892	1.180004098	1.110828280	1.026130972	.9663254237	.9215529494	.8579739798	.7963655844
1	PQ1		2.331304027	1.965348268	1.741015877	1.626077637	1.615054297	1.645371522	1.768921947	2.068216308
2	PFC		1.063418224	1.060590457	1.055791862	1.047902956	1.040819462	1.034571289	1.024063842	1.011610473
2	PQC		32.61484270	24.78471647	19.48275996	15.85468474	14.29411616	13.50621395	12.90421851	13.06215453
1	ZF1		1.058337239	1.057649357	1.057215438	1.057008220	1.056990346	1.057028678	1.057150545	1.057346942
1	ZF2		1.085779581	1.091549257	1.097864235	1.104818327	1.109483311	1.112944226	1.117956632	1.123162761
1	ZF3		1.372126027	1.390001002	1.412444421	1.440488774	1.461134869	1.477313567	1.501873147	1.528512745

(3) P 5+2 — TH = 73 — SBE = 1.045691756 — MCPER FUNCTIONS

MU		RC = 02	RC = 05	RC = 07	RC = 10	RC = 15	RC = 20	RC = 25	RC = 35	RC = 50
	AMAX DB		.0108709564	.0213327376	.0436480540	.0988323393	.1772876696	.2802872360	.5675287486	1.249387366
	AMIN DB		5.822925053	8.737194433	12.37740633	16.94739857	20.37814990	23.13882630	27.51698320	32.67001753
MU	K		.1910202D 1	.1284769D 1	.8298377D 0	.4956164D 0	.3395540D 0	.2509374D 0	.1554404D 0	.8834029D-1
1	PFR		2.343627394	1.782218864	1.362700899	1.022904664	.8371559979	.7139717234	.5516538702	.3979835373
1	PF1		1.231198292	1.181665654	1.116701073	1.034757651	.9758665592	.9314009171	.8678640218	.8059933092
1	PQ1		2.560888192	2.124233439	1.852174397	1.705391492	1.680914450	1.704328730	1.821939120	2.120478542
2	PFC		1.057339617	1.055098847	1.051117128	1.044297825	1.038005343	1.032367422	1.022753936	1.011202144
2	PQC		38.29103979	28.73366189	22.23487236	17.78452994	15.86157689	14.87598608	14.07394270	14.11970800
1	ZF1		1.052339878	1.051616318	1.051133766	1.050877636	1.050831663	1.050851464	1.050949528	1.051124430
1	ZF2		1.076089516	1.081376572	1.087332355	1.094023701	1.098558632	1.101937611	1.106845882	1.111956816
1	ZF3		1.346347579	1.362431620	1.383319689	1.410110880	1.430151248	1.445988196	1.470199819	1.496643913

(3) P 5+2 — TH = 74 — SBE = 1.040299436 — MCPER FUNCTIONS

MU		RC = 02	RC = 05	RC = 07	RC = 10	RC = 15	RC = 20	RC = 25	RC = 35	RC = 50
	AMAX DB			.0213327376	.0436480540	.0988323393	.1772876696	.2802872360	.5675287486	1.249387366
	AMIN DB			7.722779198	11.21771139	15.69903040	19.09663984	21.84101503	26.20338054	31.34601031
MU	K			.1427630D 1	.9263708D 0	.5554250D 0	.3813370D 0	.2822129D 0	.1751421D 0	.9971817D-1
1	PFR			1.887485809	1.430047543	1.064161289	.8667574116	.7370013226	.5674254238	.4032212667
1	PF1			1.182036940	1.121752932	1.043101665	.9854014291	.9413933822	.8780434854	.8159903535
1	PQ1			2.313234789	1.983267571	1.797592863	1.756670032	1.771624793	1.881832526	2.178972648
2	PFC			1.049748340	1.046505172	1.040695715	1.035169404	1.030130627	1.021405060	1.010759243
2	PQC			33.64367339	25.63012152	20.13205336	17.74794178	16.51135899	15.45517019	15.35557066
1	ZF1			1.045968222	1.045442762	1.045139140	1.045064986	1.045065992	1.045139705	1.045292394
1	ZF2			1.071839362	1.077388093	1.083773088	1.088155822	1.091439269	1.096226620	1.101227202
1	ZF3			1.335940981	1.355159287	1.380551840	1.399896546	1.415331472	1.439118125	1.465299510

(3) P 5+2 — TH = 75 — SBE = 1.035276180 — MCPER FUNCTIONS

MU		RC = 02	RC = 05	RC = 07	RC = 10	RC = 15	RC = 20	RC = 25	RC = 35	RC = 50
	AMAX DB			.0213327376	.0436480540	.0988323393	.1772876696	.2802872360	.5675287486	1.249387366
	AMIN DB			6.756222694	10.07988805	14.45197171	17.80839860	20.53271036	24.87599112	30.00637486
MU	K			.1585337D 1	.1033908D 1	.6225984D 0	.4284618D 0	.3175776D 0	.1974905D 0	.1126614D 0
1	PFR			2.006390336	1.505996189	1.110301774	.8996156839	.7624181082	.5846955069	.4193539396
1	PF1			1.181015501	1.125832801	1.051051952	.9948625524	.9514923470	.8885072300	.8263730424
1	PQ1			2.539712047	2.139264239	1.905792719	1.844604399	1.849107587	1.950033059	2.244914924
2	PFC			1.044562943	1.041975377	1.037108372	1.032318725	1.027865045	1.020018012	1.010280312
2	PQC			39.81322032	29.87039304	23.02445596	20.04660000	18.48746036	17.10458237	16.81491504
1	ZF1			1.040691781	1.040131378	1.039782922	1.039680891	1.039663017	1.039712007	1.039841881
1	ZF2			1.062940483	1.068036619	1.074068156	1.078273802	1.081446369	1.086093963	1.090967353
1	ZF3			1.310511149	1.327961688	1.351806262	1.370357995	1.385324653	1.408600873	1.434444265

$P^{(4)}_{5+4}$

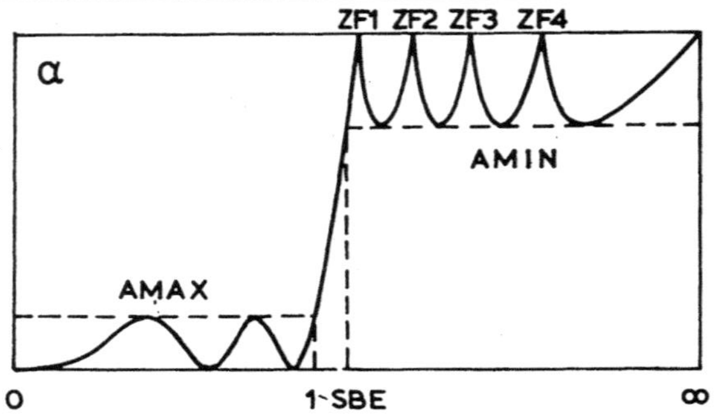

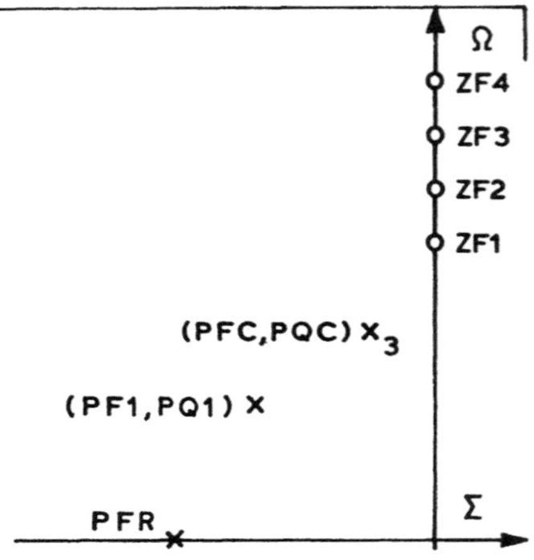

TH = 22 SBE = 2.669467163 MCPER FUNCTIONS

$P^{(4)}_{5+4}$		RC = 02	RC = 05	RC = 07	RC = 10	RC = 15	RC = 20	RC = 25	RC = 35	RC = 50
AMAX DB		.0017375255	.0108709564	.0213327376	.0436480540	.0988323393	.1772876696	.2802872360	.5675287486	1.249387366
AMIN DB		92.64057777	107.0301842	112.3446532	118.0105999	124.5209306	129.2189717	132.9400584	138.7753647	145.5630492
MU	K	.5038352D-3	.9830528D-4	.5365826D-4	.2811395D-4	.1336462D-4	.7810147D-5	.5102144D-5	.2615724D-5	.1201722D-5
1	PFR	.9794693192	.7216049930	.6368068807	.5516272070	.4598959481	.3976144895	.3506209364	.2812988001	.2081387108
1	PF1	1.141179318	.9171019851	.8458530549	.7762384818	.7041100886	.6572204583	.6231549234	.5752552864	.5281344497
1	PQ1	.6744038774	.7163370499	.7418334036	.7782978779	.8372819046	.8965040471	.9576680524	1.090907288	1.339131103
3	PFC	1.651642254	1.440617756	1.370795204	1.301388810	1.228141697	1.179647562	1.143862416	1.092501187	1.040317444
3	PQC	1.769765371	1.798397830	1.832839358	1.886109421	1.972460752	2.055348974	2.136131001	2.296674258	2.552611822
1	ZF1	2.703228044	2.704502958	2.704892619	2.705265367	2.705644529	2.705888731	2.706066187	2.706318197	2.706573511
1	ZF2	3.010344681	3.023424893	3.027332688	3.031037842	3.034776608	3.037169764	3.038902019	3.041352770	3.043825297
1	ZF3	3.915758807	3.961476624	3.974843291	3.987399818	3.999958570	4.007940218	4.013690558	4.021788187	4.029914467
1	ZF4	7.114398331	7.252396254	7.292359279	7.329718231	7.366895609	7.390420651	7.407317885	7.431038153	7.454753902

TH = 23 SBE = 2.559304665 MCPER FUNCTIONS

$P^{(4)}_{5+4}$		RC = 02	RC = 05	RC = 07	RC = 10	RC = 15	RC = 20	RC = 25	RC = 35	RC = 50
AMAX DB		.0017375255	.0108709564	.0213327376	.0436480540	.0988323393	.1772876696	.2802872360	.5675287486	1.249387366
AMIN DB		89.08055701	103.4655402	108.7790040	114.4441435	120.9537970	125.6514742	129.3723302	135.2073560	141.9948105
MU	K	.7202278D-3	.1409177D-3	.7697563D-4	.4035860D-4	.1919809D-4	.1122374D-4	.7334253D-5	.3761552D-5	.1728810D-5
1	PFR	.9863289067	.7255320252	.6400104948	.5542025166	.4618899460	.3992620556	.3520292969	.2823840538	.2089646671
1	PF1	1.146104226	.9204390168	.8487723068	.7787877164	.7063141758	.6592204258	.6250165021	.5769366060	.5296558150
1	PQ1	.6771985785	.7185149409	.7438737426	.7802366058	.8391602535	.8983790576	.9595680818	1.092908028	1.341382986
3	PFC	1.643814734	1.436711414	1.367861819	1.299287698	1.226786959	1.178713846	1.143215777	1.092214444	1.040345538
3	PQC	1.807739463	1.826182154	1.858140751	1.909329300	1.993909378	2.075892369	2.156170290	2.316307538	2.572415908
1	ZF1	2.591094040	2.592392141	2.592791583	2.593174725	2.593565435	2.593817565	2.594001013	2.594261852	2.594526476
1	ZF2	2.880253146	2.893714115	2.897749377	2.901581077	2.905453072	2.907934389	2.909731844	2.912276804	2.914846676
1	ZF3	3.733967934	3.781482201	3.795391694	3.808465471	3.821548773	3.829867581	3.835862669	3.844307599	3.852785569
1	ZF4	6.760428640	6.904477718	6.946239696	6.985294774	7.024171497	7.048776730	7.066451790	7.091266346	7.116078611

(4) P 5+4		TH = 24	SBE = 2.458593336							MCPER FUNCTIONS
		RC = 02	RC = 05	RC = 07	RC = 10	RC = 15	RC = 20	RC = 25	RC = 35	RC = 50
AMAX DB		.0017375255	.0108709564	.0213327376	.0436480540	.0988323393	.1772876696	.2802872360	.5675287486	1.249387366
AMIN DB		85.66512763	100.0447949	105.3571036	111.0213159	117.5301917	122.2274508	125.9480420	131.7827457	138.5699362
MU	K	.1014043D-2	.1989963D-3	.1087886D-3	.5708001D-4	.2717140D-4	.1589205D-4	.1038798D-4	.5329968D-5	.2450659D-5
1	PFR	.9936205077	.7296876227	.6433965101	.5569215564	.4639931369	.4009988235	.3535133190	.2835270227	.2097815191
1	PF1	1.151278368	.9239465968	.8518407627	.7814671152	.7086305922	.6613221569	.6269726797	.5787031973	.5312541900
1	PQ1	.6801945076	.7208328935	.7460410453	.7822923464	.8411486480	.9003619345	.9615760432	1.095020518	1.343758714
3	PFC	1.635629314	1.432616801	1.364784800	1.297081920	1.225363120	1.177741597	1.142534327	1.091910531	1.040372014
3	PQC	1.848856877	1.856014417	1.885242552	1.934147043	2.016784995	2.097773512	2.177494570	2.337173067	2.593437096
1	ZF1	2.488565704	2.489882415	2.490290579	2.490683239	2.491084730	2.491344359	2.491533517	2.491802828	2.492076449
1	ZF2	2.761124126	2.774939641	2.779095992	2.783048836	2.787049257	2.789616004	2.791476853	2.794113711	2.796778885
1	ZF3	3.566929734	3.616203589	3.630647282	3.644231156	3.657832733	3.666485194	3.672722739	3.681512090	3.690339271
1	ZF4	6.434096582	6.584120493	6.627670864	6.668415928	6.708988948	6.734673788	6.753126811	6.779036528	6.804946878

(4) P 5+4		TH = 25	SBE = 2.366201583							MCPER FUNCTIONS
		RC = 02	RC = 05	RC = 07	RC = 10	RC = 15	RC = 20	RC = 25	RC = 35	RC = 50
AMAX DB		.0017375255	.0108709564	.0213327376	.0436480540	.0988323393	.1772876696	.2802872360	.5675287486	1.249387366
AMIN DB		82.38228282	96.75586701	102.0668551	107.7300075	114.2379945	118.9347761	122.6550647	128.4894004	135.2762892
MU	K	.1407921D-2	.2771721D-3	.1516570D-3	.7963457D-4	.3793634D-4	.2219850D-4	.1451495D-4	.7450796D-5	.3427279D-5
1	PFR	1.001365102	.7340802243	.6469710830	.5597887680	.4662086018	.4028271675	.3550749346	.2847290986	.2106402018
1	PF1	1.156704996	.9276276448	.8550610489	.7842789972	.7110613394	.6635274560	.6290251239	.5805565483	.5329309032
1	PQ1	.6834050503	.7232979786	.7483411381	.7844699745	.8432512213	.9024564594	.9636955335	1.097248225	1.346261868
3	PFC	1.627087000	1.428332545	1.361562702	1.294770134	1.223869058	1.176714867	1.141817262	1.091588844	1.040396470
3	PQC	1.893380094	1.888032877	1.914257737	1.960655729	2.041164294	2.121060479	2.200166608	2.359327531	2.615727983
1	ZF1	2.394494356	2.395824921	2.396240707	2.396641968	2.397053440	2.397320120	2.397514693	2.397792100	2.398074388
1	ZF2	2.651652477	2.665795710	2.670066570	2.674134964	2.678258819	2.680908154	2.682830512	2.685556852	2.688315189
1	ZF3	3.412886136	3.463881017	3.478849912	3.492936358	3.507049630	3.516032057	3.522509651	3.531640386	3.540814164
1	ZF4	6.132108744	6.288017901	6.333343826	6.375771152	6.418036313	6.444799685	6.464030540	6.491036034	6.518045896

(4) P 5+4		TH = 26	SBE = 2.281172033							MCPER FUNCTIONS
		RC = 02	RC = 05	RC = 07	RC = 10	RC = 15	RC = 20	RC = 25	RC = 35	RC = 50
AMAX DB		.0017375255	.0108709564	.0213327376	.0436480540	.0988323393	.1772876696	.2802872360	.5675287486	1.249387366
AMIN DB		79.22144436	93.58809808	98.89758294	104.5595293	111.0665050	115.7627435	119.4826879	125.3166053	132.1031511
MU	K	.1929760D-2	.3811977D-3	.2087671D-3	.1097140D-3	.5230727D-4	.3062264D-4	.2003014D-4	.1028670D-4	.4733937D-5
1	PFR	1.009585725	.7387189790	.6507408643	.5628089352	.4685396495	.4047496331	.3567162122	.2859917700	.2115417139
1	PF1	1.162387264	.9314851965	.8584359168	.7872258036	.7136085321	.6658382320	.6311756008	.5824982359	.5346873627
1	PQ1	.6868450659	.7259179361	.7507803741	.7867747865	.8454724491	.9046667184	.9659304333	1.099594881	1.348896294
3	PFC	1.618189341	1.423857338	1.358194087	1.292350972	1.222303608	1.175637671	1.141063737	1.091248750	1.040418483
3	PQC	1.941602768	1.922390170	1.945310554	1.988957361	2.067131026	2.145827448	2.224254687	2.382832468	2.639345549
1	ZF1	2.307908211	2.309247703	2.309669967	2.310078876	2.310499495	2.310772756	2.310972438	2.311257549	2.311548160
1	ZF2	2.550734132	2.565177709	2.569556304	2.573734469	2.577976586	2.580705554	2.582687459	2.585500761	2.588350025
1	ZF3	3.270350238	3.323025789	3.338510466	3.353091616	3.367709683	3.377018210	3.383733320	3.393202244	3.402719864
1	ZF4	5.851682935	6.013373323	6.060459449	6.104559553	6.148511448	6.176351717	6.196359975	6.224461560	6.252572184

(4) P 5+4	TH = 27	SBE = 2.202689265						MCPER FUNCTIONS	
	RC = 02	RC = 05	RC = 07	RC = 10	RC = 15	RC = 20	RC = 25	RC = 35	RC = 50
AMAX DB	.0017375255	.0108709564	.0213327376	.0436480540	.0988323393	.1772876696	.2802872360	.5675287486	1.249387366
AMIN DB	76.17324849	90.53203752	95.83981821	101.5003978	108.0062279	112.7018511	116.4214058	122.2548493	129.0410067
MU I K	.2613638D-2	.5181609D-3	.2840541D-3	.1494115D-3	.7129381D-4	.4175974D-4	.2732482D-4	.1403998D-4	.6464331D-5
1 I PFR	1.018307693	.7436138083	.6547130406	.5659872101	.4709898324	.4067689483	.3584393662	.2873166274	.2124871207
1 I PF1	1.168328177	.9355223980	.8619682430	.7903101008	.7162744014	.6682565026	.6334259784	.5845299295	.5365250608
1 I PQ1	.6905310705	.7287012444	.7533656837	.7892125385	.8478171774	.9069971261	.9682849294	1.102064504	1.351666128
3 I PFC	1.608938502	1.419189958	1.354677534	1.289823046	1.220665571	1.174508986	1.140272874	1.090889585	1.040437608
3 I PQC	1.993853898	1.959254962	1.978537755	2.019163789	2.094776693	2.172155276	2.249833120	2.407754695	2.664351529
1 I ZF1	2.227979623	2.229322952	2.229750504	2.230166072	2.230594970	2.230874325	2.231078793	2.231371200	2.231669773
1 I ZF2	2.457428843	2.472144922	2.476624295	2.480906276	2.485261304	2.488066838	2.490106251	2.493003890	2.495941747
1 I ZF3	3.138056150	3.192370106	3.208360715	3.223428336	3.238543972	3.248174541	3.255124510	3.264928265	3.274786823
1 I ZF4	5.590453912	5.757805915	5.806634128	5.852395575	5.898027407	5.926942311	5.947727195	5.976924836	6.006137266

(4) P 5+4	TH = 28	SBE = 2.130054468						MCPER FUNCTIONS	
	RC = 02	RC = 05	RC = 07	RC = 10	RC = 15	RC = 20	RC = 25	RC = 35	RC = 50
AMAX DB	.0017375255	.0108709564	.0213327376	.0436480540	.0988323393	.1772876696	.2802872360	.5675287486	1.249387366
AMIN DB	73.22937098	87.57926721	92.88512310	98.54415948	105.0486966	109.7436255	113.4627405	119.2956494	126.0813687
MU I K	.3500843D-2	.6967292D-3	.3823415D-3	.2012989D-3	.9613892D-4	.5634351D-4	.3688170D-4	.1896049D-4	.8734314D-5
1 I PFR	1.027558859	.7487754768	.6588953791	.5693291421	.4735629648	.4088880364	.3602467662	.2887053702	.2134775591
1 I PF1	1.174530518	.9397424990	.8656610280	.7935345825	.7190612988	.6707843980	.6357782310	.5866533956	.5384455770
1 I PQ1	.6944814475	.7316571972	.7561046312	.7917894885	.8502906556	.9094524537	.9707635391	1.104661421	1.354575813
3 I PFC	1.599337329	1.414329291	1.351011654	1.287184955	1.218953715	1.173327750	1.139443754	1.090510652	1.040453373
3 I PQC	2.050502660	1.998813828	2.014089983	2.051397751	2.124201319	2.200132142	2.276982811	2.434166789	2.690812827
1 I ZF1	2.153999348	2.155341262	2.155772875	2.156194075	2.156630351	2.156915292	2.157124212	2.157423488	2.157729647
1 I ZF2	2.370930913	2.385891259	2.390464286	2.394843956	2.399306370	2.402185296	2.404280098	2.407259346	2.410283362
1 I ZF3	3.014919597	3.070827646	3.087313883	3.102859367	3.118465006	3.128413364	3.135595400	3.145730458	3.155926898
1 I ZF4	5.346399308	5.519276537	5.569825707	5.617234920	5.664538345	5.694524915	5.716085260	5.746378525	5.776693556

(4) P 5+4	TH = 29	SBE = 2.062665340						MCPER FUNCTIONS	
	RC = 02	RC = 05	RC = 07	RC = 10	RC = 15	RC = 20	RC = 25	RC = 35	RC = 50
AMAX DB	.0017375255	.0108709564	.0213327376	.0436480540	.0988323393	.1772876696	.2802872360	.5675287486	1.249387366
AMIN DB	70.38238304	84.72225684	90.02594605	95.68324608	102.1863289	106.8804771	110.5990976	116.4314056	123.2166325
MU I K	.4641016D-2	.9274283D-3	.5095028D-3	.2685147D-3	.1283628D-3	.7527250D-4	.4929245D-4	.2535486D-4	.1168524D-4
1 I PFR	1.037369911	.7542156710	.6632962785	.5728407098	.4762631418	.4111100300	.3621409481	.2901598136	.2145142413
1 I PF1	1.180996763	.9441488429	.8695173946	.7969020718	.7219717002	.6734241654	.6382344436	.5838705016	.5404505825
1 I PQ1	.6987166910	.7347959921	.7590054795	.7945124446	.8528985714	.9120378586	.9733711377	1.107390291	1.357630125
3 I PFC	1.589389415	1.409274351	1.347195099	1.284435291	1.217166775	1.172092865	1.138575424	1.090111220	1.040465282
3 I PQC	2.111964030	2.041273401	2.052133361	2.085794054	2.155514331	2.229854276	2.305791893	2.462147621	2.718801976
1 I ZF1	2.085356131	2.086691227	2.087125630	2.087551399	2.087994118	2.088284119	2.088497143	2.088802344	2.089116197
1 I ZF2	2.290545977	2.305722043	2.310381443	2.314852514	2.319416620	2.322365655	2.324513654	2.327571676	2.330679316
1 I ZF3	2.900006684	2.957462301	2.974433388	2.990447727	3.006535449	3.016797138	3.024208317	3.034670970	3.045202079
1 I ZF4	5.117780890	5.296029013	5.348274730	5.397315782	5.446280753	5.477335235	5.499669439	5.531057442	5.562475588

(4) P 5+4 — TH = 30 — SBE = 2.000000000 — MCPER FUNCTIONS

MU	P 5+4	RC = 02	RC = 05	RC = 07	RC = 10	RC = 15	RC = 20	RC = 25	RC = 35	RC = 50
	AMAX DB	.0017375255	.0108709564	.0213327376	.0436480540	.0988323393	.1772876696	.2802872360	.5675287486	1.249387366
	AMIN DB	67.62563203	81.95424382	87.25550159	92.91085401	99.41230616	104.1055793	107.8236457	113.6552803	120.4399554
MU	K	.6093426D-2	.1222958D-2	.6726442D-3	.3548657D-3	.1698135D-3	.9964067D-4	.6527813D-4	.3359727D-4	.1549405D-4
1	PFR	1.047774706	.7599470879	.6679248250	.5765283571	.4790947616	.4134382861	.3641246255	.2916818966	.2155984602
1	PF1	1.187728986	.9487448558	.8735405849	.8004155229	.7250082096	.6761781737	.6407968167	.5911832207	.5425418446
1	PQ1	.7032596864	.7381288303	.7620772624	.7973888179	.8556470916	.9147589185	.9761129886	1.110256134	1.360834197
3	PFC	1.579099179	1.404024302	1.343226575	1.281572646	1.215303458	1.170803198	1.137666893	1.039690528	1.040472814
3	PQC	2.178705329	2.086862831	2.092851292	2.122500911	2.188835542	2.261426769	2.336356443	2.491782950	2.748397653
1	ZF1	2.021520374	2.022843101	2.023278984	2.023708226	2.024156421	2.024450933	2.024667701	2.024979367	2.025299506
1	ZF2	2.215672427	2.231035440	2.235773788	2.240329820	2.244989762	2.248005517	2.250204443	2.253338298	2.256526929
1	ZF3	2.792508916	2.851463170	2.868907824	2.885381592	2.901943107	2.912513460	2.920150712	2.930937071	2.941799467
1	ZF4	4.903097557	5.086543151	5.140457455	5.191111857	5.241726450	5.273844221	5.296950194	5.329431543	5.361952986

(4) P 5+4 — TH = 31 — SBE = 1.941604026 — MCPER FUNCTIONS

MU	P 5+4	RC = 02	RC = 05	RC = 07	RC = 10	RC = 15	RC = 20	RC = 25	RC = 35	RC = 50
	AMAX DB	.0017375255	.0108709564	.0213327376	.0436480540	.0988323393	.1772876696	.2802872360	.5675287486	1.249387366
	AMIN DB	64.95314183	79.26913300	84.56766987	90.22084394	96.72047304	101.4127681	105.1302154	110.9610977	117.7451564
MU	K	.7928348D-2	.1598550D-2	.8803134D-3	.4649430D-3	.2227257D-3	.1307726D-3	.8571269D-4	.4414188D-4	.2036910D-4
1	PFR	1.058810656	.7659835351	.6727908562	.5803990329	.4820625493	.4158764034	.3662007031	.2932736905	.2167315946
1	PF1	1.194728740	.9535340321	.8777339568	.8040780228	.7281735634	.6790489181	.6434676712	.5935936368	.5447212312
1	PQ1	.7081360364	.7416680301	.7653298663	.8004266836	.8585429065	.9176216691	.9789947772	1.113264358	1.364193550
3	PFC	1.568471933	1.398578488	1.339104863	1.278595620	1.213362445	1.169457580	1.136717136	1.089247779	1.040475419
3	PQC	2.251253874	2.135836614	2.136446525	2.161681463	2.224296273	2.294964498	2.368781282	2.523166101	2.779685254
1	ZF1	1.962030966	1.963335633	1.963771650	1.964203232	1.964655903	1.964954363	1.965174501	1.965491652	1.965818152
1	ZF2	2.145786439	2.161307481	2.166117224	2.170751634	2.175501395	2.178580376	2.180827884	2.184034529	2.187301414
1	ZF3	2.691723068	2.752124398	2.770030797	2.786954128	2.803980758	2.814854883	2.822714991	2.833820972	2.845011102
1	ZF4	4.701047427	4.889496858	4.945047959	4.997294442	5.049544683	5.082720158	5.106595268	5.140167998	5.173792552

(4) P 5+4 — TH = 32 — SBE = 1.887079915 — MCPER FUNCTIONS

MU	P 5+4	RC = 02	RC = 05	RC = 07	RC = 10	RC = 15	RC = 20	RC = 25	RC = 35	RC = 50
	AMAX DB	.0017375255	.0108709564	.0213327376	.0436480540	.0988323393	.1772876696	.2802872360	.5675287486	1.249387366
	AMIN DB	62.35952944	76.66141230	81.95691220	87.60765620	94.10525259	98.79645754	102.5132148	108.3432588	115.1266310
MU	K	.1022859D-1	.2072367D-2	.1142736D-2	.6042541D-3	.2897868D-3	.1702640D-3	.1116501D-3	.5753722D-4	.2656707D-4
1	PFR	1.070519171	.7723400432	.6779050313	.5844602353	.4851715837	.4184282408	.3683722907	.2949374086	.2179151150
1	PF1	1.201996917	.9585199164	.8821009781	.8078927921	.7314706346	.6820390254	.6462494538	.5961039500	.5469907160
1	PQ1	.7133744380	.7454271564	.7687741234	.8036348483	.8615932801	.9206326459	.9820226483	1.116420794	1.367714120
3	PFC	1.557513960	1.392936453	1.334828829	1.275502831	1.211342395	1.168054808	1.135725089	1.088782141	1.040472520
3	PQC	2.330205928	2.188477843	2.183143516	2.203515508	2.262040612	2.330593161	2.403180883	2.556398708	2.812757539
1	ZF1	1.906484587	1.907765371	1.908200136	1.908632894	1.909089010	1.909390831	1.909613952	1.909936094	1.910268514
1	ZF2	2.080429794	2.096079891	2.100953364	2.105659433	2.110492847	2.113631460	2.115925130	2.119201414	2.122543713
1	ZF3	2.597034841	2.658828799	2.677184543	2.694547102	2.712029759	2.723202532	2.731282123	2.742703440	2.754217568
1	ZF4	4.510497025	4.703735371	4.760887366	4.814701649	4.868571323	4.902797865	4.927438883	4.962100394	4.996827447

TH = 33 SBE = 1.836078459 MCPER FUNCTIONS

(4) P 5+4	RC = 02	RC = 05	RC = 07	RC = 10	RC = 15	RC = 20	RC = 25	RC = 35	RC = 50
AMAX DB	.0017375255	.0108709564	.0213327376	.0436480540	.0988323393	.1772876696	.2802872360	.5675287486	1.249387366
AMIN DB	59.83993464	74.12608152	79.41819960	85.06623922	91.56157462	96.25156747	99.96755766	105.7966701	112.5792793
MU K	.1309113D-1	.2665953D-2	.1472082D-2	.7793737D-3	.3742141D-3	.2200282D-3	.1443558D-3	.7444280D-4	.3439579D-4
1 PFR	1.082946172	.7790329932	.6832789115	.5887200614	.4884273257	.4210979386	.3706427193	.2966754164	.2191505900
1 PF1	1.209533593	.9637060821	.8866452193	.8118631854	.7349024368	.6851512591	.6491447425	.5987164823	.5493523845
1 PQ1	.7190071224	.7494211684	.7724219170	.8070229267	.8648061064	.9237989313	.9852032480	1.119731732	1.371402298
3 PFC	1.546232587	1.387097977	1.330397447	1.272292923	1.209241950	1.166593651	1.134689652	1.098292749	1.040463512
3 PQC	2.416237210	2.245101959	2.233191130	2.248201481	2.302226858	2.368450456	2.439680387	2.591591571	2.847715348
1 ZF1	1.854526948	1.855777898	1.856209992	1.856642729	1.857101227	1.857405807	1.857631511	1.857958128	1.858296013
1 ZF2	2.019199929	2.034950136	2.039879576	2.044650464	2.049561218	2.052755767	2.055093105	2.058435774	2.061850543
1 ZF3	2.507905499	2.571034462	2.589826527	2.607617478	2.625546640	2.637012690	2.645308228	2.657040383	2.668874582
1 ZF4	4.330456050	4.528246083	4.586958671	4.642313226	4.697783685	4.733053500	4.758456535	4.794203522	4.830031990

TH = 34 SBE = 1.788291650 MCPER FUNCTIONS

(4) P 5+4	RC = 02	RC = 05	RC = 07	RC = 10	RC = 15	RC = 20	RC = 25	RC = 35	RC = 50
AMAX DB	.0017375255	.0108709564	.0213327376	.0436480540	.0988323393	.1772876696	.2802872360	.5675287486	1.249387366
AMIN DB	57.38996047	71.65859180	76.94695209	32.59198865	89.08481474	93.77346295	97.48860243	103.3166818	110.0984453
MU K	.1662889D-1	.3404740D-2	.1882767D-2	.9981139D-3	.4798418D-3	.2823495D-3	.1853421D-3	.9564826D-4	.4422443D-4
1 PFR	1.096142672	.7860802607	.6889250492	.5931872628	.4918356565	.4238899410	.3730155584	.2984902430	.2204396938
1 PF1	1.217337828	.9690961045	.8913703440	.8159926909	.7384721286	.6883885252	.6521562530	.6014336834	.5518084391
1 PQ1	.7250703663	.7536665885	.7762863009	.8106014284	.8681899726	.9271282072	.9885437706	1.123203963	1.375264967
3 PFC	1.534636260	1.381063101	1.325809812	1.268964577	1.207059738	1.165072846	1.133609694	1.097778700	1.040447759
3 PQC	2.510115267	2.306061084	2.286865758	2.295958711	2.345029150	2.408687412	2.478416763	2.628865606	2.884668421
1 ZF1	1.805845589	1.807060631	1.807488601	1.807920088	1.808379876	1.808686592	1.808914466	1.809245029	1.809587905
1 ZF2	1.961741733	1.977563221	1.982540784	1.987369539	1.992351182	1.995597875	1.997976313	2.001382009	2.004866200
1 ZF3	2.423860878	2.488263722	2.507478411	2.525686381	2.544052064	2.555805762	2.564313537	2.576351809	2.588501950
1 ZF4	4.160056584	4.362137821	4.422366010	4.479229817	4.536279777	4.572583810	4.598744250	4.635572626	4.672500894

TH = 35 SBE = 1.743446796 MCPER FUNCTIONS

(4) P 5+4	RC = 02	RC = 05	RC = 07	RC = 10	RC = 15	RC = 20	RC = 25	RC = 35	RC = 50
AMAX DB	.0017375255	.0108709564	.0213327376	.0436480540	.0988323393	.1772376696	.2802872360	.5675287486	1.249387366
AMIN DB	55.00562277	69.25479394	74.53898684	80.18069531	86.67074222	91.35790184	95.07209979	100.8990361	107.6798640
MU K	.2097265D-1	.4318601D-2	.2391790D-2	.1269718D-2	.6112205D-3	.3599444D-3	.2364098D-3	.1220954D-3	.5649399D-4
1 PFR	1.110165460	.7935013786	.6948570889	.5978713075	.4954029042	.4268090222	.3754946355	.3003345944	.2217842134
1 PF1	1.225407447	.9746935291	.8962800971	.8202849292	.7421830180	.6917538776	.6552868446	.6042581380	.5543612064
1 PQ1	.7316050905	.7581816960	.7803816362	.8143818556	.8717542305	.9306288142	.9920520105	1.126844826	1.379309543
3 PFC	1.522734618	1.374832163	1.321065166	1.265516523	1.204794380	1.163491101	1.132484044	1.087239056	1.040424593
3 PQC	2.612714072	2.371749050	2.344474901	2.347030012	2.390639331	2.451469900	2.519540113	2.668352920	2.923736303
1 ZF1	1.760163899	1.761336844	1.761759204	1.762188183	1.762648139	1.762956351	1.763185970	1.763519930	1.763867306
1 ZF2	1.907740752	1.923604899	1.928622675	1.933502244	1.938543195	1.941843144	1.944260044	1.947725302	1.951275762
1 ZF3	2.344482289	2.410094035	2.429716918	2.448329956	2.467121685	2.479157123	2.487873243	2.500212674	2.512674416
1 ZF4	3.998535835	4.204623680	4.266317495	4.324655796	4.383261122	4.420588947	4.447501384	4.495406205	4.523432067

(4) P 5+4	TH = 36	SBE = 1.701301617						MCPER FUNCTIONS	
	RC = 02	RC = 05	RC = 07	RC = 10	RC = 15	RC = 20	RC = 25	RC = 35	RC = 50
AMAX DB	.0017375255	.0108709564	.0213327376	.0436480540	.0988323393	.1772876696	.2802872360	.5675287486	1.249387366
AMIN DB	52.68330714	66.91089414	72.19047362	77.82850053	84.31547509	89.00098990	92.71414783	98.53982184	105.3196167
MU I K	.2627303D-1	.5442478D-2	.3019113D-2	.1605077D-2	.7737303D-3	.4560305D-3	.2996952D-3	.1549035D-3	.7172950D-4
1 I PFR	1.125077872	.8013177225	.7010898817	.6027824501	.4991358949	.4298603145	.3780840577	.3023613667	.2231860579
1 I PF1	1.233738777	.9805018319	.9013782892	.8247436506	.7460385656	.6952505245	.6585395273	.6071925725	.5570131434
1 I PQ1	.7386575593	.7629867492	.7847237467	.8183768149	.8755090767	.9343098178	.9957364217	1.130662254	1.383544026
3 I PFC	1.510538567	1.368405831	1.316162918	1.261947553	1.202444491	1.161847102	1.131311503	1.086672844	1.040393316
3 I PQC	2.725031309	2.442607235	2.406361317	2.401684663	2.439269079	2.496980360	2.563215159	2.710198040	2.965049375
1 I ZF1	1.717236145	1.718360689	1.718775923	1.719201108	1.719660082	1.719969132	1.720200057	1.720536848	1.720888219
1 I ZF2	1.856917534	1.872796005	1.877846039	1.882769280	1.887872834	1.891212059	1.893664712	1.897185964	1.900799432
1 I ZF3	2.269398948	2.336150372	2.356166218	2.375171750	2.394378520	2.406689500	2.415609878	2.428245260	2.441014040
1 I ZF4	3.845221730	4.055006722	4.118110916	4.177884956	4.238018444	4.276358142	4.304016306	4.342991685	4.382112278

(4) P 5+4	TH = 37	SBE = 1.661640141						MCPER FUNCTIONS	
	RC = 02	RC = 05	RC = 07	RC = 10	RC = 15	RC = 20	RC = 25	RC = 35	RC = 50
AMAX DB	.0017375255	.0108709564	.0213327376	.0436480540	.0988323393	.1772876696	.2802872360	.5675287486	1.249387366
AMIN DB	50.41973233	64.62341594	69.89789648	75.53185756	82.01544151	86.69914207	90.41115319	96.23543569	103.0140922
MU I K	.3270269D-1	.6817065D-2	.3788081D-2	.2016977D-2	.9737092D-3	.5744064D-3	.3777236D-3	.1953983D-3	.9055398D-4
1 I PFR	1.140950707	.8095527220	.7076396136	.6079318099	.5030419936	.4330493401	.3807882353	.3044236624	.2246472674
1 I PF1	1.242326346	.9865243729	.9066687775	.8293727309	.7500423891	.6988818344	.6619174688	.6102398624	.5597668458
1 I PQ1	.7462802020	.7681042416	.7893300977	.8226001436	.8794656437	.9381810834	.9996061831	1.134664838	1.387977052
3 I PFC	1.498060341	1.361785137	1.311102667	1.258256533	1.200008694	1.160139511	1.130090836	1.086079050	1.040353194
3 I PQC	2.848208884	2.519131370	2.472907818	2.460221832	2.491152344	2.545419769	2.609622935	2.754559288	3.008750017
1 I ZF1	1.676843299	1.677913028	1.678319592	1.678739668	1.679196483	1.679505695	1.679737476	1.680076516	1.680431358
1 I ZF2	1.809022884	1.824887714	1.829962021	1.834921717	1.840076055	1.843455490	1.845941116	1.849514691	1.853187799
1 I ZF3	2.198281641	2.266098854	2.286491562	2.305876336	2.325486578	2.338066585	2.347186928	2.360112790	2.373183804
1 I ZF4	3.699520806	3.912668002	3.977121764	4.038288532	4.099919679	4.139257720	4.167654397	4.207693409	4.247905148

(4) P 5+4	TH = 38	SBE = 1.624269245						MCPER FUNCTIONS	
	RC = 02	RC = 05	RC = 07	RC = 10	RC = 15	RC = 20	RC = 25	RC = 35	RC = 50
AMAX DB	.0017375255	.0108709564	.0213327376	.0436480540	.0988323393	.1772876696	.2802872360	.5675287486	1.249387366
AMIN DB	48.21191904	62.38916729	67.65802054	73.28749822	79.76734611	84.44904873	88.15979736	93.98254844	100.7599525
MU I K	.4045850D-1	.8489568D-2	.4725895D-2	.2520376D-2	.1218599D-2	.7195419D-3	.4734710D-3	.2451454D-3	.1137045D-3
1 I PFR	1.157863264	.8182321002	.7145239521	.6133314596	.5071291561	.4363820467	.3836119088	.3065748071	.2261700239
1 I PF1	1.251162524	.9927643395	.9121554420	.8341761656	.7541982663	.7026513424	.6654240019	.6134030409	.5626250554
1 I PQ1	.7545325785	.7735591974	.7942199996	.8270670541	.8836361027	.9422533606	1.003671273	1.138361886	1.392617954
3 I PFC	1.485313575	1.354971523	1.305884231	1.254442420	1.197485618	1.158366973	1.128820780	1.085456627	1.040303461
3 I PQC	2.983557351	2.601879491	2.544542852	2.522974541	2.546548146	2.597009886	2.658962715	2.801610359	3.054993924
1 I ZF1	1.638789531	1.639797921	1.640194243	1.640607870	1.641061322	1.641370005	1.641602178	1.641942868	1.642300642
1 I ZF2	1.763833877	1.779657548	1.784748134	1.789737005	1.794935201	1.798350692	1.800866444	1.804498568	1.808217841
1 I ZF3	2.130837402	2.199641402	2.220393917	2.240143947	2.260145479	2.272987664	2.282303460	2.295514044	2.308882238
1 I ZF4	3.560907984	3.777056478	3.842793152	3.905305120	3.968399882	4.008720992	4.037847947	4.078942540	4.120241037

(4) P 5+4 — TH = 39 — SBE = 1.589015729 — MCPER FUNCTIONS

MU	K	RC = 02	RC = 05	RC = 07	RC = 10	RC = 15	RC = 20	RC = 25	RC = 35	RC = 50
	AMAX DB	.0017375255	.0108709564	.0213327376	.0436480540	.0988323393	.1772876696	.2802872360	.5675287486	1.249387366
	AMIN DB	46.05716335	60.20521217	65.46786327	71.09240383	77.56814085	82.24764641	85.95700726	91.77807548	98.55410373
MU	K	.4976395D-1	.1051454D-1	.5864139D-2	.3132711D-2	.1517108D-2	.8966806D-3	.5904347D-3	.3059883D-3	.1420509D-3
1	PFR	1.175904563	.8273841484	.7217622120	.6189945247	.5114059870	.4398648465	.3865601782	.3088183689	.2277566625
1	PF1	1.260237122	.9992246787	.9178421563	.8391580619	.7585101380	.7065627567	.6690626325	.6166853075	.5655906693
1	PQ1	.7634825180	.7793795129	.7994148432	.8317942988	.8880337806	.9465383785	1.007942553	1.143263499	1.397476828
3	PFC	1.472313355	1.347966872	1.300507670	1.250504278	1.194873914	1.156528118	1.127500040	1.084804487	1.040243313
3	PQC	3.132585077	2.691481257	2.621746999	2.590314250	2.605743802	2.651995837	2.711454221	2.851542108	3.103951599
1	ZF1	1.602899238	1.603839649	1.604224134	1.604629949	1.605078809	1.605386254	1.605618346	1.605960070	1.606320220
1	ZF2	1.721150478	1.736905995	1.742004871	1.747015589	1.752250622	1.755697934	1.758240898	1.761907696	1.765689550
1	ZF3	2.066805010	2.136511214	2.157605438	2.177705935	2.198085916	2.211183070	2.220689569	2.234178821	2.247838870
1	ZF4	3.428917864	3.647680485	3.714627294	3.778432145	3.842952700	3.884239724	3.914087611	3.956228499	3.998608488

(4) P 5+4 — TH = 40 — SBE = 1.555723827 — MCPER FUNCTIONS

MU	K	RC = 02	RC = 05	RC = 07	RC = 10	RC = 15	RC = 20	RC = 25	RC = 35	RC = 50
	AMAX DB	.0017375255	.0108709564	.0213327376	.0436480540	.0988323393	.1772876696	.2802872360	.5675287486	1.249387366
	AMIN DB	43.95301440	58.06884583	63.32466936	68.94377988	75.41499940	80.09209215	83.79992952	89.61915100	96.39367001
MU	K	.6087148D-1	.1295481D-1	.7239362D-2	.3874249D-2	.1879401D-2	.1111957D-2	.7327124D-3	.3800921D-3	.1766168D-3
1	PFR	1.195174749	.8370400398	.7293755437	.6249352968	.5158818044	.4435046612	.3896385363	.3111581798	.2294096850
1	PF1	1.269536921	1.005908016	.9237327517	.8443226272	.7629821103	.7106199651	.6728370478	.6200930372	.5686667494
1	PQ1	.7732074638	.7855963528	.8049383706	.8368003589	.8926732931	.9510489542	1.012431861	1.147830649	1.402564608
3	PFC	1.459076272	1.340773560	1.294973317	1.246441294	1.192172258	1.154621565	1.126127296	1.084121506	1.040171910
3	PQC	3.297033177	2.788648910	2.705060560	2.662656188	2.669058645	2.710649080	2.767340152	2.904564601	3.155810055
1	ZF1	1.569014519	1.569880199	1.570251225	1.570647843	1.571090858	1.571396342	1.571627866	1.571969992	1.572331945
1	ZF2	1.680792683	1.696453636	1.701552840	1.706578042	1.711842804	1.715317625	1.717884823	1.721592316	1.725423057
1	ZF3	2.005951164	2.076468925	2.097885616	2.118320924	2.139065796	2.152410320	2.162102522	2.175864062	2.189810357
1	ZF4	3.303137282	3.524100480	3.592178260	3.657218627	3.723123113	3.765356876	3.795915149	3.839091706	3.882546954

(4) P 5+4 — TH = 41 — SBE = 1.524253087 — MCPER FUNCTIONS

MU	K	RC = 02	RC = 05	RC = 07	RC = 10	RC = 15	RC = 20	RC = 25	RC = 35	RC = 50
	AMAX DB	.0017375255	.0108709564	.0213327376	.0436480540	.0988323393	.1772876696	.2802872360	.5675287486	1.249387366
	AMIN DB	41.89725572	55.97757338	61.22588889	66.83903388	73.30529477	77.97974092	81.68590789	87.50310532	94.27597083
MU	K	.7406500D-1	.1588245D-1	.8893719D-2	.4768476D-2	.2317308D-2	.1372528D-2	.9050940D-3	.4699937D-3	.2186042D-3
1	PFR	1.215786735	.8472341892	.7373871487	.6311693613	.5205667126	.4473089708	.3928529057	.3135983591	.2311317735
1	PF1	1.279045142	1.012816561	.9298309739	.8496741552	.7676184559	.7148270415	.6767511249	.6236207909	.5718565328
1	PQ1	.7837960669	.7922446109	.8108169873	.8421056613	.8975706963	.9557991156	1.017152123	1.152725272	1.407893152
3	PFC	1.445620464	1.333394492	1.289281812	1.242252803	1.189379364	1.152645926	1.124701204	1.083406522	1.040088374
3	PQC	3.478917478	2.894190192	2.795092455	2.740465558	2.736848333	2.773270827	2.826889095	2.960909456	3.210774749
1	ZF1	1.536993026	1.537777097	1.538133024	1.538519036	1.538954933	1.539257716	1.539488177	1.539830056	1.540193225
1	ZF2	1.642598069	1.658138693	1.663230310	1.668262610	1.673549916	1.677047862	1.679636251	1.683380364	1.687256188
1	ZF3	1.948067231	2.019299341	2.041018008	2.061771515	2.082866946	2.096450823	2.106323454	2.120350559	2.134577183
1	ZF4	3.183198905	3.405922855	3.475045807	3.541259002	3.608501262	3.651660416	3.682917234	3.727117375	3.771640590

(4) P 5+4 — TH = 42 SBE = 1.494476550 — MCPER FUNCTIONS

MU		RC = 02	RC = 05	RC = 07	RC = 10	RC = 15	RC = 20	RC = 25	RC = 35	RC = 50
	AMAX DB	.0017375255	.0108709564	.0213327376	.0436480540	.0988323393	.1772876696	.2802872360	.5675287486	1.249387366
	AMIN DB	39.88789009	53.92909112	59.16915824	64.77575586	71.23657954	75.90812575	79.61246327	85.42744480	92.19850086
MU	K	.8966230D-1	.1937992D-1	.1087569D-1	.5842531D-2	.2844558D-2	.1686726D-2	.1113165D-2	.5786587D-3	.2694216D-3
1	PFR	1.237868106	.8580046663	.7458225252	.6377137424	.5254716848	.4512858704	.3962096804	.3161433404	.2329258069
1	PF1	1.288740844	1.019951989	.9361404294	.8552170071	.7724236143	.7191882523	.6808089399	.6272813260	.5751634426
1	PQ1	.7953500770	.7993634469	.8170801249	.8477328283	.9027436607	.9608042414	1.022117468	1.157810375	1.413475336
3	PFC	1.431965636	1.325833155	1.283434133	1.237938301	1.186493992	1.150599817	1.123220398	1.032658334	1.039991786
3	PQC	3.680579084	3.009023634	2.892530687	2.824264786	2.809509850	2.840196001	2.890398880	3.020832530	3.269071801
1	ZF1	1.506706119	1.507401569	1.507740732	1.508114714	1.508542198	1.508841528	1.509070417	1.509411386	1.509775167
1	ZF2	1.606419709	1.621814929	1.626891104	1.631923112	1.637225714	1.640742329	1.643348809	1.647125364	1.651042356
1	ZF3	1.892966440	1.964808638	1.986807427	2.007861484	2.029292298	2.043107057	2.053154552	2.067440128	2.081940836
1	ZF4	3.068775678	3.292794629	3.362870084	3.430187835	3.498717148	3.542778018	3.574720142	3.619930201	3.665512936

(4) P 5+4 — TH = 43 SBE = 1.466279186 — MCPER FUNCTIONS

MU		RC = 02	RC = 05	RC = 07	RC = 10	RC = 15	RC = 20	RC = 25	RC = 35	RC = 50
	AMAX DB	.0017375255	.0108709564	.0213327376	.0436480540	.0988323393	.1772876696	.2802872360	.5675287486	1.249387366
	AMIN DB	37.92312776	51.92127032	57.15228326	62.75170127	69.20656844	73.87494018	77.57727600	83.38983407	90.15891209
MU	K	.1080175D 0	.2354123D-1	.1324088D-1	.7127693D-2	.3477053D-2	.2064225D-2	.1363425D-2	.7095468D-3	.3307162D-3
1	PFR	1.261563339	.8693936712	.7547097492	.6445870690	.5306086559	.4554441326	.3997157726	.3137979010	.2347948783
1	PF1	1.298598246	1.027315309	.9426645220	.8609555885	.7774021901	.7237080623	.6850147768	.6310756077	.5785911005
1	PQ1	.8079865936	.8069969139	.8237606616	.8537069653	.9082116698	.9660812211	1.027343372	1.163150146	1.419325164
3	PFC	1.418133084	1.318093662	1.277431633	1.233497475	1.183514963	1.148481858	1.121683492	1.081875705	1.039881185
3	PQC	3.904745484	3.134196708	2.998154678	2.914642030	2.887487332	2.911797825	2.958200454	3.084617020	3.330950531
1	ZF1	1.478037276	1.478636945	1.478957662	1.479318171	1.479735928	1.480031039	1.480257840	1.480597221	1.480960992
1	ZF2	1.572124371	1.587349839	1.592402798	1.597427133	1.602737727	1.606268494	1.608889901	1.612694627	1.616648761
1	ZF3	1.840481476	1.912821971	1.935077538	1.956413364	1.978163465	1.992200146	2.002416622	2.016953176	2.031721373
1	ZF4	2.959575993	3.184398872	3.255327082	3.323675276	3.393436084	3.438372505	3.470985192	3.517139790	3.563822331

(4) P 5+4 — TH = 44 SBE = 1.439556540 — MCPER FUNCTIONS

MU		RC = 02	RC = 05	RC = 07	RC = 10	RC = 15	RC = 20	RC = 25	RC = 35	RC = 50
	AMAX DB	.0017375255	.0108709564	.0213327376	.0436480540	.0988323393	.1772876696	.2802872360	.5675287486	1.249387366
	AMIN DB	36.00137815	49.95214317	55.17322469	60.76477585	67.21312290	71.87802259	75.57817017	81.38808015	88.15499788
MU	K	.1295234D 0	.2847324D-1	.1605286D-1	.8659925D-2	.4233167D-2	.2516238D-2	.1663422D-2	.8656872D-3	.4044112D-3
1	PFR	1.287036378	.8814480825	.7640797971	.6518097632	.5359906283	.4597932785	.4033786650	.3215671954	.2367423153
1	PF1	1.308585979	1.034906702	.9494063754	.8668943194	.7825589504	.7283911406	.6893731376	.6350078219	.5821433400
1	PQ1	.8218407493	.8151946943	.8308954120	.8600559942	.9139962493	.9716486380	1.032846815	1.168760095	1.425457886
3	PFC	1.404145684	1.310180802	1.271276077	1.228930225	1.180441169	1.146290684	1.120089087	1.081057360	1.039755571
3	PQC	4.154604594	3.270907442	3.112849880	3.012261176	2.971278878	2.988493160	3.030662386	3.152577040	3.396686384
1	ZF1	1.450880731	1.451377289	1.451677861	1.452023437	1.452430137	1.452720251	1.452944436	1.453281536	1.453644663
1	ZF2	1.539590976	1.554623090	1.559645158	1.564654462	1.569965701	1.573506041	1.576139161	1.579967687	1.583954828
1	ZF3	1.790462387	1.863181402	1.885668786	1.907266361	1.929318657	1.943567766	1.953946999	1.968726608	1.983755318
1	ZF4	2.855339447	3.080450758	3.152124713	3.221423139	3.292354771	3.338137918	3.371404807	3.418586713	3.466257971

(4) P 5+4			TH = 45	SBE = 1.414213562							MCPER FUNCTIONS
			RC = 02	RC = 05	RC = 07	RC = 10	RC = 15	RC = 20	RC = 25	RC = 35	RC = 50
AMAX DB			.0017375255	.0108709564	.0213327376	.0436480540	.0988323393	.1772876696	.2802872360	.5675287486	1.249387366
AMIN DB			34.12124516	48.01989062	53.23008524	58.81302226	65.25423730	69.91534226	73.61309952	79.42011831	86.18467862
MU		K	.1546137D 0	.3429704D-1	.1938416D-1	.1048048D-1	.5134085D-2	.3055735D-2	.2021906D-2	.1054763D-2	.4927484D-3
1		PFR	1.314473639	.8942200920	.7739669170	.6594042558	.5416317921	.4643436588	.4072064706	.3244567926	.2387717020
1		PF1	1.318666269	1.042725331	.9563687405	.8730375981	.7878988185	.7332423657	.6938887516	.6390823879	.5858242204
1		PQ1	.8370689172	.8240129644	.8385256995	.8668110406	.9201212308	.9775269790	1.038646461	1.174657203	1.431890138
3		PFC	1.390027877	1.302100093	1.264969683	1.224236693	1.177271592	1.144024951	1.118435773	1.080201990	1.039613898
3		PQC	4.433894711	3.420530279	3.237625126	3.117873641	3.061444527	3.070748742	3.108196097	3.225061762	3.466584299
1		ZF1	1.425140280	1.425526213	1.425804920	1.426134090	1.426528387	1.426812716	1.427033749	1.427367862	1.427729695
1		ZF2	1.508709261	1.523525170	1.528508785	1.533495737	1.538800244	1.542346527	1.544987088	1.548834953	1.552850850
1		ZF3	1.742774764	1.815744123	1.838436605	1.860274565	1.882610871	1.897062335	1.907597731	1.922612008	1.937893849
1		ZF4	2.755833102	2.980694124	3.052999404	3.123161516	3.195197904	3.241796117	3.275699111	3.323839093	3.372536475

(4) P 5+4			TH = 46	SBE = 1.390163591							MCPER FUNCTIONS
			RC = 02	RC = 05	RC = 07	RC = 10	RC = 15	RC = 20	RC = 25	RC = 35	RC = 50
AMAX DB			.0017375255	.0108709564	.0213327376	.0436480540	.0988323393	.1772876696	.2802872360	.5675287486	1.249387366
AMIN DB			32.28152658	46.12283201	51.32109838	56.89460831	63.32802671	67.98498697	71.68013491	77.48399924	84.24598894
MU		K	.1837649D 0	.4114953D-1	.2331728D-1	.1263657D-1	.6204188D-2	.3697690D-2	.2448999D-2	.1279212D-2	.5983369D-3
1		PFR	1.344087492	.9077679376	.7844090560	.6673952338	.5475476618	.4691065448	.4112079996	.3274727186	.2408869044
1		PF1	1.328794044	1.050769122	.9635538845	.8793897549	.7934268657	.7382668295	.6985665853	.6433039729	.5896380429
1		PQ1	.8538525478	.8335154145	.8466980268	.8740068837	.9266130568	.9837388773	1.044762865	1.180860097	1.438640101
3		PFC	1.375805623	1.293857828	1.258515163	1.219417288	1.174005314	1.141683348	1.116722138	1.079308248	1.039455073
3		PQC	4.747014049	3.584647093	3.373633317	3.232332358	3.158615652	3.159088475	3.191261977	3.302460232	3.540982600
1		ZF1	1.400728258	1.400995837	1.401250941	1.401562219	1.401942755	1.402220499	1.402437838	1.402768244	1.403128118
1		ZF2	1.479378625	1.493956212	1.498893941	1.503851274	1.509141652	1.512687199	1.515333880	1.519196531	1.523236824
1		ZF3	1.697298141	1.770380906	1.793249871	1.815305382	1.837906329	1.852549447	1.863234009	1.878474064	1.894001215
1		ZF4	2.660848162	2.884898484	2.957713143	3.028645829	3.101715227	3.149093825	3.183612968	3.232689632	3.282398913

(4) P 5+4			TH = 47	SBE = 1.367327461							MCPER FUNCTIONS
			RC = 02	RC = 05	RC = 07	RC = 10	RC = 15	RC = 20	RC = 25	RC = 35	RC = 50
AMAX DB			.0017375255	.0108709564	.0213327376	.0436480540	.0988323393	.1772876696	.2802872360	.5675287486	1.249387366
AMIN DB			30.48121814	44.25941623	49.44461853	55.00781647	61.43271600	66.08515173	69.77745290	75.57787762	82.33706598
MU		K	.2174976D 0	.4918495D-1	.2794590D-1	.1518211D-1	.7471483D-2	.4459364D-2	.2956402D-2	.1546333D-2	.7242101D-3
1		PFR	1.376120297	.9221567556	.7954483540	.6758099245	.5537552328	.4740942324	.4153928359	.3306215043	.2430920980
1		PF1	1.338916015	1.059034498	.9709634576	.8859549969	.7991482986	.7434698404	.7034118522	.6476775069	.5935893671
1		PQ1	.8724027683	.8437744555	.8554648644	.8816824826	.9335011350	.9903093909	1.051218716	1.187389248	1.445727679
3		PFC	1.361506339	1.285461131	1.251915769	1.214472725	1.170641543	1.139264606	1.114946769	1.078374755	1.039277961
3		PQC	5.099154372	3.765084565	3.522196193	3.356608400	3.263506050	3.254102020	3.280376537	3.385206984	3.620257524
1		ZF1	1.377564639	1.377705889	1.377935636	1.378227525	1.378592930	1.378863282	1.379076374	1.379402342	1.379759579
1		ZF2	1.451507122	1.465824973	1.470709525	1.475630037	1.480898885	1.484439974	1.487088407	1.490961199	1.495021416
1		ZF3	1.653924580	1.726974762	1.749989551	1.772238193	1.795083119	1.809906503	1.820732798	1.836189196	1.851953360
1		ZF4	2.570197006	2.792856396	2.866050897	2.937654267	3.011678967	3.059900062	3.094913405	3.144903030	3.195608207

(4) P 5+4 — TH = 48 — SBE = 1.345632730 — MCPER FUNCTIONS

(4) P 5+4	RC = 02	RC = 05	RC = 07	RC = 10	RC = 15	RC = 20	RC = 25	RC = 35	RC = 50
AMAX DB	.0017375255	.0108709564	.0213327376	.0436480540	.0988323393	.1772876696	.2802872360	.5675287486	1.249387366
AMIN DB	28.71952302	42.42821455	47.59911258	53.15103463	59.56662995	64.21412874	67.90332554	73.70000158	80.45613882
MU I K	.2563783D 0	.5857668D-1	.3337606D-1	.1817858D-1	.8968089D-2	.5360624D-2	.3557611D-2	.1863422D-2	.8738903D-3
1 I PFR	1.410849061	.9374595703	.8071317158	.6846784232	.5602731601	.4793201608	.4197714239	.3339102392	.2453918006
1 I PF1	1.348969713	1.067516079	.9785983356	.8927373385	.8050684422	.7488569245	.7084300223	.6522031984	.5976830303
1 I PQ1	.8929658961	.8548726509	.8648855800	.8898815927	.9408182492	.9972663264	1.058039108	1.194267201	1.453174704
3 I PFC	1.347158808	1.276918003	1.245175341	1.209404051	1.167179624	1.136767507	1.113108260	1.077400101	1.039081374
3 I PQC	5.496464243	3.963959378	3.684834088	3.491810831	3.376925080	3.356454915	3.376120838	3.473738612	3.704328486
1 I ZF1	1.355576244	1.355582921	1.355785536	1.356056529	1.356405423	1.356667568	1.356875856	1.357196645	1.357550551
1 I ZF2	1.425010596	1.439047937	1.443872174	1.448748745	1.453988666	1.457520541	1.460167312	1.464045515	1.468121072
1 I ZF3	1.612557423	1.685419776	1.708547539	1.730963167	1.754030002	1.769021521	1.779981640	1.795644354	1.811636720
1 I ZF4	2.483710524	2.704381157	2.777818354	2.849985555	2.924881599	2.973703899	3.009387362	3.060263718	3.111946862

(4) P 5+4 — TH = 49 — SBE = 1.325012993 — MCPER FUNCTIONS

(4) P 5+4	RC = 02	RC = 05	RC = 07	RC = 10	RC = 15	RC = 20	RC = 25	RC = 35	RC = 50
AMAX DB	.0017375255	.0108709564	.0213327376	.0436480540	.0988323393	.1772876696	.2802872360	.5675287486	1.249387366
AMIN DB	26.99586791	40.62791484	45.78315263	51.32274794	57.72818452	62.37029824	66.05611105	71.84870322	78.60151873
MU I K	.3010190D 0	.6951910D-1	.3972765D-1	.2169590D-1	.1073078D-1	.6424305D-2	.4268181D-2	.2238914D-2	.1051464D-2
1 I PFR	1.448590777	.9537584476	.8195114745	.6940340706	.5671219641	.4847990484	.4243551669	.3373466332	.2477909073
1 I PF1	1.358882566	1.076206325	.9864584313	.8997405182	.8111927174	.7544338249	.7136268313	.6569015521	.6019241670
1 I PQ1	.9158300491	.8669044223	.8750275383	.8986534895	.9486010410	1.004640615	1.065251863	1.201518839	1.461005173
3 I PFC	1.332793070	1.268237377	1.238298348	1.204212690	1.163619069	1.134190904	1.111205226	1.076382844	1.038864080
3 I PQC	5.946248613	4.183733107	3.863301846	3.639210494	3.499793314	3.466900568	3.479150421	3.568751484	3.795164232
1 I ZF1	1.334696055	1.334559609	1.334733299	1.334981879	1.335312876	1.335565992	1.335768913	1.336083770	1.336433643
1 I ZF2	1.399811917	1.413548536	1.418305482	1.423131082	1.428334698	1.431852572	1.434494227	1.438373027	1.442459229
1 I ZF3	1.573110157	1.645620075	1.668825632	1.691380239	1.714645386	1.729792094	1.740877612	1.756735972	1.772947173
1 I ZF4	2.401235703	2.619304758	2.692839938	2.765456984	2.841133893	2.890612497	2.926839727	2.978573886	3.031214972

(4) P 5+4 — TH = 50 — SBE = 1.305407289 — MCPER FUNCTIONS

(4) P 5+4	RC = 02	RC = 05	RC = 07	RC = 10	RC = 15	RC = 20	RC = 25	RC = 35	RC = 50
AMAX DB	.0017375255	.0108709564	.0213327376	.0436480540	.0988323393	.1772876696	.2802872360	.5675287486	1.249387366
AMIN DB	25.30992656	38.85731743	43.99540993	49.52153163	55.91587887	60.55212021	64.23424525	70.02238979	76.77159021
MU I K	.3520772D 0	.8222950D-1	.4713579D-1	.2581352D-1	.1280162D-1	.7676618D-2	.5106015D-2	.2682552D-2	.1261667D-2
1 I PFR	1.489708509	.9711458397	.8326461631	.7039138914	.5743242675	.4905470483	.4291565413	.3409390866	.2502947321
1 I PF1	1.368571044	1.085095129	.9945424718	.9069678959	.8175266104	.7602064978	.7190082887	.6617633863	.6063182315
1 I PQ1	.9413330607	.8799780870	.8859674079	.9080538239	.9568905719	1.012466753	1.072887902	1.209171692	1.469245520
3 I PFC	1.318440281	1.259429159	1.231289947	1.198900474	1.159959577	1.131533730	1.109236303	1.075321517	1.038624793
3 I PQC	6.457212789	4.427279138	4.059632334	3.800268654	3.633161228	3.586294520	3.590207087	3.670710843	3.891790060
1 I ZF1	1.314862597	1.314574154	1.314717101	1.314941743	1.315253451	1.315496714	1.315693700	1.316001865	1.316346988
1 I ZF2	1.375840320	1.389256457	1.393939304	1.398707008	1.403866974	1.407366035	1.409999083	1.413873585	1.417965626
1 I ZF3	1.535505410	1.607488939	1.630734634	1.653398213	1.676836414	1.692124480	1.703326410	1.719369047	1.735789107
1 I ZF4	2.322633429	2.537476064	2.610957054	2.683902700	2.760263194	2.810349389	2.847091607	2.899651740	2.953228472

(4) P 5+4		TH = 51	SBE = 1.286759566							MCPER FUNCTIONS
		RC = 02	RC = 05	RC = 07	RC = 10	RC = 15	RC = 20	RC = 25	RC = 35	RC = 50
AMAX DB		.0017375255	.0108709564	.0213327376	.0436480540	.0988323393	.1772876696	.2802872360	.5675287486	1.249387366
AMIN DB		23.66165214	37.11533265	42.23464989	47.74604475	54.12828822	58.75812685	62.43623384	68.21953562	74.96480269
MU	K	.4102542D 0	.9695016D-1	.5575256D-1	.3062149D-1	.1522861D-1	.9147620D-2	.6091700D-2	.3205586D-2	.1509988D-2
1	PFR	1.534618229	.9897261587	.8466014155	.7143591023	.5819050687	.4965819273	.4341892255	.3446967701	.2529090544
1	PF1	1.377939972	1.094169340	1.002847735	.9144223296	.8240756346	.7661811056	.7245806850	.6667998524	.6108710221
1	PQ1	.9698719345	.8942183012	.8977927181	.9181456342	.9657329861	1.020783309	1.080981680	1.217256286	1.477924927
3	PFC	1.304132556	1.250504275	1.224156026	1.193469691	1.156201061	1.128795016	1.107200166	1.074214631	1.038362179
3	PQC	7.039760299	4.697964628	4.276189425	3.976671623	3.778231639	3.715611482	3.710132913	3.780361576	3.995296326
1	ZF1	1.296019395	1.295569737	1.295680099	1.295879270	1.296170295	1.296402876	1.296593356	1.296894061	1.297233707
1	ZF2	1.353030828	1.366107043	1.370709153	1.375412151	1.380521160	1.383996584	1.386617498	1.390482730	1.394575693
1	ZF3	1.499674017	1.570948006	1.594193580	1.616933977	1.640518175	1.655932806	1.667241544	1.683456327	1.700074615
1	ZF4	2.247776471	2.458759194	2.532026538	2.605172184	2.682111922	2.732752970	2.769978810	2.823329972	2.877817590

(4) P 5+4		TH = 52	SBE = 1.269018215							MCPER FUNCTIONS
		RC = 02	RC = 05	RC = 07	RC = 10	RC = 15	RC = 20	RC = 25	RC = 35	RC = 50
AMAX DB		.0017375255	.0108709564	.0213327376	.0436480540	.0988323393	.1772876696	.2802872360	.5675287486	1.249387366
AMIN DB		22.05131921	35.40098015	40.49972834	45.99502474	52.36405747	56.98691561	60.66064513	66.43867457	73.17966269
MU	K	.4762921D 0	.1139506D 0	.6574873D-1	.3622180D-1	.1806653D-1	.1087174D-1	.7248890D-2	.3820988D-2	.1802778D-2
1	PFR	1.583796398	1.009617616	.8614510178	.7254157043	.5898920583	.5029232728	.4394682498	.3486297167	.2556401718
1	PF1	1.386882121	1.103412226	1.011369739	.9221060254	.8308452808	.7723640045	.7303505976	.6720174544	.6155887071
1	PQ1	1.001914091	.9097690003	.9106037224	.9290005491	.9751802916	1.029633533	1.089571689	1.225806561	1.487075692
3	PFC	1.289902781	1.241474709	1.216903260	1.187923129	1.152343678	1.125973913	1.105095536	1.073060676	1.038074850
3	PQC	7.706355664	4.999751277	4.515732777	4.170372836	3.936386764	3.855965769	3.839887002	3.898491025	4.106348494
1	ZF1	1.278114481	1.277494050	1.277569956	1.277742116	1.278011062	1.278232131	1.278415532	1.278708003	1.279041437
1	ZF2	1.331323740	1.344040758	1.348555661	1.353187263	1.358238066	1.361685016	1.364290241	1.368141157	1.372230016
1	ZF3	1.465554168	1.535926570	1.559129043	1.581911817	1.605613008	1.621138363	1.632543637	1.648917606	1.665722771
1	ZF4	2.176547598	2.383032051	2.455919276	2.529128915	2.606536246	2.657675166	2.695350514	2.749454412	2.804825486

(4) P 5+4		TH = 53	SBE = 1.252135658							MCPER FUNCTIONS
		RC = 02	RC = 05	RC = 07	RC = 10	RC = 15	RC = 20	RC = 25	RC = 35	RC = 50
AMAX DB		.0017375255	.0108709564	.0213327376	.0436480540	.0988323393	.1772876696	.2802872360	.5675287486	1.249387366
AMIN DB		20.47957577	33.71339053	38.78958900	44.26728289	50.62189534	55.23714290	58.90610335	64.67839294	71.41472646
MU	K	.5509695D 0	.1335295D 0	.7731568D-1	.4272970D-1	.2137776D-1	.1288838D-1	.8604739D-2	.4543716D-2	.2147389D-2
1	PFR	1.637788184	1.030954377	.8772781411	.7371351741	.5983159878	.5095927322	.4450101690	.3527489268	.2584949614
1	PF1	1.395278237	1.112802853	1.020101877	.9300203568	.8378409545	.7787617269	.7363248933	.6774230701	.6204778543
1	PQ1	1.038010670	.9267969492	.9245156368	.9407002238	.9852912859	1.039066066	1.098701059	1.234860350	1.496733649
3	PFC	1.275784411	1.232353537	1.209539159	1.182264120	1.148387862	1.123069710	1.102921196	1.071858128	1.037761367
3	PQC	8.471964463	5.337319815	4.781497432	4.383644196	4.109221002	4.008635935	3.980565601	4.025994274	4.225699083
1	ZF1	1.261099943	1.260298870	1.260338421	1.260482021	1.260727492	1.260936221	1.261111969	1.261395427	1.261721906
1	ZF2	1.310664176	1.323002714	1.327424105	1.331977749	1.336963162	1.340376799	1.342962755	1.346794237	1.350873858
1	ZF3	1.433090607	1.502360963	1.525474535	1.548262812	1.572049896	1.587668994	1.599159805	1.615679092	1.632659003
1	ZF4	2.108837799	2.310185001	2.382518965	2.455649186	2.533404915	2.584980265	2.623068082	2.677882838	2.734107044

(4) P 5+4	TH = 54	SBE = 1.236067977						MCPER FUNCTIONS	
	RC = 02	RC = 05	RC = 07	RC = 10	RC = 15	RC = 20	RC = 25	RC = 35	RC = 50
AMAX DB	.0017375255	.0108709564	.0213327376	.0436480540	.0988323393	.1772876696	.2802872360	.5675287486	1.249387366
AMIN DB	18.94750437	32.05180940	37.10326228	42.56170059	48.90056918	53.50751816	57.17128245	62.93732283	69.66859296
MU I K	.6350952D 0	.1560174D 0	.9066745D-1	.5027531D-1	.2523331D-1	.1524260D-1	.1019040D-1	.5391005D-2	.2552329D-2
1 I PFR	1.697216100	1.053889091	.8941767885	.7495752738	.6072110983	.5166142909	.4508332621	.3570664905	.2614809494
1 I PF1	1.402997673	1.122315394	1.029034994	.9381656478	.8450678969	.7853809561	.7425107288	.6830239726	.6255454621
1 I PQ1	1.078812120	.9454960440	.9396613396	.9533380618	.9961326583	1.049135778	1.108418258	1.244459953	1.506938663
3 I PFC	1.261811241	1.223154955	1.202072123	1.176496594	1.144334358	1.120081859	1.100676010	1.070605459	1.037420238
3 I PQC	9.354584151	5.716224415	5.077292103	4.619139038	4.298580844	4.175094617	4.133426372	4.163892508	4.354201912
1 I ZF1	1.244931510	1.243939692	1.243940952	1.244054436	1.244275040	1.244470602	1.244638123	1.244911789	1.245230561
1 I ZF2	1.291001679	1.302942262	1.307263988	1.311733244	1.316646155	1.320021645	1.322584729	1.326391595	1.330456738
1 I ZF3	1.402233856	1.470193993	1.493169967	1.515924310	1.539763931	1.555458554	1.567023107	1.583672857	1.600814528
1 I ZF4	2.044544554	2.240119664	2.311721035	2.384621050	2.462598230	2.514543888	2.553004034	2.608483919	2.665527800

(4) P 5+4	TH = 55	SBE = 1.220774589						MCPER FUNCTIONS	
	RC = 02	RC = 05	RC = 07	RC = 10	RC = 15	RC = 20	RC = 25	RC = 35	RC = 50
AMAX DB	.0017375255	.0108709564	.0213327376	.0436480540	.0988323393	.1772876696	.2802872360	.5675287486	1.249387366
AMIN DB	17.45668959	30.41560473	35.43986576	40.87722649	47.19890036	51.79679846	55.45490016	61.21413566	67.93989704
MU I K	.7294994D 0	.1817785D 0	.1060429D 0	.5900529D-1	.2971392D-1	.1798590D-1	.1204161D-1	.6382713D-2	.3027438D-2
1 I PFR	1.762788708	1.078595853	.9122534974	.7628010020	.6166156225	.5240145361	.4569577643	.3615957282	.2646063921
1 I PF1	1.409899881	1.131918355	1.038156884	.9465409115	.8525310861	.7922284911	.7489155461	.6888278522	.6307989947
1 I PQ1	1.125086301	.9660925435	.9561946424	.9670212868	1.007780307	1.059904773	1.118777930	1.254652800	1.517735216
3 I PFC	1.248017157	1.213894295	1.194511487	1.170625123	1.140184264	1.117010000	1.098358932	1.069301137	1.037049918
3 I PQC	10.37587961	6.143085087	5.407621148	4.879970082	4.506613739	4.357044872	4.299917831	4.313355144	4.492829184
1 I ZF1	1.229568149	1.228375390	1.228336384	1.228418186	1.228612537	1.228794110	1.228952832	1.229215926	1.229526235
1 I ZF2	1.272289835	1.283812622	1.288028666	1.292407235	1.297240619	1.300573136	1.303109730	1.306886736	1.310932054
1 I ZF3	1.372939448	1.439374442	1.462161179	1.484839462	1.508695849	1.524446440	1.536072074	1.552836352	1.570125863
1 I ZF4	1.983570110	2.172747802	2.243431492	2.315943385	2.394007135	2.446252076	2.485041119	2.541136286	2.598962989

(4) P 5+4	TH = 56	SBE = 1.206217949						MCPER FUNCTIONS	
	RC = 02	RC = 05	RC = 07	RC = 10	RC = 15	RC = 20	RC = 25	RC = 35	RC = 50
AMAX DB	.0017375255	.0108709564	.0213327376	.0436480540	.0988323393	.1772876696	.2802872360	.5675287486	1.249387366
AMIN DB	16.00928583	28.80427796	33.79860646	39.21287463	45.51576008	50.10378347	53.75571247	59.50753609	66.22730282
MU I K	.8350244D 0	.2112132D 0	.1237082D 0	.6908476D-1	.3491134D-1	.2117710D-1	.1419933D-1	.7541711D-2	.3584103D-2
1 I PFR	1.835308833	1.105273672	.9316293458	.7768857147	.6265723743	.5318233362	.4634061403	.3663513543	.2678803692
1 I PF1	1.415836967	1.141573721	1.047451722	.9551435389	.8602351137	.7993111997	.7555470628	.6948428388	.6362464197
1 I PQ1	1.177739174	.9888514494	.9742942680	.9818734503	1.020320919	1.071443581	1.129841882	1.265492245	1.529173080
3 I PFC	1.234435855	1.204588032	1.186867569	1.164655004	1.135939066	1.113853993	1.095969036	1.067943641	1.036643811
3 I PQC	11.56193730	6.625828402	5.777836734	5.169805318	4.735828235	4.556464659	4.481715249	4.475726654	4.642692069
1 I ZF1	1.214971683	1.213567919	1.213486633	1.213535181	1.213701897	1.213868666	1.214018021	1.214269763	1.214570852
1 I ZF2	1.254485928	1.265570557	1.269675018	1.273956731	1.278703650	1.281988389	1.284494860	1.288236706	1.292256749
1 I ZF3	1.345167128	1.409856611	1.432399515	1.454956814	1.478791617	1.494577168	1.506250284	1.523111974	1.540534387
1 I ZF4	1.925819688	2.107990288	2.177566303	2.249525050	2.327532408	2.380000487	2.419071516	2.475727705	2.534296699

(4) P 5+4		TH = 57	SBE = 1.192363293						MCPER FUNCTIONS
		RC = 02	RC = 05	RC = 07	RC = 10	RC = 15	RC = 20	RC = 25	RC = 35	RC = 50
AMAX DB		.0017375255	.0108709564	.0213327376	.0436480540	.0988323393	.1772876696	.2802872360	.5675287486	1.249387366
AMIN DB		14.60807538	27.21748010	32.17878549	37.56772360	43.85006584	48.42731073	52.07250828	57.81625591	64.52949712
MU	K	.9525127D 0	.2447599D 0	.1439591D 0	.80699320D-1	.40929700D-1	.24883360D-1	.16710490D-1	.88943430D-2	.42355020D-2
1	PFR	1.915680561	1.134150533	.9524423209	.7919124487	.6371294420	.5400736856	.4702033971	.3713496686	.2713128925
1	PF1	1.420657573	1.151236038	1.056899400	.9639689257	.8681840319	.8066359567	.7624132550	.7010775244	.6418962496
1	PQ1	1.237838123	1.014084314	.9941687125	.9980374780	1.033853880	1.083832600	1.141680276	1.277038512	1.541308137
3	PFC	1.221100534	1.195253785	1.179151715	1.158592262	1.131600695	1.110613952	1.093505533	1.066531468	1.036215268
3	PQC	12.94415115	7.173989961	6.194329678	5.492988722	4.989168434	4.775661621	4.680764667	4.652559268	4.805065638
1	ZF1	1.201106406	1.199482050	1.199356430	1.199370138	1.199507850	1.199659006	1.199798431	1.200038048	1.200329156
1	ZF2	1.237550594	1.248176091	1.252163155	1.256341963	1.260995578	1.264227754	1.266700462	1.270401795	1.274391011
1	ZF3	1.318879991	1.381599891	1.403841444	1.426229947	1.450002084	1.465800023	1.477505996	1.494446697	1.511985956
1	ZF4	1.871199584	2.045776149	2.114050257	2.185284133	2.263083956	2.315693689	2.354996113	2.412154345	2.471421115

(4) P 5+4		TH = 58	SBE = 1.179178403						MCPER FUNCTIONS
		RC = 02	RC = 05	RC = 07	RC = 10	RC = 15	RC = 20	RC = 25	RC = 35	RC = 50
AMAX DB		.0017375255	.0108709564	.0213327376	.0436480540	.0988323393	.1772876696	.2802872360	.5675287486	1.249387366
AMIN DB		13.25650151	25.65503361	30.57980555	35.94091708	42.20077838	46.76625128	50.40410425	56.13904816	62.84518290
MU	K	.1082795D 1	.2828968D 0	.1671234D 0	.94057380D-1	.47887190D-1	.29181330D-1	.19628900D-1	.10470960D-1	.49968900D-2
1	PFR	2.004914113	1.165488136	.9748501180	.8079754907	.6483410095	.5488028296	.4773774620	.3766087797	.2749150325
1	PF1	1.424212210	1.160851449	1.066474771	.9730100303	.8763811639	.8142095630	.7695223301	.7075409834	.6477575876
1	PQ1	1.306637744	1.042158816	1.016062210	1.015679392	1.048493589	1.097163821	1.154373051	1.289359814	1.554203334
3	PFC	1.208043537	1.185910296	1.171376334	1.152443761	1.127171568	1.107290276	1.090967801	1.065063147	1.035747592
3	PQC	14.56025326	7.799095941	6.664770101	5.854690870	5.270106745	5.017340950	4.899337205	4.845653073	4.981419246
1	ZF1	1.187938689	1.186085123	1.185913077	1.185890352	1.185997699	1.186132447	1.186261387	1.186488112	1.186768480
1	ZF2	1.221447472	1.231592247	1.235456154	1.239526122	1.244079693	1.247254551	1.249689853	1.253345274	1.257298011
1	ZF3	1.294043517	1.354568377	1.376448219	1.398617166	1.422282664	1.438068736	1.449791830	1.466791738	1.484430559
1	ZF4	1.819615111	1.986041644	2.052816335	2.123147265	2.200580173	2.253244538	2.292723880	2.350320129	2.410235848

(4) P 5+4		TH = 59	SBE = 1.166633397						MCPER FUNCTIONS
		RC = 02	RC = 05	RC = 07	RC = 10	RC = 15	RC = 20	RC = 25	RC = 35	RC = 50
AMAX DB		.0017375255	.0108709564	.0213327376	.0436480540	.0988323393	.1772876696	.2802872360	.5675287486	1.249387366
AMIN DB		11.95865636	24.11696173	29.00118211	34.33166592	40.56689928	45.11950559	48.74933989	54.47468111	61.17307251
MU	K	.1226680D 1	.3261435D 0	.1935642D 0	.1093926D 0	.55917820D-1	.34158510D-1	.23016190D-1	.12306540D-1	.58359360D-2
1	PFR	2.104127694	1.199587387	.9990334509	.8251822415	.6602683293	.5580525856	.4849596211	.3921488689	.2786990668
1	PF1	1.426360050	1.170356711	1.076146799	.9822568487	.8848288705	.8220386409	.7768826874	.7142427930	.6538401771
1	PQ1	1.385607914	1.073510540	1.040262089	1.034992880	1.064372280	1.111542936	1.168011652	1.302533717	1.567929838
3	PFC	1.195295939	1.176577403	1.163554937	1.146217234	1.122654644	1.103883700	1.088355411	1.063537253	1.035244040
3	PQC	16.45550479	8.515146223	7.198412591	6.261099504	5.582760170	5.284689999	5.140096436	5.057105473	5.173453757
1	ZF1	1.175436596	1.173346830	1.173126236	1.173065478	1.173141112	1.173258670	1.173376579	1.173589654	1.173858528
1	ZF2	1.206142833	1.215784829	1.219519836	1.223475123	1.227922009	1.231034629	1.233429081	1.237033149	1.240943660
1	ZF3	1.270624482	1.328730493	1.350185561	1.372081215	1.395593071	1.411341215	1.423064493	1.440102266	1.457822027
1	ZF4	1.770968366	1.923729398	1.993804955	2.063049005	2.139947389	2.192573623	2.232171315	2.290136164	2.350647329

(4) TH = 60 SBE = 1.154700538 MCPER FUNCTIONS

P 5+4	MU	RC = 02	RC = 05	RC = 07	RC = 10	RC = 15	RC = 20	RC = 25	RC = 35	RC = 50
AMAX DB		.0017375255	.0108709564	.0213327376	.0436480540	.0988323393	.1772876696	.2802872360	.5675287486	1.249387366
AMIN DB		10.71920054	22.60352636	27.44255915	32.73925216	38.94746925	43.48599977	47.10707261	52.82193226	59.51188082
K	MU	.1384945D 1	.3750618D 0	.2236815D 0	.1269668D 0	.6517346D-1	.3991469D-1	.2694304D-1	.1444141D-1	.6923121D-2
PFR	1	2.214545607	1.236794749	1.025199967	.8436554374	.6729808826	.5678701408	.4929850426	.3879925010	.2826786553
PF1	1	1.426976927	1.179678252	1.085877609	.9916957961	.8935282624	.8301294993	.7845028629	.7211930494	.6601544562
PQ1	1	1.476464010	1.108657475	1.067107873	1.056204941	1.081643481	1.127091901	1.182701124	1.316648770	1.582568415
PFC	3	1.182887067	1.167275990	1.155702152	1.139921353	1.118053489	1.100395331	1.085668165	1.061952425	1.034702822
PQC	3	18.68407152	9.339228235	7.806485344	6.719661675	5.932037082	5.581484501	5.406182903	5.289372620	5.383147482
ZF1	1	1.163569491	1.161239007	1.160967733	1.160867337	1.160909928	1.161009528	1.161115872	1.161314553	1.161571184
ZF2	1	1.191605191	1.200722213	1.204322551	1.208157394	1.212491054	1.215537154	1.217886718	1.221433955	1.225296399
ZF3	1	1.248589723	1.304058628	1.325023371	1.346589027	1.369897084	1.385579312	1.397284534	1.414337162	1.432117763
ZF4	1	1.725155900	1.873787545	1.936963285	2.004931265	2.081119366	2.133608785	2.173261957	2.231520235	2.292568279

(4) TH = 61 SBE = 1.143354068 MCPER FUNCTIONS

P 5+4	MU	RC = 02	RC = 05	RC = 07	RC = 10	RC = 15	RC = 20	RC = 25	RC = 35	RC = 50
AMAX DB		.0017375255	.0108709564	.0213327376	.0436480540	.0988323393	.1772876696	.2802872360	.5675287486	1.249387366
AMIN DB		9.543192862	21.11527614	25.90373077	31.16303569	37.34156730	41.86468211	45.47617299	51.17958212	57.86031793
K	MU	.1558340D 1	.4302558D 0	.2579159D 0	.1470724D 0	.7582621D-1	.4656378D-1	.3149046D-1	.1692207D-1	.8132206D-2
PFR	1	2.337492494	1.277509494	1.053588872	.8635358036	.6865577665	.5783089340	.5014933998	.3941649942	.2868690468
PF1	1	1.425963929	1.188731347	1.095621449	1.001308981	.9024788478	.8384879629	.7923914526	.7284023798	.6667116166
PQ1	1	1.581199492	1.148217897	1.097002602	1.079582886	1.100486293	1.143952083	1.198562674	1.331806498	1.598211101
PFC	3	1.170843959	1.158027928	1.147833749	1.133565778	1.113372330	1.096826706	1.082906134	1.060307383	1.034122114
PQC	3	21.31063004	10.29229929	8.502689209	7.239393641	6.323823848	5.912222871	5.701319704	5.545346227	5.612813205
ZF1	1	1.152307685	1.149735448	1.149411379	1.149269742	1.149277979	1.149358874	1.149453132	1.149636691	1.149880343
ZF2	1	1.177804877	1.186375171	1.189835004	1.193543692	1.197757677	1.200732417	1.203033662	1.206518561	1.210327007
ZF3	1	1.227904801	1.280528782	1.300935456	1.322111508	1.345162349	1.360748621	1.372416147	1.389453798	1.407278525
ZF4	1	1.682066422	1.821168870	1.882244583	1.948742791	2.024036859	2.076284698	2.115925965	2.174396359	2.235917227

(4) TH = 62 SBE = 1.132570051 MCPER FUNCTIONS

P 5+4	MU	RC = 02	RC = 05	RC = 07	RC = 10	RC = 15	RC = 20	RC = 25	RC = 35	RC = 50
AMAX DB		.0017375255	.0108709564	.0213327376	.0436480540	.0988323393	.1772876696	.2802872360	.5675287486	1.249387366
AMIN DB		8.435818323	19.65310556	24.38467008	29.60246416	35.74831096	40.25452001	43.85552002	49.54640772	56.21708147
K	MU	.1747589D 1	.4923712D 0	.2967502D 0	.1700362D 0	.8807094D-1	.5423578D-1	.3675136D-1	.1980223D-1	.9540785D-2
PFR	1	2.474384491	1.322191896	1.084476385	.8849852317	.7010893602	.5894297127	.5105296169	.4006948612	.2912873253
PF1	1	1.423255624	1.197419523	1.105323571	1.011073358	.9116780996	.8471191528	.8005570089	.7358819478	.6735236680
PQ1	1	1.702121886	1.192932416	1.130426965	1.105443095	1.121110685	1.162288153	1.215736825	1.348123833	1.614963238
PFC	3	1.159190773	1.148855988	1.139966640	1.127161224	1.108616128	1.093179845	1.080069697	1.058600956	1.033500057
PQC	3	24.41229117	11.40018638	9.303841315	7.831280265	6.765223889	6.282297300	6.029947535	5.828450350	5.865169537
ZF1	1	1.141622133	1.138811721	1.138432808	1.138248340	1.138220937	1.138282403	1.138364071	1.138531802	1.138761750
ZF2	1	1.164713609	1.172716704	1.176030088	1.179606935	1.183694881	1.186593665	1.188842974	1.192260002	1.196008430
ZF3	1	1.208532644	1.258120195	1.277899262	1.298623328	1.321360208	1.336818317	1.348426997	1.365432859	1.383268221
ZF4	1	1.641578796	1.770829939	1.829607556	1.894438664	1.968647219	2.020542496	2.060099746	2.118694401	2.180618087

(4) P 5+4 — TH = 63 SBE = 1.122326238 — MCPER FUNCTIONS

MU	K	RC = 02	RC = 05	RC = 07	RC = 10	RC = 15	RC = 20	RC = 25	RC = 35	RC = 50
	AMAX DB	.0017375255	.0108709564	.0213327376	.0436480540	.0988323393	.1772876696	.2802872360	.5675287486	1.249387366
	AMIN DB	7.402020813	18.21832536	22.88556721	28.05708712	34.16685792	38.65449734	42.24399637	47.92117582	54.58084821
MU	K	.1953416D 1	.5620939D 0	.3407121D 0	.1962224D 0	.1021283D 0	.6307905D-1	.4283239D-1	.2314395D-1	.1118095D-1
1	PFR	2.626719332	1.371372329	1.118182147	.9081905898	.7166793347	.6013018071	.5201447667	.4076143383	.2959527038
1	PF1	1.418826685	1.205634353	1.114919092	1.020959752	.9211209284	.8560272112	.8090079037	.7436434495	.6806035074
1	PQ1	1.841894472	1.243691158	1.167957006	1.134162000	1.143764134	1.182292924	1.234387321	1.365736116	1.632945957
3	PFC	1.147948198	1.139783736	1.132118871	1.120719506	1.103790649	1.089457315	1.077159598	1.056832107	1.032834767
3	PQC	28.08099196	12.69486672	10.23070986	8.508793597	7.264866404	6.698213527	6.397396488	6.142764278	6.143430843
1	ZF1	1.131484238	1.128444993	1.128009324	1.127780467	1.127716172	1.127757509	1.127826103	1.127977324	1.128192862
1	ZF2	1.152304082	1.159721888	1.162882739	1.166322055	1.170277670	1.173095948	1.175289719	1.178633324	1.182315632
1	ZF3	1.190432310	1.236814953	1.255895615	1.276102736	1.298465552	1.313761014	1.325288084	1.342228196	1.360053752
1	ZF4	1.603560666	1.722730163	1.779015720	1.841979843	1.914904041	1.966329453	2.005725644	2.064349736	2.126599779

(4) P 5+4 — TH = 64 SBE = 1.112601940 — MCPER FUNCTIONS

MU	K	RC = 02	RC = 05	RC = 07	RC = 10	RC = 15	RC = 20	RC = 25	RC = 35	RC = 50
	AMAX DB	.0017375255	.0108709564	.0213327376	.0436480540	.0988323393	.1772876696	.2802872360	.5675287486	1.249387366
	AMIN DB	6.446072893	16.81274284	21.40687813	26.52657584	32.59640950	37.06361240	40.64048371	46.30263555	52.95026493
MU	K	.2176577D 1	.6401466D 0	.3903757D 0	.2260363D 0	.1182480D 0	.7326303D-1	.4985601D-1	.2701902D-1	.1309006D-1
1	PFR	2.796068537	1.425661161	1.155076708	.9333683004	.7334470862	.6140046744	.5303971572	.4149600246	.3008868791
1	PF1	1.412695656	1.213255826	1.124331855	1.030931741	.9307990401	.8652149512	.8177521452	.7516990950	.6879649943
1	PQ1	2.003587735	1.301567221	1.210286392	1.166189988	1.168739974	1.204193414	1.254706007	1.384800813	1.652299245
3	PFC	1.137132933	1.130835402	1.124309603	1.114253591	1.098902538	1.085662299	1.074176996	1.054999966	1.032124347
3	PQC	32.42659742	14.21611135	11.30910295	9.288571339	7.833308436	7.167875539	6.810107983	6.493179915	6.451421552
1	ZF1	1.121865785	1.118613856	1.118119761	1.117845020	1.117742619	1.117763160	1.117818219	1.117952278	1.118152721
1	ZF2	1.140549609	1.147367726	1.150369810	1.153665866	1.157482916	1.160216183	1.162350832	1.165615449	1.169225455
1	ZF3	1.173558061	1.216597558	1.234908454	1.254531389	1.276456702	1.291552658	1.302973632	1.319816703	1.337604874
1	ZF4	1.567868066	1.676830776	1.730436743	1.791332709	1.862766848	1.913598711	1.952751666	2.011302958	2.073795898

(4) P 5+4 — TH = 65 SBE = 1.103377919 — MCPER FUNCTIONS

MU	K	RC = 02	RC = 05	RC = 07	RC = 10	RC = 15	RC = 20	RC = 25	RC = 35	RC = 50
	AMAX DB	.0017375255	.0108709564	.0213327376	.0436480540	.0988323393	.1772876696	.2802872360	.5675287486	1.249387366
	AMIN DB	5.571140212	15.43874771	19.94938626	25.01075035	31.03621668	35.48087675	39.04385813	44.68951056	51.32393827
MU	K	.2417908D 1	.7272852D 0	.4463629D 0	.2599279D 0	.1367125D 0	.8498127D-1	.5796300D-1	.3151062D-1	.1531172D-1
1	PFR	2.984075553	1.485759194	1.195590223	.9607698413	.7515306966	.6276297780	.5413536533	.4227736573	.3061144592
1	PF1	1.404924920	1.220153550	1.133473421	1.040944395	.9407001560	.8746834173	.8267971394	.7600615716	.6956230296
1	PQ1	2.190746520	1.367857797	1.258254513	1.202069066	1.196387985	1.228258505	1.276918956	1.405502163	1.673135758
3	PFC	1.126757315	1.122035715	1.116559068	1.107777642	1.093959400	1.081798672	1.071123532	1.053103868	1.031366899
3	PQC	37.58106254	16.01359898	12.57129662	10.19131159	8.483563245	7.700957827	7.275923564	6.835605175	6.793722725
1	ZF1	1.112739050	1.109298143	1.108744343	1.108422325	1.108280664	1.108279779	1.108320868	1.108437149	1.108621841
1	ZF2	1.129423876	1.135632995	1.138469946	1.141616950	1.145289242	1.147933035	1.150005000	1.153185053	1.156716501
1	ZF3	1.157858939	1.197454422	1.214924543	1.233894175	1.255315297	1.270172437	1.281461008	1.298173232	1.315894089
1	ZF4	1.534346288	1.633093673	1.683841758	1.742468626	1.812200801	1.862309041	1.901131262	1.959499643	2.022144417

(4) P 5+4	TH = 66	SBE = 1.094636279						MCPER FUNCTIONS
	RC = 02	RC = 05	RC = 07	RC = 10	RC = 15	RC = 20	RC = 25	RC = 35	RC = 50
AMAX DB		.0108709564	.0213327376	.0436480540	.0988323393	.1772876696	.2802872360	.5675287486	1.249387366
AMIN DB		14.09939546	18.51427823	23.50961585	29.48558945	33.90531511	37.45298565	43.08049032	49.70042316
MU I K		.8242934D 0	.5093438D 0	.2983957D 0	.1578414D 0	.9845499D-1	.6731534D-1	.3671523D-1	.1789689D-1
1 I PFR		1.552468304	1.240222418	.9906883566	.7710905485	.6422828898	.5530912947	.4311030595	.3116634858
1 I PF1		1.226189034	1.142242296	1.050942891	.9508070716	.8844313322	.8361493775	.7637439808	.7035936398
1 I PQ1		1.444134637	1.312882021	1.242455484	1.227127953	1.254808692	1.301294247	1.428057034	1.695795582
3 I PFC		1.113409692	1.108888509	1.101307052	1.088969885	1.077871083	1.068001398	1.051143400	1.030560541
3 I PQC		18.14964104	14.05791818	11.24296256	9.231801310	8.309396890	7.804464524	7.327229396	7.175861792
1 I ZF1		1.100478745	1.099864551	1.099494039	1.099312036	1.099289136	1.099315854	1.099413780	1.099582098
1 I ZF2		1.124498093	1.127163475	1.130155554	1.133676913	1.136226810	1.138232548	1.141322459	1.144769025
1 I ZF3		1.179373285	1.195933155	1.214179039	1.235026215	1.249602728	1.260730666	1.277275526	1.294896570
1 I ZF4		1.591480063	1.639204597	1.695363493	1.763176438	1.812424644	1.850823142	1.909890146	1.971587438

(4) P 5+4	TH = 67	SBE = 1.086360377						MCPER FUNCTIONS
	RC = 02	RC = 05	RC = 07	RC = 10	RC = 15	RC = 20	RC = 25	RC = 35	RC = 50
AMAX DB		.0108709564	.0213327376	.0436480540	.0988323393	.1772876696	.2802872360	.5675287486	1.249387366
AMIN DB		12.79847436	17.10323382	22.02341096	27.94391087	32.33596700	35.86671810	41.47422060	48.07820981
MU I K		.9319785D 0	.5800360D 0	.3419905D 0	.1819959D 0	.1139372D 0	.7809960D-1	.4274503D-1	.2090526D-1
1 I PFR		1.626701786	1.289553897	1.023466590	.7923137591	.6580869248	.5656992905	.4400033059	.3175660747
1 I PF1		1.231219326	1.150523597	1.060861016	.9610965251	.8944544033	.8458140297	.7777597381	.7118940635
1 I PQ1		1.532305938	1.375415924	1.288148874	1.261467131	1.284228612	1.328151919	1.452722374	1.720352263
3 I PFC		1.104982386	1.101320093	1.094858469	1.083943767	1.073885047	1.064813419	1.049118450	1.029703427
3 I PQC		20.70270787	15.82044400	12.47631802	10.09828983	9.008047067	8.407636212	7.826873337	7.604560701
1 I ZF1		1.092137414	1.091462984	1.091043021	1.090819705	1.090774257	1.090786237	1.090865281	1.091016642
1 I ZF2		1.113944859	1.116432300	1.119263494	1.122627746	1.125079360	1.127015353	1.130009545	1.133364837
1 I ZF3		1.162342504	1.177925707	1.195376805	1.215577486	1.229829044	1.240766117	1.257104195	1.274590103
1 I ZF4		1.551948901	1.596500932	1.649997274	1.715669431	1.763914987	1.801791130	1.859429448	1.922070971

(4) P 5+4	TH = 68	SBE = 1.078534743						MCPER FUNCTIONS
	RC = 02	RC = 05	RC = 07	RC = 10	RC = 15	RC = 20	RC = 25	RC = 35	RC = 50
AMAX DB		.0108709564	.0213327376	.0436480540	.0988323393	.1772876696	.2802872360	.5675287486	1.249387366
AMIN DB		11.54053599	15.71852815	20.55267081	26.41065741	30.77189070	34.28388942	39.86929283	46.45570853
MU I K		.1051167D 1	.6592041D 0	.3913192D 0	.2095844D 0	.1317175D 0	.9053100D-1	.4973064D-1	.2440697D-1
1 I PFR		1.709493914	1.344258709	1.059506379	.8154196386	.6751854519	.5792814910	.4495381674	.3238592072
1 I PF1		1.235102152	1.158189396	1.070619633	.9715378467	.9047444546	.8557944182	.7871224247	.7205428384
1 I PQ1		1.634692391	1.447386916	1.340130026	1.300022982	1.316983289	1.357876821	1.479804765	1.747120480
3 I PFC		1.096778559	1.093876790	1.088449805	1.078892040	1.069847045	1.061563143	1.047029272	1.028793770
3 I PQC		23.77201776	17.92453160	13.93317595	11.10866552	9.815565528	9.100305633	8.395455977	8.088064062
1 I ZF1		1.084256558	1.083523223	1.083053247	1.082787794	1.082719326	1.082716245	1.082775938	1.082909805
1 I ZF2		1.103956374	1.106259784	1.108924074	1.112125030	1.114474003	1.116336755	1.119229655	1.122487224
1 I ZF3		1.146350222	1.160895323	1.177480960	1.196960220	1.210840021	1.221553924	1.237642705	1.254955075
1 I ZF4		1.514455070	1.555707260	1.606353490	1.669660345	1.716754666	1.754004068	1.811077039	1.873544755

(4) P 5+4	TH = 69	SBE = 1.071144994						MCPER FUNCTIONS	
	RC = 02	RC = 05	RC = 07	RC = 10	RC = 15	RC = 20	RC = 25	RC = 35	RC = 50
AMAX DB		.0108709564	.0213327376	.0436480540	.0988323393	.1772876696	.2802872360	.5675287486	1.249387366
AMIN DB		10.33086354	14.36314012	19.09830790	24.88542801	29.21217073	32.70331302	38.26423215	44.83123203
MU I K		.1182704D 1	.7476569D 0	.4470480D 0	.2410678D 0	.1521274D 0	.1048582D 0	.5782457D-1	.2848462D-1
1 I PFR		1.802008343	1.405118063	1.099279972	.8406664350	.6937470697	.5939594763	.4597819161	.3305857176
1 I PF1		1.237702566	1.165100052	1.080125199	.9820913630	.9152883449	.8660913353	.7968455701	.7295598855
1 I PQ1		1.754121229	1.530682502	1.399610200	1.343553032	1.353639414	1.390935401	1.509672828	1.776415934
3 I PFC		1.088822294	1.086582209	1.082100232	1.073826999	1.065764632	1.058254943	1.044876556	1.027829881
3 I PQC		27.48356478	20.45449388	15.66728642	12.29567874	10.75562103	9.901223150	9.046624734	8.636578325
1 I ZF1		1.076819044	1.076029677	1.075509693	1.075201488	1.075109610	1.075091198	1.075131134	1.075247026
1 I ZF2		1.094516738	1.096630632	1.099122009	1.102153454	1.104395452	1.106181489	1.108967535	1.112120877
1 I ZF3		1.131383396	1.144836296	1.160487416	1.179168540	1.192627402	1.203083722	1.218877413	1.235974479
1 I ZF4		1.478947352	1.516799678	1.564418631	1.625134411	1.670923292	1.707435741	1.763796849	1.825962112

(4) P 5+4	TH = 70	SBE = 1.064177772						MCPER FUNCTIONS	
	RC = 02	RC = 05	RC = 07	RC = 10	RC = 15	RC = 20	RC = 25	RC = 35	RC = 50
AMAX DB		.0108709564	.0213327376	.0436480540	.0988323393	.1772876696	.2802872360	.5675287486	1.249387366
AMIN DB		9.175349807	13.04085536	17.66171343	23.36798489	27.65593022	31.12378060	36.65748406	43.20297449
MU I K		.1327461D 1	.8462458D 0	.5099055D 0	.2769667D 0	.1755472D 0	.1213687D 0	.6720528D-1	.3323582D-1
1 I PFR		1.905545477	1.473034928	1.143343430	.8683596953	.7139708928	.6098764425	.4708215979	.3377955347
1 I PF1		1.238900811	1.171106857	1.089268495	.9927065336	.9260666212	.8767021634	.8069423438	.7389665832
1 I PQ1		1.894043980	1.627640626	1.468095906	1.392994475	1.394893539	1.427897857	1.542773489	1.808618233
3 I PFC		1.081136528	1.079460383	1.075830149	1.068762329	1.061646551	1.054894134	1.042661513	1.026810202
3 I PQC		31.99814569	23.51935292	17.74841667	13.70161310	11.85856659	10.83429294	9.797617726	9.262867286
1 I ZF1		1.069808013	1.068967428	1.068398236	1.068046956	1.067931381	1.067897426	1.067917278	1.068014782
1 I ZF2		1.085610810	1.087530747	1.089843344	1.092699048	1.094829755	1.096535626	1.099209270	1.102251834
1 I ZF3		1.117426732	1.129743438	1.144394211	1.162199492	1.175186042	1.185348253	1.200797622	1.217633960
1 I ZF4		1.445366284	1.479752399	1.524181465	1.582081274	1.626405405	1.662064851	1.717557220	1.779279836

(4) P 5+4	TH = 71	SBE = 1.057620681						MCPER FUNCTIONS	
	RC = 02	RC = 05	RC = 07	RC = 10	RC = 15	RC = 20	RC = 25	RC = 35	RC = 50
AMAX DB		.0108709564	.0213327376	.0436480540	.0988323393	.1772876696	.2802872360	.5675287486	1.249387366
AMIN DB		8.080261523	11.75634430	16.24488036	21.85831005	26.10235062	29.54406357	35.04739931	41.56898654
MU I K		.1486348D 1	.9558644D 0	.5806855D 0	.3178678D 0	.2024129D 0	.1403962D 0	.7808225D-1	.3877636D-1
1 I PFR		2.021549741	1.549049252	1.192352373	.8988626601	.7360934760	.6272021293	.4827599210	.3455472585
1 I PF1		1.238600705	1.176056306	1.097923804	1.003319815	.9370518545	.9876197402	.8174251198	.7487858244
1 I PQ1		2.058686906	1.741170292	1.547474654	1.449517262	1.441609852	1.469467744	1.579653577	1.844187860
3 I PFC		1.073742539	1.072535483	1.069661123	1.063713184	1.057502859	1.051487098	1.040385979	1.025733360
3 I PQC		37.52224976	27.26110931	20.26801551	15.38168626	13.16378687	11.93034967	10.67046070	9.983071153
1 I ZF1		1.063206736	1.062322062	1.061705538	1.061311272	1.061171842	1.061122208	1.061121739	1.061200516
1 I ZF2		1.077223948	1.078947075	1.081075375	1.083749128	1.085764244	1.087386520	1.089942232	1.092867427
1 I ZF3		1.104461591	1.115611269	1.129201134	1.146052947	1.158513920	1.168343437	1.183395681	1.199921902
1 I ZF4		1.413642100	1.444535962	1.485632184	1.540494721	1.583190395	1.617875017	1.672330930	1.733458128

(4) P 5+4 — TH = 72 — SBE = 1.051462224 — MCPER FUNCTIONS

MU	K	RC = 02	RC = 05	RC = 07	RC = 10	RC = 15	RC = 20	RC = 25	RC = 35	RC = 50
	AMAX DB		.0108709564	.0213327376	.0436480540	.0988323393	.1772876696	.2802872360	.5675287486	1.249387366
	AMIN DB		7.051882413	10.51518712	14.85054647	20.35668290	24.55070178	27.96291855	33.43221699	39.92714505
MU	K		.1660346D 1	.1077450D 1	.6602492D 0	.3644323D 0	.2332253D 0	.1623279D 0	.9070215D-1	.4524416D-1
1	PFR		2.151618996	1.634353645	1.247080308	.9326092128	.7603976071	.6461391227	.4957189636	.3539101834
1	PF1		1.236737491	1.179796200	1.105948868	1.013852272	.9482065923	.8988308993	.8283048693	.7590420402
1	PQ1		2.253248299	1.874908447	1.640129317	1.514596074	1.494871456	1.516522067	1.620988896	1.883688858
3	PFC		1.066659404	1.065831457	1.063615784	1.058696265	1.053345058	1.048041425	1.038052527	1.024598227
3	PQC		44.32317489	31.86617802	23.34719695	17.40889762	14.72304319	13.22968223	11.69365505	10.81785086
1	ZF1		1.056998545	1.056079498	1.055418932	1.054982330	1.054819060	1.054753702	1.054732782	1.054792584
1	ZF2		1.069341748	1.070867402	1.072806565	1.075292249	1.077187500	1.078722773	1.081155046	1.083956250
1	ZF3		1.092465008	1.102433063	1.114909235	1.130731469	1.142612146	1.152068440	1.166667107	1.182829553
1	ZF4		1.383693051	1.411115116	1.448761308	1.500372347	1.541272433	1.574854817	1.628095257	1.688460564

(4) P 5+4 — TH = 73 — SBE = 1.045691756 — MCPER FUNCTIONS

MU	K	RC = 02	RC = 05	RC = 07	RC = 10	RC = 15	RC = 20	RC = 25	RC = 35	RC = 50
	AMAX DB		.0108709564	.0213327376	.0436480540	.0988323393	.1772876696	.2802872360	.5675287486	1.249387366
	AMIN DB		6.096053599	9.323810712	13.48235021	18.86378488	23.00038689	26.37909921	31.81004571	38.27511622
MU	K		.1850553D 1	.1211995D 1	.7495282D 0	.4174042D 0	.2685598D 0	.1876148D 0	.1053565D 0	.5280424D-1
1	PFR		2.297519820	1.730309851	1.308439760	.9701200360	.7872235487	.6669309839	.5098449828	.3629669254
1	PF1		1.233283869	1.182183489	1.113186036	1.024207007	.9594808644	.9103145979	.8395903182	.7697611700
1	PQ1		2.484165045	2.033426978	1.749092091	1.590109062	1.556050927	1.570166347	1.667623942	1.927819567
3	PFC		1.059903502	1.059371584	1.057717670	1.053729377	1.049186235	1.044566080	1.035664609	1.023403997
3	PQC		52.75054757	37.58145794	27.14813685	19.88104472	16.60531994	14.78568100	12.90460030	11.79401455
1	ZF1		1.051166867	1.050225838	1.049526286	1.049048772	1.048861899	1.048780885	1.048739513	1.048780197
1	ZF2		1.061949831	1.063280141	1.065026431	1.067318162	1.069089310	1.070534204	1.072837563	1.075508129
1	ZF3		1.081408971	1.090199758	1.101520207	1.116240120	1.127484954	1.136525775	1.150610762	1.166351201
1	ZF4		1.355424434	1.379446466	1.413558294	1.461715129	1.500650393	1.532997846	1.584832112	1.644254119

(4) P 5+4 — TH = 74 — SBE = 1.040299436 — MCPER FUNCTIONS

MU	K	RC = 02	RC = 05	RC = 07	RC = 10	RC = 15	RC = 20	RC = 25	RC = 35	RC = 50
	AMAX DB			.0213327376	.0436480540	.0988323393	.1772876696	.2802872360	.5675287486	1.249387366
	AMIN DB			8.189301139	12.14498299	17.38083841	21.45100880	24.79137824	30.17884297	36.61031003
MU	K			.1360565D 1	.8495293D 0	.4776196D 0	.3090787D 0	.2167830D 0	.1223911D 0	.6165512D-1
1	PFR			1.838467261	1.377506451	1.012022789	.8169835037	.6898728341	.5253147228	.3728168742
1	PF1			1.183093411	1.119465106	1.034266551	.9708091821	.9220395223	.8512867797	.7809705456
1	PQ1			2.222511364	1.878254065	1.678475113	1.626908866	1.631811487	1.720627027	1.977454968
3	PFC			1.053177948	1.051990991	1.048833970	1.045041201	1.041071571	1.033226718	1.022150286
3	PQC			44.73737519	31.89058386	22.93104799	18.90397485	16.67020878	14.35313787	12.94687380
1	ZF1			1.044747234	1.044015870	1.043499894	1.043289959	1.043193497	1.043131827	1.043153375
1	ZF2			1.056174080	1.057725415	1.059817766	1.061460654	1.062811829	1.064980341	1.067514114
1	ZF3			1.078898803	1.089035578	1.102586192	1.113139679	1.121721397	1.135229058	1.150484416
1	ZF4			1.349475997	1.380009727	1.424526847	1.461327747	1.492302813	1.542528221	1.600809234

(4) P 5+4 — TH = 75 SBE = 1.035276180 — MCPEP FUNCTIONS

MU	K	RC = 02	RC = 05	RC = 07	RC = 10	RC = 15	RC = 20	RC = 25	RC = 35	RC = 50
	AMAX DB			.0213327376	.0436480540	.0988323393	.1772876696	.2802872360	.5675287486	1.249387366
	AMIN DB			7.119066252	10.84430968	15.90978660	19.90246606	23.19858452	28.53639339	34.92982346
MU	K			.1524337D 1	.9613426D 0	.5460174D 0	.3555444D 0	.2504481D 0	.1422184D 0	.7203703D-1
1	PFR			1.960586372	1.455547035	1.059077324	.8501803530	.7153252712	.5423437967	.3835807952
1	PF1			1.182428938	1.124608323	1.043890484	.9821069936	.9339610461	.8633945431	.7926986377
1	PQ1			2.449545500	2.032654596	1.782847425	1.709733831	1.703282924	1.781368213	2.033706083
3	PFC			1.047270876	1.046460314	1.044030143	1.040926637	1.037570147	1.030744584	1.020837248
3	PQC			53.78174658	37.87620263	26.74242426	21.74750098	18.98167638	16.10634706	14.32373038
1	ZF1			1.039629837	1.038876206	1.038325560	1.038093500	1.037981984	1.037900355	1.037902897
1	ZF2			1.049538148	1.050894721	1.052783060	1.054293676	1.055547855	1.057577146	1.059966475
1	ZF3			1.068513075	1.077455722	1.089778318	1.099586681	1.107664796	1.120528212	1.135230351
1	ZF4			1.321136800	1.348097057	1.388813272	1.423312385	1.452773642	1.501175383	1.558099963

(4) P 5+4 — TH = 76 SBE = 1.030613629 — MCPEP FUNCTIONS

MU	K	RC = 02	RC = 05	RC = 07	RC = 10	RC = 15	RC = 20	RC = 25	RC = 35	RC = 50
	AMAX DB			.0213327376	.0436480540	.0988323393	.1772876696	.2802872360	.5675287486	1.249387366
	AMIN DB			6.120348778	9.587413976	14.45351823	18.35509050	21.59966387	26.88028678	33.23036885
MU	K			.1704661D 1	.1086159D 1	.6236518D 0	.4088364D 0	.2893327D 0	.1653323D 0	.8424263D-1
1	PFR			2.098672373	1.544051470	1.112207228	.8874320964	.7437328599	.5611979735	.3954070581
1	PF1			1.180129108	1.128437857	1.052913710	.9932666107	.9460173817	.8759066485	.8049745889
1	PQ1			2.724058992	2.218888316	1.907390947	1.807544648	1.786978652	1.851631718	2.098004984
3	PFC			1.041668378	1.041150134	1.039341610	1.036861228	1.034076008	1.028225396	1.019465732
3	PQC			65.33096152	45.52523324	31.57295358	25.31608638	21.85746876	18.25914995	15.98915957
1	ZF1			1.034859836	1.034095931	1.033516095	1.033263378	1.033137436	1.033036422	1.033020261
1	ZF2			1.043361202	1.044526116	1.046207071	1.047581670	1.048735672	1.050619963	1.052858725
1	ZF3			1.059020024	1.066778664	1.077828420	1.086839186	1.094369078	1.106518541	1.120594132
1	ZF4			1.294347331	1.317793840	1.354581005	1.386616330	1.414419564	1.460770805	1.516104183

(4) P 5+4 — TH = 77 SBE = 1.026304108 — MCPEP FUNCTIONS

MU	K	RC = 02	RC = 05	RC = 07	RC = 10	RC = 15	RC = 20	RC = 25	RC = 35	RC = 50
	AMAX DB				.0436480540	.0988323393	.1772876696	.2802872360	.5675287486	1.249387366
	AMIN DB				8.382513665	13.01613520	16.80984049	19.99377521	25.20789898	31.50818254
MU	K				.1225301D 1	.7117082D 0	.4699707D 0	.3342876D 0	.1923280D 0	.9863122D-1
1	PFR				1.644772581	1.172539435	.9295039670	.7756489765	.5822086120	.4084802118
1	PF1				1.130785696	1.061146014	1.004152722	.9581247907	.8888058141	.8178274134
1	PQ1				2.445691344	2.057687543	1.924388870	1.886102997	1.933781734	2.172229256
3	PFC				1.036084341	1.034793080	1.032865783	1.030605530	1.025678078	1.018037476
3	PQC				55.43313698	37.79172529	29.86773134	25.49356523	20.94205127	18.03323344
1	ZF1				1.029663685	1.029062171	1.028790966	1.028651532	1.028531995	1.028497644
1	ZF2				1.038611697	1.040083771	1.041319057	1.042369870	1.044104018	1.046185651
1	ZF3				1.056998764	1.066745936	1.074913016	1.081850984	1.093214800	1.106585353
1	ZF4				1.289062570	1.321835811	1.351255224	1.377255175	1.421317514	1.474803914

$P^{(3)}_{6+0}$

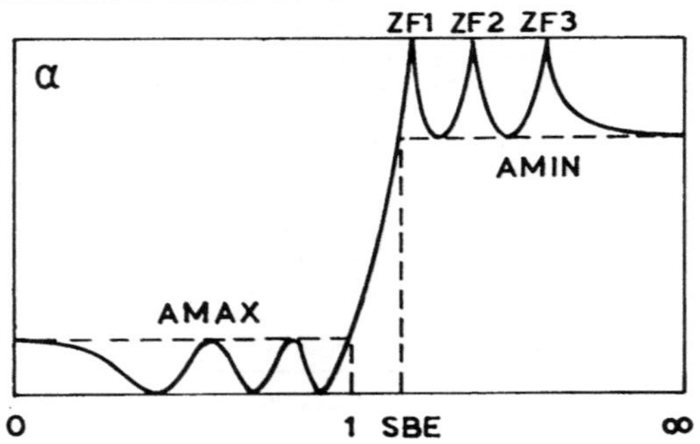

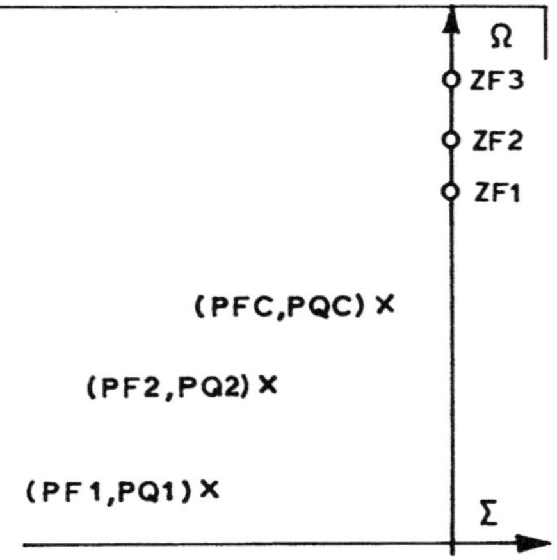

(3) P 6+0	TH = 17	SBE = 3.420303620						CAUER FUNCTIONS	
	RC = 02	RC = 05	RC = 07	RC = 10	RC = 15	RC = 20	RC = 25	RC = 35	RC = 50
AMAX DB	.0017375255	.0108709564	.0213327376	.0436480540	.0988323393	.1772876696	.2802872360	.5675287486	1.249387366
AMIN DB	89.16162702	97.12956062	100.0625831	103.1829376	106.7599471	109.3371772	111.3783770	114.5381792	118.3680770
MU I K	.3482721D-4	.1391624D-4	.9928207D-5	.6931913D-5	.4592008D-5	.3413038D-5	.2698244D-5	.1864623D-5	.1206691D-5
1 I PF1	.9149802027	.7242146358	.6603259710	.5959584713	.5268828606	.4805096506	.4460685406	.3967281294	.3479596642
1 I PQ1	.5442555728	.5591225957	.5678903626	.5804919107	.6012737191	.6227313025	.6455149929	.6971361327	.7991355942
1 I PF2	1.113814925	.9738388065	.9297658116	.8872898340	.8442839276	.8171616727	.7980474258	.7723538897	.7491346446
1 I PQ2	.9587585733	1.075527991	1.140110057	1.228579852	1.365666850	1.498456766	1.632115336	1.915059670	2.424355366
1 I PFC	1.267990915	1.158821366	1.125057966	1.092884753	1.060736828	1.040712496	1.026728387	1.008114135	.9914757193
1 I PQC	3.152906365	3.657953754	3.930991189	4.299680705	4.861807587	5.398575007	5.933277787	7.052431438	9.041058642
1 I ZF1	3.535748289	3.535748289	3.535748289	3.535748289	3.535748289	3.535748289	3.535748289	3.535748289	3.535748289
1 I ZF2	4.783909061	4.783909061	4.783909061	4.783909061	4.783909061	4.783909061	4.783909061	4.783909061	4.783909061
1 I ZF3	12.94415946	12.94415946	12.94415946	12.94415946	12.94415946	12.94415946	12.94415946	12.94415946	12.94415946

(3) P 6+0	TH = 18	SBE = 3.236067977						CAUER FUNCTIONS	
	RC = 02	RC = 05	RC = 07	RC = 10	RC = 15	RC = 20	RC = 25	RC = 35	RC = 50
AMAX DB	.0017375255	.0108709564	.0213327376	.0436480540	.0988323393	.1772876696	.2802872360	.5675287486	1.249387366
AMIN DB	86.13513489	94.10306848	97.03609098	100.1564455	103.7334550	106.3106850	108.3518848	111.5616871	115.3415849
MU I K	.4934501D-4	.1971726D-4	.1406680D-4	.9821498D-5	.6506198D-5	.4835771D-5	.3823013D-5	.2641896D-5	.1709703D-5
1 I PF1	.9188811002	.7267975901	.6625497441	.5978567190	.5284685373	.4819047839	.4473312889	.3978136057	.3488828566
1 I PQ1	.5446268362	.5594322092	.5681845560	.5807738256	.6015471721	.6230030034	.6457885122	.6974201324	.7994497811
1 I PF2	1.115222966	.9751088621	.9309872314	.8884625266	.8454056779	.8182505656	.7991128579	.7733924288	.7501390218
1 I PQ2	.9635015551	1.079899457	1.144460146	1.232988186	1.370269616	1.503310677	1.637256157	1.920868531	2.431462454
1 I PFC	1.266136312	1.157883597	1.124359647	1.092396391	1.060441120	1.040528395	1.026618501	1.008098376	.9915396745
1 I PQC	3.181245958	3.685268029	3.958634421	4.328199077	4.892183537	5.431049098	5.968023624	7.092315305	9.090479346
1 I ZF1	3.344698444	3.344698444	3.344698444	3.344698444	3.344698444	3.344698444	3.344698444	3.344698444	3.344698444
1 I ZF2	4.520147021	4.520147021	4.520147021	4.520147021	4.520147021	4.520147021	4.520147021	4.520147021	4.520147021
1 I ZF3	12.21594006	12.21594006	12.21594006	12.21594006	12.21594006	12.21594006	12.21594006	12.21594006	12.21594006

(3) P 6+0	TH = 19	SBE = 3.071553487						CAUER FUNCTIONS	
	RC = 02	RC = 05	RC = 07	RC = 10	RC = 15	RC = 20	RC = 25	RC = 35	RC = 50
AMAX DB	.0017375255	.0108709564	.0213327376	.0436480540	.0988323393	.1772876696	.2802872360	.5675287486	1.249387366
AMIN DB	83.26684258	91.23477617	94.16779866	97.28815317	100.8651626	103.4423927	105.4835925	108.6933948	112.4732926
MU I K	.6865274D-4	.2743223D-4	.1957087D-4	.1366446D-4	.9051944D-5	.6727913D-5	.5318882D-5	.3675617D-5	.2378676D-5
1 I PF1	.9230371583	.7295462123	.6649153356	.5998753874	.5301542917	.4833876953	.4486733160	.3989670405	.3498637049
1 I PQ1	.5450247607	.5597629749	.5684985552	.5810744618	.6018385440	.6232923640	.6460797098	.6977223532	.7997839953
1 I PF2	1.116703228	.9764494896	.9322777167	.8897024797	.8465925546	.8194030996	.8002408264	.7744869503	.7512029136
1 I PQ2	.9685875798	1.084575533	1.149110059	1.237697290	1.375183332	1.508490535	1.642740763	1.927063744	2.439039981
1 I PFC	1.264179825	1.156893065	1.123621747	1.091880136	1.060128363	1.040333583	1.026502141	1.008081544	.9916071573
1 I PQC	3.211659396	3.714528224	3.988231775	4.358719029	4.924676688	5.465777111	6.005174091	7.134895506	9.143296042
1 I ZF1	3.174063955	3.174063955	3.174063955	3.174063955	3.174063955	3.174063955	3.174063955	3.174063955	3.174063955
1 I ZF2	4.284259547	4.284259547	4.284259547	4.284259547	4.284259547	4.284259547	4.284259547	4.284259547	4.284259547
1 I ZF3	11.56383779	11.56383779	11.56383779	11.56383779	11.56383779	11.56383779	11.56383779	11.56383779	11.56383779

(3) P 6+0	TH = 20	SBE = 2.923804400						CAUER FUNCTIONS	
	RC = 02	RC = 05	RC = 07	RC = 10	RC = 15	RC = 20	RC = 25	RC = 35	RC = 50
AMAX DB	.0017375255	.0108709564	.0213327376	.0436480540	.0988323393	.1772876696	.2802872360	.5675287486	1.249387366
AMIN DB	80.54018488	88.50811846	91.44114095	94.56149546	98.13850493	100.7157350	102.7569348	105.9667370	109.7466349
MU I K	.9397033D-4	.3754863D-4	.2678816D-4	.1870360D-4	.1239010D-4	.9209016D-5	.7280367D-5	.5031102D-5	.3255879D-5
1 I PF1	.9274541946	.7324637334	.6674253751	.6020166022	.5319418039	.4849598122	.4500958825	.4001894820	.3509030748
1 I PQ1	.5454503845	.5601155376	.5688329142	.5813942997	.6021482538	.6235997729	.6463889581	.6980431558	.8001386113
1 I PF2	1.118254156	.9778602804	.9336370937	.8910097006	.8478447096	.8206195000	.8014316005	.7756427738	.7523266774
1 I PQ2	.9740302607	1.089566413	1.154069286	1.242716175	1.380416797	1.514005236	1.648578300	1.933655234	2.447099658
1 I PFC	1.262122157	1.155849916	1.122844324	1.091335996	1.059798534	1.040128027	1.026379273	1.008063610	.9916781456
1 I PQC	3.244233345	3.745806783	4.019853430	4.391310019	4.959358031	5.502833022	6.044806939	7.180309092	9.199614475
1 I ZF1	3.020784723	3.020784723	3.020784723	3.020784723	3.020784723	3.020784723	3.020784723	3.020784723	3.020784723
1 I ZF2	4.072065660	4.072065660	4.072065660	4.072065660	4.072065660	4.072065660	4.072065660	4.072065660	4.072065660
1 I ZF3	10.97642760	10.97642760	10.97642760	10.97642760	10.97642760	10.97642760	10.97642760	10.97642760	10.97642760

(3) P 6+0	TH = 21	SBE = 2.790428110						CAUER FUNCTIONS	
	RC = 02	RC = 05	RC = 07	RC = 10	RC = 15	RC = 20	RC = 25	RC = 35	RC = 50
AMAX DB	.0017375255	.0108709564	.0213327376	.0436480540	.0988323393	.1772876696	.2802872360	.5675287486	1.249387366
AMIN DB	77.94101623	85.90894977	88.84197226	91.96232678	95.53933624	98.11656630	100.1577661	103.3675684	107.1474662
MU I K	.1267504D-3	.5064686D-4	.3613278D-4	.2522805D-4	.1671218D-4	.1242143D-4	.9820005D-5	.6786121D-5	.4391640D-5
1 I PF1	.9321384686	.7355536223	.6700826840	.6042826425	.5338328741	.4866226635	.4516003383	.4014820524	.3520018922
1 I PQ1	.5459048415	.5604905968	.5691882334	.5817338589	.6024767543	.6239256494	.6467166583	.6983329287	.8005140320
1 I PF2	1.119874062	.9793407819	.9350651642	.8923841876	.8491622974	.8219000007	.8026854623	.7753602362	.7535106907
1 I PQ2	.9798444323	1.094883161	1.159348110	1.248054593	1.385979527	1.519864394	1.654778643	1.940653715	2.455654115
1 I PFC	1.259964058	1.154754311	1.122027446	1.090763978	1.059451611	1.039911697	1.026249865	1.008044543	.9917526160
1 I PQC	3.279062471	3.779182544	4.053575681	4.426047490	4.996304603	5.542297062	6.087006464	7.223650396	9.259549191
1 I ZF1	2.882383650	2.882383650	2.882383650	2.882383650	2.882383650	2.882383650	2.882383650	2.882383650	2.882383650
1 I ZF2	3.880180768	3.880180768	3.880180768	3.880180768	3.880180768	3.880180768	3.880180768	3.880180768	3.880180768
1 I ZF3	10.44446014	10.44446014	10.44446014	10.44446014	10.44446014	10.44446014	10.44446014	10.44446014	10.44446014

(3) P 6+0	TH = 22	SBE = 2.669467163						CAUER FUNCTIONS	
	RC = 02	RC = 05	RC = 07	RC = 10	RC = 15	RC = 20	RC = 25	RC = 35	RC = 50
AMAX DB	.0017375255	.0108709564	.0213327376	.0436480540	.0988323393	.1772876696	.2802872360	.5675287486	1.249387366
AMIN DB	75.45715956	83.42509306	86.35811555	89.47847006	93.05547952	95.63270958	97.67390941	100.8837116	104.6636095
MU K	.1687105D-3	.6741326D-4	.4809437D-4	.3357968D-4	.2224467D-4	.1653349D-4	.1307087D-4	.9032634D-5	.5845471D-5
1 PF1	.9370967054	.7388195971	.6728902840	.6066759479	.5358294282	.4883778840	.4531881264	.4028459510	.3531611463
1 PQ1	.5463893694	.5608889114	.5695651630	.5820937009	.6028245337	.6242704453	.6470632424	.6987420300	.8009106904
1 PF2	1.121561115	.9808904945	.9365617027	.8938259275	.8505454742	.8232448439	.8040027059	.7781396909	.7547553515
1 PQ2	.9860462474	1.100537768	1.164957661	1.253723086	1.391881799	1.526078383	1.661352448	1.948070736	2.464716961
1 PFC	1.257706321	1.153606420	1.121171183	1.090164093	1.059087568	1.039684557	1.026113879	1.008024310	.9918305435
1 PQC	3.316250113	3.814741229	4.089481397	4.463013299	5.035599915	5.584256157	6.131863960	7.230021532	9.323224130
1 ZF1	2.756834146	2.756834146	2.756834146	2.756834146	2.756834146	2.756834146	2.756834146	2.756834146	2.756834146
1 ZF2	3.705835670	3.705835670	3.705835670	3.705835670	3.705835670	3.705835670	3.705835670	3.705835670	3.705835670
1 ZF3	9.960367278	9.960367278	9.960367278	9.960367278	9.960367278	9.960367278	9.960367278	9.960367278	9.960367278

(3) P 6+0	TH = 23	SBE = 2.559304665						CAUER FUNCTIONS	
	RC = 02	RC = 05	RC = 07	RC = 10	RC = 15	RC = 20	RC = 25	RC = 35	RC = 50
AMAX DB	.0017375255	.0108709564	.0213327376	.0436480540	.0988323393	.1772876696	.2802872360	.5675287486	1.249387366
AMIN DB	73.07805563	81.04598905	83.97901153	87.09936604	90.67637550	93.25360556	95.29480538	98.50460761	102.2845054
MU K	.2218693D-3	.8865445D-4	.6324838D-4	.4416027D-4	.2925373D-4	.2174301D-4	.1718936D-4	.1187872D-4	.7687316D-5
1 PF1	.9423361213	.7422656379	.6758514076	.6091991262	.5379335230	.4902272194	.4548607868	.4042824571	.3543818924
1 PQ1	.5469053188	.5613113037	.5699644056	.5824744318	.6031921182	.6246346469	.6474291751	.6991210889	.8013290516
1 PF2	1.123313330	.9825088668	.9381264538	.8953348937	.8519943964	.8246542793	.8053836370	.7794815085	.7560610785
1 PQ2	.9926532854	1.106543221	1.170909981	1.259733044	1.398134699	1.532658387	1.668311195	1.955018734	2.474302841
1 PFC	1.255349787	1.152406424	1.120275613	1.089536353	1.058706380	1.039446573	1.025971279	1.008002878	.9919119016
1 PQC	3.355909038	3.852575996	4.127660528	4.502296208	5.077334429	5.628804413	6.179478222	7.334532946	9.390773286
1 ZF1	2.642462192	2.642462192	2.642462192	2.642462192	2.642462192	2.642462192	2.642462192	2.642462192	2.642462192
1 ZF2	3.546742776	3.546742776	3.546742776	3.546742776	3.546742776	3.546742776	3.546742776	3.546742776	3.546742776
1 ZF3	9.517896594	9.517896594	9.517896594	9.517896594	9.517896594	9.517896594	9.517896594	9.517896594	9.517896594

(3) P 6+0	TH = 24	SBE = 2.458593336						CAUER FUNCTIONS	
	RC = 02	RC = 05	RC = 07	RC = 10	RC = 15	RC = 20	RC = 25	RC = 35	RC = 50
AMAX DB	.0017375255	.0108709564	.0213327376	.0436480540	.0988323393	.1772876696	.2802872360	.5675287486	1.249387366
AMIN DB	70.79448685	78.76242015	81.69544261	84.81579712	88.39280657	90.97003663	93.01123646	96.22103868	100.0009365
MU K	.2885863D-3	.1153132D-3	.8226742D-4	.5743943D-4	.3805044D-4	.2828122D-4	.2235827D-4	.1545079D-4	.9998922D-5
1 PF1	.9478644514	.7458960002	.6789695081	.6118549613	.5401473529	.4921725308	.4566199613	.4057929341	.3556652543
1 PQ1	.5474541632	.5617586643	.5703867212	.5828767052	.6035800742	.6250187773	.6478149558	.6995204075	.8017696141
1 PF2	1.125128561	.9841952914	.9397591289	.8969110444	.8535092198	.8261285630	.8068285729	.7808860757	.7574283106
1 PQ2	.9996846732	1.112913580	1.177218083	1.266096765	1.404750182	1.539616457	1.675667244	1.964211088	2.484427506
1 PFC	1.252895343	1.151154518	1.119340821	1.088880772	1.058308023	1.039197708	1.025822026	1.007980210	.9919966623
1 PQC	3.398162286	3.892788043	4.168210670	4.543992418	5.121606088	5.676043654	6.229956102	7.392304028	9.462341430
1 ZF1	2.537872890	2.537872890	2.537872890	2.537872890	2.537872890	2.537872890	2.537872890	2.537872890	2.537872890
1 ZF2	3.400995771	3.400995771	3.400995771	3.400995771	3.400995771	3.400995771	3.400995771	3.400995771	3.400995771
1 ZF3	9.111837264	9.111837264	9.111837264	9.111837264	9.111837264	9.111837264	9.111837264	9.111837264	9.111837264

(3) P 6+0	TH = 25	SBE = 2.366201583						CAUER FUNCTIONS	
	RC = 02	RC = 05	RC = 07	RC = 10	RC = 15	RC = 20	RC = 25	RC = 35	RC = 50
AMAX DB	.0017375255	.0108709564	.0213327376	.0436480540	.0988323393	.1772876696	.2802872360	.5675287486	1.249387366
AMIN DB	68.59835765	76.56629075	79.49931320	82.61966769	86.19667714	88.77390720	90.81510702	94.02490925	97.80480707
MU I K	.3716055D-3	.1484860D-3	.1059337D-3	.7396336D-4	.4899662D-4	.3641704D-4	.2879020D-4	.1989549D-4	.1287537D-4
1 I PF1	.9536899796	.7497152301	.6822482719	.6146464226	.5424732568	.4942158011	.4584673985	.4073788332	.3570124277
1 I PQ1	.5480375105	.5622319576	.5708329304	.5833012262	.6039890115	.6254233985	.6482211208	.6999405630	.8022329123
1 I PF2	1.127004487	.9859491001	.9414594023	.8985543202	.8550900978	.8276679571	.8083378413	.7823537961	.7588575080
1 I PQ2	1.007161221	1.119664062	1.183896031	1.272827515	1.411741129	1.546965565	1.683433899	1.972962189	2.495107879
1 I PFC	1.250343924	1.149850909	1.118366896	1.088197366	1.057892470	1.038937923	1.025666079	1.007956267	.9920847958
1 I PQC	3.443144118	3.935487290	4.211237695	4.588206166	5.168520902	5.726084017	6.283413125	7.453463783	9.538084920
1 I ZF1	2.441894644	2.441894644	2.441894644	2.441894644	2.441894644	2.441894644	2.441894644	2.441894644	2.441894644
1 I ZF2	3.266993358	3.266993358	3.266993358	3.266993358	3.266993358	3.266993358	3.266993358	3.266993358	3.266993358
1 I ZF3	8.737811716	8.737811716	8.737811716	8.737811716	8.737811716	8.737811716	8.737811716	8.737811716	8.737811716

(3) P 6+0	TH = 26	SBE = 2.281172033						CAUER FUNCTIONS	
	RC = 02	RC = 05	RC = 07	RC = 10	RC = 15	RC = 20	RC = 25	RC = 35	RC = 50
AMAX DB	.0017375255	.0108709564	.0213327376	.0436480540	.0988323393	.1772876696	.2802872360	.5675287486	1.249387366
AMIN DB	66.48251782	74.45045060	77.38347302	80.50382750	84.08083694	86.65806700	88.69926682	91.90906904	95.68896686
MU I K	.4741045D-3	.1894425D-3	.1351532D-3	.9436450D-4	.6251125D-4	.4646187D-4	.3673133D-4	.2538321D-4	.1642675D-4
1 I PF1	.9598215721	.7537281799	.6856916302	.6175766746	.5449137253	.4963591409	.4604049586	.4090416978	.3584246842
1 I PQ1	.5486571157	.5627322278	.5713039201	.5837487547	.6044195856	.6258491145	.6486482457	.7003821099	.8027195185
1 I PF2	1.128938605	.9877695585	.9432269082	.9002646412	.8567371800	.8292727284	.8099117802	.7838850890	.7603491515
1 I PQ2	1.015105572	1.126811131	1.190959021	1.279939608	1.419121418	1.554719677	1.691625468	1.982187505	2.506362140
1 I PFC	1.247696515	1.148495819	1.117353936	1.087486155	1.057459696	1.038667181	1.025503395	1.007931011	.9921762707
1 I PQC	3.491001083	3.980793139	4.256856454	4.635050393	5.218193597	5.779044617	6.339974174	7.518151579	9.618172585
1 I ZF1	2.353536216	2.353536216	2.353536216	2.353536216	2.353536216	2.353536216	2.353536216	2.353536216	2.353536216
1 I ZF2	3.143380605	3.143380605	3.143380605	3.143380605	3.143380605	3.143380605	3.143380605	3.143380605	3.143380605
1 I ZF3	8.392115367	8.392115367	8.392115367	8.392115367	8.392115367	8.392115367	8.392115367	8.392115367	8.392115367

(3) P 6+0	TH = 27	SBE = 2.202689265						CAUER FUNCTIONS	
	RC = 02	RC = 05	RC = 07	RC = 10	RC = 15	RC = 20	RC = 25	RC = 35	RC = 50
AMAX DB	.0017375255	.0108709564	.0213327376	.0436480540	.0988323393	.1772876696	.2802872360	.5675287486	1.249387366
AMIN DB	64.44061921	72.40855151	75.34157388	78.46192833	82.03893776	84.61616781	86.65736763	89.86716985	93.64706767
MU I K	.5997483D-3	.2396472D-3	.1709705D-3	.1193723D-3	.7907753D-4	.5877486D-4	.4646561D-4	.3211009D-4	.2078005D-4
1 I PF1	.9662687123	.7579400263	.6893037738	.6206490876	.5474714086	.4986047949	.4624346197	.4107831682	.3599033741
1 I PQ1	.5493148950	.5632606059	.5718006481	.5842201098	.6048725019	.6262965739	.6490969482	.7008456425	.8032300456
1 I PF2	1.130928214	.9896558606	.9450612359	.9020419044	.8584506107	.8309431477	.8115507369	.7854803901	.7619037430
1 I PQ2	1.023542374	1.134372606	1.198423466	1.287448485	1.426905995	1.562893821	1.700257342	1.991903655	2.518209812
1 I PFC	1.244954149	1.147089484	1.116302046	1.086747160	1.057009675	1.038385441	1.025333931	1.007904399	.9922710539
1 I PQC	3.541893217	4.028835316	4.305191554	4.684647478	5.270748342	5.835054275	6.399774245	7.586517980	9.702786712
1 I ZF1	2.271953333	2.271953333	2.271953333	2.271953333	2.271953333	2.271953333	2.271953333	2.271953333	2.271953333
1 I ZF2	3.029003316	3.029003316	3.029003316	3.029003316	3.029003316	3.029003316	3.029003316	3.029003316	3.029003316
1 I ZF3	8.071591981	8.071591981	8.071591981	8.071591981	8.071591981	8.071591981	8.071591981	8.071591981	8.071591981

(3) P 6+0	TH = 28	SBE = 2.130054468						CAUER FUNCTIONS	
	RC = 02	RC = 05	RC = 07	RC = 10	RC = 15	RC = 20	RC = 25	RC = 35	RC = 50
AMAX DB	.0017375255	.0108709564	.0213327376	.0436480540	.0988323393	.1772876696	.2802872360	.5675287486	1.249387366
AMIN DB	62.46699851	70.43493005	73.36795235	76.48830676	80.06531617	82.64254622	84.68374603	87.89354825	91.67344606
MU I K	.7527488D-3	.3007831D-3	.2145865D-3	.1498251D-3	.9925084D-4	.7376879D-4	.5831935D-4	.4030163D-4	.2608121D-4
1 I PF1	.9730415402	.7623562890	.6930891671	.6238672493	.5501491257	.5009551493	.4645584836	.4126049865	.3614499314
1 I PQ1	.5500129426	.5638183165	.5723241494	.5847161742	.6053485189	.6267664733	.6495678911	.7013317978	.8037651493
1 I PF2	1.132970405	.9916071224	.9469619259	.9038859810	.8602305268	.8326794879	.8132550672	.7871401507	.7635218055
1 I PQ2	1.032498469	1.142367773	1.206307105	1.295370803	1.435110961	1.571504169	1.709346072	2.002128502	2.530671858
1 I PFC	1.242117913	1.145632153	1.115211341	1.085980409	1.056542382	1.038092662	1.025157641	1.007376388	.9923691103
1 I PQC	3.595995393	4.079754825	4.356378232	4.737130067	5.326319547	5.894252338	6.462959288	7.658725656	9.792124139
1 I ZF1	2.196422441	2.196422441	2.196422441	2.196422441	2.196422441	2.196422441	2.196422441	2.196422441	2.196422441
1 I ZF2	2.922872192	2.922872192	2.922872192	2.922872192	2.922872192	2.922872192	2.922872192	2.922872192	2.922872192
1 I ZF3	7.773535720	7.773535720	7.773535720	7.773535720	7.773535720	7.773535720	7.773535720	7.773535720	7.773535720

(3) P 6+0	TH = 29	SBE = 2.062665340						CAUER FUNCTIONS	
	RC = 02	RC = 05	RC = 07	RC = 10	RC = 15	RC = 20	RC = 25	RC = 35	RC = 50
AMAX DB	.0017375255	.0108709564	.0213327376	.0436480540	.0988323393	.1772876696	.2802872360	.5675287486	1.249387366
AMIN DB	60.55658049	68.52451089	71.45753308	74.57788744	78.15489682	80.73212685	82.77332667	85.98312888	89.76302669
MU I K	.9379312D-3	.3747783D-3	.2673766D-3	.1866834D-3	.1236674D-3	.9191654D-4	.7266640D-4	.5021617D-4	.3249740D-4
1 I PF1	.9801508941	.7669828520	.6970525647	.6272349770	.5529498735	.5034127397	.4667787831	.4145090024	.3630658774
1 I PQ1	.5507535484	.5644066869	.5728755425	.5852378997	.6058484522	.6272595607	.6500617852	.7018412580	.8043255315
1 I PF2	1.135062045	.9936223755	.9489284647	.9057967125	.8620770562	.8344820238	.8150251348	.7888648373	.7652038834
1 I PQ2	1.042003106	1.150817515	1.214629107	1.303724536	1.443753660	1.580568127	1.718909459	2.012381236	2.543770787
1 I PFC	1.239188945	1.144124094	1.114081941	1.085185929	1.056057791	1.037788801	1.024974477	1.007846932	.9924704032
1 I PQC	3.653498846	4.133704999	4.410563332	4.792641984	5.385052765	5.956789583	6.529687140	7.734950411	9.886397467
1 I ZF1	2.126319899	2.126319899	2.126319899	2.126319899	2.126319899	2.126319899	2.126319899	2.126319899	2.126319899
1 I ZF2	2.824134395	2.824134395	2.824134395	2.824134395	2.824134395	2.824134395	2.824134395	2.824134395	2.824134395
1 I ZF3	7.495613474	7.495613474	7.495613474	7.495613474	7.495613474	7.495613474	7.495613474	7.495613474	7.495613474

(3) P 6+0	TH = 30	SBE = 2.000000000						CAUER FUNCTIONS	
	RC = 02	RC = 05	RC = 07	RC = 10	RC = 15	RC = 20	RC = 25	RC = 35	RC = 50
AMAX DB	.0017375255	.0108709564	.0213327376	.0436480540	.0988323393	.1772876696	.2802872360	.5675287486	1.249387366
AMIN DB	58.70479775	66.67272644	69.60574847	72.72610275	76.30311208	78.88034210	80.92154190	84.13134411	87.91124192
MU I K	.1160807D-2	.4638352D-3	.3309120D-3	.2310441D-3	.1530539D-3	.1137582D-3	.8993379D-4	.6214881D-4	.4021961D-4
1 I PF1	.9876083568	.7718259856	.7011990290	.6307563320	.5558768373	.5059802596	.4690978887	.4164971786	.3647528262
1 I PQ1	.5515392191	.5650271561	.5734560367	.5857863128	.6063731792	.6277766397	.6505793936	.7023747542	.8049119434
1 I PF2	1.137199765	.9957005599	.9509602800	.9077739077	.8639903151	.8363510294	.8168613104	.7906549317	.7669505423
1 I PQ2	1.052088181	1.159744455	1.223410201	1.312529082	1.452852783	1.590104430	1.728966650	2.024182486	2.557530770
1 I PFC	1.236168439	1.142565588	1.112913978	1.084363754	1.055555878	1.037473815	1.024784393	1.007815984	.9925748937
1 I PQC	3.714612897	4.190852697	4.467906385	4.851339261	5.447105683	6.022829227	6.600128569	7.815382310	9.985836401
1 I ZF1	2.061105327	2.061105327	2.061105327	2.061105327	2.061105327	2.061105327	2.061105327	2.061105327	2.061105327
1 I ZF2	2.732050808	2.732050808	2.732050808	2.732050808	2.732050808	2.732050808	2.732050808	2.732050808	2.732050808
1 I ZF3	7.235802709	7.235802709	7.235802709	7.235802709	7.235802709	7.235802709	7.235802709	7.235802709	7.235802709

(3) P 6+0			TH = 31	SBE = 1.941604026					CAUER FUNCTIONS	
			RC = 02	RC = 05	RC = 07	RC = 10	RC = 15	RC = 20	RC = 25	RC = 35	RC = 50
AMAX DB			.0017375255	.0108709564	.0213327376	.0436480540	.0988323393	.1772876696	.2802872360	.5675287486	1.249387366
AMIN DB			56.90752347	64.87544963	67.80847143	70.92882558	74.50583485	77.08306484	79.12426464	82.33406684	86.11396464
MU		K	.1427657D-2	.5704630D-3	.4069831D-3	.2841572D-3	.1882384D-3	.1399094D-3	.1106081D-3	.7643577D-4	.4946543D-4
	1	PF1	.9954263044	.7768923713	.7055339497	.6344356337	.5589334020	.5086605696	.4715183174	.4185715981	.3665124896
	1	PQ1	.5523727012	.5656812850	.5740669400	.5863625207	.6069236443	.6283185735	.6511215349	.7029330700	.8055251890
	1	PF2	1.139379945	.9978405166	.9530567349	.9098173387	.8659704062	.8382867776	.8187639702	.7925109301	.7687623691
	1	PQ2	1.062788510	1.169173113	1.232672810	1.321805390	1.462428475	1.600133254	1.739538253	2.036054426	2.571977772
	1	PFC	1.233057640	1.140956935	1.111707593	1.083513922	1.055036618	1.037147660	1.024587336	1.007783495	.9926825408
	1	PQC	3.779566906	4.251379631	4.528580841	4.913391282	5.512649238	6.092548050	6.674468425	7.900226943	10.09068925
	1	ZF1	2.000308179	2.000308179	2.000308179	2.000308179	2.000308179	2.000308179	2.000308179	2.000308179	2.000308179
	1	ZF2	2.645977691	2.645977691	2.645977691	2.645977691	2.645977691	2.645977691	2.645977691	2.645977691	2.645977691
	1	ZF3	6.992341356	6.992341356	6.992341356	6.992341356	6.992341356	6.992341356	6.992341356	6.992341356	6.992341356

(3) P 6+0			TH = 32	SBE = 1.887079915					CAUER FUNCTIONS	
			RC = 02	RC = 05	RC = 07	RC = 10	RC = 15	RC = 20	RC = 25	RC = 35	RC = 50
AMAX DB			.0017375255	.0108709564	.0213327376	.0436480540	.0988323393	.1772876696	.2802872360	.5675287486	1.249387366
AMIN DB			55.16101490	63.12893738	66.06195884	69.18231280	72.75932197	75.33655193	77.37775171	80.58755390	84.36745169
MU		K	.1745618D-2	.6975143D-3	.4976248D-3	.3474436D-3	.2301621D-3	.1710694D-3	.1352423D-3	.9345925D-4	.6048218D-4
	1	PF1	1.003617961	.7821891289	.7100630641	.6382774764	.5621231644	.5114567075	.4740427408	.4207344706	.3683466827
	1	PQ1	.5532570082	.5663707685	.5747096683	.5869677187	.6075008639	.6288862898	.6516890881	.7035170452	.8061661290
	1	PF2	1.141598694	1.000040979	.9552171219	.9119267362	.8680174155	.8402895376	.8207334953	.7944333430	.7706399724
	1	PQ2	1.074142134	1.179130089	1.242441209	1.331576095	1.472502467	1.610676335	1.750646448	2.048520905	2.587139694
	1	PFC	1.229857854	1.139298454	1.110462937	1.082636474	1.054499988	1.036810292	1.024383255	1.007749413	.9927933015
	1	PQC	3.848612487	4.315483874	4.592775430	4.978982064	5.581868863	6.166137652	6.752906937	7.989706830	10.20122460
	1	ZF1	1.943516836	1.943516836	1.943516836	1.943516836	1.943516836	1.943516836	1.943516836	1.943516836	1.943516836
	1	ZF2	2.565351779	2.565351779	2.565351779	2.565351779	2.565351779	2.565351779	2.565351779	2.565351779	2.565351779
	1	ZF3	6.763687081	6.763687081	6.763687081	6.763687081	6.763687081	6.763687081	6.763687081	6.763687081	6.763687081

(3) P 6+0			TH = 33	SBE = 1.836078459					CAUER FUNCTIONS	
			RC = 02	RC = 05	RC = 07	RC = 10	RC = 15	RC = 20	RC = 25	RC = 35	RC = 50
AMAX DB			.0017375255	.0108709564	.0213327376	.0436480540	.0988323393	.1772876696	.2802872360	.5675287486	1.249387366
AMIN DB			53.46186548	61.42978264	64.36280360	67.48315730	71.06016633	73.63739624	75.67859600	78.88339816	82.66829595
MU		K	.2122788D-2	.8482246D-3	.6051455D-3	.4225150D-3	.2798928D-3	.2080320D-3	.1644638D-3	.1136528D-3	.7355043D-4
	1	PF1	1.012197455	.7877238451	.7147924802	.6422867464	.5654499464	.5143718995	.4766739948	.4229331402	.3702573301
	1	PQ1	.5541954504	.5670974483	.5753857556	.5876031979	.6081059328	.6294807865	.6522829971	.7041275805	.8068356849
	1	PF2	1.143851839	1.002300565	.9574406563	.9141017856	.8701314100	.8423595732	.8227702699	.7964226942	.7725839821
	1	PQ2	1.086190663	1.189644260	1.252741693	1.341365671	1.483098208	1.621757104	1.762315123	2.061607583	2.603046529
	1	PFC	1.226570441	1.137590482	1.109180172	1.081731459	1.053945966	1.036461566	1.024172098	1.007713684	.9929071306
	1	PQC	3.922026028	4.383381549	4.660695706	5.048311702	5.654965874	6.243805857	6.835661153	8.084062988	10.31773314
	1	ZF1	1.890369680	1.890369680	1.890369680	1.890369680	1.890369680	1.890369680	1.890369680	1.890369680	1.890369680
	1	ZF2	2.489678091	2.489678091	2.489678091	2.489678091	2.489678091	2.489678091	2.489678091	2.489678091	2.489678091
	1	ZF3	6.548483968	6.548483968	6.548483968	6.548483968	6.548483968	6.548483968	6.548483968	6.548483968	6.548483968

(3) P 6+0	TH = 34	SBE = 1.788291650						CAUER FUNCTIONS	
	RC = 02	RC = 05	RC = 07	RC = 10	RC = 15	RC = 20	RC = 25	RC = 35	RC = 50
AMAX DB	.0017375255	.0108709564	.0213327376	.0436480540	.0988323393	.1772876696	.2802872360	.5675287486	1.249387366
AMIN DB	51.80696412	59.77487365	62.70789390	65.82824722	69.40525605	71.98248589	74.02368561	77.23348775	81.01338552
MU I K	.2568336D-2	.1026257D-2	.7321588D-3	.5111962D-3	.3386392D-3	.2516956D-3	.1989829D-3	.1375073D-3	.8898785D-4
1 I PF1	1.021179885	.7935046054	.7197287016	.6464686416	.5689178097	.5174095719	.4794150901	.4253350941	.3722464727
1 I PQ1	.5551916696	.5678633278	.5760968649	.5882703539	.6087400311	.6301031366	.6529042762	.7047656420	.8075348442
1 I PF2	1.146134902	1.004617766	.9597264691	.9163421217	.8723124342	.8444971410	.8248746802	.7984795207	.7745950496
1 I PQ2	1.098979673	1.200747006	1.263602776	1.352700596	1.494241020	1.633400834	1.774570045	2.075342090	2.619730541
1 I PFC	1.223196822	1.135833374	1.107859471	1.080798929	1.053374530	1.036101734	1.023953810	1.007676253	.9930239804
1 I PQC	4.000111566	4.455308740	4.732565783	5.121597981	5.732159037	6.325778287	6.922966553	8.183556690	10.44052978
1 I ZF1	1.840547742	1.840547742	1.840547742	1.840547742	1.840547742	1.840547742	1.840547742	1.840547742	1.840547742
1 I ZF2	2.418519883	2.418519883	2.418519883	2.418519883	2.418519883	2.418519883	2.418519883	2.418519883	2.418519883
1 I ZF3	6.345535076	6.345535076	6.345535076	6.345535076	6.345535076	6.345535076	6.345535076	6.345535076	6.345535076

(3) P 6+0	TH = 35	SBE = 1.743446796						CAUER FUNCTIONS	
	RC = 02	RC = 05	RC = 07	RC = 10	RC = 15	RC = 20	RC = 25	RC = 35	RC = 50
AMAX DB	.0017375255	.0108709564	.0213327376	.0436480540	.0988323393	.1772876696	.2802872360	.5675287486	1.249387366
AMIN DB	50.19346028	58.16135898	61.09437821	64.21473100	67.79173954	70.36896928	72.41016896	75.61997105	79.39986880
MU I K	.3092623D-2	.1235754D-2	.8816193D-3	.6155502D-3	.4077679D-3	.3030760D-3	.2396026D-3	.1655775D-3	.1071535D-3
1 I PF1	1.030581384	.7995400290	.7248786545	.6508286925	.5725310713	.5205733641	.4822692232	.4277779716	.3743162751
1 I PQ1	.5562496784	.5686705885	.5768448010	.5889706965	.6094044315	.6307544958	.6535540161	.7054322668	.8082646650
1 I PF2	1.148443086	1.006990938	.9620735994	.9186473232	.8745605068	.8467024878	.8270471129	.8006043715	.7766738478
1 I PQ2	1.112559151	1.212472461	1.275055399	1.364109550	1.505958275	1.645634807	1.787438972	2.089754191	2.637226456
1 I PFC	1.219738479	1.134027509	1.106501021	1.079838943	1.052785662	1.035730451	1.023728334	1.007637061	.9931438007
1 I PQC	4.083204074	4.531523654	4.808630286	5.199078200	5.813686321	6.412300122	7.015078864	8.288471420	10.56995594
1 I ZF1	1.793768618	1.793768618	1.793768618	1.793768618	1.793768618	1.793768618	1.793768618	1.793768618	1.793768618
1 I ZF2	2.351490335	2.351490335	2.351490335	2.351490335	2.351490335	2.351490335	2.351490335	2.351490335	2.351490335
1 I ZF3	6.153779695	6.153779695	6.153779695	6.153779695	6.153779695	6.153779695	6.153779695	6.153779695	6.153779695

(3) P 6+0	TH = 36	SBE = 1.701301617						CAUER FUNCTIONS	
	RC = 02	RC = 05	RC = 07	RC = 10	RC = 15	RC = 20	RC = 25	RC = 35	RC = 50
AMAX DB	.0017375255	.0108709564	.0213327376	.0436480540	.0988323393	.1772876696	.2802872360	.5675287486	1.249387366
AMIN DB	48.61873403	56.58661747	59.51963529	62.63998732	66.21699546	68.79422505	70.83542467	74.04522670	77.82512442
MU I K	.3707348D-2	.1481389D-2	.1056862D-2	.7379053D-3	.4888214D-3	.3633195D-3	.2872293D-3	.1984900D-3	.1284529D-3
1 I PF1	1.040419196	.8058393064	.7302497171	.6553727848	.5762943213	.5238671432	.4852397891	.4303195738	.3764690332
1 I PQ1	.5573739061	.5695216090	.5776315247	.5897058608	.6101005079	.6314361090	.6542333903	.7061285690	.8090262825
1 I PF2	1.150771250	1.009418290	.9644809860	.9210169068	.8768756171	.8489758482	.8292879530	.8027978070	.7788210710
1 I PQ2	1.126984002	1.224857801	1.287133182	1.376123620	1.518279578	1.658488497	1.800951890	2.104875981	2.655571682
1 I PFC	1.216196953	1.132173287	1.105105020	1.078851566	1.052179343	1.035347771	1.023495614	1.007596049	.9932665390
1 I PQC	4.171673218	4.612309072	4.889156573	5.281011225	5.899806870	6.503638088	7.112276089	8.399115084	10.70638217
1 I ZF1	1.749781386	1.749781386	1.749781386	1.749781386	1.749781386	1.749781386	1.749781386	1.749781386	1.749781386
1 I ZF2	2.288245611	2.288245611	2.288245611	2.288245611	2.288245611	2.288245611	2.288245611	2.288245611	2.288245611
1 I ZF3	5.972274395	5.972274395	5.972274395	5.972274395	5.972274395	5.972274395	5.972274395	5.972274395	5.972274395

(3) P 6+0	TH = 37 SBE = 1.661640141								CAUER FUNCTIONS
	RC = 02	RC = 05	RC = 07	RC = 10	RC = 15	RC = 20	RC = 25	RC = 35	RC = 50
AMAX DB	.0017375255	.0108709564	.0213327376	.0436480540	.0988323393	.1772876696	.2802872360	.5675287486	1.249387366
AMIN DB	47.08037017	55.04823229	57.98124811	61.10159909	64.67860666	67.25583606	69.29703558	72.50683754	76.28673522
MU K	.4425695D-2	.1768432D-2	.1261646D-2	.8808867D-3	.5835387D-3	.4337187D-3	.3428848D-3	.2369508D-3	.1533427D-3
1 PF1	1.050711753	.8124122400	.7358497520	.6601071846	.5802124413	.5272950192	.4883303942	.4329628746	.3787071834
1 PQ1	.5585692497	.5704189858	.5784591689	.5904776191	.6108297449	.6321493186	.6549436623	.7068557460	.8098209145
1 PF2	1.153113891	1.011897871	.9669474584	.9234503210	.8792577206	.8513174415	.8315975821	.8050603981	.7810374346
1 PQ2	1.142314635	1.237943559	1.299872690	1.388776545	1.531236990	1.671993779	1.815141191	2.120742096	2.674806543
1 PFC	1.212573850	1.130271131	1.103671683	1.077836872	1.051555557	1.034953645	1.023255590	1.007553153	.9933921399
1 PQC	4.265927689	4.697975138	4.974437236	5.367679815	5.990803235	6.600082688	7.214860794	8.515822479	10.85021105
1 ZF1	1.708362360	1.708362360	1.708362360	1.708362360	1.708362360	1.708362360	1.708362360	1.708362360	1.708362360
1 ZF2	2.228479056	2.228479056	2.228479056	2.228479056	2.228479056	2.228479056	2.228479056	2.228479056	2.228479056
1 ZF3	5.800177144	5.800177144	5.800177144	5.800177144	5.800177144	5.800177144	5.800177144	5.800177144	5.800177144

(3) P 6+0	TH = 38 SBE = 1.624269245								CAUER FUNCTIONS
	RC = 02	RC = 05	RC = 07	RC = 10	RC = 15	RC = 20	RC = 25	RC = 35	RC = 50
AMAX DB	.0017375255	.0108709564	.0213327376	.0436480540	.0988323393	.1772876696	.2802872360	.5675287486	1.249387366
AMIN DB	45.57613575	53.54396829	56.47698135	59.59733086	63.17433765	65.75156678	67.79276617	71.00256802	74.78246564
MU K	.5262513D-2	.2102818D-2	.1500206D-2	.1047450D-2	.6938780D-3	.5157291D-3	.4077197D-3	.2817550D-3	.1823378D-3
1 PF1	1.061478759	.8192692900	.7416871414	.6650385658	.5842906252	.5308613623	.4915448714	.4357110327	.3810333117
1 PQ1	.5598411349	.5713655577	.5793300564	.5912878950	.6115937479	.6328955737	.6556861943	.7076150858	.8106493690
1 PF2	1.155465124	1.014427560	.9694717273	.9259469389	.8817067347	.8537274692	.8339763765	.8073927251	.7833236753
1 PQ2	1.158617623	1.251773996	1.313313744	1.402104977	1.544865264	1.686185157	1.830041909	2.137389950	2.694974554
1 PFC	1.208870841	1.128321488	1.102201237	1.076794940	1.050914292	1.034548028	1.023008203	1.007508310	.9935205452
1 PQC	4.366420180	4.788862531	5.064792957	5.459393248	6.086983886	6.701950724	7.323162689	8.638958087	11.00188051
1 ZF1	1.669311509	1.669311509	1.669311509	1.669311509	1.669311509	1.669311509	1.669311509	1.669311509	1.669311509
1 ZF2	2.171916299	2.171916299	2.171916299	2.171916299	2.171916299	2.171916299	2.171916299	2.171916299	2.171916299
1 ZF3	5.636733929	5.636733929	5.636733929	5.636733929	5.636733929	5.636733929	5.636733929	5.636733929	5.636733929

(3) P 6+0	TH = 39 SBE = 1.589015729								CAUER FUNCTIONS
	RC = 02	RC = 05	RC = 07	RC = 10	RC = 15	RC = 20	RC = 25	RC = 35	RC = 50
AMAX DB	.0017375255	.0108709564	.0213327376	.0436480540	.0988323393	.1772876696	.2802872360	.5675287486	1.249387366
AMIN DB	44.10396060	52.07175234	55.00476160	58.12510909	61.70211480	64.27934355	66.32054277	69.53034447	73.31024200
MU K	.6234505D-2	.2491222D-2	.1777305D-2	.1240922D-2	.8220425D-3	.6109882D-3	.4830286D-3	.3337973D-3	.2160170D-3
1 PF1	1.072741281	.8264216225	.7477708255	.6701740400	.5885344020	.5345708214	.4948872963	.4385674044	.3834501643
1 PQ1	.5611955845	.5723644328	.5802467204	.5921387784	.6123942554	.6336764396	.6564624554	.7084079755	.8115145523
1 PF2	1.157818653	1.017005049	.9720523735	.9285060504	.8842225337	.8562061111	.8364247041	.8097953763	.7856805505
1 PQ2	1.175966477	1.266397509	1.327499770	1.416148791	1.559202117	1.701100026	1.845691973	2.154859998	2.716122717
1 PFC	1.205089663	1.126324830	1.100693926	1.075725859	1.050255536	1.034130872	1.022753391	1.007461451	.9936516936
1 PQC	4.473653143	4.885346092	5.160575747	5.556490313	6.188686067	6.809588148	7.437541543	8.768919218	11.16186747
1 ZF1	1.632449427	1.632449427	1.632449427	1.632449427	1.632449427	1.632449427	1.632449427	1.632449427	1.632449427
1 ZF2	2.118311118	2.118311118	2.118311118	2.118311118	2.118311118	2.118311118	2.118311118	2.118311118	2.118311118
1 ZF3	5.481267434	5.481267434	5.481267434	5.481267434	5.481267434	5.481267434	5.481267434	5.481267434	5.481267434

(3) P 6+0	TH = 40	SBE = 1.555723827						CAUER FUNCTIONS	
	RC = 02	RC = 05	RC = 07	RC = 10	RC = 15	RC = 20	RC = 25	RC = 35	RC = 50
AMAX DB	.0017375255	.0108709564	.0213327376	.0436480540	.0988323393	.1772876696	.2802872360	.5675287486	1.249387366
AMIN DB	42.66192020	50.62965609	53.56266013	56.68300485	60.26000909	62.83723732	64.87843630	68.08823779	71.86813521
MU I K	.7360444D-2	.2941150D-2	.2098297D-2	.1465041D-2	.9705090D-3	.7213369D-3	.5702669D-3	.3940834D-3	.2550312D-3
1 I PF1	1.084521846	.8338811633	.7541103449	.6755211893	.5929496605	.5384283447	.4983620045	.4415355577	.3859606590
1 I PQ1	.5626392994	.5734190197	.5812119274	.5930325435	.6132331519	.6344936097	.6572740324	.7092359097	.8124164767
1 I PF2	1.160167754	1.019627832	.9746878370	.9311268543	.8868049434	.8587535216	.8389429223	.8122689470	.7891088385
1 I PQ2	1.194442518	1.281867098	1.342478193	1.430951420	1.574288533	1.716778961	1.862132498	2.173196038	2.738301856
1 I PFC	1.201232119	1.124281657	1.099150010	1.074629727	1.049579282	1.033702131	1.022491092	1.007412506	.9937855208
1 I PQC	4.588185466	4.987838975	5.262172653	5.659342708	6.296279033	6.923373294	7.558390480	8.906139572	11.33069212
1 I ZF1	1.597614759	1.597614759	1.597614759	1.597614759	1.597614759	1.597614759	1.597614759	1.597614759	1.597614759
1 I ZF2	2.067441918	2.067441918	2.067441918	2.067441918	2.067441918	2.067441918	2.067441918	2.067441918	2.067441918
1 I ZF3	5.333167413	5.333167413	5.333167413	5.333167413	5.333167413	5.333167413	5.333167413	5.333167413	5.333167413

(3) P 6+0	TH = 41	SBE = 1.524253087						CAUER FUNCTIONS	
	RC = 02	RC = 05	RC = 07	RC = 10	RC = 15	RC = 20	RC = 25	RC = 35	RC = 50
AMAX DB	.0017375255	.0108709564	.0213327376	.0436480540	.0988323393	.1772876696	.2802872360	.5675287486	1.249387366
AMIN DB	41.24822087	49.21588068	52.14887763	55.26921857	58.84622080	61.42344832	63.46464698	66.67444818	70.45434546
MU I K	.8661417D-2	.3461035D-2	.2469199D-2	.1724007D-2	.1142060D-2	.8488434D-3	.6710697D-3	.4637432D-3	.3001116D-3
1 I PF1	1.096844542	.8416606564	.7607158870	.6810881025	.5975426767	.5424392020	.5019736112	.4446192885	.3885678978
1 I PQ1	.5641797515	.5745330629	.5822287038	.5939716680	.6141124828	.6353489173	.6581226408	.7101005012	.8133572715
1 I PF2	1.162505243	1.022293184	.9773764041	.9338084488	.8894537352	.8613698256	.8415313749	.8148140375	.7906093373
1 I PQ2	1.214135907	1.298240906	1.358300884	1.446560246	1.590169106	1.733266044	1.879408107	2.192445538	2.761566992
1 I PFC	1.197300083	1.122192496	1.097569768	1.073506650	1.048885525	1.033261758	1.022221243	1.007361404	.9939219591
1 I PQC	4.710640224	5.096797418	5.370009997	5.768358924	6.410167730	7.043720553	7.686139734	9.051093272	11.50892255
1 I ZF1	1.564662002	1.564662002	1.564662002	1.564662002	1.564662002	1.564662002	1.564662002	1.564662002	1.564662002
1 I ZF2	2.019108731	2.019108731	2.019108731	2.019108731	2.019108731	2.019108731	2.019108731	2.019108731	2.019108731
1 I ZF3	5.191882466	5.191882466	5.191882466	5.191882466	5.191882466	5.191882466	5.191882466	5.191882466	5.191882466

(3) P 6+0	TH = 42	SBE = 1.494476550						CAUER FUNCTIONS	
	RC = 02	RC = 05	RC = 07	RC = 10	RC = 15	RC = 20	RC = 25	RC = 35	RC = 50
AMAX DB	.0017375255	.0108709564	.0213327376	.0436480540	.0988323393	.1772876696	.2802872360	.5675287486	1.249387366
AMIN DB	39.86118661	47.82874339	50.76173073	53.88206657	57.45906607	60.03629264	62.07749085	65.28729167	69.06718874
MU I K	.1016110D-1	.4060344D-2	.2896766D-2	.2022538D-2	.1339821D-2	.9958304D-3	.7872732D-3	.5440457D-3	.3520794D-3
1 I PF1	1.109735123	.8497737272	.7675983361	.6868834144	.6023201436	.5466090095	.5057270327	.4478226372	.3912751805
1 I PQ1	.5658252928	.5757106835	.5833003657	.5949588556	.6150344714	.6362443508	.6590101377	.7110034920	.8143386927
1 I PF2	1.164823452	1.024998147	.9801161941	.9365498219	.8921686192	.8640551140	.8441903883	.8174312519	.7931828650
1 I PQ2	1.235146817	1.315582824	1.375024672	1.463027035	1.606892434	1.750609236	1.897567297	2.212660018	2.785977768
1 I PFC	1.193295498	1.120057906	1.095953498	1.072356746	1.048174264	1.032809709	1.021943779	1.007308069	.9940609374
1 I PQC	4.841713740	5.212726230	5.484558238	5.883988686	6.530797006	7.171084559	7.821260905	9.204299451	11.69718024
1 I ZF1	1.533459623	1.533459623	1.533459623	1.533459623	1.533459623	1.533459623	1.533459623	1.533459623	1.533459623
1 I ZF2	1.973130639	1.973130639	1.973130639	1.973130639	1.973130639	1.973130639	1.973130639	1.973130639	1.973130639
1 I ZF3	5.056912989	5.056912989	5.056912989	5.056912989	5.056912989	5.056912989	5.056912989	5.056912989	5.056912989

(3) P 6+0		TH = 43	SBE = 1.466279186						CAUER FUNCTIONS	
		RC = 02	RC = 05	RC = 07	RC = 10	RC = 15	RC = 20	RC = 25	RC = 35	RC = 50
AMAX DB		.0017375255	.0108709564	.0213327376	.0436480540	.0988323393	.1772876696	.2802872360	.5675287486	1.249387366
AMIN DB		38.49924772	46.46666570	49.39964008	52.51996904	56.09696487	58.67419015	60.71538777	63.92518807	67.70508487
MU	K	.1188605D-1	.4749706D-2	.3388582D-2	.2365928D-2	.1567299D-2	.1164905D-2	.9209385D-3	.6364153D-3	.4118563D-3
1	PF1	1.123221121	.8582349527	.7747693292	.6929163497	.6072892043	.5509437570	.5096275096	.4511499079	.3940860202
1	PQ1	.5675852832	.5769564261	.5844305526	.5959970615	.6160015376	.6371820688	.6599385360	.7119467660	.8153626358
1	PF2	1.167114201	1.027739512	.9829051448	.9393498402	.8949492375	.8668094383	.8469202686	.8201211954	.7958302587
1	PQ2	1.257586816	1.333963195	1.392711924	1.480408436	1.624511549	1.768860794	1.916662850	2.233895468	2.811598922
1	PFC	1.189220379	1.117878474	1.094301516	1.071180144	1.047445503	1.032345938	1.021658635	1.007252423	.9942023808
1	PQC	4.982186177	5.336185137	5.606337577	6.006728049	6.658656415	7.305964965	7.964271831	9.366327479	11.89614609
1	ZF1	1.503888438	1.503888438	1.503888438	1.503888438	1.503888438	1.503888438	1.503888438	1.503888438	1.503888438
1	ZF2	1.929343559	1.929343559	1.929343559	1.929343559	1.929343559	1.929343559	1.929343559	1.929343559	1.929343559
1	ZF3	4.927805102	4.927805102	4.927805102	4.927805102	4.927805102	4.927805102	4.927805102	4.927805102	4.927805102

(3) P 6+0		TH = 44	SBE = 1.439556540						CAUER FUNCTIONS	
		RC = 02	RC = 05	RC = 07	RC = 10	RC = 15	RC = 20	RC = 25	RC = 35	RC = 50
AMAX DB		.0017375255	.0108709564	.0213327376	.0436480540	.0988323393	.1772876696	.2802872360	.5675287486	1.249387366
AMIN DB		37.16093075	45.12816261	48.06111963	51.18143936	54.75843026	57.33565383	59.37685065	62.58665025	66.36654668
MU	K	.1386607D-1	.5541047D-2	.3953157D-2	.2760120D-2	.1828431D-2	.1358993D-2	.1074379D-2	.7424505D-3	.4804771D-3
1	PF1	1.137331964	.8670599370	.7822413172	.6991967706	.6124574884	.5554498381	.5136806330	.4546056894	.3970041596
1	PQ1	.5694702397	.5782753119	.5856232670	.5970895210	.6170163196	.6381644185	.6609100206	.7129323633	.8164311492
1	PF2	1.169368768	1.030513795	.9857409961	.9422072373	.8977951559	.8696328055	.8497212965	.8228344728	.7985523741
1	PQ2	1.281580465	1.353459618	1.411431217	1.498766551	1.643084429	1.788077747	1.936752304	2.256212829	2.838500825
1	PFC	1.185076816	1.115654822	1.092614163	1.069976984	1.046699250	1.031870401	1.021365743	1.007194385	.9943462103
1	PQC	5.132933995	5.467796126	5.735924416	6.137125259	6.794285741	7.448911923	8.115742149	9.537802930	12.10656746
1	ZF1	1.475840214	1.475840214	1.475840214	1.475840214	1.475840214	1.475840214	1.475840214	1.475840214	1.475840214
1	ZF2	1.887598333	1.887598333	1.887598333	1.887598333	1.887598333	1.887598333	1.887598333	1.887598333	1.887598333
1	ZF3	4.804145400	4.804145400	4.804145400	4.804145400	4.804145400	4.804145400	4.804145400	4.804145400	4.804145400

(3) P 6+0		TH = 45	SBE = 1.414213562						CAUER FUNCTIONS	
		RC = 02	RC = 05	RC = 07	RC = 10	RC = 15	RC = 20	RC = 25	RC = 35	RC = 50
AMAX DB		.0017375255	.0108709564	.0213327376	.0436480540	.0988323393	.1772876696	.2802872360	.5675287486	1.249387366
AMIN DB		35.84484979	43.81183323	46.74476707	49.86507448	53.44205381	56.01928007	58.06047583	61.27027451	65.05017044
MU	K	.1613457D-1	.6447752D-2	.4600040D-2	.3211784D-2	.2127635D-2	.1581379D-2	.1250191D-2	.8639454D-3	.5591026D-3
1	PF1	1.152099082	.8762653942	.7900276314	.7057352295	.6178331521	.5601340833	.5178923730	.4531948781	.4000335900
1	PQ1	.5714920134	.5796729018	.5868829193	.5982397827	.6180816983	.6391939562	.6619269665	.7139624962	.8175464494
1	PF2	1.171577855	1.033317221	.9886212730	.9451206008	.9007058552	.8725251710	.8525937237	.8257216353	.8013500842
1	PQ2	1.307267205	1.374157877	1.431258098	1.518169585	1.662674559	1.808322437	1.957898476	2.279678539	2.866760089
1	PFC	1.180866972	1.113387606	1.090891798	1.068747420	1.045935518	1.031383057	1.021065038	1.007133873	.9944923430
1	PQC	5.294944641	5.608251996	5.873958847	6.275787527	6.938281346	7.600532371	8.276299684	9.719414422	12.32926616
1	ZF1	1.449216453	1.449216453	1.449216453	1.449216453	1.449216453	1.449216453	1.449216453	1.449216453	1.449216453
1	ZF2	1.847759065	1.847759065	1.847759065	1.847759065	1.847759065	1.847759065	1.847759065	1.847759065	1.847759065
1	ZF3	4.685556399	4.685556399	4.685556399	4.685556399	4.685556399	4.685556399	4.685556399	4.685556399	4.685556399

(3) P 6+0	TH = 46	SBE = 1.390163591						CAUER FUNCTIONS	
	RC = 02	RC = 05	RC = 07	RC = 10	RC = 15	RC = 20	RC = 25	RC = 35	RC = 50
AMAX DB	.0017375255	.0108709564	.0213327376	.0436480540	.0988323393	.1772876696	.2802872360	.5675287486	1.249387366
AMIN DB	34.54969893	42.51635217	45.44925520	48.56954624	52.14652183	54.72374004	56.76493437	59.97473182	63.75462710
MU I K	.1872902D-1	.7484838D-2	.5339951D-2	.3728402D-2	.2469869D-2	.1835748D-2	.1451287D-2	.1002913D-2	.6490358D-3
1 I PF1	1.167556024	.8858692381	.7981425573	.7125430277	.6234249229	.5650037965	.5222691106	.4619227038	.4031785715
1 I PQ1	.5736639981	.5811553674	.5882143792	.5994517466	.6192008250	.6402734704	.6629919595	.7150395671	.8187109383
1 I PF2	1.173731558	1.036145692	.9915432664	.9480883576	.9036807216	.8754864324	.8555377670	.8236334281	.8042242789
1 I PQ2	1.334803562	1.396153010	1.452275968	1.538692591	1.683351587	1.829663129	1.980170077	2.304365141	2.896460256
1 I PFC	1.176593087	1.111077519	1.089134809	1.067491620	1.045154326	1.030883865	1.020756450	1.007070798	.9946406915
1 I PQC	5.469333945	5.758326320	6.021153354	6.423388861	7.091303499	7.761497279	8.446637799	9.911921471	12.56514758
1 I ZF1	1.423927342	1.423927342	1.423927342	1.423927342	1.423927342	1.423927342	1.423927342	1.423927342	1.423927342
1 I ZF2	1.809701684	1.809701684	1.809701684	1.809701684	1.809701684	1.809701684	1.809701684	1.809701684	1.809701684
1 I ZF3	4.571692570	4.571692570	4.571692570	4.571692570	4.571692570	4.571692570	4.571692570	4.571692570	4.571692570

(3) P 6+0	TH = 47	SBE = 1.367327461						CAUER FUNCTIONS	
	RC = 02	RC = 05	RC = 07	RC = 10	RC = 15	RC = 20	RC = 25	RC = 35	RC = 50
AMAX DB	.0017375255	.0108709564	.0213327376	.0436480540	.0988323393	.1772876696	.2802872360	.5675287486	1.249387366
AMIN DB	33.27424583	41.24046192	44.17332415	47.29359352	50.87055754	53.44777169	55.48896416	58.69875997	62.47865438
MU I K	.2169141D-1	.8669158D-2	.6184916D-2	.4318375D-2	.2860699D-2	.2126236D-2	.1680938D-2	.1161614D-2	.7517393D-3
1 I PF1	1.183738563	.8958906809	.8066014151	.7196322793	.6292421487	.5700667961	.5268176729	.4657947579	.4064436558
1 I PQ1	.5760013787	.5827295757	.5896230356	.6007297069	.6203771535	.6414060084	.6641078196	.7161661894	.8199272224
1 I PF2	1.175819327	1.038994772	.9945040126	.9511087588	.9067190361	.8785164214	.8585536027	.8316202870	.8071758632
1 I PQ2	1.364365760	1.419550564	1.474577103	1.560418332	1.705192072	1.852174724	2.003642390	2.330351996	2.927692583
1 I PFC	1.172257480	1.108725292	1.087343607	1.066209768	1.044355700	1.030372786	1.020439912	1.007005072	.9947911636
1 I PQC	5.657366814	5.918885115	6.178302949	6.580679177	7.254084868	7.932550025	8.627523874	10.11616355	12.91521131
1 I ZF1	1.399890831	1.399890831	1.399890831	1.399890831	1.399890831	1.399890831	1.399890831	1.399890831	1.399890831
1 I ZF2	1.773312683	1.773312683	1.773312683	1.773312683	1.773312683	1.773312683	1.773312683	1.773312683	1.773312683
1 I ZF3	4.462236877	4.462236877	4.462236877	4.462236877	4.462236877	4.462236877	4.462236877	4.462236877	4.462236877

(3) P 6+0	TH = 48	SBE = 1.345632730						CAUER FUNCTIONS	
	RC = 02	RC = 05	RC = 07	RC = 10	RC = 15	RC = 20	RC = 25	RC = 35	RC = 50
AMAX DB	.0017375255	.0108709564	.0213327376	.0436480540	.0988323393	.1772876696	.2802872360	.5675287486	1.249387366
AMIN DB	32.01732641	39.98296588	42.91577428	46.03601506	49.61296381	52.19017262	54.23136261	57.44115628	61.22104954
MU I K	.2506881D-1	.1001963D-1	.7148440D-2	.4991134D-2	.3306373D-2	.2457488D-2	.1942817D-2	.1342586D-2	.8688554D-3
1 I PF1	1.200684785	.9063503393	.8154206486	.7270159827	.6352948523	.5753314602	.5315453721	.4698170253	.4098337115
1 I PQ1	.5785214266	.5844031861	.5911148662	.6020784026	.6216144761	.6425949061	.6652776279	.7173452107	.8211981352
1 I PF2	1.177829933	1.041859648	.9975002707	.9541798623	.9098199621	.8816148954	.8616413605	.8346828343	.8102057565
1 I PQ2	1.396152817	1.444468041	1.498263834	1.583438267	1.728280342	1.875939559	2.028398070	2.357726082	2.960556936
1 I PFC	1.167862548	1.106331697	1.085518633	1.064902061	1.043539670	1.029849784	1.020115354	1.006936600	.9949436622
1 I PQC	5.860481936	6.090900560	6.346297050	6.748494909	7.427440377	8.114516103	8.819809133	10.33307056	13.08056326
1 I ZF1	1.377031832	1.377031832	1.377031832	1.377031832	1.377031832	1.377031832	1.377031832	1.377031832	1.377031832
1 I ZF2	1.738488020	1.738488020	1.738488020	1.738488020	1.738488020	1.738488020	1.738488020	1.738488020	1.738488020
1 I ZF3	4.356897733	4.356897733	4.356897733	4.356897733	4.356897733	4.356897733	4.356897733	4.356897733	4.356897733

(3) P 6+0	TH = 49	SBE = 1.325012993						CAUER FUNCTIONS	
	RC = 02	RC = 05	RC = 07	RC = 10	RC = 15	RC = 20	RC = 25	RC = 35	RC = 50
AMAX DB	.0017375255	.0108709564	.0213327376	.0436480540	.0988323393	.1772876696	.2802872360	.5675287486	1.249387366
AMIN DB	30.77784062	38.74272207	41.67545971	44.79566289	48.37259157	50.94979336	52.99098010	56.20077093	59.98066269
MU K	.2891399D-1	.1155750D-1	.8245690D-2	.5757273D-2	.3813910D-2	.2834721D-2	.2241047D-2	.1548679D-2	.1002229D-2
1 PF1	1.218435160	.9172703495	.8246179226	.7347080988	.6415937925	.5808067768	.5364600496	.4739959193	.4133539525
1 PQ1	.5812438542	.5861847654	.5926965181	.6035030754	.6229169661	.6438438229	.6665047569	.7185797390	.8225267615
1 PF2	1.179751428	1.044735108	1.000528498	.9572995137	.9129825333	.8847815277	.8648011158	.8378216255	.8133148901
1 PQ2	1.430390223	1.471036600	1.523449922	1.607853704	1.752709484	1.901048343	2.054528046	2.386582913	2.995162819
1 PFC	1.163410771	1.103897549	1.083660353	1.063568718	1.042706278	1.029314824	1.019782707	1.006865287	.9950980847
1 PQC	6.080321436	6.275467175	6.526133418	6.927771436	7.612278710	8.308314427	9.024440057	10.56367497	13.36242983
1 ZF1	1.355281508	1.355281508	1.355281508	1.355281508	1.355281508	1.355281508	1.355281508	1.355281508	1.355281508
1 ZF2	1.705132152	1.705132152	1.705132152	1.705132152	1.705132152	1.705132152	1.705132152	1.705132152	1.705132152
1 ZF3	4.255406331	4.255406331	4.255406331	4.255406331	4.255406331	4.255406331	4.255406331	4.255406331	4.255406331

(3) P 6+0	TH = 50	SBE = 1.305407289						CAUER FUNCTIONS	
	RC = 02	RC = 05	RC = 07	RC = 10	RC = 15	RC = 20	RC = 25	RC = 35	RC = 50
AMAX DB	.0017375255	.0108709564	.0213327376	.0436480540	.0988323393	.1772876696	.2802872360	.5675287486	1.249387366
AMIN DB	29.55474930	37.51863748	40.45128236	43.57143626	47.14833864	49.72553122	51.76671370	54.97650082	58.75639061
MU K	.3328607D-1	.1330663D-1	.9493708D-2	.6628697D-2	.4391198D-2	.3263799D-2	.2580265D-2	.1783097D-2	.1153933D-2
1 PF1	1.237032576	.9286744915	.8342122293	.7427236387	.6481505314	.5865024002	.5415701239	.4783383211	.4170099691
1 PQ1	.5841912405	.5880839225	.5943754018	.6050095369	.6242892254	.6451567815	.6677929055	.7198731732	.8239164667
1 PF2	1.181571097	1.047615505	1.003584822	.9604653257	.9162056382	.8880158969	.8680328822	.8410371935	.8165042057
1 PQ2	1.467334348	1.499403049	1.550262163	1.633777132	1.778582494	1.927601225	2.082132571	2.417027604	3.031630543
1 PFC	1.158904709	1.101423707	1.081769269	1.062209974	1.041855571	1.028767875	1.019441902	1.006791032	.9952543231
1 PQC	6.318766608	6.473820994	6.718934603	7.119557667	7.809615764	8.514970523	9.242471695	10.80912594	13.66217428
1 ZF1	1.334576660	1.334576660	1.334576660	1.334576660	1.334576660	1.334576660	1.334576660	1.334576660	1.334576660
1 ZF2	1.673157185	1.673157185	1.673157185	1.673157185	1.673157185	1.673157185	1.673157185	1.673157185	1.673157185
1 ZF3	4.157514276	4.157514276	4.157514276	4.157514276	4.157514276	4.157514276	4.157514276	4.157514276	4.157514276

(3) P 6+0	TH = 51	SBE = 1.286759566						CAUER FUNCTIONS	
	RC = 02	RC = 05	RC = 07	RC = 10	RC = 15	RC = 20	RC = 25	RC = 35	RC = 50
AMAX DB	.0017375255	.0108709564	.0213327376	.0436480540	.0988323393	.1772876696	.2802872360	.5675287486	1.249387366
AMIN DB	28.34707238	36.30966283	39.24218650	42.36227600	45.93914400	48.51632455	50.55750145	53.76728372	57.54717094
MU K	.3825132D-1	.1529385D-1	.1091166D-1	.7618793D-2	.5047110D-2	.3751317D-2	.2965684D-2	.2049443D-2	.1326299D-2
1 PF1	1.256522326	.9405883229	.8442240063	.7510787603	.6549775101	.5924287138	.5468846460	.4828516237	.4208077640
1 PQ1	.5873895455	.5901114673	.5961598022	.6066042466	.6257363413	.6465382134	.6691461391	.7212292381	.8253709283
1 PF2	1.183275419	1.050494721	1.006665013	.9636746548	.9194880045	.8913174752	.8713366019	.8443300442	.8197746534
1 PQ2	1.507277727	1.529732196	1.578842275	1.661333781	1.806013608	1.955709043	2.111322430	2.449176091	3.070092591
1 PFC	1.154347011	1.098911080	1.079845915	1.060826085	1.040987607	1.028208911	1.019092869	1.006713732	.9954127629
1 PQC	6.577981217	6.687362387	6.925967420	7.325033253	8.020590460	8.735631999	9.475083250	11.07070587	13.98131589
1 ZF1	1.314859174	1.314859174	1.314859174	1.314859174	1.314859174	1.314859174	1.314859174	1.314859174	1.314859174
1 ZF2	1.642482125	1.642482125	1.642482125	1.642482125	1.642482125	1.642482125	1.642482125	1.642482125	1.642482125
1 ZF3	4.062991499	4.062991499	4.062991499	4.062991499	4.062991499	4.062991499	4.062991499	4.062991499	4.062991499

(3) P 6+0 — TH = 52 — SBE = 1.269018215 — CAUER FUNCTIONS

		RC = 02	RC = 05	RC = 07	RC = 10	RC = 15	RC = 20	RC = 25	RC = 35	RC = 50
	AMAX DB	.0017375255	.0108709564	.0213327376	.0436480540	.0988323393	.1772876696	.2802872360	.5675287486	1.249387366
	AMIN DB	27.15388835	35.11478799	38.04715367	41.16715921	44.74398240	47.32114726	49.36231691	52.57209286	56.35197671
MU	K	.4388394D-1	.1754933D-1	.1252110D-1	.8742629D-2	.5791631D-2	.4304697D-2	.3403174D-2	.2351773D-2	.1521953D-2
1	PF1	1.276952016	.9530393203	.8546752646	.7597908753	.6620881326	.5985969003	.5524133596	.4875437814	.4247537912
1	PQ1	.5908687333	.5922795993	.5980590085	.6082944041	.6272639506	.6479930119	.6705689362	.7226520236	.8268941736
1	PF2	1.184850017	1.053366127	1.009764449	.9669245761	.9228281809	.8946856145	.8747121360	.8477006495	.8231271833
1	PQ2	1.550555443	1.562209634	1.609349108	1.690663445	1.835129857	1.985494765	2.142220346	2.483156548	3.110695175
1	PFC	1.149740411	1.096360626	1.077890859	1.059417329	1.040102452	1.027637907	1.018735539	1.006633279	.9955717835
1	PQC	6.860464211	6.917683329	7.148666114	7.545528960	8.246483391	8.971586750	9.723596384	11.34984975	14.32155239
1	ZF1	1.296075538	1.296075538	1.296075538	1.296075538	1.296075538	1.296075538	1.296075538	1.296075538	1.296075538
1	ZF2	1.613032213	1.613032213	1.613032213	1.613032213	1.613032213	1.613032213	1.613032213	1.613032213	1.613032213
1	ZF3	3.971624391	3.971624391	3.971624391	3.971624391	3.971624391	3.971624391	3.971624391	3.971624391	3.971624391

(3) P 6+0 — TH = 53 — SBE = 1.252135658 — CAUER FUNCTIONS

		RC = 02	RC = 05	RC = 07	RC = 10	RC = 15	RC = 20	RC = 25	RC = 35	RC = 50
	AMAX DB	.0017375255	.0108709564	.0213327376	.0436480540	.0988323393	.1772876696	.2802872360	.5675287486	1.249387366
	AMIN DB	25.97433543	33.93303768	36.86519794	39.98509432	43.56185923	46.13900370	48.18016390	51.38993163	55.16981112
MU	K	.5026703D-1	.2010704D-1	.1434631D-1	.1001718D-1	.6636010D-2	.4932304D-2	.3899346D-2	.2694657D-2	.1743851D-2
1	PF1	1.298371375	.9660570295	.8655897287	.7688787684	.6694968606	.6050190209	.5581667707	.4924233657	.4288550009
1	PQ1	.5946635324	.5946021321	.6000834679	.6100880564	.6288783163	.6495265940	.6720662420	.7241460311	.8284906225
1	PF2	1.186279606	1.056222542	1.012878081	.9702118547	.9262245170	.8981195322	.8781592524	.8511494411	.8265627680
1	PQ2	1.597552863	1.597045045	1.641961270	1.721922629	1.866072885	2.017095185	2.174962619	2.519111034	3.153600062
1	PFC	1.145087733	1.093773359	1.075904710	1.057984010	1.039200187	1.027054844	1.018369842	1.006549560	.9957327570
1	PQC	7.169114264	7.166600177	7.388660045	7.782550922	8.488738922	9.224284480	9.989496834	11.64816798	14.68478626
1	ZF1	1.278176419	1.278176419	1.278176419	1.278176419	1.278176419	1.278176419	1.278176419	1.278176419	1.278176419
1	ZF2	1.584738339	1.584738339	1.584738339	1.584738339	1.584738339	1.584738339	1.584738339	1.584738339	1.584738339
1	ZF3	3.883214140	3.883214140	3.883214140	3.883214140	3.883214140	3.883214140	3.883214140	3.883214140	3.883214140

(3) P 6+0 — TH = 54 — SBE = 1.236067977 — CAUER FUNCTIONS

		RC = 02	RC = 05	RC = 07	RC = 10	RC = 15	RC = 20	RC = 25	RC = 35	RC = 50
	AMAX DB	.0017375255	.0108709564	.0213327376	.0436480540	.0988323393	.1772876696	.2802872360	.5675287486	1.249387366
	AMIN DB	24.80761456	32.76346781	35.69536155	38.81511630	42.39180560	44.96892360	47.01007155	50.21982861	53.99970243
MU	K	.5749357D-1	.2300523D-1	.1641466D-1	.1146157D-1	.7592936D-2	.5643569D-2	.4461659D-2	.3083249D-2	.1995331D-2
1	PF1	1.320831921	.9796732221	.8769929895	.7783627297	.6772193189	.6117081043	.5641562243	.4974996277	.4331188904
1	PQ1	.5988143647	.5970947624	.6022449689	.6119942246	.6305864153	.6511449715	.6736435309	.7257162270	.8301651366
1	PF2	1.187547938	1.059056185	1.016000393	.9735329150	.9296751411	.9016182938	.8816776138	.8546768020	.8300823486
1	PQ2	1.648715067	1.634476157	1.676880239	1.755287086	1.899001091	2.050662895	2.209701038	2.557197414	3.198986681
1	PFC	1.140391895	1.091150346	1.073888115	1.056526455	1.038280902	1.026459709	1.017995711	1.006462461	.9958950481
1	PQC	7.507309218	7.436193226	7.647806936	8.037809610	8.748991479	9.495362241	10.27446002	11.96747336	15.07315594
1	ZF1	1.261116273	1.261116273	1.261116273	1.261116273	1.261116273	1.261116273	1.261116273	1.261116273	1.261116273
1	ZF2	1.557536516	1.557536516	1.557536516	1.557536516	1.557536516	1.557536516	1.557536516	1.557536516	1.557536516
1	ZF3	3.797575240	3.797575240	3.797575240	3.797575240	3.797575240	3.797575240	3.797575240	3.797575240	3.797575240

(3) P — TH = 55 — SBE = 1.220774589 — CAUER FUNCTIONS

6+0	RC = 02	RC = 05	RC = 07	RC = 10	RC = 15	RC = 20	RC = 25	RC = 35	RC = 50
AMAX DB	.0017375255	.0108709564	.0213327376	.0436480540	.0988323393	.1772876696	.2802872360	.5675287486	1.249387366
AMIN DB	23.65299470	31.60516215	34.53671082	37.65628217	41.23287355	43.80995727	45.85108936	49.06083260	52.84069908
MU / K	.6566747D-1	.2628705D-1	.1875705D-1	.1309742D-1	.8676735D-2	.6449145D-2	.5098536D-2	.3523371D-2	.2280159D-2
1 PF1	1.344386424	.9939220554	.8889126701	.7882647010	.6852724142	.6186782463	.5703939920	.5027825692	.4375535604
1 PQ1	.6033684886	.5997753928	.6045568583	.6140230540	.6323960438	.6528548356	.6753068789	.7273681041	.8319230761
1 PF2	1.188637750	1.061858621	1.019125359	.9768838052	.9331779350	.9051807942	.8852667621	.8582830579	.8336868805
1 PQ2	1.704558399	1.674773492	1.714334078	1.790954828	1.934092154	2.086368620	2.246605136	2.597591615	3.247054607
1 PFC	1.135655916	1.088492720	1.071841768	1.055045021	1.037344701	1.025852492	1.017613078	1.006371861	.9960585134
1 PQC	7.879004446	7.728854743	7.928233007	8.313254644	9.029096997	9.786674909	10.58038154	12.30981318	15.48907264
1 ZF1	1.244853017	1.244853017	1.244853017	1.244853017	1.244853017	1.244853017	1.244853017	1.244853017	1.244853017
1 ZF2	1.531367417	1.531367417	1.531367417	1.531367417	1.531367417	1.531367417	1.531367417	1.531367417	1.531367417
1 ZF3	3.714534150	3.714534150	3.714534150	3.714534150	3.714534150	3.714534150	3.714534150	3.714534150	3.714534150

(3) P — TH = 56 — SBE = 1.206217949 — CAUER FUNCTIONS

6+0	RC = 02	RC = 05	RC = 07	RC = 10	RC = 15	RC = 20	RC = 25	RC = 35	RC = 50
AMAX DB	.0017375255	.0108709564	.0213327376	.0436480540	.0988323393	.1772876696	.2802872360	.5675287486	1.249387366
AMIN DB	22.50982091	30.45722962	33.38833229	36.50766656	40.08413135	42.66117076	44.70228234	47.91200773	51.69186472
MU / K	.7490468D-1	.3000119D-1	.2140836D-1	.1494916D-1	.9903608D-2	.7361079D-2	.5819503D-2	.4021607D-2	.2602596D-2
1 PF1	1.369088101	1.008840232	.9013786041	.7986084392	.6936744689	.6259447240	.5768933715	.5082830238	.4421677807
1 PQ1	.6083814103	.6026645217	.6070343035	.6161859933	.6343159395	.6546636552	.6770630495	.7291077540	.8337703667
1 PF2	1.189530698	1.064620710	1.022246393	.9802601587	.9367305057	.9088057363	.8889261026	.8619634665	.8373773036
1 PQ2	1.765684687	1.718246095	1.754581892	1.829149714	1.971546027	2.124403973	2.285864842	2.640490270	3.298026463
1 PFC	1.130882915	1.085801676	1.069766409	1.053540095	1.036391705	1.025233191	1.017221878	1.006277635	.9962230010
1 PQC	8.288855387	8.047347591	8.232381684	8.611116819	9.331170707	10.10033170	10.90941367	12.67750745	15.93526439
1 ZF1	1.229347718	1.229347718	1.229347718	1.229347718	1.229347718	1.229347718	1.229347718	1.229347718	1.229347718
1 ZF2	1.506175959	1.506175959	1.506175959	1.506175959	1.506175959	1.506175959	1.506175959	1.506175959	1.506175959
1 ZF3	3.633928074	3.633928074	3.633928074	3.633928074	3.633928074	3.633928074	3.633928074	3.633928074	3.633928074

(3) P — TH = 57 — SBE = 1.192363293 — CAUER FUNCTIONS

6+0	RC = 02	RC = 05	RC = 07	RC = 10	RC = 15	RC = 20	RC = 25	RC = 35	RC = 50
AMAX DB	.0017375255	.0108709564	.0213327376	.0436480540	.0988323393	.1772876696	.2802872360	.5675287486	1.249387366
AMIN DB	21.37752583	29.31880209	32.24932923	35.36835751	38.94465891	41.52164112	43.56272622	46.77242855	50.55227329
MU / K	.8533432D-1	.3420266D-1	.2440808D-1	.1704442D-1	.1129190D-1	.8393014D-2	.6635348D-2	.4585414D-2	.2967470D-2
1 PF1	1.394989444	1.024467155	.9144230264	.8094196961	.7024453717	.6335241235	.5336687989	.5140127483	.4469710639
1 PQ1	.6139186351	.6057857174	.6096946062	.6184960076	.6363559286	.6565797934	.6789195939	.7309419519	.8357135767
1 PF2	1.190207305	1.067332535	1.025356296	.9836571508	.9403301541	.9124916073	.8926548848	.8657332060	.8411545413
1 PQ2	1.832798857	1.765248486	1.797919202	1.870125755	2.011588513	2.164984730	2.327693634	2.686113857	3.352151346
1 PFC	1.126076127	1.083078479	1.067662830	1.052012096	1.035422049	1.024601813	1.016822045	1.006179653	.9963883494
1 PQC	8.742371126	8.394877224	8.563073051	8.933959135	9.657632800	10.43874016	11.26400921	13.07319489	16.41432870
1 ZF1	1.214564329	1.214564329	1.214564329	1.214564329	1.214564329	1.214564329	1.214564329	1.214564329	1.214564329
1 ZF2	1.481910928	1.481910928	1.481910928	1.481910928	1.481910928	1.481910928	1.481910928	1.481910928	1.481910928
1 ZF3	3.555603858	3.555603858	3.555603858	3.555603858	3.555603858	3.555603858	3.555603858	3.555603858	3.555603858

(3) P 6+0	TH = 58	SBE = 1.179178403						CAUER FUNCTIONS	
	RC = 02	RC = 05	RC = 07	RC = 10	RC = 15	RC = 20	RC = 25	RC = 35	RC = 50
AMAX DB	.0017375255	.0108709564	.0213327376	.0436480540	.0988323393	.1772876696	.2802872360	.5675287486	1.249387366
AMIN DB	20.25564539	28.18903290	31.11881838	34.23745224	37.81354300	40.39045147	42.43150244	45.64117504	49.42100399
MU K	.9709966D-1	.3895367D-1	.2780091D-1	.1941455D-1	.1286242D-1	.9560430D-2	.7558313D-2	.5223255D-2	.3380258D-2
1 PF1	1.422140554	1.040845058	.9280807749	.8207264186	.7116067473	.6414344861	.5907359769	.5199945281	.4519737499
1 PQ1	.6200578492	.6091661963	.6125575845	.6209678356	.6385270996	.6586126461	.6808849700	.7328782560	.8377600080
1 PF2	1.190646895	1.069983341	1.028447195	.9870694500	.9439738386	.9162366508	.8964521812	.8695773612	.8450194939
1 PQ2	1.906730860	1.816189157	1.844684463	1.914172303	2.054475543	2.208354739	2.372332236	2.734710422	3.409708895
1 PFC	1.121238908	1.080324467	1.065531875	1.050461482	1.034435887	1.023958370	1.016413517	1.006077780	.9965543871
1 PQC	9.246108163	8.775180667	8.923576874	9.284739117	10.01126390	10.80465952	11.64697461	13.49988330	16.93129592
1 ZF1	1.200469445	1.200469445	1.200469445	1.200469445	1.200469445	1.200469445	1.200469445	1.200469445	1.200469445
1 ZF2	1.458524649	1.458524649	1.458524649	1.458524649	1.458524649	1.458524649	1.458524649	1.458524649	1.458524649
1 ZF3	3.479416971	3.479416971	3.479416971	3.479416971	3.479416971	3.479416971	3.479416971	3.479416971	3.479416971

(3) P 6+0	TH = 59	SBE = 1.166633397						CAUER FUNCTIONS	
	RC = 02	RC = 05	RC = 07	RC = 10	RC = 15	RC = 20	RC = 25	RC = 35	RC = 50
AMAX DB	.0017375255	.0108709564	.0213327376	.0436480540	.0988323393	.1772876696	.2802872360	.5675287486	1.249387366
AMIN DB	19.14383972	27.06709619	29.99592702	33.11405307	36.68987258	39.26668608	41.30769309	44.51732741	48.29713602
MU K	.1103591D 0	.4432464D-1	.3163761D-1	.2209517D-1	.1463883D-1	.1088092D-1	.8602315D-2	.5944750D-2	.3847186D-2
1 PF1	1.450586822	1.058019122	.9423895000	.8325589688	.7211821487	.6496954748	.5981120221	.5262122978	.4571871039
1 PQ1	.6268916513	.6128375337	.6156460369	.6236182996	.6408420122	.6607728068	.6829686831	.7349251259	.8399178033
1 PF2	1.190827544	1.072561452	1.031510476	.9904911631	.9476581355	.9200388363	.9003168628	.8735009084	.8489730306
1 PQ2	1.988463128	1.871540998	1.895267016	1.961620339	2.100498337	2.254790619	2.420053371	2.786560026	3.471014128
1 PFC	1.116374754	1.077541062	1.063374452	1.048888748	1.033433395	1.023302885	1.015996234	1.005971377	.9967209315
1 PQC	9.807916506	9.192637197	9.317702843	9.666885394	10.39527289	11.20126559	12.06153444	13.96104182	17.48870610
1 ZF1	1.187032088	1.187032088	1.187032088	1.187032088	1.187032088	1.187032088	1.187032088	1.187032088	1.187032088
1 ZF2	1.435972684	1.435972684	1.435972684	1.435972684	1.435972684	1.435972684	1.435972684	1.435972684	1.435972684
1 ZF3	3.405230567	3.405230567	3.405230567	3.405230567	3.405230567	3.405230567	3.405230567	3.405230567	3.405230567

(3) P 6+0	TH = 60	SBE = 1.154700538						CAUER FUNCTIONS	
	RC = 02	RC = 05	RC = 07	RC = 10	RC = 15	RC = 20	RC = 25	RC = 35	RC = 50
AMAX DB	.0017375255	.0108709564	.0213327376	.0436480540	.0988323393	.1772876696	.2802872360	.5675287486	1.249387366
AMIN DB	18.04192044	25.95218730	28.87979055	31.99726334	35.57273384	38.14942514	40.19037557	43.39996063	47.17974307
MU K	.1252864D 0	.5039537D-1	.3597580D-1	.2512678D-1	.1664805D-1	.1237453D-1	.9783207D-2	.6760860D-2	.4375350D-2
1 PF1	1.480365743	1.076037523	.9573898769	.8449503662	.7311972746	.6583285635	.6058156325	.5327112792	.4626234284
1 PQ1	.6345309846	.6168365407	.6189863114	.6264666831	.6433149499	.6630722644	.6851814554	.7370920622	.8421960728
1 PF2	1.190726030	1.075054188	1.034536712	.9939157735	.9513791926	.9238958234	.9042475713	.8775036969	.8530159801
1 PQ2	2.079165149	1.931854188	1.950116851	2.012850144	2.149989654	2.304607432	2.471166668	2.841980085	3.536423244
1 PFC	1.111487317	1.074729769	1.061191531	1.047294433	1.032414768	1.022635392	1.015570140	1.005861798	.9968877881
1 PQC	10.43725442	9.652407019	9.749912888	10.08439243	10.81338032	11.63223050	12.51141041	14.46063294	18.09170236
1 ZF1	1.174223513	1.174223513	1.174223513	1.174223513	1.174223513	1.174223513	1.174223513	1.174223513	1.174223513
1 ZF2	1.414213562	1.414213562	1.414213562	1.414213562	1.414213562	1.414213562	1.414213562	1.414213562	1.414213562
1 ZF3	3.332914617	3.332914617	3.332914617	3.332914617	3.332914617	3.332914617	3.332914617	3.332914617	3.332914617

TH = 61 SBE = 1.143354068 CAUER FUNCTIONS

(3) P 6+0		RC = 02	RC = 05	RC = 07	RC = 10	RC = 15	RC = 20	RC = 25	RC = 35	RC = 50
AMAX DB		.0017375255	.0108709564	.0213327376	.0436480540	.0988323393	.1772876696	.2802872360	.5675287486	1.249387366
AMIN DB		16.94988561	24.84352445	27.76955039	30.88618340	34.46120514	37.03773928	39.07861695	42.28813865	46.06788743
MU	K	.1420711D 0	.5725637D-1	.4088109D-1	.2855557D-1	.1892081D-1	.1406414D-1	.1111909D-1	.7684101D-2	.4972853D-2
1	PF1	1.511502603	1.094951413	.9731258125	.8579365519	.7416802155	.6673572547	.6138672821	.5394981425	.4682961929
1	PQ1	.6431094660	.6212063547	.6226090073	.6295351914	.6459622246	.6655246419	.6875354282	.7393897748	.8446050454
1	PF2	1.190317805	1.077447770	1.037515571	.9973360720	.9551326771	.9278049218	.9082426874	.8815854285	.8571491201
1	PQ2	2.180237272	1.997772247	2.009756667	2.068300711	2.203331420	2.358165567	2.526025731	2.901331824	3.606340601
1	PFC	1.106580435	1.071892191	1.058984154	1.045679123	1.031380228	1.021955937	1.015135181	1.005747391	.9970547498
1	PQC	11.14559382	10.16060630	10.22546196	10.54193855	11.26992175	12.10182097	13.00091862	15.00326327	18.74564554
1	ZF1	1.162017032	1.162017032	1.162017032	1.162017032	1.162017032	1.162017032	1.162017032	1.162017032	1.162017032
1	ZF2	1.393208539	1.393208539	1.393208539	1.393208539	1.393208539	1.393208539	1.393208539	1.393208539	1.393208539
1	ZF3	3.262345083	3.262345083	3.262345083	3.262345083	3.262345083	3.262345083	3.262345083	3.262345083	3.262345083

TH = 62 SBE = 1.132570051 CAUER FUNCTIONS

(3) P 6+0		RC = 02	RC = 05	RC = 07	RC = 10	RC = 15	RC = 20	RC = 25	RC = 35	RC = 50
AMAX DB		.0017375255	.0108709564	.0213327376	.0436480540	.0988323393	.1772876696	.2802872360	.5675287486	1.249387366
AMIN DB		15.86796396	23.74035223	26.66435274	29.77990669	33.35435159	35.93068369	37.97146783	41.18090806	44.96061356
MU	K	.1609170D 0	.6501033D-1	.4642826D-1	.3243431D-1	.2149228D-1	.1597592D-1	.1263068D-1	.8728801D-2	.5648971D-2
1	PF1	1.544004732	1.114814781	.9896446354	.8715566756	.7526617329	.6768073293	.6222894438	.5465911916	.4742201864
1	PQ1	.6527888650	.6259978046	.6265498458	.6328495198	.6488025498	.6681454859	.6900444075	.7418303841	.8471562495
1	PF2	1.189576997	1.079727213	1.040435730	1.000744077	.9589137163	.9317630447	.9123002930	.8857456333	.8613731641
1	PQ2	2.293366494	2.070052153	2.074796894	2.128481392	2.260964088	2.415879175	2.585035888	2.965028118	3.681227189
1	PFC	1.101658167	1.069030033	1.056753441	1.044043456	1.030330024	1.021264577	1.014691309	1.005628498	.9972215949
1	PQC	11.94694634	10.72452988	10.75057590	11.04503400	11.76997662	12.61502061	13.53509019	15.59428322	19.45675549
1	ZF1	1.150387861	1.150387861	1.150387861	1.150387861	1.150387861	1.150387861	1.150387861	1.150387861	1.150387861
1	ZF2	1.372921373	1.372921373	1.372921373	1.372921373	1.372921373	1.372921373	1.372921373	1.372921373	1.372921373
1	ZF3	3.193403134	3.193403134	3.193403134	3.193403134	3.193403134	3.193403134	3.193403134	3.193403134	3.193403134

TH = 63 SBE = 1.122326238 CAUER FUNCTIONS

(3) P 6+0		RC = 02	RC = 05	RC = 07	RC = 10	RC = 15	RC = 20	RC = 25	RC = 35	RC = 50
AMAX DB		.0017375255	.0108709564	.0213327376	.0436480540	.0988323393	.1772876696	.2802872360	.5675287486	1.249387366
AMIN DB		14.79666974	22.64194736	25.56334807	28.67751584	32.25121940	34.82729180	36.86795569	40.07729120	43.85694105
MU	K	.1820399D 0	.7377388D-1	.5270267D-1	.3682343D-1	.2440276D-1	.1813992D-1	.1434175D-1	.9911410D-2	.6414354D-2
1	PF1	1.577853930	1.135684133	1.006997248	.8858534010	.7641755753	.6867071333	.6311068487	.5540105814	.4804116932
1	PQ1	.6637660542	.6312711279	.6308507583	.6364395566	.6518574980	.6709526204	.6927241618	.7444276657	.8498627305
1	PF2	1.188476443	1.081876209	1.043284760	1.004130945	.9627168288	.9357666555	.9164181297	.8899836419	.8656987468
1	PQ2	2.420597953	2.149589759	2.145954519	2.193986379	2.323398217	2.478226538	2.648664072	3.033543087	3.761610949
1	PFC	1.096724843	1.066145116	1.054500597	1.042388131	1.029264438	1.020561386	1.014238480	1.005504955	.9973880868
1	PQC	12.85855099	11.35293697	11.33267791	11.60020833	12.31953028	13.17768298	14.11982203	16.23994718	20.23228672
1	ZF1	1.139312975	1.139312975	1.139312975	1.139312975	1.139312975	1.139312975	1.139312975	1.139312975	1.139312975
1	ZF2	1.353318131	1.353318131	1.353318131	1.353318131	1.353318131	1.353318131	1.353318131	1.353318131	1.353318131
1	ZF3	3.125974397	3.125974397	3.125974397	3.125974397	3.125974397	3.125974397	3.125974397	3.125974397	3.125974397

(3) P 6+0	TH = 64	SBE = 1.112601940						CAUER FUNCTIONS	
	RC = 02	RC = 05	RC = 07	RC = 10	RC = 15	RC = 20	RC = 25	RC = 35	RC = 50
AMAX DB	.0017375255	.0108709564	.0213327376	.0436480540	.0988323393	.1772876696	.2802872360	.5675287486	1.249387366
AMIN DB	13.73686947	21.54762735	24.46569191	27.57807902	31.15082975	33.72656826	35.76707754	38.97627838	42.75585668
MU K	.2056632D 0	.8367941D-1	.5980196D-1	.4179228D-1	.2769864D-1	.2059072D-1	.1627969D-1	.1125087D-1	.7281271D-2
1 PF1	1.612996690	1.157617916	1.025238214	.9008732285	.7762588361	.6970879094	.6403467877	.5617795712	.4868887020
1 PQ1	.6762818446	.6370981511	.6355612567	.6403402620	.6551520677	.6739665818	.6955927878	.7471973472	.8527393152
1 PF2	1.186987806	1.083876998	1.046049009	1.007486867	.9665358465	.9398117060	.9205935485	.8942985521	.8700964057
1 PQ2	2.564427031	2.237452158	2.224076895	2.265512897	2.391228892	2.545762943	2.717450974	3.107423919	3.848099462
1 PFC	1.091785128	1.063239389	1.052226923	1.040713906	1.028183782	1.019846457	1.013776657	1.005376591	.9975539718
1 PQC	13.90178010	12.05642115	11.98067966	12.21524943	12.92567931	13.79672495	14.76206663	16.94760848	21.08074834
1 ZF1	1.128770983	1.128770983	1.128770983	1.128770983	1.128770983	1.128770983	1.128770983	1.128770983	1.128770983
1 ZF2	1.334367004	1.334367004	1.334367004	1.334367004	1.334367004	1.334367004	1.334367004	1.334367004	1.334367004
1 ZF3	3.059948208	3.059948208	3.059948208	3.059948208	3.059948208	3.059948208	3.059948208	3.059948208	3.059948208

(3) P 6+0	TH = 65	SBE = 1.103377919						CAUER FUNCTIONS	
	RC = 02	RC = 05	RC = 07	RC = 10	RC = 15	RC = 20	RC = 25	RC = 35	RC = 50
AMAX DB	.0017375255	.0108709564	.0213327376	.0436480540	.0988323393	.1772876696	.2802872360	.5675287486	1.249387366
AMIN DB	12.68986118	20.45676332	23.37054729	26.48064657	30.05217218	32.62748125	34.66779165	37.87681926	41.65630550
MU K	.2320101D 0	.9487719D-1	.6783794D-1	.4742067D-1	.3143340D-1	.2336824D-1	.1847610D-1	.1276906D-1	.8263894D-2
1 PF1	1.649331812	1.180675574	1.044425739	.9166668222	.7889523588	.7079841804	.6500394632	.5699198242	.4936711515
1 PQ1	.6906322244	.6435650732	.6407401710	.6445927725	.6587153889	.6772111578	.6986711612	.7501574735	.8558029343
1 PF2	1.185081769	1.085710228	1.048713463	1.010800953	.9703638231	.9438935645	.9248234528	.8936891910	.8745965607
1 PQ2	2.727918684	2.334919258	2.310172097	2.343884217	2.465153843	2.619136772	2.792026199	3.187305553	3.941395650
1 PFC	1.086844112	1.060314948	1.049933826	1.039021615	1.027088411	1.019119901	1.013305811	1.005243228	.9977189779
1 PQC	15.10334219	12.84789377	12.70535880	12.89951135	13.59689420	14.48037290	15.47007326	17.72596618	22.01218167
1 ZF1	1.118742016	1.118742016	1.118742016	1.118742016	1.118742016	1.118742016	1.118742016	1.118742016	1.118742016
1 ZF2	1.316038147	1.316038147	1.316038147	1.316038147	1.316038147	1.316038147	1.316038147	1.316038147	1.316038147
1 ZF3	2.995216876	2.995216876	2.995216876	2.995216876	2.995216876	2.995216876	2.995216876	2.995216876	2.995216876

(3) P 6+0	TH = 66	SBE = 1.094636279						CAUER FUNCTIONS	
	RC = 02	RC = 05	RC = 07	RC = 10	RC = 15	RC = 20	RC = 25	RC = 35	RC = 50
AMAX DB	.0017375255	.0108709564	.0213327376	.0436480540	.0988323393	.1772876696	.2802872360	.5675287486	1.249387366
AMIN DB	11.65746498	19.36879796	22.27708939	25.38424802	28.95419749	31.52895384	33.56900821	36.77781279	40.55718051
MU K	.2612924D 0	.1075375D 0	.7693882D-1	.5380066D-1	.3566893D-1	.2651870D-1	.2096764D-1	.1449137D-1	.9378664D-2
1 PF1	1.686695135	1.204916069	1.064621481	.9332893305	.8023011924	.7194341930	.6602184015	.5784617602	.5007812225
1 PQ1	.7071826478	.6507760506	.6464578722	.6492458026	.6625816106	.6807140612	.7019834972	.7533288579	.8590730185
1 PF2	1.182728379	1.087354794	1.051261594	1.014061092	.9741929291	.9480069327	.9291042303	.9031540695	.8791894883
1 PQ2	2.914862868	2.443537660	2.405447977	2.430079042	2.545996395	2.699109799	2.873127313	3.273929062	4.042317361
1 PFC	1.081907417	1.057374054	1.047622832	1.037312172	1.025978724	1.018381853	1.012825921	1.005104680	.9978828128
1 PQC	16.49689187	13.74322252	13.51985358	13.66431449	14.34335826	15.23847922	16.25369709	18.58538026	23.03851398
1 ZF1	1.109207618	1.109207618	1.109207618	1.109207618	1.109207618	1.109207618	1.109207618	1.109207618	1.109207618
1 ZF2	1.298303529	1.298303529	1.298303529	1.298303529	1.298303529	1.298303529	1.298303529	1.298303529	1.298303529
1 ZF3	2.931674928	2.931674928	2.931674928	2.931674928	2.931674928	2.931674928	2.931674928	2.931674928	2.931674928

(3) P 6+0 — TH = 67 SBE = 1.086360377 — CAUER FUNCTIONS

	RC = 02	RC = 05	RC = 07	RC = 10	RC = 15	RC = 20	RC = 25	RC = 35	RC = 50
AMAX DB	.0017375255	.0108709564	.0213327376	.0436480540	.0988323393	.1772876696	.2802872360	.5675287486	1.249387366
AMIN DB	10.64212179	18.28327050	21.18451346	24.28789004	27.85580996	30.42985418	32.46957863	35.67809576	39.45731072
MU K	.2936932D 0	.1218531D 0	.8725179D-1	.6103873D-1	.4047711D-1	.3009590D-1	.2379694D-1	.1644732D-1	.1064473D-1
1 PF1	1.724841422	1.230395663	1.085890099	.9508006722	.8163551030	.7314804337	.6709209382	.5874349768	.5082436882
1 PQ1	.7263861691	.6588578467	.6527991394	.6543574387	.6667910256	.6845077779	.7055580540	.7567356447	.8625719906
1 PF2	1.179897564	1.088787676	1.053675177	1.017253800	.9780143286	.9521457472	.9334316741	.9076913285	.8838752921
1 PQ2	3.129978043	2.565191165	2.511362920	2.525269401	2.634734830	2.786583022	2.961624007	3.368164871	4.151822027
1 PFC	1.076981345	1.054419159	1.045295606	1.035586582	1.024855168	1.017632472	1.012336979	1.004960754	.9980451618
1 PQC	18.12520226	14.76208367	14.44031804	14.52347164	15.17740982	16.08293335	17.12479945	19.53827831	24.17401328
1 ZF1	1.100150663	1.100150663	1.100150663	1.100150663	1.100150663	1.100150663	1.100150663	1.100150663	1.100150663
1 ZF2	1.281136796	1.281136796	1.281136796	1.281136796	1.281136796	1.281136796	1.281136796	1.281136796	1.281136796
1 ZF3	2.869218322	2.869218322	2.869218322	2.869218322	2.869218322	2.869218322	2.869218322	2.869218322	2.869218322

(3) P 6+0 — TH = 68 SBE = 1.078534743 — CAUER FUNCTIONS

	RC = 02	RC = 05	RC = 07	RC = 10	RC = 15	RC = 20	RC = 25	RC = 35	RC = 50
AMAX DB	.0017375255	.0108709564	.0213327376	.0436480540	.0988323393	.1772876696	.2802872360	.5675287486	1.249387366
AMIN DB	9.646992360	17.19985096	20.09204704	23.19055541	26.75585894	29.32898446	31.36828315	34.57642929	38.35544706
MU K	.3293445D 0	.1380408D 0	.9894586D-1	.6925836D-1	.4594170D-1	.3416259D-1	.2701381D-1	.1867147D-1	.1208447D-1
1 PF1	1.763423924	1.257164638	1.108298417	.9692657499	.8311691412	.7441702273	.6821887906	.5968737515	.5160863349
1 PQ1	.7488063761	.6679659134	.6598668919	.6599974557	.6713915101	.6886306459	.7094280226	.7604060166	.8663258805
1 PF2	1.176559904	1.089983762	1.055934100	1.020364035	.9818180348	.9563030639	.9378008882	.9122986753	.8886538662
1 PQ2	3.379179038	2.702193978	2.629692487	2.630870016	2.732540316	2.882629858	3.058549045	3.471042364	4.271037978
1 PFC	1.072073059	1.051452944	1.042953963	1.033845955	1.023718249	1.016871952	1.011838988	1.004311247	.9982056857
1 PQC	20.04312178	15.92911325	15.48679992	15.49398718	16.11412533	17.02820120	18.09777157	20.59968558	25.43587927
1 ZF1	1.091555261	1.091555261	1.091555261	1.091555261	1.091555261	1.091555261	1.091555261	1.091555261	1.091555261
1 ZF2	1.264513152	1.264513152	1.264513152	1.264513152	1.264513152	1.264513152	1.264513152	1.264513152	1.264513152
1 ZF3	2.807743612	2.807743612	2.807743612	2.807743612	2.807743612	2.807743612	2.807743612	2.807743612	2.807743612

(3) P 6+0 — TH = 69 SBE = 1.071144994 — CAUER FUNCTIONS

	RC = 02	RC = 05	RC = 07	RC = 10	RC = 15	RC = 20	RC = 25	RC = 35	RC = 50
AMAX DB	.0017375255	.0108709564	.0213327376	.0436480540	.0988323393	.1772876696	.2802872360	.5675287486	1.249387366
AMIN DB	8.676043296	16.11838635	18.99896829	22.09120405	25.65312973	28.22506807	30.26381640	33.47148286	37.25024563
MU K	.3682967D 0	.1563438D 0	.1122152D 0	.7860312D-1	.5216071D-1	.3879240D-1	.3067674D-1	.2120439D-1	.1372422D-1
1 PF1	1.801973127	1.285262568	1.131913988	.9887545281	.8468042652	.7575564337	.6940687381	.6068166480	.5243404733
1 PQ1	.7751462488	.6782924105	.6677870937	.6662503403	.6764403847	.6931282403	.7136326648	.7643730934	.8703651014
1 PF2	1.172687697	1.090915670	1.058016132	1.023374991	.9855927405	.9604709211	.9422061761	.9169733070	.8935248516
1 PQ2	3.669931541	2.857415263	2.762617912	2.748603349	2.840876461	2.988539298	3.165138342	3.583788039	4.401304643
1 PFC	1.067190797	1.048478358	1.040599922	1.032091520	1.022568539	1.016100518	1.011331967	1.004655951	.9983640180
1 PQC	22.32163449	17.27548145	16.68443214	16.59699808	17.17209768	18.09204184	19.19022798	21.78792394	26.84501994
1 ZF1	1.083406696	1.083406696	1.083406696	1.083406696	1.083406696	1.083406696	1.083406696	1.083406696	1.083406696
1 ZF2	1.248409242	1.248409242	1.248409242	1.248409242	1.248409242	1.248409242	1.248409242	1.248409242	1.248409242
1 ZF3	2.747147029	2.747147029	2.747147029	2.747147029	2.747147029	2.747147029	2.747147029	2.747147029	2.747147029

(3) P 6+0 — TH = 70 — SBE = 1.064177772 — CAUER FUNCTIONS

(3) P 6+0	RC = 02	RC = 05	RC = 07	RC = 10	RC = 15	RC = 20	RC = 25	RC = 35	RC = 50
AMAX DB	.0017375255	.0108709564	.0213327376	.0436480540	.0988323393	.1772876696	.2802872360	.5675287486	1.249387366
AMIN DB	7.734098774	15.03896261	17.90463263	20.98877671	24.54633386	27.11673498	29.15477034	32.36181524	36.14024751
MU I K	.4104829D 0	.1770320D 0	.1272824D 0	.8924033D-1	.5924931D-1	.4407205D-1	.3485471D-1	.2409402D-1	.1559508D-1
1 I PF1	1.839877485	1.314711574	1.156802818	1.009341878	.8633280099	.7716982558	.7066134297	.6173072494	.5330415631
1 I PQ1	.8062842536	.6900768688	.6767152634	.6732192793	.6820068463	.6980551726	.7182187832	.7686760853	.8747254403
1 I PF2	1.168256386	1.091553577	1.059896683	1.026267847	.9893256163	.9646401759	.9466409064	.9217113182	.8984875833
1 I PQ2	4.011724080	3.034447515	2.912845054	2.880586356	2.961314833	3.105872621	3.282883444	3.707375227	4.544224713
1 I PFC	1.062344128	1.045498675	1.038235688	1.030324649	1.021406689	1.015318441	1.010815957	1.004494646	.9985197617
1 I PQC	25.05349057	18.84107498	18.06507177	17.85905764	18.37448881	19.29647151	20.42393629	23.12554480	28.42708440
1 I ZF1	1.075691350	1.075691350	1.075691350	1.075691350	1.075691350	1.075691350	1.075691350	1.075691350	1.075691350
1 I ZF2	1.232803052	1.232803052	1.232803052	1.232803052	1.232803052	1.232803052	1.232803052	1.232803052	1.232803052
1 I ZF3	2.687323465	2.687323465	2.687323465	2.687323465	2.687323465	2.687323465	2.687323465	2.687323465	2.687323465

(3) P 6+0 — TH = 71 — SBE = 1.057620681 — CAUER FUNCTIONS

(3) P 6+0	RC = 02	RC = 05	RC = 07	RC = 10	RC = 15	RC = 20	RC = 25	RC = 35	RC = 50
AMAX DB	.0017375255	.0108709564	.0213327376	.0436480540	.0988323393	.1772876696	.2802872360	.5675287486	1.249387366
AMIN DB	6.826828660	13.96198643	16.80851101	19.88220306	23.43409879	26.00250462	28.03961401	31.24585135	35.02385414
MU I K	.4556785D 0	.2004014D 0	.1444024D 0	.1013654D 0	.6734340D-1	.5010427D-1	.3962956D-1	.2739728D-1	.1773402D-1
1 I PF1	1.876370677	1.345506846	1.183025813	1.031107038	.8808151803	.7866621764	.7198823505	.6283950544	.5422299839
1 I PQ1	.8433192039	.7036204917	.6868452092	.6810314803	.6881751840	.7034774531	.7232426422	.7733617896	.8794493360
1 I PF2	1.163246368	1.091865083	1.061548519	1.029021487	.9930020707	.9688003074	.9510973507	.9265100889	.9035410250
1 I PQ2	4.416700976	3.237836999	3.083766401	3.029447746	3.096122711	3.236538934	3.413601133	3.845089701	4.701732656
1 I PFC	1.057544227	1.042517571	1.035863745	1.028546877	1.020233439	1.014526041	1.010291020	1.004327105	.9986724959
1 I PQC	28.36109888	20.67756644	19.66958859	19.31391250	19.75047379	20.66907911	21.82608100	24.64059174	30.21385523
1 I ZF1	1.068396651	1.068396651	1.068396651	1.068396651	1.068396651	1.068396651	1.068396651	1.068396651	1.068396651
1 I ZF2	1.217673811	1.217673811	1.217673811	1.217673811	1.217673811	1.217673811	1.217673811	1.217673811	1.217673811
1 I ZF3	2.628165295	2.628165295	2.628165295	2.628165295	2.628165295	2.628165295	2.628165295	2.628165295	2.628165295

(3) P 6+0 — TH = 72 — SBE = 1.051462224 — CAUER FUNCTIONS

(3) P 6+0	RC = 02	RC = 05	RC = 07	RC = 10	RC = 15	RC = 20	RC = 25	RC = 35	RC = 50
AMAX DB		.0108709564	.0213327376	.0436480540	.0988323393	.1772876696	.2802872360	.5675287486	1.249387366
AMIN DB		12.88829202	15.71024393	18.77041611	22.31495744	24.88076606	26.91666921	30.12185409	33.89929695
MU I K		.2267699D 0	.1638656D 0	.1152071D 0	.7660412D-1	.5701140D-1	.4509896D-1	.3118224D-1	.2018530D-1
1 I PF1		1.377603539	1.210633408	1.054132460	.8993485284	.8025230352	.7339429771	.6401365774	.5519519935
1 I PQ1		.7193055004	.6984208786	.6898453612	.6950490886	.7094756351	.7287725121	.7784865587	.8845875471
1 I PF2		1.091815116	1.062941477	1.031612165	.9966054619	.9729391791	.9555664872	.9313631475	.9086836888
1 I PQ2		3.473403287	3.279684819	3.198488765	3.247882324	3.382897190	3.559527044	3.997617411	4.876185931
1 I PFC		1.039539217	1.033486883	1.026759936	1.019049636	1.013723695	1.009757246	1.004153093	.9988217222
1 I PQC		22.85279943	21.55111228	21.00500252	21.33725332	22.24484897	23.43100811	26.36833461	32.24515314
1 I ZF1		1.061511022	1.061511022	1.061511022	1.061511022	1.061511022	1.061511022	1.061511022	1.061511022
1 I ZF2		1.203001910	1.203001910	1.203001910	1.203001910	1.203001910	1.203001910	1.203001910	1.203001910
1 I ZF3		2.569561001	2.569561001	2.569561001	2.569561001	2.569561001	2.569561001	2.569561001	2.569561001

(3) P 6+0	TH = 73 SBE = 1.045691756							CAUER FUNCTIONS	
	RC = 02	RC = 05	RC = 07	RC = 10	RC = 15	RC = 20	RC = 25	RC = 35	RC = 50
AMAX DB		.0108709564	.0213327376	.0436480540	.0988323393	.1772876696	.2802872360	.5675287486	1.249387366
AMIN DB		11.81927769	14.60971695	17.65237612	21.18733813	23.74975488	25.78408111	28.98788930	32.76459941
MU I K		.2564697D 0	.1860005D 0	.1310332D 0	.8722342D-1	.6494000D-1	.5138022D-1	.3553084D-1	.2300223D-1
1 I PF1		1.410898939	1.239657563	1.078501676	.9190193346	.8193652568	.7488721654	.6525967142	.5622609348
1 I PQ1		.7376215298	.7117526287	.6998603961	.7027575053	.7161490586	.7348920843	.7841189228	.8902013603
1 I PF2		1.091365954	1.064042151	1.034013115	1.000116751	.9770427486	.9600377587	.9362650013	.9139135353
1 I PQ2		3.748688678	3.506127414	3.391907181	3.419905432	3.547896111	3.723443493	4.168163607	5.070488192
1 I PFC		1.036568408	1.031108277	1.024965794	1.017856257	1.012911855	1.009214764	1.003972366	.9989669593
1 I PQC		25.45716082	23.77972209	22.98904001	23.18290741	24.06872972	25.28267368	28.35369092	34.57148470
1 I ZF1		1.055023833	1.055023833	1.055023833	1.055023833	1.055023833	1.055023833	1.055023833	1.055023833
1 I ZF2		1.188768818	1.188768818	1.188768818	1.188768818	1.188768818	1.188768818	1.188768818	1.188768818
1 I ZF3		2.511393524	2.511393524	2.511393524	2.511393524	2.511393524	2.511393524	2.511393524	2.511393524

(3) P 6+0	TH = 74 SBE = 1.040299436							CAUER FUNCTIONS	
	RC = 02	RC = 05	RC = 07	RC = 10	RC = 15	RC = 20	RC = 25	RC = 35	RC = 50
AMAX DB		.0108709564	.0213327376	.0436480540	.0988323393	.1772876696	.2802872360	.5675287486	1.249387366
AMIN DB		10.75707636	13.50716524	16.52710857	20.04955594	22.60752617	24.63978244	27.84178206	31.61752903
MU I K		.2898319D 0	.2111746D 0	.1491573D 0	.9943109D-1	.7406682D-1	.5861528D-1	.4054253D-1	.2624965D-1
1 I PF1		1.445208740	1.270100018	1.104295651	.9399277569	.8372842254	.7647578157	.6658504465	.5732187684
1 I PQ1		.7592019588	.7272398425	.7113307957	.7114627108	.7236216657	.7417051271	.7903431380	.8963655570
1 I PF2		1.090477408	1.064813593	1.036194110	1.003514081	.9810947111	.9644987736	.9412084250	.9192278478
1 I PQ2		4.073600057	3.770293204	3.615114785	3.616414467	3.735269429	3.908857011	4.360117255	5.288258447
1 I PFC		1.033610728	1.028731584	1.023166710	1.016654435	1.012091061	1.008663746	1.003784676	.9991076385
1 I PQC		28.61301446	26.44935212	25.34123559	25.35052012	26.19932380	27.43814520	30.65466894	37.25779000
1 I ZF1		1.048925370	1.048925370	1.048925370	1.048925370	1.048925370	1.048925370	1.048925370	1.048925370
1 I ZF2		1.174957012	1.174957012	1.174957012	1.174957012	1.174957012	1.174957012	1.174957012	1.174957012
1 I ZF3		2.453538223	2.453538223	2.453538223	2.453538223	2.453538223	2.453538223	2.453538223	2.453538223

(3) P 6+0	TH = 75 SBE = 1.035276180							CAUER FUNCTIONS	
	RC = 02	RC = 05	RC = 07	RC = 10	RC = 15	RC = 20	RC = 25	RC = 35	RC = 50
AMAX DB		.0108709564	.0213327376	.0436480540	.0988323393	.1772876696	.2802872360	.5675287486	1.249387366
AMIN DB		9.704760356	12.40331618	15.39376286	18.89980740	21.45192323	23.48144940	26.68106119	30.45553560
MU I K		.3271613D 0	.2397917D 0	.1699464D 0	.1135036D 0	.8460652D-1	.6697728D-1	.4633903D-1	.3000704D-1
1 I PF1		1.480236440	1.301915296	1.131586766	.9621827048	.8563877738	.7817008690	.6799349931	.5848980467
1 I PQ1		.7848743603	.7454027893	.7245848188	.7213716466	.7320501107	.7493419449	.7972640707	.9031724664
1 I PF2		1.089107268	1.065215062	1.038120940	1.006772262	.9850760563	.9689349320	.9461846941	.9246230725
1 I PQ2		4.461341030	4.081703515	3.875196188	3.842873474	3.949813554	4.120249591	4.577782213	5.534067548
1 I PFC		1.030672762	1.026361063	1.021365304	1.015445495	1.011261960	1.008104421	1.003589766	.9992431468
1 I PQC		32.48895454	29.68817847	28.16309465	27.92427021	28.71429870	29.97271302	33.34736433	40.38835376
1 I ZF1		1.043206806	1.043206806	1.043206806	1.043206806	1.043206806	1.043206806	1.043206806	1.043206806
1 I ZF2		1.161549907	1.161549907	1.161549907	1.161549907	1.161549907	1.161549907	1.161549907	1.161549907
1 I ZF3		2.395860322	2.395860322	2.395860322	2.395860322	2.395860322	2.395860322	2.395860322	2.395860322

$P^{(4)}_{6+2}$

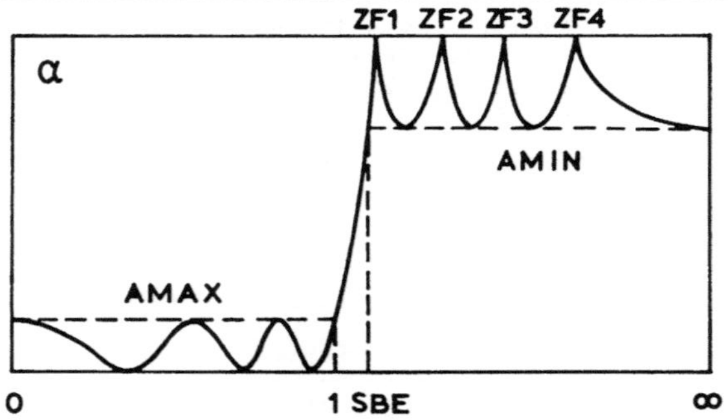

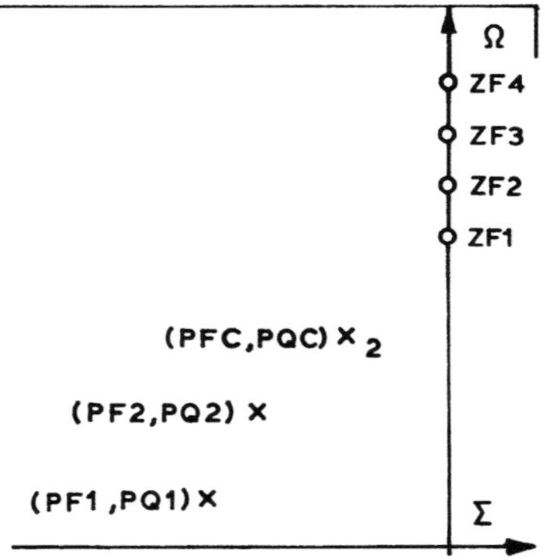

(4) P 6+2	TH = 25	SBE = 2.366201583						MCPER FUNCTIONS	
	RC = 02	RC = 05	RC = 07	RC = 10	RC = 15	RC = 20	RC = 25	RC = 35	RC = 50
AMAX DB	.0017375255	.0108709564	.0213327376	.0436480540	.0988323393	.1772876696	.2802872360	.5675287486	1.249387366
AMIN DB	89.56217446	100.2182329	104.1475228	108.3323478	113.1349882	116.5973037	119.3386354	123.6402528	128.6653541
MU I K	.3325763D-4	.9751880D-5	.6203315D-5	.3831622D-5	.2204198D-5	.1479568D-5	.1079116D-5	.6576387D-6	.3687502D-6
1 I PF1	.8656278562	.6726571857	.6087666904	.5447670844	.4765099453	.4309394146	.3972258024	.3491034981	.3016451797
1 I PQ1	.5413868814	.5553813739	.5638180228	.5760462635	.5963534082	.6174180177	.6398431723	.6907668956	.7915824555
1 I PF2	1.074776620	.9274366130	.8804475039	.8346519174	.7874596042	.7569922511	.7349912829	.7042794488	.6743617624
1 I PQ2	.9015498336	.9965054426	1.050630433	1.125588536	1.242800875	1.357083918	1.472609929	1.718317958	2.163238401
2 I PFC	1.366664076	1.239821641	1.198269785	1.157217180	1.114255323	1.086097025	1.065518097	1.036388277	1.007500698
2 I PQC	2.551347392	2.738246814	2.848540027	2.998834749	3.225070367	3.434935323	3.637172922	4.038876724	4.689493659
1 I ZF1	2.405402216	2.406169366	2.406404718	2.406631735	2.406865641	2.407018506	2.407130975	2.407293130	2.407460918
1 I ZF2	2.776864354	2.784376245	2.786671736	2.788882815	2.791158260	2.792644043	2.793736630	2.795311114	2.796939452
1 I ZF3	4.026268584	4.051882717	4.059714512	4.067259865	4.075026121	4.080097692	4.083827267	4.089201850	4.094760094
1 I ZF4	11.20311515	11.30497268	11.33622194	11.36636681	11.39742856	11.41772885	11.43266440	11.45419637	11.47647284

(4) P 6+2	TH = 26	SBE = 2.281172033						MCPER FUNCTIONS	
	RC = 02	RC = 05	RC = 07	RC = 10	RC = 15	RC = 20	RC = 25	RC = 35	RC = 50
AMAX DB	.0017375255	.0108709564	.0213327376	.0436480540	.0988323393	.1772876696	.2802872360	.5675287486	1.249387366
AMIN DB	86.74289055	97.39806473	101.3271436	105.5117908	110.3142743	113.7765014	116.5177751	120.8193194	125.8443579
MU I K	.4601034D-4	.1349263D-4	.8583073D-5	.5301643D-5	.3049905D-5	.2047269D-5	.1493177D-5	.9099846D-6	.5102489D-6
1 I PF1	.8710279794	.6761786469	.6117839676	.5473300406	.4786386715	.4328041510	.3989072891	.3505395043	.3028556783
1 I PQ1	.5418407478	.5557482529	.5641630416	.5763732219	.5966662409	.6177254517	.6401497038	.6910797899	.7919206742
1 I PF2	1.077514145	.9297580580	.8826373238	.8367157858	.7893974571	.7588517130	.7367961554	.7060120918	.6760307920
1 I PQ2	.9074913877	1.001757230	1.055779868	1.130724393	1.248061861	1.362551357	1.478329886	1.724648783	2.170782004
2 I PFC	1.362414446	1.237544647	1.196520551	1.155937674	1.113415659	1.085518481	1.065117381	1.036221208	1.007545210
2 I PQC	2.594796825	2.775342909	2.884267178	3.033644215	3.259539326	3.469681123	3.672495922	4.075874292	4.729992430
1 I ZF1	2.318427353	2.319218493	2.319461567	2.319696177	2.319938045	2.320096185	2.320212571	2.320380422	2.320554166
1 I ZF2	2.671851162	2.679622161	2.681998556	2.684288348	2.686645611	2.688185270	2.689317705	2.690949958	2.692638476
1 I ZF3	3.863789048	3.890324258	3.898444383	3.906270449	3.914328612	3.919592425	3.923464182	3.929044862	3.934817726
1 I ZF4	10.72959256	10.83507625	10.86748157	10.89875959	10.93100652	10.95209054	10.96760715	10.98998312	11.01314030

(4) P 6+2 — TH = 27 — SBE = 2.202689265 — MCPER FUNCTIONS

MU		RC = 02	RC = 05	RC = 07	RC = 10	RC = 15	RC = 20	RC = 25	RC = 35	RC = 50
	AMAX DB	.0017375255	.0108709564	.0213327376	.0436480540	.0988323393	.1772876696	.2802872360	.5675287486	1.249387366
	AMIN DB	84.02243654	94.67661011	98.60545002	102.7898961	107.5922020	111.0543291	113.7955372	118.0969988	123.1219662
	K	.6293296D-4	.1845736D-4	.1174161D-4	.7252791D-5	.4172438D-5	.2800809D-5	.2042787D-5	.1244945D-5	.6980744D-6
1	PF1	.8767145606	.6798785893	.6149521525	.5500196123	.4808713085	.4347592484	.4006698465	.3520442843	.3041238034
1	PQ1	.5423226310	.5561357552	.5645269233	.5767175983	.5969953219	.6180486079	.6404717445	.6914082932	.7922755537
1	PF2	1.080351218	.9321735303	.8849178681	.8388667173	.7914182989	.7607914521	.7386793464	.7078203841	.6777730616
1	PQ2	.9138080925	1.007316170	1.061223600	1.136147445	1.253610731	1.368313950	1.484355696	1.731313764	2.178718960
2	PFC	1.358007883	1.235178525	1.194701597	1.154606198	1.112541080	1.084915353	1.064699210	1.036046166	1.007590457
2	PQC	2.641250165	2.814824542	2.922241670	3.070597646	3.296086536	3.506493634	3.709899614	4.115021894	4.772813874
1	ZF1	2.238128622	2.238942815	2.239193379	2.239435379	2.239685022	2.239848326	2.239968551	2.240141995	2.240321594
1	ZF2	2.574713027	2.582738216	2.585194186	2.587561494	2.589999432	2.591592271	2.592764069	2.594453431	2.596201497
1	ZF3	3.713079786	3.740523154	3.748928480	3.757032597	3.765380292	3.770834993	3.774848035	3.780633665	3.786620146
1	ZF4	10.28956678	10.39861114	10.43215866	10.46455906	10.49798257	10.51984604	10.53594116	10.55915838	10.58319447

(4) P 6+2 — TH = 28 — SBE = 2.130054468 — MCPER FUNCTIONS

MU		RC = 02	RC = 05	RC = 07	RC = 10	RC = 15	RC = 20	RC = 25	RC = 35	RC = 50
	AMAX DB	.0017375255	.0108709564	.0213327376	.0436480540	.0988323393	.1772876696	.2802872360	.5675287486	1.249387366
	AMIN DB	81.39328385	92.04632960	95.97490016	100.1591195	104.9612252	108.4232394	111.1643736	115.4657419	120.4906292
	K	.8517964D-4	.2498524D-4	.1589480D-4	.9818475D-5	.5648573D-5	.3791735D-5	.2765549D-5	.1685438D-5	.9450799D-6
1	PF1	.8826979945	.6837624943	.6182756348	.5528392923	.4832105708	.4368069870	.4025154702	.3536194753	.3054508846
1	PQ1	.5428341009	.5565447923	.5649104364	.5770800462	.5973412097	.6183879967	.6408097773	.6917528643	.7926475579
1	PF2	1.083286137	.9346831245	.8872895480	.8411053450	.7935229191	.7628123266	.7406417508	.7097052569	.6795895246
1	PQ2	.9205214369	1.013196998	1.066974926	1.141869914	1.259458987	1.374383014	1.490698736	1.738324810	2.187062713
2	PFC	1.353445892	1.232723426	1.192812871	1.153222592	1.111631375	1.084287429	1.064263382	1.035862981	1.007636312
2	PQC	2.690911604	2.856832505	2.962590099	3.109810347	3.334818438	3.545475340	3.749484755	4.156420167	4.818063056
1	ZF1	2.163793624	2.164629864	2.164887671	2.165136848	2.165394066	2.165562415	2.165686398	2.165865324	2.166050674
1	ZF2	2.484609600	2.492883868	2.495418024	2.497861595	2.500379010	2.502024298	2.503234947	2.504980724	2.506787671
1	ZF3	3.572882598	3.601220517	3.609907743	3.618287109	3.626921830	3.632565989	3.636719370	3.642708738	3.648907772
1	ZF4	9.879456157	9.991990955	10.02666579	10.06017694	10.09476776	10.11740600	10.13407689	10.15813236	10.18304538

(4) P 6+2 — TH = 29 — SBE = 2.062665340 — MCPER FUNCTIONS

MU		RC = 02	RC = 05	RC = 07	RC = 10	RC = 15	RC = 20	RC = 25	RC = 35	RC = 50
	AMAX DB	.0017375255	.0108709564	.0213327376	.0436480540	.0988323393	.1772876696	.2802872360	.5675287486	1.249387366
	AMIN DB	78.84868955	89.50046866	93.42873667	97.61270133	102.4145820	105.8764696	108.6175207	112.9187841	117.9435814
	K	.1141736D-3	.3349474D-4	.2130900D-4	.1316330D-4	.7573051D-5	.5083660D-5	.3707865D-5	.2259752D-5	.1267129D-5
1	PF1	.8889894609	.6878362291	.6217591066	.5557928091	.4856593527	.4389497958	.4044462853	.3552668191	.3068383356
1	PQ1	.5433768804	.5569763549	.5653144132	.5774612714	.5977045064	.6187441671	.6411643211	.6921139954	.7930371836
1	PF2	1.086316954	.9372868817	.8897527554	.8434323079	.7957121311	.7649152276	.7426843014	.7116676845	.6814811817
1	PQ2	.9276549237	1.019415704	1.073048249	1.147905013	1.265619039	1.380770737	1.497371247	1.745694724	2.195827701
2	PFC	1.348730105	1.230179528	1.190854340	1.151786694	1.110686328	1.083634490	1.063809689	1.035671477	1.007682645
2	PQC	2.744006239	2.901520799	3.005450624	3.151407888	3.375850707	3.586737474	3.791360596	4.200178095	4.865853617
1	ZF1	2.094808365	2.095665576	2.095930363	2.096186498	2.096451070	2.096624339	2.096751991	2.096936283	2.097127271
1	ZF2	2.400816600	2.409334649	2.411945543	2.414464066	2.417059702	2.418756672	2.420005636	2.421807097	2.423672224
1	ZF3	3.442113027	3.471331164	3.480296819	3.488948480	3.497867585	3.503699693	3.507992412	3.514134238	3.520594701
1	ZF4	9.496173960	9.612124022	9.647910149	9.682519553	9.718267682	9.741675663	9.758919324	9.783809769	9.809597542

(4) P 6+2 — TH = 30 — SBE = 2.000000000 — MCPER FUNCTIONS

MU			RC = 02	RC = 05	RC = 07	RC = 10	RC = 15	RC = 20	RC = 25	RC = 35	RC = 50
		AMAX DB	.0017375255	.0108709564	.0213327376	.0436480540	.0988323393	.1772876696	.2802872360	.5675287486	1.249387366
		AMIN DB	76.38258924	87.03295054	90.96087984	95.14455932	99.94618817	103.4079338	106.1488919	110.4500379	115.4747344
MU		K	.1516598D-3	.4449923D-4	.2831105D-4	.1748928D-4	.1006215D-4	.6754657D-5	.4926692D-5	.3002604D-5	.1683694D-5
1		PF1	.8956009859	.6921060744	.6254075829	.5588841421	.4882207389	.4411902615	.4064645533	.3559881680	.3082876584
1		PQ1	.5439528626	.5574315198	.5657397561	.5778620371	.5980858605	.6191177099	.6415359336	.6924922149	.7934449633
1		PF2	1.089441453	.9399847799	.8923078570	.8458482473	.7979867714	.7671010786	.7448079691	.7137096353	.6834490814
1		PQ2	.9352342818	1.025989661	1.079459174	1.154267028	1.272104275	1.387490256	1.504386413	1.753437271	2.205029429
2		PFC	1.343862293	1.227547048	1.188825984	1.150298350	1.109705721	1.082956310	1.063337916	1.035471469	1.007729317
2		PQC	2.800782512	2.949058025	3.050974143	3.195527118	3.419309133	3.630400828	3.835645664	4.246413756	4.916308519
1		ZF1	2.030640857	2.031517892	2.031789379	2.032052208	2.032323931	2.032501985	2.032633216	2.032822748	2.033019258
1		ZF2	2.322706479	2.331462822	2.334148945	2.336741054	2.339413597	2.341161444	2.342448160	2.344304536	2.346227104
1		ZF3	3.319831420	3.349914681	3.359155107	3.368075952	3.377276658	3.383295124	3.387726123	3.394119054	3.400739754
1		ZF4	9.137045901	9.256330922	9.293211181	9.328905405	9.365800062	9.389972339	9.407785544	9.433507397	9.460167526

(4) P 6+2 — TH = 31 — SBE = 1.941604026 — MCPER FUNCTIONS

MU			RC = 02	RC = 05	RC = 07	RC = 10	RC = 15	RC = 20	RC = 25	RC = 35	RC = 50
		AMAX DB	.0017375255	.0108709564	.0213327376	.0436480540	.0988323393	.1772876696	.2802872360	.5675287486	1.249387366
		AMIN DB	73.98950763	84.63828680	88.56583805	92.74919921	97.55054692	101.0121342	103.7529884	108.0540033	113.0785873
MU		K	.1997674D-3	.5862538D-4	.3729994D-4	.2304305D-4	.1325784D-4	.8900067D-5	.6491583D-5	.3956397D-5	.2218557D-5
1		PF1	.9025455087	.6965787535	.6292264243	.5621175389	.4908980172	.4435311389	.4085726812	.3587854916	.3098004495
1		PQ1	.5445641298	.5579114583	.5661874434	.5782831683	.5984859715	.6195092609	.6419252144	.6928880896	.7938714679
1		PF2	1.092657117	.9427767234	.8949551879	.8483538032	.8003476978	.7693708353	.7470137627	.7158293225	.6854943221
1		PQ2	.9432877054	1.032937745	1.086224623	1.160971413	1.278929148	1.394555728	1.511758427	1.761567257	2.214684555
2		PFC	1.338844372	1.224826240	1.186727805	1.148757410	1.108689331	1.082252658	1.062847838	1.035262763	1.007776183
2		PQC	2.861515010	2.999628950	3.099325623	3.242317300	3.465330605	3.676596665	3.882468632	4.295255157	4.969560888
1		ZF1	1.970827899	1.971723533	1.972001421	1.972270697	1.972549322	1.972732020	1.972866732	1.973061374	1.973263280
1		ZF2	2.249732819	2.258721788	2.261481571	2.264145839	2.266893914	2.268691796	2.270015674	2.271926159	2.273905391
1		ZF3	3.205219590	3.236152075	3.245663422	3.254850174	3.264329547	3.270532693	3.275100859	3.281693464	3.288523142
1		ZF4	8.799743664	8.922278027	8.960234038	8.996998669	9.035028263	9.059958967	9.078338218	9.104887608	9.132417466

(4) P 6+2 — TH = 32 — SBE = 1.887079915 — MCPER FUNCTIONS

MU			RC = 02	RC = 05	RC = 07	RC = 10	RC = 15	RC = 20	RC = 25	RC = 35	RC = 50
		AMAX DB	.0017375255	.0108709564	.0213327376	.0436480540	.0988323393	.1772876696	.2802872360	.5675287486	1.249387366
		AMIN DB	71.66448317	82.31150183	86.23863233	90.42163918	95.22267394	98.68408489	101.4248234	105.7256925	110.7501512
MU		K	.2610813D-3	.7663460D-4	.4876053D-4	.3012437D-4	.1733270D-4	.1163579D-4	.8487090D-5	.5172677D-5	.2900631D-5
1		PF1	.9098369555	.7012614656	.6332213614	.5654975338	.4936946922	.4459753614	.4107732307	.3606608847	.3113784050
1		PQ1	.5452129761	.5584174461	.5666585370	.5787255574	.5989055939	.6199195047	.6423328083	.6933022285	.7943173092
1		PF2	1.095961096	.9456625317	.8976950442	.8509496102	.8027957876	.7717254849	.7493027298	.7180307051	.6876180529
1		PQ2	.9518461262	1.040280493	1.093362956	1.168034899	1.286109266	1.401982424	1.519502583	1.770100611	2.224810977
2		PFC	1.333678408	1.222017399	1.184559824	1.147163732	1.107636935	1.081523297	1.062339227	1.035045160	1.007823090
2		PQC	2.926507663	3.053436293	3.150685593	3.291941389	3.514064219	3.725467743	3.931969291	4.346841165	5.025754944
1		ZF1	1.914964332	1.915877258	1.916161231	1.916436680	1.916721956	1.916909148	1.917047238	1.917246851	1.917454021
1		ZF2	2.181417665	2.190633410	2.193465222	2.196200162	2.199022335	2.200869370	2.202229790	2.204193537	2.206228616
1		ZF3	3.097561860	3.129326813	3.139105026	3.148554235	3.158309183	3.164695240	3.169399396	3.176190166	3.183227488
1		ZF4	8.482230868	8.607923476	8.646935590	8.684755187	8.723907267	8.749590074	8.768531594	8.795904321	8.824301029

(4) P 6+2	TH = 33	SBE = 1.836078459						MCPER FUNCTIONS	
	RC = 02	RC = 05	RC = 07	RC = 10	RC = 15	RC = 20	RC = 25	RC = 35	RC = 50
AMAX DB	.0017375255	.0108709564	.0213327376	.0436480540	.0988323393	.1772876696	.2802872360	.5675287486	1.249387366
AMIN DB	69.40300430	80.04806914	83.97473259	88.15734610	92.95803343	96.41924861	99.15985875	103.4605658	108.4848854
MU K	.3387270D-3	.9944811D-4	.6327955D-4	.3909603D-4	.2249564D-4	.1510211D-4	.1101557D-4	.6713851D-5	.3764920D-5
1 PF1	.9174903208	.7061619209	.6373985221	.5690289678	.4966145001	.4485260535	.4130689284	.3626165745	.3130233272
1 PQ1	.5459019319	.5589508735	.5671541902	.5791901710	.5993455420	.6203491790	.6427594091	.6937352852	.7947831433
1 PF2	1.099350173	.9486419265	.9005276750	.8536362930	.8053319360	.7741660456	.7516759561	.7203139387	.6898214758
1 PQ2	.9609435217	1.048040262	1.100894112	1.175475608	1.293661499	1.409786825	1.527635371	1.779054485	2.235427940
2 PFC	1.328366633	1.219120867	1.182322086	1.145517180	1.106548307	1.080767984	1.061811845	1.034818451	1.007869880
2 PQC	2.996097425	3.110702745	3.205251839	3.344577478	3.565672519	3.777169459	3.984299648	4.401322557	5.085047044
1 ZF1	1.862694250	1.863623077	1.863912798	1.864194130	1.864485789	1.864677318	1.864818676	1.865023114	1.865235408
1 ZF2	2.117341153	2.126777645	2.129679795	2.132483860	2.135378632	2.137273898	2.138670211	2.140686332	2.142776400
1 ZF3	2.996229547	3.028809305	3.038850113	3.048558145	3.058585411	3.065152513	3.069991418	3.076978757	3.084222314
1 ZF4	8.182718873	8.311472978	8.351520224	8.390378266	8.430639475	8.457067579	8.476567292	8.504758804	8.534019216

(4) P 6+2	TH = 34	SBE = 1.788291650						MCPER FUNCTIONS	
	RC = 02	RC = 05	RC = 07	RC = 10	RC = 15	RC = 20	RC = 25	RC = 35	RC = 50
AMAX DB	.0017375255	.0108709564	.0213327376	.0436480540	.0988323393	.1772876696	.2802872360	.5675287486	1.249387366
AMIN DB	67.20095520	77.84385695	81.77000322	85.95218114	90.75248365	94.21348200	96.95394992	101.2544775	106.2786430
MU K	.4364678D-3	.1281761D-3	.8156425D-4	.5039541D-4	.2899852D-4	.1946820D-4	.1420046D-4	.8655180D-5	.4853643D-5
1 PF1	.9255217575	.7112883810	.6417644613	.5727170111	.4996614246	.4511865437	.4154626767	.3646549303	.3147371313
1 PQ1	.5466337934	.5595132574	.5676756569	.5796780564	.5998066955	.6207990789	.6432057637	.6941879627	.7952696743
1 PF2	1.102820726	.9517145179	.9034532731	.8564144607	.8079570531	.7766935650	.7541345658	.7226803766	.6921058467
1 PQ2	.9706172671	1.056241420	1.108839761	1.183313193	1.301604093	1.417986727	1.536174577	1.788447351	2.246556147
2 PFC	1.322911445	1.216137036	1.180014665	1.143817633	1.105423220	1.079986471	1.061265448	1.034582421	1.007916384
2 PQC	3.070658520	3.171673269	3.263241333	3.400420430	3.620332908	3.831871143	4.039625148	4.458863183	5.147606853
1 ZF1	1.813703760	1.814647008	1.814942120	1.815229027	1.815526787	1.815722484	1.815866997	1.816076104	1.816293375
1 ZF2	2.057132971	2.066784011	2.069754743	2.072626323	2.075592133	2.077534665	2.078966192	2.081033756	2.083177911
1 ZF3	2.900668195	2.934044130	2.944343033	2.954306059	2.964602210	2.971348390	2.976320733	2.983502956	2.990951256
1 ZF4	7.899630360	8.031343410	8.072403438	8.112282275	8.153638302	8.180804389	8.200857901	8.229863273	8.259983954

(4) P 6+2	TH = 35	SBE = 1.743446796						MCPER FUNCTIONS	
	RC = 02	RC = 05	RC = 07	RC = 10	RC = 15	RC = 20	RC = 25	RC = 35	RC = 50
AMAX DB	.0017375255	.0108709564	.0213327376	.0436480540	.0988323393	.1772876696	.2302872360	.5675287486	1.249387366
AMIN DB	65.05456938	75.69508175	79.62065660	83.80235319	88.60223047	92.06298919	94.80329992	99.10362922	104.1276243
MU K	.5588195D-3	.1641519D-3	.1044641D-3	.6454793D-4	.3714398D-4	.2493736D-4	.1819010D-4	.1108711D-4	.6217543D-5
1 PF1	.9339486754	.7166497014	.6463261935	.5765671876	.5028397155	.4539603789	.4179575668	.3667784724	.3165218533
1 PQ1	.5474116555	.5601062550	.5682243018	.5801903492	.6002900052	.6212700623	.6436726768	.6946610163	.7957776580
1 PF2	1.106368682	.9548797886	.9064719653	.8592847014	.8106720617	.7793091201	.7566797212	.7251311206	.6944724776
1 PQ2	.9809085367	1.064910556	1.117223486	1.191568981	1.309956800	1.426601367	1.545139405	1.798299124	2.258217883
2 PFC	1.317315425	1.213066350	1.177637663	1.142064978	1.104261449	1.079178508	1.060699785	1.034336844	1.007962427
2 PQC	3.150607335	3.236617727	3.324892416	3.459683729	3.678239226	3.889757511	4.098126058	4.519541288	5.213618652
1 ZF1	1.767714986	1.768671080	1.768971205	1.769263363	1.769566924	1.769766613	1.769914158	1.770127772	1.770349864
1 ZF2	2.000465286	2.010324505	2.013362001	2.016299425	2.019334643	2.021323433	2.022789465	2.024907496	2.027104792
1 ZF3	2.810386995	2.844539446	2.855091703	2.865305687	2.875867105	2.882790286	2.887894686	2.895270013	2.902921477
1 ZF4	7.631569163	7.766132660	7.808181687	7.849062485	7.891498013	7.919394233	7.939996813	7.969810721	8.000787926

(4) P 6+2 — TH = 36 SBE = 1.701301617 — MCPER FUNCTIONS

MU	(4) P 6+2	RC = 02	RC = 05	RC = 07	RC = 10	RC = 15	RC = 20	RC = 25	RC = 35	RC = 50
	AMAX DB	.0017375255	.0108709564	.0213327376	.0436480540	.0988323393	.1772876696	.2802872360	.5675287486	1.249387366
	AMIN DB	62.96038970	73.59826821	77.52321305	81.70437891	86.50378726	89.96428169	92.70441907	97.00452963	102.0283368
MU	K	.7111816D-3	.2089713D-3	.1329962D-3	.8218282D-4	.4729450D-4	.3175308D-4	.2316216D-4	.1411801D-4	.7917410D-5
1	PF1	.9427898509	.7222553803	.6510912289	.5805854016	.5061539076	.4568513402	.4205568911	.3689898833	.3183796580
1	PQ1	.5482389501	.5607316792	.5688016119	.5807282829	.6007965000	.6217630552	.6441610157	.6951552592	.7963079061
1	PF2	1.109989475	.9581370774	.9095838010	.8622475752	.8134778944	.7820138149	.7593126227	.7276675219	.6969227382
1	PQ2	.9918627636	1.074076718	1.126070977	1.200266142	1.318741023	1.435651552	1.554550604	1.808631291	2.270437156
2	PFC	1.311581342	1.209909316	1.175191213	1.140259118	1.103062768	1.078343836	1.060114598	1.034081488	1.008007826
2	PQC	3.236408099	3.305833888	3.390467289	3.522601583	3.739603547	3.951030309	4.159999029	4.583850986	5.283282808
1	ZF1	1.724481073	1.725448342	1.725753078	1.726050140	1.726359188	1.726562682	1.726713130	1.726931078	1.727157829
1	ZF2	1.947046848	1.957107715	1.960210094	1.963211625	1.966314554	1.968343551	1.969848343	1.972015819	1.974265263
1	ZF3	2.724949970	2.759858177	2.770658788	2.781119477	2.791942344	2.799040338	2.804275334	2.811841884	2.819694843
1	ZF4	7.377295116	7.514594488	7.557607240	7.599469931	7.642968590	7.671586523	7.692733094	7.723349780	7.755179433

(4) P 6+2 — TH = 37 SBE = 1.661640141 — MCPER FUNCTIONS

MU	(4) P 6+2	RC = 02	RC = 05	RC = 07	RC = 10	RC = 15	RC = 20	RC = 25	RC = 35	RC = 50
	AMAX DB	.0017375255	.0108709564	.0213327376	.0436480540	.0988323393	.1772876696	.2802872360	.5675287486	1.249387366
	AMIN DB	60.91523398	71.55021469	75.47446625	79.65504803	84.45394024	87.91414378	90.65409032	94.95396008	99.97756041
MU	K	.8999913D-3	.2645387D-3	.1683746D-3	.1040513D-3	.5988292D-4	.4020618D-4	.2932888D-4	.1787730D-4	.1002587D-4
1	PF1	.9520655473	.7281156110	.6560676130	.5847779679	.5096088424	.4598634601	.4232641588	.3712920186	.3203128479
1	PQ1	.5491194906	.5613915158	.5694092097	.5812931977	.6013272941	.6222790587	.6446717165	.6956715669	.7968612914
1	PF2	1.113677997	.9614855593	.9127887399	.8653036069	.8163754893	.7848087798	.7620345080	.7302909321	.6994580576
1	PQ2	1.003530167	1.083771686	1.135410251	1.209429878	1.327979977	1.445159816	1.564430618	1.819467051	2.283239852
2	PFC	1.305712162	1.206666502	1.172675488	1.138399971	1.101826954	1.077482196	1.059509624	1.033816111	1.008052389
2	PQC	3.328579479	3.379650866	3.460254860	3.589431310	3.804658215	4.015910175	4.225458851	4.651703937	5.356817433
1	ZF1	1.683781996	1.684758664	1.685067587	1.685369191	1.685683390	1.685890493	1.686043709	1.686265809	1.686497046
1	ZF2	1.896618052	1.906873877	1.910039194	1.913103030	1.916271905	1.918350010	1.919882787	1.922098636	1.924399186
1	ZF3	2.643968596	2.679610621	2.690654311	2.701357216	2.712437507	2.719708003	2.725072053	2.732327839	2.740880523
1	ZF4	7.135702980	7.275617460	7.319567124	7.362390351	7.406934661	7.436265287	7.457950364	7.489363650	7.522041320

(4) P 6+2 — TH = 38 SBE = 1.624269245 — MCPER FUNCTIONS

MU	(4) P 6+2	RC = 02	RC = 05	RC = 07	RC = 10	RC = 15	RC = 20	RC = 25	RC = 35	RC = 50
	AMAX DB	.0017375255	.0108709564	.0213327376	.0436480540	.0988323393	.1772876696	.2802872360	.5675287486	1.249387366
	AMIN DB	58.91616501	69.54796316	73.47145319	77.65139331	82.44971846	85.90960236	88.64933921	92.94894434	97.97231736
MU	K	.1132900D-2	.3331209D-3	.2120447D-3	.1310480D-3	.7542478D-4	.5064305D-4	.3694307D-4	.2251919D-4	.1262944D-4
1	PF1	.9617976479	.7342413403	.6612639707	.5891516437	.5132096923	.4630010419	.4260831117	.3736879205	.3223238736
1	PQ1	.5500575226	.5620879432	.5700488673	.5818865523	.6018835954	.6228191553	.6452057901	.6962108829	.7974387525
1	PF2	1.117428542	.9649242253	.9160866378	.8684532776	.8193657869	.7876951689	.7648466520	.7330027539	.7020799265
1	PQ2	1.015966357	1.094030274	1.145271912	1.219087636	1.337698873	1.455150585	1.574803745	1.830831482	2.296653907
2	PFC	1.299711060	1.203338545	1.170090699	1.136487476	1.100553789	1.076593326	1.058884592	1.033540463	1.008095914
2	PQC	3.427702260	3.458433081	3.534574005	3.660456073	3.873658157	4.084638742	4.294740453	4.723431228	5.434460247
1	ZF1	1.645421038	1.646405225	1.646717885	1.647023644	1.647342644	1.647553146	1.647708988	1.647935049	1.648170589
1	ZF2	1.848946780	1.859390721	1.862616967	1.865741240	1.868974221	1.871095290	1.872660239	1.874923341	1.877273904
1	ZF3	2.567095588	2.603448228	2.614729427	2.625669814	2.637003281	2.644443838	2.649935315	2.657878234	2.666128771
1	ZF4	6.905804682	7.048207214	7.093065379	7.136826442	7.182397762	7.212431431	7.234649160	7.266852354	7.300373232

(4) P 6+2		TH = 39	SBE = 1.589015729						MCPER FUNCTIONS
		RC = 02	RC = 05	RC = 07	RC = 10	RC = 15	RC = 20	RC = 25	RC = 35	RC = 50
AMAX DB		.0017375255	.0108709564	.0213327376	.0436480540	.0988323393	.1772876696	.2802872360	.5675287486	1.249387366
AMIN DB		56.96046457	67.58877299	71.51142791	75.69066430	80.48836744	83.94790072	86.68740752	90.98672235	96.00984596
MU	K	.1418982D-2	.4174086D-3	.2657227D-3	.1642354D-3	.9453261D-4	.6347533D-4	.4630519D-4	.2822695D-4	.1583098D-4
1	PF1	.9720098030	.7406443334	.6666895543	.5937136655	.5169619861	.4662686805	.4290177418	.3761808310	.3244153441
1	PQ1	.5510577840	.5628233556	.5707225238	.5825099363	.6024667153	.6233845174	.6457643294	.6967742254	.7980413003
1	PF2	1.121234754	.9684518589	.9194772317	.8716970154	.8224497248	.7906741584	.7677503657	.7358044420	.7047898989
1	PQ2	1.029233041	1.104890687	1.155689431	1.229269345	1.347925121	1.465650371	1.585696324	1.842751721	2.310709500
2	PFC	1.293581431	1.199926159	1.167437100	1.134521590	1.099243057	1.075676958	1.058239227	1.033254286	1.008138192
2	PQC	3.534428338	3.542584824	3.613777306	3.735987998	3.946883517	4.157481047	4.368101167	4.799285512	5.516470692
1	ZF1	1.609221804	1.610211517	1.610527440	1.610836948	1.611160377	1.611374059	1.611532377	1.611762196	1.612001848
1	ZF2	1.803824879	1.814449949	1.817735051	1.820917823	1.824213001	1.826375838	1.827972110	1.830281293	1.832680722
1	ZF3	2.494019644	2.531058344	2.542571169	2.553744037	2.565326192	2.572934233	2.578551415	2.586679244	2.595125651
1	ZF4	6.686714288	6.831471442	6.877208048	6.921882852	6.968461325	6.999187720	7.021931812	7.054917724	7.089276596

(4) P 6+2		TH = 40	SBE = 1.555723827						MCPER FUNCTIONS
		RC = 02	RC = 05	RC = 07	RC = 10	RC = 15	RC = 20	RC = 25	RC = 35	RC = 50
AMAX DB		.0017375255	.0108709564	.0213327376	.0436480540	.0988323393	.1772876696	.2802872360	.5675287486	1.249387366
AMIN DB		55.04561063	65.67009805	69.59183855	73.77030430	78.56732624	82.02647547	84.76573030	89.06472711	94.08757749
MU	K	.1768966D-2	.5205891D-3	.3314424D-3	.2048730D-3	.1179325D-3	.7919107D-4	.5777148D-4	.3521792D-4	.1975246D-4
1	PF1	.9827275920	.7473372450	.6723542976	.5984717884	.5208716381	.4696712854	.4320723112	.3787742079	.3265900393
1	PQ1	.5521255742	.5636003882	.5714323041	.5831650841	.6030780788	.6239764160	.6463485173	.6973626935	.7986700240
1	PF2	1.125089559	.9720670106	.9229601222	.8750351848	.8256282327	.7937469435	.7707469953	.7386975042	.7075895948
1	PQ2	1.043398826	1.116394911	1.166699473	1.240007690	1.358688567	1.476687982	1.597136940	1.855257164	2.325439271
2	PFC	1.287326896	1.196430136	1.164714995	1.132502295	1.097894548	1.074732827	1.057573245	1.032957312	1.008179002
2	PQC	3.649491260	3.632555527	3.698255357	3.816371753	4.024642670	4.234728257	4.445823302	4.879543435	5.603132324
1	ZF1	1.575025688	1.576018820	1.576337504	1.576650332	1.576977797	1.577194429	1.577355063	1.577538429	1.577831992
1	ZF2	1.761065177	1.771864247	1.775206065	1.778445333	1.781800719	1.784004079	1.785630789	1.787984825	1.790431918
1	ZF3	2.424460983	2.462159732	2.473897964	2.485298028	2.497124132	2.504896932	2.510637999	2.518948385	2.527538560
1	ZF4	6.477635222	6.624607126	6.671190416	6.716753415	6.764317908	6.795726010	6.818989731	6.852750631	6.887941847

(4) P 6+2		TH = 41	SBE = 1.524253087						MCPER FUNCTIONS
		RC = 02	RC = 05	RC = 07	RC = 10	RC = 15	RC = 20	RC = 25	RC = 35	RC = 50
AMAX DB		.0017375255	.0108709564	.0213327376	.0436480540	.0988323393	.1772876696	.2802872360	.5675287486	1.249387366
AMIN DB		53.16925737	63.78956648	67.71030713	71.88793017	76.68420710	80.14293628	82.88191549	87.18056442	92.20311591
MU	K	.2195519D-2	.6464303D-3	.4116088D-3	.2544506D-3	.1464838D-3	.9836785D-4	.7176360D-4	.4374937D-4	.2453828D-4
1	PF1	.9939787016	.7543336986	.6782688745	.6034343302	.5249449803	.4732141067	.4352513740	.3814717413	.3288509227
1	PQ1	.5532668358	.5644219470	.5721805402	.5838538912	.6037192366	.6245962308	.6469596356	.6979774757	.7993260990
1	PF2	1.128985103	.9757679696	.9265347549	.8784680748	.8289022263	.7969147359	.7738379206	.7416335015	.7104807018
1	PQ2	1.058540166	1.128589178	1.178342270	1.251338418	1.370021749	1.488294769	1.609156654	1.868379698	2.340878564
2	PFC	1.280951316	1.192851357	1.161924739	1.130429600	1.096508062	1.073760663	1.056886361	1.032649264	1.008218114
2	PQC	3.773718640	3.728845884	3.788441739	3.901988662	4.107275673	4.316700796	4.528217083	4.964508398	5.694755523
1	ZF1	1.542689708	1.543684031	1.544004947	1.544320641	1.544651730	1.544871068	1.545033852	1.545270541	1.545517802
1	ZF2	1.720498927	1.731464733	1.734861064	1.738154750	1.741568282	1.743810865	1.745467088	1.747864693	1.750358190
1	ZF3	2.358167539	2.396498764	2.408455823	2.420077497	2.432142536	2.440077214	2.445940240	2.454430692	2.463262407
1	ZF4	6.277849342	6.426889640	6.474286107	6.520710249	6.569238300	6.601316353	6.625092495	6.659620078	6.695637526

(4) P 6+2 — TH = 42, SBE = 1.494476550 — MCPER FUNCTIONS

MU	K	RC = 02	RC = 05	RC = 07	RC = 10	RC = 15	RC = 20	RC = 25	RC = 35	RC = 50
	AMAX DB	.0017375255	.0108709564	.0213327376	.0436480540	.0988323393	.1772876696	.2802872360	.5675287486	1.249387366
	AMIN DB	51.32921766	61.94496287	65.86461165	70.04131434	74.83677756	78.29504790	81.03372602	85.33199490	90.35421982
	K	.2713558D-2	.7993774D-3	.5090605D-3	.3147272D-3	.1812012D-3	.1216880D-3	.8877971D-4	.5412529D-4	.3035911D-4
1	PF1	1.005793123	.7616483750	.6844447656	.6086102206	.5291887974	.4769027630	.4385597996	.3842773728	.3312011567
1	PQ1	.5544882496	.5652912432	.5729697963	.5845784323	.6043918789	.6252454614	.6475990748	.6986198584	.8000107946
1	PF2	1.132912680	.9795527328	.9302003985	.8819958858	.8322726009	.8001787594	.7770245538	.7447640490	.7134649776
1	PQ2	1.074742457	1.141524486	1.190662042	1.263300693	1.381960197	1.500504895	1.621789270	1.882153956	2.357065696
2	PFC	1.274458799	1.189190794	1.159066745	1.128303543	1.095083406	1.072760197	1.056178281	1.032329858	1.008255289
2	PQC	3.908046821	3.832014961	3.884818774	3.993261452	4.195158224	4.403751925	4.615624016	5.054513703	5.791680587
1	ZF1	1.512084651	1.513077811	1.513400401	1.513718485	1.514052763	1.514274551	1.514439309	1.514679086	1.514929822
1	ZF2	1.681973622	1.693098769	1.696547343	1.699893301	1.703362833	1.705643288	1.707328057	1.709767888	1.712306468
1	ZF3	2.294911701	2.333846149	2.346015068	2.357852439	2.370151111	2.378244617	2.384227564	2.392895441	2.401916333
1	ZF4	6.086707577	6.237663399	6.285837742	6.333094414	6.382562169	6.415297643	6.439578500	6.474863850	6.511700927

(4) P 6+2 — TH = 43, SBE = 1.466279186 — MCPER FUNCTIONS

MU	K	RC = 02	RC = 05	RC = 07	RC = 10	RC = 15	RC = 20	RC = 25	RC = 35	RC = 50
	AMAX DB	.0017375255	.0108709564	.0213327376	.0436480540	.0988323393	.1772876696	.2802872360	.5675287486	1.249387366
	AMIN DB	49.52344754	60.13421242	64.05267022	68.22836898	73.02294449	76.48071422	79.21906381	83.51691799	88.53878653
	K	.3340624D-2	.9846670D-3	.6271429D-3	.3877766D-3	.2232815D-3	.1499562D-3	.1094074D-3	.6670434D-4	.3741629D-4
1	PF1	1.018203367	.7692971090	.6908943308	.6140090556	.5336103659	.4807432732	.4420027999	.3871953159	.3336441177
1	PQ1	.5557973478	.5662118323	.5738028972	.5853409822	.6050978505	.6259257402	.6482683460	.6992912357	.8007254834
1	PF2	1.136862660	.9834189700	.9339561217	.8856187145	.8357402226	.8035402467	.7803083375	.7479408167	.7165442526
1	PQ2	1.092101323	1.155257199	1.203707482	1.275937492	1.394542773	1.513355647	1.635071627	1.896617608	2.374042272
2	PFC	1.267853718	1.185449522	1.156141487	1.126124195	1.093620398	1.071731162	1.055448710	1.031998798	1.008290274
2	PQC	4.053538280	3.942688500	3.987924214	4.090659741	4.288706227	4.496271852	4.708420742	5.149926164	5.894281260
1	ZF1	1.483093475	1.484082992	1.484406667	1.484726639	1.485063648	1.485287616	1.485454165	1.485696782	1.485950758
1	ZF2	1.645351115	1.656628084	1.660126565	1.663522574	1.667045879	1.669362796	1.671075104	1.673555755	1.676138032
1	ZF3	2.234487514	2.273994124	2.286367521	2.298414326	2.310941015	2.319190121	2.325290829	2.334133334	2.343340891
1	ZF4	5.903621859	6.056333815	6.105248889	6.153307867	6.203690021	6.237069570	6.261846909	6.297880465	6.335530040

(4) P 6+2 — TH = 44, SBE = 1.439556540 — MCPER FUNCTIONS

MU	K	RC = 02	RC = 05	RC = 07	RC = 10	RC = 15	RC = 20	RC = 25	RC = 35	RC = 50
	AMAX DB	.0017375255	.0108709564	.0213327376	.0436480540	.0988323393	.1772876696	.2802872360	.5675287486	1.249387366
	AMIN DB	47.75003262	58.35536687	62.27252691	66.44713173	71.24073987	74.69796401	77.43595552	81.73335766	86.75483771
	K	.4097306D-2	.1208458D-2	.7697925D-3	.4760400D-3	.2741341D-3	.1841204D-3	.1343390D-3	.8190909D-4	.4594710D-4
1	PF1	1.031244701	.7772969990	.6976308911	.6196411573	.5382174983	.4847420908	.4455859583	.3902300792	.3361834149
1	PQ1	.5572026468	.5671876595	.5746829605	.5861440395	.6058391680	.6266388469	.6489690930	.6999931206	.8014716514
1	PF2	1.140824402	.9873639866	.9378007669	.8893365371	.8393059204	.8070004337	.7836907416	.7512155294	.7197204320
1	PQ2	1.110724115	1.169849749	1.217532318	1.289296065	1.407812052	1.526887790	1.649043940	1.911811692	2.391853523
2	PFC	1.261140713	1.181628723	1.153149510	1.123891665	1.092118868	1.070673293	1.054697347	1.031655780	1.008322806
2	PQC	4.211402343	4.061568636	4.098359033	4.194706406	4.388381066	4.594692472	4.807023466	5.251150233	6.002968764
1	ZF1	1.455609936	1.456593195	1.456917337	1.457238668	1.457577926	1.457803792	1.457971936	1.458217134	1.458474104
1	ZF2	1.610505997	1.621927143	1.625473128	1.628916894	1.632491660	1.634843572	1.636582365	1.639102368	1.641726890
1	ZF3	2.176708263	2.216754028	2.229324074	2.241573676	2.254322432	2.262723715	2.268939894	2.277954058	2.287345616
1	ZF4	5.728058149	5.882360340	5.931977116	5.980806506	6.032076242	6.066085666	6.091350699	6.128122216	6.166576596

(4) P 6+2	TH = 45	SBE = 1.414213562						MCPER FUNCTIONS	
	RC = 02	RC = 05	RC = 07	RC = 10	RC = 15	RC = 20	RC = 25	RC = 35	RC = 50
AMAX DB	.0017375255	.0108709564	.0213327376	.0436480540	.0988323393	.1772876696	.2802872360	.5675287486	1.249387366
AMIN DB	46.00717599	56.60659186	60.52233909	64.69575303	69.48830801	72.94493811	75.68253974	79.97944964	85.00050662
MU I K	.5007733D-2	.1477986D-2	.9416360D-3	.5823879D-3	.3354166D-3	.2252958D-3	.1643891D-3	.1002369D-3	.5623085D-4
1 I PF1	1.044955410	.7856665270	.7046688190	.6255176419	.5430185910	.4889061434	.4493152626	.3933864919	.3388229097
1 I PQ1	.5587138057	.5682231127	.5756134347	.5869903538	.6066180402	.6273867244	.6497031073	.7007271573	.8022509101
1 I PF2	1.144786177	.9913846809	.9417329205	.8931491907	.8429704754	.8105605541	.7871732613	.7545899671	.7229954986
1 I PQ2	1.130731693	1.185371437	1.232195951	1.303428458	1.421814764	1.541145965	1.663750179	1.927780984	2.410548708
2 I PFC	1.254324708	1.177729695	1.150091430	1.121606100	1.090578663	1.069586325	1.053923888	1.031300492	1.008352612
2 I PQC	4.383019963	4.189445322	4.216796540	4.305984997	4.494695715	4.699492843	4.911893080	5.358632765	6.118196434
1 I ZF1	1.429537387	1.430511638	1.430835595	1.431157730	1.431498732	1.431726197	1.431895734	1.432143238	1.432402942
1 I ZF2	1.577324190	1.588881749	1.592472766	1.595961917	1.599585745	1.601971124	1.603735305	1.606293122	1.608958367
1 I ZF3	2.121404382	2.161954205	2.174712588	2.187157945	2.200122459	2.208672289	2.215001510	2.224184181	2.233756906
1 I ZF4	5.559530387	5.715250435	5.765527957	5.815094143	5.867223069	5.901847269	5.927590625	5.965083132	6.004340026

(4) P 6+2	TH = 46	SBE = 1.390163591						MCPER FUNCTIONS	
	RC = 02	RC = 05	RC = 07	RC = 10	RC = 15	RC = 20	RC = 25	RC = 35	RC = 50
AMAX DB	.0017375255	.0108709564	.0213327376	.0436480540	.0988323393	.1772876696	.2802872360	.5675287486	1.249387366
AMIN DB	44.29318770	54.88615561	58.80036597	62.97248461	67.76389404	71.21987797	73.95705543	78.25342980	83.27402645
MU I K	.6100151D-2	.1801740D-2	.1148105D-2	.7101920D-3	.4090772D-3	.2747933D-3	.2005152D-3	.1222724D-3	.6859598D-4
1 I PF1	1.059377073	.7944256940	.7120236399	.6316504943	.5480226785	.4932428754	.4531971416	.3966697322	.3415667376
1 I PQ1	.5603418154	.5693230841	.5765981425	.5878829573	.6074368903	.6281714981	.6504723441	.7014351356	.8030650090
1 I PF2	1.148735074	.9954774987	.9457508808	.8970563520	.8467346095	.8142218329	.7907574127	.7580659648	.7263715152
1 I PQ2	1.152260520	1.201899372	1.247764201	1.318392125	1.436602298	1.556179153	1.679238512	1.944574426	2.430181557
2 I PFC	1.247410922	1.173753860	1.146967944	1.119267694	1.088999647	1.068470001	1.053128027	1.030932610	1.008379403
2 I PQC	4.569973477	4.327209799	4.343993070	4.425148394	4.608221841	4.811205537	5.023541073	5.472868500	6.240465050
1 I ZF1	1.404787747	1.405750096	1.406073186	1.406395540	1.406737755	1.406966508	1.407137222	1.407386746	1.407648913
1 I ZF2	1.545701732	1.557387813	1.561021323	1.564553410	1.568223810	1.570641064	1.572429485	1.575023509	1.577727881
1 I ZF3	2.068421647	2.109438177	2.122376069	2.135009706	2.148183278	2.156877802	2.163317484	2.172665314	2.182416193
1 I ZF4	5.397595220	5.554554316	5.605449663	5.655717256	5.708675338	5.743898277	5.770109965	5.808323724	5.848362203

(4) P 6+2	TH = 47	SBE = 1.367327461						MCPER FUNCTIONS	
	RC = 02	RC = 05	RC = 07	RC = 10	RC = 15	RC = 20	RC = 25	RC = 35	RC = 50
AMAX DB	.0017375255	.0108709564	.0213327376	.0436480540	.0988323393	.1772876696	.2802872360	.5675287486	1.249387366
AMIN DB	42.60647551	53.19241871	57.10495833	61.27566912	66.06583350	69.52111506	72.25783147	76.55362366	81.57371985
MU I K	.7407578D-2	.2189672D-2	.1395571D-2	.8634089D-3	.4974029D-3	.3341521D-3	.2438420D-3	.1487027D-3	.8342842D-4
1 I PF1	1.074554868	.8035961697	.7197121447	.6380526525	.5532394942	.4977602962	.4572385057	.4000853593	.3444193330
1 I PQ1	.5620992244	.5704930420	.5776413313	.5888252005	.6082983819	.6289954971	.6512789413	.7022990067	.8039158506
1 I PF2	1.152656911	.9996383823	.9498526215	.9010575135	.8505989721	.8179854788	.7944447290	.7616454125	.7298506274
1 I PQ2	1.175465155	1.219519563	1.264310168	1.334250618	1.452231281	1.572041198	1.695561798	1.962245609	2.450810781
2 I PFC	1.240404879	1.169702777	1.143779839	1.116876691	1.087381703	1.067324067	1.052309455	1.030551803	1.008402880
2 I PQC	4.774082502	4.475870553	4.480800584	4.552928960	4.729598097	4.930424021	5.142536405	5.594406424	6.370328990
1 I ZF1	1.381280589	1.382228000	1.382549504	1.382871466	1.383214337	1.383444048	1.383615715	1.383866959	1.384131301
1 I ZF2	1.515543708	1.527350295	1.531023688	1.534596179	1.538310569	1.540758040	1.542569506	1.545198056	1.547939880
1 I ZF3	2.017619598	2.059063061	2.072171074	2.084985046	2.098360560	2.107195684	2.113743087	2.123252515	2.133178338
1 I ZF4	5.241847402	5.399860370	5.451328620	5.502260400	5.556015902	5.591820556	5.618489937	5.657406395	5.698222850

(4) P 6+2			TH = 48	SBE = 1.345632730						MCPER FUNCTIONS	
			RC = 02	RC = 05	RC = 07	RC = 10	RC = 15	RC = 20	RC = 25	RC = 35	RC = 50
AMAX DB			.0017375255	.0108709564	.0213327376	.0436480540	.0988323393	.1772876696	.2802872360	.5675287486	1.249387366
AMIN DB			40.94553691	51.52382488	55.43454910	59.60373066	64.39254277	67.84706145	70.58327704	74.87843681	79.99798928
MU		K	.8968569D-2	.2653437D-2	.1691502D-2	.1046679D-2	.6030771D-3	.4051790D-3	.2956897D-3	.1803342D-3	.1011814D-3
1		PF1	1.090537897	.8132014601	.7277525172	.6447381014	.5586795381	.5024670349	.4614467926	.4036393485	.3473854565
1		PQ1	.5640004117	.5717391155	.5787477321	.5898207944	.6092054492	.6298612781	.6521252409	.7031409019	.8048055073
1		PF2	1.156536134	1.003862714	.9540357514	.9051519567	.8545641251	.8218526749	.7982367554	.7653302542	.7334350665
1		PQ2	1.200521231	1.238328195	1.281915237	1.351074402	1.468764248	1.588791415	1.712778170	1.980853329	2.472500649
2		PFC	1.233312422	1.165578145	1.140527993	1.114433389	1.085724738	1.066148280	1.051467863	1.030157730	1.008422729
2		PQC	4.997447448	4.636572280	4.628181560	4.690150477	4.859539824	5.057811288	5.269513514	5.723857147	6.508403361
1		ZF1	1.358942352	1.359871639	1.360190804	1.360511732	1.360854674	1.361035001	1.361257385	1.361510032	1.361776251
1		ZF2	1.486763333	1.498682269	1.502392861	1.506003145	1.509758844	1.512234807	1.514068069	1.516729385	1.519506906
1		ZF3	1.968870167	2.010698179	2.023966323	2.036952180	2.050522070	2.059493441	2.066145647	2.075812884	2.085910233
1		ZF4	5.091915761	5.250791129	5.302785323	5.354342194	5.408861606	5.445229920	5.472345671	5.511951415	5.553535510

(4) P 6+2			TH = 49	SBE = 1.325012993						MCPER FUNCTIONS	
			RC = 02	RC = 05	RC = 07	RC = 10	RC = 15	RC = 20	RC = 25	RC = 35	RC = 50
AMAX DB			.0017375255	.0108709564	.0213327376	.0436480540	.0988323393	.1772876696	.2802872360	.5675287486	1.249387366
AMIN DB			39.30895245	49.87889250	53.78764482	57.95516613	62.74251034	66.19620091	68.93187284	73.22634603	78.24530817
MU		K	.1082810D-1	.3206678D-2	.2044644D-2	.1265440D-2	.7292467D-3	.4899931D-3	.3576073D-3	.2181136D-3	.1223868D-3
1		PF1	1.107379522	.8232670947	.7361644748	.6517219784	.5643541523	.5073724003	.4658300183	.4073381312	.3504702257
1		PQ1	.5660619159	.5730681936	.5799226281	.5908738586	.6101613312	.6307716540	.6530138131	.7040231530	.8057362405
1		PF2	1.160355718	1.008145256	.9582974693	.9093387223	.8586305255	.8258245689	.8021350431	.7691224871	.7371271517
1		PQ2	1.227629018	1.258433132	1.300670248	1.368941794	1.486270414	1.606495285	1.730951687	2.000462223	2.495321660
2		PFC	1.226139723	1.161381819	1.137213391	1.111938148	1.084028683	1.064942403	1.050602941	1.029750039	1.008438622
2		PQC	5.242502509	4.810618541	4.787226684	4.837742263	4.998850463	5.194109956	5.405181667	5.861901509	6.655372289
1		ZF1	1.337705644	1.338613469	1.338929509	1.339248730	1.339591132	1.339821713	1.339994568	1.340248287	1.340516068
1		ZF2	1.459281132	1.471304106	1.475049142	1.478694520	1.482488748	1.484991405	1.486845160	1.489537403	1.492348780
1		ZF3	1.922056476	1.964223847	1.977641484	1.990790229	2.004546441	2.013649425	2.020403331	2.030224339	2.040489568
1		ZF4	4.947459657	5.106999737	5.159470853	5.211611786	5.266859758	5.303772595	5.331322673	5.371603378	5.413944007

(4) P 6+2			TH = 50	SBE = 1.305407289						MCPER FUNCTIONS	
			RC = 02	RC = 05	RC = 07	RC = 10	RC = 15	RC = 20	RC = 25	RC = 35	RC = 50
AMAX DB			.0017375255	.0108709564	.0213327376	.0436480540	.0988323393	.1772876696	.2802872360	.5675287486	1.249387366
AMIN DB			37.69538033	48.25620700	52.16281773	56.32853720	61.11428875	64.56708087	67.30216295	71.59589116	76.61421272
MU		K	.1303860D-1	.3865357D-2	.2465239D-2	.1526065D-2	.8796007D-3	.5910796D-3	.4314116D-3	.2631513D-3	.1476690D-3
1		PF1	1.125137723	.8338208346	.7449694283	.6590206923	.5702756074	.5124864494	.4703968344	.4111886397	.3536791496
1		PQ1	.5683028352	.5744880431	.5811719359	.5919889784	.6111696115	.6317297262	.6539474848	.7049483161	.8067105219
1		PF2	1.164097066	1.012480081	.9626345161	.9136165761	.8627985063	.8299022611	.8061411420	.7730241598	.7409292934
1		PQ2	1.257017716	1.279955690	1.320676878	1.387940062	1.504826570	1.625225269	1.750153106	2.021143504	2.519351303
2		PFC	1.218893299	1.157115819	1.133837127	1.109391394	1.082293503	1.063706212	1.049714382	1.029328372	1.008450216
2		PQC	5.512080533	4.999498934	4.959175939	4.996755895	5.148435014	5.340154149	5.550335950	6.009300646	6.811998587
1		ZF1	1.317508630	1.318391501	1.318703592	1.319020403	1.319361624	1.319592082	1.319765150	1.320019594	1.320288607
1		ZF2	1.433024240	1.445142764	1.448919410	1.452597094	1.456426965	1.458954445	1.460827332	1.463548577	1.466391877
1		ZF3	1.877071793	1.919530311	1.933086100	1.946388148	1.960322103	1.969551763	1.976404058	1.986374534	1.996803752
1		ZF4	4.808165861	4.968166832	5.021063749	5.073745735	5.129685009	5.167122099	5.195093702	5.236034078	5.279119311

(4) P 6+2 — TH = 51 — SBE = 1.286759566 — MCPER FUNCTIONS

MU		RC = 02	RC = 05	RC = 07	RC = 10	RC = 15	RC = 20	RC = 25	RC = 35	RC = 50
	AMAX DB	.0017375255	.0108709564	.0213327376	.0436480540	.0988323393	.1772876696	.2802872360	.5675287486	1.249387366
	AMIN DB	36.10355241	46.65441382	50.55869853	54.72246289	59.50648705	62.95830482	65.69274719	69.98566745	75.00329418
	K	.1566110D-1	.4648141D-2	.2965276D-2	.1836018D-2	.1058463D-2	.7113523D-3	.5192334D-3	.3167500D-3	.1777605D-3
1	PF1	1.143875444	.8448929061	.7541906601	.6666520569	.5764571982	.5178200644	.4751565934	.4151983572	.3570181673
1	PQ1	.5707453166	.5760074487	.5825023002	.5931712702	.6122342646	.6327389212	.6549293716	.7059191996	.8077310592
1	PF2	1.167739902	1.016860498	.9670431144	.9179839709	.8670682538	.8340867905	.8102565925	.7770373707	.7448439960
1	PQ2	1.288950655	1.303032734	1.342049253	1.408166708	1.524518133	1.645061741	1.770460761	2.042975812	2.544674938
2	PFC	1.211580025	1.152782343	1.130400418	1.106793626	1.080519192	1.062439497	1.048801879	1.028892360	1.008457153
2	PQC	5.809492844	5.204921834	5.145443870	5.168385149	5.309315985	5.496883539	5.705870198	6.166907813	6.979135073
1	ZF1	1.298294495	1.299148763	1.299456043	1.299769710	1.300109080	1.300339021	1.300512030	1.300766837	1.301036735
1	ZF2	1.407925775	1.420131151	1.423936490	1.427643601	1.431506120	1.434056474	1.435947072	1.438695304	1.441568501
1	ZF3	1.833818607	1.876516810	1.890198654	1.903643783	1.917746331	1.927097402	1.934044558	1.944159907	1.954748956
1	ZF4	4.673745800	4.833997779	4.887267256	4.940445251	4.997036589	5.034976478	5.063356008	5.104939743	5.148756771

(4) P 6+2 — TH = 52 — SBE = 1.269018215 — MCPER FUNCTIONS

MU		RC = 02	RC = 05	RC = 07	RC = 10	RC = 15	RC = 20	RC = 25	RC = 35	RC = 50
	AMAX DB	.0017375255	.0108709564	.0213327376	.0436480540	.0988323393	.1772876696	.2802872360	.5675287486	1.249387366
	AMIN DB	34.53227180	45.07221192	48.97396958	53.13561263	57.91776370	61.36352513	64.10227399	68.39431829	73.41119167
	K	.1876664D-1	.5576856D-2	.3558783D-2	.2204039D-2	.1270901D-2	.8542279D-3	.6235716D-3	.3804382D-3	.2135209D-3
1	PF1	1.163660931	.8565162592	.7638535253	.6746354415	.5829133531	.5233850393	.4801194215	.4193753751	.3604936923
1	PQ1	.5734151549	.5776363792	.5839212065	.5944264597	.6133597101	.6338030349	.6559629157	.7069388952	.8088008239
1	PF2	1.171262167	1.021278974	.9715189118	.9224390038	.8714397821	.8383791191	.8144829154	.7811542647	.7488738602
1	PQ2	1.323731613	1.327819175	1.364915871	1.429730987	1.545440368	1.666094090	1.791961597	2.066046194	2.571386816
2	PFC	1.204207145	1.148383780	1.126904613	1.104145425	1.078705786	1.061142065	1.047865133	1.028441624	1.008459058
2	PQC	6.138627995	5.428854057	5.347649995	5.353989852	5.482652373	5.665359982	5.872792310	6.335682331	7.157737879
1	ZF1	1.280010965	1.280832818	1.281134390	1.281444147	1.281780966	1.282009978	1.282182646	1.282437434	1.282707857
1	ZF2	1.383924297	1.396207569	1.400038600	1.403772161	1.407664220	1.410235416	1.412142242	1.414915350	1.417816316
1	ZF3	1.792207827	1.835090754	1.848885739	1.862463038	1.876724414	1.886191276	1.893229523	1.903484836	1.914229270
1	ZF4	4.543933129	4.704220222	4.757806867	4.811433747	4.868635857	4.907055854	4.935828870	4.978038571	5.022573649

(4) P 6+2 — TH = 53 — SBE = 1.252135658 — MCPER FUNCTIONS

MU		RC = 02	RC = 05	RC = 07	RC = 10	RC = 15	RC = 20	RC = 25	RC = 35	RC = 50
	AMAX DB	.0017375255	.0108709564	.0213327376	.0436480540	.0988323393	.1772876696	.2802872360	.5675287486	1.249387366
	AMIN DB	32.98041233	43.50834798	47.40735856	51.56669967	56.34681988	59.79643633	62.52943353	66.82052841	71.83658521
	K	.2243775D-1	.6677019D-2	.4262183D-2	.2640371D-2	.1522857D-2	.1023713D-2	.7473574D-3	.4560092D-3	.2559592D-3
1	PF1	1.184568029	.8687268576	.7739856769	.6829919423	.5896597574	.5291941788	.4852963015	.4237284568	.3641126612
1	PQ1	.5763425328	.5793861884	.5854371153	.5957609734	.6145508751	.6349262824	.6570519291	.7080108144	.8099230848
1	PF2	1.174639919	1.025727038	.9760569093	.9269793676	.8759129042	.8427801132	.8188216006	.7854070301	.7530215860
1	PQ2	1.361712546	1.354490961	1.389421877	1.452755687	1.567699825	1.688421995	1.814752371	2.090451252	2.599591264
2	PFC	1.196782292	1.143922724	1.123351202	1.101447461	1.076853363	1.059813741	1.046903845	1.027975780	1.008455539
2	PQC	6.504074630	5.673569145	5.567655588	5.555124565	5.669762359	5.846787350	6.052242435	6.516706091	7.348382158
1	ZF1	1.262609883	1.263395348	1.263690278	1.263995324	1.264328864	1.264556516	1.264728547	1.264982920	1.265253488
1	ZF2	1.360963332	1.373315229	1.377168856	1.380925788	1.384844157	1.387434078	1.389355582	1.392151358	1.395077858
1	ZF3	1.752158078	1.795166999	1.809061328	1.822759140	1.837168912	1.846745558	1.853870867	1.864260896	1.875155953
1	ZF4	4.418481581	4.578581898	4.632428146	4.686454651	4.744224112	4.783100228	4.812251408	4.855068536	4.900306920

(4) P 6+2		TH = 54	SBE = 1.236067977						MCPER FUNCTIONS	
		RC = 02	RC = 05	RC = 07	RC = 10	RC = 15	RC = 20	RC = 25	RC = 35	RC = 50
	AMAX DB	.0017375255	.0108709564	.0213327376	.0436480540	.0988323393	.1772876696	.2802872360	.5675287486	1.249387366
	AMIN DB	31.44692038	41.96161090	45.85763258	50.01447435	54.79239303	58.24076855	60.97295118	65.26301722	70.27818906
MU	K	.2677035D-1	.7978467D-2	.5094697D-2	.3157012D-2	.1821295D-2	.1224508D-2	.8940307D-3	.5455683D-3	.3062602D-3
1	PF1	1.206676393	.8815639985	.7846173207	.6917445747	.5967134936	.5352614098	.4906991672	.4282671114	.3678825909
1	PQ1	.5795629385	.5812698556	.5870596215	.5971820469	.6158132693	.6361133570	.6582006442	.7091337312	.8111014456
1	PF2	1.177847226	1.030195188	.9806513849	.9316022964	.8804871982	.8472905222	.8232740931	.7897678933	.7572899751
1	PQ2	1.403303092	1.383248663	1.415731780	1.477379255	1.591416037	1.712156940	1.838941063	2.116298483	2.629404061
2	PFC	1.189313503	1.139401988	1.119741835	1.098700503	1.074962049	1.058454374	1.045917730	1.027494432	1.008446187
2	PQC	6.911275222	5.941706466	5.807608408	5.773573247	5.872150559	6.042535259	6.245514685	6.711203198	7.553780709
1	ZF1	1.246046837	1.246791774	1.247079090	1.247378595	1.247708095	1.247933939	1.248105025	1.248358567	1.248628884
1	ZF2	1.338990957	1.351401809	1.355274835	1.359051949	1.362993273	1.365599713	1.367534274	1.370350404	1.373300091
1	ZF3	1.713595077	1.756667206	1.770646131	1.784451993	1.798999003	1.808679008	1.815887067	1.826406190	1.837446767
1	ZF4	4.297163064	4.456848683	4.510894766	4.565269450	4.623560638	4.662867527	4.692380616	4.735785418	4.781711299

(4) P 6+2		TH = 55	SBE = 1.220774589						MCPER FUNCTIONS	
		RC = 02	RC = 05	RC = 07	RC = 10	RC = 15	RC = 20	RC = 25	RC = 35	RC = 50
	AMAX DB	.0017375255	.0108709564	.0213327376	.0436480540	.0988323393	.1772876696	.2802872360	.5675287486	1.249387366
	AMIN DB	29.93081954	40.43082700	44.32359248	48.47772056	53.25325062	56.70028118	59.43158112	63.72053232	68.73474522
MU	K	.3187565D-1	.9516092D-2	.6078835D-2	.3768027D-2	.2174390D-2	.1462130D-2	.1067630D-2	.6515885D-3	.3658160D-3
1	PF1	1.230071563	.8950706680	.7957815012	.7009184932	.6040932010	.5416019096	.4963410109	.4330016767	.3718116421
1	PQ1	.5831183098	.5833022778	.5887996467	.5986978540	.6171530718	.6373694999	.6594137733	.7103263309	.8123398895
1	PF2	1.180856078	1.034672785	.9852958087	.9363045039	.8851619698	.8519109539	.8278417768	.7942491139	.7616819333
1	PQ2	1.448982317	1.414321814	1.444032715	1.503758319	1.616723529	1.737423989	1.864648540	2.143707853	2.660954055
2	PFC	1.181809236	1.134824622	1.116078328	1.095905428	1.073032025	1.057063841	1.044906509	1.026997180	1.008430575
2	PQC	7.366719568	6.236343986	6.069997383	6.011391366	6.091540940	6.254167609	6.454083191	6.920563449	7.773806157
1	ZF1	1.230280821	1.230980925	1.231259616	1.231552716	1.231877387	1.232100955	1.232270774	1.232523052	1.232792704
1	ZF2	1.317959443	1.330419077	1.334308192	1.338102182	1.342062972	1.344683628	1.346629553	1.349463611	1.352434017
1	ZF3	1.676451101	1.719519281	1.733567021	1.747467595	1.762139900	1.771916383	1.779202574	1.789844766	1.801025387
1	ZF4	4.179765987	4.338802842	4.392986760	4.447655940	4.506420943	4.546131841	4.575989604	4.619961041	4.666557469

(4) P 6+2		TH = 56	SBE = 1.206217949						MCPER FUNCTIONS	
		RC = 02	RC = 05	RC = 07	RC = 10	RC = 15	RC = 20	RC = 25	RC = 35	RC = 50
	AMAX DB	.0017375255	.0108709564	.0213327376	.0436480540	.0988323393	.1772876696	.2802872360	.5675287486	1.249387366
	AMIN DB	28.43121910	38.91485562	42.80406754	46.95524493	51.72818406	55.17375667	57.90410002	62.19184318	67.20501698
MU	K	.3788254D-1	.1133071D-1	.7240968D-2	.4489911D-2	.2591736D-2	.1743059D-2	.1272902D-2	.7769764D-3	.4362638D-3
1	PF1	1.254844826	.9092939315	.8075144233	.7105412401	.6118192583	.5482322508	.5022360062	.4379434146	.3759086925
1	PQ1	.5870584678	.5855006262	.5906696706	.6003176611	.6185772349	.6387005814	.6606965784	.7115797684	.8136428302
1	PF2	1.183636302	1.039147927	.9899827472	.9410821138	.8899362079	.8566418456	.8325259556	.7988529767	.7662004726
1	PQ2	1.499313319	1.447974177	1.474538376	1.532070721	1.643774211	1.764363899	1.892010518	2.172813649	2.694385056
2	PFC	1.174278392	1.130193932	1.112362682	1.093063233	1.071063534	1.055642047	1.043869918	1.026483613	1.008408254
2	PQC	7.878189818	6.561088398	6.357719870	6.270957348	6.329916773	6.483477100	6.679633500	7.146370526	8.010517498
1	ZF1	1.215273939	1.215924735	1.216193756	1.216479554	1.216798575	1.217019380	1.217187598	1.217438161	1.217706715
1	ZF2	1.297824942	1.310322546	1.314224314	1.318031749	1.322008373	1.324640840	1.326596357	1.329445801	1.332434333
1	ZF3	1.640664515	1.683656871	1.697756533	1.711737534	1.726522337	1.736387923	1.743747291	1.754506086	1.765820869
1	ZF4	4.066093798	4.224241453	4.278498938	4.333406632	4.392595169	4.432681825	4.462865995	4.507381663	4.554630474

(4) P 6+2 — TH = 57 — SBE = 1.192363293 — MCPER FUNCTIONS

MU		RC = 02	RC = 05	RC = 07	RC = 10	RC = 15	RC = 20	RC = 25	RC = 35	RC = 50
	AMAX DB	.0017375255	.0108709564	.0213327376	.0436480540	.0988323393	.1772876696	.2802872360	.5675287486	1.249387366
	AMIN DB	26.94732729	37.41258535	41.29791045	45.44587612	50.21600254	53.65999421	56.38930069	60.67573466	65.68778240
MU	K	.4494006D-1	.1347010D-1	.8612009D-2	.5342028D-2	.3084607D-2	.2074915D-2	.1515427D-2	.9251524D-3	.5195303D-3
1	PF1	1.281092728	.9242853601	.8198558143	.7206430293	.6199139917	.5551705695	.5083996486	.4431046205	.3801834215
1	PQ1	.5914429243	.5878847812	.5926840119	.6020520127	.6200936081	.6401131981	.6620549532	.7129027354	.8150151724
1	PF2	1.186155491	1.043607331	.9947037577	.9459305814	.8948085348	.8614834310	.8373278312	.8035817831	.7708487130
1	PQ2	1.554961485	1.484510183	1.507493785	1.562519165	1.672740246	1.793135638	1.921179906	2.203766678	2.729858083
2	PFC	1.166730339	1.125513498	1.108597100	1.090175044	1.069056889	1.054188937	1.042807707	1.025953318	1.008378757
2	PQC	8.455072629	6.920187472	6.674164878	6.555035809	6.589569398	6.732527181	6.924100657	7.390436051	8.265692042
1	ZF1	1.200991136	1.201587981	1.201846248	1.202123817	1.202436337	1.202653875	1.202820142	1.203068520	1.203335524
1	ZF2	1.278547226	1.291071177	1.294982019	1.298799330	1.302788003	1.305429772	1.307393025	1.310255187	1.313259115
1	ZF3	1.606179363	1.649018931	1.663152418	1.677198537	1.692082111	1.702028883	1.709456109	1.720324559	1.731767175
1	ZF4	3.955963698	4.112974985	4.167239456	4.222327326	4.281886657	4.322319264	4.352810484	4.397846532	4.445728257

(4) P 6+2 — TH = 58 — SBE = 1.179178403 — MCPER FUNCTIONS

MU		RC = 02	RC = 05	RC = 07	RC = 10	RC = 15	RC = 20	RC = 25	RC = 35	RC = 50
	AMAX DB	.0017375255	.0108709564	.0213327376	.0436480540	.0988323393	.1772876696	.2802872360	.5675287486	1.249387366
	AMIN DB	25.47847093	35.92293099	39.80399244	43.94845662	48.71552697	52.15780354	54.88598573	59.17100052	64.18182773
MU	K	.5322019D-1	.1599018D-1	.1022823D-1	.6347127D-2	.3666263D-2	.2466663D-2	.1801776D-2	.1100145D-2	.6178864D-3
1	PF1	1.308916073	.9401014910	.8328493298	.7312570691	.6284019152	.5624367567	.5148489183	.4484987491	.3846464061
1	PQ1	.5963431698	.5904778703	.5948591711	.6039129558	.6217110855	.6416147879	.6634955207	.7143015406	.8164623824
1	PF2	1.188378964	1.048036185	.9994492703	.9508446042	.8997771481	.8664357010	.8422484766	.8084378400	.7756298832
1	PQ2	1.616717440	1.524282841	1.543181112	1.595335631	1.703817488	1.823919399	1.952329593	2.236736880	2.767554014
2	PFC	1.159174940	1.120787193	1.104784000	1.087242137	1.067012481	1.052704494	1.041719648	1.025405874	1.008341594
2	PQC	9.108759298	7.318671036	7.023316719	6.866855746	6.873158125	7.003703362	7.189715622	7.654840930	8.541364042
1	ZF1	1.187399959	1.187938039	1.188184434	1.188452815	1.188757955	1.188971701	1.189135657	1.189381359	1.189646343
1	ZF2	1.260089470	1.272627115	1.276543288	1.280366751	1.284363526	1.287011970	1.288981017	1.291853095	1.294869547
1	ZF3	1.572945025	1.615549337	1.629697251	1.643792062	1.658759673	1.668779126	1.676268491	1.687239121	1.698802754
1	ZF4	3.849205546	4.004826009	4.059028515	4.114235797	4.174110628	4.214857746	4.245635509	4.291166548	4.339660328

(4) P 6+2 — TH = 59 — SBE = 1.166633397 — MCPER FUNCTIONS

MU		RC = 02	RC = 05	RC = 07	RC = 10	RC = 15	RC = 20	RC = 25	RC = 35	RC = 50
	AMAX DB	.0017375255	.0108709564	.0213327376	.0436480540	.0988323393	.1772876696	.2802872360	.5675287486	1.249387366
	AMIN DB	24.02412326	34.44483134	38.32119887	42.46183767	47.22558374	50.66599834	53.39296087	57.67643668	62.68594058
MU	K	.6292074D-1	.1895651D-1	.1213221D-1	.7531962D-2	.4352320D-2	.2928870D-2	.2139695D-2	.1306707D-2	.7340117D-3
1	PF1	1.338418173	.9568043169	.8465430067	.7424199312	.6373100055	.5700526811	.5216024672	.4541405598	.3893092332
1	PQ1	.6018455856	.5933069340	.5972142516	.6059143125	.6234397847	.6432137680	.6650257500	.7157827045	.8179905715
1	PF2	1.190269746	1.052418000	1.004208456	.9558180212	.9048397548	.8714983591	.8472888047	.8134234460	.7805473219
1	PQ2	1.685526065	1.567703531	1.581926819	1.630786767	1.737229662	1.856920239	1.985655807	2.271916462	2.807676759
2	PFC	1.151622589	1.116019214	1.100926043	1.084265946	1.064930788	1.051188752	1.040605534	1.024840857	1.008296254
2	PQC	9.853161792	7.762529151	7.409884994	7.210207883	7.183784262	7.299776351	7.479062218	7.941985859	8.839871690
1	ZF1	1.174470335	1.174944674	1.175178045	1.175436250	1.175733098	1.175942512	1.176103780	1.176346299	1.176608773
1	ZF2	1.242418073	1.254955456	1.258873022	1.262698747	1.266699497	1.269351869	1.271324673	1.274203725	1.277229676
1	ZF3	1.540915912	1.583196552	1.597338085	1.611463955	1.626499767	1.636582753	1.644128101	1.655192360	1.666870158
1	ZF4	3.745660927	3.899628030	3.953697174	4.008960598	4.069092975	4.110121454	4.141164042	4.187163047	4.236246527

(4) P 6+2 — TH = 60 SBE = 1.154700538 MCPER FUNCTIONS

MU		RC = 02	RC = 05	RC = 07	RC = 10	RC = 15	RC = 20	RC = 25	RC = 35	RC = 50
	AMAX DB	.0017375255	.0108709564	.0213327376	.0436480540	.0988323393	.1772876696	.2802872360	.5675287486	1.249387366
	AMIN DB	22.58394247	32.97724817	36.84842505	40.98487356	45.74499820	49.18338960	51.90902811	56.19083414	61.19890268
MU	K	.7426820D-1	.2244593D-1	.1437404D-1	.8928044D-2	.5161193D-2	.3474006D-2	.2538333D-2	.1550452D-2	.8710736D-3
1	PF1	1.369701993	.9744617921	.8609897678	.7541719703	.6466680209	.5780424453	.5286808362	.4600462863	.3941846293
1	PQ1	.6080551641	.5964037581	.5997714792	.6080720133	.6252912621	.6449197012	.6666540954	.7173535727	.8196065953
1	PF2	1.191788606	1.056734441	1.008969081	.9608436986	.9099934944	.8766707679	.8524495323	.8185408751	.7856044769
1	PQ2	1.762523369	1.615254200	1.624110483	1.669180488	1.773233434	1.892372479	2.021382172	2.309523664	2.850457062
2	PFC	1.144084245	1.111214103	1.097026149	1.081248083	1.062812389	1.049641798	1.039465188	1.024257842	1.008242204
2	PQC	10.70538225	8.258938951	7.839468784	7.589566702	7.525083240	7.623980246	7.795147434	8.254653405	9.163914647
1	ZF1	1.162174383	1.162579843	1.162799005	1.163046017	1.163333637	1.163538158	1.163696352	1.163935159	1.164194614
1	ZF2	1.225502523	1.238024044	1.241938841	1.245762744	1.249763150	1.252416570	1.254390993	1.257273924	1.260306185
1	ZF3	1.510051221	1.551913329	1.566026146	1.580164131	1.595251109	1.605387777	1.612982476	1.624132678	1.635915701
1	ZF4	3.645182390	3.797224421	3.851086252	3.906339953	3.966669155	4.007944042	4.039228451	4.085666659	4.135315881

(4) P 6+2 — TH = 61 SBE = 1.143354068 MCPER FUNCTIONS

MU		RC = 02	RC = 05	RC = 07	RC = 10	RC = 15	RC = 20	RC = 25	RC = 35	RC = 50
	AMAX DB	.0017375255	.0108709564	.0213327376	.0436480540	.0988323393	.1772876696	.2802872360	.5675287486	1.249387366
	AMIN DB	21.15782398	31.51916669	35.38457255	39.51641593	44.27258795	47.70877836	50.43297831	54.71297141	59.71948219
MU	K	.8752030D-1	.2654860D-1	.1701263D-1	.1057254D-1	.6114636D-2	.4116812D-2	.3008507D-2	.1838025D-2	.1032823D-2
1	PF1	1.402865735	.9931483373	.8762479778	.7665578049	.6565088690	.5864326854	.5361067088	.4662338325	.3992866114
1	PQ1	.6151002795	.5998059186	.6025568496	.6104045084	.6272787763	.6467434971	.6683901648	.7190224516	.8213181715
1	PF2	1.192894154	1.060965151	1.013717343	.9659133995	.9152348504	.8819518884	.8577311362	.8237323560	.7908049039
1	PQ2	1.849083594	1.667502665	1.670175783	1.710874130	1.812124635	1.930545082	2.059764635	2.349807315	2.896157078
2	PFC	1.136571489	1.106376774	1.093087522	1.078190356	1.060657974	1.048063787	1.038298467	1.023656402	1.008178884
2	PQC	11.68658661	8.816555729	8.318765666	8.010244591	7.901340061	7.980110082	8.141488823	8.596084826	9.516624923
1	ZF1	1.150486223	1.150817523	1.151021260	1.151256037	1.151533465	1.151732515	1.151887234	1.152121785	1.152377689
1	ZF2	1.209315301	1.221803295	1.225710892	1.229528675	1.233524204	1.236175650	1.238149445	1.241033000	1.244068207
1	ZF3	1.480314720	1.521656462	1.535716573	1.549846302	1.564966105	1.575145835	1.582782728	1.594008992	1.605889150
1	ZF4	3.547632843	3.697467453	3.751045329	3.806220729	3.866683137	3.908167588	3.939669455	3.986516243	4.036705526

(4) P 6+2 — TH = 62 SBE = 1.132570051 MCPER FUNCTIONS

MU		RC = 02	RC = 05	RC = 07	RC = 10	RC = 15	RC = 20	RC = 25	RC = 35	RC = 50
	AMAX DB	.0017375255	.0108709564	.0213327376	.0436480540	.0988323393	.1772876696	.2802872360	.5675287486	1.249387366
	AMIN DB	19.74597013	30.06959819	33.92854621	38.05530800	42.80715569	46.24094814	48.96358334	53.24160638	58.24642537
MU	K	.1029678D 0	.3137040D-1	.2011744D-1	.1250935D-1	.7238394D-2	.4874753D-2	.3563041D-2	.2177307D-2	.1223711D-2
1	PF1	1.437996235	1.012945309	.8923820498	.7796268631	.6668690345	.5952529190	.5439052062	.4727230040	.4046306651
1	PQ1	.6231388209	.6035581061	.6056009383	.6129332760	.6294176106	.6486976571	.6702449239	.7207987714	.8231340224
1	PF2	1.193543039	1.065087555	1.018437687	.9710176363	.9205595480	.8873402067	.8631338021	.8291800475	.7961522634
1	PQ2	1.946879651	1.725121952	1.720644266	1.756284588	1.854245930	1.971748243	2.101097501	2.393052399	2.945075940
2	PFC	1.129096578	1.101512553	1.089113683	1.075094795	1.058468360	1.046454943	1.037105268	1.023036113	1.008105711
2	PQC	12.82315090	9.445889566	8.855839959	8.478588075	8.317635119	8.372644401	8.522223944	8.970075814	9.901654805
1	ZF1	1.139381819	1.139633551	1.139820620	1.140042095	1.140308343	1.140501331	1.140652161	1.140881890	1.141133697
1	ZF2	1.193831821	1.206266052	1.210161697	1.213968810	1.217954694	1.220600990	1.222571796	1.225452544	1.228487144
1	ZF3	1.451674566	1.492386568	1.506368190	1.520467741	1.535600602	1.545811931	1.553483283	1.564775461	1.576743450
1	ZF4	3.452885101	3.600217402	3.653431801	3.708457477	3.768986430	3.810641598	3.842335117	3.889557881	3.940259689

(4) P 6+2 — TH = 63 — SBE = 1.122326238 — MCPER FUNCTIONS

MU	K	RC = 02	RC = 05	RC = 07	RC = 10	RC = 15	RC = 20	RC = 25	RC = 35	RC = 50
	AMAX DB	.0017375255	.0108709564	.0213327376	.0436480540	.0988323393	.1772876696	.2802872360	.5675287486	1.249387366
	AMIN DB	18.34898110	28.62758548	32.47925199	36.60037874	41.34748146	44.77865655	47.49958741	51.77546733	56.77844737
MU	K	.1209347D 0	.3703571D-1	.2377045D-1	.1479044D-1	.8563000D-2	.5768557D-2	.4217165D-2	.2577666D-2	.1449031D-2
1	PF1	1.475159363	1.033941381	.9094630954	.7934340052	.6777890761	.6045359534	.5521042355	.4795357788	.4102339524
1	PQ1	.6323660854	.6077138133	.6089399243	.6156834591	.6317254717	.6507965752	.6722309455	.7226932854	.8250640454
1	PF2	1.193690269	1.069076654	1.023112608	.9761455020	.9259624353	.8928336503	.8686573633	.8347060078	.8016503163
1	PQ2	2.057960996	1.788914946	1.776132743	1.805901018	1.899996360	2.016341549	2.145720869	2.439586861	2.997556550
2	PFC	1.121672519	1.096627210	1.085108494	1.071963670	1.056244506	1.044815578	1.035885536	1.022396555	1.008022076
2	PQC	14.14817266	10.15979689	9.460469998	9.002229797	8.780029860	8.806900517	8.942248493	9.381096733	10.32328677
1	ZF1	1.128838812	1.129005483	1.129174618	1.129381704	1.129635764	1.129822081	1.129968597	1.130192923	1.130440062
1	ZF2	1.179030410	1.191387456	1.195266009	1.199057618	1.203028825	1.205666626	1.207631952	1.210506277	1.213536515
1	ZF3	1.424103149	1.464067911	1.477943312	1.491989075	1.507113665	1.517344205	1.525041649	1.536338749	1.548434472
1	ZF4	3.360821571	3.505341748	3.558110006	3.612911511	3.673437141	3.715222063	3.747079892	3.794643906	3.845828699

(4) P 6+2 — TH = 64 — SBE = 1.112601940 — MCPER FUNCTIONS

MU	K	RC = 02	RC = 05	RC = 07	RC = 10	RC = 15	RC = 20	RC = 25	RC = 35	RC = 50
	AMAX DB	.0017375255	.0108709564	.0213327376	.0436480540	.0988323393	.1772876696	.2802872360	.5675287486	1.249387366
	AMIN DB	16.96797128	27.19221254	31.03559624	35.15043703	39.89231443	43.32062614	46.03969744	50.31324289	55.31422194
MU	K	.1417756D 0	.4369074D-1	.2806856D-1	.1747745D-1	.1012475D-1	.6822895D-2	.4989019D-2	.3050267D-2	.1715098D-2
1	PF1	1.514386428	1.056232751	.9275696016	.8080402277	.6893142059	.6143183657	.5607348990	.4866966287	.4161155595
1	PQ1	.6430249310	.6123375051	.6126168914	.6186846653	.6342229861	.6530569103	.6743627159	.7247183124	.8271195215
1	PF2	1.193289734	1.072904803	1.027722421	.9812844783	.9314373450	.8984294874	.8743012283	.8403721574	.8073029160
1	PQ2	2.184854279	1.859846054	1.837375458	1.860300886	1.949843276	2.064744139	2.194029855	2.489790035	3.053993931
2	PFC	1.114313154	1.091727002	1.081076198	1.068799523	1.053987534	1.043146101	1.034639272	1.021737317	1.007927343
2	PQC	15.70347288	10.97412798	10.14460189	9.590414974	9.295805277	9.289232878	9.407392097	9.834444842	10.78657194
1	ZF1	1.118836375	1.118912466	1.119062381	1.119253972	1.119494817	1.119673842	1.119815609	1.120033934	1.120275819
1	ZF2	1.164892313	1.177144848	1.181000704	1.184771635	1.188722840	1.191348615	1.193305835	1.196169915	1.199191821
1	ZF3	1.397576926	1.436668255	1.450407584	1.464374109	1.479467394	1.489703741	1.497418212	1.508808305	1.520920788
1	ZF4	3.271334040	3.412714428	3.464950401	3.519450041	3.579899076	3.621770537	3.653763696	3.701631956	3.753268028

(4) P 6+2 — TH = 65 — SBE = 1.103377919 — MCPER FUNCTIONS

MU	K	RC = 02	RC = 05	RC = 07	RC = 10	RC = 15	RC = 20	RC = 25	RC = 35	RC = 50
	AMAX DB	.0017375255	.0108709564	.0213327376	.0436480540	.0988323393	.1772876696	.2802872360	.5675287486	1.249387366
	AMIN DB	15.60471342	25.76261967	29.59648687	33.70426606	38.44036376	41.86553417	44.58257235	48.85357080	53.85236976
MU	K	.1658687D 0	.5150733D-1	.3312651D-1	.2064366D-1	.1196690D-1	.8067209D-2	.5900263D-2	.3608456D-2	.2029465D-2
1	PF1	1.555655472	1.079923044	.9467881070	.8235134571	.7014949668	.6246410690	.5698319801	.4942329004	.4222967905
1	PQ1	.6554187983	.6175074286	.6166834992	.6219719332	.6369343219	.6554980485	.6766570149	.7268880351	.8293133700
1	PF2	1.192294969	1.076541481	1.032245010	.9864202157	.9369769332	.9041242092	.8800642937	.8461802333	.8131139990
1	PQ2	2.330693738	1.939082242	1.905252570	1.920170410	2.004337432	2.117447468	2.246436148	2.544103127	3.114845588
2	PFC	1.107033253	1.086818721	1.077021457	1.065605197	1.051698748	1.041447034	1.033366546	1.021058001	1.007820949
2	PQC	17.54225735	11.90858815	10.92294824	10.25442850	9.873771129	9.827288809	9.924644065	10.33643373	11.29750604
1	ZF1	1.109355053	1.109335115	1.109464510	1.109639492	1.109866080	1.110037180	1.110173754	1.110385464	1.110621488
1	ZF2	1.151401717	1.163517692	1.167344678	1.171089365	1.175014906	1.177624921	1.179571259	1.182421053	1.185430418
1	ZF3	1.372076233	1.410158752	1.423729850	1.437589681	1.452626759	1.462854391	1.470576052	1.481996178	1.494163471
1	ZF4	3.184323502	3.322215174	3.373828775	3.427945323	3.488240848	3.530153233	3.562250988	3.610384036	3.662437334

(4) P 6+2 — TH = 66 — SBE = 1.094636279 — MCPER FUNCTIONS

MU	P 6+2	RC = 02	RC = 05	RC = 07	RC = 10	RC = 15	RC = 20	RC = 25	RC = 35	RC = 50
	AMAX DB	.0017375255	.0108709564	.0213327376	.0436480540	.0988323393	.1772876696	.2802872360	.5675287486	1.249387366
	AMIN DB	14.26181037	24.33802664	28.16083750	32.26061793	36.99028865	40.41200109	43.12681073	47.39502491	52.39144508
MU	K	.1936018D 0	.6068742D-1	.3908032D-1	.2437637D-1	.1414118D-1	.9536740D-2	.6976851D-2	.4268239D-2	.2401197D-2
1	PF1	1.598866357	1.105122717	.9672138221	.8399294336	.7143880244	.6355499830	.5794345217	.5021752720	.4288015202
1	PQ1	.6699283153	.6233192823	.6212021449	.6255872831	.6398879774	.6581426855	.6791333909	.7292188712	.8316604624
1	PF2	1.190660265	1.079953055	1.036655554	.9915362824	.9425724908	.9099133899	.8859448407	.8521317334	.8190875695
1	PQ2	2.499390355	2.028046708	1.980827137	1.986330863	2.064132201	2.175031480	2.303632582	2.603042379	3.180644450
2	PFC	1.099848623	1.081909750	1.072949401	1.062383877	1.049379664	1.039719030	1.032067506	1.020358227	1.007701904
2	PQC	19.73266548	12.98789494	11.81378657	11.00815951	10.52467094	10.43034151	10.50244610	10.89466898	11.86325495
1	ZF1	1.100376606	1.100255409	1.100362979	1.100520228	1.100731511	1.100894045	1.101024974	1.101229441	1.101458982
1	ZF2	1.138545779	1.150487526	1.154278782	1.157991192	1.161885026	1.164475318	1.166407832	1.169239060	1.172231422
1	ZF3	1.347585029	1.384513863	1.397882059	1.411605549	1.426559465	1.436762630	1.444480792	1.455916841	1.468125908
1	ZF4	3.099699929	3.233728916	3.284625511	3.338273830	3.398334998	3.440240110	3.472409841	3.520765572	3.573199485

(4) P 6+2 — TH = 67 — SBE = 1.086360377 — MCPER FUNCTIONS

MU	P 6+2	RC = 02	RC = 05	RC = 07	RC = 10	RC = 15	RC = 20	RC = 25	RC = 35	RC = 50
	AMAX DB	.0017375255	.0108709564	.0213327376	.0436480540	.0988323393	.1772876696	.2802872360	.5675287486	1.249387366
	AMIN DB	12.94288749	22.91776681	26.72757584	30.81820914	35.54068740	38.95857746	41.67093683	45.93610022	50.92992011
MU	K	.2253490D 0	.7146801D-1	.4609154D-1	.2877992D-1	.1670958D-1	.1127382D-1	.8249985D-2	.5048879D-2	.2841214D-2
1	PF1	1.643809008	1.131947646	.9889511086	.8573726823	.7280570945	.6470968327	.5895865195	.5105583024	.4355566189
1	PQ1	.6870322738	.6298910469	.6262487878	.6295808987	.6431177864	.6610175629	.6818147582	.7317299369	.8341780125
1	PF2	1.188342220	1.083102541	1.040926219	.9966138758	.9482137216	.9157915204	.8919404093	.8582278470	.8252276805
1	PQ2	2.695851387	2.128489752	2.065393566	2.059772745	2.130008316	2.238185301	2.366111708	2.667215752	3.252015177
2	PFC	1.092776207	1.077008130	1.068865676	1.059139124	1.047032033	1.037962895	1.030742396	1.019637639	1.007569788
2	PQC	22.36251443	14.24334964	12.84003583	11.86885610	11.26171824	11.10972825	11.15107584	11.51832468	12.49244694
1	ZF1	1.091883837	1.091656587	1.091741031	1.091879427	1.092074351	1.092227676	1.092352501	1.092549086	1.092771505
1	ZF2	1.126314622	1.138037931	1.141785765	1.145459309	1.149314953	1.151881304	1.153796872	1.156604991	1.159575609
1	ZF3	1.324090497	1.359711311	1.372839196	1.386394298	1.401235848	1.411397432	1.419100463	1.430537050	1.442773639
1	ZF4	3.017381792	3.147145252	3.197224863	3.250315419	3.310057090	3.351903946	3.384110989	3.432644426	3.485419553

(4) P 6+2 — TH = 68 — SBE = 1.078534743 — MCPER FUNCTIONS

MU	P 6+2	RC = 02	RC = 05	RC = 07	RC = 10	RC = 15	RC = 20	RC = 25	RC = 35	RC = 50
	AMAX DB	.0017375255	.0108709564	.0213327376	.0436480540	.0988323393	.1772876696	.2802872360	.5675287486	1.249387366
	AMIN DB	11.65278753	21.50133650	25.29565832	29.37571730	34.09008527	37.50372888	40.21338415	44.47519527	49.46616671
MU	K	.2614331D 0	.8412657D-1	.5435219D-1	.3397928D-1	.1974675D-1	.1332949D-1	.9757325D-2	.5973656D-2	.3362727D-2
1	PF1	1.690125288	1.160516474	1.012113672	.8759375520	.7425740281	.6593401015	.6003377550	.5194210959	.4428924691
1	PQ1	.7073337766	.6373693950	.6319166742	.6340138229	.6466642157	.6641544112	.6847281589	.7344436319	.8368860562
1	PF2	1.185301814	1.085949382	1.045025818	1.001631492	.9538884823	.9217518094	.8980476479	.8644693701	.8315384071
1	PQ2	2.926265189	2.242583303	2.160541828	2.141700647	2.202906039	2.307733975	2.434689693	2.737344263	3.329694932
2	PFC	1.085834158	1.072122633	1.064776508	1.055874932	1.044657881	1.036179608	1.029391570	1.018895914	1.007423754
2	PQC	25.54565615	15.71499588	14.03072551	12.85814620	12.10131489	11.87943365	11.88315504	12.21862334	13.19555507
1	ZF1	1.083860396	1.083523058	1.083583095	1.083701526	1.083879045	1.084022513	1.084140773	1.084323329	1.084543475
1	ZF2	1.114701254	1.126154538	1.129850241	1.133477670	1.137288134	1.139826040	1.141721336	1.144501514	1.147445345
1	ZF3	1.301582447	1.335732061	1.348579244	1.361931289	1.376628790	1.386730176	1.394405395	1.405825710	1.418074214
1	ZF4	2.937295079	3.062358002	3.111514275	3.163952494	3.223284779	3.265019353	3.297226829	3.345889862	3.398963744

(4) P 6+2		TH = 69 SBE = 1.071144994								MCPER FUNCTIONS
		RC = 02	RC = 05	RC = 07	RC = 10	RC = 15	RC = 20	RC = 25	RC = 35	RC = 50
AMAX DB		.0017375255	.0108709564	.0213327376	.0436480540	.0988323393	.1772876696	.2802872360	.5675287486	1.249387366
AMIN DB		10.39773372	20.08846523	23.86409388	27.93178018	32.63692119	36.04581867	38.75247623	43.01059129	47.99843450
MU	K	.3020740D 0	.9898668D-1	.6409074D-1	.4012463D-1	.2334285D-1	.1576555D-1	.1154453D-1	.7070831D-2	.3981789D-2
1	PF1	1.737267146	1.190946023	1.036824236	.8957292829	.7580200796	.6723461730	.6117448053	.5288081138	.4505435969
1	PQ1	.7315922593	.6459382654	.6383213008	.6389606051	.6505760572	.6675911693	.6879057401	.7373863366	.8398081216
1	PF2	1.181507069	1.088449270	1.048919454	1.006564547	.9595824732	.9277859443	.9042621294	.8708565992	.8380238114
1	PQ2	3.198470788	2.373048269	2.268243493	2.233592874	2.283967479	2.384673413	2.510287404	2.814289582	3.414560065
2	PFC	1.079041870	1.067262851	1.060688776	1.052595778	1.042259544	1.034370357	1.028015516	1.018132773	1.007263028
2	PQC	29.43052275	17.45461953	15.42302611	14.00343684	13.06402575	12.75687587	12.71433127	13.00938385	13.98540366
1	ZF1	1.076290569	1.075840308	1.075874697	1.075972076	1.076131157	1.076264124	1.076375359	1.076554232	1.076760442
1	ZF2	1.103701347	1.114825044	1.118458689	1.122031961	1.125789663	1.128294293	1.130165767	1.132912856	1.135824524
1	ZF3	1.280052399	1.312560333	1.325083188	1.338194628	1.352713668	1.362734574	1.370368132	1.381753776	1.393997056
1	ZF4	2.859371440	2.979264829	3.027383711	3.079069141	3.137896820	3.179461740	3.211630337	3.260371420	3.313698227

(4) P 6+2		TH = 70 SBE = 1.064177772								MCPER FUNCTIONS
		RC = 02	RC = 05	RC = 07	RC = 10	RC = 15	RC = 20	RC = 25	RC = 35	RC = 50
AMAX DB		.0017375255	.0108709564	.0213327376	.0436480540	.0988323393	.1772876696	.2802872360	.5675287486	1.249387366
AMIN DB		9.185404926	18.67921448	22.43198096	26.48499859	31.17953321	34.58308762	37.28640388	41.54042716	46.52482451
MU	K	.3473200D 0	.1164231D 0	.7557897D-1	.4739691D-1	.2760726D-1	.1865716D-1	.1366721D-1	.8374881D-2	.4718009D-2
1	PF1	1.784457265	1.223343832	1.063213339	.9168650224	.7744873803	.6861907054	.6238722720	.5387701714	.4586494526
1	PQ1	.7607618945	.6558304205	.6456070938	.6445132196	.6549126581	.6713735779	.6913860256	.7405836237	.8429719212
1	PF2	1.176936244	1.090554026	1.052568127	1.011384938	.9652788722	.9338838039	.9105781276	.8773892000	.8446878951
1	PQ2	3.522438214	2.523328016	2.390968428	2.337281640	2.374592932	2.470216610	2.594021330	2.899090125	3.507660837
2	PFC	1.072419901	1.062439305	1.056610093	1.049306691	1.039839720	1.032536567	1.026614879	1.017347989	1.007086808
2	PQC	34.21167575	19.52997002	17.06509418	15.33985860	14.17591938	13.76398058	13.66420459	13.90779994	14.87784756
1	ZF1	1.069159044	1.068594815	1.068602386	1.068677663	1.068817297	1.068939132	1.069042883	1.069211922	1.069409025
1	ZF2	1.093312796	1.104039273	1.107599463	1.111109590	1.114806252	1.117272405	1.119116254	1.121824751	1.124698515
1	ZF3	1.259492282	1.290183634	1.302335042	1.315165176	1.329468330	1.339386623	1.346963371	1.358294161	1.370513356
1	ZF4	2.783545008	2.897766945	2.944725009	2.995550223	3.053771996	3.095106160	3.127193875	3.175957672	3.229487842

(4) P 6+2		TH = 71 SBE = 1.057620681								MCPER FUNCTIONS
		RC = 02	RC = 05	RC = 07	RC = 10	RC = 15	RC = 20	RC = 25	RC = 35	RC = 50
AMAX DB		.0017375255	.0108709564	.0213327376	.0436480540	.0988323393	.1772876696	.2802872360	.5675287486	1.249387366
AMIN DB		8.024847245	17.27411416	20.99856382	25.03394533	29.71614302	33.11363038	35.81319794	40.06266901	45.04325696
MU	K	.3969700D 0	.1368656D 0	.8913983D-1	.5601479D-1	.3267329D-1	.2209625D-1	.1619348D-1	.9928109D-2	.5595477D-2
1	PF1	1.830662901	1.257796461	1.091416674	.9394746458	.7920806388	.7009602861	.6367942851	.5493656724	.4672553815
1	PQ1	.7960366704	.6673431336	.6539564977	.6507862913	.6597469000	.6755572900	.6952155877	.7440910130	.8464104791
1	PF2	1.171581357	1.092211621	1.055928341	1.016060550	.9709578984	.9400331087	.9169883424	.8840660419	.8515345377
1	PQ2	3.910896281	2.697827252	2.531845162	2.455062439	2.476516937	2.565855593	2.687258159	2.993008857	3.610267110
2	PFC	1.065989752	1.057663561	1.052548908	1.046013330	1.037401525	1.030679946	1.025190492	1.016541408	1.006894269
2	PQC	40.14556583	22.03077735	19.02011723	16.91300930	15.47044036	14.92867011	14.75760566	14.93549935	15.89269573
1	ZF1	1.062450674	1.061773963	1.061753657	1.061805842	1.061925062	1.062035150	1.062130967	1.062289524	1.062476844
1	ZF2	1.083534998	1.093789242	1.097262838	1.100699698	1.104326229	1.106748276	1.108560410	1.111224420	1.114054137
1	ZF3	1.239892674	1.268592798	1.280321918	1.292826594	1.306873104	1.316664594	1.324167926	1.335421673	1.347595975
1	ZF4	2.709747481	2.817768863	2.863431255	2.913280415	2.970787925	3.011826021	3.043787848	3.092514802	3.146194609

(4) P 6+2	TH = 72	SBE = 1.051462224						MCPER FUNCTIONS	
	RC = 02	RC = 05	RC = 07	RC = 10	RC = 15	RC = 20	RC = 25	RC = 35	RC = 50
AMAX DB	.0017375255	.0108709564	.0213327376	.0436480540	.0988323393	.1772876696	.2802872360	.5675287486	1.249387366
AMIN DB	6.926139639	15.87434786	19.56331620	23.57718396	28.24483981	31.63536784	34.33069649	38.57507294	43.55143145
MU I K	.4504982D 0	.1607987D 0	.1051560D 0	.6624312D-1	.3870419D-1	.2619580D-1	.1920725D-1	.1178274D-1	.6643982D-2
1 I PF1	1.874597855	1.294351720	1.121570092	.9637011254	.8109190737	.7167544307	.6505963559	.5606621529	.4764138458
1 I PQ1	.8389024969	.6808596160	.6636024756	.6579242497	.6651692305	.6802107024	.6994512758	.7479361293	.8501634247
1 I PF2	1.165451592	1.093366403	1.058951710	1.020554682	.9765962894	.9462189950	.9234835606	.8908849897	.8585674133
1 I PQ2	4.380162921	2.902244801	2.694884265	2.589845381	2.591912433	2.673445651	2.791688590	3.097597510	3.723929325
2 I PFC	1.059773448	1.052948377	1.048514625	1.042722080	1.034948556	1.028802532	1.023743415	1.015712964	1.006684564
2 I PQC	47.57237969	25.07744659	21.37215565	18.78288960	16.99106729	16.28696188	16.02638711	16.12001699	17.05498645
1 I ZF1	1.056150269	1.055365948	1.055316882	1.055345070	1.055442968	1.055540719	1.055628169	1.055775605	1.055952471
1 I ZF2	1.074367797	1.084069253	1.087441081	1.090793199	1.094339544	1.096711365	1.098487369	1.101100554	1.103879633
1 I ZF3	1.221240657	1.247782021	1.259034129	1.271165427	1.284910848	1.294549045	1.301960724	1.313112975	1.325219360
1 I ZF4	2.637901205	2.739178181	2.783396159	2.832143166	2.888819715	2.929491604	2.961279141	3.009904951	3.063675993

(4) P 6+2	TH = 73	SBE = 1.045691756						MCPER FUNCTIONS	
	RC = 02	RC = 05	RC = 07	RC = 10	RC = 15	RC = 20	RC = 25	RC = 35	RC = 50
AMAX DB		.0108709564	.0213327376	.0436480540	.0988323393	.1772876696	.2802872360	.5675287486	1.249387366
AMIN DB		14.48199824	18.12606422	22.11330282	26.76356453	30.14601463	32.83650488	37.07513877	42.04677721
MU I K		.1887557D 0	.1240786D 0	.7840339D-1	.4590096D-1	.3109562D-1	.2281260D-1	.1400371D-1	.7900619D-2
1 I PF1		1.332992462	1.153800944	.9897000143	.8311385573	.7336879944	.6653776714	.5727382262	.4861959774
1 I PQ1		.6968784339	.6748458930	.6661112567	.6712932015	.6854188104	.7041632310	.7521806731	.8542787898
1 I PF2		1.093959651	1.061584632	1.024825405	.9821666733	.9524234917	.9300522334	.8978426384	.8657898800
1 I PQ2		3.144042700	2.885294153	2.745367979	2.723535739	2.795321700	2.909428757	3.214784142	3.850561069
2 I PFC		1.048307855	1.044517738	1.039440158	1.032484980	1.026906758	1.022274977	1.014862707	1.006456831
2 I PQC		28.83380217	24.23472311	21.02965062	18.79515517	17.88597875	17.51197860	17.49637936	18.39677765
1 I ZF1		1.049359683	1.049281235	1.049284651	1.049360397	1.049445262	1.049523928	1.049659628	1.049825375
1 I ZF2		1.074875978	1.078128539	1.081382845	1.084837811	1.087152713	1.088887800	1.091443327	1.094164679
1 I ZF3		1.227748836	1.238465317	1.250171235	1.263567037	1.273022883	1.280322834	1.291346575	1.303359490
1 I ZF4		2.661905273	2.704513429	2.752019546	2.807738401	2.847968320	2.879529269	2.927984247	2.981782820

(4) P 6+2	TH = 74	SBE = 1.040299436						MCPER FUNCTIONS	
	RC = 02	RC = 05	RC = 07	RC = 10	RC = 15	RC = 20	RC = 25	RC = 35	RC = 50
AMAX DB		.0108709564	.0213327376	.0436480540	.0988323393	.1772876696	.2802872360	.5675287486	1.249387366
AMIN DB		13.10036033	16.68716367	20.64097288	25.27009647	28.64304120	31.32794696	35.56005189	40.52638984
MU I K		.2213003D 0	.1464340D 0	.9288623D-1	.5451238D-1	.3696987D-1	.2713955D-1	.1667237D-1	.9411969D-2
1 I PF1		1.373599267	1.188213751	1.017637309	.8528938895	.7518940739	.6812539503	.5856860614	.4966435759
1 I PQ1		.7160540454	.6880799602	.6755851887	.6782631993	.6912885427	.7094390286	.7568934935	.8588154175
1 I PF2		1.093930647	1.063768067	1.028824852	.9876368119	.9586248739	.9366799467	.9049339737	.8732048318
1 I PQ2		3.433115761	3.109934612	2.926499583	2.874931972	2.934461907	3.043162027	3.346994596	3.992552766
2 I PFC		1.043757608	1.040570002	1.036175754	1.030015626	1.024995523	1.020786838	1.013990829	1.006210197
2 I PQC		33.52605685	27.76362008	23.76215428	20.95960165	19.78835362	19.26910319	19.11261069	19.95970746
1 I ZF1		1.043744681	1.043636619	1.043614668	1.043667546	1.043739024	1.043808521	1.043931900	1.044085878
1 I ZF2		1.066208522	1.069321766	1.072463317	1.075814371	1.078064993	1.079753950	1.082244418	1.084900397
1 I ZF3		1.208493991	1.218612597	1.229836778	1.242829908	1.252071464	1.259237536	1.270102838	1.281993835
1 I ZF4		2.585862707	2.626676090	2.672786949	2.727409105	2.767114613	2.798392137	2.846600389	2.900356719

(4) P 6+2	TH = 75	SBE = 1.035276180							MCPER FUNCTIONS
	RC = 02	RC = 05	RC = 07	RC = 10	RC = 15	RC = 20	RC = 25	RC = 35	RC = 50
AMAX DB		.0108709564	.0213327376	.0436480540	.0988323393	.1772876696	.2802872360	.5675287486	1.249387366
AMIN DB		11.73432084	15.24775124	19.15904220	23.76204537	27.12362989	29.80200484	34.02660930	38.98695005
MU I K		.2589906D 0	.1728273D 0	.1101661D 0	.6484817D-1	.4403708D-1	.3235190D-1	.1989159D-1	.1123705D-1
1 I PF1		1.415899520	1.224867363	1.047684474	.8763610076	.7715274714	.6983610129	.5996145732	.5078717115
1 I PQ1		.7392527887	.7038249721	.6866576363	.6862654256	.6979562765	.7153894742	.7621607750	.8638462678
1 I PF2		1.093218465	1.065437559	1.032498440	.9929686852	.9647968575	.9433487508	.9121519319	.8808144981
1 I PQ2		3.782760231	3.377979146	3.139685273	3.050731862	3.094722273	3.196341081	3.497324157	4.152931088
2 I PFC		1.039314919	1.036684618	1.032938190	1.027546115	1.023072288	1.019281055	1.013097709	1.005943792
2 I PQC		39.47151970	32.17650967	27.13100065	23.58938998	22.07881739	21.37130128	21.02916152	21.79873335
1 I ZF1		1.038510922	1.038373584	1.038325929	1.038355373	1.038413034	1.038473018	1.038583531	1.038725120
1 I ZF2		1.058068396	1.061019637	1.064031358	1.067264384	1.069442587	1.071079705	1.073497074	1.076079405
1 I ZF3		1.190021103	1.199476694	1.210158249	1.222690664	1.231682754	1.238690446	1.249364050	1.261101337
1 I ZF4		2.510964018	2.549775656	2.594317600	2.647688809	2.686779379	2.717711260	2.765589643	2.819226923

(4) P 6+2	TH = 76	SBE = 1.030613629							MCPER FUNCTIONS
	RC = 02	RC = 05	RC = 07	RC = 10	RC = 15	RC = 20	RC = 25	RC = 35	RC = 50
AMAX DB		.0108709564	.0213327376	.0436480540	.0988323393	.1772876696	.2802872360	.5675287486	1.249387366
AMIN DB		10.39077824	13.81009209	17.66668624	22.23685525	25.58462478	28.25524433	32.47112437	37.42461735
MU I K		.3023160D 0	.2039367D 0	.1308175D 0	.7729604D-1	.5257373D-1	.3865786D-1	.2379270D-1	.1345145D-1
1 I PF1		1.459402067	1.263739702	1.080008620	.9017387205	.7927687676	.7168592571	.6146535735	.5199721661
1 I PQ1		.7676303074	.7227782044	.6997429884	.6955437993	.7055986359	.7221568735	.7690919496	.8694630395
1 I PF2		1.091764738	1.066523671	1.035784085	.9981173852	.9709075893	.9500363045	.9194868248	.8886201717
1 I PQ2		4.211096507	3.701900183	3.393606746	3.257090131	3.281180413	3.373482242	3.669785362	4.335587175
2 I PFC		1.034998854	1.032876452	1.029738120	1.025083003	1.021141191	1.017760175	1.012133961	1.005656763
2 I PQC		47.12286451	37.78342545	31.34883040	26.83079018	24.87429726	23.91934368	23.33058995	23.98772036
1 I ZF1		1.033648704	1.033483230	1.033409906	1.033415554	1.033459059	1.033509237	1.033606399	1.033735016
1 I ZF2		1.050459359	1.053223475	1.056085923	1.059184967	1.061281699	1.062860699	1.065196190	1.067695888
1 I ZF3		1.172335972	1.181062016	1.191135566	1.203143755	1.211847559	1.218669696	1.229114523	1.240662433
1 I ZF4		2.437121214	2.473700961	2.516476760	2.568423608	2.606798735	2.637316243	2.684772996	2.738206147

(4) P 6+2	TH = 77	SBE = 1.026304108							MCPER FUNCTIONS
	RC = 02	RC = 05	RC = 07	RC = 10	RC = 15	RC = 20	RC = 25	RC = 35	RC = 50
AMAX DB		.0108709564	.0213327376	.0436480540	.0988323393	.1772876696	.2802872360	.5675287486	1.249387366
AMIN DB		9.079038453	12.37804340	16.16364219	20.69183043	24.02247637	26.68372219	30.38930222	35.83488877
MU I K		.3515994D 0	.2404904D 0	.1555313D 0	.9234396D-1	.6293267D-1	.4632484D-1	.2854532D-1	.1615309D-1
1 I PF1		1.503320721	1.304675241	1.114754627	.9292491502	.8158289645	.7369392692	.6309592432	.5330680575
1 I PQ1		.8027387726	.7458863083	.7154014706	.7064234791	.7144483536	.7299270993	.7748282321	.8757827721
1 I PF2		1.089517622	1.066953101	1.038611470	1.003029782	.9769183715	.9567147691	.9269255768	.8966218331
1 I PQ2		4.743196432	4.098964151	3.700189206	3.502348982	3.500651860	3.580604687	3.869674388	4.545611016
2 I PFC		1.030830292	1.029162259	1.026587767	1.022633977	1.019207195	1.016227348	1.011250505	1.005348295
2 I PQC		57.13789940	45.03545707	36.72281063	30.89233027	28.33983080	27.05440128	26.13341679	26.62801572
1 I ZF1		1.029148460	1.028957097	1.028858662	1.028840433	1.028869562	1.028909712	1.028993116	1.029108234
1 I ZF2		1.043387004	1.045937101	1.048628366	1.051575365	1.053580510	1.055094447	1.057338434	1.059745706
1 I ZF3		1.155445377	1.163376574	1.172772703	1.184187253	1.192559853	1.199166207	1.209340770	1.220659110
1 I ZF4		2.364240016	2.398336244	2.439120478	2.489445252	2.526991864	2.557018222	2.603951131	2.657085197

(4) P 6+2	TH = 78	SBE = 1.022340595						MCPER FUNCTIONS	
	RC = 02	RC = 05	RC = 07	RC = 10	RC = 15	RC = 20	RC = 25	RC = 35	RC = 50
AMAX DB		.0108709564	.0213327376	.0436480540	.0988323393	.1772876696	.2802872360	.5675287486	1.249387366
AMIN DB		7.811060616	10.95764010	14.65056678	19.12420205	22.43318445	25.08287065	29.27607435	34.21240737
MU I K		.4068618D 0	.2832161D 0	.1851278D 0	.1106089D 0	.7556850D-1	.5570016D-1	.3437133D-1	.1947061D-1
1 I PF1		1.546497762	1.347310242	1.152014113	.9591353097	.8409544658	.7588288225	.6487214345	.5473101618
1 I PQ1		.8466760285	.7744514232	.7344040224	.7193464805	.7248181543	.7389476411	.7825552506	.8829575099
1 I PF2		1.086437023	1.066650846	1.040901530	1.007642926	.9827820384	.9633493656	.9344507003	.9048176262
1 I PQ2		5.414335668	4.593552082	4.076175248	3.798070339	3.762487354	3.825904291	4.104127665	4.789794602
2 I PFC		1.026831780	1.025560903	1.023501208	1.020208088	1.017276277	1.014686479	1.010298651	1.005017642
2 I PQC		70.49247378	54.60454664	43.70817144	36.07916637	32.71444413	30.97931447	29.60317457	29.86200560
1 I ZF1		1.025000583	1.024787032	1.024664779	1.024622975	1.024637666	1.024667656	1.024736997	1.024838165
1 I ZF2		1.036857992	1.039166777	1.041662630	1.044437177	1.046339390	1.047780545	1.049922427	1.052226556
1 I ZF3		1.139355178	1.146431600	1.155078036	1.165823348	1.173817218	1.180174072	1.190031765	1.201075021
1 I ZF4		2.292211438	2.323557724	2.362092581	2.410566666	2.447155563	2.476603824	2.522897722	2.575625712

(4) P 6+2	TH = 79	SBE = 1.018716695						MCPER FUNCTIONS	
	RC = 02	RC = 05	RC = 07	RC = 10	RC = 15	RC = 20	RC = 25	RC = 35	RC = 50
AMAX DB		.0108709564	.0213327376	.0436480540	.0988323393	.1772876696	.2802872360	.5675287486	1.249387366
AMIN DB		6.601361752	9.557770527	13.12956942	17.53126649	20.81224714	23.44735421	27.62537543	32.55069715
MU I K		.4676618D 0	.3327450D 0	.2205573D 0	.1328730D 0	.9107258D-1	.6724071D-1	.4156533D-1	.2357573D-1
1 I PF1		1.587349908	1.390972895	1.191773609	.9916528492	.8684317103	.7828014840	.6681735461	.5623857411
1 I PQ1		.9022956489	.8102884585	.7578320441	.7349270339	.7371377333	.7495553884	.7915222781	.8911888480
1 I PF2		1.082500769	1.065543867	1.042566439	1.011882180	.9884408814	.9698964683	.9420388975	.9132031146
1 I PQ2		6.275109964	5.220856889	4.545657109	4.160692860	4.079843829	4.120820729	4.382994324	5.077412448
2 I PFC		1.023027154	1.022093506	1.020494700	1.017816055	1.015355672	1.013142420	1.009330225	1.004664177
2 I PQC		88.66928010	67.51836653	52.99942495	42.85144972	38.35477316	35.99419614	33.98180445	33.89530072
1 I ZF1		1.021195274	1.020965028	1.020821262	1.020756710	1.020757112	1.020776928	1.020832036	1.020918902
1 I ZF2		1.030878871	1.032920906	1.035195414	1.037774634	1.039561165	1.040920931	1.042948274	1.045138185
1 I ZF3		1.124067639	1.130240653	1.138064628	1.148058966	1.155621457	1.161691087	1.171179335	1.181895672
1 I ZF4		2.220897920	2.249227296	2.285220305	2.331575968	2.367056905	2.395826959	2.441350218	2.493550155

(4) P 6+2	TH = 80	SBE = 1.015426612						MCPER FUNCTIONS	
	RC = 02	RC = 05	RC = 07	RC = 10	RC = 15	RC = 20	RC = 25	RC = 35	RC = 50
AMAX DB			.0213327376	.0436480540	.0988323393	.1772876696	.2802872360	.5675287486	1.249387366
AMIN DB			8.190834246	11.60497952	15.91064901	19.15463415	21.77089437	25.92983636	30.84178695
MU I K			.3894559D 0	.2628761D 0	.1601281D 0	.1102220D 0	.8155588D-1	.5052522D-1	.2870190D-1
1 I PF1			1.434562450	1.233831125	1.027050437	.8985898038	.8091868560	.6896060607	.5800311507
1 I PQ1			.8559616333	.7872340854	.7540401635	.7520128608	.7622208094	.8020724936	.9007505495
1 I PF2			1.063566678	1.043510552	1.015659091	.9938239871	.9763010554	.9496591155	.9217702052
1 I PQ2			6.032961793	5.144287758	4.614289092	4.471786205	4.481790593	4.720254171	5.421476161
2 I PFC			1.018783468	1.017587040	1.015470643	1.013454191	1.011601231	1.008347734	1.004237452
2 I PQC			85.39679000	65.69184720	51.92757756	45.81171063	42.55855402	39.63490947	39.03452876
1 I ZF1			1.017483053	1.017321422	1.017235679	1.017222220	1.017232004	1.017272880	1.017345225
1 I ZF2			1.027209367	1.029236244	1.031594924	1.033251476	1.034520229	1.036421402	1.038482702
1 I ZF3			1.114817919	1.121750313	1.130906555	1.137979409	1.143719505	1.152778722	1.163108738
1 I ZF4			2.175180996	2.208307396	2.252228192	2.286423066	2.314397361	2.358996363	2.410527611

$P^{(5)}_{6+4}$

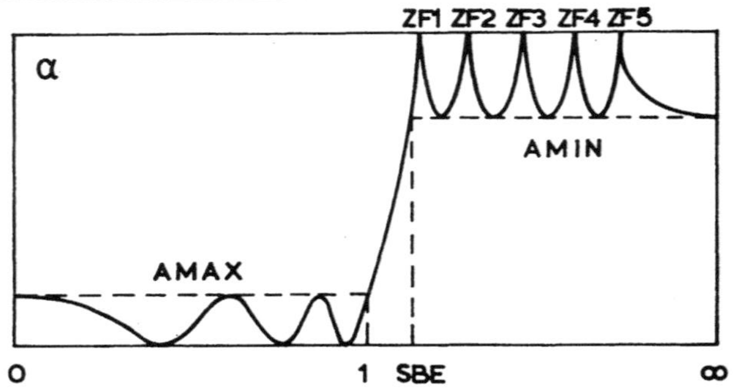

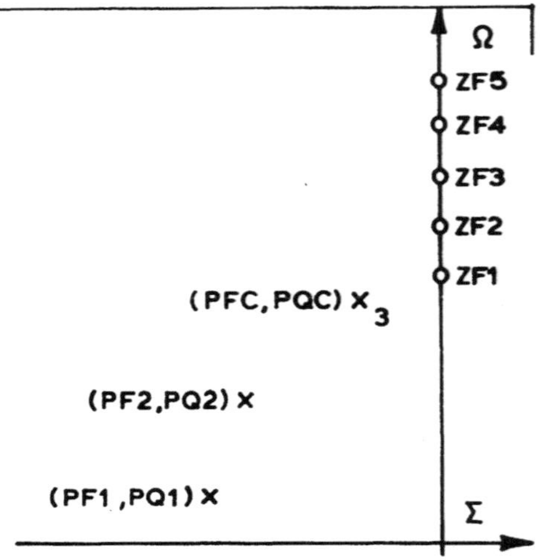

(5) P 6+4	TH = 30	SBE = 2.000000000							MCPER FUNCTIONS
	RC = 02	RC = 05	RC = 07	RC = 10	RC = 15	RC = 20	RC = 25	RC = 35	RC = 50
AMAX DB	.0017375255	.0108709564	.0213327376	.0436480540	.0988323393	.1772876696	.2802872360	.5675287486	1.249387366
AMIN DB	90.95011146	104.2976820	109.2353508	114.5049519	120.5675864	124.9485279	128.4226901	133.8301644	140.2480691
MU I K	.2834617D-4	.6096996D-5	.3453285D-5	.1882576D-5	.9367435D-6	.5656836D-6	.3791975D-6	.2022981D-6	.9718440D-7
1 I PF1	.8495069070	.6516918010	.5868111787	.5221015072	.4533892334	.4076844899	.3739548752	.3259142254	.2785837130
1 I PQ1	.5403549858	.5537168273	.5619243492	.5738990366	.5938899905	.6146979504	.6368930791	.6873801280	.7874797462
1 I PF2	1.062857212	.9090912657	.8597677856	.8114321202	.7611759893	.7283431759	.7043432715	.6702160364	.6358129860
1 I PQ2	.8835963765	.9663009257	1.014985194	1.083107444	1.190477016	1.295708744	1.402419684	1.630101440	2.043944921
3 I PFC	1.425372055	1.294972932	1.250399063	1.205326821	1.156876457	1.124260463	1.099894856	1.064444143	1.027781629
3 I PQC	2.434108310	2.467018120	2.512939064	2.585027999	2.702644710	2.815731677	2.925911584	3.144585991	3.492345669
1 I ZF1	2.018176886	2.019025387	2.019296933	2.019563642	2.019843191	2.020028395	2.020165878	2.020365846	2.020574851
1 I ZF2	2.180150973	2.188690635	2.191374056	2.193990751	2.196715602	2.198511834	2.199841007	2.201768354	2.203775981
1 I ZF3	2.625160041	2.653642681	2.662453832	2.670987679	2.679816511	2.685605910	2.689874973	2.696043581	2.702443249
1 I ZF4	3.890610309	3.965092245	3.988037541	4.010205511	4.033075789	4.048034209	4.059043686	4.074919411	4.091348030
1 I ZF5	10.92739374	11.19902441	11.28282603	11.36379394	11.44730234	11.50189301	11.54205155	11.59992221	11.65975205

(5) P 6+4 — TH = 31 — SBE = 1.941604026 — MCPER FUNCTIONS

MU	P 6+4	RC = 02	RC = 05	RC = 07	RC = 10	RC = 15	RC = 20	RC = 25	RC = 35	RC = 50
	AMAX DB	.0017375255	.0108709564	.0213327376	.0436480540	.0988323393	.1772876696	.2802872360	.5675287486	1.249387366
	AMIN DB	87.96758975	101.3089495	106.2451351	111.5134873	117.5750185	121.9553380	125.4290922	130.8860509	137.2535118
MU	K	.3995954D-4	.8601071D-5	.4872404D-5	.2656597D-5	.1322054D-5	.7984231D-6	.5352361D-6	.2855601D-6	.1371906D-6
1	PF1	.8560641924	.6558887450	.5903861959	.5251201780	.4558795533	.4098550387	.3759039730	.3275670780	.2799642302
1	PQ1	.5408754905	.5541237205	.5623029652	.5742539209	.5942252131	.6150241604	.6372156204	.6877046296	.7878237958
1	PF2	1.066580976	.9121836498	.8626614846	.8141365388	.7636917787	.7307418008	.7066601908	.6724241865	.6379233930
1	PQ2	.8905474014	.9722084208	1.020701819	1.088730954	1.196146014	1.301529884	1.408449356	1.636664692	2.051600134
3	PFC	1.418697954	1.291292474	1.247531958	1.203188951	1.155427079	1.123223275	1.099140414	1.064064776	1.027747471
3	PQC	2.495637161	2.515029993	2.557657388	2.627027371	2.742428419	2.854490306	2.964201941	3.132792609	3.531599436
1	ZF1	1.958868220	1.959724301	1.959999709	1.960270801	1.960555524	1.960744461	1.960884864	1.961089295	1.961303223
1	ZF2	2.112716684	2.121401855	2.124138807	2.126811154	2.129597494	2.131436190	2.132797754	2.134773509	2.136833333
1	ZF3	2.536136311	2.565312166	2.574350999	2.583111439	2.592181068	2.598131927	2.602521873	2.608367913	2.615455138
1	ZF4	3.744965665	3.821468387	3.845075384	3.867898641	3.891460599	3.906879606	3.918232061	3.934608125	3.951561485
1	ZF5	10.49178501	10.77084737	10.85713358	10.94057661	11.02670773	11.08304794	11.12450981	11.18428122	11.24610219

(5) P 6+4 — TH = 32 — SBE = 1.887079915 — MCPER FUNCTIONS

MU	P 6+4	RC = 02	RC = 05	RC = 07	RC = 10	RC = 15	RC = 20	RC = 25	RC = 35	RC = 50
	AMAX DB	.0017375255	.0108709564	.0213327376	.0436480540	.0988323393	.1772876696	.2802872360	.5675287486	1.249387366
	AMIN DB	85.07114012	98.40558683	103.3401212	108.6070830	114.6673860	119.0470131	122.5203128	127.9766977	134.3436644
MU	K	.5577544D-4	.1201491D-4	.6807599D-5	.3712324D-5	.1847697D-5	.1115962D-5	.7481426D-6	.3991766D-6	.1917859D-6
1	PF1	.8629597835	.6602875615	.5941296680	.5282783915	.4584828573	.4121229384	.3779397938	.3292926920	.2814049286
1	PQ1	.5414285908	.5545529903	.5627016106	.5746269099	.5945769392	.6153660826	.6375534633	.6880442187	.7881835522
1	PF2	1.070422662	.9153884041	.8656631288	.8169438784	.7663048391	.7332339158	.7090678579	.6747193164	.6401173029
1	PQ2	.8979457127	.9784569480	1.026737661	1.094658683	1.202112055	1.307649939	1.414784224	1.643553611	2.059627991
3	PFC	1.411818742	1.287486938	1.244564454	1.200973885	1.153923357	1.122145874	1.098355649	1.063668463	1.027708942
3	PQC	2.561850178	2.566315448	2.605320560	2.671700581	2.784657905	2.895578045	3.004755302	3.223204449	3.573065397
1	ZF1	1.903483921	1.904345462	1.904624196	1.904899201	1.905188665	1.905381079	1.905524227	1.905732888	1.905951522
1	ZF2	2.049653136	2.058470631	2.061257589	2.063982443	2.066827308	2.068706687	2.070099414	2.072121929	2.074232399
1	ZF3	2.452621170	2.482465750	2.491725766	2.500707014	2.510012006	2.516121078	2.520629679	2.527150186	2.533922222
1	ZF4	3.607884508	3.686349478	3.710605128	3.734073091	3.758317782	3.774192548	3.785884920	3.802757621	3.820232740
1	ZF5	10.08103359	10.36727774	10.45599824	10.54187857	10.63060363	10.68867974	10.73143731	10.79310182	10.85690984

(5) P 6+4			TH = 33	SBE = 1.836078459						MCPER FUNCTIONS	
			RC = 02	RC = 05	RC = 07	RC = 10	RC = 15	RC = 20	RC = 25	RC = 35	RC = 50
AMAX DB			.0017375255	.0108709564	.0213327376	.0436480540	.0988323393	.1772876696	.2802872360	.5675287486	1.249387366
AMIN DB			82.25520813	95.58198015	100.5146811	105.7800991	111.8390382	116.2178964	119.6906915	125.1464390	131.5128570
MU		K	.7713289D-4	.1663033D-4	.9424665D-5	.5140378D-5	.2558869D-5	.1545629D-5	.1036252D-5	.5529401D-6	.2656790D-6
1		PF1	.8702095881	.6648962133	.5980478675	.5315810574	.4612028953	.4144913011	.3800650365	.3310932492	.2829075505
1		PQ1	.5420165778	.5550058752	.5631213071	.5750188544	.5949458794	.6157243561	.6379072049	.6883994508	.7885595626
1		PF2	1.074380073	.9187061368	.8687737531	.8198554439	.7690166369	.7358210431	.7115678158	.6771029777	.6423962595
1		PQ2	.9058229505	.9850665741	1.033110364	1.100906414	1.208389549	1.314082771	1.421437978	1.650782128	2.068043686
3		PFC	1.404737392	1.283556426	1.241496253	1.198681106	1.152364698	1.121027681	1.097540019	1.063254744	1.027665694
3		PQC	2.633159288	2.621121700	2.656139789	2.719229648	2.829490552	2.939139168	3.047707890	3.265948797	3.616865741
1		ZF1	1.851670598	1.852535407	1.852816909	1.853095341	1.853389095	1.853584719	1.853730430	1.853943077	1.854166190
1		ZF2	1.990569187	1.999505519	2.002338855	2.005112971	2.008013291	2.009931505	2.011354118	2.013421677	2.015581179
1		ZF3	2.374121995	2.404610022	2.414084491	2.423280548	2.432815257	2.439079161	2.443704093	2.450395970	2.457349942
1		ZF4	3.478602455	3.558967017	3.583857339	3.607958668	3.632876495	3.649201838	3.661230842	3.678596198	3.696589862
1		ZF5	9.692931958	9.986087397	10.07718724	10.16546333	10.25675051	10.31654730	10.36059198	10.42414086	10.48993102

(5) P 6+4			TH = 34	SBE = 1.788291650						MCPER FUNCTIONS	
			RC = 02	RC = 05	RC = 07	RC = 10	RC = 15	RC = 20	RC = 25	RC = 35	RC = 50
AMAX DB			.0017375255	.0108709564	.0213327376	.0436480540	.0988323393	.1772876696	.2802872360	.5675287486	1.249387366
AMIN DB			79.51473975	92.83301217	97.76368246	103.0273903	109.0848185	113.4628250	116.9350611	122.3901025	128.7559125
MU		K	.1057458D-3	.2282177D-4	.1293647D-4	.7057168D-5	.3513655D-5	.2122554D-5	.1423138D-5	.7594425D-6	.3649256D-6
1		PF1	.8778308218	.6697232603	.6021475232	.5350334331	.4640436750	.4162634487	.3822825796	.3329710746	.2844739504
1		PQ1	.5426419932	.5554837309	.5635631685	.5754306782	.5953328032	.6160996712	.6382774892	.6887709241	.7889524151
1		PF2	1.078450561	.9221373721	.8719943693	.8228725591	.7718286864	.7335047645	.7141616740	.6795767951	.6447618831
1		PQ2	.9142140588	.9920592342	1.039839152	1.107491299	1.214994105	1.320843372	1.428425407	1.658365265	2.076863577
3		PFC	1.397457238	1.279501143	1.238327108	1.196310119	1.150750505	1.119868102	1.096692965	1.062823141	1.027617364
3		PQC	2.710025179	2.679722166	2.710347896	2.769814589	2.377098752	2.985331384	3.093208394	3.311164299	3.663133293
1		ZF1	1.803116569	1.803982385	1.804266077	1.804547432	1.804845008	1.805043565	1.805191648	1.805408029	1.805635384
1		ZF2	1.935119917	1.944161321	1.947037304	1.949857334	1.952809936	1.954765071	1.956216243	1.958327062	1.960533912
1		ZF3	2.300204493	2.331309869	2.340991826	2.350396472	2.360155031	2.366570247	2.371309088	2.378169101	2.385302091
1		ZF4	3.356446283	3.438643434	3.464153456	3.488876000	3.514456676	3.531227035	3.543589143	3.561442873	3.579951620
1		ZF5	9.325537408	9.625312378	9.718731692	9.809358057	9.903172285	9.964672862	10.00999505	10.07541836	10.14318492

(5) P 6+4		TH = 35 SBE = 1.743446796								MCPER FUNCTIONS
		RC = 02	RC = 05	RC = 07	RC = 10	RC = 15	RC = 20	RC = 25	RC = 35	RC = 50
AMAX DB		.0017375255	.0108709564	.0213327376	.0436480540	.0988323393	.1772876696	.2802872360	.5675287486	1.249387366
AMIN DB		76.84512377	90.15400415	95.08243015	100.3442479	106.4000061	110.7770712	114.2486893	119.7029502	126.0680882
MU	K	.1437950D-3	.3106703D-4	.1761483D-4	.9611421D-5	.4786298D-5	.2891655D-5	.1938945D-5	.1034791D-5	.4972738D-6
1	PF1	.8858421288	.6747779099	.6064358580	.5386411506	.4670094814	.4195429290	.3845954943	.3349236456	.2861061025
1	PQ1	.5433076620	.5559880426	.5640284102	.5758633861	.5957385440	.6164927741	.6386650112	.6891592826	.7893627417
1	PF2	1.082630968	.9256825325	.8753259560	.8259965613	.7747425478	.7412867223	.7168511099	.6821424683	.6472158741
1	PQ2	.9231576917	.9994589349	1.046945003	1.114431996	1.221942655	1.327947972	1.435762500	1.666319232	2.086105284
3	PFC	1.389981999	1.275321398	1.235056829	1.193860445	1.149080177	1.118666536	1.095813913	1.062373160	1.027563569
3	PQC	2.792964111	2.742419746	2.768201919	2.823675503	2.927671582	3.034327306	3.141419301	3.359002143	3.712012577
1	ZF1	1.757545908	1.758410401	1.758695688	1.758979442	1.759280355	1.759481556	1.759631814	1.759851666	1.760083017
1	ZF2	1.883000022	1.892132476	1.895047278	1.897909777	1.900911383	1.902901455	1.904379812	1.906532038	1.908784483
1	ZF3	2.230484391	2.262180163	2.272062398	2.281669184	2.291645500	2.298208365	2.303058589	2.310033364	2.317392047
1	ZF4	3.240820988	3.324779141	3.350892841	3.376223537	3.402456085	3.419665498	3.432356920	3.450694426	3.469714527
1	ZF5	8.977134373	9.283215239	9.378889011	9.471815992	9.568113748	9.631304403	9.677893392	9.745179945	9.814916176

(5) P 6+4		TH = 36 SBE = 1.701301617								MCPER FUNCTIONS
		RC = 02	RC = 05	RC = 07	RC = 10	RC = 15	RC = 20	RC = 25	RC = 35	RC = 50
AMAX DB		.0017375255	.0108709564	.0213327376	.0436480540	.0988323393	.1772876696	.2802872360	.5675287486	1.249387366
AMIN DB		74.24214229	87.54066616	92.46661695	97.72635005	103.7802662	108.1562928	111.6272293	117.0806293	123.4450259
MU	K	.1940407D-3	.4197268D-4	.2380505D-4	.1299219D-4	.6471228D-5	.3910077D-5	.2622035D-5	.1399486D-5	.6725874D-6
1	PF1	.8942637154	.6800700729	.6109206299	.5424102469	.4701048982	.4222335322	.3870070593	.3369686039	.2878061095
1	PQ1	.5440167304	.5565204405	.5645183607	.5763180716	.5961640056	.6169044722	.6390705218	.6895652209	.7897912225
1	PF2	1.086917556	.9293419179	.8787694461	.8292287955	.7777598256	.7441686192	.7196378703	.6848017752	.6497600163
1	PQ2	.9326966805	1.007291986	1.054450826	1.121748823	1.229253580	1.335414158	1.443466559	1.674661541	2.095787808
3	PFC	1.382315809	1.271017625	1.231685290	1.191331635	1.147353113	1.117422366	1.094902272	1.061904285	1.027503913
3	PQC	2.882555868	2.809550585	2.829986091	2.881054955	2.981416708	3.086316092	3.192518403	3.409627389	3.763661036
1	ZF1	1.714713483	1.715574257	1.715860523	1.716146134	1.716449882	1.716653429	1.716805657	1.717028706	1.717263796
1	ZF2	1.833938314	1.843147567	1.846097264	1.848998687	1.852045915	1.854068874	1.855572990	1.857764698	1.860060914
1	ZF3	2.164620521	2.196878832	2.206953883	2.216756125	2.226943867	2.233650570	2.238609545	2.245795560	2.253276735
1	ZF4	3.131199012	3.216841752	3.243541992	3.269467010	3.296339516	3.313981586	3.326998258	3.345814597	3.365342035
1	ZF5	8.646203003	8.958253746	9.056111581	9.151285134	9.250034226	9.314884327	9.362728216	9.431865437	9.503563593

(5) P 6+4	TH = 37	SBE = 1.661640141						MCPER FUNCTIONS	
	RC = 02	RC = 05	RC = 07	RC = 10	RC = 15	RC = 20	RC = 25	RC = 35	RC = 50
AMAX DB	.0017375255	.0108709564	.0213327376	.0436480540	.0988323393	.1772876696	.2802872360	.5675287486	1.249387366
AMIN DB	71.70192816	84.98905407	89.91228023	95.16971840	101.2216066	105.5964898	109.0666760	114.5191279	120.8827084
MU I K	.2599582D-3	.5630504D-4	.3194376D-4	.1743870D-4	.8687997D-5	.5250196D-5	.3521001D-5	.1879506D-5	.9033677D-6
1 I PF1	.9031174996	.6856104253	.6156101773	.5463471971	.4733348316	.4250393098	.3895207759	.3390937671	.2895762116
1 I PQ1	.5447727096	.5570827162	.5650344733	.5767959265	.5966101692	.6173356397	.6394948326	.6899894879	.7902385897
1 I PF2	1.091305938	.9331156829	.8823257136	.8325706072	.7808821668	.7471522187	.7225237724	.6875565742	.6523961802
1 I PQ2	.9428785705	1.015587263	1.062381686	1.129463931	1.236946859	1.343261007	1.451556325	1.683411120	2.105931653
3 I PFC	1.374463238	1.266590392	1.228212442	1.188723270	1.145568716	1.116134970	1.093957437	1.061415984	1.027437977
3 I PQC	2.979453037	2.881488415	2.896015270	2.942220704	3.038562555	3.141505348	3.246700507	3.463220480	3.818250387
1 I ZF1	1.674400792	1.675255389	1.675541999	1.675828910	1.676134972	1.676340555	1.676494540	1.676720501	1.676959063
1 I ZF2	1.787693103	1.796964700	1.799945279	1.802881986	1.805971353	1.808025077	1.809553478	1.811782671	1.814120762
1 I ZF3	2.102309019	2.135101062	2.145361203	2.155351968	2.165744562	2.172591137	2.177656123	2.184999701	2.192649927
1 I ZF4	3.027111213	3.114357050	3.141625509	3.168129893	3.195629755	3.213697626	3.227035184	3.246325045	3.266355495
1 I ZF5	8.331392821	8.649054618	8.749020513	8.846381979	8.947531327	9.014023185	9.063108803	9.134082631	9.207733839

(5) P 6+4	TH = 38	SBE = 1.624269245						MCPER FUNCTIONS	
	RC = 02	RC = 05	RC = 07	RC = 10	RC = 15	RC = 20	RC = 25	RC = 35	RC = 50
AMAX DB	.0017375255	.0108709564	.0213327376	.0436480540	.0988323393	.1772876696	.2802872360	.5675287486	1.249387366
AMIN DB	69.22092808	82.49553237	87.41576466	92.67068094	98.72034050	103.0939672	106.5633285	112.0147386	118.3774219
MU I K	.3459024D-3	.7502800D-4	.4258060D-4	.2325230D-4	.1158732D-4	.7003282D-5	.4697141D-5	.2507628D-5	.1205394D-5
1 I PF1	.9124272752	.6914104767	.6205134701	.5504589518	.4767045368	.4279645960	.3921403861	.3413071427	.2914187971
1 I PQ1	.5455795266	.5576768418	.5655783408	.5772982513	.5970781013	.6177872246	.6399388222	.6904328927	.7907056322
1 I PF2	1.095790991	.9370038107	.8859955576	.8360233342	.7841112578	.7502393451	.7255107057	.6904088074	.6551263271
1 I PQ2	.9537562409	1.024376507	1.070765034	1.137601507	1.245044233	1.351509236	1.460052121	1.692538460	2.116558964
3 I PFC	1.366429324	1.262040414	1.224638318	1.186034968	1.143726394	1.114803715	1.092978789	1.060907706	1.027365327
3 I PQC	3.084391908	2.958649574	2.966638873	3.007468829	3.099360789	3.200123266	3.304179374	3.519978246	3.875968165
1 I ZF1	1.636412467	1.637258368	1.637544668	1.637832303	1.638140142	1.638347441	1.638502963	1.638731540	1.638973297
1 I ZF2	1.744048322	1.753367634	1.756374992	1.759343252	1.762471170	1.764553472	1.766104629	1.768369239	1.770747232
1 I ZF3	2.043278460	2.076574411	2.087011639	2.097183740	2.107774357	2.114756676	2.119924816	2.127422120	2.135237739
1 I ZF4	2.928139272	3.016901387	3.044718489	3.071786298	3.099899966	3.118386286	3.132040051	3.151797728	3.172326533
1 I ZF5	8.031500497	8.354391414	8.456383544	8.555869426	8.659368840	8.727477575	8.777790391	8.850585166	8.926179328

(5) P 6+4			TH = 39	SBE = 1.589015729						MCPER FUNCTIONS
			RC = 02	RC = 05	RC = 07	RC = 10	RC = 15	RC = 20	RC = 25	RC = 35	RC = 50
AMAX DB			.0017375255	.0108709564	.0213327376	.0436480540	.0988323393	.1772876696	.2802872360	.5675287486	1.249387366
AMIN DB			66.79587056	80.05674174	84.97368972	90.22583923	96.27305359	100.6453015	104.1137579	109.5640247	115.9257235
MU		K	.4573055D-3	.9934887D-4	.5640473D-4	.3081116D-4	.1535845D-4	.9283996D-5	.6227477D-5	.3325054D-5	.1598504D-5
	1	PF1	.9222188950	.6974826466	.6256401650	.5547529775	.4802196467	.4310140299	.3948698916	.3436119428	.2933364136
	1	PQ1	.5464415835	.5583049923	.5661517115	.5778264673	.5975689621	.6182602556	.6404034424	.6908963095	.7911932010
	1	PF2	1.100366776	.9410060836	.8897796847	.8395882975	.7874488214	.7534318823	.7286006326	.6933605030	.6579525133
	1	PQ2	.9653886220	1.033694665	1.079630987	1.146187995	1.253569393	1.360181374	1.468976014	1.702215756	2.127693690
	3	PFC	1.358219596	1.257368567	1.220963044	1.183266393	1.141825561	1.113427963	1.091965696	1.060378880	1.027285506
	3	PQC	3.198205283	3.041498832	3.042245434	3.077127326	3.164089152	3.262421075	3.365189938	3.580119350	3.937019471
	1	ZF1	1.600573305	1.601407935	1.601693255	1.601981021	1.602290082	1.602498765	1.602655596	1.602886485	1.603131147
	1	ZF2	1.702810231	1.712162486	1.715192441	1.718183431	1.721351212	1.723459833	1.725032170	1.727330052	1.729745901
	1	ZF3	1.987285730	2.021054681	2.031660701	2.042006681	2.052788225	2.059901994	2.065170309	2.072817336	2.080794704
	1	ZF4	2.833909264	2.924095251	2.952440095	2.980054283	3.008767254	3.027664145	3.041629094	3.061848459	3.082870603
	1	ZF5	7.745451022	8.073165820	8.177096327	8.278638071	8.384433025	8.454131430	8.505655456	8.580253804	8.657779501

(5) P 6+4			TH = 40	SBE = 1.555723827						MCPER FUNCTIONS
			RC = 02	RC = 05	RC = 07	RC = 10	RC = 15	RC = 20	RC = 25	RC = 35	RC = 50
AMAX DB			.0017375255	.0108709564	.0213327376	.0436480540	.0988323393	.1772876696	.2802872360	.5675287486	1.249387366
AMIN DB			64.42373813	77.66957066	82.58292124	87.83203994	93.87657553	98.24731295	101.7147777	107.1637919	113.5244120
MU		K	.6009151D-3	.1307740D-3	.7427693D-4	.4058803D-4	.2023817D-4	.1223586D-4	.8208449D-5	.4383393D-5	.2107557D-5
	1	PF1	.9325204736	.7038403494	.6310006683	.5592373025	.4838862046	.4341925816	.3977135749	.3460116004	.2953317806
	1	PQ1	.5473638270	.5589695704	.5667565072	.5783821298	.5980840162	.6187558507	.6408897257	.6913806843	.7917022147
	1	PF2	1.105026433	.9451220500	.8936786885	.8432667902	.7908966127	.7567317741	.7317955895	.6964137786	.6608768941
	1	PQ2	.9778415264	1.043580282	1.089012641	1.155252353	1.262548189	1.369301947	1.478351990	1.712317089	2.139361758
	3	PFC	1.349840100	1.252575909	1.217186852	1.180417256	1.139865647	1.112007074	1.090917512	1.059828918	1.027198040
	3	PQC	3.321837603	3.130556160	3.123267869	3.151560248	3.233054720	3.328675846	3.429990832	3.643879496	4.001628954
	1	ZF1	1.566725757	1.567546487	1.567830138	1.568117427	1.568427138	1.568636864	1.568794767	1.569027650	1.569274919
	1	ZF2	1.663804631	1.673174948	1.676223240	1.679243050	1.682436905	1.684569520	1.686161406	1.688490341	1.690941919
	1	ZF3	1.934112531	1.968322408	1.979088617	1.989600736	2.000565832	2.007806579	2.013171967	2.020964536	2.029099657
	1	ZF4	2.744086199	2.835597811	2.864448089	2.892590440	2.921887192	2.941186213	2.955456959	2.976131429	2.997641512
	1	ZF5	7.472281664	7.804391738	7.910166539	8.013690316	8.121721719	8.192980116	8.245697816	8.322080518	8.401524903

(5) P 6+4 — TH = 41 — SBE = 1.524253087 — MCPER FUNCTIONS

MU	K	RC = 02	RC = 05	RC = 07	RC = 10	RC = 15	RC = 20	RC = 25	RC = 35	RC = 50
	AMAX DB	.0017375255	.0108709564	.0213327376	.0436480540	.0988323393	.1772876696	.2802872360	.5675287486	1.249387366
	AMIN DB	62.10174289	75.33113067	80.24054642	85.48634971	91.52795475	95.89703955	99.36341944	104.8110631	111.1705029
MU	K	.7850781D-3	.1711762D-3	.9726860D-4	.5317194D-4	.2652176D-4	.1603792D-4	.1076042D-4	.5747075D-5	.2763598D-5
1	PF1	.9433626110	.7104980883	.6366062052	.5639205669	.4877106993	.4375055799	.4006760234	.3485097873	.2974078032
1	PQ1	.5483518300	.5596732356	.5673948444	.5789669437	.5986246430	.6192752265	.6413987935	.6918870422	.7922336664
1	PF2	1.109762077	.9493509871	.8976930271	.8470600647	.7944564141	.7601410222	.7350976882	.6995708442	.6639017286
1	PQ2	.9911886174	1.054075953	1.098946424	1.164826340	1.272008874	1.378897700	1.488206164	1.722918616	2.151591267
3	PFC	1.341297431	1.247663686	1.213310089	1.177487331	1.137846095	1.110540402	1.089833583	1.059257211	1.027102433
3	PQC	3.456362857	3.226404628	3.210189597	3.231172502	3.306597655	3.399193711	3.498867285	3.711520955	4.070043069
1	ZF1	1.534727770	1.535531923	1.535813201	1.536099386	1.536409161	1.536619575	1.536778306	1.537012856	1.537262423
1	ZF2	1.626874475	1.636247895	1.639310195	1.642349832	1.645570876	1.647725088	1.649334841	1.651692533	1.654177636
1	ZF3	1.883562400	1.918179869	1.929097340	1.939767559	1.950908538	1.958271607	1.963730832	1.971664579	1.979953333
1	ZF4	2.658369367	2.751102256	2.780434176	2.809085231	2.838949156	2.858641261	2.873212026	2.894334528	2.916326736
1	ZF5	7.211128238	7.547181695	7.654700302	7.760126794	7.870330761	7.943116857	7.997009041	8.075154909	8.156503592

(5) P 6+4 — TH = 42 — SBE = 1.494476550 — MCPER FUNCTIONS

MU	K	RC = 02	RC = 05	RC = 07	RC = 10	RC = 15	RC = 20	RC = 25	RC = 35	RC = 50
	AMAX DB	.0017375255	.0108709564	.0213327376	.0436480540	.0988323393	.1772876696	.2802872360	.5675287486	1.249387366
	AMIN DB	59.82730533	73.03873439	77.94385186	83.18603309	89.22443623	93.59171527	97.05690971	102.5030557	108.8612057
MU	K	.1020081D-2	.2228760D-3	.1267090D-3	.6929443D-4	.3457627D-4	.2091290D-4	.1403313D-4	.7496304D-5	.3605286D-5
1	PF1	.9547786408	.7174715607	.6424688966	.5688120794	.4917001053	.4409587441	.4037621550	.3511104349	.2995675874
1	PQ1	.5494118872	.5604189376	.5680690581	.5795827808	.5991923504	.6198197087	.6419318652	.6924164947	.7927886304
1	PF2	1.114564687	.9536918589	.9018229960	.8509693187	.7981300300	.7636616846	.7335091159	.7028340056	.6670293848
1	PQ2	1.005512538	1.065228832	1.109472512	1.174944849	1.281982370	1.388997833	1.498567004	1.734048787	2.164412719
3	PFC	1.332598754	1.242633361	1.209333232	1.174476453	1.135766370	1.109027303	1.088713241	1.058663134	1.026998166
3	PQC	3.603005875	3.329699649	3.303551665	3.316415399	3.385095530	3.474313569	3.572134446	3.783331955	4.142532655
1	ZF1	1.504450952	1.505235803	1.505513990	1.505798429	1.506107662	1.506318402	1.506477710	1.506713588	1.506965130
1	ZF2	1.591877823	1.601239352	1.604311265	1.607366656	1.610610907	1.612784252	1.614410139	1.616794212	1.619310553
1	ZF3	1.835458155	1.870448523	1.881507978	1.892327944	1.903636826	1.911117367	1.916667054	1.924737420	1.933175497
1	ZF4	2.576488343	2.670331806	2.700120009	2.729258988	2.759672322	2.779747821	2.794612406	2.816175345	2.838543416
1	ZF5	6.961213303	7.300735187	7.409890547	7.517134744	7.629442355	7.703721093	7.758766317	7.838652552	7.921889483

(5) P 6+4	TH = 43	SBE = 1.466279186						MCPER FUNCTIONS	
	RC = 02	RC = 05	RC = 07	RC = 10	RC = 15	RC = 20	RC = 25	RC = 35	RC = 50
AMAX DB	.0017375255	.0108709564	.0213327376	.0436480540	.0988323393	.1772876696	.2802872360	.5675287486	1.249387366
AMIN DB	57.59803554	70.78987632	75.69030406	80.92853293	86.96344182	91.32875007	94.79265068	100.2371620	106.5939042
MU I K	.1318555D-2	.2887395D-3	.1642422D-3	.8986144D-4	.4485676D-4	.2713704D-4	.1821241D-4	.9730651D-5	.4680635D-5
1 I PF1	.9668049037	.7247777753	.6486018457	.5739218795	.4958619268	.4445582185	.4069772479	.3538177560	.3018144569
1 I PQ1	.5505511288	.5612099550	.5687817297	.5802317002	.5997887885	.6203907440	.6424902690	.6929702492	.7933682709
1 I PF2	1.119423979	.9581432689	.9060686992	.8549956778	.8019192790	.7672958736	.7420321359	.7062056670	.6702623442
1 I PQ2	1.020906234	1.077091239	1.120635294	1.185646286	1.292502583	1.399634307	1.509465598	1.745738595	2.177859263
3 I PFC	1.323751832	1.237486624	1.205256900	1.171384538	1.133625963	1.107467137	1.087555812	1.058046041	1.026884700
3 I PQC	3.763167728	3.441179838	3.403961081	3.407793124	3.468968340	3.554411357	3.650141215	3.859630703	4.219395885
1 I ZF1	1.475778980	1.476541766	1.476816125	1.477098164	1.477406236	1.477616926	1.477776553	1.478013408	1.478266593
1 I ZF2	1.558686085	1.568020726	1.571097802	1.574164799	1.577428181	1.579618127	1.581258357	1.583666361	1.586211572
1 I ZF3	1.789639710	1.824966807	1.836158598	1.847119619	1.858588100	1.866181056	1.871817684	1.880019904	1.888602790
1 I ZF4	2.498199557	2.593036289	2.623253759	2.652858482	2.683802228	2.704250740	2.719402499	2.741397714	2.764334908
1 I ZF5	6.721835962	7.064328641	7.175006992	7.283977992	7.398315068	7.474048469	7.530224923	7.611824977	7.696932314

(5) P 6+4	TH = 44	SBE = 1.439556540						MCPER FUNCTIONS	
	RC = 02	RC = 05	RC = 07	RC = 10	RC = 15	RC = 20	RC = 25	RC = 35	RC = 50
AMAX DB	.0017375255	.0108709564	.0213327376	.0436480540	.0988323393	.1772876696	.2802872360	.5675287486	1.249387366
AMIN DB	55.41171697	68.58221566	73.47753218	78.71145294	84.74255265	89.10571233	92.56820230	98.01093133	104.3661384
MU I K	.1695954D-2	.3722967D-3	.2118963D-3	.1159918D-3	.5792584D-4	.3505213D-4	.2352826D-4	.1257342D-4	.6049132D-5
1 I PF1	.9794810479	.7324351832	.6550192341	.5792608077	.5002042473	.4483106114	.4103269723	.3566362596	.3041519724
1 I PQ1	.5517776544	.5620499403	.5695357187	.5809159706	.6004157670	.6209899139	.6430754537	.6935496192	.7939738508
1 I PF2	1.124328275	.9627034075	.9104300149	.8591401769	.8058259866	.7710457529	.7456690876	.7096883349	.6736032079
1 I PQ2	1.037474508	1.089721342	1.132483915	1.196973005	1.303606756	1.410842108	1.520935945	1.758021357	2.191966983
3 I PFC	1.314765054	1.232225413	1.201081866	1.168211582	1.131424395	1.105859270	1.086360616	1.057405268	1.026761471
3 I PQC	3.938456145	3.561679795	3.512100576	3.505870272	3.558684321	3.639905008	3.733274672	3.940769201	4.300961650
1 I ZF1	1.448606236	1.449344155	1.449613939	1.449892908	1.450199182	1.450409440	1.450569117	1.450806539	1.451061072
1 I ZF2	1.527182502	1.536475298	1.539553035	1.542627421	1.545905766	1.548109714	1.549762448	1.552191350	1.554763474
1 I ZF3	1.745962184	1.781588237	1.792902316	1.803995344	1.815614770	1.823314868	1.829034760	1.837363355	1.846036824
1 I ZF4	2.423283328	2.518989180	2.549607157	2.579653961	2.611197807	2.631918215	2.647350018	2.669768744	2.693167798
1 I ZF5	6.492363018	6.837306735	6.949387488	7.059988307	7.176275172	7.253422181	7.310704579	7.393990999	7.480943980

(5) P 6+4 — TH = 45 SBE = 1.414213562 — MCPER FUNCTIONS

MU	K	RC = 02	RC = 05	RC = 07	RC = 10	RC = 15	RC = 20	RC = 25	RC = 35	RC = 50
	AMAX DB	.0017375255	.0108709564	.0213327376	.0436480540	.0988323393	.1772876696	.2802872360	.5675287486	1.249387366
	AMIN DB	53.26629198	66.41356114	71.30331265	76.53254218	82.55949348	86.92031303	90.38126652	95.82205415	102.1755390
MU	K	.2171128D-2	.4778834D-3	.2721663D-3	.1490640D-3	.7447754D-4	.4508005D-4	.3026472D-4	.1617697D-4	.7784318D-5
1	PF1	.9928503585	.7404638238	.6617364291	.5848405836	.5047357843	.4522230385	.4138174269	.3595708289	.3065839520
1	PQ1	.5531006932	.5629429717	.5703341995	.5816380974	.6010752733	.6216189499	.6436890019	.6941560353	.7946067422
1	PF2	1.129264365	.9673699920	.9149065569	.8634037371	.8098519746	.7749135328	.7494223858	.7132846210	.6770547022
1	PQ2	1.055335851	1.103183960	1.145072905	1.208971807	1.315335877	1.422659663	1.533015301	1.770935531	2.206775222
3	PFC	1.305647454	1.226851938	1.196809080	1.164957679	1.129161223	1.104203076	1.085126966	1.056740133	1.026627893
3	PQC	4.130722093	3.692145237	3.628740095	3.611280677	3.654766728	3.731260190	3.821965192	4.027137664	4.387593457
1	ZF1	1.422836619	1.423546836	1.423811285	1.424086500	1.424390327	1.424599757	1.424759211	1.424796927	1.425252354
1	ZF2	1.497260827	1.506496900	1.509570755	1.512648247	1.515937301	1.518152583	1.519815930	1.522264116	1.524859612
1	ZF3	1.704294270	1.740179769	1.751605660	1.762821263	1.774582610	1.782384348	1.788183663	1.796634424	1.805492522
1	ZF4	2.351541296	2.447985044	2.478972934	2.509436583	2.541378819	2.562539217	2.578243417	2.601076237	2.624929332
1	ZF5	6.272221265	6.619074860	6.732430505	6.844557905	6.962709152	7.041225478	7.099586938	7.184529213	7.273316009

(5) P 6+4 — TH = 46 SBE = 1.390163591 — MCPER FUNCTIONS

MU	K	RC = 02	RC = 05	RC = 07	RC = 10	RC = 15	RC = 20	RC = 25	RC = 35	RC = 50
	AMAX DB	.0017375255	.0108709564	.0213327376	.0436480540	.0988323393	.1772876696	.2802872360	.5675287486	1.249387366
	AMIN DB	51.15984936	64.28185751	69.16555543	74.38968113	80.41211859	84.77039163	88.22967305	93.66834783	100.0700627
MU	K	.2766990D-2	.6108114D-3	.3481146D-3	.1907726D-3	.9536611D-4	.5774048D-4	.3877183D-4	.2072920D-4	.9976929D-5
1	PF1	1.006960118	.7488854885	.6687701047	.5906738939	.5094659516	.4563031710	.4174551794	.3626266519	.3091144948
1	PQ1	.5545307951	.5638936144	.5711807040	.5824008524	.6017694946	.6222797513	.6443326458	.6947910585	.7952684381
1	PF2	1.134217356	.9721402007	.9194976317	.8677871411	.8139990498	.7789014667	.7532945200	.7169972455	.6806196846
1	PQ2	1.074624613	1.117551503	1.158462908	1.221694521	1.327735140	1.435129227	1.545744559	1.784520080	2.222326948
3	PFC	1.296408736	1.221368699	1.192439678	1.161623031	1.126836050	1.102497944	1.083854173	1.056049937	1.026483354
3	PQC	4.342103935	3.833650950	3.754750362	3.724737794	3.757801762	3.828997011	3.916692416	4.119169626	4.479693928
1	ZF1	1.398382512	1.399062163	1.399320505	1.399591272	1.399891988	1.400100187	1.400259135	1.400496713	1.400752716
1	ZF2	1.468824183	1.477988775	1.481054172	1.484130428	1.487425852	1.489649737	1.491321749	1.493786026	1.496402764
1	ZF3	1.664516828	1.700620371	1.712147134	1.723475471	1.735369323	1.743266952	1.749141676	1.757708650	1.766696685
1	ZF4	2.282794184	2.379837321	2.411162599	2.442016193	2.474423619	2.495921254	2.511889651	2.535126448	2.559425159
1	ZF5	6.060890744	6.409092563	6.523588601	6.637132923	6.757057188	6.836895133	6.896306562	6.982871455	7.073463020

(5) P 6+4 — TH = 47 — SBE = 1.367327461 — MCPER FUNCTIONS

MU	K	RC = 02	RC = 05	RC = 07	RC = 10	RC = 15	RC = 20	RC = 25	RC = 35	RC = 50
	AMAX DB	.0017375255	.0108709564	.0213327376	.0436480540	.0988323393	.1772876696	.2802872360	.5675287486	1.249387366
	AMIN DB	49.09061344	62.18517344	67.06229165	72.28086914	78.29839908	82.65390323	86.11136645	91.54774359	97.89747918
	K	.3511311D-2	.7775733D-3	.4434916D-3	.2431961D-3	.1216410D-3	.7367240D-4	.4948023D-4	.2646140D-4	.1273873D-4
1	PF1	1.021861993	.7577239038	.6761383765	.5967744906	.5144049281	.4605592908	.4212473115	.3658093562	.3117480070
1	PQ1	.5560800600	.5649069922	.5720791714	.5832073096	.6025008430	.6229744055	.6450082846	.6954563948	.7959605657
1	PF2	1.139170519	.9770105980	.9242021883	.8722910030	.8182689904	.7830118438	.7572880527	.7209290404	.6843011509
1	PQ2	1.095493577	1.132905054	1.172721519	1.235198667	1.340854492	1.448297366	1.559168698	1.798319894	2.238669170
3	PFC	1.287059297	1.215778509	1.187975002	1.158207958	1.124448527	1.100743282	1.082541549	1.055333964	1.026327217
3	PQC	4.575080990	3.987422183	3.891118969	3.847046928	3.868447862	3.933697848	4.017992190	4.217347389	4.577710007
1	ZF1	1.375163886	1.375810080	1.376061535	1.376327147	1.376624076	1.376830639	1.376988783	1.377225829	1.377482030
1	ZF2	1.441784062	1.450862577	1.453914918	1.456985545	1.460282919	1.462512609	1.464191288	1.466668881	1.469304139
1	ZF3	1.626521649	1.662799787	1.674415985	1.685846774	1.697863292	1.705350802	1.711796735	1.720474213	1.729586736
1	ZF4	2.216879852	2.314376390	2.346004504	2.377219383	2.410067214	2.431888433	2.448112230	2.471742133	2.496477383
1	ZF5	5.857898825	6.206867800	6.322362707	6.437207723	6.558807459	6.639915755	6.700345721	6.788497095	6.880867009

(5) P 6+4 — TH = 48 — SBE = 1.345632730 — MCPER FUNCTIONS

MU	K	RC = 02	RC = 05	RC = 07	RC = 10	RC = 15	RC = 20	RC = 25	RC = 35	RC = 50
	AMAX DB	.0017375255	.0108709564	.0213327376	.0436480540	.0988323393	.1772876696	.2802872360	.5675287486	1.249387366
	AMIN DB	47.05693470	60.12169068	64.99166257	70.20421312	76.21641133	80.56890691	84.02439442	89.45327473	95.80585933
	K	.4437652D-2	.9860875D-3	.5628814D-3	.3088797D-3	.1545893D-3	.9366011D-4	.6291878D-4	.3365784D-4	.1620716D-4
1	PF1	1.037612455	.7670049356	.6838609537	.6031573018	.5195637359	.4650003509	.4252014703	.3691249974	.3144892313
1	PQ1	.5577624142	.5659888711	.5730340064	.5840608853	.6032719837	.6237052108	.6457180044	.6961539124	.7966849013
1	PF2	1.144105137	.9819770515	.9290187630	.8769157350	.8226635295	.7872469824	.7614056167	.7247829524	.6881022416
1	PQ2	1.118117038	1.149335658	1.187924280	1.249548240	1.354749248	1.462215508	1.573337293	1.813883758	2.255853421
3	PFC	1.277610244	1.210084522	1.183416622	1.154712914	1.121998365	1.098938518	1.081188409	1.054591481	1.026158821
3	PQC	4.832538760	4.154860256	4.038969509	3.979119722	3.987446657	4.046016536	4.126464706	4.322211447	4.682139017
1	ZF1	1.353107511	1.353717334	1.353961114	1.354220856	1.354513309	1.354717795	1.354874858	1.355110967	1.355366977
1	ZF2	1.416059457	1.425037503	1.428072173	1.431132731	1.434427560	1.436660196	1.438343491	1.440831540	1.443482507
1	ZF3	1.590210392	1.626617457	1.638311113	1.649833597	1.661962491	1.670033597	1.676046339	1.684828338	1.694059628
1	ZF4	2.153651579	2.251447883	2.283342159	2.314887805	2.348149567	2.370279750	2.386749513	2.410760839	2.435922848
1	ZF5	5.662814979	6.011951899	6.128297123	6.244319909	6.367491159	6.449814800	6.511229401	6.600928043	6.695047337

(5) P 6+4	TH = 49	SBE = 1.325012993						MCPER FUNCTIONS	
	RC = 02	RC = 05	RC = 07	RC = 10	RC = 15	RC = 20	RC = 25	RC = 35	RC = 50
AMAX DB	.0017375255	.0108709564	.0213327376	.0436480540	.0988323393	.1772876696	.2802872360	.5675287486	1.249387366
AMIN DB	45.05728185	58.08969440	62.95190955	68.15791729	74.16432651	78.51355514	81.96689709	87.39806576	93.74331414
MU I K	.5586450D-2	.1245992D-2	.7118763D-3	.3909346D-3	.1957869D-3	.1186649D-3	.7973613D-4	.4266745D-4	.2055106D-4
1 I PF1	1.054273221	.7767568187	.6919593095	.6098385560	.5249543280	.4696360446	.4293259246	.3725801135	.3173432797
1 I PQ1	.5595939452	.5671457586	.5740501467	.5849653859	.6040858687	.6244747034	.6464641013	.6968856605	.7974433878
1 I PF2	1.149000340	.9870346399	.9339454159	.8816615084	.8271843367	.7916092204	.7656499121	.7288620458	.6920262498
1 I PQ2	1.142694485	1.166945823	1.204155829	1.264814611	1.369480815	1.476940581	1.588305107	1.829765411	2.273936307
3 I PFC	1.268073408	1.204290253	1.178766355	1.151138500	1.119485343	1.097083111	1.079794074	1.053321739	1.025977475
3 I PQC	5.117848765	4.337573328	4.199584455	4.121991375	4.115635903	4.166689193	4.242784055	4.434363598	4.793535724
1 I ZF1	1.332146268	1.332716786	1.332952094	1.333205242	1.333492519	1.333694507	1.333850177	1.334034936	1.334340352
1 I ZF2	1.391576093	1.400439519	1.403451902	1.406497914	1.409785633	1.412018292	1.413704102	1.416199666	1.418863435
1 I ZF3	1.555493642	1.591981573	1.603740126	1.615343038	1.627573533	1.635721649	1.641796592	1.650676833	1.660020883
1 I ZF4	2.092976562	2.190911217	2.223032758	2.254876687	2.288524112	2.310947607	2.327653216	2.352033410	2.377611633
1 I ZF5	5.475246164	5.823935123	5.940975114	6.058045937	6.182678124	6.266158191	6.328520923	6.419724351	6.515561331

(5) P 6+4	TH = 50	SBE = 1.305407289						MCPER FUNCTIONS	
	RC = 02	RC = 05	RC = 07	RC = 10	RC = 15	RC = 20	RC = 25	RC = 35	RC = 50
AMAX DB	.0017375255	.0108709564	.0213327376	.0436480540	.0988323393	.1772876696	.2802872360	.5675287486	1.249387366
AMIN DB	43.09023543	56.08756446	60.94136495	66.14027377	72.14040097	76.48608397	79.93709716	85.36532244	91.70803477
MU I K	.7006292D-2	.1568996D-2	.8972878D-3	.4931583D-3	.2471610D-3	.1498635D-3	.1007268D-3	.5391801D-4	.2597755D-4
1 I PF1	1.071911718	.7870104132	.7004568734	.6168359232	.5305896873	.4744768827	.4336296307	.3761917732	.3203156702
1 I PQ1	.5615933089	.5683850217	.5751331433	.5859250635	.6049457748	.6252856884	.6472491082	.6976538918	.7982381540
1 I PF2	1.153832950	.9921775501	.9389796594	.8865282094	.8318329951	.7961009041	.7700237012	.7330695053	.6960766284
1 I PQ2	1.169455035	1.185851306	1.221511266	1.281077590	1.385117537	1.492535748	1.604132770	1.846524172	2.292980135
3 I PFC	1.258461360	1.198399607	1.174026284	1.147485485	1.116909318	1.095176555	1.078357876	1.053023977	1.025782463
3 I PQC	5.434966706	4.537413545	4.374432616	4.276841193	4.253964848	4.296547022	4.367709502	4.554431480	4.912520597
1 I ZF1	1.312218541	1.312746797	1.312972833	1.313219657	1.313500049	1.313699101	1.313853069	1.314036052	1.314340463
1 I ZF2	1.368265756	1.377000696	1.379986180	1.383013140	1.386289118	1.388518820	1.390204990	1.392705046	1.395378614
1 I ZF3	1.522290095	1.558808254	1.570618512	1.582290032	1.594610833	1.602829053	1.608961363	1.617933256	1.627383745
1 I ZF4	2.034734584	2.132638290	2.164945883	2.197053541	2.231056454	2.253756502	2.270687103	2.295422666	2.321405737
1 I ZF5	5.294832719	5.642442739	5.760015024	5.877997253	6.003972971	6.088546454	6.151818078	6.244430340	6.342000397

(5) P 6+4		TH = 51 SBE = 1.286759566								MCPER FUNCTIONS
		RC = 02	RC = 05	RC = 07	RC = 10	RC = 15	RC = 20	RC = 25	RC = 35	RC = 50
AMAX DB		.0017375255	.0108709564	.0213327376	.0436480540	.0988323393	.1772876696	.2802872360	.5675287486	1.249387366
AMIN DB		41.15448287	54.11376761	58.95844393	64.14965406	70.14296736	74.48480399	77.93329072	83.35832248	89.69828301
MU	K	.8755397D-2	.1969299D-2	.1127399D-2	.6201793D-3	.3110653D-3	.1886947D-3	.1268631D-3	.6793348D-4	.3274054D-4
1	PF1	1.090601543	.7977994917	.7093792473	.6241686735	.5364839385	.4795342822	.4381223042	.3799376318	.3234123691
1	PQ1	.5637822279	.5697150262	.5762892552	.5869446801	.6058553487	.6261412749	.6480758250	.6984610877	.7990715373
1	PF2	1.158577336	.9973989634	.9441183779	.8915153877	.8366109754	.8007243748	.7745298027	.7374036381	.7002569989
1	PQ2	1.198662805	1.206183224	1.240097750	1.298426667	1.401735668	1.509071269	1.620887577	1.864225683	2.313053647
3	PFC	1.248787420	1.192416908	1.169198786	1.143754815	1.114270235	1.093218383	1.076879160	1.052197419	1.025573041
3	PQC	5.788553779	4.756522098	4.565202237	4.445017232	4.403512532	4.436531509	4.502098829	4.683327307	5.039789522
1	ZF1	1.293267682	1.293750700	1.293966662	1.294204425	1.294479214	1.294674886	1.294826837	1.295057610	1.295310592
1	ZF2	1.346065702	1.354658608	1.357612600	1.360615980	1.363875527	1.366099234	1.367783561	1.370285002	1.372965271
1	ZF3	1.490525832	1.527020822	1.538868911	1.550596623	1.562995872	1.571276949	1.577461548	1.586518166	1.596068446
1	ZF4	1.978816834	2.076512356	2.108962368	2.141297024	2.175623218	2.198581881	2.215725828	2.240302248	2.267177909
1	ZF5	5.121244714	5.467131543	5.585066830	5.703816873	5.831011687	5.916611316	5.980749709	6.074321181	6.173986583

(5) P 6+4		TH = 52 SBE = 1.269018215								MCPER-FUNCTIONS
		RC = 02	RC = 05	RC = 07	RC = 10	RC = 15	RC = 20	RC = 25	RC = 35	RC = 50
AMAX DB		.0017375255	.0108709564	.0213327376	.0436480540	.0988323393	.1772876696	.2802872360	.5675287486	1.249387366
AMIN DB		39.24881524	52.16685044	57.00163694	62.18450111	68.17042635	72.50809196	75.95383871	81.37540681	87.71238250
MU	K	.1090333D-1	.2464095D-2	.1412271D-2	.7776335D-3	.3903720D-3	.2369166D-3	.1593339D-3	.8535514D-4	.4115105D-4
1	PF1	1.110422927	.8091610599	.7187544495	.6318578566	.5426524756	.4848206650	.4428145038	.3838559946	.3266398381
1	PQ1	.5661861029	.5711453043	.5775255605	.5880295850	.6068186595	.6270449182	.6489473549	.6993099879	.7999461098
1	PF2	1.163205267	1.002690930	.9493577368	.8966221980	.8415196054	.8054819522	.7791710842	.7418828756	.7045711594
1	PQ2	1.230623426	1.228090580	1.260036406	1.316962487	1.419420536	1.526625509	1.638644411	1.882942754	2.334232865
3	PFC	1.239065659	1.186346926	1.164286556	1.139947643	1.111568140	1.091208181	1.075357295	1.051341280	1.025348434
3	PQC	6.184127373	4.997384137	4.773841116	4.628065977	4.565509717	4.587712559	4.646924214	4.821761693	5.176125261
1	ZF1	1.275241538	1.275676323	1.275881404	1.276110367	1.276377830	1.276569669	1.276719283	1.276947403	1.277198522
1	ZF2	1.324918135	1.333355810	1.336273747	1.339249007	1.342487382	1.344702002	1.346382233	1.348881868	1.351565641
1	ZF3	1.460133690	1.496549169	1.508420476	1.520191318	1.532656552	1.540992869	1.547224418	1.556358475	1.566001542
1	ZF4	1.925124854	2.022427016	2.054973305	2.087495932	2.122111050	2.145309125	2.162653925	2.188055595	2.214810619
1	ZF5	4.954178695	5.297686749	5.415809080	5.535176346	5.663458609	5.750012670	5.814972684	5.910399848	6.011169527

(5) P 6+4 — TH = 53 — SBE = 1.252135658 — MCPER FUNCTIONS

MU		RC = 02	RC = 05	RC = 07	RC = 10	RC = 15	RC = 20	RC = 25	RC = 35	RC = 50
	AMAX DB	.0017375255	.0108709564	.0213327376	.0436480540	.0988323393	.1772876696	.2802872360	.5675287486	1.249387366
	AMIN DB	37.37212602	50.24543316	55.06950282	60.24332199	66.22123900	70.55438280	73.99715887	79.41497134	85.74871029
	K	.1353299D-1	.3074173D-2	.1764107D-2	.9723753D-3	.4885827D-3	.2966749D-3	.1995915D-3	.1069674D-3	.5158988D-4
1	PF1	1.131463138	.8211357145	.7286131897	.6399265065	.5491121058	.4903495721	.4477177248	.3879458877	.3300050878
1	PQ1	.5688347695	.5726867557	.5788500909	.5891858044	.6078402596	.6280004681	.6498671450	.7002036244	.8008647079
1	PF2	1.167685796	1.008044233	.9546930804	.9018473332	.8465600340	.8103759153	.7839504517	.7464957748	.7090230945
1	PQ2	1.265692019	1.251743277	1.281464588	1.336798583	1.438267889	1.545286115	1.657486833	1.902756368	2.356602074
3	PFC	1.229310902	1.180194906	1.159292632	1.136065339	1.108803194	1.089145592	1.073791671	1.050454766	1.025107839
3	PQC	6.628249258	5.262896057	5.002605434	4.827768269	4.741365255	4.751310216	4.803291201	4.970759496	5.322411026
1	ZF1	1.258092036	1.258475566	1.258668961	1.258888383	1.259147790	1.259335341	1.259482293	1.259707309	1.259956120
1	ZF2	1.304769744	1.313039374	1.315916733	1.318859331	1.322071747	1.324274138	1.325947972	1.328442529	1.331126510
1	ZF3	1.431052691	1.467329193	1.479208309	1.491008525	1.503526627	1.511910168	1.518183050	1.527386871	1.537115339
1	ZF4	1.873569597	1.970285342	2.002879161	2.035548325	2.070415724	2.093832656	2.111364902	2.137075035	2.164195150
1	ZF5	4.793354774	5.133819215	5.251946167	5.371773055	5.501003727	5.588435888	5.654169200	5.750894416	5.853223740

(5) P 6+4 — TH = 54 — SBE = 1.236067977 — MCPER FUNCTIONS

MU		RC = 02	RC = 05	RC = 07	RC = 10	RC = 15	RC = 20	RC = 25	RC = 35	RC = 50
	AMAX DB	.0017375255	.0108709564	.0213327376	.0436480540	.0988323393	.1772876696	.2802872360	.5675287486	1.249387366
	AMIN DB	35.52341218	48.34820400	53.16066257	58.32468111	64.29391946	68.62216214	72.06171803	77.47545906	83.80568876
	K	.1674285D-1	.3824633D-2	.2197692D-2	.1212735D-2	.6099638D-3	.3705885D-3	.2494101D-3	.1337294D-3	.6452315D-4
1	PF1	1.153816813	.8337680421	.7389891793	.6483998727	.5558812142	.4961357922	.4528445070	.3922171402	.3335157395
1	PQ1	.5717634371	.5743518897	.5802719900	.5904201486	.6089252575	.6290122257	.6508390350	.7011453611	.8018304675
1	PF2	1.171985167	1.013448232	.9601188174	.9071889476	.8517331888	.8154084781	.7888708376	.7512510180	.7136169838
1	PQ2	1.304282978	1.277335751	1.304538587	1.358063445	1.458385508	1.565151411	1.677508349	1.923756847	2.330254970
3	PFC	1.219538719	1.173966602	1.154220427	1.132109521	1.105975690	1.087030328	1.072181714	1.049537076	1.024850426
3	PQC	7.128761882	5.556448390	5.254119546	5.046183011	4.932697969	4.928720820	4.972461461	5.131428757	5.479646636
1	ZF1	1.241774807	1.242104032	1.242284937	1.242494077	1.242744697	1.242927499	1.243071459	1.243292912	1.243538961
1	ZF2	1.285571297	1.293660473	1.296492783	1.299398186	1.302579819	1.304766789	1.306431880	1.308918006	1.311598797
1	ZF3	1.403227545	1.439302315	1.451172973	1.462988052	1.475545196	1.483967516	1.490275811	1.499541306	1.509347380
1	ZF4	1.824070575	1.919999089	1.952589004	1.985360746	2.020441357	2.044055147	2.061760455	2.087760990	2.115230785
1	ZF5	4.638514024	4.975262957	5.093205875	5.213327796	5.343360287	5.431589419	5.498044374	5.596005644	5.699846168

(5) P 6+4			TH = 55 SBE = 1.220774589								MCPER FUNCTIONS
			RC = 02	RC = 05	RC = 07	RC = 10	RC = 15	RC = 20	RC = 25	RC = 35	RC = 50
AMAX DB			.0017375255	.0108709564	.0213327376	.0436480540	.0988323393	.1772876696	.2802872360	.5675287486	1.249387366
AMIN DB			33.70177832	46.47391433	51.27379352	56.42719379	62.38702803	66.70995910	70.14602473	75.55535244	81.88177780
MU		K	.2064957D-1	.4745744D-2	.2730928D-2	.1508830D-2	.7597113D-3	.4618512D-3	.3109559D-3	.1668140D-3	.8052136D-4
1		PF1	1.177586124	.8471070610	.7499194816	.6573056834	.5629799519	.5021955098	.4582085570	.3966804770	.3371800944
1		PQ1	.5750133615	.5761551196	.5818017063	.5917403403	.6100794032	.6300350103	.6518673144	.7021389402	.8028468636
1		PF2	1.176066749	1.018890707	.9656282922	.9126445687	.8570397275	.8205817619	.7939351852	.7561524127	.7183572127
1		PQ2	1.346882063	1.305091369	1.329436888	1.380902994	1.479895113	1.586332046	1.698813924	1.946045221	2.405296010
3		PFC	1.209765413	1.167668309	1.149073756	1.128082077	1.103086063	1.084862182	1.070526889	1.048587407	1.024575331
3		PQC	7.695086712	5.882028485	5.531449565	5.285700630	5.141375398	5.121548661	5.155880269	5.305033460	5.648967841
1		ZF1	1.226248859	1.226520693	1.226688307	1.226886426	1.227127526	1.227305115	1.227445751	1.227663174	1.227905997
1		ZF2	1.267277278	1.275174016	1.277956870	1.280820561	1.283966561	1.286134875	1.287788832	1.290263091	1.292937194
1		ZF3	1.376608203	1.412415038	1.424260052	1.436074671	1.448656268	1.457108456	1.463445918	1.472764546	1.482639989
1		ZF4	1.776555090	1.871488009	1.904019816	1.936847539	1.972099726	1.995886826	2.013749759	2.040021256	2.067824088
1		ZF5	4.489416161	4.821772895	4.939337185	5.059582614	5.190262637	5.279202637	5.346324089	5.445454804	5.550754013

(5) P 6+4			TH = 56 SBE = 1.206217949								MCPER FUNCTIONS
			RC = 02	RC = 05	RC = 07	RC = 10	RC = 15	RC = 20	RC = 25	RC = 35	RC = 50
AMAX DB			.0017375255	.0108709564	.0213327376	.0436480540	.0988323393	.1772876696	.2802872360	.5675287486	1.249387366
AMIN DB			31.90644459	44.62137449	49.40762410	54.54952027	60.49916458	64.81633931	68.24862205	73.65316591	79.97546709
MU		K	.2539088D-1	.5873964D-2	.3385469D-2	.1872941D-2	.9441517D-3	.5743585D-3	.3868734D-3	.2076547D-3	.1002828D-3
1		PF1	1.202880697	.8612067089	.7614449068	.6666744465	.5704304509	.5085464743	.4638248896	.4013476265	.3410072144
1		PQ1	.5786358180	.5781131215	.5834512240	.5931551674	.6113091893	.6312242388	.6529567883	.7031385362	.8039177584
1		PF2	1.179891023	1.024357670	.9712136417	.9182109973	.8624799794	.8258977610	.7991464297	.7612038889	.7232483819
1		PQ2	1.394061411	1.335267765	1.356364100	1.405483545	1.502934640	1.608952952	1.721521766	1.969734861	2.431841997
3		PFC	1.200008001	1.161306892	1.143856869	1.123985191	1.100134917	1.082641039	1.068826711	1.047604956	1.024281663
3		PQC	8.338603067	6.244348482	5.838194470	5.549108843	5.369561144	5.331644489	5.355209811	5.493020990	5.831669540
1		ZF1	1.211476279	1.211687591	1.211841119	1.212027482	1.212258331	1.212430242	1.212567218	1.212780140	1.213019264
1		ZF2	1.249845571	1.257538317	1.260267385	1.263084878	1.266190373	1.268336758	1.269977148	1.272436027	1.275099841
1		ZF3	1.351149453	1.386618573	1.398419770	1.410217730	1.422808365	1.431281009	1.437641039	1.447003773	1.456939870
1		ZF4	1.730957534	1.824679238	1.857095895	1.889930241	1.925309647	1.949244862	1.967248850	1.993770383	2.021888266
1		ZF5	4.345837471	4.673122827	4.790108283	4.910298848	5.041464292	5.131023901	5.198753055	5.298981732	5.405682759

(5) P 6+4	TH = 57	SBE = 1.192363293						MCPER FUNCTIONS	
	RC = 02	RC = 05	RC = 07	RC = 10	RC = 15	RC = 20	RC = 25	RC = 35	RC = 50
AMAX DB	.0017375255	.0108709564	.0213327376	.0436480540	.0988323393	.1772876696	.2802872360	.5675287486	1.249387366
AMIN DB	30.13675975	42.78945037	47.56092900	52.69035993	58.62896204	62.93989807	66.36808048	71.76743850	78.08526839
MU I K	.3112877D-1	.7253164D-2	.4187488D-2	.2319968D-2	.1170987D-2	.7128614D-3	.4803922D-3	.2580050D-3	.1246627D-3
1 I PF1	1.229817132	.8761263778	.7736104562	.6765397914	.5782570716	.5152081951	.4697099894	.4062314429	.3450070141
1 I PQ1	.5826889591	.5802452746	.5852343453	.5946746671	.6126219732	.6324360193	.6541128576	.7042988203	.8050474555
1 I PF2	1.183415623	1.029833183	.9768656351	.9238841923	.8680538792	.8313583028	.8045074742	.7664094949	.7282953186
1 I PQ2	1.446498286	1.368163376	1.385555736	1.431995397	1.527660974	1.633155691	1.745765452	1.994953403	2.460023956
3 I PFC	1.190284186	1.154889825	1.138574490	1.119821375	1.097123041	1.080366892	1.067080752	1.046588927	1.023968498
3 I PQC	9.073131851	6.649005813	6.178599599	5.839674039	5.619773058	5.561152637	5.572369819	5.697055486	6.029233838
1 I ZF1	1.197421975	1.197569580	1.197708230	1.197882110	1.198101979	1.198267747	1.198400726	1.198608669	1.198843613
1 I ZF2	1.233237172	1.240714809	1.243385847	1.246152695	1.249212806	1.251333952	1.252958301	1.255398209	1.258048030
1 I ZF3	1.326810557	1.361868500	1.373606656	1.385370816	1.397954180	1.406437325	1.412812940	1.422210230	1.432197749
1 I ZF4	1.687218740	1.779506746	1.811748311	1.844537055	1.879996448	1.904052816	1.922180070	1.948929105	1.977342593
1 I ZF5	4.207568979	4.529103582	4.645304762	4.765255362	4.896736182	4.986813815	5.055093053	5.156343060	5.264384390

(5) P 6+4	TH = 58	SBE = 1.179178403						MCPER FUNCTIONS	
	RC = 02	RC = 05	RC = 07	RC = 10	RC = 15	RC = 20	RC = 25	RC = 35	RC = 50
AMAX DB	.0017375255	.0108709564	.0213327376	.0436480540	.0988323393	.1772876696	.2802872360	.5675287486	1.249387366
AMIN DB	28.39222091	40.97706086	45.73252480	50.84844581	56.77508011	61.07925358	64.50299086	69.89672631	76.20970776
MU I K	.3805300D-1	.8936078D-2	.5168610D-2	.2867991D-2	.1449593D-2	.8831558D-3	.5954571D-3	.3200101D-3	.1547087D-3
1 I PF1	1.258517928	.8919314928	.7864658223	.6869388604	.5864866862	.5222021664	.4758819986	.4113460492	.3491903675
1 I PQ1	.5872451727	.5825742043	.5871670344	.5963103495	.6140261219	.6337272635	.6553416122	.7054750357	.8062407660
1 I PF2	1.186595447	1.035299147	.9825734962	.9296591401	.8737608870	.8369649991	.8100211614	.7717733917	.7335030866
1 I PQ2	1.504998627	1.404125475	1.417284047	1.460657176	1.554253244	1.659101267	1.771696487	2.021845050	2.499989364
3 I PFC	1.180612322	1.148425220	1.133231848	1.115593497	1.094051433	1.078039858	1.065288658	1.045538532	1.023634887
3 I PQC	9.915556788	7.102685814	6.557698963	6.161242642	5.894955204	5.812569026	5.809587414	5.919058547	6.243364128
1 I ZF1	1.184053432	1.184134078	1.184257069	1.184417744	1.184625912	1.184785073	1.184913717	1.185116201	1.185346479
1 I ZF2	1.217415942	1.224667776	1.227276637	1.229988446	1.232998293	1.235090862	1.236696660	1.239113930	1.241745953
1 I ZF3	1.303554922	1.338124475	1.349779250	1.361491457	1.374050276	1.382533377	1.388917184	1.398338905	1.408369057
1 I ZF4	1.645285385	1.735910849	1.767914443	1.800602378	1.836091489	1.860240160	1.878471586	1.905423847	1.934111904
1 I ZF5	4.074414829	4.389521332	4.504727975	4.624246940	4.755865069	4.846368637	4.915121348	5.017310613	5.126625763

(5) P 6+4 — TH = 59 SBE = 1.166633397 — MCPER FUNCTIONS

6+4	RC = 02	RC = 05	RC = 07	RC = 10	RC = 15	RC = 20	RC = 25	RC = 35	RC = 50
AMAX DB	.0017375255	.0108709564	.0213327376	.0436480540	.0988323393	.1772876696	.2802872360	.5675287436	1.249387366
AMIN DB	26.67250198	39.18317632	43.92126611	49.02253937	54.93619893	59.23304006	62.65195716	68.03959495	74.34731757
MU I K	.4638472D-1	.1098604D-1	.6367027D-2	.3538939D-2	.1791390D-2	.1092315D-2	.7368891D-3	.3962965D-3	.1917053D-3
1 I PF1	1.289109503	.9086941311	.8000659484	.6979127551	.5951490075	.5295521273	.4823609336	.4167070022	.3535692316
1 I PQ1	.5923915854	.5851264575	.5892678402	.5980754695	.6155311877	.6351058214	.6566499438	.7067230875	.8075030851
1 I PF2	1.189382852	1.040735087	.9883247067	.9355297051	.8795998970	.8427191886	.8156902375	.7772998430	.7388769977
1 I PQ2	1.570526771	1.443560124	1.451865201	1.491721148	1.582916829	1.686973529	1.799487371	2.050573338	2.521904822
3 I PFC	1.171011366	1.141921863	1.127834716	1.111304820	1.090921324	1.075660196	1.063450154	1.044453003	1.023279847
3 I PQC	10.88662662	7.613419279	6.981494886	6.518368176	6.198567423	6.088813017	6.069458638	6.161259241	6.476026750
1 I ZF1	1.171340507	1.171350865	1.171457421	1.171604182	1.171799937	1.171952028	1.172076002	1.172272543	1.172497660
1 I ZF2	1.202348387	1.209364125	1.211906769	1.214559207	1.217513925	1.219574554	1.221159257	1.223550150	1.226160464
1 I ZF3	1.281349793	1.315349964	1.326899841	1.338540862	1.351056821	1.359528695	1.365912851	1.375348264	1.385408650
1 I ZF4	1.605109432	1.693837782	1.725537559	1.758066393	1.793531751	1.817741863	1.836056955	1.863186281	1.892126141
1 I ZF5	3.946190877	4.254196058	4.368193532	4.487082813	4.618652095	4.709468833	4.778629236	4.881669940	4.992187116

(5) P 6+4 — TH = 60 SBE = 1.154700538 — MCPER FUNCTIONS

6+4	RC = 02	RC = 05	RC = 07	RC = 10	RC = 15	RC = 20	RC = 25	RC = 35	RC = 50
AMAX DB	.0017375255	.0108709564	.0213327376	.0436480540	.0988323393	.1772876696	.2802872360	.5675287486	1.249387366
AMIN DB	24.97749376	37.40681839	42.12604224	47.21142532	53.11101274	57.39990082	60.81358914	66.19461158	72.49662810
MU I K	.5638003D-1	.1347904D-1	.7828848D-2	.4359420D-2	.2210291D-2	.1348978D-2	.9105851D-3	.4900828D-3	.2372294D-3
1 I PF1	1.321718927	.9264936663	.8144716523	.7095070481	.6042769678	.5372843638	.4891689373	.4223314844	.3581567904
1 I PQ1	.5982344066	.5879333456	.5915584206	.5999853592	.6171481215	.6365806430	.6580456809	.7080496505	.8088404836
1 I PF2	1.191727956	1.046117927	.9941047896	.9414884601	.8855691292	.8486218681	.8215173096	.7829932031	.7444226219
1 I PQ2	1.644243182	1.486944567	1.489668151	1.525479747	1.613888228	1.716983305	1.829335330	2.081324467	2.555959254
3 I PFC	1.161500831	1.135389248	1.122389456	1.106959038	1.087734212	1.073228327	1.061565069	1.043331592	1.022902371
3 I PQC	12.01199661	8.190912266	7.457186459	6.916471721	6.534697588	6.393317035	6.355024926	6.426255933	6.729502224
1 I ZF1	1.159255231	1.159191884	1.159281237	1.159413389	1.159596028	1.159740595	1.159859561	1.160049678	1.160269136
1 I ZF2	1.188003456	1.194773175	1.197245677	1.199834487	1.202729234	1.204754542	1.206315577	1.208676284	1.211260877
1 I ZF3	1.260165972	1.293512014	1.304934244	1.316483693	1.328937357	1.337386131	1.343762306	1.353199998	1.363280561
1 I ZF4	1.566647597	1.653239298	1.684566443	1.716874707	1.752259469	1.776498009	1.794874748	1.822152936	1.851319939
1 I ZF5	3.822723474	4.122960130	4.235529916	4.353585313	4.484911454	4.575927748	4.645420708	4.749218965	4.860860690

(5) P 6+4		TH = 61 RC = 02	SBE = 1.143354068 RC = 05	RC = 07	RC = 10	RC = 15	RC = 20	RC = 25	MCPER FUNCTIONS RC = 35	RC = 50
AMAX DB		.0017375255	.0108709564	.0213327376	.0436480540	.0988323393	.1772876696	.2802872360	.5675287486	1.249387366
AMIN DB		23.30735885	35.64706150	40.34577456	45.41390675	51.29822339	55.57848094	58.98649454	64.36033660	70.65615874
MU	K	.68333250D-1	.1650619D-1	.9609732D-2	.5361727D-2	.2723258D-2	.1663704D-2	.1123764D-2	.6053174D-3	.2932190D-3
1	PF1	1.356468784	.9454174188	.8297503162	.7217723660	.6139071592	.5454280609	.4963305742	.4282385290	.3629676233
1	PQ1	.6049038612	.5910320069	.5940641941	.6020578391	.6188895326	.6381619758	.6595377517	.7094622982	.8102598180
1	PF2	1.193579067	1.051421754	.9998970717	.9475264928	.8916660036	.8546736103	.8275047923	.7888578995	.7501457980
1	PQ2	1.727552557	1.534842793	1.531125698	1.562273641	1.647441067	1.749373458	1.861466851	2.114311345	2.592367779
3	PFC	1.152100720	1.128837608	1.116903054	1.102560315	1.084491886	1.070744861	1.059633345	1.042173587	1.022501423
3	PQC	13.32358740	8.846971658	7.993462265	7.362046032	6.908203326	6.730139134	6.669868798	6.717093247	7.006448681
1	ZF1	1.147771636	1.147631069	1.147702464	1.147819323	1.147988161	1.148124756	1.148238382	1.148421591	1.148634890
1	ZF2	1.174352377	1.180866472	1.183265031	1.185786039	1.188616007	1.190602605	1.192137369	1.194464017	1.197018777
1	ZF3	1.239977555	1.272581048	1.283851600	1.295287869	1.307658600	1.316071653	1.322430988	1.331858817	1.341947782
1	ZF4	1.529860833	1.614072321	1.644955076	1.676978042	1.712221822	1.736453485	1.754868221	1.782264855	1.811632274
1	ZF5	3.703848424	3.995657007	4.106577199	4.223588613	4.354469157	4.445565379	4.515311220	4.619766730	4.732449449

(5) P 6+4		TH = 62 RC = 02	SBE = 1.132570051 RC = 05	RC = 07	RC = 10	RC = 15	RC = 20	RC = 25	MCPER FUNCTIONS RC = 35	RC = 50
AMAX DB		.0017375255	.0108709564	.0213327376	.0436480540	.0988323393	.1772876696	.2802872360	.5675287486	1.249387366
AMIN DB		21.66260549	33.90303679	38.57941473	43.62880030	49.49653369	53.76741973	57.16927097	62.53531471	68.82440838
MU	K	.8257902D-1	.2017661D-1	.1177685D-1	.6585064D-2	.3350991D-2	.2049411D-2	.1385276D-2	.7468515D-3	.3620592D-3
1	PF1	1.393469427	.9655612696	.8459766416	.7347650566	.6240803448	.5540157142	.5038731754	.4344492939	.3680179032
1	PQ1	.6125605339	.5944667534	.5968151585	.6043137276	.6207700082	.6398616048	.6611363825	.7109696590	.8117688624
1	PF2	1.194883266	1.056617587	1.005682424	.9536331873	.8978869923	.8608744658	.8336548434	.7948984095	.7560526435
1	PQ2	1.822165427	1.587925212	1.576748362	1.602501790	1.683893502	1.784425108	1.896143257	2.149778533	2.631376426
3	PFC	1.142831455	1.122277946	1.111383169	1.098113333	1.081196471	1.068210622	1.057655060	1.040978320	1.022075945
3	PQC	14.86136634	9.596058733	8.600878523	7.862917303	7.324892335	7.104105494	7.018234537	7.037358583	7.309980962
1	ZF1	1.136865585	1.136644195	1.136696883	1.136797785	1.136952154	1.137080336	1.137188296	1.137364119	1.137570755
1	ZF2	1.161368495	1.167617621	1.169938563	1.172387693	1.175148122	1.177092616	1.178598487	1.180837139	1.133407853
1	ZF3	1.220761673	1.252530685	1.263624205	1.274924400	1.287190266	1.295554171	1.301887236	1.311292264	1.321377070
1	ZF4	1.494713825	1.576298622	1.606662346	1.638331956	1.673370658	1.697557694	1.715985023	1.743467284	1.773006132
1	ZF5	3.589410097	3.872140032	3.981185846	4.096937565	4.227161890	4.318212225	4.388126532	4.493132222	4.606765870

(5) P 6+4		TH = 63 SBE = 1.122326238							MCPER FUNCTIONS	
		RC = 02	RC = 05	RC = 07	RC = 10	RC = 15	RC = 20	RC = 25	RC = 35	RC = 50
AMAX DB		.0017375255	.0108709564	.0213327376	.0436480540	.0988323393	.1772876696	.2802872360	.5675287486	1.249387366
AMIN DB		20.04418428	32.17393910	36.82594409	41.85493162	47.70464040	51.96534258	55.36049703	60.71806525	66.99984504
MU	K	.9949260D-1	.2462085D-1	.1441129D-1	.8077062D-2	.4118774D-2	.2521929D-2	.1705985D-2	.9206546D-3	.4466916D-3
1	PF1	1.432807662	.9870301792	.8632334614	.7485479476	.6348420551	.5630836131	.5118272456	.4409873234	.3733256302
1	PQ1	.6214035308	.5982907910	.5998469256	.6067774787	.6228065091	.6416931484	.6628533408	.7125816062	.8133764676
1	PF2	1.195587181	1.061673137	1.011438982	.9597959754	.9042274471	.8672238457	.8399692853	.8011192303	.7621495623
1	PQ2	1.930177318	1.646993727	1.627141929	1.646634061	1.723617458	1.822465345	1.933667591	2.188008321	2.673267893
3	PFC	1.133713797	1.115722064	1.105938176	1.093623336	1.077850457	1.065626675	1.055630454	1.039745176	1.021624858
3	PQC	16.67569151	10.45601539	9.292351800	8.428583714	7.791753737	7.520992258	7.405181570	7.391304097	7.643770006
1	ZF1	1.126514623	1.126208730	1.126241973	1.126326273	1.126465527	1.126584868	1.126686841	1.126854805	1.127054277
1	ZF2	1.149027142	1.155002139	1.157241920	1.159615199	1.162301393	1.164200392	1.165674731	1.167921395	1.170403752
1	ZF3	1.202498243	1.233337567	1.244227359	1.255367238	1.267504933	1.275805390	1.282102130	1.291470556	1.301537788
1	ZF4	1.461174461	1.539884531	1.569651792	1.600896619	1.635662266	1.659764322	1.678176953	1.705709416	1.735388225
1	ZF5	3.479260654	3.752271331	3.859215600	3.973486597	4.102835927	4.193708209	4.263701623	4.369143260	4.483630791

(5) P 6+4		TH = 64 SBE = 1.112601940							MCPER FUNCTIONS	
		RC = 02	RC = 05	RC = 07	RC = 10	RC = 15	RC = 20	RC = 25	RC = 35	RC = 50
AMAX DB		.0017375255	.0108709564	.0213327376	.0436480540	.0988323393	.1772876696	.2802872360	.5675287486	1.249387366
AMIN DB		18.45361193	30.45903860	35.08437472	40.09113112	45.92122685	50.17085216	53.55872267	58.90707146	65.18089423
MU	K	.1194867D 0	.2999494D-1	.1761089D-1	.9895630D-2	.5057532D-2	.3100683D-2	.2099249D-2	.1134087D-2	.5507510D-3
1	PF1	1.474530676	1.009938508	.8816125909	.7631912088	.6462432839	.5726724097	.5202269436	.4478790174	.3789109096
1	PQ1	.6316809640	.6025684338	.6032020396	.6094779860	.6250188620	.6436724240	.6647022336	.7143094907	.8150927562
1	PF2	1.195637998	1.066552589	1.017141844	.9660000565	.9106813970	.8737203811	.8464495091	.8075248399	.7684432522
1	PQ2	2.054170734	1.713013952	1.683029822	1.695227195	1.767050219	1.863876871	1.974393181	2.229328257	2.718368663
3	PFC	1.124768758	1.109182590	1.100277211	1.089096184	1.074456752	1.062994367	1.053559949	1.038473612	1.021147063
3	PQC	18.83040656	11.44902541	10.08380688	9.070657284	8.317257537	7.987759609	7.836781108	7.784002171	8.012168758
1	ZF1	1.116697818	1.116303711	1.116316782	1.116383856	1.116507373	1.116617462	1.116713136	1.116872777	1.117064585
1	ZF2	1.137305505	1.142997319	1.145152529	1.147446098	1.150053431	1.151903560	1.153343715	1.155544350	1.157983943
1	ZF3	1.185169703	1.214981220	1.225639235	1.236593158	1.248577911	1.256799682	1.263049369	1.272366447	1.282401754
1	ZF4	1.429213262	1.504800659	1.533891379	1.564636605	1.599057182	1.623031138	1.641399749	1.668944155	1.698728739
1	ZF5	3.373259306	3.635920782	3.740534428	3.853098669	3.981346103	4.071901640	4.141879649	4.247635425	4.362872302

(5) P 6+4		TH = 65	SBE = 1.103377919						MCPER FUNCTIONS	
		RC = 02	RC = 05	RC = 07	RC = 10	RC = 15	RC = 20	RC = 25	RC = 35	RC = 50
AMAX DB		.0017375255	.0108709564	.0213327376	.0436480540	.0988323393	.1772876696	.2802872360	.5675287486	1.249387366
AMIN DB		16.89312348	28.75769849	33.35375288	38.33623014	44.14495508	48.38251881	51.76245852	57.10076856	63.36592591
MU	K	.1430026D 0	.3648506D-1	.2149376D-1	.1211124D-1	.6205149D-2	.3809553D-2	.2581529D-2	.1396245D-2	.6787404D-3
1	PF1	1.518623926	1.034409985	.9012156847	.7787733236	.6583413023	.5828277914	.5291106536	.4551539724	.3847962837
1	PQ1	.6437033589	.6073779782	.6069316723	.6124496044	.6274303792	.6458179043	.6666988794	.7161664284	.8169293610
1	PF2	1.194984758	1.071216398	1.022762751	.9722280824	.9172413120	.8803617556	.8530963572	.8141196456	.7749407079
1	PQ2	2.197346774	1.787156956	1.745281878	1.748945173	1.814709098	1.909110128	2.018734398	2.274120549	2.767057830
3	PFC	1.116017488	1.102672998	1.094710219	1.084538408	1.071018726	1.060315362	1.051444186	1.037163173	1.020641453
3	PQC	21.40693657	12.60289876	10.99503695	9.803446145	8.911745956	8.512855922	8.320370580	8.221545331	8.420373057
1	ZF1	1.107395609	1.106909628	1.106901811	1.106951059	1.107058248	1.107158689	1.107247763	1.107398628	1.107582279
1	ZF2	1.126182513	1.131582118	1.133649476	1.135859598	1.138383532	1.140181434	1.141584750	1.143735269	1.146127603
1	ZF3	1.168760723	1.197443908	1.207840766	1.218581655	1.230387151	1.238513988	1.244705164	1.253955123	1.263943129
1	ZF4	1.398802746	1.471021649	1.499353301	1.529520733	1.563520036	1.587319830	1.605612916	1.633127928	1.662981113
1	ZF5	3.271271513	3.522965087	3.625017528	3.735644275	3.862554814	3.952648219	4.022510928	4.128451014	4.244324650

(5) P 6+4		TH = 66	SBE = 1.094636279						MCPER FUNCTIONS	
		RC = 02	RC = 05	RC = 07	RC = 10	RC = 15	RC = 20	RC = 25	RC = 35	RC = 50
AMAX DB		.0017375255	.0108709564	.0213327376	.0436480540	.0988323393	.1772876696	.2802872360	.5675287486	1.249387366
AMIN DB		15.36585084	27.06940149	31.63316575	36.58905797	42.37445744	46.59386983	49.97016338	55.29753014	61.55323965
MU	K	.1704934D 0	.4431287D-1	.2620244D-1	.1480973D-1	.7608116D-2	.4677960D-2	.3173159D-2	.1718397D-2	.8362536D-3
1	PF1	1.564981819	1.060577087	.9221550359	.7953821745	.6712006129	.5936012796	.5385216668	.4628455596	.3910071303
1	PQ1	.6578606733	.6128154601	.6110978196	.6157334606	.6300686434	.6481512922	.6688617725	.7181676563	.8188997198
1	PF2	1.193580008	1.075621115	1.028269754	.9784598022	.9238978254	.8871445038	.8599099807	.8209079189	.7816492216
1	PQ2	2.363695050	1.870853868	1.814951719	1.808585465	1.867210180	1.958698724	2.067180262	2.322833900	2.819778167
3	PFC	1.107481158	1.096207623	1.089147999	1.079957265	1.067540268	1.057591689	1.049284056	1.035813511	1.020106912
3	PQC	24.50971757	13.95280449	12.05085751	10.64472988	9.587950054	9.106616955	8.864886106	8.711305853	8.874629196
1	ZF1	1.098589637	1.098008316	1.097978910	1.098009758	1.098100061	1.098190480	1.098272667	1.098414317	1.098589325
1	ZF2	1.115638711	1.120737049	1.122713398	1.124836463	1.127272559	1.129014911	1.130378731	1.132475013	1.134815503
1	ZF3	1.153257873	1.180710508	1.190815541	1.201314864	1.212913165	1.220927741	1.227048159	1.236214108	1.246138316
1	ZF4	1.369916646	1.438525931	1.466013799	1.495521925	1.529019424	1.552595880	1.570779591	1.598220520	1.628101843
1	ZF5	3.173167948	3.413286908	3.512546380	3.621000464	3.746331047	3.835810052	3.905451945	4.011438006	4.127827145

(5) P 6+4	TH = 67	SBE = 1.086360377						MCPER FUNCTIONS	
	RC = 02	RC = 05	RC = 07	RC = 10	RC = 15	RC = 20	RC = 25	RC = 35	RC = 50
AMAX DB	.0017375255	.0108709564	.0213327376	.0436480540	.0988323393	.1772876696	.2802872360	.5675287486	1.249387366
AMIN DB	13.87601658	25.39378853	29.92175320	34.84843992	40.60832747	44.81837752	48.18023305	53.49565258	59.74104733
MU I K	.2023947D 0	.5374160D-1	.3190894D-1	.1809581D-1	.9323600D-2	.5742237D-2	.3899315D-2	.2114547D-2	.1030262D-2
1 I PF1	1.613370644	1.088579466	.9445542162	.8131162374	.6848940671	.6050511805	.5485089990	.4709915527	.3975721456
1 I PQ1	.6746436511	.6189996102	.6157761768	.6193791499	.6329665114	.6506982487	.6712126699	.7203309809	.8210194427
1 I PF2	1.191381859	1.079719280	1.033626863	.9846716667	.9306394100	.8940637695	.8668896633	.8278937117	.7885763774
1 I PQ2	2.558213030	1.965867934	1.893325753	1.875113158	1.925292485	2.013279203	2.120311826	2.375998546	2.877050111
3 I PFC	1.099180797	1.089801677	1.083602252	1.075360807	1.064025848	1.054825787	1.047080744	1.034424412	1.019542332
3 I PQC	28.27340221	15.54363295	13.28267137	11.61680346	10.36167929	9.781795454	9.481301598	9.262276356	9.382504388
1 I ZF1	1.090262567	1.089582858	1.089531178	1.089543083	1.089615980	1.089696028	1.089771056	1.089903072	1.090068964
1 I ZF2	1.105656135	1.110444087	1.112326387	1.114358917	1.116702848	1.118386369	1.119708046	1.121745941	1.124029921
1 I ZF3	1.138649193	1.164768382	1.174549718	1.184777489	1.196138972	1.204022809	1.210059370	1.219123204	1.228965886
1 I ZF4	1.342528965	1.407295478	1.433853000	1.462617099	1.495527832	1.518828478	1.536866439	1.564134951	1.594050323
1 I ZF5	3.078822978	3.306774089	3.403007846	3.509049894	3.632549403	3.721254661	3.790564341	3.896449010	4.013223042

(5) P 6+4	TH = 68	SBE = 1.078534743						MCPER FUNCTIONS	
	RC = 02	RC = 05	RC = 07	RC = 10	RC = 15	RC = 20	RC = 25	RC = 35	RC = 50
AMAX DB	.0017375255	.0108709564	.0213327376	.0436480540	.0988323393	.1772876696	.2802872360	.5675287486	1.249387366
AMIN DB	12.42911850	23.73071424	28.21872667	33.11319749	38.84511017	43.03944556	46.39097960	51.69333685	57.92745301
MU I K	.2390805D 0	.6508251D-1	.3882073D-1	.2209735D-1	.1142206D-1	.7047381D-2	.4791274D-2	.2602155D-2	.1269484D-2
1 I PF1	1.663384770	1.118560883	.9685483827	.8320858643	.6995041744	.6172437223	.5591283765	.4796349012	.4045239314
1 I PQ1	.6946701908	.6260784365	.6210599377	.6234469548	.6361634114	.6534893279	.6737773385	.7226773436	.8233067742
1 I PF2	1.188356494	1.083459373	1.038793708	.9908363877	.9374519976	.9011130154	.8740336072	.8350807478	.7957300379
1 I PQ2	2.787188980	2.074390576	1.981988124	1.949705842	1.989849462	2.073616667	2.178824614	2.434245552	2.939489599
3 I PFC	1.091137101	1.083471246	1.078085622	1.070757943	1.060480582	1.052020573	1.044835774	1.032995824	1.018946614
3 I PQC	32.87243510	17.43325066	14.73061611	12.74789968	11.25275286	10.55427203	10.18321541	9.885521758	9.953243113
1 I ZF1	1.082397879	1.081617490	1.081542880	1.081535332	1.081590348	1.081659703	1.081727319	1.081849305	1.082005623
1 I ZF2	1.096218160	1.100686592	1.102471911	1.104410560	1.106658124	1.108279584	1.109556486	1.111531829	1.113754560
1 I ZF3	1.124923657	1.149607248	1.159031938	1.168956755	1.180050062	1.187783465	1.193722155	1.202664446	1.212406523
1 I ZF4	1.316612756	1.377315569	1.402854767	1.430787081	1.463021569	1.485990461	1.503843593	1.530987383	1.560788708
1 I ZF5	2.988112364	3.203318955	3.296293310	3.399679877	3.521089112	3.608853975	3.677713876	3.783340173	3.900358372

(5) P 6+4	TH = 69	SBE = 1.071144994						MCPER FUNCTIONS	
	RC = 02	RC = 05	RC = 07	RC = 10	RC = 15	RC = 20	RC = 25	RC = 35	RC = 50
AMAX DB	.0017375255	.0108709564	.0213327376	.0436480540	.0988323393	.1772876696	.2802872360	.5675287486	1.249387366
AMIN DB	11.03206148	22.08032558	26.52339842	31.38215155	37.08329163	41.26039348	44.60061813	49.88866711	56.11042894
MU I K	.2807999D 0	.7870163D-1	.4718784D-1	.2697071D-1	.1399057D-1	.8649287D-2	.5888018D-2	.3203072D-2	.1564871D-2
1 I PF1	1.714399941	1.150663862	.9942839790	.8524146071	.7151246326	.6302544193	.5704434307	.4888246770	.4118997168
1 I PQ1	.7187172521	.6342380390	.6270648625	.6280107790	.6397070361	.6565611857	.6765865184	.7252315444	.8257831794
1 I PF2	1.184481071	1.086785892	1.043725191	.9969224502	.9443185338	.9082836733	.8813386688	.8424722856	.8031183201
1 I PQ2	3.058567004	2.199170723	2.082907769	2.033813309	2.061970491	2.140638335	2.243557876	2.498331831	3.007831072
3 I PFC	1.083370169	1.077233289	1.072611738	1.066158512	1.056910308	1.049179499	1.042551066	1.031527892	1.018318691
3 I PQC	38.53381245	19.69703559	16.44654771	14.07415243	12.28627379	11.44402152	10.98764393	10.59478633	10.59824223
1 I ZF1	1.074979644	1.074097518	1.073999359	1.073971892	1.074008600	1.074066974	1.074126949	1.074238534	1.074384840
1 I ZF2	1.087309314	1.091449234	1.093134734	1.094976293	1.097123423	1.098679655	1.099909177	1.101817795	1.103974469
1 I ZF3	1.112070459	1.135219045	1.144253245	1.153842363	1.164634380	1.172196372	1.178022198	1.186822083	1.196442995
1 I ZF4	1.292138617	1.348574522	1.373006558	1.400016541	1.431480761	1.454058302	1.471684625	1.498597062	1.528281813
1 I ZF5	2.900909807	3.102817669	3.192297847	3.292781432	3.411833008	3.498483270	3.566769341	3.671970016	3.789080681

(5) P 6+4	TH = 70	SBE = 1.064177772						MCPER FUNCTIONS	
	RC = 02	RC = 05	RC = 07	RC = 10	RC = 15	RC = 20	RC = 25	RC = 35	RC = 50
AMAX DB	.0017375255	.0108709564	.0213327376	.0436480540	.0988323393	.1772876696	.2802872360	.5675287486	1.249387366
AMIN DB	9.693170704	20.44317176	24.83522596	29.65413034	35.32128809	39.47943888	42.80724268	48.07958516	54.28778687
MU I K	.3275982D 0	.9502577D-1	.5731109D-1	.3290739D-1	.1713703D-1	.1061764D-1	.7238321D-2	.3944761D-2	.1930237D-2
1 I PF1	1.765531807	1.185020908	1.021917391	.8742404877	.7318621082	.6441697129	.5825271547	.4986172384	.4197422511
1 I PQ1	.7477585716	.6437144946	.6339361120	.6331620697	.6436555800	.6599581647	.6796771754	.7280231756	.8284740952
1 I PF2	1.179746909	1.089639586	1.048371190	1.002893579	.9512184576	.9155647222	.8888000328	.8500709397	.8107495571
1 I PQ2	3.382419354	2.343690745	2.198556504	2.129237941	2.142996262	2.215478053	2.315533219	2.569172966	3.082957466
3 I PFC	1.075899160	1.071105618	1.067195253	1.061573353	1.053321674	1.046306634	1.040228998	1.030021000	1.017657540
3 I PQC	45.55418968	22.43427036	18.49824110	15.64234428	13.49439636	12.47644468	11.91610986	11.40732000	11.23169269
1 I ZF1	1.067992273	1.067009228	1.066886966	1.066839163	1.066857193	1.066904336	1.066956466	1.067057314	1.067193197
1 I ZF2	1.078915050	1.082717930	1.084300853	1.086042249	1.088085025	1.089572937	1.090752508	1.092590236	1.094675990
1 I ZF3	1.100078107	1.121597780	1.130207001	1.139426460	1.149882326	1.157250598	1.162947526	1.171582591	1.181060151
1 I ZF4	1.269072858	1.321063403	1.344299281	1.370293953	1.400889351	1.423012130	1.440366562	1.466936300	1.496497036
1 I ZF5	2.817082010	3.005169628	3.090919429	3.188248306	3.304666452	3.390020033	3.457601370	3.562198172	3.679237647

(5) P 6+4 — TH = 71 SBE = 1.057620681 — MCPER FUNCTIONS

MU	K	RC = 02	RC = 05	RC = 07	RC = 10	RC = 15	RC = 20	RC = 25	RC = 35	RC = 50
	AMAX DB	.0017375255	.0108709564	.0213327376	.0436480540	.0988323393	.1772876696	.2802872360	.5675287486	1.249387366
	AMIN DB	8.422004114	18.82035522	23.15387802	27.92798517	33.55743498	37.69467674	41.00880121	46.26385957	52.45714298
MU	K	.3792275D 0	.1145466D 0	.6955144D-1	.4014216D-1	.2099560D-1	.1303966D-1	.8903483D-2	.4861911D-2	.2383103D-2
1	PF1	1.815612976	1.221740700	1.051611875	.8977170420	.7498382903	.6590889504	.5954636878	.5030776761	.4281009185
1	PQ1	.7830082581	.6548099838	.6418575557	.6390151272	.6480807318	.6637343950	.6830941529	.7310378427	.8314099059
1	PF2	1.174162579	1.091957921	1.052676302	1.008708161	.9581270908	.9229421780	.8964108111	.8578784539	.8186322382
1	PQ2	3.771562317	2.512408366	2.332071265	2.238244444	2.234593652	2.299536077	2.396006262	2.647936815	3.165939805
3	PFC	1.068741867	1.065106866	1.061851873	1.057014384	1.049722228	1.043406749	1.037872485	1.028475821	1.016962206
3	PQC	54.32309347	25.77726454	20.97538848	17.51380950	14.91881501	13.68423134	12.99615677	12.34501961	12.17145871
1	ZF1	1.061420273	1.060339800	1.060192983	1.060124490	1.060123540	1.060159245	1.060203357	1.060293170	1.060418248
1	ZF2	1.071021468	1.074479782	1.075957447	1.077595736	1.079530398	1.080946988	1.082074078	1.083836773	1.085846698
1	ZF3	1.088933328	1.108739348	1.116888797	1.125703611	1.135786772	1.142937647	1.148488543	1.156934702	1.166244939
1	ZF4	1.247375442	1.294775666	1.316727141	1.341611563	1.371235162	1.392835801	1.409869944	1.436130499	1.465404318
1	ZF5	2.736482106	2.910276845	2.992058140	3.085975971	3.199476162	3.283342710	3.350081131	3.453883964	3.570675523

(5) P 6+4 — TH = 72 SBE = 1.051462224 — MCPER FUNCTIONS

MU	K	RC = 02	RC = 05	RC = 07	RC = 10	RC = 15	RC = 20	RC = 25	RC = 35	RC = 50
	AMAX DB	.0017375255	.0108709564	.0213327376	.0436480540	.0988323393	.1772876696	.2802872360	.5675287486	1.249387366
	AMIN DB	7.228885434	17.21373475	21.47933164	26.20261791	31.78997653	35.90405566	39.20306506	44.43904803	50.61587493
MU	K	.4350666D 0	.1378203D 0	.8433997D-1	.4896312D-1	.2573364D-1	.1602497D-1	.1096091D-1	.5998568D-2	.2945820D-2
1	PF1	1.863205770	1.260887090	1.083531708	.9230138323	.7691922262	.6751267746	.6093505146	.5202816212	.4370331441
1	PQ1	.8259705392	.6679147895	.6510645882	.6457143850	.6530717318	.6679566175	.6868923764	.7344687797	.8346272228
1	PF2	1.167756407	1.093675861	1.056579697	1.014318625	.9650149253	.9303984756	.9041615429	.8658954059	.8267749216
1	PQ2	4.242371629	2.711092965	2.487480022	2.363711802	2.338858473	2.394560574	2.486536688	2.735852257	3.258090178
3	PFC	1.061914198	1.059256434	1.056598378	1.052494675	1.046120521	1.040485419	1.035485063	1.026893381	1.016231832
3	PQC	65.35503792	29.90453644	23.99829486	19.77006964	16.61432801	15.11000977	14.26349151	13.43603033	13.14030099
1	ZF1	1.055248051	1.054077218	1.053905552	1.053816103	1.053795948	1.053820059	1.053856015	1.053934541	1.054048471
1	ZF2	1.063615009	1.066723019	1.068092819	1.069625192	1.071448151	1.072790514	1.073862654	1.075546209	1.077475366
1	ZF3	1.078619847	1.096641297	1.104296343	1.112670770	1.122343100	1.129251515	1.134638098	1.142869473	1.151986462
1	ZF4	1.226997910	1.269706704	1.290287464	1.313965377	1.342509971	1.363517002	1.380178936	1.406008225	1.434976147
1	ZF5	2.658941688	2.818043217	2.895615397	2.985860567	3.096148921	3.178329307	3.244078837	3.346884772	3.463237315

(5) P 6+4 — TH = 73 — SBE = 1.045691756 — MCPER FUNCTIONS

MU	K	RC = 02	RC = 05	RC = 07	RC = 10	RC = 15	RC = 20	RC = 25	RC = 35	RC = 50
AMAX DB		.0017375255	.0108709564	.0213327376	.0436480540	.0988323393	.1772876696	.2802872360	.5675287486	1.249387366
AMIN DB		6.124121283	15.62619064	19.81201269	24.47702670	30.01705753	34.10535037	37.38759196	42.60245078	48.76106809
MU	K	.4940762D 0	.1654590D 0	.1021879D 0	.5972397D-1	.3156074D-1	.1971208D-1	.1350891D-1	.7411011D-2	.3647091D-2
1	PF1	1.906666912	1.302448258	1.117831959	.9503159054	.7900829120	.6924160060	.6243011887	.5323175296	.4466061846
1	PQ1	.8784971115	.6835374192	.6618619582	.6534445299	.6587409556	.6727080359	.6911398417	.7382190179	.8381705892
1	PF2	1.160577674	1.094727115	1.060015133	1.019670806	.9718467903	.9379117178	.9120395748	.8741208269	.8351861065
1	PQ2	4.815887584	2.947298424	2.670020922	2.509347073	2.458458727	2.502760474	2.589084780	2.834789544	3.361033701
3	PFC	1.055429636	1.053574415	1.051452636	1.048028532	1.042526223	1.037549135	1.033070995	1.025275129	1.015465692
3	PQC	79.33517992	35.06011597	27.73068938	22.52111522	18.65403268	16.81018306	15.76506944	14.71703276	14.26760830
1	ZF1	1.049459810	1.048210162	1.048013604	1.047903051	1.047863559	1.047875978	1.047903681	1.047970722	1.048073206
1	ZF2	1.056682151	1.059436931	1.060696359	1.062120143	1.063827989	1.065093339	1.066108125	1.067708492	1.069551928
1	ZF3	1.069117227	1.085302530	1.092429328	1.100327253	1.109549243	1.116188777	1.121391578	1.129380385	1.138276065
1	ZF4	1.207881684	1.245853239	1.264980468	1.287355148	1.314709639	1.335047422	1.351281491	1.376601333	1.405187595
1	ZF5	2.584262252	2.728373479	2.801493115	2.887797760	2.994570126	3.074855759	3.139462000	3.241054107	3.356760632

(5) P 6+4 — TH = 74 — SBE = 1.040299436 — MCPER FUNCTIONS

MU	K	RC = 02	RC = 05	RC = 07	RC = 10	RC = 15	RC = 20	RC = 25	RC = 35	RC = 50
AMAX DB			.0108709564	.0213327376	.0436480540	.0988323393	.1772876696	.2802872360	.5675287486	1.249387366
AMIN DB			14.06195615	18.15299640	22.75037976	28.23671963	32.29613033	35.55968119	40.75105238	46.88944768
MU	K		.1981081D 0	.1236944D 0	.7285863D-1	.3874039D-1	.2427691D-1	.1667308D-1	.9171648D-2	.4524052D-2
1	PF1		1.346293073	1.154641545	.9798213154	.8126920378	.7111111089	.6404487242	.5452895924	.4568994380
1	PQ1		.7023459475	.6746488235	.6624447734	.6652317221	.6780936562	.6959217282	.7424043431	.8420947841
1	PF2		1.095045988	1.062911237	1.024703306	.9785808835	.9454547620	.9200282843	.8825517051	.8438740503
1	PQ2		3.231036032	2.886600081	2.679991376	2.596837925	2.626964524	2.706146732	2.946372108	3.476807893
3	PFC		1.048081473	1.046433595	1.043631565	1.038950246	1.034605434	1.030635392	1.023623029	1.014663240
3	PQC		41.58221709	32.39893075	25.91781264	21.13705202	18.86059575	17.56362604	16.23657342	15.59189554
1	ZF1		1.042727893	1.042506783	1.042375145	1.042316303	1.042317001	1.042336402	1.042391822	1.042482617
1	ZF2		1.052611795	1.053758496	1.055071165	1.056660688	1.057846368	1.058801481	1.060314690	1.062067459
1	ZF3		1.074722886	1.081289231	1.088674682	1.097405744	1.103748682	1.108747039	1.116463480	1.125107466
1	ZF4		1.223212477	1.240808950	1.261784358	1.287834275	1.307422973	1.323169586	1.347395167	1.376016422
1	ZF5		2.641171537	2.709592723	2.791681494	2.894622110	2.972794015	3.036093354	3.136239279	3.251075057

(5) P 6+4	TH = 75	SBE = 1.035276180						MCPER FUNCTIONS	
	RC = 02	RC = 05	RC = 07	RC = 10	RC = 15	RC = 20	RC = 25	RC = 35	RC = 50
AMAX DB		.0108709564	.0213327376	.0436480540	.0988323393	.1772876696	.2802872360	.5675287486	1.249387366
AMIN DB		12.52700498	16.50428754	21.02213165	26.44690672	30.47372451	33.71631882	38.88144790	44.99729171
MU I K		.2364012D 0	.1495497D 0	.8889829D-1	.4760523D-1	.2994427D-1	.2061503D-1	.1137438D-1	.5625167D-2
1 I PF1		1.392112351	1.194036246	1.011735250	.8372266530	.7313923232	.6579498309	.5593214859	.4680074577
1 I PQ1		.7252247913	.6899543455	.6730292876	.6727294059	.6842478241	.7013461652	.7471074004	.8464679908
1 I PF2		1.094570039	1.065192182	1.029346901	.9851676471	.9529941041	.9281061048	.8911323366	.8528465071
1 I PQ2		3.575744742	3.146459128	2.882066130	2.758509318	2.770849856	2.840948052	3.074385079	3.609002482
3 I PFC		1.042798664	1.041561251	1.039320763	1.035404884	1.031663055	1.028184354	1.021939671	1.013324167
3 I PQC		49.94650225	38.32134318	30.17088720	24.20028210	21.36509145	19.74448130	18.06002075	17.16448492
1 I ZF1		1.037620109	1.037375359	1.037222901	1.037144942	1.037133869	1.037144975	1.037138713	1.037267639
1 I ZF2		1.046238771	1.047270655	1.048469862	1.049938069	1.051041573	1.051934792	1.053356979	1.055014165
1 I ZF3		1.064902570	1.070879057	1.077716912	1.085915819	1.091933285	1.096705357	1.104117551	1.112476934
1 I ZF4		1.201780933	1.217777839	1.237260178	1.261888434	1.280644077	1.295839534	1.319873839	1.347443237
1 I ZF5		2.556337659	2.619813886	2.697402546	2.796182152	2.872009723	2.933828301	3.032278512	3.145998365

(5) P 6+4	TH = 76	SBE = 1.030613629						MCPER FUNCTIONS	
	RC = 02	RC = 05	RC = 07	RC = 10	RC = 15	RC = 20	RC = 25	RC = 35	RC = 50
AMAX DB		.0108709564	.0213327376	.0436480540	.0988323393	.1772876696	.2802872360	.5675287486	1.249387366
AMIN DB		11.02945517	14.96920092	19.29220368	24.64548707	28.63518314	31.85411056	36.98974820	43.08031749
MU I K		.2808842D 0	.1805261D 0	.1084900D 0	.5857680D-1	.3700333D-1	.2554433D-1	.1414206D-1	.7014297D-2
1 I PF1		1.439345070	1.235997253	1.046258380	.8639212475	.7534705112	.6769902199	.5745612544	.4800439453
1 I PQ1		.7533526542	.7084885964	.6856169367	.6814775587	.6913450750	.7075524762	.7524335017	.8513762367
1 I PF2		1.093243716	1.066778978	1.033524091	.9915484767	.9604885154	.9362452946	.9000034692	.8621103547
1 I PQ2		3.999711456	3.462166119	3.124237328	2.949491092	2.939277697	2.997725993	3.222454882	3.757960190
3 I PFC		1.037747186	1.036856581	1.035114556	1.031903962	1.028732107	1.025725140	1.020228403	1.012948478
3 I PQC		60.83255467	45.95395347	35.58064805	28.03669598	24.46875921	22.42601109	20.27712149	19.05506045
1 I ZF1		1.032876785	1.032610133	1.032437472	1.032340523	1.032318029	1.032320911	1.032352992	1.032419947
1 I ZF2		1.040309765	1.041225201	1.042308838	1.043652984	1.044671976	1.045501200	1.046828635	1.048385385
1 I ZF3		1.055841363	1.061202961	1.067459902	1.075085410	1.080747596	1.085270438	1.092344375	1.100383537
1 I ZF4		1.181552795	1.195893525	1.213793366	1.236881363	1.254716037	1.269292394	1.292545611	1.319451756
1 I ZF5		2.473763729	2.532052487	2.604846783	2.699120081	2.772359351	2.832511715	2.928997271	3.041334829

(5) P 6+4	TH = 77 SBE = 1.026304108								MCPER FUNCTIONS
	RC = 02	RC = 05	RC = 07	RC = 10	RC = 15	RC = 20	RC = 25	RC = 35	RC = 50
AMAX DB		.0108709564	.0213327376	.0436480540	.0988323393	.1772876696	.2802872360	.5675287486	1.249387366
AMIN DB		9.579902810	13.25285616	17.56125876	22.83030521	26.77723917	29.96919947	35.07145660	41.13353126
MU I K		.3318982D 0	.2174489D 0	.1324150D 0	.7219128D-1	.4582875D-1	.3173511D-1	.1763710D-1	.8776542D-2
1 I PF1		1.487095148	1.280350033	1.083565622	.8930382613	.7775926606	.6977912398	.5911877422	.4931471213
1 I PQ1		.7883095705	.7312162538	.7007752646	.6918017926	.6996161053	.7147232322	.7585190272	.8569297495
1 I PF2		1.091023034	1.067591649	1.037148907	.9976542563	.9678873713	.9444103731	.9090011635	.8716710605
1 I PQ2		4.528188536	3.851115661	3.418427042	3.177965835	3.138798663	3.182152206	3.394389809	3.931071950
3 I PFC		1.032948002	1.032341418	1.031032853	1.028462995	1.025824270	1.023266373	1.018493502	1.012036596
3 I PQC		75.22824740	55.96343755	42.58474399	32.92465623	28.37901555	25.77621528	23.01384173	21.36027972
1 I ZF1		1.028487981	1.028202314	1.028010582	1.027895333	1.027861585	1.027856393	1.027876943	1.027931916
1 I ZF2		1.034817237	1.035615359	1.036581671	1.037799308	1.038731656	1.039494926	1.040724049	1.042175610
1 I ZF3		1.047537578	1.052265693	1.057911515	1.064923232	1.070199742	1.074449441	1.081149016	1.088829456
1 I ZF4		1.162517747	1.175162836	1.191398063	1.212827276	1.229649498	1.243534499	1.265893417	1.292029163
1 I ZF5		2.393325534	2.446197433	2.513892878	2.603295273	2.673686554	2.731973843	2.826203506	2.936864737

(5) P 6+4	TH = 78 SBE = 1.022340595								MCPER FUNCTIONS
	RC = 02	RC = 05	RC = 07	RC = 10	RC = 15	RC = 20	RC = 25	RC = 35	RC = 50
AMAX DB		.0108709564	.0213327376	.0436480540	.0988323393	.1772876696	.2802872360	.5675287486	1.249387366
AMIN DB		8.191535950	11.66277900	15.83111231	20.99928569	24.89627515	28.05716578	33.12130601	39.15102507
MU I K		.3894244D 0	.2611326D 0	.1616011D 0	.8913242D-1	.5690969D-1	.3954956D-1	.2207673D-1	.1102678D-1
1 I PF1		1.534053930	1.326678702	1.123769420	.9248651213	.8040477413	.7206181299	.6094191731	.5074870798
1 I PQ1		.8322245931	.7594639370	.7192867243	.7041459952	.7093718655	.7231023197	.7655438360	.8632722838
1 I PF2		1.087881120	1.067552625	1.040127217	1.003403736	.9751286060	.9525561235	.9181552653	.8815319127
1 I PQ2		5.196625128	4.337851182	3.781393055	3.455307621	3.378434678	3.401982927	3.597199186	4.133222692
3 I PFC		1.028421348	1.028038243	1.027097065	1.025099362	1.022953021	1.020818273	1.016740383	1.011089483
3 I PQC		94.59809966	69.34650605	51.83711676	39.27633031	33.40022846	30.03977063	26.45151036	24.21753549
1 I ZF1		1.024443666	1.024143379	1.023934446	1.023801842	1.023757260	1.023744238	1.023753503	1.023796590
1 I ZF2		1.029753953	1.030435106	1.031282882	1.032371945	1.033215753	1.033911282	1.035038753	1.036380515
1 I ZF3		1.039986744	1.044071820	1.049081182	1.055440772	1.060301144	1.064253065	1.070540204	1.077820408
1 I ZF4		1.144658227	1.155591522	1.170091396	1.189745650	1.205461004	1.218578128	1.239925553	1.265166641
1 I ZF5		2.314870787	2.362125078	2.424409048	2.508552791	2.575817435	2.632024903	2.723681310	2.832342035

(5) P 6+4		TH = 79	SBE = 1.018716695							MCPER FUNCTIONS
		RC = 02	RC = 05	RC = 07	RC = 10	RC = 15	RC = 20	RC = 25	RC = 35	RC = 50
AMAX DB			.0108709564	.0213327376	.0436480540	.0988323393	.1772876696	.2802872360	.5675287486	1.249387366
AMIN DB			6.879823357	10.10955116	14.10532953	19.15062582	22.98830821	26.11290647	31.13304017	37.12569647
K	MU		.4529068D 0	.3122644D 0	.1971213D 0	.1102729D 0	.7088994D-1	.4947145D-1	.2775543D-1	.1392243D-1
PF1	1		1.578456033	1.374214666	1.166858863	.9597042440	.8331720606	.7457901161	.6295247371	.5232760609
PQ1	1		.8879825535	.7950783086	.7422502643	.7191286607	.7210409334	.7330228105	.7737505312	.8705951929
PF2	1		1.083814030	1.066591696	1.042357881	1.008701815	.9821362132	.9606250253	.9274373311	.8916929017
PQ2	1		6.055748234	4.957755398	4.237266897	3.797731989	3.670927926	3.668091433	3.839909943	4.372485470
PFC	3		1.024186103	1.023969862	1.023330060	1.021832477	1.020133895	1.018392946	1.014975858	1.010108822
PQC	3		121.1642539	87.63128594	64.34283150	47.72051467	39.99266207	35.58382601	30.85889650	27.82774051
ZF1	1		1.020733583	1.020424905	1.020201665	1.020053157	1.019998354	1.019977860	1.019976229	1.020007656
ZF2	1		1.025112699	1.025679005	1.026407889	1.027366826	1.028120498	1.028746707	1.029769461	1.030997019
ZF3	1		1.033180104	1.036624620	1.040979392	1.046652229	1.051066688	1.054695856	1.060530301	1.067366187
ZF4	1		1.127946314	1.137182037	1.149892740	1.167661509	1.182173687	1.194442373	1.214651611	1.238860107
ZF5	1		2.238201707	2.279689696	2.336247995	2.414718174	2.478554200	2.532447809	2.621182264	2.727481707

(5) P 6+4		TH = 80	SBE = 1.015426612							MCPER FUNCTIONS
		RC = 02	RC = 05	RC = 07	RC = 10	RC = 15	RC = 20	RC = 25	RC = 35	RC = 50
AMAX DB				.0213327376	.0436480540	.0988323393	.1772876696	.2802872360	.5675287486	1.249387366
AMIN DB				8.607362134	12.39005903	17.28313859	21.04901774	24.13049524	29.09911512	35.04885214
K	MU			.3712205D 0	.2401580D 0	.1367235D 0	.8862354D-1	.6215488D-1	.3507876D-1	.1768306D-1
PF1	1			1.421707653	1.212602919	.9978494289	.8653520739	.7736923203	.6518404342	.5407831277
PQ1	1			.8406622019	.7712403258	.7376330495	.7352293421	.7449513712	.7834739511	.8791593628
PF2	1			1.064652702	1.043735283	1.013437875	.9888172155	.9685439366	.9368077767	.9021490724
PQ2	1			5.763105158	4.821767349	4.229042625	4.034809946	3.996168718	4.135398502	4.660241503
PFC	3			1.020158915	1.019756053	1.018683967	1.017384779	1.016004727	1.013208470	1.009097245
PQC	3			113.2296290	81.69707229	59.25014780	48.87707081	42.97882904	36.64824152	32.49462240
ZF1	1			1.017038391	1.016805106	1.016642848	1.016578695	1.016551233	1.016539273	1.016559417
ZF2	1			1.021341973	1.021952920	1.022780922	1.023443241	1.023998819	1.024914137	1.026023372
ZF3	1			1.029924629	1.033616884	1.038574328	1.042514859	1.045796554	1.051138377	1.057481383
ZF4	1			1.119930485	1.130822445	1.146605624	1.159818072	1.171154228	1.190088722	1.213111257
ZF5	1			2.198707448	2.249238464	2.321590051	2.381666381	2.432988039	2.518413281	2.621945957

(5) P 6+4	TH = 81	SBE = 1.012465126						MCPER FUNCTIONS	
	RC = 02	RC = 05	RC = 07	RC = 10	RC = 15	RC = 20	RC = 25	RC = 35	RC = 50
AMAX DB			.0213327376	.0436480540	.0988323393	.1772876696	.2802872360	.5675287486	1.249387366
AMIN DB			7.174191573	10.69512158	15.39684411	19.07386521	22.10303064	27.01028408	32.90963157
MU I K			.4378148D 0	.2919066D 0	.1698861D 0	.1112517D 0	.7849617D-1	.4461550D-1	.2262134D-1
1 I PF1			1.467304687	1.260404520	1.039536729	.9010200729	.8047886647	.6767909860	.5603556862
1 I PQ1			.8999355758	.8085615217	.7609561901	.7528198779	.7595617530	.7951908255	.8893307299
1 I PF2			1.061701687	1.044153962	1.017484499	.9950580534	.9762197872	.9462118974	.9128880562
1 I PQ2			6.833409642	5.589470012	4.785396509	4.497978746	4.409647854	4.502657932	5.013064948
3 I PFC			1.016627212	1.016400358	1.015677817	1.014726243	1.013670587	1.011448920	1.008058644
3 I PQC			150.0818936	106.5277631	75.49764380	61.22949937	53.14833224	44.47803592	38.69583835
1 I ZF1			1.013975121	1.013737752	1.013564878	1.013492597	1.013458855	1.013437345	1.013446770
1 I ZF2			1.017418987	1.017914885	1.018612242	1.019182503	1.019666478	1.020472099	1.021459274
1 I ZF3			1.023967852	1.027003452	1.031225917	1.034667804	1.037578443	1.042385917	1.048186313
1 I ZF4			1.103822720	1.112899690	1.126614516	1.138433014	1.148749985	1.166263259	1.187929075
1 I ZF5			2.118930370	2.163170484	2.228928979	2.284878148	2.333339079	2.415018993	2.515323289

(5) P 6+4	TH = 82	SBE = 1.009827573						MCPER FUNCTIONS	
	RC = 02	RC = 05	RC = 07	RC = 10	RC = 15	RC = 20	RC = 25	RC = 35	RC = 50
AMAX DB				.0436480540	.0988323393	.1772376696	.2802872360	.5675287486	1.249387366
AMIN DB				9.035276847	13.49396000	17.05840043	20.02249847	24.85500902	30.69414105
MU I K				.3533753D 0	.2114959D 0	.1403072D 0	.9974131D-1	.5718071D-1	.2919396D-1
1 I PF1				1.309096409	1.084848773	.9406328102	.8396328615	.7049203779	.5824519980
1 I PQ1				.8576685165	.7910643714	.7751439484	.7778619031	.8096038562	.9016404691
1 I PF2				1.043516151	1.020697184	1.000720434	.9835339862	.9555742226	.9233862942
1 I PQ2				6.626948974	5.523984674	5.104203196	4.944995864	4.970858921	5.456057712
3 I PFC				1.013288949	1.012840475	1.012181903	1.011410628	1.009710616	1.006998612
3 I PQC				143.3599181	99.27810547	79.06551096	67.66327648	55.45286165	47.21930977
1 I ZF1				1.010992536	1.010813498	1.010734792	1.010695706	1.010665696	1.010665192
1 I ZF2				1.014291163	1.014859809	1.015338028	1.015749886	1.016444148	1.017306057
1 I ZF3				1.021146216	1.024627239	1.027551220	1.030069673	1.034302669	1.039508267
1 I ZF4				1.096139115	1.107729985	1.118066700	1.127276989	1.143213171	1.163332020
1 I ZF5				2.077768924	2.136439308	2.187849049	2.233120577	2.310555242	2.407096793

$P^{(3)}_{7+0}$

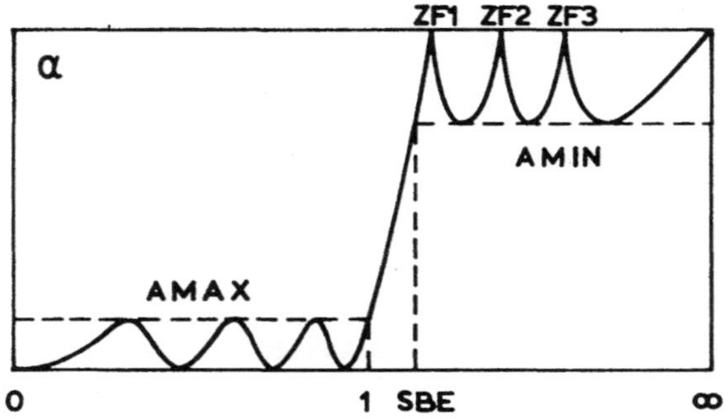

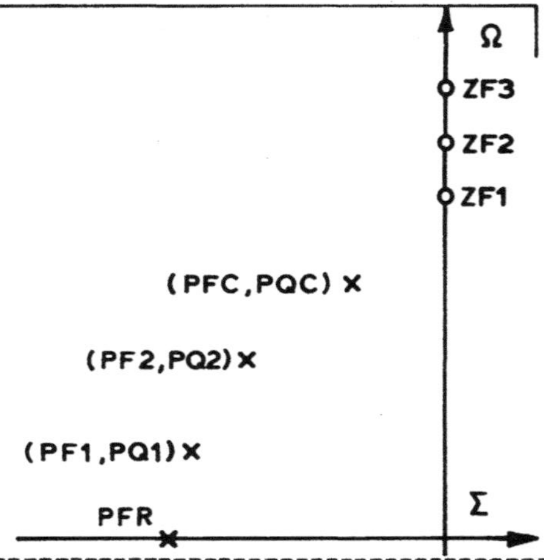

(3) P 7+0	TH = 25	SBE = 2.366201583						CAUER FUNCTIONS	
	RC = 02	RC = 05	RC = 07	RC = 10	RC = 15	RC = 20	RC = 25	RC = 35	RC = 50
AMAX DB	.0017375255	.0108709564	.0213327376	.0436480540	.0988323393	.1772876696	.2802872360	.5675287486	1.249387366
AMIN DB	87.70122696	95.66916056	98.60218306	101.7225376	105.2995470	107.8767771	109.9179769	113.1277792	116.9076770
MU I K	.6501131D-3	.2597719D-3	.1853280D-3	.1293968D-3	.8571817D-4	.6371056D-4	.5036762D-4	.3480658D-4	.2252508D-4
1 I PFR	.7604272031	.5879877245	.5285018847	.4672174594	.3993137114	.3519204271	.3153577026	.2599228683	.1990251102
1 I PF1	.8750850156	.7365708685	.6914380137	.6469913167	.6008218896	.5709715193	.5495335551	.5201193231	.4928387173
1 I PQ1	.6706382542	.7239857347	.7543157910	.7966557108	.8637732667	.9302051280	.9981957853	1.144936153	1.415399774
1 I PF2	1.066820088	.9711882297	.9411213126	.9122152584	.8830571591	.8647436666	.8518803998	.8346582451	.8191622498
1 I PQ2	1.312787617	1.507377980	1.613353305	1.757065897	1.977194622	2.188231838	2.399053720	2.841674470	3.630823097
1 I PFC	1.181342250	1.108269988	1.085425343	1.063549994	1.041588833	1.027858103	1.018245919	1.005421723	.9939305865
1 I PQC	4.415237328	5.163519456	5.566102335	6.108019731	6.931402122	7.715296170	8.494536653	10.12197202	13.00642668
1 I ZF1	2.421357053	2.421357053	2.421357053	2.421357053	2.421357053	2.421357053	2.421357053	2.421357053	2.421357053
1 I ZF2	2.970712942	2.970712942	2.970712942	2.970712942	2.970712942	2.970712942	2.970712942	2.970712942	2.970712942
1 I ZF3	5.243629654	5.243629654	5.243629654	5.243629654	5.243629654	5.243629654	5.243629654	5.243629654	5.243629654

(3) P 7+0	TH = 26	SBE = 2.281172033						CAUER FUNCTIONS	
	RC = 02	RC = 05	RC = 07	RC = 10	RC = 15	RC = 20	RC = 25	RC = 35	RC = 50
AMAX DB	.0017375255	.0108709564	.0213327376	.0436480540	.0988323393	.1772876696	.2802872360	.5675287486	1.249387366
AMIN DB	85.23274673	93.20068032	96.13370282	99.25405733	102.8310668	105.4082969	107.4494967	110.6592989	114.4391967
MU I K	.8293881D-3	.3314066D-3	.2364340D-3	.1650792D-3	.1093558D-3	.8127936D-4	.6425698D-4	.4440483D-4	.2873659D-4
1 I PFR	.7651928801	.5911585729	.5312184874	.4695101277	.4011829988	.3535195940	.3167610205	.2610474337	.1998646268
1 I PF1	.8790117519	.7395276943	.6941171404	.6494140631	.6029953764	.5729926423	.5514493261	.5218958668	.4944914118
1 I PQ1	.6723991815	.7255862333	.7558950896	.7982385270	.8654009394	.9319005830	.9999730808	1.146911803	1.417774162
1 I PF2	1.067821249	.9723328908	.9422944209	.9134089289	.8842652159	.8659576598	.8530971909	.8358770678	.8203812750
1 I PQ2	1.323502562	1.517919222	1.624093288	1.768218299	1.989152078	2.201068216	2.412827873	2.857526535	3.650557794
1 I PFC	1.179530880	1.107307548	1.084696238	1.063030746	1.041267299	1.027653755	1.018120723	1.005398489	.9939952389
1 I PQC	4.469587918	5.218651676	5.622883090	6.167632497	6.996075005	7.785264964	8.570042898	10.20955222	13.11625881
1 I ZF1	2.333899616	2.333899616	2.333899616	2.333899616	2.333899616	2.333899616	2.333899616	2.333899616	2.333899616
1 I ZF2	2.859592270	2.859592270	2.859592270	2.859592270	2.859592270	2.859592270	2.859592270	2.859592270	2.859592270
1 I ZF3	5.038750333	5.038750333	5.038750333	5.038750333	5.038750333	5.038750333	5.038750333	5.038750333	5.038750333

(3) P 7+0	TH = 27	SBE = 2.202689265						CAUER FUNCTIONS	
	RC = 02	RC = 05	RC = 07	RC = 10	RC = 15	RC = 20	RC = 25	RC = 35	RC = 50
AMAX DB	.0017375255	.0108709564	.0213327376	.0436480540	.0988323393	.1772876696	.2802872360	.5675287486	1.249387366
AMIN DB	82.85053101	90.81846460	93.75148709	96.87184161	100.4488511	103.0260811	105.0672810	108.2770832	112.0569810
MU I K	.1049123D-2	.4192083D-3	.2990741D-3	.2088147D-3	.1383281D-3	.1028132D-3	.8128100D-4	.5616929D-4	.3634997D-4
1 I PFR	.7702075536	.5944889215	.5340702155	.4719156467	.4031433588	.3551961859	.3182319875	.2622259024	.2007441813
1 I PF1	.8831193962	.7426204075	.6969192528	.6519478900	.6052683803	.5751062222	.5534526766	.5237535507	.4962195256
1 I PQ1	.6742619731	.7272739474	.7575589053	.7999046149	.8671128103	.9336828332	1.001840705	1.148986906	1.420267064
1 I PF2	1.068849213	.9735177531	.9435107048	.9146481391	.8855207378	.8672201062	.8543630319	.8371456186	.8216505252
1 I PQ2	1.334854689	1.529060827	1.635437076	1.779990110	2.001765744	2.214603860	2.427348554	2.874231969	3.671348543
1 I PFC	1.177653573	1.106308700	1.083939219	1.062491375	1.040933118	1.027441257	1.017990443	1.005374158	.9940622611
1 I PQC	4.527266926	5.277049959	5.682994731	6.230710427	7.064473688	7.859242632	8.649858955	10.30210657	13.23230156
1 I ZF1	2.253155726	2.253155726	2.253155726	2.253155726	2.253155726	2.253155726	2.253155726	2.253155726	2.253155726
1 I ZF2	2.756829163	2.756829163	2.756829163	2.756829163	2.756829163	2.756829163	2.756829163	2.756829163	2.756829163
1 I ZF3	4.848897170	4.848897170	4.848897170	4.848897170	4.848897170	4.848897170	4.848897170	4.848897170	4.848897170

(3) P 7+0	TH = 28	SBE = 2.130054468						CAUER FUNCTIONS	
	RC = 02	RC = 05	RC = 07	RC = 10	RC = 15	RC = 20	RC = 25	RC = 35	RC = 50
AMAX DB	.0017375255	.0108709564	.0213327376	.0436480540	.0988323393	.1772876696	.2802872360	.5675287486	1.249387366
AMIN DB	80.54797249	88.51590606	91.44892856	94.56928307	98.14629254	100.7235226	102.7647224	105.9745247	109.7544225
MU I K	.1316673D-2	.5261159D-3	.3753447D-3	.2620671D-3	.1736049D-3	.1290329D-3	.1020095D-3	.7049373D-4	.4562003D-4
1 I PFR	.7754798498	.5979835374	.5370609188	.4744370846	.4051971523	.3569521499	.3197722688	.2634595633	.2016647066
1 I PF1	.8874112054	.7458514855	.6998465916	.6545948162	.6076427029	.5773139271	.5555451825	.5256938294	.4980244055
1 I PQ1	.6762313893	.7290523281	.7593103991	.8016569191	.8689116871	.9355546647	1.003801480	1.151164454	1.422881924
1 I PF2	1.069902338	.9747421595	.9447697476	.9159326749	.8868236898	.8685310702	.8556780517	.8384641064	.8229702744
1 I PQ2	1.346875547	1.540829440	1.647410544	1.792406980	2.015061866	2.228866133	2.442644526	2.891822973	3.693234448
1 I PFC	1.175710952	1.105273617	1.083154378	1.061931916	1.040586291	1.027220597	1.017855061	1.005348706	.9941316323
1 I PQC	4.588451606	5.338874896	5.746597056	6.297415917	7.136768750	7.937410068	8.734177338	10.39985430	13.35482536
1 I ZF1	2.178408537	2.178408537	2.178408537	2.178408537	2.178408537	2.178408537	2.178408537	2.178408537	2.178408537
1 I ZF2	2.661529103	2.661529103	2.661529103	2.661529103	2.661529103	2.661529103	2.661529103	2.661529103	2.661529103
1 I ZF3	4.672456623	4.672456623	4.672456623	4.672456623	4.672456623	4.672456623	4.672456623	4.672456623	4.672456623

(3) P 7+0	TH = 29	SBE = 2.062665340						CAUER FUNCTIONS	
	RC = 02	RC = 05	RC = 07	RC = 10	RC = 15	RC = 20	RC = 25	RC = 35	RC = 50
AMAX DB	.0017375255	.0108709564	.0213327376	.0436480540	.0988323393	.1772876696	.2802872360	.5675287486	1.249387366
AMIN DB	78.31914991	86.28708346	89.22010595	92.34046047	95.91746993	98.49469999	100.5358998	103.7457020	107.5255999
MU I K	.1640461D-2	.6554947D-3	.4676469D-3	.3265129D-3	.2162967D-3	.1607638D-3	.1270950D-3	.8782906D-4	.5683860D-4
1 I PFR	.7810190478	.6016475235	.5401947123	.4770777174	.4073468965	.3587895604	.3213836374	.2647497378	.2026271942
1 I PF1	.8918905823	.7492235337	.7029015162	.6573569700	.6101202446	.5796175178	.5577285080	.5277182390	.4999074743
1 I PQ1	.6783125938	.7309250981	.7611529758	.8034986064	.8708005845	.9375190664	1.005858430	1.153447654	1.425622430
1 I PF2	1.070978879	.9760054046	.9460710979	.9172623004	.8881740256	.8698906118	.8570423792	.8398327446	.8243408060
1 I PQ2	1.359599371	1.553253859	1.660041621	1.805496570	2.029068719	2.243884491	2.458746739	2.910334177	3.716257544
1 I PFC	1.173703666	1.104202480	1.082341811	1.061352409	1.040226818	1.026991760	1.017714557	1.005322111	.9942033304
1 I PQC	4.653335028	5.404300769	5.813863304	6.367924852	7.213144776	8.019962911	8.823206178	10.50303230	13.48412231
1 I ZF1	2.109040189	2.109040189	2.109040189	2.109040189	2.109040189	2.109040189	2.109040189	2.109040189	2.109040189
1 I ZF2	2.572920998	2.572920998	2.572920998	2.572920998	2.572920998	2.572920998	2.572920998	2.572920998	2.572920998
1 I ZF3	4.508037507	4.508037507	4.508037507	4.508037507	4.508037507	4.508037507	4.508037507	4.508037507	4.508037507

(3) P 7+0 TH = 30 SBE = 2.000000000 CAUER FUNCTIONS

MU		RC = 02	RC = 05	RC = 07	RC = 10	RC = 15	RC = 20	RC = 25	RC = 35	RC = 50
	AMAX DB	.0017375255	.0108709564	.0213327376	.0436480540	.0988323393	.1772876696	.2802872360	.5675287486	1.249387366
	AMIN DB	76.15873438	84.12666790	87.05969039	90.18004490	93.75705436	96.33428442	98.37548425	101.5852865	105.3651843
	K	.2030102D-2	.8111875D-3	.5787222D-3	.4040661D-3	.2676714D-3	.1989484D-3	.1572825D-3	.1086902D-3	.7033888D-4
1	PFR	.7868351315	.6054863430	.5434759955	.4798410419	.4095952747	.3607106269	.3230679808	.2660980349	.2036326983
1	PF1	.8965610765	.7527392890	.7060865092	.6602365931	.6127030097	.5820188522	.5600044093	.5293284017	.5018702346
1	PQ1	.6805111910	.7328962746	.7630903030	.8054330825	.8727827398	.9395792449	1.008014798	1.155839939	1.428492532
1	PF2	1.072076983	.9773067308	.9474142677	.9186367556	.8895716867	.8712987851	.8584561415	.8412517510	.8257624117
1	PQ2	1.373063337	1.566365221	1.673360470	1.819288718	2.043816770	2.259690652	2.475688504	2.929802830	3.740463022
1	PFC	1.171632397	1.103095481	1.081501623	1.060752895	1.039854701	1.026754732	1.017568910	1.005294349	.9942773324
1	PQC	4.722127633	5.473516819	5.884981386	6.442427828	7.293801597	8.107112840	8.917170578	10.61189662	13.62050807
1	ZF1	2.044515315	2.044515315	2.044515315	2.044515315	2.044515315	2.044515315	2.044515315	2.044515315	2.044515315
1	ZF2	2.490336612	2.490336612	2.490336612	2.490336612	2.490336612	2.490336612	2.490336612	2.490336612	2.490336612
1	ZF3	4.354433926	4.354433926	4.354433926	4.354433926	4.354433926	4.354433926	4.354433926	4.354433926	4.354433926

(3) P 7+0 TH = 31 SBE = 1.941604026 CAUER FUNCTIONS

MU		RC = 02	RC = 05	RC = 07	RC = 10	RC = 15	RC = 20	RC = 25	RC = 35	RC = 50
	AMAX DB	.0017375255	.0108709564	.0213327376	.0436480540	.0988323393	.1772876696	.2802872360	.5675287486	1.249387366
	AMIN DB	74.06191095	82.02984441	84.96286689	88.08322140	91.66023087	94.23746093	96.27866075	99.43846298	103.2683608
	K	.2496550D-2	.9975706D-3	.7116927D-3	.4969066D-3	.3291731D-3	.2446599D-3	.1934206D-3	.1336635D-3	.8650034D-4
1	PFR	.7929388487	.6095058456	.5469094713	.4827307914	.4119451476	.3627177028	.3248273081	.2675058562	.2046823395
1	PF1	.9014263846	.7564016240	.7094041801	.6632360452	.6153931101	.5845198839	.5623747388	.5320260288	.5039142728
1	PQ1	.6828332677	.7349701948	.7651263335	.8074640121	.8748616307	.9417386411	1.010274061	1.158344987	1.431496453
1	PF2	1.073194688	.9786453259	.9487987289	.9200557540	.8910166007	.8727556376	.8599194639	.8427213468	.8272353921
1	PQ2	1.387307839	1.580197214	1.687399680	1.833815625	2.059338858	2.276318783	2.493505686	2.950269013	3.765899477
1	PFC	1.169497859	1.101952825	1.080633924	1.060133422	1.039469944	1.026509501	1.017418101	1.005265393	.9943536138
1	PQC	4.795058998	5.546728676	5.960155253	6.521131497	7.378955666	8.199089004	9.016314123	10.72672424	13.76432383
1	ZF1	1.984367739	1.984367739	1.984367739	1.984367739	1.984367739	1.984367739	1.984367739	1.984367739	1.984367739
1	ZF2	2.413193977	2.413193977	2.413193977	2.413193977	2.413193977	2.413193977	2.413193977	2.413193977	2.413193977
1	ZF3	4.210595375	4.210595375	4.210595375	4.210595375	4.210595375	4.210595375	4.210595375	4.210595375	4.210595375

(3) P 7+0 TH = 32 SBE = 1.887079915 CAUER FUNCTIONS

MU		RC = 02	RC = 05	RC = 07	RC = 10	RC = 15	RC = 20	RC = 25	RC = 35	RC = 50
	AMAX DB	.0017375255	.0108709564	.0213327376	.0436480540	.0988323393	.1772876696	.2802872360	.5675287486	1.249387366
	AMIN DB	72.02431261	79.99224598	82.92526846	86.04562296	89.62263242	92.19986248	94.24106231	97.45086453	101.2307623
	K	.3052248D-2	.1219616D-2	.8701056D-3	.6075112D-3	.4024424D-3	.2991178D-3	.2364734D-3	.1634151D-3	.1057541D-3
1	PFR	.7993417746	.6137122968	.5505001690	.4857509518	.4143995650	.3648132950	.3266637580	.2689749022	.2057773085
1	PF1	.9064903496	.7602135513	.7128572708	.6663578091	.6181927702	.5871226924	.5648414496	.5343129259	.5060412624
1	PQ1	.6852854400	.7371515434	.7672653292	.8095953393	.8770409939	.9440009489	1.012639951	1.160966742	1.434638720
1	PF2	1.074329911	.9800203185	.9502239111	.9215189812	.8925086797	.8742612085	.8614324684	.8442417566	.8287600558
1	PQ2	1.402376806	1.594786302	1.702194483	1.849112059	2.075670402	2.293805704	2.512236916	2.971775868	3.792619190
1	PFC	1.167300796	1.100774726	1.079738831	1.059494037	1.039072550	1.026256054	1.017262107	1.005235219	.9944321489
1	PQC	4.872379816	5.624159956	6.039606438	6.604260085	7.468841602	8.296139632	9.120900566	10.84781486	13.91593868
1	ZF1	1.928189667	1.928189667	1.928189667	1.928189667	1.928189667	1.928189667	1.928189667	1.928189667	1.928189667
1	ZF2	2.340983918	2.340983918	2.340983918	2.340983918	2.340983918	2.340983918	2.340983918	2.340983918	2.340983918
1	ZF3	4.075602454	4.075602454	4.075602454	4.075602454	4.075602454	4.075602454	4.075602454	4.075602454	4.075602454

(3) P 7+0 — TH = 33 SBE = 1.836078459 — CAUER FUNCTIONS

MU		RC = 02	RC = 05	RC = 07	RC = 10	RC = 15	RC = 20	RC = 25	RC = 35	RC = 50
	AMAX DB	.0017375255	.0108709564	.0213327376	.0436480540	.0988323393	.1772876696	.2802872360	.5675287486	1.249387366
	AMIN DB	70.04196439	78.00989763	80.94292009	84.06327459	87.64028405	90.21751411	92.25871393	95.46851616	99.24841397
	K	.3711302D-2	.1482961D-2	.1057983D-2	.7386875D-3	.4893395D-3	.3637046D-3	.2875337D-3	.1987004D-3	.1285890D-3
1	PFR	.8060563849	.6181124105	.5542534686	.4889057803	.4169617792	.3670000740	.3285796077	.2705069286	.2069188713
1	PF1	.9117569599	.7641782278	.7164486600	.6696044954	.6211043321	.5898294376	.5674066002	.5366909968	.5082529694
1	PQ1	.6878749060	.7394453847	.7695118880	.8118313120	.8793248468	.9463701347	1.015116470	1.163709431	1.437924179
1	PF2	1.075480445	.9814307750	.9516891981	.9230260914	.8940478186	.8758155280	.8629952728	.8458132069	.8303367192
1	PQ2	1.418318057	1.610171989	1.717782992	1.865215584	2.092849619	2.312191118	2.531923829	2.994369861	3.820678435
1	PFC	1.165041988	1.099561412	1.078816468	1.058834796	1.038662525	1.025994376	1.017100907	1.005203798	.9945129106
1	PQC	4.954364144	5.706054056	6.123575776	6.692057083	7.563713913	8.398533820	9.231215712	10.97549307	14.07575214
1	ZF1	1.875622844	1.875622844	1.875622844	1.875622844	1.875622844	1.875622844	1.875622844	1.875622844	1.875622844
1	ZF2	2.273259020	2.273259020	2.273259020	2.273259020	2.273259020	2.273259020	2.273259020	2.273259020	2.273259020
1	ZF3	3.948646987	3.948646987	3.948646987	3.948646987	3.948646987	3.948646987	3.948646987	3.948646987	3.948646987

(3) P 7+0 — TH = 34 SBE = 1.788291650 — CAUER FUNCTIONS

MU		RC = 02	RC = 05	RC = 07	RC = 10	RC = 15	RC = 20	RC = 25	RC = 35	RC = 50
	AMAX DB	.0017375255	.0108709564	.0213327376	.0436480540	.0988323393	.1772876696	.2802872360	.5675287486	1.249387366
	AMIN DB	68.11123578	76.07916882	79.01219126	82.13254575	85.70955520	88.28678526	90.32798508	93.53778731	97.31768512
	K	.4489676D-2	.1793983D-2	.1279874D-2	.8936130D-3	.5919691D-3	.4399846D-3	.3478384D-3	.2403740D-3	.1555581D-3
1	PFR	.8130961354	.6227133842	.5581751274	.4921998253	.4196352592	.3692808858	.3305772818	.2721038031	.2081083735
1	PF1	.9172303470	.7682989587	.7201813681	.6729788479	.6241302607	.5926424154	.5700723597	.5391622486	.5105512560
1	PQ1	.6906095048	.7418571965	.7718709739	.8141765075	.8817175104	.9488504605	1.017707916	1.166577585	1.441358023
1	PF2	1.076643957	.9828756950	.9531939245	.9245767056	.8956338931	.8774186152	.8646079897	.8474359264	.8319657063
1	PQ2	1.435183699	1.626397107	1.734206473	1.882166818	2.110917780	2.331517864	2.552611325	3.018101067	3.850137818
1	PFC	1.162722249	1.098313123	1.077866970	1.058155756	1.038239878	1.025724456	1.016934481	1.005171104	.9945958704
1	PQC	5.041311933	5.792676175	6.212325342	6.784787150	7.663848922	8.506563540	9.347569517	11.11011070	14.24419700
1	ZF1	1.826351274	1.826351274	1.826351274	1.826351274	1.826351274	1.826351274	1.826351274	1.826351274	1.826351274
1	ZF2	2.209624550	2.209624550	2.209624550	2.209624550	2.209624550	2.209624550	2.209624550	2.209624550	2.209624550
1	ZF3	3.829015661	3.829015661	3.829015661	3.829015661	3.829015661	3.829015661	3.829015661	3.829015661	3.829015661

(3) P 7+0 — TH = 35 SBE = 1.743446796 — CAUER FUNCTIONS

MU		RC = 02	RC = 05	RC = 07	RC = 10	RC = 15	RC = 20	RC = 25	RC = 35	RC = 50
	AMAX DB	.0017375255	.0108709564	.0213327376	.0436480540	.0988323393	.1772876696	.2802872360	.5675287486	1.249387366
	AMIN DB	66.22879995	74.19673269	77.12975510	80.25010957	83.82711902	86.40434907	88.44554889	91.65535112	95.43524893
	K	.5405403D-2	.2159889D-2	.1540921D-2	.1075877D-2	.7127087D-3	.5297251D-3	.4187845D-3	.2894013D-3	.1872861D-3
1	PFR	.8204755520	.6275229390	.5622713100	.4956379483	.4224237068	.3716587641	.3326593634	.2737675138	.2093472461
1	PF1	.9229147820	.7725792023	.7240585628	.6764837496	.6272731498	.5955640381	.5728410136	.5417287972	.5129380864
1	PQ1	.6934977831	.7443929100	.7743479504	.8166358622	.8842236359	.9514465084	1.020418906	1.169576070	1.444945821
1	PF2	1.077817974	.9843540069	.9547373721	.9261704084	.8972667577	.8790704769	.8662707251	.8491101440	.8336473480
1	PQ2	1.453030577	1.643508141	1.751509644	1.900009715	2.129919488	2.351832203	2.574347863	3.043023489	3.881062659
1	PFC	1.160342426	1.097030113	1.076890478	1.057456980	1.037804617	1.025446282	1.016762804	1.005137105	.9946809983
1	PQC	5.133551915	5.884315592	6.306140627	6.882738247	7.769546940	8.620545885	9.470298446	11.25204940	14.42174254
1	ZF1	1.780095181	1.780095181	1.780095181	1.780095181	1.780095181	1.780095181	1.780095181	1.780095181	1.780095181
1	ZF2	2.149730934	2.149730934	2.149730934	2.149730934	2.149730934	2.149730934	2.149730934	2.149730934	2.149730934
1	ZF3	3.716076458	3.716076458	3.716076458	3.716076458	3.716076458	3.716076458	3.716076458	3.716076458	3.716076458

(3) P 7+0	TH = 36	SBE = 1.701301617						CAUER FUNCTIONS	
	RC = 02	RC = 05	RC = 07	RC = 10	RC = 15	RC = 20	RC = 25	RC = 35	RC = 50
AMAX DB	.0017375255	.0108709564	.0213327376	.0436480540	.0988323393	.1772876696	.2802872360	.5675287486	1.249387366
AMIN DB	64.39159869	72.35953097	75.29255334	78.41290779	81.98991722	84.56714727	86.60834709	89.81814931	93.59804713
MU I K	.6478823D-2	.2588805D-2	.1846921D-2	.1289527D-2	.8542404D-3	.6349194D-3	.5019478D-3	.3468714D-3	.2244779D-3
1 I PFR	.8282103317	.6325493631	.5665486213	.4992253489	.4253310743	.3741369441	.3348286051	.2755001768	.2106370111
1 I PF1	.9288146713	.7770225742	.7280835646	.6801222281	.6305357278	.5985968456	.5757149692	.5443928732	.5154155322
1 I PQ1	.6965490708	.7470589529	.7769486175	.8192147035	.8868482331	.9541632081	1.023254403	1.172710110	1.448693545
1 I PF2	1.078999884	.9858645635	.9563187662	.9278067447	.8989462437	.8807711057	.8679835772	.8508360890	.8353819817
1 I PQ2	1.471920789	1.661555591	1.769741011	1.918791885	2.149902986	2.373184132	2.597185783	3.069195413	3.913523411
1 I PFC	1.157903408	1.095712648	1.075887142	1.056738535	1.037356753	1.025159841	1.016585856	1.005101773	.9947682626
1 I PQC	5.231444874	5.981288237	6.405333000	6.986224041	7.881134690	8.740825593	9.599768124	11.40172366	14.60889808
1 I ZF1	1.736605987	1.736605987	1.736605987	1.736605987	1.736605987	1.736605987	1.736605987	1.736605987	1.736605987
1 I ZF2	2.093267485	2.093267485	2.093267485	2.093267485	2.093267485	2.093267485	2.093267485	2.093267485	2.093267485
1 I ZF3	3.609267353	3.609267353	3.609267353	3.609267353	3.609267353	3.609267353	3.609267353	3.609267353	3.609267353

(3) P 7+0	TH = 37	SBE = 1.661640141						CAUER FUNCTIONS	
	RC = 02	RC = 05	RC = 07	RC = 10	RC = 15	RC = 20	RC = 25	RC = 35	RC = 50
AMAX DB	.0017375255	.0108709564	.0213327376	.0436480540	.0988323393	.1772876696	.2802872360	.5675287486	1.249387366
AMIN DB	62.59681205	70.56474365	73.49776596	76.61812037	80.19512979	82.77235983	84.81355965	88.02336137	91.80325968
MU I K	.7732846D-2	.3089888D-2	.2204406D-2	.1539125D-2	.1019585D-2	.7578126D-3	.5991035D-3	.4140109D-3	.2679272D-3
1 I PFR	.8363174551	.6378015611	.5710141428	.5029675913	.4283615833	.3767188781	.3370879418	.2773040464	.2119792884
1 I PF1	.9349345495	.7816328515	.7322598520	.6838974626	.6339208648	.6017435119	.5786967625	.5471568277	.5179857780
1 I PQ1	.6997735663	.7498622985	.7796792539	.8219187864	.8895967032	.9570058674	1.026219746	1.175985316	1.452607609
1 I PF2	1.080186924	.9874061364	.9579372711	.9294852169	.9006721562	.8825204783	.8697466346	.8526139895	.8371699508
1 I PQ2	1.491922254	1.680594388	1.788953248	1.938564951	2.170920513	2.395627737	2.621181674	3.096679808	3.947596134
1 I PFC	1.155406117	1.094361013	1.074857122	1.056000495	1.036896300	1.024865123	1.016403613	1.005065076	.9948576302
1 I PQC	5.335387379	6.083939608	6.510242482	7.095586618	7.998968059	8.867777892	9.736376305	11.55958408	14.80621702
1 I ZF1	1.695662098	1.695662098	1.695662098	1.695662098	1.695662098	1.695662098	1.695662098	1.695662098	1.695662098
1 I ZF2	2.039957148	2.039957148	2.039957148	2.039957148	2.039957148	2.039957148	2.039957148	2.039957148	2.039957148
1 I ZF3	3.508086841	3.508086841	3.508086841	3.508086841	3.508086841	3.508086841	3.508086841	3.508086841	3.508086841

(3) P 7+0	TH = 38	SBE = 1.624269245						CAUER FUNCTIONS	
	RC = 02	RC = 05	RC = 07	RC = 10	RC = 15	RC = 20	RC = 25	RC = 35	RC = 50
AMAX DB	.0017375255	.0108709564	.0213327376	.0436480540	.0988323393	.1772876696	.2802872360	.5675287486	1.249387366
AMIN DB	60.84183200	68.80976260	71.74278482	74.86313918	78.44014857	81.01737861	83.05857842	86.26838063	90.04827845
MU I K	.9193251D-2	.3673436D-2	.2620724D-2	.1829800D-2	.1212141D-2	.9009311D-3	.7122486D-3	.4921999D-3	.3185273D-3
1 I PFR	.8448153137	.6432891081	.5756754734	.5068706349	.4315197465	.3794082519	.3394405043	.2791815248	.2133758024
1 I PF1	.9412790716	.7864139773	.7365910679	.6878127901	.6374315790	.6050068521	.5817890651	.5500231390	.5206511283
1 I PQ1	.7031824328	.7528105208	.7825466632	.8247543339	.8924748742	.9599802066	1.029320684	1.179407727	1.456694900
1 I PF2	1.081376175	.9889774112	.9595919858	.9312052811	.9024442721	.8843185535	.8715599749	.8544440712	.8390116040
1 I PQ2	1.513109380	1.700684351	1.809203622	1.959384951	2.193028687	2.419221588	2.646396781	3.125544764	3.983363018
1 I PFC	1.152851517	1.092975504	1.073800589	1.055242940	1.036423274	1.024562117	1.016216052	1.005026930	.9949490662
1 I PQC	5.445816046	6.192648073	6.621240889	7.211199549	8.123435186	9.001811698	9.880556232	11.72612113	15.01430139
1 I ZF1	1.657065355	1.657065355	1.657065355	1.657065355	1.657065355	1.657065355	1.657065355	1.657065355	1.657065355
1 I ZF2	1.989552079	1.989552079	1.989552079	1.989552079	1.989552079	1.989552079	1.989552079	1.989552079	1.989552079
1 I ZF3	3.412085962	3.412085962	3.412085962	3.412085962	3.412085962	3.412085962	3.412085962	3.412085962	3.412085962

(3) P 7+0	TH = 39	SBE = 1.589015729						CAUER FUNCTIONS	
	RC = 02	RC = 05	RC = 07	RC = 10	RC = 15	RC = 20	RC = 25	RC = 35	RC = 50
AMAX DB	.0017375255	.0108709564	.0213327376	.0436480540	.0988323393	.1772876696	.2802872360	.5675287486	1.249387366
AMIN DB	59.12423942	67.09216856	70.02519064	73.14554494	76.72255429	79.29978431	81.34098411	84.55078632	88.33068413
MU I K	.1088901D-1	.4351027D-2	.3104135D-2	.2167319D-2	.1435729D-2	.1067114D-2	.8436279D-3	.5829896D-3	.3772819D-3
1 I PFR	.8537238529	.6490223105	.5805407744	.5109408667	.4348103911	.3822090033	.3418896349	.2811351735	.2148233902
1 I PF1	.9478530018	.7913700648	.7410810252	.6918717128	.6410710448	.6083898299	.5849946913	.5529944203	.5234140140
1 I PQ1	.7067879092	.7559118563	.7855582268	.8277280826	.8954890411	.9630923958	1.032563412	1.182983839	1.460962826
1 I PF2	1.082564553	.9905769814	.9612819392	.9329663431	.9042623368	.8861652697	.8734236629	.8563265566	.8409072945
1 I PQ2	1.535563803	1.721890710	1.830554472	1.981312790	2.216288949	2.444029185	2.672897456	3.155863992	4.020912969
1 I PFC	1.150240615	1.091556439	1.072717723	1.054465953	1.035937694	1.024250814	1.016023151	1.004737453	.9950425336
1 I PQC	5.563212428	6.307828628	6.738735404	7.333471361	8.254959966	9.143373241	10.03278042	11.90186937	15.23380689
1 I ZF1	1.620638033	1.620638033	1.620638033	1.620638033	1.620638033	1.620638033	1.620638033	1.620638033	1.620638033
1 I ZF2	1.941829903	1.941829903	1.941829903	1.941829903	1.941829903	1.941829903	1.941829903	1.941829903	1.941829903
1 I ZF3	3.320861549	3.320861549	3.320861549	3.320861549	3.320861549	3.320861549	3.320861549	3.320861549	3.320861549

(3) P 7+0	TH = 40	SBE = 1.555723827						CAUER FUNCTIONS	
	RC = 02	RC = 05	RC = 07	RC = 10	RC = 15	RC = 20	RC = 25	RC = 35	RC = 50
AMAX DB	.0017375255	.0108709564	.0213327376	.0436480540	.0988323393	.1772876696	.2802872360	.5675287486	1.249387366
AMIN DB	57.44178391	65.40971093	68.34273281	71.46308700	75.04009630	77.61732630	79.65852610	82.86832330	86.64822610
MU I K	.1285266D-1	.5135663D-2	.3663915D-2	.2558160D-2	.1694639D-2	.1259551D-2	.9957623D-3	.6881222D-3	.4453184D-3
1 I PFR	.8630647342	.6550122736	.5856188197	.5151851390	.4382386851	.3851253428	.3444389041	.2831677258	.2163390103
1 I PF1	.9546612000	.7965054007	.7457337130	.6960779051	.6448426002	.6118955655	.5883166062	.5560734273	.5262770007
1 I PQ1	.7106034348	.7591752741	.7887219626	.8308473329	.8986460110	.9663490982	1.035954614	1.186720653	1.465419358
1 I PF2	1.083748804	.9922033424	.9630060848	.9347677541	.9061260613	.8880605425	.8753377488	.8582616634	.8428573794
1 I PQ2	1.559375248	1.744284704	1.853073753	2.004414748	2.240768055	2.470119454	2.700755675	3.187717381	4.060342273
1 I PFC	1.147574456	1.090104148	1.071608719	1.053669628	1.035439580	1.023931206	1.015824886	1.004946459	.9951379937
1 I PQC	5.688108630	6.429937191	6.863172638	7.462849491	8.394006017	9.292950161	10.19356493	12.08741219	15.46544868
1 I ZF1	1.586220293	1.586220293	1.586220293	1.586220293	1.586220293	1.586220293	1.586220293	1.586220293	1.586220293
1 I ZF2	1.896590529	1.896590529	1.896590529	1.896590529	1.896590529	1.896590529	1.896590529	1.896590529	1.896590529
1 I ZF3	3.234050487	3.234050487	3.234050487	3.234050487	3.234050487	3.234050487	3.234050487	3.234050487	3.234050487

(3) P 7+0	TH = 41	SBE = 1.524253087						CAUER FUNCTIONS	
	RC = 02	RC = 05	RC = 07	RC = 10	RC = 15	RC = 20	RC = 25	RC = 35	RC = 50
AMAX DB	.0017375255	.0108709564	.0213327376	.0436480540	.0988323393	.1772876696	.2802872360	.5675287486	1.249387366
AMIN DB	55.79236599	63.76028998	66.69331158	69.81366562	73.39067483	75.96790480	78.00910459	81.21890678	84.99880458
MU I K	.1512073D-1	.6041934D-2	.4310472D-2	.3009588D-2	.1993685D-2	.1481819D-2	.1171481D-2	.8095525D-3	.5239021D-3
1 I PFR	.8728615184	.6612709781	.5909190521	.5196108100	.4418101661	.3881617765	.3470921296	.2852821007	.2179097518
1 I PF1	.9617086042	.8018244492	.7505533030	.7004352215	.6487497557	.6155273447	.5917579349	.5592630664	.5292427960
1 I PQ1	.7146437934	.7626105550	.7920465918	.8341200069	.9019531532	.9697575166	1.039501508	1.190625722	1.470073086
1 I PF2	1.084925493	.9938548844	.9647632949	.9366088068	.9080351187	.8900042625	.8773022655	.8602496033	.8448622187
1 I PQ2	1.584642505	1.767944246	1.876835649	2.028763055	2.266538635	2.497567309	2.730049616	3.221191620	4.101755329
1 I PFC	1.144854133	1.088618983	1.070473780	1.052854064	1.034928957	1.023603285	1.015621234	1.004903961	.9952354058
1 I PQC	5.821093770	6.559475490	6.995043261	7.599824769	8.541081191	9.451076164	10.36347427	12.28338725	15.71000794
1 I ZF1	1.553668001	1.553668001	1.553668001	1.553668001	1.553668001	1.553668001	1.553668001	1.553668001	1.553668001
1 I ZF2	1.853653435	1.853653435	1.853653435	1.853653435	1.853653435	1.853653435	1.853653435	1.853653435	1.853653435
1 I ZF3	3.151324814	3.151324814	3.151324814	3.151324814	3.151324814	3.151324814	3.151324814	3.151324814	3.151324814

(3) P 7+0	TH = 42	SBE = 1.494476550						CAUER FUNCTIONS	
	RC = 02	RC = 05	RC = 07	RC = 10	RC = 15	RC = 20	RC = 25	RC = 35	RC = 50
AMAX DB	.0017375255	.0108709564	.0213327376	.0436480540	.0988323393	.1772876696	.2802872360	.5675287486	1.249387366
AMIN DB	54.17402146	62.14194111	65.07496230	68.19531612	71.77232522	74.34955516	76.39075492	79.60055710	83.38045489
MU I K	.1773415D-1	.7086203D-2	.5055481D-2	.3529756D-2	.2338268D-2	.1737932D-2	.1373956D-2	.9494731D-3	.6144517D-3
1 I PFR	.8831398736	.6678113648	.5964516465	.5242257901	.4455307741	.3913231306	.3498533966	.2874814180	.2195428450
1 I PF1	.9690002095	.8073318544	.7555441554	.7049477050	.6527962031	.6192886277	.5953219709	.5625664045	.5323142593
1 I PQ1	.7189252775	.7662283808	.7955416142	.8375547126	.9054184557	.9733254475	1.043211899	1.194707198	1.474933272
1 I PF2	1.086090995	.9955298861	.9665523550	.9384887296	.9099891405	.8919962916	.8793172268	.8622905799	.8469221745
1 I PQ2	1.611474556	1.792954691	1.901921271	2.054436544	2.293679819	2.526454286	2.760864303	3.256380908	4.145265487
1 I PFC	1.142080784	1.087101315	1.069313126	1.052019368	1.034405854	1.023267046	1.015412174	1.004359924	.9953347269
1 I PQC	5.962821460	6.696996691	7.134887304	7.744936552	8.696742716	9.618336320	10.54312688	12.49049261	15.96833922
1 I ZF1	1.522850865	1.522850865	1.522850865	1.522850865	1.522850865	1.522850865	1.522850865	1.522850865	1.522850865
1 I ZF2	1.812855345	1.812855345	1.812855345	1.812855345	1.812855345	1.812855345	1.812855345	1.812855345	1.812855345
1 I ZF3	3.072387516	3.072387516	3.072387516	3.072387516	3.072387516	3.072387516	3.072387516	3.072387516	3.072387516

(3) P 7+0	TH = 43	SBE = 1.466279186						CAUER FUNCTIONS	
	RC = 02	RC = 05	RC = 07	RC = 10	RC = 15	RC = 20	RC = 25	RC = 35	RC = 50
AMAX DB	.0017375255	.0108709564	.0213327376	.0436480540	.0988323393	.1772876696	.2802872360	.5675287486	1.249387366
AMIN DB	52.58490734	60.55282082	63.48584144	66.60619496	70.18320389	72.76043377	74.80163351	78.01143566	81.79133344
MU I K	.2073883D-1	.8286815D-2	.5912029D-2	.4127802D-2	.2734440D-2	.2032389D-2	.1606745D-2	.1110342D-2	.7185580D-3
1 I PFR	.8939278131	.6746474315	.6022275795	.5290385928	.4494068869	.3946145804	.3527270812	.2897690152	.2212406734
1 I PF1	.9765410418	.8130324431	.7607108253	.7096195951	.6569858257	.6231830596	.5990121871	.5659366784	.5354944109
1 I PQ1	.7234658776	.7700404369	.7992173931	.8411608168	.9090505891	.9770613402	1.047094238	1.198373897	1.480009919
1 I PF2	1.087241488	.9972265064	.9683719568	.9404066820	.9119877129	.8940364606	.8813826247	.8643847872	.8490376095
1 I PQ2	1.639991868	1.819409712	1.928419451	2.081521393	2.322277957	2.556869261	2.793292348	3.293387748	4.190995987
1 I PFC	1.139255594	1.085551536	1.068126987	1.051165656	1.033870303	1.022922483	1.015197680	1.004314306	.9954359118
1 I PQC	6.114018469	6.843111845	7.283300230	7.898778590	8.861603062	9.795373107	10.73320145	12.70949367	16.24137890
1 I ZF1	1.493650827	1.493650827	1.493650827	1.493650827	1.493650827	1.493650827	1.493650827	1.493650827	1.493650827
1 I ZF2	1.774048220	1.774048220	1.774048220	1.774048220	1.774048220	1.774048220	1.774048220	1.774048220	1.774048220
1 I ZF3	2.996968914	2.996968914	2.996968914	2.996968914	2.996968914	2.996968914	2.996968914	2.996968914	2.996968914

(3) P 7+0	TH = 44	SBE = 1.439556540						CAUER FUNCTIONS	
	RC = 02	RC = 05	RC = 07	RC = 10	RC = 15	RC = 20	RC = 25	RC = 35	RC = 50
AMAX DB	.0017375255	.0108709564	.0213327376	.0436480540	.0988323393	.1772876696	.2802872360	.5675287486	1.249387366
AMIN DB	51.02328946	58.99119423	61.92421404	65.04456712	68.62157583	71.19880562	73.24000533	76.44980745	80.22970521
MU I K	.2418624D-1	.9664328D-2	.6894782D-2	.4813964D-2	.3188984D-2	.2370232D-2	.1873833D-2	.1294913D-2	.8380035D-3
1 I PFR	.9052559659	.6817943409	.6082587099	.5340583928	.4534453609	.3980416808	.3557178761	.2921484665	.2230057862
1 I PF1	.9843361257	.8189312256	.7660580685	.7144553372	.6613227082	.6272144809	.6028322462	.5695273060	.5387864432
1 I PQ1	.7282855002	.7740595281	.8030852529	.8449485274	.9128589783	.9809743651	1.051157685	1.203435362	1.485313840
1 I PF2	1.088372944	.9989427764	.9702206914	.9423617484	.9140303718	.8961245653	.8834984262	.8665324078	.8512088861
1 I PQ2	1.670327894	1.847412296	1.956427647	2.110111965	2.352427424	2.588909264	2.827434781	3.332323851	4.239081018
1 I PFC	1.136379800	1.083970059	1.066915612	1.050293053	1.033322339	1.022569596	1.014977733	1.004767070	.9955389127
1 I PQC	6.275494800	6.998497332	7.440939926	8.062005762	9.036336667	9.982893328	10.93444415	12.94123126	16.53015483
1 I ZF1	1.465960679	1.465960679	1.465960679	1.465960679	1.465960679	1.465960679	1.465960679	1.465960679	1.465960679
1 I ZF2	1.737097533	1.737097533	1.737097533	1.737097533	1.737097533	1.737097533	1.737097533	1.737097533	1.737097533
1 I ZF3	2.924823532	2.924823532	2.924823532	2.924823532	2.924823532	2.924823532	2.924823532	2.924823532	2.924823532

(3) P 7+0	TH = 45 SBE = 1.414213562								CAUER FUNCTIONS
	RC = 02	RC = 05	RC = 07	RC = 10	RC = 15	RC = 20	RC = 25	RC = 35	RC = 50
AMAX DB	.0017375255	.0108709564	.0213327376	.0436480540	.0988323393	.1772876696	.2802872360	.5675287486	1.249387366
AMIN DB	49.48753126	57.45542380	60.38844246	63.50879494	67.08580332	69.66303300	71.70423265	74.91403473	78.69393246
MU K	.2813401D-1	.1124178D-1	.8020174D-2	.5599717D-2	.3709502D-2	.2757110D-2	.2179687D-2	.1506274D-2	.9747855D-3
1 PFR	.9171578873	.6892685434	.6145578674	.5392950899	.4576535762	.4016104020	.3588308189	.2946236036	.2248409132
1 PF1	.9923904457	.8250333966	.7715908477	.7194595912	.6658111478	.6313869392	.6067860125	.5731918982	.5421937318
1 PQ1	.7334062206	.7782997116	.8071575886	.8489289875	.9168538840	.9850744890	1.055412185	1.208101939	1.490856743
1 PF2	1.089481117	1.000676591	.9720970417	.9443529322	.9161165986	.8982603627	.8856645707	.8687336102	.8534363649
1 PQ2	1.702630813	1.877075893	1.986052983	2.140311772	2.384231547	2.622680404	2.863401999	3.373311156	4.289666930
1 PFC	1.133454690	1.082357323	1.065679264	1.049401692	1.032762004	1.022208383	1.014752309	1.004718172	.9956436795
1 PQC	6.448155459	7.163903472	7.608534763	8.235341827	9.221687664	10.18167604	11.14767687	13.18663069	16.83579730
1 ZF1	1.439682860	1.439682860	1.439682860	1.439682860	1.439682860	1.439682860	1.439682860	1.439682860	1.439682860
1 ZF2	1.701880767	1.701880767	1.701880767	1.701880767	1.701880767	1.701880767	1.701880767	1.701880767	1.701880767
1 ZF3	2.855727391	2.855727391	2.855727391	2.855727391	2.855727391	2.855727391	2.855727391	2.855727391	2.855727391

(3) P 7+0	TH = 46 SBE = 1.390163591								CAUER FUNCTIONS
	RC = 02	RC = 05	RC = 07	RC = 10	RC = 15	RC = 20	RC = 25	RC = 35	RC = 50
AMAX DB	.0017375255	.0108709564	.0213327376	.0436480540	.0988323393	.1772876696	.2802872360	.5675287486	1.249387366
AMIN DB	47.97608373	55.94395918	58.87697625	61.99732788	65.57433581	68.15156533	70.19276491	73.40256692	77.13246462
MU K	.3264672D-1	.1304496D-1	.9306614D-2	.6497914D-2	.4304508D-2	.3199352D-2	.2529310D-2	.1747881D-2	.1131142D-2
1 PFR	.9296704164	.6970879168	.6211389537	.5447593826	.4620394868	.4053271680	.3620713248	.2971385391	.2267489802
1 PF1	1.000708899	.8313443334	.7773143376	.7246372413	.6704556664	.6357047020	.6108775652	.5759342714	.5457198483
1 PQ1	.7388525751	.7827764491	.8114479926	.8531143816	.9210464950	.9893725622	1.059868553	1.212984860	1.496651322
1 PF2	1.090561532	1.002425699	.9739993743	.9463791485	.9182458146	.9004435668	.8878809661	.8709885467	.8557204031
1 PQ2	1.737065543	1.908525736	2.017413438	2.172234578	2.417803659	2.658298928	2.901314844	3.416483001	4.342913596
1 PFC	1.130481608	1.080713788	1.064418223	1.048491720	1.032189344	1.021838846	1.014521386	1.004667571	.9957501591
1 PQC	6.633014232	7.340164505	7.786892930	8.419588381	9.418478775	10.39259170	11.37380671	13.44671230	17.15955169
1 ZF1	1.414728409	1.414728409	1.414728409	1.414728409	1.414728409	1.414728409	1.414728409	1.414728409	1.414728409
1 ZF2	1.668286117	1.668286117	1.668286117	1.668286117	1.668286117	1.668286117	1.668286117	1.668286117	1.668286117
1 ZF3	2.789475645	2.789475645	2.789475645	2.789475645	2.789475645	2.789475645	2.789475645	2.789475645	2.789475645

(3) P 7+0	TH = 47 SBE = 1.367327461								CAUER FUNCTIONS
	RC = 02	RC = 05	RC = 07	RC = 10	RC = 15	RC = 20	RC = 25	RC = 35	RC = 50
AMAX DB	.0017375255	.0108709564	.0213327376	.0436480540	.0988323393	.1772876696	.2802872360	.5675287486	1.249387366
AMIN DB	46.48747635	54.45532801	57.38834286	60.50869332	64.08570062	66.66292992	68.70412940	71.91393132	75.69382897
MU K	.3779673D-1	.1510280D-1	.1077473D-1	.7522959D-2	.4983543D-2	.3704049D-2	.2928308D-2	.2023609D-2	.1309579D-2
1 PFR	.9428340866	.7052719242	.6280170575	.5504628498	.4666116778	.4091989013	.3654452220	.2998776927	.2237331277
1 PF1	1.009296236	.8378695936	.7832339299	.7299934058	.6752610226	.6401722697	.6151117116	.5809084618	.5493685744
1 PQ1	.7446519030	.7875067824	.8159713984	.8575180572	.9254490331	.9938804161	1.064538565	1.218096336	1.592711361
1 PF2	1.091609477	1.004187693	.9759259300	.9484392171	.9204173753	.9026738442	.8901474852	.8732973504	.8580613530
1 PQ2	1.773816116	1.941900370	2.050639217	2.206005666	2.453268308	2.695892426	2.941305841	3.461985448	4.398995983
1 PFC	1.127461955	1.079039945	1.063132788	1.047563292	1.031604411	1.021460999	1.014284945	1.004615222	.9958592956
1 PQC	6.831209892	7.528210220	7.976913261	8.615635221	9.627621602	10.61656269	11.61383698	13.72260356	17.50279302
1 ZF1	1.391016063	1.391016063	1.391016063	1.391016063	1.391016063	1.391016063	1.391016063	1.391016063	1.391016063
1 ZF2	1.636211350	1.636211350	1.636211350	1.636211350	1.636211350	1.636211350	1.636211350	1.636211350	1.636211350
1 ZF3	2.725880525	2.725880525	2.725880525	2.725880525	2.725880525	2.725880525	2.725880525	2.725880525	2.725880525

(3) P 7+0	TH = 48	SBE = 1.345632730							CAUER FUNCTIONS
	RC = 02	RC = 05	RC = 07	RC = 10	RC = 15	RC = 20	RC = 25	RC = 35	RC = 50
AMAX DB	.0017375255	.0108709564	.0213327376	.0436480540	.0988323393	.1772876696	.2802872360	.5675287486	1.249387366
AMIN DB	45.02030878	52.98812751	55.92113929	59.04148811	62.61849454	65.19572354	67.23692287	70.44672467	74.22662226
MU I K	.4366513D-1	.1744770D-1	.1244764D-1	.8690990D-2	.5757299D-2	.4279148D-2	.3382963D-2	.2337799D-2	.1512907D-2
1 I PFR	.9566936029	.7138417952	.6352085845	.5564180440	.4713794300	.4132330719	.3689587923	.3026658208	.2307967305
1 I PF1	1.018156996	.8446149093	.7893552374	.7355334467	.6802322255	.6447943910	.6194915029	.5849687408	.5531439171
1 I PQ1	.7508347450	.7925095348	.8207442470	.8621546648	.9300748734	.9986109747	1.069435073	1.223449672	1.509051850
1 I PF2	1.092619989	1.005960000	.9778748151	.9505318540	.9226305643	.9049508086	.8924639614	.8756601327	.8604595595
1 I PQ2	1.813088440	1.977353418	2.085874340	2.241763293	2.490762649	2.735601218	2.983520608	3.509978808	4.458105929
1 I PFC	1.124397193	1.077336310	1.061823277	1.046616579	1.031007261	1.021074820	1.014042965	1.004561079	.9959680301
1 I PQC	7.044025334	7.729079534	8.179597860	8.824472400	9.850128557	10.85467552	11.86887983	14.01555306	17.86704271
1 I ZF1	1.368471449	1.368471449	1.368471449	1.368471449	1.368471449	1.368471449	1.368471449	1.368471449	1.368471449
1 I ZF2	1.605562808	1.605562808	1.605562808	1.605562808	1.605562808	1.605562808	1.605562808	1.605562808	1.605562808
1 I ZF3	2.664769530	2.664769530	2.664769530	2.664769530	2.664769530	2.664769530	2.664769530	2.664769530	2.664769530

(3) P 7+0	TH = 49	SBE = 1.325012993							CAUER FUNCTIONS
	RC = 02	RC = 05	RC = 07	RC = 10	RC = 15	RC = 20	RC = 25	RC = 35	RC = 50
AMAX DB	.0017375255	.0108709564	.0213327376	.0436480540	.0988323393	.1772876696	.2802872360	.5675287486	1.249387366
AMIN DB	43.57324338	51.54101669	54.47402423	57.59437080	61.17137602	63.74860460	65.78980374	68.99960537	72.77950287
MU I K	.5034280D-1	.2011596D-1	.1435124D-1	.1002010D-1	.6637757D-2	.4933554D-2	.3900317D-2	.2695316D-2	.1744275D-2
1 I PFR	.9712983944	.7228207325	.6427314057	.5626385974	.4763527921	.4174377535	.3726188169	.3055680492	.2329434203
1 I PF1	1.027295420	.8515861798	.7956840971	.7412629801	.6853745490	.6495760778	.6240232500	.5891696316	.5570501250
1 I PQ1	.7574353115	.7978055450	.8257846779	.8670403173	.9349386810	1.003578382	1.074572119	1.229059381	1.515689122
1 I PF2	1.093587845	1.007739865	.9798439900	.9526556631	.9248845863	.9072740157	.8948301841	.8780769793	.8629153584
1 I PQ2	1.855113560	2.015055640	2.123278474	2.279660381	2.530438041	2.777579953	3.028119483	3.560639387	4.520454197
1 I PFC	1.121288846	1.075603433	1.060490029	1.045651761	1.030397959	1.020680347	1.013795426	1.004505095	.9960793001
1 I PQC	7.272910254	7.943936424	8.396066860	9.047204286	10.08712674	11.10809503	12.14017101	14.32694671	18.25398805
1 I ZF1	1.347026393	1.347026393	1.347026393	1.347026393	1.347026393	1.347026393	1.347026393	1.347026393	1.347026393
1 I ZF2	1.576254545	1.576254545	1.576254545	1.576254545	1.576254545	1.576254545	1.576254545	1.576254545	1.576254545
1 I ZF3	2.605983841	2.605983841	2.605983841	2.605983841	2.605983841	2.605983841	2.605983841	2.605983841	2.605983841

(3) P 7+0	TH = 50	SBE = 1.305407289							CAUER FUNCTIONS
	RC = 02	RC = 05	RC = 07	RC = 10	RC = 15	RC = 20	RC = 25	RC = 35	RC = 50
AMAX DB	.0017375255	.0108709564	.0213327376	.0436480540	.0988323393	.1772876696	.2802872360	.5675287486	1.249387366
AMIN DB	42.14499831	50.11270920	53.04571091	56.16605439	59.74305796	62.32028596	64.36148484	67.57128623	71.35118361
MU I K	.5793165D-1	.2314831D-1	.1651460D-1	.1153056D-1	.7638356D-2	.5677255D-2	.4488264D-2	.3101619D-2	.2007213D-2
1 I PFR	.9867032591	.7322341504	.6506050276	.5691393419	.4815426636	.4218216869	.3764326274	.3085399108	.2351771114
1 I PF1	1.036715352	.8587894595	.8022265718	.7471878858	.6906935465	.6545226228	.6287115418	.5935159277	.5610917075
1 I PQ1	.7644920341	.8034179390	.8311127503	.8721927746	.9400565686	1.008798149	1.079965085	1.234941332	1.522641004
1 I PF2	1.094507545	1.009524345	.9818312583	.9548091260	.9271785593	.9096429567	.8972458937	.8805479467	.8554290739
1 I PQ2	1.900151522	2.055197337	2.163029075	2.319866473	2.572461903	2.821999447	3.075279402	3.614161501	4.586272828
1 I PFC	1.118138506	1.073841892	1.059133408	1.044669035	1.029776576	1.020277586	1.013542311	1.004447220	.9961920397
1 I PQC	7.519508164	8.174088676	8.627575744	9.285066013	10.33987414	11.37813092	12.42908697	14.65832663	18.66550478
1 I ZF1	1.326618303	1.326618303	1.326618303	1.326618303	1.326618303	1.326618303	1.326618303	1.326618303	1.326618303
1 I ZF2	1.548207550	1.548207550	1.548207550	1.548207550	1.548207550	1.548207550	1.548207550	1.548207550	1.548207550
1 I ZF3	2.549376921	2.549376921	2.549376921	2.549376921	2.549376921	2.549376921	2.549376921	2.549376921	2.549376921

(3) P 7+0	TH = 51	SBE = 1.286759566							CAUER FUNCTIONS
	RC = 02	RC = 05	RC = 07	RC = 10	RC = 15	RC = 20	RC = 25	RC = 35	RC = 50
AMAX DB	.0017375255	.0108709564	.0213327376	.0436480540	.0988323393	.1772876696	.2802872360	.5675287486	1.249387366
AMIN DB	40.73434123	48.70196665	51.63496039	54.75529963	58.33230094	60.90952815	62.95072666	66.16052773	69.94042494
MU I K	.6654602D-1	.2659043D-1	.1897030D-1	.1324514D-1	.8774171D-2	.6521456D-2	.5155664D-2	.3562825D-2	.2305683D-2
1 I PFR	1.002969118	.7421099481	.6588507867	.5759364470	.4869608885	.4263943518	.3804081649	.3117373872	.2375020288
1 I PF1	1.046420118	.8662309425	.8089889501	.7533143177	.6961950673	.6596396178	.6335617645	.5980127133	.5652734545
1 I PQ1	.7720482181	.8093724462	.8367507002	.8776316572	.9454462783	1.014287320	1.085630949	1.241112905	1.529926994
1 I PF2	1.095373300	1.011310290	.9838342546	.9569905918	.9295115067	.9120570515	.8997107756	.8830730582	.8680010153
1 I PQ2	1.948495945	2.097991161	2.205323882	2.362570006	2.617019864	2.869049823	3.125196067	3.670759802	4.655817864
1 I PFC	1.114947835	1.072052304	1.057753801	1.043669615	1.029143193	1.019866552	1.013283603	1.004387405	.9963061793
1 I PQC	7.785688711	8.421010059	8.875535768	9.539442831	10.60977867	11.66624711	12.73716498	15.01141319	19.10368350
1 I ZF1	1.307189627	1.307189627	1.307189627	1.307189627	1.307189627	1.307189627	1.307189627	1.307189627	1.307189627
1 I ZF2	1.521349077	1.521349077	1.521349077	1.521349077	1.521349077	1.521349077	1.521349077	1.521349077	1.521349077
1 I ZF3	2.494813278	2.494813278	2.494813278	2.494813278	2.494813278	2.494813278	2.494813278	2.494813278	2.494813278

(3) P 7+0	TH = 52	SBE = 1.269018215							CAUER FUNCTIONS
	RC = 02	RC = 05	RC = 07	RC = 10	RC = 15	RC = 20	RC = 25	RC = 35	RC = 50
AMAX DB	.0017375255	.0108709564	.0213327376	.0436480540	.0988323393	.1772876696	.2802872360	.5675287486	1.249387366
AMIN DB	39.34008355	47.30759229	50.24057514	53.36090860	56.93790683	59.51513296	61.55633096	64.76613160	68.54602858
MU I K	.7631426D-1	.3049362D-1	.2175494D-1	.1518938D-1	.1006212D-1	.7478736D-2	.5912460D-2	.4085809D-2	.2644132D-2
1 I PFR	1.020163902	.7524788272	.6674920741	.5830475776	.4926203630	.4311660487	.3845540456	.3150169571	.2399227415
1 I PF1	1.056412382	.8739169416	.8159777437	.7596487126	.7018852737	.6649329735	.6385796228	.6026653853	.5696004593
1 I PQ1	.7801528185	.8156977700	.8427232389	.8833786931	.9511273909	1.020064667	1.091587972	1.247593179	1.537568455
1 I PF2	1.096179020	1.013094330	.9858504303	.9591982655	.9318823477	.9145156413	.9022244539	.8856522991	.8706314745
1 I PQ2	2.000479489	2.143675434	2.250383848	2.407980978	2.664318268	2.918937992	3.178086470	3.730671977	4.729372494
1 I PFC	1.111718571	1.070235320	1.056351620	1.042650729	1.028497896	1.019447268	1.013019286	1.004325598	.9964216450
1 I PQC	8.073586524	8.686366644	9.141538118	9.811892928	10.89842050	11.97408451	13.06612661	15.38813085	19.57086049
1 I ZF1	1.288687371	1.288687371	1.288687371	1.288687371	1.288687371	1.288687371	1.288687371	1.288687371	1.288687371
1 I ZF2	1.495612045	1.495612045	1.495612045	1.495612045	1.495612045	1.495612045	1.495612045	1.495612045	1.495612045
1 I ZF3	2.442167364	2.442167364	2.442167364	2.442167364	2.442167364	2.442167364	2.442167364	2.442167364	2.442167364

(3) P 7+0	TH = 53	SBE = 1.252135658							CAUER FUNCTIONS
	RC = 02	RC = 05	RC = 07	RC = 10	RC = 15	RC = 20	RC = 25	RC = 35	RC = 50
AMAX DB	.0017375255	.0108709564	.0213327376	.0436480540	.0988323393	.1772876696	.2802872360	.5675287486	1.249387366
AMIN DB	37.96107508	45.92842502	48.86139306	51.98171865	55.55871267	58.13593733	60.17713466	63.38693470	67.16683136
MU I K	.8738059D-1	.3491550D-1	.2490962D-1	.1739200D-1	.1152123D-1	.8563227D-2	.6769825D-2	.4678292D-2	.3027558D-2
1 I PFR	1.038363597	.7633746597	.6765545948	.5904920758	.4985351577	.4361479926	.3888796371	.3184356505	.2424441986
1 I PF1	1.066693975	.8818538604	.8231996822	.7661977995	.7077706586	.6704089410	.6437711629	.6074796779	.5740781437
1 I PQ1	.7888613652	.8224260230	.8490579035	.8894580063	.9571215693	1.026150911	1.097856915	1.254403145	1.545588852
1 I PF2	1.096918294	1.014872858	.9878770397	.9614301949	.9342898872	.9170179800	.9047864847	.8882856112	.8733207222
1 I PQ2	2.056480412	2.192518062	2.298456581	2.456334068	2.714587105	2.971900552	3.234191835	3.794161890	4.807250718
1 I PFC	1.108452526	1.068391634	1.054927308	1.041615626	1.027840785	1.019019760	1.012749348	1.004261743	.9965383588
1 I PQC	8.385648169	8.972048205	9.427382638	10.10417448	11.20757850	12.30348786	13.41790552	15.79063859	20.06965408
1 I ZF1	1.271062679	1.271062679	1.271062679	1.271062679	1.271062679	1.271062679	1.271062679	1.271062679	1.271062679
1 I ZF2	1.470934506	1.470934506	1.470934506	1.470934506	1.470934506	1.470934506	1.470934506	1.470934506	1.470934506
1 I ZF3	2.391322596	2.391322596	2.391322596	2.391322596	2.391322596	2.391322596	2.391322596	2.391322596	2.391322596

(3) P 7+0			TH = 54	SBE = 1.236067977							CAUER FUNCTIONS
			RC = 02	RC = 05	RC = 07	RC = 10	RC = 15	RC = 20	RC = 25	RC = 35	RC = 50
AMAX DB			.0017375255	.0108709564	.0213327376	.0436480540	.0988323393	.1772876696	.2802872360	.5675287486	1.249387366
AMIN DB			36.59619928	44.56333368	47.49628161	50.61659651	54.19358483	56.77080749	58.81200389	62.02180313	65.80169937
MU		K	.9990715D-1	.3992086D-1	.2848058D-1	.1988525D-1	.1317288D-1	.9790820D-2	.7740323D-2	.5348955D-2	.3461578D-2
1		PFR	1.057653485	.7748349173	.6860666685	.5982911704	.5047206599	.4413524202	.3933951450	.3220011108	.2450717717
1		PF1	1.077265682	.8900481574	.8306617034	.7729686075	.7138580646	.6760741352	.6491427985	.6124516898	.5787122851
1		PQ1	.7982370746	.8295932405	.8557854677	.8958964540	.9634528417	1.032568988	1.104460287	1.261565949	1.554014016
1		PF2	1.097584377	1.016642015	.9899111230	.9636842573	.9367328048	.9195632251	.9073963477	.8909728875	.8760690036
1		PQ2	2.116930462	2.244821184	2.349820430	2.507892320	2.768083456	3.028197192	3.293781059	3.861523254	4.889801626
1		PFC	1.105151601	1.066521980	1.053481337	1.040563573	1.027171967	1.018584060	1.012473777	1.004195786	.9956562378
1		PQC	8.724689188	9.280205851	9.735112138	10.41827788	11.53926166	12.65653756	13.79468028	16.22136586	20.60300749
1		ZF1	1.254270452	1.254270452	1.254270452	1.254270452	1.254270452	1.254270452	1.254270452	1.254270452	1.254270452
1		ZF2	1.447259176	1.447259176	1.447259176	1.447259176	1.447259176	1.447259176	1.447259176	1.447259176	1.447259176
1		ZF3	2.342170480	2.342170480	2.342170480	2.342170480	2.342170480	2.342170480	2.342170480	2.342170480	2.342170480

(3) P 7+0			TH = 55	SBE = 1.220774589							CAUER FUNCTIONS
			RC = 02	RC = 05	RC = 07	RC = 10	RC = 15	RC = 20	RC = 25	RC = 35	RC = 50
AMAX DB			.0017375255	.0108709564	.0213327376	.0436480540	.0988323393	.1772876696	.2802872360	.5675287486	1.249387366
AMIN DB			35.24436892	43.21121144	46.14413213	49.26443257	52.84141316	55.41863312	57.45982827	60.66962642	64.44952208
MU		K	.1140765D 0	.4558265D-1	.3251984D-1	.2270548D-1	.1504112D-1	.1117941D-1	.8838096D-2	.6107572D-2	.3952517D-2
1		PFR	1.078129617	.7869011736	.6960595801	.6064682204	.5111937358	.4467927138	.3981117121	.3257216663	.2478113030
1		PF1	1.088126995	.8985063000	.8383709390	.7799684718	.7201547038	.6819355598	.6547013378	.6176179140	.5835090484
1		PQ1	.8083521890	.8372399901	.8629404269	.9027240220	.9701479332	1.039344347	1.111423136	1.269107181	1.562872448
1		PF2	1.098170169	1.018397665	.9919494895	.9659581435	.9392096420	.9221504273	.9100534379	.8937139651	.8788765346
1		PQ2	2.182324456	2.300926724	2.404789318	2.562951496	2.825095552	3.088119690	3.357154761	3.933083945	4.977414403
1		PFC	1.101817781	1.064627141	1.052014214	1.039494860	1.026491562	1.018140205	1.012192564	1.004127669	.9967751941
1		PQC	9.093963806	9.613297363	10.06705356	10.75646422	11.89574647	13.03558755	14.19891350	16.68305531	21.17423971
1		ZF1	1.238269015	1.238269015	1.238269015	1.238269015	1.238269015	1.238269015	1.238269015	1.238269015	1.238269015
1		ZF2	1.424533018	1.424533018	1.424533018	1.424533018	1.424533018	1.424533018	1.424533018	1.424533018	1.424533018
1		ZF3	2.294609832	2.294609832	2.294609832	2.294609832	2.294609832	2.294609832	2.294609832	2.294609832	2.294609832

(3) P 7+0			TH = 56	SBE = 1.206217949							CAUER FUNCTIONS
			RC = 02	RC = 05	RC = 07	RC = 10	RC = 15	RC = 20	RC = 25	RC = 35	RC = 50
AMAX DB			.0017375255	.0108709564	.0213327376	.0436480540	.0988323393	.1772876696	.2802872360	.5675287486	1.249387366
AMIN DB			33.90452228	41.87097043	44.80385432	47.92413520	51.50110536	54.07832166	56.11951512	59.32931180	63.10920667
MU		K	.1300944D 0	.5198306D-1	.3708606D-1	.2589363D-1	.1715310D-1	.1274914D-1	.1007908D-1	.6965157D-2	.4507503D-2
1		PFR	1.099900582	.7996196955	.7065679898	.6150489981	.5179729202	.4524835438	.4030415338	.3296064119	.2506691603
1		PF1	1.099275810	.9072347063	.8463346943	.7872050381	.7266681782	.6880006346	.6604540143	.6229552712	.5884750197
1		PQ1	.8192896009	.8454120963	.8705615724	.9099742922	.9772366543	1.046505304	1.118773288	1.277055206	1.572195685
1		PF2	1.098668196	1.020135382	.9939886974	.9682493409	.9417187886	.9247785189	.9127570556	.8965086182	.8817434959
1		PQ2	2.253231948	2.361223040	2.463718533	2.621845262	2.885947589	3.151995648	3.424650066	4.009211081	5.070524229
1		PFC	1.098453145	1.062707948	1.050526481	1.038409801	1.025799704	1.017688241	1.011905702	1.004057333	.9968951344
1		PQC	9.497250629	9.974142098	10.42586766	11.12131170	12.27962186	13.44331069	14.63339849	17.17881375	21.78710609
1		ZF1	1.223019816	1.223019816	1.223019816	1.223019816	1.223019816	1.223019816	1.223019816	1.223019816	1.223019816
1		ZF2	1.402706864	1.402706864	1.402706864	1.402706864	1.402706864	1.402706864	1.402706864	1.402706864	1.402706864
1		ZF3	2.248546061	2.248546061	2.248546061	2.248546061	2.248546061	2.248546061	2.248546061	2.248546061	2.248546061

(3) P 7+0	TH = 57 SBE = 1.192363293								CAUER FUNCTIONS
	RC = 02	RC = 05	RC = 07	RC = 10	RC = 15	RC = 20	RC = 25	RC = 35	RC = 50
AMAX DB	.0017375255	.0108709564	.0213327376	.0436480540	.0988323393	.1772876696	.2802872360	.5675287486	1.249387366
AMIN DB	32.57562004	40.54153632	43.47437056	46.59462506	50.17158115	52.74879252	54.78998370	57.99977840	61.77967221
MU I K	.1481929D 0	.5921487D-1	.4224542D-1	.2949592D-1	.1953941D-1	.1452279D-1	.1148127D-1	.7934140D-2	.5134581D-2
1 I PFR	1.123089633	.8130421411	.7176304146	.6240620200	.5250786361	.4584410344	.4081979909	.3336653040	.2536523006
1 I PF1	1.110708066	.9162396703	.8545604189	.7946862644	.7334065012	.6942772245	.6664085201	.6284811470	.5936172455
1 I PQ1	.8311448334	.8541615098	.8786926763	.9176849955	.9847523604	1.054083467	1.126541746	1.285441558	1.582018710
1 I PF2	1.099070590	1.021850418	.9960250325	.9705551148	.9442584671	.9274463008	.9155063951	.8993565494	.8846700277
1 I PQ2	2.330311543	2.426152949	2.527011663	2.684951380	2.951005456	3.220194118	3.496646275	4.090317041	5.169619254
1 I PFC	1.095059873	1.060765289	1.049018721	1.037308737	1.025096539	1.017228218	1.011613188	1.003984715	.9970159594
1 I PQC	9.938958665	10.36598788	10.81460928	11.51577157	12.69384320	13.88275324	15.10131533	17.71217328	22.44587092
1 I ZF1	1.208487159	1.208487159	1.208487159	1.208487159	1.208487159	1.208487159	1.208487159	1.208487159	1.208487159
1 I ZF2	1.381735085	1.381735085	1.381735085	1.381735085	1.381735085	1.381735085	1.381735085	1.381735085	1.381735085
1 I ZF3	2.203890539	2.203890539	2.203890539	2.203890539	2.203890539	2.203890539	2.203890539	2.203890539	2.203890539

(3) P 7+0	TH = 58 SBE = 1.179178403								CAUER FUNCTIONS
	RC = 02	RC = 05	RC = 07	RC = 10	RC = 15	RC = 20	RC = 25	RC = 35	RC = 50
AMAX DB	.0017375255	.0108709564	.0213327376	.0436480540	.0988323393	.1772876696	.2802872360	.5675287486	1.249387366
AMIN DB	31.25664281	39.22184297	42.15461036	45.27482935	48.85176647	51.42897121	53.47015932	56.67995134	60.45984374
MU I K	.1686345D 0	.6738292D-1	.4807272D-1	.3356456D-1	.2223466D-1	.1652605D-1	.1306499D-1	.9028569D-2	.5842840D-2
1 I PFR	1.147837249	.8272263885	.7292897976	.6335389361	.5325334520	.4646829566	.4135958040	.3379092701	.2567683437
1 I PF1	1.122417308	.9255272692	.8630556690	.8024204182	.7403781191	.7007736717	.6725730429	.6342034327	.5989432762
1 I PQ1	.8440284698	.8635473523	.8873833106	.9258986715	.9927324950	1.062114222	1.134763159	1.294301395	1.592380453
1 I PF2	1.099369062	1.023537685	.9980544850	.9728724872	.9468267155	.9301524283	.9183005329	.9022573808	.8876562230
1 I PQ2	2.414328568	2.496223451	2.595128975	2.752699151	3.020683600	3.293132315	3.573571631	4.176366628	5.275248883
1 I PFC	1.091640250	1.058800108	1.047491558	1.036192035	1.024382229	1.016760197	1.011315024	1.003909753	.9971375636
1 I PQC	10.42425941	10.79259293	11.23680084	11.94323623	13.14179785	14.35740084	15.60629866	18.28716493	23.15539478
1 I ZF1	1.194637964	1.194637964	1.194637964	1.194637964	1.194637964	1.194637964	1.194637964	1.194637964	1.194637964
1 I ZF2	1.361575290	1.361575290	1.361575290	1.361575290	1.361575290	1.361575290	1.361575290	1.361575290	1.361575290
1 I ZF3	2.160560015	2.160560015	2.160560015	2.160560015	2.160560015	2.160560015	2.160560015	2.160560015	2.160560015

(3) P 7+0	TH = 59 SBE = 1.166633397								CAUER FUNCTIONS
	RC = 02	RC = 05	RC = 07	RC = 10	RC = 15	RC = 20	RC = 25	RC = 35	RC = 50
AMAX DB	.0017375255	.0108709564	.0213327376	.0436480540	.0988323393	.1772876696	.2802872360	.5675287486	1.249387366
AMIN DB	29.94658971	37.91082705	40.84350454	43.96367576	47.54058739	50.11778321	52.15896718	55.36375561	59.14864609
MU I K	.1917162D 0	.7660589D-1	.5465263D-1	.3815868D-1	.2527801D-1	.1878803D-1	.1485325D-1	.1026435D-1	.6642573D-2
1 I PFR	1.174304263	.8422375253	.7415941828	.6435149887	.5403623837	.4712289546	.4192512146	.3423503366	.2600256575
1 I PF1	1.134394163	.9351032450	.8718280547	.8104160696	.7475919328	.7074988293	.6789563057	.6401305716	.6044612151
1 I PQ1	.8580691518	.8736371827	.8966898334	.9346634573	1.001219239	1.070637324	1.143476390	1.303674046	1.603324364
1 I PF2	1.099554876	1.025191722	1.000072723	.9751982138	.9494213681	.9328953944	.9211384132	.9052106429	.8907021200
1 I PQ2	2.506177027	2.572017600	2.668597579	2.825578410	3.095453283	3.371283686	3.655911430	4.269385642	5.388033671
1 I PFC	1.088196676	1.056813413	1.045945664	1.035060095	1.023656953	1.016284246	1.011011212	1.003332381	.9972598344
1 I PQC	10.95925254	11.25832701	11.69652266	12.40762251	13.62738498	14.87125865	16.15251989	18.90840806	23.92124037
1 I ZF1	1.181441556	1.181441556	1.181441556	1.181441556	1.181441556	1.181441556	1.181441556	1.181441556	1.181441556
1 I ZF2	1.342188062	1.342188062	1.342188062	1.342188062	1.342188062	1.342188062	1.342188062	1.342188062	1.342188062
1 I ZF3	2.118476085	2.118476085	2.118476085	2.118476085	2.118476085	2.118476085	2.118476085	2.118476085	2.118476085

(3) P 7+0	TH = 60	SBE = 1.154700538							CAUER FUNCTIONS
	RC = 02	RC = 05	RC = 07	RC = 10	RC = 15	RC = 20	RC = 25	RC = 35	RC = 50
AMAX DB	.0017375255	.0108709564	.0213327376	.0436480540	.0988323393	.1772876696	.2802872360	.5675287486	1.249387366
AMIN DB	28.64447823	36.60742254	39.53997927	42.66008633	46.23696370	48.81414753	50.85532596	54.06510955	57.84499747
MU I K	.2177748D 0	.8701839D-1	.6208117D-1	.4334532D-1	.2871387D-1	.2134176D-1	.1687214D-1	.1165950D-1	.7545451D-2
1 I PFR	1.202675676	.8581490358	.7545975206	.6540295574	.5485932507	.4781008109	.4251821966	.3470017789	.2634334577
1 I PF1	1.146625706	.9449728577	.8808851731	.8186820763	.7550573200	.7144620983	.6855676116	.6462716108	.6101797738
1 I PQ1	.8734173048	.8845085383	.9066765830	.9440340380	1.010260284	1.079697597	1.152775132	1.313603654	1.614899105
1 I PF2	1.099618821	1.026806668	1.002075061	.9775287575	.9520400339	.9356735119	.9240188326	.9082157626	.8938076934
1 I PQ2	2.606907098	2.654209066	2.748023844	2.904150462	3.175852573	3.455187648	3.744217795	4.368471187	5.508677213
1 I PFC	1.084731671	1.054806283	1.044381762	1.033913353	1.022920912	1.015800445	1.010701762	1.003752530	.9973826519
1 I PQC	11.55117545	11.76829684	12.19852453	12.91347416	14.15511344	15.42894952	16.74478773	19.58121927	24.74980128
1 I ZF1	1.168869467	1.168869467	1.168869467	1.168869467	1.168869467	1.168869467	1.168869467	1.168869467	1.168869467
1 I ZF2	1.323536714	1.323536714	1.323536714	1.323536714	1.323536714	1.323536714	1.323536714	1.323536714	1.323536714
1 I ZF3	2.077564703	2.077564703	2.077564703	2.077564703	2.077564703	2.077564703	2.077564703	2.077564703	2.077564703

(3) P 7+0	TH = 61	SBE = 1.143354068							CAUER FUNCTIONS
	RC = 02	RC = 05	RC = 07	RC = 10	RC = 15	RC = 20	RC = 25	RC = 35	RC = 50
AMAX DB	.0017375255	.0108709564	.0213327376	.0436480540	.0988323393	.1772876696	.2802872360	.5675287486	1.249387366
AMIN DB	27.34934575	35.31055518	38.24294981	41.36297074	44.93980214	47.51696987	49.55814085	52.76791796	56.54780243
MU I K	.2471932D 0	.9877337D-1	.7046748D-1	.4920067D-1	.3259272D-1	.2422474D-1	.1915134D-1	.1323454D-1	.8564738D-2
1 I PFR	1.233165354	.8750442337	.7683606342	.6651268087	.5572570987	.4853227611	.4314087067	.3518782964	.2670019248
1 I PF1	1.159094696	.9551406991	.8902345191	.8272275607	.7627841547	.7216734664	.6924168926	.6526362590	.6161083356
1 I PQ1	.8902497990	.8962508238	.9174173320	.9540727982	1.019909773	1.089345771	1.162558797	1.324139946	1.627159365
1 I PF2	1.099551175	1.028376222	1.004056434	.9798602596	.9546800717	.9384848925	.9269404217	.9112720496	.8969728449
1 I PQ2	2.717759812	2.743580154	2.834108663	2.989061458	3.262498524	3.545461418	3.839121535	4.474304148	5.637980495
1 I PFC	1.081247885	1.052779871	1.042800631	1.032752281	1.022174323	1.015308885	1.010386689	1.003670133	.9975058879
1 I PQC	12.20867023	12.32850317	12.74836454	13.46608882	14.73022261	16.03583478	17.38867183	20.31174615	25.64945988
1 I ZF1	1.156895267	1.156895267	1.156895267	1.156895267	1.156895267	1.156895267	1.156895267	1.156895267	1.156895267
1 I ZF2	1.305587077	1.305587077	1.305587077	1.305587077	1.305587077	1.305587077	1.305587077	1.305587077	1.305587077
1 I ZF3	2.037755733	2.037755733	2.037755733	2.037755733	2.037755733	2.037755733	2.037755733	2.037755733	2.037755733

(3) P 7+0	TH = 62	SBE = 1.132570051							CAUER FUNCTIONS
	RC = 02	RC = 05	RC = 07	RC = 10	RC = 15	RC = 20	RC = 25	RC = 35	RC = 50
AMAX DB	.0017375255	.0108709564	.0213327376	.0436480540	.0988323393	.1772876696	.2802872360	.5675287486	1.249387366
AMIN DB	26.06025342	34.01913669	36.95131388	40.07121926	43.64798897	46.22513511	48.26629610	51.47606450	55.25594435
MU I K	.2804074D 0	.1120451D 0	.7993585D-1	.5581152D-1	.3697204D-1	.2747969D-1	.2172461D-1	.1501280D-1	.9715539D-2
1 I PFR	1.266021839	.8930180026	.7829523859	.6768564761	.5663887042	.4929218673	.4379529811	.3569962217	.2707423419
1 I PF1	1.171778654	.9656104609	.8998833705	.8360618743	.7707828272	.7291435502	.6995147635	.6592349518	.6222570271
1 I PQ1	.9087758234	.9089676433	.9289970714	.9648512250	1.030229428	1.099639504	1.173033382	1.335339162	1.640166846
1 I PF2	1.099341671	1.029893610	1.006011351	.9821885066	.9573385633	.9413274233	.9299016252	.9143786804	.9001973916
1 I PQ2	2.840211151	2.841044225	2.927666352	3.081058875	3.356102095	3.642814480	3.941346617	4.589164382	5.776859339
1 I PFC	1.077748113	1.050735413	1.041203111	1.031577392	1.021417432	1.014809670	1.010066013	1.003585115	.9976294054
1 I PQC	12.94212693	12.94603859	13.35258324	14.07167660	15.35883261	16.69816413	18.09065577	21.10713258	26.62578206
1 I ZF1	1.145494408	1.145494408	1.145494408	1.145494408	1.145494408	1.145494408	1.145494408	1.145494408	1.145494408
1 I ZF2	1.288307305	1.288307305	1.288307305	1.288307305	1.288307305	1.288307305	1.288307305	1.288307305	1.288307305
1 I ZF3	1.998982518	1.998982518	1.998982518	1.998982518	1.998982518	1.998982518	1.998982518	1.998982518	1.998982518

(3) P 7+0	TH = 63	SBE = 1.122326238						CAUER FUNCTIONS	
	RC = 02	RC = 05	RC = 07	RC = 10	RC = 15	RC = 20	RC = 25	RC = 35	RC = 50
AMAX DB	.0017375255	.0108709564	.0213327376	.0436480540	.0988323393	.1772876696	.2802872360	.5675287486	1.249387366
AMIN DB	24.77629331	32.73205899	35.66394455	38.78369494	42.36038191	44.93749910	46.97864668	50.18840341	53.96827705
MU K	.3179150D 0	.1270324D 0	.9062818D-1	.6327695D-1	.4191747D-1	.3115542D-1	.2463052D-1	.1702094D-1	.1101510D-1
1 PFR	1.301535586	.9121789235	.7984510935	.6892748012	.5760271796	.5009284650	.4448398887	.3623737670	.2746672585
1 PF1	1.184648750	.9763846421	.9098386390	.8451945468	.7790642585	.7368836392	.7068725812	.6660789258	.6286368004
1 PQ1	.9292443457	.9227796982	.9415142128	.9764516307	1.041289936	1.110644612	1.184212678	1.347265175	1.653991444
1 PF2	1.098979449	1.031351537	1.007933862	.9845088932	.9600122816	.9441987398	.9329006786	.9175346807	.9034810531
1 PQ2	2.976028625	2.947673853	3.029648227	3.181011965	3.457486587	3.748066393	4.051727945	4.712449343	5.926365752
1 PFC	1.074235300	1.048674236	1.039590111	1.030389249	1.020650506	1.014302918	1.009739762	1.003497402	.9977530577
1 PQC	13.76412890	13.62933899	14.01892370	14.73755946	16.04813242	17.42326290	18.85832806	21.97572443	27.69175928
1 ZF1	1.134644082	1.134644082	1.134644082	1.134644082	1.134644082	1.134644082	1.134644082	1.134644082	1.134644082
1 ZF2	1.271667702	1.271667702	1.271667702	1.271667702	1.271667702	1.271667702	1.271667702	1.271667702	1.271667702
1 ZF3	1.961181489	1.961181489	1.961181489	1.961181489	1.961181489	1.961181489	1.961181489	1.961181489	1.961181489

(3) P 7+0	TH = 64	SBE = 1.112601940						CAUER FUNCTIONS	
	RC = 02	RC = 05	RC = 07	RC = 10	RC = 15	RC = 20	RC = 25	RC = 35	RC = 50
AMAX DB	.0017375255	.0108709564	.0213327376	.0436480540	.0988323393	.1772876696	.2802872360	.5675287486	1.249387366
AMIN DB	23.49659995	31.44818820	34.37968262	37.49922513	41.07580111	43.65287944	45.69400903	48.90375011	52.68361543
MU K	.3602862D 0	.1439630D 0	.1027069D 0	.7171038D-1	.4750415D-1	.3530775D-1	.2791323D-1	.1928946D-1	.1248318D-1
1 PFR	1.340048037	.9326518924	.8149462631	.7024456779	.5862167041	.5093767009	.4520973568	.3680313204	.2787906850
1 PF1	1.197668460	.9874641815	.9201066779	.8546352136	.7876399118	.7449057424	.7145025099	.6731803024	.6352595280
1 PQ1	.9519536620	.9378284182	.9550833286	.9889692818	1.053177649	1.122436585	1.196169679	1.359990849	1.668712688
1 PF2	1.098453017	1.032742139	1.009817510	.9868163803	.9626976553	.9470961959	.9359355815	.9207389040	.9068234364
1 PQ2	3.127344494	3.064736439	3.141172245	3.289937315	3.567610560	3.862168882	4.171233371	4.845697076	6.087714217
1 PFC	1.070712561	1.046597764	1.037962615	1.029188463	1.019873844	1.013788764	1.009407970	1.003406919	.9978766873
1 PQC	14.69003616	14.38850604	14.75661204	15.47242419	16.80661744	18.21976783	19.70062207	22.92732768	28.85811154
1 ZF1	1.124323097	1.124323097	1.124323097	1.124323097	1.124323097	1.124323097	1.124323097	1.124323097	1.124323097
1 ZF2	1.255640566	1.255640566	1.255640566	1.255640566	1.255640566	1.255640566	1.255640566	1.255640566	1.255640566
1 ZF3	1.924291780	1.924291780	1.924291780	1.924291780	1.924291780	1.924291780	1.924291780	1.924291780	1.924291780

(3) P 7+0	TH = 65	SBE = 1.103377919						CAUER FUNCTIONS	
	RC = 02	RC = 05	RC = 07	RC = 10	RC = 15	RC = 20	RC = 25	RC = 35	RC = 50
AMAX DB	.0017375255	.0108709564	.0213327376	.0436480540	.0988323393	.1772876696	.2802872360	.5675287486	1.249387366
AMIN DB	22.22036814	30.16635869	33.09732835	36.21659191	39.79301894	42.37004513	44.41115059	47.62087065	51.40072481
MU K	.4081751D 0	.1630984D 0	.1163586D 0	.8124206D-1	.5381836D-1	.4000083D-1	.3162343D-1	.2185339D-1	.1414243D-1
1 PFR	1.381963079	.9545813676	.8325407227	.7164420534	.5970074136	.5183051841	.4597568833	.3739918024	.2831283268
1 PF1	1.210791949	.9988479951	.9306930339	.8643935169	.7965217954	.7532226364	.7224175932	.6805521830	.6421381118
1 PQ1	.9772637272	.9542805474	.9698385890	1.002515052	1.065971684	1.135102446	1.208988330	1.373599713	1.684421491
1 PF2	1.097750194	1.034056929	1.011655274	.9891054471	.9653907279	.9500168283	.9390040668	.9239900081	.9102240182
1 PQ2	3.296751467	3.193739709	3.263560615	3.409031091	3.687596585	3.986233448	4.300991155	4.990114826	6.262314340
1 PFC	1.067183196	1.044507534	1.036321687	1.027975704	1.019087780	1.013267361	1.009070683	1.003313585	.9980001253
1 PQC	15.73875769	15.23572451	15.57671844	16.28664611	17.64439205	19.09792657	20.62812042	23.97353485	30.13867003
1 ZF1	1.114511760	1.114511760	1.114511760	1.114511760	1.114511760	1.114511760	1.114511760	1.114511760	1.114511760
1 ZF2	1.240200048	1.240200048	1.240200048	1.240200048	1.240200048	1.240200048	1.240200048	1.240200048	1.240200048
1 ZF3	1.888254859	1.888254859	1.888254859	1.888254859	1.888254859	1.888254859	1.888254859	1.888254859	1.888254859

(3) P 7+0 — TH = 66 SBE = 1.094636279 — CAUER FUNCTIONS

(3) P 7+0	RC = 02	RC = 05	RC = 07	RC = 10	RC = 15	RC = 20	RC = 25	RC = 35	RC = 50
AMAX DB	.0017375255	.0108709564	.0213327376	.0436480540	.0988323393	.1772876696	.2802872360	.5675287486	1.249387366
AMIN DB	20.94687932	28.98536730	31.81563247	34.93452144	38.51074845	41.08770461	43.12877766	46.33346949	50.11830866
MU K	.4623356D 0	.1847399D 0	.1317982D 0	.9202203D-1	.6095949D-1	.4530851D-1	.3581953D-1	.2475311D-1	.1601898D-1
1 PFR	1.427761599	.9781354260	.8513532713	.7313476545	.6084564890	.5277577813	.4678541614	.3802810972	.2876978676
1 PF1	1.223962128	1.010532391	.9416021264	.8744789693	.8057224553	.7618479141	.7306318332	.6882087577	.6492866098
1 PQ1	1.005612214	.9723339894	.9859381097	1.017218756	1.079796554	1.148743072	1.222765689	1.338188026	1.701222315
1 PF2	1.096858057	1.035286734	1.013439513	.9913700363	.9680871111	.9529573161	.9421035660	.9272364273	.9136821251
1 PQ2	3.487429063	3.336490358	3.398387933	3.539710050	3.818767606	4.121566183	4.442324508	5.147114928	6.451811709
1 PFC	1.063650710	1.042405203	1.034668488	1.026751706	1.018292680	1.012738884	1.008727956	1.003217319	.9981231898
1 PQC	16.93378507	16.18580809	16.49262611	17.19270733	18.57355853	20.06998162	21.65344422	25.12814164	31.54986434
1 ZF1	1.105191779	1.105191779	1.105191779	1.105191779	1.105191779	1.105191779	1.105191779	1.105191779	1.105191779
1 ZF2	1.225322028	1.225322028	1.225322028	1.225322028	1.225322028	1.225322028	1.225322028	1.225322028	1.225322028
1 ZF3	1.853014166	1.853014166	1.853014166	1.853014166	1.853014166	1.853014166	1.853014166	1.853014166	1.853014166

(3) P 7+0 — TH = 67 SBE = 1.086360377 — CAUER FUNCTIONS

(3) P 7+0	RC = 02	RC = 05	RC = 07	RC = 10	RC = 15	RC = 20	RC = 25	RC = 35	RC = 50
AMAX DB	.0017375255	.0108709564	.0213327376	.0436480540	.0988323393	.1772876696	.2802872360	.5675287486	1.249387366
AMIN DB	19.67553980	27.60396794	30.53328640	33.65167187	37.22762995	39.80449197	41.84552143	45.05517531	48.83499431
MU K	.5236393D 0	.2092356D 0	.1492741D 0	.1042237D 0	.6904245D-1	.5131623D-1	.4056904D-1	.2803526D-1	.1814303D-1
1 PFR	1.478020113	1.003510876	.8715219942	.7472591350	.6206295026	.5377845939	.4764298456	.3869285810	.2925193132
1 PF1	1.237108329	1.022510330	.9528368331	.8849007691	.8152549521	.7707960332	.7391602768	.6961654305	.6567203841
1 PQ1	1.037535623	.9922253273	1.003569504	1.033233373	1.094775466	1.163476101	1.237614633	1.403867342	1.719235853
1 PF2	1.095762888	1.036421619	1.015161896	.9936034905	.9707819304	.9559139341	.9452311672	.9306263394	.9171969090
1 PQ2	3.703312257	3.495170454	3.547543441	3.683664227	3.962693379	4.269712064	4.596795435	5.318360234	6.658139481
1 PFC	1.060118842	1.040292564	1.033004278	1.025517275	1.017488955	1.012203532	1.008379857	1.003118037	.9982456847
1 PQC	18.30459246	17.25692109	17.52064725	18.20574286	19.60872214	21.15066728	22.79175607	26.40768387	33.11134914
1 ZF1	1.096346166	1.096346166	1.096346166	1.096346166	1.096346166	1.096346166	1.096346166	1.096346166	1.096346166
1 ZF2	1.210984003	1.210984003	1.210984003	1.210984003	1.210984003	1.210984003	1.210984003	1.210984003	1.210984003
1 ZF3	1.818514748	1.818514748	1.818514748	1.818514748	1.818514748	1.818514748	1.818514748	1.818514748	1.818514748

(3) P 7+0 — TH = 68 SBE = 1.078534743 — CAUER FUNCTIONS

(3) P 7+0	RC = 02	RC = 05	RC = 07	RC = 10	RC = 15	RC = 20	RC = 25	RC = 35	RC = 50
AMAX DB	.0017375255	.0108709564	.0213327376	.0436480540	.0988323393	.1772876696	.2802872360	.5675287486	1.249387366
AMIN DB	18.40593545	26.32086719	29.24891175	32.36661942	35.94221542	38.51895066	40.55992144	43.76952421	47.54931606
MU K	.5930977D 0	.2369898D 0	.1690746D 0	.1180486D 0	.7820063D-1	.5812310D-1	.4595034D-1	.3175402D-1	.2054962D-1
1 PFR	1.533434770	1.030939754	.8932084490	.7642887683	.6336020927	.5484431705	.4855304976	.3939677714	.2976154144
1 PF1	1.250143538	1.034770483	.9643979532	.8956675516	.8251328156	.7800823611	.7480191099	.7044389621	.6644562749
1 PQ1	1.073697299	1.014239599	1.022957045	1.050740460	1.111059529	1.179439599	1.253667271	1.420767739	1.738602388
1 PF2	1.094450110	1.037450806	1.016813323	.9957984801	.9734697627	.9588824982	.9483835679	.9340076284	.9207673193
1 PQ2	3.949319175	3.672438084	3.713312403	3.842925340	4.121250361	4.432511883	4.766260958	5.505322176	6.883584141
1 PFC	1.056591597	1.038171559	1.031330433	1.024273298	1.016677063	1.011661529	1.008026466	1.003015653	.9983673983
1 PQC	19.88855676	18.47154423	18.68084146	19.34426150	20.76765410	22.35785996	24.06141676	27.83213662	34.84681824
1 ZF1	1.087959159	1.087959159	1.087959159	1.087959159	1.087959159	1.087959159	1.087959159	1.087959159	1.087959159
1 ZF2	1.197164987	1.197164987	1.197164987	1.197164987	1.197164987	1.197164987	1.197164987	1.197164987	1.197164987
1 ZF3	1.784702884	1.784702884	1.784702884	1.784702884	1.784702884	1.784702884	1.784702884	1.784702884	1.784702884

(3) P 7+0	TH = 69	SBE = 1.071144994						CAUER	FUNCTIONS
	RC = 02	RC = 05	RC = 07	RC = 10	RC = 15	RC = 20	RC = 25	RC = 35	RC = 50
AMAX DB	.0017375255	.0108709564	.0213327376	.0436480540	.0988323393	.1772876696	.2802872360	.5675287486	1.249387366
AMIN DB	17.13790883	25.03472147	27.96104898	31.07784259	34.65295016	37.22951436	39.27040595	42.47993977	46.25969498
MU I K	.6718904D 0	.2684737D 0	.1915361D 0	.1337312D 0	.8858953D-1	.6584471D-1	.5205482D-1	.3597251D-1	.2327962D-1
1 I PFR	1.594852483	1.060697669	.9166030022	.7825678564	.6474620697	.5598000233	.4952097674	.4014371344	.3030121900
1 I PF1	1.262961151	1.047296036	.9762835088	.9067870527	.8353699680	.7897232140	.7572257597	.7130476357	.6725128050
1 I PQ1	1.114924980	1.038723146	1.044370998	1.069957144	1.128828138	1.196796766	1.271079289	1.439041914	1.759486037
1 I PF2	1.092904251	1.038362582	1.018383830	.9979469178	.9761445624	.9618583011	.9515570181	.9374278390	.9243920697
1 I PQ2	4.231662228	3.871561571	3.898483754	4.019956595	4.296699824	4.612176191	4.952945945	5.711855740	7.130870196
1 I PFC	1.053073289	1.036044303	1.029648462	1.023020752	1.015857516	1.011113135	1.007667883	1.002910078	.9984881015
1 I PQC	21.73362692	19.85778372	19.99811667	20.63110940	22.07217233	23.71343905	25.48485264	29.42583503	36.78507420
1 I ZF1	1.080016141	1.080016141	1.080016141	1.080016141	1.080016141	1.080016141	1.080016141	1.080016141	1.080016141
1 I ZF2	1.183845426	1.183845426	1.183845426	1.183845426	1.183845426	1.183845426	1.183845426	1.183845426	1.183845426
1 I ZF3	1.751525696	1.751525696	1.751525696	1.751525696	1.751525696	1.751525696	1.751525696	1.751525696	1.751525696

(3) P 7+0	TH = 70	SBE = 1.064177772						CAUER	FUNCTIONS
	RC = 02	RC = 05	RC = 07	RC = 10	RC = 15	RC = 20	RC = 25	RC = 35	RC = 50
AMAX DB	.0017375255	.0108709564	.0213327376	.0436480540	.0988323393	.1772876696	.2802872360	.5675287486	1.249387366
AMIN DB	15.87166665	23.74413725	26.66814562	29.78370376	33.35815090	35.93448380	37.97526830	41.18470885	44.96441451
MU I K	.7613989D 0	.3042395D 0	.2170523D 0	.1515468D 0	.1003913D 0	.7461647D-1	.5898950D-1	.4076473D-1	.2638091D-1
1 I PFR	1.663311559	1.093114609	.9419317021	.8022510891	.6623120895	.5719325447	.5055298805	.4093810974	.3087395804
1 I PF1	1.275431214	1.060063155	.9884878344	.9182656555	.8459806013	.7997358841	.7667990031	.7220114488	.6809104243
1 I PQ1	1.162261638	1.066100698	1.068139941	1.091145288	1.148295980	1.215741997	1.290035545	1.458870433	1.792080177
1 I PF2	1.091108913	1.039144181	1.019862485	1.000039859	.9787995753	.9648360369	.9547472539	.9408841236	.9280695984
1 I PQ2	4.558278750	4.096600797	4.106494309	4.217772093	4.491790954	4.811382503	5.159538583	5.939297407	7.403270477
1 I PFC	1.049568602	1.033913102	1.027960017	1.021760721	1.015030887	1.010558641	1.007304225	1.002301224	.9986075459
1 I PQC	23.90208838	21.45117120	21.50373084	22.09477485	23.54932648	25.24444095	27.08971559	31.21870201	38.96145094
1 I ZF1	1.072503581	1.072503581	1.072503581	1.072503581	1.072503581	1.072503581	1.072503581	1.072503581	1.072503581
1 I ZF2	1.171007119	1.171007119	1.171007119	1.171007119	1.171007119	1.171007119	1.171007119	1.171007119	1.171007119
1 I ZF3	1.718930728	1.718930728	1.718930728	1.718930728	1.718930728	1.718930728	1.718930728	1.718930728	1.718930728

(3) P 7+0	TH = 71	SBE = 1.057620681						CAUER	FUNCTIONS
	RC = 02	RC = 05	RC = 07	RC = 10	RC = 15	RC = 20	RC = 25	RC = 35	RC = 50
AMAX DB	.0017375255	.0108709564	.0213327376	.0436480540	.0988323393	.1772876696	.2802872360	.5675287486	1.249387366
AMIN DB	14.60792769	22.44767603	25.36854433	28.48242806	32.05597967	34.63199883	36.67263807	39.88195211	43.66159055
MU I K	.8632508D 0	.3449374D 0	.2460873D 0	.1718191D 0	.1138206D 0	.8459788D-1	.6688050D-1	.4621781D-1	.2990987D-1
1 I PFR	1.740095031	1.128589112	.9694652281	.8235221812	.6782730844	.5849314555	.5165635309	.4178513317	.3148322727
1 I PF1	1.287396188	1.073039023	1.001000383	.9301077781	.8569789908	.8101386474	.7767590810	.7313523363	.6896718016
1 I PQ1	1.217035026	1.096898400	1.094667221	1.114623620	1.169722230	1.236508814	1.310757363	1.430468537	1.806614445
1 I PF2	1.089046795	1.039781664	1.021237259	1.002067386	.9814272360	.9678097114	.9579494178	.9443731787	.9317980203
1 I PQ2	4.939435096	4.352655732	4.341624675	4.440097714	4.709898719	5.033404700	5.389317112	6.191594692	7.704751592
1 I PFC	1.046082661	1.031780483	1.026266922	1.020494404	1.014197819	1.009998384	1.006935633	1.002638998	.9987254612
1 I PQC	26.47595406	23.29717655	23.23737092	23.77118126	25.23301515	26.98462786	28.91045579	33.24790748	41.41973104
1 I ZF1	1.065408967	1.065408967	1.065408967	1.065408967	1.065408967	1.065408967	1.065408967	1.065408967	1.065408967
1 I ZF2	1.158633155	1.158633155	1.158633155	1.158633155	1.158633155	1.158633155	1.158633155	1.158633155	1.158633155
1 I ZF3	1.686865484	1.686865484	1.686865484	1.686865484	1.686865484	1.686865484	1.686865484	1.686865484	1.686865484

(3) P 7+0		TH = 72 SBE = 1.051462224								CAUER FUNCTIONS
		RC = 02	RC = 05	RC = 07	RC = 10	RC = 15	RC = 20	RC = 25	RC = 35	RC = 50
AMAX DB		.0017375255	.0108709564	.0213327376	.0436480540	.0988323393	.1772876696	.2802872360	.5675287486	1.249387366
AMIN DB		13.34812288	21.14386722	24.06047170	27.17207884	30.74441200	33.32000417	35.36044568	38.56958749	42.34913442
MU	K	.9793747D 0	.3913382D 0	.2791908D 0	.1949321D 0	.1291317D 0	.9597793D-1	.7587722D-1	.5243500D-1	.3393332D-1
1	PFR	1.826800982	1.167607047	.9995306857	.8466012498	.6954887188	.5989039678	.5283963190	.4269083971	.3213307539
1	PF1	1.298666363	1.086179308	1.013804154	.9423150427	.8683792174	.8209507388	.7871278147	.7410944309	.6988221776
1	PQ1	1.280953824	1.131775265	1.124453251	1.140784040	1.193422802	1.259380377	1.333512150	1.504095082	1.833363902
1	PF2	1.086699782	1.040259766	1.022494882	1.004018465	.9840190460	.9707725345	.9611579635	.9478911682	.9355750698
1	PQ2	5.388586173	4.646212432	4.609269540	4.691591178	4.955210921	5.282287779	5.646320431	6.472979828	8.040168245
1	PFC	1.042621144	1.029649226	1.024571192	1.019223142	1.013359039	1.009432749	1.006562274	1.002573308	.9988415532
1	PQC	29.56481196	25.45477593	25.25007864	25.70619035	27.16622725	28.97665288	30.99048639	35.56014417	44.21476814
1	ZF1	1.058720761	1.058720761	1.058720761	1.058720761	1.058720761	1.058720761	1.058720761	1.058720761	1.058720761
1	ZF2	1.146707862	1.146707862	1.146707862	1.146707862	1.146707862	1.146707862	1.146707862	1.146707862	1.146707862
1	ZF3	1.655276904	1.655276904	1.655276904	1.655276904	1.655276904	1.655276904	1.655276904	1.655276904	1.655276904

(3) P 7+0		TH = 73 SBE = 1.045691756								CAUER FUNCTIONS
		RC = 02	RC = 05	RC = 07	RC = 10	RC = 15	RC = 20	RC = 25	RC = 35	RC = 50
AMAX DB		.0017375255	.0108709564	.0213327376	.0436480540	.0988323393	.1772876696	.2802872360	.5675287486	1.249387366
AMIN DB		12.09465977	19.83123352	22.74202936	25.85052953	29.42119825	31.99620686	34.03637809	37.24528447	41.02470629
MU	K	.1112071D 1	.4443611D 0	.3170187D 0	.2213437D 0	.1466279D 0	.1089821D 0	.8615791D-1	.5953948D-1	.3853099D-1
1	PFR	1.925435627	1.210766819	1.032527352	.8717545968	.7141312516	.6139779261	.5411299312	.4366238768	.3282826726
1	PF1	1.309015229	1.099424890	1.026873609	.9548851436	.8801947567	.8321922783	.7979287175	.7512643680	.7083897947
1	PQ1	1.356241957	1.171566815	1.158126196	1.170113854	1.219786904	1.284703596	1.358626277	1.530064444	1.862661173
1	PF2	1.084049165	1.040561734	1.023620663	1.005880786	.9865654264	.9737167903	.9643665405	.9514336300	.9393980303
1	PQ2	5.923616760	4.985633046	4.916316841	4.978147993	5.232986022	5.563087499	5.935581286	6.788706185	8.415526985
1	PFC	1.039190419	1.027522411	1.022875068	1.017948430	1.012515368	1.008862179	1.006184348	1.002454060	.9989555000
1	PQC	33.31745090	28.00161254	27.60844354	27.95915914	29.40420196	31.27509875	33.38521267	38.21480065	47.41613178
1	ZF1	1.052428347	1.052428347	1.052428347	1.052428347	1.052428347	1.052428347	1.052428347	1.052428347	1.052428347
1	ZF2	1.135216760	1.135216760	1.135216760	1.135216760	1.135216760	1.135216760	1.135216760	1.135216760	1.135216760
1	ZF3	1.624110757	1.624110757	1.624110757	1.624110757	1.624110757	1.624110757	1.624110757	1.624110757	1.624110757

(3) P 7+0		TH = 74 SBE = 1.040299436								CAUER FUNCTIONS
		RC = 02	RC = 05	RC = 07	RC = 10	RC = 15	RC = 20	RC = 25	RC = 35	RC = 50
AMAX DB		.0017375255	.0108709564	.0213327376	.0436480540	.0988323393	.1772876696	.2802872360	.5675287486	1.249387366
AMIN DB		10.85126037	18.50833637	21.41118990	24.51543150	28.08381585	30.65802307	32.69782305	35.90640598	39.68565591
MU	K	.1264107D 1	.5051112D 0	.3603594D 0	.2516044D 0	.1666739D 0	.1238814D 0	.9793687D-1	.6767934D-1	.4379870D-1
1	PFR	2.038536716	1.258813647	1.068947994	.8993078679	.7344093608	.6303073050	.5548863463	.4470831866	.3357446217
1	PF1	1.318175389	1.112697638	1.040171900	.9678102948	.8924378778	.8438841170	.8091870920	.7618916444	.7194064249
1	PQ1	1.445828228	1.217346622	1.196484893	1.203226608	1.249299773	1.312908409	1.386502583	1.558762618	1.894912790
1	PF2	1.081076069	1.040669131	1.024598284	1.007640554	.9890555394	.9766336786	.9675678526	.9549953618	.9432636481
1	PQ2	6.568662600	5.381863000	5.271689725	5.307334612	5.549914839	5.882204789	6.263450506	7.145375510	8.838349662
1	PFC	1.035797738	1.025403463	1.021181052	1.016671952	1.011667740	1.008287183	1.005802092	1.002331163	.9990669491
1	PQC	37.93940954	31.04161576	30.40073334	30.60809365	32.01896893	33.95082295	36.16634963	41.28346599	51.11326506
1	ZF1	1.046521998	1.046521998	1.046521998	1.046521998	1.046521998	1.046521998	1.046521998	1.046521998	1.046521998
1	ZF2	1.124146540	1.124146540	1.124146540	1.124146540	1.124146540	1.124146540	1.124146540	1.124146540	1.124146540
1	ZF3	1.593310913	1.593310913	1.593310913	1.593310913	1.593310913	1.593310913	1.593310913	1.593310913	1.593310913

(3) P 7+0 — TH = 75 — SBE = 1.035276180 — CAUER FUNCTIONS

MU		RC = 02	RC = 05	RC = 07	RC = 10	RC = 15	RC = 20	RC = 25	RC = 35	RC = 50
	AMAX DB	.0017375255	.0108709564	.0213327376	.0436480540	.0988323393	.1772876696	.2802872360	.5675287486	1.249387366
	AMIN DB	9.623370190	17.17385260	20.06580238	23.16417800	26.72941012	29.30251055	31.34179762	34.54993363	38.32894600
	K	.1438834D 1	.5749286D 0	.4101689D 0	.2863816D 0	.1897119D 0	.1410045D 0	.1114739D 0	.77034090-1	.49852630-1
1	PFR	2.169337163	1.312686830	1.109408187	.9296640365	.7565787533	.6430796243	.5698134854	.4583893298	.3437845120
1	PF1	1.325836034	1.125894919	1.053647146	.9810750847	.9051187616	.8560475597	.8209300945	.7730090403	.7289080275
1	PQ1	1.553616909	1.270514708	1.240560044	1.240905630	1.282574480	1.344534565	1.417644605	1.590669361	1.930621568
1	PF2	1.077762191	1.040561633	1.025409559	1.009282255	.9914770685	.9795131197	.9707534834	.9585702793	.9471680246
1	PQ2	7.356826972	5.849473049	5.687137335	5.689014217	5.914638020	6.247861416	6.638056438	7.551399390	9.318184088
1	PFC	1.032451494	1.023296233	1.019491959	1.015395612	1.010817224	1.007708354	1.005415791	1.002204527	.9991755124
1	PQC	43.72000808	34.71650780	33.74605950	33.75728072	35.10601105	37.09730091	39.42820371	44.88145573	55.42292989
1	ZF1	1.040992839	1.040992839	1.040992839	1.040992839	1.040992839	1.040992839	1.040992839	1.040992839	1.040992839
1	ZF2	1.113485040	1.113485040	1.113485040	1.113485040	1.113485040	1.113485040	1.113485040	1.113485040	1.113485040
1	ZF3	1.562818460	1.562818460	1.562818460	1.562818460	1.562818460	1.562818460	1.562818460	1.562818460	1.562818460

(3) P 7+0 — TH = 76 — SBE = 1.030613629 — CAUER FUNCTIONS

MU		RC = 02	RC = 05	RC = 07	RC = 10	RC = 15	RC = 20	RC = 25	RC = 35	RC = 50
	AMAX DB	.0017375255	.0108709564	.0213327376	.0436480540	.0988323393	.1772876696	.2802872360	.5675287486	1.249387366
	AMIN DB	8.418611367	15.82669998	18.70361523	21.79386530	25.35472012	27.92628183	29.96485551	33.17236973	36.95105155
	K	.1640357D 1	.6554534D 0	.4676174D 0	.3264923D 0	.2162830D 0	.1607537D 0	.1270870D 0	.8782352D-1	.5683501D-1
1	PFR	2.321981599	1.373585923	1.154687289	.9633284170	.7809568100	.6675261274	.5860929381	.4706680026	.3524847818
1	PF1	1.331643563	1.138882507	1.067227416	.9946534872	.9182442046	.8687038954	.8331867346	.7846531155	.7399355738
1	PQ1	1.684878134	1.332927051	1.291703232	1.284166827	1.320397306	1.380269604	1.452690758	1.626389257	1.970417804
1	PF2	1.074090996	1.040216805	1.026034138	1.010788349	.9938159468	.9823435106	.9739136781	.9621512382	.9511064808
1	PQ2	8.334306518	6.408230626	6.178415587	6.136271473	6.338502853	6.670791374	7.069969982	8.017663759	9.867336098
1	PFC	1.029161554	1.021205082	1.017810981	1.014121576	1.009965046	1.007126384	1.005025788	1.002074065	.9992807620
1	PQC	51.07491428	39.22362660	37.80842718	37.54886844	38.79427699	40.84010494	43.29702491	49.12748280	60.50019800
1	ZF1	1.035832822	1.035832822	1.035832822	1.035832822	1.035832822	1.035832822	1.035832822	1.035832822	1.035832822
1	ZF2	1.103221259	1.103221259	1.103221259	1.103221259	1.103221259	1.103221259	1.103221259	1.103221259	1.103221259
1	ZF3	1.532570586	1.532570586	1.532570586	1.532570586	1.532570586	1.532570586	1.532570586	1.532570586	1.532570586

(3) P 7+0 — TH = 77 — SBE = 1.026304108 — CAUER FUNCTIONS

MU		RC = 02	RC = 05	RC = 07	RC = 10	RC = 15	RC = 20	RC = 25	RC = 35	RC = 50
	AMAX DB	.0017375255	.0108709564	.0213327376	.0436480540	.0988323393	.1772876696	.2802872360	.5675287486	1.249387366
	AMIN DB	7.247205128	14.46623863	17.32232973	20.40125332	23.95598538	26.52539127	28.56296485	31.76960705	35.54782518
	K	.1873760D 1	.7487162D 0	.5341535D 0	.3729481D 0	.2470574D 0	.1836269D 0	.1451698D 0	.1003197D 0	.6492193D-1
1	PFR	2.501812139	1.443064869	1.205786820	1.000944138	.8079432025	.6889360324	.6039507339	.4840746721	.3619468242
1	PF1	1.335207566	1.151485439	1.080813910	1.008504653	.9318157001	.8818736210	.8459877574	.7968647928	.7515360991
1	PQ1	1.846818054	1.407089533	1.351719457	1.334351335	1.363794000	1.421003907	1.492463654	1.666696229	2.015103752
1	PF2	1.070049542	1.039609898	1.026449171	1.012138911	.9960560156	.9851114174	.9770370662	.9657298069	.9550733848
1	PQ2	9.566782778	7.085531087	6.767096560	6.666814463	6.836697987	7.165270620	7.573190691	8.558508651	10.50194386
1	PFC	1.025939680	1.019135011	1.016141766	1.012852330	1.009112627	1.006542082	1.004632493	1.001939699	.9993822238
1	PQC	60.61482338	44.84432679	42.81888790	42.18092987	43.26163684	45.35144713	47.94529772	54.20836128	66.55509203
1	ZF1	1.031034712	1.031034712	1.031034712	1.031034712	1.031034712	1.031034712	1.031034712	1.031034712	1.031034712
1	ZF2	1.093345368	1.093345368	1.093345368	1.093345368	1.093345368	1.093345368	1.093345368	1.093345368	1.093345368
1	ZF3	1.502499142	1.502499142	1.502499142	1.502499142	1.502499142	1.502499142	1.502499142	1.502499142	1.502499142

CAUER FUNCTIONS

(3) P 7+0 — TH = 78 — SBE = 1.022340595

MU	K	RC = 02	RC = 05	RC = 07	RC = 10	RC = 15	RC = 20	RC = 25	RC = 35	RC = 50
	AMAX DB	.0017375255	.0108709564	.0213327376	.0436480540	.0988323393	.1772876696	.2802872360	.5675287486	1.249387366
	AMIN DB	6.122215884	13.09258840	15.91970695	18.98273167	22.52882763	25.09518577	27.13134422	30.33675191	34.11431332
MU	K	.2145413D 1	.8572634D 0	.6115939D 0	.4270173D 0	.2828753D 0	.2102488D 0	.1662162D 0	.1148638D 0	.7433416D-1
1	PFR	2.715744948	1.523168184	1.264015269	1.043343558	.8380495734	.7126769340	.6236726992	.4988046080	.3722972288
1	PF1	1.336114788	1.163476353	1.094271665	1.022566917	.9458265682	.8955751778	.8593653160	.8096900374	.7637640614
1	PQ1	2.049424017	1.496455523	1.423069579	1.393265853	1.414129056	1.467912584	1.538043066	1.712598938	2.065718544
1	PF2	1.065631062	1.038713697	1.026628925	1.013311174	.9981785900	.9878011807	.9801103052	.9692959738	.9590619263
1	PQ2	11.14955088	7.920269556	7.483423563	7.305162534	7.430003629	7.750694003	8.166649554	9.193212214	11.24359673
1	PFC	1.022800011	1.017091832	1.014488536	1.011590754	1.008261629	1.005956409	1.004236409	1.001801363	.9994793701
1	PQC	73.25936887	51.99070935	49.11160261	47.93653872	48.75948151	50.87317298	53.61423706	60.37702551	73.87851889
1	ZF1	1.026592071	1.026592071	1.026592071	1.026592071	1.026592071	1.026592071	1.026592071	1.026592071	1.026592071
1	ZF2	1.083848770	1.083848770	1.083848770	1.083848770	1.083848770	1.083848770	1.083848770	1.083848770	1.083848770
1	ZF3	1.472528748	1.472528748	1.472528748	1.472528748	1.472528748	1.472528748	1.472528748	1.472528748	1.472528748

CAUER FUNCTIONS

(3) P 7+0 — TH = 79 — SBE = 1.018716695

MU	K	RC = 02	RC = 05	RC = 07	RC = 10	RC = 15	RC = 20	RC = 25	RC = 35	RC = 50
	AMAX DB		.0108709564	.0213327376	.0436480540	.0988323393	.1772876696	.2802872360	.5675287486	1.249387366
	AMIN DB		11.70712041	14.49376591	17.53430333	21.06810073	23.63010441	25.66423848	28.86787884	32.64449961
MU	K		.9843359D 0	.7022507D 0	.4903142D 0	.3248060D 0	.2414140D 0	.1908545D 0	.1318902D 0	.8535275D-1
1	PFR		1.616631590	1.331114137	1.091624637	.8719433644	.7392247689	.6456269144	.5151074623	.3836967994
1	PF1		1.174560891	1.107416811	1.036749167	.9602576083	.9098228949	.8733522719	.8231806337	.7766831259
1	PQ1		1.605894999	1.509188690	1.463401633	1.473258925	1.522580874	1.590877042	1.765439641	2.123635682
1	PF2		1.037498508	1.026544357	1.014278973	1.000161901	.9903944017	.9831176135	.9728377604	.9630638209
1	PQ2		8.969211465	8.370969198	8.086176090	8.147580358	8.454068568	8.876572633	9.948309217	12.12185028
1	PFC		1.015082411	1.012856239	1.010340222	1.007414009	1.005370513	1.003838149	1.001659007	.9995716115
1	PQC		61.28536130	57.18473912	55.23229024	55.65327886	57.75439641	60.65043211	67.99479996	82.88404792
1	ZF1		1.022499263	1.022499263	1.022499263	1.022499263	1.022499263	1.022499263	1.022499263	1.022499263
1	ZF2		1.074724174	1.074724174	1.074724174	1.074724174	1.074724174	1.074724174	1.074724174	1.074724174
1	ZF3		1.442574204	1.442574204	1.442574204	1.442574204	1.442574204	1.442574204	1.442574204	1.442574204

CAUER FUNCTIONS

(3) P 7+0 — TH = 80 — SBE = 1.015426612

MU	K	RC = 02	RC = 05	RC = 07	RC = 10	RC = 15	RC = 20	RC = 25	RC = 35	RC = 50
	AMAX DB		.0108709564	.0213327376	.0436480540	.0988323393	.1772876696	.2802872360	.5675287486	1.249387366
	AMIN DB		10.31320108	13.04313700	16.05161058	19.56770031	22.12340497	24.15460537	27.35568182	31.13093693
MU	K		.1134132D 1	.8091194D 0	.5649303D 0	.3742350D 0	.2781525D 0	.2198988D 0	.1519612D 0	.9834175D-1
1	PFR		1.727183362	1.409449914	1.147267579	.9105144518	.7692091375	.6702975356	.5333090885	.3963539710
1	PF1		1.184360043	1.119999116	1.050918234	.9750703991	.9246236188	.8879808341	.8373950409	.7903685516
1	PQ1		1.742457048	1.615003668	1.548288125	1.543776753	1.587202843	1.652956919	1.827048619	2.190714224
1	PF2		1.035932464	1.026162708	1.015012053	1.001980364	.9928692602	.9860401463	.9763407026	.9670689082
1	PQ2		10.31787198	9.494532782	9.060974863	9.031584093	9.314105021	9.740337504	10.86134617	13.17827451
1	PFC		1.013115001	1.011250770	1.009104742	1.006572109	1.004785781	1.003438475	1.001512611	.9996582855
1	PQC		73.70324048	67.80775220	64.70273736	64.49223385	66.51563159	69.56795824	77.59415224	94.17786515
1	ZF1		1.018751463	1.018751463	1.018751463	1.018751463	1.018751463	1.018751463	1.018751463	1.018751463
1	ZF2		1.065965738	1.065965738	1.065965738	1.065965738	1.065965738	1.065965738	1.065965738	1.065965738
1	ZF3		1.412536877	1.412536877	1.412536877	1.412536877	1.412536877	1.412536877	1.412536877	1.412536877

$P^{(4)}_{7+2}$

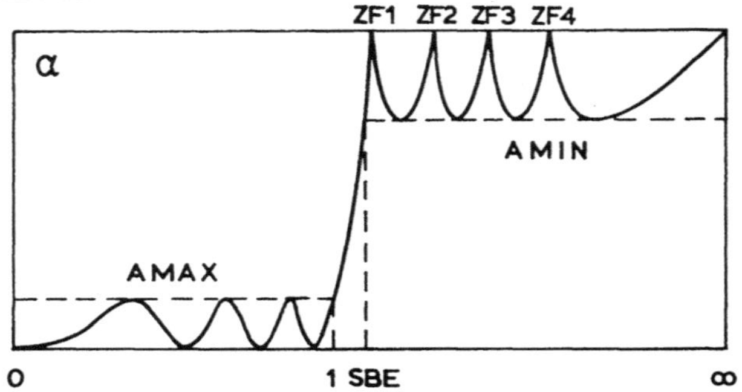

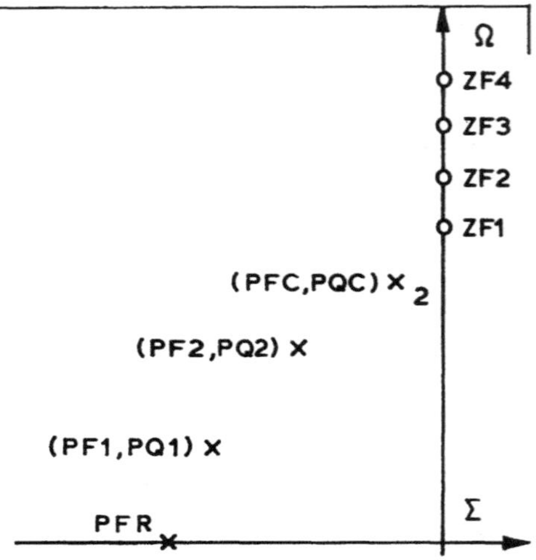

(4) P 7+2			TH = 32	SBE = 1.887079915						MCPER FUNCTIONS
			RC = 02	RC = 05	RC = 07	RC = 10	RC = 15	RC = 20	RC = 25	RC = 35	RC = 50
AMAX DB			.0017375255	.0108709564	.0213327376	.0436480540	.0988323393	.1772876696	.2802872360	.5675287486	1.249387366
AMIN DB			90.69983292	100.9848421	104.7786517	108.8200869	113.4595213	116.8053355	119.4553690	123.6162103	128.4832717
MU		K	.4390242D-3	.1356901D-3	.8794894D-4	.5540015D-4	.3258102D-4	.2221392D-4	.1639965D-4	.1018205D-4	.5828368D-5
	1	PFR	.7205627723	.5475827320	.4888083241	.4287907678	.3629976495	.3175770447	.2828531581	.2308003934	.1745642476
	1	PF1	.8343727815	.6914928606	.6450608677	.5993094502	.5516474000	.5206582683	.4982444864	.4671041958	.4374190081
	1	PQ1	.6604153638	.7103341691	.7391802384	.7796976983	.8442556669	.9083752496	.9741317411	1.116307793	1.378801810
	1	PF2	1.038101338	.9349426142	.9018001827	.8693898748	.8358544037	.8140997964	.7983145796	.7761154983	.7541844106
	1	PQ2	1.210321620	1.364851994	1.451165586	1.569278396	1.751563937	1.927282621	2.103474826	2.474957059	3.141057670
	2	PFC	1.251767276	1.166836244	1.138536263	1.110326187	1.080520995	1.060817631	1.046327377	1.025675078	1.005013402
	2	PQC	3.472426707	3.739882873	3.895648819	4.106779242	4.422990252	4.715140597	4.995898042	5.551952884	6.449610204
	1	ZF1	1.909690052	1.910195241	1.910355206	1.910512286	1.910677427	1.910787408	1.910869479	1.910989689	1.911116632
	1	ZF2	2.117429574	2.122264157	2.123792865	2.125293611	2.126871269	2.127922042	2.128706259	2.129855092	2.131068562
	1	ZF3	2.740066715	2.755449278	2.760332068	2.765134275	2.770191760	2.773565278	2.776085548	2.779731312	2.783689435
	1	ZF4	4.962559564	5.005802181	5.019623202	5.033257023	5.047656701	5.057283602	5.064486200	5.075063297	5.086265733

(4) P 7+2			TH = 33 SBE = 1.836078459 RC = 02	RC = 05	RC = 07	RC = 10	RC = 15	RC = 20	RC = 25	RC = 35	MCPER FUNCTIONS RC = 50
AMAX DB			.0017375255	.0108709564	.0213327376	.0436480540	.0988323393	.1772876696	.2802872360	.5675287486	1.249387366
AMIN DB			88.15349744	98.43744553	102.2309853	106.2721860	110.9114058	114.2570950	116.9070447	121.0677775	125.9347427
MU		K	.5681639D-3	.1757321D-3	.1139288D-3	.7178127D-4	.4222471D-4	.2879352D-4	.2125956D-4	.1320164D-4	.7558818D-5
1		PFR	.7263463502	.5513720217	.4920389361	.4315034931	.3651964336	.3194496523	.2844901808	.2321036621	.1755287452
1		PF1	.8394789888	.6952816179	.6484764585	.6023819496	.5543873018	.5231949890	.5006405085	.4693134051	.4394597979
1		PQ1	.6623820136	.7120650644	.7408690826	.7813698034	.8459500148	.9101201033	.9759432113	1.118288557	1.381132438
1		PF2	1.040127562	.9369984606	.9038454650	.8714178213	.8378590027	.8160875451	.8002899899	.7780747844	.7561318348
1		PQ2	1.222474651	1.376290698	1.462633720	1.580981154	1.763854602	1.940273015	2.117237261	2.490467946	3.159881761
2		PFC	1.248071211	1.164733605	1.136885410	1.109089823	1.079685735	1.060227741	1.045908274	1.025485696	1.005038262
2		PQC	3.542640260	3.802849667	3.957328839	4.167916035	4.484652990	4.778064760	5.060446990	5.620431996	6.525517037
1		ZF1	1.857697628	1.858214583	1.858378442	1.858539415	1.858708714	1.858821500	1.858905682	1.859029009	1.859159277
1		ZF2	2.056572692	2.061530752	2.063099347	2.064639649	2.066259319	2.067338319	2.068143732	2.069323822	2.070570575
1		ZF3	2.654393500	2.670180369	2.675195298	2.680129246	2.685327382	2.688795804	2.691387540	2.695138974	2.699209964
1		ZF4	4.795410777	4.839754549	4.853947035	4.867956062	4.882761138	4.892664098	4.900075781	4.910963731	4.922500132

(4) P 7+2			TH = 34 SBE = 1.788291650 RC = 02	RC = 05	RC = 07	RC = 10	RC = 15	RC = 20	RC = 25	RC = 35	MCPER FUNCTIONS RC = 50
AMAX DB			.0017375255	.0108709564	.0213327376	.0436480540	.0988323393	.1772876696	.2802872360	.5675287486	1.249387366
AMIN DB			85.67378771	95.95656088	99.74980190	103.7907427	108.4297248	111.7752757	114.4251324	118.5857450	123.4526036
MU		K	.7302234D-3	.2260286D-3	.1465714D-3	.9236939D-4	.5434874D-4	.3706703D-4	.2737154D-4	.1700000D-4	.9735422D-5
1		PFR	.7324143541	.5553365450	.4954162546	.4343373344	.3674917677	.3214036496	.2861978494	.2334626422	.1765341177
1		PF1	.8448013844	.6992280344	.6520333449	.6055808131	.5572392111	.5258350104	.5031338322	.4716119910	.4415828599
1		PQ1	.6644600105	.7138854357	.7426428353	.7831237879	.8477251804	.9119468322	.9778387125	1.120359798	1.383568024
1		PF2	1.042198187	.9391145154	.9059539284	.8735110686	.8399304043	.8181427656	.8023332429	.7801023315	.7581479199
1		PQ2	1.235346620	1.388359068	1.474719216	1.593300569	1.776779005	1.953924041	2.131692703	2.506749266	3.179628565
2		PFC	1.244274077	1.162568594	1.135184324	1.107814842	1.078823550	1.059618306	1.045474867	1.025289197	1.005062854
2		PQC	3.617497881	3.869699811	4.022730757	4.232664966	4.549881004	4.844576699	5.128639320	5.692721742	6.605538874
1		ZF1	1.808969625	1.809497851	1.809665473	1.809830218	1.810003561	1.810119079	1.810205320	1.810331693	1.810465214
1		ZF2	1.999424802	2.004503462	2.006111111	2.007690187	2.009351089	2.010457822	2.011284075	2.012494919	2.013774451
1		ZF3	2.573701735	2.589885237	2.595030226	2.600094034	2.605431020	2.608993247	2.611655681	2.615561745	2.619694608
1		ZF4	4.637597253	4.683015490	4.697572927	4.711951645	4.727157251	4.737333544	4.744952540	4.756149140	4.768017673

(4) P 7+2	TH = 35 SBE = 1.743446796								MCPER FUNCTIONS
	RC = 02	RC = 05	RC = 07	RC = 10	RC = 15	RC = 20	RC = 25	RC = 35	RC = 50
AMAX DB	.0017375255	.0108709564	.0213327376	.0436480540	.0988323393	.1772876696	.2802872360	.5675287486	1.249387366
AMIN DB	83.25644743	93.53792255	97.33083342	101.3714871	106.0102064	109.3556045	112.0053585	116.1658382	121.0325789
MU I K	.9324453D-3	.2888508D-3	.1873557D-3	.1181003D-3	.6950599D-4	.4741255D-4	.3501539D-4	.2175142D-4	.1245880D-4
1 I PFR	.7387800087	.5594832930	.4989458514	.4372966760	.3698869793	.3234417556	.2879784719	.2343790993	.1775816243
1 I PF1	.8503462136	.7033365358	.6557354393	.6089094874	.5602061317	.5285810710	.5057270153	.4740022746	.4437903055
1 I PQ1	.6666561338	.7157998882	.7445056259	.7849634151	.8495846639	.9138588474	.9798216485	1.122525040	1.386112508
1 I PF2	1.044310717	.9412901203	.9081253282	.8756697068	.8420689814	.8202659829	.8044449600	.7821988783	.7602335027
1 I PQ2	1.248983463	1.401092744	1.487455645	1.606268863	1.790368906	1.968267904	2.146874265	2.523836796	3.200339733
2 I PFC	1.240377346	1.160341558	1.133433151	1.106501250	1.077934358	1.058989212	1.045027032	1.025085459	1.005087076
2 I PQC	3.697349389	3.940696875	4.092098737	4.301255149	4.618892569	4.914890992	5.200689220	5.769040699	6.690058764
1 I ZF1	1.763229193	1.763768162	1.763939407	1.764107798	1.764285060	1.764403234	1.764491480	1.764620822	1.764757519
1 I ZF2	1.945671070	1.950867320	1.952513147	1.954130178	1.955831491	1.956965434	1.957812152	1.959053219	1.960364998
1 I ZF3	2.497566648	2.514138753	2.519411621	2.524603318	2.530077262	2.533732139	2.536464466	2.540474062	2.544717750
1 I ZF4	4.488319712	4.534784363	4.549699887	4.564442484	4.580043479	4.590490223	4.598314656	4.609817573	4.622016290

(4) P 7+2	TH = 36 SBE = 1.701301617								MCPER FUNCTIONS
	RC = 02	RC = 05	RC = 07	RC = 10	RC = 15	RC = 20	RC = 25	RC = 35	RC = 50
AMAX DB	.0017375255	.0108709564	.0213327376	.0436480540	.0988323393	.1772876696	.2802872360	.5675287486	1.249387366
AMIN DB	80.89756707	91.17761113	94.97015792	99.01049493	103.6489245	106.9941540	109.6437948	113.8041279	118.6707386
MU I K	.1183450D-2	.3669065D-3	.2380457D-3	.1500905D-3	.8835653D-4	.6028168D-4	.4452533D-4	.2766417D-4	.1584865D-4
1 I PFR	.7454576355	.5638197839	.5026337055	.4403862139	.3723856262	.3255668737	.2898345111	.2363549155	.1786726057
1 I PF1	.8561200398	.7076117963	.6595868772	.6123716178	.5632912421	.5314360692	.5084227645	.4764867125	.4460843696
1 I PQ1	.6689778103	.7178134256	.7464619302	.7868927547	.8515322419	.9158598243	.9818956835	1.124788073	1.388770119
1 I PF2	1.046462432	.9435245273	.9103593623	.8778937958	.8442750992	.8224577268	.8066257753	.7843651862	.7623894506
1 I PQ2	1.263435485	1.414530516	1.500879475	1.619920978	1.804658679	1.983339427	2.162817720	2.541769142	3.222060167
2 I PFC	1.236382588	1.158052880	1.131632053	1.105149062	1.077018077	1.058340345	1.044564642	1.024874356	1.005110818
2 I PQC	3.782581720	4.016130370	4.165700419	4.373937242	4.691926016	4.989241641	5.276830058	5.849626720	6.779179999
1 I ZF1	1.720230375	1.720779520	1.720954238	1.721126140	1.721307191	1.721427939	1.721518131	1.721650361	1.721790153
1 I ZF2	1.895031794	1.900342486	1.902025573	1.903679696	1.905420557	1.906581162	1.907447948	1.908718680	1.910062145
1 I ZF3	2.425610805	2.442563105	2.447961566	2.453279083	2.458888000	2.462634314	2.465435683	2.469547656	2.473901066
1 I ZF4	4.346867778	4.394349383	4.409615770	4.424716119	4.440707079	4.451421227	4.459449111	4.471255873	4.483782705

(4) P 7+2	TH = 37	SBE = 1.661640141						MCPER FUNCTIONS	
	RC = 02	RC = 05	RC = 07	RC = 10	RC = 15	RC = 20	RC = 25	RC = 35	RC = 50
AMAX DB	.0017375255	.0108709564	.0213327376	.0436480540	.0988323393	.1772876696	.2802872360	.5675287486	1.249387366
AMIN DB	78.59354493	88.87201437	92.66416049	96.70414895	101.3422597	104.6873034	107.3368196	111.4969914	116.3634589
MU I K	.1493463D-2	.4634133D-3	.3007385D-3	.1896685D-3	.1116860D-3	.7621209D-4	.5629943D-4	.3498642D-4	.2004764D-4
1 I PFR	.7524627623	.5683541092	.5064862385	.4436109817	.3749915142	.3277821060	.2917685962	.2378920977	.1798084905
1 I PF1	.8621297548	.7120587515	.6635920297	.6159710604	.5664979057	.5344030724	.5112239453	.4790679053	.4484674181
1 I PQ1	.6714331881	.7199314911	.7485166051	.7889162123	.8535719935	.9179537270	.9840647655	1.127152974	1.391545404
1 I PF2	1.048650369	.9458168900	.9126556647	.8801833603	.8465491127	.8247185303	.8088763347	.7866020392	.7646166624
1 I PQ2	1.278757868	1.428714654	1.515030367	1.634294844	1.819695571	1.999176297	2.179561750	2.560588009	3.244838313
2 I PFC	1.232291467	1.155702977	1.129781217	1.103758302	1.076074631	1.057671590	1.044087571	1.024655759	1.005133967
2 I PQC	3.873623628	4.096318788	4.243829605	4.450985847	4.769241911	5.067884153	5.357316411	5.934738944	6.873228196
1 I ZF1	1.679753929	1.680312645	1.680490676	1.680665945	1.680850645	1.680973881	1.681065958	1.681200990	1.681343792
1 I ZF2	1.847257662	1.852679510	1.854398894	1.856089203	1.857868706	1.859055395	1.859941831	1.861241640	1.862616198
1 I ZF3	2.357497718	2.374821407	2.380343066	2.385784233	2.391526038	2.395362511	2.398232030	2.402445163	2.406907129
1 I ZF4	4.212607975	4.261075614	4.276685260	4.292136905	4.308512104	4.319490437	4.327719674	4.339827660	4.352680410

(4) P 7+2	TH = 38	SBE = 1.624269245						MCPER FUNCTIONS	
	RC = 02	RC = 05	RC = 07	RC = 10	RC = 15	RC = 20	RC = 25	RC = 35	RC = 50
AMAX DB	.0017375255	.0108709564	.0213327376	.0436480540	.0988323393	.1772876696	.2802872360	.5675287486	1.249387366
AMIN DB	76.34105341	86.61779339	90.40949932	94.44910486	99.08686513	102.4317049	105.0810841	109.2410786	114.1073888
MU I K	.1874584D-2	.5821844D-3	.3779210D-3	.2384099D-3	.1404270D-3	.9584224D-4	.7081047D-4	.4401301D-4	.2522535D-4
1 I PFR	.7598122454	.5730949858	.5105103529	.4469763786	.3777087176	.3300907697	.2937835360	.2394927877	.1809908013
1 I PF1	.8683825881	.7166826122	.6677555170	.6197118939	.5698296826	.5374853283	.5141335914	.4817486068	.4509419564
1 I PQ1	.6740312206	.7221600132	.7506749267	.7910385622	.8557083286	.9201448345	.9863331518	1.129624133	1.394443252
1 I PF2	1.050871305	.9481662536	.9150137982	.8825383845	.8488913631	.8270489273	.8111972949	.7889102433	.7669160686
1 I PQ2	1.295011231	1.443691275	1.529951506	1.649431685	1.835489986	2.015819346	2.197148230	2.580338485	3.268726492
2 I PFC	1.228105755	1.153292304	1.127880851	1.102329011	1.075103945	1.056982833	1.043595686	1.024429534	1.005156403
2 I PQC	3.970951121	4.181613073	4.326809316	4.532702240	4.851125532	5.151097889	5.442426361	6.024660058	6.972503621
1 I ZF1	1.641603820	1.642171459	1.642352635	1.642531117	1.642719319	1.642844951	1.642938847	1.643076590	1.643222311
1 I ZF2	1.802125762	1.807655334	1.809410008	1.811135554	1.812952746	1.814164910	1.815070558	1.816398822	1.817803849
1 I ZF3	2.292926474	2.310612324	2.316254668	2.321817209	2.327689712	2.331615003	2.334551730	2.338864742	2.343434035
1 I ZF4	4.084973608	4.134394846	4.150339754	4.166135895	4.182889296	4.194128414	4.202556782	4.214963216	4.228139548

(4) P 7+2			TH = 39 RC = 02	SBE = 1.589015729 RC = 05	RC = 07	RC = 10	RC = 15	RC = 20	RC = 25	MCPER FUNCTIONS RC = 35	RC = 50
AMAX DB			.0017375255	.0108709564	.0213327376	.0436480540	.0988323393	.1772876696	.2802872360	.5675287486	1.249387366
AMIN DB			74.13700946	84.41185312	88.20307628	92.24226178	96.87963764	100.2242537	102.8734826	107.0332825	111.8994201
MU		K	.2341087D-2	.7277267D-3	.4725338D-3	.2981794D-3	.1756835D-3	.1199283D-3	.8861858D-4	.5509341D-4	.3158287D-4
	1	PFR	.7675244084	.5780518141	.5147134750	.4504882013	.3805416015	.3324964138	.2958823332	.2411592721	.1822211622
	1	PF1	.8748861155	.7214888790	.6720822225	.6235984335	.5732903413	.5406862758	.5171549160	.4845317337	.4535106379
	1	PQ1	.6767817620	.7245054578	.7529426337	.7932649844	.8579460195	.9224377707	.9887054377	1.132206283	1.397468925
	1	PF2	1.053121734	.9505715439	.9174332460	.8849588066	.8513021746	.8294494508	.8135893213	.7912906260	.7692886317
	1	PQ2	1.312262276	1.459510772	1.545689981	1.665376357	1.852115803	2.033312863	2.215622541	2.601069375	3.293781273
	2	PFC	1.223827334	1.150821358	1.125931192	1.100861239	1.074105950	1.056273958	1.043088856	1.024195542	1.005178003
	2	PQC	4.075093735	4.272400602	4.414995281	4.619417483	4.937889677	5.239188740	5.532464095	6.119698862	7.077333839
	1	ZF1	1.605604251	1.606180122	1.606364260	1.606545794	1.606737340	1.606865271	1.606960917	1.607101274	1.607249818
	1	ZF2	1.759436202	1.765069923	1.766858830	1.768618617	1.770472495	1.771709496	1.772633893	1.773989960	1.775424797
	1	ZF3	2.231627187	2.249665520	2.255425913	2.261107438	2.267108337	2.271121034	2.274123981	2.278535521	2.283210843
	1	ZF4	3.963456203	4.013797041	4.030068803	4.046202280	4.063327515	4.074823823	4.083448974	4.096150916	4.109648350

(4) P 7+2			TH = 40 RC = 02	SBE = 1.555723827 RC = 05	RC = 07	RC = 10	RC = 15	RC = 20	RC = 25	MCPER FUNCTIONS RC = 35	RC = 50
AMAX DB			.0017375255	.0108709564	.0213327376	.0436480540	.0988323393	.1772876696	.2802872360	.5675287486	1.249387366
AMIN DB			71.97854882	82.25131643	86.04201092	90.08073641	94.71769123	98.06206216	100.7111264	104.8707132	109.7366617
MU		K	.2909771D-2	.9053544D-3	.5880467D-3	.3711782D-3	.2187597D-3	.1493638D-3	.1103858D-3	.6864087D-4	.3935808D-4
	1	PFR	.7756191993	.5832347432	.5191036034	.4541526790	.3834948464	.3350028388	.2980682001	.2428939944	.1835013067
	1	PF1	.8816482667	.7264833578	.6765773079	.6276352452	.5768838725	.5440095575	.5202913229	.4874203766	.4561762747
	1	PQ1	.6796956773	.7269748864	.7553259759	.7956011053	.8602902369	.9248375373	.9911865888	1.134904531	1.400628094
	1	PF2	1.055397849	.9530315558	.9199134031	.8874445122	.8537818502	.8319206299	.8160530868	.7937440361	.7717353464
	1	PQ2	1.330584532	1.476228287	1.562297205	1.682177742	1.869610736	2.051704947	2.235033921	2.622833568	3.320063879
	2	PFC	1.219458198	1.148290682	1.123932503	1.099355056	1.073080583	1.055544852	1.042566947	1.023953640	1.005198637
	2	PQC	4.186641828	4.369109754	4.508779944	4.711495983	5.029877871	5.332492163	5.627762857	6.220193171	7.198076695
	1	ZF1	1.571597135	1.572180499	1.572367408	1.572551819	1.572746543	1.572876670	1.572973993	1.573116862	1.573268127
	1	ZF2	1.719009241	1.724743386	1.726565421	1.728358404	1.730247918	1.731509082	1.732451742	1.733834923	1.735298877
	1	ZF3	2.173357128	2.191737790	2.197613468	2.203411466	2.209538343	2.213636961	2.216705085	2.221213729	2.225993711
	1	ZF4	3.847598235	3.898823054	3.915412832	3.931876106	3.949366460	3.961116167	3.969935615	3.982929949	3.996745852

(4) P 7+2			TH = 41	SBE = 1.524253087						MCPER FUNCTIONS
			RC = 02	RC = 05	RC = 07	RC = 10	RC = 15	RC = 20	RC = 25	RC = 35	RC = 50
AMAX DB			.0017375255	.0108709564	.0213327376	.0436480540	.0988323393	.1772876696	.2802872360	.5675287486	1.249387366
AMIN DB			69.86300327	80.13350134	83.92361772	87.96184012	92.59833442	95.94243723	98.59132136	102.7506750	107.6164166
MU		K	.3600368D-2	.1121321D-2	.7285447D-3	.4599992D-3	.2711929D-3	.1852024D-3	.1368932D-3	.8514305D-4	.4883184D-4
1		PFR	.7841183691	.5886547437	.5236893625	.4579765128	.3865734758	.3376141182	.3003445765	.2446995677	.1848330864
1		PF1	.8886773303	.7316721757	.6812462300	.6318271613	.5806145035	.5474590334	.5235464199	.4904178119	.4589418481
1		PQ1	.6827849685	.7295760233	.7578317702	.7980530453	.8627465904	.9273495514	.9937819766	1.137724391	1.403926874
1		PF2	1.057695518	.9555449401	.9224535664	.8899953280	.8563306675	.8344629867	.8185892695	.7962713421	.7742572403
1		PQ2	1.350059206	1.493904257	1.579829406	1.699889180	1.888026746	2.071047905	2.255435868	2.645688449	3.347640658
2		PFC	1.215000466	1.145700865	1.121885077	1.097810544	1.072027786	1.054795402	1.042029821	1.023703680	1.005218169
2		PQC	4.306255057	4.472215175	4.608597053	4.809339560	5.127468022	5.431376642	5.728688319	6.326513125	7.305123706
1		ZF1	1.539439941	1.540030009	1.540219484	1.540406590	1.540604315	1.540736528	1.540835451	1.540980724	1.541134603
1		ZF2	1.680682837	1.686513530	1.688367537	1.690192622	1.692116667	1.693401289	1.694361700	1.695771270	1.697263608
1		ZF3	2.117897434	2.136609762	2.142597820	2.148509655	2.154759966	2.158942943	2.162075146	2.166679394	2.171562590
1		ZF4	3.736986913	3.789058420	3.805956927	3.822742068	3.840590461	3.852589558	3.861600676	3.874884102	3.889015674

(4) P 7+2			TH = 42	SBE = 1.494476550						MCPER FUNCTIONS
			RC = 02	RC = 05	RC = 07	RC = 10	RC = 15	RC = 20	RC = 25	RC = 35	RC = 50
AMAX DB			.0017375255	.0108709564	.0213327376	.0436480540	.0988323393	.1772876696	.2802872360	.5675287486	1.249387366
AMIN DB			67.78788053	78.05590083	81.84538588	85.88305881	90.51905006	93.86285997	96.51154734	100.6706463	105.5361618
MU		K	.4436002D-2	.1382969D-2	.8988280D-3	.5676913D-3	.3347917D-3	.2286843D-3	.1690601D-3	.1051744D-3	.6033513D-4
1		PFR	.7930456743	.5943236899	.5284800626	.4619669194	.3897828872	.3403346222	.3027151492	.2465787889	.1862184808
1		PF1	.8959819573	.7370617982	.6860947574	.6361792976	.5844867133	.5510387956	.5269240327	.4935275143	.4618105199
1		PQ1	.6860629191	.7323173309	.7604674622	.8006274707	.8653211729	.9299796875	.9964974192	1.140671827	1.407371869
1		PF2	1.060010258	.9581101900	.9250529241	.8926110133	.8589488740	.8370770339	.8211985507	.7988734324	.7768553737
1		PQ2	1.370776160	1.512605036	1.598348167	1.718568968	1.907420496	2.091398702	2.276886582	2.669696367	3.376583602
2		PFC	1.210456383	1.143052547	1.119789242	1.096227807	1.070947508	1.054025498	1.041477341	1.023445512	1.005236457
2		PQC	4.434672284	4.582243859	4.714926957	4.913392124	5.231076612	5.536247654	5.835642414	6.439064949	7.428903921
1		ZF1	1.509003846	1.509599779	1.509791602	1.509981207	1.510181746	1.510315930	1.510416369	1.510563932	1.510720312
1		ZF2	1.644310549	1.650233759	1.652118530	1.653974570	1.655931989	1.657239325	1.658216948	1.659652144	1.661172096
1		ZF3	2.065050281	2.084083069	2.090180454	2.096203354	2.102574424	2.106840114	2.110035240	2.114733508	2.119718392
1		ZF4	3.631248855	3.684128027	3.701325510	3.718424170	3.736623140	3.748867396	3.758067406	3.771636427	3.786080697

(4) P 7+2 — TH = 43 SBE = 1.466279186 — MCPER FUNCTIONS

MU	K	RC = 02	RC = 05	RC = 07	RC = 10	RC = 15	RC = 20	RC = 25	RC = 35	RC = 50
	AMAX DB	.0017375255	.0108709564	.0213327376	.0436480540	.0988323393	.1772876696	.2802872360	.5675287486	1.249387366
	AMIN DB	65.75084648	76.01616499	79.80496142	83.84203494	88.47747736	91.82096767	94.46944033	98.62826128	103.4935302
	K	.5443729D-2	.1698908D-2	.1104527D-2	.6978332D-3	.4116803D-3	.2812669D-3	.2079675D-3	.1294104D-3	.7425736D-4
1	PFR	.8024271089	.6002544528	.5334857681	.4661316800	.3931288860	.3431690444	.3051838736	.2485346545	.1876596080
1	PF1	.9035711609	.7426590466	.6911289896	.6406970712	.5385052506	.5547531841	.5304282196	.4967531712	.4647956465
1	PQ1	.6895442622	.7352080965	.7632411981	.8033316537	.8680206125	.9327343242	.9993392263	1.143753297	1.410970220
1	PF2	1.062337212	.9607256255	.9277105440	.8952912523	.8616366811	.8397632700	.8238816126	.8015512135	.7795308400
1	PQ2	1.392835049	1.532403606	1.617921065	1.738280927	1.927853876	2.112819468	2.299449479	2.694925160	3.406970937
2	PFC	1.205828330	1.140346424	1.117645361	1.094606971	1.069839709	1.053235030	1.040909370	1.023178978	1.005253355
2	PQC	4.572723176	4.699782201	4.828302719	5.024145063	5.341163509	5.647552199	5.949067735	6.558295254	7.559888317
1	ZF1	1.480172147	1.480773050	1.480966990	1.481158887	1.481362041	1.481498073	1.481599943	1.481749675	1.481908434
1	ZF2	1.609759730	1.615771269	1.617685542	1.619571336	1.621560913	1.622890183	1.623884452	1.625344471	1.626891223
1	ZF3	2.014636456	2.033977915	2.040181418	2.046312466	2.052801479	2.057148149	2.060404978	2.065195595	2.070280553
1	ZF4	3.530045507	3.583691533	3.601177751	3.618581152	3.637122833	3.649607780	3.658993743	3.672844654	3.687598462

(4) P 7+2 — TH = 44 SBE = 1.439556540 — MCPER FUNCTIONS

MU	K	RC = 02	RC = 05	RC = 07	RC = 10	RC = 15	RC = 20	RC = 25	RC = 35	RC = 50
	AMAX DB	.0017375255	.0108709564	.0213327376	.0436480540	.0988323393	.1772876696	.2802872360	.5675287486	1.249387366
	AMIN DB	63.74970927	74.01208498	77.80013114	81.83655149	86.47139577	89.81453775	92.46277637	96.62129430	101.4862943
	K	.6655147D-2	.2079213D-2	.1352238D-2	.8546189D-3	.5043500D-3	.3446600D-3	.2548838D-3	.1586443D-3	.9105604D-4
1	PFR	.8122911703	.6064610046	.5387173736	.4704791960	.3966177243	.3461224317	.3077549981	.2505703788	.1891587371
1	PF1	.9114543120	.7484711165	.6963553761	.6453862202	.5926751512	.5586068044	.5340632890	.5000986979	.4678707920
1	PQ1	.6932453747	.7382585309	.7661619056	.8061735404	.8708521299	.9356203979	1.002314251	1.146975802	1.414729653
1	PF2	1.064671123	.9633893773	.9304253608	.8980356442	.8643942588	.8425221762	.8266391349	.8043056088	.7822847658
1	PQ2	1.416346634	1.553380393	1.638622376	1.759095042	1.949394599	2.135378077	2.323193767	2.721448756	3.438887793
2	PFC	1.201118827	1.137583252	1.115453837	1.092948180	1.068704356	1.052423894	1.040325768	1.022903920	1.005268706
2	PQC	4.721341854	4.825484189	4.949317198	5.142143462	5.458237506	5.765784022	6.069452585	6.684695971	7.698594832
1	ZF1	1.452838888	1.453443810	1.453639621	1.453833590	1.454039149	1.454176899	1.454280107	1.454431880	1.454592893
1	ZF2	1.576909964	1.583005487	1.584947944	1.586862236	1.588882696	1.590233080	1.591243400	1.592727396	1.594300091
1	ZF3	1.966493258	1.986130975	1.992437219	1.998673346	2.005277337	2.009703158	2.013020404	2.017901607	2.023084933
1	ZF4	3.433069180	3.487439408	3.505203611	3.522902520	3.541778622	3.554499540	3.564068348	3.578197225	3.593257213

(4) P 7+2	TH = 45 SBE = 1.414213562							MCPER FUNCTIONS	
	RC = 02	RC = 05	RC = 07	RC = 10	RC = 15	RC = 20	RC = 25	RC = 35	RC = 50
AMAX DB	.0017375255	.0108709564	.0213327376	.0436480540	.0988323393	.1772876696	.2802872360	.5675287486	1.249387366
AMIN DB	61.78240506	72.04157871	75.82880827	79.86451760	84.49871067	87.84147336	90.48945712	94.64764506	99.51235214
MU K	.8107112D-2	.2535662D-2	.1649673D-2	.1042957D-2	.6157188D-3	.4208670D-3	.3112958D-3	.1938064D-3	.1112680D-3
1 PFR	.8226691637	.6129585374	.5441866900	.4750185503	.4002561436	.3492002172	.3104330913	.2526894126	.1907183010
1 PF1	.9196411305	.7545055966	.7017807376	.6502528248	.5970017591	.5626045469	.5378338169	.5035682551	.4710697445
1 PQ1	.6971845025	.7414798822	.7692393861	.8091618273	.8738236040	.9386454632	1.005429946	1.150346943	1.418658546
1 PF2	1.067006304	.9660993685	.9331961620	.9008436935	.8672217282	.8453542110	.8294717920	.8071375575	.7851183108
1 PQ2	1.441434310	1.575624205	1.660533910	1.781088209	1.972116877	2.159148806	2.348195095	2.749347846	3.472426952
2 PFC	1.196330547	1.134763850	1.113215115	1.091251609	1.067541428	1.051591988	1.039726396	1.022620171	1.005282351
2 PQC	4.881582994	4.960080964	5.078631289	5.267993316	5.582862725	5.891489628	6.197336781	6.818810009	7.845594137
1 ZF1	1.426907671	1.427515599	1.427713020	1.427908830	1.428116571	1.428255902	1.428360351	1.428514030	1.428677162
1 ZF2	1.545651723	1.551826723	1.553795987	1.555737462	1.557787468	1.559158106	1.560183850	1.561690933	1.563288667
1 ZF3	1.920472690	1.940393580	1.946799008	1.953136980	1.959852822	1.964355868	1.967732174	1.972702098	1.977981988
1 ZF4	3.340039618	3.395089495	3.413120405	3.431105111	3.450306898	3.463258805	3.473007170	3.487409852	3.502772454

(4) P 7+2	TH = 46 SBE = 1.390163591							MCPER FUNCTIONS	
	RC = 02	RC = 05	RC = 07	RC = 10	RC = 15	RC = 20	RC = 25	RC = 35	RC = 50
AMAX DB	.0017375255	.0108709564	.0213327376	.0436480540	.0988323393	.1772876696	.2802872360	.5675287486	1.249387366
AMIN DB	59.84698529	70.10267796	73.88901953	77.92395557	82.55744030	85.89979036	88.54749686	92.70532576	97.56971414
MU K	.9842555D-2	.3082003D-2	.2005844D-2	.1268587D-2	.7492000D-3	.5122330D-3	.3789446D-3	.2359863D-3	.1355223D-3
1 PFR	.8335955537	.6197635985	.5499065421	.4797595780	.4040514236	.3524082578	.3132230723	.2548954663	.1923409117
1 PF1	.9281416686	.7607704882	.7074122883	.6553033298	.6014907478	.5667516072	.5417446674	.5071662667	.4743865332
1 PQ1	.7013820245	.7448845658	.7724844214	.8123060503	.8769436462	.9418177609	1.008694433	1.153874985	1.422765994
1 PF2	1.069336608	.9688532952	.9360215728	.9037147979	.8701191548	.8482598053	.8323802490	.8100480125	.7880326679
1 PQ2	1.468235896	1.599233318	1.683745952	1.804345073	1.996102201	2.184213087	2.374536312	2.778710659	3.507689717
2 PFC	1.191466315	1.131889106	1.110929685	1.089517454	1.066350917	1.050739215	1.039111115	1.022327564	1.005294120
2 PQC	5.054640912	5.104392011	5.216983526	5.402369909	5.715666033	6.025275231	6.333318370	6.961237783	8.001516267
1 ZF1	1.402290623	1.402900481	1.403099236	1.403296639	1.403506327	1.403647093	1.403752680	1.403908123	1.404073233
1 ZF2	1.515885183	1.522134992	1.524129628	1.526096910	1.528175061	1.529565047	1.530605556	1.532134789	1.533756612
1 ZF3	1.876439879	1.896630135	1.903130996	1.909567406	1.916391801	1.920970037	1.924403969	1.929460646	1.934835188
1 ZF4	3.250701014	3.306384030	3.324669814	3.342930110	3.362448376	3.375626009	3.385550454	3.400222531	3.415883953

(4) P 7+2		TH = 47	SBE = 1.367327461							MCPER FUNCTIONS
		RC = 02	RC = 05	RC = 07	RC = 10	RC = 15	RC = 20	RC = 25	RC = 35	RC = 50
AMAX DB		.0017375255	.0108709564	.0213327376	.0436480540	.0988323393	.1772876696	.2802872360	.5675287486	1.249387366
AMIN DB		57.94160513	68.19351660	71.97889338	76.01298910	80.64570398	83.98760553	86.63501062	90.79244918	95.65649105
MU	K	.1191142D-1	.3734269D-2	.2431267D-2	.1538209D-2	.9087832D-3	.6214998D-3	.4598664D-3	.2864594D-3	.1645558D-3
1	PFR	.8451083710	.6268942443	.5558908797	.4847129449	.4080114362	.3557528760	.3161302450	.2571925337	.1940293768
1	PF1	.9369662867	.7672742246	.7132576586	.6605445694	.6061481443	.5710535097	.5458010144	.5108974410	.4778254472
1	PQ1	.7058607603	.7484863154	.7759088953	.8156166848	.8802216861	.9451462964	1.012116570	1.157568933	1.427061885
1	PF2	1.071655401	.9716486056	.9389000387	.9066482366	.8730865395	.8512393567	.8353651577	.8130379385	.7910290627
1	PQ2	1.496905727	1.624316731	1.708358356	1.828959007	2.021440229	2.210660366	2.402308310	2.809633840	3.544786879
2	PFC	1.186529126	1.128959982	1.108598089	1.087745948	1.065132828	1.049865486	1.038479788	1.022025924	1.005303837
2	PQC	5.241872274	5.259338305	5.365201330	5.546027598	5.857345667	6.167814832	6.478061397	7.112644762	8.167058286
1	ZF1	1.378907493	1.379518139	1.379717936	1.379916671	1.380128055	1.380270104	1.380376721	1.380533776	1.380700714
1	ZF2	1.487519210	1.493838992	1.495857503	1.497849154	1.499953981	1.501362365	1.502416948	1.503967344	1.505612252
1	ZF3	1.834271708	1.854716742	1.861309079	1.867840328	1.874769797	1.879421075	1.882911117	1.888052463	1.893519636
1	ZF4	3.164819399	3.221087031	3.239615279	3.258140435	3.277965478	3.291363278	3.301460124	3.316396922	3.332353130

(4) P 7+2		TH = 48	SBE = 1.345632730							MCPER FUNCTIONS
		RC = 02	RC = 05	RC = 07	RC = 10	RC = 15	RC = 20	RC = 25	RC = 35	RC = 50
AMAX DB		.0017375255	.0108709564	.0213327376	.0436480540	.0988323393	.1772876696	.2802872360	.5675287486	1.249387366
AMIN DB		56.06451296	66.31232000	70.09664929	74.12983254	78.76171132	82.10312571	84.75020338	88.90721789	93.77088329
MU	K	.1437178D-1	.4511144D-2	.2938206D-2	.1859645D-2	.1099127D-2	.7518711D-3	.5564414D-3	.3467171D-3	.1992305D-3
1	PFR	.8572496840	.6343702161	.5621549036	.4898902380	.4121447079	.3592409074	.3191603366	.2595849203	.1957867184
1	PF1	.9461256177	.7740256907	.7193249203	.6659837932	.6109803552	.5755161321	.5500083658	.5147667928	.4813910568
1	PQ1	.7106463327	.7523003572	.7795259342	.8191052617	.8836680680	.9486409284	1.015706045	1.161438608	1.431556987
1	PF2	1.073955526	.9744824765	.9418298075	.9096431557	.8761238094	.8542932224	.8384271519	.8161083094	.7941087534
1	PQ2	1.527617121	1.650995625	1.734481815	1.855033237	2.048229810	2.238589095	2.431611014	2.842223459	3.583839845
2	PFC	1.181522145	1.125977518	1.106220921	1.085937352	1.063887181	1.048970714	1.037832278	1.021715074	1.005311319
2	PQC	5.444823232	5.425957814	5.524214218	5.699811258	6.008681298	6.319859607	6.632304943	7.273770216	8.342993147
1	ZF1	1.356684865	1.357295088	1.357495616	1.357695407	1.357908224	1.358051393	1.358158925	1.358317433	1.358486041
1	ZF2	1.460470455	1.466855210	1.468896035	1.470910553	1.473040515	1.474466301	1.475534229	1.477104750	1.478771685
1	ZF3	1.793855615	1.814539997	1.821219630	1.827841920	1.834872785	1.839594834	1.843139383	1.848363193	1.853929852
1	ZF4	3.082180361	3.138982014	3.157739716	3.176518454	3.196640050	3.210252148	3.220517500	3.235714061	3.251960765

(4) P 7+2 — TH = 49 SBE = 1.325012993 — MCPER FUNCTIONS

MU	7+2	RC = 02	RC = 05	RC = 07	RC = 10	RC = 15	RC = 20	RC = 25	RC = 35	RC = 50
	AMAX DB	.0017375255	.0108709564	.0213327376	.0436480540	.0988323393	.1772876696	.2802872360	.5675287486	1.249387366
	AMIN DB	54.21404087	64.45739516	68.24058793	72.27278101	76.90375236	80.24463793	82.89136016	87.04791429	91.91117092
MU	K	.1729104D-1	.5434378D-2	.3540948D-2	.2242013D-2	.1325668D-2	.9070873D-3	.6714496D-3	.4185025D-3	.2405546D-3
1	PFR	.8700661486	.6422131425	.5687152097	.4953040670	.4164604889	.3628797542	.3223195405	.2620772743	.1976161941
1	PF1	.9556305188	.7810342424	.7256226118	.6716286948	.6159941953	.5801457331	.5543725902	.5187796690	.4850882366
1	PQ1	.7157675950	.7563436123	.7833500700	.8227845008	.8872941630	.9523124703	1.019473465	1.165494745	1.436263049
1	PF2	1.076229274	.9773517884	.9448089090	.9126985529	.8792308075	.8574217122	.8415668413	.8192601057	.7972730300
1	PQ2	1.560565288	1.679405067	1.762239334	1.882682145	2.076580172	2.268107867	2.462554507	2.876596171	3.624981915
2	PFC	1.176448724	1.122942842	1.103798836	1.084091965	1.062614015	1.048054825	1.037168451	1.021394831	1.005316374
2	PQC	5.665262001	5.605423859	5.695069391	5.864669742	6.170545813	6.482248891	6.796873666	7.445437412	8.530179984
1	ZF1	1.335555464	1.336163981	1.336364913	1.336565468	1.336779439	1.336923557	1.337031882	1.337191675	1.337361787
1	ZF2	1.434662579	1.441107137	1.443168651	1.445204464	1.447357949	1.448800088	1.449880597	1.451470149	1.453157996
1	ZF3	1.755088547	1.775995938	1.782758448	1.789467761	1.796596134	1.801386550	1.804983909	1.810287847	1.815933718
1	ZF4	3.002587039	3.059869991	3.078843511	3.097863977	3.118271353	3.132091545	3.142521283	3.157972350	3.174504984

(4) P 7+2 — TH = 50 SBE = 1.305407289 — MCPER FUNCTIONS

MU	7+2	RC = 02	RC = 05	RC = 07	RC = 10	RC = 15	RC = 20	RC = 25	RC = 35	RC = 50
	AMAX DB	.0017375255	.0108709564	.0213327376	.0436480540	.0988323393	.1772876696	.2802872360	.5675287486	1.249387366
	AMIN DB	52.38859591	62.62712173	66.40908208	70.44020129	75.07018832	78.41050019	81.05683678	85.21289135	90.07570438
MU	K	.2074750D-1	.6529282D-2	.4256134D-2	.2695941D-2	.1594748D-2	.1091514D-2	.8081369D-3	.5038517D-3	.2897064D-3
1	PFR	.8836096509	.6504467715	.5755899543	.5009681816	.4209688334	.3666774458	.3256145663	.2646746218	.1995213209
1	PF1	.9654920059	.7883097244	.7321597654	.6774874419	.6211969180	.5849489827	.5588999461	.5229417756	.4889221912
1	PQ1	.7212571369	.7606349334	.7873974286	.8266684651	.8911124971	.9561728078	1.023430472	1.169749103	1.441192912
1	PF2	1.078468344	.9802530982	.9478351324	.9158132610	.8824072808	.8606250796	.8447848066	.8224943111	.8005232136
1	PQ2	1.595970796	1.709696003	1.791767946	1.912032784	2.106612292	2.299336740	2.495260334	2.912880549	3.668359760
2	PFC	1.171312406	1.119857172	1.101332552	1.082210129	1.061313389	1.047117751	1.036488175	1.021065012	1.005318800
2	PQC	5.905218151	5.799066949	5.878950190	6.041671755	6.343919172	6.655923045	6.972690140	7.628565503	8.729576119
1	ZF1	1.315457549	1.316063004	1.316263992	1.316465002	1.316679834	1.316824719	1.316933710	1.317094611	1.317266051
1	ZF2	1.410025557	1.416524577	1.418605087	1.420660553	1.422835871	1.424293264	1.425385548	1.426992981	1.428700565
1	ZF3	1.717876045	1.738989117	1.745829824	1.752621911	1.759843676	1.764699914	1.768348286	1.773729874	1.779461544
1	ZF4	2.925858358	2.983567700	3.002742747	3.021992487	3.042674290	3.056696023	3.067285784	3.082985778	3.099799481

(4) P 7+2 — TH = 51 SBE = 1.286759566 — MCPER FUNCTIONS

MU	(4) P 7+2	RC = 02	RC = 05	RC = 07	RC = 10	RC = 15	RC = 20	RC = 25	RC = 35	RC = 50
	AMAX DB	.0017375255	.0108709564	.0213327376	.0436480540	.0988323393	.1772876696	.2802872360	.5675287486	1.249387366
	AMIN DB	50.58665203	60.81994359	64.60056813	68.63052323	73.25944307	76.59913286	79.24505124	83.40056400	88.26289585
MU	K	.2483193D-1	.7825297D-2	.5103135D-2	.3233811D-2	.1913762D-2	.1310244D-2	.9702915D-3	.6051428D-3	.3480628D-3
1	PFR	.8979380624	.6590972397	.5827990441	.5068976045	.4256806894	.3706427078	.3290526948	.2673824071	.2015059016
1	PF1	.9757211670	.7958624875	.7389459341	.6835687092	.6265962500	.5899329956	.5635971144	.5272592088	.4928984839
1	PQ1	.7271518874	.7651953806	.7916859498	.8307727394	.8951368998	.9602350323	1.027589868	1.174214582	1.446360640
1	PF2	1.080663814	.9831826097	.9509060022	.9189859282	.8856528672	.8639035127	.8480815910	.8258119085	.8039606555
1	PQ2	1.634083685	1.742037606	1.823220725	1.943226650	2.138460490	2.332408762	2.529863013	2.951219627	3.714135119
2	PFC	1.166116941	1.116721826	1.098822857	1.080292226	1.059985383	1.046159435	1.035791322	1.020725425	1.005318390
2	PQC	6.167030243	6.008400851	6.077198047	6.232024663	6.529904751	6.841938622	7.160789364	7.824183496	8.942251095
1	ZF1	1.296334378	1.296935335	1.297136013	1.297337153	1.297552533	1.297697995	1.297807517	1.297969339	1.298141921
1	ZF2	1.386495071	1.393043040	1.395140783	1.397214189	1.399409569	1.400881061	1.401984276	1.403608378	1.405334459
1	ZF3	1.682131428	1.703431788	1.710345734	1.717216089	1.724526885	1.729446249	1.733143728	1.738600337	1.744415242
1	ZF4	2.851827464	2.909906043	2.929267648	2.948733574	2.969677838	2.983894186	2.994639351	3.010582350	3.027671948

(4) P 7+2 — TH = 52 SBE = 1.269018215 — MCPER FUNCTIONS

MU	(4) P 7+2	RC = 02	RC = 05	RC = 07	RC = 10	RC = 15	RC = 20	RC = 25	RC = 35	RC = 50
	AMAX DB	.0017375255	.0108709564	.0213327376	.0436480540	.0988323393	.1772876696	.2802872360	.5675287486	1.249387366
	AMIN DB	48.80674267	59.03436091	62.81353811	66.84223180	71.46999506	74.80901056	77.45447557	81.60940093	86.47121106
MU	K	.2964962D-1	.9356658D-2	.6104495D-2	.3870047D-2	.2291329D-2	.1569218D-2	.1162335D-2	.7251526D-3	.4172330D-3
1	PFR	.9131161310	.6681933831	.5903643552	.5131087847	.4306080027	.3747850408	.3326418421	.2702065380	.2035740557
1	PF1	.9863290491	.8037034025	.7459912206	.6898817130	.6322004277	.5951053676	.5684712341	.5317384888	.4970230690
1	PQ1	.7334938345	.7700485469	.7962356448	.8351146383	.8993826757	.9645135976	1.031965764	1.178905370	1.451781665
1	PF2	1.082806105	.9861361412	.9540187508	.9222149977	.8889670812	.8672571235	.8514576932	.8292138760	.8072867355
1	PQ2	1.675188420	1.776620041	1.856769156	1.976421751	2.172274292	2.367471755	2.566511790	2.991767693	3.762486775
2	PFC	1.160866292	1.113538231	1.096270613	1.078338688	1.058630103	1.045179833	1.035077770	1.020375877	1.005314923
2	PQC	6.453403873	6.235153888	6.291338710	6.437096877	6.729748721	7.041486278	7.362335875	8.033446668	9.169403165
1	ZF1	1.278133729	1.278728673	1.278928654	1.279129579	1.279345182	1.279491019	1.279600930	1.279763476	1.279937005
1	ZF2	1.364011979	1.370603206	1.372716349	1.374805904	1.377019491	1.378503871	1.379617126	1.381256621	1.382999892
1	ZF3	1.647775091	1.669243187	1.676225110	1.683168957	1.690564159	1.695543786	1.699288348	1.704817185	1.710712599
1	ZF4	2.780340340	2.838728706	2.858261187	2.877929550	2.899123663	2.913527306	2.924422982	2.940602699	2.957962677

(4) P 7+2	TH = 53 SBE = 1.252135658							MCPER FUNCTIONS	
	RC = 02	RC = 05	RC = 07	RC = 10	RC = 15	RC = 20	RC = 25	RC = 35	RC = 50
AMAX DB	.0017375255	.0108709564	.0213327376	.0436480540	.0988323393	.1772876696	.2802872360	.5675287486	1.249387366
AMIN DB	47.04745392	57.26892275	61.04653217	65.07385939	69.70036960	73.03865444	75.68362814	79.83791686	84.69916145
MU I K	.3532265D-1	.1116317D-1	.7286451D-2	.4621447D-2	.2737501D-2	.1875366D-2	.1389425D-2	.8671237D-3	.4990971D-3
1 I PFR	.9292165361	.6777671002	.5983099872	.5196197750	.4357638360	.3791148106	.3363906321	.2731534374	.2057302538
1 I PF1	.9973265128	.8118438716	.7533063053	.6964362489	.6380182381	.6004742167	.5735299423	.5363965981	.5013023272
1 I PQ1	.7403308900	.7752209413	.8010688976	.8397134498	.9038668046	.9690244992	1.036573748	1.183337106	1.457472965
1 I PF2	1.084884938	.9891090908	.9571702899	.9254986843	.8923492967	.8706859351	.8549135578	.8327011810	.8108028606
1 I PQ2	1.719609835	1.813657751	1.892605926	2.011795046	2.208220615	2.404690394	2.605372692	3.034702371	3.813612844
2 I PFC	1.155564652	1.110307926	1.093676767	1.076350000	1.057247680	1.044178916	1.034347402	1.020016173	1.005308171
2 I PQC	6.767482802	6.481306666	6.523113732	6.658444607	6.944863117	7.255912058	7.578644045	8.257656001	9.412378743
1 I ZF1	1.260807482	1.261394816	1.261593693	1.261794041	1.262009520	1.262155521	1.262265671	1.262428732	1.262603002
1 I ZF2	1.342521834	1.349150451	1.351277083	1.353380919	1.355610771	1.357106766	1.358229126	1.359882672	1.361641753
1 I ZF3	1.614733878	1.636348898	1.643393209	1.650405476	1.657880171	1.662917022	1.666706516	1.672304614	1.678277633
1 I ZF4	2.711254579	2.769890921	2.789577860	2.809434213	2.830864881	2.845448080	2.856489081	2.872398838	2.890523319

(4) P 7+2	TH = 54 SBE = 1.236067977							MCPER FUNCTIONS	
	RC = 02	RC = 05	RC = 07	RC = 10	RC = 15	RC = 20	RC = 25	RC = 35	RC = 50
AMAX DB	.0017375255	.0108709564	.0213327376	.0436480540	.0988323393	.1772876696	.2802872360	.5675287486	1.249387366
AMIN DB	45.30741815	55.52221985	59.29813126	63.32397852	67.94913144	71.28662474	73.93106612	78.08466499	82.94529671
MU I K	.4199252D-1	.1329113D-1	.8679539D-2	.5507582D-2	.3263994D-2	.2236773D-2	.1657586D-2	.1034843D-2	.5958526D-3
1 I PFR	.9463211458	.6878537759	.6066625583	.5264504360	.4411625055	.3836433520	.3403084795	.2762301027	.2079793576
1 I PF1	1.008724045	.8202958340	.7609024747	.7032427309	.6440590631	.6060482271	.5787814180	.5412110237	.5057431060
1 I PQ1	.7477179340	.7807424434	.8062108216	.8445907204	.9086081753	.9737854850	1.041431086	1.189027068	1.463453254
1 I PF2	1.086889308	.9920963981	.9603571767	.9288349484	.8957987290	.8741898687	.8584495648	.8362747748	.8144104625
1 I PQ2	1.767720354	1.853393371	1.930948243	2.049545326	2.246486332	2.444248660	2.646630926	3.080217065	3.867733449
2 I PFC	1.150216456	1.107032577	1.091042353	1.074326706	1.055838281	1.043156670	1.033600108	1.019646111	1.005297895
2 I PQC	7.112936624	6.749137819	6.774518453	6.897844001	7.176853425	7.486742777	7.811202234	8.498281238	9.672695432
1 I ZF1	1.244311245	1.244889287	1.245086630	1.245286021	1.245501012	1.245646954	1.245757185	1.245920542	1.246095335
1 I ZF2	1.321974470	1.328634421	1.330772559	1.332888724	1.335132806	1.336639081	1.337769562	1.339435743	1.341209179
1 I ZF3	1.582940528	1.604680296	1.611781050	1.618856342	1.626405309	1.631496151	1.635328290	1.640992488	1.647040021
1 I ZF4	2.644438283	2.703258375	2.723082580	2.743111747	2.764764954	2.779519523	2.790700355	2.807333054	2.825215770

(4) P 7+2		TH = 55 SBE = 1.220774589								MCPER FUNCTIONS
		RC = 02	RC = 05	RC = 07	RC = 10	RC = 15	RC = 20	RC = 25	RC = 35	RC = 50
AMAX DB		.0017375255	.0108709564	.0213327376	.0436480540	.0988323393	.1772876696	.2802872360	.5675287486	1.249387366
AMIN DB		43.58530819	53.79287777	57.56695015	61.59119467	66.21487758	69.55151350	72.19537823	76.34822970	81.20819731
MU	K	.4982332D-1	.1579436D-1	.1031931D-1	.6551253D-2	.3884477D-2	.2662876D-2	.1973847D-2	.1232737D-2	.7100688D-3
1	PFR	.9645225208	.6984927788	.6154515500	.5336226738	.4468197397	.3883830877	.3444056861	.2794441741	.2103266644
1	PF1	1.020531520	.8290717653	.7687916480	.7103122340	.6503329279	.6118366993	.5842344319	.5462198049	.5103527641
1	PQ1	.7557180822	.7866468445	.8116896807	.8497705915	.9136278588	.9788163006	1.046556951	1.194494396	1.469743215
1	PF2	1.088807441	.9950925036	.9635755794	.9322214670	.8993144141	.8777687277	.8620660181	.8399355853	.8181109951
1	PQ2	1.819948766	1.896102411	1.972041788	2.089896633	2.287281314	2.486352715	2.690493715	3.128528821	3.925093848
2	PFC	1.144826395	1.103713987	1.088368504	1.072269414	1.054402102	1.042113100	1.032835786	1.019265489	1.005283844
2	PQC	7.494069485	7.041279748	7.047848062	7.157329947	7.427551744	7.735716441	8.061701693	8.756988408	9.952069435
1	ZF1	1.228604022	1.229171002	1.229366362	1.229564393	1.229778515	1.229924162	1.230034307	1.230197731	1.230372819
1	ZF2	1.302323629	1.309008666	1.311156244	1.313282702	1.315538884	1.317054035	1.318191604	1.319868930	1.321655185
1	ZF3	1.552333190	1.574174045	1.581324911	1.588457487	1.596075168	1.601216559	1.605088907	1.610815838	1.616934587
1	ZF4	2.579769088	2.638706222	2.658649699	2.678835737	2.700696704	2.715613983	2.726928817	2.743776914	2.761911175

(4) P 7+2		TH = 56 SBE = 1.206217949								MCPER FUNCTIONS
		RC = 02	RC = 05	RC = 07	RC = 10	RC = 15	RC = 20	RC = 25	PC = 35	RC = 50
AMAX DB		.0017375255	.0108709564	.0213327376	.0436480540	.0988323393	.1772876696	.2802872360	.5675287486	1.249387366
AMIN DB		41.87983199	52.07955012	55.85163048	59.87413926	64.49623015	67.83193740	70.47517748	74.62721923	79.48646726
MU	K	.5900529D-1	.1873550D-1	.1224716D-1	.7779043D-2	.4614896D-2	.3164692D-2	.2346423D-2	.1465977D-2	.8447518D-3
1	PFR	.9839257235	.7097280487	.6247097112	.5411607181	.4527528631	.3933476669	.3486935513	.2828040124	.2127779598
1	PF1	1.032757898	.8381846674	.7769864029	.7176565377	.6568505544	.6178496052	.5898984005	.5514215862	.5151392230
1	PQ1	.7644042330	.7929724957	.8175373915	.8552801974	.9189494315	.9841389777	1.051972696	1.200260350	1.476365768
1	PF2	1.090626759	.9980913029	.9668212369	.9356556022	.9028951846	.8814221796	.8657631315	.8436345087	.8219059314
1	PQ2	1.876790992	1.942098900	2.016165448	2.133102337	2.330842030	2.531234325	2.737193636	3.179980709	3.985968128
2	PFC	1.139399429	1.100354102	1.085656461	1.070178803	1.052939382	1.041048231	1.032054347	1.018874103	1.005265758
2	PQC	7.915955776	7.360786960	7.345753783	7.439243164	7.699056851	8.004818905	8.332071305	9.035672813	10.25244833
1	ZF1	1.213647913	1.214201976	1.214394877	1.214591127	1.214803978	1.214949083	1.215058969	1.215222216	1.215397358
1	ZF2	1.283526626	1.290230302	1.292385173	1.294519797	1.296785848	1.298308402	1.299451970	1.301138871	1.302936325
1	ZF3	1.522854998	1.544771665	1.551965890	1.559149631	1.566830103	1.572018376	1.575928336	1.581714407	1.587900850
1	ZF4	2.517133283	2.576118205	2.596162123	2.616488281	2.638541419	2.653612246	2.665054899	2.682110365	2.700489029

(4) P 7+2 — TH = 57 SBE = 1.192363293 — MCPER FUNCTIONS

MU		RC = 02	RC = 05	RC = 07	RC = 10	RC = 15	RC = 20	RC = 25	RC = 35	RC = 50
	AMAX DB	.0017375255	.0108709564	.0213327376	.0436480540	.0988323393	.1772876696	.2802872360	.5675287486	1.249387366
	AMIN DB	40.18972789	50.38091191	54.15083389	58.17146261	62.79182919	66.12653054	68.76909394	72.92025844	77.77872675
MU	K	.6975913D-1	.2218746D-1	.1451134D-1	.9221960D-2	.5473874D-2	.3755095D-2	.2784918D-2	.1740616D-2	.1003421D-2
1	PFR	1.004650506	.7216087919	.6344735350	.5490914464	.4589810107	.3985521261	.3531845004	.2863187903	.2153395777
1	PF1	1.045410841	.8476480460	.7854999965	.7252881725	.6636234192	.6240976485	.5957834481	.5568256772	.5201110234
1	PQ1	.7738609673	.7997630870	.8237901229	.8611501392	.9245993561	.9897781740	1.057702168	1.206348611	1.483346384
1	PF2	1.092333851	1.001086099	.9700894159	.9391343648	.9065396438	.8851497367	.8695410130	.8475224000	.8257967589
1	PQ2	1.938823367	1.991742216	2.063637023	2.179450027	2.377435860	2.579154925	2.786992601	3.234545824	4.050663556
2	PFC	1.133940809	1.096955030	1.082907578	1.068055630	1.051450402	1.039962115	1.031255710	1.018471745	1.005243364
2	PQC	8.384610671	7.711220334	7.671311784	7.746287699	7.993782880	8.296328267	8.624519585	9.336498793	10.57605045
1	ZF1	1.199407852	1.199947052	1.200136999	1.200331026	1.200542182	1.200686486	1.200795929	1.200958745	1.201133686
1	ZF2	1.265544064	1.272259717	1.274419646	1.276560218	1.278833801	1.280362212	1.281510635	1.283205455	1.285012397
1	ZF3	1.494453692	1.516419138	1.523649514	1.530877889	1.538614832	1.543846072	1.547790870	1.553632245	1.559882617
1	ZF4	2.456425019	2.515385862	2.535510512	2.555959192	2.578188058	2.593402730	2.604966639	2.622220931	2.640836364

(4) P 7+2 — TH = 58 SBE = 1.179178403 — MCPER FUNCTIONS

MU		RC = 02	RC = 05	RC = 07	RC = 10	RC = 15	RC = 20	RC = 25	RC = 35	RC = 50
	AMAX DB	.0017375255	.0108709564	.0213327376	.0436480540	.0988323393	.1772876696	.2802872360	.5675287486	1.249387366
	AMIN DB	38.51376053	48.69565286	52.46323508	56.48182683	61.10032548	64.43393710	67.07576736	71.22598131	76.08360462
MU	K	.8234100D-1	.2623518D-1	.1716810D-1	.1091620D-1	.6483173D-2	.4449141D-2	.3300566D-2	.2063741D-2	.1190199D-2
1	PFR	1.026833974	.7341903103	.6447838222	.5574447673	.4655253793	.4040130764	.3578922345	.2899985971	.2180184698
1	PF1	1.058496232	.8574758710	.7943463816	.7332204675	.6706638174	.6305923329	.6019004751	.5624421210	.5252773911
1	PQ1	.7841868963	.8070685902	.8304890184	.8674150505	.9306074361	.9957615763	1.063772090	1.212785642	1.490713448
1	PF2	1.093914436	1.004069550	.9733748630	.9426543749	.9102461358	.8889507321	.8733996473	.8514500625	.8297849755
1	PQ2	2.006719120	2.045445397	2.114820129	2.229267404	2.427366254	2.630410484	2.840186599	3.292832062	4.119525763
2	PFC	1.128456090	1.093519049	1.080123340	1.065900739	1.049935490	1.038854829	1.030439812	1.018058209	1.005216375
2	PQC	8.907205990	8.096751730	8.028108253	8.081601526	8.314518793	8.612868949	8.941585727	9.661947906	10.92541247
1	ZF1	1.185851369	1.186373669	1.186560141	1.186751481	1.186960502	1.187103733	1.187212538	1.187374655	1.187549131
1	ZF2	1.248339571	1.255060308	1.257222972	1.259367175	1.261645842	1.263178484	1.264330558	1.266031553	1.267846175
1	ZF3	1.467081285	1.489066567	1.496325389	1.503591419	1.511378079	1.516648103	1.520624776	1.526517350	1.532827621
1	ZF4	2.397545606	2.456407804	2.476592565	2.497145275	2.519532518	2.534880759	2.546558954	2.564002969	2.582347010

(4) P 7+2	TH = 59	SBE = 1.166633397						MCPER FUNCTIONS	
	RC = 02	RC = 05	RC = 07	RC = 10	RC = 15	RC = 20	RC = 25	RC = 35	RC = 50
AMAX DB	.0017375255	.0108709564	.0213327376	.0436480540	.0988323393	.1772876696	.2802872360	.5675287486	1.249387366
AMIN DB	36.85071775	47.02247072	50.78751472	54.80389849	59.42037299	62.75280375	65.39383951	69.54302325	74.39973062
MU I K	.9704836D-1	.3097770D-1	.2028313D-1	.1290406D-1	.7668251D-2	.5264454D-2	.3906527D-2	.2443658D-2	.1409926D-2
1 I PFR	1.050633840	.7475349940	.6556863527	.5662540734	.4724095229	.4097489236	.3628319055	.2938545621	.2208222867
1 I PF1	1.072017562	.8676825156	.8035402157	.7414675987	.6779849317	.6373460377	.6082612350	.5682917699	.5306483101
1 I PQ1	.7954975821	.8149464060	.8376810677	.8741142780	.9370073574	1.002120381	1.070212506	1.219601109	1.498498699
1 I PF2	1.095353335	1.007033611	.9766717518	.9462118166	.9140127111	.8928242944	.8773388747	.8554682337	.8338720831
1 I PQ2	2.081268995	2.103685338	2.170132622	2.282929424	2.480978971	2.685337362	2.897111402	3.355087825	4.192944921
2 I PFC	1.122951157	1.090048627	1.077305370	1.063715068	1.048395033	1.037726483	1.029606605	1.017633286	1.005184492
2 I PQC	9.492345528	8.522294729	8.420345099	8.448843812	8.664501519	8.957477984	9.286202018	10.01437767	11.30344725
1 I ZF1	1.172948373	1.173451646	1.173634100	1.173822269	1.174028690	1.174170562	1.174278529	1.174439665	1.174613395
1 I ZF2	1.231879573	1.238598246	1.240761228	1.242906643	1.245187829	1.246722996	1.247877452	1.249592783	1.251403175
1 I ZF3	1.440693781	1.462667874	1.469946892	1.477243110	1.485072262	1.490376592	1.494381967	1.500321346	1.506687195
1 I ZF4	2.340402869	2.399089066	2.419312361	2.439949669	2.462476972	2.477947892	2.489732967	2.507357003	2.526420915

(4) P 7+2	TH = 60	SBE = 1.154700538						MCPER FUNCTIONS	
	RC = 02	RC = 05	RC = 07	RC = 10	RC = 15	RC = 20	RC = 25	RC = 35	RC = 50
AMAX DB	.0017375255	.0108709564	.0213327376	.0436480540	.0988323393	.1772876696	.2802872360	.5675287486	1.249387366
AMIN DB	35.19940864	45.36006430	49.12235207	53.13634096	57.75062101	61.08177170	63.72194619	67.87001301	72.72572719
MU I K	.1142270D 0	.3653059D-1	.2393315D-1	.1523502D-1	.9058925D-2	.6221693D-2	.4618238D-2	.2890118D-2	.1668285D-2
1 I PFR	1.076232437	.7617135172	.6672326890	.5755567799	.4796597023	.4157801279	.3680203219	.2978989985	.2237594728
1 I PF1	1.085975165	.8782826652	.8130968572	.7500446368	.6856009085	.6443721016	.6148784216	.5743563732	.5362346077
1 I PQ1	.8079291952	.8234627692	.8454201683	.8812927043	.9438373407	1.008889867	1.077057316	1.226828382	1.506737742
1 I PF2	1.096634445	1.009969472	.9799736245	.9498023887	.9178370887	.8967693171	.8813583676	.8595775715	.8380595805
1 I PQ2	2.163407241	2.167015353	2.230056914	2.340867018	2.538669639	2.744319392	2.958149439	3.421708904	4.271363174
2 I PFC	1.117432239	1.086546436	1.074455445	1.061499662	1.046829474	1.036577222	1.028756062	1.017196768	1.005147405
2 I PQC	10.15041908	8.993669210	8.852972219	8.852303500	9.047506453	9.333686762	9.661770647	10.39859357	11.71351452
1 I ZF1	1.160670964	1.161152993	1.161330861	1.161515354	1.161718691	1.161858905	1.161965822	1.162125682	1.162298371
1 I ZF2	1.216133090	1.222842270	1.225003050	1.227147151	1.229428168	1.230964066	1.232119571	1.233827298	1.235651440
1 I ZF3	1.415250917	1.437180529	1.444470902	1.451789306	1.459653209	1.464987049	1.469017723	1.474999195	1.481415985
1 I ZF4	2.284910572	2.343340513	2.363579764	2.384281245	2.406929265	2.422511318	2.434395396	2.452189102	2.471463526

(4) P 7+2 — TH = 61 SBE = 1.143354068 — MCPER FUNCTIONS

MU	K	RC = 02	RC = 05	RC = 07	RC = 10	RC = 15	RC = 20	RC = 25	RC = 35	RC = 50
	AMAX DB	.0017375255	.0108709564	.0213327376	.0436480540	.0988323393	.1772876696	.2802872360	.5675287486	1.249387366
	AMIN DB	33.55866330	43.70712645	47.46641729	51.47780632	56.08970585	59.41946821	62.05870857	66.20556393	71.06020070
MU	K	.1342790D 0	.4302896D-1	.2820798D-1	.1796705D-1	.1069017D-1	.7345109D-2	.5453824D-2	.3414581D-2	.1971962D-2
1	PFR	1.103841689	.7768062879	.6794811429	.5853949688	.4873053009	.4221295107	.3734761923	.3021455732	.2268393775
1	PF1	1.100365257	.8892911884	.8230323472	.7589675942	.6935269400	.6516849164	.6217657664	.5806786755	.5420480507
1	PQ1	.8216431300	.8326944797	.8537684242	.8990017488	.9511409288	1.016110090	1.084344918	1.234505144	1.515470567
1	PF2	1.097740715	1.012867491	.9832733281	.9534212479	.9217166125	.9007844245	.8854576032	.8637786354	.8423489552
1	PQ2	2.254244619	2.236080798	2.295152720	2.403577792	2.600893000	2.807796506	3.023738163	3.493146805	4.355283599
2	PFC	1.111905941	1.083015370	1.071575512	1.059255685	1.045239328	1.035407233	1.027888179	1.016748451	1.005104785
2	PQC	10.89406153	9.517810160	9.331854248	9.297035404	9.467960256	9.745622566	10.07225885	10.81693805	12.15950789
1	ZF1	1.148993257	1.149451736	1.149624429	1.149804717	1.150004465	1.150142708	1.150248353	1.150406627	1.150577969
1	ZF2	1.201071565	1.207763499	1.209919452	1.212059598	1.214337629	1.215872372	1.217027520	1.218735597	1.220561355
1	ZF3	1.390715951	1.412565318	1.419857559	1.427189564	1.435079913	1.440438113	1.444490437	1.450508938	1.456971684
1	ZF4	2.230987903	2.289078292	2.309309870	2.330054049	2.352802353	2.368483288	2.380457986	2.398410312	2.417885211

(4) P 7+2 — TH = 62 SBE = 1.132570051 — MCPER FUNCTIONS

MU	K	RC = 02	RC = 05	RC = 07	RC = 10	RC = 15	RC = 20	RC = 25	RC = 35	RC = 50
	AMAX DB	.0017375255	.0108709564	.0213327376	.0436480540	.0988323393	.1772876696	.2802872360	.5675287486	1.249387366
	AMIN DB	31.92733516	42.06233664	45.81836320	49.82692666	54.43624172	57.76449734	60.40272399	64.54826458	69.40173193
MU	K	.1576730D 0	.5063089D-1	.3321285D-1	.2116819D-1	.1260307D-1	.8663217D-2	.6434616D-2	.4030546D-2	.2328839D-2
1	PFR	1.133709308	.7929052159	.6924979472	.5958161651	.4953793215	.4288226211	.3792204145	.3066095078	.2300723869
1	PF1	1.115178724	.9007229562	.8333633687	.7682534664	.7017793534	.6593000319	.6289381493	.5872625304	.5481014574
1	PQ1	.8368318742	.8427310527	.8627977470	.8973005918	.9589679452	1.023826721	1.092118987	1.242674112	1.524742785
1	PF2	1.098654133	1.015717122	.9865629422	.9570629462	.9256482010	.9048679321	.8896358315	.8680718665	.8467416723
1	PQ2	2.355110697	2.311638610	2.366072907	2.471639261	2.668174268	2.876275303	3.094380269	3.569918898	4.445281107
2	PFC	1.106379265	1.079458569	1.068667703	1.056984429	1.043625188	1.034216749	1.027002978	1.016288132	1.005056292
2	PQC	11.73875360	10.10303477	9.863982451	9.789032105	9.931082476	10.19813558	10.52231764	11.27440101	12.64596265
1	ZF1	1.137891227	1.138323761	1.138490667	1.138666203	1.138861834	1.138997782	1.139101922	1.139258288	1.139427957
1	ZF2	1.186668711	1.193335268	1.195483654	1.197617084	1.199889174	1.201420778	1.202574089	1.204280355	1.206105472
1	ZF3	1.367055476	1.388786141	1.396070056	1.403406440	1.411314309	1.416691337	1.420761384	1.426811468	1.433314803
1	ZF4	2.178559000	2.236223331	2.256422502	2.277186793	2.300013780	2.315780590	2.327836981	2.345936116	2.365600714

(4) P 7+2 — TH = 63 SBE = 1.122326238 — MCPER FUNCTIONS

MU	K	RC = 02	RC = 05	RC = 07	RC = 10	RC = 15	RC = 20	RC = 25	RC = 35	RC = 50
	AMAX DB	.0017375255	.0108709564	.0213327376	.0436480540	.0988323393	.1772876696	.2802872360	.5675287486	1.249387366
	AMIN DB	30.30430679	40.42435332	44.17681645	48.18230455	52.78881085	56.11542991	58.75255562	62.89666834	67.74886554
MU	K	.1849550D 0	.5952174D-1	.3907131D-1	.2491837D-1	.1484600D-1	.1020961D-1	.7585757D-2	.4753936D-2	.2748224D-2
1	PFR	1.166126581	.8101158847	.7063586833	.6068742756	.5039189846	.4358881743	.3852764215	.3113078185	.2334700801
1	PF1	1.130399602	.9125925949	.8441071750	.7779202678	.7103757070	.6672342729	.6364117240	.5941230306	.5544088258
1	PQ1	.8537265300	.8536774068	.8725918428	.9062576858	.9673756675	1.032092068	1.100429414	1.251383919	1.534605520
1	PF2	1.099355713	1.018506836	.9898337011	.9607213597	.9296282911	.9090178012	.8938920384	.8724575626	.8512391626
1	PQ2	2.467608528	2.394581975	2.443583364	2.545725330	2.741123206	2.950342101	3.170656279	3.652620883	4.542015777
2	PFC	1.100859643	1.075879439	1.065734354	1.054687331	1.041987731	1.033006058	1.026100514	1.015815615	1.005001567
2	PQC	12.70361544	10.75938715	10.45774629	10.33544286	10.44306479	10.69695903	11.01743107	11.77625837	13.17819036
1	ZF1	1.127342570	1.127746680	1.127907163	1.128077378	1.128268346	1.128401657	1.128504052	1.128658171	1.128825829
1	ZF2	1.172900383	1.179532984	1.181670940	1.183794764	1.186057811	1.187584192	1.188734105	1.190436279	1.192258364
1	ZF3	1.344239270	1.365809833	1.373074457	1.380405296	1.388321082	1.393710981	1.397794525	1.403870324	1.410408460
1	ZF4	2.127552545	2.184700874	2.204841737	2.225602373	2.248485198	2.264324061	2.276452626	2.294685932	2.314528647

(4) P 7+2 — TH = 64 SBE = 1.112601940 — MCPER FUNCTIONS

MU	K	RC = 02	RC = 05	RC = 07	RC = 10	RC = 15	RC = 20	RC = 25	RC = 35	RC = 50
	AMAX DB	.0017375255	.0108709564	.0213327376	.0436480540	.0988323393	.1772876696	.2802872360	.5675287486	1.249387366
	AMIN DB	28.68850095	38.79180586	42.54036792	46.54250258	51.14595247	54.47079212	57.10672109	61.24928179	66.10009831
MU	K	.2167628D 0	.6991922D-1	.4592870D-1	.2931182D-1	.1747602D-1	.1202395D-1	.8936953D-2	.5603534D-2	.3241135D-2
1	PFR	1.201438215	.8285602380	.7211500359	.6186307343	.5129664529	.4433585814	.3916705972	.3162596030	.2370454163
1	PF1	1.146003179	.9249141493	.8552814771	.7879870540	.7193348930	.6755058694	.6442040601	.6012766567	.5609854834
1	PQ1	.8726065378	.8656572539	.8832436985	.9159526301	.9764302747	1.040966333	1.109333465	1.260690187	1.545117498
1	PF2	1.099825499	1.021224034	.9930759053	.9643896084	.9336527727	.9132315861	.8982249035	.8769358499	.8558428072
1	PQ2	2.593686000	2.485971724	2.528588100	2.626626998	2.820451685	3.030679149	3.253240153	3.741942180	4.646249277
2	PFC	1.095354972	1.072281680	1.062778032	1.052365993	1.040327736	1.031775509	1.025180878	1.015330709	1.004940239
2	PQC	13.81246553	11.49908733	11.12328465	10.94485482	11.01129978	11.24891274	11.56410546	12.32874700	13.76244807
1	ZF1	1.117326569	1.117699694	1.117853101	1.118017407	1.118203143	1.118333466	1.118433862	1.118585383	1.118750675
1	ZF2	1.159744470	1.166334001	1.168458526	1.170569716	1.172820464	1.174339425	1.175484295	1.177179967	1.178996489
1	ZF3	1.322240167	1.343606020	1.350839541	1.358154147	1.366067501	1.371463848	1.375556327	1.381651509	1.388218192
1	ZF4	2.077901390	2.134440044	2.154495458	2.175227417	2.198141896	2.214038114	2.226228694	2.244582636	2.264591001

MU	(4) P 7+2	TH = 65	SBE = 1.103377919							MCPER FUNCTIONS
		RC = 02	RC = 05	RC = 07	RC = 10	RC = 15	RC = 20	RC = 25	RC = 35	RC = 50
	AMAX DB	.0017375255	.0108709564	.0213327376	.0436480540	.0988323393	.1772876696	.2802872360	.5675287486	1.249387366
	AMIN DB	27.07889927	37.16328631	40.90756217	44.90603163	49.50615032	52.82905287	55.46367953	59.60455152	64.45386576
MU	K	.2538419D 0	.8207960D-1	.5395638D-1	.3445974D-1	.2056058D-1	.1415319D-1	.1052339D-1	.6601820D-2	.3820649D-2
1	PFR	1.240054874	.8483799304	.7369719618	.6311559090	.5225697143	.4512705920	.3984327802	.3214863876	.2408129587
1	PF1	1.161953610	.9377006289	.8669042737	.7984739273	.7286772453	.6841346023	.6523343050	.6087414512	.5678482613
1	PQ1	.8938123531	.8788174100	.8948837178	.9264785194	.9862086462	1.050519165	1.118897208	1.270656847	1.556345869
1	PF2	1.100042578	1.023854955	.9962788246	.9680599659	.9377169142	.9175063730	.9026327490	.8815066484	.8605539198
1	PQ2	2.735729897	2.587076665	2.622161260	2.715278569	2.906995784	3.118084971	3.342918804	3.838685121	4.758865261
2	PFC	1.089873649	1.068669317	1.059801553	1.050022198	1.038646093	1.030525520	1.024244207	1.014333236	1.004871916
2	PQC	15.09524868	12.33712219	11.87294455	11.62765778	11.64467565	11.86216426	12.17011143	12.93928743	14.40615315
1	ZF1	1.107823984	1.108163486	1.108309143	1.108466935	1.108646852	1.108773820	1.108871957	1.109020511	1.109183068
1	ZF2	1.147180818	1.153717506	1.155825451	1.157920828	1.160155855	1.161665084	1.162803175	1.164489796	1.166298075
1	ZF3	1.301033968	1.322146990	1.329336674	1.336623516	1.344523267	1.349919126	1.354015614	1.360123329	1.366711791
1	ZF4	2.029542247	2.085373432	2.105314935	2.125991852	2.148912366	2.164850287	2.177092028	2.195552087	2.215712665

MU	(4) P 7+2	TH = 66	SBE = 1.094636279							MCPER FUNCTIONS
		RC = 02	RC = 05	RC = 07	RC = 10	RC = 15	RC = 20	RC = 25	RC = 35	RC = 50
	AMAX DB	.0017375255	.0108709564	.0213327376	.0436480540	.0988323393	.1772876696	.2802872360	.5675287486	1.249387366
	AMIN DB	25.47457207	35.53734092	39.27688593	43.27133765	47.86781847	51.18860906	53.82181665	57.96084901	62.80852679
MU	K	.2970655D 0	.9630526D-1	.6335690D-1	.4049381D-1	.2417969D-1	.1665304D-1	.1238685D-1	.7775190D-2	.4502330D-2
1	PFR	1.282469247	.8697405373	.7539403963	.6445308424	.5327836680	.4596660806	.4055968782	.3270125476	.2447891473
1	PF1	1.178200959	.9509633986	.8789936037	.8094020136	.7384246533	.6931419665	.6608233685	.6165372184	.5750156990
1	PQ1	.9177621248	.8933333286	.9076337124	.9379449086	.9968006154	1.060831602	1.129197283	1.281357768	1.568367950
1	PF2	1.099985112	1.026384578	.9994305901	.9717237561	.9418152760	.9218387094	.9071134820	.8861696322	.8653737200
1	PQ2	2.896691178	2.699425922	2.725588339	2.812791161	3.001743855	3.213500072	3.440616705	3.943789042	4.880894899
2	PFC	1.084424618	1.065046730	1.056808015	1.047657931	1.036943820	1.029256593	1.023290685	1.014323028	1.004796192
2	PQC	16.58998326	13.29203152	12.72188656	12.39652209	12.35395957	12.54656708	12.84479635	13.61677038	15.11815827
1	ZF1	1.098816942	1.099120114	1.099257331	1.099407986	1.099581479	1.099704714	1.099800320	1.099945525	1.100104961
1	ZF2	1.135191168	1.141664431	1.143752476	1.145828698	1.148044404	1.149541463	1.150670942	1.152345818	1.154143009
1	ZF3	1.280599371	1.301407602	1.308539699	1.315786321	1.323660394	1.329048262	1.333143423	1.339256255	1.345859156
1	ZF4	1.982415423	2.037436712	2.057234419	2.077828485	2.100727865	2.116690807	2.128972098	2.147522679	2.167820962

(4) P 7+2			TH = 67 SBE = 1.086360377							MCPER FUNCTIONS
			RC = 02	RC = 05	RC = 07	RC = 10	RC = 15	RC = 20	RC = 25	RC = 35	RC = 50
AMAX DB			.0017375255	.0108709564	.0213327376	.0436480540	.0988323393	.1772876696	.2802872360	.5675287486	1.249387366
AMIN DB			23.87472432	33.91246180	37.64675525	41.63678651	46.22928486	49.54776867	52.17942752	56.31645296	61.16234576
MU		K	.3474575D 0	.1129540D 0	.7437036D-1	.4757039D-1	.2842850D-1	.1958985D-1	.1457715D-1	.9155361D-2	.5304765D-2
1		PFR	1.329276793	.8928368952	.7721906577	.6588494257	.5436714686	.4685930163	.4132016234	.3328658235	.2489926308
1		PF1	1.194677514	.9647113624	.8915671919	.8207933990	.7486006750	.7025513507	.6696941340	.6246857596	.5825082857
1		PQ1	.9449738416	.9094162729	.9216620311	.9504815919	1.008311819	1.071998505	1.140323126	1.292378800	1.581273256
1		PF2	1.099630398	1.028796509	1.002518073	.9753712366	.9459416113	.9262245203	.9116645245	.8909241807	.8703033157
1		PQ2	3.080253721	2.824877579	2.840419788	2.920495972	3.105872490	3.318039763	3.547427201	4.058360829	5.013549113
2		PFC	1.079017423	1.061418695	1.053800828	1.045275407	1.035222077	1.027969321	1.022320558	1.013799935	1.004712642
2		PQC	18.34544773	14.38696598	13.68889298	13.26703163	13.15230171	13.31410317	13.59949191	14.37192858	15.90910689
1		ZF1	1.090288837	1.090552914	1.090680988	1.090823868	1.090990317	1.091109428	1.091202223	1.091343683	1.091499596
1		ZF2	1.123759124	1.130157370	1.132222003	1.134275549	1.136468140	1.137950459	1.139069389	1.140729668	1.142512755
1		ZF3	1.260917930	1.281365206	1.288424855	1.295617785	1.303453106	1.308824850	1.312912902	1.319022797	1.325632160
1		ZF4	1.936464621	1.990568263	2.010190749	2.030672597	2.053521992	2.069492150	2.081800549	2.100424874	2.120845169

(4) P 7+2			TH = 68 SBE = 1.078534743							MCPER FUNCTIONS
			RC = 02	RC = 05	RC = 07	RC = 10	RC = 15	RC = 20	RC = 25	RC = 35	RC = 50
AMAX DB			.0017375255	.0108709564	.0213327376	.0436480540	.0988323393	.1772876696	.2802872360	.5675287486	1.249387366
AMIN DB			22.27876501	32.28707920	36.01550143	40.00064667	44.58877215	47.90473078	50.53469617	54.66952851	59.51347132
MU		K	.4062211D 0	.1324503D 0	.8728228D-1	.5587574D-1	.3342056D-1	.2304289D-1	.1715381D-1	.1078024D-1	.6250231D-2
1		PFR	1.381202671	.9178999349	.7918817686	.6742211361	.5553062074	.4781066692	.4212915082	.3390779574	.2534446749
1		PF1	1.211293260	.9789498724	.9046419499	.8326710082	.7592306490	.7123982379	.6789716996	.6332111469	.5903487465
1		PQ1	.9760950122	.9273227124	.9371652189	.9642434665	1.020867336	1.084131661	1.152379784	1.305320336	1.595166063
1		PF2	1.098954964	1.031072868	1.005526749	.9789914651	.9500887498	.9306590117	.9162827323	.8957693215	.8753436600
1		PQ2	3.291064187	2.965709779	2.968541537	3.040000735	3.220793109	3.433036471	3.664652763	4.183714020	5.158259681
2		PFC	1.073662267	1.057790426	1.050783754	1.042877102	1.033482194	1.026664408	1.021334141	1.013263830	1.004620822
2		PQC	20.42493240	15.65112907	14.79745989	14.25853240	14.05590557	14.17946873	14.44805164	15.21782507	16.79189792
1		ZF1	1.082224244	1.082446415	1.082564631	1.082699090	1.082857860	1.082972447	1.083062142	1.083199447	1.083351419
1		ZF2	1.112870148	1.119180521	1.121218009	1.123245159	1.125410632	1.126875493	1.127981827	1.129624487	1.131390268
1		ZF3	1.241974049	1.261999599	1.268970720	1.276095366	1.283877757	1.289224548	1.293299204	1.299397405	1.306004543
1		ZF4	1.891636805	1.944708813	1.964122960	1.984461527	2.007230259	2.023188598	2.035510752	2.054190735	2.074716035

(4) P 7+2		TH = 69 SBE = 1.071144994								MCPER FUNCTIONS
		RC = 02	RC = 05	RC = 07	RC = 10	RC = 15	RC = 20	RC = 25	RC = 35	RC = 50
	AMAX DB	.0017375255	.0108709564	.0213327376	.0436480540	.0988323393	.1772876696	.2802872360	.5675287486	1.249387366
	AMIN DB	20.68641027	30.65955532	34.38135558	38.36106901	42.94437496	46.25756190	48.88567143	53.01810246	57.85991117
MU	K	.4747718D 0	.1552996D 0	.1024334D 0	.6563262D-1	.3929192D-1	.2710723D-1	.2018833D-1	.1269540D-1	.7365534D-2
1	PFR	1.439136899	.9452055138	.8132019984	.6907745183	.5677730312	.4882711237	.4299179571	.3456854855	.2581696687
1	PF1	1.227930375	.9936792701	.9182332769	.8450583966	.7703417953	.7226804257	.6886836550	.6421400470	.5985623842
1	PQ1	1.011942815	.9473667754	.9543817609	.9794168632	1.034616379	1.097363753	1.165491510	1.318890583	1.610168633
1	PF2	1.097934722	1.033194161	1.008440551	.9825721463	.9542484632	.9351365549	.9209642974	.9007036605	.8804955160
1	PQ2	3.535048934	3.124743113	3.112268898	3.173264273	3.348212993	3.560094915	3.793857292	4.321419386	5.316732201
2	PFC	1.068370089	1.054167626	1.047760946	1.040465784	1.031725687	1.025342683	1.020331832	1.012714609	1.004520272
2	PQC	22.91154606	17.12176940	16.07729270	15.39528476	15.08493061	15.16085880	15.40756842	16.17050160	17.78229979
1	ZF1	1.074608830	1.074786260	1.074893898	1.075019281	1.075169727	1.075279380	1.075365679	1.075498407	1.075646007
1	ZF2	1.102511577	1.108719637	1.110725989	1.112722797	1.114856922	1.116301454	1.117393025	1.119014862	1.120759933
1	ZF3	1.223754989	1.243292999	1.250158175	1.257198711	1.264912771	1.270225006	1.274279421	1.280356382	1.286951813
1	ZF4	1.847882147	1.899801087	1.918971895	1.939134258	1.961789644	1.977715777	1.990037327	2.003753435	2.029365267

(4) P 7+2		TH = 70 SBE = 1.064177772								MCPER FUNCTIONS
		RC = 02	RC = 05	RC = 07	RC = 10	RC = 15	RC = 20	RC = 25	RC = 35	RC = 50
	AMAX DB	.0017375255	.0108709564	.0213327376	.0436480540	.0988323393	.1772876696	.2802872360	.5675287486	1.249387366
	AMIN DB	19.09783419	29.02818104	32.74243193	36.71606336	41.29403282	44.60416750	47.23023766	51.36003323	56.19950148
MU	K	.5547775D 0	.1821061D 0	.1202321D 0	.7710850D-1	.4620629D-1	.3189742D-1	.2376690D-1	.1495586D-1	.8683084D-2
1	PFR	1.504180456	.9750859512	.8363760446	.7086616596	.5811718445	.4991611980	.4391408057	.3527307350	.2631957599
1	PF1	1.244436650	1.008892934	.9323540863	.8579794199	.7819632939	.7334582657	.6988603985	.6515021043	.6071774913
1	PQ1	1.053558908	.9699369540	.9736037026	.9962278812	1.049738412	1.111853518	1.179806409	1.333459759	1.626425365
1	PF2	1.096545208	1.035139145	1.011241694	.9860994561	.9584113061	.9396505507	.9257046323	.9057252986	.8857594052
1	PQ2	3.819856297	3.305507336	3.274473256	3.322696202	3.490216214	3.701164888	3.936934879	4.473371117	5.491015083
2	PFC	1.063152646	1.050556544	1.044737001	1.038044554	1.029954292	1.024005120	1.019314126	1.012152209	1.004410514
2	PQC	25.91582044	18.84697414	17.56638380	16.70805011	16.26472438	16.28103256	16.49934488	17.24985105	18.89977250
1	ZF1	1.067429270	1.067559132	1.067655472	1.067771122	1.067912593	1.068016896	1.068099495	1.068227213	1.068369995
1	ZF2	1.092672686	1.098762002	1.100732922	1.102695189	1.104793482	1.106214644	1.107289153	1.108886768	1.110607509
1	ZF3	1.206250908	1.225230055	1.231970395	1.238909635	1.246538610	1.251805823	1.255832529	1.261877819	1.268451164
1	ZF4	1.805154054	1.855789460	1.874679807	1.894630977	1.917138125	1.933010172	1.945315644	1.964046731	1.984724978

(4) P 7+2		TH = 71	SBE = 1.057620681							MCPER FUNCTIONS
		RC = 02	RC = 05	RC = 07	RC = 10	RC = 15	RC = 20	RC = 25	RC = 35	RC = 50
	AMAX DB	.0017375255	.0108709564	.0213327376	.0436480540	.0988323393	.1772876696	.2802872360	.5675287486	1.249387366
	AMIN DB	17.51388607	27.39117829	31.09671042	35.06347105	39.63549749	42.94225740	45.56607922	49.69297430	54.52986950
MU	K	.6482066D 0	.2135944D 0	.1411695D 0	.9062604D-1	.5436158D-1	.3755216D-1	.2799399D-1	.1762837D-1	.1024227D-1
1	PFR	1.577705864	1.007945261	.8616744476	.7280640082	.5956207909	.5108649039	.4490301899	.3602630913	.2685556580
1	PF1	1.260617872	1.024574669	.9470134515	.8714577265	.7941263240	.7447549203	.7095355003	.6613303966	.6162258479
1	PQ1	1.102284885	.9955188154	.9951922999	1.014953505	1.066451245	1.127792516	1.195502516	1.349465554	1.644108149
1	PF2	1.094761959	1.036884686	1.013910489	.9895578387	.9625664299	.9441932669	.9304982300	.9108317292	.8911355456
1	PQ2	4.155481599	3.512471816	3.458755501	3.491292149	3.649372423	3.858638568	4.096201403	4.641874558	5.683590586
2	PFC	1.058022623	1.046964045	1.041717015	1.035616895	1.028169999	1.022652868	1.018281633	1.011576604	1.004291051
2	PQC	29.58675502	20.88964916	19.31394496	18.23630957	17.62752959	17.56878076	17.75022459	18.48080763	20.16858320
1	ZF1	1.060673163	1.060752689	1.060837018	1.060942280	1.061074123	1.061172657	1.061251249	1.061373514	1.061511018
1	ZF2	1.083344763	1.089296418	1.091227245	1.093150480	1.095208176	1.096602743	1.097657753	1.099227534	1.100920089
1	ZF3	1.189454898	1.207797880	1.214392877	1.221212131	1.228737761	1.233948523	1.237939354	1.243941549	1.250481420
1	ZF4	1.763409296	1.812619621	1.831189941	1.850892614	1.873214172	1.889008593	1.901281267	1.920004379	1.940727072

(4) P 7+2		TH = 72	SBE = 1.051462224							MCPER FUNCTIONS
		RC = 02	RC = 05	RC = 07	RC = 10	RC = 15	RC = 20	RC = 25	RC = 35	RC = 50
	AMAX DB	.0017375255	.0108709564	.0213327376	.0436480540	.0988323393	.1772876696	.2802872360	.5675287486	1.249387366
	AMIN DB	15.93639709	25.74671205	29.44201937	33.40093286	37.96629320	41.26930326	43.89063648	48.01432877	52.84838714
MU	K	.7573836D 0	.2506381D 0	.1658399D 0	.1065765D 0	.6399841D-1	.4424026D-1	.3299689D-1	.2079441D-1	.1209121D-1
1	PFR	1.661436837	1.044279502	.8894260851	.7492000304	.6112607950	.5234866382	.4596689879	.3693406265	.2742876623
1	PF1	1.276229415	1.040695200	.9622147248	.8855159927	.8068640347	.7566066315	.7207461183	.6716619810	.6257433315
1	PQ1	1.159866991	1.024726322	1.019599398	1.035936643	1.085021852	1.145414141	1.212795852	1.367020327	1.663423394
1	PF2	1.092561074	1.038405610	1.016425131	.9929297703	.9667013610	.9487556428	.9353384951	.9160197128	.8966237746
1	PQ2	4.555161619	3.751369245	3.669687189	3.682819960	3.828885232	4.035482495	4.274518251	4.829764089	5.897496537
2	PFC	1.052993751	1.043397685	1.038706657	1.033186725	1.026375090	1.021287273	1.017235103	1.010987826	1.004161371
2	PQC	34.12808001	23.33328984	21.38461794	20.03141915	19.21498806	19.06097952	19.19444745	19.89499583	21.61934295
1	ZF1	1.054328943	1.054355499	1.054427118	1.054521350	1.054642918	1.054735261	1.054809536	1.054925900	1.055057656
1	ZF2	1.074521222	1.080313217	1.082198851	1.084078223	1.086090245	1.087454787	1.088487703	1.090025811	1.091686072
1	ZF3	1.173363022	1.190986121	1.197413489	1.204092408	1.211494761	1.216636564	1.220582568	1.226529125	1.233022992
1	ZF4	1.722608206	1.770238234	1.788446106	1.807860339	1.829956191	1.845647579	1.857869340	1.876559486	1.897302554

(4) P 7+2 — TH = 73 SBE = 1.045691756 — MCPER FUNCTIONS

MU	K	RC = 02	RC = 05	RC = 07	RC = 10	RC = 15	RC = 20	RC = 25	RC = 35	RC = 50
	AMAX DB	.0017375255	.0108709564	.0213327376	.0436480540	.0988323393	.1772876696	.2802872360	.5675287486	1.249387366
	AMIN DB	14.36860125	24.09291894	27.77602024	31.72585154	36.28366780	39.58248564	42.20105099	46.32119255	51.15211272
	K	.8850567D 0	.2942950D 0	.1949658D 0	.1254372D 0	.7541102D-1	.5216856D-1	.3893173D-1	.2455407D-1	.1428914D-1
1	PFR	1.757552628	1.084704306	.9200349738	.7723354279	.6282615671	.5371513761	.4711560154	.3770322205	.2804369948
1	PF1	1.290967731	1.057207478	.9779529193	.9001747869	.8202114038	.7690529877	.7325334700	.6825385525	.6357706685
1	PQ1	1.228602394	1.058345694	1.047397101	1.059606788	1.105781095	1.165005782	1.231951241	1.386370715	1.694621363
1	PF2	1.089920086	1.039674557	1.018761454	.9961954850	.9708017360	.9533270539	.9402175374	.9212851223	.9022234513
1	PQ2	5.036670930	4.029658391	3.913151771	3.902080489	4.032797955	4.235419435	4.475462114	5.040563989	6.136491564
2	PFC	1.048080946	1.039865807	1.035712242	1.030759468	1.024572189	1.019909924	1.016175449	1.010385977	1.004020949
2	PQC	39.82253752	26.29050857	23.86463604	22.16118043	21.08108750	20.80551628	20.87628059	21.53305427	23.29115937
1	ZF1	1.048385772	1.048356984	1.048415221	1.048497801	1.048608460	1.048694196	1.048763844	1.048873852	1.048999382
1	ZF2	1.066197724	1.071804298	1.073639105	1.075469384	1.077430298	1.078761162	1.079769236	1.081271567	1.082895149
1	ZF3	1.157974288	1.174787069	1.181022563	1.187538961	1.194796241	1.199855372	1.203746712	1.209623819	1.216057849
1	ZF4	1.682714909	1.728592611	1.746392197	1.765474992	1.787301380	1.802862724	1.815013871	1.833643742	1.854380719

(4) P 7+2 — TH = 74 SBE = 1.040299436 — MCPER FUNCTIONS

MU	K	RC = 02	RC = 05	RC = 07	RC = 10	RC = 15	RC = 20	RC = 25	RC = 35	RC = 50
	AMAX DB	.0017375255	.0108709564	.0213327376	.0436480540	.0988323393	.1772876696	.2802872360	.5675287486	1.249387366
	AMIN DB	12.81568852	22.42796340	26.09619814	30.03534856	34.58453225	37.87862751	40.49409595	44.61028202	49.43771671
	K	.1034476D 1	.3458532D 0	.2294311D 0	.1477941D 0	.8896182D-1	.6159221D-1	.4599136D-1	.2903117D-1	.1690943D-1
1	PFR	1.868823747	1.129992584	.9540031919	.7977969715	.6468296594	.5520102653	.4836102694	.3864203616	.2870575522
1	PF1	1.304463179	1.074040375	.9942110565	.9154508891	.8342049154	.7821371696	.7449433660	.6940072406	.6463543701
1	PQ1	1.311544637	1.097397881	1.079319662	1.086508885	1.129144064	1.186925521	1.253297094	1.407820685	1.708008768
1	PF2	1.086819251	1.040661839	1.020892670	.9993326525	.9748509840	.9578950234	.9451259182	.9266227498	.9079333330
1	PQ2	5.624224634	4.357198413	4.196836717	4.155279726	4.266284055	4.463183456	4.703562039	5.278711693	6.405283142
2	PFC	1.043300459	1.036377656	1.032740834	1.028337129	1.022764323	1.018522690	1.015103782	1.009771242	1.003869249
2	PQC	47.06865094	29.91590169	26.87103073	24.71659923	23.29820312	22.86554005	22.85382172	23.44797248	25.23470801
1	ZF1	1.042833419	1.042747361	1.042791589	1.042861924	1.042961064	1.043039785	1.043104501	1.043207702	1.043326521
1	ZF2	1.058372274	1.063763185	1.065540877	1.067316364	1.069220330	1.070513612	1.071493896	1.072956087	1.074538302
1	ZF3	1.143290498	1.159195809	1.165213036	1.171542694	1.178631025	1.183592411	1.187418251	1.193210656	1.199569532
1	ZF4	1.643697452	1.687630388	1.704971686	1.723676442	1.745187489	1.760587863	1.772646888	1.791186526	1.811888184

(4) P 7+2		TH = 75 SBE = 1.035276180								MCPER FUNCTIONS
		RC = 02	RC = 05	RC = 07	RC = 10	RC = 15	RC = 20	RC = 25	RC = 35	RC = 50
AMAX DB		.0017375255	.0108709564	.0213327376	.0436480540	.0988323393	.1772876696	.2802872360	.5675287486	1.249387366
AMIN DB		11.28548359	20.75013964	24.39986449	28.32621502	32.86538501	36.15410947	38.76608715	42.87784071	47.70138571
MU	K	.1209489D 1	.4068905D 0	.2703245D 0	.1743718D 0	.1051005D 0	.7282847D-1	.5441587D-1	.3438017D-1	.2004379D-1
1	PFR	1.998786313	1.181126879	.9919626342	.8259915392	.6672194524	.5682482053	.4971766542	.3966049036	.2942142434
1	PF1	1.316276703	1.091090253	1.010955058	.9313548128	.8488819483	.7959061332	.7580268016	.7061215833	.6575479113
1	PQ1	1.412793386	1.143229119	1.116323748	1.117344447	1.155638712	1.211625464	1.277245966	1.431749541	1.733966055
1	PF2	1.083243411	1.041335342	1.022789073	1.002316000	.9788299401	.9624448686	.9500523344	.9320260631	.9137514171
1	PQ2	6.351307207	4.747252291	4.530958433	4.450570928	4.536065522	4.724884800	4.964636900	5.549873556	6.709848208
2	PFC	1.038669997	1.032943505	1.029800360	1.025928394	1.020954995	1.017127778	1.014021454	1.009143918	1.003705727
2	PQC	56.43723240	34.42591222	30.56571028	27.82211022	25.96363772	25.32577176	25.20461654	25.70998436	27.51670540
1	ZF1	1.037662098	1.037517592	1.037547244	1.037604791	1.037691836	1.037763152	1.037822637	1.037918584	1.038030208
1	ZF2	1.051045239	1.056185123	1.057898614	1.059613041	1.061453751	1.062705268	1.063654614	1.065071997	1.066607828
1	ZF3	1.129315856	1.144210418	1.149980629	1.156097079	1.162990263	1.167837301	1.171585672	1.177276480	1.183543175
1	ZF4	1.605527572	1.647299223	1.664127060	1.682402841	1.703546923	1.718754111	1.730697416	1.749113794	1.769747700

(4) P 7+2		TH = 76 SBE = 1.030613629								MCPER FUNCTIONS
		RC = 02	RC = 05	RC = 07	RC = 10	RC = 15	RC = 20	RC = 25	RC = 35	RC = 50
AMAX DB		.0017375255	.0108709564	.0213327376	.0436480540	.0988323393	.1772876696	.2802872360	.5675287486	1.249387366
AMIN DB		9.789183374	19.05804831	22.68418386	26.59485795	31.12221686	34.40476039	37.01276690	41.11951713	45.93869623
MU	K	.1414658D 1	.4793517D 0	.3189973D 0	.2060744D 0	.1243903D 0	.8627548D-1	.6450701D-1	.4079556D-1	.2380791D-1
1	PFR	2.151962264	1.239373032	1.034719740	.8574327855	.6897484101	.5860943067	.5120338457	.4077081942	.3019861635
1	PF1	1.325904331	1.108209731	1.028126578	.9478871421	.8642797026	.8104106625	.7718405958	.7189427206	.6694132332
1	PQ1	1.537904611	1.197644825	1.159676965	1.153031350	1.185947016	1.239685015	1.304323698	1.458637289	1.762971584
1	PF2	1.079184581	1.041660487	1.024417720	1.005116867	.9827163735	.9669592603	.9549832225	.9374368957	.9196747362
1	PQ2	7.264936281	5.218014834	4.929355497	4.798863006	4.851029291	5.028543456	5.266286341	5.861401069	7.057893404
2	PFC	1.034208803	1.029574827	1.026899752	1.023538749	1.019149273	1.015727804	1.012930112	1.008504441	1.003529844
2	PQC	68.75963870	40.13027349	35.17754907	31.65144483	29.21168327	28.30210505	28.03415789	28.41391540	30.22657584
1	ZF1	1.032862272	1.032659322	1.032673924	1.032718204	1.032792633	1.032856178	1.032910149	1.032998408	1.033102357
1	ZF2	1.044219191	1.049067201	1.050708426	1.052354850	1.054125450	1.055330701	1.056245734	1.057613305	1.059097374
1	ZF3	1.116056157	1.129832213	1.135324106	1.141198395	1.147867635	1.152581993	1.156239635	1.161810067	1.167965584
1	ZF4	1.568179624	1.607546503	1.623799179	1.641589721	1.662310654	1.677288661	1.689090199	1.707346704	1.727876661

(4) P — TH = 77 SBE = 1.026304108 — MCPER FUNCTIONS

7+2			RC = 02	RC = 05	RC = 07	RC = 10	RC = 15	RC = 20	RC = 25	RC = 35	RC = 50
AMAX DB			.0017375255	.0108709564	.0213327376	.0436480540	.0988323393	.1772876696	.2802872360	.5675287486	1.249387366
AMIN DB			8.341975359	17.35089390	20.94624588	24.83724586	29.35039041	32.62571443	35.22915062	39.33020167	44.14444566
MU		K	.1655414D 1	.5656509D 0	.3771409D 0	.2440397D 0	.1475432D 0	.1024380D 0	.7664799D-1	.4852488D-1	.2834936D-1
1		PFR	2.334132507	1.306385259	1.083319626	.8927792433	.7148186951	.6058366158	.5284053110	.4198822159	.3104709891
1		PF1	1.332794620	1.125192903	1.045632923	.9650330981	.8804333849	.8257051715	.7864480461	.7325408702	.6820226838
1		PQ1	1.694474453	1.263111671	1.211090928	1.194792873	1.220965523	1.271859427	1.335211705	1.489101142	1.795636182
1		PF2	1.074645268	1.041600336	1.025742110	1.007702676	.9864844042	.9714176692	.9599022542	.9429950462	.9256990884
1		PQ2	8.432192371	5.794999339	5.411181264	5.215057852	5.223158886	5.384889176	5.618622770	6.223005973	7.459531438
2		PFC	1.029937578	1.026284484	1.024049119	1.021175624	1.017348901	1.014325880	1.011831762	1.007853423	1.003341069
2		PQC	85.26828441	47.48318777	41.03805774	36.45297548	33.23175510	31.95662275	31.48910018	31.69051240	33.48665017
1		ZF1	1.028424406	1.028164822	1.028164036	1.028194663	1.028256025	1.028311470	1.028359663	1.028439823	1.028535627
1		ZF2	1.037898422	1.042408513	1.043968226	1.045538884	1.047231879	1.048386000	1.049263092	1.050575480	1.052002009
1		ZF3	1.103517357	1.116066047	1.121245595	1.126846033	1.133259626	1.137821011	1.141373192	1.146802300	1.152825350
1		ZF4	1.531627930	1.568319076	1.583926554	1.601168911	1.621404345	1.636113254	1.647744077	1.665799840	1.686185182

(4) P — TH = 78 SBE = 1.022340595 — MCPER FUNCTIONS

7+2			RC = 02	RC = 05	RC = 07	RC = 10	RC = 15	RC = 20	RC = 25	RC = 35	RC = 50
AMAX DB			.0017375255	.0108709564	.0213327376	.0436480540	.0988323393	.1772876696	.2802872360	.5675287486	1.249387366
AMIN DB			6.963200107	15.62897377	19.18321940	23.04886276	27.54448694	30.81122135	33.40932026	37.50380492	42.31242143
MU		K	.1938281D 1	.6688086D 0	.4468929D 0	.2897163D 0	.1754708D 0	.1219638D 0	.9133148D-1	.5788717D-1	.3385877D-1
1		PFR	2.552672311	1.384358053	1.139139801	.9328899572	.7429484992	.6278433226	.5465751035	.4333187525	.3197911976
1		PF1	1.336384015	1.141756267	1.063332877	.9827544312	.8973731944	.8418470416	.8019195268	.7469971564	.6954615637
1		PQ1	1.892980024	1.343069583	1.272926423	1.244293841	1.261896806	1.309152506	1.370809910	1.523949398	1.832753758
1		PF2	1.069642295	1.041115969	1.026721893	1.010036319	.9901037754	.9757956605	.9647896864	.9485377534	.9318186782
1		PQ2	9.950425780	6.514871671	6.003602916	5.720001152	5.668983968	5.808595564	6.035393726	6.647787062	7.928302035
2		PFC	1.025878178	1.023086966	1.021259963	1.018847578	1.015562448	1.012925724	1.010728859	1.007191707	1.003138900
2		PQC	107.8259563	57.16923416	48.64087769	42.59153591	38.29808660	36.52183571	35.77845307	35.72445341	37.46823883
1		ZF1	1.024338718	1.024026915	1.024010602	1.024027317	1.024075260	1.024122325	1.024164510	1.024236191	1.024323404
1		ZF2	1.032087977	1.036210347	1.037677899	1.039164041	1.040771180	1.041868893	1.042704126	1.043955551	1.045318321
1		ZF3	1.091703347	1.102920632	1.107751002	1.113042912	1.119165901	1.123551797	1.126982096	1.132246422	1.138113032
1		ZF4	1.495841555	1.529562961	1.544444509	1.561067178	1.580747086	1.595142202	1.606569851	1.624378880	1.644573549

(4) P 7+2		TH = 79	SBE = 1.018716695						MCPER FUNCTIONS
		RC = 02	RC = 05	RC = 07	RC = 10	RC = 15	RC = 20	RC = 25	RC = 35	RC = 50
AMAX DB			.0108709564	.0213327376	.0436480540	.0988323393	.1772876696	.2802872360	.5675287486	1.249387366
AMIN DB			13.89446255	17.39265363	21.22469085	25.69811201	28.95439112	31.54614102	35.63294950	40.43507829
MU	K		.7926362D 0	.5309839D 0	.3449706D 0	.2093549D 0	.1456957D 0	.1091999D 0	.6929959D-1	.4058603D-1
1	PFR		1.476248543	1.204029930	.9789076978	.7748186400	.6525941109	.5669110968	.4482642411	.3301030965
1	PF1		1.157515033	1.081016809	1.000977237	.9151193484	.8588951181	.8183328828	.7624058859	.7098315206
1	PQ1		1.442425280	1.348518861	1.303855154	1.310391847	1.352928792	1.412333197	1.564263277	1.875377553
1	PF2		1.040167324	1.027312673	1.012075440	.9935389376	.9800639823	.9696215090	.9540989965	.9380256287
1	PQ2		7.431805201	6.746251491	6.343664024	6.211912056	6.320247051	6.535759123	7.153841813	8.482761240
2	PFC		1.019998669	1.018545443	1.016564531	1.013795485	1.011531805	1.009624424	1.006520433	1.002922886
2	PQC		70.25262868	58.74471738	50.62004773	44.81990095	42.34125170	41.20880953	40.78399666	42.41784516
1	ZF1		1.020238890	1.020207207	1.020209931	1.020244235	1.020282710	1.020318694	1.020381567	1.020459776
1	ZF2		1.030476372	1.031839519	1.033231223	1.034743362	1.035778910	1.036568038	1.037752260	1.039044563
1	ZF3		1.090408837	1.094850522	1.099796026	1.105589828	1.109775187	1.113065238	1.118138400	1.123821435
1	ZF4		1.491222929	1.505284164	1.521204506	1.540249065	1.554279731	1.565467419	1.582977440	1.602928775

(4) P 7+2		TH = 80	SBE = 1.015426612						MCPER FUNCTIONS
		RC = 02	RC = 05	RC = 07	RC = 10	RC = 15	RC = 20	RC = 25	RC = 35	RC = 50
AMAX DB			.0108709564	.0213327376	.0436480540	.0988323393	.1772876696	.2802872360	.5675287486	1.249387366
AMIN DB			12.15262866	15.57303792	19.35926430	23.80365073	27.04684476	29.63086428	33.70853097	38.50307500
MU	K		.9419927D 0	.6329464D 0	.4122425D 0	.2507528D 0	.1747486D 0	.1311051D 0	.8331701D-1	.4886450D-1
1	PFR		1.586105899	1.280524789	1.032386833	.8113439251	.6807279035	.5899001984	.4650421206	.3416103203
1	PF1		1.171955678	1.098379072	1.019573518	.9336738456	.8769066815	.8357733129	.7788773746	.7252551655
1	PQ1		1.568354192	1.442711208	1.376803621	1.368779559	1.405092725	1.461464402	1.611525468	1.924938838
1	PF2		1.038714879	1.027466079	1.013771640	.9967478844	.9841873700	.9743683088	.9596585433	.9443093046
1	PQ2		8.628391381	7.693843238	7.130538421	6.886127093	6.949595413	7.147212882	7.766890307	9.149045714
2	PFC		1.017038202	1.015920706	1.014338058	1.012055831	1.010149542	1.008522194	1.005841134	1.002692659
2	PQC		88.44685636	72.55906466	61.40758358	53.42988078	49.94019714	48.24538756	47.27181879	48.70165390
1	ZF1		1.016794379	1.016747930	1.016736835	1.016757462	1.016787228	1.016816878	1.016370677	1.016939520
1	ZF2		1.025212789	1.026457605	1.027743593	1.029150534	1.030117602	1.030856011	1.031966273	1.033180856
1	ZF3		1.078547848	1.082559239	1.087117159	1.092539177	1.096496073	1.099625262	1.104477451	1.109946015
1	ZF4		1.453241606	1.466371130	1.481491844	1.499808414	1.513416347	1.524321817	1.541472678	1.561119760

(4) P 7+2	TH = 81	SBE = 1.012465126						MCPER FUNCTIONS	
	RC = 02	RC = 05	RC = 07	RC = 10	RC = 15	RC = 20	RC = 25	RC = 35	RC = 50
AMAX DB		.0108709564	.0213327376	.0436480540	.0988323393	.1772876696	.2802872360	.5675287486	1.249387366
AMIN DB		10.41362250	13.72480607	17.44688126	21.85196777	25.07822942	27.65255771	31.71907367	36.50458860
MU I K		.1123151D 1	.7574161D 0	.4947761D 0	.3017547D 0	.2106260D 0	.1581991D 0	.1006927D 0	.5914823D-1
1 I PFR		1.719564960	1.372176215	1.095495833	.8537862075	.7131181417	.6162036107	.4840876511	.3545847121
1 I PF1		1.184408352	1.114980484	1.038333093	.9530067060	.8959316250	.8543321223	.7965414219	.7418824868
1 I PQ1		1.731643241	1.562758689	1.468065502	1.440452182	1.468386649	1.520606885	1.667830147	1.983438956
1 I PF2		1.036722688	1.027130387	1.015069651	.9996806633	.9881229500	.9789937215	.9651906243	.9506553508
1 I PQ2		10.23514222	8.954816943	8.149212149	7.743410168	7.741197777	7.910602347	8.524735685	9.965188538
2 I PFC		1.014226687	1.013403284	1.012181775	1.010352864	1.008785556	1.007426832	1.005155868	1.002447986
2 I PQC		114.6240902	92.09583921	76.38010828	65.14819929	60.15573895	57.62311240	55.81647434	56.88466828
1 I ZF1		1.013687200	1.013627257	1.013602881	1.013610037	1.013631098	1.013654356	1.013698909	1.013758095
1 I ZF2		1.020428300	1.021539387	1.022706898	1.023997227	1.024888860	1.025571508	1.026600475	1.027729481
1 I ZF3		1.067358959	1.070897778	1.075023823	1.080027100	1.083724330	1.086669440	1.091266806	1.096485484
1 I ZF4		1.415557307	1.427623614	1.441828046	1.459306822	1.472423688	1.482997577	1.499718993	1.518990328

(4) P 7+2	TH = 82	SBE = 1.009827573						MCPER FUNCTIONS	
	RC = 02	RC = 05	RC = 07	RC = 10	RC = 15	RC = 20	RC = 25	RC = 35	RC = 50
AMAX DB		.0108709564	.0213327376	.0436480540	.0988323393	.1772876696	.2802872360	.5675287486	1.249387366
AMIN DB		8.694869153	11.85207557	15.48215015	19.83206229	23.03553531	25.59726489	29.64975341	34.42426151
MU I K		.1344348D 1	.9105845D 0	.5969716D 0	.3652300D 0	.2554037D 0	.1920786D 0	.1224749D 0	.7207146D-1
1 I PFR		1.884596215	1.484083388	1.171349077	.9039406731	.7509964245	.6467472067	.5060044468	.3693999816
1 I PF1		1.194027865	1.130199586	1.056920683	.9730326478	.9160014158	.8741030555	.8155514585	.7598999914
1 I PQ1		1.949021114	1.719929591	1.585221153	1.530544363	1.546910959	1.593321968	1.736242608	2.053774669
1 I PF2		1.034163488	1.026252258	1.015906656	1.002277472	.9918180702	.9834522748	.9706620364	.9570442835
1 I PQ2		12.46704672	10.66701855	9.510298210	8.865735058	8.764713195	8.889218757	9.485248473	10.98875929
2 I PFC		1.011587969	1.011013571	1.010111834	1.008697954	1.007448027	1.006344224	1.004467408	1.002188849
2 I PQC		153.8301069	120.8741121	98.00069338	81.70540227	74.38958421	70.56097044	67.44667566	67.87963170
1 I ZF1		1.010911140	1.010839956	1.010803374	1.010797607	1.010810132	1.010827042	1.010862301	1.010911639
1 I ZF2		1.016133713	1.017095009	1.018129862	1.019290815	1.020099328	1.020720697	1.021660384	1.022695280
1 I ZF3		1.056866590	1.059892875	1.063540457	1.068073481	1.071476151	1.074210942	1.078514837	1.083442724
1 I ZF4		1.378098698	1.388949185	1.402095506	1.418603184	1.431146983	1.441330388	1.457538621	1.476348779

(4) P 7+2	TH = 83	SBE = 1.007509825						MCPER FUNCTIONS	
	RC = 02	RC = 05	RC = 07	RC = 10	RC = 15	RC = 20	RC = 25	RC = 35	RC = 50
AMAX DB		.0108709564	.0213327376	.0436480540	.0988323393	.1772876696	.2802872360	.5675287486	1.249387366
AMIN DB		7.023724578	9.965513600	13.46121997	17.73073610	20.90212279	23.44672788	27.48085344	32.24151220
MU I K		.1616682D 1	.1100909D 1	.7249460D 0	.4452269D 0	.3120316D 0	.2350217D 0	.1501673D 0	.8854818D-1
1 I PFR		2.092671733	1.623763729	1.264572079	.9644573876	.7961657239	.6828768461	.5316609580	.3865887161
1 I PF1		1.199799249	1.143174651	1.074810608	.9935708111	.9371080456	.8951734649	.8360891257	.7795441151
1 I PQ1		2.247391239	1.932480373	1.740484738	1.647201202	1.647086585	1.685131853	1.821451379	2.140321119
1 I PF2		1.031026530	1.024780550	1.016212179	1.004466255	.9952073011	.9876862997	.9760293378	.9634493532
1 I PQ2		15.69620519	13.09913003	11.40224362	10.39010179	10.13496936	10.18635317	10.74158777	12.31124848
2 I PFC		1.009148551	1.008775307	1.008147567	1.007105055	1.006147206	1.005281897	1.003779523	1.001915567
2 I PQC		215.4426213	165.4317600	130.7983103	106.2258508	95.13723056	89.20631751	83.94755294	83.24721231
1 I ZF1		1.008459710	1.008380911	1.008333988	1.008316328	1.008320712	1.008331458	1.008357540	1.008396979
1 I ZF2		1.012340843	1.013137496	1.014024613	1.015042086	1.015759000	1.016313040	1.017154750	1.018086263
1 I ZF3		1.047095945	1.049577547	1.052699906	1.056706862	1.059775994	1.062270725	1.066236765	1.070826202
1 I ZF4		1.340772482	1.350238316	1.362153397	1.377524046	1.389393541	1.399114259	1.414707038	1.432951552

(4) P 7+2	TH = 84	SBE = 1.005508280						MCPER FUNCTIONS	
	RC = 02	RC = 05	RC = 07	RC = 10	RC = 15	RC = 20	RC = 25	RC = 35	RC = 50
AMAX DB			.0213327376	.0436480540	.0988323393	.1772876696	.2802872360	.5675287486	1.249387366
AMIN DB			8.086669944	11.38436472	15.53246808	18.65633469	21.17637709	25.18520227	29.92767340
MU I K			.1340310D 1	.8874726D 0	.5476555D 0	.3848559D 0	.2904037D 0	.1860122D 0	.1099485D 0
1 I PFR			1.802634590	1.382309268	1.039428158	.8513906032	.7266420460	.5623655964	.4069450733
1 I PF1			1.152746883	1.091189167	1.014273658	.9591630499	.9176042004	.8583681899	.8011215695
1 I PQ1			2.231532753	1.954676906	1.804121148	1.779569473	1.805103637	1.931037931	2.250052010
1 I PF2			1.022673726	1.015909477	1.006159809	.9982082245	.9916213765	.9812345336	.9698331530
1 I PQ2			16.73673075	14.16874600	12.56163824	12.05436429	11.98203000	12.45359305	14.08731653
2 I PFC			1.006715979	1.006312292	1.005591539	1.004896139	1.004249642	1.003097407	1.001629000
2 I PQC			238.8528265	183.7746170	144.8109061	127.1994591	117.6419233	108.6557922	105.8554431
1 I ZF1			1.006244892	1.006190639	1.006162819	1.006159767	1.006164720	1.006181972	1.006211652
1 I ZF2			1.009682146	1.010407033	1.011265962	1.011882026	1.012362144	1.013096421	1.013914484
1 I ZF3			1.039990227	1.042545087	1.045967182	1.048659507	1.050880437	1.054457360	1.058652256
1 I ZF4			1.311349988	1.321825956	1.335848310	1.346914251	1.356080715	1.370929036	1.388476261

$P^{(5)}_{7+4}$

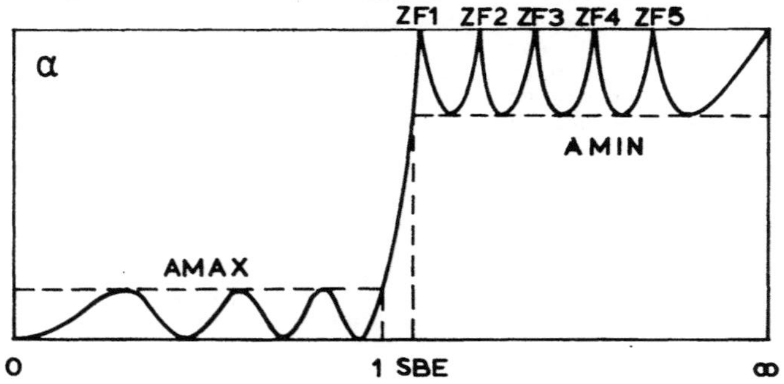

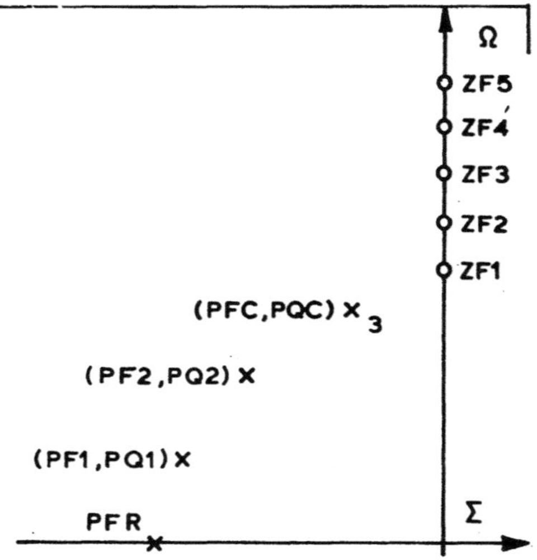

$P^{(5)}_{7+4}$	TH = 37	SBE = 1.661640141						MCPER FUNCTIONS	
	RC = 02	RC = 05	RC = 07	RC = 10	RC = 15	RC = 20	RC = 25	RC = 35	RC = 50
AMAX DB	.0017375255	.0108709564	.0213327376	.0436480540	.0988323393	.1772876696	.2802872360	.5675287486	1.249387366
AMIN DB	91.52354234	104.1292941	108.7977908	113.7835313	119.5243429	123.6763443	126.9715153	132.1536290	138.2129832
MU I K	.3922081D-3	.9423015D-4	.5549505D-4	.3150639D-4	.1640471D-4	.1022759D-4	.7027627D-5	.3893402D-5	.1950434D-5
1 I PFR	.7089128991	.5316570940	.4721803987	.4118506266	.3462220271	.3012594110	.2670995971	.2162890293	.1620260577
1 I PF1	.8224416352	.6737876766	.6256714594	.5782873879	.5288732670	.4966484463	.4732442333	.4404812536	.4087258030
1 I PQ1	.6573699030	.7047135931	.7325390335	.7718517248	.8347789808	.8974660923	.9618631238	1.101310913	1.359131605
1 I PF2	1.029781971	.9208765103	.8854344692	.8504374993	.8137112831	.7894706148	.7715837162	.7458120290	.7192471816
1 I PQ2	1.182110459	1.313958478	1.390049534	1.495278541	1.659000648	1.817664474	1.977266536	2.314867264	2.922599581
3 I PFC	1.291679319	1.205813248	1.175805142	1.145090901	1.111631328	1.088829730	1.071640179	1.046374812	1.019897234
3 I PQC	3.302195212	3.345334105	3.406472588	3.502666041	3.659760090	3.810812362	3.957936755	4.249764820	4.713452814
1 I ZF1	1.672684633	1.673257818	1.673446830	1.673635784	1.673837894	1.673974397	1.674077223	1.674229279	1.674391700
1 I ZF2	1.769730023	1.775397059	1.777239315	1.779070667	1.781019634	1.782330894	1.783316247	1.784769979	1.786318836
1 I ZF3	2.024199944	2.042445672	2.048311484	2.054114886	2.060263187	2.064384852	2.067474733	2.072022519	2.076854704
1 I ZF4	2.663253601	2.707292661	2.721443988	2.735436004	2.750246619	2.760166029	2.767596655	2.778523499	2.790119778
1 I ZF5	4.862567291	4.974408054	5.010516149	5.046276963	5.084179188	5.109584506	5.128622702	5.156624742	5.186343819

(5) P 7+4	TH = 38 SBE = 1.624269245 RC = 02	RC = 05	RC = 07	RC = 10	RC = 15	RC = 20	RC = 25	MCPER FUNCTIONS RC = 35	RC = 50
AMAX DB	.0017375255	.0108709564	.0213327376	.0436480540	.0988323393	.1772876696	.2802872360	.5675287486	1.249387366
AMIN DB	88.78191114	101.3805724	106.0472584	111.0314211	116.7707858	120.9219431	124.2165461	129.3979133	135.4566188
MU I K	.5193261D-3	.1250415D-3	.7368892D-4	.4186180D-4	.2181046D-4	.1360346D-4	.9350102D-5	.5182337D-5	.2597297D-5
1 I PFR	.7157658358	.5360563731	.4759069253	.4149592871	.3487231088	.3033779015	.2689433305	.2177460709	.1630943381
1 I PF1	.8286421063	.6783148366	.6297300162	.5819170226	.5320884194	.4996106081	.4760309747	.4430340518	.4110650219
1 I PQ1	.6596162506	.7066216029	.7343788113	.7736508919	.8365760739	.8992969064	.9637469225	1.103340217	1.361474401
1 I PF2	1.032590156	.9236513128	.8881701598	.8531257381	.8163437961	.7920652193	.7741509904	.7483430395	.7217484121
1 I PQ2	1.196251787	1.326726263	1.402665987	1.507957273	1.672081252	1.831308318	1.991566098	2.330701379	2.941401806
3 I PFC	1.286045795	1.202487629	1.173148581	1.143055107	1.110204010	1.087779505	1.070855304	1.045951924	1.019819749
3 I PQC	3.398088915	3.423788795	3.480724884	3.573538314	3.728062318	3.878125859	4.025006033	4.317507746	4.783904407
1 I ZF1	1.634790734	1.635366225	1.635556809	1.635747681	1.635952196	1.636090515	1.636194804	1.636349169	1.636514233
1 I ZF2	1.727261984	1.732988437	1.734854612	1.736711863	1.738690642	1.740023228	1.741025261	1.742504605	1.744082032
1 I ZF3	1.970237203	1.988780811	1.994750953	2.000661731	2.006928340	2.011131959	2.014284679	2.018927150	2.023862825
1 I ZF4	2.583101676	2.627952818	2.642389476	2.656674649	2.671807169	2.681949664	2.689548989	2.700730475	2.712603518
1 I ZF5	4.701639245	4.815530150	4.852385708	4.888924211	4.927688898	4.953693044	4.973190334	5.001882966	5.032353961

(5) P 7+4	TH = 39 SBE = 1.589015729 RC = 02	RC = 05	RC = 07	RC = 10	RC = 15	RC = 20	RC = 25	MCPER FUNCTIONS RC = 35	RC = 50
AMAX DB	.0017375255	.0108709564	.0213327376	.0436480540	.0988323393	.1772876696	.2802872360	.5675237486	1.249387366
AMIN DB	86.10058812	98.69148170	103.3561830	108.3386157	114.0763938	118.2266254	121.5206052	126.7011636	132.7591475
MU I K	.6831800D-3	.1648675D-3	.9722539D-4	.5526869D-4	.2881483D-4	.1797995D-4	.1236212D-4	.6854375D-5	.3437145D-5
1 I PFR	.7229657363	.5406602950	.4798024922	.4182056752	.3513324530	.3055867813	.2708649396	.2192638209	.1642065652
1 I PF1	.8351067305	.6830290306	.6339544875	.5856934916	.5354322297	.5026904731	.4789278900	.4456870715	.4134954667
1 I PQ1	.6619978665	.7086316055	.7363133879	.7755395742	.8384594908	.9012137419	.9657178598	1.105461413	1.363921182
1 I PF2	1.035449120	.9265003215	.8909840313	.8558947658	.8190586938	.7947427764	.7768014492	.7509574019	.7243331334
1 I PQ2	1.211285605	1.340225720	1.415983705	1.521320134	1.685846427	1.845652205	2.006588858	2.347320169	2.961117026
3 I PFC	1.280285115	1.199075428	1.170419844	1.140961572	1.108734107	1.086696594	1.070044946	1.045513697	1.019736953
3 I PQC	3.501261996	3.507631362	3.559914528	3.648975690	3.800619715	3.949541431	4.096096414	4.389217819	4.858383409
1 I ZF1	1.599038342	1.599614963	1.599806799	1.599999300	1.600205941	1.600345905	1.600451538	1.600608045	1.600775591
1 I ZF2	1.687135858	1.692913941	1.694801817	1.696682917	1.698689493	1.700042145	1.701059963	1.702563678	1.704168456
1 I ZF3	1.919094383	1.937919903	1.943989791	1.950003694	1.956384426	1.960667384	1.963881087	1.968615679	1.973652343
1 I ZF4	2.506879702	2.552507163	2.567219889	2.581790072	2.597236934	2.607596044	2.615363027	2.626795149	2.638941306
1 I ZF5	4.548239907	4.664078558	4.701656893	4.738952422	4.778561599	4.805154676	4.825104724	4.854480236	4.885596953

(5) P 7+4 — TH = 40 — SBE = 1.555723827 — MCPER FUNCTIONS

(5) P 7+4		RC = 02	RC = 05	RC = 07	RC = 10	RC = 15	RC = 20	RC = 25	RC = 35	RC = 50
	AMAX DB	.0017375255	.0108709564	.0213327376	.0436480540	.0988323393	.1772876696	.2802872360	.5675287486	1.249387366
	AMIN DB	83.47615534	96.05855327	100.7210825	105.7016217	111.4376626	115.5868805	118.8801779	124.0598503	130.1170439
MU	K	.8932155D-3	.2160668D-3	.1275098D-3	.7253372D-4	.3784252D-4	.2362376D-4	.1624790D-4	.9013833D-5	.4521881D-5
1	PFR	.7305325665	.5454788940	.4838749708	.4215958889	.3540546112	.3078897263	.2728675168	.2208446120	.1653643750
1	PF1	.8418454031	.6879370296	.6383507755	.5896219180	.5389090883	.5058919968	.4819386396	.4484435881	.4160200704
1	PQ1	.6645248258	.7107499991	.7383483755	.7775227734	.8404337690	.9032209427	.9677802092	1.107678828	1.366476650
1	PF2	1.038354927	.9294227560	.8938759732	.8587449961	.8218568289	.7975043431	.7795362848	.7536564684	.7270028282
1	PQ2	1.227280431	1.354506241	1.430048019	1.535409463	1.700336610	1.860736219	2.022375278	2.364766083	2.981792958
3	PFC	1.274400365	1.195577326	1.167619169	1.138810229	1.107221361	1.085580675	1.069208761	1.045059794	1.019648556
3	PQC	3.612403143	3.597315276	3.644441093	3.729333302	3.877749525	4.025356237	4.171492962	4.465167492	4.937157326
1	ZF1	1.565270194	1.565846728	1.566039485	1.566233315	1.566441794	1.566583224	1.566690075	1.566848551	1.567018410
1	ZF2	1.649180964	1.655002689	1.656909977	1.658812803	1.660845088	1.662216499	1.663249171	1.664775963	1.666406819
1	ZF3	1.870566260	1.889657213	1.895822102	1.901934717	1.908425220	1.912784793	1.916057544	1.920881578	1.926016614
1	ZF4	2.434301461	2.480667642	2.495646689	2.510493301	2.526246535	2.536818544	2.544748973	2.556427499	2.568842908
1	ZF5	4.401821251	4.519499677	4.557774619	4.595805218	4.636239733	4.663411172	4.683807196	4.713857315	4.745813076

(5) P 7+4 — TH = 41 — SBE = 1.524253087 — MCPER FUNCTIONS

(5) P 7+4		RC = 02	RC = 05	RC = 07	RC = 10	RC = 15	RC = 20	RC = 25	RC = 35	RC = 50
	AMAX DB	.0017375255	.0108709564	.0213327376	.0436480540	.0988323393	.1772876696	.2802872360	.5675287486	1.249387366
	AMIN DB	80.90543995	93.47856026	98.13871602	103.1171854	108.8513272	112.9994366	116.2919877	121.4706910	127.5270202
MU	K	.1161040D-2	.2815522D-3	.1662797D-3	.9465563D-4	.4942009D-4	.3086582D-4	.2123617D-4	.1178700D-4	.5916069D-5
1	PFR	.7384881460	.5505230382	.4881328643	.4251365000	.3568944769	.3102906828	.2749543786	.2224909422	.1665695173
1	PF1	.8488685641	.6930460213	.6429251515	.5937077479	.5425236638	.5092193856	.4850671158	.4513070838	.4186419502
1	PQ1	.6672082987	.7129838024	.7404899117	.7796059429	.8425038416	.9053232226	.9699386021	1.109997140	1.369145882
1	PF2	1.041303202	.9324176718	.8968457745	.8616767932	.8247390173	.8003509921	.7823567186	.7564416377	.7297590395
1	PQ2	1.244312165	1.369622091	1.444908608	1.550271547	1.715595903	1.876604005	2.038969363	2.383085221	3.003481333
3	PFC	1.268394886	1.191994101	1.164746855	1.136601047	1.105665525	1.084431425	1.068346402	1.044589871	1.019554254
3	PQC	3.732287045	3.693345666	3.734748338	3.815004423	3.959802026	4.105897755	4.251509390	4.545655996	5.020519370
1	ZF1	1.533344499	1.533919691	1.534113026	1.534307874	1.534517889	1.534660600	1.534768537	1.534928802	1.535100797
1	ZF2	1.613243465	1.619100649	1.621024990	1.622947349	1.625003176	1.626391987	1.627438545	1.628987069	1.630642675
1	ZF3	1.824467937	1.843807310	1.850062286	1.856269035	1.862864784	1.867298137	1.870627915	1.875538593	1.880769263
1	ZF4	2.365109254	2.412174621	2.427409725	2.442523728	2.458574929	2.469354862	2.477445345	2.489365802	2.502046375
1	ZF5	4.261889869	4.381294381	4.420238213	4.458980568	4.500220031	4.527958549	4.548793297	4.579509149	4.612196763

(5) P 7+4	TH = 42 SBE = 1.494476550								MCPER FUNCTIONS
	RC = 02	RC = 05	RC = 07	RC = 10	RC = 15	RC = 20	RC = 25	RC = 35	RC = 50
AMAX DB	.0017375255	.0108709564	.0213327376	.0436480540	.0988323393	.1772876696	.2802872360	.5675287486	1.249387366
AMIN DB	78.38549002	90.94849323	95.60605896	100.5822691	106.3143374	110.4612362	113.7529723	118.9306172	124.9860019
MU I K	.1500863D-2	.3649075D-3	.2156767D-3	.1228670D-3	.6419829D-4	.4011550D-4	.2761006D-4	.1533271D-4	.7699784D-5
1 I PFR	.7468563616	.5558045150	.4925853701	.4288345986	.3598573166	.3127938914	.2771290849	.2242054892	.1678238650
1 I PF1	.8561872167	.6983636370	.6476842807	.5979567732	.5462809229	.5126771148	.4883174595	.4542812632	.4213644218
1 I PQ1	.6700606948	.7153407281	.7427447187	.7817950374	.8446750789	.9075257038	.9721980643	1.112421417	1.371934370
1 I PF2	1.044289088	.9354839409	.8998931094	.8646904626	.8277060828	.8032838087	.7852640007	.7593143550	.7326033719
1 I PQ2	1.262465058	1.385632994	1.460620005	1.565957069	1.731672474	1.893303158	2.056419045	2.402327708	3.026238314
3 I PFC	1.262272285	1.188326638	1.161803267	1.134334030	1.104066366	1.083248526	1.067457517	1.044103570	1.019453732
3 I PQC	3.861787365	3.796286452	3.831330142	3.906425442	4.047164718	4.191527543	4.336491555	4.631012550	5.108791461
1 I ZF1	1.503133099	1.503705655	1.503899214	1.504094757	1.504305996	1.504449795	1.504558683	1.504720549	1.504894496
1 I ZF2	1.579184360	1.585068635	1.587007599	1.588947226	1.591024355	1.592429158	1.593488595	1.595057448	1.596736419
1 I ZF3	1.780632438	1.800202667	1.806542639	1.812838773	1.819535063	1.824039239	1.827423939	1.832418336	1.837741774
1 I ZF4	2.299070546	2.346793525	2.362273881	2.377645748	2.393986054	2.404968657	2.413215612	2.425373270	2.438314676
1 I ZF5	4.128000495	4.249011556	4.288594955	4.328024331	4.370047048	4.398340608	4.419606327	4.450978401	4.484390128

(5) P 7+4	TH = 43 SBE = 1.466279186								MCPER FUNCTIONS
	RC = 02	RC = 05	RC = 07	RC = 10	RC = 15	RC = 20	PC = 25	RC = 35	RC = 50
AMAX DB	.0017375255	.0108709564	.0213327376	.0436480540	.0988323393	.1772876696	.2802872360	.5675287486	1.249387366
AMIN DB	75.91355326	88.46553857	93.12028156	98.09402881	103.8238359	107.9694144	111.2602614	116.4367516	122.4911058
MU I K	.1930025D-2	.4705286D-3	.2783312D-3	.1586841D-3	.8297896D-4	.5187772D-4	.3571903D-4	.1984656D-4	.9972054D-5
1 I PFR	.7556634102	.5613361275	.4972424500	.4326978441	.3629488049	.3154039141	.2793954615	.2259911263	.1691294253
1 I PF1	.8638129445	.7038979793	.6526352493	.6023751568	.5501861528	.5162699490	.4916940797	.4573700699	.4241910142
1 I PQ1	.6730958303	.7178292664	.7451201715	.7840965695	.8469533358	.9098339596	.9745640572	1.114957156	1.374848058
1 I PF2	1.047307206	.9386202297	.9030175219	.8677862406	.8307588261	.8063038874	.7882594071	.7622761125	.7355374936
1 I PQ2	1.281832831	1.402604802	1.477242180	1.582521623	1.748619022	1.910885659	2.074776610	2.422548132	3.050124957
3 I PFC	1.256036453	1.184575932	1.158788844	1.132009222	1.102423666	1.082031660	1.066541751	1.043600527	1.019346661
3 I PQC	4.001892030	3.906768660	3.934737385	4.004081617	4.140267139	4.282645567	4.426821486	4.721600032	5.202327668
1 I ZF1	1.474519889	1.475088478	1.475281894	1.475477798	1.475689937	1.475834624	1.475944321	1.476107592	1.476283300
1 I ZF2	1.546877767	1.552780590	1.554731680	1.556686240	1.558782350	1.560201683	1.561272955	1.562860678	1.564561570
1 I ZF3	1.738908649	1.758691591	1.765111285	1.771491874	1.778283814	1.782855734	1.786293158	1.791368219	1.796781423
1 I ZF4	2.235975092	2.284311958	2.300026182	2.315645873	2.332265930	2.343445656	2.351845296	2.364235150	2.377432804
1 I ZF5	3.999750444	4.122242547	4.162434520	4.202524697	4.245307604	4.274143371	4.295831775	4.327849883	4.361977392

(5) P 7+4			TH = 44	SBE = 1.439556540							MCPER FUNCTIONS
			RC = 02	RC = 05	RC = 07	RC = 10	RC = 15	RC = 20	RC = 25	RC = 35	RC = 50
AMAX DB			.0017375255	.0108709564	.0213327376	.0436480540	.0988323393	.1772876696	.2802872360	.5675287486	1.249387366
AMIN DB			73.48705805	86.02705927	90.67872957	95.64979493	101.3771391	105.5212793	108.8111575	113.9853892	120.0396203
MU		K	.2469613D-2	.6037917D-3	.3574659D-3	.2039675D-3	.1067474D-3	.6677350D-4	.4599322D-4	.2556958D-4	.1285503D-4
1		PFR	.7649380776	.5671318038	.5021149093	.4367345216	.3661750637	.3181256646	.2817576241	.2278509397	.1704883511
1		PF1	.8717579245	.7096576526	.6577855936	.6069694593	.5542449850	.5200029640	.4952016731	.4605777049	.4271254864
1		PQ1	.6763291223	.7204587811	.7476243759	.7865176734	.8493450055	.9122540635	.9770425236	1.117610330	1.377893392
1		PF2	1.050351610	.9418249748	.9062184078	.8709642823	.8338980233	.8094123280	.7913442382	.7653284494	.7385631385
1		PQ2	1.302519977	1.420610274	1.494841207	1.600026296	1.766493296	1.929408374	2.094099187	2.443306032	3.075207742
3		PFC	1.249691577	1.180743105	1.155704105	1.129626716	1.100737228	1.080780517	1.065598747	1.043080365	1.019232698
3		PQC	4.153721327	4.025500133	4.045585971	4.108513763	4.239586446	4.379695192	4.522922015	4.817819194	5.301518147
1		ZF1	1.447399445	1.447962702	1.448155599	1.448351518	1.448564222	1.448709590	1.448819948	1.448984422	1.449161691
1		ZF2	1.516209433	1.522122099	1.524082749	1.526049835	1.528162532	1.529594880	1.530676901	1.532281978	1.534003286
1		ZF3	1.699159535	1.719136441	1.725630392	1.732090320	1.738972821	1.743609277	1.747097132	1.752249665	1.757749492
1		ZF4	2.175632451	2.224537210	2.240473308	2.256330233	2.273220171	2.284591159	2.293139478	2.305756232	2.319205272
1		ZF5	3.876774794	4.000616356	4.041384180	4.082107389	4.125625976	4.154990275	4.177092516	4.209745744	4.244580078

(5) P 7+4			TH = 45	SBE = 1.414213562							MCPER FUNCTIONS
			RC = 02	RC = 05	RC = 07	RC = 10	RC = 15	RC = 20	RC = 25	RC = 35	RC = 50
AMAX DB			.0017375255	.0108709564	.0213327376	.0436480540	.0988323393	.1772876696	.2802872360	.5675287486	1.249387366
AMIN DB			71.10359664	83.63057777	88.27890696	93.24705516	98.97171959	103.1142942	106.4031179	111.5769796	117.6289878
MU		K	.3145197D-2	.7712531D-3	.4570174D-3	.2609937D-3	.1367112D-3	.8556486D-4	.5896083D-4	.3279807D-4	.1649902D-4
1		PFR	.7747120586	.5732067203	.5072144861	.4409536065	.3695427067	.3209644421	.2842200063	.2297882488	.1719029543
1		PF1	.8800349357	.7156517943	.6631433313	.6117466685	.5584634219	.5238815709	.4988452474	.4639086476	.4301718454
1		PQ1	.6797778146	.7232396199	.7502662572	.7890661784	.8518570807	.9147926449	.9796399407	1.120387430	1.381077368
1		PF2	1.053415738	.9450963570	.9094949957	.8742246482	.8371244180	.8126102313	.7945198162	.7684729519	.7416821074
1		PQ2	1.324643279	1.439729960	1.513490028	1.618538345	1.785358698	1.948933626	2.114449304	2.466166457	3.101559166
3		PFC	1.243242160	1.176829412	1.152549658	1.127186654	1.099006872	1.079494793	1.064628146	1.042542701	1.019111488
3		PQC	4.318549404	4.153276944	4.164566204	4.220326035	4.345653887	4.483168974	4.625262131	4.920113530	5.406793672
1		ZF1	1.421675851	1.422232375	1.422424363	1.422619942	1.422832865	1.422978697	1.423089565	1.423255029	1.423433652
1		ZF2	1.487075445	1.492989101	1.494956680	1.496933817	1.499060623	1.500504420	1.501596064	1.503216919	1.504957074
1		ZF3	1.661260603	1.681412084	1.687974629	1.694508583	1.701476352	1.706174000	1.709709891	1.714936559	1.720519717
1		ZF4	2.117869841	2.167294105	2.183439436	2.199522422	2.216671817	2.228227871	2.236920634	2.249758680	2.263453949
1		ZF5	3.758742213	3.883795482	3.925104642	3.966431496	4.010659741	4.040538005	4.063044637	4.096321304	4.131852832

(5) P 7+4	TH = 46 SBE = 1.390163591								MCPER FUNCTIONS
	RC = 02	RC = 05	RC = 07	RC = 10	RC = 15	RC = 20	RC = 25	RC = 35	RC = 50
AMAX DB	.0017375255	.0108709564	.0213327376	.0436480540	.0988323393	.1772876696	.2802872360	.5675287486	1.249387366
AMIN DB	68.76091011	81.27376055	85.91846044	90.88343867	96.60519011	100.7460625	104.0337394	109.2061111	115.2567889
MU K	.3987696D-2	.9808892D-3	.5817838D-3	.3325418D-3	.1743473D-3	.1091847D-3	.7526897D-4	.4189542D-4	.2108862D-4
1 PFR	.7850203258	.5795774433	.5125539531	.4453648365	.3730588888	.3239259695	.2867873900	.2318066281	.1733757207
1 PF1	.8886573613	.7218901080	.6687169949	.6167142310	.5628478657	.5279115434	.5026301464	.4673676777	.4333343669
1 PQ1	.6834612431	.7261832421	.7530556629	.7917506925	.8544972230	.9174569516	.9823633794	1.123295534	1.384407595
1 PF2	1.056492361	.9484322718	.9128463250	.8775672900	.8404387125	.8158986934	.7977874816	.7717112522	.7448962709
1 PQ2	1.348333584	1.460053241	1.533269340	1.638131964	1.805284975	1.969529844	2.135895526	2.489700598	3.129258431
3 PFC	1.236693035	1.172836252	1.149326207	1.124689236	1.097232449	1.078174193	1.063629590	1.041987139	1.018982658
3 PQC	4.497829911	4.290996821	4.292453785	4.340195021	4.459062338	4.593615360	4.734363191	5.028974898	5.518630856
1 ZF1	1.397261657	1.397810019	1.398000697	1.398195570	1.398408354	1.398554428	1.398665647	1.398831882	1.399011642
1 ZF2	1.459381110	1.465286769	1.467258585	1.469243225	1.471381589	1.472835210	1.473935311	1.475570306	1.477327675
1 ZF3	1.625098566	1.645404556	1.652029822	1.658632284	1.665679812	1.670435166	1.674016595	1.679313907	1.684976947
1 ZF4	2.062530272	2.112423130	2.128764364	2.145061624	2.162459461	2.174194029	2.183026752	2.196080145	2.210016172
1 ZF5	3.645351325	3.771472307	3.813286435	3.855185861	3.900096150	3.930472873	3.953373815	3.987261421	4.023479791

(5) P 7+4	TH = 47 SBE = 1.367327461								MCPER FUNCTIONS
	RC = 02	RC = 05	RC = 07	RC = 10	RC = 15	RC = 20	RC = 25	RC = 35	RC = 50
AMAX DB	.0017375255	.0108709564	.0213327376	.0436480540	.0988323393	.1772876696	.2802872360	.5675287486	1.249387366
AMIN DB	66.45687493	78.95440429	83.59516545	88.55670213	94.27528985	98.41431307	101.7007438	106.8714961	112.9207280
MU K	.5034399D-2	.1242381D-2	.7375989D-3	.4219973D-3	.2214577D-3	.1387726D-3	.9570893D-4	.5330655D-4	.2685006D-4
1 PFR	.7959015566	.5862620895	.5181472331	.4499787937	.3767313625	.3270164362	.2894649408	.2339099319	.1749093265
1 PF1	.8976391822	.7283828988	.6745156681	.6218800874	.5674051501	.5320990465	.5065620769	.4709599001	.4366176169
1 PQ1	.6874011458	.7293023656	.7560034801	.7945806986	.3572738426	.9202549221	.9852205723	1.126342362	1.387892360
1 PF2	1.059573536	.9518302976	.9162712221	.8809920332	.8438415581	.8192787996	.8011485900	.7750450288	.7482075703
1 PQ2	1.373737894	1.481679520	1.554268617	1.658889179	1.826349001	1.991272309	2.158513178	2.514486520	3.158392213
3 PFC	1.230049382	1.168765180	1.146034564	1.122134727	1.095413835	1.076818436	1.062602723	1.041413278	1.018845924
3 PQC	4.693226711	4.439675018	4.430122764	4.468880415	4.580475102	4.711646526	4.850806135	5.144950038	5.637558196
1 ZF1	1.374076992	1.374615731	1.374804691	1.374998479	1.375210754	1.375356840	1.375468246	1.375635026	1.375815698
1 ZF2	1.433039976	1.438928538	1.440901834	1.442891363	1.445038652	1.446500421	1.447607769	1.449255203	1.451028086
1 ZF3	1.590570177	1.611009894	1.617691785	1.624357022	1.631478575	1.636287998	1.639912357	1.645276662	1.651015969
1 ZF4	2.009470925	2.059778811	2.076301890	2.092800974	2.110435612	2.122341759	2.131309694	2.144572134	2.158743108
1 ZF5	3.536327530	3.663365936	3.705646750	3.748085920	3.793648969	3.824507653	3.847792152	3.882277325	3.919171414

(5) P 7+4	TH = 48	SBE = 1.345632730						MCPER FUNCTIONS	
	RC = 02	RC = 05	RC = 07	RC = 10	RC = 15	RC = 20	PC = 25	RC = 35	RC = 50
AMAX DB	.0017375255	.0108709564	.0213327376	.0436480540	.0988323393	.1772876696	.2802872360	.5675287486	1.249387366
AMIN DB	64.18949085	76.67042326	81.30691352	86.26471684	91.97987128	96.11688710	99.40196495	104.5709587	110.6186199
MU I K	.6330166D-2	.1567451D-2	.9315422D-3	.5334759D-3	.2802376D-3	.1757186D-3	.1212465D-3	.6757512D-4	.3406018D-4
1 I PFR	.8073986278	.5932805101	.5240095298	.4548069980	.3805685410	.3302425468	.2922582464	.2361023223	.1765066572
1 I PF1	.9069949592	.7351411088	.6805490238	.6272527095	.5721425754	.5364506691	.5106471397	.4746907733	.4400264776
1 I PQ1	.6916220287	.7326111381	.7591217723	.7975666648	.8601961898	.9231952677	.9882199910	1.129536354	1.391540703
1 I PF2	1.062650546	.9552876608	.9197682738	.8844985586	.8473335435	.8227516175	.8046045074	.7784760048	.7516180201
1 I PQ2	1.401021819	1.504719622	1.576587299	1.680900874	1.848635695	2.014244018	2.182385188	2.540609991	3.189055552
3 I PFC	1.223316744	1.164617922	1.142675653	1.119523461	1.093550941	1.075427254	1.061547192	1.040820706	1.018700584
3 I PQC	4.906650824	4.600463162	4.578560837	4.607237573	4.710636234	4.837947511	4.975239908	5.268648167	5.764163079
1 I ZF1	1.352048772	1.352576402	1.352763223	1.352955540	1.353166925	1.353312785	1.353424209	1.353591297	1.353772648
1 I ZF2	1.407972978	1.413835243	1.415807206	1.417798944	1.419952448	1.421420630	1.422533974	1.424192082	1.425978711
1 I ZF3	1.557581216	1.578133115	1.584865304	1.591587354	1.598776960	1.603636659	1.607301220	1.612728700	1.618540484
1 I ZF4	1.958561739	2.009228286	2.025918378	2.042606137	2.060465264	2.072535648	2.081633761	2.095098563	2.109498312
1 I ZF5	3.431420226	3.559219435	3.601926683	3.644870941	3.691055715	3.722378812	3.746035404	3.781103843	3.818661705

(5) P 7+4	TH = 49	SBE = 1.325012993						MCPER FUNCTIONS	
	RC = 02	RC = 05	RC = 07	RC = 10	RC = 15	RC = 20	RC = 25	RC = 35	RC = 50
AMAX DB	.0017375255	.0108709564	.0213327376	.0436480540	.0988323393	.1772876696	.2802872360	.5675287486	1.249387366
AMIN DB	61.95687007	74.41983796	79.05170064	84.00545708	89.71688836	93.85172660	97.13533651	102.3024219	108.3483782
MU I K	.7928846D-2	.1970273D-2	.1172188D-2	.6719722D-3	.3533561D-3	.2217146D-3	.1530586D-3	.8536453D-4	.4305718D-4
1 I PFR	.8195591919	.6006545021	.5301574790	.4598620130	.3845795710	.3336115756	.2951733614	.2383883011	.1781708279
1 I PF1	.9167397999	.7421763545	.6868273661	.6328411418	.5770679464	.5409734603	.5148918630	.4785661389	.4435661741
1 I PQ1	.6961515982	.7361253366	.7624239370	.8007201725	.8632744594	.9262875665	.9913709348	1.132886755	1.395362501
1 I PF2	1.065713846	.9588011970	.9233357977	.8880863815	.8509151821	.8263181886	.8081566050	.7820059467	.7551297095
1 I PQ2	1.430372491	1.529297425	1.600336174	1.704267994	1.872239077	2.038536670	2.207603046	2.568165439	3.221352872
3 I PFC	1.216501045	1.160396388	1.139250528	1.116855854	1.091643717	1.074000395	1.060462650	1.040209004	1.018546522
3 I PQC	5.140305067	4.774671690	4.738887512	4.756232355	4.850382680	4.973286964	5.108391333	5.400749839	5.899099951
1 I ZF1	1.331110014	1.331625026	1.331809278	1.331999726	1.332209831	1.332355219	1.332466486	1.332633637	1.332815426
1 I ZF2	1.384107683	1.389934377	1.391902140	1.393893341	1.396050270	1.397523078	1.398641121	1.400308072	1.402106606
1 I ZF3	1.526045603	1.546687323	1.553463237	1.560235900	1.567487334	1.572393351	1.576095264	1.581581921	1.587462206
1 I ZF4	1.909684168	1.960650067	1.977491519	1.994354058	2.012424645	2.024651487	2.033874442	2.047534511	2.062156468
1 I ZF5	3.330400353	3.458797404	3.501888805	3.545301598	3.592075225	3.623844083	3.647860547	3.683496963	3.721705768

(5) P 7+4			TH = 50	SBE = 1.305407289						MCPER FUNCTIONS
			RC = 02	RC = 05	RC = 07	RC = 10	RC = 15	RC = 20	RC = 25	RC = 35	RC = 50
AMAX DB			.0017375255	.0108709564	.0213327376	.0436480540	.0988323393	.1772876696	.2802872360	.5675287486	1.249387366
AMIN DB			59.75722739	72.20076461	76.82761675	81.77698934	87.48438569	91.61686317	94.89888132	100.0638969	106.1080037
MU		K	.9894936D-2	.2467943D-2	.1469906D-2	.8435382D-3	.4440539D-3	.2788179D-3	.1925773D-3	.1074831D-3	.5425373D-4
1		PFR	.8324363510	.6084080518	.5366093204	.4651575671	.3887744141	.3371314295	.2982168576	.2407727450	.1799052077
1		PF1	.9268893066	.7495009645	.6933616730	.6386550459	.5821896151	.5456749681	.5193032400	.4825922566	.4472423054
1		PQ1	.7010212742	.7398626001	.7659248900	.8040540644	.8665199123	.9295423725	.9946836325	1.136403707	1.399368575
1		PF2	1.068753011	.9623673092	.9269718094	.8917548281	.8545868970	.8299795187	.8118062534	.7856366628	.7587448043
1		PQ2	1.462002025	1.555551776	1.625638999	1.729102932	1.897263486	2.064251822	2.234267921	2.597257062	3.255399151
3		PFC	1.209608608	1.156102682	1.135760380	1.114132404	1.089692155	1.072537629	1.059348756	1.039577747	1.018383204
3		PQC	5.396738256	4.963796725	4.912375741	4.916958733	5.000658631	5.118529784	5.251076710	5.542017341	6.043099862
1		ZF1	1.311199229	1.311700092	1.311881339	1.312069510	1.312277932	1.312422596	1.312533525	1.312700488	1.312882463
1		ZF2	1.361377634	1.367159429	1.369120072	1.371107928	1.373265419	1.374741005	1.375862407	1.377536304	1.379344829
1		ZF3	1.495884624	1.516592932	1.523405735	1.530222554	1.537529329	1.542477527	1.546213812	1.551755458	1.557700073
1		ZF4	1.862730101	1.913932951	1.930909235	1.947931865	1.966200118	1.978575173	1.987917306	2.001765101	2.016602282
1		ZF5	3.233058237	3.361883844	3.405315031	3.449157833	3.496485520	3.528680318	3.553043634	3.589231684	3.628077654

(5) P 7+4			TH = 51	SBE = 1.286759566						MCPER FUNCTIONS
			RC = 02	RC = 05	RC = 07	RC = 10	RC = 15	RC = 20	RC = 25	RC = 35	RC = 50
AMAX DB			.0017375255	.0108709564	.0213327376	.0436480540	.0988323393	.1772876696	.2802872360	.5675287486	1.249387366
AMIN DB			57.58887138	70.01140565	74.63283585	79.57746234	85.28048834	89.41040787	92.69070088	97.85347267	103.8955742
MU		K	.1230554D-1	.3081061D-2	.1837213D-2	.1055496D-2	.5562602D-3	.3495263D-3	.2415434D-3	.1349147D-3	.6815280D-4
1		PFR	.8460894465	.6165676159	.5433850967	.4707086931	.3931639410	.3408107185	.3013958806	.2432609466	.1817134461
1		PF1	.9374594994	.7571280168	.7001636437	.6447047487	.5875165267	.5505632838	.5238887702	.4867759417	.4510608792
1		PQ1	.7062668021	.7438427026	.7696412795	.8075826162	.8699450152	.9329713402	.9981693598	1.140098369	1.403570792
1		PF2	1.071756672	.9659819218	.9306739862	.8955030087	.8583490047	.8337365671	.8155548150	.7893700004	.7624655485
1		PQ2	1.496151652	1.583638747	1.652634399	1.755531166	1.923825017	2.091502224	2.262491956	2.628000102	3.291321281
3		PFC	1.202646176	1.151739122	1.132206549	1.111353710	1.087696298	1.071038753	1.058205180	1.038926501	1.018210182
3		PQC	5.678911339	5.169552367	5.100477815	5.090659779	5.162532540	5.274652091	5.404215510	5.693306921	6.196981663
1		ZF1	1.292259885	1.292745053	1.292922848	1.293108328	1.293314656	1.293458335	1.293568740	1.293735254	1.293917154
1		ZF2	1.339721763	1.345449301	1.347399861	1.349381503	1.351536613	1.353013075	1.354136453	1.355815328	1.357631856
1		ZF3	1.467026241	1.487776968	1.494619547	1.501473799	1.508829144	1.513815202	1.517582742	1.523174989	1.529179553
1		ZF4	1.817600905	1.868975058	1.886068724	1.903235907	1.921687214	1.934201734	1.943657031	1.957684537	1.972729502
1		ZF5	3.139201673	3.268280266	3.312004732	3.356236977	3.404081913	3.436681599	3.461377897	3.498100117	3.537568463

(5) P 7+4			TH = 52	SBE = 1.269018215						MCPER FUNCTIONS	
			RC = 02	RC = 05	RC = 07	RC = 10	RC = 15	RC = 20	RC = 25	RC = 35	RC = 50
		AMAX DB	.0017375255	.0108709564	.0213327376	.0436480540	.0988323393	.1772876696	.2802872360	.5675287486	1.249387366
		AMIN DB	55.45019632	67.85004078	72.46560700	77.40509784	83.10339244	87.23054162	90.50896578	95.66930632	101.7092343
MU		K	.1525267D-1	.3834405D-2	.2289204D-2	.1316694D-2	.6947343D-3	.4368692D-3	.3020702D-3	.1688556D-3	.8536690D-4
	1	PFR	.8605849909	.6251624482	.5505068821	.4765318872	.3977600387	.3446588373	.3047182156	.2458586601	.1835995037
	1	PF1	.9484667071	.7650713758	.7072457465	.6510012952	.5930582713	.5556470905	.5286565058	.4911241076	.4550283498
	1	PQ1	.7119289847	.7480878760	.7735917384	.8113217379	.8735636038	.9365873702	1.001840575	1.143983033	1.407982202
	1	PF2	1.074712467	.9696404304	.9344396271	.8993297884	.8622016958	.8375902339	.8194036360	.7932078432	.7662942656
	1	PQ2	1.533096679	1.613734308	1.681478105	1.783693174	1.952053186	2.120413378	2.292399775	2.560522338	3.329259630
	3	PFC	1.195620922	1.147308259	1.128590542	1.108520475	1.095656243	1.069503591	1.057031603	1.038254831	1.018026987
	3	PQC	5.990278521	5.393909716	5.304856507	5.278752783	5.337217412	5.442758993	5.568846609	5.855583248	6.361665194
	1	ZF1	1.274239939	1.274707852	1.274881743	1.275064108	1.275267919	1.275410347	1.275520036	1.275685831	1.275867388
	1	ZF2	1.319083976	1.324747798	1.326685266	1.328657769	1.330807485	1.332282864	1.333406787	1.335098604	1.336911068
	1	ZF3	1.439404502	1.460172474	1.467037418	1.473922092	1.481318940	1.486338340	1.490133870	1.495772117	1.501832026
	1	ZF4	1.774206586	1.825682996	1.842875610	1.860170905	1.878789779	1.891434478	1.900996547	1.915195236	1.930440056
	1	ZF5	3.048654225	3.177804019	3.221773072	3.266352073	3.314675339	3.347657563	3.372672072	3.409909801	3.449984647

(5) P 7+4			TH = 53	SBE = 1.252135658						MCPER FUNCTIONS	
			RC = 02	RC = 05	RC = 07	RC = 10	RC = 15	RC = 20	RC = 25	RC = 35	RC = 50
		AMAX DB	.0017375255	.0108709564	.0213327376	.0436480540	.0988323393	.1772876696	.2802872360	.5675287486	1.249387366
		AMIN DB	53.33967496	65.71501871	70.32424576	75.25818180	80.95135621	85.07550616	88.35190651	93.50961354	99.54713647
MU		K	.1884593D-1	.4757741D-2	.2844057D-2	.1637815D-2	.8652348D-3	.5445180D-3	.3767215D-3	.2107593D-3	.1066414D-3
	1	PFR	.8759977698	.6342249786	.5579990483	.4826452941	.4025757338	.3486860581	.3081923613	.2485721543	.1855676867
	1	PF1	.9599274179	.7733457263	.7146212699	.6575565051	.5988251396	.5609357170	.5336151030	.4956448133	.4591516614
	1	PQ1	.7180545639	.7526231896	.7777971804	.8152892087	.8773910724	.9404047777	1.005711076	1.148071282	1.412617189
	1	PF2	1.077606983	.9733376470	.9382656076	.9032337535	.8661450131	.8415413451	.8233540359	.7971521069	.7702333600
	1	PQ2	1.573152470	1.646037504	1.712345607	1.813746689	1.982092891	2.151125367	2.324130250	2.694965815	3.369369877
	3	PFC	1.188540474	1.142812886	1.124914042	1.105633520	1.083572151	1.067931999	1.055827726	1.037562293	1.017833137
	3	PQC	6.334887315	5.639144235	5.527422686	5.482859464	5.526095080	5.624105810	5.746147624	6.029936559	6.538186897
	1	ZF1	1.257091412	1.257540501	1.257710030	1.257888849	1.258089713	1.258230616	1.258339392	1.258504191	1.258685125
	1	ZF2	1.299412199	1.305003167	1.306924502	1.308884888	1.311026121	1.312498403	1.313621396	1.315304047	1.317130300
	1	ZF3	1.412959004	1.433717967	1.440597551	1.447505330	1.454936297	1.459984311	1.463804410	1.469483830	1.475594247
	1	ZF4	1.732465052	1.783971115	1.801243202	1.818649207	1.837419220	1.850184235	1.859846280	1.874207070	1.889643285
	1	ZF5	2.961253712	3.090286810	3.134449518	3.179330399	3.228090869	3.261431908	3.286748903	3.324482207	3.365146509

(5) P 7+4	TH = 54	SBE = 1.236067977						MCPER FUNCTIONS	
	RC = 02	RC = 05	RC = 07	RC = 10	RC = 15	RC = 20	RC = 25	RC = 35	RC = 50
AMAX DB	.0017375255	.0108709564	.0213327376	.0436480540	.0988323393	.1772876696	.2802872360	.5675287486	1.249387366
AMIN DB	51.25585201	63.60474946	68.20712618	73.13505620	78.82269149	82.94359538	86.21780476	91.37265997	97.40768186
MU I K	.2321568D-1	.5886794D-2	.3523646D-2	.2031741D-2	.1074724D-2	.6769186D-3	.4686058D-3	.2623907D-3	.1328827D-3
1 I PFR	.8924121521	.6437912579	.5658885734	.4890689201	.4076253350	.3529036377	.3118276164	.2514082729	.1876226869
1 I PF1	.9718580783	.7819666043	.7223043749	.6643830340	.6048281859	.5664391979	.5387738799	.5003463168	.4634382972
1 I PQ1	.7246972858	.7574770034	.7822811498	.8195049537	.8814445962	.9444394894	1.009796184	1.152378157	1.417491639
1 I PF2	1.080425710	.9770677408	.9421483311	.9072131742	.8701788272	.8455906357	.8274072964	.8012047346	.7742853183
1 I PQ2	1.616681703	1.680774245	1.745435308	1.845869378	2.014106732	2.183795006	2.357838567	2.731488882	3.411825143
3 I PFC	1.181412934	1.138256068	1.121178930	1.102693796	1.081444254	1.066323875	1.054593264	1.036848445	1.017628129
3 I PQC	6.717502661	5.907893568	5.770381009	5.704842484	5.730745408	5.820123498	5.937458032	6.217603119	6.727718353
1 I ZF1	1.240770021	1.241198710	1.241363414	1.241538252	1.241735729	1.241874827	1.241982488	1.242146005	1.242326030
1 I ZF2	1.280658966	1.286167696	1.288069825	1.290015068	1.292144662	1.293611775	1.294732318	1.296413623	1.298241437
1 I ZF3	1.387634436	1.408356973	1.415243132	1.422166375	1.429623739	1.434695415	1.438536498	1.444252023	1.450407857
1 I ZF4	1.692301455	1.743760858	1.761091838	1.778590122	1.797493846	1.810368690	1.820123477	1.834636688	1.850255266
1 I ZF5	2.876850849	3.005573375	3.049876527	3.095012146	3.144166387	3.177841072	3.203443814	3.241651398	3.282886861

(5) P 7+4	TH = 55	SBE = 1.220774589						MCPER FUNCTIONS	
	RC = 02	RC = 05	RC = 07	RC = 10	RC = 15	RC = 20	RC = 25	RC = 35	RC = 50
AMAX DB	.0017375255	.0108709564	.0213327376	.0436480540	.0988323393	.1772876696	.2802872360	.5675287486	1.249387366
AMIN DB	49.19733831	61.51769694	66.11267312	71.03411100	76.71575547	80.83314695	84.10498487	89.25675243	95.28901143
MU I K	.2851678D-1	.7264395D-2	.4354265D-2	.2514002D-2	.1331616D-2	.8394522D-3	.5814897D-3	.3258924D-3	.1651930D-3
1 I PFR	.9099236568	.6539014783	.5742054023	.4958248820	.4129245979	.3573239410	.3156341774	.2543745037	.1897696271
1 I PF1	.9842748267	.7909504210	.7303101482	.6714944396	.6110792964	.5721683408	.5441428806	.5052376348	.4678963334
1 I PQ1	.7319191949	.7626815075	.7870702375	.8239913709	.8857433926	.9487092744	1.014112958	1.156920360	1.422623147
1 I PF2	1.083152990	.9808241738	.9460836740	.9112659626	.8743028080	.8497387305	.8315646476	.8053676909	.7784527092
1 I PQ2	1.664103217	1.718201860	1.780972286	1.880262021	2.048277745	2.218598385	2.393698673	2.770263579	3.456818494
3 I PFC	1.174246888	1.133641153	1.117387298	1.099702399	1.079272866	1.064679161	1.053327961	1.036112842	1.017411444
3 I PQC	7.143761823	6.203228482	6.036285705	5.946849819	5.952981603	6.032449305	6.144306966	6.419989723	6.931588423
1 I ZF1	1.225234845	1.225641558	1.225800972	1.225971385	1.226165029	1.226302037	1.226408375	1.226570317	1.226749135
1 I ZF2	1.262780061	1.268197353	1.270077182	1.272004211	1.274118940	1.275578760	1.276695286	1.278372993	1.280200056
1 I ZF3	1.363380163	1.384037608	1.390921914	1.397852633	1.405328311	1.410418461	1.414276766	1.420023069	1.426218963
1 I ZF4	1.653647613	1.704980181	1.722348304	1.739919344	1.758938271	1.771911792	1.781751614	1.796406921	1.812198207
1 I ZF5	2.795308035	2.923520301	2.967908363	3.013249239	3.062751405	3.096733040	3.122603716	3.161252834	3.203049820

I (5) P 7+4		TH = 56	SBE = 1.206217949						MCPER FUNCTIONS	
		RC = 02	RC = 05	RC = 07	RC = 10	RC = 15	RC = 20	RC = 25	RC = 35	RC = 50
AMAX DB		.0017375255	.0108709564	.0213327376	.0436480540	.0988323393	.1772876696	.2802872360	.5675287486	1.249387366
AMIN DB		47.16280568	59.45237203	64.03935490	68.95377654	74.62894269	78.74253411	82.01180554	87.16023047	93.18949739
MU	K	.3493290D-1	.8941856D-2	.5367507D-2	.3103304D-2	.1646074D-2	.1038630D-2	.7199365D-3	.4038642D-3	.2049125D-3
1	PFR	.9286408334	.6646005879	.5829828704	.5029376964	.4184909160	.3619605846	.3196232537	.2574790590	.1920141136
1	PF1	.9971931447	.8003144784	.7386546554	.6789052525	.6175912655	.5781348004	.5497329479	.5103285104	.4725345019
1	PQ1	.7397922161	.7682733696	.7921945764	.8287737190	.8903090295	.9532340156	1.018680446	1.161716486	1.428031251
1	PF2	1.085771986	.9845996309	.9500669254	.9153896254	.8785163924	.8539861208	.8358272516	.8096429544	.7827381844
1	PQ2	1.715902858	1.758614586	1.819212822	1.917152321	2.084812675	2.255733883	2.431906168	2.811503463	3.504565896
3	PFC	1.167051435	1.128971801	1.113541469	1.096660580	1.077058389	1.062997853	1.052031586	1.035355039	1.017182543
3	PQC	7.620368956	6.528740459	6.328109076	6.211368957	6.194893116	6.262963895	6.368446810	6.638703201	7.151309807
1	ZF1	1.210448029	1.210831193	1.210984848	1.211150391	1.211339750	1.211474376	1.211579179	1.211739245	1.211916551
1	ZF2	1.245734695	1.251051464	1.252905881	1.254811588	1.256908166	1.258358514	1.259469411	1.261141193	1.262965108
1	ZF3	1.340149865	1.360712213	1.367585849	1.374515681	1.382001208	1.387104388	1.390975964	1.396747446	1.402977753
1	ZF4	1.616441499	1.667563049	1.684945328	1.702568435	1.721682904	1.734743226	1.744659868	1.759446246	1.775399912
1	ZF5	2.716498256	2.843994963	2.888410045	2.933904284	2.983706006	3.017966287	3.044085940	3.083172298	3.125489723

I (5) P 7+4		TH = 57	SBE = 1.192363293						MCPER FUNCTIONS	
		RC = 02	RC = 05	RC = 07	RC = 10	RC = 15	RC = 20	RC = 25	PC = 35	RC = 50
AMAX DB		.0017375255	.0108709564	.0213327376	.0436480540	.0988323393	.1772876696	.2802872360	.5675287486	1.249387366
AMIN DB		45.15098253	57.40732580	61.98567612	66.89251586	72.56067716	76.67015754	79.93665151	85.08145787	91.10748450
MU	K	.4268165D-1	.1098061D-1	.6601310D-2	.3822174D-2	.2030369D-2	.1282329D-2	.8894739D-3	.4994615D-3	.2536701D-3
1	PFR	.9486875339	.6759390185	.5922582017	.5104346200	.4243435449	.3668286032	.3238071995	.2607309682	.1943622970
1	PF1	1.010627404	.8100769710	.7473549921	.6866310510	.6243778789	.5843511624	.5555558041	.5156294895	.4773622607
1	PQ1	.7484000989	.7742945152	.7976884359	.8338805795	.8951657911	.9580360311	1.023519983	1.166787301	1.433737709
1	PF2	1.088264647	.9883859458	.9540927210	.9195812106	.8828187478	.8583331388	.8401961843	.8140325084	.7871444775
1	PQ2	1.772646884	1.802350235	1.860449856	1.956799474	2.123945875	2.295425772	2.472681754	2.855416960	3.555309701
3	PFC	1.159836194	1.124252006	1.109644020	1.093569769	1.074801326	1.061280005	1.050703944	1.034574597	1.016940872
3	PQC	8.155341126	6.888650476	6.649326119	6.501293518	6.458898102	6.513836564	6.611894000	6.875586131	7.388611077
1	ZF1	1.196374523	1.196732569	1.196879997	1.197040222	1.197224839	1.197356785	1.197459838	1.197617718	1.197793198
1	ZF2	1.229485118	1.234692424	1.236518313	1.238399556	1.240474638	1.241913282	1.243016894	1.244680350	1.246498631
1	ZF3	1.317901220	1.338337026	1.345190758	1.352110944	1.359597439	1.364707936	1.368588629	1.374379397	1.380638163
1	ZF4	1.580626783	1.631448989	1.648821126	1.666474368	1.685663480	1.698797953	1.708782651	1.723688318	1.739793300
1	ZF5	2.640304095	2.766874569	2.811256395	2.856849616	2.906899891	2.941408814	2.967757278	3.007244923	3.050070148

(5) P 7+4 — TH = 58 SBE = 1.179178403 — MCPER FUNCTIONS

MU	K	RC = 02	RC = 05	RC = 07	RC = 10	RC = 15	RC = 20	RC = 25	RC = 35	RC = 50
	AMAX DB	.0017375255	.0108709564	.0213327376	.0436480540	.0988323393	.1772876696	.2802872360	.5675287486	1.249387366
	AMIN DB	43.16065042	55.38114300	59.95017054	64.84881728	70.50940434	74.61443715	77.87792522	83.01881403	89.04133128
MU	K	.5202058D-1	.1345418D-1	.8101218D-2	.4697736D-2	.2499325D-2	.1580079D-2	.1096799D-2	.6165146D-3	.3134474D-3
1	PFR	.9702056724	.6879735520	.6020730992	.5183460497	.4305038634	.3719446444	.3281996691	.2641401859	.1968209422
1	PF1	1.024590284	.8202569700	.7564293316	.6946885417	.6314540071	.5908310369	.5616241431	.5211520077	.4823898736
1	PQ1	.7578408215	.7807930734	.8035909372	.8393444108	.9003411139	.9631404552	1.028655543	1.172156070	1.439766825
1	PF2	1.090611698	.9921740196	.9581549696	.9238372484	.8872087293	.8627799267	.8446724131	.8185383305	.7916744035
1	PQ2	1.834998604	1.849798317	1.905019620	1.999499695	2.165944007	2.337928532	2.516275356	2.902261364	3.609322806
3	PFC	1.152611328	1.119486118	1.105697802	1.090431584	1.072502297	1.059525742	1.049344878	1.033771081	1.016685855
3	PQC	8.758321622	7.287945025	7.004019733	6.820005613	6.747807909	6.787580652	6.876978897	7.132760324	7.645475521
1	ZF1	1.182981841	1.183313210	1.183453945	1.183608402	1.183787814	1.183916781	1.184017864	1.184173243	1.184346573
1	ZF2	1.213996360	1.219085444	1.220879688	1.222733302	1.224783487	1.226208146	1.227302772	1.228955427	1.230765500
1	ZF3	1.296595619	1.316871900	1.323696045	1.330597396	1.338075540	1.343187348	1.347072791	1.352876635	1.359157576
1	ZF4	1.546152424	1.596582694	1.613919009	1.631579132	1.650820661	1.664015798	1.674059190	1.689071541	1.705315979
1	ZF5	2.566616844	2.692045305	2.736331183	2.781966448	2.832211524	2.866937288	2.893493107	2.938354318	2.976663027

(5) P 7+4 — TH = 59 SBE = 1.166633397 — MCPER FUNCTIONS

MU	K	RC = 02	RC = 05	RC = 07	RC = 10	RC = 15	RC = 20	RC = 25	RC = 35	RC = 50
	AMAX DB	.0017375255	.0108709564	.0213327376	.0436480540	.0988323393	.1772876696	.2802872360	.5675287486	1.249387366
	AMIN DB	41.19064156	53.37243561	57.93139408	62.82118675	68.47358312	72.57380371	75.83403822	80.97068515	86.98940089
MU	K	.6325419D-1	.1645052D-1	.9921902D-2	.5762650D-2	.3070841D-2	.1943416D-2	.1350029D-2	.7596746D-3	.3866552D-3
1	PFR	.9933585965	.7007683568	.6124744468	.5267059965	.4369956826	.3773271980	.3328157982	.2677177182	.1993975113
1	PF1	1.039092022	.8308743813	.7658969654	.7030956423	.6388357066	.5975891628	.5679517335	.5269084885	.4876285026
1	PQ1	.7682295833	.7878245300	.8099469211	.8452022157	.9058661090	.9685756938	1.034114154	1.177948937	1.446145837
1	PF2	1.092792636	.9959537360	.9622467729	.9281536840	.8916848322	.8673264018	.8492567709	.8231623790	.7963308560
1	PQ2	1.903739222	1.901410026	1.953309734	2.045592938	2.211111721	2.383532055	2.562971084	2.952322640	3.666913621
3	PFC	1.145387555	1.114678875	1.101705966	1.087247861	1.070162048	1.057735266	1.047954281	1.032944068	1.016416899
3	PQC	9.440981952	7.732547340	7.397012396	7.171478365	7.064906897	7.087121847	7.166407098	7.412680072	7.924188525
1	ZF1	1.170239851	1.170542996	1.170676574	1.170824815	1.170998558	1.171124243	1.171223131	1.171375688	1.171546536
1	ZF2	1.199236002	1.204198316	1.205957811	1.207780612	1.209802452	1.211210796	1.212294692	1.213933996	1.215733195
1	ZF3	1.276197913	1.296280050	1.303064446	1.309937313	1.317397311	1.322504112	1.326389704	1.332200078	1.338496557
1	ZF4	1.512972310	1.562913679	1.580187037	1.597829378	1.617099672	1.630341084	1.640433165	1.655538702	1.671909864
1	ZF5	2.495335687	2.619401551	2.663526356	2.709144096	2.759527345	2.794436258	2.821176606	2.861381766	2.905147837

(5) TH = 60 SBE = 1.154700538 MCPER FUNCTIONS

MU	P 7+4	RC = 02	RC = 05	RC = 07	RC = 10	RC = 15	RC = 20	RC = 25	RC = 35	RC = 50
	AMAX DB	.0017375255	.0108709564	.0213327376	.0436480540	.0988323393	.1772876696	.2802872360	.5675287486	1.249387366
	AMIN DB	39.23983781	51.37983654	55.92791765	60.80814006	66.45167745	70.54669017	73.80340225	78.93545496	84.95005161
MU	K	.7674218D-1	.2007484D-1	.1212899D-1	.7056251D-2	.3766549D-2	.2386304D-2	.1659007D-2	.9345937D-3	.4762296D-3
1	PFR	1.018335233	.7143962364	.6235151495	.5355526492	.4438456106	.3829968655	.3376724169	.2714757705	.2021002590
1	PF1	1.054139463	.8419498693	.7757783327	.7118715686	.6465403325	.6046415254	.5745535355	.5329124561	.4930903127
1	PQ1	.7797025553	.7954531410	.8168080043	.8514963500	.9117761910	.9743739713	1.039926408	1.183895398	1.452905371
1	PF2	1.094785750	.9997138696	.9663603368	.9325258025	.8962451363	.8719722159	.8539499259	.8279065766	.8011168045
1	PQ2	1.979794113	1.957710576	2.005769157	2.095471108	2.259798574	2.432567962	2.613093239	3.005926206	3.728432036
3	PFC	1.138176166	1.109835431	1.097671991	1.084020669	1.067781470	1.055908871	1.046532100	1.032093147	1.016133396
3	PQC	10.21754170	8.229534551	7.834032151	7.560404386	7.414051871	7.415882960	7.483335311	7.718197778	8.227395738
1	ZF1	1.158120584	1.158393972	1.158519934	1.158661514	1.158829124	1.158951219	1.159047687	1.159197095	1.159365122
1	ZF2	1.185173969	1.190001207	1.191722871	1.193511666	1.195501671	1.196891326	1.197962705	1.199586035	1.201371604
1	ZF3	1.256676195	1.276527832	1.283261799	1.290096042	1.297527591	1.302622733	1.306503624	1.312313617	1.318618613
1	ZF4	1.481044932	1.530395980	1.547577713	1.565176113	1.584449994	1.597722315	1.607852381	1.623036634	1.639520836
1	ZF5	2.426366962	2.548845174	2.592741333	2.638279278	2.688741074	2.723797441	2.750698035	2.791215495	2.835410854

(5) TH = 61 SBE = 1.143354068 MCPER FUNCTIONS

MU	P 7+4	RC = 02	RC = 05	RC = 07	RC = 10	RC = 15	RC = 20	RC = 25	RC = 35	RC = 50
	AMAX DB	.0017375255	.0108709564	.0213327376	.0436480540	.0988323393	.1772876696	.2802872360	.5675287486	1.249387366
	AMIN DB	37.30717150	49.40199321	53.93831985	58.80819483	64.44214756	68.53152249	71.78441978	76.91149492	82.92162671
MU	K	.9290908D-1	.2445300D-1	.1480127D-1	.8625931D-2	.4612600D-2	.2925665D-2	.2035675D-2	.1148146D-2	.5857490D-3
1	PFR	1.045355218	.7289401395	.6352551438	.5449290481	.4510834865	.3889766803	.3427883032	.2754279231	.2049383465
1	PF1	1.069734851	.8535047337	.7860950331	.7210369211	.6545866624	.6120054881	.5814458340	.5391786644	.4987885931
1	PQ1	.7924216133	.8037536785	.8242338762	.8582755079	.9181118402	.9805719898	1.046127055	1.190328845	1.460079992
1	PF2	1.096568162	1.003441990	.9704868748	.9369481443	.9008872427	.8767167070	.8587523461	.8327727909	.8060352889
1	PQ2	2.064266221	2.019314557	2.062920505	2.149588170	2.312407486	2.485417309	2.667013624	3.063443960	3.794276650
3	PFC	1.130989033	1.104961386	1.093599710	1.080752337	1.065361616	1.054046954	1.045078349	1.031217927	1.015834717
3	PQC	11.10544640	8.787415246	8.321923449	7.992357999	7.799796858	7.777889879	7.831465915	8.052645410	8.558174931
1	ZF1	1.146598061	1.146840176	1.146958072	1.147092549	1.147253563	1.147371761	1.147465580	1.147611506	1.147776366
1	ZF2	1.171782343	1.176466476	1.178147258	1.179898854	1.181853501	1.183222050	1.184279084	1.185883743	1.187652832
1	ZF3	1.238001601	1.257584550	1.264256857	1.271041805	1.278434055	1.283510523	1.287381596	1.293183906	1.299489987
1	ZF4	1.450333087	1.498987891	1.516047723	1.533574434	1.552825083	1.566111896	1.576268484	1.591515924	1.608098443
1	ZF5	2.359623492	2.480284879	2.523882364	2.569275472	2.619753057	2.654919068	2.681954072	2.722750005	2.767344467

MU	(5) P 7+4	TH = .62	SBE = 1.132570051						MCPER FUNCTIONS	
		RC = 02	RC = 05	RC = 07	RC = 10	RC = 15	RC = 20	RC = 25	RC = 35	RC = 50
	AMAX DB	.0017375255	.0108709564	.0213327376	.0436480540	.0988323393	.1772876696	.2802872360	.5675287486	1.249387366
	AMIN DB	35.39162896	47.43756127	51.96117936	56.81986193	62.44344068	66.52670981	69.77547389	74.89715369	80.90244345
MU	K	.1122556D 0	.2973560D-1	.1803342D-1	.1052883D-1	.5640637D-2	.3582021D-2	.2494543D-2	.1408701D-2	.7195805D-3
1	PFR	1.074675290	.7444950017	.6477626206	.5548838966	.4587429006	.3952924895	.3481844830	.2795893379	.2079219751
1	PF1	1.085874296	.8655607232	.7968698144	.7306137719	.6629950314	.6196999406	.5886463891	.5457232443	.5047378971
1	PQ1	.8065803583	.8128136007	.8322938990	.8655959308	.9249195320	.9872117307	1.052755745	1.197187241	1.467708867
1	PF2	1.098115895	1.007124362	.9746165022	.9414144104	.9056082010	.8815588448	.8636642566	.8377628108	.8110894135
1	PQ2	2.158478876	2.086945197	2.125375401	2.208472686	2.369405184	2.542520051	2.725160500	3.125302860	3.864903563
3	PFC	1.123838626	1.100062822	1.089493347	1.077445482	1.062903723	1.052150033	1.043593116	1.030318040	1.015520220
3	PQC	12.12625931	9.416487172	8.868917118	8.474001668	8.227550879	8.177904978	8.215165652	8.419936288	8.920125418
1	ZF1	1.135648141	1.135857484	1.135966872	1.136093812	1.136247770	1.136361764	1.136452704	1.136594813	1.136756151
1	ZF2	1.159035204	1.163568502	1.165205394	1.166916610	1.168832348	1.170177338	1.171218161	1.172801381	1.174551045
1	ZF3	1.220148131	1.239422287	1.246021114	1.252745528	1.260087041	1.265137432	1.268993280	1.274780184	1.281079474
1	ZF4	1.420803616	1.468651734	1.485557709	1.502983296	1.522182137	1.535465884	1.545636709	1.560930653	1.577595625
1	ZF5	2.295023974	2.413635619	2.456861944	2.502042332	2.552469677	2.587705282	2.614847206	2.655885442	2.700846536

MU	(5) P 7+4	TH = 63	SBE = 1.122326238						MCPER FUNCTIONS	
		RC = 02	RC = 05	RC = 07	RC = 10	RC = 15	RC = 20	RC = 25	PC = 35	RC = 50
	AMAX DB	.0017375255	.0108709564	.0213327376	.0436480540	.0988323393	.1772876696	.2802872360	.5675287486	1.249387366
	AMIN DB	33.49225789	45.48519811	49.99506703	54.84163642	60.45398088	64.53063383	67.77491723	72.89074550	78.89078109
MU	K	.1353717D 0	.3610288D-1	.2193918D-1	.1283388D-1	.6888986D-2	.4380288D-2	.3053260D-2	.1726463D-2	.8830617D-3
1	PFR	1.106597304	.7611700035	.6611155148	.5654725427	.4668618188	.4019734119	.3538845899	.2839770067	.2110625457
1	PF1	1.102545859	.8781397650	.8081265257	.7406257447	.6717874796	.6277454656	.5961746081	.5525638745	.5109542048
1	PQ1	.8224118339	.8227357692	.8410690973	.8735229015	.9322528774	.9943414339	1.059857913	1.204513929	1.475836560
1	PF2	1.099403973	1.010745841	.9787381208	.9459173560	.9104044257	.8864971659	.8686855906	.8428783176	.8162823374
1	PQ2	2.264031161	2.161458707	2.193853761	2.272743463	2.431335159	2.604386758	2.788029645	3.191995466	3.940837145
3	PFC	1.116738013	1.095146338	1.085357549	1.074103040	1.060409232	1.050218758	1.042076579	1.029393150	1.015189246
3	PQC	13.30684535	10.12930234	9.484979066	9.013359937	8.703779078	8.621596461	8.639615822	8.824693312	9.317480163
1	ZF1	1.125248382	1.125423470	1.125523917	1.125642896	1.125789345	1.125898829	1.125986659	1.126124612	1.126282071
1	ZF2	1.146908472	1.151283536	1.152873587	1.154541261	1.156414525	1.157733467	1.158756176	1.160315120	1.162042322
1	ZF3	1.203092495	1.222015761	1.228528660	1.235180692	1.242459397	1.247475892	1.251310797	1.257074114	1.263358254
1	ZF4	1.392427160	1.439353666	1.456072073	1.473365318	1.492481891	1.505743786	1.515915665	1.531238168	1.547968477
1	ZF5	2.232492424	2.343818042	2.391598266	2.436495140	2.486802800	2.522065582	2.549285166	2.590527020	2.635819792

(5) P 7+4 — TH = 64 — SBE = 1.112601940 — MCPER FUNCTIONS

MU	K	RC = 02	RC = 05	RC = 07	RC = 10	RC = 15	RC = 20	RC = 25	RC = 35	RC = 50
	AMAX DB	.0017375255	.0108709564	.0213327376	.0436480540	.0988323393	.1772876696	.2802872360	.5675287486	1.249387366
	AMIN DB	31.60818022	43.54355652	48.03853737	52.87198759	58.47215802	62.54163698	65.78105975	70.89053731	76.88486739
	K	.1629525D 0	.4377071D-1	.2665544D-1	.1562438D-1	.8404133D-2	.5350763D-2	.3733320D-2	.2113895D-2	.1082728D-2
1	PFR	1.141478361	.7790913614	.6754033337	.5767581757	.4754833408	.4090523903	.3599152985	.2886100476	.2143728500
1	PF1	1.119727133	.8912635811	.8198900200	.7510980852	.6809879131	.6361645263	.6040517416	.5597199790	.5174551122
1	PQ1	.8401985012	.8336418801	.8506546521	.8821326019	.9401740323	1.002016804	1.067485876	1.212358631	1.484514006
1	PF2	1.100406556	1.014289765	.9828392939	.9504486709	.9152716003	.8915296983	.8738159301	.8481208504	.8216172625
1	PQ2	2.382870208	2.243875333	2.269208222	2.343129259	2.498833878	2.671613215	2.856198101	3.264093010	4.022683330
3	PFC	1.109700869	1.090219088	1.081197423	1.070729299	1.057879816	1.048253938	1.040529020	1.028442959	1.014841125
3	PQC	14.68095844	10.94127754	10.18226482	9.620117304	9.236261364	9.115755161	9.111003926	9.272411801	9.755247474
1	ZF1	1.115377911	1.115517274	1.115608366	1.115718969	1.115857465	1.115962135	1.116046627	1.116180083	1.116333299
1	ZF2	1.135379785	1.139589563	1.141129890	1.142750894	1.144578111	1.145868490	1.146871151	1.148402912	1.150104521
1	ZF3	1.186813964	1.205342197	1.211756055	1.218323207	1.225526354	1.230500680	1.234308587	1.240039645	1.246299758
1	ZF4	1.365177933	1.411063511	1.427558821	1.444686620	1.463688447	1.476908369	1.487067149	1.502398886	1.519176038
1	ZF5	2.171957681	2.285757986	2.328014722	2.372554310	2.422669269	2.457914298	2.485180396	2.526584466	2.572171264

(5) P 7+4 — TH = 65 — SBE = 1.103377919 — MCPER FUNCTIONS

MU	K	RC = 02	RC = 05	RC = 07	RC = 10	RC = 15	RC = 20	RC = 25	RC = 35	RC = 50
	AMAX DB	.0017375255	.0108709564	.0213327376	.0436480540	.0988323393	.1772876696	.2802872360	.5675287486	1.249387366
	AMIN DB	29.73861320	41.61127821	46.09011950	50.90934816	56.49631538	60.55800927	63.79215495	68.89473432	74.88286345
	K	.1958158D 0	.5299781D-1	.3234701D-1	.1900103D-1	.1024254D-1	.6530346D-2	.4560959D-2	.2586246D-2	.1326596D-2
1	PFR	1.179743683	.7984058030	.6907294157	.5888132948	.4846566227	.4165668628	.3663068494	.2935100645	.2178673002
1	PF1	1.137382248	.9049531549	.8321859868	.7620577117	.6906222754	.6449816767	.6123011079	.5672129558	.5242600541
1	PQ1	.8602852429	.8456768314	.8611630502	.8915144446	.9487554536	1.010302504	1.075700190	1.220778664	1.493799705
1	PF2	1.101097118	1.017737849	.9869061107	.9549988454	.9202045660	.8966538737	.8790544369	.8534917635	.8270974182
1	PQ2	2.517386501	2.335419311	2.352455357	2.420493798	2.572651247	2.744898779	2.930341405	3.342261707	4.111146151
3	PFC	1.102741472	1.085288822	1.077018584	1.067324937	1.055317409	1.046256557	1.038950841	1.027467217	1.014475175
3	PQC	16.29139139	11.87150373	10.97571793	10.30615698	9.834428155	9.668574640	9.636770429	9.769668140	10.23939164
1	ZF1	1.106017314	1.106119498	1.106200832	1.106302661	1.106432770	1.106532329	1.106613256	1.106741873	1.106890479
1	ZF2	1.124428376	1.128466181	1.129953981	1.131525233	1.133302838	1.134562117	1.135542764	1.137044373	1.138717164
1	ZF3	1.171294242	1.189381221	1.195682224	1.202151307	1.209265416	1.214188815	1.217963306	1.223652898	1.229879538
1	ZF4	1.339033509	1.383754618	1.399989427	1.416916693	1.435769134	1.448925524	1.459055988	1.474376123	1.491180106
1	ZF5	2.113352964	2.224386003	2.266039429	2.310144907	2.359990419	2.395170100	2.422449541	2.463971494	2.509811724

(5) P 7+4		TH = 66 SBE = 1.094636279							MCPER FUNCTIONS	
		RC = 02	RC = 05	RC = 07	RC = 10	RC = 15	RC = 20	RC = 25	RC = 35	RC = 50
AMAX DB		.0017375255	.0108709564	.0213327376	.0436480540	.0988323393	.1772876696	.2802872360	.5675287486	1.249387366
AMIN DB		27.88290230	39.68698775	44.14830742	48.95210210	54.52473571	58.57797325	61.80638419	66.90146339	72.88284635
MU	K	.2349233D 0	.6409452D-1	.3921263D-1	.2308588D-1	.1247293D-1	.7964073D-2	.5568256D-2	.3162218D-2	.1624528D-2
1	PFR	1.221903101	.8192849299	.7072137446	.6017215266	.4944380128	.4245595849	.3730936895	.2987015835	.2215622082
1	PF1	1.155458145	.9192279973	.8450406875	.7735332361	.7007187305	.6542237987	.6209483506	.5750664442	.5313905648
1	PQ1	.8830964650	.8590143352	.8727281002	.9017740212	.9580821037	1.019274026	1.084571330	1.229840463	1.503761200
1	PF2	1.101448671	1.021070078	.9909230411	.9595570200	.9251971937	.9018664233	.8843997692	.8589921751	.8327260398
1	PQ2	2.670540860	2.437570809	2.444815951	2.505867837	2.653676663	2.825069648	3.011255369	3.427283293	4.207048467
3	PFC	1.095874703	1.080363931	1.072827195	1.063897066	1.052724241	1.044227807	1.037342587	1.026465735	1.014090707
3	PQC	18.19291932	12.94383110	11.88386614	11.08606302	10.50980060	10.29001669	10.22592960	10.32438969	10.77706536
1	ZF1	1.097148529	1.097212088	1.097283282	1.097375956	1.097497260	1.097591417	1.097668554	1.097791995	1.097935622
1	ZF2	1.114034969	1.117894493	1.119327057	1.120845534	1.122569979	1.123795606	1.124752249	1.126220676	1.127861335
1	ZF3	1.156517344	1.174114770	1.180288373	1.186645461	1.193656270	1.198519460	1.202253724	1.207892064	1.214075172
1	ZF4	1.313974626	1.357403753	1.373338730	1.390028292	1.408694401	1.421764138	1.431849915	1.447135953	1.463945080
1	ZF5	2.056615483	2.164636915	2.205604787	2.249196208	2.298691615	2.333755520	2.361012962	2.402605295	2.448655148

(5) P 7+4		TH = 67 SBE = 1.086360377							MCPER FUNCTIONS	
		RC = 02	RC = 05	RC = 07	RC = 10	RC = 15	RC = 20	RC = 25	RC = 35	RC = 50
AMAX DB		.0017375255	.0108709564	.0213327376	.0436480540	.0988323393	.1772876696	.2802872360	.5675287486	1.249387366
AMIN DB		26.04057155	37.76928696	42.21154942	46.99857098	52.55562520	56.59966684	59.82183865	64.90875383	70.88278889
MU	K	.2814057D 0	.7743358D-1	.4749228D-1	.2802700D-1	.1517909D-1	.9707031D-2	.6794539D-2	.3864807D-2	.1988685D-2
1	PFR	1.268572305	.8419307435	.7249964862	.6155798903	.5048924562	.4330796424	.3803152606	.3042125884	.2254761264
1	PF1	1.173880005	.9341051466	.8584805571	.7855549367	.7113078530	.6639203665	.6300217334	.5833066376	.5388705887
1	PQ1	.9091587932	.8738641988	.8855100976	.9130368675	.9682542439	1.029020049	1.094181820	1.239621481	1.514476913
1	PF2	1.101434044	1.024264603	.9948727799	.9641108172	.9302422361	.9071632569	.8898499835	.8646229045	.8385063430
1	PQ2	2.846035459	2.552134163	2.547767502	2.600490658	2.742972493	2.913108645	3.099883877	3.520081116	4.311358188
3	PFC	1.089116028	1.075453490	1.068630022	1.060449277	1.050102873	1.042169110	1.035704965	1.025438396	1.013687030
3	PQC	20.45637453	14.18834031	12.92989294	11.97795680	11.27657412	10.99229336	10.88949109	10.94620750	11.37691187
1	ZF1	1.088754755	1.088778250	1.088838941	1.088922101	1.089034201	1.089122673	1.089195800	1.089313732	1.089452009
1	ZF2	1.104181692	1.107857005	1.109231729	1.110694480	1.112362254	1.113551671	1.114482298	1.115914460	1.117519582
1	ZF3	1.142469477	1.159527016	1.165557913	1.171788306	1.178680719	1.183473852	1.187160654	1.192737327	1.198866174
1	ZF4	1.289984980	1.331990996	1.347584859	1.363997369	1.382437735	1.395396014	1.405419470	1.420647097	1.437437823
1	ZF5	2.001686112	2.106449393	2.146647051	2.189641267	2.238701814	2.273596496	2.300794266	2.342406031	2.388618177

(5) P 7+4 — TH = 68 — SBE = 1.078534743 — MCPER FUNCTIONS

MU		RC = 02	RC = 05	RC = 07	RC = 10	RC = 15	RC = 20	RC = 25	RC = 35	RC = 50
	AMAX DB	.0017375255	.0108709564	.0213327376	.0436480540	.0988323393	.1772876696	.2802872360	.5675287486	1.249387366
	AMIN DB	24.21139913	35.85675052	40.27823675	45.04699849	50.58709502	54.62112314	57.83649811	62.91451490	68.88053586
MU	K	.3365899D 0	.9346324D-1	.5747633D-1	.3400454D-1	.1846349D-1	.1182679D-1	.8288157D-2	.4722372D-2	.2434112D-2
1	PFR	1.320500376	.8665827082	.7442424728	.6305016494	.5160952509	.4421837123	.3880169763	.3100751819	.2296302677
1	PF1	1.192545707	.9495978140	.8725316189	.7981546604	.7224228208	.6741037431	.6395524802	.5919626518	.5467268489
1	PQ1	.9391314640	.8904818713	.8997025418	.9254533172	.9793910069	1.039645437	1.104628915	1.250212585	1.526038462
1	PF2	1.101026232	1.027297645	.9987360813	.9686461526	.9353311564	.9125393186	.8954024170	.8703843958	.8444414886
1	PQ2	3.048546766	2.681328566	2.663113438	2.705864372	2.841817574	3.010193734	3.197354706	3.621753623	4.425221793
3	PFC	1.082481486	1.070567310	1.064434483	1.056986696	1.047456247	1.040082160	1.034038876	1.024385173	1.013263454
3	PQC	23.17435417	15.64336290	14.14309746	13.00455792	12.15240097	11.79051002	11.64102014	11.64692248	12.04946189
1	ZF1	1.080820359	1.080802358	1.080852208	1.080925516	1.081028035	1.081110553	1.081179459	1.081291551	1.081424113
1	ZF2	1.094851994	1.098337538	1.099651938	1.101056103	1.102663739	1.103814392	1.104716978	1.106109743	1.107675838
1	ZF3	1.129138924	1.145604303	1.151476411	1.157564597	1.164322628	1.169035251	1.172666894	1.178170806	1.184233931
1	ZF4	1.267051030	1.307499679	1.322709179	1.338803027	1.356975617	1.369795806	1.379737934	1.394380831	1.411627555
1	ZF5	1.948509122	2.049765552	2.089105927	2.131416498	2.179953126	2.214621917	2.241719824	2.283296330	2.329619563

(5) P 7+4 — TH = 69 — SBE = 1.071144994 — MCPER FUNCTIONS

MU		RC = 02	RC = 05	RC = 07	RC = 10	RC = 15	RC = 20	RC = 25	RC = 35	RC = 50
	AMAX DB	.0017375255	.0108709564	.0213327376	.0436480540	.0988323393	.1772876696	.2802872360	.5675287486	1.249387366
	AMIN DB	22.39552926	33.94792377	38.34669147	43.09553258	48.61713951	52.64024671	55.84820592	60.91650903	66.87377572
MU	K	.4020327D 0	.1127234D 0	.6951675D-1	.4123810D-1	.2245174D-1	.1440648D-1	.1010874D-1	.5770010D-2	.2979491D-2
1	PFR	1.378605651	.8935268695	.7651469470	.6466199355	.5281342607	.4519376443	.3962514455	.3163264103	.2340490278
1	PF1	1.211319199	.9657135572	.8872186440	.8113656185	.7340996007	.6848095078	.6495751656	.6010669598	.5549892901
1	PQ1	.9738473858	.9091811035	.9155409642	.9392048293	.9916350154	1.051275096	1.116028055	1.261721118	1.538553592
1	PF2	1.100198815	1.030143412	1.002491584	.9731470243	.9404539287	.9179884168	.9010535456	.8762766231	.8505345387
1	PQ2	3.284046568	2.827910105	2.793075467	2.823825796	2.951764344	3.117748370	3.305026212	3.733617779	4.550007517
3	PFC	1.075987661	1.065715981	1.060248713	1.053515040	1.044787729	1.037968962	1.032345443	1.023306148	1.012819301
3	PQC	26.46931112	17.35829339	15.56091041	14.19464976	13.15945331	12.70353405	12.49739131	12.44112892	12.80766142
1	ZF1	1.073330797	1.073269879	1.073308571	1.073371719	1.073464308	1.073540617	1.073605097	1.073711032	1.073837514
1	ZF2	1.086030575	1.089321153	1.090572876	1.091915695	1.093459797	1.094569144	1.095441656	1.096791852	1.098315350
1	ZF3	1.116515937	1.132335095	1.138031549	1.143961173	1.150567894	1.155188903	1.158757196	1.164176511	1.170161656
1	ZF4	1.245161774	1.283916323	1.298696264	1.314427507	1.332287506	1.344941003	1.354781291	1.369810932	1.386485764
1	ZF5	1.897031978	1.994530575	2.032924170	2.074461267	2.122380379	2.156763164	2.183718295	2.225200760	2.271579603

(5) P 7+4		TH = 70 SBE = 1.064177772								MCPER FUNCTIONS
		RC = 02	RC = 05	RC = 07	RC = 10	RC = 15	RC = 20	RC = 25	RC = 35	RC = 50
AMAX DB		.0017375255	.0108709564	.0213327376	.0436480540	.0988323393	.1772876696	.2802872360	.5675287486	1.249387366
AMIN DB		20.59363567	32.04132442	36.41515396	41.14220496	46.64361050	50.65478519	53.85463894	58.91231964	64.86000648
MU	K	.4795574D 0	.1358654D 0	.84041250D-1	.4999625D-1	.2729838D-1	.1754874D-1	.12330100D-1	.7051321D-2	.36481120D-2
1	PFR	1.444022588	.9231077487	.7879429965	.6640923996	.5411127349	.4624184662	.4050800175	.3230093015	.2387606397
1	PF1	1.230022794	.9824518264	.9025639566	.8252220285	.7463771125	.6960768173	.6601281648	.6106559086	.5636916130
1	PQ1	1.014369093	.9303509429	.9333146739	.9545123311	1.005158418	1.064058992	1.128517324	1.274274842	1.552149950
1	PF2	1.098926491	1.032774028	1.006115633	.9775952776	.9455988068	.9235030229	.9067988146	.8822989746	.8567883995
1	PQ2	3.560249672	2.995338223	2.940418435	2.956642831	3.074714843	3.237508142	3.424548883	3.857266051	4.687361931
3	PFC	1.069651647	1.060910929	1.056081626	1.050040687	1.042101174	1.035831875	1.030626053	1.022201536	1.012353907
3	PQC	30.50516564	19.39756125	17.23171817	15.58511567	14.32588729	13.75518447	13.47981628	13.34705942	13.66758245
1	ZF1	1.066272530	1.066167296	1.066194545	1.066247256	1.066329596	1.066399459	1.066459322	1.066558792	1.066678838
1	ZF2	1.077703328	1.080794075	1.081980912	1.083259749	1.084737002	1.085802529	1.086642934	1.087947357	1.089424613
1	ZF3	1.104592604	1.119709932	1.125213094	1.130966938	1.137404433	1.141922026	1.145418243	1.150740322	1.156634357
1	ZF4	1.224308514	1.261230605	1.275533897	1.290856206	1.308355856	1.320811928	1.330528227	1.345413651	1.361986147
1	ZF5	1.847205185	1.940692342	1.978047191	2.018717474	2.065920678	2.099953632	2.126720123	2.168045284	2.214419524

(5) P 7+4		TH = 71 SBE = 1.057620681								MCPER FUNCTIONS
		RC = 02	RC = 05	RC = 07	RC = 10	RC = 15	RC = 20	RC = 25	RC = 35	RC = 50
AMAX DB		.0017375255	.0108709564	.0213327376	.0436480540	.0988323393	.1772876696	.2802872360	.5675287486	1.249387366
AMIN DB		18.80715621	30.13545113	34.48177043	39.18490736	44.66418631	48.66229501	51.85327110	56.89931194	62.83649403
MU	K	.5712967D 0	.1636771D 0	.1015710D 0	.6060843D-1	.3319418D-1	.2138080D-1	.1504399D-1	.8620714D-2	.4469149D-2
1	PFR	1.518163113	.9557440313	.8129112894	.6831072544	.5551529386	.4737169541	.4145747564	.3301741846	.2437980034
1	PF1	1.248428530	.9998006749	.9185857517	.8397585283	.7592973481	.7079487959	.6712541702	.6207703359	.5728719237
1	PQ1	1.062065600	.9544788478	.9533825917	.9716473636	1.020170878	1.078178752	1.142263304	1.288027078	1.566979965
1	PF2	1.097185740	1.035159496	1.009582086	.9819703423	.9507520541	.9290740302	.9126324319	.8894501095	.8632057476
1	PQ2	3.887245925	3.188006132	3.108621576	3.107145507	3.213023155	3.371610129	3.557947574	3.994642127	4.839284486
3	PFC	1.063491004	1.056164466	1.051942983	1.046570749	1.039400987	1.033673676	1.028882401	1.021071713	1.011866643
3	PQC	35.50417940	21.84633049	19.21888213	17.22381218	15.68789035	14.97588529	14.61526595	14.38774672	14.64939370
1	ZF1	1.059632945	1.059482047	1.059497597	1.059539629	1.059611440	1.059674642	1.059729708	1.059822423	1.059935688
1	ZF2	1.069857279	1.072743640	1.073863533	1.075075887	1.076483086	1.077502314	1.078308589	1.079564017	1.080991321
1	ZF3	1.093362713	1.107721393	1.113012886	1.118572866	1.124822184	1.129223820	1.132638661	1.137849994	1.143638834
1	ZF4	1.204484555	1.239435327	1.253213092	1.268077726	1.285166173	1.297391797	1.306960162	1.321667717	1.338104587
1	ZF5	1.798982178	1.888201081	1.924422663	1.964129125	2.010512931	2.044128235	2.070657002	2.111756663	2.158060815

(5) P 7+4 — TH = 72 SBE = 1.051462224 — MCPER FUNCTIONS

MU		K	RC = 02	RC = 05	RC = 07	RC = 10	RC = 15	RC = 20	RC = 25	RC = 35	RC = 50
		AMAX DB	.0017375255	.0108709564	.0213327376	.0436480540	.0988323393	.1772876696	.2802872360	.5675287486	1.249387366
		AMIN DB	17.03862239	28.22880348	32.54458200	37.22136407	42.67633445	46.66009935	49.84132852	54.87458454	60.80022065
		K	.6797402D 0	.1971128D 0	.1227430D 0	.7348029D-1	.4037572D-1	.2606107D-1	.1836500D-1	.1054642D-1	.5479339D-2
1		PFR	1.602796797	.9919495000	.8403929777	.7038912178	.5704008859	.4859409642	.4248209911	.3378803885	.2491997501
1		PF1	1.266249122	1.017732362	.9352957370	.8550092628	.7729054089	.7204729496	.6830007847	.6314563104	.5825735301
1		PQ1	1.118718721	.9821825742	.9761948985	.9909471768	1.036930291	1.093856461	1.157468847	1.303153493	1.583227275
1		PF2	1.094955673	1.037267682	1.012862135	.9862489409	.9558976279	.9346904662	.9185471181	.8947277783	.8697380346
1		PQ2	4.278403314	3.411564866	3.302117113	3.278906899	3.369635574	3.522714474	3.707732947	4.148142967	5.008227634
3		PFC	1.057523694	1.051489840	1.047843468	1.043113154	1.036692205	1.031497618	1.027116544	1.019917253	1.011356922
3		PQC	41.77178815	24.81882070	21.60655837	19.17369286	17.29259173	16.40500142	15.93847026	15.59264627	15.77870695
1		ZF1	1.053400259	1.053202461	1.053206092	1.053237243	1.053298288	1.053354639	1.053404748	1.053490436	1.053596589
1		ZF2	1.062480540	1.065158234	1.066209291	1.067352814	1.068686879	1.069657389	1.070427529	1.071630734	1.073004318
1		ZF3	1.082821576	1.096364058	1.101424829	1.106772010	1.112813139	1.117085497	1.120409045	1.125495177	1.131163694
1		ZF4	1.185684823	1.218526405	1.231728136	1.246083960	1.262707116	1.274666799	1.284061333	1.298554392	1.314819147
1		ZF5	1.752319178	1.837009024	1.872000116	1.910641885	1.956097358	1.989222854	2.015461289	2.056261792	2.102424476

(5) P 7+4 — TH = 73 SBE = 1.045691756 — MCPER FUNCTIONS

MU		K	RC = 02	RC = 05	RC = 07	RC = 10	RC = 15	RC = 20	RC = 25	RC = 35	RC = 50
		AMAX DB	.0017375255	.0108709564	.0213327376	.0436480540	.0988323393	.1772876696	.2802872360	.5675287486	1.249387366
		AMIN DB	15.29210707	26.31992114	30.60151753	35.24910042	40.67726580	44.64523620	47.81573379	52.83490904	58.74782044
		K	.8077881D 0	.2373323D 0	.1483384D 0	.8911341D-1	.4913772D-1	.3178785D-1	.2243703D-1	.1291451D-1	.6725219D-2
1		PFR	1.700155167	1.032361307	.8708070304	.7267200997	.5870325917	.4992198105	.4359206542	.3461984540	.2550116224
1		PF1	1.283129555	1.036197488	.9526958384	.8710064897	.7872494040	.7337015868	.6954211966	.6427660241	.5928459270
1		PQ1	1.186670978	1.014254840	1.002323103	1.012835495	1.055757380	1.111366574	1.174383546	1.319911200	1.601115283
1		PF2	1.092219148	1.039064358	1.015924115	.9904047651	.9610168093	.9403391460	.9245338004	.9011285989	.8765398626
1		PQ2	4.751674202	3.673387042	3.526628490	3.476497148	3.548285540	3.694172457	3.877054885	4.320758470	5.197232969
3		PFC	1.051767999	1.046901286	1.043794766	1.039676740	1.033980583	1.029307512	1.025330968	1.018738970	1.010824223
3		PQC	49.73361352	28.47068503	24.50828478	21.51883853	19.20227696	18.09419857	17.49478060	16.99994232	17.08847634
1		ZF1	1.047563420	1.047317703	1.047309234	1.047329350	1.047379438	1.047428781	1.047473791	1.047552208	1.047650936
1		ZF2	1.055562249	1.058027253	1.059007752	1.060080269	1.061338275	1.062257714	1.062989751	1.064137514	1.065453568
1		ZF3	1.072965794	1.085634475	1.090444896	1.095559534	1.101371392	1.105500331	1.108722015	1.113667471	1.119199394
1		ZF4	1.167905341	1.198502852	1.211076655	1.224870208	1.240970636	1.252626244	1.261818914	1.276057556	1.292110117
1		ZF5	1.707174924	1.787070095	1.820730527	1.858202598	1.902614931	1.935173730	1.961065337	2.001486952	2.047430147

(5) TH = 74 SBE = 1.040299436 MCPER FUNCTIONS

P 7+4	MU		RC = 02	RC = 05	RC = 07	RC = 10	RC = 15	RC = 20	RC = 25	RC = 35	RC = 50
AMAX DB			.0017375255	.0108709564	.0213327376	.0436480540	.0988323393	.1772876696	.2802872360	.5675287486	1.249387366
AMIN DB			13.57380357	24.40745468	28.65039459	33.26540707	38.66387827	42.61439311	45.77303558	50.77665345	56.67549730
	MU	K	.9588125D 0	.2857486D 0	.1793191D 0	.1081309D 0	.5984942D-1	.3881080D-1	.2744189D-1	.1583425D-1	.8266147D-2
	1	PFR	1.813066125	1.077778640	.9046738493	.7519331203	.6052624519	.5137100998	.4479967143	.3552130571	.2612882911
	1	PF1	1.298641318	1.055117210	.9707736065	.8877784871	.8023801168	.7476922183	.7085749442	.6547588760	.6037460248
	1	PQ1	1.269032313	1.051726924	1.032502518	1.037851569	1.077055899	1.131052299	1.193318065	1.338551123	1.670918729
	1	PF2	1.088964238	1.040513310	1.018733340	.9944081180	.9660877687	.9460042537	.9305812356	.9076477737	.8834598222
	1	PQ2	5.331507695	3.983243269	3.789659023	3.705846837	3.753769682	3.890263014	4.069917410	4.516265929	5.410119630
	3	PFC	1.046242400	1.042414071	1.039809650	1.036271360	1.031272700	1.027107819	1.023528661	1.017537972	1.010268114
	3	PQC	59.99134176	33.01780401	28.07991764	24.37345926	21.50061487	20.11237353	19.34434838	18.65988912	18.62172888
	1	ZF1	1.042111971	1.041817718	1.041797017	1.041805993	1.041844995	1.041887206	1.041927002	1.041997932	1.042088947
	1	ZF2	1.049092502	1.051341062	1.052249463	1.053248986	1.054428185	1.055294295	1.055986305	1.057075438	1.058330125
	1	ZF3	1.063792918	1.075531108	1.080071129	1.084932755	1.090493205	1.094463740	1.097572296	1.102360506	1.107738312
	1	ZF4	1.151142457	1.179366766	1.191259707	1.204435340	1.219952177	1.231262758	1.240223209	1.254163859	1.269960102
	1	ZF5	1.663510043	1.738339599	1.770565887	1.806758755	1.850006749	1.881916758	1.907400732	1.947356917	1.992995077

(5) TH = 75 SBE = 1.035276180 MCPER FUNCTIONS

P 7+4	MU		RC = 02	RC = 05	RC = 07	RC = 10	RC = 15	RC = 20	RC = 25	RC = 35	RC = 50
AMAX DB			.0017375255	.0108709564	.0213327376	.0436480540	.0988323393	.1772876696	.2802872360	.5675287486	1.249387366
AMIN DB			11.89271578	22.49028794	26.68893637	31.26730095	36.63268668	40.56382450	43.70931873	48.69568366	54.57891915
	MU	K	.1136729D 1	.3440888D 0	.2168750D 0	.1313113D 0	.7297626D-1	.4744636D-1	.3361092D-1	.1944539D-1	.1017843D-1
	1	PFR	1.945124740	1.129216285	.9426479393	.7799525914	.6253546608	.5296036333	.4611991547	.3650269313	.2680957822
	1	PF1	1.312282457	1.074372964	.9894958152	.9053464380	.8183502997	.7625078825	.7225287636	.6675027913	.6153396998
	1	PQ1	1.369969917	1.095960840	1.067693398	1.066691594	1.101341128	1.153348528	1.214664114	1.359435124	1.642979614
	1	PF2	1.085186153	1.041576543	1.021251952	.9982255221	.9710850519	.9516668338	.9366755431	.9142787298	.8905492772
	1	PQ2	6.051684434	4.354309453	4.101212928	3.974777700	3.992346713	4.116532203	4.291485692	4.739505965	5.651749905
	3	PFC	1.040965400	1.038044537	1.035902064	1.032907998	1.028576077	1.024903754	1.021713215	1.016315724	1.009688284
	3	PQC	73.40823830	38.76546644	32.53957359	27.89564398	24.30206834	22.55305105	21.56836019	20.63975716	20.43557312
	1	ZF1	1.037035889	1.036693178	1.036660174	1.036657969	1.036685817	1.036720817	1.036755311	1.036818576	1.036901621
	1	ZF2	1.043062247	1.045090966	1.045925913	1.046850664	1.047948517	1.048759153	1.049409276	1.050436645	1.051626120
	1	ZF3	1.055300949	1.066054282	1.070303640	1.074891201	1.080177106	1.083973400	1.086956842	1.091570063	1.096774865
	1	ZF4	1.135391741	1.161123294	1.172281879	1.184781998	1.199650938	1.210572555	1.219267908	1.232862944	1.248354168
	1	ZF5	1.621285707	1.690773934	1.721458744	1.756257896	1.798213305	1.829386654	1.854397364	1.893793889	1.939032864

(5) TH = 76 SBE = 1.030613629 MCPER FUNCTIONS

MU	P 7+4	RC = 02	RC = 05	RC = 07	RC = 10	RC = 15	RC = 20	RC = 25	RC = 35	RC = 50
	AMAX DB	.0017375255	.0108709564	.0213327376	.0436480540	.0988323393	.1772876696	.2802872360	.5675287486	1.249387366
	AMIN DB	10.26136757	20.56774422	24.71481825	29.25148495	34.57973501	38.48924527	41.62008745	46.58723476	52.45307917
MU	K	.1346091D 1	.4144712D 0	.2624847D 0	.1596328D 0	.8910919D-1	.5809876D-1	.4124084D-1	.2392322D-1	.1256105D-1
1	PFR	2.100906358	1.187979801	.9855638500	.8113114536	.6476390488	.5471383018	.4757131809	.3757662231	.2755147767
1	PF1	1.323488038	1.093793037	1.008799554	.9237198089	.8352133683	.7782173045	.7373575059	.6810758394	.6277037714
1	PQ1	1.495115971	1.148785489	1.109170772	1.100269008	1.129280887	1.178814614	1.238922335	1.383010179	1.667729538
1	PF2	1.080889670	1.042214636	1.023438823	1.001819293	.9759789726	.9573041677	.9427996226	.9210126563	.8978075799
1	PQ2	6.959576107	4.804701437	4.474884338	4.293795883	4.272326997	4.380291387	4.548534294	4.996732569	5.928411605
3	PFC	1.035955253	1.033810133	1.032087212	1.029598895	1.025899322	1.022701428	1.019888937	1.015074138	1.009084589
3	PQC	91.24203650	46.15480524	38.19916378	32.30890640	27.76648021	25.54576143	24.27857333	23.03133412	22.60722504
1	ZF1	1.032325368	1.031935428	1.031890131	1.031876775	1.031893482	1.031921238	1.031950376	1.032005844	1.032080699
1	ZF2	1.037463136	1.039269180	1.040029515	1.040877944	1.041892150	1.042645305	1.043251767	1.044214316	1.045334753
1	ZF3	1.047487586	1.057206085	1.061144609	1.065436669	1.070424010	1.074029385	1.076874997	1.081294235	1.086305662
1	ZF4	1.120646372	1.143780564	1.154151405	1.165916844	1.180070210	1.190555793	1.198950436	1.212147746	1.227280063
1	ZF5	1.580461003	1.644330312	1.673361708	1.706646922	1.747173610	1.777515942	1.801982304	1.840716166	1.885451888

(5) TH = 77 SBE = 1.026304108 MCPER FUNCTIONS

MU	P 7+4	RC = 02	RC = 05	RC = 07	RC = 10	RC = 15	RC = 20	RC = 25	RC = 35	RC = 50
	AMAX DB	.0017375255	.0108709564	.0213327376	.0436480540	.0988323393	.1772876696	.2802872360	.5675287486	1.249387366
	AMIN DB	8.696315092	18.63992602	22.72576808	27.21431205	32.50048620	36.38569209	39.50011140	44.44573993	50.29211056
MU	K	.1592218D 1	.4995049D 0	.3179974D 0	.1943341D 0	.1090048D 0	.7128888D-1	.5071581D-1	.2951759D-1	.1554371D-1
1	PFR	2.286227980	1.255772365	1.034501917	.8466925636	.6725334988	.5666134127	.4917707169	.3875879508	.2836451824
1	PF1	1.331656259	1.113135331	1.028579854	.9428895021	.8530211321	.7948947251	.7531450756	.6955682235	.6409285567
1	PQ1	1.652144841	1.212701913	1.158659467	1.139804353	1.161755977	1.208182291	1.266746082	1.409853261	1.695721702
1	PF2	1.076091988	1.042387322	1.025249544	1.005147082	.9807348910	.9628890061	.9489324185	.9278379049	.9052325865
1	PQ2	8.122650840	5.359872435	4.929545670	4.677309001	4.604966514	4.691366859	4.850117994	5.296457261	6.248381739
3	PFC	1.031229564	1.029729428	1.028381654	1.026357693	1.023252293	1.020507993	1.018060991	1.013815671	1.008457104
3	PQC	115.3554293	55.83871657	45.51569745	37.93693259	32.12233324	29.27403959	27.63230966	25.96261453	25.24331100
1	ZF1	1.027970567	1.027536426	1.027478962	1.027454575	1.027460239	1.027480778	1.027504548	1.027552139	1.027618633
1	ZF2	1.032287298	1.033868805	1.034553584	1.035324389	1.036252925	1.036946761	1.037507892	1.038402682	1.039450306
1	ZF3	1.040349154	1.048990234	1.052598250	1.056573292	1.061237357	1.064634350	1.067328646	1.071533645	1.076329716
1	ZF4	1.106894943	1.127349552	1.136880284	1.147850862	1.161217799	1.171217039	1.179272424	1.192014922	1.206728540
1	ZF5	1.540988274	1.598966430	1.626226909	1.657871281	1.696824104	1.726233679	1.750078372	1.788036455	1.832153292

(5) P 7+4		TH = 78 SBE = 1.022340595								MCPER FUNCTIONS
		RC = 02	RC = 05	RC = 07	RC = 10	RC = 15	RC = 20	RC = 25	RC = 35	RC = 50
	AMAX DB	.0017375255	.0108709564	.0213327376	.0436480540	.0988323393	.1772876696	.2802872360	.5675287486	1.249387366
	AMIN DB	7.218085076	16.70826437	20.71976214	25.15176607	30.38968358	34.24734026	37.34321906	42.26459834	48.08903416
MU	K	.1881415D 1	.6024174D 0	.3857395D 0	.2369959D 0	.1336407D 0	.8769495D-1	.6253854D-1	.3652293D-1	.1929847D-1
1	PFR	2.508467535	1.334848528	1.090884086	.8869859688	.7005764019	.5884117460	.5096668725	.4006906231	.2926126122
1	PF1	1.336194139	1.132065875	1.048672569	.9628176961	.8718199753	.8126191159	.7699852900	.7110847353	.6551212172
1	PQ1	1.851601544	1.291199578	1.218542231	1.186962769	1.199951589	1.242427561	1.298997982	1.440722953	1.727677924
1	PF2	1.070825654	1.042054429	1.026636543	1.008161406	.9853123528	.9683886148	.9550479848	.9347392014	.9128201345
1	PQ2	9.638624771	6.056463544	5.492042954	5.145549369	5.005862289	5.063260507	5.208604366	5.649856235	6.622780183
3	PFC	1.026804727	1.025822107	1.024803380	1.023199590	1.020646289	1.018331846	1.016235576	1.012543464	1.007806201
3	PQC	148.5581853	68.80975311	55.17745413	45.26147327	37.70506273	34.00482534	31.85679020	29.61659202	28.49467332
1	ZF1	1.023961353	1.023488670	1.023419337	1.023384154	1.023378983	1.023392397	1.023410833	1.023450531	1.023508549
1	ZF2	1.027527027	1.028883808	1.029492322	1.030184483	1.031025643	1.031658522	1.032172790	1.032997035	1.033968161
1	ZF3	1.033879166	1.041411844	1.044670747	1.048307600	1.052623286	1.055793760	1.058322576	1.062291726	1.066848737
1	ZF4	1.094118707	1.111843803	1.120484375	1.130599711	1.143106569	1.152565871	1.160240317	1.172465420	1.186693812
1	ZF5	1.502805697	1.554640015	1.580005356	1.609873975	1.647097292	1.675463826	1.698602291	1.735659661	1.779028326

(5) P 7+4		TH = 79 SBE = 1.018716695								MCPER FUNCTIONS
		RC = 02	RC = 05	RC = 07	RC = 10	RC = 15	RC = 20	RC = 25	RC = 35	RC = 50
	AMAX DB		.0108709564	.0213327376	.0436480540	.0988323393	.1772876696	.2802872360	.5675287486	1.249387366
	AMIN DB		14.77638469	18.69538901	23.05948422	28.24117775	32.06725814	35.14201527	40.03585357	45.83540650
MU	K		.7272244D 0	.4686615D 0	.2896549D 0	.1642948D 0	.1082107D 0	.7737524D-1	.4535763D-1	.2405675D-1
1	PFR		1.428237990	1.156616360	.9333744173	.7324748651	.6130321256	.5297841508	.4153297518	.3025777950
1	PF1		1.150133385	1.068830912	.9834227352	.8916445170	.8314722798	.7879824486	.7277477806	.6704102121
1	PQ1		1.389255422	1.292189599	1.244070743	1.245499823	1.282881830	1.336849510	1.476638062	1.764559662
1	PF2		1.041177378	1.027549458	1.010809184	.9896640611	.9737635778	.9611142794	.9416965958	.9205633181
1	PQ2		6.948681674	6.201643440	5.727717383	5.497204420	5.515006802	5.641315482	6.072699974	7.066903307
3	PFC		1.022108922	1.021371895	1.020141509	1.018094282	1.016182852	1.014420142	1.011261511	1.007132648
3	PQC		86.62288877	68.25403518	55.02244370	45.02254905	40.13828405	37.29009256	34.26256905	32.58077252
1	ZF1		1.019785110	1.019704471	1.019658881	1.019643211	1.019649673	1.019662866	1.019694731	1.019744227
1	ZF2		1.024308975	1.024840813	1.025453625	1.026206071	1.026776600	1.027242637	1.027993759	1.028884841
1	ZF3		1.034477080	1.037370121	1.040648555	1.044590826	1.047516155	1.049864690	1.053575090	1.057867516
1	ZF4		1.097278928	1.104983435	1.114184127	1.125755113	1.134617661	1.141866196	1.153505265	1.167174201
1	ZF5		1.511307958	1.534646124	1.562594316	1.597919953	1.625123082	1.647462215	1.683479919	1.725954760

(5) P 7+4 — TH = 80 — SBE = 1.015426612 — MCPER FUNCTIONS

MU		RC = 02	RC = 05	RC = 07	RC = 10	RC = 15	RC = 20	RC = 25	RC = 35	RC = 50
	AMAX DB		.0108709564	.0213327376	.0436480540	.0988323393	.1772876696	.2802872360	.5675287486	1.249387366
	AMIN DB		12.85142207	16.65250326	20.93287239	26.04771446	29.83707349	32.88748657	37.74973658	43.52081615
MU	K		.8789595D 0	.5705386D 0	.3549649D 0	.2026590D 0	.1340304D 0	.9612194D-1	.5658213D-1	.3013502D-1
1	PFR		1.540074730	1.234304915	.9874644434	.7691784398	.6411389721	.5526291077	.4318411910	.3137506559
1	PF1		1.166741962	1.088693839	1.004556602	.9125071335	.8515349383	.8072511899	.7457010908	.6869513304
1	PQ1		1.514142030	1.384499338	1.314469629	1.300709703	1.331409717	1.381928801	1.519001411	1.807678909
1	PF2		1.039721580	1.027935950	1.013031352	.9937346495	.9789662810	.9670915910	.9486840393	.9284514695
1	PQ2		8.119237181	7.117656391	6.467322551	6.111553748	6.074259290	6.173238124	6.587440887	7.602384003
3	PFC		1.018611577	1.018108262	1.017202277	1.015611178	1.014072634	1.012623667	1.009974885	1.006437748
3	PQC		111.7984614	86.46860386	68.39855964	54.87199040	48.29591137	44.45306302	40.31053546	37.83157733
1	ZF1		1.016419021	1.016328050	1.016272664	1.016246994	1.016246770	1.016254878	1.016279062	1.016320076
1	ZF2		1.020139857	1.020595005	1.021128141	1.021790965	1.022298042	1.022714691	1.023390378	1.024198078
1	ZF3		1.028192572	1.030705980	1.033607544	1.037152097	1.039813482	1.041966461	1.045394008	1.049394439
1	ZF4		1.083671680	1.090401022	1.098630372	1.109188601	1.117394594	1.124168861	1.135146639	1.148173047
1	ZF5		1.468924487	1.490095166	1.515966288	1.549210752	1.575117944	1.596554352	1.631376479	1.672791882

(5) P 7+4 — TH = 81 — SBE = 1.012465126 — MCPER FUNCTIONS

MU		RC = 02	RC = 05	RC = 07	RC = 10	RC = 15	RC = 20	RC = 25	RC = 35	RC = 50
	AMAX DB		.0108709564	.0213327376	.0436480540	.0988323393	.1772876696	.2802872360	.5675287486	1.249387366
	AMIN DB		10.94589967	14.59336668	18.76741529	23.80068667	27.54651406	30.56843731	35.39399740	41.13214008
MU	K		.1064001D 1	.6962563D 0	.4364305D 0	.2510089D 0	.1667777D 0	.1200054D 0	.7097136D-1	.3797478D-1
1	PFR		1.676086231	1.327591185	1.051493338	.8119957199	.6736403368	.5788898321	.4506774906	.3264121147
1	PF1		1.181127524	1.107744846	1.025971305	.9343805813	.8728791672	.8279147866	.7651141950	.7049360259
1	PQ1		1.676775618	1.502810363	1.403115285	1.368953822	1.390705311	1.436569850	1.569802629	1.858878141
1	PF2		1.037660159	1.027743269	1.014762678	.9974592248	.9839389730	.9729304528	.9556674202	.9364686909
1	PQ2		9.698864783	8.333069899	7.431678364	6.898472537	6.782655721	6.841700557	7.227255430	8.260842265
3	PFC		1.015352498	1.015035117	1.014402810	1.013214138	1.012014927	1.010857009	1.008690038	1.005723532
3	PQC		148.5835320	112.7157707	87.34992895	68.55924813	59.48435564	54.18201993	48.41035430	44.76301074
1	ZF1		1.013383835	1.013284146	1.013219896	1.013184939	1.013178415	1.013181676	1.013198438	1.013231114
1	ZF2		1.016372645	1.016751687	1.017205296	1.017778106	1.018220986	1.018587346	1.019185635	1.019906901
1	ZF3		1.022564488	1.024689081	1.027198280	1.030322516	1.032701471	1.034643411	1.037763021	1.041442195
1	ZF4		1.071038330	1.076764133	1.083970685	1.093439838	1.100927001	1.107175290	1.117409371	1.129700034
1	ZF5		1.427437152	1.446293262	1.469916261	1.500876913	1.525340566	1.545758160	1.579207841	1.619373316

(5) P 7+4 — TH = 82 SBE = 1.009827573 — MCPER FUNCTIONS

MU	K	RC = 02	RC = 05	RC = 07	RC = 10	RC = 15	RC = 20	RC = 25	RC = 35	RC = 50
	AMAX DB		.0108709564	.0213327376	.0436480540	.0988323393	.1772876696	.2802872360	.5675287486	1.249387366
	AMIN DB		9.080129387	12.52457217	16.55938148	21.48987695	25.18276856	28.17066234	32.95289437	38.65240554
MU	K		.1290572D 1	.8522295D 0	.5387561D 0	.3124634D 0	.2087048D 0	.1507415D 0	.8962114D-1	.4820680D-1
1	PFR		1.844330889	1.441683615	1.128666930	.8627960597	.7118158773	.6095298509	.4724668981	.3409492462
1	PF1		1.192349318	1.125261552	1.047268936	.9571689269	.8955541107	.8501001566	.7861875926	.7246031953
1	PQ1		1.894045982	1.658515455	1.517642751	1.455350177	1.464810093	1.504244128	1.631966355	1.920836127
1	PF2		1.034979523	1.026921166	1.015932055	1.000761673	.9886112577	.9785688292	.9626017381	.9445916880
1	PQ2		11.90285223	10.00022332	8.729706469	7.936808068	7.705656639	7.704881558	8.043225296	9.090377038
3	PFC		1.012354413	1.012176584	1.011766289	1.010922941	1.010026015	1.009133353	1.007415201	1.004993030
3	PQC		204.4673732	152.1030388	115.3070975	88.33595921	75.41882583	67.88864696	59.64359562	54.22087010
1	ZF1		1.010672918	1.010567088	1.010495393	1.010452157	1.010439863	1.010438612	1.010448346	1.010472951
1	ZF2		1.013003969	1.013308427	1.013683305	1.014166351	1.014544725	1.014860221	1.015379598	1.016011783
1	ZF3		1.017597083	1.019330563	1.021436541	1.024120967	1.026200064	1.027915711	1.030701756	1.034028668
1	ZF4		1.059391977	1.064102303	1.070243677	1.078550586	1.085255129	1.090922639	1.100323061	1.111773116
1	ZF5		1.386778270	1.403172004	1.424359485	1.452809403	1.475662727	1.494929310	1.526803058	1.565496307

(5) P 7+4 — TH = 83 SBE = 1.007509825 — MCPER FUNCTIONS

MU	K	RC = 02	RC = 05	RC = 07	RC = 10	RC = 15	RC = 20	RC = 25	RC = 35	RC = 50
	AMAX DB		.0108709564	.0213327376	.0436480540	.0988323393	.1772876696	.2802872360	.5675287486	1.249387366
	AMIN DB		7.284644516	10.46011292	14.30731011	19.10328011	22.72959475	25.67569857	30.40560301	36.05896968
MU	K		.1569586D 1	.1047059D 1	.6683895D 0	.3913995D 0	.2630168D 0	.1907951D 0	.1141255D 0	.6175961D-1
1	PFR		2.056338467	1.584224529	1.223732083	.9242895770	.7575371488	.6459493013	.4981140028	.3579148677
1	PF1		1.199314494	1.140261457	1.067828421	.9806582890	.9195590312	.8739259844	.8091581192	.7462562576
1	PQ1		2.193035332	1.870048731	1.670362041	1.568036029	1.560072717	1.590354436	1.709985314	1.997617660
1	PF2		1.031687061	1.025427026	1.016463814	1.003552786	.9928968304	.9839282469	.9694273243	.9527864665
1	PQ2		15.10351207	12.38216726	10.54753060	9.358985221	8.951799981	8.858322825	9.118100147	10.16785007
3	PFC		1.009639646	1.009558048	1.009318298	1.008760417	1.008125284	1.007468806	1.006160938	1.004250662
3	PQC		293.3999306	214.1896557	158.6582342	118.3362634	99.20975650	88.10722880	75.92206200	67.67532117
1	ZF1		1.008279326	1.008171279	1.008094303	1.008044212	1.008026871	1.008021568	1.008024827	1.008041783
1	ZF2		1.010030561	1.010263444	1.010561337	1.010955704	1.011269814	1.011534287	1.011973826	1.012514832
1	ZF3		1.013290560	1.014640640	1.016339605	1.018569873	1.020333871	1.021808909	1.024235976	1.027178357
1	ZF4		1.048738410	1.052445726	1.057494467	1.064573218	1.070431522	1.075461018	1.083930148	1.094421436
1	ZF5		1.346847532	1.360645188	1.379194109	1.404875559	1.425926620	1.443888881	1.473948300	1.510905049

(5) P 7+4 — TH = 84 SBE = 1.005508280 — MCPER FUNCTIONS

(5) P 7+4	RC = 02	RC = 05	RC = 07	RC = 10	RC = 15	RC = 20	RC = 25	RC = 35	RC = 50
AMAX DB			.0213327376	.0436480540	.0988323393	.1772876696	.2802872360	.5675287486	1.249387366
AMIN DB			8.425813404	12.01498976	16.62726261	20.16609024	23.05888109	27.72358715	33.32044642
MU I K			.1292651D 1	.8344197D 0	.4941463D 0	.3344229D 0	.2438280D 0	.1468873D 0	.80052350-1
1 I PFR			1.766754805	1.343991098	1.000727259	.8136698194	.6902788079	.5289822799	.3781349241
1 I PF1			1.151459605	1.086701532	1.004432113	.9447917470	.8994761312	.8343027190	.7702886468
1 I PQ1			2.168767269	1.882254782	1.720722635	1.687060433	1.703804032	1.811156372	2.095733426
1 I PF2			1.023234461	1.016281178	1.005728481	.9966892304	.9889083616	.9760632944	.9610030960
1 I PQ2			15.96345465	13.22546775	11.40309928	10.71355922	10.46965368	10.59484444	11.62400291
3 I PFC			1.007205644	1.007086872	1.006752939	1.006335922	1.005883183	1.004940920	1.003502835
3 I PQC			318.2164294	230.2503989	166.7733628	136.9590295	119.7559159	100.8930996	87.87877763
1 I ZF1			1.006090971	1.006011968	1.005957069	1.005935665	1.005926927	1.005924469	1.005934390
1 I ZF2			1.007615355	1.007839468	1.008147396	1.008398211	1.008612039	1.008971595	1.009420092
1 I ZF3			1.010626474	1.011925144	1.013695044	1.015132607	1.016354775	1.018398996	1.020924133
1 I ZF4			1.041821730	1.045774188	1.051572717	1.056524275	1.060857465	1.068290520	1.077689871
1 I ZF5			1.318589794	1.334289539	1.356906896	1.375929958	1.392405894	1.420364982	1.455263338

(5) P 7+4 — TH = 85 SBE = 1.003819838 — MCPER FUNCTIONS

(5) P 7+4	RC = 02	RC = 05	RC = 07	RC = 10	RC = 15	RC = 20	RC = 25	RC = 35	RC = 50
AMAX DB			.0213327376	.0436480540	.0988323393	.1772876696	.2802872360	.5675287486	1.249387366
AMIN DB			6.464530099	9.697100210	14.04774672	17.46508920	20.28621424	24.86598545	30.39115035
MU I K			.1606353D 1	.1050186D 1	.6302312D 0	.4301426D 0	.3155259D 0	.1917024D 0	.1053634D 0
1 I PFR			2.007332945	1.501202367	1.098968672	.8848641066	.7459577513	.5672520212	.4029177661
1 I PF1			1.157268693	1.102474975	1.027723259	.9709473145	.9267403981	.8619350606	.7972230918
1 I PQ1			2.611514834	2.191676306	1.938299309	1.864731439	1.860397373	1.948224741	2.226375630
1 I PF2			1.020346468	1.015313952	1.007168928	.9998564324	.9933791751	.9823985986	.9691669896
1 I PQ2			21.71366253	17.44211241	14.53677957	13.36352068	12.85955861	12.74219607	13.70020763
3 I PFC			1.005145314	1.005102317	1.004930963	1.004685808	1.004401068	1.003773065	1.002758855
3 I PQC			506.7186850	358.5986912	251.7184870	201.9282311	173.4022373	142.2474209	120.5119190
1 I ZF1			1.004320040	1.004243741	1.004187008	1.004162900	1.004151558	1.004144401	1.004148141
1 I ZF2			1.005362758	1.005518544	1.005743984	1.005933468	1.006097762	1.006378254	1.006733988
1 I ZF3			1.007288961	1.008209154	1.009525030	1.010631355	1.011592243	1.013233553	1.015310007
1 I ZF4			1.032248905	1.035138062	1.039628985	1.043621567	1.047201814	1.053488702	1.061646505
1 I ZF5			1.276803950	1.289460941	1.308676756	1.325399957	1.340167018	1.365671727	1.398106655

$P^{(4)}_{8+0}$

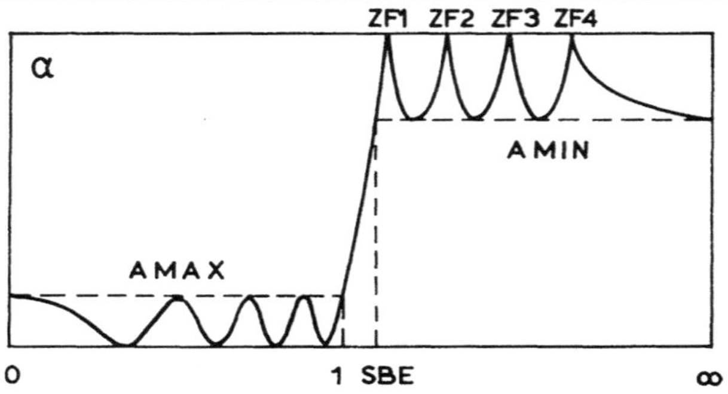

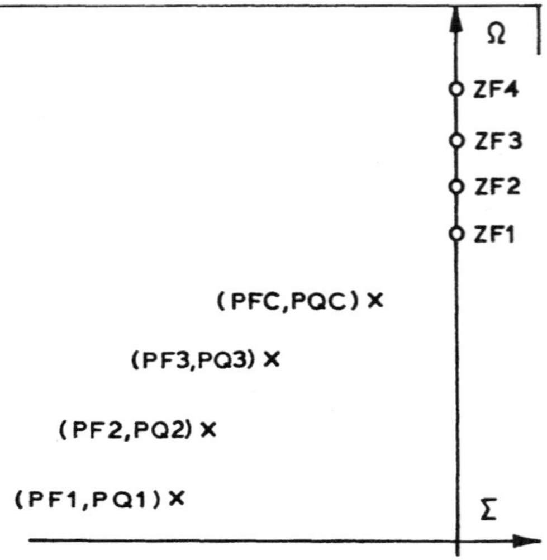

$P^{(4)}_{8+0}$	TH = 32	SBE = 1.887079915						CAUER FUNCTIONS	
	RC = 02	RC = 05	RC = 07	RC = 10	RC = 15	RC = 20	RC = 25	RC = 35	RC = 50
AMAX DB	.0017375255	.0108709564	.0213327376	.0436480540	.0988323393	.1772876696	.2802872360	.5675287486	1.249387366
AMIN DB	88.88762301	96.85555661	99.78857911	102.9089336	106.4859431	109.0631782	111.1043730	114.3141752	118.0940730
MU I K	.3594337D-4	.1436224D-4	.1024639D-4	.7154072D-5	.4739176D-5	.3522422D-5	.2784719D-5	.1924382D-5	.1245364D-5
1 I PF1	.7101114363	.5673756952	.5189019994	.4697116393	.4165277021	.3805964364	.3537954640	.3152352733	.2769402492
1 I PQ1	.5404221784	.5552166452	.5639357328	.5764629768	.5971157429	.6184354385	.6410698950	.6923474549	.7936590010
1 I PF2	.8790468429	.7789919518	.7469876918	.7158958490	.6841568944	.6640007880	.6497290915	.6304596372	.6129489111
1 I PQ2	.8773814255	.9849569715	1.044387544	1.125751150	1.251752359	1.373748881	1.496508266	1.756308356	2.223815720
1 I PF3	1.042092246	.9748171730	.9536224848	.9332417001	.9126889258	.8997881326	.8907323232	.8786172876	.8677279365
1 I PQ3	1.793281546	2.087814240	2.246578642	2.460607352	2.786370790	3.097004672	3.406153027	4.052621177	5.200068562
1 I PFC	1.127209984	1.076347837	1.060327981	1.044934847	1.029429255	1.019708275	1.012891029	1.003780267	.9956015781
1 I PQC	6.075105973	7.143901095	7.716253494	8.484842934	9.649795193	10.75670513	11.85558955	14.14748750	18.20366085
1 I ZF1	1.918380750	1.918380750	1.918380750	1.918380750	1.918380750	1.918380750	1.918380750	1.918380750	1.918380750
1 I ZF2	2.215314882	2.215314882	2.215314882	2.215314882	2.215314882	2.215314882	2.215314882	2.215314882	2.215314882
1 I ZF3	3.214661927	3.214661927	3.214661927	3.214661927	3.214661927	3.214661927	3.214661927	3.214661927	3.214661927
1 I ZF4	8.951402272	8.951402272	8.951402272	8.951402272	8.951402272	8.951402272	8.951402272	8.951402272	8.951402272

(4) P 8+0	TH = 33 SBE = 1.836078459							CAUER FUNCTIONS	
	RC = 02	RC = 05	RC = 07	RC = 10	RC = 15	RC = 20	RC = 25	RC = 35	RC = 50
AMAX DB	.0017375255	.0108709564	.0213327376	.0436480540	.0988323393	.1772876696	.2802872360	.5675287486	1.249387366
AMIN DB	86.62208202	94.59001561	97.52303811	100.6433926	104.2204021	106.7976322	108.8388320	112.0486342	115.8285320
MU K	.4665475D-4	.1864229D-4	.1329989D-4	.9286036D-5	.6151484D-5	.4572128D-5	.3614585D-5	.2497861D-5	.1616491D-5
1 PF1	.7151680132	.5709731222	.5220733185	.4724823613	.4188982227	.3827141972	.3557331574	.3169255314	.2783970873
1 PQ1	.5407944649	.5555431509	.5642513826	.5767707274	.5974197198	.6187410397	.6413801459	.6926733230	.7940233399
1 PF2	.8826973852	.7820676586	.7498892309	.7186328304	.6867306499	.6664733849	.6521312340	.6327681889	.6151739363
1 PQ2	.8820872065	.9895382018	1.049032356	1.130546832	1.256858363	1.379202018	1.502336283	1.762975581	2.232066053
1 PF3	1.042758921	.9757868414	.9546698262	.9343562803	.9138642258	.9009980446	.8919649654	.8798783666	.8690126920
1 PQ3	1.813811744	2.108934298	2.268430562	2.483652522	2.811487227	3.124257045	3.435622667	4.086897673	5.243161029
1 PFC	1.125555280	1.075438829	1.059631432	1.044432796	1.029113816	1.019505172	1.012764604	1.003753648	.9956619901
1 PQC	6.169181440	7.242757351	7.819273476	8.594270395	9.769956619	10.88772097	11.99775993	14.31363750	18.41346270
1 ZF1	1.866186240	1.866186240	1.866186240	1.866186240	1.866186240	1.866186240	1.866186240	1.866186240	1.866186240
1 ZF2	2.152117016	2.152117016	2.152117016	2.152117016	2.152117016	2.152117016	2.152117016	2.152117016	2.152117016
1 ZF3	3.116609880	3.116609880	3.116609880	3.116609880	3.116609880	3.116609880	3.116609880	3.116609880	3.116609880
1 ZF4	8.665147716	8.665147716	8.665147716	8.665147716	8.665147716	8.665147716	8.665147716	8.665147716	8.665147716

(4) P 8+0	TH = 34 SBE = 1.788291650							CAUER FUNCTIONS	
	RC = 02	RC = 05	RC = 07	RC = 10	RC = 15	RC = 20	RC = 25	RC = 35	RC = 50
AMAX DB	.0017375255	.0108709564	.0213327376	.0436480540	.0988323393	.1772876696	.2802872360	.5675287486	1.249387366
AMIN DB	84.41553477	92.38346836	95.31649085	98.43684537	102.0138548	104.5910849	106.6322847	109.8420869	113.6219848
MU K	.6014829D-4	.2403403D-4	.1714650D-4	.1197175D-4	.7930622D-5	.5894483D-5	.4659998D-5	.3220295D-5	.2084015D-5
1 PF1	.7204524626	.5747270373	.5253811282	.4753711425	.4213687549	.3849207618	.3577517604	.3186859730	.2799141167
1 PQ1	.5411862241	.5558855135	.5645820148	.5770927705	.5977375101	.6190603414	.6417041748	.6930134817	.7944034783
1 PF2	.8864753457	.7852544988	.7528965269	.7214703233	.6893995678	.6690377607	.6546227700	.6351629300	.6174822590
1 PQ2	.8870453504	.9943511415	1.053907904	1.135576705	1.262209504	1.384914202	1.508439122	1.769954137	2.240698290
1 PF3	1.043432687	.9767801363	.9557451121	.9355025814	.9150746965	.9022451153	.8932360827	.8811795816	.8703389811
1 PQ3	1.835498752	2.131194756	2.291447247	2.507911160	2.837910311	3.152916630	3.466606130	4.122922845	5.288438529
1 PFC	1.123855055	1.074503533	1.058914421	1.043915762	1.028788780	1.019295782	1.012634178	1.003726041	.9957240752
1 PQC	6.268786268	7.347226945	7.928083188	8.709789462	9.896742718	11.02591773	12.14769106	14.48880880	18.63460208
1 ZF1	1.817267780	1.817267780	1.817267780	1.817267780	1.817267780	1.817267780	1.817267780	1.817267780	1.817267780
1 ZF2	2.092769684	2.092769684	2.092769684	2.092769684	2.092769684	2.092769684	2.092769684	2.092769684	2.092769684
1 ZF3	3.024288523	3.024288523	3.024288523	3.024288523	3.024288523	3.024288523	3.024288523	3.024288523	3.024288523
1 ZF4	8.395142656	8.395142656	8.395142656	8.395142656	8.395142656	8.395142656	8.395142656	8.395142656	8.395142656

(4) P 8+0	TH = 35	SBE = 1.743446796						CAUER FUNCTIONS	
	RC = 02	RC = 05	RC = 07	RC = 10	RC = 15	RC = 20	RC = 25	RC = 35	RC = 50
AMAX DB	.0017375255	.0108709564	.0213327376	.0436480540	.0988323393	.1772876696	.2802872360	.5675287486	1.249387366
AMIN DB	82.26417913	90.23211271	93.16513520	96.28548972	99.86249919	102.4397292	104.4809291	107.6907313	111.4706291
MU I K	.7705326D-4	.3078891D-4	.2196561D-4	.1533647D-4	.1015956D-4	.7551158D-5	.5969714D-5	.4125375D-5	.2669737D-5
1 I PF1	.7259731693	.5786426486	.5288298265	.4783816697	.4239423239	.3872187629	.3598536378	.3205186105	.2814930337
1 I PQ1	.5415984820	.5562444494	.5649282722	.5774296882	.5980696463	.6193938533	.6420424814	.6933684318	.7947999520
1 I PF2	.8903818376	.7885539789	.7560111372	.7244099090	.6921652305	.6716954885	.6572052615	.6376454027	.6198753989
1 I PQ2	.8922694988	.9994069814	1.059024926	1.140851281	1.267816377	1.390896366	1.514828185	1.777256645	2.249727590
1 I PF3	1.044112130	.9777963556	.9568478377	.9366802755	.9163201762	.9035292810	.8945456770	.8825210190	.8717069623
1 I PQ3	1.858410349	2.154657577	2.315690496	2.533446193	2.865706254	3.183053710	3.499178259	4.160781978	5.336006332
1 I PFC	1.122109848	1.073542130	1.058177058	1.043383799	1.028454166	1.019080104	1.012499743	1.003697430	.9957878140
1 I PQC	6.374269210	7.457643660	8.043020749	8.831746794	10.03052739	11.17169644	12.30581321	14.67349669	18.86769602
1 I ZF1	1.771347283	1.771347283	1.771347283	1.771347283	1.771347283	1.771347283	1.771347283	1.771347283	1.771347283
1 I ZF2	2.036944353	2.036944353	2.036944353	2.036944353	2.036944353	2.036944353	2.036944353	2.036944353	2.036944353
1 I ZF3	2.937204784	2.937204784	2.937204784	2.937204784	2.937204784	2.937204784	2.937204784	2.937204784	2.937204784
1 I ZF4	8.139979496	8.139979496	8.139979496	8.139979496	8.139979496	8.139979496	8.139979496	8.139979496	8.139979496

(4) P 8+0	TH = 36	SBE = 1.701301617						CAUER FUNCTIONS	
	RC = 02	RC = 05	RC = 07	RC = 10	RC = 15	RC = 20	RC = 25	RC = 35	RC = 50
AMAX DB	.0017375255	.0108709564	.0213327376	.0436480540	.0988323393	.1772876696	.2802872360	.5675287486	1.249387366
AMIN DB	80.16451994	88.13245351	91.06547600	94.18583051	97.76283998	100.3400700	102.3812699	105.5910721	109.3709699
MU I K	.9812372D-4	.3920824D-4	.2797217D-4	.1953028D-4	.1293773D-4	.9616045D-5	.7602151D-5	.5253472D-5	.3399785D-5
1 I PF1	.7317391066	.5827255116	.5324241009	.4815178687	.4266221482	.3896109995	.3620413028	.3224255815	.2831356394
1 I PQ1	.5420323546	.5566207333	.5652908483	.5777821074	.5984167012	.6197421230	.6423956023	.6937387100	.7952133350
1 I PF2	.8944179494	.7919676446	.7592346715	.7274532294	.6950292876	.6744482114	.6598803422	.6402172225	.6223549499
1 I PQ2	.8977744683	1.004717819	1.064395014	1.146381889	1.273690390	1.397160270	1.521515730	1.784896664	2.259170229
1 I PF3	1.044795764	.9788347535	.9579774635	.9378890090	.9176004859	.9048504660	.8958937420	.8839027615	.8731167949
1 I PQ3	1.882620377	2.179390011	2.341227311	2.560325777	2.894946701	3.214744288	3.533419952	4.200567208	5.385978112
1 I PFC	1.120320221	1.072554810	1.057419457	1.042836967	1.028109993	1.018858140	1.012361292	1.003667800	.9958531859
1 I PQC	6.486011841	7.574371447	8.164454473	8.960525924	10.17171720	11.32549296	12.47259376	14.86823896	19.11341425
1 I ZF1	1.728177672	1.728177672	1.728177672	1.728177672	1.728177672	1.728177672	1.728177672	1.728177672	1.728177672
1 I ZF2	1.984349062	1.984349062	1.984349062	1.984349062	1.984349062	1.984349062	1.984349062	1.984349062	1.984349062
1 I ZF3	2.854920275	2.854920275	2.854920275	2.854920275	2.854920275	2.854920275	2.854920275	2.854920275	2.854920275
1 I ZF4	7.898406208	7.898406208	7.898406208	7.898406208	7.898406208	7.898406208	7.898406208	7.898406208	7.898406208

(4) P 8+0	TH = 37	SBE = 1.661640141						CAUER FUNCTIONS	
	RC = 02	RC = 05	RC = 07	RC = 10	RC = 15	RC = 20	RC = 25	RC = 35	RC = 50
AMAX DB	.0017375255	.0108709564	.0213327376	.0436480540	.0988323393	.1772876696	.2802872360	.5675287486	1.249387366
AMIN DB	78.11333431	86.08126786	89.01429035	92.13464486	95.71165432	98.28988439	100.3300842	103.5398864	107.3197843
MU I K	.1242606D-3	.4965198D-4	.3542301D-4	.2473249D-4	.1638390D-4	.1217743D-4	.9627107D-5	.6652819D-5	.4305373D-5
1 I PF1	.7377598833	.5869815561	.5361689490	.4847839211	.4294116524	.3921004485	.3643174270	.3244091574	.2848438466
1 I PQ1	.5424890576	.5570152033	.5656704919	.5781507037	.5987792909	.6201057391	.6427641135	.6941248921	.7956442428
1 I PF2	.8985847338	.7954970793	.7625687908	.7306019882	.6979934581	.6772976452	.6626497203	.6428330811	.6249225830
1 I PQ2	.9035763703	1.010296746	1.070030694	1.152180755	1.279843838	1.403718575	1.528514949	1.792888775	2.269043698
1 I PF3	1.045482024	.9798945376	.9591334132	.9391284003	.9189154277	.9062085909	.8972802618	.8853248873	.8745686374
1 I PQ3	1.908209384	2.205465132	2.368130411	2.588623807	2.925709249	3.248070631	3.569418741	4.242378168	5.438476728
1 I PFC	1.118486763	1.071541773	1.056641741	1.042275331	1.027756283	1.018629894	1.012218817	1.003637132	.9959201692
1 I PQC	6.604432232	7.697807647	8.292786051	9.096540486	10.32075474	11.48778148	12.64854110	15.07362022	19.37248451
1 I ZF1	1.687538690	1.687538690	1.687538690	1.687538690	1.687538690	1.687538690	1.687538690	1.687538690	1.687538690
1 I ZF2	1.934723479	1.934723479	1.934723479	1.934723479	1.934723479	1.934723479	1.934723479	1.934723479	1.934723479
1 I ZF3	2.777043898	2.777043898	2.777043898	2.777043898	2.777043898	2.777043898	2.777043898	2.777043898	2.777043898
1 I ZF4	7.669305265	7.669305265	7.669305265	7.669305265	7.669305265	7.669305265	7.669305265	7.669305265	7.669305265

(4) P 8+0	TH = 38	SBE = 1.624269245						CAUER FUNCTIONS	
	RC = 02	RC = 05	RC = 07	RC = 10	RC = 15	RC = 20	RC = 25	RC = 35	RC = 50
AMAX DB	.0017375255	.0108709564	.0213327376	.0436480540	.0988323393	.1772876696	.2802872360	.5675287486	1.249387366
AMIN DB	76.10764151	84.07557502	87.00859751	90.12895202	93.70596148	96.28319155	98.32439137	101.5341936	105.3140914
MU I K	.1565373D-3	.6254913D-4	.4462417D-4	.3115677D-4	.2063963D-4	.1534053D-4	.1212776D-4	.8380893D-5	.5423697D-5
1 I PF1	.7440457959	.5914171143	.5400697022	.4881842835	.4323144830	.3946902774	.3666848524	.3264717524	.2866196875
1 I PQ1	.5429699171	.5574287676	.5660680129	.5785362065	.5991580792	.6204853346	.6431486343	.6945275964	.7960933352
1 I PF2	.9028831969	.7991439012	.7660152069	.7338579515	.7010595319	.6802455805	.6655151814	.6456357498	.6275800505
1 I PQ2	.9096927457	1.016157942	1.075945514	1.158261084	1.286289979	1.410584929	1.535840048	1.801248669	2.279366804
1 I PF3	1.046169267	.9809748659	.9603150706	.9403980379	.9202647831	.9076035218	.8987052095	.8867874686	.8760626477
1 I PQ3	1.935265356	2.232962439	2.396478814	2.618420489	2.958078038	3.283121882	3.607269430	4.286322707	5.493635105
1 I PFC	1.116610088	1.070503232	1.055844040	1.041698960	1.027393059	1.018395369	1.012072310	1.003605409	.9959887410
1 I PQC	6.729989131	7.828386641	8.428454158	9.240242842	10.47812241	11.65907857	12.83420877	15.29027677	19.64569848
1 I ZF1	1.649233379	1.649233379	1.649233379	1.649233379	1.649233379	1.649233379	1.649233379	1.649233379	1.649233379
1 I ZF2	1.887834746	1.887834746	1.887834746	1.887834746	1.887834746	1.887834746	1.887834746	1.887834746	1.887834746
1 I ZF3	2.703225615	2.703225615	2.703225615	2.703225615	2.703225615	2.703225615	2.703225615	2.703225615	2.703225615
1 I ZF4	7.451675889	7.451675889	7.451675889	7.451675889	7.451675889	7.451675889	7.451675889	7.451675889	7.451675889

(4) P 8+0		TH = 39 SBE = 1.589015729								CAUER FUNCTIONS
		RC = 02	RC = 05	RC = 07	RC = 10	RC = 15	RC = 20	RC = 25	RC = 35	RC = 50
AMAX DB		.0017375255	.0108709564	.0213327376	.0436480540	.0988323393	.1772876696	.2802872360	.5675287486	1.249387366
AMIN DB		74.14467663	82.11261010	85.04563258	88.16598709	91.74299655	94.32022661	96.36142644	99.57122866	103.3511265
MU	K	.1962303D-3	.7840965D-4	.5593947D-4	.3905716D-4	.2587320D-4	.1923042D-4	.1520298D-4	.1050603D-4	.6798979D-5
1	PF1	.7506078859	.5960389531	.5441320520	.4917237083	.4353345247	.3973838586	.3691466031	.3286159347	.2884653226
1	PQ1	.5434763816	.5578624110	.5664842874	.5789394033	.5995537813	.6208815913	.6435498306	.6949474873	.7965613210
1	PF2	.9073142843	.8029097603	.7695756811	.7372229484	.7042293719	.6832938858	.6684785911	.6484860829	.6303291890
1	PQ2	.9161427170	1.022316784	1.082154143	1.164637154	1.293043128	1.417774049	1.543506344	1.809993248	2.290159792
1	PF3	1.046855766	.9820748438	.9615217777	.9416974780	.9216483114	.9090351684	.9001685461	.8882905714	.8775989816
1	PQ3	1.963884538	2.261968525	2.426358491	2.649802989	2.992144401	3.319994747	3.647074816	4.332517697	5.551597210
1	PFC	1.114690839	1.069439410	1.055026492	1.041107927	1.027020347	1.018154568	1.011921765	1.003572613	.9960588772
1	PQC	6.863186722	7.966583999	8.571938530	9.392127199	10.64434675	11.83994770	13.03020036	15.51890199	19.93391851
1	ZF1	1.613085092	1.613085092	1.613085092	1.613085092	1.613085092	1.613085092	1.613085092	1.613085092	1.613085092
1	ZF2	1.843473955	1.843473955	1.843473955	1.843473955	1.843473955	1.843473955	1.843473955	1.843473955	1.843473955
1	ZF3	2.633151179	2.633151179	2.633151179	2.633151179	2.633151179	2.633151179	2.633151179	2.633151179	2.633151179
1	ZF4	7.244619029	7.244619029	7.244619029	7.244619029	7.244619029	7.244619029	7.244619029	7.244619029	7.244619029

(4) P 8+0		TH = 40 SBE = 1.555723827								CAUER FUNCTIONS
		RC = 02	RC = 05	RC = 07	RC = 10	RC = 15	RC = 20	RC = 25	RC = 35	RC = 50
AMAX DB		.0017375255	.0108709564	.0213327376	.0436480540	.0988323393	.1772876696	.2802872360	.5675287486	1.249387366
AMIN DB		72.22186755	80.18980093	83.12282341	86.24317791	89.82018737	92.39741743	94.43861726	97.64841948	101.4283173
MU	K	.2448537D-3	.9783854D-4	.6980055D-4	.4873502D-4	.3228424D-4	.2399546D-4	.1897008D-4	.1310929D-4	.8483677D-5
1	PF1	.7574580041	.6008543087	.5483620778	.4954072668	.4384759185	.4001847848	.3717058992	.3309444369	.2903830495
1	PQ1	.5440100358	.5583172030	.5669202650	.5793611457	.5999671693	.6212952439	.6439684199	.6953852792	.7970489611
1	PF2	.9118788668	.8067963347	.7732520233	.7406988714	.7075049157	.6864445099	.6715418985	.6514330212	.6331719242
1	PQ2	.9229471607	1.028789970	1.088672482	1.171324418	1.300118758	1.425301829	1.551530362	1.819140735	2.301444476
1	PF3	1.047539708	.9831935215	.9627528313	.9430262421	.9230657473	.9105033823	.9016702194	.8898342538	.8791777925
1	PQ3	1.994172377	2.292577851	2.457863096	2.682866154	3.028007613	3.358794259	3.688946501	4.381089937	5.612519156
1	PFC	1.112729687	1.068350544	1.054189242	1.040502311	1.026638177	1.017907499	1.011767175	1.003538725	.9961305525
1	PQC	7.004580077	8.112921189	8.723764605	9.552734271	10.82000322	12.03100436	13.23717489	15.76025257	20.23808520
1	ZF1	1.578934960	1.578934960	1.578934960	1.578934960	1.578934960	1.578934960	1.578934960	1.578934960	1.578934960
1	ZF2	1.801453159	1.801453159	1.801453159	1.801453159	1.801453159	1.801453159	1.801453159	1.801453159	1.801453159
1	ZF3	2.566537652	2.566537652	2.566537652	2.566537652	2.566537652	2.566537652	2.566537652	2.566537652	2.566537652
1	ZF4	7.047324587	7.047324587	7.047324587	7.047324587	7.047324587	7.047324587	7.047324587	7.047324587	7.047324587

(4) P 8+0	TH = 41	SBE = 1.524253087						CAUER FUNCTIONS	
	RC = 02	RC = 05	RC = 07	RC = 10	RC = 15	RC = 20	RC = 25	RC = 35	RC = 50
AMAX DB	.0017375255	.0108709564	.0213327376	.0436480540	.0988323393	.1772876696	.2802872360	.5675287486	1.249387366
AMIN DB	70.33681451	78.30474778	81.23777024	84.35812474	87.93513420	90.51236426	92.55356408	95.76336631	99.54326412
MU I K	.3042000D-3	.1215521D-3	.8671845D-4	.6054716D-4	.4010913D-4	.2981136D-4	.2356795D-4	.1628665D-4	.1053991D-4
1 I PF1	.7646088812	.6058709264	.5527662790	.4992403748	.4417430816	.4030968857	.3743661721	.3331601689	.2923753138
1 I PQ1	.5445726159	.5587943063	.5673769756	.5798023556	.6003990769	.6217270845	.6444051751	.6958417415	.7975570743
1 I PF2	.9165777231	.8108053258	.7770460897	.7442876767	.7108881780	.6896994845	.6747071395	.6544785962	.6361102747
1 I PQ2	.9301289029	1.035595654	1.095517793	1.178339622	1.307533610	1.433185451	1.559929959	1.828710799	2.313244381
1 I PF3	1.048219188	.9843298895	.9640074803	.9443838144	.9245167992	.9120080056	.9032101616	.8914185649	.8807992304
1 I PQ3	2.026244594	2.324893604	2.491094810	2.717713335	3.065775718	3.399634656	3.733005799	4.432177178	5.676570437
1 I PFC	1.110727332	1.067236883	1.053332447	1.039882198	1.026246580	1.017654167	1.011608533	1.003503726	.9962037401
1 I PQC	7.154781397	8.267970953	8.884508795	9.722656585	11.00572180	12.23292188	13.45585300	16.01515547	20.55922601
1 I ZF1	1.546639728	1.546639728	1.546639728	1.546639728	1.546639728	1.546639728	1.546639728	1.546639728	1.546639728
1 I ZF2	1.761602820	1.761602820	1.761602820	1.761602820	1.761602820	1.761602820	1.761602820	1.761602820	1.761602820
1 I ZF3	2.503129581	2.503129581	2.503129581	2.503129581	2.503129581	2.503129581	2.503129581	2.503129581	2.503129581
1 I ZF4	6.859060505	6.859060505	6.859060505	6.859060505	6.859060505	6.859060505	6.859060505	6.859060505	6.859060505

(4) P 8+0	TH = 42	SBE = 1.494476550						CAUER FUNCTIONS	
	RC = 02	RC = 05	RC = 07	RC = 10	RC = 15	RC = 20	RC = 25	RC = 35	RC = 50
AMAX DB	.0017375255	.0108709564	.0213327376	.0436480540	.0988323393	.1772876696	.2802872360	.5675287486	1.249387366
AMIN DB	68.48727221	76.45520530	79.38822775	82.50858224	86.08559169	88.66282175	90.70402157	93.91382379	97.69372161
MU I K	.3763885D-3	.1503972D-3	.1072973D-3	.7491536D-4	.4962727D-4	.3688577D-4	.2916077D-4	.2015157D-4	.1304109D-4
1 I PF1	.7720742064	.6110971025	.5573516097	.5032288203	.4451407300	.4061242472	.3771310805	.3355662320	.2944447200
1 I PQ1	.5451660281	.5592949875	.5678555383	.5802640322	.6008504061	.6221779694	.6448609305	.6963177030	.7980865419
1 I PF2	.9214115208	.8149384538	.7809597812	.7479913845	.7143812523	.6930609276	.6779764402	.6576249340	.6391463568
1 I PQ2	.9377129416	1.042753601	1.102708835	1.185700940	1.315305820	1.441443512	1.568724445	1.838724698	2.325584911
1 I PF3	1.048892207	.9854828760	.9652849219	.9457696390	.9260011467	.9135488585	.9047882886	.8930435438	.8824634405
1 I PQ3	2.060228407	2.359028691	2.526165276	2.754457318	3.105566480	3.442640353	3.779384773	4.485929268	5.743935331
1 I PFC	1.108684508	1.066098692	1.052456272	1.039247678	1.025845590	1.017394582	1.011445834	1.003467596	.9962784118
1 I PQC	7.314467190	8.432363448	9.054804515	9.902544521	11.20219319	12.44643805	13.68702395	16.28451587	20.89846498
1 I ZF1	1.516069897	1.516069897	1.516069897	1.516069897	1.516069897	1.516069897	1.516069897	1.516069897	1.516069897
1 I ZF2	1.723769619	1.723769619	1.723769619	1.723769619	1.723769619	1.723769619	1.723769619	1.723769619	1.723769619
1 I ZF3	2.442695718	2.442695718	2.442695718	2.442695718	2.442695718	2.442695718	2.442695718	2.442695718	2.442695718
1 I ZF4	6.679163411	6.679163411	6.679163411	6.679163411	6.679163411	6.679163411	6.679163411	6.679163411	6.679163411

(4) P 8+0	TH = 43	SBE = 1.466279186							CAUER FUNCTIONS
	RC = 02	RC = 05	RC = 07	RC = 10	RC = 15	RC = 20	RC = 25	RC = 35	RC = 50
AMAX DB	.0017375255	.0108709564	.0213327376	.0436480540	.0988323393	.1772876696	.2802872360	.5675287486	1.249387366
AMIN DB	66.67113372	74.63906654	77.57208896	80.69244344	84.26945288	86.84668294	88.88788276	92.09768499	95.87758280
MU I K	.4639202D-3	.1853731D-3	.1322500D-3	.9233744D-4	.6116843D-4	.4546381D-4	.3594230D-4	.2483795D-4	.1607389D-4
1 I PF1	.7798687158	.6165417332	.5621255173	.5073787951	.4486739027	.4092712319	.3800045292	.3380659333	.2965940444
1 I PQ1	.5457923684	.5598206274	.5683571703	.5807472601	.6013221335	.6226488242	.6453365871	.6968140582	.7986383134
1 I PF2	.9263807951	.8191974510	.7849950399	.7518120787	.7179863127	.6965310462	.6813520203	.6608742600	.6422823898
1 I PQ2	.9457267008	1.050285365	1.110266027	1.193428120	1.323455066	1.450096165	1.577934736	1.849205430	2.338493529
1 I PF3	1.049556666	.9866513421	.9665842988	.9471831167	.9275184379	.9151257374	.9064044972	.8947092175	.8841705624
1 I PQ3	2.096263936	2.395106853	2.563196684	2.793221374	3.147508450	3.487947062	3.828227394	4.542509460	5.814814470
1 I PFC	1.106601980	1.064936247	1.051560893	1.038598848	1.025435247	1.017128754	1.011279073	1.003430314	.9963545378
1 I PQC	7.484386546	8.606793288	9.235349072	10.09311323	11.41017606	12.67236265	13.93155364	16.56932620	21.25703389
1 I ZF1	1.487108131	1.487108131	1.487108131	1.487108131	1.487108131	1.487108131	1.487108131	1.487108131	1.487108131
1 I ZF2	1.687814574	1.687814574	1.687814574	1.687814574	1.687814574	1.687814574	1.687814574	1.687814574	1.687814574
1 I ZF3	2.385026199	2.385026199	2.385026199	2.385026199	2.385026199	2.385026199	2.385026199	2.385026199	2.385026199
1 I ZF4	6.507030559	6.507030559	6.507030559	6.507030559	6.507030559	6.507030559	6.507030559	6.507030559	6.507030559

(4) P 8+0	TH = 44	SBE = 1.439556540							CAUER FUNCTIONS
	RC = 02	RC = 05	RC = 07	RC = 10	RC = 15	RC = 20	RC = 25	RC = 35	RC = 50
AMAX DB	.0017375255	.0108709564	.0213327376	.0436480540	.0988323393	.1772876696	.2802872360	.5675287486	1.249387366
AMIN DB	64.88641620	72.85434862	75.78737100	78.90772546	82.48473490	85.06196495	87.10316477	90.31296699	94.09286481
MU I K	.5697433D-3	.2276578D-3	.1624170D-3	.1134002D-3	.7512133D-4	.5583439D-4	.4414096D-4	.3050364D-4	.1974044D-4
1 I PF1	.7880082901	.6222143670	.5670959860	.5116969291	.4523479889	.4125425023	.3829906892	.3406628033	.2988262485
1 I PQ1	.5464539466	.5603727346	.5688831981	.5812532180	.6018153180	.6231406510	.6458331198	.6973317731	.7992134135
1 I PF2	.9314859237	.8235840551	.7891538464	.7557519064	.7217056162	.7001121394	.6848361978	.6642289033	.6455207010
1 I PQ2	.9542003209	1.058214483	1.118211631	1.201542652	1.332002725	1.459165284	1.587583514	1.860177917	2.351999968
1 I PF3	1.050210363	.9878340790	.9679046947	.9486236018	.9290682872	.9167384127	.9080596640	.8964155999	.8859207291
1 I PQ3	2.134505820	2.433263951	2.602322989	2.834140467	3.191742185	3.535703046	3.879690865	4.602095881	5.889426633
1 I PFC	1.104480548	1.063749845	1.050646496	1.037935814	1.025015592	1.016856696	1.011108246	1.003391359	.9964320868
1 I PQC	7.665370707	8.792027642	9.426911583	10.29515053	11.63050515	12.91158609	14.19039369	16.87067652	21.63623487
1 I ZF1	1.459647879	1.459647879	1.459647879	1.459647879	1.459647879	1.459647879	1.459647879	1.459647879	1.459647879
1 I ZF2	1.653611413	1.653611413	1.653611413	1.653611413	1.653611413	1.653611413	1.653611413	1.653611413	1.653611413
1 I ZF3	2.329930103	2.329930103	2.329930103	2.329930103	2.329930103	2.329930103	2.329930103	2.329930103	2.329930103
1 I ZF4	6.342112863	6.342112863	6.342112863	6.342112863	6.342112863	6.342112863	6.342112863	6.342112863	6.342112863

(4) P 8+0		TH = 45 SBE = 1.414213562								CAUER FUNCTIONS
		RC = 02	RC = 05	RC = 07	RC = 10	RC = 15	RC = 20	RC = 25	RC = 35	RC = 50
AMAX DB		.0017375255	.0108709564	.0213327376	.0436480540	.0988323393	.1772876696	.2802872360	.5675287486	1.249387366
AMIN DB		63.13124803	71.09917986	74.03220219	77.15255662	80.72956604	83.30679608	85.34799590	88.55779812	92.33769594
MU	K	.6973288D-3	.2786384D-3	.1987879D-3	.1387945D-3	.9194364D-4	.6833767D-4	.5402568D-4	.3733448D-4	.2416102D-4
1	PF1	.7965100641	.6281252648	.5722715841	.5161903287	.4561687581	.4159430462	.3860940203	.3433606142	.3011444948
1	PQ1	.5471533128	.5609529601	.5694350692	.5817831893	.6023311096	.6236545365	.6463515843	.6978718929	.7998129491
1	PF2	.9367270996	.8281000004	.7934382152	.7598130772	.7255415041	.7038066020	.6884313923	.6676913021	.6488637318
1	PQ2	.9631669918	1.066566709	1.126569952	1.210067959	1.340972057	1.468674642	1.597695414	1.871669195	2.366136457
1	PF3	1.050850987	.9890297984	.9692451304	.9500903985	.9306502717	.9183866257	.9097506423	.8981626896	.8877140649
1	PQ3	2.175125076	2.473649431	2.643691318	2.877362619	3.238421633	3.586070562	3.933947129	4.664883213	5.968010775
1	PFC	1.102321049	1.062539797	1.049713281	1.037258688	1.024586670	1.016578420	1.010933348	1.003352209	.9965110257
1	PQC	7.858344170	8.988915565	9.630342077	10.50952604	11.86410078	13.16508928	14.46459196	17.18976626	22.03770493
1	ZF1	1.433592176	1.433592176	1.433592176	1.433592176	1.433592176	1.433592176	1.433592176	1.433592176	1.433592176
1	ZF2	1.621045169	1.621045169	1.621045169	1.621045169	1.621045169	1.621045169	1.621045169	1.621045169	1.621045169
1	ZF3	2.277233339	2.277233339	2.277233339	2.277233339	2.277233339	2.277233339	2.277233339	2.277233339	2.277233339
1	ZF4	6.183908856	6.183908856	6.183908856	6.183908856	6.183908856	6.183908856	6.183908856	6.183908856	6.183908856

(4) P 8+0		TH = 46 SBE = 1.390163591								CAUER FUNCTIONS
		RC = 02	RC = 05	RC = 07	RC = 10	RC = 15	RC = 20	RC = 25	RC = 35	RC = 50
AMAX DB		.0017375255	.0108709564	.0213327376	.0436480540	.0988323393	.1772876696	.2802872360	.5675287486	1.249387366
AMIN DB		61.40385721	69.37178817	72.30481042	75.42516481	79.00217420	81.57940424	83.62060406	86.83040627	90.61030408
MU	K	.8507602D-3	.3399465D-3	.2425267D-3	.1693331D-3	.1121738D-3	.8337384D-4	.6591281D-4	.4554909D-4	.2947710D-4
1	PF1	.8053925499	.6342854653	.5776615179	.5208666201	.4601423943	.4194782052	.3893192959	.3461634006	.3035521636
1	PQ1	.5478932894	.5615631135	.5700143659	.5823385732	.6028707590	.6241916607	.6468931259	.6984355493	.8004381177
1	PF2	.9421042993	.8327470086	.7978501897	.7639978615	.7294964043	.7076169274	.6921401295	.6712640091	.6523140439
1	PQ2	.9726633363	1.075370268	1.135367579	1.219029617	1.350388409	1.478650119	1.608297232	1.883708645	2.380937988
1	PF3	1.051476112	.9902371373	.9706045594	.9515827569	.9322639284	.9200700868	.9114802604	.8999504681	.8895506848
1	PQ3	2.218311256	2.516428014	2.687463577	2.923050482	3.287715714	3.639227497	3.991184580	4.731084602	6.050828336
1	PFC	1.100124360	1.061306433	1.048761459	1.036567590	1.024148533	1.016293944	1.010754376	1.003311341	.9965913194
1	PQC	8.064337613	9.198398795	9.846582018	10.73720164	12.11197960	13.43395503	14.75530451	17.52791788	22.46293262
1	ZF1	1.408852614	1.408852614	1.408852614	1.408852614	1.408852614	1.408852614	1.408852614	1.408852614	1.408852614
1	ZF2	1.590010949	1.590010949	1.590010949	1.590010949	1.590010949	1.590010949	1.590010949	1.590010949	1.590010949
1	ZF3	2.226776810	2.226776810	2.226776810	2.226776810	2.226776810	2.226776810	2.226776810	2.226776810	2.226776810
1	ZF4	6.031959418	6.031959418	6.031959418	6.031959418	6.031959418	6.031959418	6.031959418	6.031959418	6.031959418

(4) P 8+0	TH = 47 SBE = 1.367327461								CAUER FUNCTIONS
	RC = 02	RC = 05	RC = 07	RC = 10	RC = 15	RC = 20	RC = 25	RC = 35	RC = 50
AMAX DB	.0017375255	.0108709564	.0213327376	.0436480540	.0988323393	.1772876696	.2802872360	.5675287486	1.249387366
AMIN DB	59.70256079	67.67049049	70.60351262	73.72386694	77.30087630	79.87810633	81.91930614	85.12910835	88.90900616
MU K	.1034837D-2	.4135000D-3	.2950016D-3	.2059713D-3	.1364445D-3	.1014132D-3	.8017421D-4	.5540444D-4	.3585500D-4
1 PF1	.8146757742	.6407068600	.5832756911	.5257339968	.4642755325	.4231537065	.3926716310	.3490754826	.3060528720
1 PQ1	.5486770067	.5622051828	.5706228216	.5829208982	.6034356288	.6247533070	.6474589886	.6990239701	.8010902161
1 PF2	.9476172475	.8375267768	.8023918364	.7683085895	.7335728324	.7115457113	.6959650461	.6749496970	.6558743258
1 PQ2	.9827298533	1.084656153	1.144633647	1.228455597	1.360279455	1.489119935	1.619418166	1.896328245	2.396442614
1 PF3	1.052083196	.9914546431	.9719818630	.9530998688	.9339087504	.9217884712	.9132473185	.9017788979	.8914306922
1 PQ3	2.264274949	2.561781639	2.733818295	2.971383136	3.339810142	3.695369245	4.051610027	4.800933842	6.138165883
1 PFC	1.097891396	1.060050102	1.047791255	1.035862650	1.023701233	1.016003287	1.010571330	1.003269233	.9966729310
1 PQC	8.284502992	9.421524282	10.07667648	10.97924359	12.37526714	13.71938115	15.06380954	17.88659248	22.91377713
1 ZF1	1.385348426	1.385348426	1.385348426	1.385348426	1.385348426	1.385348426	1.385348426	1.385348426	1.385348426
1 ZF2	1.560412873	1.560412873	1.560412873	1.560412873	1.560412873	1.560412873	1.560412873	1.560412873	1.560412873
1 ZF3	2.178414801	2.178414801	2.178414801	2.178414801	2.178414801	2.178414801	2.178414801	2.178414801	2.178414801
1 ZF4	5.885843183	5.885843183	5.885843183	5.885843183	5.885843183	5.885843183	5.885843183	5.885843183	5.885843183

(4) P 8+0	TH = 48 SBE = 1.345632730								CAUER FUNCTIONS
	RC = 02	RC = 05	RC = 07	RC = 10	RC = 15	RC = 20	RC = 25	RC = 35	RC = 50
AMAX DB	.0017375255	.0108709564	.0213327376	.0436480540	.0988323393	.1772876696	.2802872360	.5675287486	1.249387366
AMIN DB	58.02575525	65.99368310	68.92670506	72.04705929	75.62406960	78.20129861	80.24249842	83.45230062	87.23219843
MU K	.1255198D-2	.5015519D-3	.3578201D-3	.2498314D-3	.1654995D-3	.1230085D-3	.9724675D-4	.6720243D-4	.4349007D-4
1 PF1	.8243814340	.6474022759	.5891247726	.5308012730	.4685753017	.4269756980	.3961565136	.3521014916	.3086504951
1 PQ1	.5495079454	.5628813558	.5712623381	.5835318367	.6040272058	.6253408734	.6480505263	.6996384892	.8017706518
1 PF2	.9532653764	.8424409647	.8070652375	.7727476478	.7377733936	.7155956554	.6999088941	.6787511651	.6595474007
1 PQ2	.9934114306	1.094458468	1.154400146	1.238376550	1.370675459	1.500114917	1.631090081	1.909562862	2.412691778
1 PF3	1.052669570	.9926807728	.9733758451	.9546408629	.9355841834	.9235414169	.9150515858	.9036479196	.8933541770
1 PQ3	2.313250699	2.609911708	2.782952763	3.022558167	3.394909507	3.754710862	4.115450947	4.874687877	6.230338130
1 PFC	1.095623118	1.058771173	1.046802910	1.035144007	1.023244833	1.015706469	1.010384209	1.003225859	.9967558215
1 PQC	8.520131254	9.659458780	10.32178830	11.23683665	12.65521228	14.02269564	15.39152334	18.26740788	23.39224047
1 ZF1	1.363005702	1.363005702	1.363005702	1.363005702	1.363005702	1.363005702	1.363005702	1.363005702	1.363005702
1 ZF2	1.532163135	1.532163135	1.532163135	1.532163135	1.532163135	1.532163135	1.532163135	1.532163135	1.532163135
1 ZF3	2.132013580	2.132013580	2.132013580	2.132013580	2.132013580	2.132013580	2.132013580	2.132013580	2.132013580
1 ZF4	5.745172499	5.745172499	5.745172499	5.745172499	5.745172499	5.745172499	5.745172499	5.745172499	5.745172499

(4) P 8+0	TH = 49 SBE = 1.325012993								CAUER FUNCTIONS
	RC = 02	RC = 05	RC = 07	RC = 10	RC = 15	RC = 20	RC = 25	RC = 35	RC = 50
AMAX DB	.0017375255	.0108709564	.0213327376	.0436480540	.0988323393	.1772876696	.2802872360	.5675287486	1.249387366
AMIN DB	56.37190760	64.33983279	67.27285450	70.39320860	73.97021784	76.54744783	78.58864762	81.79844981	85.57834761
MU I K	.1518464D-2	.6067480D-3	.4328698D-3	.3022314D-3	.2002115D-3	.1488085D-3	.1176434D-3	.8129756D-4	.5261173D-4
1 I PF1	.8345330712	.6543855695	.5952202706	.5360779446	.4730493711	.4309507883	.3997798390	.3552463991	.3113491896
1 I PQ1	.5503899858	.5635940461	.5719350068	.5841732224	.6046471155	.6259558859	.6486692145	.7002805582	.8024809540
1 I PF2	.9590477799	.8474911788	.8118724822	.7773174770	.7421007837	.7197695704	.7039745461	.6826713457	.6633362338
1 I PQ2	1.004757941	1.104814819	1.164702275	1.248826132	1.381609585	1.511668802	1.643347824	1.923450577	2.429730707
1 I PF3	1.053232434	.9939138857	.9747852266	.9562048004	.9372896209	.9253285200	.9168927974	.9055574496	.8953212142
1 I PQ3	2.365500412	2.661041694	2.835085498	3.076794059	3.453239678	3.817489548	4.182958073	4.952629682	6.327691411
1 I PFC	1.093320531	1.057470040	1.045796680	1.034411811	1.022779397	1.015403515	1.010193012	1.003181196	.9968399495
1 I PQC	8.772673172	9.913505890	10.58321462	11.51130042	12.95320408	14.34537431	15.74001897	18.67215955	23.90054301
1 I ZF1	1.341756690	1.341756690	1.341756690	1.341756690	1.341756690	1.341756690	1.341756690	1.341756690	1.341756690
1 I ZF2	1.505181184	1.505181184	1.505181184	1.505181184	1.505181184	1.505181184	1.505181184	1.505181184	1.505181184
1 I ZF3	2.087450162	2.087450162	2.087450162	2.087450162	2.087450162	2.087450162	2.087450162	2.087450162	2.087450162
1 I ZF4	5.609589875	5.609589875	5.609589875	5.609589875	5.609589875	5.609589875	5.609589875	5.609589875	5.609589875

(4) P 8+0	TH = 50 SBE = 1.305407289								CAUER FUNCTIONS
	RC = 02	RC = 05	RC = 07	RC = 10	RC = 15	RC = 20	RC = 25	RC = 35	RC = 50
AMAX DB	.0017375255	.0108709564	.0213327376	.0436480540	.0988323393	.1772876696	.2802872360	.5675287486	1.249387366
AMIN DB	54.73954718	62.70746853	65.64048988	68.76084379	72.33785293	74.91508288	76.95628266	80.16608484	83.94598263
MU I K	.1832410D-2	.7321947D-3	.5223667D-3	.3647185D-3	.2416058D-3	.1795750D-3	.1419665D-3	.9810604D-4	.6348935D-4
1 I PF1	.8451562700	.6616717319	.6015746184	.5415742559	.4777060034	.4350860912	.4035479497	.3585155491	.3141534204
1 I PQ1	.5513274660	.5643459220	.5726431325	.5848470701	.6052971385	.6266000127	.6493166646	.7009517595	.8032227872
1 I PF2	.9649631620	.8526789549	.8168156565	.7820205668	.7465577900	.7240703800	.7081650004	.6867133114	.6672439414
1 I PQ2	1.016824939	1.115766767	1.175578843	1.259841371	1.393118246	1.523818586	1.656229575	1.938033067	2.447609844
1 I PF3	1.053768852	.9951522367	.9762086389	.9577906635	.9390244002	.9271493312	.9187706508	.9075073774	.8973318607
1 I PQ3	2.421317359	2.715420180	2.890459113	3.134332973	3.515050593	3.883967513	4.254408390	5.035071577	6.430607685
1 I PFC	1.090984689	1.056147118	1.044772838	1.033666222	1.022304997	1.015094454	1.009997743	1.003135218	.9969252713
1 I PQC	9.043763994	10.18512607	10.86240615	11.80410851	13.27079146	14.68906131	16.11104784	19.10284505	24.44115333
1 I ZF1	1.321539187	1.321539187	1.321539187	1.321539187	1.321539187	1.321539187	1.321539187	1.321539187	1.321539187
1 I ZF2	1.479393007	1.479393007	1.479393007	1.479393007	1.479393007	1.479393007	1.479393007	1.479393007	1.479393007
1 I ZF3	2.044611217	2.044611217	2.044611217	2.044611217	2.044611217	2.044611217	2.044611217	2.044611217	2.044611217
1 I ZF4	5.478764844	5.478764844	5.478764844	5.478764844	5.478764844	5.478764844	5.478764844	5.478764844	5.478764844

(4) P 8+0	TH = 51 RC = 02	SBE = 1.286759566 RC = 05	RC = 07	RC = 10	RC = 15	RC = 20	RC = 25	CAUER FUNCTIONS RC = 35	RC = 50
AMAX DB	.0017375255	.0108709564	.0213327376	.0436480540	.0988323393	.1772876696	.2802872360	.5675287486	1.249387366
AMIN DB	53.12725795	61.09517380	64.02819463	67.14854827	70.72555727	73.30278716	75.34398691	78.55378907	82.33368686
MU I K	.2206160D-2	.8815386D-3	.6289126D-3	.4391093D-3	.2908855D-3	.2162025D-3	.1709231D-3	.1181165D-3	.7643912D-4
1 I PF1	.8562788807	.6692770085	.6082012698	.5473012766	.4825541142	.4393892755	.4074676784	.3619146942	.3170679901
1 I PQ1	.5523252496	.5651399411	.5733892610	.5855555981	.6059792293	.6272750815	.6499946395	.7016538215	.8039979667
1 I PF2	.9710097771	.8580057367	.8218968309	.7868594516	.7511472922	.7285011246	.7124833864	.6908802826	.6712737998
1 I PQ2	1.029674477	1.127360352	1.187072743	1.271463103	1.405241510	1.536604926	1.669777250	1.953356034	2.466380348
1 I PF3	1.054275745	.9963939697	.9776446178	.9593973755	.9407877968	.9290033513	.9206848018	.9094975618	.8993861536
1 I PQ3	2.481030904	2.773324401	2.949343662	3.195443977	3.580619485	3.954435308	4.330108600	5.122359074	6.539509158
1 I PFC	1.088616699	1.054802849	1.043731675	1.032907414	1.021821711	1.014779315	1.009798405	1.003087899	.9970117406
1 I PQC	9.335252681	10.47596017	11.16098998	12.11691098	13.60970612	15.05559322	16.50656511	19.56169243	25.01682292
1 I ZF1	1.302295998	1.302295998	1.302295998	1.302295998	1.302295998	1.302295998	1.302295998	1.302295998	1.302295998
1 I ZF2	1.454730488	1.454730488	1.454730488	1.454730488	1.454730488	1.454730488	1.454730488	1.454730488	1.454730488
1 I ZF3	2.003392115	2.003392115	2.003392115	2.003392115	2.003392115	2.003392115	2.003392115	2.003392115	2.003392115
1 I ZF4	5.352391179	5.352391179	5.352391179	5.352391179	5.352391179	5.352391179	5.352391179	5.352391179	5.352391179

(4) P 8+0	TH = 52 RC = 02	SBE = 1.269018215 RC = 05	RC = 07	RC = 10	RC = 15	RC = 20	RC = 25	CAUER FUNCTIONS RC = 35	RC = 50
AMAX DB	.0017375255	.0108709564	.0213327376	.0436480540	.0988323393	.1772876696	.2802872360	.5675287486	1.249387366
AMIN DB	51.53367126	59.50157923	62.43459933	65.55495258	69.13196136	71.70919119	73.75039091	76.96019304	80.74009080
MU I K	.2650431D-2	.1059061D-2	.7555619D-3	.5275363D-3	.3494636D-3	.2597410D-3	.2053433D-3	.1419026D-3	.9183230D-4
1 I PF1	.8679312732	.6772190342	.6151148074	.5532709880	.4876033391	.4438686222	.4115463989	.3654500367	.3200980730
1 I PQ1	.5533888063	.5659793897	.5741762112	.5863012546	.6066955379	.6279830989	.6507050722	.7023886357	.8048084760
1 I PF2	.9771853636	.8634728517	.8271180462	.7918367032	.7558722619	.7330649644	.7169329704	.6951756359	.6754292561
1 I PQ2	1.043376063	1.139646713	1.199231495	1.283736464	1.418023567	1.550072600	1.684036973	1.969469704	2.486104678
1 I PF3	1.054749881	.9976371090	.9790915957	.9610237433	.9425790192	.9308900263	.9226348604	.9115278278	.9014841069
1 I PQ3	2.545012086	2.835064414	3.012040549	3.260426822	3.650254651	4.029215688	4.410399155	5.214875338	6.654863654
1 I PFC	1.086217721	1.053437702	1.042673504	1.032135573	1.021329625	1.014458135	1.009595003	1.003039213	.9970993083
1 I PQC	9.649236793	10.78785734	11.48079631	12.45156080	13.97188966	15.44702715	16.92875932	20.05119365	25.63062692
1 I ZF1	1.283974463	1.283974463	1.283974463	1.283974463	1.283974463	1.283974463	1.283974463	1.283974463	1.283974463
1 I ZF2	1.431130849	1.431130849	1.431130849	1.431130849	1.431130849	1.431130849	1.431130849	1.431130849	1.431130849
1 I ZF3	1.963696070	1.963696070	1.963696070	1.963696070	1.963696070	1.963696070	1.963696070	1.963696070	1.963696070
1 I ZF4	5.230184416	5.230184416	5.230184416	5.230184416	5.230184416	5.230184416	5.230184416	5.230184416	5.230184416

(4) P 8+0	TH = 53	SBE = 1.252135658						CAUER FUNCTIONS	
	RC = 02	RC = 05	RC = 07	RC = 10	RC = 15	RC = 20	RC = 25	RC = 35	RC = 50
AMAX DB	.0017375255	.0108709564	.0213327376	.0436480540	.0988323393	.1772876696	.2802872360	.5675287486	1.249387366
AMIN DB	49.95745892	57.92535567	60.85837473	63.97872742	67.55573591	70.13296563	72.17416530	75.38396739	79.16386513
MU I K	.3177804D-2	.1269791D-2	.9059021D-3	.6325045D-3	.4189992D-3	.3114237D-3	.2462021D-3	.1701381D-3	.1101049D-3
1 I PF1	.8801466261	.6855169868	.6223310658	.5594963810	.4928641103	.4485330880	.4157920811	.3691282752	.3232492527
1 I PQ1	.5545243079	.5668679293	.5750071124	.5870867480	.6074484350	.6287262728	.6514500864	.7031582764	.8056564866
1 I PF2	.9834870668	.8690814829	.8324812958	.7969549231	.7607357628	.7377651833	.7215171615	.6996029129	.6797139384
1 I PQ2	1.058007795	1.152682795	1.212107883	1.296711471	1.431513272	1.564271040	1.699059613	1.986429398	2.506847248
1 I PF3	1.055187872	.9988795517	.9805478939	.9626685009	.9443972025	.9328087419	.9246203857	.9135979628	.9036257083
1 I PQ3	2.613680275	2.900987998	3.078887134	3.329616364	3.724299854	4.108668142	4.495658961	5.313046389	6.777190875
1 I PFC	1.083788978	1.052052176	1.041598656	1.031350900	1.020828833	1.014130953	1.009387546	1.002989130	.9971879222
1 I PQC	9.988104277	11.12290804	11.82389025	12.81014513	14.35952540	15.86567413	17.38008746	20.57414384	26.28601191
1 I ZF1	1.266526031	1.266526031	1.266526031	1.266526031	1.266526031	1.266526031	1.266526031	1.266526031	1.266526031
1 I ZF2	1.408536155	1.408536155	1.408536155	1.408536155	1.408536155	1.408536155	1.408536155	1.408536155	1.408536155
1 I ZF3	1.925433386	1.925433386	1.925433386	1.925433386	1.925433386	1.925433386	1.925433386	1.925433386	1.925433386
1 I ZF4	5.111879641	5.111879641	5.111879641	5.111879641	5.111879641	5.111879641	5.111879641	5.111879641	5.111879641

(4) P 8+0	TH = 54	SBE = 1.236067977						CAUER FUNCTIONS	
	RC = 02	RC = 05	RC = 07	RC = 10	RC = 15	RC = 20	RC = 25	RC = 35	RC = 50
AMAX DB	.0017375255	.0108709564	.0213327376	.0436480540	.0988323393	.1772876696	.2802872360	.5675287486	1.249387366
AMIN DB	48.39732662	56.36520744	59.29822501	62.41857691	65.99558498	68.57281455	70.61401415	73.82381618	77.60371389
MU I K	.3803064D-2	.1519636D-2	.1084148D-2	.7569569D-3	.5014421D-3	.3726999D-3	.2946451D-3	.2036147D-3	.1317693D-3
1 I PF1	.8929612547	.6941917624	.6298672719	.5659915676	.4983477430	.4533923789	.4202133560	.3729566564	.3265275652
1 I PQ1	.5557387423	.5678096518	.5758854481	.5879150826	.6082405409	.6295070386	.6522320210	.7039650238	.8065443811
1 I PF2	.9899113522	.8748326356	.8379885060	.8022167327	.7657409489	.7426051921	.7262395181	.7041658293	.6841316686
1 I PQ2	1.073657709	1.166532196	1.225760698	1.310443694	1.445764774	1.579254957	1.714901412	2.004296203	2.528680204
1 I PF3	1.055586161	1.000119058	.9820117137	.9643302764	.9462414019	.9347588178	.9266408807	.9157077114	.9058109159
1 I PQ3	2.687511123	2.971486460	3.150262156	3.403387773	3.803139504	4.193194194	4.586310895	5.417347197	6.907069711
1 I PFC	1.081331751	1.050646803	1.040507490	1.030553612	1.020319438	1.013797814	1.009176044	1.002937624	.9972775273
1 I PQC	10.35458380	11.48348364	12.19260965	13.19502249	14.77507628	16.31413839	17.86331631	21.13368770	26.98685250
1 I ZF1	1.249905884	1.249905884	1.249905884	1.249905884	1.249905884	1.249905884	1.249905884	1.249905884	1.249905884
1 I ZF2	1.386892868	1.386892868	1.386892868	1.386892868	1.386892868	1.386892868	1.386892868	1.386892868	1.386892868
1 I ZF3	1.888520781	1.888520781	1.888520781	1.888520781	1.888520781	1.888520781	1.888520781	1.888520781	1.888520781
1 I ZF4	4.997229501	4.997229501	4.997229501	4.997229501	4.997229501	4.997229501	4.997229501	4.997229501	4.997229501

(4) P 8+0	TH = 55 SBE = 1.220774589								CAUER FUNCTIONS
	RC = 02	RC = 05	RC = 07	RC = 10	RC = 15	RC = 20	RC = 25	RC = 35	RC = 50
AMAX DB	.0017375255	.0108709564	.0213327376	.0436480540	.0988323393	.1772876696	.2802872360	.5675287486	1.249387366
AMIN DB	46.85200744	54.81986570	57.75288117	60.87323195	64.45023942	67.02746879	69.06866829	72.27847024	76.05836790
MU I K	.4543595D-2	.1815544D-2	.1295257D-2	.9043539D-3	.5990844D-3	.4452732D-3	.3520194D-3	.2432632D-3	.1574279D-3
1 I PF1	.9064149872	.7032661755	.6377422053	.5727719082	.5040665342	.4584570322	.4248195874	.3769430360	.3299395479
1 I PQ1	.5570400511	.5688091439	.5768151073	.5887895998	.6090747593	.6303280887	.6530534582	.7048113896	.8074747799
1 I PF2	.9964539065	.8807270992	.8436415116	.8076247608	.7708910623	.7475885313	.7311037534	.7088682861	.6886864755
1 I PQ2	1.090425384	1.181266156	1.240255612	1.324995049	1.460838252	1.595085064	1.731624718	2.023137740	2.551683315
1 I PF3	1.055941016	1.001353240	.9834811274	.9660075879	.9481105849	.9367395011	.9286957866	.9178567714	.9080396541
1 I PQ3	2.767046114	3.047001542	3.226592173	3.482162694	3.887204758	4.283243654	4.682828301	5.528308843	7.045146832
1 I PFC	1.078847392	1.049222149	1.039400386	1.029743941	1.019801550	1.013458768	1.008960508	1.002884666	.9973680646
1 I PQC	10.75180568	11.87228370	12.58961033	13.60886709	15.22132987	16.79536426	18.38157168	21.73337464	27.73751828
1 I ZF1	1.234072610	1.234072610	1.234072610	1.234072610	1.234072610	1.234072610	1.234072610	1.234072610	1.234072610
1 I ZF2	1.366151453	1.366151453	1.366151453	1.366151453	1.366151453	1.366151453	1.366151453	1.366151453	1.366151453
1 I ZF3	1.852880781	1.852880781	1.852880781	1.852880781	1.852880781	1.852880781	1.852880781	1.852880781	1.852880781
1 I ZF4	4.886002414	4.886002414	4.886002414	4.886002414	4.886002414	4.886002414	4.886002414	4.886002414	4.886002414

(4) P 8+0	TH = 56 SBE = 1.206217949								CAUER FUNCTIONS
	RC = 02	RC = 05	RC = 07	RC = 10	RC = 15	RC = 20	RC = 25	RC = 35	RC = 50
AMAX DB	.0017375255	.0108709564	.0213327376	.0436480540	.0988323393	.1772876696	.2802872360	.5675287486	1.249387366
AMIN DB	45.32025558	53.28808198	56.22109447	59.34144367	62.91845030	65.49567937	67.53687874	70.74668057	74.52657817
MU I K	.5419849D-2	.2165688D-2	.1545060D-2	.1078767D-2	.7146238D-3	.5311486D-3	.4199099D-3	.2901790D-3	.1877894D-3
1 I PF1	.9205515939	.7127651891	.6459763818	.5798541577	.5100338761	.4637385127	.4296209558	.3810959461	.3334922952
1 I PQ1	.5584372945	.5698715636	.5778004452	.5897140272	.6099543172	.6311924085	.6539172561	.7057001485	.8084505717
1 I PF2	1.003109524	.8867654020	.8494420277	.8131816290	.7761894301	.7527188743	.7361137429	.7137143794	.6933826090
1 I PQ2	1.108423864	1.196964732	1.255666209	1.340434724	1.476800782	1.611828921	1.749298835	2.043029069	2.575945013
1 I PF3	1.056248521	1.002579555	.9849540674	.9676988344	.9500036232	.9387499586	.9307844761	.9200447877	.9103118092
1 I PQ3	2.852904106	3.128033661	3.308359225	3.566416567	3.976980760	4.379322017	4.785742661	5.646526970	7.192146822
1 I PFC	1.076337327	1.047778817	1.038277755	1.028922140	1.019275294	1.013113869	1.008740954	1.002830224	.9974594717
1 I PQC	11.18337607	12.29239309	13.01792052	14.05472203	15.70145231	17.31269218	18.93839719	22.37722449	23.54295373
1 I ZF1	1.218987898	1.218987898	1.218987898	1.218987898	1.218987898	1.218987898	1.218987898	1.218987898	1.218987898
1 I ZF2	1.346266034	1.346266034	1.346266034	1.346266034	1.346266034	1.346266034	1.346266034	1.346266034	1.346266034
1 I ZF3	1.818441185	1.818441185	1.818441185	1.818441185	1.818441185	1.818441185	1.818441185	1.818441185	1.818441185
1 I ZF4	4.777980943	4.777980943	4.777980943	4.777980943	4.777980943	4.777980943	4.777980943	4.777980943	4.777980943

(4) P 8+0 — TH = 57 SBE = 1.192363293 — CAUER FUNCTIONS

MU		RC = 02	RC = 05	RC = 07	RC = 10	RC = 15	RC = 20	RC = 25	RC = 35	RC = 50
	AMAX DB	.0017375255	.0108709564	.0213327376	.0436480540	.0988323393	.1772876696	.2802872360	.5675287486	1.249387366
	AMIN DB	43.80084006	51.76862155	54.70162986	57.82197683	61.39898228	63.97621093	66.01741011	69.22721176	73.00710928
MU	K	.6455918D-2	.2579698D-2	.1840427D-2	.1284994D-2	.8512378D-3	.6326878D-3	.5001837D-3	.3456523D-3	.2236890D-3
1	PF1	.9354192802	.7227161786	.6545922648	.5872566327	.5162643856	.4692493204	.4346285534	.3854246740	.3371935229
1	PQ1	.5599408528	.5710027317	.5788463552	.5906925356	.6108828109	.6321033171	.6548265871	.7066343733	.8094749494
1	PF2	1.009871977	.8929477588	.8553916160	.8188899334	.7816394594	.7580000292	.7412735304	.7187084133	.6982245564
1	PQ2	1.127781986	1.213718194	1.272075212	1.356840282	1.493727356	1.629561937	1.768001031	2.064053752	2.601563617
1	PF3	1.056504566	1.003795289	.9864283146	.9694022853	.9519192834	.9407892690	.9329062458	.9222713465	.9126272244
1	PQ3	2.945795352	3.215151818	3.396110020	3.656687364	4.073015259	4.481999258	4.895652716	5.772671824	7.348884183
1	PFC	1.073803056	1.046317452	1.037140036	1.028088483	1.018740802	1.012763181	1.008517402	1.002774269	.9975516820
1	PQC	11.65346784	12.74735122	13.48100654	14.53606345	16.21905316	17.86992595	19.53782482	23.06980636	29.40877391
1	ZF1	1.204616277	1.204616277	1.204616277	1.204616277	1.204616277	1.204616277	1.204616277	1.204616277	1.204616277
1	ZF2	1.327194075	1.327194075	1.327194075	1.327194075	1.327194075	1.327194075	1.327194075	1.327194075	1.327194075
1	ZF3	1.785134575	1.785134575	1.785134575	1.785134575	1.785134575	1.785134575	1.785134575	1.785134575	1.785134575
1	ZF4	4.672960313	4.672960313	4.672960313	4.672960313	4.672960313	4.672960313	4.672960313	4.672960313	4.672960313

(4) P 8+0 — TH = 58 SBE = 1.179178403 — CAUER FUNCTIONS

MU		RC = 02	RC = 05	RC = 07	RC = 10	RC = 15	RC = 20	RC = 25	RC = 35	RC = 50
	AMAX DB	.0017375255	.0108709564	.0213327376	.0436480540	.0988323393	.1772876696	.2802872360	.5675287486	1.249387366
	AMIN DB	42.29253847	50.26025680	53.19325921	56.31360306	59.89060683	62.46783490	64.50903380	67.71883523	71.49873262
MU	K	.7680210D-2	.3068931D-2	.2189460D-2	.1528691D-2	.1012674D-2	.7526763D-3	.5950429D-3	.4112049D-3	.2661113D-3
1	PF1	.9510712524	.7331492371	.6636145091	.5949994054	.5227740545	.4750031172	.4398544940	.3899393524	.3410516411
1	PQ1	.5615626709	.5722092414	.5799583544	.5917298078	.6118642615	.6330645156	.6557849827	.7076174765	.8105514519
1	PF2	1.016733867	.8992740100	.8614916450	.8247522231	.7872446308	.7634359405	.7465873350	.7238549122	.7032170598
1	PQ2	1.148647194	1.231628699	1.289575936	1.374298967	1.511702084	1.648368549	1.787817721	2.086305100	2.628648770
1	PF3	1.056704833	1.004997544	.9879014865	.9711160683	.9538562171	.9428564138	.9350603084	.9245359680	.9149856940
1	PQ3	3.046538645	3.309005556	3.490466976	3.753586080	4.175928925	4.591920344	5.013235351	5.907500205	7.516277634
1	PFC	1.071246168	1.044838743	1.035987702	1.027243263	1.018198222	1.012406773	1.008289873	1.002716770	.9976446248
1	PQC	12.16693257	13.24123638	13.98285282	15.05687832	16.77826386	18.47141402	20.18445999	23.81633358	30.34137963
1	ZF1	1.190924874	1.190924874	1.190924874	1.190924874	1.190924874	1.190924874	1.190924874	1.190924874	1.190924874
1	ZF2	1.308896105	1.308896105	1.308896105	1.308896105	1.308896105	1.308896105	1.308896105	1.308896105	1.308896105
1	ZF3	1.752897875	1.752897875	1.752897875	1.752897875	1.752897875	1.752897875	1.752897875	1.752897875	1.752897875
1	ZF4	4.570747051	4.570747051	4.570747051	4.570747051	4.570747051	4.570747051	4.570747051	4.570747051	4.570747051

(4) P 8+0	TH = 59	SBE = 1.166633397						CAUER FUNCTIONS	
	RC = 02	RC = 05	RC = 07	RC = 10	RC = 15	RC = 20	RC = 25	RC = 35	RC = 50
AMAX DB	.0017375255	.0108709564	.0213327376	.0436480540	.0988323393	.1772876696	.2802872360	.5675287486	1.249387366
AMIN DB	40.79413062	48.76176025	51.69475438	54.81509383	58.39209526	60.96932251	63.01052103	66.22032212	70.00021934
MU I K	.9126273D-2	.3646800D-2	.2601730D-2	.1816541D-2	.1203359D-2	.8944043D-3	.7070888D-3	.4886342D-3	.3162198D-3
1 I PF1	.9675663679	.7440975292	.6730702441	.6031045277	.5295804227	.4810148712	.4453120400	.3946510633	.3450753397
1 I PQ1	.5633165602	.5734985903	.5811426868	.5928311194	.6129031794	.6340801451	.6567963865	.7086532600	.8116840133
1 I PF2	1.023686457	.9057435485	.8677432431	.8307709734	.7930034894	.7690306902	.7520595577	.7291586347	.7083651357
1 I PQ2	1.171188969	1.250812300	1.308274041	1.392909258	1.530819634	1.668343621	1.808845878	2.109887653	2.657323135
1 I PF3	1.056844786	1.006183227	.9893710225	.9728381570	.9558129488	.9449502670	.9372457833	.9268330391	.9173869573
1 I PQ3	3.156082403	3.410339492	3.592141605	3.857809391	4.286427761	4.709817866	5.139258660	6.051869803	7.695367222
1 I PFC	1.068668342	1.043343429	1.034821262	1.026386802	1.017647714	1.012044723	1.008058394	1.002557593	.9977382243
1 I PQC	12.72943954	13.77876921	14.52805955	15.62175923	17.38383322	19.12214815	20.88358491	24.62277888	31.34809689
1 I ZF1	1.177883204	1.177883204	1.177883204	1.177883204	1.177883204	1.177883204	1.177883204	1.177883204	1.177883204
1 I ZF2	1.291335465	1.291335465	1.291335465	1.291335465	1.291335465	1.291335465	1.291335465	1.291335465	1.291335465
1 I ZF3	1.721671956	1.721671956	1.721671956	1.721671956	1.721671956	1.721671956	1.721671956	1.721671956	1.721671956
1 I ZF4	4.471157722	4.471157722	4.471157722	4.471157722	4.471157722	4.471157722	4.471157722	4.471157722	4.471157722

(4) P 8+0	TH = 60	SBE = 1.154700538						CAUER FUNCTIONS	
	RC = 02	RC = 05	RC = 07	RC = 10	RC = 15	RC = 20	RC = 25	RC = 35	RC = 50
AMAX DB	.0017375255	.0108709564	.0213327376	.0436480540	.0988323393	.1772876696	.2802872360	.5675287486	1.249387366
AMIN DB	39.30439207	47.27189731	50.20487984	53.32521312	56.90221125	59.47943735	61.52063534	64.73043597	68.51033294
MU I K	.1083379D-1	.4329175D-2	.3088560D-2	.2156450D-2	.1428530D-2	.1061764D-2	.8393986D-3	.5800671D-3	.3753906D-3
1 I PF1	.9849698852	.7555977021	.6829894026	.6115962917	.5367027797	.4873010261	.4510157497	.3995719591	.3492761866
1 I PQ1	.5652185709	.5748793402	.5824064463	.5940024366	.6140046421	.6351548542	.6578652173	.7097459728	.8128770202
1 I PF2	1.030719472	.9123552364	.8741472418	.8369485536	.7989346333	.7747884966	.7576947881	.7346245889	.7136740964
1 I PQ2	1.195603061	1.271401385	1.328289638	1.412782749	1.551186950	1.689594123	1.831194704	2.134918943	2.687724418
1 I PF3	1.056919661	1.007349031	.9908341689	.9745663559	.9577878641	.9470695836	.9394616861	.9291771043	.9198306913
1 I PQ3	3.275530800	3.520011076	3.701950800	3.970155025	4.405318137	4.836527299	5.274597716	6.206756473	7.887334923
1 I PFC	1.066071354	1.041832301	1.033641264	1.025519445	1.017089456	1.011677117	1.007822997	1.002597006	.9978323997
1 I PQC	13.34764943	14.36544051	15.12196303	16.23602086	18.04124417	19.82788413	21.64128468	25.49601475	32.43734671
1 I ZF1	1.165462976	1.165462976	1.165462976	1.165462976	1.165462976	1.165462976	1.165462976	1.165462976	1.165462976
1 I ZF2	1.274478085	1.274478085	1.274478085	1.274478085	1.274478085	1.274478085	1.274478085	1.274478085	1.274478085
1 I ZF3	1.691401271	1.691401271	1.691401271	1.691401271	1.691401271	1.691401271	1.691401271	1.691401271	1.691401271
1 I ZF4	4.374017752	4.374017752	4.374017752	4.374017752	4.374017752	4.374017752	4.374017752	4.374017752	4.374017752

(4) P 8+0	TH = 61	SBE = 1.143354068							CAUER FUNCTIONS
	RC = 02	RC = 05	RC = 07	RC = 10	RC = 15	RC = 20	RC = 25	RC = 35	RC = 50
AMAX DB	.0017375255	.0108709564	.0213327376	.0436480540	.0988323393	.1772876696	.2802872360	.5675287486	1.249387366
AMIN DB	37.82208758	45.78941854	48.72238481	51.84270945	55.41970298	57.99692746	60.03812470	63.24792468	67.02782130
MU I K	.1284978D-1	.5134866D-2	.3663370D-2	.2557788D-2	.1694396D-2	.1259371D-2	.9956204D-3	.6830243D-3	.4452551D-3
1 I PF1	1.003354326	.7676903656	.6934051047	.6205015344	.5441624003	.4938796981	.4569816498	.4047154035	.3536637428
1 I PQ1	.5672874559	.5763613126	.5837577278	.5952505337	.6151743875	.6362938805	.6589964437	.7109003799	.8141353810
1 I PF2	1.037820873	.9191073045	.8807041097	.8432871882	.8050266986	.7807137132	.7634978106	.7402580478	.7191495738
1 I PQ2	1.222116701	1.293547637	1.349759840	1.434046408	1.572925322	1.712241132	1.854987645	2.161531599	2.720007776
1 I PF3	1.056924447	1.008491418	.9922879616	.9762982848	.9597791943	.9492129869	.9417069172	.9315522557	.9223165023
1 I PQ3	3.406176374	3.639012436	3.820836805	4.091540532	4.533525092	4.973005561	5.420253716	6.373275177	8.093529577
1 I PFC	1.063457091	1.040306209	1.032448299	1.024641570	1.016523641	1.011304051	1.007583717	1.002534676	.9979270645
1 I PQC	14.02943313	15.00767005	15.77078469	16.90584370	18.75685778	20.59529009	22.46460226	26.44398568	33.61885335
1 I ZF1	1.153637924	1.153637924	1.153637924	1.153637924	1.153637924	1.153637924	1.153637924	1.153637924	1.153637924
1 I ZF2	1.258292282	1.258292282	1.258292282	1.258292282	1.258292282	1.258292282	1.258292282	1.258292282	1.258292282
1 I ZF3	1.662033527	1.662033527	1.662033527	1.662033527	1.662033527	1.662033527	1.662033527	1.662033527	1.662033527
1 I ZF4	4.279160316	4.279160316	4.279160316	4.279160316	4.279160316	4.279160316	4.279160316	4.279160316	4.279160316

(4) P 8+0	TH = 62	SBE = 1.132570051							CAUER FUNCTIONS
	RC = 02	RC = 05	RC = 07	RC = 10	RC = 15	RC = 20	RC = 25	RC = 35	RC = 50
AMAX DB	.0017375255	.0108709564	.0213327376	.0436480540	.0988323393	.1772876696	.2802872360	.5675287486	1.249387366
AMIN DB	36.34596429	44.31305129	47.24599479	50.36630734	53.94329441	56.52051663	58.56171283	61.77151189	65.55140803
MU I K	.1523007D-1	.6086217D-2	.4342104D-2	.3031689D-2	.2008331D-2	.1492706D-2	.1180088D-2	.8155008D-3	.5277516D-3
1 I PF1	1.022800469	.7804206549	.7043541095	.6298499956	.5519828203	.5007709066	.4632274365	.4100961368	.3582506962
1 I PQ1	.5695452545	.5779558282	.5852058095	.5965831356	.6164189257	.6375031487	.6601956743	.7121218451	.8154646070
1 I PF2	1.044976602	.9259972349	.8874138722	.8497889093	.8112883403	.7868108241	.7694736101	.7460645678	.7247975464
1 I PQ2	1.250995111	1.317425673	1.372841887	1.456845328	1.596172886	1.736422271	1.880364816	2.189875889	2.754348724
1 I PF3	1.056853871	1.009606602	.9937292066	.9780313603	.9617850013	.9513789540	.9439802485	.9339627222	.9248439169
1 I PQ3	3.549541043	3.768497451	3.949891859	4.223026384	4.672114767	5.120353712	5.577377378	6.552705527	8.315497272
1 I PFC	1.060827555	1.038766065	1.031243007	1.023753586	1.015950483	1.010925634	1.007340597	1.002470668	.9980221262
1 I PQC	14.78414951	15.71300553	16.48181711	17.63845302	19.53809176	21.43213037	23.36173019	27.47592098	34.90390105
1 I ZF1	1.142383649	1.142383649	1.142383649	1.142383649	1.142383649	1.142383649	1.142383649	1.142383649	1.142383649
1 I ZF2	1.242748584	1.242748584	1.242748584	1.242748584	1.242748584	1.242748584	1.242748584	1.242748584	1.242748584
1 I ZF3	1.633519375	1.633519375	1.633519375	1.633519375	1.633519375	1.633519375	1.633519375	1.633519375	1.633519375
1 I ZF4	4.186425270	4.186425270	4.186425270	4.186425270	4.186425270	4.186425270	4.186425270	4.186425270	4.186425270

(4) P 8+0	TH = 63	SBE = 1.122326238						CAUER FUNCTIONS	
	RC = 02	RC = 05	RC = 07	RC = 10	RC = 15	RC = 20	RC = 25	RC = 35	RC = 50
AMAX DB	.0017375255	.0108709564	.0213327376	.0436480540	.0988323393	.1772876696	.2802872360	.5675287486	1.249387366
AMIN DB	34.87474494	42.84149053	45.77440217	48.89469779	52.47167583	55.04889489	57.09008962	60.29988741	64.07978287
MU K	.1804109D-1	.7209837D-2	.5143750D-2	.3591411D-2	.2379119D-2	.1768297D-2	.1397962D-2	.9660634D-3	.6251883D-3
1 PF1	1.043398490	.7938388921	.7158773463	.6396747391	.5601901627	.5079968468	.4697727139	.4157304702	.3630505205
1 PQ1	.5720180293	.5796760040	.5867613788	.5980090921	.6177456757	.6387893888	.6614692656	.7134164303	.8168709108
1 PF2	1.052170273	.9330216219	.8942760170	.8564554990	.8177232076	.7930844379	.7756273774	.7520500066	.7306243689
1 PQ2	1.282549678	1.343237528	1.397716973	1.481346090	1.621087680	1.762294653	1.907485948	2.220122790	2.790946636
1 PF3	1.056702383	1.010690523	.9951544594	.9797627761	.9638031587	.9535657992	.9462803083	.9364075561	.9274123710
1 PQ3	3.707428131	3.909815552	4.090388835	4.365844567	4.822322107	5.279844923	5.747297727	6.746523171	8.555018601
1 PFC	1.058184875	1.037212854	1.030026079	1.022855940	1.015370214	1.010541985	1.007093686	1.002404947	.9981174855
1 PQC	15.62300127	16.49037378	17.26365801	18.44234312	20.39364369	22.34749474	24.34224947	28.60259951	36.30565322
1 ZF1	1.131677478	1.131677478	1.131677478	1.131677478	1.131677478	1.131677478	1.131677478	1.131677478	1.131677478
1 ZF2	1.227819563	1.227819563	1.227819563	1.227819563	1.227819563	1.227819563	1.227819563	1.227819563	1.227819563
1 ZF3	1.605812133	1.605812133	1.605812133	1.605812133	1.605812133	1.605812133	1.605812133	1.605812133	1.605812133
1 ZF4	4.095658114	4.095658114	4.095658114	4.095658114	4.095658114	4.095658114	4.095658114	4.095658114	4.095658114

(4) P 8+0	TH = 64	SBE = 1.112601940						CAUER FUNCTIONS	
	RC = 02	RC = 05	RC = 07	RC = 10	RC = 15	RC = 20	RC = 25	RC = 35	RC = 50
AMAX DB	.0017375255	.0108709564	.0213327376	.0436480540	.0988323393	.1772876696	.2802872360	.5675287486	1.249387366
AMIN DB	33.40712088	41.37338874	44.30625580	47.42652774	51.00349313	53.58070776	55.62190045	58.83169645	62.61159097
MU K	.2136210D-1	.8537497D-2	.6090981D-2	.4252787D-2	.2817250D-2	.2093942D-2	.1655408D-2	.1143971D-2	.7403217D-3
1 PF1	1.065249275	.8080013686	.7280205444	.6500126533	.5688135238	.5155322130	.4766392752	.4216365161	.3680781618
1 PQ1	.5747368087	.5815371226	.5884368108	.5995385916	.6191631315	.6401602800	.6628244542	.7147910150	.8183613241
1 PF2	1.059382820	.9401760065	.9012893790	.8632884197	.8243349125	.7995392781	.7819645127	.7582205443	.7366368056
1 PQ2	1.317148313	1.371218242	1.424594987	1.507740912	1.647851389	1.790038499	1.936533986	2.252467746	2.830029016
1 PF3	1.056464134	1.011738827	.9965599999	.9814894798	.9658313316	.9557716556	.9486055642	.9388856789	.9300211980
1 PQ3	3.881988957	4.064554255	4.243819629	4.521434284	4.985585305	5.452959190	5.931557774	6.955438608	8.814154693
1 PFC	1.055531323	1.035647637	1.028798267	1.021949118	1.014783095	1.010153240	1.006843041	1.002337478	.9982130357
1 PQC	16.55949499	17.35040053	18.12650696	19.32756083	21.33377216	23.35208679	25.41742966	29.83668168	37.83955206
1 ZF1	1.121498343	1.121498343	1.121498343	1.121498343	1.121498343	1.121498343	1.121498343	1.121498343	1.121498343
1 ZF2	1.213479695	1.213479695	1.213479695	1.213479695	1.213479695	1.213479695	1.213479695	1.213479695	1.213479695
1 ZF3	1.578867522	1.578867522	1.578867522	1.578867522	1.578867522	1.578867522	1.578867522	1.578867522	1.578867522
1 ZF4	4.006708977	4.006708977	4.006708977	4.006708977	4.006708977	4.006708977	4.006708977	4.006708977	4.006708977

(4) P 8+0 — TH = 65 SBE = 1.103377919 — CAUER FUNCTIONS

	RC = 02	RC = 05	RC = 07	RC = 10	RC = 15	RC = 20	RC = 25	RC = 35	RC = 50
AMAX DB	.0017375255	.0108709564	.0213327376	.0436480540	.0988323393	.1772876696	.2802872360	.5675287486	1.249387366
AMIN DB	31.94174540	39.90734458	42.84014922	45.96038800	49.53733569	52.11454413	54.15573394	57.36552745	61.14542064
MU K	.2528790D-1	.1010724D-1	.7210951D-2	.5034781D-2	.3335287D-2	.2478979D-2	.1959807D-2	.1354327D-2	.8764537D-3
1 PF1	1.088465902	.8229712722	.7408349822	.6609050468	.5778854338	.5235545826	.4838514387	.4278344631	.3733502619
1 PQ1	.5777388013	.5835570973	.5902465159	.6011834274	.6206810675	.6416246278	.6642695176	.7162534431	.8199438418
1 PF2	1.066592091	.9474546788	.9084520033	.8702887278	.8311269907	.8061801685	.7884906282	.7645827051	.7428420686
1 PQ2	1.355228695	1.401642858	1.453720439	1.536252800	1.676673981	1.819861598	1.967719521	2.287135267	2.871856737
1 PF3	1.056132956	1.012746834	.9979418064	.9832081468	.9678669529	.9579944535	.9509543037	.9413958640	.9326696141
1 PQ3	4.075808890	4.234593163	4.411943670	4.691486880	5.163588988	5.641426752	6.131959073	7.184445589	9.095304502
1 PFC	1.052869321	1.034071563	1.027560385	1.021033653	1.014189409	1.009759548	1.006588726	1.002268224	.9983086620
1 PQC	17.61004205	18.30582145	19.08254505	20.30606685	22.37065404	24.45858958	26.60060884	31.19312927	39.52382318
1 ZF1	1.111826664	1.111826664	1.111826664	1.111826664	1.111826664	1.111826664	1.111826664	1.111826664	1.111826664
1 ZF2	1.199705228	1.199705228	1.199705228	1.199705228	1.199705228	1.199705228	1.199705228	1.199705228	1.199705228
1 ZF3	1.552643416	1.552643416	1.552643416	1.552643416	1.552643416	1.552643416	1.552643416	1.552643416	1.552643416
1 ZF4	3.919431580	3.919431580	3.919431580	3.919431580	3.919431580	3.919431580	3.919431580	3.919431580	3.919431580

(4) P 8+0 — TH = 66 SBE = 1.094636279 — CAUER FUNCTIONS

	RC = 02	RC = 05	RC = 07	RC = 10	RC = 15	RC = 20	RC = 25	RC = 35	RC = 50
AMAX DB	.0017375255	.0108709564	.0213327376	.0436480540	.0988323393	.1772876696	.2802872360	.5675287486	1.249387366
AMIN DB	30.47722740	38.44189002	41.37460722	44.49479955	48.07172244	50.64892220	52.69010800	55.89999301	59.67978934
MU K	.2993220D-1	.1196480D-1	.8536299D-2	.5960189D-2	.3948334D-2	.2934634D-2	.2320035D-2	.1603264D-2	.1037554D-2
1 PF1	1.113175331	.8388197918	.7543783843	.6723983634	.5874424098	.5319448791	.4914364509	.4343469047	.3788854256
1 PQ1	.5810689758	.5857570622	.5922073779	.6029573288	.6223107926	.6431925819	.6658139723	.7178127026	.8216275969
1 PF2	1.073772361	.9548504415	.9157609785	.8774569692	.8381028510	.8130120141	.7952115477	.7711433810	.7492478600
1 PQ2	1.397315373	1.434835293	1.485379902	1.567141995	1.707799477	1.852004847	2.001286294	2.324384637	2.916730502
1 PF3	1.055702332	1.013709514	.9992955238	.9849151514	.9699071967	.9602318961	.9533246117	.9439367176	.9353567026
1 PQ3	4.292019730	4.422173167	4.596849777	4.878002874	5.358318824	5.847282875	6.350617857	7.432881965	9.401276709
1 PFC	1.050201460	1.032485875	1.026313322	1.020110127	1.013589472	1.009361077	1.006330816	1.002197148	.9984042407
1 PQC	18.79475204	19.37201538	20.14642544	21.39220084	23.51884258	25.68213355	27.90767923	32.68974087	41.38011828
1 ZF1	1.102644247	1.102644247	1.102644247	1.102644247	1.102644247	1.102644247	1.102644247	1.102644247	1.102644247
1 ZF2	1.186474069	1.186474069	1.186474069	1.186474069	1.186474069	1.186474069	1.186474069	1.186474069	1.186474069
1 ZF3	1.527099609	1.527099609	1.527099609	1.527099609	1.527099609	1.527099609	1.527099609	1.527099609	1.527099609
1 ZF4	3.833682188	3.833682188	3.833682188	3.833682188	3.833682188	3.833682188	3.833682188	3.833682188	3.833682188

(4) P 8+0 — TH = 67 — SBE = 1.086360377 — CAUER FUNCTIONS

		RC = 02	RC = 05	RC = 07	RC = 10	RC = 15	RC = 20	RC = 25	RC = 35	RC = 50
	AMAX DB	.0017375255	.0108709564	.0213327376	.0436480540	.0988323393	.1772876696	.2802872360	.5675287486	1.249387366
	AMIN DB	29.01212625	36.97547575	39.90807033	43.02819750	46.60508561	49.18227319	51.22345336	54.43323847	58.21312719
MU	K	.3543184D-1	.1416531D-1	.1010640D-1	.7056513D-2	.4674614D-2	.3474452D-2	.2746802D-2	.1898183D-2	.1228411D-2
1	PF1	1.139520266	.8556274385	.7687159993	.6845450443	.5975256231	.5407879299	.4994249729	.4411992372	.3847045433
1	PQ1	.5847821406	.5881621265	.5943393100	.6048763794	.6240654691	.6448759085	.6674688204	.7194791469	.8234230779
1	PF2	1.080893765	.9623543256	.9232122350	.8847930497	.8452657118	.8200397740	.8021333037	.7779098576	.7558624210
1	PQ2	1.444042097	1.471179657	1.519911480	1.600714128	1.741513215	1.886749202	2.037518073	2.364517032	2.964998901
1	PF3	1.055165374	1.014621444	1.000616430	.9866065329	.9719489475	.9624814313	.9557143445	.9465066562	.9380813945
1	PQ3	4.534448117	4.629986022	4.801035807	5.083365068	5.572131221	6.072937598	6.590036392	7.704506963	9.735380772
1	PFC	1.047530520	1.030891923	1.025058045	1.019179180	1.012983633	1.008958015	1.006069398	1.002124214	.9984996383
1	PQC	20.13849267	20.56770379	21.33591366	22.60328600	24.79586073	27.04090116	29.35771385	34.34784160	43.43434169
1	ZF1	1.093934193	1.093934193	1.093934193	1.093934193	1.093934193	1.093934193	1.093934193	1.093934193	1.093934193
1	ZF2	1.173765681	1.173765681	1.173765681	1.173765681	1.173765681	1.173765681	1.173765681	1.173765681	1.173765681
1	ZF3	1.502197588	1.502197588	1.502197588	1.502197588	1.502197588	1.502197588	1.502197588	1.502197588	1.502197588
1	ZF4	3.749318496	3.749318496	3.749318496	3.749318496	3.749318496	3.749318496	3.749318496	3.749318496	3.749318496

(4) P 8+0 — TH = 68 — SBE = 1.078534743 — CAUER FUNCTIONS

		RC = 02	RC = 05	RC = 07	RC = 10	RC = 15	RC = 20	RC = 25	RC = 35	RC = 50
	AMAX DB	.0017375255	.0108709564	.0213327376	.0436480540	.0988323393	.1772876696	.2802872360	.5675287486	1.249387366
	AMIN DB	27.54494859	35.50645437	38.43887670	41.55891234	45.13575160	47.71292208	49.75409433	52.96387255	56.74375761
MU	K	.4195199D-1	.1677557D-1	.1196895D-1	.8357077D-2	.5536208D-2	.4114849D-2	.3253084D-2	.2248052D-2	.1454830D-2
1	PF1	1.167661180	.8734856308	.7839219048	.6974045751	.6081817116	.5501231450	.5078516731	.4484201433	.3908311830
1	PQ1	.5889457109	.5908023524	.5966659685	.6069595498	.6259605145	.6466883322	.6692468585	.7212647732	.8253423971
1	PF2	1.087921626	.9699552457	.9308002979	.8922960756	.8526185206	.8272684250	.8092621296	.7848898408	.7626945868
1	PQ2	1.496181287	1.511134873	1.557716941	1.637330617	1.778151068	1.924424470	2.076747350	2.407884486	3.017068539
1	PF3	1.054514786	1.015476774	1.001899393	.9882779571	.9739887645	.9647402193	.9581210997	.9491038803	.9408424466
1	PQ3	4.807813923	4.861291584	5.027512312	5.310433212	5.807843188	6.321265419	6.853194566	8.002599554	10.10154333
1	PFC	1.044859484	1.029291176	1.023795615	1.018241515	1.012372277	1.008550574	1.005804569	1.002049386	.9985947102
1	PQC	21.67232393	21.91587883	22.67273284	23.96042359	26.22297775	28.55691665	30.97378359	36.19318079	45.71772427
1	ZF1	1.085680812	1.085680812	1.085680812	1.085680812	1.085680812	1.085680812	1.085680812	1.085680812	1.085680812
1	ZF2	1.161560994	1.161560994	1.161560994	1.161560994	1.161560994	1.161560994	1.161560994	1.161560994	1.161560994
1	ZF3	1.477900311	1.477900311	1.477900311	1.477900311	1.477900311	1.477900311	1.477900311	1.477900311	1.477900311
1	ZF4	3.666198440	3.666198440	3.666198440	3.666198440	3.666198440	3.666198440	3.666198440	3.666198440	3.666198440

(4) P 8+0	TH = 69	SBE = 1.071144994						CAUER FUNCTIONS	
	RC = 02	RC = 05	RC = 07	RC = 10	RC = 15	RC = 20	RC = 25	RC = 35	RC = 50
AMAX DB	.0017375255	.0108709564	.0213327376	.0436480540	.0988323393	.1772876696	.2802872360	.5675287486	1.249387366
AMIN DB	26.07414873	34.03306095	36.96524084	40.08514766	43.66191814	46.23906455	48.28022566	51.43999417	55.26987408
MU K	.4969270D-1	.1987682D-1	.1418202D-1	.9902449D-2	.6560004D-2	.4875810D-2	.3854683D-2	.2663792D-2	.1723877D-2
1 PF1	1.197778411	.8924986038	.8000805935	.7110447656	.6194637752	.5599953509	.5167559532	.4560421848	.3972920716
1 PQ1	.5936434362	.5937140327	.5992156774	.6092293818	.6280141149	.6486459706	.6711630684	.7231835709	.8273996293
1 PF2	1.094815654	.9776395810	.9385179850	.8999641554	.8601638503	.8347029123	.8166044452	.7920914868	.7697538511
1 PQ2	1.554683437	1.555253768	1.599277466	1.677422073	1.818111258	1.965420548	2.119366422	2.454901301	3.073416877
1 PF3	1.053742828	1.016269182	1.003138830	.9899246713	.9760228408	.9670050944	.9605421812	.9517263438	.9436384158
1 PQ3	5.117998967	5.120073063	5.279939005	5.562668002	6.068849454	6.595721880	7.143668733	8.331087414	10.50445857
1 PFC	1.042191567	1.027685239	1.022527192	1.017297909	1.011755837	1.008138991	1.005536445	1.001972625	.9986892998
1 PQC	23.43546558	23.44505006	24.18369313	25.48954979	27.82623780	30.25708932	32.78403503	38.25711346	48.26823448
1 ZF1	1.077869552	1.077869552	1.077869552	1.077869552	1.077869552	1.077869552	1.077869552	1.077869552	1.077869552
1 ZF2	1.149842327	1.149842327	1.149842327	1.149842327	1.149842327	1.149842327	1.149842327	1.149842327	1.149842327
1 ZF3	1.454171986	1.454171986	1.454171986	1.454171986	1.454171986	1.454171986	1.454171986	1.454171986	1.454171986
1 ZF4	3.584178884	3.584178884	3.584178884	3.584178884	3.584178884	3.584178884	3.584178884	3.584178884	3.584178884

(4) P 8+0	TH = 70	SBE = 1.064177772						CAUER FUNCTIONS	
	RC = 02	RC = 05	RC = 07	RC = 10	RC = 15	RC = 20	RC = 25	RC = 35	RC = 50
AMAX DB	.0017375255	.0108709564	.0213327376	.0436480540	.0988323393	.1772876696	.2802872360	.5675287486	1.249387366
AMIN DB	24.59813513	32.55339053	35.48522833	38.60495335	42.18162677	44.75873921	46.79988459	50.00963941	53.78951204
MU K	.5889701D-1	.2356842D-1	.1681662D-1	.1174228D-1	.7778908D-2	.5781800D-2	.4570943D-2	.3158770D-2	.2044205D-2
1 PF1	1.230074168	.9127857155	.8172889145	.7255433261	.6314326057	.5704558199	.5261828476	.4641025374	.4041176895
1 PQ1	.5989804871	.5969413840	.6020226431	.6117128793	.6302478870	.6507678908	.6732351164	.7252519664	.8296112394
1 PF2	1.101528986	.9853906622	.9463560317	.9077941513	.8679037682	.8423480847	.8241668351	.7995234317	.7770504394
1 PQ2	1.620730484	1.604208293	1.645174310	1.721505788	1.861869688	2.010201953	2.165841688	2.506058703	3.134608707
1 PF3	1.052841274	1.016991816	1.004328650	.9915414531	.9780469547	.9692725207	.9629745578	.9543717179	.9464676276
1 PQ3	5.472416240	5.411245328	5.562806676	5.844295446	6.359276959	6.900496883	7.465787695	8.694713487	10.94978473
1 PFC	1.039530240	1.026075865	1.021254053	1.016349219	1.011134791	1.007723533	1.005265156	1.001893897	.9987832370
1 PQC	25.47803711	25.19094258	25.90222123	27.22285955	29.63783947	32.17460916	34.82312967	40.57817465	51.13245503
1 ZF1	1.070486927	1.070486927	1.070486927	1.070486927	1.070486927	1.070486927	1.070486927	1.070486927	1.070486927
1 ZF2	1.138593321	1.138593321	1.138593321	1.138593321	1.138593321	1.138593321	1.138593321	1.138593321	1.138593321
1 ZF3	1.430977845	1.430977845	1.430977845	1.430977845	1.430977845	1.430977845	1.430977845	1.430977845	1.430977845
1 ZF4	3.503114139	3.503114139	3.503114139	3.503114139	3.503114139	3.503114139	3.503114139	3.503114139	3.503114139

(4) P 8+0		TH = 71	SBE = 1.057620681						CAUER FUNCTIONS	
		RC = 02	RC = 05	RC = 07	RC = 10	RC = 15	RC = 20	RC = 25	RC = 35	RC = 50
AMAX DB		.0017375255	.0108709564	.0213327376	.0436480540	.0988323393	.1772876696	.2802872360	.5675287486	1.249387366
AMIN DB		23.11528684	31.06537198	33.99672556	37.11619321	40.69272922	43.26979356	45.31091668	48.52065211	52.30051444
MU	K	.6986114D-1	.2797251D-1	.1996015D-1	.1393768D-1	.9233440D-2	.6862942D-2	.5425680D-2	.3749449D-2	.2426466D-2
1	PF1	1.264774104	.9344842360	.8356584616	.7409898243	.6441582202	.5815635566	.5361841454	.4726439068	.4113430141
1	PQ1	.6050904978	.6005388193	.6051285675	.6144426812	.6326877416	.6530768336	.6754839995	.7274893953	.8319966314
1	PF2	1.108007037	.9931881371	.9543026266	.9157813673	.8758396655	.8502086060	.8319560138	.8071948241	.7845953948
1	PQ2	1.695809161	1.658823311	1.696116270	1.770208830	1.910000047	2.059326811	2.216732292	2.561943917	3.201317522
1	PF3	1.051801366	1.017637239	1.005462192	.9931225493	.9800564139	.9715385398	.9654148147	.9570373483	.9493281406
1	PQ3	5.880525172	5.740938371	5.881683080	6.160527713	6.684191294	7.240718933	7.824840072	9.099258646	11.44440404
1	PFC	1.036879266	1.024464980	1.019977608	1.015396397	1.010509676	1.007304504	1.004990852	1.001813165	.9988763362
1	PQC	27.86493678	27.19884462	27.87046172	29.20075249	31.69801198	34.35083969	37.13419233	43.20420573	54.36811251
1	ZF1	1.063520463	1.063520463	1.063520463	1.063520463	1.063520463	1.063520463	1.063520463	1.063520463	1.063520463
1	ZF2	1.127798885	1.127798885	1.127798885	1.127798885	1.127798885	1.127798885	1.127798885	1.127798885	1.127798885
1	ZF3	1.408283913	1.408283913	1.408283913	1.408283913	1.408283913	1.408283913	1.408283913	1.408283913	1.408283913
1	ZF4	3.422854258	3.422854258	3.422854258	3.422854258	3.422854258	3.422854259	3.422854258	3.422854258	3.422854258

(4) P 8+0		TH = 72	SBE = 1.051462224						CAUER FUNCTIONS	
		RC = 02	RC = 05	RC = 07	RC = 10	RC = 15	RC = 20	RC = 25	RC = 35	RC = 50
AMAX DB		.0017375255	.0108709564	.0213327376	.0436480540	.0988323393	.1772876696	.2802872360	.5675287486	1.249387366
AMIN DB		21.62398643	29.56673763	32.49740328	35.61650520	39.19284593	41.76984190	43.81093338	47.02064126	50.80048395
MU	K	.8294700D-1	.3324016D-1	.2372083D-1	.1656436D-1	.1097382D-1	.8156576D-2	.6448420D-2	.4456233D-2	.2883869D-2
1	PF1	1.302127890	.9577527218	.8553185277	.7574881322	.6577217903	.5933869193	.5468198071	.4817156861	.4190084573
1	PQ1	.6121454593	.6045740421	.6085848226	.6174586264	.6353650262	.6556001632	.6779348913	.7299190471	.8345788595
1	PF2	1.114186106	1.001007183	.9623428259	.9239191534	.8839720384	.8582888375	.8399787751	.8151153567	.7924006779
1	PQ2	1.781813417	1.720122517	1.752976728	1.824298960	1.963200573	2.113472031	2.272714770	2.623265286	3.274353585
1	PF3	1.050613758	1.018197352	1.006532155	.9946616046	.9820459880	.9737987083	.9678590954	.9597202041	.9522177020
1	PQ3	6.354563076	6.116889116	6.243550330	6.517864248	7.049876537	7.622731073	8.227353835	9.551840239	11.99677134
1	PFC	1.034242738	1.022854705	1.018699414	1.014440502	1.009881096	1.006882246	1.004713707	1.001730396	.9989683952
1	PQC	30.68143286	29.52690709	30.14221089	31.47453401	34.05760598	36.83792392	39.77148822	46.19526365	58.04753714
1	ZF1	1.056958636	1.056958636	1.056958636	1.056958636	1.056958636	1.056958636	1.056958636	1.056958636	1.056958636
1	ZF2	1.117445151	1.117445151	1.117445151	1.117445151	1.117445151	1.117445151	1.117445151	1.117445151	1.117445151
1	ZF3	1.386056754	1.386056754	1.386056754	1.386056754	1.386056754	1.386056754	1.386056754	1.386056754	1.386056754
1	ZF4	3.343243012	3.343243012	3.343243012	3.343243012	3.343243012	3.343243012	3.343243012	3.343243012	3.343243012

(4) P 8+0	TH = 73	SBE = 1.045691756						CAUER FUNCTIONS	
	RC = 02	RC = 05	RC = 07	RC = 10	RC = 15	RC = 20	RC = 25	RC = 35	RC = 50
AMAX DB	.0017375255	.0108709564	.0213327376	.0436480540	.0988323393	.1772876696	.2802872360	.5675287486	1.249387366
AMIN DB	20.12267933	28.05498838	30.98467235	34.10325223	37.67931415	40.25621253	42.29725882	45.50692736	49.28675415
MU I K	.9859753D-1	.3955948D-1	.2823361D-1	.1971684D-1	.1306274D-1	.9709332D-2	.7676037D-2	.5304612D-2	.3432909D-2
1 I PF1	1.342407691	.9827750866	.8764197795	.7751595157	.6722180953	.6060056863	.5581597717	.4913754336	.4271610633
1 I PQ1	.6203698473	.6091323275	.6124554279	.6208098753	.6383180569	.6583711355	.6806182650	.7325688488	.8373855630
1 I PF2	1.119991693	1.008817515	.9704578146	.9321983980	.8923002010	.8665926799	.8482419179	.8232952983	.8004792855
1 I PQ2	1.881189856	1.789391869	1.816844392	1.884726673	2.022330165	2.173467141	2.334616091	2.690835833	3.354701318
1 I PF3	1.049268468	1.018663318	1.007530511	.9961515772	.9840098289	.9760480236	.9703030320	.9624168161	.9551336951
1 I PQ3	6.910602921	6.548992390	6.657275653	6.924509029	7.464220622	8.054469517	8.681478974	10.06131970	12.61738865
1 I PFC	1.031625130	1.021247384	1.017421202	1.013482715	1.009249728	1.006457149	1.004433923	1.001645558	.9990591921
1 I PQC	34.04138357	32.25086596	32.78708579	34.11022985	36.78173175	39.70243211	42.80416247	49.62770575	62.26247170
1 I ZF1	1.050790836	1.050790836	1.050790836	1.050790836	1.050790836	1.050790836	1.050790836	1.050790836	1.050790836
1 I ZF2	1.107519446	1.107519446	1.107519446	1.107519446	1.107519446	1.107519446	1.107519446	1.107519446	1.107519446
1 I ZF3	1.364263198	1.364263198	1.364263198	1.364263198	1.364263198	1.364263198	1.364263198	1.364263198	1.364263198
1 I ZF4	3.264115451	3.264115451	3.264115451	3.264115451	3.264115451	3.264115451	3.264115451	3.264115451	3.264115451

(4) P 8+0	TH = 74	SBE = 1.040299436						CAUER FUNCTIONS	
	RC = 02	RC = 05	RC = 07	RC = 10	RC = 15	RC = 20	RC = 25	RC = 35	RC = 50
AMAX DB	.0017375255	.0108709564	.0213327376	.0436480540	.0988323393	.1772876696	.2802872360	.5675287486	1.249387366
AMIN DB	18.60997555	26.52735404	29.45562987	32.57346061	36.14912237	38.72588063	40.76686206	43.97647411	47.75627090
MU I K	.1173549D 0	.4716635D-1	.3366809D-1	.2351402D-1	.1557915D-1	.1157993D-1	.9154969D-2	.6326686D-2	.4094364D-2
1 I PF1	1.385902711	1.009765477	.8991388460	.7941465665	.6877586760	.6195137214	.5702862907	.5016907835	.4358560584
1 I PQ1	.6300611414	.6143225502	.6168211943	.6245578315	.6415942064	.6614306139	.6835714068	.7354727833	.8404502122
1 I PF2	1.125336452	1.016582133	.9786239585	.9406068726	.9008219081	.8751233586	.8567521405	.8317455223	.8088453907
1 I PQ2	1.997148168	1.868268796	1.889094017	1.952683316	2.088457940	2.240340510	2.403458653	2.765868797	3.443569873
1 I PF3	1.047754821	1.019025456	1.008448400	.9975846363	.9859413742	.9782808333	.9727416607	.9651232021	.9580730748
1 I PQ3	7.570114586	7.050090633	7.134280482	7.390959267	7.937255396	8.545992227	9.197520339	10.63886830	13.31946220
1 I PFC	1.029031358	1.019645620	1.016144905	1.012524365	1.008616339	1.006029658	1.004151733	1.001558623	.9991484834
1 I PQC	38.09958760	35.47093893	35.89656566	37.19407681	39.95496459	43.03056278	46.32155971	53.59989865	67.13087639
1 I ZF1	1.045007317	1.045007317	1.045007317	1.045007317	1.045007317	1.045007317	1.045007317	1.045007317	1.045007317
1 I ZF2	1.098010270	1.098010270	1.098010270	1.098010270	1.098010270	1.098010270	1.098010270	1.098010270	1.098010270
1 I ZF3	1.342870031	1.342870031	1.342870031	1.342870031	1.342870031	1.342870031	1.342870031	1.342870031	1.342870031
1 I ZF4	3.185294883	3.185294883	3.185294883	3.185294883	3.185294883	3.185294883	3.185294883	3.185294883	3.185294883

(4) P 8+0		TH = 75 SBE = 1.035276180								CAUER FUNCTIONS
		RC = 02	RC = 05	RC = 07	RC = 10	RC = 15	RC = 20	RC = 25	RC = 35	RC = 50
AMAX DB		.0017375255	.0108709564	.0213327376	.0436480540	.0988323393	.1772876696	.2802872360	.5675287486	1.249387366
AMIN DB		17.08481894	24.98074953	27.90699326	31.02374225	34.59882598	37.17538183	39.21626955	42.42580000	46.20555343
MU	K	.1398811D 0	.5635890D-1	.4023929D-1	.2810690D-1	.1862339D-1	.1384302D-1	.1094426D-1	.7563277D-2	.4894658D-2
1	PF1	1.432906572	1.038974024	.9236840640	.8146182547	.7044759384	.6340224578	.5832969852	.5127419513	.4451588835
1	PQ1	.6416201740	.6202858377	.6217856049	.6287802401	.6452528005	.6648294339	.6868404915	.7386726855	.8438137970
1	PF2	1.130117731	1.024255709	.9868115792	.9491283735	.9095328521	.8838831306	.8655158863	.8404775291	.8175145112
1	PQ2	2.133971744	1.958870161	1.971486906	2.029682960	2.162932484	2.315383637	2.480522681	2.849540523	3.542462732
1	PF3	1.046061420	1.019273128	1.009276014	.9989520403	.9878332306	.9804907257	.9751693186	.9678347746	.9610322881
1	PQ3	8.362321477	7.637130174	7.689511318	7.930852822	8.481928391	9.110231127	9.788693862	11.29876731	14.11982879
1	PFC	1.026466865	1.018052315	1.014872688	1.011566947	1.007981800	1.005600284	1.003867411	1.001469565	.9992360006
1	PQC	43.07079242	39.32212691	39.59294081	40.84059691	43.68894916	46.93571227	50.44094715	58.24060297	72.80675195
1	ZF1	1.039599165	1.039599165	1.039599165	1.039599165	1.039599165	1.039599165	1.039599165	1.039599165	1.039599165
1	ZF2	1.088907294	1.088907294	1.088907294	1.088907294	1.088907294	1.088907294	1.088907294	1.088907294	1.088907294
1	ZF3	1.321843626	1.321843626	1.321843626	1.321843626	1.321843626	1.321843626	1.321843626	1.321843626	1.321843626
1	ZF4	3.106589056	3.106589056	3.106589056	3.106589056	3.106589056	3.106589056	3.106589056	3.106589056	3.106589056

(4) P 8+0		TH = 76 SBE = 1.030613629								CAUER FUNCTIONS
		RC = 02	RC = 05	RC = 07	RC = 10	RC = 15	RC = 20	RC = 25	RC = 35	RC = 50
AMAX DB		.0017375255	.0108709564	.0213327376	.0436480540	.0988323393	.1772876696	.2802872360	.5675287486	1.249387366
AMIN DB		15.54676279	23.41172911	26.33501965	29.45019518	33.02443774	35.60069896	37.64145028	40.85086194	44.63055225
MU	K	.1669790D 0	.6751706D-1	.4822242D-1	.3368916D-1	.2232431D-1	.1659453D-1	.1311981D-1	.9066860D-2	.5867761D-2
1	PF1	1.483692040	1.070693430	.9503026713	.8367764801	.7225285653	.6496655173	.5973089088	.5246250727	.4551479055
1	PQ1	.6555968817	.6272082610	.6274833369	.6335770574	.6493692167	.6686317237	.6904834824	.7422207346	.8475271567
1	PF2	1.134214630	1.031782500	.9949833460	.9577415851	.9184259854	.8928728805	.8745391238	.8495034567	.8265037128
1	PQ2	2.297486387	2.063979981	2.066316852	2.117680208	2.247480621	2.400242375	2.567434445	2.943578992	3.653275328
1	PF3	1.044176138	1.019394598	1.010002440	1.000243986	.9896770293	.9826703935	.9775795156	.9705462266	.9640071751
1	PQ3	9.327846669	8.332895086	8.342881495	8.562215843	9.115230152	9.764085931	10.47222291	12.05956368	15.04028982
1	PFC	1.023937729	1.016470729	1.013606988	1.010612155	1,007347108	1.005169614	1.003581275	1.001378369	.9993214467
1	PQC	49.25974064	43.98999543	44.04289801	45.20475813	48.13377097	51.56974829	55.31898601	63.72128690	79.49464053
1	ZF1	1.034558271	1.034558271	1.034558271	1.034558271	1.034558271	1.034558271	1.034558271	1.034558271	1.034558271
1	ZF2	1.080201379	1.080201379	1.080201379	1.080201379	1.080201379	1.080201379	1.080201379	1.080201379	1.080201379
1	ZF3	1.301149510	1.301149510	1.301149510	1.301149510	1.301149510	1.301149510	1.301149510	1.301149510	1.301149510
1	ZF4	3.027785234	3.027785234	3.027785234	3.027785234	3.027785234	3.027785234	3.027785234	3.027785234	3.027785234

(4) P 8+0	TH = 77 SBE = 1.026304108							CAUER FUNCTIONS	
	RC = 02	RC = 05	RC = 07	RC = 10	RC = 15	RC = 20	RC = 25	RC = 35	RC = 50
AMAX DB	.0017375255	.0108709564	.0213327376	.0436480540	.0988323393	.1772876696	.2802872360	.5675287486	1.249387366
AMIN DB	13.99641107	21.81644396	24.73540776	27.84827504	31.42128277	33.99711135	36.03766234	39.24689951	43.02649711
MU / K	.1996087D 0	.8112931D-1	.5797351D-1	.4051224D-1	.2684948D-1	.1995926D-1	.1578036D-1	.1090574D-1	.7057894D-2
1 PF1	1.538463958	1.105266070	.9792897663	.8608646426	.7421087544	.6666049347	.6124640415	.5374567285	.4659180972
1 PQ1	.6727606560	.6353399001	.6340929061	.6390790521	.6540408141	.6729196716	.6945742722	.7461329889	.8516542624
1 PF2	1.137484604	1.039093599	1.003092137	.9664185490	.9274905930	.9020915559	.8838270246	.8538360690	.8358318583
1 PQ2	2.495784588	2.187331763	2.176627336	2.219243771	2.344351578	2.497049338	2.666293856	3.050141265	3.778435013
1 PF3	1.042086210	1.019376867	1.010615487	1.001449424	.9914632469	.9848114637	.9799647737	.9732513874	.9669928436
1 PQ3	10.52451020	9.168678188	9.121469156	9.309347066	9.859882285	10.53005395	11.27096658	12.94577958	16.10957758
1 PFC	1.021450815	1.014904539	1.012350575	1.009661920	1.006713406	1.004738332	1.003293702	1.001285024	.9994044915
1 PQC	57.11009537	49.73555261	49.47872712	50.50079385	53.49541755	57.14024363	61.16921670	70.27486925	87.47158006
1 ZF1	1.029877308	1.029877308	1.029877308	1.029877308	1.029877308	1.029877308	1.029877308	1.029877308	1.029877308
1 ZF2	1.071884612	1.071884612	1.071884612	1.071884612	1.071884612	1.071884612	1.071884612	1.071884612	1.071884612
1 ZF3	1.280751817	1.280751817	1.280751817	1.280751817	1.280751817	1.280751817	1.280751817	1.280751817	1.280751817
1 ZF4	2.948643682	2.948643682	2.948643682	2.948643682	2.948643682	2.948643682	2.948643682	2.948643682	2.948643682

(4) P 8+0	TH = 78 SBE = 1.022340595							CAUER FUNCTIONS	
	RC = 02	RC = 05	RC = 07	RC = 10	RC = 15	RC = 20	RC = 25	RC = 35	RC = 50
AMAX DB	.0017375255	.0108709564	.0213327376	.0436480540	.0988323393	.1772876696	.2802872360	.5675287486	1.249387366
AMIN DB	12.43611053	20.19061482	23.10318005	26.21262732	29.78380361	32.35899021	34.39924387	37.60822207	41.38768205
MU / K	.2388881D 0	.9782937D-1	.6995858D-1	.4890673D-1	.3241976D-1	.2410186D-1	.1905627D-1	.1317009D-1	.8523459D-2
1 PF1	1.597275588	1.143090651	1.010999310	.8871789431	.7634520526	.6850396980	.6289368600	.5513802050	.4775861372
1 PQ1	.6942107391	.6450242632	.6418559384	.6454607442	.6593957356	.6778005419	.6992087410	.7506445157	.8562769497
1 PF2	1.139759723	1.046103272	1.011078147	.9751225689	.9367110021	.9115353647	.8933834930	.8684887071	.8455199147
1 PQ2	2.740374213	2.334045377	2.306543889	2.337819271	2.456533021	2.608621131	2.779863747	3.172051574	3.921106742
1 PF3	1.039778468	1.019205478	1.011101456	1.002555826	.9931809785	.9869042818	.9823164226	.9759430384	.9699835088
1 PQ3	12.03682945	10.18852374	10.06296968	10.20574758	10.74693755	11.43872702	12.21590551	13.99051043	17.36633276
1 PFC	1.019013989	1.013357932	1.011106608	1.008718463	1.006082019	1.004307236	1.003005138	1.001189533	.9994847668
1 PQC	67.28679544	56.93576674	56.23249256	57.03224221	60.06341135	63.93761657	68.28950696	78.22501667	97.12131552
1 ZF1	1.025549716	1.025549716	1.025549716	1.025549716	1.025549716	1.025549716	1.025549716	1.025549716	1.025549716
1 ZF2	1.063950369	1.063950369	1.063950369	1.063950369	1.063950369	1.063950369	1.063950369	1.063950369	1.063950369
1 ZF3	1.260612590	1.260612590	1.260612590	1.260612590	1.260612590	1.260612590	1.260612590	1.260612590	1.260612590
1 ZF4	2.868888847	2.868888847	2.868888847	2.868888847	2.868888847	2.868888847	2.868888847	2.868888847	2.868888847

(4) P 8+0 — TH = 79 — SBE = 1.018716695 — CAUER FUNCTIONS

(4) P 8+0	RC = 02	RC = 05	RC = 07	RC = 10	RC = 15	RC = 20	RC = 25	RC = 35	RC = 50
AMAX DB	.0017375255	.0108709564	.0213327376	.0436480540	.0988323393	.1772876696	.2802872360	.5675287486	1.249387366
AMIN DB	10.87100667	18.52954431	21.43254530	24.53686609	28.10529295	30.67951509	32.71932197	35.92791093	39.70716406
MU I K	.2860551D 0	.1184467D 0	.8479549D-1	.5931393D-1	.3933103D-1	.2924316D-1	.2312245D-1	.1598102D-1	.1034289D-1
1 I PF1	1.659885917	1.184626073	1.045857195	.9160833673	.7868509461	.7052177019	.6469449960	.5665743502	.4902976432
1 I PQ1	.7215530198	.6567450766	.6511074453	.6529604616	.6656063568	.6834173003	.7045138736	.7557170492	.8615019300
1 I PF2	1.140843060	1.052704009	1.018864902	.9838052848	.9460647495	.9211966154	.9032104623	.8784751743	.8555913377
1 I PQ2	3.048052543	2.511325314	2.461799167	2.478139703	2.588084763	2.738761010	2.911859057	3.313087087	4.085504092
1 I PF3	1.037239889	1.018864292	1.011444874	1.003548892	.9948176472	.9889376328	.9846243341	.9786126732	.9729722835
1 I PQ3	13.99212019	11.45620761	11.22130811	11.29883110	11.81991725	12.53274115	13.35005852	15.23949868	18.86375382
1 I PFC	1.016636418	1.011835721	1.009878730	1.007784353	1.005454492	1.003877265	1.002716119	1.001091916	.9995618590
1 I PQC	80.81995210	66.15207471	64.79314702	65.24163465	68.25608701	72.37936152	77.10662624	88.03351083	108.9894072
1 I ZF1	1.021569693	1.021569693	1.021569693	1.021569693	1.021569693	1.021569693	1.021569693	1.021569693	1.021569693
1 I ZF2	1.056393427	1.056393427	1.056393427	1.056393427	1.056393427	1.056393427	1.056393427	1.056393427	1.056393427
1 I ZF3	1.240690862	1.240690862	1.240690862	1.240690862	1.240690862	1.240690862	1.240690862	1.240690862	1.240690862
1 I ZF4	2.788197113	2.788197113	2.788197113	2.788197113	2.788197113	2.788197113	2.788197113	2.788197113	2.788197113

(4) P 8+0 — TH = 80 — SBE = 1.015426612 — CAUER FUNCTIONS

(4) P 8+0	RC = 02	RC = 05	RC = 07	RC = 10	RC = 15	RC = 20	RC = 25	RC = 35	RC = 50
AMAX DB	.0017375255	.0108709564	.0213327376	.0436480540	.0988323393	.1772876696	.2802872360	.5675287486	1.249387366
AMIN DB	9.310582635	16.82821884	19.71674878	22.81327990	26.37751949	28.95027120	30.98939663	34.19739177	37.97632927
MU I K	.3423504D 0	.1440751D 0	.1033148D 0	.7233292D-1	.4798705D-1	.3568506D-1	.2821826D-1	.1950430D-1	.1262361D-1
1 I PF1	1.725527580	1.230387088	1.084375630	.9480295688	.8126739848	.7274528653	.6667646158	.5832664384	.5042377138
1 I PQ1	.7571888857	.6712032979	.6623250780	.6619125225	.6729105449	.6899652448	.7106619326	.7615507866	.8674715073
1 I PF2	1.140506239	1.058759734	1.026353659	.9924024945	.9555199185	.9310619988	.9133068264	.8888095052	.8660725579
1 I PQ2	3.444060415	2.729619035	2.650593717	2.646890184	2.744672693	2.892739634	3.067403963	3.478426852	4.277375337
1 I PF3	1.034458701	1.018335240	1.011628146	1.004412176	.9963586239	.9908983752	.9868765724	.9812501792	.9759508978
1 I PQ3	16.58883286	13.06719806	12.67613427	12.65779720	13.14164875	13.87326113	14.73488533	16.75775707	20.67712225
1 I PFC	1.014329014	1.010343501	1.008671184	1.006862597	1.004832650	1.003449540	1.002427289	1.000992217	.9996353018
1 I PQC	99.36712911	78.25137034	75.90614151	75.79617060	78.69784318	83.08506388	88.25154663	100.3798753	123.8754929
1 I ZF1	1.017932203	1.017932203	1.017932203	1.017932203	1.017932203	1.017932203	1.017932203	1.017932203	1.017932203
1 I ZF2	1.049210112	1.049210112	1.049210112	1.049210112	1.049210112	1.049210112	1.049210112	1.049210112	1.049210112
1 I ZF3	1.220941391	1.220941391	1.220941391	1.220941391	1.220941391	1.220941391	1.220941391	1.220941391	1.220941391
1 I ZF4	2.706179283	2.706179283	2.706179283	2.706179283	2.706179283	2.706179283	2.706179283	2.706179283	2.706179283

(4) P 8+0	TH = 81 SBE = 1.012465126								CAUER FUNCTIONS
	RC = 02	RC = 05	RC = 07	RC = 10	RC = 15	RC = 20	RC = 25	RC = 35	RC = 50
AMAX DB	.0017375255	.0108709564	.0213327376	.0436480540	.0988323393	.1772876696	.2802872360	.5675287486	1.249387366
AMIN DB	7.770725693	15.08160154	17.94793467	21.03243948	24.59019166	27.16066144	29.19872866	32.40580133	36.18424838
MU I K	.4087556D 0	.1761651D 0	.1266494D 0	.8879286D-1	.5895090D-1	.4384973D-1	.3467876D-1	.2397231D-1	.1551628D-1
1 I PF1	1.792554984	1.280919742	1.127165983	.9835830035	.8413932131	.7521502775	.6887532498	.6017514337	.5196467866
1 I PQ1	.8047964270	.6894489210	.6762125551	.6728010034	.6816466161	.6977190714	.7178933551	.7683532590	.8743805697
1 I PF2	1.138490220	1.064096389	1.033415367	1.000828028	.9650311612	.9411099759	.9236667720	.8995055299	.8769936053
1 I PQ2	3.967578879	3.004623704	2.885069431	2.853831266	2.934459243	3.078090483	3.253784160	3.675385561	4.504793320
1 I PF3	1.031426418	1.017598087	1.011631152	1.005126595	.9977867224	.9927709473	.9890589204	.9838434047	.9789093183
1 I PQ3	20.14860411	15.17014991	14.54965877	14.38740921	14.80608118	15.55110456	16.46121708	18.64080755	22.91650764
1 I PFC	1.012105048	1.008887871	1.007488980	1.005956756	1.004218669	1.003025410	1.002139440	1.000890510	.9997045653
1 I PQC	125.7151474	94.63088190	90.75609048	89.74285751	92.35739756	97.00975121	102.6927136	116.3015289	142.9949392
1 I ZF1	1.014632987	1.014632987	1.014632987	1.014632987	1.014632987	1.014632987	1.014632987	1.014632987	1.014632987
1 I ZF2	1.042398542	1.042398542	1.042398542	1.042398542	1.042398542	1.042398542	1.042398542	1.042398542	1.042398542
1 I ZF3	1.201312874	1.201312874	1.201312874	1.201312874	1.201312874	1.201312874	1.201312874	1.201312874	1.201312874
1 I ZF4	2.622354687	2.622354687	2.622354687	2.622354687	2.622354687	2.622354687	2.622354687	2.622354687	2.622354687

(4) P 8+0	TH = 82 SBE = 1.009827573								CAUER FUNCTIONS
	RC = 02	RC = 05	RC = 07	RC = 10	RC = 15	RC = 20	RC = 25	RC = 35	RC = 50
AMAX DB		.0108709564	.0213327376	.0436480540	.0988323393	.1772876696	.2802872360	.5675287486	1.249387366
AMIN DB		13.28531897	16.11708856	19.18267665	22.73016832	25.29701957	27.33340708	30.53901464	34.31668240
MU I K		.2166377D 0	.1563672D 0	.1098667D 0	.7302837D-1	.5434368D-1	.4298626D-1	.2972003D-1	.1923826D-1
1 I PF1		1.336730315	1.174941722	1.023456164	.8736242828	.7798442343	.7133848179	.6224218962	.5368454095
1 I PQ1		.7131169173	.6938478159	.6863539602	.6923135908	.7070790072	.7265554689	.7764208969	.8825047423
1 I PF2		1.068488755	1.039878834	1.008964491	.9745335562	.9513066702	.9342770958	.9105760597	.8883889186
1 I PQ2		3.360944467	3.183964772	3.113795956	3.169663869	3.305991517	3.481745780	3.914670349	4.779500147
1 I PF3		1.016630290	1.011430752	1.005669793	.9990815102	.9945366875	.9911542227	.9863775536	.9818352105
1 I PQ3		18.00777521	17.03822385	16.65346401	16.95982431	17.70685362	18.66891096	21.03475282	25.74932899
1 I PFC		1.007476770	1.006338126	1.005071114	1.003615186	1.002606525	1.001853555	1.000786923	.9997690431
1 I PQC		117.6624019	111.3224658	108.8064239	110.8097953	115.6946928	121.9854453	137.4544353	168.2779218
1 I ZF1		1.011668599	1.011668599	1.011668599	1.011668599	1.011668599	1.011668599	1.011668599	1.011668599
1 I ZF2		1.035958965	1.035958965	1.035958965	1.035958965	1.035958965	1.035958965	1.035958965	1.035958965
1 I ZF3		1.181745323	1.181745323	1.181745323	1.181745323	1.181745323	1.181745323	1.181745323	1.181745323
1 I ZF4		2.536111391	2.536111391	2.536111391	2.536111391	2.536111391	2.536111391	2.536111391	2.536111391

(4) P 8+0	TH = 83	SBE = 1.007509825						CAUER FUNCTIONS	
	RC = 02	RC = 05	RC = 07	RC = 10	RC = 15	RC = 20	RC = 25	RC = 35	RC = 50
AMAX DB		.0108709564	.0213327376	.0436480540	.0988323393	.1772876696	.2802872360	.5675287486	1.249387366
AMIN DB		11.43714447	14.21423126	17.24940993	20.78026198	23.34122035	25.37486802	28.57308351	32.35447805
MU I K		.2680049D 0	.1946653D 0	.1372554D 0	.9140857D-1	.6806737D-1	.5385879D-1	.3724739D-1	.2411438D-1
1 I PF1		1.398110709	1.228489089	1.068546970	.9101861195	.8112577032	.7413054914	.6458164658	.5562747366
1 I PQ1		.7448721603	.7169612903	.7037182940	.7056811481	.7186539947	.7371716940	.7861947877	.8922495849
1 I PF2		1.071642051	1.045512894	1.016648783	.9839327456	.9615991451	.9451127114	.9220313533	.9002984046
1 I PQ2		3.839182642	3.577705657	3.450465556	3.469467767	3.593811818	3.767873469	4.212665332	5.119330005
1 I PF3		1.015407163	1.011000268	1.006015299	1.000218347	.9961728681	.9931414427	.9888343097	.9847131580
1 I PQ3		22.00003460	20.47567000	19.73313068	19.84405719	20.56962372	21.58450324	24.17442951	29.44296314
1 I PFC		1.00[illegible]	1.005225981	1.004210922	1.003025456	1.002194944	1.001570881	1.000681652	.9998280351
1 I PQC		1[illegible]	141.1247848	136.0039237	136.7689266	141.7727209	148.7715629	166.6315542	202.9608018
1 I ZF1		1[illegible]	1.009036462	1.009036462	1.009036462	1.009036462	1.009036462	1.009036462	1.009036462
1 I ZF2		1[illegible]	1.029894294	1.029894294	1.029894294	1.029894294	1.029894294	1.029894294	1.029894294
1 I ZF3		1[illegible]	1.162166004	1.162166004	1.162166004	1.162166004	1.162166004	1.162166004	1.162166004
1 I ZF4		2[illegible]	2.446642165	2.446642165	2.446642165	2.446642165	2.446642165	2.446642165	2.446642165

(4) P 8+0	TH = 84	SBE = 1.005508280						CAUER FUNCTIONS	
	RC = 02	RC = 05	RC = 07	RC = 10	RC = 15	RC = 20	RC = 25	RC = 35	RC = 50
AMAX DB		.0108709564	.0213327376	.0436480540	.0988323393	.1772876696	.2802872360	.5675287486	1.249387366
AMIN DB		9.540051956	12.22928563	15.21433734	18.71736298	21.26839804	23.29741963	26.49658979	30.27082871
MU I K		.3334244D 0	.2446446D 0	.1734935D 0	.1159129D 0	.8641320D-1	.6841149D-1	.4733371D-1	.3065198D-1
1 I PF1		1.464738900	1.288546746	1.119969872	.9521904999	.8473988165	.7734270130	.6727027395	.5785665648
1 I PQ1		.7893015101	.7484951625	.7268079347	.7230049282	.7334225142	.7505740700	.7983660605	.9042424755
1 I PF2		1.073166704	1.049998877	1.023648496	.9930883121	.9719038331	.9561287417	.9338761345	.9127688234
1 I PQ2		4.510730130	4.118941684	3.904076129	3.865842829	3.970165884	4.139274142	4.595376963	5.552894352
1 I PF3		1.013902844	1.010309075	1.006131436	1.001167008	.9976512712	.9949942577	.9911905157	.9875234754
1 I PQ3		27.92829407	25.47172112	24.12308910	23.88299186	24.53802972	25.59928474	28.46180995	34.45164550
1 I PFC		1.004829825	1.004161808	1.003382785	1.002453597	1.001793297	1.001293034	1.000575010	.9998807255
1 I PQC		204.9678384	186.9329190	177.0316736	175.2689295	180.0780645	187.8694415	208.8845719	252.8568405
1 I ZF1		1.006734965	1.006734965	1.006734965	1.006734965	1.006734965	1.006734965	1.006734965	1.006734965
1 I ZF2		1.024210938	1.024210938	1.024210938	1.024210938	1.024210938	1.024210938	1.024210938	1.024210938
1 I ZF3		1.142482851	1.142482851	1.142482851	1.142482851	1.142482851	1.142482851	1.142482851	1.142482851
1 I ZF4		2.352835354	2.352835354	2.352835354	2.352835354	2.352835354	2.352835354	2.352835354	2.352835354

$P^{(5)}_{8+2}$

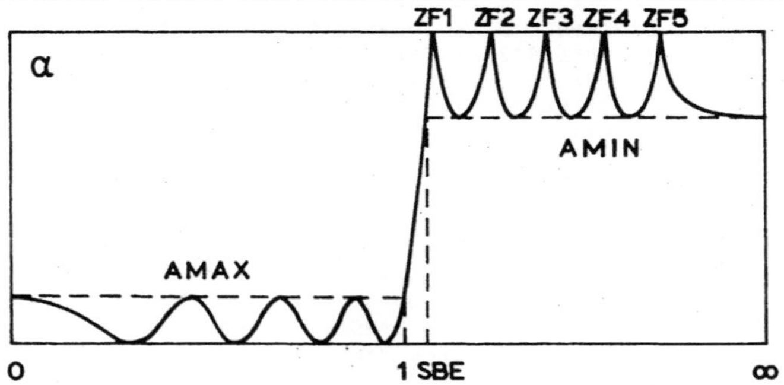

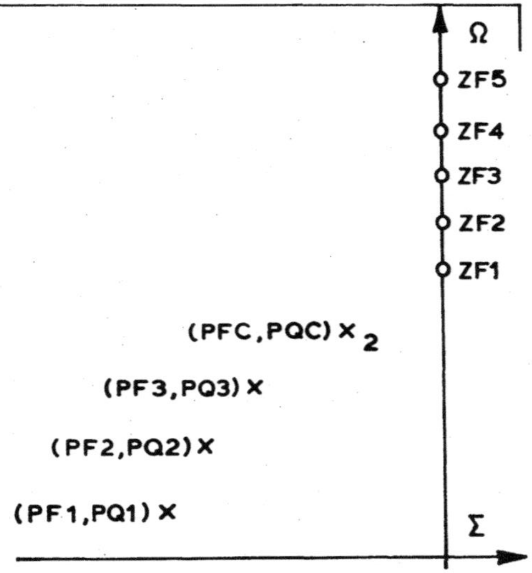

(5) P 8+2		TH = 40	SBE = 1.555723827						MCPER FUNCTIONS	
		RC = 02	RC = 05	RC = 07	RC = 10	RC = 15	RC = 20	RC = 25	RC = 35	RC = 50
	AMAX DB	.0017375255	.0108709564	.0213327376	.0436480540	.0988323393	.1772876696	.2802872360	.5675287486	1.249387366
	AMIN DB	88.41594114	98.42047796	102.1117930	106.0446211	110.5602025	113.8174021	116.3978449	120.4510834	125.1968382
MU	K	.3794923D-4	.1199433D-4	.7841702D-5	.4986191D-5	.2964762D-5	.2037651D-5	.1513937D-5	.9493926D-6	.5497410D-6
1	PF1	.6868711709	.5396294661	.4903514917	.4407236342	.3875037426	.3518091287	.3253209970	.2874036018	.2499184945
1	PQ1	.5397107292	.5540349915	.5625926001	.5749470463	.5953906629	.6165429414	.6390271728	.6900155188	.7908381673
1	PF2	.8534595165	.7463226094	.7118533545	.6781434493	.6433351304	.6208756178	.6047029685	.5822837908	.5608291630
1	PQ2	.8636800798	.9623503546	1.017930783	1.094564514	1.213925693	1.329937434	1.446934639	1.695037756	2.142346995
1	PF3	1.023310092	.9496917897	.9257751319	.9022541051	.8777598205	.8617671305	.8500991711	.8335730496	.8170592127
1	PQ3	1.666301629	1.894010044	2.020639283	2.193234316	2.458307532	2.712687560	2.966914448	3.501008269	4.454805478
2	PFC	1.173159499	1.116233183	1.096943269	1.077547605	1.056866962	1.043084477	1.032889352	1.018267123	1.003522151
2	PQC	4.864373596	5.223987194	5.436987579	5.727001039	6.162614755	6.565660149	6.953206420	7.720974917	8.960444465
1	ZF1	1.568953718	1.569305818	1.569420046	1.569533741	1.569655093	1.569737054	1.569798862	1.569890452	1.569988630
1	ZF2	1.688256844	1.691583868	1.692662936	1.693737346	1.694884788	1.695660283	1.696245393	1.697112973	1.698043649
1	ZF3	2.022721324	2.032887923	2.036206215	2.039520403	2.043070998	2.045477016	2.047295633	2.049997161	2.052901302
1	ZF4	2.986826264	3.011607725	3.019779407	3.027979458	3.036805034	3.042808106	3.047356954	3.054131033	3.061433616
1	ZF5	8.374136166	8.461395851	8.490372241	8.519543781	8.551040816	8.572520512	8.588824922	8.613146898	8.639417047

(5) P 8+2 — TH = 41 SBE = 1.524253087 — MCPER FUNCTIONS

MU	K	RC = 02	RC = 05	RC = 07	RC = 10	RC = 15	RC = 20	RC = 25	RC = 35	RC = 50
AMAX DB		.0017375255	.0108709564	.0213327376	.0436480540	.0988323393	.1772876696	.2802872360	.5675287486	1.249387366
AMIN DB		86.06293462	96.06610437	99.75705536	103.6895592	108.2048361	111.4618542	114.0421730	118.0952480	122.8408543
	K	.4975689D-4	.1572877D-4	.1028365D-4	.6539161D-5	.3888286D-5	.2672436D-5	.1985598D-5	.1245196D-5	.7210366D-6
1	PF1	.6932840868	.5441089222	.4942770025	.4441325468	.3904003827	.3543840267	.3276674308	.2894368871	.2516566471
1	PQ1	.5401475758	.5544038906	.5629446714	.5752856813	.5957199030	.6168700654	.6393560822	.6903555561	.7912110191
1	PF2	.8584918563	.7505064763	.7157811961	.6818299134	.6467825192	.6241748079	.6078987502	.5853415298	.5637620337
1	PQ2	.8692755619	.9675833598	1.023163021	1.099887694	1.219497041	1.335812367	1.453148632	1.702027289	2.150819483
1	PF3	1.024843321	.9515473827	.9277075375	.9042511220	.8798145172	.8638554657	.8522107850	.8357172119	.8192380618
1	PQ3	1.690970925	1.918109339	2.045112588	2.218534428	2.485249348	2.741426620	2.997568623	3.535892261	4.497552697
2	PFC	1.169836678	1.114254526	1.095363965	1.076343670	1.056035834	1.042486791	1.032456899	1.018060932	1.003531702
2	PQC	4.987821107	5.337906245	5.549689684	5.839837538	6.277650666	6.683890026	7.075126471	7.851279856	9.105931494
1	ZF1	1.536930740	1.537288537	1.537404714	1.537520389	1.537643894	1.537727331	1.537790261	1.537883532	1.537983530
1	ZF2	1.651412009	1.654799969	1.655899260	1.656994030	1.658163462	1.658953966	1.659550481	1.660435101	1.661384235
1	ZF3	1.973461102	1.983820195	1.987203648	1.990584053	1.994206852	1.996662538	1.998519095	2.001277615	2.004243865
1	ZF4	2.905615239	2.930843285	2.939173544	2.947538068	2.956546433	2.962677129	2.967324396	2.974247684	2.981714505
1	ZF5	8.130343055	8.219105647	8.248633608	8.278385341	8.310535690	8.332476199	8.349138303	8.374005972	8.400880896

(5) P 8+2 — TH = 42 SBE = 1.494476550 — MCPER FUNCTIONS

MU	K	RC = 02	RC = 05	RC = 07	RC = 10	RC = 15	RC = 20	RC = 25	RC = 35	RC = 50
AMAX DB		.0017375255	.0108709564	.0213327376	.0436480540	.0988323393	.1772876696	.2802872360	.5675287486	1.249387366
AMIN DB		83.75462246	93.75629905	97.44685233	101.3790019	105.8939461	109.1507659	111.7309492	115.7838455	120.5292896
	K	.6490361D-4	.2052036D-4	.1341706D-4	.8531981D-5	.5073442D-5	.3487078D-5	.2590911D-5	.1624329D-5	.9408828D-6
1	PF1	.6999886986	.5487807821	.4983682233	.4476830213	.3934153673	.3570630621	.3301080891	.2915510424	.2534633215
1	PQ1	.5406086259	.5547911938	.5633137469	.5756401738	.5960640909	.6172117567	.6396994408	.6907102612	.7915996875
1	PF2	.8636909762	.7548339360	.7198448570	.6856446027	.6503503875	.6275895764	.6112066496	.5885067186	.5667980891
1	PQ2	.8751895061	.9730893458	1.028660871	1.105474128	1.225336720	1.341965549	1.459653448	1.709338655	2.159676142
1	PF3	1.026388723	.9534399778	.9296831357	.9062966419	.8819225106	.8659998704	.8543803388	.8379217857	.8214796630
1	PQ3	1.717139365	1.943575974	2.070944733	2.245210118	2.513624769	2.771673790	3.029815285	3.572563027	4.542459071
2	PFC	1.166447313	1.112231390	1.093747860	1.075110635	1.055183730	1.041873462	1.032012695	1.017348478	1.003540397
2	PQC	5.119802424	5.459177486	5.669509126	5.959650617	6.399645060	6.809170327	7.204241787	7.989162449	9.259755590
1	ZF1	1.506622955	1.506986101	1.507104129	1.507221693	1.507347259	1.507432112	1.507496122	1.507591011	1.507692768
1	ZF2	1.616463990	1.619910750	1.621029612	1.622144111	1.623334885	1.624139974	1.624747580	1.625548787	1.626615912
1	ZF3	1.926573685	1.937119766	1.940566768	1.944011874	1.947705351	1.950209732	1.952103525	1.954918043	1.957945402
1	ZF4	2.828065890	2.853725297	2.862210050	2.870735397	2.879923108	2.886179363	2.890923633	2.897994241	2.905623558
1	ZF5	7.897104728	7.987312843	8.017377354	8.047696152	8.080487828	8.102882243	8.119897461	8.145305018	8.172779761

(5) P 8+2 — TH = 43 SBE = 1.466279186 — MCPER FUNCTIONS

MU	K	RC = 02	RC = 05	RC = 07	RC = 10	RC = 15	RC = 20	RC = 25	RC = 35	RC = 50
	AMAX DB	.0017375255	.0108709564	.0213327376	.0436480540	.0988323393	.1772876696	.2802872360	.5675287486	1.249387366
	AMIN DB	81.48839458	91.48844221	95.17856154	99.11032454	103.6249057	106.8815091	109.4615445	113.5142459	118.2595128
	K	.8425201D-4	.2664268D-4	.1742095D-4	.1107857D-4	.6588017D-5	.4528189D-5	.3364517D-5	.2110026D-5	.1221868D-5
1	PF1	.7069996603	.5536537285	.5026323780	.4513810249	.3965535158	.3598503830	.3326466665	.2937491711	.2553410944
1	PQ1	.5410955693	.5551980057	.5637007970	.5760113844	.5964239967	.6175687408	.6400579510	.6910803234	.7920048922
1	PF2	.8690597795	.7593083164	.7240476541	.6895907758	.6540418962	.6311230059	.6146296871	.5917822857	.5699401668
1	PQ2	.8814443709	.9788854037	1.034440320	1.111339054	1.231459642	1.348412038	1.466464525	1.716988485	2.168936315
1	PF3	1.027943467	.9553683879	.9317011738	.9083902934	.8840837795	.8682005271	.8566081511	.8401872660	.8237846678
1	PQ3	1.744919912	1.970504322	2.098226857	2.273351074	2.543524535	2.803522681	3.063751882	3.611127702	4.589651448
2	PFC	1.162992901	1.110164275	1.092095245	1.073848633	1.054310665	1.041244452	1.031556676	1.017629674	1.003548142
2	PQC	5.261066089	5.588395045	5.797006015	6.086975153	6.529117019	6.942017590	7.341072573	8.135160256	9.422496965
1	ZF1	1.477913940	1.478282062	1.478401837	1.478521193	1.478648723	1.478734929	1.478799974	1.478896416	1.478999863
1	ZF2	1.583283763	1.586787074	1.587924817	1.589058382	1.590269812	1.591089036	1.591707402	1.592624719	1.593609339
1	ZF3	1.881895964	1.892623247	1.896132103	1.899640311	1.903402859	1.905954909	1.907885196	1.910754661	1.913842074
1	ZF4	2.753922624	2.779997449	2.788632410	2.797314746	2.806678180	2.813057822	2.817897599	2.825113536	2.832903505
1	ZF5	7.673674836	7.765268924	7.795854336	7.826726488	7.860146949	7.882988026	7.900351549	7.926292386	7.954362220

(5) P 8+2 — TH = 44 SBE = 1.439556540 — MCPER FUNCTIONS

MU	K	RC = 02	RC = 05	RC = 07	RC = 10	RC = 15	RC = 20	RC = 25	RC = 35	RC = 50
	AMAX DB	.0017375255	.0108709564	.0213327376	.0436480540	.0988323393	.1772876696	.2802872360	.5675287486	1.249387366
	AMIN DB	79.26179772	89.26007023	92.94971659	96.88105818	101.3952435	104.6516111	107.2314853	111.2839740	116.0290478
	K	.1088705D-3	.3443471D-4	.2251719D-4	.1432013D-4	.8516043D-5	.5853552D-5	.4349364D-5	.2727729D-5	.1579602D-5
1	PF1	.7143328290	.5587371107	.5070772333	.4552329635	.3998199945	.3627504329	.3352871181	.2960345925	.2572927197
1	PQ1	.5416102724	.5556255355	.5641068811	.5764002507	.5968004571	.6179418046	.6404323736	.6914664884	.7924274112
1	PF2	.8746011479	.7639330602	.7283930403	.6936718411	.6578603636	.6347783381	.6181710426	.5951713188	.5731912616
1	PQ2	.8880649189	.9849902436	1.040518833	1.117499087	1.237882030	1.355168193	1.473598621	1.724994812	2.178620952
1	PF3	1.029504520	.9573313145	.9337608147	.9105316456	.8862982671	.8704575968	.8588945293	.8425141488	.8261537395
1	PQ3	1.774437502	1.998998139	2.127058978	2.303055553	2.575047884	2.837075564	3.099484790	3.651703109	4.639267997
2	PFC	1.159475027	1.108053723	1.090406438	1.072557814	1.053416662	1.040599728	1.031088777	1.017404430	1.003554842
2	PQC	5.412449472	5.726219055	5.932801418	6.222403084	6.666639870	7.083001618	7.486192154	8.289864975	9.594793469
1	ZF1	1.450697986	1.451070686	1.451192096	1.451313139	1.451442529	1.451530022	1.451596053	1.451693978	1.451799045
1	ZF2	1.551754195	1.555311690	1.556467589	1.557619518	1.558850879	1.559683763	1.560312539	1.561245459	1.562247049
1	ZF3	1.839279819	1.850182215	1.853751141	1.857320769	1.861150694	1.863749328	1.865715325	1.868639628	1.871784977
1	ZF4	2.682953104	2.709426653	2.718207325	2.727042619	2.736577965	2.743078702	2.748012409	2.755371573	2.763320242
1	ZF5	7.459374129	7.552292441	7.583382451	7.614793648	7.648829772	7.672109918	7.689816695	7.716285384	7.744943792

(5) P 8+2 — TH = 45 SBE = 1.414213562 — MCPER FUNCTIONS

MU	K	RC = 02	RC = 05	RC = 07	RC = 10	RC = 15	RC = 20	RC = 25	RC = 35	RC = 50
	AMAX DB	.0017375255	.0108709564	.0213327376	.0436480540	.0988323393	.1772876696	.2802872360	.5675287486	1.249387366
	AMIN DB	77.07251956	87.06885962	90.75799098	94.68887370	99.20262806	102.4587387	105.0384373	109.0906945	113.8355580
	K	.1400793D-3	.4431564D-4	.2898014D-4	.1843131D-4	.1096146D-4	.7534650D-5	.5598583D-5	.3511278D-5	.2033397D-5
1	PF1	.7220053860	.5640410081	.5117111500	.4592457231	.4032203489	.3657679763	.3380336825	.2984108606	.2593211432
1	PQ1	.5421548015	.5560751091	.5645331576	.5768077949	.5971943820	.6183318026	.6408235341	.6918695641	.7928680860
1	PF2	.8803179177	.7687117234	.7328846075	.6978913620	.6618092726	.6385589827	.6218340649	.5986770738	.5765545353
1	PQ2	.8950785080	.9914243845	1.046915520	1.123972365	1.244621562	1.362251822	1.481073959	1.733377218	2.188752776
1	PF3	1.031068634	.9593273377	.9358611291	.9127202022	.8885658764	.8727712159	.8612397661	.8449029292	.8285875531
1	PQ3	1.805830611	2.029171717	2.157551070	2.334431402	2.608303546	2.872444362	3.137130331	3.694416850	4.691459466
2	PFC	1.155895368	1.105900313	1.088681785	1.071238343	1.052501752	1.039939260	1.030608937	1.017172657	1.003560396
2	PQC	5.574891705	5.873384663	6.077585516	6.366590891	6.812848161	7.232752234	7.640233653	8.453929162	9.777347641
1	ZF1	1.424878912	1.425255762	1.425378688	1.425501308	1.425632447	1.425721156	1.425788121	1.425887456	1.425994066
1	ZF2	1.521768733	1.525377920	1.526551209	1.527720761	1.528971288	1.529817330	1.530456146	1.531404135	1.532422138
1	ZF3	1.798590458	1.809661563	1.813288681	1.816917954	1.820813469	1.823457544	1.825458422	1.828434391	1.831638492
1	ZF4	2.614945670	2.641800457	2.650722117	2.659706134	2.669409381	2.676028800	2.681054772	2.688554939	2.696660242
1	ZF5	7.253582925	7.347761469	7.379339109	7.411274419	7.445912483	7.469623739	7.487668468	7.514657748	7.543899406

(5) P 8+2 — TH = 46 SBE = 1.390163591 — MCPER FUNCTIONS

MU	K	RC = 02	RC = 05	RC = 07	RC = 10	RC = 15	RC = 20	RC = 25	RC = 35	RC = 50
	AMAX DB	.0017375255	.0108709564	.0213327376	.0436480540	.0988323393	.1772876696	.2802872360	.5675287486	1.249387366
	AMIN DB	74.91837433	84.91261261	88.60118372	92.53156720	97.04485260	100.3006836	102.8801912	106.9321964	111.6768310
	K	.1795070D-3	.5680275D-4	.3714846D-4	.2362771D-4	.1405262D-4	.9659748D-5	.7177785D-5	.4501841D-5	.2607105D-5
1	PF1	.7300359753	.5695763014	.5165431388	.4634267139	.4067605377	.3689081282	.3408909066	.3008817849	.2614295203
1	PQ1	.5427314493	.5565481831	.5649808955	.5772351339	.5976067628	.6187396646	.6412323293	.6922904267	.7933278280
1	PF2	.8862128512	.7736479732	.7375260889	.7022530635	.6658922797	.6424685270	.6256222822	.6023029866	.5800333281
1	PQ2	.9025154282	.9982103711	1.053651329	1.130778731	1.251697535	1.369682331	1.488910381	1.742156998	2.199356467
1	PF3	1.032632329	.9613549062	.9380010868	.9149553941	.8908864656	.8751414919	.8636441369	.8473540996	.8310867937
1	PQ3	1.839253074	2.061151213	2.189824287	2.367597216	2.643410867	2.909751783	3.176815928	3.739408538	4.746390605
2	PFC	1.152255696	1.103704671	1.086921662	1.069890409	1.051565979	1.039263023	1.030117094	1.016934260	1.003564696
2	PQC	5.749448873	6.030712525	6.232127070	6.520268274	6.968445724	7.391967092	7.803897698	8.628073963	9.970934826
1	ZF1	1.400369033	1.400749574	1.400873890	1.400997969	1.401130738	1.401220589	1.401288432	1.401389099	1.401497172
1	ZF2	1.493230256	1.496888518	1.498078389	1.499264783	1.500533667	1.501392336	1.502040800	1.503003291	1.504037121
1	ZF3	1.759704991	1.770938058	1.774621387	1.778308435	1.782267655	1.784955963	1.786990846	1.790018240	1.793278841
1	ZF4	2.549707090	2.576924800	2.585982492	2.595110778	2.604977701	2.611713257	2.616829737	2.624468557	2.632723306
1	ZF5	7.055734554	7.151107061	7.183154670	7.215598521	7.250824177	7.274958201	7.293335314	7.320838073	7.350656837

(5) P 8+2	TH = 47	SBE = 1.367327461						MCPER FUNCTIONS	
	RC = 02	RC = 05	RC = 07	RC = 10	RC = 15	RC = 20	RC = 25	RC = 35	RC = 50
AMAX DB	.0017375255	.0108709564	.0213327376	.0436480540	.0988323393	.1772876696	.2802872360	.5675287486	1.249387366
AMIN DB	72.79728986	82.78924407	86.47720622	90.40704699	94.91982254	98.17534962	100.7546495	104.8063807	109.5507663
MU I K	.2291583D-3	.7253336D-4	.4743945D-4	.3017503D-4	.1794770D-4	.1233765D-4	.9167851D-5	.5759174D-5	.3330134D-5
1 I PF1	.7384448579	.5753547528	.5215829247	.4677839210	.4104469716	.3721763866	.3438636746	.3034514532	.2636212338
1 I PQ1	.5433427652	.5570463611	.5654514874	.5776834898	.5980386817	.6191664031	.6416597350	.6927300282	.7938076260
1 I PF2	.8922886038	.7787455844	.7423213614	.7067608372	.6701132235	.6465107459	.6295394122	.6060526839	.5836311708
1 I PQ2	.9104092916	1.005373023	1.060749268	1.137939930	1.259131044	1.377480910	1.497129530	1.751357348	2.210458875
1 I PF3	1.034191883	.9634123262	.9401795472	.9172365723	.8932598417	.8775684994	.8661078962	.8498681472	.8336521558
1 I PQ3	1.874876206	2.095076183	2.224012385	2.402683674	2.680501105	2.949132614	3.218681428	3.786331214	4.804241795
2 I PFC	1.148557887	1.101467471	1.085126480	1.068514217	1.050609394	1.038570997	1.029613191	1.016689146	1.003567634
2 I PQC	5.937311939	6.199121068	6.397284473	6.684248248	7.134215034	7.561420710	7.977961325	8.813098048	10.17641253
1 I ZF1	1.377088261	1.377472004	1.377597573	1.377722985	1.377857261	1.377948172	1.378016837	1.378118752	1.378228202
1 I ZF2	1.466050086	1.469754673	1.470960275	1.472162686	1.473449073	1.474319810	1.474977508	1.475953902	1.477002936
1 I ZF3	1.722511169	1.733899089	1.737636543	1.741379392	1.745400325	1.748131590	1.750199552	1.753277058	1.756592832
1 I ZF4	2.487060603	2.514622054	2.523810570	2.533078442	2.543104591	2.549953598	2.555158732	2.562933720	2.571345595
1 I ZF5	6.865309622	6.961807510	6.994306719	7.027242873	7.063041121	7.087589173	7.106292828	7.134301582	7.164690968

(5) P 8+2	TH = 48	SBE = 1.345632730						MCPER FUNCTIONS	
	RC = 02	RC = 05	RC = 07	RC = 10	RC = 15	RC = 20	RC = 25	RC = 35	RC = 50
AMAX DB	.0017375255	.0108709564	.0213327376	.0436480540	.0988323393	.1772876696	.2802872360	.5675287486	1.249387366
AMIN DB	70.70729569	80.69676960	84.38407030	88.31332152	92.82554319	96.08074001	98.65981412	102.7112476	107.4553627
MU I K	.2914978D-3	.9229146D-4	.6036657D-4	.3840024D-4	.2284141D-4	.1570229D-4	.1166835D-4	.7318762D-5	.4238692D-5
1 I PF1	.7472540882	.5813890960	.5268410183	.4723259609	.4142865577	.3755786687	.3469572395	.3061242583	.2658999156
1 I PQ1	.5439915922	.5575714117	.5659464647	.5781542027	.5984913215	.6196131235	.6421068150	.6931894045	.7943085544
1 I PF2	.8985476837	.7840084340	.7472744461	.7114187482	.6744761338	.6506896129	.6335893746	.6099299961	.5873517976
1 I PQ2	.9187974859	1.012939722	1.068234657	1.145479836	1.266945201	1.385670737	1.505755057	1.761003582	2.222089264
1 I PF3	1.035743316	.9654977494	.9423952493	.9195629996	.8956857548	.8800522747	.8686312732	.8524455514	.8362843410
1 I PQ3	1.912891274	2.131101356	2.260263355	2.439835083	2.719718915	2.990735220	3.262880623	3.836852969	4.865210908
2 I PFC	1.144803925	1.099189437	1.083296686	1.067110001	1.049632063	1.037863170	1.029097172	1.016437217	1.003569095
2 I PQC	6.139827980	6.379640890	6.574018673	6.859438911	7.311028101	7.741974984	8.163288330	9.009887965	10.39473123
1 I ZF1	1.354963320	1.355349741	1.355476420	1.355603031	1.355738680	1.355830566	1.355899990	1.356003066	1.356113803
1 I ZF2	1.440147115	1.443895140	1.445115576	1.446333135	1.447636124	1.448518335	1.449184830	1.450174493	1.451238073
1 I ZF3	1.686906300	1.698441574	1.702230948	1.706027516	1.710108060	1.712880931	1.714980993	1.718107222	1.721476763
1 I ZF4	2.426844206	2.454729302	2.464043173	2.473445702	2.483626389	2.490586014	2.495877840	2.503786367	2.512347909
1 I ZF5	6.681830991	6.779383327	6.812315036	6.845726569	6.882081731	6.907034655	6.926058716	6.954565593	6.985518758

(5) P 8+2		TH = 49 SBE = 1.325012993							MCPER FUNCTIONS	
		RC = 02	RC = 05	RC = 07	RC = 10	RC = 15	RC = 20	RC = 25	RC = 35	RC = 50
AMAX DB		.0017375255	.0108709564	.0213327376	.0436480540	.0988323393	.1772876696	.2802872360	.5675287486	1.249387366
AMIN DB		68.64651221	78.63329448	82.31987720	86.24848842	90.76010872	94.01494695	96.59377596	100.6448861	105.3887075
MU	K	.3695510D-3	.1170403D-3	.7656074D-4	.4870523D-4	.2897307D-4	.1991832D-4	.1480169D-4	.9284440D-5	.5377304D-5
1	PF1	.7564877130	.5876931381	.5323287967	.4770621450	.4182867477	.3791213520	.3501772589	.3089049262	.2682694698
1	PQ1	.5446811082	.5581252904	.5664675148	.5786487446	.5989659778	.6200810339	.6425747305	.6936696847	.7948317824
1	PF2	.9049924059	.7894404943	.7523895092	.7162310402	.6789852421	.6550093118	.6377763033	.6139389713	.5911991608
1	PQ2	.9277217035	1.020940743	1.076135421	1.153424714	1.275165371	1.394277217	1.514812855	1.771123373	2.234279578
1	PF3	1.037282371	.9676091597	.9446468005	.9219338409	.8981638913	.8825928111	.8712144681	.8550867810	.8389840574
1	PQ3	1.953512400	2.169398704	2.298741324	2.479211156	2.761224070	3.034723251	3.309582993	3.889658798	4.929515442
2	PFC	1.140995909	1.096871350	1.081432766	1.065678017	1.048634063	1.037139534	1.028568983	1.016178375	1.003568957
2	PQC	6.358525440	6.573431715	6.763408359	7.046857250	7.499859195	7.934591436	8.360841314	9.219430177	10.62694696
1	ZF1	1.333927056	1.334315596	1.334443230	1.334570898	1.334707778	1.334800550	1.334870668	1.334974810	1.335086739
1	ZF2	1.415447045	1.419235478	1.420469804	1.421701593	1.423020233	1.423913292	1.424588121	1.425590383	1.426667811
1	ZF3	1.652796286	1.664470995	1.668309965	1.672158050	1.676295982	1.679109032	1.681240157	1.684413636	1.687835452
1	ZF4	2.368909149	2.397096837	2.406530313	2.416062314	2.426392597	2.433459848	2.438836293	2.446875575	2.455584175
1	ZF5	6.504859355	6.603392819	6.636737179	6.670606453	6.707502148	6.732850350	6.752188378	6.781185097	6.812694818

(5) P 8+2		TH = 50 SBE = 1.305407289							MCPER FUNCTIONS	
		RC = 02	RC = 05	RC = 07	RC = 10	RC = 15	RC = 20	RC = 25	RC = 35	RC = 50
AMAX DB		.0017375255	.0108709564	.0213327376	.0436480540	.0988323393	.1772876696	.2802872360	.5675287486	1.249387366
AMIN DB		66.61314064	76.59700360	80.28280740	84.21072423	88.72169199	91.97614108	94.55470408	98.60546336	103.3489659
MU	K	.4670281D-3	.1479619D-3	.9679649D-4	.6158342D-4	.3663662D-4	.2518796D-4	.1871823D-4	.1174159D-4	.6800670D-5
1	PF1	.7661719988	.5942818749	.5380585946	.4820025518	.4224555934	.3828113206	.3535298349	.3117985495	.2707340991
1	PQ1	.5454148760	.5587101642	.5670165015	.5791687363	.5994640725	.6205714579	.6430647520	.6941721018	.7953785847
1	PF2	.9116248366	.7950458234	.7576708615	.7212021414	.6836449912	.6594742489	.6421045605	.6180338901	.5951774467
1	PQ2	.9372285603	1.029409637	1.084482425	1.161803527	1.283819454	1.403328244	1.524331326	1.781747031	2.247064761
1	PF3	1.038804500	.9697443597	.9469326644	.9243481541	.9006938659	.8851900522	.8738576468	.8577922906	.8417520165
1	PQ3	1.996979977	2.210159848	2.339628764	2.520989081	2.805193447	3.081277637	3.358975730	3.945452755	4.997394982
2	PFC	1.137136060	1.094514051	1.079535248	1.064218549	1.047615489	1.036400092	1.028028575	1.015912521	1.003567097
2	PQC	6.595144300	6.781802447	6.966667860	7.247645358	7.701799765	8.140345559	8.571695772	9.442325110	10.87423593
1	ZF1	1.313917830	1.314307892	1.314436317	1.314564890	1.314702851	1.314796412	1.314867154	1.314972264	1.315085283
1	ZF2	1.391881710	1.395707377	1.396954600	1.398199652	1.399532937	1.400436184	1.401118857	1.402133009	1.403223548
1	ZF3	1.620094783	1.631900560	1.635786668	1.639683945	1.643876911	1.646728629	1.648889720	1.652108888	1.655581396
1	ZF4	2.313118611	2.341586833	2.351133871	2.360789884	2.371264550	2.378436265	2.383895137	2.392062224	2.400915111
1	ZF5	6.333989335	6.433428190	6.467164577	6.501473216	6.538892328	6.564625757	6.584270993	6.613748842	6.645807494

(5) P — TH = 51 — SBE = 1.286759566 — MCPER FUNCTIONS

MU	K	RC = 02	RC = 05	RC = 07	RC = 10	RC = 15	RC = 20	RC = 25	RC = 35	RC = 50
8+2										
AMAX DB		.0017375255	.0108709564	.0213327376	.0436480540	.0988323393	.1772876696	.2802872360	.5675287486	1.249387366
AMIN DB		64.60545368	74.58615192	78.27111112	82.19827490	86.70853491	89.96256190	92.54083633	96.59121495	101.3343716
MU	K	.5884740D-3	.1865058D-3	.1220238D-3	.7764013D-4	.4619269D-4	.3175937D-4	.2360251D-4	.1480505D-4	.8575934D-5
1	PF1	.7763356900	.6011716220	.5440438082	.4871581080	.4268018094	.3866560168	.3570215596	.3148106238	.2732933347
1	PQ1	.5461969021	.5593284403	.5675954875	.5797159656	.5999871692	.6210858480	.6435782718	.6946980048	.7959503534
1	PF2	.9184467288	.8008285529	.7631229552	.7263366697	.6884600465	.6640890669	.6465787515	.6223692821	.5992910931
1	PQ2	.9473703249	1.038383681	1.093309868	1.170648280	1.292938202	1.412854519	1.534341695	1.792907827	2.260483113
1	PF3	1.040304848	.9719009546	.9492511474	.9268048784	.9032752136	.8878438855	.8765609355	.8605525168	.8445889321
1	PQ3	2.043564713	2.253598879	2.383129073	2.565365931	2.851823340	3.130598886	3.411266075	4.004460434	5.069114044
2	PFC	1.133226732	1.092118444	1.077604710	1.062731914	1.046576451	1.035644854	1.027475903	1.015639554	1.003563382
2	PQC	6.851672295	7.006234989	7.185168347	7.463089592	7.918076011	8.360443671	8.797056620	9.681303613	11.13791165
1	ZF1	1.294878983	1.295269928	1.295398970	1.295528288	1.295667169	1.295761418	1.295832711	1.295938683	1.296052685
1	ZF2	1.369388496	1.373248069	1.374507142	1.375764440	1.377111310	1.378024045	1.378714045	1.379739337	1.380842204
1	ZF3	1.588722466	1.600650459	1.604581105	1.608525111	1.612770623	1.615659407	1.617849299	1.621112501	1.624634020
1	ZF4	2.259346535	2.288072174	2.297726421	2.307500695	2.318114242	2.325387080	2.330926058	2.339217823	2.348212055
1	ZF5	6.168846023	6.269112074	6.303219068	6.337947934	6.375872581	6.401980708	6.421926058	6.451875867	6.484475401

(5) P — TH = 52 — SBE = 1.269018215 — MCPER FUNCTIONS

MU	K	RC = 02	RC = 05	RC = 07	RC = 10	RC = 15	RC = 20	RC = 25	RC = 35	RC = 50
8+2										
AMAX DB		.0017375255	.0108709564	.0213327376	.0436480540	.0988323393	.1772876696	.2802872360	.5675287486	1.249387366
AMIN DB		62.62178673	72.59905555	76.28309930	80.20944675	84.71893943	87.97250873	90.55047026	94.60043607	99.34321753
MU	K	.7394532D-3	.2344484D-3	.1534069D-3	.9761750D-4	.5808353D-4	.3993692D-4	.2968086D-4	.1861994D-4	.1078547D-4
1	PF1	.7870103052	.6083801642	.5502990127	.4925406809	.4313348437	.3906635001	.3606595664	.3179470897	.2759670695
1	PQ1	.5470317056	.5599828000	.5682067623	.5802924102	.6005369916	.6216258020	.6441168194	.6952488722	.7965486119
1	PF2	.9254594459	.8067928722	.7687503813	.7316394375	.6934353071	.6688586590	.6512037406	.6267999438	.6035448083
1	PQ2	.9582057781	1.047904397	1.102655734	1.179994427	1.302555592	1.422889908	1.544878362	1.804642357	2.274576700
1	PF3	1.041778233	.9740763364	.9516003841	.9293028223	.9059073798	.8905541347	.8793244143	.8633978739	.8474955166
1	PQ3	2.093572430	2.299955682	2.429469597	2.612561484	2.901332155	3.182909769	3.466684039	4.066931353	5.144965366
2	PFC	1.129270415	1.089685503	1.075641775	1.061218459	1.045517077	1.034873841	1.026910927	1.015359371	1.003557676
2	PQC	7.130388596	7.248412658	7.420463040	7.694643279	8.150069646	8.596242805	9.038277680	9.936246311	11.41944503
1	ZF1	1.276758362	1.277149512	1.277278984	1.277408875	1.277548508	1.277643336	1.277715102	1.277821825	1.277936696
1	ZF2	1.347909809	1.351799804	1.353069628	1.354338096	1.355697433	1.356618917	1.357315698	1.358351336	1.359465705
1	ZF3	1.558606371	1.570647204	1.574619636	1.578607765	1.582903185	1.585827338	1.588044796	1.591350274	1.594919019
1	ZF4	2.207476590	2.236435420	2.246190192	2.256076666	2.266823287	2.274193713	2.279810339	2.288223465	2.297355912
1	ZF5	6.009081903	6.110094465	6.144549827	6.179678988	6.218090491	6.244562287	6.264800301	6.295212423	6.328344337

(5) P 8+2 — TH = 53 SBE = 1.252135658 — MCPER FUNCTIONS

MU		RC = 02	RC = 05	RC = 07	RC = 10	RC = 15	RC = 20	RC = 25	RC = 35	RC = 50
	AMAX DB	.0017375255	.0108709564	.0213327376	.0436480540	.0988323393	.1772876696	.2802872360	.5675287486	1.249387366
	AMIN DB	60.66052961	70.63408332	74.31713512	78.24259791	82.75125893	86.00433212	88.58195445	92.63147272	97.37384742
	K	.9267733D-3	.2939651D-3	.1923726D-3	.1224250D-3	.7285126D-4	.5009373D-4	.3723079D-4	.2335750D-4	.1353031D-4
1	PF1	.7982304763	.6159269253	.5568400967	.4981631838	.4360649582	.3948425148	.3644515879	.3212143798	.2787455967
1	PQ1	.5479244014	.5606762380	.5688528730	.5809002621	.6011154437	.6221930812	.6446820782	.6958263286	.7971750310
1	PF2	.9326638720	.8129430085	.7745578622	.7371154557	.6985759172	.6737881842	.6559846687	.6313809584	.6079435924
1	PQ2	.9698012332	1.058018163	1.112562329	1.189881348	1.312709258	1.433471853	1.555979317	1.816990973	2.289391831
1	PF3	1.043219127	.9762676651	.9539783217	.9318406502	.9085897102	.8933205515	.8821481108	.8662987492	.8504724791
1	PQ3	2.147349793	2.349499855	2.478905202	2.662821530	2.953963567	3.238458455	3.525485584	4.133144817	5.225273745
2	PFC	1.125269747	1.087216280	1.073647125	1.059678572	1.044437516	1.034087085	1.026333611	1.015071869	1.003549837
2	PQC	7.433916758	7.510254246	7.674317328	7.943953768	8.399342552	8.849274269	9.296884724	10.20920646	11.72048803
1	ZF1	1.259507902	1.259898533	1.260028239	1.260158521	1.260298725	1.260394019	1.260466174	1.260573531	1.260689149
1	ZF2	1.327392616	1.331309390	1.332588807	1.333867316	1.335237938	1.336167392	1.336870376	1.337915520	1.339040515
1	ZF3	1.529679322	1.541823056	1.545834357	1.549863847	1.554206377	1.557164097	1.559407810	1.562753698	1.566367769
1	ZF4	2.157401256	2.186567884	2.196416147	2.206408432	2.217281991	2.224746265	2.230437934	2.238968899	2.248236231
1	ZF5	5.854374107	5.956049968	5.990830613	6.026339318	6.065218164	6.092042074	6.112564931	6.143429210	6.177084524

(5) P 8+2 — TH = 54 SBE = 1.236067977 — MCPER FUNCTIONS

MU		RC = 02	RC = 05	RC = 07	RC = 10	RC = 15	RC = 20	RC = 25	RC = 35	RC = 50
	AMAX DB	.0017375255	.0108709564	.0213327376	.0436480540	.0988323393	.1772876696	.2802872360	.5675287486	1.249387366
	AMIN DB	58.72011864	68.68964855	72.37162575	76.29613006	80.80388990	84.05642544	86.63368013	90.68271332	95.42464710
	K	.1158762D-2	.3677203D-3	.2406682D-3	.1531770D-3	.9116025D-4	.6268718D-4	.4659250D-4	.2923239D-4	.1693432D-4
1	PF1	.8100343386	.6238331640	.5636844158	.5040396962	.4410033204	.3992025652	.3684060227	.3246194727	.2816396524
1	PQ1	.5488807995	.5614121093	.5695366615	.5815419583	.6017246344	.6227896322	.6452759055	.6964321626	.7978314467
1	PF2	.9400603069	.8192832020	.7805502438	.7427699368	.7038872791	.6788830840	.6609269724	.6361177173	.6124927607
1	PQ2	.9822317512	1.068776937	1.123076901	1.200352894	1.323440993	1.444641861	1.567686623	1.829998271	2.304979607
1	PF3	1.044621640	.9784718501	.9563827019	.9344168678	.9113214391	.8961428059	.8850319914	.8692654970	.8535205210
1	PQ3	2.205291183	2.402535355	2.531722479	2.716421774	3.009990234	3.297522198	3.587956343	4.203408357	5.310400517
2	PFC	1.121227525	1.084711907	1.071621501	1.058112678	1.043337940	1.033284632	1.025743929	1.014776943	1.003539715
2	PQC	7.765289269	7.793955046	7.948744917	8.212894787	8.667666189	9.121271700	9.574602869	10.50193707	12.04290174
1	ZF1	1.243083253	1.243472597	1.243602327	1.243732807	1.243873390	1.243969028	1.244041485	1.244149350	1.244265588
1	ZF2	1.307788040	1.311727781	1.313015576	1.314302934	1.315683595	1.316620195	1.317328770	1.318382532	1.319517224
1	ZF3	1.501879423	1.514115507	1.518162584	1.522230502	1.526617174	1.529606549	1.531875125	1.535259436	1.538916814
1	ZF4	2.109021009	2.138368818	2.148303168	2.158394524	2.169388535	2.176942694	2.182706642	2.191351704	2.200750376
1	ZF5	5.704421951	5.806675337	5.841757309	5.877623954	5.916949761	5.944113677	5.964913166	5.996218909	6.030388134

(5) P	TH = 55	SBE = 1.220774589						MCPER FUNCTIONS	
8+2	RC = 02	RC = 05	RC = 07	RC = 10	RC = 15	RC = 20	RC = 25	RC = 35	RC = 50
AMAX DB	.0017375255	.0108709564	.0213327376	.0436480540	.0988323393	.1772876696	.2802872360	.5675287486	1.249387366
AMIN DB	56.79902892	66.76420111	70.44501438	74.36848038	78.87526383	82.12721677	84.70407307	88.75258054	93.49403642
MU I K	.1445601D-2	.4589760D-3	.3004341D-3	.1912388D-3	.1138248D-3	.7827790D-4	.5818303D-4	.3650657D-4	.2114941D-4
1 I PF1	.8224639809	.6321221983	.5708509691	.5101856013	.4461621088	.4037540039	.3725320103	.3281699542	.2846554651
1 I PQ1	.5499075241	.5621941840	.5702613082	.5822202153	.6023669055	.6234176112	.6459003560	.6970683484	.7985198814
1 I PF2	.9476483422	.8258176747	.7867324824	.7486082966	.7093750646	.6841490997	.6660364043	.6410159441	.6171979698
1 I PQ2	.9955825970	1.080239106	1.134252375	1.211458045	1.334797332	1.456446069	1.580046970	1.843713670	2.321396562
1 I PF3	1.045979498	.9806855279	.9588110425	.9370298052	.9141016765	.8990204757	.8879759535	.8722984327	.8566403324
1 I PQ3	2.267846979	2.459406052	2.588244759	2.773672462	3.069718181	3.360411673	3.654416012	4.278069853	5.400748812
2 I PFC	1.117146713	1.082173604	1.069565708	1.056521248	1.042218547	1.032466542	1.025141861	1.014474491	1.003527154
2 I PQC	8.128026680	8.102036523	8.246051454	8.503605363	8.957056832	9.414204603	9.873389292	10.81542318	12.38878969
1 I ZF1	1.227443451	1.227830688	1.227960220	1.228090693	1.228231454	1.228327308	1.228399973	1.228508214	1.228624937
1 I ZF2	1.289050987	1.293009713	1.294304609	1.295599561	1.296988948	1.297931823	1.298645343	1.299706783	1.300850189
1 I ZF3	1.475149607	1.487466829	1.491546395	1.495649630	1.500077289	1.503096286	1.505388243	1.508808866	1.512507396
1 I ZF4	2.062243595	2.091744676	2.101757316	2.111940629	2.123048231	2.130688077	2.136521370	2.145276551	2.154802782
1 I ZF5	5.558944722	5.661687255	5.697045702	5.733247798	5.772999275	5.800490511	5.821558003	5.853293951	5.887967057

(5) P	TH = 56	SBE = 1.206217949						MCPER FUNCTIONS	
8+2	RC = 02	RC = 05	RC = 07	RC = 10	RC = 15	RC = 20	RC = 25	RC = 35	RC = 50
AMAX DB	.0017375255	.0108709564	.0213327376	.0436480540	.0988323393	.1772876696	.2802872360	.5675287486	1.249387366
AMIN DB	54.89576688	64.85621964	68.53577226	72.45811353	76.96383917	80.21516087	82.79158545	86.83952319	91.58046112
MU I K	.1799748D-2	.5717274D-3	.3742927D-3	.2382837D-3	.1418430D-3	.9755330D-4	.7251381D-4	.4550130D-4	.2636191D-4
1 I PF1	.8355659657	.6408196642	.5783606026	.5166177449	.4515546336	.4085081310	.3768395187	.3318740874	.2877998122
1 I PQ1	.5510121585	.5630267130	.5710303831	.5829380699	.6030448648	.6240794138	.6465577080	.6977370704	.7992425686
1 I PF2	.9554267156	.8325505921	.7931096288	.7546361542	.7150452283	.6895922914	.6713190557	.6460817205	.6220652463
1 I PQ2	1.009950991	1.092470498	1.146148228	1.223251668	1.346830253	1.468935905	1.593112336	1.858192086	2.338705408
1 I PF3	1.047286024	.9829050399	.9612606155	.9396775995	.9169293932	.9019530346	.8909798145	.8753978252	.8598325875
1 I PQ3	2.335533606	2.520502361	2.648838090	2.834923884	3.133492005	3.427476120	3.725223535	4.357515496	5.496769750
2 I PFC	1.113030457	1.079602688	1.067480624	1.054904801	1.041079562	1.031632891	1.024527395	1.014164408	1.003511991
2 I PQC	8.526235193	8.437406798	8.568887424	8.818536857	9.269818025	9.730318674	10.19547249	11.15492061	12.76053768
1 I ZF1	1.212550618	1.212934881	1.213063977	1.213194227	1.213334951	1.213430884	1.213503659	1.213612137	1.213729201
1 I ZF2	1.271139828	1.275113382	1.276414037	1.277715263	1.279111990	1.280060221	1.280778001	1.281846124	1.282997204
1 I ZF3	1.449437241	1.461823665	1.465932229	1.470067471	1.474532763	1.477579219	1.479892979	1.483347664	1.487085052
1 I ZF4	2.016983376	2.046608467	2.056691181	2.066958940	2.078172872	2.085893949	2.091793470	2.100654535	2.110304291
1 I ZF5	5.417679675	5.520820330	5.556429470	5.592943610	5.633098515	5.660903773	5.682230207	5.714384497	5.749550879

(5) P 8+2		TH = 57 SBE = 1.192363293							MCPER FUNCTIONS	
		RC = 02	RC = 05	RC = 07	RC = 10	RC = 15	RC = 20	RC = 25	RC = 35	RC = 50
AMAX DB		.0017375255	.0108709564	.0213327376	.0436480540	.0988323393	.1772876696	.2802872360	.5675287486	1.249387366
AMIN DB		53.00886284	62.96420368	66.64239084	70.56351372	75.06809331	78.31873100	80.89468775	84.94200804	89.68238455
MU	K	.2236438D-2	.7108694D-3	.4654580D-3	.2963632D-3	.1764393D-3	.1213566D-3	.9021227D-4	.5661084D-4	.3280052D-4
1	PF1	.8493919331	.6499538159	.5862362436	.5233546182	.4571954753	.4134773099	.3813394434	.3357408939	.2910800847
1	PQ1	.5522034214	.5639145061	.5718479075	.5836989283	.6037614250	.6247777088	.6472504955	.6984407524	.8000019812
1	PF2	.9633931407	.8394860153	.7996868057	.7608593296	.7209040201	.6952190580	.6767813804	.6513215153	.6271010189
1	PQ2	1.025448233	1.105545591	1.158831519	1.235795428	1.359597984	1.482168872	1.606940756	1.873494718	2.356975911
1	PF3	1.048534116	.9851264071	.9637284247	.9423581736	.9198034050	.9049398389	.8940433013	.8785638885	.8630979393
1	PQ3	2.408945789	2.586269247	2.713918413	2.900572955	3.201701080	3.499109463	3.800783272	4.442181765	5.598969776
2	PFC	1.108882096	1.077000582	1.065367203	1.053263909	1.039921241	1.030783774	1.023900532	1.013846589	1.003494056
2	PQC	8.964727797	8.803434707	8.920312697	9.160510134	9.608591996	10.07218451	10.54339963	11.52000245	13.16086164
1	ZF1	1.198369700	1.198750069	1.198878480	1.199008276	1.199148736	1.199244604	1.199317385	1.199425951	1.199543204
1	ZF2	1.254016111	1.258000154	1.259305158	1.260611269	1.262013873	1.262966488	1.263687808	1.264761559	1.265919208
1	ZF3	1.424693768	1.437136680	1.441270523	1.445434250	1.449933603	1.453005213	1.455339097	1.458825444	1.462599241
1	ZF4	1.973160752	2.002879166	2.013023293	2.023367562	2.034680131	2.042477710	2.048440145	2.057402581	2.067171555
1	ZF5	5.280380206	5.383825266	5.419658362	5.456460179	5.496995275	5.525100615	5.546676466	5.579236598	5.614885036

(5) P 8+2		TH = 58 SBE = 1.179178403							MCPER FUNCTIONS	
		RC = 02	RC = 05	RC = '07	RC = 10	RC = 15	RC = 20	RC = 25	RC = 35	RC = 50
AMAX DB		.0017375255	.0108709564	.0213327376	.0436480540	.0988323393	.1772876696	.2802872360	.5675287486	1.249387366
AMIN DB		51.13686359	61.08666572	64.76337367	68.68317655	73.18651432	76.43641075	79.01186045	83.05851147	87.79827938
MU	K	.2774322D-2	.8824025D-3	.5778716D-3	.3679944D-3	.2191161D-3	.1507230D-3	.1120487D-3	.7031928D-4	.4074610D-4
1	PF1	.8639993024	.6595558740	.5945031744	.5304165699	.4631006463	.4186751002	.3860437236	.3397302467	.2945043615
1	PQ1	.5534913828	.5648630265	.5727184263	.5845066232	.6045198495	.6255154784	.6479815450	.6991320916	.8008003650
1	PF2	.9715441058	.8466278426	.8064691801	.7672838383	.7269579978	.7010361584	.6824302213	.6567422166	.6323121546
1	PQ2	1.042202281	1.119548964	1.172378130	1.249158872	1.373165977	1.496209471	1.621597237	1.889689981	2.376285917
1	PF3	1.049716231	.9873453025	.9662111796	.9450692147	.9227223543	.9079801121	.8971660371	.8817967730	.8664370131
1	PQ3	2.488771584	2.657215916	2.783960221	2.971071136	3.274786996	3.575757632	3.881552363	4.532560651	5.707919380
2	PFC	1.104705176	1.074368823	1.063226486	1.051599205	1.038743876	1.029919309	1.023261283	1.013520932	1.003473170
2	PQC	9.449175654	9.204041058	9.303875720	9.532785464	9.976422268	10.44275683	10.92009401	11.91461547	13.59286539
1	ZF1	1.184868232	1.185243731	1.185371193	1.185500292	1.185640248	1.185735899	1.185808575	1.185917072	1.186034355
1	ZF2	1.237644298	1.241634306	1.242942181	1.244251714	1.245658653	1.246614626	1.247338720	1.248416985	1.249580034
1	ZF3	1.400874400	1.413360232	1.417515393	1.421703851	1.426233459	1.429327767	1.431679982	1.435195431	1.439003019
1	ZF4	1.930701637	1.960481188	1.970677595	1.981089981	1.992493032	2.000362088	2.006383908	2.015442904	2.025326490
1	ZF5	5.146814194	5.250467194	5.286496515	5.323560646	5.364451649	5.392842455	5.414657709	5.447610503	5.483729122

(5) P 8+2	TH = 59	SBE = 1.166633397						MCPER FUNCTIONS	
	RC = 02	RC = 05	RC = 07	RC = 10	RC = 15	RC = 20	RC = 25	RC = 35	RC = 50
AMAX DB	.0017375255	.0108709564	.0213327376	.0436480540	.0988323393	.1772876696	.2802872360	.5675287486	1.249387366
AMIN DB	49.27832490	59.22212305	62.89722815	66.81560059	71.31759248	74.56668541	77.14158546	81.18751087	85.92661887
MU I K	.3436242D-2	.1093689D-2	.7163720D-3	.4562680D-3	.2717192D-3	.1869243D-3	.1389699D-3	.8722168D-4	.5054394D-4
1 I PF1	.8794520914	.6696604324	.6031893491	.5378260531	.4692877780	.4241164125	.3909654754	.3440029781	.2980814964
1 I PQ1	.5548877304	.5658785040	.5736470968	.5853654833	.6053238073	.6262960664	.6487540191	.6999640982	.8016422775
1 I PF2	.9798746382	.8539797379	.8134619262	.7739158822	.7332140389	.7070507340	.6882728391	.6623511667	.6377059987
1 I PQ2	1.060360925	1.134577049	1.186874250	1.263420717	1.387608063	1.511130309	1.637154841	1.906854612	2.396722577
1 I PF3	1.050824358	.9895570221	.9687052680	.9478081489	.9256846896	.9110729279	.9003475268	.8850965538	.8698503994
1 I PQ3	2.575810963	2.733927617	2.859507071	3.046934011	3.353252524	3.657927360	3.968049572	4.629209406	5.824263533
2 I PFC	1.100503468	1.071709070	1.061059607	1.049911386	1.037547797	1.029039633	1.022609674	1.013187337	1.003449147
2 I PQC	9.986298653	9.643811813	9.723711334	9.939148497	10.37683039	10.84544681	11.32892515	12.34214868	14.06011080
1 I ZF1	1.172016123	1.172385723	1.172511955	1.172640100	1.172779298	1.172874571	1.172947026	1.173055287	1.173172431
1 I ZF2	1.221991542	1.225982797	1.227291991	1.228603408	1.230013056	1.230971302	1.231697362	1.232778960	1.233946168
1 I ZF3	1.377937843	1.390452099	1.394624349	1.398833534	1.403389336	1.406503719	1.408872351	1.412414164	1.416252745
1 I ZF4	1.889536988	1.919343911	1.929582962	1.940054586	1.951539467	1.959474655	1.965552101	1.974702518	1.984695785
1 I ZF5	5.016762458	5.120524121	5.156720908	5.194020951	5.235242479	5.263903417	5.285947544	5.319279097	5.355855318

(5) P 8+2	TH = 60	SBE = 1.154700538						MCPER FUNCTIONS	
	RC = 02	RC = 05	RC = 07	RC = 10	RC = 15	RC = 20	RC = 25	RC = 35	RC = 50
AMAX DB	.0017375255	.0108709564	.0213327376	.0436480540	.0988323393	.1772876696	.2802872360	.5675287486	1.249387366
AMIN DB	47.43180381	57.36908919	61.04245680	64.95927859	69.45981132	72.70803304	75.28233701	79.32747548	84.06586769
MU I K	.4250204D-2	.1353772D-2	.8869051D-3	.5649839D-3	.3365189D-3	.2315252D-3	.1721405D-3	.1080504D-3	.6261907D-4
1 I PF1	.8958218743	.6803059345	.6123257666	.5456079142	.4757763387	.4298176886	.3961191479	.3484210060	.3018212180
1 I PQ1	.5564061012	.5669680738	.5746397937	.5862804156	.6061774380	.6271232342	.6495714692	.7007901432	.8025296347
1 I PF2	.9883780265	.8615450419	.8206701794	.7807618354	.7396793522	.7132703324	.6943169434	.6681562022	.6432904199
1 I PQ2	1.080095689	1.150740255	1.202418181	1.278670413	1.403007840	1.527013409	1.653695980	1.925074984	2.418383788
1 I PF3	1.051850004	.9917564518	.9712067243	.9505721137	.9286886423	.9142171908	.9035871409	.8884632187	.8733386460
1 I PQ3	2.670998930	2.817080120	2.941184423	3.128752935	3.437672484	3.746196838	4.060865968	4.732761696	5.948734233
2 I PFC	1.096280983	1.069023121	1.058867803	1.048201222	1.036333379	1.028144915	1.021945745	1.012845703	1.003421792
2 I PQC	10.58410711	10.12813944	10.18466239	10.38401674	10.81391058	11.28421108	11.77379489	12.80651739	14.56670346
1 I ZF1	1.159785468	1.160148080	1.160272787	1.160399707	1.160537877	1.160632602	1.160704714	1.160812565	1.160929389
1 I ZF2	1.207027478	1.211015064	1.212323950	1.213635632	1.215046275	1.216005648	1.216732819	1.217816499	1.218986552
1 I ZF3	1.355846055	1.368373226	1.372558048	1.376783680	1.381361337	1.384492991	1.386875993	1.390441244	1.394307821
1 I ZF4	1.849602380	1.879401248	1.889672769	1.900194232	1.911751755	1.919747384	1.925876445	1.935112792	1.945210455
1 I ZF5	4.890017338	4.993785501	5.030119927	5.067628389	5.109153906	5.138068885	5.160330802	5.194026440	5.231046933

(5) P 8+2			TH = 61 SBE = 1.143354068							MCPER FUNCTIONS	
			RC = 02	RC = 05	RC = 07	RC = 10	RC = 15	RC = 20	RC = 25	RC = 35	RC = 50
AMAX DB			.0017375255	.0108709564	.0213327376	.0436480540	.0988323393	.1772876696	.2802872360	.5675287486	1.249387366
AMIN DB			45.59585078	55.52606483	59.19754799	63.11268804	67.61163813	70.85891475	73.43257207	77.47685669	82.21447217
MU		K	.5250582D-2	.1673774D-2	.1096788D-2	.6988204D-3	.4163112D-3	.2864536D-3	.2129960D-3	.1337079D-3	.7749548D-4
1		PF1	.9131889043	.6915352333	.6219469075	.5537897314	.4825878902	.4357971131	.4015207058	.3530474804	.3057342477
1		PQ1	.5580624961	.5681399453	.5757032391	.5872570060	.6070854302	.6280012296	.6504378968	.7016640150	.8034667655
1		PF2	.9970454946	.8693266645	.8280989769	.7878282234	.7463614870	.7197029324	.7005707265	.6741656969	.6490738614
1		PQ2	1.101606685	1.168165570	1.219122519	1.295010031	1.419460341	1.543951802	1.671313972	1.944448686	2.441379932
1		PF3	1.052784172	.9939380330	.9737111960	.9533579264	.9317322012	.9174116140	.9068840958	.8918966538	.8769022474
1		PQ3	2.775434504	2.907457590	3.029715416	3.217209306	3.528707007	3.841228674	4.160677906	4.843941118	6.082165681
2		PFC	1.092041992	1.066312924	1.056652425	1.046469563	1.035101043	1.027235351	1.021269556	1.012495935	1.003390901
2		PQC	11.25221090	10.66340074	10.69243205	10.87257231	11.29244720	11.76366199	12.25924371	13.31226672	15.11739809
1		ZF1	1.148150372	1.148504851	1.148627721	1.148753129	1.148889991	1.148983987	1.149055625	1.149162882	1.149279197
1		ZF2	1.192724041	1.196702837	1.198009707	1.199319952	1.200729782	1.201689073	1.202416451	1.203500888	1.204672390
1		ZF3	1.334564033	1.347087502	1.351280063	1.355517567	1.360112435	1.363258361	1.365653541	1.369239093	1.373130454
1		ZF4	1.810837623	1.840591258	1.850884503	1.861445843	1.873066243	1.881116247	1.887292640	1.896609040	1.906805426
1		ZF5	4.766381360	4.870050879	4.906492004	4.944180252	4.985982002	5.015134127	5.037602166	5.071646390	5.109097018

(5) P 8+2			TH = 62 SBE = 1.132570051							MCPER FUNCTIONS	
			RC = 02	RC = 05	RC = 07	RC = 10	RC = 15	RC = 20	RC = 25	RC = 35	RC = 50
AMAX DB			.0017375255	.0108709564	.0213327376	.0436480540	.0988323393	.1772876696	.2802872360	.5675287486	1.249387366
AMIN DB			43.76900144	53.69152803	57.36096594	61.27428098	65.77151350	69.01776437	71.59071975	75.63407744	80.37084957
MU		K	.6479626D-2	.2067396D-2	.1355039D-2	.8635469D-3	.5145461D-3	.3540885D-3	.2633080D-3	.1653089D-3	.9582030D-4
1		PF1	.9316434337	.7033962546	.6320912509	.5624022151	.4897463899	.4420748616	.4071878444	.3578969564	.3098324369
1		PQ1	.5598758039	.5694036106	.5768451596	.5883016402	.6080531165	.6289348683	.6513578287	.7025899864	.8044579778
1		PF2	1.005865817	.8773269519	.8357531834	.7951216946	.7532683413	.7263569698	.7070429001	.6803886115	.6550653988
1		PQ2	1.125128701	1.186999758	1.237116835	1.312556562	1.437074067	1.562051446	1.690114922	1.965086421	2.465835957
1		PF3	1.053617341	.9960957239	.9762139051	.9561620499	.9348130822	.9206546944	.9102374331	.8953966271	.8805416341
1		PQ3	2.890417361	3.005974814	3.125940406	3.313092155	3.627117814	3.943785763	4.268262892	4.963577688	6.225512755
2		PFC	1.087791046	1.063580588	1.054414949	1.044717347	1.033851267	1.026311172	1.020581188	1.012137942	1.003356261
2		PQC	12.00221815	11.25718242	11.25377600	11.41092873	11.81806170	12.28920509	12.79058326	13.86470021	15.71772896
1		ZF1	1.137086799	1.137431941	1.137552648	1.137676242	1.137811496	1.137904576	1.137975602	1.138082069	1.138197676
1		ZF2	1.179055304	1.183019976	1.184323041	1.185630059	1.187037169	1.187995099	1.188721726	1.189805517	1.190976988
1		ZF3	1.314059632	1.326561573	1.330756694	1.335001168	1.339608273	1.342765255	1.345170264	1.348772753	1.352685448
1		ZF4	1.773186413	1.802855788	1.813159397	1.823750052	1.835422939	1.843520845	1.849739992	1.859130145	1.869419157
1		ZF5	4.645665976	4.749128608	4.785644330	4.823482533	4.865531470	4.894902993	4.917564861	4.951941290	4.939807045

(5) P 8+2		TH = 63 SBE = 1.122326238								MCPER FUNCTIONS
		RC = 02	RC = 05	RC = 07	RC = 10	RC = 15	RC = 20	RC = 25	RC = 35	RC = 50
	AMAX DB	.0017375255	.0108709564	.0213327376	.0436480540	.0988323393	.1772876696	.2802872360	.5675287486	1.249387366
	AMIN DB	41.94976811	51.86392355	55.53113974	59.44247278	63.93784004	67.18297694	69.75516978	73.79752052	78.53337636
MU	K	.7989353D-2	.2551548D-2	.1672796D-2	.1066293D-2	.6354889D-3	.4373722D-3	.3252681D-3	.2042321D-3	.1183944D-3
1	PF1	.9512872659	.7159427826	.6428018881	.5714796819	.4972785478	.4486733959	.4131402429	.3629855989	.3141289306
1	PQ1	.5618684674	.5707701038	.5780744812	.5894216538	.6090865883	.6299296331	.6523364070	.7035728967	.8055081370
1	PF2	1.014824864	.8855475238	.8436373974	.8026489806	.7604081658	.7332413642	.7137427353	.6868345479	.6612748051
1	PQ2	1.150938869	1.207413321	1.256550983	1.331444743	1.455973467	1.581433565	1.710220004	1.987114318	2.491893896
1	PF3	1.054339452	.9982229574	.9787096067	.9589805531	.9379286956	.9239446839	.9136459939	.8989627692	.8842571593
1	PQ3	3.017494540	3.113705046	3.230841333	3.417319996	3.733788358	4.054750880	4.384519130	5.092628101	6.379873639
2	PFC	1.083533002	1.060828405	1.052156989	1.042945613	1.032584589	1.025372648	1.019880744	1.011771638	1.003317648
2	PQC	12.84825456	11.91856949	11.87674704	12.00634218	12.39739791	12.86721187	13.37406237	14.47004059	16.37417263
1	ZF1	1.126572434	1.126906975	1.127025173	1.127146637	1.127279970	1.127371933	1.127442204	1.127547675	1.127662364
1	ZF2	1.165997338	1.169942325	1.171239709	1.172541619	1.173943999	1.174899216	1.175624080	1.176705738	1.177875606
1	ZF3	1.294303406	1.306764668	1.310956792	1.315202976	1.319816983	1.322981570	1.325393886	1.329009699	1.332940019
1	ZF4	1.736596016	1.766140148	1.776442107	1.787050868	1.798765174	1.806904068	1.813161064	1.822518210	1.832993284
1	ZF5	4.527690356	4.630834615	4.667391612	4.705348674	4.747614387	4.777186648	4.800029382	4.834720687	4.872985623

(5) P 8+2		TH = 64 SBE = 1.112601940								MCPER FUNCTIONS
		RC = 02	RC = 05	RC = 07	RC = 10	RC = 15	RC = 20	RC = 25	RC = 35	RC = 50
	AMAX DB	.0017375255	.0108709564	.0213327376	.0436480540	.0988323393	.1772876696	.2802872360	.5675287486	1.249387366
	AMIN DB	40.13663098	50.04165104	53.70645112	57.61562967	62.10896955	65.35289538	67.92425950	71.96551549	76.70037499
MU	K	.9843929D-2	.3147150D-2	.2063847D-2	.1315887D-2	.7844252D-3	.5399521D-3	.4015938D-3	.2521879D-3	.1462114D-3
1	PF1	.9722355839	.7292353992	.6541272546	.5810606211	.5052142531	.4556178138	.4193998674	.3683314246	.3186383599
1	PQ1	.5640673407	.5722523243	.5794015706	.5906255147	.6101928377	.6309917950	.6533794990	.7046182498	.8066227611
1	PF2	1.023905073	.8939890737	.8517558324	.8104168434	.7677895642	.7403655456	.7206801054	.6935138104	.6677126255
1	PQ2	1.179366452	1.229605455	1.277599199	1.351830550	1.476301990	1.602237518	1.731768267	2.010676740	2.519715942
1	PF3	1.054939890	1.000312596	.9811925422	.9618090675	.9410761077	.9272795562	.9171083906	.9025945513	.8880490828
1	PQ3	3.158520564	3.231915208	3.345572354	3.530968166	3.849748914	4.175151013	4.510489761	5.232200788	6.546517765
2	PFC	1.079273048	1.058058865	1.049880314	1.041155510	1.031301617	1.024420094	1.019168358	1.011396944	1.003274829
2	PQC	13.80764619	12.65851749	12.57100897	12.66748123	13.03835744	13.50523877	14.01707710	15.13563245	17.09435238
1	ZF1	1.116586550	1.116909167	1.117024496	1.117143498	1.117274580	1.117365217	1.117434580	1.117538839	1.117652387
1	ZF2	1.153528076	1.157447583	1.158737319	1.160032142	1.161427674	1.162378750	1.163100778	1.164178727	1.165345325
1	ZF3	1.275268475	1.287668453	1.291851611	1.296093855	1.300709029	1.303877514	1.306294424	1.309919673	1.313863626
1	ZF4	1.701016980	1.730392815	1.740680408	1.751295364	1.763039289	1.771211773	1.777501362	1.787018234	1.797472293
1	ZF5	4.412280222	4.514991183	4.551554847	4.589598337	4.632048962	4.661802328	4.684812247	4.719800078	4.758447233

(5) P 8+2 — TH = 65 SBE = 1.103377919 — MCPER FUNCTIONS

MU	K	RC = 02	RC = 05	RC = 07	RC = 10	RC = 15	RC = 20	RC = 25	RC = 35	RC = 50
	AMAX DB	.0017375255	.0108709564	.0213327376	.0436480540	.0988323393	.1772876696	.2802872360	.5675287486	1.249387366
	AMIN DB	38.32802919	48.22305174	51.88522068	55.79205455	60.28318860	63.52579687	66.09625915	70.13632383	74.87009894
	K	.1212268D-1	.3880140D-2	.2545300D-2	.1623294D-2	.9679225D-3	.6663619D-3	.4956636D-3	.3113034D-3	.1805074D-3
1	PF1	.9946191021	.7433426093	.6661220089	.5911883736	.5135870849	.4629362695	.4259913323	.3739545928	.3233770733
1	PQ1	.5665048009	.5738654461	.5808385378	.5919230528	.6113799309	.6321285623	.6544938314	.7057323349	.8078081364
1	PF2	1.033084809	.9026511234	.8601121687	.8184320051	.7754214870	.7477394814	.7278655325	.7004374746	.6743902628
1	PQ2	1.210805413	1.253810281	1.300465243	1.373895546	1.498225864	1.624624352	1.754920082	2.035939756	2.549488214
1	PF3	1.055407471	1.002356882	.9836563874	.9646427370	.9442519973	.9306569695	.9206229737	.9062912590	.8919175533
1	PQ3	3.315735579	3.362110751	3.471498699	3.655303344	3.976208114	4.306187860	4.647393173	5.383587168	6.726920550
2	PFC	1.075016730	1.055274678	1.047586863	1.039348311	1.030003037	1.023453874	1.018444193	1.011013789	1.003227560
2	PQC	14.90182733	13.49033760	13.34824342	13.40477470	13.75040168	14.21230708	14.72843861	15.87019999	17.88729746
1	ZF1	1.107109899	1.107419212	1.107531295	1.107647486	1.107775971	1.107865062	1.107933359	1.108036178	1.108148349
1	ZF2	1.141627213	1.145515190	1.146795216	1.148080872	1.149467321	1.150412748	1.151130806	1.152203374	1.153364930
1	ZF3	1.256930411	1.269246913	1.273414679	1.277646904	1.282257076	1.285425465	1.287844048	1.291474544	1.295427823
1	ZF4	1.666402874	1.695565155	1.705824911	1.716433399	1.728194343	1.736392492	1.742709027	1.752277801	1.762803200
1	ZF5	4.299266701	4.401425751	4.437960112	4.476056176	4.518658296	4.548572091	4.571734744	4.606999647	4.646010957

(5) P 8+2 — TH = 66 SBE = 1.094636279 — MCPER FUNCTIONS

MU	K	RC = 02	RC = 05	RC = 07	RC = 10	RC = 15	RC = 20	RC = 25	RC = 35	RC = 50
	AMAX DB	.0017375255	.0108709564	.0213327376	.0436480540	.0988323393	.1772876696	.2802872360	.5675287486	1.249387366
	AMIN DB	36.52235197	46.40639345	50.06569214	53.96997080	58.45870197	61.69987509	64.26935499	68.30812185	73.04071542
	K	.1492390D-1	.4782779D-2	.3138451D-2	.2002172D-2	.1194167D-2	.8222545D-3	.6116912D-3	.3842323D-3	.2228252D-3
1	PF1	1.018586593	.7583421980	.6788480933	.6019119514	.5224349285	.4706604790	.4329423357	.3798777545	.3283634141
1	PQ1	.5692202062	.5756274379	.5823996183	.5933257449	.6126572256	.6333482644	.6556871567	.7069223748	.8090714599
1	PF2	1.042337623	.9115317198	.8687093658	.8267010554	.7833132168	.7553737017	.7353102375	.7076174653	.6813200758
1	PQ2	1.245730754	1.280304774	1.325388897	1.397852350	1.521938830	1.648781216	1.779861439	2.063095435	2.581425414
1	PF3	1.055730436	1.004347382	.9860941945	.9674761622	.9474526060	.9340742224	.9241877933	.9100519626	.8958625862
1	PQ3	3.491868024	3.506093420	3.610246400	3.791827505	4.114592902	4.449276362	4.796661225	5.548300940	6.922806964
2	PFC	1.070769989	1.052478798	1.045278767	1.037525428	1.028689621	1.022474413	1.017708451	1.010622116	1.003175585
2	PQC	16.15756124	14.43033534	14.22268234	14.23086433	14.54494274	14.99926412	15.51871802	16.68417768	18.76377457
1	ZF1	1.098124604	1.098419179	1.098527622	1.098640639	1.098766166	1.098853480	1.098920542	1.099021682	1.099132232
1	ZF2	1.130276101	1.134126225	1.135394378	1.136668682	1.138043691	1.138981872	1.139694759	1.140760175	1.141914806
1	ZF3	1.239267154	1.251476241	1.255621691	1.259837346	1.264435864	1.267599853	1.270016958	1.273648179	1.277606132
1	ZF4	1.632710048	1.661611168	1.671828804	1.682417341	1.694181835	1.702397146	1.708734553	1.718346790	1.728935254
1	ZF5	4.188485174	4.289969681	4.326437316	4.364550584	4.407269120	4.437321545	4.460621645	4.496142971	4.535499172

(5) P 8+2	TH = 67	SBE = 1.086360377						MCPER FUNCTIONS	
	RC = 02	RC = 05	RC = 07	RC = 10	RC = 15	RC = 20	RC = 25	RC = 35	RC = 50
AMAX DB	.0017375255	.0108709564	.0213327376	.0436480540	.0988323393	.1772876696	.2802872360	.5675287486	1.249387366
AMIN DB	34.71793044	44.58985332	48.24601415	52.14750347	56.63361331	59.87322111	62.44162954	66.47898073	71.21028531
MU I K	.1836976D-1	.5895319D-2	.3869896D-2	.2469590D-2	.1473395D-2	.1014703D-2	.7549506D-3	.4742976D-3	.2750969D-3
1 I PF1	1.044307838	.7743228753	.6923760229	.6132870341	.5318007235	.4788263348	.4402841884	.3861264756	.3336180565
1 I PQ1	.5722618306	.5775597334	.5841016595	.5948470774	.6140356427	.6346605822	.6569684596	.7081967091	.8104210139
1 I PF2	1.051631369	.9206270613	.8775494263	.8352303291	.7914743418	.7632793224	.7430261929	.7150666438	.6885154924
1 I PQ2	1.284719973	1.309418935	1.352654242	1.423951566	1.547668106	1.674926927	1.806809326	2.092367208	2.615776639
1 I PF3	1.055896443	1.006274928	.9884983279	.9703033366	.9506736812	.9375282035	.9278005545	.9138754310	.8998840383
1 I PQ3	3.690271114	3.666036367	3.763766601	3.942334458	4.266599670	4.606093853	4.959987887	5.728127942	7.136206645
2 I PFC	1.066539195	1.049674450	1.042958368	1.035688430	1.027362244	1.021482198	1.016961376	1.010221877	1.003118637
2 I PQC	17.60860231	15.49866134	15.21181272	15.16120058	15.43585506	15.87925424	16.40069486	17.59013876	19.73671606
1 I ZF1	1.089614069	1.089892418	1.089996813	1.090106276	1.090228469	1.090313765	1.090379416	1.090478625	1.090587294
1 I ZF2	1.119457672	1.123263319	1.124517333	1.125777983	1.127139067	1.128068317	1.128774761	1.129831144	1.130976850
1 I ZF3	1.222258941	1.234334760	1.238450423	1.242642436	1.247222120	1.250377057	1.252789279	1.256416336	1.260373927
1 I ZF4	1.599897419	1.628487247	1.638647604	1.649201818	1.660955440	1.669178774	1.675530509	1.685177088	1.695819642
1 I ZF5	4.079774126	4.180456997	4.216818916	4.254912387	4.297710469	4.327878508	4.351299864	4.387055662	4.426736177

(5) P 8+2	TH = 68	SBE = 1.078534743						MCPER FUNCTIONS	
	RC = 02	RC = 05	RC = 07	RC = 10	RC = 15	RC = 20	RC = 25	RC = 35	RC = 50
AMAX DB	.0017375255	.0108709564	.0213327376	.0436480540	.0988323393	.1772876696	.2802872360	.5675287486	1.249387366
AMIN DB	32.91303102	42.77149785	46.42421905	50.32265715	54.80590248	58.04380033	60.61103832	64.64684306	69.37673938
MU I K	.2261249D-1	.7268171D-2	.4772974D-2	.3046963D-2	.1818465D-2	.1252593D-2	.9320690D-3	.5856766D-3	.3397528D-3
1 I PF1	1.071977045	.7913862806	.7067864607	.6253771886	.5417333799	.4874746570	.4480524596	.3927297543	.3391644158
1 I PQ1	.5756894607	.5796881016	.5859647484	.5965030080	.6155280109	.6360768380	.6583482170	.7095650245	.8118663846
1 I PF2	1.060927148	.9299310340	.8866330980	.8440257470	.7999147130	.7714680638	.7510261791	.7227989076	.6959911414
1 I PQ2	1.328481570	1.341548988	1.382600328	1.452490662	1.575681971	1.703319011	1.836018548	2.124016652	2.652832678
1 I PF3	1.055892568	1.008129553	.9908603912	.9731175742	.9539104093	.9410153323	.9314585648	.9177603401	.9039815768
1 I PQ3	3.915106557	3.844582965	3.934419607	4.108983314	4.434260346	4.778643379	5.139391771	5.925190036	7.369524313
2 I PFC	1.062331188	1.046865163	1.040628249	1.033839061	1.026021896	1.020477795	1.016203265	1.009813045	1.003056438
2 I PQC	19.29798977	16.72045819	16.33732047	16.21483559	16.44015224	16.86833956	17.38994670	18.60335669	20.82177880
1 I ZF1	1.081562890	1.081823475	1.081923400	1.082028918	1.082147383	1.082230409	1.082294466	1.082391481	1.082497997
1 I ZF2	1.109156375	1.112910581	1.114148075	1.115392653	1.116737192	1.117655727	1.118354379	1.119399735	1.120534387
1 I ZF3	1.205888266	1.217802862	1.221880659	1.226041386	1.230594469	1.233735323	1.236138974	1.239756569	1.243708337
1 I ZF4	1.567926278	1.596151952	1.606238921	1.616743471	1.628470758	1.636692273	1.643051268	1.652722312	1.663409201
1 I ZF5	3.972973943	4.072723027	4.108938542	4.146973450	4.189812264	4.220071584	4.243597019	4.279563917	4.319546728

(5) P 8+2 — TH = 69 — SBE = 1.071144994 — MCPER FUNCTIONS

MU	K	RC = 02	RC = 05	RC = 07	RC = 10	RC = 15	RC = 20	RC = 25	RC = 35	RC = 50
	AMAX DB	.0017375255	.0108709564	.0213327376	.0436480540	.0988323393	.1772876696	.2802872360	.5675287486	1.249387366
	AMIN DB	31.10585213	40.94925979	44.59819774	48.49328973	52.97339839	56.20942488	58.77538215	62.80949469	67.53784986
MU	K	.2784245D-1	.8964726D-2	.5889658D-2	.3761279D-2	.2245588D-2	.1547137D-2	.1151412D-2	.7236445D-3	.4198629D-3
1	PF1	1.101816737	.8096494383	.7221721570	.6382553723	.5522889079	.4966521192	.4562877732	.3997206574	.3450291526
1	PQ1	.5795779290	.5820437926	.5880130282	.5983125623	.6171495070	.6376103646	.6598387268	.7110386440	.8134187366
1	PF2	1.070178062	.9394346352	.8959594955	.8530926068	.8086443794	.7799522622	.7593238430	.7308293031	.7037630069
1	PQ2	1.377893434	1.377174686	1.415635078	1.483825454	1.606299540	1.734262741	1.867790430	2.158352109	2.692935295
1	PF3	1.055705318	1.009900419	.9931711464	.9759114261	.9571573396	.9445314905	.9351586723	.9217047230	.9081546431
1	PQ3	4.171595252	4.044978377	4.125086057	4.294395093	4.620028794	4.969336186	5.337297369	6.142027839	7.625630655
2	PFC	1.058153328	1.044054803	1.038291263	1.031979270	1.024669695	1.019461859	1.015434470	1.009395611	1.002988696
2	PQC	21.28125902	18.12742749	17.62637087	17.41549208	17.57889363	17.98632821	18.50563357	19.74254903	22.03808206
1	ZF1	1.073956774	1.074198018	1.074293036	1.074394203	1.074508533	1.074589027	1.074651299	1.074745846	1.074849924
1	ZF2	1.099358118	1.103053535	1.104272008	1.105497967	1.106823196	1.107729129	1.108418563	1.109450773	1.110572109
1	ZF3	1.190139854	1.201862962	1.205894151	1.210015313	1.214533381	1.217654694	1.220045775	1.223648156	1.227588165
1	ZF4	1.536760115	1.564565793	1.574562233	1.585000720	1.596685056	1.604894138	1.611252745	1.620937530	1.631658122
1	ZF5	3.867925674	3.966602952	4.002629504	4.040565154	4.083403771	4.113728595	4.137339852	4.173492914	4.213754415

(5) P 8+2 — TH = 70 — SBE = 1.064177772 — MCPER FUNCTIONS

MU	K	RC = 02	RC = 05	RC = 07	RC = 10	RC = 15	RC = 20	RC = 25	RC = 35	RC = 50
	AMAX DB	.0017375255	.0108709564	.0213327376	.0436480540	.0988323393	.1772876696	.2802872360	.5675287486	1.249387366
	AMIN DB	29.29452710	39.12091088	42.76566975	46.65708080	51.13374635	54.36772046	56.93227348	60.96453075	65.69119612
MU	K	.3429838D-1	.1106508D-1	.7273049D-2	.4646714D-2	.2775318D-2	.1912555D-2	.1423595D-2	.8948978D-3	.5193262D-3
1	PF1	1.134082045	.8292477826	.7386403543	.6520058029	.5635318231	.5064124010	.4650367980	.4071371117	.3512427998
1	PQ1	.5840219876	.5846650636	.5902757756	.6002986120	.6189182264	.6392769800	.6614545299	.7126308967	.8150911609
1	PF2	1.079327728	.9491252538	.9055256194	.8624353088	.8176734932	.7887448702	.7679337572	.7391741538	.7118486085
1	PQ2	1.434055221	1.416882320	1.452253605	1.518385149	1.639903461	1.768122844	1.902484075	2.195739801	2.736489129
1	PF3	1.055320650	1.011575736	.9954204221	.9786765855	.9604082944	.9480719409	.9388971930	.9257064103	.9124024082
1	PQ3	4.466364465	4.271247102	4.339315873	4.501781312	4.826895117	5.181100435	5.556641872	6.381709107	7.907981001
2	PFC	1.054013548	1.041247616	1.035950567	1.030111231	1.023306913	1.018435143	1.014655414	1.008969593	1.002915110
2	PQC	23.63100892	19.76000193	19.11337041	18.79302390	18.87841488	19.25789377	19.77155509	21.03087596	23.40919394
1	ZF1	1.066782472	1.067002762	1.067092428	1.067188825	1.067298600	1.067376291	1.067436579	1.067528370	1.067629713
1	ZF2	1.090050240	1.093679065	1.094875887	1.096080547	1.097383546	1.098274882	1.098953584	1.099970401	1.101076010
1	ZF3	1.175000668	1.186499486	1.190474588	1.194547206	1.199021123	1.202116968	1.204491130	1.208072040	1.211993825
1	ZF4	1.506364475	1.533691022	1.543578661	1.553933524	1.565557026	1.573742195	1.580092118	1.589778976	1.600521651
1	ZF5	3.764469707	3.861930175	3.897723120	3.935516692	3.978311855	4.008674820	4.032352453	4.068665017	4.109179837

(5) P 8+2 — TH = 71 — SBE = 1.057620681 — MCPER FUNCTIONS

MU		RC = 02	RC = 05	RC = 07	RC = 10	RC = 15	RC = 20	RC = 25	RC = 35	RC = 50
	AMAX DB	.0017375255	.0108709564	.0213327376	.0436480540	.0988323393	.1772876696	.2802872360	.5675287486	1.249387366
	AMIN DB	27.47713796	37.28402994	40.92414724	44.81149327	49.23436817	52.51608570	55.07909539	59.10931400	63.83412255
	K	.4228079D-1	.1367094D-1	.8990682D-2	.5746790D-2	.3433852D-2	.2366986D-2	.1762160D-2	.1107986D-2	.6431227D-3
1	PF1	1.169065215	.8503388998	.7563157897	.6667263030	.5755369116	.5168176340	.4743534900	.4150228958	.3578405474
1	PQ1	.5891431324	.5875992399	.5927888393	.6024889050	.6208559291	.6410956055	.6632129574	.7143575948	.8168991226
1	PF2	1.088308497	.9589857662	.9153257423	.8720569961	.8270121726	.7978594403	.7768714786	.7478512070	.7202672152
1	PQ2	1.498360910	1.461395683	1.493062698	1.556692290	1.676956649	1.805338854	1.940531083	2.236618307	2.783977155
1	PF3	1.054724009	1.013142684	.9975970097	.9814037780	.9636562643	.9516312312	.9426698236	.9297627092	.9167237191
1	PQ3	4.807936410	4.528436071	4.581530652	4.735117531	5.058538918	5.417526403	5.801017414	6.647972534	8.220772207
2	PFC	1.049920414	1.038448277	1.033609664	1.028237380	1.021934992	1.017398524	1.013866600	1.008535039	1.002835364
2	PQC	26.44350910	21.67040446	20.84242842	20.38544200	20.37202263	20.71411025	21.21759543	22.49729919	24.96446822
1	ZF1	1.060027709	1.060225408	1.060309267	1.060400468	1.060505255	1.060579862	1.060637960	1.060726698	1.060824995
1	ZF2	1.081221494	1.084775384	1.085947781	1.087128317	1.088406004	1.089280631	1.089946997	1.090946031	1.092033348
1	ZF3	1.160459932	1.171698870	1.175607596	1.179621914	1.184041746	1.187105668	1.189458173	1.193010787	1.196907287
1	ZF4	1.476706846	1.503491428	1.513250754	1.523503148	1.535046523	1.543195336	1.549527554	1.559203747	1.569955769
1	ZF5	3.662444361	3.758534483	3.794046814	3.831653114	3.874358984	3.904730968	3.928454210	3.964897694	4.005638497

(5) P 8+2 — TH = 72 — SBE = 1.051462224 — MCPER FUNCTIONS

MU		RC = 02	RC = 05	RC = 07	RC = 10	RC = 15	RC = 20	RC = 25	RC = 35	RC = 50
	AMAX DB	.0017375255	.0108709564	.0213327376	.0436480540	.0988323393	.1772876696	.2802872360	.5675287486	1.249387366
	AMIN DB	25.65174847	35.43596534	39.07089125	42.95372625	47.42241280	50.65164177	53.21295063	57.24092322	61.96368626
	K	.5216901D-1	.1691226D-1	.1112898D-1	.7117274D-2	.4254802D-2	.2933715D-2	.2184502D-2	.1373896D-2	.7976561D-3
1	PF1	1.207099820	.8731071786	.7753444734	.6825312668	.5883914689	.5279402383	.4843006699	.4234288996	.3648632374
1	PQ1	.5950993133	.5909055419	.5955965897	.6049174449	.6229890300	.6430890802	.6651348485	.7162376557	.8188610420
1	PF2	1.097039336	.9689933966	.9253506194	.8819590797	.8366703073	.8073100831	.7861536029	.7568798013	.7290401003
1	PQ2	1.572600944	1.511618430	1.538814000	1.599389586	1.718024621	1.846445520	1.982455054	2.281517669	2.835981004
1	PF3	1.053900388	1.014587310	.9996885464	.9840826336	.9668932840	.9552030798	.9464715374	.9338703669	.9211170338
1	PQ3	5.207427924	4.822952901	4.857302999	4.999381104	5.319537797	5.683063175	6.074863214	6.945421213	8.569152906
2	PFC	1.045883201	1.035661947	1.031272449	1.026360453	1.020555580	1.016353015	1.013068630	1.008092038	1.002749132
2	PQC	29.84843709	23.92703211	22.87085528	22.24177110	22.10236603	22.39458985	22.88172907	24.17845973	26.74088339
1	ZF1	1.053681123	1.053854582	1.053932175	1.054017747	1.054117105	1.054188341	1.054244036	1.054329413	1.054424341
1	ZF2	1.072862049	1.076332003	1.077477045	1.078630473	1.079879592	1.080735274	1.081387603	1.082366316	1.083432607
1	ZF3	1.146509195	1.157449594	1.161280760	1.165226161	1.169581081	1.172606039	1.174931712	1.178448568	1.182312054
1	ZF4	1.447756571	1.473932139	1.483542260	1.493671915	1.505114297	1.513213230	1.519517902	1.529169487	1.539916849
1	ZF5	3.561684359	3.656239902	3.691421886	3.728793033	3.771360878	3.801710793	3.825457403	3.862001073	3.902938312

(5) P 8+2 — TH = 73 — SBE = 1.045691756 — MCPER FUNCTIONS

MU	K	RC = 02	RC = 05	RC = 07	RC = 10	RC = 15	RC = 20	RC = 25	RC = 35	RC = 50
	AMAX DB	.0017375255	.0108709564	.0213327376	.0436480540	.0988323393	.1772876696	.2802872360	.5675287486	1.249387366
	AMIN DB	23.81646960	33.57379106	37.20285831	41.08065641	45.54469463	48.77116936	51.33059770	55.35608829	60.07659143
MU	K	.6444311D-1	.2095610D-1	.1379930D-1	.8830132D-2	.5281597D-2	.3642852D-2	.2713127D-2	.1706851D-2	.9912208D-3
1	PF1	1.248563681	.8977696043	.7958984798	.6995554527	.6021981711	.5398652780	.4949520463	.4324147381	.3723586405
1	PQ1	.6020989860	.5946590258	.5987546039	.6076263706	.6253499322	.6452852507	.6672455064	.7182939257	.8209990595
1	PF2	1.105423295	.9791182681	.9355864671	.8921406085	.8466572834	.8171113881	.7957978086	.7662810576	.7381908469
1	PQ2	1.659108532	1.568692944	1.590449711	1.647276553	1.763805764	1.892100329	2.028897810	2.331084962	2.893208069
1	PF3	1.052834421	1.015894434	1.001681381	.9867015387	.9701102859	.9587802388	.9502964582	.9380254654	.9255803406
1	PQ3	5.679571712	5.163044946	5.173748670	5.300882346	5.615655608	5.983289345	6.383729173	7.279786916	8.959509579
2	PFC	1.041911976	1.032894337	1.028943270	1.024483533	1.019170556	1.015299794	1.012262215	1.007640725	1.002656079
2	PQC	34.02351938	26.62086020	25.27422711	24.42615157	24.12481459	24.35051278	24.81285335	26.12131650	28.78561461
1	ZF1	1.047732209	1.047879785	1.047950653	1.048030161	1.048123642	1.048191212	1.048244287	1.048325985	1.048417209
1	ZF2	1.064963523	1.068339722	1.069454303	1.070577469	1.071794576	1.072628945	1.073265431	1.074221123	1.075263474
1	ZF3	1.133142411	1.143742245	1.147483675	1.151348587	1.155626774	1.158605066	1.160898243	1.164371161	1.168193146
1	ZF4	1.419484830	1.444979415	1.454417898	1.464402949	1.475721707	1.483756015	1.490022371	1.499634028	1.510361267
1	ZF5	3.462019176	3.554862164	3.589660848	3.626745873	3.669123675	3.699418214	3.723164282	3.759774976	3.800876600

(5) P 8+2 — TH = 74 — SBE = 1.040299436 — MCPER FUNCTIONS

MU	K	RC = 02	RC = 05	RC = 07	RC = 10	RC = 15	RC = 20	RC = 25	RC = 35	RC = 50
	AMAX DB	.0017375255	.0108709564	.0213327376	.0436480540	.0988323393	.1772876696	.2802872360	.5675287486	1.249387366
	AMIN DB	21.96958013	31.69425576	35.31663472	39.18876435	43.64761516	46.87102836	49.42836945	53.45110749	58.16910546
MU	K	.7971150D-1	.2601880D-1	.1714621D-1	.1097897D-1	.6570815D-2	.4533656D-2	.3377392D-2	.2125419D-2	.1234652D-2
1	PF1	1.293878469	.9245829787	.8181820662	.7179588810	.6170788020	.5526935233	.5063948473	.4420508512	.3803831131
1	PQ1	.6104218164	.5989561803	.6023334301	.6106685576	.6279788503	.6477184513	.6695759906	.7205542865	.8233400578
1	PF2	1.113344500	.9893215494	.9460136332	.9025974296	.8569815950	.8272782883	.8058228828	.7760780981	.7477457178
1	PQ2	1.760973158	1.634083680	1.649166658	1.701360423	1.815173038	1.943121237	2.080655214	2.386119077	2.956528203
1	PF3	1.051510539	1.017047535	1.003560424	.9892474618	.9732969231	.9623543279	.9541377069	.9422232919	.9301110577
1	PQ3	6.244238147	5.559491204	5.540087692	5.647734387	5.954248155	6.325292167	6.734643829	7.658297316	9.399862326
2	PFC	1.038017698	1.030151788	1.026626995	1.022610104	1.017782077	1.014240234	1.011448208	1.007181296	1.002555857
2	PQC	39.21703354	29.87499877	28.15387220	27.02384913	26.51236430	26.64900405	27.07486488	28.38692415	31.15969740
1	ZF1	1.042171261	1.042291340	1.042355030	1.042428040	1.042515195	1.042573802	1.042629035	1.042706728	1.042793902
1	ZF2	1.057519038	1.060790636	1.061871446	1.062961011	1.064142460	1.064953004	1.065571730	1.066501531	1.067516832
1	ZF3	1.120356068	1.130569609	1.134208036	1.137979823	1.142168344	1.145091528	1.147345989	1.150765978	1.154537112
1	ZF4	1.391864674	1.416600449	1.425843110	1.435659893	1.446830395	1.454783945	1.461000157	1.470554989	1.481244963
1	ZF5	3.363271254	3.454205627	3.488564171	3.525308482	3.567440445	3.597643760	3.621363477	3.658005261	3.699236361

(5) P 8+2 — TH = 75 SBE = 1.035276180 — MCPER FUNCTIONS

MU		RC = 02	RC = 05	RC = 07	RC = 10	RC = 15	RC = 20	RC = 25	RC = 35	RC = 50
	AMAX DB	.0017375255	.0108709564	.0213327376	.0436480540	.0988323393	.1772876696	.2802872360	.5675287486	1.249387366
	AMIN DB	20.10973878	29.79372534	33.40835544	37.27404105	41.72706242	44.94705428	47.50206790	51.52174045	56.23695045
	K	.9874453D-1	.3238275D-1	.2135906D-1	.1368667D-1	.8196848D-2	.5657796D-2	.4215961D-2	.2654074D-2	.1542242D-2
1	PF1	1.343502186	.9538528851	.8424395412	.7379332311	.6331791550	.5665454851	.5187332894	.4524212710	.3890037826
1	PQ1	.6204497612	.6039230358	.6064239673	.6141112887	.6309263522	.6504315501	.6721648943	.7230531665	.8259170504
1	PF2	1.120664633	.9995530640	.9566048475	.9133210564	.8676502957	.8378258420	.8162487152	.7862962960	.7577341073
1	PQ2	1.882358496	1.709697596	1.716507915	1.762928261	1.873232500	2.000539332	2.138726985	2.447618847	3.027024276
1	PF3	1.049913199	1.018028634	1.005308970	.9917057473	.9764413580	.9659156316	.9579872128	.9464581781	.9347059082
1	PQ3	6.928754780	6.026623771	5.968464456	6.050533653	6.344845679	6.718209068	7.136637977	8.090196014	9.900422707
2	PFC	1.034212342	1.027441364	1.024329092	1.020744120	1.016392625	1.013175938	1.010627622	1.006714022	1.002448116
2	PQC	45.78322556	33.85929474	31.64819509	30.15025957	29.36292088	29.37959382	29.75265193	31.05595519	33.94335394
1	ZF1	1.036989315	1.037080347	1.037136421	1.037202507	1.037282891	1.037342237	1.037389403	1.037462760	1.037545531
1	ZF2	1.050523316	1.053678156	1.054721653	1.055774066	1.056915993	1.057700049	1.058298980	1.059199833	1.060184767
1	ZF3	1.108149341	1.117926811	1.121447762	1.125112608	1.129197283	1.132056082	1.134264980	1.137622125	1.141332077
1	ZF4	1.364871155	1.388763151	1.397783794	1.407406594	1.418401930	1.426256997	1.432410017	1.441889303	1.452522918
1	ZF5	3.265254058	3.354059483	3.387916231	3.424260898	3.466086797	3.496160103	3.519825463	3.556459205	3.597781570

(5) P 8+2 — TH = 76 SBE = 1.030613629 — MCPER FUNCTIONS

MU		RC = 02	RC = 05	RC = 07	RC = 10	RC = 15	RC = 20	RC = 25	RC = 35	RC = 50
	AMAX DB	.0017375255	.0108709564	.0213327376	.0436480540	.0988323393	.1772876696	.2802872360	.5675287486	1.249387366
	AMIN DB	18.23634789	27.86812105	31.47360381	35.33186757	39.77827951	42.99442256	45.54682609	49.56306736	54.27516033
	K	.1225131D 0	.4041978D-1	.2668823D-1	.1711617D-1	.1025855D-1	.7084005D-2	.5280301D-2	.3325421D-2	.1933045D-2
1	PF1	1.397907390	.9859446979	.8689654380	.7597102919	.6506755725	.5815668139	.5320932238	.4636273313	.3983014753
1	PQ1	.6327146319	.6097271820	.6111453261	.6180415445	.6342569716	.6534788346	.6750608366	.7258336451	.8287711015
1	PF2	1.127218901	1.009748178	.9673228987	.9242971319	.8786682174	.8487688882	.8270962368	.7969635597	.7681890995
1	PQ2	2.028986021	1.798062167	1.794496752	1.833651308	1.939407176	2.065673863	2.204387376	2.516850975	3.106063202
1	PF3	1.048027264	1.018818174	1.006908501	.9940598716	.9795300027	.9694528499	.9618354785	.9507232978	.9393607602
1	PQ3	7.771529635	6.583873209	6.475175181	6.523368396	6.800008083	7.174018234	7.601506133	8.587494780	10.47439667
2	PFC	1.030509044	1.024770967	1.022055730	1.018890088	1.015005070	1.012108789	1.009801674	1.006239266	1.002332500
2	PQC	54.23951745	38.81324335	35.95025541	33.96474623	32.81038049	32.66398504	32.96111501	34.23696276	37.24391625
1	ZF1	1.032178097	1.032238643	1.032286682	1.032345438	1.032418616	1.032473406	1.032517282	1.032585969	1.032663975
1	ZF2	1.043972805	1.046997059	1.047999424	1.049010899	1.050109199	1.050863940	1.051440911	1.052309562	1.053260584
1	ZF3	1.096524281	1.105811504	1.109199172	1.112741950	1.116707201	1.119491393	1.121647172	1.124930501	1.128567812
1	ZF4	1.338481583	1.361435942	1.370206012	1.379606742	1.390397370	1.398134392	1.404209755	1.413592649	1.424148521
1	ZF5	3.167769989	3.254192985	3.287480139	3.323360910	3.364815235	3.394716283	3.418296711	3.454879543	3.496251105

(5) P 8+2	TH = 77	SBE = 1.026304108						MCPER FUNCTIONS	
	RC = 02	RC = 05	RC = 07	RC = 10	RC = 15	RC = 20	RC = 25	RC = 35	RC = 50
AMAX DB	.0017375255	.0108709564	.0213327376	.0436480540	.0988323393	.1772876696	.2802872360	.5675287486	1.249387366
AMIN DB	16.35016239	25.91285960	29.50728815	33.35685871	37.79569047	41.00746743	43.55692334	47.56930033	52.27788919
MU I K	.1522271D 0	.5062407D-1	.3346845D-1	.2148607D-1	.1288889D-1	.8904850D-2	.6639782D-2	.4183454D-2	.2432795D-2
1 I PF1	1.457532406	1.021296773	.8981177040	.7835732624	.6697838110	.5979356469	.5466284710	.4757927286	.4083747118
1 I PQ1	.6479722929	.6165960279	.6166566008	.6225738108	.6380544629	.6569301720	.6783260337	.7289504441	.8319540374
1 I PF2	1.132811682	1.019823707	.9781175137	.9355033138	.8900368523	.8601215126	.8383872677	.8081106454	.7791481637
1 I PQ2	2.208887585	1.902595357	1.885837432	1.915739695	2.015560087	2.140241603	2.279287757	2.595448065	3.195397900
1 I PF3	1.045838603	1.019394896	1.008338460	.9962911502	.9825472006	.9729527854	.9656712846	.9550104091	.9440704219
1 I PQ3	8.827864335	7.258172080	7.082555236	7.085352970	7.336612812	7.708721840	8.144941332	9.166088195	11.13916782
2 I PFC	1.026922275	1.022149470	1.019813901	1.017053167	1.013622747	1.011041008	1.008971824	1.005757512	1.002208657
2 I PQC	65.36154740	45.08302665	41.33586619	38.69251883	37.04196198	36.67122982	36.85911599	38.07907485	41.20792212
1 I ZF1	1.027729948	1.027758755	1.027798379	1.027849425	1.027914984	1.027964931	1.028005297	1.028068982	1.028141859
1 I ZF2	1.037865853	1.040743560	1.041700640	1.042667119	1.043717424	1.044439845	1.044992555	1.045825536	1.046738859
1 I ZF3	1.085486022	1.094224118	1.097461234	1.100865356	1.104694035	1.107392329	1.109486618	1.112683950	1.116235857
1 I ZF4	1.312675921	1.334587532	1.343075654	1.352223441	1.362776750	1.370374012	1.376355594	1.385618745	1.396072779
1 I ZF5	3.070608072	3.154349322	3.186991028	3.222336938	3.263347725	3.293030113	3.316491969	3.352976592	3.394350712

(5) P 8+2	TH = 78	SBE = 1.022340595						MCPER FUNCTIONS	
	RC = 02	RC = 05	RC = 07	RC = 10	RC = 15	RC = 20	RC = 25	RC = 35	RC = 50
AMAX DB	.0017375255	.0108709564	.0213327376	.0436480540	.0988323393	.1772876696	.2802872360	.5675287486	1.249387366
AMIN DB	14.45428457	23.92281075	27.50349178	31.34265748	35.77266508	38.97943573	41.52553338	45.53352550	50.23814896
MU I K	.1893589D 0	.6365895D-1	.4215270D-1	.2709363D-1	.1626922D-1	.1124678D-1	.8389254D-2	.5288393D-2	.3076752D-2
1 I PF1	1.522683338	1.060435427	.9303347653	.8098720394	.6907712682	.6158728010	.5625296276	.4890705711	.4193452731
1 I PQ1	.6673205571	.6248453340	.6231749922	.6278619167	.6424296462	.6608771607	.6820415456	.7324742854	.8355323669
1 I PF2	1.137212365	1.029672487	.9889211085	.9469063261	.9017527310	.8718962190	.8501442147	.8197715168	.7906540354
1 I PQ2	2.433604884	2.028027029	1.994224889	2.012178247	2.104179638	2.226520507	2.365609404	2.685553725	3.297317367
1 I PF3	1.043335083	1.019735733	1.009575989	.9983783842	.9854748297	.9763999499	.9694813119	.9593095220	.9488283734
1 I PQ3	10.17955485	8.087812234	7.821971288	7.763032471	7.977851059	8.344167776	8.788274200	9.847451194	11.91808894
2 I PFC	1.023468057	1.019586891	1.017611570	1.015239301	1.012249553	1.009975227	1.008139838	1.005269396	1.002076249
2 I PQC	80.34596170	53.18345235	48.21002747	44.66037069	42.32618938	41.64205343	41.67172050	42.79209284	46.04013513
1 I ZF1	1.023637742	1.023633868	1.023664747	1.023707744	1.023765302	1.023810135	1.023846780	1.023905139	1.023972524
1 I ZF2	1.032202906	1.034915419	1.035822656	1.036739766	1.037737413	1.038424314	1.038950313	1.039743928	1.040615502
1 I ZF3	1.075042955	1.083168190	1.086235878	1.089483141	1.093156334	1.095756221	1.097779719	1.100877479	1.104329704
1 I ZF4	1.287437322	1.308186708	1.316358055	1.325218688	1.335498432	1.342931672	1.348801379	1.357918447	1.368243292
1 I ZF5	2.973541208	3.054237623	3.086147132	3.120878491	3.161365687	3.190777919	3.214083845	3.250417604	3.291742141

(5) TH = 79 SBE = 1.018716695 MCPER FUNCTIONS

MU	P 8+2	RC = 02	RC = 05	RC = 07	RC = 10	RC = 15	RC = 20	RC = 25	RC = 35	RC = 50
	AMAX DB	.0017375255	.0108709564	.0213327376	.0436480540	.0988323393	.1772876696	.2802872360	.5675287486	1.249387366
	AMIN DB	12.55573154	21.89230623	25.45529680	29.28166246	33.70119532	36.90214481	39.44437286	43.44734211	48.14744168
MU	K	.2356207D 0	.8042382D-1	.5336238D-1	.3434922D-1	.2065096D-1	.1428541D-1	.1066059D-1	.6724080D-2	.3914064D-2
1	PF1	1.593351048	1.103989896	.9661573432	.8390441349	.7139741731	.6356562214	.5800365933	.5036534287	.4313661438
1	PQ1	.6923896319	.6349252099	.6310046045	.6341175482	.6475324742	.6654424949	.6863152068	.7364984162	.8395931048
1	PF2	1.140152619	1.039156082	.9996429170	.9584577786	.9138050361	.8841026404	.8623895169	.8319837026	.8027558434
1	PQ2	2.720149716	2.181078731	2.124838129	2.127097501	2.208667677	2.327602096	2.466297911	2.790043688	3.414873997
1	PF3	1.040508155	1.019815745	1.010595647	1.000297432	.9882918013	.9797760572	.9732496524	.9636084610	.9536264097
1	PQ3	11.95125524	9.128861093	8.738734947	8.594282846	8.756426442	9.110947368	9.561225136	10.66130721	12.84327967
2	PFC	1.020164191	1.017094598	1.015457864	1.013455379	1.010890075	1.008914593	1.007307860	1.004775752	1.001934958
2	PQC	101.0987029	63.90561886	57.18664975	52.35727738	49.05974526	47.92926002	47.72693856	48.67943513	52.03448812
1	ZF1	1.019894768	1.019857779	1.019879657	1.019914325	1.019963550	1.020003023	1.020035749	1.020088470	1.020150009
1	ZF2	1.026986698	1.029512094	1.030364428	1.031227423	1.032167415	1.032815388	1.033312066	1.034062371	1.034887864
1	ZF3	1.065206778	1.072650791	1.075528433	1.078598843	1.082095656	1.084583240	1.086525579	1.089508571	1.092845063
1	ZF4	1.262752690	1.282202144	1.290017540	1.298552729	1.308518269	1.315760163	1.321497526	1.330438563	1.340602911
1	ZF5	2.876322263	2.953522276	2.984597662	3.018623069	3.058496172	3.087580340	3.110688351	3.146811988	3.188028049

(5) TH = 80 SBE = 1.015426612 MCPER FUNCTIONS

MU	P 8+2	RC = 02	RC = 05	RC = 07	RC = 10	RC = 15	RC = 20	RC = 25	RC = 35	RC = 50
	AMAX DB	.0017375255	.0108709564	.0213327376	.0436480540	.0988323393	.1772876696	.2802872360	.5675287486	1.249387366
	AMIN DB	10.66776050	19.81527284	23.35459168	27.16466341	31.57144060	34.76549546	37.30319846	41.30034388	45.99522996
MU	K	.2928276D 0	.1021495D 0	.6796267D-1	.4382953D-1	.2638931D-1	.1826944D-1	.1364081D-1	.8609597D-2	.5014625D-2
1	PF1	1.668892606	1.152702002	1.006255452	.8716435391	.7398222681	.6576419655	.5994568509	.5197880532	.4446331640
1	PQ1	.7256538830	.6474969327	.6405838813	.6416402522	.6535712461	.6707947108	.6912940168	.7411487030	.8442526997
1	PF2	1.141327606	1.048094932	1.010160746	.9700881266	.9261720247	.8967455067	.8751446655	.8447886570	.8155105788
1	PQ2	3.094293519	2.371602255	2.285156449	2.266379366	2.333803427	2.447794603	2.585435502	2.912874404	3.552239510
1	PF3	1.037355392	1.019608153	1.011369096	1.002020686	.9909734184	.9830593621	.9769571616	.9678922782	.9584541570
1	PQ3	14.33934884	10.46628481	9.900507945	9.634893614	9.719882678	10.05318280	10.50642585	11.64997869	13.96015592
2	PFC	1.017030487	1.014685576	1.013363314	1.011709443	1.009549755	1.007862893	1.006478520	1.004277671	1.001784512
2	PQC	130.7646625	78.51259558	69.23163722	62.54172175	57.84931944	56.06770738	55.51896104	56.19435878	59.62645791
1	ZF1	1.016494553	1.016424857	1.016437580	1.016463722	1.016504350	1.016538253	1.016566885	1.016613682	1.016669038
1	ZF2	1.022222334	1.024534947	1.025326691	1.026130377	1.027007345	1.027612747	1.028077321	1.028780109	1.029554884
1	ZF3	1.055992210	1.062683082	1.065348192	1.068219797	1.071517109	1.073876914	1.075726489	1.078577631	1.081780245
1	ZF4	1.238612906	1.256602246	1.264016904	1.272183220	1.281788509	1.288807978	1.294389616	1.303120232	1.313087884
1	ZF5	2.778676975	2.851808346	2.881925908	2.915137673	2.954292217	2.982982099	3.005844298	3.041690203	3.082730414

(5) P 8+2			TH = 81	SBE = 1.012465126						MCPER FUNCTIONS
			RC = 02	RC = 05	RC = 07	RC = 10	RC = 15	RC = 20	RC = 25	RC = 35	RC = 50
AMAX DB			.0017375255	.0108709564	.0213327376	.0436480540	.0988323393	.1772876696	.2802872360	.5675287486	1.249387366
AMIN DB			8.812954366	17.68564308	21.19190020	24.98035380	29.37107340	32.55675878	35.08906531	39.07935013	43.76814993
MU		K	.3625370D 0	.1305323D 0	.8717762D-1	.5636147D-1	.3399745D-1	.2355928D-1	.1760138D-1	.1111815D-1	.6480261D-2
1		PF1	1.747526990	1.207416915	1.051459161	.9083806716	.7688750623	.6822955393	.6211929529	.5377976295	.4594026815
1		PQ1	.7709469888	.6635664171	.6525670643	.6508681291	.6608444682	.6771723442	.6971842697	.7465998466	.8496712368
1		PF2	1.140405592	1.056254987	1.020309192	.9816977439	.9388155408	.9098213877	.8884284866	.8582320240	.8289850391
1		PQ2	3.596250560	2.614571078	2.486370287	2.438689597	2.486522826	2.593294036	2.728856915	3.059654951	3.715284583
1		PF3	1.033884341	1.019084582	1.011864812	1.003516436	.9934905412	.9862237793	.9805805857	.9721424493	.9632983983
1		PQ3	17.66490149	12.23418115	11.41238489	10.97021202	10.93989133	11.23679850	11.68718275	12.87579445	15.33508124
2		PFC	1.014088894	1.012374763	1.011340166	1.010010963	1.008235108	1.006824733	1.005655070	1.003776589	1.001624713
2		PQC	174.7423193	99.11400754	85.93338775	76.44176479	69.66192435	66.90159310	65.82240774	66.04058388	69.48595377
1		ZF1	1.013430610	1.013329987	1.013333555	1.013351087	1.013382952	1.013411126	1.013435520	1.013476146	1.013525008
1		ZF2	1.017917067	1.019987507	1.020712206	1.021450852	1.022258988	1.022817925	1.023247424	1.023898207	1.024617301
1		ZF3	1.047415999	1.053281018	1.055709186	1.058357906	1.061430115	1.063644857	1.065388636	1.068088627	1.071136725
1		ZF4	1.215011769	1.231355088	1.238316774	1.246064142	1.255256282	1.262017541	1.267416430	1.275896664	1.285625250
1		ZF5	2.680288606	2.748621059	2.777624997	2.809891929	2.848204054	2.876422244	2.898982928	2.934472605	2.975258649

(5) P 8+2			TH = 82	SBE = 1.009827573						MCPER FUNCTIONS
			RC = 02	RC = 05	RC = 07	RC = 10	RC = 15	RC = 20	RC = 25	RC = 35	RC = 50
AMAX DB			.0017375255	.0108709564	.0213327376	.0436480540	.0988323393	.1772876696	.2802872360	.5675287486	1.249387366
AMIN DB			7.026446373	15.49836151	18.95633731	22.71468641	27.08431302	30.25949735	32.78519480	36.76722347	41.44879686
MU		K	.4453256D 0	.1679121D 0	.1127673D 0	.7315865D-1	.4423687D-1	.3069200D-1	.2294776D-1	.1450904D-1	.8463698D-2
1		PF1	1.825654072	1.269021930	1.102785841	.9501771169	.8018763836	.7102402036	.6457853136	.5581166871	.4760193449
1		PQ1	.8343191556	.6847267270	.6679710050	.6624677901	.6697963017	.6849254270	.7042858107	.7531026392	.8560760546
1		PF2	1.137052281	1.063329720	1.029862512	.9931433867	.9516723565	.9233133347	.9022541087	.8723636474	.8432584535
1		PQ2	4.290845775	2.933743954	2.745949688	2.657330946	2.677295307	2.773357836	2.905219053	3.238631202	3.912568008
1		PF3	1.030117860	1.018215767	1.012047878	1.004748109	.9958084820	.9892376822	.9840913514	.9763357387	.9681421081
1		PQ3	22.47533014	14.65472663	13.44554390	12.73692782	12.52935188	12.76459817	13.20152964	14.43437253	17.06896631
2		PFC	1.011363358	1.010179483	1.009402790	1.008371204	1.006954023	1.005805781	1.004841580	1.003274406	1.001455478
2		PQC	242.7273938	129.4282683	110.0467809	96.15155229	86.11799020	81.83003469	79.91108835	79.36295787	82.69153861
1		ZF1	1.010696100	1.010568506	1.010563139	1.010572140	1.010595207	1.010617568	1.010637626	1.010671890	1.010713992
1		ZF2	1.014079413	1.015875805	1.016526084	1.017193311	1.017926313	1.018434607	1.018825863	1.019419850	1.020077957
1		ZF3	1.039494683	1.044466219	1.046631203	1.049030728	1.051849483	1.053899819	1.055523111	1.058050033	1.060919977
1		ZF4	1.191941757	1.206428433	1.212874863	1.220144321	1.228861468	1.235322666	1.240507100	1.248639831	1.258129012
1		ZF5	2.580763032	2.643375805	2.671061898	2.702217581	2.739535289	2.767188701	2.789381442	2.824421681	2.864860637

(5) TH = 83 SBE = 1.007509825 MCPER FUNCTIONS

MU	P 8+2	RC = 02	RC = 05	RC = 07	RC = 10	RC = 15	RC = 20	RC = 25	RC = 35	RC = 50
	AMAX DB		.0108709564	.0213327376	.0436480540	.0988323393	.1772876696	.2802872360	.5675287486	1.249387366
	AMIN DB		13.25163289	16.63596734	20.35005658	24.69045841	27.85187009	30.36917235	34.34096941	39.01376397
MU	K		.2174795D 0	.1472996D 0	.9604996D-1	.5827430D-1	.4049547D-1	.3030689D-1	.1918455D-1	.1120242D-1
1	PF1		1.338255595	1.161439603	.9982374589	.8398384791	.7423342145	.6739816330	.5813483817	.4949619447
1	PQ1		.7136215121	.6884552092	.6775036651	.6811188608	.6945909740	.7130537226	.7610323825	.8638031879
1	PF2		1.068916653	1.038509358	1.004217048	.9646400263	.9371817721	.9166234899	.8872369440	.8584261012
1	PQ2		3.368818381	3.092620371	2.943783723	2.922709798	3.002484868	3.127972305	3.462480834	4.157123905
1	PF3		1.016973199	1.011880080	1.005673336	.9978855172	.9920622230	.9874538400	.9804425445	.9729630195
1	PQ3		18.11839532	16.29554601	15.16670566	14.67604362	14.80560424	15.20935895	16.47999332	19.32298762
2	PFC		1.008119990	1.007568229	1.006803730	1.005716184	1.004813110	1.004043225	1.002773675	1.001276919
2	PQC		176.4512194	146.6720511	125.4778957	110.1062125	103.3162179	100.0077083	98.13479663	101.0803098
1	ZF1		1.008136101	1.008122356	1.008123133	1.008137555	1.008154118	1.008169810	1.008197600	1.008232746
1	ZF2		1.012208734	1.012776213	1.013364892	1.014015888	1.014469060	1.014818691	1.015350784	1.015942239
1	ZF3		1.036266970	1.038141175	1.040262995	1.042796979	1.044661234	1.046147417	1.048476238	1.051140741
1	ZF4		1.181789774	1.187645037	1.194365374	1.202533569	1.208644768	1.213576830	1.221405442	1.230494276
1	ZF5		2.535331718	2.561421632	2.591244804	2.627373231	2.654345671	2.676088588	2.710565383	2.750544022

(5) TH = 84 SBE = 1.005508280 MCPER FUNCTIONS

MU	P 8+2	RC = 02	RC = 05	RC = 07	RC = 10	RC = 15	RC = 20	RC = 25	RC = 35	RC = 50
	AMAX DB		.0108709564	.0213327376	.0436480540	.0988323393	.1772876696	.2802872360	.5675287486	1.249387366
	AMIN DB		10.95165260	14.21925793	17.86439561	22.16159455	25.30384463	27.80992921	31.76851025	36.43029254
MU	K		.2834114D 0	.1945526D 0	.1278734D 0	.7796870D-1	.5430099D-1	.4069149D-1	.2579731D-1	.1508292D-1
1	PF1		1.415212829	1.228710494	1.054130844	.8841743507	.7797996096	.7068552232	.6083637852	.5169234961
1	PQ1		.7548828150	.7168932569	.6977803329	.6959550721	.7070404450	.7242174901	.7709809986	.8733751310
1	PF2		1.072489314	1.045815279	1.014611818	.9775527675	.9513483252	.9315171767	.9029067545	.8746044040
1	PQ2		3.990436634	3.576765998	3.335013597	3.250717098	3.304780288	3.419272835	3.751833720	4.469929163
1	PF3		1.015332730	1.011320791	1.006243132	.9996708574	.9946489078	.9906228340	.9844243858	.9777314026
1	PQ3		23.36339624	20.51147523	18.68131147	17.71459521	17.65701391	17.98920107	19.27805435	22.37114601
2	PFC		1.006220126	1.005856935	1.005325107	1.004533675	1.003855712	1.003266702	1.002277873	1.001089455
2	PQC		254.4503029	206.0565525	171.9234607	147.1968239	136.0424272	130.2942728	126.0165018	128.0132964
1	ZF1		1.006028645	1.006007617	1.006000816	1.006007005	1.006017926	1.006029314	1.006050635	1.006078726
1	ZF2		1.008998321	1.009473774	1.009975990	1.010537495	1.010930748	1.011235162	1.011699956	1.012218736
1	ZF3		1.028719152	1.030275000	1.032088820	1.034303670	1.035957568	1.037287790	1.039389761	1.041817033
1	ZF4		1.157405738	1.162575996	1.168658702	1.176186894	1.181886903	1.186520101	1.193924893	1.202587790
1	ZF5		2.423513317	2.447615648	2.475796586	2.510471937	2.536611058	2.557798662	2.591566927	2.630942269

(5) P 8+2	TH = 85	SBE = 1.003819838						MCPER FUNCTIONS	
	RC = 02	RC = 05	RC = 07	RC = 10	RC = 15	RC = 20	RC = 25	RC = 35	RC = 50
AMAX DB		.0108709564	.0213327376	.0436480540	.0988323393	.1772876696	.2802872360	.5675287486	1.249387366
AMIN DB		8.621880061	11.69932830	15.23062425	19.45890878	22.57235547	25.06235787	29.00281677	33.65012400
MU I K		.3706005D 0	.2600361D 0	.1731685D 0	.1064277D 0	.7436734D-1	.5583186D-1	.3546983D-1	.2077274D-1
1 I PF1		1.498187996	1.305567295	1.119833442	.9369071820	.8244473008	.7460191430	.6404862811	.5429617493
1 I PQ1		.8171592154	.7586376265	.7265993862	.7163395272	.7237952320	.7390328473	.7839483882	.8856598597
1 I PF2		1.073371265	1.051167906	1.023864797	.9901380910	.9656654381	.9468741005	.9194234693	.8919382257
1 I PQ2		4.934604515	4.294352495	3.900112517	3.712349623	3.723580234	3.818522758	4.142851864	4.887275928
1 I PF3		1.013281997	1.010329841	1.006401253	1.001101865	.9969359798	.9935395704	.9882298848	.9824064224
1 I PQ3		31.92097239	27.21453275	24.12380460	22.29780350	21.88957209	22.07007095	23.32514860	26.71896380
2 I PFC		1.004507998	1.004293749	1.003955927	1.003421894	1.002945277	1.002520893	1.001791854	1.000894027
2 I PQC		395.6671443	311.0825710	251.9194407	209.2966116	189.8534809	179.4587081	170.4860197	170.2455144
1 I ZF1		1.004241939	1.004215581	1.004202370	1.004201116	1.004206752	1.004214025	1.004229048	1.004250131
1 I ZF2		1.006259542	1.006633764	1.007041043	1.007505017	1.007833266	1.008088681	1.008480503	1.008920267
1 I ZF3		1.021866490	1.023079833	1.024554943	1.026413582	1.027830078	1.028982946	1.030824871	1.032977502
1 I ZF4		1.133238312	1.137608717	1.142940774	1.149712721	1.154923870	1.159199240	1.166091559	1.174231678
1 I ZF5		2.306560540	2.328118381	2.354200898	2.387041774	2.412134509	2.432622519	2.465487285	2.504069726

(5) P 8+2	TH = 86	SBE = 1.002441898						MCPER FUNCTIONS	
	RC = 02	RC = 05	RC = 07	RC = 10	RC = 15	RC = 20	RC = 25	RC = 35	RC = 50
AMAX DB			.0213327376	.0436480540	.0988323393	.1772876696	.2802872360	.5675287486	1.249387366
AMIN DB			9.084923005	12.41819386	16.52669119	19.59231045	22.05689124	25.97021771	30.59709304
MU I K			.3513612D 0	.2393813D 0	.1491645D 0	.1048056D 0	.7891425D-1	.5029087D-1	.2952197D-1
1 I PF1			1.391380252	1.197520905	1.000986274	.8790903643	.7940433662	.6798661127	.5748107190
1 I PQ1			.8246182045	.7706365444	.7462792203	.7477881745	.7598817176	.8017895132	.9022356087
1 I PF2			1.053701961	1.031258454	1.001933459	.9798548522	.9625486296	.9368174490	.9106113561
1 I PQ2			5.449267991	4.783445838	4.411575828	4.345601272	4.403551457	4.705727690	5.478494816
1 I PF3			1.008875203	1.006085318	1.002100366	.9988428282	.9961253093	.9917879070	.9869297190
1 I PQ3			39.00426726	33.40799764	29.86752038	28.74063392	28.58362672	29.66437458	33.41012518
2 I PFC			1.002909132	1.002722305	1.002400991	1.002097465	1.001817959	1.001322625	1.000692473
2 I PQC			520.6477400	407.0119840	325.7277595	288.5323931	268.1908299	248.9947242	243.2302991
1 I ZF1			1.002742932	1.002725312	1.002717975	1.002718970	1.002722505	1.002731639	1.002745975
1 I ZF2			1.004275164	1.004579491	1.004937745	1.005195772	1.005398319	1.005711364	1.006065572
1 I ZF3			1.016616510	1.017725565	1.019189725	1.020339168	1.021290535	1.022833962	1.024667423
1 I ZF4			1.112669028	1.117104494	1.122965438	1.127584366	1.131423560	1.137685399	1.145172965
1 I ZF5			2.200637502	2.223922580	2.254334120	2.278056182	2.297630352	2.329308743	2.366828685

$P^{(6)}_{8+4}$

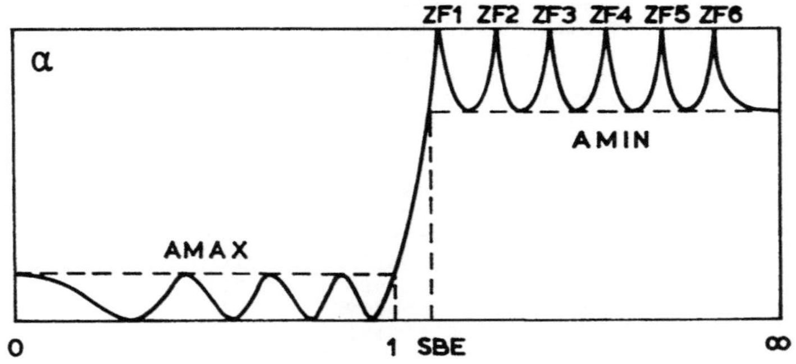

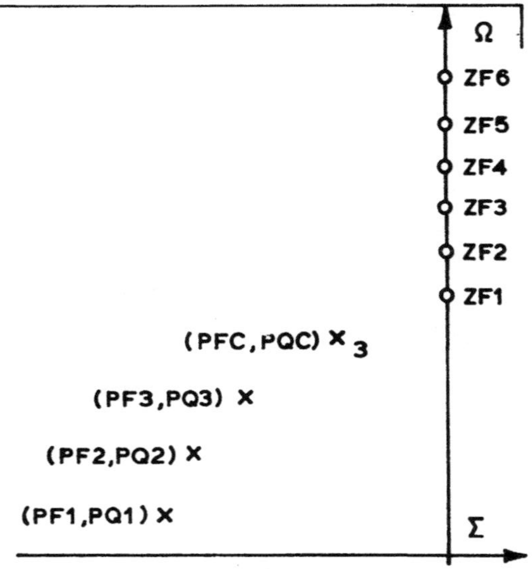

(6) P 8+4	TH = 44	SBE = 1.439556540						MCPER FUNCTIONS	
	RC = 02	RC = 05	RC = 07	RC = 10	RC = 15	RC = 20	RC = 25	RC = 35	RC = 50
AMAX DB	.0017375255	.0108709564	.0213327376	.0436480540	.0988323393	.1772876696	.2802872360	.5675287486	1.249387366
AMIN DB	90.65388197	102.6976711	107.1618436	111.9316966	117.4270473	121.4038434	124.5616173	129.5314925	135.3516004
MU I K	.2932958D-4	.7330210D-5	.4384376D-5	.2531717D-5	.1344769D-5	.8507615D-6	.5914515D-6	.3337532D-6	.1707733D-6
1 I PF1	.6740088005	.5233706983	.4734240129	.4233571632	.3699309605	.3342527457	.3078558559	.2701759565	.2330039632
1 I PQ1	.5390788061	.5530733798	.5615154335	.5737437505	.5940309173	.6150552631	.6374231553	.6881850465	.7886220077
1 I PF2	.8411957375	.7286062242	.6922683603	.6565878371	.6194824615	.5953038831	.5777125402	.5529362890	.5285041209
1 I PQ2	.8543259807	.9469046337	.9998484367	1.073224199	1.187975091	1.299801270	1.412748839	1.652593548	2.085585226
1 I PF3	1.016509091	.9379200225	.9119108518	.8859941863	.8585080673	.8401710122	.8265189812	.8066252190	.7857644959
1 I PQ3	1.608997772	1.799066193	1.908228853	2.058575006	2.291329907	2.515868129	2.740982985	3.215446265	4.066088355
3 I PFC	1.202576094	1.145087764	1.124607908	1.103424755	1.080081177	1.064005460	1.051791750	1.033683982	1.014493709
3 I PQC	4.553786964	4.595196042	4.672041293	4.796007736	5.001337193	5.200242402	5.394692839	5.781549084	6.397879554
1 I ZF1	1.446472474	1.446869217	1.447003260	1.447139153	1.447286839	1.447388089	1.447465227	1.447580768	1.447706268
1 I ZF2	1.506661787	1.510546991	1.511843467	1.513151482	1.514566905	1.515534169	1.516269618	1.517369107	1.518560935
1 I ZF3	1.659761452	1.672006688	1.676057888	1.680130604	1.684523269	1.687517422	1.689790226	1.693182517	1.696853005
1 I ZF4	2.016413153	2.044597387	2.053940620	2.063340000	2.073483097	2.080398893	2.085649076	2.093485318	2.101963196
1 I ZF5	3.002189142	3.063753380	3.084312006	3.105056995	3.127504864	3.142841601	3.154498997	3.171917752	3.190783439
1 I ZF6	8.451348599	8.659469386	8.729347473	8.800024530	8.876669659	8.929121982	8.969032116	9.028724174	9.093438878

(6) P 8+4 — TH = 45 SBE = 1.414213562 — MCPER FUNCTIONS

MU	K	RC = 02	RC = 05	RC = 07	RC = 10	RC = 15	RC = 20	RC = 25	RC = 35	RC = 50
	AMAX DB	.0017375255	.0108709564	.0213327376	.0436480540	.0988323393	.1772876696	.2802872360	.5675287486	1.249387366
	AMIN DB	88.04121237	100.0768657	104.5388540	109.3067535	114.8002617	118.7759547	121.9329719	126.9018443	132.7210338
MU	K	.3962227D-4	.9911896D-5	.5930036D-5	.3425014D-5	.1819646D-5	.1151336D-5	.8004817D-6	.4517600D-6	.2311790D-6
1	PF1	.6813126953	.5283823453	.4777899565	.4271253491	.3731105079	.3370645937	.3104074763	.2723718830	.2348651842
1	PQ1	.5395544769	.5534625523	.5618828808	.5740931844	.5943661723	.6153850553	.6377520233	.6885204180	.7889834813
1	PF2	.8471693407	.7335090001	.6968488119	.6608646039	.6234581627	.5990923389	.5813696902	.5564164728	.5318206701
1	PQ2	.8604622323	.9524529991	1.005331479	1.078733841	1.193658722	1.305730138	1.418964388	1.659483163	2.093785818
1	PF3	1.018653818	.9404071504	.9144717441	.8886136597	.8611769680	.8428678909	.8292353510	.8093701627	.7885434469
1	PQ3	1.636411743	1.824782380	1.933965941	2.084769731	2.318720280	2.544693258	2.771392755	3.249440717	4.106855634
3	PFC	1.197719480	1.142071298	1.122152680	1.101504482	1.078700909	1.062968967	1.051001987	1.033238151	1.014385243
3	PQC	4.710432079	4.727282184	4.798327098	4.917778035	5.119968722	5.317999521	5.512644765	5.901584819	6.523647810
1	ZF1	1.420801361	1.421197636	1.421332040	1.421468529	1.421617101	1.421719991	1.421796862	1.421913452	1.422040224
1	ZF2	1.478154358	1.482057885	1.483363472	1.484682105	1.486110568	1.487087637	1.487831013	1.488943094	1.490149546
1	ZF3	1.624388974	1.636757528	1.640855298	1.644977745	1.649427339	1.652462261	1.654767083	1.658208883	1.661935270
1	ZF4	1.966755814	1.995279540	2.004751221	2.014287462	2.024586635	2.031613729	2.036951002	2.044921336	2.053549737
1	ZF5	2.918095012	2.980396065	3.001248923	3.022313369	3.045131146	3.060734380	3.072601492	3.090344654	3.109575963
1	ZF6	8.196010065	8.406543538	8.477422750	8.549201925	8.627138152	8.680528257	8.721179895	8.782023803	8.848041955

(6) P 8+4 — TH = 46 SBE = 1.390163591 — MCPER FUNCTIONS

MU	K	RC = 02	RC = 05	RC = 07	RC = 10	RC = 15	RC = 20	RC = 25	RC = 35	RC = 50
	AMAX DB	.0017375255	.0108709564	.0213327376	.0436480540	.0988323393	.1772876696	.2802872360	.5675287486	1.249387366
	AMIN DB	85.47194537	97.49876757	101.9583831	106.7241600	112.2156653	116.1901589	119.3463533	124.3141350	130.1323253
MU	K	.5326019D-4	.1333711D-4	.7981433D-5	.4610967D-5	.2450286D-5	.1550572D-5	.1078158D-5	.6085458D-6	.3114467D-6
1	PF1	.6889701242	.5336194816	.4823479944	.4310558406	.3764241171	.3399934184	.3130642414	.2746571244	.2368011720
1	PQ1	.5400589897	.5538725583	.5622692520	.5744599606	.5947174577	.6157302512	.6380959992	.6888708522	.7893608491
1	PF2	.8533481580	.7385858273	.7015927658	.6652945646	.6275765412	.6030168050	.5851581253	.5600215168	.5352560895
1	PQ2	.8669797892	.9583113568	1.011110730	1.084531587	1.199629810	1.311952520	1.425482948	1.666701538	2.102369666
1	PF3	1.020810000	.9429408264	.9170875544	.8912950552	.8639139491	.8456363381	.8320256174	.8121920729	.7914023193
1	PQ3	1.665646175	1.852065411	1.961228140	2.112474746	2.347645208	2.575103050	2.803451130	3.285241453	4.149744816
3	PFC	1.192782472	1.138994019	1.119644981	1.099540713	1.077287221	1.061906003	1.050190998	1.032778766	1.014271161
3	PQC	4.879715834	4.869088992	4.933631776	5.047989110	5.246567518	5.443500022	5.638231951	6.029215116	6.657186556
1	ZF1	1.396435928	1.396830964	1.396965510	1.397102388	1.397251641	1.397354242	1.397432550	1.397550058	1.397677967
1	ZF2	1.451055545	1.454972060	1.456285158	1.457612884	1.459052940	1.460038719	1.460789298	1.461912938	1.463132963
1	ZF3	1.590656705	1.603136952	1.607277761	1.611446560	1.615949639	1.619023108	1.621358328	1.624847336	1.628627292
1	ZF4	1.919222185	1.948060848	1.957654022	1.967320703	1.977769655	1.984904126	1.990325708	1.998426301	2.007201503
1	ZF5	2.837351075	2.900332345	2.921464254	2.942834647	2.966009692	2.981871931	2.993943663	3.012004550	3.031595460
1	ZF6	7.950430764	8.163180807	8.235009222	8.307844982	8.387030667	8.441334553	8.482711797	8.544687454	8.611992019

(6) P 8+4			TH = 47 SBE = 1.367327461							MCPER FUNCTIONS
			RC = 02	RC = 05	RC = 07	RC = 10	RC = 15	RC = 20	RC = 25	RC = 35	RC = 50
AMAX DB			.0017375255	.0108709564	.0213327376	.0436480540	.0988323393	.1772876696	.2802872360	.5675287486	1.249387366
AMIN DB			82.94368488	94.96093074	99.41797106	104.1814433	109.6707735	113.6439642	116.7992645	121.7658610	127.5829652
MU		K	.7125507D-4	.1786296D-4	.1069305D-4	.6179137D-5	.3284440D-5	.2078748D-5	.1445562D-5	.8160316D-6	.4176878D-6
1		PF1	.6970022245	.5390941710	.4871080381	.4351567196	.3798782225	.3430447021	.3158309950	.2770356975	.2388152116
1		PQ1	.5405947203	.5543048711	.5626758193	.5748451878	.5950857460	.6160917539	.6384559471	.6892371312	.7897549603
1		PF2	.8597369153	.7438419264	.7065053482	.6698826771	.6318423240	.6070818369	.5890822697	.5637556546	.5388144244
1		PQ2	.8739101625	.9645023275	1.017206925	1.090636785	1.205906835	1.318486750	1.432323048	1.674268180	2.111358841
1		PF3	1.022973100	.9455193324	.9197572905	.8940380069	.8667192084	.8484768746	.8348905157	.8150919599	.7943423696
1		PQ3	1.696859680	1.881039360	1.990132424	2.141802233	2.378214828	2.607208645	2.837271756	3.322969834	4.194894813
3		PFC	1.187768154	1.135856959	1.117085396	1.097533693	1.075840102	1.060816441	1.049358596	1.032305591	1.014151216
3		PQC	5.062978487	5.021553927	5.078798049	5.187405005	5.381830579	5.577405876	5.772095905	6.165062005	6.799115354
1		ZF1	1.373296407	1.373689410	1.373823867	1.373960921	1.374110643	1.374213718	1.374292466	1.374410751	1.374539658
1		ZF2	1.425280157	1.429204172	1.430523124	1.431858364	1.433308205	1.434301855	1.435058880	1.436193002	1.437425503
1		ZF3	1.558465731	1.571045632	1.575225813	1.579437446	1.583990416	1.587100108	1.589464033	1.592997837	1.596828915
1		ZF4	1.873683444	1.902811304	1.912518670	1.922309042	1.932901149	1.940138864	1.945641826	1.953868635	1.962786703
1		ZF5	2.759751320	2.823353581	2.844748583	2.866410694	2.889929667	2.906042984	2.918313939	2.936685463	2.956629570
1		ZF6	7.714001489	7.928764313	8.001487602	8.075332147	8.155723431	8.210915742	8.253001758	8.316087823	8.384660638

(6) P 8+4			TH = 48 SBE = 1.345632730							MCPER FUNCTIONS
			RC = 02	RC = 05	RC = 07	RC = 10	RC = 15	RC = 20	RC = 25	RC = 35	RC = 50
AMAX DB			.0017375255	.0108709564	.0213327376	.0436480540	.0988323393	.1772876696	.2802872360	.5675287486	1.249387366
AMIN DB			80.45416336	92.46103414	96.91528197	101.6762542	107.1632239	111.1350003	114.2893299	119.2546394	125.0705641
MU		K	.9490560D-4	.2382036D-4	.1426382D-4	.8244936D-5	.4383680D-5	.2774917D-5	.1929894D-5	.1089602D-5	.5577914D-6
1		PF1	.7054320694	.5448195073	.4920808260	.4394367272	.3834797711	.3462243576	.3187129571	.2795119269	.2409108361
1		PQ1	.5411643248	.5547611197	.5631039846	.5752500842	.5954721024	.6164705510	.6388328105	.6896203123	.7901667390
1		PF2	.8663403001	.7492827258	.7115919260	.6746341588	.6362605045	.6112922561	.5931468111	.5676233780	.5424999710
1		PQ2	.8812885151	.9710509332	1.023642940	1.097070730	1.212510080	1.325352918	1.439504969	1.682204408	2.120777421
1		PF3	1.025138194	.9481407436	.9224798076	.8968420452	.8695928874	.8513899919	.8378307698	.8180708469	.7973648886
1		PQ3	1.730229777	1.911842069	2.020808448	2.172876260	2.410550672	2.641132496	2.872979713	3.362759432	4.242458029
3		PFC	1.182679826	1.132661252	1.114474582	1.095483712	1.074359562	1.059700163	1.048504596	1.031818382	1.014025152
3		PQC	5.261746275	5.185734704	5.234774019	5.336882923	5.526537639	5.720456236	5.914951941	6.309817699	6.950122267
1		ZF1	1.351309812	1.351699961	1.351834093	1.351971101	1.352121069	1.352224477	1.352303563	1.352422482	1.352552239
1		ZF2	1.400750293	1.404676172	1.405999266	1.407340381	1.408798437	1.409798777	1.410561460	1.411704939	1.412948767
1		ZF3	1.527725688	1.540392772	1.544608510	1.548859315	1.553458427	1.556601913	1.558992770	1.562568843	1.566448475
1		ZF4	1.830021973	1.859412033	1.869225923	1.879132889	1.889861176	1.897197781	1.902779034	1.911127787	1.920184561
1		ZF5	2.685107410	2.749268717	2.770910045	2.792848885	2.816697702	2.833053716	2.845518175	2.864192813	2.884483303
1		ZF6	7.486164079	7.702728191	7.776289552	7.851092742	7.932643458	7.988697416	8.031474386	8.095648206	8.165469915

(6) P 8+4		TH = 49	SBE = 1.325012993						MCPER FUNCTIONS	
		RC = 02	RC = 05	RC = 07	RC = 10	RC = 15	RC = 20	RC = 25	RC = 35	RC = 50
AMAX DB		.0017375255	.0108709564	.0213327376	.0436480540	.0988323393	.1772876696	.2802872360	.5675287486	1.249387366
AMIN DB		78.00122933	89.99686904	94.44809107	99.20635337	104.6907634	108.6610055	111.8142820	116.7781949	122.5928394
MU	K	.1258747D-3	.3163418D-4	.1894940D-4	.1095676D-4	.5827226D-5	.3689349D-5	.2566173D-5	.1449073D-5	.7419216D-6
1	PF1	.7142848887	.5508097246	.4972780103	.4439053319	.3872362749	.3495387712	.3217157613	.2820904745	.2430918500
1	PQ1	.5417707806	.5552431095	.5635552965	.5756759912	.5958776964	.6168677241	.6392276218	.6900212362	.7905971928
1	PF2	.8731629143	.7549138612	.7168581149	.6795544990	.6408363590	.6156531681	.5973567194	.5716294562	.5463172961
1	PQ2	.8891541852	.9779849145	1.030444066	1.103856899	1.219461854	1.332573080	1.447050952	1.690533563	2.130651728
1	PF3	1.027299943	.9508029087	.9252537919	.8997065850	.8725350615	.8543761451	.8408470878	.8211297670	.8004712001
1	PQ3	1.765955700	1.944627054	2.053400271	2.205834361	2.444787142	2.677009821	2.910712959	3.404757018	4.292602021
3	PFC	1.177521013	1.129408141	1.111813274	1.093391110	1.072845637	1.058557064	1.047628818	1.031316895	1.013892702
3	PQC	5.477762996	5.362828276	5.402629415	5.497387252	5.681563322	5.873478593	6.067599717	6.464254441	7.110973285
1	ZF1	1.330409252	1.330795702	1.330929263	1.331065995	1.331215978	1.331319572	1.331398890	1.331518292	1.331648747
1	ZF2	1.377394606	1.381316567	1.382642037	1.383987331	1.385451867	1.386457775	1.387225294	1.388376956	1.389630909
1	ZF3	1.498353903	1.511095251	1.515342582	1.519628742	1.524270085	1.527444825	1.529860758	1.533476450	1.537401937
1	ZF4	1.788130253	1.817754187	1.827666537	1.837682631	1.848539751	1.855970656	1.861626936	1.870093122	1.879284200
1	ZF5	2.613246907	2.677902505	2.699772537	2.721972324	2.746136116	2.762725956	2.775377860	2.794347626	2.814977246
1	ZF6	7.266406115	7.484552305	7.558892411	7.634601686	7.717263261	7.774150602	7.817599672	7.882837208	7.953887185

(6) P 8+4		TH = 50	SBE = 1.305407289						MCPER FUNCTIONS	
		RC = 02	RC = 05	RC = 07	RC = 10	RC = 15	RC = 20	RC = 25	RC = 35	RC = 50
AMAX DB		.0017375255	.0108709564	.0213327376	.0436480540	.0988323393	.1772876696	.2802872360	.5675287486	1.249387366
AMIN DB		75.58283598	87.56632717	92.01427293	96.76959965	102.2512357	106.2198147	109.3719493	114.3343475	120.1476034
MU	K	.1662870D-3	.4184886D-4	.2507762D-4	.1450508D-4	.7716817D-5	.4886628D-5	.3399402D-5	.1919913D-5	.9831501D-6
1	PF1	.7235883202	.5570803235	.5027122561	.4485728069	.3911558693	.3529948513	.3248454970	.2847763737	.2453623566
1	PQ1	.5424174349	.5557528459	.5640314694	.5761243888	.5963038139	.6172844606	.6396415126	.6904410366	.7910474219
1	PF2	.8802092143	.7607411723	.7223097860	.6846494728	.6455754643	.6201699815	.6017172672	.5757789568	.5502712579
1	PQ2	.8975513011	.9853351003	1.037638330	1.111021241	1.226786742	1.340171504	1.454985439	1.699281248	2.141010586
1	PF3	1.029452572	.9535034275	.9280777427	.9026309107	.8755457300	.8574357459	.8439401562	.8242697605	.8036626608
1	PQ3	1.804261696	1.979565736	2.088068357	2.240829371	2.481073231	2.714990273	2.950624007	3.449124794	4.345511396
3	PFC	1.172295479	1.126098988	1.109102291	1.091256283	1.071298395	1.057387057	1.046731091	1.030800881	1.013753590
3	PQC	5.713027991	5.554193415	5.583574855	5.670006147	5.847891461	6.037401890	6.230935598	6.629236021	7.282523336
1	ZF1	1.310533324	1.310915205	1.311047943	1.311184160	1.311333919	1.311437547	1.311516984	1.311636712	1.311767707
1	ZF2	1.355147654	1.359059779	1.360385801	1.361733521	1.363202740	1.364213045	1.364984545	1.366143164	1.367405988
1	ZF3	1.470274653	1.483076873	1.487351675	1.491669216	1.496348705	1.499552044	1.501991109	1.505643641	1.509612150
1	ZF4	1.747909898	1.777737967	1.787740298	1.797857657	1.808835867	1.816356225	1.822084088	1.830662935	1.839983654
1	ZF5	2.544011699	2.609093933	2.631174159	2.653618268	2.678081337	2.694895619	2.707728544	2.726984957	2.747945984
1	ZF6	7.054256236	7.273757589	7.348814524	7.425374829	7.509096190	7.566787092	7.610888318	7.677164070	7.749420340

(6) P 8+4 — TH = 51   SBE = 1.286759566   MCPER FUNCTIONS

MU	K	RC = 02	RC = 05	RC = 07	RC = 10	RC = 15	RC = 20	RC = 25	RC = 35	RC = 50
	AMAX DB	.0017375255	.0108709564	.0213327376	.0436480540	.0988323393	.1772876696	.2802872360	.5675287486	1.249387366
	AMIN DB	73.19703067	85.16738992	89.61179041	94.36393905	99.84257072	103.8093478	106.9602449	111.9210013	117.7327515
MU	K	.2188510D-3	.5516079D-4	.3306819D-4	.1913388D-4	.1018290D-4	.6449597D-5	.4487327D-5	.2534836D-5	.1298262D-5
1	PF1	.7333726966	.5636482137	.5083973541	.4534503174	.3952473802	.3566000839	.3281087568	.2875750676	.2477267885
1	PQ1	.5431080620	.5562925624	.5645344053	.5765969136	.5967518725	.6177220666	.6400757254	.6908809015	.7915186302
1	PF2	.8874834405	.7667706981	.7279530717	.6899251543	.6504837167	.6248484295	.6062340520	.5800772693	.5543670301
1	PQ2	.9065295048	.9931358374	1.045256864	1.118592501	1.234511904	1.348174951	1.463335345	1.708475608	2.151885630
1	PF3	1.031589835	.9562396277	.9309499518	.9056141606	.8786248054	.8605691542	.8471106332	.8274918707	.8069406586
1	PQ3	1.845400943	2.016850055	2.124991923	2.278031567	2.519574515	2.755239892	2.992881863	3.496042136	4.401390017
3	PFC	1.167007240	1.122735284	1.106342545	1.089079685	1.069717933	1.056190070	1.045811252	1.030270089	1.013607529
3	PQC	5.969842050	5.761377699	5.778984787	5.855971211	6.026632035	6.213271984	6.405967215	6.805731313	7.465729184
1	ZF1	1.291625578	1.292002000	1.292133652	1.292269109	1.292418397	1.292521898	1.292601341	1.292721231	1.292852598
1	ZF2	1.333949325	1.337845565	1.339170264	1.340518597	1.341990634	1.343004123	1.343778714	1.344943012	1.346213394
1	ZF3	1.443418503	1.456267710	1.460565696	1.464910477	1.469623847	1.472853005	1.475313167	1.478999625	1.483008175
1	ZF4	1.709270804	1.739271770	1.749355157	1.759565499	1.770656630	1.778261324	1.784057127	1.792743587	1.802189006
1	ZF5	2.477256626	2.542694849	2.564965827	2.587636754	2.612382523	2.629411310	2.642418446	2.661952494	2.683236701
1	ZF6	6.849279990	7.069901915	7.145611125	7.222964832	7.307692302	7.366155313	7.410887604	7.478174527	7.551613683

(6) P 8+4 — TH = 52   SBE = 1.269018215   MCPER FUNCTIONS

MU	K	RC = 02	RC = 05	RC = 07	RC = 10	RC = 15	RC = 20	RC = 25	RC = 35	RC = 50
	AMAX DB	.0017375255	.0108709564	.0213327376	.0436480540	.0988323393	.1772876696	.2802872360	.5675287486	1.249387366
	AMIN DB	70.84194523	82.79811815	87.23868441	91.98739423	97.46277351	101.4275992	104.5771558	109.5361334	115.3462518
MU	K	.2870138D-3	.7245929D-4	.4345760D-4	.2515535D-4	.1339249D-4	.8484379D-5	.5903944D-5	.3335749D-5	.1708785D-5
1	PF1	.7436713740	.5705318775	.5143483478	.4585500209	.3995203992	.3603625945	.3315126908	.2904924525	.2501899415
1	PQ1	.5438469309	.5568647530	.5650662203	.5770953798	.5972234378	.6181819815	.6405316281	.6913421360	.7920121374
1	PF2	.8949895308	.7730086670	.7337943702	.6953879310	.6555673525	.6296945930	.6109130208	.5845301304	.5586101278
1	PQ2	.9161448092	1.001425495	1.053334344	1.126602599	1.242667411	1.356612995	1.472130377	1.718147655	2.163311647
1	PF3	1.033704991	.9590085383	.9338684821	.9086553094	.8817720999	.8637766681	.8503591420	.8307971401	.8103066118
1	PQ3	1.889660210	2.056695555	2.164371693	2.317631178	2.560475484	2.797943370	3.037674286	3.545707973	4.460463552
3	PFC	1.161660581	1.119318658	1.103535048	1.086861841	1.068104388	1.054966053	1.044869146	1.029724266	1.013454221
3	PQC	6.250863148	5.986149893	5.990424914	6.056680941	6.219041273	6.402269978	6.593830693	6.994830248	7.661664620
1	ZF1	1.273634050	1.274004100	1.274134398	1.274268840	1.274417400	1.274520611	1.274599940	1.274719822	1.274851387
1	ZF2	1.313744329	1.317618517	1.318939963	1.320287041	1.321759965	1.322775376	1.323552131	1.324720776	1.325997346
1	ZF3	1.417721726	1.430603518	1.434920227	1.439287930	1.444030728	1.447282800	1.449761926	1.453479250	1.457524709
1	ZF4	1.672130401	1.702271436	1.712426480	1.722721072	1.733916501	1.741600122	1.747460012	1.756248739	1.765813616
1	ZF5	2.412848255	2.478568740	2.501010054	2.523889362	2.548900324	2.566133099	2.579307222	2.599109327	2.620707942
1	ZF6	6.651076143	6.872576426	6.948870663	7.026957492	7.112634692	7.171836650	7.217177711	7.285447123	7.360044235

(6) P 8+4			TH = 53 SBE = 1.252135658							MCPER FUNCTIONS	
			RC = 02	RC = 05	RC = 07	RC = 10	RC = 15	RC = 20	RC = 25	RC = 35	RC = 50
AMAX DB			.0017375255	.0108709564	.0213327376	.0436480540	.0988323393	.1772876696	.2802872360	.5675287486	1.249387366
AMIN DB			68.51578692	80.45664263	84.89306411	89.63805469	95.10991458	99.07262776	102.2207330	107.1777839	112.9861345
MU		K	.3751549D-3	.9487851D-4	.5693074D-4	.3296835D-4	.1755918D-4	.1112676D-4	.7743964D-5	.4376337D-5	.2242298D-5
1		PF1	.7545211092	.5777515564	.5205816803	.4638851802	.4039853697	.3642912196	.3350650674	.2935349270	.2527570143
1		PQ1	.5446388870	.5574722102	.5656292744	.5776218035	.5977202436	.6186657959	.6410107294	.6918261764	.7925293938
1		PF2	.9027310164	.7794614837	.7398403479	.7010445171	.6608329691	.6347149251	.6157604963	.5891436523	.5630064363
1		PQ2	.9264606158	1.010247054	1.061909489	1.135087076	1.251286647	1.365518404	1.481403405	1.728331634	2.175326989
1		PF3	1.035790771	.9618068623	.9368311429	.9117531487	.8849873116	.8670585134	.8536862621	.8341866054	.8137619666
1		PQ3	1.937365433	2.099345032	2.206433154	2.359841333	2.603982286	2.843306713	3.085210432	3.598343491	4.522982454
3		PFC	1.156260066	1.115850891	1.100680922	1.084603347	1.066457937	1.053714978	1.043904631	1.029163157	1.013293358
3		PQC	6.559174507	6.230539016	6.219685163	6.273728810	6.426545643	6.605734029	6.795811109	7.197762738	7.871538398
1		ZF1	1.256510838	1.256873585	1.257002253	1.257135418	1.257282985	1.257385735	1.257464826	1.257584523	1.257716107
1		ZF2	1.294481754	1.298327613	1.299643825	1.300987720	1.302459534	1.303475559	1.304253514	1.305425119	1.306706444
1		ZF3	1.393125796	1.406025225	1.410356014	1.414742138	1.419509710	1.422781651	1.425277505	1.429022484	1.433101557
1		ZF4	1.636412995	1.666659585	1.676876381	1.687246009	1.698536626	1.706293453	1.712213350	1.721098675	1.730777446
1		ZF5	2.350663800	2.416589652	2.439179865	2.462248140	2.487505803	2.504931426	2.518264869	2.538324846	2.560228516
1		ZF6	6.459273381	6.681402265	6.758211536	6.836968478	6.923536215	6.983442172	7.029368438	7.098589931	7.174318451

(6) P 8+4			TH = 54 SBE = 1.236067977							MCPER FUNCTIONS	
			RC = 02	RC = 05	RC = 07	RC = 10	RC = 15	RC = 20	RC = 25	RC = 35	RC = 50
AMAX DB			.0017375255	.0108709564	.0213327376	.0436480540	.0988323393	.1772876696	.2902872360	.5675287486	1.249387366
AMIN DB			66.21682991	78.14115496	82.57309780	87.31406732	92.78212022	96.74254732	99.88908138	104.8440462	110.6504821
MU		K	.4888307D-3	.1238632D-3	.7436098D-4	.4308208D-4	.2295588D-4	.1455032D-4	.1012852D-4	.5725293D-5	.2934107D-5
1		PF1	.7659624937	.5853294654	.5271153615	.4694702935	.4086536856	.3683955876	.3387743438	.2967094486	.2554336526
1		PQ1	.5454894491	.5581180706	.5662262074	.5781784310	.5982442145	.6191752711	.6415146978	.6923346077	.7930719960
1		PF2	.9107108977	.7861357094	.7460979394	.7069019669	.6662875482	.6399162785	.6207832061	.5939243540	.5675622436
1		PQ2	.9375489305	1.019648811	1.071025665	1.144085610	1.260406764	1.374927578	1.491190891	1.739065458	2.187974031
1		PF3	1.037839353	.9646309457	.9398354638	.9149062650	.8882700084	.8704148314	.8570925196	.8376512915	.8173081955
1		PQ3	1.988888407	2.145072874	2.251430408	2.404901556	2.650325939	2.891560364	3.135723954	3.654195312	4.589225427
3		PFC	1.150810563	1.112333924	1.097781411	1.082304883	1.064778805	1.052436847	1.042917582	1.028586509	1.013124621
3		PQC	6.898368131	6.496881698	6.468819492	6.508937053	6.650770605	6.825185418	7.013366907	7.415921196	8.096715539
1		ZF1	1.240211734	1.240566230	1.240692985	1.240824603	1.240970904	1.241073018	1.241151741	1.241271070	1.241402483
1		ZF2	1.276114662	1.279925819	1.281234765	1.282573493	1.284042132	1.285057414	1.285835570	1.287008689	1.288293275
1		ZF3	1.369576931	1.382478474	1.386818508	1.391218357	1.396005839	1.399294463	1.401804700	1.405573961	1.409683181
1		ZF4	1.602049189	1.632365027	1.642633131	1.653068066	1.664444238	1.672268215	1.678243796	1.687219705	1.697006457
1		ZF5	2.290590156	2.356641225	2.379357835	2.402594633	2.428079460	2.445686128	2.459170754	2.479477766	2.501676509
1		ZF6	6.273527362	6.496027644	6.573279164	6.652640401	6.740036560	6.800609693	6.847096264	6.917237603	6.994069259

(6) P 8+4	TH = 55	SBE = 1.220774589						MCPER FUNCTIONS	
	RC = 02	RC = 05	RC = 07	RC = 10	RC = 15	RC = 20	RC = 25	RC = 35	RC = 50
AMAX DB	.0017375255	.0108709564	.0213327376	.0436480540	.0988323393	.1772876696	.2802872360	.5675287486	1.249387366
AMIN DB	63.94340725	75.84989883	80.27700386	85.01362735	90.47756326	94.43551701	97.58035067	102.5330572	108.3374195
MU K	.6350818D-3	.1612520D-3	.9686119D-4	.5614598D-4	.2993104D-4	.1897685D-4	.1321242D-4	.7470456D-5	.3829385D-5
1 PF1	.7780404542	.5932900394	.5339691608	.4753212445	.4135378041	.3726862117	.3426497462	.3000235971	.2582260000
1 PQ1	.5464049269	.5588058678	.5668599808	.5787677725	.5987974927	.6197123632	.6420453826	.6928691829	.7936417064
1 PF2	.9189314938	.7930380361	.7525743437	.7129676878	.6719384801	.6453059344	.6259883152	.5988791956	.5722842762
1 PQ2	.9494918211	1.029685197	1.080731587	1.153642635	1.270069235	1.384881061	1.501533389	1.750391212	2.201299719
1 PF3	1.039842332	.9674767444	.9428786647	.9181130158	.8916196095	.8738456645	.8605783767	.8412222048	.8209467938
1 PQ3	2.044654879	2.194190242	2.299650762	2.453081901	2.699766155	2.942962892	3.189476660	3.713539230	4.659503514
3 PFC	1.145317250	1.108769877	1.094837889	1.079967218	1.063067267	1.051131690	1.041907887	1.027994068	1.012947680
3 PQC	7.272647955	6.787879893	6.740194209	6.764397520	6.893575240	7.062359882	7.248159176	7.650887469	8.338742725
1 ZF1	1.224695892	1.225041174	1.225165726	1.225295522	1.225440275	1.225541569	1.225619793	1.225738562	1.225869610
1 ZF2	1.258599735	1.262369733	1.263669335	1.265000856	1.266464190	1.267477324	1.268254639	1.269427769	1.270714055
1 ZF3	1.347025698	1.359913223	1.364257464	1.368666138	1.373468443	1.376770410	1.379292571	1.383082570	1.387218285
1 ZF4	1.568975369	1.599322250	1.609630639	1.620120602	1.631572136	1.639456843	1.645483522	1.654543630	1.664432074
1 ZF5	2.232523044	2.298615843	2.321435234	2.344819025	2.370510372	2.388285576	2.401912742	2.422455253	2.444938410
1 ZF6	6.093518061	6.316125212	6.393743356	6.473640182	6.561799613	6.623001137	6.670021708	6.741048722	6.818953405

(6) P 8+4	TH = 56	SBE = 1.206217949						MCPER FUNCTIONS	
	RC = 02	RC = 05	RC = 07	RC = 10	RC = 15	RC = 20	RC = 25	RC = 35	RC = 50
AMAX DB	.0017375255	.0108709564	.0213327376	.0436480540	.0988323393	.1772876696	.2802872360	.5675287486	1.249387366
AMIN DB	61.69390321	73.58116158	78.00304215	82.73496941	88.19445395	92.14973215	95.29272584	100.2429879	106.0451050
MU K	.8228200D-3	.2093832D-3	.1258485D-3	.7298801D-4	.3892936D-4	.2468956D-4	.1719348D-4	.9724127D-5	.4985914D-5
1 PF1	.7908048324	.6016602186	.5411648297	.4814554735	.4186513753	.3771745972	.3467013614	.3034856493	.2611407564
1 PQ1	.5473925635	.5595395964	.5675339274	.5793926412	.5993824693	.6202792500	.6426048395	.6934318460	.7942404750
1 PF2	.9273942636	.8001752509	.7592770170	.7192494519	.6777935889	.6508916346	.6313834604	.6040156167	.5771797393
1 PQ2	.9623831760	1.040417773	1.091082152	1.163808058	1.280320486	1.395424148	1.512476137	1.762355742	2.215356205
1 PF3	1.041790693	.9703397887	.9459576245	.9213715023	.8950353654	.8773509408	.8641442187	.8448703254	.8246792766
1 PQ3	2.105154380	2.247051296	2.351420221	2.504687891	2.752595864	2.997805369	3.246762845	3.776684623	4.734164906
3 PFC	1.139785643	1.105161059	1.091851872	1.077591222	1.061323661	1.049799574	1.040875458	1.027385584	1.012762193
3 PQC	7.686957987	7.106671620	7.036546966	7.042521350	7.157094182	7.319245437	7.502086938	7.904465209	8.599378733
1 ZF1	1.209925531	1.210260625	1.210382679	1.210510370	1.210653285	1.210753572	1.210831158	1.210949170	1.211079650
1 ZF2	1.241896956	1.245619275	1.246907408	1.248229628	1.249685460	1.250694989	1.251470387	1.252641963	1.253928321
1 ZF3	1.325426658	1.338283385	1.342626582	1.347038964	1.351850771	1.355162582	1.357694084	1.361501094	1.365659459
1 ZF4	1.537133253	1.567470957	1.577807993	1.588342112	1.599858212	1.607796843	1.613869750	1.623007269	1.632990707
1 ZF5	2.176366244	2.242413872	2.265311262	2.288819372	2.314695419	2.332625899	2.346386420	2.367152143	2.389908322
1 ZF6	5.918947373	6.141389674	6.219295937	6.299656675	6.388511072	6.450300148	6.497826930	6.569703401	6.648649048

(6) P 8+4	TH = 57 SBE = 1.192363293								MCPER FUNCTIONS
	RC = 02	RC = 05	RC = 07	RC = 10	RC = 15	RC = 20	RC = 25	RC = 35	RC = 50
AMAX DB	.0017375255	.0108709564	.0213327376	.0436480540	.0988323393	.1772876696	.2802872360	.5675287486	1.249387366
AMIN DB	59.46674589	71.33326582	75.74950539	80.47635872	85.93103088	89.88341507	93.02441786	97.97203408	103.7717201
MU I K	.1063317D-2	.2712294D-3	.1631266D-3	.9466339D-4	.5051827D-4	.3205009D-4	.2232436D-4	.1262985D-4	.6477598D-5
1 I PF1	.8043110574	.6104697781	.5487263584	.4878921772	.4240093924	.3818733641	.3509402433	.3071046633	.2641852446
1 I PQ1	.5484607093	.5603237881	.5682518106	.5800562002	.6000018215	.6208783633	.6431953599	.6940247596	.7948704655
1 I PF2	.9360995911	.8075541903	.7662136586	.7257554055	.6838611593	.6566816155	.6369767890	.6093415789	.5822563613
1 I PQ2	.9763308360	1.051916398	1.102139441	1.174638123	1.291212657	1.406607597	1.524069750	1.775011349	2.230201592
1 I PF3	1.043674783	.9732151444	.9490688451	.9246795397	.8985163346	.8809304558	.8677903397	.8486065975	.8285071742
1 I PQ3	2.170952236	2.304060705	2.407110094	2.560066440	2.809146638	3.056416588	3.307914446	3.843979695	4.813600638
3 I PFC	1.134221611	1.101509987	1.088825036	1.075177872	1.059548385	1.048440604	1.039820230	1.026760809	1.012567805
3 I PQC	8.147142535	7.456918195	7.361059180	7.346099706	7.443788674	7.598128312	7.777329901	8.178718980	8.880631092
1 I ZF1	1.195865673	1.196189597	1.196308852	1.196434148	1.196574929	1.196674016	1.196750822	1.196867873	1.196997572
1 I ZF2	1.225969328	1.229637397	1.230911897	1.232222670	1.233668736	1.234673156	1.235445518	1.236613913	1.237899646
1 I ZF3	1.304738051	1.317546516	1.321883192	1.326293936	1.331109675	1.334427659	1.336965790	1.340785892	1.344962853
1 I ZF4	1.506469493	1.536755668	1.547109052	1.557675824	1.569245046	1.577230373	1.583344334	1.592552040	1.602623334
1 I ZF5	2.122030909	2.187942972	2.210892362	2.234500913	2.260538594	2.278610282	2.292494395	2.313470233	2.336487253
1 I ZF6	5.749536932	5.971535636	6.049648589	6.130398511	6.219876288	6.282209921	6.330213567	6.402901104	6.482853565

(6) P 8+4	TH = 58 SBE = 1.179178403								MCPER FUNCTIONS
	RC = 02	RC = 05	RC = 07	RC = 10	RC = 15	RC = 20	RC = 25	RC = 35	RC = 50
AMAX DB	.0017375255	.0108709564	.0213327376	.0436480540	.0988323393	.1772876696	.2802872360	.5675287486	1.249387366
AMIN DB	57.26039999	69.10456118	73.51471052	78.23608222	83.68555186	87.63480572	90.77365427	95.71840655	101.5154601
MU I K	.1370819D-2	.3505677D-3	.2109913D-3	.1225169D-3	.6542179D-4	.4152023D-4	.2892793D-4	.1637117D-4	.8398989D-5
1 I PF1	.8186209276	.6197517126	.5566802756	.4946525396	.4296283657	.3867963906	.3553785357	.3108905768	.2673674887
1 I PQ1	.5496190347	.5611636034	.5690178950	.5807620175	.6006585563	.6215124275	.6438195058	.6946503315	.7955340853
1 I PF2	.9450465302	.8151816809	.7733921902	.7324940769	.6901499649	.6626846454	.6427770012	.6148656134	.5875224450
1 I PQ2	.9914591947	1.064260659	1.113973909	1.186196434	1.302804502	1.418488474	1.536371042	1.788416611	2.245900811
1 I PF3	1.045484289	.9760973712	.9522084124	.9280346235	.9020613575	.8845838512	.8715169261	.8524319191	.8324320272
1 I PQ3	2.242704337	2.365682770	2.467144982	2.619612985	2.869795190	3.119169315	3.373307208	3.915817735	4.898251374
3 I PFC	1.128631393	1.097819403	1.085759227	1.072728269	1.057741915	1.047054935	1.038742164	1.026119502	1.012364153
3 I PQC	8.660147927	7.842912467	7.717445503	7.678378489	7.756509101	7.901649043	8.076400566	8.476021769	9.184800486
1 I ZF1	1.182483905	1.182795671	1.182911823	1.183034429	1.183172772	1.183270460	1.183346340	1.183462219	1.183590920
1 I ZF2	1.210782616	1.214389839	1.215648502	1.216945632	1.218379603	1.219377361	1.220145528	1.221309054	1.222590391
1 I ZF3	1.284921524	1.297663532	1.301987969	1.306391487	1.311205323	1.314525630	1.317067539	1.320896608	1.325087889
1 I ZF4	1.476935322	1.507125360	1.517482090	1.528069330	1.539679533	1.547703876	1.553853385	1.563123581	1.573275113
1 I ZF5	2.069434948	2.135117495	2.158091615	2.181775459	2.207950379	2.226148350	2.240145668	2.261317653	2.284582474
1 I ZF6	5.585026135	5.806295635	5.884530882	5.965592126	6.055618289	6.118451223	6.166900737	6.240358649	6.321281545

(6) P 8+4	TH = 59	SBE = 1.166633397						MCPER FUNCTIONS	
	RC = 02	RC = 05	RC = 07	RC = 10	RC = 15	RC = 20	RC = 25	RC = 35	RC = 50
AMAX DB	.0017375255	.0108709564	.0213327376	.0436480540	.0988323393	.1772876696	.2802872360	.5675287486	1.249387366
AMIN DB	55.07335978	66.89341583	71.29698998	76.01243944	81.45628450	85.40215213	88.53866945	93.48032130	99.27452374
MU I K	.1763324D-2	.4521986D-3	.2723645D-3	.1582625D-3	.8456405D-4	.5368988D-4	.3741679D-4	.2118283D-4	.1087111D-4
1 I PF1	.8338035234	.6295426850	.5650559972	.5017600012	.4355265247	.3919589788	.3600296152	.3148543213	.2706963029
1 I PQ1	.5508787946	.5620649426	.5698370321	.5815141333	.6013560623	.6221845040	.6444801509	.6953112619	.7962340211
1 I PF2	.9542324995	.8230644628	.7808207262	.7394743800	.6966692962	.6689100649	.6487933964	.6205968735	.5929869237
1 I PQ2	1.007912389	1.077541578	1.126665823	1.198555186	1.315162466	1.431131167	1.549444007	1.802637352	2.262526668
1 I PF3	1.047208221	.9789804775	.9553719542	.9314338923	.9056690269	.8883105923	.8753240374	.8563471286	.8364553795
1 I PQ3	2.321175426	2.432452559	2.532012500	2.683780138	2.934971231	3.186487816	3.443368102	3.992644619	4.988615532
3 I PFC	1.123021625	1.094092291	1.082656478	1.070243648	1.055904808	1.045642770	1.037641255	1.025461429	1.012150860
3 I PQC	9.234278331	8.269714025	8.110065084	8.043150837	8.098572045	8.232871388	8.402208111	8.799113056	9.514534837
1 I ZF1	1.169750169	1.170048786	1.170161527	1.170281145	1.170416739	1.170512824	1.170587627	1.170702116	1.170829592
1 I ZF2	1.196305122	1.199844896	1.201085486	1.202366731	1.203786216	1.204775708	1.205538481	1.206695382	1.207971481
1 I ZF3	1.265941882	1.278598460	1.282904688	1.287295135	1.292100953	1.295419538	1.297962230	1.301795920	1.305997008
1 I ZF4	1.448486251	1.478533156	1.488879478	1.499474271	1.511112560	1.519167751	1.525346943	1.534671422	1.544895051
1 I ZF5	2.018502475	2.083857930	2.106828184	2.130560832	2.156847191	2.175155596	2.189255063	2.210608284	2.234106941
1 I ZF6	5.425170314	5.645418326	5.723688469	5.804979948	5.895475958	5.958760568	6.007623216	6.081808363	6.163662934

(6) P 8+4	TH = 60	SBE = 1.154700538						MCPER FUNCTIONS	
	RC = 02	RC = 05	RC = 07	RC = 10	RC = 15	RC = 20	RC = 25	RC = 35	RC = 50
AMAX DB	.0017375255	.0108709564	.0213327376	.0436480540	.0988323393	.1772876696	.2802872360	.5675287486	1.249387366
AMIN DB	52.90414214	64.69820793	69.09468269	73.80373318	79.24149646	83.18370040	86.31769442	91.25598911	97.04710292
MU I K	.2263565D-2	.5822233D-3	.3509667D-3	.2040861D-3	.1091252D-3	.6931305D-4	.4831870D-4	.2736532D-4	.1404898D-4
1 I PF1	.8499362749	.6398835524	.5738862352	.5092405742	.4417240546	.3973780485	.3649082575	.3190079549	.2741813970
1 I PQ1	.5522531555	.5630345808	.5707147627	.5823171389	.6020981729	.6228980458	.6451805293	.6960105769	.7969732813
1 I PF2	.9636529206	.8312090925	.7885075312	.7467056131	.7034289903	.6753678297	.6550359244	.6265451942	.5986594260
1 I PQ2	1.025858240	1.091863703	1.140307007	1.211796655	1.328361980	1.444608589	1.563360991	1.817747811	2.280161077
1 I PF3	1.048834892	.9818578720	.9585545922	.9348740854	.9093376547	.8921099400	.8792115839	.8603529907	.8405787711
1 I PQ3	2.407261893	2.504989599	2.602275176	2.753088227	3.005167012	3.258856991	3.518584296	4.074967865	5.085259046
3 I PFC	1.117399361	1.090331902	1.079519030	1.067725392	1.054037714	1.044204375	1.036517535	1.024786368	1.011927539
3 I PQC	9.879522686	8.743319318	8.544060875	8.444872410	8.473855886	8.595367571	8.758137213	9.151170230	9.872895600
1 I ZF1	1.157636566	1.157921048	1.158030067	1.158146394	1.158278924	1.158373197	1.158446770	1.158559644	1.158685661
1 I ZF2	1.182507478	1.185973218	1.187193470	1.188456544	1.189859091	1.190838664	1.191594805	1.192743263	1.194012203
1 I ZF3	1.247766877	1.260318224	1.264600001	1.268971256	1.273762645	1.277075259	1.279615581	1.283449314	1.287655442
1 I ZF4	1.421081790	1.450936056	1.461257409	1.471846057	1.483498726	1.491576064	1.497778687	1.507148684	1.517435702
1 I ZF5	1.969163305	2.034090417	2.057026821	2.080780375	2.107150869	2.125552875	2.139742714	2.161261244	2.184978759
1 I ZF6	5.269739065	5.488666809	5.566881405	5.648318724	5.739202352	5.802888519	5.852129731	5.926996374	6.009741310

(6) P 8+4			TH = 61 SBE = 1.143354068							MCPER FUNCTIONS
			RC = 02	RC = 05	RC = 07	RC = 10	RC = 15	RC = 20	RC = 25	RC = 35	RC = 50
AMAX DB			.0017375255	.0108709564	.0213327376	.0436480540	.0988323393	.1772876696	.2802872360	.5675287486	1.249387366
AMIN DB			50.75127951	62.51731658	66.90612454	71.60825950	77.03944508	80.97768403	84.10894610	89.04360444	94.83137114
MU		K	.2900254D-2	.7484007D-3	.4515374D-3	.2627768D-3	.1406137D-3	.8935437D-4	.6230927D-4	.3530366D-4	.1813140D-4
1		PF1	.8671062124	.6508199854	.5832074793	.5171232139	.4482433751	.4030723657	.3700308337	.3233648183	.2778334987
1		PQ1	.5537576075	.5640803338	.5716574433	.5831762746	.6028892406	.6236569618	.6459242948	.6967516867	.7977552455
1		PF2	.9733007874	.8396218199	.7964609629	.7541974508	.7104394597	.6820685573	.6615152416	.6327211582	.6045503485
1		PQ2	1.045493159	1.107347668	1.155002964	1.226014997	1.342489021	1.459003644	1.578204099	1.833832031	2.298896551
1		PF3	1.050351908	.9847223123	.9617508915	.9383514971	.9130652335	.8959809202	.8831793017	.8644501783	.8448037285
1		PQ3	2.502020409	2.584014820	2.678585115	2.828138247	3.080948989	3.336833497	3.599514072	4.163367626	5.188827200
3		PFC	1.111772092	1.086541773	1.076349348	1.065175050	1.052141387	1.042740081	1.035371080	1.024094109	1.011693790
3		PQC	10.60797596	9.270877310	9.025535363	8.888806129	8.886920219	8.993324414	9.148145972	9.535897014	10.26343956
1		ZF1	1.146117190	1.146386556	1.146491543	1.146604274	1.146733419	1.146825666	1.146897850	1.147008880	1.147133195
1		ZF2	1.169362460	1.172747626	1.173945254	1.175187834	1.176570933	1.177538886	1.178287114	1.179425243	1.180685027
1		ZF3	1.230367013	1.242792445	1.247043242	1.251388896	1.256159126	1.259461303	1.261995935	1.265824880	1.270031011
1		ZF4	1.394685219	1.424294689	1.434575655	1.445143618	1.456796084	1.464886296	1.471105674	1.480511818	1.490852897
1		ZF5	1.921352504	1.985746299	2.008617425	2.032362495	2.058788224	2.077265939	2.091533598	2.113200406	2.137120704
1		ZF6	5.118514686	5.335817066	5.413882588	5.495377943	5.586563120	5.650598106	5.700181366	5.775681000	5.859272257

(6) P 8+4			TH = 62 SBE = 1.132570051							MCPER FUNCTIONS
			RC = 02	RC = 05	RC = 07	RC = 10	RC = 15	RC = 20	RC = 25	RC = 35	RC = 50
AMAX DB			.0017375255	.0108709564	.0213327376	.0436480540	.0988323393	.1772876696	.2802872360	.5675287486	1.249387366
AMIN DB			48.61331300	60.34911236	64.72963831	69.42429717	74.84836625	78.78231256	81.91061558	86.84133350	92.62547133
MU		K	.3709662D-2	.9606040D-3	.5801203D-3	.3378976D-3	.1809596D-3	.1150494D-3	.8025447D-4	.4549182D-4	.2337364D-4
1		PF1	.8854114367	.6624032010	.5930605657	.5254402562	.4551094675	.4090628104	.3754155401	.3279397183	.2816644980
1		PQ1	.5554104840	.5652112624	.5726723991	.5840975466	.6037342284	.6244656945	.6467155909	.6975384473	.7985837250
1		PF2	.9831661565	.8483084333	.8046893956	.7619599277	.7177117207	.6890235771	.6682427727	.6391361707	.6106709384
1		PQ2	1.067048292	1.124133340	1.170875486	1.241318453	1.357642021	1.474410992	1.594066907	1.850985549	2.318837985
1		PF3	1.051746172	.9875658491	.9649548028	.9418619234	.9168493936	.8999222884	.8872267227	.8686392525	.8491317535
1		PQ3	2.606704190	2.670371670	2.761702192	2.909627853	3.162972183	3.421059409	3.686800206	4.258510116	5.300059035
3		PFC	1.106147779	1.082725756	1.073150147	1.062594354	1.050216698	1.041250298	1.034202016	1.023384458	1.011449204
3		PQC	11.43438671	9.860965072	9.561773999	9.381205280	9.343156179	9.431676493	9.576888461	9.957633725	10.69032048
1		ZF1	1.135167970	1.135421249	1.135521895	1.135630722	1.135756157	1.135846164	1.135916796	1.136025745	1.136148110
1		ZF2	1.156844822	1.160142948	1.161315650	1.162535380	1.163896466	1.164851053	1.165590048	1.166715894	1.167964445
1		ZF3	1.213715385	1.225993276	1.230206257	1.234519592	1.239261595	1.242548635	1.245074080	1.248893139	1.253093943
1		ZF4	1.369263373	1.398573118	1.408797360	1.419329191	1.430965927	1.439059114	1.445288114	1.454720374	1.465105507
1		ZF5	1.875009983	1.938761723	1.961534634	1.985240261	2.011690622	2.030225014	2.044557102	2.066353963	2.090459769
1		ZF6	4.971290720	5.186656489	5.264476276	5.345938359	5.437335009	5.501663316	5.551550050	5.627631218	5.712021817

(6) P 8+4	TH = 63	SBE = 1.122326238						MCPER FUNCTIONS	
	RC = 02	RC = 05	RC = 07	RC = 10	RC = 15	RC = 20	RC = 25	RC = 35	RC = 50
AMAX DB	.0017375255	.0108709564	.0213327376	.0436480540	.0988323393	.1772876696	.2802872360	.5675287486	1.249387366
AMIN DB	46.48878536	58.19194709	62.56352277	67.25009609	72.66646230	76.59575900	79.72085536	84.64730111	90.42750237
MU I K	.4737626D-2	.1231410D-2	.7444300D-3	.4340055D-3	.2326360D-3	.1479831D-3	.1032660D-3	.5856457D-4	.3010405D-4
1 I PF1	.9049628477	.6746908327	.6034913537	.5342279380	.4623502647	.4153726953	.3810826716	.3327491452	.2856876179
1 I PQ1	.5572336253	.5664379264	.5737681149	.5850878720	.6046388206	.6253293154	.6475591363	.6983752364	.7994630347
1 I PF2	.9932355435	.8572740640	.8132011195	.7700034088	.7252574202	.6962449827	.6752307791	.6458025434	.6170333882
1 I PQ2	1.090797288	1.142383736	1.188065869	1.257832058	1.373934193	1.490939212	1.611056544	1.869317440	2.340104832
1 I PF3	1.053003883	.9903797674	.9681596011	.9454006040	.9206873539	.9039324879	.8913531403	.8729206381	.8535643093
1 I PQ3	2.722809303	2.765052655	2.852516813	2.998371280	3.251997968	3.512279108	3.781186466	4.361164131	5.419805020
3 I PFC	1.100534870	1.078888045	1.069924415	1.059985237	1.048264652	1.039735523	1.033010528	1.022657246	1.011193364
3 I PQC	12.37687550	10.52394305	10.16153165	9.929547077	9.848977149	9.916274486	10.04986926	10.42149582	11.15841627
1 I ZF1	1.124766527	1.125002768	1.125098763	1.125203379	1.125324779	1.125412325	1.125481242	1.125587867	1.125708025
1 I ZF2	1.144931147	1.148135870	1.149281339	1.150475833	1.151812292	1.152751719	1.153480122	1.154591667	1.155826825
1 I ZF3	1.197787528	1.209895253	1.214063257	1.218337222	1.223043573	1.226310527	1.228823095	1.232626885	1.236816720
1 I ZF4	1.344786470	1.373738657	1.383888859	1.394368142	1.405972595	1.414058183	1.420289170	1.429736793	1.440155228
1 I ZF5	1.830080119	1.893077280	1.915717467	1.939351031	1.965793601	1.984364418	1.998746637	2.020654026	2.044926755
1 I ZF6	4.827870580	5.040982469	5.118456681	5.199790567	5.291304428	5.355867650	5.406017098	5.482625191	5.567764993

(6) P 8+4	TH = 64	SBE = 1.112601940						MCPER FUNCTIONS	
	RC = 02	RC = 05	RC = 07	RC = 10	RC = 15	RC = 20	RC = 25	RC = 35	RC = 50
AMAX DB	.0017375255	.0108709564	.0213327376	.0436480540	.0988323393	.1772876696	.2802872360	.5675287486	1.249387366
AMIN DB	44.37623408	56.04414275	60.40604070	65.08386460	70.49188849	74.41614593	77.53776515	82.45957601	88.23550397
MU I K	.6042105D-2	.1576859D-2	.9543287D-3	.5569379D-3	.2988172D-3	.1901922D-3	.1327736D-3	.7533923D-4	.3874582D-4
1 I PF1	.9258861799	.6877479691	.6145515327	.5435270189	.4699971153	.4220281458	.3870549474	.3378115313	.2899176175
1 I PQ1	.5592532312	.5677727032	.5749544703	.5861552573	.6056095587	.6262536400	.6484603291	.6992670459	.8003980803
1 I PF2	1.003491211	.8665229411	.8220042076	.7783385450	.7330888580	.7037456879	.6824924332	.6527335891	.6236509445
1 I PQ2	1.117066197	1.162289920	1.206738927	1.275701001	1.391496405	1.508713461	1.629296239	1.888952827	2.362833740
1 I PF3	1.054110561	.9931545241	.9713578169	.9489621564	.9245758652	.9080096034	.8955575699	.8772945954	.8581028039
1 I PQ3	2.852134378	2.869232941	2.952078592	3.095324335	3.348916308	3.611360305	3.883538119	4.472221508	5.549048882
3 I PFC	1.094942335	1.075033203	1.066675441	1.057349861	1.046286399	1.038196357	1.031796870	1.021912326	1.010925841
3 I PQC	13.45788687	11.27241726	10.83540323	10.54283320	10.41206281	10.45409983	10.57364002	10.93354942	11.67348958
1 I ZF1	1.114892055	1.115110324	1.115201364	1.115301465	1.115418501	1.115503364	1.115570399	1.115674453	1.115792142
1 I ZF2	1.133599710	1.136704803	1.137820731	1.138987586	1.140296759	1.141219191	1.141935603	1.143030761	1.144250286
1 I ZF3	1.182561295	1.194475166	1.198590685	1.202817880	1.207480767	1.210722419	1.213218219	1.217001047	1.221173932
1 I ZF4	1.321227951	1.349761724	1.359819531	1.370228804	1.381783316	1.389849996	1.396074788	1.405526232	1.415966400
1 I ZF5	1.786511422	1.848637675	1.871108990	1.894636118	1.921036534	1.939622203	1.954039273	1.976036249	2.000455879
1 I ZF6	4.688066239	4.898601044	4.975626596	5.056733625	5.148266057	5.213002714	5.263371789	5.340448825	5.426284286

(6) P 8+4	TH = 65	SBE = 1.103377919						MCPER FUNCTIONS	
	RC = 02	RC = 05	RC = 07	RC = 10	RC = 15	RC = 20	RC = 25	RC = 35	RC = 50
AMAX DB	.0017375255	.0108709564	.0213327376	.0436480540	.0988323393	.1772876696	.2802872360	.5675287486	1.249387366
AMIN DB	42.27418468	53.90397913	58.25540553	62.92375517	68.32273792	72.24152981	75.35937577	80.27615432	86.04743965
MU I K	.7696456D-2	.2017442D-2	.1222446D-2	.7141875D-3	.3835863D-3	.2443000D-3	.1706205D-3	.9687067D-4	.4984574D-4
1 I PF1	.9483243974	.7016484002	.6262995924	.5533835269	.4780853415	.4290585549	.3933579009	.3431475612	.2943710348
1 I PQ1	.5615009715	.5692301904	.5762430371	.5873090206	.6066540093	.6272453705	.6494253749	.7002195948	.8013944647
1 I PF2	1.013910326	.8760580848	.8311063437	.7869762067	.7412190048	.7115394855	.6900419002	.6599437294	.6305380320
1 I PQ2	1.146246212	1.184077186	1.227088022	1.295094815	1.410480752	1.527878763	1.648928483	1.910035964	2.387181805
1 I PF3	1.055051074	.9958796823	.9745411620	.9525405025	.9285111459	.9121513051	.8998387017	.8817611862	.8627485699
1 I PQ3	2.996858340	2.984313323	3.061632828	3.201616019	3.454773748	3.719320447	3.994867626	4.592722689	5.688934781
3 I PFC	1.089379691	1.071166203	1.063406843	1.054690638	1.044283261	1.036633514	1.030561374	1.021149585	1.010646200
3 I PQC	14.70546448	12.12184654	11.59630777	11.23198017	11.04167436	11.05354091	11.15605155	11.50103558	12.24239202
1 I ZF1	1.105525199	1.105724588	1.105810376	1.105905660	1.106018005	1.106099962	1.106164945	1.106266178	1.106381128
1 I ZF2	1.122830355	1.125829760	1.126913851	1.128050651	1.129329841	1.130233404	1.130936388	1.132013009	1.133214575
1 I ZF3	1.168016742	1.179711951	1.183767106	1.187939756	1.192550956	1.195761808	1.198236727	1.201992569	1.206142164
1 I ZF4	1.298564350	1.326615724	1.336561673	1.346882355	1.358368070	1.366403730	1.372613553	1.382056406	1.392505842
1 I ZF5	1.744256223	1.805391431	1.827656017	1.851040483	1.877362305	1.895939830	1.910375403	1.932439480	1.956984410
1 I ZF6	4.551696970	4.759325560	4.835796051	4.916573692	5.008021460	5.072866815	5.123409952	5.200894329	5.287368220

(6) P 8+4	TH = 66	SBE = 1.094636279						MCPER FUNCTIONS	
	RC = 02	RC = 05	RC = 07	RC = 10	RC = 15	RC = 20	RC = 25	RC = 35	RC = 50
AMAX DB	.0017375255	.0108709564	.0213327376	.0436480540	.0988323393	.1772876696	.2802872360	.5675287486	1.249387366
AMIN DB	40.18114460	51.76968014	56.10976626	60.76784823	66.15702425	70.06988311	73.18363080	78.09494053	83.86117716
MU I K	.9793609D-2	.2579384D-2	.1564990D-2	.9153930D-3	.4922081D-3	.3136937D-3	.2191889D-3	.1245240D-3	.6411227D-4
1 I PF1	.9724405059	.7164761201	.6388019942	.5638496599	.4866549120	.4364971334	.4000203506	.3487805454	.2990664800
1 I PQ1	.5640154422	.5708277190	.5776474528	.5885600704	.6077809741	.6283122728	.6504614444	.7012394684	.8024586187
1 I PF2	1.024463978	.8858809219	.8405145991	.7959273898	.7496615108	.7196411063	.6978944282	.6674486162	.6377103980
1 I PQ2	1.178810223	1.208012922	1.249341435	1.316212658	1.431065026	1.548604119	1.670118962	1.932734075	2.413330601
1 I PF3	1.055809692	.9985438426	.9777004476	.9561287854	.9324888086	.9163547860	.9041948464	.8863202340	.8675028404
1 I PQ3	3.159642713	3.111975656	3.182667316	3.318588926	3.570808985	3.837360159	4.116367102	4.723888909	5.840801422
3 I PFC	1.083857029	1.067292459	1.060122609	1.052010265	1.042256748	1.035047839	1.029304464	1.020368949	1.010354000
3 I PQC	16.15497855	13.09134907	12.46012743	12.01033166	11.74906555	11.72475183	11.80658057	12.13265968	12.87332667
1 I ZF1	1.096647954	1.096827587	1.096907832	1.096998003	1.097105333	1.097184162	1.097246923	1.097345080	1.097457015
1 I ZF2	1.112604387	1.115492249	1.116542231	1.117646556	1.118893036	1.119775817	1.120463904	1.121519770	1.122700969
1 I ZF3	1.154136033	1.165586595	1.169573116	1.173683046	1.178233899	1.181408147	1.183857838	1.187580315	1.191699885
1 I ZF4	1.276775177	1.304276951	1.314090403	1.324302714	1.335699480	1.343691137	1.349876561	1.359297463	1.369742708
1 I ZF5	1.703270393	1.763290616	1.785308837	1.808512457	1.834717015	1.853261860	1.867698430	1.889805428	1.914452320
1 I ZF6	4.418588127	4.622975345	4.698780963	4.779122654	4.870377691	4.935263545	4.985932530	5.063758749	5.150809853

(6) P 8+4 — TH = 67 SBE = 1.086360377 — MCPER FUNCTIONS

MU	K	RC = 02	RC = 05	RC = 07	RC = 10	RC = 15	RC = 20	RC = 25	RC = 35	RC = 50
	AMAX DB	.0017375255	.0108709564	.0213327376	.0436480540	.0988323393	.1772876696	.2802872360	.5675287486	1.249387366
	AMIN DB	38.09559812	49.63939827	53.96719027	58.61413349	63.99266170	67.89907350	71.00836530	75.91372541	81.67446563
MU	K	.1245145D-1	.3296325D-2	.2002813D-2	.1172987D-2	.6314906D-3	.4027600D-3	.2815670D-3	.1600714D-3	.8246634D-4
1	PF1	.9984208374	.7323271473	.6521345938	.5749848821	.4957512610	.4443815781	.4070749726	.3547368744	.3040249910
1	PQ1	.5668441027	.5725860151	.5791839001	.5899212592	.6090007525	.6294633971	.6515768708	.7023342912	.8035979627
1	PF2	1.035116016	.8959908037	.8502351436	.8052030846	.7584307022	.7280662797	.7060664450	.6752652706	.6451852798
1	PQ2	1.215334564	1.234416728	1.273770500	1.339290023	1.453458371	1.571087688	1.693061496	1.957242156	2.441491198
1	PF3	1.056370153	1.001134572	.9808254955	.9597192765	.9365037744	.9206166870	.9086238711	.8909712758	.8723667185
1	PQ3	3.343767826	3.254255159	3.316973043	3.447851317	3.698498503	3.966906001	4.249449728	4.867163115	6.006225267
3	PFC	1.078385047	1.063417874	1.056827131	1.049311749	1.040208587	1.033440330	1.028026671	1.019570393	1.010048797
3	PQC	17.85149539	14.20478568	13.44656043	12.89433802	12.54802472	12.48012355	12.53675718	12.83896844	13.57618845
1	ZF1	1.088243572	1.088402607	1.088477030	1.088561800	1.088663795	1.088739273	1.088799642	1.088894466	1.089003107
1	ZF2	1.102904468	1.105675174	1.106688805	1.107758246	1.108969267	1.109829323	1.110501009	1.111533843	1.112692182
1	ZF3	1.140903361	1.152082064	1.155991265	1.160029875	1.164511251	1.167642763	1.170062628	1.173744976	1.177827364
1	ZF4	1.255842824	1.282724516	1.292383592	1.302466465	1.313752730	1.321686449	1.327837330	1.337221875	1.347648373
1	ZF5	1.663513089	1.722290600	1.744020963	1.767003477	1.793049715	1.811535675	1.825954466	1.848078351	1.872801942
1	ZF6	4.288569939	4.489374356	4.564401752	4.644196719	4.735145848	4.800000315	4.850744090	4.928942450	5.016405205

(6) P 8+4 — TH = 68 SBE = 1.078534743 — MCPER FUNCTIONS

MU	K	RC = 02	RC = 05	RC = 07	RC = 10	RC = 15	RC = 20	RC = 25	RC = 35	RC = 50
	AMAX DB	.0017375255	.0108709564	.0213327376	.0436480540	.0988323393	.1772876696	.2802872360	.5675287486	1.249387366
	AMIN DB	36.01600367	47.51119675	51.82564328	56.46048824	61.82744178	65.72683956	68.83128071	73.73016010	79.48490879
MU	K	.1581976D-1	.4211533D-2	.2562818D-2	.1503057D-2	.8102666D-3	.5171994D-3	.3617728D-3	.2058221D-3	.1061096D-3
1	PF1	1.026478853	.7493117442	.6663843822	.5868572666	.5054262913	.4527548901	.4145590008	.3610465737	.3092704673
1	PQ1	.5700458800	.5745300612	.5808717233	.5914078349	.6103254761	.6307093570	.6527813982	.7035129432	.8048211088
1	PF2	1.045821697	.9063844036	.8602728730	.8148140948	.7675415588	.7368317915	.7145756641	.6834122408	.6529815996
1	PQ2	1.256527931	1.263673554	1.300700114	1.364607341	1.477908498	1.595563371	1.717984311	1.983789051	2.471910505
1	PF3	1.056715759	1.003638330	.9839050410	.9633032708	.9405501754	.9249330106	.9131231235	.8957135053	.8773411417
1	PQ3	3.553316415	3.413635753	3.466723739	3.591345019	3.839615718	4.109665692	4.395803093	5.024262491	6.187075791
3	PFC	1.072975077	1.059548880	1.053525249	1.046598455	1.038140749	1.031812162	1.026728651	1.018753948	1.009730148
3	PQC	19.85307084	15.49223330	14.58027334	13.90446860	13.45559746	13.33490928	13.36073080	13.63284681	14.36301064
1	ZF1	1.080296474	1.080434108	1.080502443	1.080581535	1.080677884	1.080749792	1.080807598	1.080898831	1.081003894
1	ZF2	1.093714531	1.096362743	1.097337833	1.098369997	1.099542800	1.100378162	1.101031913	1.102039376	1.103172281
1	ZF3	1.128304874	1.139183235	1.143005993	1.146964233	1.151366501	1.154448796	1.156833964	1.160469008	1.164506598
1	ZF4	1.235752481	1.261940302	1.271421816	1.281352812	1.292505506	1.300366313	1.306471726	1.315804350	1.326196332
1	ZF5	1.624946519	1.682349823	1.703748905	1.726467854	1.752312145	1.770711202	1.785092054	1.807204744	1.831977641
1	ZF6	4.161476319	4.358349768	4.432481898	4.511614936	4.602139555	4.666886800	4.717651246	4.795947491	4.883951597

(6) P 8+4		TH = 69 SBE = 1.071144994							MCPER FUNCTIONS	
		RC = 02	RC = 05	RC = 07	RC = 10	RC = 15	RC = 20	RC = 25	RC = 35	RC = 50
AMAX DB		.0017375255	.0108709564	.0213327376	.0436480540	.0988323393	.1772876696	.2802872360	.5675287486	1.249387366
AMIN DB		33.94079500	45.38302911	49.68296598	54.30465164	59.65900548	63.55076181	66.64991516	71.54172512	77.28993288
MU	K	.2008909D-1	.5380821D-2	.3279833D-2	.1926493D-2	.1040039D-2	.6644494D-3	.4650549D-3	.2647974D-3	.1366166D-3
1	PF1	1.056859470	.7675571349	.6816516276	.5995451517	.5157396123	.4616663853	.4225150910	.3677439890	.3148302066
1	PQ1	.5736947182	.5766902310	.5827342338	.5930380253	.6117695365	.6320626861	.6540864966	.7047858344	.8061381145
1	PF2	1.056526108	.9170549608	.8706309292	.8247707928	.7770096659	.7459555368	.7234411976	.6919097828	.6611201938
1	PQ2	1.303270301	1.296250962	1.330522441	1.392501153	1.504710978	1.622309265	1.745158067	2.012645204	2.504879320
1	PF3	1.056829496	1.006040398	.9869266276	.9668709686	.9446212412	.9292990190	.9176893421	.9005457041	.8824268364
1	PQ3	3.793424523	3.593177348	3.634581408	3.751434937	3.996308509	4.267700228	4.557458675	5.197246603	6.385586917
3	PFC	1.067639115	1.055692492	1.050222300	1.043874139	1.036055487	1.030164708	1.025411203	1.017919717	1.009397612
3	PQC	22.23539397	16.99201619	15.89247856	15.06645220	14.49306257	14.30806540	14.29602841	14.53018040	15.24855723
1	ZF1	1.072792173	1.072907647	1.072969645	1.073042793	1.073133195	1.073201318	1.073256394	1.073343779	1.073444978
1	ZF2	1.085019701	1.087540393	1.088474810	1.089467339	1.090599166	1.091407846	1.092042102	1.093021802	1.094126616
1	ZF3	1.116328610	1.126876859	1.130603589	1.134471931	1.138784929	1.141811149	1.144156459	1.147736575	1.151721256
1	ZF4	1.216492066	1.241908942	1.251188339	1.260943554	1.271937970	1.279709759	1.285757914	1.295021782	1.305362126
1	ZF5	1.587535735	1.643429593	1.664451953	1.686862545	1.712458496	1.730740652	1.745061887	1.767133042	1.791925477
1	ZF6	4.037143661	4.229730494	4.302846396	4.381197597	4.471173315	4.535733255	4.586460942	4.664875858	4.753245819

(6) P 8+4		TH = 70 SBE = 1.064177772							MCPER FUNCTIONS	
		RC = 02	RC = 05	RC = 07	RC = 10	RC = 15	RC = 20	RC = 25	RC = 35	RC = 50
AMAX DB		.0017375255	.0108709564	.0213327376	.0436480540	.0988323393	.1772876696	.2802872360	.5675287486	1.249387366
AMIN DB		31.86838966	43.25271537	47.53684630	52.14419404	57.48481008	61.36822795	64.46160752	69.34569310	75.08674799
MU	K	.2550237D-1	.6876449D-2	.4199114D-2	.2470531D-2	.1335856D-2	.8542571D-3	.5983009D-3	.3409694D-3	.1760608D-3
1	PF1	1.089843851	.7872108519	.6980525321	.6131392011	.5267600824	.4711729527	.4309923982	.3748686392	.3207355707
1	PQ1	.5778844845	.5791038036	.5847997727	.5948338018	.6133501402	.6335382956	.6555057665	.7061652543	.8075608070
1	PF2	1.067162340	.9279913326	.8813100798	.8350827873	.7868511278	.7554565617	.7326836756	.7007800675	.6696240825
1	PQ2	1.356666058	1.332722096	1.363715017	1.423378755	1.534221375	1.651658650	1.774906266	2.044132665	2.540742697
1	PF3	1.056694191	1.008324804	.9898764930	.9704113421	.9487091671	.9337091144	.9223185497	.9054661584	.8876242632
1	PQ3	4.070629380	3.796688057	3.823838114	3.931028578	4.171202100	4.443519213	4.736883414	5.388606868	6.604450395
3	PFC	1.062389849	1.051856366	1.046924177	1.041143009	1.033955372	1.028499575	1.024075301	1.017067887	1.009050757
3	PQC	25.09843689	18.75354735	17.42312486	16.41298757	15.68726803	15.42339693	15.36458392	15.55075037	16.25111897
1	ZF1	1.065717205	1.065809807	1.065865235	1.065932194	1.066016362	1.066080491	1.066132672	1.066215954	1.066313000
1	ZF2	1.076806218	1.079194713	1.080086402	1.081036988	1.082125096	1.082905093	1.083518274	1.084467769	1.085541757
1	ZF3	1.104964445	1.115151521	1.118772162	1.122540574	1.126753571	1.129716458	1.132016434	1.135533518	1.139456639
1	ZF4	1.198052162	1.222617819	1.231669119	1.241223095	1.252032755	1.259698181	1.265676345	1.274853209	1.285123289
1	ZF5	1.551248453	1.605493891	1.626091988	1.648146950	1.673445176	1.691578277	1.705816539	1.727913315	1.752592869
1	ZF6	3.915409631	4.103345556	4.175320066	4.252764484	4.342060678	4.406348634	4.456978532	4.535427481	4.624082080

(6) P 8+4 — TH = 71 SBE = 1.057620681 — MCPER FUNCTIONS

MU	K	RC = 02	RC = 05	RC = 07	RC = 10	RC = 15	RC = 20	RC = 25	RC = 35	RC = 50
	AMAX DB	.0017375255	.0108709564	.0213327376	.0436480540	.0988323393	.1772876696	.2802872360	.5675287486	1.249387366
	AMIN DB	29.79720971	41.11791438	45.38478626	49.97647991	55.30208868	59.17639043	62.26345373	67.13908320	72.87230078
MU	K	.3236976D-1	.8792336D-2	.5379733D-2	.3170852D-2	.1717495D-2	.1099463D-2	.7705970D-3	.4395880D-3	.2271878D-3
1	PF1	1.125754420	.8084448766	.7157225480	.6277449877	.5385677457	.4813406355	.4400479288	.3824662874	.3270228230
1	PQ1	.5827358517	.5818170122	.5871031336	.5968219908	.6150880352	.6351540729	.6570554653	.7076658247	.8091031988
1	PF2	1.077649371	.9391768024	.8923079183	.8457584752	.7970824293	.7653550882	.7423253706	.7100474176	.6785187904
1	PQ2	1.418117576	1.373796650	1.400864985	1.457737669	1.566871249	1.684014433	1.807618888	2.078638157	2.579913424
1	PF3	1.056292708	1.010474254	.9927394470	.9739119851	.9528049611	.9381566972	.9270059253	.9104725584	.8929335488
1	PQ3	4.393359563	4.028960619	4.038609222	4.133737856	4.367537414	4.640208517	4.937101991	5.601385508	6.846939446
3	PFC	1.057240682	1.048048857	1.043637388	1.038409773	1.031843339	1.026818642	1.022722115	1.016198750	1.008689169
3	PQC	28.57613809	20.84136738	19.22398727	17.98613822	17.07248725	16.71114134	16.59415611	16.71945961	17.39359808
1	ZF1	1.059059061	1.059128130	1.059176779	1.059237321	1.059314985	1.059374921	1.059424048	1.059502975	1.059595584
1	ZF2	1.069061384	1.071313381	1.072160383	1.073066779	1.074108459	1.074857770	1.075448281	1.076365084	1.077405438
1	ZF3	1.094204040	1.103997624	1.107501621	1.111159547	1.115261212	1.118153077	1.120401909	1.123847334	1.127699658
1	ZF4	1.180425961	1.204057092	1.212852846	1.222178483	1.232775003	1.240315369	1.246209764	1.255279812	1.265459320
1	ZF5	1.516054898	1.568509197	1.588633291	1.610282708	1.635230587	1.653180107	1.667310194	1.689196956	1.713928230
1	ZF6	3.796111924	3.979022303	4.049725656	4.126132884	4.214612173	4.278538455	4.329005591	4.407397975	4.496249657

(6) P 8+4 — TH = 72 SBE = 1.051462224 — MCPER FUNCTIONS

MU	K	RC = 02	RC = 05	RC = 07	RC = 10	RC = 15	RC = 20	RC = 25	RC = 35	RC = 50
	AMAX DB	.0017375255	.0108709564	.0213327376	.0436480540	.0988323393	.1772876696	.2802872360	.5675287486	1.249387366
	AMIN DB	27.72572375	38.97609157	43.22406201	47.79862247	53.10780043	56.97211421	60.05225274	64.91860474	70.64321588
MU	K	.4108789D-1	.1125111D-1	.6899171D-2	.4074449D-2	.2211108D-2	.1417080D-2	.9940022D-3	.5676358D-3	.2936562D-3
1	PF1	1.164959572	.8314607825	.7348205510	.6434862638	.5512562922	.4922466405	.4497482599	.3905903024	.3337341917
1	PQ1	.5884061102	.5848878586	.5896874926	.5990351322	.6170084788	.6369316713	.6587551982	.7093050911	.8107820307
1	PF2	1.087889647	.9505875798	.9036178276	.8568044388	.8077202209	.7756725092	.7523903216	.7197385807	.6878327291
1	PQ2	1.489428793	1.420363263	1.442701981	1.496191880	1.603189575	1.719868383	1.843770506	2.116630359	2.622889783
1	PF3	1.055608191	1.012470069	.9954987400	.9773589430	.9568982644	.9426339972	.9317456509	.9155618749	.8983544005
1	PQ3	4.772637645	4.296102225	4.284100889	4.364101464	4.589359899	4.861603747	5.161862443	5.839335944	7.117074724
3	PFC	1.052205758	1.044279091	1.040369127	1.035679710	1.029722748	1.025124100	1.021353052	1.015312724	1.008312457
3	PQC	32.85076168	23.34098712	21.36310100	19.84074228	18.69304083	18.21019358	18.02031302	18.06803904	18.70500920
1	ZF1	1.052806131	1.052851057	1.052892745	1.052946667	1.053017580	1.053073136	1.053119056	1.053193385	1.053281272
1	ZF2	1.061773498	1.063885109	1.064685574	1.065545620	1.066538213	1.067254844	1.067821083	1.068702672	1.069706515
1	ZF3	1.084040790	1.093407381	1.096783683	1.100320022	1.104298390	1.107111088	1.109302598	1.112667180	1.116438833
1	ZF4	1.163609204	1.186219748	1.194731011	1.203799483	1.214152427	1.221547560	1.227343261	1.236284942	1.246351684
1	ZF5	1.481927689	1.532444328	1.552042377	1.573233510	1.597774901	1.615503707	1.629498363	1.651236356	1.675880590
1	ZF6	3.679086989	3.856584364	3.925881662	4.001115288	4.088632880	4.152102296	4.202337353	4.280575963	4.369530110

(6) P 8+4	TH = 73 SBE = 1.045691756								MCPER FUNCTIONS
	RC = 02	RC = 05	RC = 07	RC = 10	RC = 15	RC = 20	RC = 25	RC = 35	RC = 50
AMAX DB	.0017375255	.0108709564	.0213327376	.0436480540	.0988323393	.1772876696	.2802872360	.5675287486	1.249387366
AMIN DB	25.65252459	36.82448177	41.05167513	45.60742753	50.89856847	54.75191129	57.82443886	62.68058653	68.39572234
MU K	.5216435D-1	.1441371D-1	.8859643D-2	.5243589D-2	.2851488D-2	.1829803D-2	.1284630D-2	.7344643D-3	.3803767D-3
1 PF1	1.207876940	.8564961407	.7555341300	.6605091419	.5649362123	.5039819183	.4601717481	.3993034078	.3409192355
1 PQ1	.5951033958	.5883900431	.5926070720	.6015143313	.6191425401	.6388975682	.6606288409	.7111043080	.8126174861
1 PF2	1.097766326	.9621909075	.9152276356	.8682246326	.8187809976	.7864313372	.7629044510	.7298830426	.6975976574
1 PQ2	1.572953530	1.473548511	1.490143519	1.539507747	1.643831837	1.759827151	1.883944702	2.158683099	2.670279283
1 PF3	1.054624370	1.014292134	.9981359241	.9807365198	.9609771406	.9471318715	.9365307253	.9207302070	.9038859998
1 PQ3	5.223106681	4.606002855	4.566986724	4.627895638	4.841781440	5.112529730	5.415864048	6.107142460	7.419850504
3 PFC	1.047299974	1.040557039	1.037127357	1.032958746	1.027597441	1.023418509	1.019969799	1.014410381	1.007920267
3 PQC	38.17461389	26.36750439	23.93124545	22.04935802	20.60706681	19.97128828	19.68925831	19.63746374	20.22259140
1 ZF1	1.046947645	1.046967879	1.047002451	1.047049580	1.047113522	1.047164524	1.047207093	1.047276590	1.047359479
1 ZF2	1.054931816	1.056899588	1.057651797	1.058463435	1.059404355	1.060086336	1.060626700	1.061470532	1.062434927
1 ZF3	1.074469772	1.083374819	1.086611883	1.090014978	1.093857424	1.096582322	1.098709952	1.101983896	1.105664310
1 ZF4	1.147600120	1.169101678	1.177298005	1.186078688	1.196155426	1.203383542	1.209064355	1.217854184	1.227783846
1 ZF5	1.448841760	1.497270286	1.516287827	1.536964913	1.561039833	1.578507910	1.592337589	1.613384549	1.638399167
1 ZF6	3.564168727	3.735849293	3.803599771	3.877516667	3.963919551	4.026830801	4.076759628	4.154739883	4.243693945

(6) P 8+4	TH = 74 SBE = 1.040299436								MCPER FUNCTIONS
	RC = 02	RC = 05	RC = 07	RC = 10	RC = 15	RC = 20	RC = 25	RC = 35	RC = 50
AMAX DB	.0017375255	.0108709564	.0213327376	.0436480540	.0988323393	.1772876696	.2802872360	.5675287486	1.249387366
AMIN DB	23.57646695	34.66004688	38.86429320	43.39932321	48.67060153	52.51185796	55.57599594	60.42088704	66.12556007
MU K	.6624859D-1	.1849259D-1	.1139686D-1	.6761357D-2	.3685275D-2	.2368139D-2	.1664180D-2	.9526989D-3	.4939944D-3
1 PF1	1.254972966	.8838325009	.7780863411	.6789874931	.5797388948	.5166545192	.4714114035	.4086799589	.3486366220
1 PQ1	.6031076908	.5924185569	.5959308797	.6043108283	.6215288376	.6410845050	.6627057892	.7130894953	.8146341498
1 PF2	1.107140195	.9739426642	.9271178620	.8800192852	.8302806253	.7976550820	.7738956640	.7405133863	.7078492393
1 PQ2	1.671811856	1.534800512	1.544358702	1.588653991	1.689620206	1.804648015	1.928867499	2.205506964	2.722830862
1 PF3	1.053325945	1.015918861	1.000630708	.9840270606	.9650278273	.9516395617	.9413527365	.9259725941	.9095268667
1 PQ3	5.764560884	4.969013461	4.895948410	4.932576559	5.131350249	5.399137039	5.705076398	6.410726208	7.761548188
3 PFC	1.042538992	1.036893602	1.033920896	1.030253540	1.025471826	1.021704863	1.018574375	1.013492479	1.007512296
3 PQC	44.90359528	30.07858633	27.05162690	24.70958536	22.89205247	22.06164094	21.66193566	21.48143946	21.99483644
1 ZF1	1.041473625	1.041468682	1.041496018	1.041536213	1.041592994	1.041639289	1.041678376	1.041742821	1.041820442
1 ZF2	1.048526511	1.050347447	1.051049832	1.051811131	1.052697889	1.053343287	1.053856188	1.054659708	1.055581666
1 ZF3	1.065487684	1.073895794	1.076981608	1.080239247	1.083932460	1.086560415	1.088617190	1.091790059	1.095367902
1 ZF4	1.132399345	1.152701779	1.160551257	1.169011676	1.178777237	1.185814797	1.191363134	1.199975466	1.209741344
1 ZF5	1.416774343	1.462960124	1.481340135	1.501444151	1.524988410	1.542152536	1.555785125	1.577094818	1.601432387
1 ZF6	3.451187129	3.616625761	3.682681781	3.755131168	3.840257082	3.902502011	3.952045027	4.029654032	4.118496481

(6) P 8+4		TH = 75	SBE = 1.035276180						MCPER FUNCTIONS	
		RC = 02	RC = 05	RC = 07	RC = 10	RC = 15	RC = 20	RC = 25	RC = 35	RC = 50
AMAX DB		.0017375255	.0108709564	.0213327376	.0436480540	.0988323393	.1772876696	.2802872360	.5675287486	1.249387366
AMIN DB		21.49690479	32.47942950	36.65817675	41.17027092	46.41959338	50.24748810	53.30234670	58.13477839	63.82785884
MU	K	.8416950D-1	.2376996D-1	.1469235D-1	.8739497D-2	.4775516D-2	.3073446D-2	.2162134D-2	.1239542D-2	.6435867D-3
1	PF1	1.306754508	.9138052966	.8027443905	.6991299963	.5958220154	.5303940178	.4835786831	.4198089496	.3569564758
1	PQ1	.6128024013	.5970978003	.5997480642	.6074901284	.6242162824	.6435334820	.6650226800	.7152928849	.8168623117
1	PF2	1.115846301	.9857843203	.9392594245	.8921834096	.8422336490	.8093680175	.7853939096	.7516657039	.7186277254
1	PQ2	1.790212073	1.606009838	1.606859459	1.644872641	1.741600057	1.855288912	1.979453984	2.257993165	2.781479364
1	PF3	1.051699066	1.017327174	1.002960807	.9872107047	.9690344406	.9561443984	.9462015834	.9312927793	.9152746873
1	PQ3	6.424279693	5.398949716	5.282468275	5.287923272	5.466583841	5.729384313	6.037194957	6.757679758	8.150179071
3	PFC	1.037939238	1.033300696	1.030759527	1.027571594	1.023350961	1.019986663	1.017169192	1.012560011	1.007088312
3	PQC	53.55025358	34.69449907	30.89467927	27.95515523	25.65313326	24.57186775	24.02011990	23.67154256	24.08592203
1	ZF1	1.036374829	1.036344301	1.036364319	1.036397478	1.036446950	1.036488406	1.036523893	1.036583084	1.036655183
1	ZF2	1.042548640	1.044220214	1.044871377	1.045580559	1.046410793	1.047017731	1.047501606	1.048262265	1.049138763
1	ZF3	1.057092768	1.064968010	1.067890138	1.070989570	1.074519549	1.077040882	1.079019391	1.082080020	1.085543164
1	ZF4	1.118009812	1.137022082	1.144491410	1.152597220	1.162014158	1.168835717	1.174232449	1.182639229	1.192211909
1	ZF5	1.385705009	1.429488822	1.447171549	1.466639946	1.489584715	1.506398086	1.519798574	1.540820248	1.564929819
1	ZF6	3.339966881	3.498710171	3.562915823	3.633738013	3.717414110	3.778876766	3.827948216	3.905063612	3.993672646

(6) P 8+4		TH = 76	SBE = 1.030613629						MCPER FUNCTIONS	
		RC = 02	RC = 05	RC = 07	RC = 10	RC = 15	RC = 20	RC = 25	RC = 35	RC = 50
AMAX DB		.0017375255	.0108709564	.0213327376	.0436480540	.0988323393	.1772876696	.2802872360	.5675287486	1.249387366
AMIN DB		19.41409237	30.27890576	34.42908984	38.91565132	44.14059207	47.95365394	50.99820787	55.81679423	61.49697929
MU	K	.1069782D 0	.3062349D-1	.1899090D-1	.1132967D-1	.6208267D-2	.4002371D-2	.2818964D-2	.1618677D-2	.8416878D-3
1	PF1	1.363744524	.9468160039	.8298308559	.7211894502	.6133767339	.5453574372	.4968085758	.4297980476	.3659635312
1	PQ1	.6247227413	.6025936369	.6041757537	.6111371445	.6272671302	.6462965763	.6676258110	.7177549385	.8193397744
1	PF2	1.123690440	.9976390554	.9516106170	.9047047716	.8546522846	.8215947796	.7974311716	.7633900641	.7299787933
1	PQ2	1.933940068	1.689688394	1.679634123	1.709781886	1.801121155	1.912979881	2.036874625	2.317275599	2.847408146
1	PF3	1.049731974	1.018492537	1.005101795	.9902651063	.9729786230	.9606314414	.9510651289	.9366529112	.9211260902
1	PQ3	7.240674393	5.914614390	5.742019264	5.706993950	5.858754499	6.113744981	6.422305837	7.157897567	8.596123010
3	PFC	1.033517889	1.029791357	1.027654105	1.024921368	1.021240671	1.018268015	1.015757143	1.011614253	1.006648181
3	PQC	64.86979987	40.52984091	35.70128681	31.97316173	29.03585930	27.62655955	26.87567927	26.30498022	26.58235594
1	ZF1	1.031642697	1.031586281	1.031598942	1.031625008	1.031667067	1.031703580	1.031735369	1.031789126	1.031855467
1	ZF2	1.036990115	1.038510283	1.039109023	1.039764488	1.040535993	1.041102675	1.041556000	1.042271279	1.043099276
1	ZF3	1.049284713	1.056591046	1.059336708	1.062264687	1.065616743	1.068021225	1.069913599	1.072850098	1.076185497
1	ZF4	1.104436588	1.122067909	1.129122550	1.136837554	1.145865831	1.152443881	1.157668182	1.165838658	1.175185654
1	ZF5	1.355615769	1.396833185	1.413755930	1.432522310	1.454793611	1.471205390	1.484335474	1.505013185	1.528836495
1	ZF6	3.230325831	3.381882465	3.444071612	3.513096295	3.595137390	3.655692812	3.704199821	3.780688329	3.868930270

(6) P 8+4	TH = 77 SBE = 1.026304108								MCPER FUNCTIONS
	RC = 02	RC = 05	RC = 07	RC = 10	RC = 15	RC = 20	RC = 25	RC = 35	RC = 50
AMAX DB	.0017375255	.0108709564	.0213327376	.0436480540	.0988323393	.1772876696	.2802872360	.5675287486	1.249387366
AMIN DB	17.32984851	28.05434564	32.17219111	36.63011658	41.82782694	45.62434071	48.65739698	53.46052632	59.12630024
MU I K	.1359902D 0	.3956241D-1	.2462581D-1	.1473989D-1	.8102306D-2	.5233388D-2	.3690882D-2	.2123116D-2	.1105821D-2
1 I PF1	1.426427549	.9833466939	.8597382289	.7454752196	.6326374338	.5617373131	.5112665389	.4417791068	.3757614416
1 I PQ1	.6396312551	.6091317479	.6093708177	.6153639476	.6307626565	.6494410122	.6705746131	.7205272229	.8221144126
1 I PF2	1.130445857	1.009406783	.9641131040	.9175610984	.8675449494	.8343597029	.8100413394	.7757010370	.7419545947
1 I PQ2	2.111119869	1.789240888	1.765347919	1.785528848	1.869957026	1.979327491	2.102651787	2.384820664	2.922139198
1 I PF3	1.047415850	1.019389027	1.007026960	.9931651196	.9768391198	.9650830341	.9559287657	.9420731610	.9270763549
1 I PQ3	8.269137378	6.542170833	6.295900314	6.207585957	6.323076715	6.566269074	6.873880161	7.624513175	9.113071859
3 I PFC	1.029292839	1.026379838	1.024616699	1.022312423	1.019147679	1.016553745	1.014341692	1.010656837	1.006191897
3 I PQC	80.00340251	48.04533960	41.82031080	37.03162580	33.24639747	31.40089618	30.38505224	29.51662771	29.60320066
1 I ZF1	1.027269286	1.027186834	1.027192148	1.027211117	1.027245713	1.027277213	1.027305228	1.027353402	1.027413771
1 I ZF2	1.031843683	1.033210896	1.033756225	1.034356586	1.035067353	1.035592086	1.036013398	1.036680829	1.037457289
1 I ZF3	1.042064502	1.048766378	1.051322572	1.054065436	1.057224229	1.059501080	1.061298974	1.064098653	1.067292300
1 I ZF4	1.091686601	1.107848052	1.114452470	1.121738713	1.130335620	1.136640437	1.141669618	1.149570019	1.158655336
1 I ZF5	1.326491224	1.364971767	1.381068602	1.399062324	1.420580422	1.436535193	1.449352794	1.469624596	1.493097067
1 I ZF6	3.122073176	3.265900844	3.325894351	3.392938231	3.473144442	3.532657070	3.580498424	3.656213984	3.743941222

(6) P 8+4	TH = 78 SBE = 1.022340595								MCPER FUNCTIONS
	RC = 02	RC = 05	RC = 07	RC = 10	RC = 15	RC = 20	RC = 25	RC = 35	RC = 50
AMAX DB	.0017375255	.0108709564	.0213327376	.0436480540	.0988323393	.1772876696	.2802872360	.5675287486	1.249387366
AMIN DB	15.24862984	25.80119933	29.88190419	34.30739706	39.47447581	43.25241540	46.27256965	51.05834723	56.70792790
MU I K	.1728099D 0	.5127906D-1	.3205566D-1	.1925884D-1	.1062371D-1	.6876687D-2	.4857038D-2	.2799514D-2	.1460843D-2
1 I PF1	1.495140399	1.023976524	.8929477166	.7723700695	.6538951782	.5797728866	.5271581495	.4549158539	.3864787953
1 I PQ1	.6586377811	.6170263697	.6155480122	.6203215416	.6348106477	.6530551915	.6739467637	.7236766075	.8252478823
1 I PF2	1.135850905	1.020957746	.9766865656	.9307162089	.8809141072	.8476857501	.8232598691	.7886782697	.7546150716
1 I PQ2	2.333423628	1.909388844	1.867652541	1.875022858	1.950485247	2.056471513	2.178803880	2.462560566	3.007666124
1 I PF3	1.044746011	1.019989499	1.008707185	.9958824437	.9805912666	.9694782499	.9607748662	.9475312268	.9331190245
1 I PQ3	9.591696178	7.318960885	6.974155205	6.814532704	6.880555241	7.106224697	7.410303918	8.175331451	9.719461519
3 I PFC	1.025282632	1.023081723	1.021660731	1.019755589	1.017079765	1.014849545	1.012927007	1.009689839	1.005719636
3 I PQC	100.7240057	57.93530093	49.77129461	43.52506085	38.58454743	36.14759349	34.77259579	33.49328393	33.31629029
1 I ZF1	1.023247186	1.023138801	1.023136839	1.023148766	1.023175911	1.023202368	1.023226562	1.023269037	1.023323252
1 I ZF2	1.027102890	1.028316116	1.028807290	1.029351403	1.029999663	1.030480890	1.030868809	1.031486012	1.032207933
1 I ZF3	1.035434185	1.041497395	1.043851082	1.046394883	1.049344503	1.051482406	1.053176994	1.055826333	1.058863178
1 I ZF4	1.079768200	1.094374970	1.100493011	1.107310965	1.115431106	1.121430602	1.126239933	1.133833108	1.142616740
1 I ZF5	1.298318644	1.333884826	1.349086212	1.366231898	1.386910552	1.402347654	1.414806324	1.434603242	1.457652234
1 I ZF6	3.015006825	3.150494953	3.208096732	3.272960193	3.351113710	3.409435254	3.456499776	3.531781108	3.618329443

(6) P 8+4	TH = 79	SBE = 1.018716695						MCPER FUNCTIONS	
	RC = 02	RC = 05	RC = 07	RC = 10	RC = 15	RC = 20	RC = 25	RC = 35	RC = 50
AMAX DB	.0017375255	.0108709564	.0213327376	.0436480540	.0988323393	.1772876696	.2802872360	.5675287486	1.249387366
AMIN DB	13.17919833	23.51454902	27.55177070	31.94004618	37.07234548	40.82927860	43.83485406	48.60102320	54.23228945
MU K	.2193007D 0	.6672254D-1	.4191905D-1	.2529285D-1	.1400821D-1	.9089418D-2	.6430686D-2	.3714915D-2	.1942610D-2
1 PF1	1.569868940	1.069398117	.9300532491	.8023531837	.6775166075	.5997659695	.5447428372	.4694148587	.3982777145
1 PQ1	.6833929198	.6267267189	.6230088714	.6262183150	.6395573810	.6572578925	.6778459301	.7272915649	.8298211446
1 PF2	1.139609187	1.032124242	.9892214708	.9441145858	.8947530818	.8615928017	.8371230916	.8023670902	.7680296334
1 PQ2	2.618044312	2.056854888	1.991679934	1.982303118	2.045968271	2.147327185	2.268067153	2.553097068	3.106656102
1 PF3	1.041723616	1.020265859	1.010110860	.9983852277	.9842063644	.9737921933	.9655820742	.9530116811	.9392453836
1 PQ3	11.33344887	8.299869408	7.820380610	7.563443148	7.560959015	7.760720302	8.057309227	8.835088150	10.44071464
3 PFC	1.021506351	1.019914039	1.018801146	1.017263162	1.015045967	1.013162154	1.011518113	1.008715897	1.005231813
3 PQC	129.8716967	71.28103631	60.35333726	52.05270168	45.49973712	42.24223420	40.36962495	38.53058463	37.96486105
1 ZF1	1.019569399	1.019435616	1.019426517	1.019431529	1.019451312	1.019472741	1.019493099	1.019529805	1.019577724
1 ZF2	1.022762036	1.023820826	1.024257369	1.024744370	1.025328643	1.025764979	1.026118236	1.026682957	1.027347413
1 ZF3	1.029396495	1.034789389	1.036927762	1.039258473	1.041982577	1.043969729	1.045551717	1.048036345	1.050900227
1 ZF4	1.068690375	1.081665005	1.087260447	1.093569348	1.101164716	1.106824329	1.111386863	1.118631878	1.127069211
1 ZF5	1.271087726	1.303554339	1.317786598	1.334003494	1.353749014	1.368601704	1.380649877	1.399894619	1.422437824
1 ZF6	2.908908646	3.035356890	3.090348172	3.152810499	3.228671048	3.285637559	3.331801916	3.405469243	3.491654342

(6) P 8+4	TH = 80	SBE = 1.015426612						MCPER FUNCTIONS	
	RC = 02	RC = 05	RC = 07	RC = 10	RC = 15	RC = 20	RC = 25	RC = 35	RC = 50
AMAX DB	.0017375255	.0108709564	.0213327376	.0436480540	.0988323393	.1772876696	.2802872360	.5675287486	1.249387366
AMIN DB	11.13704296	21.18930935	25.17430216	29.51910236	34.61142394	38.34437012	41.33332922	46.07715938	51.68754890
MU K	.2774264D 0	.8720362D-1	.5511691D-1	.3342296D-1	.1859640D-1	.1209989D-1	.8576963D-2	.4967548D-2	.2603890D-2
1 PF1	1.649897448	1.120427872	.9717911210	.8360319061	.7039710714	.6221039905	.5643539332	.4855416189	.4113664895
1 PQ1	.7164036902	.6388946700	.6321894039	.6333500087	.6452066590	.6622127834	.6824138945	.7314919372	.8329429555
1 PF2	1.141394959	1.042689899	1.001569219	.9576736580	.9090412843	.8760949181	.8516669143	.8168290831	.7822793176
1 PQ2	2.990996211	2.241509154	2.144859957	2.113139912	2.161006751	2.255973251	2.374248990	2.660023668	3.222767047
1 PF3	1.038358131	1.020189561	1.011203884	1.000637637	.9876509113	.9779951077	.9703243816	.9584950951	.9454437409
1 PQ3	13.69151373	9.568379501	8.899975482	8.507037954	8.407796639	8.569056975	8.851986647	9.639149204	11.31288288
3 PFC	1.017983403	1.016895367	1.016054597	1.014849134	1.013056825	1.011499583	1.010121102	1.007738366	1.004729171
3 PQC	172.1524700	89.83221764	74.84270609	63.55884069	54.69017480	50.26283393	47.68248797	45.03812130	43.91308695
1 ZF1	1.016229155	1.016071259	1.016055254	1.016053562	1.016066159	1.016082635	1.016099181	1.016130104	1.016171634
1 ZF2	1.018816077	1.019720714	1.020102451	1.020531805	1.021050957	1.021441234	1.021758707	1.022268870	1.022873058
1 ZF3	1.023954193	1.028649511	1.030560374	1.032664194	1.035146244	1.036970452	1.038430120	1.040734828	1.043408416
1 ZF4	1.058461429	1.069738586	1.074775953	1.080534343	1.087554560	1.092837190	1.097123593	1.103975302	1.112016412
1 ZF5	1.244789421	1.273964083	1.287148663	1.302349790	1.321059832	1.335254221	1.346834253	1.365439534	1.387382877
1 ZF6	2.803534486	2.920128972	2.972259940	3.032072329	3.105370646	3.160798379	3.205924017	3.278274541	3.363387099

(6) P 8+4 — TH = 81, SBE = 1.012465126 — MCPER FUNCTIONS

MU		RC = 02	RC = 05	RC = 07	RC = 10	RC = 15	RC = 20	RC = 25	RC = 35	RC = 50
	AMAX DB	.0017375255	.0108709564	.0213327376	.0436480540	.0988323393	.1772876696	.2802872360	.5675287486	1.249387366
	AMIN DB	9.147476594	18.82074760	22.74087966	27.03364281	32.07923818	35.78444706	38.75425799	43.47237964	49.05874120
	K	.3488399D 0	.1145414D 0	.7293836D-1	.4449568D-1	.2489076D-1	.1624717D-1	.1154216D-1	.6704726D-2	.3524219D-2
1	PF1	1.733262567	1.177994894	1.019073576	.8741856167	.7338704642	.6472944066	.5864288228	.5036448931	.4260187520
1	PQ1	.7615523666	.6545399330	.6437421712	.6421504583	.6520514419	.6681522514	.6878502942	.7364446642	.8377634085
1	PF2	1.140868241	1.052375861	1.013528529	.9712726412	.9237369412	.8911959167	.8669244744	.8321325097	.7974596124
1	PQ2	3.492798244	2.478374414	2.338334924	2.276061118	2.302302868	2.388299673	2.502814104	2.788455859	3.361165656
1	PF3	1.034670753	1.019732421	1.011949803	1.002599395	.9908856494	.9820512190	.9749699015	.9639568416	.9516984192
1	PQ3	16.98802263	11.25671598	10.31498744	9.726401684	9.487030808	9.590284782	9.849707694	10.63983730	12.38882115
3	PFC	1.014733141	1.014045944	1.013439643	1.012529468	1.011124687	1.009871418	1.008743403	1.006761525	1.004212908
3	PQC	235.6642984	116.5499921	95.37063740	79.59791584	67.28756291	61.13666686	57.51729599	53.68938336	51.72774421
1	ZF1	1.013219651	1.013040214	1.013017655	1.013009575	1.013015267	1.013026933	1.013039742	1.013064932	1.013100042
1	ZF2	1.015260435	1.016012284	1.016339378	1.016710927	1.017164243	1.017507566	1.017788321	1.018242092	1.018783401
1	ZF3	1.019108968	1.023086638	1.024758950	1.026622759	1.028846392	1.030495250	1.031822541	1.033931343	1.036396112
1	ZF4	1.049086731	1.058620379	1.063066132	1.068232711	1.074625516	1.079491569	1.083469995	1.089878578	1.097467410
1	ZF5	1.219412600	1.245099784	1.257152284	1.271243274	1.288805242	1.302258890	1.313305800	1.331172131	1.352406957
1	ZF6	2.698592261	2.804386458	2.853363952	2.910239046	2.980667219	3.034346609	3.078275343	3.149076808	3.232875708

(6) P 8+4 — TH = 82, SBE = 1.009827573 — MCPER FUNCTIONS

MU		RC = 02	RC = 05	RC = 07	RC = 10	RC = 15	RC = 20	RC = 25	RC = 35	RC = 50
	AMAX DB	.0017375255	.0108709564	.0213327376	.0436480540	.0988323393	.1772876696	.2802872360	.5675287486	1.249387366
	AMIN DB	7.248610540	16.40567354	20.24182924	24.47020775	29.45991014	33.13249501	36.07991916	40.76806905	46.32643948
	K	.4340797D 0	.1512573D 0	.9725424D-1	.5977087D-1	.3365151D-1	.2204831D-1	.1570377D-1	.9153697D-2	.4827008D-2
1	PF1	1.816038726	1.243071511	1.073018018	.9178257035	.7680291268	.6760177398	.6115558952	.5241949072	.4426037083
1	PQ1	.8249539673	.6752659179	.6586845096	.6532818472	.6605290200	.6754184242	.6944462918	.7423902858	.8434965140
1	PF2	1.137705169	1.060823279	1.024826447	.9847360733	.9387657679	.9068821024	.8829219285	.8483522967	.8136841940
1	PQ2	4.188612618	2.791313435	2.589536434	2.484197484	2.480012005	2.553143679	2.661905813	2.945946689	3.529414370
1	PF3	1.030698778	1.018868055	1.012310235	1.004225338	.9938643762	.9859172098	.9794792025	.9693654153	.9579882965
1	PQ3	21.77167859	13.58467233	12.23239193	11.35211785	10.90320159	10.91704851	11.13672123	11.91783203	13.74920446
3	PFC	1.011774215	1.011387743	1.010976971	1.010322422	1.009264095	1.008289194	1.007394136	1.005790865	1.003684848
3	PQC	334.9608424	156.7205323	125.6915249	102.8670462	85.22275034	76.42794152	71.22241189	65.58834420	62.33327929
1	ZF1	1.010533704	1.010337394	1.010308817	1.010294800	1.010293998	1.010301075	1.010310281	1.010329871	1.010358612
1	ZF2	1.012090663	1.012692842	1.012965860	1.013279890	1.013667160	1.013962971	1.014206319	1.014602190	1.015078288
1	ZF3	1.014859643	1.018111106	1.019535749	1.021147776	1.023097383	1.024558569	1.025743253	1.027639542	1.029875807
1	ZF4	1.040565221	1.048339312	1.052163608	1.056698556	1.062410683	1.066818294	1.070454329	1.076364355	1.083438259
1	ZF5	1.194936507	1.216949256	1.227778215	1.240655717	1.256944589	1.269564618	1.280004373	1.297017043	1.317416241
1	ZF6	2.593694997	2.687611801	2.733081346	2.786677258	2.853873963	2.905560537	2.948107961	3.017089124	3.099291433

(6) P 8+4	TH = 83	SBE = 1.007509825						MCPER FUNCTIONS	
	RC = 02	RC = 05	RC = 07	RC = 10	RC = 15	RC = 20	RC = 25	RC = 35	RC = 50
AMAX DB		.0108709564	.0213327376	.0436480540	.0988323393	.1772876696	.2802872360	.5675287486	1.249387366
AMIN DB		13.94499261	17.66699280	21.81210601	26.73273175	30.36602279	33.28675414	37.93935840	43.46460905
MU I K		.2007938D 0	.1308128D 0	.8116984D-1	.4606419D-1	.3031788D-1	.2166019D-1	.1267746D-1	.6710727D-2
1 I PF1		1.316459162	1.134944356	.9682738673	.8075560863	.7092123436	.6405507082	.5478460799	.4616360784
1 I PQ1		.7037396116	.6786817244	.6678049537	.6713230501	.6845379731	.7026453521	.7496901368	.8504599332
1 I PF2		1.067572290	1.035091618	.9978089440	.9540028235	.9231100104	.8996708281	.8655689824	.8310899376
1 I PQ2		3.220222423	2.927179404	2.758813011	2.710287561	2.764406032	2.864234326	3.144162580	3.739079537
1 I PF3		1.017574386	1.012245858	1.005465085	.9965324538	.9895401702	.9838029943	.9746800059	.9642846253
1 I PQ3		16.93965110	14.94177956	13.60676769	12.83131945	12.70268207	12.85461198	13.60406477	15.52345472
3 I PFC		1.008944491	1.008689609	1.008248936	1.007492271	1.006766911	1.006084602	1.004833497	1.003147708
3 I PQC		220.4084677	172.8653900	138.3589042	112.0051678	98.94420623	91.19525392	82.67336516	77.32753475
1 I ZF1		1.007958048	1.007924272	1.007904956	1.007898231	1.007901045	1.007906852	1.007921077	1.007943594
1 I ZF2		1.009760486	1.009980495	1.010237827	1.010559458	1.010807625	1.011013191	1.011350091	1.011759048
1 I ZF3		1.013734170	1.014905081	1.016255875	1.017917486	1.019179246	1.020211219	1.021878073	1.023865139
1 I ZF4		1.038928245	1.042107620	1.045974643	1.050953331	1.054858909	1.058115682	1.063467752	1.069954393
1 I ZF5		1.189502171	1.199007942	1.210557461	1.225432728	1.237113176	1.246860308	1.262885121	1.282297586
1 I ZF6		2.569153212	2.610673520	2.660568974	2.724096084	2.773496163	2.814441386	2.881277300	2.961542940

(6) P 8+4	TH = 84	SBE = 1.005508280						MCPER FUNCTIONS	
	RC = 02	RC = 05	RC = 07	RC = 10	RC = 15	RC = 20	RC = 25	RC = 35	RC = 50
AMAX DB		.0108709564	.0213327376	.0436480540	.0988323393	.1772876696	.2802872360	.5675287486	1.249387366
AMIN DB		11.44891063	15.00757389	19.03874342	23.86995851	27.45427003	30.34227351	34.95171408	40.43695259
MU I K		.2676421D 0	.1776729D 0	.1117025D 0	.6404748D-1	.4239225D-1	.3040089D-1	.1788193D-1	.9509384D-2
1 I PF1		1.398239061	1.206259303	1.027249803	.8539998949	.7482156215	.6745857674	.5755457963	.4838634563
1 I PQ1		.7446411520	.7066254887	.6875241116	.6855664012	.6963686084	.7131609484	.7589163741	.8591497566
1 I PF2		1.072040097	1.043816829	1.010118106	.9692423729	.9397849896	.9171539451	.8838652394	.8498436977
1 I PQ2		3.836098474	3.401701170	3.136600420	3.020451817	3.045300950	3.130836960	3.402182403	4.008863343
1 I PF3		1.015838188	1.011718538	1.006263055	.9988249577	.9928547822	.9878788106	.9798468622	.9705476389
1 I PQ3		22.05575651	18.98341162	16.89794510	15.58544079	15.21876890	15.25180651	15.92525700	17.93272334
3 I PFC		1.006741558	1.006603135	1.006333073	1.005829757	1.005321713	1.004828961	1.003898734	1.002605494
3 I PQC		328.3587427	251.2961759	196.1004926	154.5450934	134.1369809	122.0418657	108.6064134	99.68323465
1 I ZF1		1.005897591	1.005859907	1.005836206	1.005824345	1.005823344	1.005826048	1.005835266	1.005851832
1 I ZF2		1.007214027	1.007382792	1.007584918	1.007842083	1.008043017	1.008210835	1.008488282	1.008828734
1 I ZF3		1.009966965	1.010882835	1.011966714	1.013329362	1.014381297	1.015251084	1.016671845	1.018388377
1 I ZF4		1.030422889	1.032944494	1.036114059	1.040309610	1.043668936	1.046507542	1.051235179	1.057054342
1 I ZF5		1.162748179	1.170823090	1.180916301	1.194217585	1.204835563	1.213789672	1.228666692	1.246909067
1 I ZF6		2.448150482	2.485160438	2.530815804	2.590119837	2.636866905	2.675935675	2.740223693	2.818130648

(6) P 8+4 — TH = 85 SBE = 1.003819838 — MCPER FUNCTIONS

(6) P 8+4	RC = 02	RC = 05	RC = 07	RC = 10	RC = 15	RC = 20	RC = 25	RC = 35	RC = 50
AMAX DB		.0108709564	.0213327376	.0436480540	.0988323393	.1772876696	.2802872360	.5675287486	1.249387366
AMIN DB		8.946793880	12.26110771	16.12556122	20.83332279	24.35338478	27.19956014	31.75475642	37.19022543
MU I K		.3569935D 0	.2437500D 0	.1562147D 0	.9085187D-1	.6058021D-1	.4365379D-1	.2583820D-1	.1381939D-1
1 I PF1		1.486507860	1.287999459	1.096912271	.9095768612	.7950104683	.7154252309	.6087364626	.5104307041
1 I PQ1		.8067109637	.7479138600	.7157511943	.7052831017	.7124133482	.7272260155	.7710395380	.8703939011
1 I PF2		1.073507442	1.050308126	1.021110701	.9841454816	.9567219236	.9352977598	.9033166503	.8701514470
1 I PQ2		4.775529238	4.109255413	3.686308593	3.460582040	3.437756010	3.499315604	3.753629906	4.371290797
1 I PF3		1.013663245	1.010695886	1.006559484	1.000664462	.9957793748	.9916260900	.9847935920	.9767209866
1 I PQ3		30.45812126	25.46464880	22.04515443	19.78282594	18.99108309	18.80400104	19.30888228	21.38753513
3 I PFC		1.004805576	1.004745783	1.004602541	1.004301240	1.003974844	1.003645203	1.002998980	1.002064136
3 I PQC		527.8622874	393.6188872	298.4807722	227.9520845	193.7418066	173.5589708	151.0411640	135.4927147
1 I ZF1		1.004151338	1.004111829	1.004085099	1.004069186	1.004064982	1.004064990	1.004069718	1.004080763
1 I ZF2		1.005052786	1.005173144	1.005322472	1.005517335	1.005672149	1.005802796	1.006021105	1.006292497
1 I ZF3		1.006818507	1.007485439	1.008302693	1.009360520	1.010194872	1.010894485	1.012053826	1.013478633
1 I ZF4		1.022859112	1.024727610	1.027182244	1.030552376	1.033322720	1.035703261	1.039735380	1.044795832
1 I ZF5		1.136669535	1.143202970	1.151695348	1.163236124	1.172645982	1.180686387	1.194220220	1.211063391
1 I ZF6		2.323383389	2.355170400	2.395867842	2.450213768	2.493826987	2.530662113	2.591880034	2.666881393

(6) P 8+4 — TH = 86 SBE = 1.002441898 — MCPER FUNCTIONS

(6) P 8+4	RC = 02	RC = 05	RC = 07	RC = 10	RC = 15	RC = 20	RC = 25	RC = 35	RC = 50
AMAX DB		.0108709564	.0213327376	.0436480540	.0988323393	.1772876696	.2802872360	.5675287486	1.249387366
AMIN DB		6.503281401	9.443652151	13.04663602	17.56852811	20.99754431	23.78668606	28.26999410	33.64079568
MU I K		.4729725D 0	.3371455D 0	.2226733D 0	.1323042D 0	.8915029D-1	.6466447D-1	.3859227D-1	.2079506D-1
1 I PF1		1.574814269	1.379422583	1.179616699	.9775118519	.8526763511	.7658814949	.6497658777	.5432205824
1 I PQ1		.9075377299	.8135810429	.7592019807	.7344743954	.7355812562	.7471889604	.7878681421	.8857051842
1 I PF2		1.071135769	1.053622696	1.029953570	.9981433738	.9735682569	.9539154806	.9239680436	.8922700391
1 I PQ2		6.329517936	5.254154034	4.551872304	4.133087536	4.026071539	4.044344329	4.264157534	4.888860406
1 I PF3		1.011084385	1.009160970	1.006294087	1.001958759	.9982103122	.9949385707	.9894196876	.9827219702
1 I PQ3		45.72725882	36.96091383	30.91935598	26.79777891	25.16976319	24.53830626	24.66097104	26.74181367
3 I PFC		1.003163486	1.003148308	1.003089224	1.002936662	1.002753007	1.002556603	1.002151123	1.001532559
3 I PQC		942.2419462	684.9379806	503.2251690	370.3601263	306.8677954	269.7127364	228.3103296	199.0215806
1 I ZF1		1.002714146	1.002676129	1.002648455	1.002630035	1.002623463	1.002621330	1.002622289	1.002628446
1 I ZF2		1.003276138	1.003352701	1.003453003	1.003589088	1.003699841	1.003794631	1.003955232	1.004159183
1 I ZF3		1.004292062	1.004727622	1.005287991	1.006043604	1.006657372	1.007181809	1.008067684	1.009181341
1 I ZF4		1.016267285	1.017515940	1.019259289	1.021776816	1.023920992	1.025804868	1.029067103	1.033266455
1 I ZF5		1.111217332	1.116114880	1.122847559	1.132406818	1.140430939	1.147407949	1.159351642	1.174500713
1 I ZF6		2.192916363	2.218623214	2.253383843	2.301737310	2.341543208	2.375648184	2.433074538	2.504416964

$P^{(4)}_{9+0}$

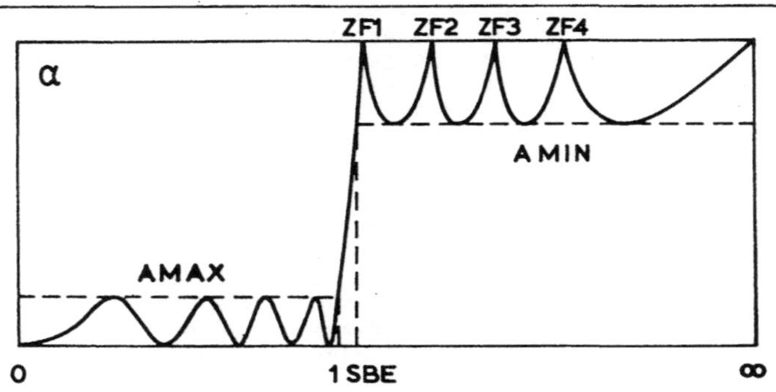

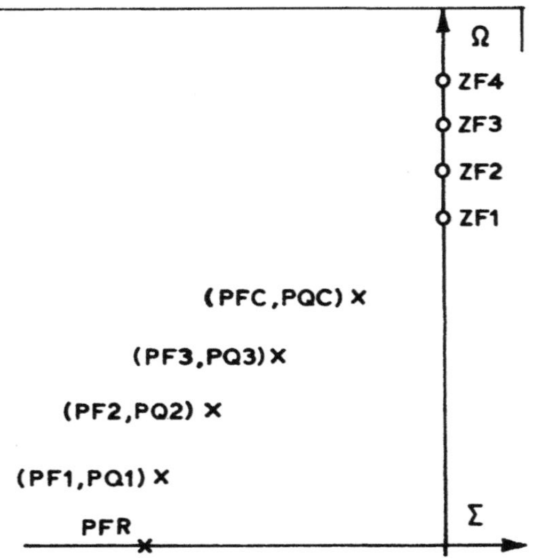

(4) P 9+0	TH = 38	SBE = 1.624269245						CAUER FUNCTIONS	
	RC = 02	RC = 05	RC = 07	RC = 10	RC = 15	RC = 20	RC = 25	RC = 35	RC = 50
AMAX DB	.0017375255	.0108709564	.0213327376	.0436480540	.0988323393	.1772876696	.2802872360	.5675287486	1.249387366
AMIN DB	91.37345438	99.34138798	102.2744105	105.3947650	108.9717745	111.5490045	113.5902043	116.8000066	120.5799044
MU I K	.3515866D-3	.1404868D-3	.1002269D-3	.6997884D-4	.4635710D-4	.3445520D-4	.2723923D-4	.1882369D-4	.1218175D-4
1 I PFR	.6215973271	.4835639048	.4354776412	.3856931934	.3302550227	.2914006708	.2613416154	.2156406085	.1652824557
1 I PF1	.7202434452	.6102105993	.5740196161	.5382100873	.5008328142	.4765687703	.4590952053	.4350558345	.4126939037
1 I PQ1	.6571327612	.7098573447	.7397343705	.7813930195	.8473709340	.9126382260	.9794169898	1.123506160	1.389027563
1 I PF2	.8958397141	.8228636727	.7997084084	.7773436268	.7546762283	.7403819629	.7303144835	.7167994117	.7046028480
1 I PQ2	1.169397457	1.345532780	1.441054957	1.570382434	1.768210169	1.957694232	2.146883947	2.543895887	3.251401052
1 I PF3	1.027777878	.9792967018	.9639795344	.9492354789	.9343549425	.9250097783	.9184482091	.9096684684	.9017760394
1 I PQ3	2.422031395	2.844729544	3.070870728	3.374549846	3.834961519	4.272600182	4.707210873	5.614025357	7.219688718
1 I PFC	1.091866872	1.055324313	1.043758696	1.032620066	1.021374708	1.014311595	1.009352291	1.002716802	.9967526109
1 I PQC	8.205228977	9.690939772	10.48291217	11.54419030	13.14959622	14.67271287	16.18330799	19.33083337	24.89558153
1 I ZF1	1.643915866	1.643915866	1.643915866	1.643915866	1.643915866	1.643915866	1.643915866	1.643915866	1.643915866
1 I ZF2	1.824497221	1.824497221	1.824497221	1.824497221	1.824497221	1.824497221	1.824497221	1.824497221	1.824497221
1 I ZF3	2.365290161	2.365290161	2.365290161	2.365290161	2.365290161	2.365290161	2.365290161	2.365290161	2.365290161
1 I ZF4	4.291506701	4.291506701	4.291506701	4.291506701	4.291506701	4.291506701	4.291506701	4.291506701	4.291506701

(4) P 9+0 — TH = 39 SBE = 1.589015729 — CAUER FUNCTIONS

MU		RC = 02	RC = 05	RC = 07	RC = 10	RC = 15	RC = 20	RC = 25	RC = 35	RC = 50
	AMAX DB	.0017375255	.0108709564	.0213327376	.0436480540	.0988323393	.1772876696	.2802872360	.5675287486	1.249387366
	AMIN DB	89.16511883	97.13305243	100.0660749	103.1864294	106.7634389	109.3406690	111.3818688	114.5916710	118.3715688
MU	K	.4406448D-3	.1760727D-3	.1256148D-3	.8770475D-4	.5809954D-4	.4318284D-4	.3413903D-4	.2359130D-4	.1526744D-4
1	PFR	.6270422212	.4873833716	.4388070590	.3885510719	.3326260585	.2934517051	.2631556700	.2171098466	.1663899126
1	PF1	.7253916874	.6142333262	.5777081057	.5415842346	.5038954144	.4794377579	.4618288003	.4376084793	.4150835399
1	PQ1	.6588000591	.7114279507	.7413027553	.7829835124	.8490265644	.9143763458	.9812491874	1.125558289	1.391510932
1	PF2	.8991023636	.8258687557	.8026281368	.7801796805	.7574264574	.7430776358	.7329715628	.7194044922	.7071608451
1	PQ2	1.179355613	1.355715319	1.451563807	1.581434350	1.780216783	1.970692577	2.160915930	2.560176769	3.271820792
1	PF3	1.028211067	.9801044022	.9648903074	.9502389595	.9354457354	.9261522875	.9196256268	.9108908183	.9030370750
1	PQ3	2.457805468	2.882722913	3.110600925	3.416892916	3.881617568	4.323581228	4.762617182	5.678910471	7.301773511
1	PFC	1.090392505	1.054495648	1.043118740	1.032155036	1.021079658	1.014119976	1.009231784	1.002689517	.9968072586
1	PQC	8.358296745	9.855968491	10.65634623	11.72993901	13.35530484	14.89822165	16.42895839	19.61941188	25.26170097
1	ZF1	1.607957479	1.607957479	1.607957479	1.607957479	1.607957479	1.607957479	1.607957479	1.607957479	1.607957479
1	ZF2	1.782265986	1.782265986	1.782265986	1.782265986	1.782265986	1.782265986	1.782265986	1.782265986	1.782265986
1	ZF3	2.305597626	2.305597626	2.305597626	2.305597626	2.305597626	2.305597626	2.305597626	2.305597626	2.305597626
1	ZF4	4.174652287	4.174652287	4.174652287	4.174652287	4.174652287	4.174652287	4.174652287	4.174652287	4.174652287

(4) P 9+0 — TH = 40 SBE = 1.555723827 — CAUER FUNCTIONS

MU		RC = 02	RC = 05	RC = 07	RC = 10	RC = 15	RC = 20	RC = 25	RC = 35	RC = 50
	AMAX DB	.0017375255	.0108709564	.0213327376	.0436480540	.0988323393	.1772876696	.2802872360	.5675287486	1.249387366
	AMIN DB	87.00195850	94.96989210	97.90291459	101.0232691	104.6002786	107.1775086	109.2187085	112.4285107	116.2084085
MU	K	.5497035D-3	.2196503D-3	.1567042D-3	.1094115D-3	.7247905D-4	.5387050D-4	.4258837D-4	.2943072D-4	.1904610D-4
1	PFR	.6327290554	.4913650028	.4422758990	.3915270789	.3350938450	.2955857684	.2650427551	.2186378013	.1675413405
1	PF1	.7307417593	.6184111300	.5815380242	.5450870869	.5070742230	.4824152484	.4646655380	.4402571407	.4175628107
1	PQ1	.6605527398	.7130735515	.7429444144	.7846467802	.8507563841	.9161913416	.9831616856	1.127699268	1.394100658
1	PF2	.9024536620	.8289627302	.8056360930	.7831030348	.7602627880	.7458585316	.7357131933	.7220931581	.7098014962
1	PQ2	1.189847040	1.366415133	1.462597876	1.593030117	1.792805129	1.984314649	2.175616713	2.577226701	3.293197377
1	PF3	1.028642805	.9809263189	.9658197022	.9512650631	.9365629560	.9273234960	.9208332851	.9121454033	.9043320727
1	PQ3	2.495625751	2.922805366	3.152489601	3.461510393	3.930751547	4.377251025	4.820931437	5.747179088	7.388114144
1	PFC	1.088885246	1.053647383	1.042463366	1.031678590	1.020777202	1.013923449	1.009108113	1.002661386	.9968631300
1	PQC	8.520558533	10.03058570	10.83975680	11.92627246	13.57262709	15.13638909	16.68834330	19.92404032	25.64808798
1	ZF1	1.573989479	1.573989479	1.573989479	1.573989479	1.573989479	1.573989479	1.573989479	1.573989479	1.573989479
1	ZF2	1.742285078	1.742285078	1.742285078	1.742285078	1.742285078	1.742285078	1.742285078	1.742285078	1.742285078
1	ZF3	2.248906639	2.248906639	2.248906639	2.248906639	2.248906639	2.248906639	2.248906639	2.248906639	2.248906639
1	ZF4	4.063382474	4.063382474	4.063382474	4.063382474	4.063382474	4.063382474	4.063382474	4.063382474	4.063382474

(4) P 9+0	TH = 41 SBE = 1.524253087								CAUER FUNCTIONS
	RC = 02	RC = 05	RC = 07	RC = 10	RC = 15	RC = 20	RC = 25	RC = 35	RC = 50
AMAX DB	.0017375255	.0108709564	.0213327376	.0436480540	.0988323393	.1772876696	.2802872360	.5675287486	1.249387366
AMIN DB	84.88127369	92.84920728	95.78222977	98.90258429	102.4795938	105.0568238	107.0980236	110.3078259	114.0877237
MU I K	.6827648D-3	.2728189D-3	.1946361D-3	.1358957D-3	.9002334D-4	.6691040D-4	.5289732D-4	.3655473D-4	.2365640D-4
1 I PFR	.6386689375	.4955155470	.4458897844	.3946258364	.3376620526	.2978059499	.2670055508	.2202265884	.1687382974
1 I PF1	.7363003362	.6227489890	.5855138630	.5487226875	.5103728492	.4855045864	.4676085810	.4430047415	.4201344267
1 I PQ1	.6623958325	.7147980927	.7446630614	.7863863772	.8525638803	.9180867391	.9851581030	1.129933009	1.396801320
1 I PF2	.9058933572	.8321462451	.8087331351	.7861147192	.7631863952	.7487259037	.7385406790	.7248667711	.7125262096
1 I PQ2	1.200904679	1.377661752	1.474186398	1.605199267	1.806006016	1.998592958	2.191020763	2.595084724	3.315578721
1 I PF3	1.029071936	.9817617840	.9667672000	.9523134111	.9377063634	.9285232473	.9220710861	.9134322017	.9056610780
1 I PQ3	2.535630432	2.965110336	3.196672256	3.508542710	3.982513936	4.433771176	4.882327149	5.819031151	7.478959944
1 I PFC	1.087345579	1.052779704	1.041792695	1.031190800	1.020467369	1.013722025	1.008981277	1.002632398	.9969202064
1 I PQC	8.692678949	10.21545342	11.03382316	12.13390183	13.80233311	15.38804782	16.96236100	20.24576053	26.05604940
1 I ZF1	1.541869206	1.541869206	1.541869206	1.541869206	1.541869206	1.541869206	1.541869206	1.541869206	1.541869206
1 I ZF2	1.704391854	1.704391854	1.704391854	1.704391854	1.704391854	1.704391854	1.704391854	1.704391854	1.704391854
1 I ZF3	2.194996871	2.194996871	2.194996871	2.194996871	2.194996871	2.194996871	2.194996871	2.194996871	2.194996871
1 I ZF4	3.957280512	3.957280512	3.957280512	3.957280512	3.957280512	3.957280512	3.957280512	3.957280512	3.957280512

(4) P 9+0	TH = 42 SBE = 1.494476550								CAUER FUNCTIONS
	RC = 02	RC = 05	RC = 07	RC = 10	RC = 15	RC = 20	RC = 25	RC = 35	RC = 50
AMAX DB	.0017375255	.0108709564	.0213327376	.0436480540	.0988323393	.1772876696	.2802872360	.5675287486	1.249387366
AMIN DB	82.80053837	90.76847195	93.70149445	96.82184896	100.3988584	102.9760885	105.0172883	108.2270905	112.0069884
MU I K	.8445553D-3	.3374671D-3	.2407577D-3	.1680980D-3	.1113556D-3	.8276573D-4	.6543206D-4	.4521687D-4	.2926211D-4
1 I PFR	.6448738562	.4998422539	.4496547473	.3978522958	.3403346085	.3001155522	.2690469207	.2218784663	.1699824458
1 I PF1	.7420744839	.6272521793	.5896403835	.5524953231	.5137951200	.4887093188	.4706612832	.4458543814	.4228012624
1 I PQ1	.6643348270	.7166058578	.7464627208	.7882061498	.8544528207	.9200663438	.9872423431	1.132263725	1.399617849
1 I PF2	.9094210930	.8354199292	.8119201212	.7892157804	.7661984866	.7516810465	.7414553658	.7277267460	.7153364518
1 I PQ2	1.212564507	1.389487295	1.486361130	1.617973846	1.819852833	2.013562715	2.207165398	2.613793076	3.339016632
1 I PF3	1.029497248	.9826100906	.9677322490	.9533835981	.9388756963	.9297513685	.9233389175	.9147511816	.9070241299
1 I PQ3	2.577971154	3.009783745	3.243296920	3.558143144	4.037068829	4.493317799	4.946993348	5.894684376	7.574582271
1 I PFC	1.085774009	1.051892807	1.041106856	1.030691742	1.020150196	1.013515717	1.008851278	1.002602542	.9969784680
1 I PQC	8.875390623	10.41129911	11.23929096	12.35360719	14.04526690	15.65410988	17.25199512	20.58571285	26.48701502
1 I ZF1	1.511467704	1.511467704	1.511467704	1.511467704	1.511467704	1.511467704	1.511467704	1.511467704	1.511467704
1 I ZF2	1.668439251	1.668439251	1.668439251	1.668439251	1.668439251	1.668439251	1.668439251	1.668439251	1.668439251
1 I ZF3	2.143668934	2.143668934	2.143668934	2.143668934	2.143668934	2.143668934	2.143668934	2.143668934	2.143668934
1 I ZF4	3.855968875	3.855968875	3.855968875	3.855968875	3.855968875	3.855968875	3.855968875	3.855968875	3.855968875

(4) P 9+0	TH = 43	SBE = 1.466279186							CAUER FUNCTIONS
	RC = 02	RC = 05	RC = 07	RC = 10	RC = 15	RC = 20	RC = 25	RC = 35	RC = 50
AMAX DB	.0017375255	.0108709564	.0213327376	.0436480540	.0988323393	.1772876696	.2802872360	.5675287486	1.249387366
AMIN DB	80.75738222	88.72531579	91.65833829	94.77869280	98.35570227	100.9329323	102.9741322	106.1839344	109.9638322
MU K	.1040648D-2	.4158217D-3	.2966580D-3	.2071278D-3	.1372107D-3	.1019827D-3	.8062438D-4	.5571553D-4	.3605632D-4
1 PFR	.6513567684	.5043529204	.4535772656	.4012117672	.3431157184	.3025181091	.2711699274	.2235958489	.1712755610
1 PF1	.7480716824	.6319262952	.5939226359	.5564095415	.5173450957	.4920332092	.4738272033	.4488093494	.4255663678
1 PQ1	.6663757251	.7185015034	.7483477607	.7901102652	.8564272814	.9221342689	.9894186217	1.134695962	1.402555562
1 PF2	.9130363955	.8387843852	.8151979061	.7924072808	.7693003016	.7547252956	.7444586532	.7306745521	.7182337503
1 PQ2	1.224865879	1.401926749	1.499156620	1.631388676	1.834381821	2.029262110	2.224091068	2.633397515	3.363567204
1 PF3	1.029917474	.9834704904	.9687142622	.9544751894	.9400706706	.9310076680	.9246366519	.9161022998	.9084212593
1 PQ3	2.622814625	3.056985434	3.292525573	3.610479251	4.094595439	4.556083131	5.015136277	5.974376180	7.675276909
1 PFC	1.084171070	1.050986903	1.040405987	1.030181497	1.019825718	1.013304542	1.008718120	1.002571806	.9970378941
1 PQC	9.069502726	10.61892353	11.45698028	12.58624565	14.30235505	15.93557600	17.55832455	20.94514743	26.94255172
1 ZF1	1.482668125	1.482668125	1.482668125	1.482668125	1.482668125	1.482668125	1.482668125	1.482668125	1.482668125
1 ZF2	1.634293972	1.634293972	1.634293972	1.634293972	1.634293972	1.634293972	1.634293972	1.634293972	1.634293972
1 ZF3	2.094741945	2.094741945	2.094741945	2.094741945	2.094741945	2.094741945	2.094741945	2.094741945	2.094741945
1 ZF4	3.759104669	3.759104669	3.759104669	3.759104669	3.759104669	3.759104669	3.759104669	3.759104669	3.759104669

(4) P 9+0	TH = 44	SBE = 1.439556540							CAUER FUNCTIONS
	RC = 02	RC = 05	RC = 07	RC = 10	RC = 15	RC = 20	RC = 25	RC = 35	RC = 50
AMAX DB	.0017375255	.0108709564	.0213327376	.0436480540	.0988323393	.1772876696	.2802872360	.5675287486	1.249387366
AMIN DB	78.74957450	86.71750805	89.65053054	92.77088506	96.34789452	98.92512458	100.9663244	104.1761266	107.9560245
MU K	.1277606D-2	.5105052D-3	.3642076D-3	.2542912D-3	.1684538D-3	.1252043D-3	.9898271D-4	.6840207D-4	.4426641D-4
1 PFR	.6581316971	.5090559422	.4576643039	.4047099513	.3460098914	.3050174060	.2733778493	.2253313179	.1726195408
1 PF1	.7542998528	.6367772715	.5983659793	.5604701700	.5210270870	.4954802533	.4771101198	.4518731377	.4284329817
1 PQ1	.6685251002	.7204900995	.7503229282	.7921032455	.8584916776	.9242949649	.9916914974	1.137234632	1.405620198
1 PF2	.9167386579	.8422401831	.8185673363	.7956902952	.7724931113	.7578600280	.7475519845	.7337117150	.7212196948
1 PQ2	1.237851922	1.415018283	1.512610511	1.645481651	1.849632369	2.045732619	2.241841685	2.653947680	3.389291252
1 PF3	1.030331287	.9843421912	.9697126148	.9555877194	.9412909778	.9322919343	.9259641450	.9174955006	.9098524887
1 PQ3	2.670344460	3.106890798	3.344535766	3.665734509	4.155289812	4.622277338	5.086981323	6.058365874	7.781366764
1 PFC	1.082537318	1.050062210	1.039690234	1.029660152	1.019493977	1.013088515	1.008581805	1.002540179	.9970984625
1 PQC	9.275910788	10.83920986	11.68779466	12.83276070	14.57461662	16.23354620	17.88253476	21.32543727	27.42437957
1 ZF1	1.455364364	1.455364364	1.455364364	1.455364364	1.455364364	1.455364364	1.455364364	1.455364364	1.455364364
1 ZF2	1.601834928	1.601834928	1.601834928	1.601834928	1.601834928	1.601834928	1.601834928	1.601834928	1.601834928
1 ZF3	2.048051420	2.048051420	2.048051420	2.048051420	2.048051420	2.048051420	2.048051420	2.048051420	2.048051420
1 ZF4	3.666375662	3.666375662	3.666375662	3.666375662	3.666375662	3.666375662	3.666375662	3.666375662	3.666375662

(4) P 9+0	TH = 45	SBE = 1.414213562						CAUER FUNCTIONS	
	RC = 02	RC = 05	RC = 07	RC = 10	RC = 15	RC = 20	RC = 25	RC = 35	RC = 50
AMAX DB	.0017375255	.0108709564	.0213327376	.0436480540	.0988323393	.1772876696	.2802872360	.5675287486	1.249387366
AMIN DB	76.77500955	84.74294308	87.67596556	90.79632008	94.37332954	96.95055960	98.99175943	102.2015617	105.9814595
MU K	.1563147D-2	.6246015D-3	.4456069D-3	.3111245D-3	.2061027D-3	.1531871D-3	.1211050D-3	.8368971D-4	.5415982D-4
1 PFR	.6652138428	.5139603714	.4619233597	.4083529760	.3490219669	.3076175022	.2756741997	.2272376378	.1740164154
1 PF1	.7607673850	.6418114075	.6029761043	.5646823359	.5248456734	.4990546964	.4805140474	.4550494572	.4314045452
1 PQ1	.6707901645	.7225771750	.7523933907	.7941900040	.8606507988	.9265532548	.9940659066	1.139885048	1.408817959
1 PF2	.9205271245	.8457878530	.8220292455	.7990659089	.7757782165	.7610866623	.7507368529	.7368398180	.7242959407
1 PQ2	1.251569978	1.428803608	1.526763881	1.660294072	1.865647356	2.063019358	2.260464989	2.675497497	3.416254801
1 PF3	1.030737297	.9852243542	.9707266422	.9567206886	.9425362832	.9336039342	.9273212338	.9189007138	.9113178302
1 PQ3	2.720763303	3.159692657	3.399522465	3.724110181	4.219366780	4.692130583	5.162775212	6.146937163	7.893204940
1 PFC	1.080873335	1.049118964	1.038959753	1.029127801	1.019155016	1.012867657	1.008442337	1.002507649	.9971601501
1 PQC	9.495608053	11.07313415	11.93273160	13.09419297	14.86317450	16.54923189	18.22593078	21.72809307	27.93439032
1 ZF1	1.429459865	1.429459865	1.429459865	1.429459865	1.429459865	1.429459865	1.429459865	1.429459865	1.429459865
1 ZF2	1.570951885	1.570951885	1.570951885	1.570951885	1.570951885	1.570951885	1.570951885	1.570951885	1.570951885
1 ZF3	2.003447449	2.003447449	2.003447449	2.003447449	2.003447449	2.003447449	2.003447449	2.003447449	2.003447449
1 ZF4	3.577496845	3.577496845	3.577496845	3.577496845	3.577496845	3.577496845	3.577496845	3.577496845	3.577496845

(4) P 9+0	TH = 46	SBE = 1.390163591						CAUER FUNCTIONS	
	RC = 02	RC = 05	RC = 07	RC = 10	RC = 15	RC = 20	RC = 25	RC = 35	RC = 50
AMAX DB	.0017375255	.0108709564	.0213327376	.0436480540	.0988323393	.1772876696	.2802872360	.5675287486	1.249387366
AMIN DB	74.83169376	82.79962725	85.73264973	88.85300424	92.43001371	95.00724377	97.04844359	100.2582458	104.0381436
MU K	.1906339D-2	.7617340D-3	.5434408D-3	.3794325D-3	.2513530D-3	.1868196D-3	.1476939D-3	.1020640D-3	.6605072D-4
1 PFR	.6726197095	.5190759820	.4663625148	.4121474367	.3521571459	.3103227557	.2780627485	.2291677723	.1754683600
1 PF1	.7674831678	.6470353932	.6077590575	.5690514888	.5288057239	.5027610526	.4840432554	.4583422540	.4344847183
1 PQ1	.6731788458	.7247687692	.7545647820	.7963758882	.8629098482	.9289143724	.9965472021	1.142652968	1.412155559
1 PF2	.9244008716	.8494278767	.8255844489	.8025352136	.7791569474	.7644066587	.7540148014	.7400605041	.7274642110
1 PQ2	1.266072120	1.443328387	1.541661639	1.675871030	1.882473534	2.081171482	2.280012962	2.698105645	3.444529651
1 PF3	1.031134047	.9861160911	.9717556363	.9578735620	.9438062233	.9349434106	.9287077352	.9203478539	.9128172848
1 PQ3	2.774295269	3.215603400	3.457700173	3.785827459	4.287062195	4.765895386	5.242788515	6.240400990	8.011178244
1 PFC	1.079179736	1.048157411	1.038214708	1.028584544	1.018808883	1.012641987	1.008299721	1.002474203	.9972229322
1 PQC	9.729698645	11.32177733	12.19289461	13.37169252	15.16926851	16.88396986	18.58995210	22.15478032	28.47466857
1 ZF1	1.404866590	1.404866590	1.404866590	1.404866590	1.404866590	1.404866590	1.404866590	1.404866590	1.404866590
1 ZF2	1.541544289	1.541544289	1.541544289	1.541544289	1.541544289	1.541544289	1.541544289	1.541544289	1.541544289
1 ZF3	1.960793107	1.960793107	1.960793107	1.960793107	1.960793107	1.960793107	1.960793107	1.960793107	1.960793107
1 ZF4	3.492207426	3.492207426	3.492207426	3.492207426	3.492207426	3.492207426	3.492207426	3.492207426	3.492207426

(4) P 9+0	TH = 47	SBE = 1.367327461						CAUER FUNCTIONS	
	RC = 02	RC = 05	RC = 07	RC = 10	RC = 15	RC = 20	RC = 25	RC = 35	RC = 50
AMAX DB	.0017375255	.0108709564	.0213327376	.0436480540	.0988323393	.1772876696	.2802872360	.5675287486	1.249387366
AMIN DB	72.91773368	80.88566709	83.81868957	86.93904408	90.51605354	93.09328360	95.13448343	98.34428565	102.1241835
MU K	.2317827D-2	.9261563D-3	.6607439D-3	.4613340D-3	.3056082D-3	.2271452D-3	.1795740D-3	.1240947D-3	.8030794D-4
1 PFR	.6803672473	.5244133433	.4709904936	.4161004413	.3554210246	.3131378514	.2805475449	.2311749020	.1769777069
1 PF1	.7744566215	.6524563380	.6127212679	.5735834258	.5329124191	.5066041257	.4877022872	.4617557282	.4376773964
1 PQ1	.6756998770	.7270714912	.7568432550	.7986667275	.8652744878	.9313840063	.9991411971	1.145544635	1.415640274
1 PF2	.9283587877	.8531606783	.8292337366	.8060993037	.7826306616	.7678215186	.7573874230	.7433754775	.7307262991
1 PQ2	1.281415743	1.458642704	1.557352968	1.692261842	1.900161969	2.100242634	2.300542305	2.721836078	3.474194004
1 PF3	1.031520014	.9870164607	.9727988436	.9590457653	.9451004037	.9363100805	.9301234436	.9218268179	.9143508404
1 PQ3	2.831188775	3.274857458	3.519305365	3.851129913	4.358635491	4.843849328	5.327318519	6.339098792	8.135711191
1 PFC	1.077457162	1.047177815	1.037455276	1.028030489	1.018455628	1.012411529	1.008153965	1.002439828	.9972867829
1 PQC	9.979412888	11.58633919	12.46950721	13.66653314	15.49427048	17.23923833	18.97618988	22.60733894	29.04751619
1 ZF1	1.381504113	1.381504113	1.381504113	1.381504113	1.381504113	1.381504113	1.381504113	1.381504113	1.381504113
1 ZF2	1.513520241	1.513520241	1.513520241	1.513520241	1.513520241	1.513520241	1.513520241	1.513520241	1.513520241
1 ZF3	1.919963071	1.919963071	1.919963071	1.919963071	1.919963071	1.919963071	1.919963071	1.919963071	1.919963071
1 ZF4	3.410268222	3.410268222	3.410268222	3.410268222	3.410268222	3.410268222	3.410268222	3.410268222	3.410268222

(4) P 9+0	TH = 48	SBE = 1.345632730						CAUER FUNCTIONS	
	RC = 02	RC = 05	RC = 07	RC = 10	RC = 15	RC = 20	RC = 25	RC = 35	RC = 50
AMAX DB	.0017375255	.0108709564	.0213327376	.0436480540	.0988323393	.1772876696	.2802872360	.5675287486	1.249387366
AMIN DB	71.03132509	78.99925841	81.93228088	85.05263538	88.62964484	91.20687490	93.24807472	96.45787695	100.2377748
MU K	.2810099D-2	.1122858D-2	.8010761D-3	.5593145D-3	.3705148D-3	.2753874D-3	.2177129D-3	.1504506D-3	.9736416D-4
1 PFR	.6884760164	.5299839031	.4758167286	.4202196621	.3588196332	.3160678325	.2831329453	.2332624455	.1785469609
1 PF1	.7816977317	.6580818026	.6178695765	.5782843182	.5371712762	.5105890326	.4914959829	.4652943546	.4409867301
1 PQ1	.6783628979	.7294925868	.7592355419	.8010688878	.8677508886	.9339683495	1.001854215	1.148566335	1.419280003
1 PF2	.9323995504	.8569866139	.8329778664	.8097592724	.7862007421	.7713327843	.7608563618	.7467865057	.7340840719
1 PQ2	1.297664242	1.474801594	1.573891834	1.709520545	1.918768539	2.120291464	2.322114969	2.746758623	3.505333184
1 PF3	1.031893602	.9879244658	.9738554612	.9602366829	.9464183962	.9377036326	.9315681291	.9233374837	.9159184706
1 PQ3	2.891719832	3.337714159	3.584599296	3.920286325	4.434372625	4.926298157	5.416692524	6.443406241	8.267270605
1 PFC	1.075706288	1.046180453	1.036681644	1.027465750	1.018095308	1.012176309	1.008005074	1.002404513	.9973516745
1 PQC	10.24612521	11.86815464	12.76392907	13.98012890	15.83970173	17.61667555	19.38640677	23.08780607	29.65548045
1 ZF1	1.359298837	1.359298837	1.359298837	1.359298837	1.359298837	1.359298837	1.359298837	1.359298837	1.359298837
1 ZF2	1.486795599	1.486795599	1.486795599	1.486795599	1.486795599	1.486795599	1.486795599	1.486795599	1.486795599
1 ZF3	1.880842406	1.880842406	1.880842406	1.880842406	1.880842406	1.880842406	1.880842406	1.880842406	1.880842406
1 ZF4	3.331459354	3.331459354	3.331459354	3.331459354	3.331459354	3.331459354	3.331459354	3.331459354	3.331459354

(4) P 9+0	TH = 49 RC = 02	SBE = 1.325012993 RC = 05	RC = 07	RC = 10	RC = 15	RC = 20	RC = 25	RC = 35	CAUER FUNCTIONS RC = 50
AMAX DB	.0017375255	.0108709564	.0213327376	.0436480540	.0988323393	.1772876696	.2802872360	.5675287486	1.249387366
AMIN DB	69.17074311	77.13867627	80.07169872	83.19205322	86.76906267	89.34629273	91.38749255	94.59729478	98.37719259
MU K	.3397796D-2	.1357690D-2	.9686113D-3	.6762882D-3	.4480034D-3	.3329813D-3	.2632448D-3	.1819155D-3	.1177267D-3
1 PFR	.6969673733	.5358000818	.4808514349	.4245133930	.3623594797	.3191181357	.2858236422	.2354340814	.1801788151
1 PF1	.7892170864	.6639198324	.6232112690	.5831607419	.5415881763	.5147212287	.4954295035	.4689629053	.4444171460
1 PQ1	.6811785737	.7320400163	.7617490230	.8035893338	.8703457892	.9366741562	1.004693148	1.151726954	1.423083336
1 PF2	.9365216012	.8609059592	.8368175557	.8135162056	.7898685956	.7749420386	.7644233130	.7502954215	.7375394723
1 PQ2	1.314887805	1.491865670	1.591337578	1.727706472	1.938354506	2.141382217	2.344798780	2.772949661	3.538040461
1 PF3	1.032253139	.9888390493	.9749246336	.9614456542	.9477597364	.9391237246	.9330415347	.9248797076	.9175201326
1 PQ3	2.956195865	3.404461031	3.653871252	3.993593950	4.514589473	5.013579371	5.511271634	6.553737542	8.406370900
1 PFC	1.073927823	1.045165621	1.035894013	1.026890451	1.017727984	1.011936354	1.007853059	1.002368243	.9974175779
1 PQC	10.53137510	12.16871259	13.07767492	14.31405335	16.20725333	18.01810142	19.82255996	23.59844231	30.30138667
1 ZF1	1.338183298	1.338183298	1.338183298	1.338183298	1.338183298	1.338183298	1.338183298	1.338183298	1.338183298
1 ZF2	1.461293193	1.461293193	1.461293193	1.461293193	1.461293193	1.461293193	1.461293193	1.461293193	1.461293193
1 ZF3	1.843325504	1.843325504	1.843325504	1.843325504	1.843325504	1.843325504	1.843325504	1.843325504	1.843325504
1 ZF4	3.255578234	3.255578234	3.255578234	3.255578234	3.255578234	3.255578234	3.255578234	3.255578234	3.255578234

(4) P 9+0	TH = 50 RC = 02	SBE = 1.305407289 RC = 05	RC = 07	RC = 10	RC = 15	RC = 20	RC = 25	RC = 35	CAUER FUNCTIONS RC = 50
AMAX DB	.0017375255	.0108709564	.0213327376	.0436480540	.0988323393	.1772876696	.2802872360	.5675287486	1.249387366
AMIN DB	67.33433277	75.30226570	78.23528813	81.35564262	84.93265206	87.50988212	89.55108194	92.76088417	96.54078198
MU K	.4098080D-2	.1637509D-2	.1168242D-2	.8156708D-3	.5403367D-3	.4016085D-3	.3174994D-3	.2194082D-3	.1419901D-3
1 PFR	.7058646851	.5418753802	.4861056942	.4289906152	.3660475985	.3222946311	.2886246984	.2376937750	.1818761697
1 PF1	.7970259144	.6699789959	.6287541105	.5882197100	.5461693951	.5190065368	.4995083582	.4727664751	.4479733708
1 PQ1	.6841587323	.7347225432	.7643918055	.8062357006	.8730665612	.9395088062	1.007665517	1.155033047	1.427059629
1 PF2	.9407231179	.8649188967	.8407534714	.8173711768	.7936356493	.7786509035	.7680900240	.7539041241	.7410945230
1 PQ2	1.333164323	1.509901826	1.609755587	1.746884899	1.958987174	2.163585409	2.368668136	2.800492910	3.572417983
1 PF3	1.032596875	.9897590902	.9760054489	.9626719700	.9491239200	.9405699808	.9345433742	.9264533226	.9191557656
1 PQ3	3.024960203	3.475417650	3.727442319	4.071382303	4.599635766	5.106066356	5.611455154	6.670550408	8.553580184
1 PFC	1.072122512	1.044133633	1.035092596	1.026304724	1.017353721	1.011691696	1.007697928	1.002331004	.9974844623
1 PQC	10.83689183	12.48967815	13.41243658	14.67006196	16.59880980	18.44554262	20.28682804	24.14176247	30.98837614
1 ZF1	1.318095555	1.318095555	1.318095555	1.318095555	1.318095555	1.318095555	1.318095555	1.318095555	1.318095555
1 ZF2	1.436942132	1.436942132	1.436942132	1.436942132	1.436942132	1.436942132	1.436942132	1.436942132	1.436942132
1 ZF3	1.807315141	1.807315141	1.807315141	1.807315141	1.807315141	1.807315141	1.807315141	1.807315141	1.807315141
1 ZF4	3.182437779	3.182437779	3.182437779	3.182437779	3.182437779	3.182437779	3.182437779	3.182437779	3.182437779

(4) P 9+0		TH = 51	SBE = 1.286759566						CAUER FUNCTIONS	
		RC = 02	RC = 05	RC = 07	RC = 10	RC = 15	RC = 20	RC = 25	RC = 35	RC = 50
AMAX DB		.0017375255	.0108709564	.0213327376	.0436480540	.0988323393	.1772876696	.2802872360	.5675287486	1.249387366
AMIN DB		65.52050043	73.48843301	76.42145541	79.54180988	83.11881932	85.69604937	87.73724919	90.94705141	94.72694923
MU	K	.4931060D-2	.1970351D-2	.1405700D-2	.9814649D-3	.6501661D-3	.4832399D-3	.3820347D-3	.2640053D-3	.1708511D-3
1	PFR	.7151935765	.5482245018	.4915915508	.4336610718	.3698916065	.3256036674	.2915415853	.2400458067	.1836421526
1	PF1	.8051361264	.6762684256	.6345063852	.5934687100	.5509216362	.5234511784	.5037384345	.4767105097	.4516604572
1	PQ1	.6873165245	.7375498373	.7671728143	.8090163753	.8759212859	.9424803786	1.010779551	1.158493910	1.431219093
1	PF2	.9450019826	.8690254996	.8447862199	.8213252398	.7975033476	.7824610396	.7718582935	.7576145819	.7447513294
1	PQ2	1.352580464	1.528984069	1.629218075	1.767127810	1.980740646	2.186978609	2.393804819	2.829480318	3.608577862
1	PF3	1.032922975	.9906833994	.9770969345	.9639148690	.9505104000	.9420419883	.9360733289	.9280581353	.9208252882
1	PQ3	3.098397335	3.550940129	3.805669784	4.154017565	4.689899651	5.204173197	5.717685669	6.794351812	8.709527325
1	PFC	1.070291141	1.043084823	1.034277622	1.025708713	1.016972590	1.011442369	1.007539694	1.002292784	.9975522947
1	PQC	11.16462360	12.83291859	13.77010883	15.05011846	17.01647670	18.90126199	20.78164228	24.72057126	31.71995039
1	ZF1	1.298978659	1.298978659	1.298978659	1.298978659	1.298978659	1.298978659	1.298978659	1.298978659	1.298978659
1	ZF2	1.413677194	1.413677194	1.413677194	1.413677194	1.413677194	1.413677194	1.413677194	1.413677194	1.413677194
1	ZF3	1.772721658	1.772721658	1.772721658	1.772721658	1.772721658	1.772721658	1.772721658	1.772721658	1.772721658
1	ZF4	3.111864836	3.111864836	3.111864836	3.111864836	3.111864836	3.111864836	3.111864836	3.111864836	3.111864836

(4) P 9+0		TH = 52	SBE = 1.269018215						CAUER FUNCTIONS	
		RC = 02	RC = 05	RC = 07	RC = 10	RC = 15	RC = 20	RC = 25	RC = 35	RC = 50
AMAX DB		.0017375255	.0108709564	.0213327376	.0436480540	.0988323393	.1772876696	.2802872360	.5675287486	1.249387366
AMIN DB		63.72770548	71.69563754	74.62865989	77.74901433	81.32602375	83.90325380	85.94445362	89.15425584	92.93415365
MU	K	.5920302D-2	.2365632D-2	.1687703D-2	.1178361D-2	.7805988D-3	.5801849D-3	.4586764D-3	.3169636D-3	.2051263D-3
1	PFR	.7249822152	.5548634933	.4973221216	.4385353520	.3738997659	.3290521233	.2945802257	.2424948052	.1854801428
1	PF1	.8135603585	.6827978630	.6404769396	.5989157440	.5558520687	.5280618094	.5081260322	.4808008374	.4554838139
1	PQ1	.6906666110	.7405325933	.7701018973	.8119405920	.8789188405	.9455977365	1.014044267	1.162119176	1.435572892
1	PF2	.9493557475	.8732257156	.8489163342	.8253794208	.8014731436	.7863741438	.7757299725	.7614288340	.7485120831
1	PQ2	1.373232918	1.549194474	1.649804993	1.788514767	2.003696698	2.211647332	2.420298934	2.860013098	3.646643407
1	PF3	1.033229520	.9916107148	.9781980527	.9651735334	.9519185824	.9435392941	.9376310448	.9296939235	.9225285961
1	PQ3	3.176939118	3.631426377	3.888952270	4.241907724	4.785813017	5.308360270	5.830454976	6.925704690	8.874910162
1	PFC	1.068434535	1.042019546	1.033449333	1.025102569	1.016584668	1.011188411	1.007378371	1.002253568	.9976210405
1	PQC	11.51677216	13.20053413	14.15281988	15.45642573	17.46261320	19.38779305	21.30972340	25.33800529	32.50002301
1	ZF1	1.280780173	1.280780173	1.280780173	1.280780173	1.280780173	1.280780173	1.280780173	1.280780173	1.280780173
1	ZF2	1.391438289	1.391438289	1.391438289	1.391438289	1.391438289	1.391438289	1.391438289	1.391438289	1.391438289
1	ZF3	1.739462222	1.739462222	1.739462222	1.739462222	1.739462222	1.739462222	1.739462222	1.739462222	1.739462222
1	ZF4	3.043698777	3.043698777	3.043698777	3.043698777	3.043698777	3.043698777	3.043698777	3.043698777	3.043698777

(4) P 9+0	TH = 53	SBE = 1.252135658							CAUER FUNCTIONS
	RC = 02	RC = 05	RC = 07	RC = 10	RC = 15	RC = 20	RC = 25	RC = 35	RC = 50
AMAX DB	.0017375255	.0108709564	.0213327376	.0436480540	.0988323393	.1772876696	.2802872360	.5675287486	1.249387366
AMIN DB	61.95445251	69.92238379	72.85540606	75.97576046	79.55276987	82.12999991	84.17119973	87.38100194	91.16089976
MU I K	.7093425D-2	.2834388D-2	.2022126D-2	.1411856D-2	.9352763D-3	.6951499D-3	.5495642D-3	.3797767D-3	.2457726D-3
1 I PFR	.7352616430	.5618099060	.5033117213	.4436249884	.3780810567	.3326474655	.2977470437	.2450457849	.1873937973
1 I PF1	.8223120167	.6895777086	.6466752300	.6045693740	.5609683689	.5328455591	.5126779008	.4850437040	.4594492379
1 I PQ1	.6942253822	.7436826699	.7731899473	.8150185405	.8820689990	.9488706250	1.017469572	1.165919410	1.440133261
1 I PF2	.9537815956	.8775193463	.8531442601	.8295347093	.8055465189	.7903919479	.7797069634	.7653489922	.7523790661
1 I PQ2	1.395229876	1.570624305	1.671605079	1.811133930	2.027945795	2.237686084	2.448249988	2.892202924	3.686750561
1 I PF3	1.033514494	.9925396964	.9793076958	.9664470842	.9533478225	.9450614014	.9392161288	.9313604330	.9242655594
1 I PQ3	3.261072123	3.717322284	3.977735776	4.335508604	4.887857728	5.419140795	5.950311004	7.065235757	9.050505086
1 I PFC	1.066553565	1.040938179	1.032607991	1.024486457	1.016190041	1.010929864	1.007213975	1.002213341	.9976906627
1 I PQC	11.89583411	13.59489427	14.56296742	15.89146232	17.93987040	19.90798065	21.87412489	25.99758236	33.33298065
1 I ZF1	1.263451752	1.263451752	1.263451752	1.263451752	1.263451752	1.263451752	1.263451752	1.263451752	1.263451752
1 I ZF2	1.370169978	1.370169978	1.370169978	1.370169978	1.370169978	1.370169978	1.370169978	1.370169978	1.370169978
1 I ZF3	1.707460178	1.707460178	1.707460178	1.707460178	1.707460178	1.707460178	1.707460178	1.707460178	1.707460178
1 I ZF4	2.977790247	2.977790247	2.977790247	2.977790247	2.977790247	2.977790247	2.977790247	2.977790247	2.977790247

(4) P 9+0	TH = 54	SBE = 1.236067977							CAUER FUNCTIONS
	RC = 02	RC = 05	RC = 07	RC = 10	RC = 15	RC = 20	RC = 25	RC = 35	RC = 50
AMAX DB	.0017375255	.0108709564	.0213327376	.0436480540	.0988323393	.1772876696	.2802872360	.5675287486	1.249387366
AMIN DB	60.19928371	68.16721383	71.10023600	74.22059035	77.79759972	80.37482975	82.41602956	85.62583177	89.40572958
MU I K	.8482805D-2	.3389556D-2	.2418197D-2	.1688395D-2	.1118468D-2	.8313080D-3	.6572067D-3	.4541631D-3	.2939118D-3
1 I PFR	.7460661617	.5690829824	.5095760078	.4489425681	.3824452594	.3363978156	.3010490199	.2477041888	.1893870812
1 I PF1	.8314053240	.6966190759	.6531113756	.6104387719	.5662787671	.5378100749	.5174012817	.4894458128	.4635629526
1 I PQ1	.6980112173	.7470132511	.7764490434	.8182614937	.8853825489	.9523097834	1.021066369	1.169906228	1.444913634
1 I PF2	.9582762974	.8819060258	.8574703397	.8337920472	.8097249287	.7945162157	.7837912196	.7693772428	.7563546544
1 I PQ2	1.418692757	1.593375326	1.694717094	1.835083245	2.053588269	2.265199572	2.477768147	2.926173316	3.729049564
1 I PF3	1.033775787	.9934689205	.9804246808	.9677345765	.9547974200	.9466077649	.9408281448	.9330573740	.9260360200
1 I PQ3	3.351346386	3.809129036	4.072520797	4.435330965	4.996572945	5.537088514	6.077865935	7.213644675	9.237178264
1 I PFC	1.064649152	1.039841125	1.031753875	1.023860555	1.015788799	1.010666774	1.007046522	1.002172090	.9977611219
1 I PQC	12.30465032	14.01868113	15.00326142	16.35802581	18.45123669	20.46502908	22.47828415	26.70325979	34.22375495
1 I ZF1	1.246948772	1.246948772	1.246948772	1.246948772	1.246948772	1.246948772	1.246948772	1.246948772	1.246948772
1 I ZF2	1.349821053	1.349821053	1.349821053	1.349821053	1.349821053	1.349821053	1.349821053	1.349821053	1.349821053
1 I ZF3	1.676644474	1.676644474	1.676644474	1.676644474	1.676644474	1.676644474	1.676644474	1.676644474	1.676644474
1 I ZF4	2.914000048	2.914000048	2.914000048	2.914000048	2.914000048	2.914000048	2.914000048	2.914000048	2.914000048

(4) P 9+0	TH = 55 SBE = 1.220774589							CAUER FUNCTIONS	
	RC = 02	RC = 05	RC = 07	RC = 10	RC = 15	RC = 20	RC = 25	RC = 35	RC = 50
AMAX DB	.0017375255	.0108709564	.0213327376	.0436480540	.0988323393	.1772876696	.2802872360	.5675287486	1.249387366
AMIN DB	58.46077147	66.42869988	69.36172189	72.48207614	76.05908547	78.63631549	80.67751529	83.88731750	87.66721531
MU K	.1012642D-1	.4046311D-2	.2886743D-2	.2015535D-2	.1335180D-2	.9923810D-3	.7845461D-3	.5421611D-3	.3508597D-3
1 PFR	.7574337848	.5767038715	.5161321483	.4545018607	.3870030493	.3403120264	.3044937570	.2504759368	.1914643026
1 PF1	.8408553680	.7039338517	.6597962170	.6165337755	.5717921000	.5429635711	.5223039560	.4940143690	.4678316486
1 PQ1	.7020447916	.7505390358	.7798926170	.8216819553	.8888714259	.9559270766	1.024846696	1.174092433	1.449928799
1 PF2	.9628361623	.8863851951	.8618947937	.8381523170	.8140098446	.7987487399	.7879847445	.7735158483	.7604413223
1 PQ2	1.443758275	1.617561344	1.719251270	1.860471835	2.080735704	2.294304119	2.508975710	2.962061260	3.773706888
1 PF3	1.034011186	.9943968739	.9815477433	.9690349937	.9562666142	.9481777866	.9424666098	.9347844177	.9278397873
1 PQ3	3.448385858	3.907411791	4.173870757	4.541948893	5.112563771	5.662846763	6.213805777	7.371714843	9.435898835
1 PFC	1.062722265	1.038728814	1.030887282	1.023225053	1.015381041	1.010399191	1.006876035	1.002129798	.9978323765
1 PQC	12.74646548	14.47494126	15.47677529	16.85928447	19.00009181	21.06255923	23.12608336	27.45950366	35.17790802
1 ZF1	1.231229998	1.231229998	1.231229998	1.231229998	1.231229998	1.231229998	1.231229998	1.231229998	1.231229998
1 ZF2	1.330344157	1.330344157	1.330344157	1.330344157	1.330344157	1.330344157	1.330344157	1.330344157	1.330344157
1 ZF3	1.646949140	1.646949140	1.646949140	1.646949140	1.646949140	1.646949140	1.646949140	1.646949140	1.646949140
1 ZF4	2.852198127	2.852198127	2.852198127	2.852198127	2.852198127	2.852198127	2.852198127	2.852198127	2.852198127

(4) P 9+0	TH = 56 SBE = 1.206217949							CAUER FUNCTIONS	
	RC = 02	RC = 05	RC = 07	RC = 10	RC = 15	RC = 20	RC = 25	RC = 35	RC = 50
AMAX DB	.0017375255	.0108709564	.0213327376	.0436480540	.0988323393	.1772876696	.2802872360	.5675287486	1.249387366
AMIN DB	56.73751098	64.70543685	67.63845862	70.75881276	74.33582202	76.91305201	78.95425180	82.16405400	85.94395180
MU K	.1206884D-1	.4822464D-2	.3440470D-2	.2402150D-2	.1591291D-2	.1182737D-2	.9350356D-3	.6461570D-3	.4181606D-3
1 PFR	.7694067690	.5846958799	.5229990137	.4603179663	.3917661064	.3443997701	.3080895524	.2533674820	.1936301522
1 PF1	.8506781492	.7115347620	.6667413813	.6228649515	.5775178688	.5483148855	.5273942980	.4987571300	.4722625316
1 PQ1	.7063494419	.7542764616	.7835356466	.8252938330	.8925488722	.9597356473	1.028823869	1.178492171	1.455195074
1 PF2	.9674569839	.8909560735	.8664176995	.8426163271	.8184027221	.8030913379	.7922895900	.7777671497	.7646416470
1 PQ2	1.470580895	1.643310034	1.745331019	1.887421634	2.109512557	2.325129319	2.542008824	3.000019100	3.820907504
1 PF3	1.034218371	.9953219470	.9826755312	.9703472415	.9577545783	.9497708107	.9441309887	.9365411914	.9296766356
1 PQ3	3.552900977	4.012810030	4.282422048	4.656009771	5.236511500	5.797139193	6.358901687	7.540326155	9.647754508
1 PFC	1.060773932	1.037601706	1.030008536	1.022580158	1.014966876	1.010127171	1.006702535	1.002086451	.9979043818
1 PQC	13.22500015	14.96714789	15.98700722	17.39883909	19.59027143	21.70467699	23.82192204	28.27137135	36.20173414
1 ZF1	1.216257282	1.216257282	1.216257282	1.216257282	1.216257282	1.216257282	1.216257282	1.216257282	1.216257282
1 ZF2	1.311695448	1.311695448	1.311695448	1.311695448	1.311695448	1.311695448	1.311695448	1.311695448	1.311695448
1 ZF3	1.618312832	1.618312832	1.618312832	1.618312832	1.618312832	1.618312832	1.618312832	1.618312832	1.618312832
1 ZF4	2.792262667	2.792262667	2.792262667	2.792262667	2.792262667	2.792262667	2.792262667	2.792262667	2.792262667

(4) P 9+0	TH = 57 SBE = 1.192363293								CAUER FUNCTIONS
	RC = 02	RC = 05	RC = 07	RC = 10	RC = 15	RC = 20	RC = 25	RC = 35	RC = 50
AMAX DB	.0017375255	.0108709564	.0213327376	.0436480540	.0988323393	.1772876696	.2802872360	.5675287486	1.249387366
AMIN DB	55.02811283	62.99603497	65.92905640	69.04941035	72.62641951	75.20364946	77.24484924	80.45465143	84.23454922
MU K	.1436243D-1	.5738934D-2	.4094304D-2	.2858659D-2	.1893703D-2	.1407507D-2	.1112732D-2	.7689539D-3	.4976287D-3
1 PFR	.7820322417	.5930847656	.5301974043	.4664074876	.3967472419	.3486716397	.3118454849	.2563358751	.1958897489
1 PF1	.8608906281	.7194354439	.6739593544	.6294436648	.5834663062	.5538735416	.5326813354	.5036924623	.4768633759
1 PQ1	.7109516037	.7582439683	.7873948871	.8291126417	.8964296212	.9637500947	1.033012664	1.183121112	1.460730516
1 PF2	.9721339790	.8956176256	.8710389675	.8471847954	.8229049966	.8075458473	.7967078544	.7821335675	.7689583131
1 PQ2	1.499335778	1.670765102	1.773094952	1.916069330	2.140058072	2.357819995	2.577019511	3.040216761	3.870857537
1 PF3	1.034394905	.9962424268	.9838065973	.9716701408	.9592604134	.9513861177	.9458206894	.9383272745	.9315462990
1 PQ3	3.665703856	4.126049966	4.398896033	4.778246178	5.369185829	5.940782558	6.514023444	7.720470163	9.873970073
1 PFC	1.058805240	1.036460290	1.029117979	1.021926092	1.014546422	1.009850776	1.006526050	1.002042033	.9979770904
1 PQC	13.74453849	15.49927621	16.53795414	17.98079745	20.22614480	22.39605519	24.57080419	29.14461051	37.30238171
1 ZF1	1.201995304	1.201995304	1.201995304	1.201995304	1.201995304	1.201995304	1.201995304	1.201995304	1.201995304
1 ZF2	1.293834302	1.293834302	1.293834302	1.293834302	1.293834302	1.293834302	1.293834302	1.293834302	1.293834302
1 ZF3	1.590678413	1.590678413	1.590678413	1.590678413	1.590678413	1.590678413	1.590678413	1.590678413	1.590678413
1 ZF4	2.734079263	2.734079263	2.734079263	2.734079263	2.734079263	2.734079263	2.734079263	2.734079263	2.734079263

(4) P 9+0	TH = 58 SBE = 1.179178403								CAUER FUNCTIONS
	RC = 02	RC = 05	RC = 07	RC = 10	RC = 15	RC = 20	RC = 25	RC = 35	RC = 50
AMAX DB	.0017375255	.0108709564	.0213327376	.0436480540	.0988323393	.1772876696	.2802872360	.5675287486	1.249387366
AMIN DB	53.33119549	61.29911215	64.23213306	67.35248674	70.92949576	73.50672566	75.54792542	78.75772758	82.53762536
MU K	.1706875D-1	.6820324D-2	.4865796D-2	.3397318D-2	.2250534D-2	.1672724D-2	.1322404D-2	.9138482D-3	.5913971D-3
1 PFR	.7953629467	.6018990817	.5377503129	.4727887301	.4019605450	.3531392673	.3157715126	.2595388394	.1982486927
1 PF1	.8715107699	.7276505261	.6814635616	.6362821578	.5896484503	.5596498207	.5381748184	.5087994067	.4816425855
1 PQ1	.7158813374	.7624623127	.7914891417	.8331557446	.9005301158	.9679866837	1.037429525	1.187996670	1.466555157
1 PF2	.9768617184	.9003685241	.8757583128	.8518583304	.8275180720	.8121141195	.8012416799	.7866176035	.7733941179
1 PQ2	1.530222325	1.700088879	1.802699291	1.946568656	2.172528546	2.392538511	2.614178059	3.082844382	3.923787406
1 PF3	1.034538232	.9971564886	.9849393919	.9730024210	.9607831417	.9530229182	.9475350564	.9401421926	.9334484671
1 PQ3	3.787726736	4.247959495	4.524113487	4.909490172	5.511459487	6.094701983	6.680155549	7.913268167	10.11592948
1 PFC	1.056817342	1.035305094	1.028215982	1.021263095	1.014119806	1.009570073	1.006346608	1.001996529	.9980504516
1 PQC	14.31003569	16.07589480	17.13420144	18.60986449	20.91270857	23.14203265	25.37844335	30.08577833	38.48800005
1 ZF1	1.188411333	1.188411333	1.188411333	1.188411333	1.188411333	1.188411333	1.188411333	1.188411333	1.188411333
1 ZF2	1.276723042	1.276723042	1.276723042	1.276723042	1.276723042	1.276723042	1.276723042	1.276723042	1.276723042
1 ZF3	1.563992587	1.563992587	1.563992587	1.563992587	1.563992587	1.563992587	1.563992587	1.563992587	1.563992587
1 ZF4	2.677540167	2.677540167	2.677540167	2.677540167	2.677540167	2.677540167	2.677540167	2.677540167	2.677540167

(4) P 9+0	TH = 59	SBE = 1.166633397						CAUER FUNCTIONS	
	RC = 02	RC = 05	RC = 07	RC = 10	RC = 15	RC = 20	RC = 25	RC = 35	RC = 50
AMAX DB	.0017375255	.0108709564	.0213327376	.0436480540	.0988323393	.1772876696	.2802872360	.5675287486	1.249387366
AMIN DB	51.64537745	59.61328607	62.54630624	65.66665951	69.24366832	71.82089815	73.86209787	77.07190000	80.85179777
MU I K	.2026028D-1	.8095595D-2	.5775607D-2	.4032551D-2	.2671342D-2	.1985491D-2	.1569668D-2	.1084720D-2	.7019771D-3
1 I PFR	.8094581345	.6111705825	.5456832361	.4794819374	.4074215555	.3578154615	.3198785887	.2628348571	.2007131263
1 I PF1	.8825575854	.7361957158	.6892684578	.6433936387	.5960762296	.5656548423	.5438852975	.5141177519	.4866092635
1 I PQ1	.7211729650	.7669549436	.7958395859	.8374426380	.9048687658	.9724635932	1.042092806	1.193138251	1.472691292
1 I PF2	.9816340495	.9052071066	.8805752231	.8566374083	.8322433084	.8167980124	.8058932488	.7912218418	.7779519761
1 I PQ2	1.563468486	1.731465421	1.834320756	1.979093140	2.207100034	2.429467518	2.653675876	3.128115430	3.979955543
1 I PF3	1.034645668	.9980621875	.9860722540	.9743427111	.9623216986	.9546803458	.9492733646	.9419854117	.9353827798
1 I PQ3	3.920044565	4.379486324	4.659012051	5.050690515	5.664325841	6.259949350	6.858416567	8.119992945	10.37520233
1 I PFC	1.054811462	1.034136679	1.027302944	1.020591428	1.013687169	1.009285138	1.006164243	1.001949924	.9981244112
1 I PQC	14.92725037	16.70227746	17.78103242	19.29145202	21.65570076	23.94873438	26.25139003	31.10238602	39.76791708
1 I ZF1	1.175475012	1.175475012	1.175475012	1.175475012	1.175475012	1.175475012	1.175475012	1.175475012	1.175475012
1 I ZF2	1.260326702	1.260326702	1.260326702	1.260326702	1.260326702	1.260326702	1.260326702	1.260326702	1.260326702
1 I ZF3	1.538205558	1.538205558	1.538205558	1.538205558	1.538205558	1.538205558	1.538205558	1.538205558	1.538205558
1 I ZF4	2.622543592	2.622543592	2.622543592	2.622543592	2.622543592	2.622543592	2.622543592	2.622543592	2.622543592

(4) P 9+0	TH = 60	SBE = 1.154700538						CAUER FUNCTIONS	
	RC = 02	RC = 05	RC = 07	RC = 10	RC = 15	RC = 20	RC = 25	RC = 35	RC = 50
AMAX DB	.0017375255	.0108709564	.0213327376	.0436480540	.0988323393	.1772876696	.2802872360	.5675287486	1.249387366
AMIN DB	49.96926903	57.93716588	60.87018494	63.99053764	67.56754613	70.14477585	72.18597552	75.39577761	79.17567536
MU I K	.2402248D-1	.9598894D-2	.6848100D-2	.4781370D-2	.3167392D-2	.2354184D-2	.1861146D-2	.1236146D-2	.8323297D-3
1 I PFR	.8243846322	.6209347058	.5540245399	.4865095672	.4131474651	.3627143684	.3241787963	.2662332710	.2032898060
1 I PF1	.8940511631	.7450878955	.6973896276	.6507923817	.6027625593	.5719006571	.5498242135	.5196481193	.4917732925
1 I PQ1	.7268658443	.7717484532	.8004701552	.8419952918	.9094662549	.9772012093	1.047023065	1.198567552	1.479163811
1 I PF2	.9864440086	.9101313266	.8854889213	.8615223474	.8370870063	.8215993819	.8106647790	.7959489499	.7826349263
1 I PQ2	1.599336003	1.765104255	1.868160029	2.013839400	2.243971581	2.468813252	2.695728878	3.176270427	4.039652827
1 I PF3	1.034714395	.9989574489	.9872034014	.9756895306	.9638749244	.9563574492	.9510348117	.9438563318	.9373488208
1 I PQ3	4.063902778	4.521720077	4.804667464	5.202933589	5.829920221	6.437725516	7.050082508	8.342094974	10.65357578
1 I PFC	1.052788902	1.032955653	1.026379295	1.019911372	1.013248663	1.008996051	1.005978994	1.001902201	.9981989114
1 I PQC	15.60290890	17.38454076	18.48456266	20.03181320	22.46174013	24.82321832	27.19718724	32.20307503	41.15285581
1 I ZF1	1.163158171	1.163158171	1.163158171	1.163158171	1.163158171	1.163158171	1.163158171	1.163158171	1.163158171
1 I ZF2	1.244612806	1.244612806	1.244612806	1.244612806	1.244612806	1.244612806	1.244612806	1.244612806	1.244612806
1 I ZF3	1.513270730	1.513270730	1.513270730	1.513270730	1.513270730	1.513270730	1.513270730	1.513270730	1.513270730
1 I ZF4	2.568993074	2.568993074	2.568993074	2.568993074	2.568993074	2.568993074	2.568993074	2.568993074	2.568993074

(4) P 9+0	TH = 61	SBE = 1.143354068						CAUER FUNCTIONS	
	RC = 02	RC = 05	RC = 07	RC = 10	RC = 15	RC = 20	RC = 25	RC = 35	RC = 50
AMAX DB	.0017375255	.0108709564	.0213327376	.0436480540	.0988323393	.1772876696	.2802872360	.5675287486	1.249387366
AMIN DB	48.30146359	56.26934324	59.20236070	62.32271254	65.89972058	68.47695014	70.51814973	73.72795176	77.50784947
MU I K	.2845628D-1	.1137055D-1	.8112044D-2	.5663861D-2	.3751994D-2	.2788692D-2	.2204655D-2	.1523528D-2	.9859517D-3
1 I PFR	.8402181368	.6312311466	.5628058960	.4938966194	.4191573555	.3678516617	.3286855063	.2698944029	.2059861857
1 I PF1	.9060126896	.7543452279	.7058438975	.6584938398	.6097214505	.5784003526	.5560039991	.5254020604	.4971454269
1 I PQ1	.7330053184	.7768731225	.8054080120	.8468385585	.9143459068	.9822224763	1.052243409	1.204308918	1.486000599
1 I PF2	.9912837213	.9151386962	.8904983226	.8665132776	.8420353885	.8265200705	.8155585181	.8008016793	.7874461360
1 I PQ2	1.638126856	1.801244939	1.904445964	2.051031123	2.283369120	2.510809482	2.740581584	3.227581412	4.103207895
1 I PF3	1.034741447	.9998400574	.9883309206	.9770412794	.9654415550	.9580531833	.9528185097	.9457542787	.9393461117
1 I PQ3	4.220751742	4.675919419	4.962319559	5.367468925	6.006545909	6.629407359	7.256615241	8.581234265	10.95309331
1 I PFC	1.050751048	1.031762664	1.025445499	1.019223232	1.012804456	1.008702905	1.005790902	1.001853344	.9982738901
1 I PQC	16.34491074	18.12981446	19.25190599	20.83820829	23.33849777	25.77365574	28.22456148	33.39783314	42.65519967
1 I ZF1	1.151434651	1.151434651	1.151434651	1.151434651	1.151434651	1.151434651	1.151434651	1.151434651	1.151434651
1 I ZF2	1.229551181	1.229551181	1.229551181	1.229551181	1.229551181	1.229551181	1.229551181	1.229551181	1.229551181
1 I ZF3	1.489144435	1.489144435	1.489144435	1.489144435	1.489144435	1.489144435	1.489144435	1.489144435	1.489144435
1 I ZF4	2.516796865	2.516796865	2.516796865	2.516796865	2.516796865	2.516796865	2.516796865	2.516796865	2.516796865

(4) P 9+0	TH = 62	SBE = 1.132570051						CAUER FUNCTIONS	
	RC = 02	RC = 05	RC = 07	RC = 10	RC = 15	RC = 20	RC = 25	RC = 35	RC = 50
AMAX DB	.0017375255	.0108709564	.0213327376	.0436480540	.0988323393	.1772876696	.2802872360	.5675287486	1.249387366
AMIN DB	46.64052813	54.60838263	57.54139775	60.66174834	64.23875572	66.81598505	68.85718454	72.06698647	75.84688413
MU I K	.3368109D-1	.1345828D-1	.9601483D-2	.6703793D-2	.4440891D-2	.3300720D-2	.2609448D-2	.1803261D-2	.1166981D-2
1 I PFR	.8570447881	.6421045463	.5720628013	.5016710250	.4254724807	.3732447665	.3334135651	.2736796944	.2088105156
1 I PF1	.9184644521	.7639872721	.7146494619	.6665147733	.6169681362	.5851681746	.5624381980	.5313921693	.5027373990
1 I PQ1	.7396438863	.7823635853	.8106841118	.8520006660	.9195341269	.9875533177	1.057779908	1.210389767	1.493233006
1 I PF2	.9961442899	.9202262208	.8956019840	.8716101046	.8471045788	.8315618946	.8205767359	.8057828650	.7923889087
1 I PQ2	1.680191231	1.840162622	1.943440708	2.090923905	2.325550198	2.555722303	2.788512045	3.282357335	4.170993546
1 I PF3	1.034723708	1.000707645	.9894527544	.9783962264	.9670202106	.9597663993	.9546234757	.9476784963	.9413741035
1 I PQ3	4.392289756	4.843545585	5.133404291	5.545740595	6.198706001	6.836580898	7.479697272	8.839319276	11.27610198
1 I PFC	1.048699378	1.030558415	1.024502058	1.018527341	1.012354731	1.008405799	1.005600015	1.001803337	.9983492807
1 I PQC	17.16258713	18.94645410	20.09138118	21.71911053	24.29490952	26.80955452	29.34365889	34.69826141	44.28931984
1 I ZF1	1.140280153	1.140280153	1.140280153	1.140280153	1.140280153	1.140280153	1.140280153	1.140280153	1.140280153
1 I ZF2	1.215113786	1.215113786	1.215113786	1.215113786	1.215113786	1.215113786	1.215113786	1.215113786	1.215113786
1 I ZF3	1.465785677	1.465785677	1.465785677	1.465785677	1.465785677	1.465785677	1.465785677	1.465785677	1.465785677
1 I ZF4	2.465867373	2.465867373	2.465867373	2.465867373	2.465867373	2.465867373	2.465867373	2.465867373	2.465867373

(4) P 9+0	TH = 63	SBE = 1.122326238						CAUER FUNCTIONS	
	RC = 02	RC = 05	RC = 07	RC = 10	RC = 15	RC = 20	RC = 25	RC = 35	RC = 50
AMAX DB	.0017375255	.0108709564	.0213327376	.0436480540	.0988323393	.1772876696	.2802872360	.5675287486	1.249387366
AMIN DB	44.98499288	52.95281067	55.88582236	59.00617114	62.58317754	65.16040653	67.20160586	70.41140766	74.19130525
MU I K	.3983851D-1	.1591866D-1	.1135678D-1	.7929350D-2	.5252755D-2	.3904142D-2	.3086495D-2	.2132925D-2	.1380323D-2
1 I PFR	.8749630977	.6536053251	.5818352040	.5098641136	.4321166041	.3789131271	.3383795169	.2776518729	.2117719584
1 I PF1	.9314298136	.7740351088	.7238260236	.6748733956	.6245192152	.5922196678	.5691416008	.5376322130	.5085620438
1 I PQ1	.7468426581	.7882596402	.8163338929	.8575138172	.9250609350	.9932231435	1.063662097	1.216841100	1.500896423
1 I PF2	1.001015665	.9253903230	.9007980460	.8768124677	.8522905759	.8367266283	.8257217153	.8108954252	.7974666897
1 I PQ2	1.725937455	1.882174899	1.985445986	2.133811166	2.370809738	2.603855967	2.839837853	3.340950600	4.243434515
1 I PF3	1.034657900	1.001557679	.9905666893	.9797524962	.9686093842	.9614958331	.9564486211	.9496281360	.9434321683
1 I PQ3	4.580517192	5.026304120	5.319593499	5.739426073	6.407142758	7.061082131	7.721274572	9.118554785	11.62531058
1 I PFC	1.046635472	1.029343662	1.023549514	1.017824061	1.011899686	1.008104845	1.005406387	1.001752166	.9984250118
1 I PQC	18.06702959	19.84430837	21.01277098	22.68446362	25.34144133	27.94203739	30.56633933	36.11790624	46.07198213
1 I ZF1	1.129672099	1.129672099	1.129672099	1.129672099	1.129672099	1.129672099	1.129672099	1.129672099	1.129672099
1 I ZF2	1.201274549	1.201274549	1.201274549	1.201274549	1.201274549	1.201274549	1.201274549	1.201274549	1.201274549
1 I ZF3	1.443155911	1.443155911	1.443155911	1.443155911	1.443155911	1.443155911	1.443155911	1.443155911	1.443155911
1 I ZF4	2.416120614	2.416120614	2.416120614	2.416120614	2.416120614	2.416120614	2.416120614	2.416120614	2.416120614

(4) P 9+0	TH = 64	SBE = 1.112601940						CAUER FUNCTIONS	
	RC = 02	RC = 05	RC = 07	RC = 10	RC = 15	RC = 20	RC = 25	RC = 35	RC = 50
AMAX DB	.0017375255	.0108709564	.0213327376	.0436480540	.0988323393	.1772876696	.2802872360	.5675287486	1.249387366
AMIN DB	43.33333987	51.30110408	54.23411077	57.35445689	60.93146187	63.50869037	65.54988947	68.75969106	72.53958854
MU I K	.4709687D-1	.1881895D-1	.1342593D-1	.9374034D-2	.6209777D-2	.4615455D-2	.3648838D-2	.2521532D-2	.1631810D-2
1 I PFR	.8940863293	.6657906928	.5921682603	.5185111755	.4391164027	.3848785270	.3436018697	.2818251498	.2148807289
1 I PF1	.9449331509	.7845114774	.7333949516	.6835895396	.6323928188	.5995718386	.5761304041	.5441372839	.5146334432
1 I PQ1	.7546731770	.7946072538	.8223981238	.8634149225	.9309606164	.9992654689	1.069923579	1.223698120	1.509030964
1 I PF2	1.005886497	.9306267541	.9060841639	.8821196912	.8575942232	.8420159843	.8309957406	.8161423593	.8026830738
1 I PQ2	1.775844478	1.927650307	2.030810873	2.180031442	2.419487116	2.655560028	2.894923497	3.403765076	4.321016955
1 I PF3	1.034540568	1.002387446	.9916703399	.9811080552	.9702074278	.9632400928	.9582927395	.9516022461	.9455195892
1 I PQ3	4.787805210	5.226197256	5.522844643	5.950484709	6.633886576	7.305047775	7.983609741	9.421501215	12.00386152
1 I PFC	1.044561022	1.028119221	1.022588460	1.017113785	1.011439544	1.007800165	1.005210078	1.001699813	.9985010064
1 I PQC	19.07151091	20.83505803	22.02764971	23.74600649	26.49042295	29.18419158	31.90654557	37.67267567	48.02285648
1 I ZF1	1.119589501	1.119589501	1.119589501	1.119589501	1.119589501	1.119589501	1.119589501	1.119589501	1.119589501
1 I ZF2	1.188009239	1.188009239	1.188009239	1.188009239	1.188009239	1.188009239	1.188009239	1.188009239	1.188009239
1 I ZF3	1.421218829	1.421218829	1.421218829	1.421218829	1.421218829	1.421218829	1.421218829	1.421218829	1.421218829
1 I ZF4	2.367475691	2.367475691	2.367475691	2.367475691	2.367475691	2.367475691	2.367475691	2.367475691	2.367475691

(4) P 9+0 — TH = 65 — SBE = 1.103377919 — CAUER FUNCTIONS

9+0	RC = 02	RC = 05	RC = 07	RC = 10	RC = 15	RC = 20	RC = 25	RC = 35	RC = 50
AMAX DB	.0017375255	.0108709564	.0213327376	.0436480540	.0988323393	.1772876696	.2802872360	.5675287486	1.249387366
AMIN DB	41.68399015	49.65167610	52.58467549	55.70501773	59.28202064	61.85924841	63.90044718	67.11024848	70.89014580
MU I K	.5565678D-1	.2223931D-1	.1586610D-1	.1107778D-1	.7338412D-2	.5454319D-2	.4312018D-2	.2979823D-2	.1928393D-2
1 I PFR	.9145454632	.6787258880	.6031132575	.5276521444	.4465019556	.3911654741	.3491014155	.2862154593	.2181482598
1 I PF1	.9589997377	.7954409218	.7433794578	.6926848483	.6406088028	.6072433441	.5834223962	.5509239781	.5209670972
1 I PQ1	.7632197227	.8014598070	.8289239524	.8697465029	.9372725190	1.005718669	1.076602773	1.231000987	1.517682308
1 I PF2	1.010743969	.9359304907	.9114574271	.8875307246	.8630161711	.8474315898	.8364010827	.8215267457	.8080418123
1 I PQ2	1.830477690	1.977018955	2.079941468	2.229977438	2.471974921	2.711238179	2.954189436	3.471265959	4.404300114
1 I PF3	1.034368076	1.003194035	.9927611329	.9824606948	.9718125367	.9649976439	.9601544927	.9535997592	.9476355491
1 I PQ3	5.016983815	5.445590183	5.745463532	6.181218693	6.881317389	7.570978791	8.269348497	9.751148737	12.41542059
1 I PFC	1.042477843	1.026885979	1.021619534	1.016396943	1.010974544	1.007491896	1.005011156	1.001646263	.9985771816
1 I PQC	20.19202974	21.93264941	23.14980190	24.91768636	27.75647137	30.55151175	33.38077050	39.38136560	50.16516028
1 I ZF1	1.110012855	1.110012855	1.110012855	1.110012855	1.110012855	1.110012855	1.110012855	1.110012855	1.110012855
1 I ZF2	1.175295336	1.175295336	1.175295336	1.175295336	1.175295336	1.175295336	1.175295336	1.175295336	1.175295336
1 I ZF3	1.399940170	1.399940170	1.399940170	1.399940170	1.399940170	1.399940170	1.399940170	1.399940170	1.399940170
1 I ZF4	2.319854268	2.319854268	2.319854268	2.319854268	2.319854268	2.319854268	2.319854268	2.319854268	2.319854268

(4) P 9+0 — TH = 66 — SBE = 1.094636279 — CAUER FUNCTIONS

9+0	RC = 02	RC = 05	RC = 07	RC = 10	RC = 15	RC = 20	RC = 25	RC = 35	RC = 50
AMAX DB	.0017375255	.0108709564	.0213327376	.0436480540	.0988323393	.1772876696	.2802872360	.5675287486	1.249387366
AMIN DB	40.03528929	48.00286088	50.93584960	54.05618617	57.63318606	60.21041277	62.25161105	65.46141192	69.24130902
MU I K	.6575811D-1	.2627560D-1	.1874569D-1	.1308832D-1	.8670284D-2	.6444242D-2	.5094621D-2	.3520641D-2	.2278384D-2
1 I PFR	.9364929182	.6924857063	.6147287466	.5373324323	.4543073395	.3978016702	.3549016187	.2908407462	.2215874034
1 I PF1	.9736555504	.8068499474	.7538047945	.7021829926	.6491889721	.6152547145	.5910371747	.5580106057	.5275801248
1 I PQ1	.7725822460	.8088796571	.8359662138	.8765578116	.9440420399	1.012626911	1.083743819	1.238795743	1.526902722
1 I PF2	1.015573603	.9412956137	.9169142645	.8930440742	.8685568336	.8529749582	.8419399812	.8270517365	.8135468205
1 I PQ2	1.890509171	2.030785924	2.133313056	2.284107359	2.528729891	2.771359252	3.018123388	3.543992016	4.493930805
1 I PF3	1.034136585	1.003974323	.9938362881	.9838080130	.9734227324	.9667667926	.9620323950	.9556194774	.9497791171
1 I PQ3	5.271455797	5.687295697	5.990184163	6.434350401	7.152242351	7.861820612	8.581603544	10.11101061	12.86428985
1 I PFC	1.040387884	1.025644895	1.020643436	1.015674006	1.010504954	1.007180189	1.004809698	1.001591502	.9986534479
1 I PQC	21.44802237	23.15385433	24.39575554	26.21618893	29.15703296	32.06246624	35.00865379	41.26633088	52.52647795
1 I ZF1	1.100924033	1.100924033	1.100924033	1.100924033	1.100924033	1.100924033	1.100924033	1.100924033	1.100924033
1 I ZF2	1.163111923	1.163111923	1.163111923	1.163111923	1.163111923	1.163111923	1.163111923	1.163111923	1.163111923
1 I ZF3	1.379287542	1.379287542	1.379287542	1.379287542	1.379287542	1.379287542	1.379287542	1.379287542	1.379287542
1 I ZF4	2.273180055	2.273180055	2.273180055	2.273180055	2.273180055	2.273180055	2.273180055	2.273180055	2.273180055

(4) P 9+0	TH = 67 SBE = 1.086360377 RC = 02	RC = 05	RC = 07	RC = 10	RC = 15	RC = 20	RC = 25	RC = 35	CAUER FUNCTIONS RC = 50
AMAX DB	.0017375255	.0108709564	.0213327376	.0436480540	.0988323393	.1772876696	.2802872360	.5675287486	1.249387366
AMIN DB	38.38549104	46.35289533	49.28586844	52.40619672	55.98319218	58.56041734	60.60161490	63.81141515	67.59131192
MU I K	.7768864D-1	.3104280D-1	.2214674D-1	.1546294D-1	.1024334D-1	.7613424D-2	.6018940D-2	.4159393D-2	.2691752D-2
1 I PFR	.9601072699	.7071564036	.6270819468	.5476039585	.4625713585	.4048185835	.3610290897	.2957213170	.2252126757
1 I PF1	.9889269674	.8187671832	.7646984754	.7121099225	.6581573424	.6236286136	.5989964044	.5654174403	.5344915040
1 I PQ1	.7828801500	.8169401143	.8435890766	.8839062383	.9513218550	1.020041315	1.091397717	1.247135461	1.536752332
1 I PF2	1.020359028	.9467151654	.9224503315	.8986577190	.8742163342	.8586474536	.8476146218	.8327205522	.8192021861
1 I PQ2	1.956743866	2.089548326	2.191485520	2.342959220	2.590286669	2.836471031	3.087294488	3.622570924	4.590661496
1 I PF3	1.033842048	1.004724949	.9948927969	.9851473930	.9750358416	.9685456668	.9639247949	.9576600554	.9519492334
1 I PQ3	5.555345750	5.954683356	6.260271295	6.713121442	7.449995020	8.181065392	8.924061338	10.50524180	13.35555095
1 I PFC	1.038293247	1.024397015	1.019660928	1.014945490	1.010031068	1.006865211	1.004605791	1.001535515	.9987297086
1 I PQC	22.86330358	24.51900056	25.78549041	27.66162591	30.71308446	33.73922799	36.81375126	43.35434948	55.13981487
1 I ZF1	1.092306192	1.092306192	1.092306192	1.092306192	1.092306192	1.092306192	1.092306192	1.092306192	1.092306192
1 I ZF2	1.151439588	1.151439588	1.151439588	1.151439588	1.151439588	1.151439588	1.151439588	1.151439588	1.151439588
1 I ZF3	1.359230256	1.359230256	1.359230256	1.359230256	1.359230256	1.359230256	1.359230256	1.359230256	1.359230256
1 I ZF4	2.227378269	2.227378269	2.227378269	2.227378269	2.227378269	2.227378269	2.227378269	2.227378269	2.227378269

(4) P 9+0	TH = 68 SBE = 1.078534743 RC = 02	RC = 05	RC = 07	RC = 10	RC = 15	RC = 20	RC = 25	RC = 35	CAUER FUNCTIONS RC = 50
AMAX DB	.0017375255	.0108709564	.0213327376	.0436480540	.0988323393	.1772876696	.2802872360	.5675287486	1.249387366
AMIN DB	36.73273859	44.69989773	47.63284797	50.75316409	54.33015307	56.90737596	58.94857247	62.15337180	65.93826808
MU I K	.9179494D-1	.3667939D-1	.2616802D-1	.1827062D-1	.1210327D-1	.8995830D-2	.7111828D-2	.4914633D-2	.3180506D-2
1 I PFR	.9855992928	.7228380872	.6402504979	.5585264282	.4713384473	.4122521560	.3675141686	.3008302714	.2290405556
1 I PF1	1.004840314	.8312235481	.7760905212	.7224941559	.6675404478	.6323901479	.6073241227	.5731670171	.5417223587
1 I PQ1	.7942572092	.8257279654	.8518681340	.8918590843	.9591734665	1.028021403	1.099623741	1.256081673	1.547300698
1 I PF2	1.025081724	.9521809804	.9280603760	.9043690114	.8799944411	.8644502476	.8534271076	.8385364723	.8250121771
1 I PQ2	2.030153816	2.154017244	2.255123070	2.407169103	2.657275283	2.907217772	3.162371211	3.707738636	4.695373132
1 I PF3	1.033480186	1.005442293	.9959273982	.9864759794	.9766494738	.9703321935	.9658298538	.9597199815	.9541446920
1 I PQ3	5.873697282	6.251821771	6.559653635	7.021420819	7.778563272	8.532883624	9.301119334	10.33379125	13.89524870
1 I PFC	1.036196203	1.023143478	1.018672845	1.014211963	1.009553209	1.006547150	1.004399531	1.001478287	.9988058588
1 I PQC	24.46732524	26.05293575	27.34334843	29.27843596	32.45004893	35.60862811	38.82453742	45.67774681	58.04496701
1 I ZF1	1.084143692	1.084143692	1.084143692	1.084143692	1.084143692	1.084143692	1.084143692	1.084143692	1.084143692
1 I ZF2	1.140260342	1.140260342	1.140260342	1.140260342	1.140260342	1.140260342	1.140260342	1.140260342	1.140260342
1 I ZF3	1.339739170	1.339739170	1.339739170	1.339739170	1.339739170	1.339739170	1.339739170	1.339739170	1.339739170
1 I ZF4	2.182375073	2.182375073	2.182375073	2.182375073	2.182375073	2.182375073	2.182375073	2.182375073	2.182375073

(4) P 9+0			TH = 69 SBE = 1.071144994								CAUER FUNCTIONS
			RC = 02	RC = 05	RC = 07	RC = 10	RC = 15	RC = 20	RC = 25	RC = 35	RC = 50
AMAX DB			.0017375255	.0108709564	.0213327376	.0436480540	.0988323393	.1772876696	.2802872360	.5675287486	1.249387366
AMIN DB			35.07504325	43.04184239	45.97475903	49.09505731	52.67203676	55.24925632	57.29045128	60.50024927	64.28014484
MU		K	.1084962D 0	.4335287D-1	.3092906D-1	.2159479D-1	.1430535D-1	.1063254D-1	.8405760D-2	.5808806D-2	.3759170D-2
	1	PFR	1.013219783	.7396477491	.6543246738	.5701689382	.4806597990	.4201436842	.3743916506	.3063440360	.2330898555
	1	PF1	1.021421192	.8442524116	.7880137309	.7333671125	.6773677034	.6415672347	.6160471066	.5812844910	.5492963055
	1	PQ1	.8068880480	.8353467317	.8608930877	.9004958293	.9676691671	1.036636946	1.108491232	1.265706181	1.558628803
	1	PF2	1.029720707	.9576834839	.9337380769	.9101745565	.8858904884	.8703842661	.8593794239	.8445028237	.8309812509
	1	PQ2	2.111923448	2.225046252	2.325019779	2.477494668	2.730443588	2.984362632	3.244144285	3.800364117	4.809104211
	1	PF3	1.033046477	1.006122450	.9969365505	.9877906501	.9782609938	.9721240739	.9677455216	.9617975542	.9563641191
	1	PQ3	6.232737271	6.583666245	6.893098730	7.363952599	8.142756067	8.922295396	9.718064340	11.41759937	14.49062900
	1	PFC	1.034099207	1.021885530	1.017680102	1.013474049	1.009071739	1.006226215	1.004191028	1.001419807	.9988817847
	1	PQC	26.29688231	27.78631481	29.09926177	31.09657948	34.39900671	37.70341401	41.07572658	48.27587507	61.29032053
	1	ZF1	1.076422021	1.076422021	1.076422021	1.076422021	1.076422021	1.076422021	1.076422021	1.076422021	1.076422021
	1	ZF2	1.129557539	1.129557539	1.129557539	1.129557539	1.129557539	1.129557539	1.129557539	1.129557539	1.129557539
	1	ZF3	1.320786539	1.320786539	1.320786539	1.320786539	1.320786539	1.320786539	1.320786539	1.320786539	1.320786539
	1	ZF4	2.138096974	2.138096974	2.138096974	2.138096974	2.138096974	2.138096974	2.138096974	2.138096974	2.138096974

(4) P 9+0			TH = 70 SBE = 1.064177772								CAUER FUNCTIONS
			RC = 02	RC = 05	RC = 07	RC = 10	RC = 15	RC = 20	RC = 25	RC = 35	RC = 50
AMAX DB			.0017375255	.0108709564	.0213327376	.0436480540	.0988323393	.1772876696	.2802872360	.5675287486	1.249387366
AMIN DB			33.41026004	41.37652910	44.30939627	47.42966827	51.00663369	53.58384833	55.62504102	58.83483704	62.61473155
MU		K	.1283018D 0	.5126680D-1	.3657507D-1	.2553685D-1	.1691675D-1	.1257348D-1	.9940207D-2	.6869186D-2	.4445396D-2
	1	PFR	1.043269805	.7577231581	.6694102022	.5826120116	.4905947877	.4285409254	.3817016949	.3121430327	.2373821827
	1	PF1	1.038693510	.8578897385	.8005039793	.7447634999	.6876718368	.6511910426	.6251953132	.5897980711	.5572398754
	1	PQ1	.8209867796	.8459209224	.8707712306	.9099110561	.9768945633	1.045970312	1.118081886	1.276093345	1.570831572
	1	PF2	1.034252164	.9632114497	.9394758501	.9160700655	.8919032782	.8764501223	.8654733934	.8506229639	.8371140621
	1	PQ2	2.203510326	2.303668949	2.402133010	2.554845812	2.810686388	3.068816718	3.333556325	3.901481228	4.933088222
	1	PF3	1.032536133	1.006761202	.9979164005	.9890879840	.9798674907	.9739187526	.9696695080	.9638908560	.9586059494
	1	PQ3	6.640234951	6.956309355	7.266445477	7.746457934	8.548423488	9.355395853	10.18130697	11.94885717	15.15045046
	1	PFC	1.032004929	1.020624537	1.016683706	1.012732439	1.008587057	1.005902641	1.003980408	1.001360063	.9989573622
	1	PQC	28.39845889	29.75734241	31.09040030	33.15314327	36.59831619	40.06393037	43.61003572	51.19708394	64.93524561
	1	ZF1	1.069127723	1.069127723	1.069127723	1.069127723	1.069127723	1.069127723	1.069127723	1.069127723	1.069127723
	1	ZF2	1.119315818	1.119315818	1.119315818	1.119315818	1.119315818	1.119315818	1.119315818	1.119315818	1.119315818
	1	ZF3	1.302345879	1.302345879	1.302345879	1.302345879	1.302345879	1.302345879	1.302345879	1.302345879	1.302345879
	1	ZF4	2.094470158	2.094470158	2.094470158	2.094470158	2.094470158	2.094470158	2.094470158	2.094470158	2.094470158

(4) P 9+0	TH = 71	SBE = 1.057620681						CAUER FUNCTIONS	
	RC = 02	RC = 05	RC = 07	RC = 10	RC = 15	RC = 20	RC = 25	RC = 35	RC = 50
AMAX DB	.0017375255	.0108709564	.0213327376	.0436480540	.0988323393	.1772876696	.2802872360	.5675287486	1.249387366
AMIN DB	31.73606019	39.70154612	42.63434019	45.75457335	49.33151804	51.90872543	53.94991476	57.15970785	60.93960082
MU I K	.1518344D 0	.6066994D-1	.4328351D-1	.3022071D-1	.2001955D-1	.1487965D-1	.1176340D-1	.8129103D-2	.5260751D-2
1 I PFR	1.076114298	.7772279106	.6856319027	.5959502073	.5012127814	.4374995037	.3894909776	.3183125232	.2419425216
1 I PF1	1.056678074	.8721741966	.8136005375	.7567217604	.6984894020	.6612965242	.6348024164	.5987395514	.5655830314
1 I PQ1	.8368186864	.8576016587	.8816320227	.9202182709	.9869518510	1.056119521	1.128492706	1.287343049	1.584021109
1 I PF2	1.038649015	.9687517077	.9452646121	.9220501767	.8980309610	.8826480349	.8717106194	.8569002593	.8434154715
1 I PQ2	2.306727790	2.391148991	2.487627744	2.640325118	2.899083707	3.161677147	3.431740582	4.012330370	5.068802422
1 I PF3	1.031944078	1.007353975	.9988627452	.9903642236	.9814657413	.9757133829	.9715992482	.9659977219	.9608683969
1 I PQ3	7.105997250	7.377320263	7.686917442	8.176012986	9.002750828	9.839655863	10.69869385	12.54135073	15.88539782
1 I PFC	1.029916273	1.019362001	1.015684771	1.011987898	1.008099611	1.005576689	1.003767813	1.001299045	.9990324561
1 I PQC	30.83150551	32.01416573	33.36342655	35.49452592	39.09581363	42.74039565	46.48056854	54.50138091	69.05332325
1 I ZF1	1.062248344	1.062248344	1.062248344	1.062248344	1.062248344	1.062248344	1.062248344	1.062248344	1.062248344
1 I ZF2	1.109521047	1.109521047	1.109521047	1.109521047	1.109521047	1.109521047	1.109521047	1.109521047	1.109521047
1 I ZF3	1.284391822	1.284391822	1.284391822	1.284391822	1.284391822	1.284391822	1.284391822	1.284391822	1.284391822
1 I ZF4	2.051419738	2.051419738	2.051419738	2.051419738	2.051419738	2.051419738	2.051419738	2.051419738	2.051419738

(4) P 9+0	TH = 72	SBE = 1.051462224						CAUER FUNCTIONS	
	RC = 02	RC = 05	RC = 07	RC = 10	RC = 15	RC = 20	RC = 25	RC = 35	RC = 50
AMAX DB	.0017375255	.0108709564	.0213327376	.0436480540	.0988323393	.1772876696	.2802872360	.5675287486	1.249387366
AMIN DB	30.04990051	38.01422477	40.94691038	44.06708592	47.64399985	50.22119647	52.26238082	55.47216957	59.25206023
MU I K	.1798592D 0	.7186806D-1	.5127254D-1	.3579869D-1	.2371464D-1	.1762605D-1	.1393462D-1	.9629528D-2	.6231751D-2
1 I PFR	1.112200379	.7983580740	.7031384378	.6102955063	.5125954809	.4470847181	.3978141699	.3248936880	.2467999758
1 I PF1	1.075390561	.8871471969	.8273464098	.7692845876	.7098613968	.6719230641	.6449064629	.6081449655	.5743598100
1 I PQ1	.8547162533	.8705742129	.8936331833	.9315549582	.9979641243	1.067202237	1.139839865	1.299574550	1.598330902
1 I PF2	1.042880387	.9742887870	.9510934897	.9281082338	.9042708857	.8889777231	.8780924137	.8633380552	.8498905542
1 I PQ2	2.423859224	2.489047771	2.582936184	2.735281977	2.996951736	3.264277594	3.540072318	4.134413527	5.218032204
1 I PF3	1.031264925	1.007895807	.9997709902	.9916152311	.9830521668	.9775047848	.9735318636	.9681157013	.9631494206
1 I PQ3	7.642563611	7.856211206	8.163551372	8.661434528	9.514657311	10.38432866	11.27992912	13.20592520	16.70863809
1 I PFC	1.027836413	1.018099579	1.014684526	1.011241278	1.007609899	1.005248656	1.003553403	1.001236746	.9991069184
1 I PQC	33.67309752	34.61821331	35.97762620	38.17946022	41.95184372	45.79603151	49.75408793	58.26407625	73.73675905
1 I ZF1	1.055772373	1.055772373	1.055772373	1.055772373	1.055772373	1.055772373	1.055772373	1.055772373	1.055772373
1 I ZF2	1.100160286	1.100160286	1.100160286	1.100160286	1.100160286	1.100160286	1.100160286	1.100160286	1.100160286
1 I ZF3	1.266899981	1.266899981	1.266899981	1.266899981	1.266899981	1.266899981	1.266899981	1.266899981	1.266899981
1 I ZF4	2.008868884	2.008868884	2.008868884	2.008868884	2.008868884	2.008868884	2.008868884	2.008868884	2.008868884

(4) P 9+0		TH = 73 SBE = 1.045691756								CAUER FUNCTIONS
		RC = 02	RC = 05	RC = 07	RC = 10	RC = 15	RC = 20	RC = 25	RC = 35	RC = 50
AMAX DB		.0017375255	.0108709564	.0213327376	.0436480540	.0988323393	.1772876696	.2802872360	.5675287486	1.249387366
AMIN DB		28.34899032	36.31158314	39.24410704	42.36419665	45.94106471	48.51824528	50.55942220	53.76920447	57.54909169
MU	K	.2133233D 0	.8523965D-1	.6081217D-1	.4245930D-1	.2812693D-1	.2090551D-1	.1652726D-1	.1142117D-1	.7391215D-2
1	PFR	1.152082379	.8213510465	.7221086038	.6257817660	.5248399782	.4573738951	.4067358580	.3319350238	.2519887257
1	PF1	1.094838614	.9028528137	.8417886722	.7824995236	.7218340089	.6831152733	.6555506848	.6180554027	.5836091253
1	PQ1	.8751015468	.8850682709	.9069689233	.9440893657	1.010081123	1.079361074	1.152263822	1.312931564	1.613921362
1	PF2	1.046910969	.9798044767	.9569494619	.9342360101	.9106194118	.8954382755	.8846197044	.8699396366	.8565446067
1	PQ2	2.557818806	2.599317427	2.689839168	2.841386099	3.105911700	3.378256395	3.660237881	4.269568115	5.382957247
1	PF3	1.030492951	1.008381300	1.000636098	.9928364364	.9846227817	.9792893959	.9754641130	.9702420140	.9654466825
1	PQ3	8.266198740	8.405089716	8.707792967	9.213841091	10.09534523	11.00100843	11.93715251	13.95611981	17.63657997
1	PFC	1.025768828	1.016839106	1.013684342	1.010493529	1.007118482	1.004918877	1.003337364	1.001173162	.9991805868
1	PQC	37.02468510	37.64893676	39.00933160	41.28326452	45.24350504	49.31143638	53.51558218	62.58085842	79.10251866
1	ZF1	1.049689195	1.049689195	1.049689195	1.049689195	1.049689195	1.049689195	1.049689195	1.049689195	1.049689195
1	ZF2	1.091221749	1.091221749	1.091221749	1.091221749	1.091221749	1.091221749	1.091221749	1.091221749	1.091221749
1	ZF3	1.249846811	1.249846811	1.249846811	1.249846811	1.249846811	1.249846811	1.249846811	1.249846811	1.249846811
1	ZF4	1.966737792	1.966737792	1.966737792	1.966737792	1.966737792	1.966737792	1.966737792	1.966737792	1.966737792

(4) P 9+0		TH = 74 SBE = 1.040299436								CAUER FUNCTIONS
		RC = 02	RC = 05	RC = 07	RC = 10	RC = 15	RC = 20	RC = 25	RC = 35	RC = 50
AMAX DB		.0017375255	.0108709564	.0213327376	.0436480540	.0988323393	.1772876696	.2802872360	.5675287486	1.249387366
AMIN DB		26.63025825	34.59025574	37.52253710	40.64249783	44.21929710	46.79645358	48.83761936	52.04739193	55.82727399
MU	K	.2534082D 0	.1012568D 0	.7223920D-1	.5043769D-1	.3341217D-1	.2483380D-1	.1963284D-1	.1356729D-1	.8780075D-2
1	PFR	1.196456627	.8464975645	.7427597921	.6425706693	.5380628110	.4684594981	.4163330714	.3394941770	.2575492814
1	PF1	1.115017631	.9193375044	.8569787848	.7964196435	.7344595233	.6949239729	.6667845125	.6285180358	.5933757875
1	PQ1	.8985180288	.9013721386	.9218812541	.9580297587	1.023487015	1.092770760	1.165936208	1.327589096	1.630987223
1	PF2	1.050700224	.9852772797	.9628169109	.9404233621	.9170716702	.9020279802	.8912929160	.8767081744	.8633831522
1	PQ2	2.712382155	2.724430950	2.810579275	2.960729042	3.227984440	3.505649801	3.794329094	4.420067605	5.566268639
1	PF3	1.029622060	1.008804560	1.001452531	.9940227759	.9861731327	.9810632111	.9773923351	.9723734949	.9677574980
1	PQ3	8.998340354	9.039587683	9.334340358	9.847443232	10.75906955	11.70441095	12.68574563	14.80905137	18.68992852
1	PFC	1.023717347	1.015582624	1.012685744	1.009745714	1.006625992	1.004587730	1.003119906	1.001108291	.9992532825
1	PQC	41.02208465	41.21067813	42.55829261	44.90393594	49.07070806	53.39081022	57.87475282	67.57499012	85.30101054
1	ZF1	1.043989048	1.043989048	1.043989048	1.043989048	1.043989048	1.043989048	1.043989048	1.043989048	1.043989048
1	ZF2	1.082694798	1.082694798	1.082694798	1.082694798	1.082694798	1.082694798	1.082694798	1.082694798	1.082694798
1	ZF3	1.233209462	1.233209462	1.233209462	1.233209462	1.233209462	1.233209462	1.233209462	1.233209462	1.233209462
1	ZF4	1.924942428	1.924942428	1.924942428	1.924942428	1.924942428	1.924942428	1.924942428	1.924942428	1.924942428

(4) P 9+0	TH = 75 SBE = 1.035276180								CAUER FUNCTIONS
	RC = 02	RC = 05	RC = 07	RC = 10	RC = 15	RC = 20	RC = 25	RC = 35	RC = 50
AMAX DB	.0017375255	.0108709564	.0213327376	.0436480540	.0988323393	.1772876696	.2802872360	.5675287486	1.249387366
AMIN DB	24.89032314	32.84640458	35.77831968	38.89808578	42.47478114	45.05190126	47.09305019	50.30280811	54.08268238
MU I K	.3016005D 0	.1205134D 0	.8597738D-1	.6002974D-1	.3976637D-1	.2955660D-1	.2336654D-1	.1614747D-1	.1044984D-1
1 I PFR	1.246210735	.8741582622	.7653595619	.6608597995	.5524054286	.4804532992	.4266986674	.3476403879	.2635301477
1 I PF1	1.135904680	.9366494938	.8729728279	.8111043340	.7477974311	.7074074242	.6786648563	.6395874329	.6037118117
1 I PQ1	.9256767135	.9198527819	.9386758082	.9736372694	1.038411133	1.107647964	1.181069242	1.343762729	1.649767591
1 I PF2	1.054201400	.9906817243	.9686770541	.9466577871	.9236212572	.9087441064	.8981118130	.8836466529	.8704119444
1 I PQ2	2.892524463	2.867567816	2.948021035	3.095967092	3.365722741	3.649021984	3.944974546	4.588761084	5.771330931
1 I PF3	1.028645757	1.009159139	1.002214178	.9951686181	.9876982238	.9828217101	.9793123782	.9745065281	.9700787735
1 I PQ3	9.867759246	9.780212662	10.06236284	10.58067877	11.52423770	12.51348574	13.54547838	15.78666742	19.89517754
1 I PFC	1.021686202	1.014332413	1.011690448	1.008999033	1.006133146	1.004255649	1.002901274	1.001042139	.9993248080
1 I PQC	45.85062498	45.44283598	46.75705041	49.17106874	53.56499951	58.17099678	62.97542173	73.40771744	92.52862086
1 I ZF1	1.038662991	1.038662991	1.038662991	1.038662991	1.038662991	1.038662991	1.038662991	1.038662991	1.038662991
1 I ZF2	1.074569932	1.074569932	1.074569932	1.074569932	1.074569932	1.074569932	1.074569932	1.074569932	1.074569932
1 I ZF3	1.216965624	1.216965624	1.216965624	1.216965624	1.216965624	1.216965624	1.216965624	1.216965624	1.216965624
1 I ZF4	1.883392962	1.883392962	1.883392962	1.883392962	1.883392962	1.883392962	1.883392962	1.883392962	1.883392962

(4) P 9+0	TH = 76 SBE = 1.030613629 .								CAUER FUNCTIONS
	RC = 02	RC = 05	RC = 07	RC = 10	RC = 15	RC = 20	RC = 25	RC = 35	RC = 50
AMAX DB	.0017375255	.0108709564	.0213327376	.0436480540	.0988323393	.1772876696	.2802872360	.5675287486	1.249387366
AMIN DB	23.12547885	31.07560591	34.00696343	37.12643316	40.70297029	43.28003502	45.32115832	48.53089391	52.31075632
MU I K	.3597891D 0	.1437644D 0	.1025653D 0	.7161146D-1	.4743862D-1	.3525905D-1	.2787472D-1	.1926285D-1	.1246596D-1
1 I PFR	1.302494818	.9047869599	.7902417772	.6808938108	.5680416939	.4934920784	.4379459379	.3564578058	.2699900733
1 I PF1	1.157449627	.9548376077	.8898315701	.8266201596	.7619157962	.7206328818	.6912577426	.6513272537	.6146781228
1 I PQ1	.9575246627	.9409846998	.9577444307	.9912440944	1.055143047	1.124265040	1.197928865	1.361721537	1.670559850
1 I PF2	1.057360283	.9959874855	.9745072163	.9529238537	.9302598348	.9155826174	.9050752928	.8907577708	.8776369650
1 I PQ2	3.104928424	3.032884787	3.105882774	3.250526596	3.522400278	3.811650000	4.115525908	4.779270688	6.002410759
1 I PF3	1.027557111	1.009437956	1.002914271	.9962676739	.9891924265	.9845597690	.9812195132	.9766369646	.9724069301
1 I PQ3	10.91386578	10.65435992	10.91730240	11.43788041	12.41501578	13.45303507	14.54217415	16.91756113	21.28676621
1 I PFC	1.019680095	1.013091031	1.010700382	1.008254843	1.005640757	1.003923129	1.002681750	1.000974715	.9993949443
1 I PQC	51.76873152	50.53528967	51.78506551	54.25921865	58.90271683	63.83492230	69.00948417	80.29367048	101.0462189
1 I ZF1	1.033702867	1.033702867	1.033702867	1.033702867	1.033702867	1.033702867	1.033702867	1.033702867	1.033702867
1 I ZF2	1.066838803	1.066838803	1.066838803	1.066838803	1.066838803	1.066838803	1.066838803	1.066838803	1.066838803
1 I ZF3	1.201093353	1.201093353	1.201093353	1.201093353	1.201093353	1.201093353	1.201093353	1.201093353	1.201093353
1 I ZF4	1.841991777	1.841991777	1.841991777	1.841991777	1.841991777	1.841991777	1.841991777	1.841991777	1.841991777

(4) P 9+0	TH = 77 SBE = 1.026304108								CAUER FUNCTIONS
	RC = 02	RC = 05	RC = 07	RC = 10	RC = 15	RC = 20	RC = 25	RC = 35	RC = 50
AMAX DB	.0017375255	.0108709564	.0213327376	.0436480540	.0988323393	.1772876696	.2802872360	.5675287486	1.249387366
AMIN DB	21.33171192	29.27270460	32.20320511	35.32221923	38.89851306	41.47549262	43.51657650	46.72627776	50.50612194
MU I K	.4304029D 0	.1719802D 0	.1226951D 0	.8566623D-1	.5674912D-1	.4217913D-1	.3334553D-1	.2304345D-1	.1491258D-1
1 I PFR	1.366826788	.9389641481	.8178295960	.7029802056	.5851883920	.5077455680	.4502150115	.3660500734	.2770011509
1 I PF1	1.179562122	.9739491929	.9076202065	.8430417931	.7768929448	.7346785790	.7046404370	.6638124775	.6263468140
1 I PQ1	.9953492299	.9653925252	.9795981962	1.011279858	1.074054226	1.142969670	1.216853444	1.381806385	1.693739299
1 I PF2	1.060113611	1.001158246	.9802798800	.9592024544	.9369766048	.9225377885	.9121811079	.8980438048	.8850644160
1 I PQ2	3.358768278	3.225920994	3.289080438	3.428906143	3.702287740	3.997792980	4.310328381	4.996277209	6.265006626
1 I PF3	1.026348724	1.009633205	1.003545283	.9973128876	.9906493671	.9862715490	.9831083255	.9787600190	.9747378077
1 I PQ3	12.19191978	11.69938884	11.93360645	12.45178789	13.46373326	14.55612891	15.71018779	18.23966420	22.91027045
1 I PFC	1.017704281	1.011861370	1.009717734	1.007514688	1.005149759	1.003590741	1.002461662	1.000906040	.9994634474
1 I PQC	59.14577861	56.75246322	57.89059842	60.40847035	65.32511734	70.63209228	76.23814363	88.52423626	111.2071768
1 I ZF1	1.029101282	1.029101282	1.029101282	1.029101282	1.029101282	1.029101282	1.029101282	1.029101282	1.029101282
1 I ZF2	1.059494254	1.059494254	1.059494254	1.059494254	1.059494254	1.059494254	1.059494254	1.059494254	1.059494254
1 I ZF3	1.185570868	1.185570868	1.185570868	1.185570868	1.185570868	1.185570868	1.185570868	1.185570868	1.185570868
1 I ZF4	1.800630864	1.800630864	1.800630864	1.800630864	1.800630864	1.800630864	1.800630864	1.800630864	1.800630864

(4) P 9+0	TH = 78 SBE = 1.022340595								CAUER FUNCTIONS
	RC = 02	RC = 05	RC = 07	RC = 10	RC = 15	RC = 20	RC = 25	RC = 35	RC = 50
AMAX DB	.0017375255	.0108709564	.0213327376	.0436480540	.0988323393	.1772876696	.2802872360	.5675287486	1.249387366
AMIN DB	19.50478932	27.43162578	30.36079423	33.47909990	37.05501535	39.63186245	41.67288500	44.88253286	48.66234867
MU I K	.5166053D 0	.2064249D 0	.1472689D 0	.1028237D 0	.6811501D-1	.5062690D-1	.4002408D-1	.2765867D-1	.1789931D-1
1 I PFR	1.441251923	.9774463421	.8486690309	.7275121577	.6041202967	.5234277813	.4636819472	.3765468102	.2846531820
1 I PF1	1.202092378	.9940265150	.9264074750	.8604529399	.7928195662	.7496362956	.7189042363	.6771323823	.6388041955
1 I PQ1	1.040941255	.9939156070	1.004916951	1.034310798	1.095629945	1.164213639	1.238281061	1.404456508	1.719787548
1 I PF2	1.062387014	1.006150190	.9859614275	.9654698100	.9437576003	.9296016901	.9194254865	.9055064179	.8927006992
1 I PQ2	3.666957982	3.454221506	3.504251843	3.637133466	3.911067145	4.213094267	4.535123516	5.245343907	6.566336727
1 I PF3	1.025012706	1.009736248	1.004098802	.9982962998	.9920617865	.9879503574	.9849725782	.9808701392	.9770665415
1 I PQ3	13.78151370	12.96747139	13.15899710	13.66744299	14.71458558	15.86780884	17.09619617	19.80435232	24.82725512
1 I PFC	1.015764674	1.010646714	1.008745003	1.006780339	1.004661228	1.003259150	1.002241397	1.000836142	.9995300442
1 I PQC	68.52497552	64.47208298	65.42566707	67.95706840	73.17011287	78.91078894	85.02518915	98.50407630	123.5010349
1 I ZF1	1.024851586	1.024851586	1.024851586	1.024851586	1.024851586	1.024851586	1.024851586	1.024851586	1.024851586
1 I ZF2	1.052530377	1.052530377	1.052530377	1.052530377	1.052530377	1.052530377	1.052530377	1.052530377	1.052530377
1 I ZF3	1.170376315	1.170376315	1.170376315	1.170376315	1.170376315	1.170376315	1.170376315	1.170376315	1.170376315
1 I ZF4	1.759188333	1.759188333	1.759188333	1.759188333	1.759188333	1.759188333	1.759188333	1.759188333	1.759188333

(4) P 9+0 — TH = 79 SBE = 1.018716695 — CAUER FUNCTIONS

(4) P 9+0	RC = 02	RC = 05	RC = 07	RC = 10	RC = 15	RC = 20	RC = 25	RC = 35	RC = 50
AMAX DB	.0017375255	.0108709564	.0213327376	.0436480540	.0988323393	.1772876696	.2802872360	.5675287486	1.249387366
AMIN DB	17.64048904	25.54513059	28.47220097	31.58939011	35.16470904	37.74134727	39.78227313	42.99183679	46.77160786
MU K	.6225820D 0	.2487711D 0	.1774797D 0	.1239170D 0	.8208817D-1	.6101253D-1	.4823464D-1	.3333258D-1	.2157119D-1
1 PFR	1.528591427	1.021240931	.8834793301	.7550024415	.6251923556	.5408135970	.4785719968	.3881130177	.2930599814
1 PF1	1.224802599	1.015100582	.9462636185	.8789470995	.8098013252	.7656147251	.7341582083	.6913346058	.6521549983
1 PQ1	1.096858794	1.027709005	1.034625940	1.061099744	1.120517629	1.188596181	1.262790469	1.430249183	1.749334714
1 PF2	1.064092346	1.010909981	.9915104382	.9716961145	.9505847152	.9367634736	.9268026045	.9131463364	.9005523768
1 PQ2	4.048209058	3.728329389	3.760582059	3.883478184	4.156472679	4.465200383	4.797667229	5.536564098	6.916081776
1 PF3	1.023540690	1.009737482	1.004565376	.9992088769	.9934213617	.9895884688	.9868050365	.9829608373	.9793874036
1 PQ3	15.80091431	14.53350466	14.66137136	15.14843186	16.22952317	17.45095337	18.76518346	21.68290431	27.12288429
1 PFC	1.013867993	1.009450826	1.007785060	1.006053840	1.004176416	1.002929133	1.002021410	1.000765066	.9995944273
1 PQC	80.73206284	74.24993599	74.90393285	77.39503284	82.92404935	89.17039654	95.89093304	110.8101194	138.6238435
1 ZF1	1.020947861	1.020947861	1.020947861	1.020947861	1.020947861	1.020947861	1.020947861	1.020947861	1.020947861
1 ZF2	1.045942611	1.045942611	1.045942611	1.045942611	1.045942611	1.045942611	1.045942611	1.045942611	1.045942611
1 ZF3	1.155487466	1.155487466	1.155487466	1.155487466	1.155487466	1.155487466	1.155487466	1.155487466	1.155487466
1 ZF4	1.717523623	1.717523623	1.717523623	1.717523623	1.717523623	1.717523623	1.717523623	1.717523623	1.717523623

(4) P 9+0 — TH = 80 SBE = 1.015426612 — CAUER FUNCTIONS

(4) P 9+0	RC = 02	RC = 05	RC = 07	RC = 10	RC = 15	RC = 20	RC = 25	RC = 35	RC = 50
AMAX DB	.0017375255	.0108709564	.0213327376	.0436480540	.0988323393	.1772876696	.2802872360	.5675287486	1.249387366
AMIN DB	15.73511392	23.60450039	26.52821391	29.64361490	33.21797802	35.79428148	37.83505235	41.04448103	44.82418039
MU K	.7539729D 0	.3012722D 0	.2149354D 0	.1500687D 0	.9941221D-1	.7388873D-1	.5841417D-1	.4036715D-1	.2612361D-1
1 PFR	1.632841454	1.071723465	.9232310915	.7861354941	.6488733818	.5602637880	.4951795316	.4009631328	.3023687523
1 PF1	1.247324329	1.037180538	.9672562081	.8986278149	.8279620961	.7827439516	.7505342941	.7067308120	.6665283146
1 PQ1	1.166868862	1.068407217	1.070018566	1.092701152	1.149602631	1.216931472	1.291164564	1.459960356	1.783224108
1 PF2	1.065124231	1.015371970	.9968753290	.9778436468	.9574343414	.9440083619	.9343038334	.9209632028	.9086261017
1 PQ2	4.530562133	4.063425644	4.071152256	4.179604682	4.449317580	4.764747505	5.108711243	5.879584891	7.327554838
1 PF3	1.021923901	1.009626180	1.004934315	1.000040292	.9947184740	.9911768926	.9885972367	.9850244671	.9816935920
1 PQ3	18.43238222	16.50855958	16.54040357	16.98728323	18.09797786	19.39591814	20.81024711	23.97704170	29.91829716
1 PFC	1.012021953	1.008278054	1.006841235	1.005337576	1.003696792	1.002601607	1.001802247	1.000692877	.9996562487
1 PQC	97.07200456	86.93280465	87.10064089	89.45592622	95.30964112	102.1498672	109.6031351	126.2906402	157.5959352
1 ZF1	1.017384918	1.017384918	1.017384918	1.017384918	1.017384918	1.017384918	1.017384918	1.017384918	1.017384918
1 ZF2	1.039727883	1.039727883	1.039727883	1.039727883	1.039727883	1.039727883	1.039727883	1.039727883	1.039727883
1 ZF3	1.140881327	1.140881327	1.140881327	1.140881327	1.140881327	1.140881327	1.140881327	1.140881327	1.140881327
1 ZF4	1.675470708	1.675470708	1.675470708	1.675470708	1.675470708	1.675470708	1.675470708	1.675470708	1.675470708

(4) P 9+0		TH = 81 SBE = 1.012465126								CAUER FUNCTIONS
		RC = 02	RC = 05	RC = 07	RC = 10	RC = 15	RC = 20	RC = 25	RC = 35	RC = 50
AMAX DB		.0017375255	.0108709564	.0213327376	.0436480540	.0988323393	.1772876696	.2802872360	.5675287486	1.249387366
AMIN DB		13.78655795	21.59913623	24.51737745	27.62985890	31.20266013	33.77841634	35.81893381	39.02814180	42.80772389
MU	K	.9185454D 0	.3670320D 0	.2618501D 0	.1828248D 0	.1211113D 0	.9001671D-1	.7116446D-1	.4917824D-1	.3182571D-1
1	PFR	1.759835285	1.130828552	.9692720262	.8218503179	.6757991094	.5822641653	.5138990635	.4153927720	.3127745136
1	PF1	1.269094933	1.060235252	.9894419445	.9196076661	.8474479011	.8011814872	.7681944257	.7233048103	.6820862417
1	PQ1	1.256717988	1.118401021	1.112961363	1.130617847	1.184131729	1.250357839	1.324492588	1.494664956	1.822615175
1	PF2	1.065355654	1.019454253	1.001990997	.9838640661	.9642754022	.9513161820	.9419166388	.9289544815	.9169284932
1	PQ2	5.157733742	4.482165212	4.455242977	4.542518449	4.805209611	5.126999478	5.483626210	6.291291756	7.819618141
1	PF3	1.020153406	1.009390303	1.005193452	1.000778643	.9959419035	.9927050625	.9903391779	.9870519238	.9839769453
1	PQ3	21.96917994	19.06386181	18.94798695	19.32359576	20.45365771	21.83709142	23.36937737	26.83688234	33.39161000
1	PFC	1.010235550	1.007133480	1.005917439	1.004634358	1.003224096	1.002277671	1.001584567	1.000619668	.9997151112
1	PQC	119.7045084	103.8656321	103.2338899	105.2816755	111.4425222	118.9844299	127.3375364	146.2394657	181.9682781
1	ZF1	1.014158306	1.014158306	1.014158306	1.014158306	1.014158306	1.014158306	1.014158306	1.014158306	1.014158306
1	ZF2	1.033884824	1.033884824	1.033884824	1.033884824	1.033884824	1.033884824	1.033884824	1.033884824	1.033884824
1	ZF3	1.126533611	1.126533611	1.126533611	1.126533611	1.126533611	1.126533611	1.126533611	1.126533611	1.126533611
1	ZF4	1.632828178	1.632828178	1.632828178	1.632828178	1.632828178	1.632828178	1.632828178	1.632828178	1.632828178

(4) P 9+0		TH = 82 SBE = 1.009827573								CAUER FUNCTIONS
		RC = 02	RC = 05	RC = 07	RC = 10	RC = 15	RC = 20	RC = 25	RC = 35	RC = 50
AMAX DB		.0017375255	.0108709564	.0213327376	.0436480540	.0988323393	.1772876696	.2802872360	.5675287486	1.249387366
AMIN DB		11.79642718	19.51607622	22.42519454	25.53279625	29.10298331	31.67782299	33.71791597	36.92675418	40.70613977
MU	K	.1127285D 1	.4504401D 0	.3213557D 0	.2243718D 0	.1486339D 0	.1104730D 0	.8733660D-1	.6035401D-1	.3905811D-1
1	PFR	1.918382454	1.201374788	1.023538686	.8634784389	.7068594299	.6074894624	.5352756100	.4317637978	.3245432325
1	PF1	1.289264201	1.084160767	1.012850720	.9420043804	.8684314924	.8211205264	.7873407014	.7413245588	.6990368961
1	PQ1	1.375539740	1.181329608	1.166253497	1.177072321	1.225924085	1.290522812	1.364342379	1.535895566	1.869154678
1	PF2	1.064632692	1.023052942	1.006773887	.9896944079	.9710664154	.9586591592	.9496229134	.9371150412	.9254659194
1	PQ2	6.001150717	5.019852686	4.942479418	4.998024548	5.247563374	5.574704464	5.945219631	6.795716513	8.419976227
1	PF3	1.018220697	1.009016286	1.005328827	1.001410074	.9970784153	.9941604121	.9920188971	.9890322287	.9862275448
1	PQ3	26.90816341	22.47607899	22.12632065	22.37741231	23.50513381	24.98285849	26.65570394	30.49313850	37.81542025
1	PFC	1.008519478	1.006023119	1.005018315	1.003947542	1.002760427	1.001958656	1.001369178	1.000545567	.9997705580
1	PQC	152.4078630	127.2953037	125.3134580	126.7362274	133.1261831	141.4991443	150.9774516	172.7195606	214.2042671
1	ZF1	1.011264333	1.011264333	1.011264333	1.011264333	1.011264333	1.011264333	1.011264333	1.011264333	1.011264333
1	ZF2	1.028414081	1.028414081	1.028414081	1.028414081	1.028414081	1.028414081	1.028414081	1.028414081	1.028414081
1	ZF3	1.112417981	1.112417981	1.112417981	1.112417981	1.112417981	1.112417981	1.112417981	1.112417981	1.112417981
1	ZF4	1.589344134	1.589344134	1.589344134	1.589344134	1.589344134	1.589344134	1.589344134	1.589344134	1.589344134

(4) P 9+0	TH = 83	SBE = 1.007509825							CAUER FUNCTIONS
	RC = 02	RC = 05	RC = 07	RC = 10	RC = 15	RC = 20	RC = 25	RC = 35	RC = 50
AMAX DB	.0017375255	.0108709564	.0213327376	.0436480540	.0988323393	.1772876696	.2802872360	.5675287486	1.249387366
AMIN DB	9.774092633	17.33949537	20.23298922	23.33219688	26.89787652	29.47113415	31.51049408	34.71869358	38.49773971
MU I K	.1396248D 1	.5579122D 0	.3980290D 0	.2779054D 0	.1840969D 0	.1368312D 0	.1081745D 0	.7475408D-1	.4837712D-1
1 I PFR	2.122305985	1.287648003	1.088932568	.9129860479	.7433495401	.6369131203	.5600907572	.4506637943	.3380508030
1 I PF1	1.306562145	1.108721272	1.037454290	.9659304143	.8911170523	.8428013417	.8082303431	.7610605908	.7176548430
1 I PQ1	1.538565098	1.263009828	1.234296138	1.235504644	1.277752788	1.339914838	1.413067216	1.585936670	1.925277162
1 I PF2	1.062769108	1.026033575	1.011114467	.9952509172	.9777509298	.9659984637	.9573963465	.9454354304	.9342441077
1 I PQ2	7.184251169	5.734526446	5.580785252	5.587269050	5.813221511	6.143384152	6.528926443	7.429959127	9.171169895
1 I PF3	1.016119074	1.008488814	1.005324294	1.001918273	.9981121836	.9955277772	.9936218662	.9909519335	.9884331418
1 I PQ3	34.14545387	27.21904527	26.48389342	26.51490039	27.59528876	29.17342720	31.01559373	35.31860179	43.62807426
1 I PFC	1.006886782	1.004954223	1.004149480	1.003281203	1.002308347	1.001646203	1.001157090	1.000470760	.9998220584
1 I PQC	202.2553535	161.2181993	156.8624527	157.0461686	163.4474124	172.7976575	183.7120498	209.2057389	258.4346306
1 I ZF1	1.008700105	1.008700105	1.008700105	1.008700105	1.008700105	1.008700105	1.008700105	1.008700105	1.008700105
1 I ZF2	1.023318786	1.023318786	1.023318786	1.023318786	1.023318786	1.023318786	1.023318786	1.023318786	1.023318786
1 I ZF3	1.098504920	1.098504920	1.098504920	1.098504920	1.098504920	1.098504920	1.098504920	1.098504920	1.098504920
1 I ZF4	1.544692190	1.544692190	1.544692190	1.544692190	1.544692190	1.544692190	1.544692190	1.544692190	1.544692190

(4) P 9+0	TH = 84	SBE = 1.005508280							CAUER FUNCTIONS
	RC = 02	RC = 05	RC = 07	RC = 10	RC = 15	RC = 20	RC = 25	RC = 35	RC = 50
AMAX DB	.0017375255	.0108709564	.0213327376	.0436480540	.0988323393	.1772876696	.2802872360	.5675287486	1.249387366
AMIN DB	7.743962532	15.05045124	17.91630057	21.00054221	24.55815211	27.12857180	29.16661578	32.37366319	36.15210445
MU I K	.1749900D 1	.6992242D 0	.4988447D 0	.3482953D 0	.2307263D 0	.1714887D 0	.1355738D 0	.9368833D-1	.6063043D-1
1 I PFR	2.394241809	1.396517940	1.170033702	.9734279040	.7872493900	.6720084463	.5895193502	.4729134581	.3538529130
1 I PF1	1.319127962	1.133438625	1.063102943	.9914674372	.9157430351	.8665269990	.8311983809	.7828756237	.7383148525
1 I PQ1	1.772672239	1.373325935	1.324434509	1.311555309	1.344035830	1.402500964	1.474391619	1.648369355	1.994768853
1 I PF2	1.059543569	1.028217820	1.014865256	1.000419085	.9842500672	.9732785225	.9651980607	.9538994331	.9432674213
1 I PQ2	8.935391827	6.728267906	6.452912502	6.380115191	6.563772199	6.891886809	7.293103377	8.254663034	10.14230045
1 I PF3	1.013846651	1.007790688	1.005161039	1.002283778	.9990239651	.9967885214	.9951301020	.9927942325	.9905782956
1 I PQ3	45.42673265	34.16648823	32.76059563	32.38885515	33.32665570	35.00173540	37.04946600	41.95513241	51.58002683
1 I PFC	1.005353891	1.003935726	1.003317864	1.002640398	1.001871062	1.001342383	1.000949592	1.000395516	.9998689888
1 I PQC	283.6531710	213.3321566	204.5520315	202.2304122	208.0872175	218.5484324	231.3367941	261.9723344	322.0794226
1 I ZF1	1.006463601	1.006463601	1.006463601	1.006463601	1.006463601	1.006463601	1.006463601	1.006463601	1.006463601
1 I ZF2	1.018605302	1.018605302	1.018605302	1.018605302	1.018605302	1.018605302	1.018605302	1.018605302	1.018605302
1 I ZF3	1.084759944	1.084759944	1.084759944	1.084759944	1.084759944	1.084759944	1.084759944	1.084759944	1.084759944
1 I ZF4	1.498431087	1.498431087	1.498431087	1.498431087	1.498431087	1.498431087	1.498431087	1.498431087	1.498431087

(4) P 9+0	TH = 85 RC = 02	SBE = 1.003819838 RC = 05	RC = 07	RC = 10	RC = 15	RC = 20	RC = 25	CAUER FUNCTIONS RC = 35	RC = 50
AMAX DB		.0108709564	.0213327376	.0436480540	.0988323393	.1772876696	.2802872360	.5675287486	1.249387366
AMIN DB		12.62775759	15.44274828	18.49906879	22.04150844	24.60657376	26.64213149	29.84701490	33.62429732
MU I K		.8901083D 0	.6350263D 0	.4433779D 0	.2937132D 0	.2183042D 0	.1725846D 0	.1192647D 0	.7718218D-1
1 I PFR		1.539753462	1.274571377	1.049872297	.8417849299	.7151486591	.6254389838	.4998266774	.3728205839
1 I PF1		1.157377647	1.089391782	1.018604188	.9425766001	.8926842900	.8566932005	.8072749330	.7615514399
1 I PQ1		1.530449912	1.449914767	1.415195094	1.432662769	1.485068061	1.554633526	1.729201474	2.083921180
1 I PF2		1.029362267	1.017820769	1.005036566	.9904495417	.9804170271	.9729688824	.9624795672	.9525374403
1 I PQ2		8.197147061	7.714502886	7.505384169	7.610677818	7.925569847	8.341483501	9.376661454	11.45418782
1 I PF3		1.006903034	1.004817065	1.002483015	.9997898632	.9979191976	.9965207855	.9945375676	.9926430066
1 I PQ3		45.09327018	42.42831078	41.27344036	41.85494736	43.59381095	45.89012075	51.60384149	63.06603279
1 I PFC		1.002978984	1.002532276	1.002031571	1.001452692	1.001049884	1.000748382	1.000320232	.9999106056
1 I PQC		300.1034870	282.3655242	274.6786643	278.5492000	290.1231074	305.4072968	343.4373405	419.7276477
1 I ZF1		1.004553792	1.004553792	1.004553792	1.004553792	1.004553792	1.004553792	1.004553792	1.004553792
1 I ZF2		1.014284415	1.014284415	1.014284415	1.014284415	1.014284415	1.014284415	1.014284415	1.014284415
1 I ZF3		1.071140563	1.071140563	1.071140563	1.071140563	1.071140563	1.071140563	1.071140563	1.071140563
1 I ZF4		1.449931781	1.449931781	1.449931781	1.449931781	1.449931781	1.449931781	1.449931781	1.449931781

(4) P 9+0	TH = 86 RC = 02	SBE = 1.002441898 RC = 05	RC = 07	RC = 10	RC = 15	RC = 20	RC = 25	CAUER FUNCTIONS RC = 35	RC = 50
AMAX DB		.0108709564	.0213327376	.0436480540	.0988323393	.1772876696	.2802872360	.5675287486	1.249387366
AMIN DB		10.05278033	12.76984870	15.77095507	19.28294739	21.83718953	23.86770789	27.06318781	30.84312500
MU I K		.1157842D 1	.8260343D 0	.5767404D 0	.3820585D 0	.2839673D 0	.2244959D 0	.1551380D 0	.1003976D 0
1 I PFR		1.739334551	1.416799730	1.151504939	.9126861043	.7704939521	.6711106462	.5336577321	.3964323075
1 I PF1		1.178711431	1.115360763	1.047078515	.9718776093	.9217673800	.8853338525	.8349985597	.7881733172
1 I PQ1		1.771493371	1.637264451	1.565951622	1.558285556	1.600404416	1.665576142	1.839489050	2.204178330
1 I PF2		1.029124013	1.019681953	1.008861632	.9961751251	.9872858239	.9806144237	.9711282184	.9620499444
1 I PQ2		10.56552601	9.694596214	9.229083798	9.178908138	9.454254909	9.878830108	11.00450019	13.34049412
1 I PF3		1.005806724	1.004266894	1.002486995	1.000379401	.9988893787	.9977639845	.9961532824	.9946003896
1 I PQ3		64.11526024	58.81809729	55.98627294	55.68101892	57.35609060	59.93873606	66.78480772	80.98766470
1 I PFC		1.002099048	1.001804358	1.001463266	1.001058741	1.000772335	1.000555789	1.000245520	.9999460080
1 I PQC		462.2183283	424.0277568	403.6112618	401.4104791	413.4871928	432.1071632	481.4646965	583.8609963
1 I ZF1		1.002970852	1.002970852	1.002970852	1.002970852	1.002970852	1.002970852	1.002970852	1.002970852
1 I ZF2		1.010373446	1.010373446	1.010373446	1.010373446	1.010373446	1.010373446	1.010373446	1.010373446
1 I ZF3		1.057590660	1.057590660	1.057590660	1.057590660	1.057590660	1.057590660	1.057590660	1.057590660
1 I ZF4		1.398232859	1.398232859	1.398232859	1.398232859	1.398232859	1.398232859	1.398232859	1.398232859

$P^{(5)}_{9+2}$

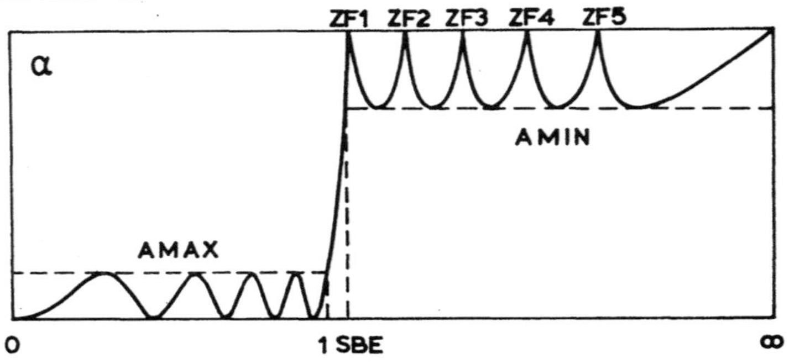

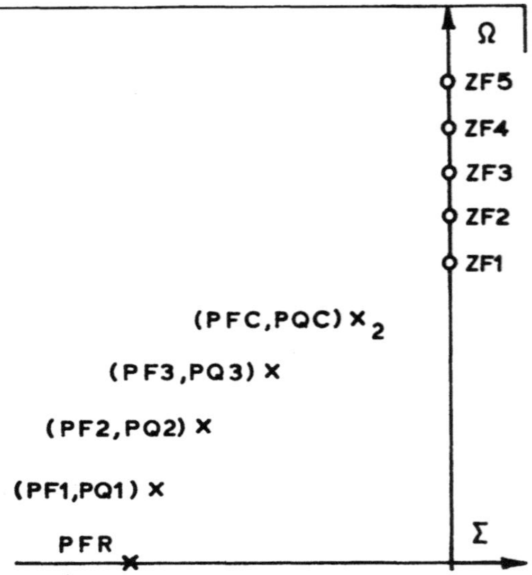

	TH = 46 SBE = 1.390163591								MCPER FUNCTIONS
(5) P 9+2	RC = 02	RC = 05	RC = 07	RC = 10	RC = 15	RC = 20	RC = 25	RC = 35	RC = 50
AMAX DB	.0017375255	.0108709564	.0213327376	.0436480540	.0988323393	.1772876696	.2802872360	.5675287486	1.249387366
AMIN DB	89.65634763	99.44324080	103.0546649	106.9026218	111.3211312	114.5086769	117.0342937	121.0025270	125.6520867
MU I K	.4021816D-3	.1316871D-3	.8719372D-4	.5618596D-4	.3391457D-4	.2355973D-4	.1765137D-4	.1121256D-4	.6587009D-5
1 I PFR	.6086060284	.4658298204	.4169769356	.3668937121	.3117399027	.2734932084	.2441482416	.1999654942	.1519300105
1 I PF1	.7050765431	.5887343410	.5507553469	.5132713811	.4741879071	.4437858051	.4304379943	.4050272272	.3809925354
1 I PQ1	.6560936716	.7067932900	.7359104083	.7767009172	.8415393103	.9058248571	.9716792159	1.113916397	1.376243850
1 I PF2	.8779600762	.7980533579	.7723092563	.7471251345	.7210999616	.7042764870	.6921298298	.6752064600	.6588320395
1 I PQ2	1.153474309	1.312906933	1.401136740	1.521478582	1.706674175	1.884776132	2.063019468	2.437856976	3.107200559
1 I PF3	1.014181020	.9608238962	.9433144361	.9260016717	.9078618013	.8959463918	.8872104426	.8747616829	.8622077609
1 I PQ3	2.260509052	2.582787093	2.760798561	3.002582620	3.372600461	3.726661934	4.079810516	4.820209876	6.139501781
2 I PFC	1.123934486	1.083809314	1.070083131	1.056210605	1.041336062	1.031372603	1.023974962	1.013321477	1.002522090
2 I PQC	6.589907795	7.084778078	7.375441340	7.770138769	8.361707858	8.908220154	9.433224521	10.47237004	12.14836579
1 I ZF1	1.398837151	1.399081219	1.399161775	1.399242744	1.399330118	1.399389736	1.399435038	1.399502742	1.399576106
1 I ZF2	1.475954956	1.478236956	1.478990894	1.479749376	1.480568814	1.481128570	1.481554270	1.482191063	1.482881887
1 I ZF3	1.682260382	1.689057571	1.691320504	1.693605739	1.696084262	1.697782932	1.699077723	1.701019067	1.703130816
1 I ZF4	2.207417229	2.222820004	2.228008369	2.233276960	2.239022756	2.242978643	2.246003223	2.250552143	2.255517778
1 I ZF5	4.023904000	4.061692942	4.074541238	4.087646341	4.102001976	4.111921903	4.119525187	4.130988836	4.143538033

(5) P 9+2	TH = 47	SBE = 1.367327461						MCPER FUNCTIONS	
	RC = 02	RC = 05	RC = 07	RC = 10	RC = 15	RC = 20	RC = 25	RC = 35	RC = 50
AMAX DB	.0017375255	.0108709564	.0213327376	.0436480540	.0988323393	.1772876696	.2802872360	.5675287486	1.249387366
AMIN DB	87.32086195	97.10627089	100.7172858	104.5648716	108.9830256	112.1703556	114.6958232	118.6638568	123.3132320
MU K	.5121689D-3	.1677995D-3	.1111263D-3	.7162169D-4	.4324090D-4	.3004286D-4	.2251110D-4	.1430190D-4	.8403372D-5
1 PFR	.6154954581	.4705649875	.4210775304	.3703901498	.3146192310	.2759706780	.2463301265	.2017205835	.1532420390
1 PF1	.7117211617	.5938463943	.5554180244	.5175135252	.4780147771	.4523549223	.4338266875	.4081738229	.3839184240
1 PQ1	.6580962111	.7086068273	.7376971390	.7784874640	.8433692110	.9077233856	.9736615703	1.116103102	1.378842326
1 PF2	.6825582711	.8022474690	.7763691236	.7510536476	.7248942172	.7079857552	.6957791924	.6787754589	.6623286991
1 PQ2	1.165606500	1.324803503	1.413236005	1.534008303	1.720046992	1.899067325	2.078287473	2.455279901	3.128625602
1 PF3	1.015314418	.9624226954	.9450398065	.9278414187	.9098114926	.8979638681	.8892760217	.8768946935	.8644097592
1 PQ3	2.304284176	2.626847632	2.805986212	3.049754652	3.423347766	3.781155592	4.138217433	4.887130979	6.222064406
2 PFC	1.121044823	1.082034879	1.068650933	1.055105609	1.040562038	1.030809258	1.023562514	1.013118287	1.002521576
2 PQC	6.786921839	7.271224952	7.561453747	7.957939546	8.554865799	9.107894468	9.640002644	10.69468421	12.39801737
1 ZF1	1.375634948	1.375881966	1.375963551	1.376045573	1.376134105	1.376194524	1.376240441	1.376309070	1.376383450
1 ZF2	1.449609301	1.451923235	1.452687955	1.453457399	1.454288810	1.454856824	1.455288851	1.455935182	1.456636463
1 ZF3	1.648248520	1.655145212	1.657442625	1.659763360	1.662281159	1.664007212	1.665323131	1.667296571	1.669443904
1 ZF4	2.156270398	2.171888378	2.177155732	2.182507702	2.188347934	2.192370885	2.195447828	2.200077210	2.205132928
1 ZF5	3.920356163	3.958648571	3.971689979	3.985002839	3.999598010	4.009690419	4.017429580	4.029103838	4.041891046

(5) P 9+2	TH = 48	SBE = 1.345632730						MCPER FUNCTIONS	
	RC = 02	RC = 05	RC = 07	RC = 10	RC = 15	RC = 20	RC = 25	RC = 35	RC = 50
AMAX DB	.0017375255	.0108709564	.0213327376	.0436480540	.0988323393	.1772876696	.2802872360	.5675287486	1.249387366
AMIN DB	85.01937948	94.80317580	98.41374609	102.2609285	106.6786963	109.8657921	112.3910974	116.3589141	121.0080987
MU K	.6498594D-3	.2130403D-3	.1411158D-3	.9096850D-4	.5493333D-4	.3817220D-4	.2860564D-4	.1817698D-4	.1068219D-4
1 PFR	.6227128630	.4755109770	.4253570207	.3740362168	.3176194375	.2785509159	.2486017583	.2035470458	.1546068608
1 PF1	.7186391117	.5991623766	.5602648557	.5219215393	.4819898361	.4560613667	.4373451715	.4114491822	.3869550157
1 PQ1	.6602133474	.7105147314	.7395741127	.7803616981	.8452863660	.9097108021	.9757355449	1.118389112	1.381556887
1 PF2	.8872757043	.8065626491	.7805491629	.7551009251	.7288053395	.7118104814	.6995429350	.6824572945	.6659367306
1 PQ2	1.178465067	1.337360062	1.425990509	1.547200906	1.734110588	1.914085521	2.094323685	2.473566213	3.151096918
1 PF3	1.016444897	.9640436677	.9467943070	.9297166034	.9118026603	.9000264945	.8913893106	.8790788907	.8666662943
1 PQ3	2.350903927	2.673600774	2.853881926	3.099700294	3.477020586	3.838750624	4.199917427	4.957775384	6.309159764
2 PFC	1.118109264	1.080227972	1.067191348	1.053978518	1.039771708	1.030233524	1.023140593	1.012909825	1.002520030
2 PQC	6.998558168	7.470661116	7.760166343	8.158311856	8.760690906	9.320488362	9.860028894	10.93104370	12.66322101
1 ZF1	1.353585608	1.353835324	1.353917861	1.354000865	1.354090482	1.354151653	1.354198148	1.354267651	1.354342992
1 ZF2	1.424513351	1.426857543	1.427632514	1.428412398	1.429255234	1.429831139	1.430269217	1.430924684	1.431635997
1 ZF3	1.615723556	1.622715534	1.625046127	1.627401104	1.629956874	1.631709453	1.633045866	1.635050483	1.637232247
1 ZF4	2.107165367	2.122988680	2.128332189	2.133764901	2.139696962	2.143785339	2.146913476	2.151621696	2.156765897
1 ZF5	3.820657207	3.859429009	3.872656810	3.886171325	3.901000195	3.911261557	3.919134158	3.931015780	3.944038009

(5) P 9+2 — TH = 49 — SBE = 1.325012993 — MCPER FUNCTIONS

MU		RC = 02	RC = 05	RC = 07	RC = 10	RC = 15	RC = 20	RC = .25	RC = 35	RC = 50
	AMAX DB	.0017375255	.0108709564	.0213327376	.0436480540	.0988323393	.1772876696	.2802872360	.5675287486	1.249387366
	AMIN DB	82.74981775	92.53186301	96.14195047	99.98869495	104.4060433	107.5928845	110.1180136	114.0855947	118.7345514
MU	K	.8217392D-3	.2695570D-3	.1785891D-3	.1151493D-3	.6955110D-4	.4833725D-4	.3622739D-4	.2302410D-4	.1353327D-4
1	PFR	.6302782473	.4806793659	.4298249076	.3778396070	.3207465363	.2812389286	.2509674443	.2054482368	.1560269103
1	PF1	.7258429362	.6046913152	.5653038829	.5265027183	.4861193699	.4599109051	.4409988563	.4148312457	.3901068312
1	PQ1	.6624539429	.7125235067	.7415473184	.7823292258	.8472961419	.9117924382	.9779065368	1.120780130	1.384394047
1	PF2	.8921124874	.8110007116	.7848515461	.7592694161	.7328359967	.7157534435	.7034239009	.6862548841	.6696591062
1	PQ2	1.192107866	1.350624758	1.439446704	1.561102073	1.748911187	1.929878440	2.111177828	2.492770716	3.174679659
1	PF3	1.017569828	.9656854266	.9485769205	.9316265521	.9138349653	.9021341367	.8935503167	.8813144718	.8689777382
1	PQ3	2.400613580	2.723263394	2.904699070	3.152633385	3.533839410	3.899677310	4.265152347	5.032412286	6.401110412
2	PFC	1.115129251	1.078389177	1.065704758	1.052829556	1.038965164	1.029645425	1.022709187	1.012696036	1.002517374
2	PQC	7.226265212	7.684289273	7.972729805	8.372370635	8.980281745	9.547105909	10.09442322	11.18261759	12.94525111
1	ZF1	1.332622085	1.332874226	1.332957635	1.333041545	1.333132167	1.333194039	1.333241073	1.333311395	1.333387636
1	ZF2	1.400594230	1.402966903	1.403751564	1.404541335	1.405395011	1.405978419	1.406422256	1.407086432	1.407807324
1	ZF3	1.584597356	1.591680161	1.594042561	1.596430447	1.599022806	1.600801001	1.602157236	1.604192055	1.606407336
1	ZF4	2.059977310	2.075995576	2.081412253	2.086922927	2.092944065	2.097096137	2.100274230	2.105059566	2.110290553
1	ZF5	3.724563342	3.763789545	3.777196737	3.790906535	3.805962995	3.816389608	3.824393093	3.836478664	3.849732758

(5) P 9+2 — TH = 50 — SBE = 1.305407289 — MCPER FUNCTIONS

MU		RC = 02	RC = 05	RC ≠ 07	RC = 10	RC = 15	RC = 20	RC = 25	RC = 35	RC = 50
	AMAX DB	.0017375255	.0108709564	.0213327376	.0436480540	.0988323393	.1772876696	.2802872360	.5675287486	1.249387366
	AMIN DB	80.51018193	90.29032688	93.89989028	97.74615936	102.1630524	105.3496173	107.8745552	111.8418804	116.4906007
MU	K	.1035729D-2	.3399746D-3	.2252912D-3	.1452929D-3	.8777861D-4	.6101486D-4	.4573442D-4	.2907146D-4	.1709114D-4
1	PFR	.6382135344	.4860827575	.4344915142	.3818086643	.3240070369	.2840401292	.2534318377	.2074277783	.1575048126
1	PF1	.7333460646	.6104428879	.5705437283	.5312645725	.4904101166	.4639097185	.4447935403	.4133523074	.3933787133
1	PQ1	.6648278521	.7146403292	.7436233480	.7843962038	.8494044189	.9139741281	.9801804462	1.123282387	1.387360908
1	PF2	.8970684982	.8155634496	.7892784731	.7635616369	.7369889573	.7198175374	.7074250621	.6901712867	.6734989506
1	PQ2	1.206599270	1.364650827	1.453655852	1.575762128	1.764499613	1.946498494	2.128904474	2.512953485	3.199445171
1	PF3	1.018686416	.9673464714	.9503865356	.9335705165	.9159080141	.9042866185	.8957590152	.8836016137	.8713444547
1	PQ3	2.453687551	2.776076785	2.958674635	3.208791059	3.594048343	3.964190326	4.334189459	5.111338851	6.498272043
2	PFC	1.112106313	1.076519120	1.064191578	1.051658964	1.038142508	1.029044995	1.022268285	1.012476367	1.002513527
2	PQC	7.471683214	7.913462603	8.200436238	8.601365500	9.214867224	9.788980656	10.34443580	11.45070928	13.24552667
1	ZF1	1.312682832	1.312937105	1.313021301	1.313106035	1.313197577	1.313260095	1.313307628	1.313378708	1.313455788
1	ZF2	1.377785071	1.380184341	1.380978095	1.381777167	1.382641065	1.383231562	1.383680850	1.384353278	1.385083269
1	ZF3	1.554789026	1.561957940	1.564350692	1.566770076	1.569397558	1.571200405	1.572575748	1.574639735	1.576887457
1	ZF4	2.014591433	2.030793734	2.036280434	2.041866136	2.047973441	2.052187377	2.055414117	2.060274743	2.065590714
1	ZF5	3.631849698	3.671504359	3.685083643	3.698982067	3.714259725	3.724847708	3.732979389	3.745265315	3.758747943

(5) P 9+2			TH = 51	SBE = 1.286759566						MCPER FUNCTIONS
			RC = 02	RC = 05	RC = 07	RC = 10	RC = 15	RC = 20	RC = 25	RC = 35	RC = 50
AMAX DB			.0017375255	.0108709564	.0213327376	.0436480540	.0988323393	.1772876696	.2802872360	.5675287436	1.249387366
AMIN DB			78.29855450	88.07663808	91.68563295	95.53138627	99.94778528	103.1340504	105.6587807	109.6258234	114.2742921
MU	K		.1301492D-2	.4274993D-3	.2833544D-3	.1827795D-3	.1104527D-3	.7678824D-4	.5756471D-4	.3659328D-4	.2152048D-4
1	PFR		.6465428065	.4917348995	.4393680774	.3859524534	.3274079984	.2869603807	.2559999744	.2094395850	.1590434024
1	PF1		.7411628853	.6164274811	.5759936471	.5362153749	.4948693090	.4630644406	.4487354458	.4220090471	.3967758572
1	PQ1		.6673460622	.7168731338	.7458094727	.7865694078	.8516176538	.9162622685	.9825637356	1.125902695	1.390465229
1	PF2		.9021433419	.8202526214	.7938321640	.7679801686	.7412670920	.7240057809	.7115495252	.6942097107	.6774595501
1	PQ2		1.222011091	1.379497277	1.468674656	1.591236631	1.780931877	1.964003375	2.147563640	2.534180513	3.225471737
1	PF3		1.019791687	.9690251783	.9522219382	.9355476664	.9180213526	.9064837172	.8980153442	.8859404702	.8737667964
1	PQ3		2.510433743	2.832310118	3.016072528	3.268436941	3.657918297	4.032572008	4.407324821	5.194333831	6.601037797
2	PFC		1.109042067	1.074618471	1.062652256	1.050467008	1.037303858	1.028432274	1.021817884	1.012252264	1.002508400
2	PQC		7.736675709	8.159708180	8.444740808	8.846701016	9.465825885	10.04749458	10.61146594	11.73677572	13.56563185
1	ZF1		1.293711270	1.293967361	1.294052252	1.294137722	1.294230096	1.294293200	1.294341188	1.294412963	1.294490815
1	ZF2		1.356024430	1.358448307	1.359250522	1.360058271	1.360931734	1.361528880	1.361983291	1.362663491	1.363402069
1	ZF3		1.526224215	1.533474243	1.535895805	1.538345191	1.541006243	1.542832718	1.544226411	1.546318468	1.548597487
1	ZF4		1.970902000	1.987276844	1.992830248	1.998487882	2.004678280	2.008952139	2.012226140	2.017160119	2.022559160
1	ZF5		3.542308416	3.582364606	3.596108375	3.610188473	3.625680636	3.636425915	3.644682969	3.657165466	3.670873113

(5) P 9+2			TH = 52	SBE = 1.269018215						MCPER FUNCTIONS
			RC = 02	RC = 05	RC = 07	RC = 10	RC = 15	RC = 20	RC = 25	RC = 35	RC = 50
AMAX DB			.0017375255	.0108709564	.0213327376	.0436480540	.0988323393	.1772876696	.2802872360	.5675287486	1.249387366
AMIN DB			76.11308538	85.88893377	89.49731208	93.34250602	97.75836925	100.9443092	103.4688144	107.4355610	112.0837463
MU	K		.1630805D-2	.5360437D-3	.3553816D-3	.2292941D-3	.1385959D-3	.9637012D-4	.7225368D-4	.4594600D-4	.2702270D-4
1	PFR		.6552925804	.4976508192	.4444668534	.3902808419	.3309570892	.2900060445	.2586773135	.2116378963	.1606457458
1	PF1		.7493088249	.6226562543	.5816635861	.5413639638	.4995047205	.4723822009	.4528312591	.4258075678	.4003038432
1	PQ1		.6700208566	.7192307167	.7481137322	.7888563114	.8539429513	.9186638875	.9850634982	1.128648523	1.393715501
1	PF2		.9073363056	.8250699346	.7985148498	.7725276532	.7456733751	.7283213183	.7158005380	.6983735229	.6815443629
1	PQ2		1.238423663	1.395229670	1.484565998	1.607587076	1.798269848	1.982456730	2.167221479	2.556524456	3.252845446
1	PF3		1.020882481	.9707197904	.9540818028	.9375570816	.9201744599	.9087251581	.9003192003	.8883311678	.8762451024
1	PQ3		2.571198677	2.892264502	3.077187429	3.331864877	3.725750710	4.105136147	4.484887208	5.283412136	6.709843258
2	PFC		1.105938230	1.072687953	1.061087276	1.049253981	1.036449345	1.027807310	1.021357982	1.012022173	1.002501906
2	PQC		8.023367258	8.424754780	8.707287390	9.109960675	9.734708672	10.32420043	10.89708434	12.04245013	13.90734007
1	ZF1		1.275655314	1.275912886	1.275998375	1.276084488	1.276177596	1.276241224	1.276289621	1.276362024	1.276440577
1	ZF2		1.335255785	1.337702160	1.338512164	1.339327931	1.340210262	1.340813592	1.341272776	1.341960236	1.342706859
1	ZF3		1.498834501	1.506160352	1.508609091	1.511086893	1.513779866	1.515628882	1.517040121	1.519159079	1.521468179
1	ZF4		1.928811461	1.945346752	1.950963359	1.956689651	1.962959889	1.967291619	1.970611409	1.975616687	1.981096761
1	ZF5		3.455746953	3.496176723	3.510077048	3.524331556	3.540031214	3.550929515	3.559308972	3.571984054	3.585913005

(5) P 9+2		TH = 53 SBE = 1.252135658								MCPER FUNCTIONS
		RC = 02	RC = 05	RC = 07	RC = 10	RC = 15	RC = 20	RC = 25	RC = 35	RC = 50
	AMAX DB	.0017375255	.0108709564	.0213327376	.0436480540	.0988323393	.1772876696	.2802872360	.5675287486	1.249387366
	AMIN DB	73.95198270	83.72540810	87.33311797	91.17770543	95.59298769	98.77857508	101.3028362	105.2692563	109.9171397
MU	K	.2038016D-2	.6703796D-3	.4445486D-3	.2868939D-3	.1734566D-3	.1206311D-3	.9045545D-4	.5753190D-4	.3384408D-4
1	PFR	.6644921279	.5038469791	.4498012393	.3948045927	.3346626567	.2931840362	.2614697847	.2138773111	.1623151655
1	PF1	.7578004340	.6291412120	.5875642490	.5467198517	.5043247177	.4768706722	.4570881760	.4297544359	.4039686752
1	PQ1	.6728660080	.7217228530	.7505450374	.7912651775	.8563881471	.9211867248	.9876875368	1.131528074	1.397121042
1	PF2	.9126463078	.8300170270	.8033287612	.7772067900	.7502108862	.7327674252	.7201814953	.7026662573	.6857570303
1	PQ2	1.255927125	1.411921057	1.501399785	1.624881693	1.816582037	2.001928949	2.187951088	2.580065493	3.281661185
1	PF3	1.021955436	.9724284069	.9559646822	.9395977435	.9223667409	.9110106078	.9026704329	.8907738012	.8787796948
1	PQ3	2.636373590	2.956277759	3.142349321	3.399403326	3.797881902	4.182232435	4.567242695	5.377329211	6.825172275
2	PFC	1.102796621	1.070728342	1.059497164	1.048020201	1.035579116	1.027170163	1.020888587	1.011786540	1.002493949
2	PQC	8.334188861	8.710566098	8.989939112	9.392935375	10.02326591	10.62084813	11.20305941	12.36956873	14.27264239
1	ZF1	1.258466958	1.258725649	1.258811630	1.258898286	1.258992028	1.259056113	1.259104869	1.259177828	1.259257008
1	ZF2	1.315427072	1.317893715	1.318710798	1.319533882	1.320424342	1.321033362	1.321496949	1.322191125	1.322945217
1	ZF3	1.472556853	1.479952917	1.482427101	1.484931635	1.487654779	1.489525181	1.490953112	1.493097730	1.495435615
1	ZF4	1.888229696	1.904912689	1.910588801	1.916380293	1.922726928	1.927114351	1.930478368	1.935552760	1.941111700
1	ZF5	3.371986568	3.412760913	3.426809525	3.441230853	3.457130664	3.468177497	3.476676235	3.489539701	3.503686031

(5) P 9+2		TH = 54 SBE = 1.236067977								MCPER FUNCTIONS
		RC = 02	RC = 05	RC = 07	RC = 10	RC = 15	RC = 20	RC = 25	RC = 35	RC = 50
	AMAX DB	.0017375255	.0108709564	.0213327376	.0436480540	.0988323393	.1772876696	.2802872360	.5675287486	1.249387366
	AMIN DB	71.81350376	81.58430318	85.19128849	89.03521852	93.44987097	96.63507612	99.15907253	103.1251396	107.7726947
MU	K	.2540591D-2	.8363245D-3	.5547287D-3	.3580886D-3	.2165589D-3	.1506342D-3	.1129689D-3	.7186579D-4	.4228558D-4
1	PFR	.6741738508	.5103414570	.4553859117	.3995354710	.3385338058	.2965018892	.2643838411	.2162128283	.1640552691
1	PF1	.7666554814	.6358952832	.5937071693	.5522932912	.5093383192	.4815381251	.4615139524	.4338567286	.4077768221
1	PQ1	.6758970068	.7243604362	.7531132912	.7938051651	.8589619043	.9238393233	.9904444544	1.134550378	1.400692092
1	PF2	.9180718395	.8350954441	.8082761154	.7820203288	.7548828105	.7373475127	.7246959467	.7070916246	.6901013888
1	PQ2	1.274622926	1.429653056	1.519253959	1.643196394	1.835944506	2.022498078	2.209833438	2.604892324	3.312023797
1	PF3	1.023006980	.9741489714	.9578689970	.9416685256	.9245975185	.9133396678	.9050688384	.8932634289	.8813708747
1	PQ3	2.706401699	3.024730087	3.211928844	3.471420517	3.874688200	4.264251681	4.654800042	5.477087474	6.947563778
2	PFC	1.099619168	1.068740473	1.057882487	1.046766018	1.034693338	1.026520902	1.020409708	1.011545312	1.002484433
2	PQC	8.671932926	9.019380583	9.294814893	9.697657412	10.33347941	10.93941624	11.53138851	12.72020242	14.66378112
1	ZF1	1.242101897	1.242361322	1.242447683	1.242534775	1.242629041	1.242693512	1.242742576	1.242816014	1.242895741
1	ZF2	1.296490290	1.298974846	1.299798255	1.300627914	1.301525718	1.302139902	1.302607500	1.303307814	1.304068763
1	ZF3	1.447333151	1.454793474	1.457291263	1.459820743	1.462572198	1.464462760	1.465906472	1.468075429	1.470440720
1	ZF4	1.849073324	1.865890589	1.871622300	1.877475332	1.883894715	1.888335521	1.891742108	1.896883291	1.902518789
1	ZF5	3.290860971	3.331949789	3.346138067	3.360718281	3.376810549	3.388001201	3.396615933	3.409663353	3.424022910

(5) P 9+2	TH = 55 SBE = 1.220774589 RC = 02	RC = 05	RC = 07	RC = 10	RC = 15	RC = 20	RC = 25	MCPER FUNCTIONS RC = 35	RC = 50
AMAX DB	.0017375255	.0108709564	.0213327376	.0436480540	.0988323393	.1772876696	.2802872360	.5675287486	1.249387366
AMIN DB	69.69594638	79.46390016	83.07010020	86.91331766	91.32728743	94.51207821	97.03578771	101.0014701	105.6486713
MU I K	.3159795D-2	.1040969D-2	.6906453D-3	.4459403D-3	.2697631D-3	.1876773D-3	.1407697D-3	.8957045D-4	.5271501D-4
1 I PFR	.6843737188	.5171541544	.4612369886	.4044863682	.3425804906	.2999678279	.2674265210	.2186498929	.1658699815
1 I PF1	.7758930546	.6429324114	.6001047927	.5580953497	.5145552622	.4863934892	.4661169618	.4381220956	.4117352667
1 I PQ1	.6791313321	.7271556405	.7558295292	.7964864533	.8616738254	.9266311368	.9933437601	1.137725401	1.404439939
1 I PF2	.9236108978	.8403066124	.8133590996	.7869710632	.7596924399	.7420651319	.7293476031	.7116535232	.6945814833
1 I PQ2	1.294625630	1.448517139	1.538215668	1.662615873	1.856441937	2.044250887	2.232958463	2.631103339	3.344049429
1 I PF3	1.024033321	.9758792590	.9597930234	.9437681827	.9268660244	.9157118666	.9075141532	.8958150673	.8840189188
1 I PQ3	2.781786923	3.098050787	3.286343644	3.548330565	3.956591984	4.351631963	4.748017028	5.583192879	7.077619778
2 I PFC	1.096407922	1.066725246	1.056243864	1.045491816	1.033792195	1.025859610	1.019921366	1.011298437	1.002473254
2 I PQC	9.039820148	9.353759462	9.624333376	10.02644124	10.66760091	11.28214941	11.88433521	13.09569457	15.08328978
1 I ZF1	1.226519205	1.226778948	1.226865569	1.226952984	1.227047659	1.227112440	1.227161755	1.227235593	1.227315781
1 I ZF2	1.278401144	1.280901122	1.281730060	1.282565506	1.283469822	1.284088614	1.284559806	1.285265644	1.286032799
1 I ZF3	1.423109756	1.430628018	1.433147459	1.435699985	1.438477774	1.440387190	1.441845715	1.444037604	1.446428832
1 I ZF4	1.811265107	1.828202481	1.833985661	1.839896357	1.846384619	1.850876355	1.854323749	1.859529252	1.865238846
1 I ZF5	3.212215110	3.253587160	3.267906112	3.282636917	3.298913576	3.310243090	3.318970356	3.332197055	3.346765445

(5) P 9+2	TH = 56 SBE = 1.206217949 RC = 02	RC = 05	RC = 07	RC = 10	RC = 15	RC = 20	RC = 25	MCPER FUNCTIONS RC = 35	RC = 50
AMAX DB	.0017375255	.0108709564	.0213327376	.0436480540	.0988323393	.1772876696	.2802872360	.5675287486	1.249387366
AMIN DB	67.59764032	77.36251053	80.96785953	84.81030466	89.22353446	92.40787608	94.93127455	98.89654088	103.5433574
MU I K	.3921509D-2	.1292957D-2	.8580583D-3	.5541843D-3	.3353394D-3	.2333451D-3	.1750493D-3	.1114067D-3	.6558188D-4
1 I PFR	.6951317860	.5243070392	.4673722160	.4096714435	.3468136187	.3035908516	.2706055186	.2211944492	.1677635821
1 I PF1	.7855336698	.6502676536	.6067705692	.5641379934	.5199860768	.4914464222	.4709062607	.4425587693	.4158515597
1 I PQ1	.6825887764	.7301221136	.7587060859	.7993203873	.8645345832	.9295726547	.9963959920	1.141064171	1.408377050
1 I PF2	.9292609096	.8456518084	.8185798518	.7920618210	.7646431713	.7469239783	.7341403442	.7163560502	.6992015814
1 I PQ2	1.316065069	1.468616146	1.558382661	1.683234914	1.878168886	2.067284128	2.257426341	2.658807981	3.377867103
1 I PF3	1.025030426	.9776168625	.9617348809	.9458953392	.9291713892	.9181266512	.9100060463	.8984136850	.8867240740
1 I PQ3	2.863104388	3.176726307	3.366065941	3.630600740	4.044068864	4.444865915	4.847407949	5.696213083	7.216014811
2 I PFC	1.093165061	1.064683627	1.054581963	1.044198013	1.032875896	1.025186382	1.019423588	1.011045864	1.002460306
2 I PQC	9.441581348	9.716644951	9.981266021	10.38193261	11.02819822	11.65160343	12.26447392	13.50170617	15.53404078
1 I ZF1	1.211681033	1.211940651	1.212027404	1.212115021	1.212209982	1.212274992	1.212324499	1.212398649	1.212479209
1 I ZF2	1.261118721	1.263631494	1.264465117	1.265305517	1.266215462	1.266838270	1.267312612	1.268023323	1.268795994
1 I ZF3	1.399837142	1.407406621	1.409945637	1.412519190	1.415321211	1.417248091	1.418720398	1.420933719	1.423349317
1 I ZF4	1.774733405	1.791775943	1.797606222	1.803570477	1.810123512	1.814663571	1.818149896	1.823417085	1.829198149
1 I ZF5	3.135904074	3.177526935	3.191967180	3.206839898	3.223292486	3.234755652	3.243591805	3.256992848	3.271765418

(5) P 9+2	TH = 57 SBE = 1.192363293								MCPER FUNCTIONS
	RC = 02	RC = 05	RC = 07	RC = 10	RC = 15	RC = 20	RC = 25	RC = 35	RC = 50
AMAX DB	.0017375255	.0108709564	.0213327376	.0436480540	.0988323393	.1772876696	.2802872360	.5675287486	1.249387366
AMIN DB	65.51693864	75.27846730	78.88289392	82.72450194	87.13692962	90.32078432	92.84384560	96.80866088	101.4550602
MU I K	.4857241D-2	.1602826D-2	.1063993D-2	.6873792D-3	.4160600D-3	.2895729D-3	.2172632D-3	.1383045D-3	.8143577D-4
1 I PFR	.7064928014	.5318244297	.4738111858	.4151062907	.3512451733	.3073808325	.2739292647	.2238530019	.1697407479
1 I PF1	.7955993888	.6579172924	.6137190565	.5704341819	.5256421719	.4967073885	.4758916623	.4471757323	.4201338817
1 I PQ1	.6862918337	.7332752049	.7617567904	.8023196501	.8675560747	.9326755485	.9996128611	1.144578921	1.412517237
1 I PF2	.9350186441	.8511321226	.8239404378	.7972954519	.7697385059	.7519278957	.7390782266	.7212035133	.7039661896
1 I PQ2	1.339088925	1.490066091	1.579864936	1.705159933	1.901231262	2.091706010	2.283348997	2.688128350	3.413620565
1 I PF3	1.025994019	.9793591775	.9636925172	.9480484765	.9315126319	.9205833773	.9125441103	.9010641954	.8894865524
1 I PQ3	2.951013152	3.261309897	3.451631599	3.718760156	4.137656234	4.544509399	4.953552536	5.816787077	7.363507126
2 I PFC	1.089892900	1.062616662	1.052897510	1.042885069	1.031944672	1.024501329	1.018916410	1.010787543	1.002445477
2 I PQC	9.881558231	10.11143121	10.36880164	10.76716810	11.41821104	12.05069948	12.67474372	13.93827014	16.01930266
1 I ZF1	1.197552348	1.197811367	1.197898116	1.197985806	1.198080920	1.198146075	1.198195711	1.198270082	1.198350917
1 I ZF2	1.244605215	1.247128011	1.247965425	1.248809893	1.249724531	1.250350728	1.250827749	1.251542642	1.252320093
1 I ZF3	1.377469552	1.385083099	1.387639481	1.390231911	1.393055922	1.394998786	1.396483777	1.398716931	1.401155227
1 I ZF4	1.739411693	1.756543622	1.762416372	1.768429831	1.775043278	1.779628885	1.783152143	1.788478209	1.794327936
1 I ZF5	3.061792105	3.103632124	3.118183872	3.133189424	3.149809060	3.161400394	3.170341592	3.183911762	3.198883581

(5) P 9+2	TH = 58 SBE = 1.179178403								MCPER FUNCTIONS
	RC = 02	RC = 05	RC = 07	RC = 10	RC = 15	RC = 20	RC = 25	RC = 35	RC = 50
AMAX DB	.0017375255	.0108709564	.0213327376	.0436480540	.0988323393	.1772876696	.2802872360	.5675287486	1.249387366
AMIN DB	63.45220911	73.21011606	76.81354236	80.65424340	85.06580146	88.24912823	90.77182389	94.73615008	99.38209710
MU I K	.6005365D-2	.1983428D-2	.1317023D-2	.8510904D-3	.5153110D-3	.3587257D-3	.2691905D-3	.1714006D-3	.1009489D-3
1 I PFR	.7185069349	.5397333282	.4805755916	.4208081310	.3558883549	.3113486280	.2774070222	.2266326871	.1718066030
1 I PF1	.8061139442	.6658989609	.6209660377	.5769979759	.5315359314	.5021877493	.4810838213	.4519326951	.4245911144
1 I PQ1	.6902661697	.7366322372	.7649371992	.8054984659	.8707516022	.9359528443	1.003007421	1.148233261	1.416875842
1 I PF2	.9408801135	.8567484157	.8294428229	.8026748138	.7749820456	.7570808800	.7441654911	.7262004446	.7088800707
1 I PQ2	1.363865864	1.512998329	1.602786716	1.728510810	1.925748089	2.117637954	2.310851884	2.719201102	3.451470466
1 I PF3	1.026919556	.9811033854	.9656636929	.9502259186	.9338886469	.9230812985	.9151278526	.9037654492	.8923065252
1 I PQ3	3.046271733	3.352433272	3.543651042	3.813410206	4.237963529	4.651191887	5.067106609	5.945636670	7.520952016
2 I PFC	1.086593905	1.060525476	1.051191292	1.041553489	1.030998783	1.023804578	1.018399877	1.010523427	1.002428650
2 I PQC	10.36482822	10.54205131	10.79062525	11.18564770	11.84101880	12.48279006	13.11851347	14.40935699	16.54280929
1 I ZF1	1.184100698	1.184358612	1.184445212	1.184532837	1.184627965	1.184693173	1.184742871	1.184817366	1.184898374
1 I ZF2	1.228825667	1.231355562	1.232195822	1.233043422	1.233961760	1.234590679	1.235069881	1.235788219	1.236569669
1 I ZF3	1.355964709	1.363614708	1.366186102	1.368795119	1.371638734	1.373596003	1.375092505	1.377343736	1.379802992
1 I ZF4	1.705238133	1.722442791	1.728353108	1.734411149	1.741080369	1.745708571	1.749266632	1.754548575	1.760563963
1 I ZF5	2.989751692	3.031773934	3.046426964	3.061555844	3.078333205	3.090046934	3.099089121	3.112822901	3.127989736

(5) P 9+2	TH = 59 SBE = 1.166633397								MCPER FUNCTIONS
	RC = 02	RC = 05	RC = 07	RC = 10	RC = 15	RC = 20	RC = 25	RC = 35	RC = 50
AMAX DB	.0017375255	.0108709564	.0213327376	.0436480540	.0988323393	.1772876696	.2802872360	.5675287486	1.249387366
AMIN DB	61.40182516	71.15580578	74.75814853	78.59786507	83.00848010	86.19123432	88.71353346	92.67732918	97.32278555
MU K	.7412639D-2	.2450451D-2	.1627618D-2	.1052117D-2	.6372313D-3	.4436952D-3	.3330071D-3	.2120864D-3	.1249443D-3
1 PFR	.7312306459	.5480638149	.4876895287	.4267960400	.3607577464	.3155062137	.2810489958	.2295413544	.1739667760
1 PF1	.8171028712	.6742317836	.6285286535	.5838446596	.5376808253	.5078998647	.4864943306	.4569902346	.4292329220
1 PQ1	.6945411916	.7402128327	.7684448746	.8088728413	.8741360887	.9394191269	1.006594266	1.152192384	1.421469962
1 PF2	.9468404580	.8625012684	.8350888391	.8082027541	.7803774891	.7623870819	.7494065711	.7313516139	.7139482627
1 PQ2	1.390589332	1.537562150	1.627288806	1.753423089	1.951853600	2.145216681	2.340076104	2.752179696	3.491596942
1 PF3	1.027802221	.9828464357	.9676459644	.9524258157	.9362981904	.9256195545	.9177566836	.9065202251	.8951841151
1 PQ3	3.149757144	3.450820789	3.642822488	3.915237171	4.345684570	4.765628944	5.188814881	6.083580271	7.689317788
2 PFC	1.083270702	1.058411286	1.049464168	1.040203825	1.030038518	1.023096278	1.017874050	1.010253471	1.002409703
2 PQC	10.89736022	11.01308463	11.25101512	11.64142380	12.30052352	12.95173930	13.59966129	14.92045470	17.10884366
1 ZF1	1.171296000	1.171552268	1.171638563	1.171725978	1.171820971	1.171886135	1.171935823	1.172010339	1.172091413
1 ZF2	1.213747743	1.216281656	1.217123762	1.217973503	1.218894487	1.219525421	1.220006273	1.220727274	1.221511893
1 ZF3	1.335283548	1.342961878	1.345545775	1.348168938	1.351029609	1.352999598	1.354506360	1.356773944	1.359252151
1 ZF4	1.672155189	1.689414964	1.695357646	1.701455359	1.708175416	1.712843063	1.716433652	1.721868265	1.727846099
1 ZF5	2.919662748	2.961830935	2.976574573	2.991816826	3.008742113	3.020572154	3.029711046	3.043602599	3.058956892

(5) P 9+2	TH = 60 SBE = 1.154700538								MCPER FUNCTIONS
	RC = 02	RC = 05	RC = 07	RC = 10	RC = 15	RC = 20	RC = 25	RC = 35	RC = 50
AMAX DB	.0017375255	.0108709564	.0213327376	.0436480540	.0988323393	.1772876696	.2802872360	.5675287486	1.249387366
AMIN DB	59.36415663	69.11387905	72.71504603	76.55369524	80.96328728	84.14542036	86.66728929	90.63050945	95.27543342
MU K	.9136087D-2	.3023057D-2	.2008572D-2	.1298775D-2	.7868855D-3	.5480213D-3	.4113775D-3	.2620657D-3	.1544303D-3
1 PFR	.7447277274	.5568495154	.4951798493	.4330912156	.3658695082	.3198668373	.2848664614	.2325876538	.1762274678
1 PF1	.8285936456	.6829365338	.6364255524	.5909908805	.5440915359	.5138572118	.4921358326	.4622093868	.4340698447
1 PQ1	.6991507445	.7440393058	.7721197175	.8124608551	.8777263338	.9430907819	1.010389771	1.156323301	1.426318703
1 PF2	.9528938149	.8683909218	.8408801453	.8138820873	.7859286245	.7678508094	.7548061003	.7366620447	.7191761000
1 PQ2	1.419482216	1.563927935	1.653531451	1.780050616	1.979699753	2.174596709	2.371180936	2.787237085	3.534202699
1 PF3	1.028636903	.9845850264	.9696366642	.9546461273	.9387398652	.9281971567	.9204299051	.9093252199	.8981193894
1 PQ3	3.262488411	3.557306756	3.749948067	4.025027545	4.461612545	4.888637328	5.319526425	6.231549504	7.869704978
2 PFC	1.079926093	1.056275408	1.047717069	1.038836684	1.029064200	1.022376596	1.017338998	1.009977632	1.002388509
2 PQC	11.48621028	11.52989015	11.75496286	12.13921100	12.80125184	13.46202162	14.12267185	15.47466644	17.72234036
1 ZF1	1.159110347	1.159364394	1.159450220	1.159537268	1.159631968	1.159696986	1.159746589	1.159821015	1.159902041
1 ZF2	1.199341529	1.201876218	1.202719112	1.203569943	1.204492457	1.205124656	1.205606593	1.206329428	1.207116332
1 ZF3	1.315389987	1.323087977	1.325681700	1.328316402	1.331191410	1.333172319	1.334688003	1.336969940	1.339465104
1 ZF4	1.640109272	1.657405542	1.663375069	1.669507230	1.676272864	1.680976593	1.684597280	1.690031132	1.696117964
1 ZF5	2.851411842	2.893688298	2.908511389	2.923856577	2.940919489	2.952859425	2.962090496	2.976133636	2.991670481

(5) P 9+2		TH = 61	SBE = 1.143354068						MCPER FUNCTIONS	
		RC = 02	RC = 05	RC = 07	RC = 10	RC = 15	RC = 20	RC = 25	RC = 35	RC = 50
	AMAX DB	.0017375255	.0108709564	.0213327376	.0436480540	.0988323393	.1772876696	.2802872360	.5675287486	1.249387366
	AMIN DB	57.33755980	67.08266178	70.68255295	74.52004391	78.92852572	82.10998463	84.63138660	88.59398198	93.23832798
MU	K	.1124533D-1	.3724681D-2	.2475544D-2	.1601244D-2	.9704775D-3	.6760417D-3	.5075676D-3	.3234235D-3	.1906443D-3
1	PFR	.7590705709	.5661281591	.5030765854	.4397172946	.3712416071	.3244452016	.2888719174	.2357811087	.1785955302
1	PF1	.8406158236	.6920358115	.6446770619	.5984548090	.5507841040	.5200745204	.4980221478	.4676542646	.4391134079
1	PQ1	.7041339678	.7481371395	.7760443709	.8162830056	.8815413208	.9469862859	1.014412373	1.160695128	1.431443501
1	PF2	.9590331677	.8744172066	.8468181796	.8197155679	.7916393207	.7734765290	.7603689209	.7421370298	.7245692373
1	PQ2	1.450802574	1.592290980	1.681697792	1.808568736	2.009459266	2.205953359	2.404346886	2.824568937	3.579516696
1	PF3	1.029418186	.9863155820	.9716328800	.9568846018	.9412121029	.9308129735	.9231466959	.9121810366	.9011123495
1	PQ3	3.385655817	3.672856751	3.865953596	4.143686776	4.586658256	5.021153380	5.460213480	6.390609376	8.063369643
2	PFC	1.076563068	1.054119264	1.045951012	1.037452732	1.028076188	1.021645725	1.016794810	1.009695873	1.002364932
2	PQC	12.13976952	12.09877332	12.30832318	12.68452265	13.34848136	14.01884372	14.69275650	16.07783086	18.38901143
1	ZF1	1.147517842	1.147769055	1.147854236	1.147940753	1.148034992	1.148099754	1.148149191	1.148223412	1.148304267
1	ZF2	1.185579353	1.188111407	1.188953971	1.189804780	1.190727638	1.191360305	1.191842730	1.192566515	1.193354765
1	ZF3	1.296250716	1.303959100	1.306559788	1.309203244	1.312089683	1.314079586	1.315602761	1.317896965	1.320406896
1	ZF4	1.609050439	1.626363498	1.632354002	1.638515052	1.645320651	1.650056869	1.653705053	1.659234465	1.665326592
1	ZF5	2.784891493	2.827237070	2.842127951	2.857565130	2.874754823	2.886797881	2.896116344	2.910304507	2.926017617

(5) P 9+2		TH = 62	SBE = 1.132570051						MCPER FUNCTIONS	
		RC = 02	RC = 05	RC = 07	RC = 10	RC = 15	RC = 20	RC = 25	RC = 35	RC = 50
	AMAX DB	.0017375255	.0108709564	.0213327376	.0436480540	.0988323393	.1772876696	.2802872360	.5675287486	1.249387366
	AMIN DB	55.32036672	65.06045197	68.65895798	72.49519133	76.90246756	80.08319434	82.60408916	86.56600595	91.20972415
MU	K	.1382545D-1	.4584024D-2	.3047718D-2	.1972004D-2	.1195619D-2	.8330806D-3	.6255870D-3	.3987415D-3	.2351066D-3
1	PFR	.7743417067	.5759422491	.5114134530	.4467007300	.3768940899	.3292576807	.2930792650	.2391325993	.1810785587
1	PF1	.8532011819	.7015542433	.6533053824	.6062563218	.5577760967	.5265679301	.5041684233	.4733371955	.4443762477
1	PQ1	.7095363559	.7525355659	.7802447084	.8203626312	.8856025862	.9511265543	1.018682910	1.165329427	1.436868486
1	PF2	.9652501728	.8805794597	.8529041020	.8257058563	.7975135147	.7792688654	.7661000915	.7477821497	.7301336751
1	PQ2	1.484850756	1.622876205	1.711998099	1.839178183	2.041329298	2.239486403	2.439779369	2.864397534	3.627798597
1	PF3	1.030140332	.9880342300	.9736314313	.9591387545	.9437131444	.9334657119	.9259060962	.9150871715	.9041629212
1	PQ3	3.520657565	3.798594015	3.991912998	4.272262338	4.721872537	5.164255541	5.611994458	6.561982909	8.271751739
2	PFC	1.073184824	1.051944398	1.044167102	1.036052702	1.027074884	1.020903884	1.016241590	1.009408160	1.002338834
2	PQC	12.86808047	12.72719643	12.91800198	13.28384162	13.94839821	14.62829635	15.31600262	16.73617135	19.11550213
1	ZF1	1.136494441	1.136742168	1.136826519	1.136912328	1.137005927	1.137070318	1.137119504	1.137193396	1.137273952
1	ZF2	1.172435624	1.174961452	1.175802507	1.176652116	1.177574064	1.178206353	1.178688629	1.179412425	1.180201021
1	ZF3	1.277835025	1.285543884	1.288148477	1.290797708	1.293692467	1.295689303	1.297218435	1.299522669	1.302045018
1	ZF4	1.578932106	1.596241089	1.602246332	1.608430344	1.615269914	1.620034777	1.623707670	1.629278694	1.635422134
1	ZF5	2.719999504	2.762373507	2.777319977	2.792837656	2.810142710	2.822281734	2.831682515	2.846008734	2.861891410

(5) P 9+2 — TH = 63 — SBE = 1.122326238 — MCPER FUNCTIONS

MU		RC = 02	RC = 05	RC = 07	RC = 10	RC = 15	RC = 20	RC = 25	RC = 35	RC = 50
	AMAX DB	.0017375255	.0108709564	.0213327376	.0436480540	.0988323393	.1772876696	.2802872360	.5675287486	1.249387366
	AMIN DB	53.31087360	63.04550742	66.64250849	70.47737541	74.88334163	78.06327277	80.58361642	84.54479566	89.18783146
MU	K	.1698066D-1	.5636293D-2	.3748646D-2	.2426387D-2	.1471663D-2	.1025684D-2	.7703672D-3	.4911630D-3	.2896891D-3
1	PFR	.7906356958	.5863398728	.5202284611	.4540712443	.3828494080	.3343225768	.2975040217	.2426537220	.1836850030
1	PF1	.8663838493	.7115187072	.6623348099	.6144172131	.5650868023	.5333551733	.5105913069	.4702738814	.4498722590
1	PQ1	.7154110824	.7572682800	.7847504328	.8247264227	.8899346682	.9555353633	1.023225032	1.170250614	1.442620930
1	PF2	.9715349604	.8868764257	.8591387261	.8318554777	.8035551953	.7852325991	.7720048935	.7536032908	.7358757884
1	PQ2	1.521978304	1.655943947	1.744674966	1.872109853	2.075535947	2.275424505	2.477713193	2.906976497	3.679344159
1	PF3	1.030797266	.9897367752	.9756288432	.9614058428	.9462410175	.9361538980	.9287069892	.9180429985	.9072709423
1	PQ3	3.669146089	3.935832420	4.129078686	4.411972349	4.868473952	5.319192149	5.776162290	6.747081416	8.496509955
2	PFC	1.069794785	1.049752483	1.042366549	1.034637400	1.026060738	1.020151324	1.015679460	1.009114464	1.002310067
2	PQC	13.68324481	13.42404593	13.59219458	13.94483590	14.60329514	15.29754467	15.99956050	17.45698240	19.90958500
1	ZF1	1.126017814	1.126261363	1.126344686	1.126429601	1.126522372	1.126586268	1.126635113	1.126708545	1.126788664
1	ZF2	1.159886683	1.162402516	1.163240813	1.164087977	1.165007684	1.165638697	1.166120148	1.166842958	1.167630836
1	ZF3	1.260114637	1.267813344	1.270418566	1.273070380	1.275970126	1.277971685	1.279505131	1.281816995	1.284349241
1	ZF4	1.549710797	1.566993599	1.573006938	1.579207591	1.586074723	1.590864109	1.594558720	1.600167112	1.606357576
1	ZF5	2.656638332	2.698998434	2.713987714	2.729573822	2.746982195	2.759209614	2.768687333	2.783144200	2.799189292

(5) P 9+2 — TH = 64 — SBE = 1.112601940 — MCPER FUNCTIONS

MU		RC = 02	RC = 05	RC = 07	RC = 10	RC = 15	RC = 20	RC = 25	RC = 35	RC = 50
	AMAX DB	.0017375255	.0108709564	.0213327376	.0436480540	.0988323393	.1772876696	.2802872360	.5675287486	1.249387366
	AMIN DB	51.30732791	61.03603194	64.63139655	68.46477759	72.86931915	76.04838495	78.56812911	82.52850606	87.17079948
MU	K	.2083872D-1	.6924759D-2	.4607289D-2	.2983257D-2	.1810129D-2	.1261915D-2	.9479858D-3	.6045874D-3	.3567009D-3
1	PFR	.8080614701	.5973756879	.5295646476	.4618623748	.3891328097	.3396604299	.3021635772	.2463578285	.1864242986
1	PF1	.8802004219	.7219585868	.6717919898	.6229614394	.5727374578	.5404557880	.5173091492	.4854810867	.4556167684
1	PQ1	.7218206650	.7623743230	.7895958126	.8294050514	.8945656547	.9602398621	1.028065699	1.175486464	1.448731791
1	PF2	.9778759040	.8933061391	.8655224362	.8381667705	.8097683811	.7913726621	.7780888386	.7596066667	.7418023598
1	PQ2	1.562599181	1.691797183	1.780009758	1.907630686	2.112339790	2.314030677	2.518418061	2.952596582	3.734491819
1	PF3	1.031382559	.9914186707	.9776213173	.9636828386	.9487935118	.9388758541	.9315480804	.9210477516	.9104361479
1	PQ3	3.833087105	4.086117941	4.278919649	4.564241322	5.027883305	5.487415972	5.954219608	6.947542007	8.739564763
2	PFC	1.066396619	1.047545339	1.040550675	1.033207714	1.025034249	1.019388329	1.015108568	1.008814765	1.002278479
2	PQC	14.59995343	14.19997536	14.34068984	14.67663280	15.33682264	16.03506893	16.75187931	18.24886431	20.78040376
1	ZF1	1.116067221	1.116305859	1.116387944	1.116471768	1.116563510	1.116626782	1.116675190	1.116748023	1.116827560
1	ZF2	1.147910678	1.150412557	1.151246781	1.152090181	1.153006236	1.153635020	1.154114930	1.154835692	1.155621718
1	ZF3	1.243063580	1.250740730	1.253343066	1.255994042	1.258895200	1.260899110	1.262435106	1.264752021	1.267291452
1	ZF4	1.521345921	1.538579107	1.544593465	1.550804005	1.557691834	1.562501320	1.566214433	1.571855624	1.578088484
1	ZF5	2.594714478	2.637016621	2.652035322	2.667677161	2.685176140	2.697483934	2.707032880	2.721612504	2.737812371

(5) P 9+2 — TH = 65 SBE = 1.103377919 — MCPER FUNCTIONS

			RC = 02	RC = 05	RC = 07	RC = 10	RC = 15	RC = 20	RC = 25	RC = 35	RC = 50
		AMAX DB	.0017375255	.0108709564	.0213327376	.0436480540	.0988323393	.1772876696	.2802872360	.5675287486	1.249387366
		AMIN DB	49.30791396	59.03015979	62.62374317	66.45550686	70.85849761	74.03662136	76.55571282	80.51521623	85.15670125
MU		K	.2555670D-1	.8502727D-2	.5659353D-2	.3665885D-2	.2225237D-2	.1551735D-2	.1165951D-2	.7438281D-3	.4389976D-3
1		PFR	.8267452517	.6091121310	.5394709780	.4701121377	.3957728128	.3452943897	.3070775015	.2502598137	.1893070241
1		PF1	.8946900472	.7329060551	.6817062084	.6319154023	.5807515138	.5478913689	.5243422427	.4919773594	.4616267390
1		PQ1	.7288390796	.7678991838	.7948205965	.8344339444	.8995278560	.9652712032	1.033235792	1.181068714	1.455236367
1		PF2	.9842593551	.8998657845	.8720550879	.8446418242	.8161570924	.7976941292	.7843576729	.7657388395	.7479206148
1		PQ2	1.607204067	1.730790574	1.818330657	1.946051003	2.152042742	2.355609029	2.562205361	3.001592809	3.793630797
1		PF3	1.031889416	.9930749866	.9796046993	.9659663962	.9513681499	.9416296725	.9344278738	.9241005050	.9136581545
1		PQ3	4.014835581	4.251281308	4.443169629	4.730745231	5.201766964	5.670627490	6.147922751	7.165274410	9.003152048
2		PFC	1.062994263	1.045324945	1.038720926	1.031764626	1.023995978	1.018615223	1.014529085	1.008509050	1.002243909
2		PQC	15.63618289	15.06784902	15.17526224	15.49017094	16.14430990	16.85097149	17.58300745	19.12202118	21.73878210
1		ZF1	1.106623395	1.106856348	1.106936974	1.107019498	1.107109999	1.107172508	1.107220377	1.107292465	1.107371267
1		ZF2	1.136487452	1.138971222	1.139799983	1.140638225	1.141549133	1.142174677	1.142652281	1.143369866	1.144152835
1		ZF3	1.226658067	1.234301406	1.236897083	1.239543545	1.242442273	1.244445988	1.245982634	1.248301827	1.250845526
1		ZF4	1.493799566	1.510958272	1.516966098	1.523179301	1.530080467	1.534905300	1.538633452	1.544302515	1.550572768
1		ZF5	2.534137902	2.576336184	2.591370256	2.607054459	2.624630605	2.637010266	2.646624365	2.661318330	2.677664793

(5) P 9+2 — TH = 66 SBE = 1.094636279 — MCPER FUNCTIONS

			RC = 02	RC = 05	RC = 07	RC = 10	RC = 15	RC = 20	RC = 25	RC = 35	RC = 50
		AMAX DB	.0017375255	.0108709564	.0213327376	.0436480540	.0988323393	.1772876696	.2802872360	.5675287486	1.249387366
		AMIN DB	47.31073663	57.02593790	60.61758012	64.44758137	68.84888211	72.02597917	74.54435924	78.50291045	83.14351421
MU		K	.3132816D-1	.1043603D-1	.6948964D-2	.4503055D-2	.2734586D-2	.1907476D-2	.1433564D-2	.9148512D-3	.5401207D-3
1		PFR	.8468342272	.6216209121	.5500034508	.4788638396	.4028017821	.3512506689	.3122679192	.2543764824	.1923450919
1		PF1	.9098944552	.7443963951	.6921097277	.6413082789	.5891549467	.5556858629	.5317131045	.4987832942	.4679210132
1		PQ1	.7365544687	.7738961842	.8004711575	.8398542512	.9048586393	.9706653187	1.038770864	1.187033812	1.462175104
1		PF2	.9906693354	.9065515293	.8787358876	.8512824016	.8227253150	.8042022059	.7908173813	.7721867433	.7542382625
1		PQ2	1.656378715	1.773341920	1.860022894	1.987733693	2.194996640	2.400513181	2.609436622	3.054353329	3.857211131
1		PF3	1.032310660	.9947003756	.9815744429	.9682528172	.9539621547	.9444131852	.9373446442	.9272001486	.9169364403
1		PQ3	4.217234494	4.433505461	4.623888585	4.913468789	5.392091746	5.870829583	6.359337294	7.402519901	9.289890464
2		PFC	1.059591951	1.043093461	1.036878890	1.030309218	1.022946552	1.017832376	1.013941212	1.008197315	1.002206189
2		PQC	16.81412046	16.04332229	16.11018211	16.39865656	17.04317919	17.75737193	18.50497851	20.08864286	22.79761848
1		ZF1	1.097668444	1.097894895	1.097973828	1.098054830	1.098143866	1.098205467	1.098252690	1.098323877	1.098401781
1		ZF2	1.125598436	1.128059739	1.128881572	1.129713183	1.130617359	1.131238588	1.131713077	1.132426284	1.133204911
1		ZF3	1.210876401	1.218472741	1.221057702	1.223695699	1.226587866	1.228588641	1.230123895	1.232442379	1.234987197
1		ZF4	1.467036314	1.484094147	1.490037376	1.496295505	1.503202100	1.508037162	1.511776619	1.517468227	1.523770455
1		ZF5	2.474821429	2.516867975	2.531902660	2.547615131	2.565254218	2.577696709	2.587369489	2.602168806	2.618653089

(5) P 9+2 — TH = 67 — SBE = 1.086360377 — MCPER FUNCTIONS

MU	K	RC = 02	RC = 05	RC = 07	RC = 10	RC = 15	RC = 20	RC = 25	RC = 35	RC = 50
	AMAX DB	.0017375255	.0108709564	.0213327376	.0436480540	.0988323393	.1772876696	.2802872360	.5675287486	1.249387366
	AMIN DB	45.31380269	55.02130509	58.61082891	62.43890705	66.83836371	70.01434040	72.53194410	76.48945607	81.12909787
	K	.3839223D-1	.1280623D-1	.8530830D-2	.5530474D-2	.3360027D-2	.2344460D-2	.1762384D-2	.1125075D-2	.6644763D-3
1	PFR	.8685012189	.6349848772	.5612264672	.4881670797	.4102566381	.3575590980	.3177599670	.2587268852	.1955519804
1	PF1	.9258579066	.7564683569	.7030381705	.6511724077	.5979766274	.5638659197	.5394468115	.5059218458	.4745206040
1	PQ1	.7450726454	.7804282383	.8066019366	.8457140564	.9106014715	.9764638896	1.044712076	1.193423846	1.469594595
1	PF2	.9970871806	.9133583238	.8855632460	.8580898437	.8294769529	.8109022096	.7974741871	.7787777095	.7607635416
1	PQ2	1.710827811	1.819946784	1.905541185	2.033105861	2.241614057	2.449156850	2.660534132	3.111330524	3.925756228
1	PF3	1.032638717	.9962890346	.9835255691	.9705380097	.9565724112	.9472239298	.9402964054	.9303453616	.9202703227
1	PQ3	4.443744624	4.635412852	4.823541901	5.114778972	5.601195088	6.090397291	6.590908749	7.661926170	9.602866796
2	PFC	1.056194242	1.040853243	1.035026309	1.028842690	1.021886672	1.017040205	1.013345185	1.007879569	1.002165144
2	PQC	18.16140745	17.14560861	17.16288703	17.41816105	18.04848615	18.76892093	19.53231207	21.16339920	23.97239559
1	ZF1	1.089185754	1.089404843	1.089481836	1.089561085	1.089648419	1.089708958	1.089755422	1.089825543	1.089902376
1	ZF2	1.115226563	1.117660832	1.118474191	1.119297613	1.120193379	1.120809154	1.121279666	1.121987217	1.122760132
1	ZF3	1.195698899	1.203234028	1.205803903	1.208429179	1.211310334	1.213305214	1.214836873	1.217151422	1.219693962
1	ZF4	1.441023083	1.457951997	1.463922006	1.470116763	1.477020281	1.481860050	1.485606777	1.491315164	1.497643485
1	ZF5	2.416680166	2.458524971	2.473544744	2.489270599	2.506957543	2.519453243	2.529177797	2.544072842	2.560685517

(5) P 9+2 — TH = 68 — SBE = 1.078534743 — MCPER FUNCTIONS

MU	K	RC = 02	RC = 05	RC = 07	RC = 10	RC = 15	RC = 20	RC = 25	RC = 35	RC = 50
	AMAX DB	.0017375255	.0108709564	.0213327376	.0436480540	.0988323393	.1772876696	.2802872360	.5675287486	1.249387366
	AMIN DB	43.31499951	53.01406777	56.60127590	60.42725237	64.82469383	67.99944612	70.51620132	74.47257738	79.11116761
	K	.4704538D-1	.1571470D-1	.1047300D-1	.6792600D-2	.4128789D-2	.2881786D-2	.2166826D-2	.1383757D-2	.8175661D-3
1	PFR	.8919506829	.6493003513	.5732145451	.4980789964	.4181797341	.3642538102	.3235823572	.2633327329	.1989430183
1	PF1	.9426270093	.7691645547	.7145309637	.6615437415	.6072487588	.5724613111	.5475714037	.5134187051	.4814490445
1	PQ1	.7545216808	.7875701069	.8132772836	.8520699173	.9168072340	.9827155631	1.051107371	1.200287704	1.477548828
1	PF2	1.003491124	.9202796596	.8925346003	.8650649510	.8364157681	.8177995440	.8043345499	.7855794930	.7675052717
1	PQ2	1.771406293	1.871197358	1.955427187	2.082673708	2.292382077	2.502027297	2.715994395	3.173055043	3.999878692
1	PF3	1.032865608	.9978346623	.9854526200	.9728174425	.9591954237	.9500591086	.9432808727	.9335345793	.9236589318
1	PQ3	4.698617187	4.860179663	5.045103552	5.337520382	5.831875683	6.332167688	6.845553495	7.946643353	9.045745205
2	PFC	1.052806055	1.038606873	1.033165104	1.027366372	1.020817127	1.016239189	1.012741278	1.007555835	1.002120590
2	PQC	19.71283207	18.39850547	18.35487522	18.56841261	19.17863218	19.90347582	20.68267082	22.36408648	25.28194485
1	ZF1	1.081159906	1.081370731	1.081445526	1.081522777	1.081608160	1.081667474	1.081713058	1.081781940	1.081857521
1	ZF2	1.105356192	1.107758639	1.108561898	1.109375484	1.110261061	1.110870172	1.111335791	1.112036327	1.112802071
1	ZF3	1.181107828	1.188566411	1.191116488	1.193724451	1.196589792	1.198575585	1.200101268	1.202408401	1.204944986
1	ZF4	1.415728974	1.432499149	1.438436704	1.444609175	1.451500448	1.456338957	1.460098584	1.465807494	1.472155510
1	ZF5	2.359630889	2.401221632	2.416210137	2.431933628	2.449652404	2.462191055	2.471959997	2.486940442	2.503671352

(5) P 9+2	TH = 69	SBE = 1.071144994						MCPER FUNCTIONS	
	RC = 02	RC = 05	RC = 07	RC = 10	RC = 15	RC = 20	RC = 25	RC = 35	RC = 50
AMAX DB	.0017375255	.0108709564	.0213327376	.0436480540	.0988323393	.1772876696	.2802872360	.5675287486	1.249387366
AMIN DB	41.31207029	51.00187098	54.58654274	58.41021824	62.80545373	65.97886568	68.49469197	72.44982438	77.08726315
MU I K	.5765643D-1	.1928792D-1	.1286048D-1	.8344997D-2	.5074931D-2	.3543361D-2	.2664943D-2	.1702496D-2	.1006288D-2
1 I PFR	.9174265020	.6646801176	.5860544850	.5086658343	.4266199540	.3713740957	.3297680787	.2682189096	.2025357369
1 I PF1	.9602503382	.7825319041	.7266318493	.6724623813	.6170073994	.5815054344	.5561183688	.5213027560	.4887328136
1 I PQ1	.7650579765	.7954113196	.8205738279	.8589888336	.9235358970	.9894774980	1.058012958	1.207682539	1.486100749
1 I PF2	1.009855815	.9273072782	.8996461958	.8722078368	.8435453024	.8248996626	.8114051569	.7926003001	.7744729124
1 I PQ2	1.839161021	1.927807098	2.010330085	2.137041766	2.347880011	2.559702559	2.776405360	3.240153729	4.080300491
1 I PF3	1.032982941	.9993304126	.9873496078	.9750860921	.9618272650	.9529155426	.9462954197	.9367659556	.9271011784
1 I PQ3	4.987126438	5.111687026	5.292192022	5.585140330	6.087511861	6.599556946	7.126776977	8.260448555	10.32290849
2 I PFC	1.049432706	1.036357180	1.031297394	1.025881742	1.019738804	1.015429872	1.012129812	1.007226151	1.002072335
2 I PQC	21.51266662	19.83178455	19.71290973	19.87385776	20.45631505	21.18299919	21.97773260	23.71248291	26.74882297
1 I ZF1	1.073576598	1.073778217	1.073850543	1.073925538	1.074008708	1.074066627	1.074111206	1.074178666	1.074252802
1 I ZF2	1.095973040	1.098338651	1.099130093	1.099932104	1.100805610	1.101406772	1.101866526	1.102558598	1.103315616
1 I ZF3	1.167087372	1.174452835	1.176978023	1.179563715	1.182408050	1.184381305	1.185898438	1.188194380	1.190721027
1 I ZF4	1.391125148	1.407704844	1.413600044	1.419740638	1.426609767	1.431440553	1.435188338	1.440910965	1.447271698
1 I ZF5	2.303591404	2.344873224	2.359813188	2.375517626	2.393251172	2.405821807	2.415627219	2.430681961	2.447520133

(5) P 9+2	TH = 70	SBE = 1.064177772						MCPER FUNCTIONS	
	RC = 02	RC = 05	RC = 07	RC = 10	RC = 15	RC = 20	RC = 25	RC = 35	RC = 50
AMAX DB	.0017375255	.0108709564	.0213327376	.0436480540	.0988323393	.1772876696	.2802872360	.5675287486	1.249387366
AMIN DB	39.30258556	48.98216358	52.56405066	56.38520165	60.77801752	63.94995939	66.46476681	70.41853501	75.05471053
MU I K	.7068618D-1	.2368440D-1	.1579985D-1	.1025743D-1	.6241255D-2	.4359248D-2	.3279442D-2	.2095394D-2	.1239332D-2
1 I PFR	.9452222345	.6812572454	.5998481341	.5200049328	.4356340969	.3789654760	.3363552779	.2734141158	.2063503100
1 I PF1	.9787777522	.7966220965	.7393894717	.6839732075	.6272930915	.5910359223	.5651232324	.5296066307	.4964018555
1 I PQ1	.7768744143	.8040600061	.8285835653	.8665507968	.9308586748	.9968173463	1.065495211	1.215675635	1.495324257
1 I PF2	1.016151747	.9344308176	.9068928179	.8795177427	.8508687776	.8322080186	.8186929077	.7998488170	.7816766299
1 I PQ2	1.915386943	1.990643245	2.071035274	2.196938035	2.408802445	2.622873821	2.842468721	3.313372758	4.167878911
1 I PF3	1.032981902	1.000768843	.9892099562	.9773383821	.9644635183	.9557896155	.9493370266	.9400373181	.9305957168
1 I PQ3	5.315887578	5.394723825	5.569251651	5.861854938	6.372218060	6.896714633	7.438829198	8.607909178	10.73964288
2 I PFC	1.046079955	1.034107278	1.029425520	1.024390449	1.018652699	1.014612876	1.011511157	1.006890580	1.002020179
2 I PQC	23.61794421	21.48310187	21.27066358	21.36510410	21.90981413	22.63476985	23.44436289	25.23549490	28.40148282
1 I ZF1	1.066422574	1.066614009	1.066683582	1.066756054	1.066836736	1.066893080	1.066936522	1.067002366	1.067074855
1 I ZF2	1.087064131	1.089387647	1.090165467	1.090954067	1.091813506	1.092405358	1.092858210	1.093540279	1.094286911
1 I ZF3	1.153623608	1.160878015	1.163372806	1.165930859	1.168748568	1.170705546	1.172211336	1.174491997	1.177004380
1 I ZF4	1.367184710	1.383540109	1.389382324	1.395480701	1.402316974	1.407133023	1.410873806	1.416592730	1.422958548
1 I ZF5	2.248479859	2.289395069	2.304268215	2.319935866	2.337665975	2.350256843	2.360090212	2.375207275	2.392140829

(5) P 9+2	TH = 71	SBE = 1.057620681						MCPER FUNCTIONS	
	RC = 02	RC = 05	RC = 07	RC = 10	RC = 15	RC = 20	RC = 25	RC = 35	RC = 50
AMAX DB	.0017375255	.0108709564	.0213327376	.0436480540	.0988323393	.1772876696	.2802872360	.5675287486	1.249387366
AMIN DB	37.28391037	46.95215593	50.53097693	54.34935123	58.73950702	61.90983291	64.47352035	68.37578881	73.01057543
MU K	.8671295D-1	.2910378D-1	.1942552D-1	.1261792D-1	.7681824D-2	.5367440D-2	.4039041D-2	.2582428D-2	.1527702D-2
1 PFR	.9756947744	.6991900694	.6147159590	.5321872800	.4452886457	.3870810708	.3433883757	.2739516823	.2104101094
1 PF1	.9982592603	.8114921006	.7528580523	.6961266300	.6381516232	.6010953863	.5746262805	.5383673946	.5044902217
1 PQ1	.7902114443	.8136479843	.8374179307	.8748521307	.9388608325	1.004815825	1.073633133	1.224346818	1.505306770
1 PF2	1.022344594	.9416373795	.9142674605	.8869928085	.8583889654	.8397299979	.8262048886	.8073342395	.7891273742
1 PQ2	2.001703940	2.060770379	2.138501832	2.263247259	2.475989586	2.692374777	2.915029140	3.393607858	4.263640379
1 PF3	1.032853259	1.002141857	.9910264339	.9795681136	.9670992084	.9586772099	.9524022197	.9433461140	.9341408994
1 PQ3	5.693298036	5.715262567	5.881797953	6.172872700	6.691054396	7.228729735	7.786911981	8.994599891	11.20238283
2 PFC	1.042754053	1.031860595	1.027552081	1.022894333	1.017559943	1.013788913	1.010885750	1.006549204	1.001963915
2 PQC	26.10312960	23.40066317	23.07100142	23.08090990	23.57475299	24.29303783	25.11621314	26.96671433	30.27485868
1 ZF1	1.059685563	1.059865801	1.059932327	1.060001995	1.060079903	1.060134484	1.060176649	1.060240676	1.060311304
1 ZF2	1.078617748	1.080893654	1.081655951	1.082429201	1.083272462	1.083853556	1.084298409	1.084968833	1.085703310
1 ZF3	1.140704507	1.147828419	1.150286847	1.152811442	1.155596429	1.157533070	1.159024487	1.161285417	1.163778828
1 ZF4	1.343882617	1.359977640	1.365755434	1.371800429	1.378592231	1.383385909	1.387114061	1.392821165	1.399183700
1 ZF5	2.194213988	2.234701723	2.249488655	2.265100624	2.282807806	2.295406277	2.305258428	2.320424857	2.337440881

(5) P 9+2	TH = 72	SBE = 1.051462224						MCPER FUNCTIONS	
	RC = 02	RC = 05	RC = 07	RC = 10	RC = 15	RC = 20	RC = 25	RC = 35	RC = 50
AMAX DB	.0017375255	.0108709564	.0213327376	.0436480540	.0988323393	.1772876696	.2802872360	.5675287486	1.249387366
AMIN DB	35.25316682	44.90876779	48.48420103	52.29951216	56.68673567	59.85528067	62.36773385	66.31834952	70.95160527
MU K	.1064664D 0	.3579887D-1	.2390788D-1	.1553819D-1	.9465323D-2	.6616244D-2	.4980265D-2	.3185617D-2	.1885415D-2
1 PFR	1.009282830	.7186687532	.6308017184	.5453208311	.4556620507	.3957833578	.3509195023	.2848706107	.2147424132
1 PF1	1.018743207	.8272046730	.7670981593	.7089794812	.6496349551	.6117323311	.5846734517	.5476273969	.5130368717
1 PQ1	.8053724136	.8243376132	.8472132413	.8840099277	.9476453885	1.013570096	1.082521576	1.233791601	1.516152555
1 PF2	1.028394413	.9489109987	.9217609137	.8946297794	.8661080191	.8474708260	.8339483327	.8150663009	.7968369674
1 PQ2	2.100163711	2.139509576	2.213913239	2.337055595	2.550467786	2.769220694	2.995113025	3.481944314	4.368825085
1 PF3	1.032587371	1.003440641	.9927910789	.9817683834	.9697287213	.9615736303	.9554869991	.9466893462	.9377347220
1 PQ3	6.130160612	6.080840516	6.236754533	6.524699865	7.050311438	7.601909705	8.177458425	9.427394914	11.71903973
2 PFC	1.039461804	1.029620926	1.025679967	1.021395456	1.016461811	1.012958798	1.010254097	1.006202141	1.001903325
2 PQC	29.06690125	25.64700622	25.16919647	25.07097258	25.49655509	26.20132155	27.03593356	28.94856654	32.41304488
1 ZF1	1.053354220	1.053522217	1.053585392	1.053651966	1.053726802	1.053779425	1.053820167	1.053882164	1.053950707
1 ZF2	1.070623403	1.072845909	1.073590680	1.074346535	1.075171384	1.075740186	1.076175871	1.076832903	1.077553337
1 ZF3	1.128319961	1.135292286	1.137707872	1.140192690	1.142938328	1.144850217	1.146323960	1.148560315	1.151029619
1 ZF4	1.321195596	1.336991692	1.342692745	1.348672274	1.355406976	1.360169953	1.363879317	1.369565701	1.375915742
1 ZF5	2.140710259	2.180706016	2.195386082	2.210922171	2.228585500	2.241177954	2.251038961	2.266240690	2.283325106

(5) P 9+2	TH = 73	SBE = 1.045691756						MCPER FUNCTIONS	
	RC = 02	RC = 05	RC = 07	RC = 10	RC = 15	RC = 20	RC = 25	RC = 35	RC = 50
AMAX DB	.0017375255	.0108709564	.0213327376	.0436480540	.0988323393	.1772876696	.2802872360	.5675287486	1.249387366
AMIN DB	33.20719257	42.84856368	46.42023738	50.23215718	54.61613813	57.78271468	60.29380377	64.24259273	68.87415638
MU I K	.1308727D 0	.4409188D-1	.2946431D-1	.1916095D-1	.1167959D-1	.8167478D-2	.6149884D-2	.3935601D-2	.2330448D-2
1 I PFR	1.046532319	.7399240672	.6482786654	.5595348820	.4668477190	.4051464658	.3590103628	.2912169236	.2193793225
1 I PF1	1.040273432	.8438288383	.7821775773	.7225960843	.6618023599	.6230022898	.5953174521	.5574353373	.5220866821
1 I PQ1	.8227451192	.8363311683	.8581380855	.8941680259	.9573380688	1.023198264	1.092275523	1.244125342	1.527987132
1 I PF2	1.034254710	.9562319858	.9293612462	.9024236354	.8740272535	.8554354437	.8419305609	.8230552993	.8048182063
1 I PQ2	2.213401121	2.228520096	2.298746920	2.419711482	2.633504438	2.854661165	3.083980777	3.579710663	4.484946706
1 I PF3	1.032174205	1.004655592	.9944951126	.9839314894	.9723457089	.9644735111	.9585867511	.9500634952	.9413747576
1 I PQ3	6.640580229	6.501095402	6.642923613	6.925564453	7.457903399	8.024164418	8.618518156	9.914867840	12.29945010
2 I PFC	1.036210628	1.027392473	1.023812403	1.019896135	1.015359757	1.012123468	1.009616791	1.005849543	1.001838185
2 I PQC	32.64220002	28.30446945	27.63754937	27.39990664	27.73392491	28.41565135	29.25828856	31.23532273	34.87223947
1 I ZF1	1.047418066	1.047572760	1.047632273	1.047695454	1.047766908	1.047817368	1.047856537	1.047916282	1.047982504
1 I ZF2	1.063071822	1.065234832	1.065959972	1.066696273	1.067500345	1.068055227	1.068480501	1.069122282	1.069826656
1 I ZF3	1.116461826	1.123259648	1.125625351	1.128063518	1.130762592	1.132644915	1.134097384	1.136303870	1.138743453
1 I ZF4	1.299102093	1.314557991	1.320168998	1.326069951	1.332733790	1.337456949	1.341140769	1.346796636	1.353124010
1 I ZF5	2.087882902	2.127317937	2.141869056	2.157307593	2.174904512	2.187476205	2.197335287	2.212556989	2.229694380

(5) P 9+2	TH = 74	SBE = 1.040299436						MCPER FUNCTIONS	
	RC = 02	RC = 05	RC = 07	RC = 10	RC = 15	RC = 20	RC = 25	RC = 35	RC = 50
AMAX DB	.0017375255	.0108709564	.0213327376	.0436480540	.0988323393	.1772876696	.2802872360	.5675287486	1.249387366
AMIN DB	31.14249690	40.76767155	44.33515011	48.14329882	52.52368066	55.69807393	58.19765036	62.14441379	66.77410101
MU I K	.1611150D 0	.5439639D-1	.3637425D-1	.2366983D-1	.1443779D-1	.1010085D-1	.7608242D-2	.4871299D-2	.2885036D-2
1 I PFR	1.088131888	.7632393205	.6673579066	.5749859193	.4789579856	.4152592100	.3677346979	.2980454384	.2243589610
1 I PF1	1.062884886	.8614402712	.7981722749	.7370495309	.6747218319	.6349692457	.6066191598	.5678476218	.5316917328
1 I PQ1	.8428326943	.8498838891	.8704035180	.9055051929	.9680940388	1.033845472	1.103035859	1.255488842	1.540963156
1 I PF2	1.039871302	.9635761071	.9370531510	.9103671136	.8921468559	.8636283361	.8501588938	.8313121191	.8130849813
1 I PQ2	2.344852849	2.329914223	2.394871297	2.512910040	2.726683618	2.950252505	3.183198296	3.688551941	4.613873602
1 I PF3	1.031603377	1.005776244	.9961288397	.9860488196	.9749429744	.9673707060	.9616961433	.9534644239	.9450580742
1 I PQ3	7.243281848	6.988532901	7.111655487	7.386015761	7.923920699	8.505543750	9.120292929	10.46784797	12.95599711
2 I PFC	1.033008643	1.025179916	1.021952999	1.018398988	1.014255436	1.011284008	1.008974527	1.005491607	1.001768266
2 I PQC	37.01145618	31.48326389	30.57215504	30.15303145	30.36387867	31.00923705	31.85462651	33.99740372	37.72507489
1 I ZF1	1.041867449	1.042007765	1.042063298	1.042122779	1.042190533	1.042238620	1.042276058	1.042333320	1.042396377
1 I ZF2	1.055954940	1.058052017	1.058755307	1.059469777	1.060250571	1.060739806	1.061203351	1.061827899	1.062514061
1 I ZF3	1.105124006	1.111722404	1.114030554	1.116414577	1.119059212	1.120906711	1.122333967	1.124504790	1.126908495
1 I ZF4	1.277582240	1.292653646	1.298160208	1.303968326	1.310546254	1.315219585	1.318870417	1.324484943	1.330778371
1 I ZF5	2.035642774	2.074443284	2.088841735	2.104159357	2.121665444	2.134200343	2.144045739	2.159270627	2.176444031

(5) P 9+2 — TH = 75 SBE = 1.035276180 — MCPER FUNCTIONS

MU	K	RC = 02	RC = 05	RC = 07	RC = 10	RC = 15	RC = 20	RC = 25	RC = 35	RC = 50
	AMAX DB	.0017375255	.0108709564	.0213327376	.0436480540	.0988323393	.1772876696	.2802872360	.5675287486	1.249387366
	AMIN DB	29.05521910	38.66167952	42.22444371	46.02837622	50.40474519	53.56670702	56.07459951	60.01910839	64.64670698
MU	K	.1987168D 0	.6724766D-1	.4499981D-1	.2930314D-1	.1788697D-1	.1252004D-1	.9433868D-2	.6043404D-2	.3582472D-2
1	PFR	1.134963562	.7889668727	.6882998677	.5918655797	.4921294747	.4262291691	.3771815817	.3054221347	.2297270725
1	PF1	1.086596882	.8801214550	.8151674464	.7524232124	.6884718460	.6477074327	.6186494183	.5789301048	.5419129715
1	PQ1	.8662988038	.8653224917	.8842773949	.9182465371	.9801072168	1.045692280	1.114977300	1.268055973	1.555268429
1	PF2	1.045180949	.9709135516	.9448171112	.9184500906	.8904655021	.8720532996	.8586405249	.8398482448	.8216524164
1	PQ2	2.499080351	2.446422005	2.504683900	2.618812307	2.832012221	3.057958933	3.294736648	3.810531378	4.757941154
1	PF3	1.030864205	1.006791178	.9976815348	.9881107202	.9775123353	.9702581542	.9648089950	.9568872602	.9487811337
1	PQ3	7.963593283	7.559649906	7.657820973	7.919789322	8.461422025	9.059005070	9.695901044	11.10020925	13.70449856
2	PFC	1.029864758	1.022988479	1.020105816	1.016906976	1.013150746	1.010441671	1.008328124	1.005128536	1.001693330
2	PQC	42.43024740	35.33366645	34.10304241	33.44497958	33.48917222	34.07932828	34.91942799	37.02765687	41.06690613
1	ZF1	1.036693488	1.036818353	1.036869585	1.036925057	1.036988787	1.037034283	1.037069828	1.037124369	1.037185205
1	ZF2	1.049265914	1.051290226	1.051969330	1.052659568	1.053414439	1.053936197	1.054336608	1.054941815	1.055607468
1	ZF3	1.094302562	1.100674405	1.102916642	1.105238339	1.107819921	1.109626836	1.111024560	1.113153359	1.115514414
1	ZF4	1.256617860	1.271257081	1.276643570	1.282343297	1.288818806	1.293431275	1.297040892	1.302602068	1.308848986
1	ZF5	1.983895999	2.021982009	2.036202148	2.051373532	2.068762204	2.081242783	2.091061593	2.106271190	2.123461843

(5) P 9+2 — TH = 76 SBE = 1.030613629 — MCPER FUNCTIONS

MU	K	RC = 02	RC = 05	RC = 07	RC = 10	RC = 15	RC = 20	RC = 25	RC = 35	RC = 50
	AMAX DB	.0017375255	.0108709564	.0213327376	.0436480540	.0988323393	.1772876696	.2802872360	.5675287486	1.249387366
	AMIN DB	26.94109980	36.52550344	40.08292073	43.88210689	48.25397701	51.41321789	53.91922713	57.86121538	62.48647904
MU	K	.2456563D 0	.8334493D-1	.5581481D-1	.3637312D-1	.2222011D-1	.1556119D-1	.1172997D-1	.7518612D-2	.4459650D-2
1	PFR	1.188176483	.8175514489	.7114303282	.6104116838	.5065304704	.4381882664	.3874599194	.3134273709	.2355391824
1	PF1	1.111402758	.8999614001	.8332585745	.7688126484	.7031435728	.6613036431	.6314913536	.5907603621	.5528223952
1	PQ1	.8940353320	.8830720052	.9001049521	.9326797422	.9936234083	1.058966431	1.128319462	1.282044314	1.571136996
1	PF2	1.050109684	.9782076157	.9526283221	.9266587727	.8989798419	.8807131204	.8673823353	.8486757593	.8305370330
1	PQ2	2.682258911	2.581632979	2.631312259	2.740217333	2.952072041	3.180296908	3.421113793	3.948274439	4.920110270
1	PF3	1.029945808	1.007687928	.9991413092	.9901063387	.9800444561	.9731277141	.9679181177	.9603262505	.9525396640
1	PQ3	8.836507779	8.236618417	8.301257327	8.545083532	9.087595114	9.701533153	10.36248857	11.83001327	14.56542380
2	PFC	1.026788778	1.020824019	1.018275436	1.015423473	1.012047868	1.009597918	1.007678546	1.004760804	1.001613137
2	PQC	49.26509781	40.06492359	38.40975791	37.43282084	37.24953645	37.75754026	38.58012765	40.75072516	45.02515250
1	ZF1	1.031888034	1.031996395	1.032043010	1.032094162	1.032153538	1.032196224	1.032229710	1.032281283	1.032339033
1	ZF2	1.042999166	1.044943410	1.045595863	1.046259337	1.046985488	1.047487828	1.047873618	1.048457237	1.049099927
1	ZF3	1.083995862	1.090111599	1.092278792	1.094529212	1.097038301	1.098798305	1.100161749	1.102241515	1.104552449
1	ZF4	1.236192517	1.250347976	1.255597376	1.261171684	1.267526594	1.272065991	1.275625251	1.281119691	1.287306033
1	ZF5	1.932542350	1.969826126	1.983840024	1.998837533	2.016079673	2.028486655	2.038264643	2.053438482	2.070625499

(5) P 9+2 — TH = 77 — SBE = 1.026304108 — MCPER FUNCTIONS

MU	K	RC = 02	RC = 05	RC = 07	RC = 10	RC = 15	RC = 20	RC = 25	RC = 35	RC = 50
	AMAX DB	.0017375255	.0108709564	.0213327376	.0436480540	.0988323393	.1772876696	.2802872360	.5675287486	1.249387366
	AMIN DB	24.79548798	34.35321534	37.90449370	41.69828919	46.06508097	49.22125905	51.72515060	55.66430597	60.28694609
	K	.3045273D 0	.1036115D 0	.6944597D-1	.4529349D-1	.2769327D-1	.1940520D-1	.1463377D-1	.9385677D-2	.5570718D-2
1	PFR	1.249296806	.8495638111	.7371633448	.6309238633	.5223712651	.4513005619	.3987047056	.3221603397	.2418635804
1	PF1	1.137253980	.9210545340	.8525523859	.7863276541	.7188436915	.6758602214	.6452434135	.6034307014	.5645059623
1	PQ1	.9272664075	.9036956359	.9183390529	.9491786998	1.008959250	1.073959719	1.143342691	1.297730247	1.588864856
1	PF2	1.054570752	.9854130078	.9604552829	.9349746202	.9076838012	.8896091261	.8763906251	.8578073180	.8397569445
1	PQ2	2.902939087	2.740360063	2.778913478	2.880816641	3.090242830	3.320545814	3.565600802	4.105177468	5.104196047
1	PF3	1.028837249	1.008452884	1.000494958	.9920234360	.9825286437	.9759699593	.9710151151	.9637745752	.9563284965
1	PQ3	9.911556481	9.049879394	9.068975696	9.286493002	9.825501170	10.45581527	11.14288608	12.68120542	15.56586563
2	PFC	1.023791539	1.018693136	1.016467062	1.013952337	1.010949330	1.008754456	1.007026942	1.004388672	1.001527450
2	PQC	58.05579933	45.97543268	43.74601154	42.33663786	41.83914190	42.22581941	43.01224292	45.23740335	49.77353837
1	ZF1	1.027443632	1.027534477	1.027576165	1.027622689	1.027677385	1.027717039	1.027748297	1.027796650	1.027851043
1	ZF2	1.037150438	1.039006737	1.039629938	1.040263973	1.040958450	1.041439317	1.041808904	1.042368545	1.042985650
1	ZF3	1.074204785	1.080032213	1.082114380	1.084283714	1.086709940	1.088416066	1.089739992	1.091762984	1.094015520
1	ZF4	1.216291640	1.229907229	1.235000940	1.240431109	1.246645316	1.251098059	1.254596759	1.260009462	1.266119390
1	ZF5	1.881473251	1.917857019	1.931633963	1.946427189	1.963490660	1.975802686	1.985524031	2.000639272	2.017799231

(5) P 9+2 — TH = 78 — SBE = 1.022340595 — MCPER FUNCTIONS

MU	K	RC = 02	RC = 05	RC = 07	RC = 10	RC = 15	RC = 20	RC = 25	RC = 35	RC = 50
	AMAX DB	.0017375255	.0108709564	.0213327376	.0436480540	.0988323393	.1772876696	.2802872360	.5675287486	1.249387366
	AMIN DB	22.61343051	32.13781955	35.68193302	39.46953419	43.83054347	46.98324895	49.48474339	53.42069503	58.04036956
	K	.3787693D 0	.1292818D 0	.8673292D-1	.5661944D-1	.3465072D-1	.2429550D-1	.1833009D-1	.1176430D-1	.6987438D-2
1	PFR	1.320395616	.8857506475	.7660348493	.6537862356	.5399190286	.4657733871	.4110859336	.3317453797	.2487855300
1	PF1	1.164035680	.9434980748	.8731674434	.8050948672	.7356979950	.6914990044	.6600234173	.6170522237	.5770675510
1	PQ1	.9677129343	.9279556424	.9395858592	.9682388205	1.026530231	1.091052795	1.160411227	1.315470963	1.608832715
1	PF2	1.058462059	.9924736291	.9682579276	.9433728990	.9165676202	.8987405536	.8856707190	.8672560796	.8493320866
1	PQ2	3.173270064	2.929203612	2.953133766	3.045580940	3.251038349	3.483063720	3.732529788	4.285717765	5.315206092
1	PF3	1.027527769	1.009071183	1.001727783	.9938481587	.9849525915	.9787739189	.9740901297	.9672241111	.9601413563
1	PQ3	11.26082468	10.04226419	9.998634280	10.17802368	10.70677245	11.35281823	12.06814832	13.68620233	16.74228215
2	PFC	1.020885066	1.016603301	1.014686627	1.012498009	1.009858072	1.007913299	1.006374679	1.004012710	1.001436032
2	PQC	69.62187869	53.50257232	50.47995171	48.47283073	47.53460529	47.74189745	48.46339500	50.72730242	55.55445071
1	ZF1	1.023353472	1.023425862	1.023462330	1.023503932	1.023553627	1.023590027	1.023618889	1.023663769	1.023714526
1	ZF2	1.031716902	1.033476651	1.034067841	1.034669616	1.035329299	1.035786516	1.036138221	1.036671342	1.037260065
1	ZF3	1.064932978	1.070437007	1.072423227	1.074500713	1.076832653	1.078477210	1.079755922	1.081713453	1.093893390
1	ZF4	1.196902749	1.209916939	1.214834527	1.220099885	1.226151039	1.230501941	1.233928624	1.239242670	1.245258233
1	ZF5	1.830569327	1.865941893	1.879447700	1.894002839	1.910851842	1.923045038	1.932691995	1.947722936	1.964829344

(5) P 9+2 — TH = 79 SBE = 1.018716695 — MCPER FUNCTIONS

MU	9+2	RC = 02	RC = 05	RC = 07	RC = 10	RC = 15	RC = 20	RC = 25	RC = 35	RC = 50
	AMAX DB	.0017375255	.0108709564	.0213327376	.0436480540	.0988323393	.1772876696	.2802872360	.5675287486	1.249387366
	AMIN DB	20.38994051	29.87095947	33.40652306	37.18689412	41.54124417	44.68997652	47.18873534	51.12103632	55.73733453
MU	K	.4730002D 0	.1620305D 0	.1088172D 0	.7110744D-1	.4356242D-1	.3056487D-1	.2307178D-1	.1431345D-1	.8808244D-2
1	PFR	1.404352676	.9271106737	.7987533191	.6795012304	.5595197626	.4818736827	.4248216129	.3423411497	.2564133572
1	PF1	1.191528888	.9673866412	.8952338508	.8252605895	.7538560483	.7083665870	.6759740511	.6317604146	.5906344587
1	PQ1	1.017861299	.9569092946	.9646759519	.9905313975	1.046893357	1.110752678	1.180008091	1.335737376	1.631539867
1	PF2	1.061663028	.9993196229	.9759851016	.9518206925	.9256165059	.9081036417	.8952263820	.8770355674	.8592844926
1	PQ2	3.511036029	3.157455612	3.161836761	3.241366775	3.440628463	3.673775130	3.927767709	4.495928185	5.559855831
1	PF3	1.026007157	1.009526603	1.002823385	.9955647604	.9373020588	.9815267484	.9771315153	.9706651232	.9639705857
1	PQ3	12.99265357	11.27578137	11.14419615	11.26795681	11.77592367	12.43588385	13.18157426	14.88996537	18.14499777
2	PFC	1.018082768	1.014563023	1.012940946	1.011065635	1.008777544	1.007076834	1.005723405	1.003633581	1.001338664
2	PQC	85.25196874	63.30821276	59.16250636	56.31004402	54.74154148	54.68133260	55.29280407	57.56538303	62.71524914
1	ZF1	1.019611351	1.019664462	1.019695443	1.019731849	1.019776238	1.019809170	1.019835469	1.019876624	1.019923466
1	ZF2	1.026697300	1.028350954	1.028907202	1.029473733	1.030095325	1.030526591	1.030858636	1.031362532	1.031919887
1	ZF3	1.056187178	1.061329614	1.063207933	1.065181743	1.067406797	1.068981265	1.070208128	1.072090339	1.074197896
1	ZF4	1.178015840	1.190360436	1.195079312	1.200156891	1.206019992	1.210251963	1.213593674	1.218739847	1.224690539
1	ZF5	1.779697352	1.813928986	1.827125012	1.841403996	1.857998194	1.870045585	1.879598051	1.894515461	1.911538040

(5) P 9+2 — TH = 80 SBE = 1.015426612 — MCPER FUNCTIONS

MU	9+2	RC = 02	RC = 05	RC = 07	RC = 10	RC = 15	RC = 20	RC = 25	RC = 35	RC = 50
	AMAX DB	.0017375255	.0108709564	.0213327376	.0436480540	.0988323393	.1772876696	.2802872360	.5675287486	1.249387366
	AMIN DB	18.12063856	27.54253470	31.06758563	34.83933557	39.18590003	42.33003296	44.82563756	48.75373948	53.36616010
MU	K	.5935121D 0	.2041684D 0	.1372784D 0	.8980653D-1	.5508164D-1	.3867646D-1	.2921117D-1	.1877691D-1	.1117067D-1
1	PFR	1.505283051	.9750146718	.8362787529	.7087417057	.5816317426	.4999527184	.4401973768	.3541543667	.2648875266
1	PF1	1.219352328	.9928018044	.9188910219	.8469937412	.7734972474	.7266414791	.6932704799	.6477230236	.6053652375
1	PQ1	1.081418137	.9920657963	.9947791252	1.016990274	1.070814359	1.133751424	1.202789363	1.359165022	1.657656274
1	PF2	1.064030774	1.005863371	.9835710855	.9602741188	.9348087038	.9176903046	.9050589389	.8871594112	.8696386208
1	PQ2	3.943184927	3.438606912	3.416302493	3.477903940	3.667679907	3.900951443	4.159471191	4.744148869	5.847380368
1	PF3	1.024266361	1.009801466	1.003763424	.9971552585	.9895604592	.9842132991	.9801254076	.9740358536	.9678067736
1	PQ3	15.27610681	12.84326754	12.58552650	12.62701061	13.09751217	13.76748801	14.54531268	16.35665441	19.84542188
2	PFC	1.015399679	1.012582053	1.011237896	1.009661224	1.007711825	1.006247920	1.005075120	1.003252135	1.001235148
2	PQC	107.0565251	76.43219539	70.64855036	66.56719293	64.07524258	63.60986151	64.03892287	66.26743897	71.76960659
1	ZF1	1.016211620	1.016244808	1.016270076	1.016301045	1.016339848	1.016369111	1.016392689	1.016429871	1.016472520
1	ZF2	1.022092148	1.023628927	1.024147097	1.024675224	1.025255248	1.025658130	1.025968633	1.026440435	1.026963237
1	ZF3	1.047977622	1.052716987	1.054474326	1.056331455	1.058435692	1.059930608	1.061098551	1.062895653	1.064913293
1	ZF4	1.159624017	1.171222358	1.175717343	1.180581437	1.186228323	1.190321976	1.193563956	1.198620242	1.204382429
1	ZF5	1.728706397	1.761640930	1.774482604	1.788441851	1.804735144	1.816605848	1.826040750	1.840810968	1.857714680

(5) P 9+2			TH = 81 SBE = 1.012465126							MCPER FUNCTIONS
			RC = 02	RC = 05	RC = 07	RC = 10	RC = 15	RC = 20	RC = 25	RC = 35	RC = 50
AMAX DB			.0017375255	.0108709564	.0213327376	.0436480540	.0988323393	.1772876696	.2802872360	.5675287486	1.249387366
AMIN DB			15.80315435	25.14021089	28.65180912	32.41297290	36.75024494	39.88896987	42.38088880	46.30410325	50.91202216
MU		K	.7490213D 0	.2589495D 0	.1743480D 0	.1142028D 0	.7013612D-1	.4928914D-1	.3724999D-1	.2396603D-1	.1427123D-1
1		PFR	1.629249021	1.031402317	.8799505958	.7424343286	.6068783679	.5204848694	.4575970860	.3674611128	.2743946383
1		PF1	1.246873520	1.019792249	.9442813771	.8704883370	.7948387189	.7465440129	.7121311413	.6651514768	.6214611903
1		PQ1	1.164111053	1.035654167	1.031597638	1.048955786	1.099377841	1.161021502	1.229671948	1.386635756	1.688105615
1		PF2	1.065395746	1.011993945	.9909306886	.9686743340	.9441126684	.9274861424	.9151659166	.8976408839	.8804217423
1		PQ2	4.512229427	3.792977837	3.733291454	3.769474066	3.944768609	4.176512696	4.439339418	5.042266980	6.190865461
1		PF3	1.022298518	1.009876588	1.004527358	.9985990069	.9917083260	.9868155493	.9830551481	.9774719639	.9716382460
1		PQ3	18.38691667	14.88939316	14.44546840	14.36268754	14.76858007	15.44090359	16.25170490	18.18097745	21.94830106
2		PFC	1.012852765	1.010671657	1.009586666	1.008291855	1.006665787	1.005430012	1.004432274	1.002369464	1.001125330
2		PQC	138.6605079	94.58277433	86.32314559	80.39418750	76.50757062	75.41410755	75.54094765	77.62739447	83.50606732
1		ZF1	1.013149121	1.013162019	1.013181407	1.013206746	1.013239725	1.013265138	1.013285348	1.013318826	1.013357013
1		ZF2	1.017904009	1.019311508	1.019788218	1.020274588	1.020809373	1.021181307	1.021468279	1.021904945	1.022389804
1		ZF3	1.040318508	1.044610013	1.046232082	1.047958230	1.049926223	1.051331047	1.052432048	1.054131538	1.056046735
1		ZF4	1.141724466	1.152488842	1.156731565	1.161353135	1.166751788	1.170684935	1.173810195	1.178701122	1.184297273
1		ZF5	1.677422860	1.708865295	1.721299856	1.734888401	1.750827143	1.762485217	1.771775652	1.786359311	1.803102986

(5) P 9+2			TH = 82 SBE = 1.009827573							MCPEP FUNCTIONS
			RC = 02	RC = 05	RC = 07	RC = 10	RC = 15	RC = 20	RC = 25	RC = 35	RC = 50
AMAX DB			.0017375255	.0108709564	.0213327376	.0436480540	.0988323393	.1772876696	.2802872360	.5675287486	1.249387366
AMIN DB			13.44003487	22.64882709	26.14229220	29.88992018	34.21577660	37.34800558	39.83554027	43.75297723	48.35559741
MU		K	.9518405D 0	.3310694D 0	.2232588D 0	.1464565D 0	.9007847D-1	.6336524D-1	.4792203D-1	.3086394D-1	.1839834D-1
1		PFR	1.785486542	1.099119885	.9317045907	.7818990936	.6361354234	.5441311213	.4775526714	.3826412907	.2851899297
1		PF1	1.273076629	1.048336145	.9715355195	.8959639591	.8181455501	.7683503276	.7328334686	.6843179478	.6391838349
1		PQ1	1.275161332	1.091104411	1.077709280	1.088425836	1.134176612	1.193978283	1.261982176	1.419414975	1.724203802
1		PF2	1.065557552	1.017569230	.9979521295	.9769416232	.9534827845	.9374673696	.9255388944	.9094920725	.8916643902
1		PQ2	5.288480047	4.252568028	4.138793367	4.137916730	4.290875274	4.518309387	4.784798313	5.407362376	6.609535820
1		PF3	1.020100766	1.009731304	1.005092144	.9998721597	.9937226016	.9893118298	.9859004909	.9808057550	.9754503446
1		PQ3	22.79895942	17.65082252	16.92201391	16.64594026	16.94148027	17.60142855	18.44381646	20.50858227	24.61357550
2		PFC	1.010461298	1.008844961	1.007998073	1.006965963	1.005645320	1.004627342	1.003797913	1.002436991	1.001009128
2		PQC	186.6497003	120.7221114	108.5509916	99.72594062	93.65069481	91.55185871	91.17004841	92.93418191	99.19116972
1		ZF1	1.010419080	1.010411769	1.010425196	1.010444783	1.010471762	1.010493178	1.010510895	1.010539461	1.010572936
1		ZF2	1.014137806	1.015401527	1.015833099	1.016274137	1.016759813	1.017099101	1.017359448	1.017757766	1.018201082
1		ZF3	1.033228462	1.037024332	1.038495574	1.040075039	1.041889694	1.043192654	1.044217709	1.045306036	1.047603992
1		ZF4	1.124319854	1.134147900	1.138105919	1.142451755	1.147565358	1.151312307	1.154301070	1.158996705	1.164394413
1		ZF5	1.625643604	1.655340682	1.667303533	1.680460108	1.695980402	1.707383130	1.716497098	1.730847259	1.747381598

(5) P 9+2 — TH = 83 — SBE = 1.007509825 — MCPER FUNCTIONS

MU	9+2	RC = 02	RC = 05	RC = 07	RC = 10	RC = 15	RC = 20	RC = 25	RC = 35	RC = 50
	AMAX DB	.0017375255	.0108709564	.0213327376	.0436480540	.0988323393	.1772876696	.2802872360	.5675287486	1.249387366
	AMIN DB	11.04447604	20.04979457	23.51717505	27.24651584	31.55776173	34.68194857	37.16413712	41.07459973	45.67036921
MU	K	.1219775D 1	.4275091D 0	.2888421D 0	.1898088D 0	.1169456D 0	.8235701D-1	.6233632D-1	.4019478D-1	.2398974D-1
1	PFR	1.988585497	1.182531597	.9944609760	.8290957686	.6706830131	.5718486507	.5008296131	.4002372404	.2976353012
1	PF1	1.296378599	1.078268934	1.000739307	.9236604007	.8437436488	.7924124994	.7557375165	.7055318939	.6598825777
1	PQ1	1.430125891	1.163966244	1.137205738	1.138520673	1.177656514	1.234773020	1.301721902	1.459395129	1.767902803
1	PF2	1.064283311	1.022404462	1.004486359	.9849663418	.9628526497	.9475958900	.9361599729	.9197223772	.9034008516
1	PQ2	6.394870637	4.870611499	4.675294106	4.618366009	4.736075809	4.954372199	5.223043159	5.868146377	7.132924082
1	PF3	1.017677428	1.009343724	1.005431968	1.000946999	.9955756728	.9916757434	.9886364781	.9840650422	.9792243718
1	PQ3	29.37928809	21.53867282	20.35329895	19.76422581	19.86862208	20.48762532	21.35507427	23.57508987	28.09787809
2	PFC	1.008247288	1.007117421	1.006485002	1.005693721	1.004657641	1.003845170	1.003175876	1.002106597	1.000386577
2	PQC	263.8154883	160.3408822	141.6395010	128.0291270	118.3462381	114.5666171	113.3014289	114.3976180	120.9762589
1	ZF1	1.008016939	1.007990250	1.007997759	1.008011575	1.008032465	1.008049788	1.008064419	1.008088403	1.008116949
1	ZF2	1.010801097	1.011904054	1.012286438	1.012678320	1.013110812	1.013412628	1.013646154	1.014002748	1.014400714
1	ZF3	1.026730765	1.029981479	1.031285075	1.032700680	1.034343067	1.035530993	1.036469957	1.037931744	1.039595531
1	ZF4	1.107420001	1.116190135	1.119825613	1.123857104	1.128642699	1.132173267	1.135002160	1.139467183	1.144627286
1	ZF5	1.573125034	1.600735443	1.612144024	1.624792257	1.639815641	1.650910861	1.659809312	1.673869608	1.690133137

(5) P 9+2 — TH = 84 — SBE = 1.005508280 — MCPER FUNCTIONS

MU	9+2	RC = 02	RC = 05	RC = 07	RC = 10	RC = 15	RC = 20	RC = 25	RC = 35	RC = 50
	AMAX DB	.0017375255	.0108709564	.0213327376	.0436480540	.0988323393	.1772876696	.2802872360	.5675287486	1.249387366
	AMIN DB	8.650699900	17.32087751	20.74771753	24.45047972	28.74189880	31.85568049	34.33111200	38.23290205	42.82136899
MU	K	.1579356D 1	.5590432D 0	.3786051D 0	.2493224D 0	.1539330D 0	.1085488D 0	.8224040D-1	.5310246D-1	.3173858D-1
1	PFR	2.262483957	1.288699671	1.072867248	.8870892784	.7124880980	.6050925897	.5285832494	.4210600274	.3122677083
1	PF1	1.314411654	1.109141917	1.031861239	.9538167071	.8720344482	.8191880026	.7813227561	.7294332960	.6810409061
1	PQ1	1.656684492	1.263799327	1.217007044	1.204409369	1.233791454	1.286859900	1.352079498	1.509558523	1.822249793
1	PF2	1.061314529	1.026255196	1.010330440	.9925944108	.9721240319	.9578110235	.9469957002	.9313357005	.9156696236
1	PQ2	8.061014841	5.742012124	5.417052960	5.271101003	5.331063744	5.531449447	5.799101671	6.467743364	7.809035824
1	PF3	1.015045592	1.008691479	1.005518089	1.001791098	.9972340270	.9938746042	.9912317842	.9872214617	.9829359761
1	PQ3	39.83943304	27.32353487	25.36088158	24.23644892	23.99777210	24.51824179	25.39220581	27.78697118	32.83971743
2	PFC	1.006235868	1.005507468	1.005063020	1.004487597	1.003711757	1.003090161	1.002571107	1.001730813	1.000757902
2	PQC	397.0893183	224.5112543	194.1070868	172.0305210	156.0045186	149.2440380	146.3667258	146.0933055	152.7846365
1	ZF1	1.005938057	1.005894120	1.005895934	1.005904106	1.005918952	1.005932156	1.005943656	1.005962952	1.005986401
1	ZF2	1.007904010	1.008826863	1.009155561	1.009494178	1.009869210	1.010131614	1.010335039	1.010646389	1.010995009
1	ZF3	1.020852778	1.023510462	1.024628502	1.025861612	1.027310839	1.028368989	1.029210396	1.030528106	1.032038201
1	ZF4	1.091042751	1.098610086	1.101877738	1.105548988	1.109955450	1.113233499	1.115874373	1.120065621	1.124940473
1	ZF5	1.519561065	1.544613526	1.555356513	1.567396547	1.581822686	1.592544371	1.601178010	1.614874028	1.630792212

(5) P 9+2	TH = 85	SBE = 1.003819838						MCPER FUNCTIONS	
	RC = 02	RC = 05	RC = 07	RC = 10	RC = 15	RC = 20	RC = 25	RC = 35	RC = 50
AMAX DB		.0108709564	.0213327376	.0436480540	.0988323393	.1772876696	.2802872360	.5675287486	1.249387366
AMIN DB		14.43769352	17.79583340	21.45619796	25.71838310	28.81778477	31.28420174	35.17472061	39.75325201
MU I K		.7431183D 0	.5048174D 0	.3333293D 0	.2063300D 0	.1457335D 0	.1105423D 0	.7149611D-1	.4280504D-1
1 I PFR		1.429846676	1.174842363	.9610060148	.7647758881	.6462196432	.5626690555	.4463981380	.3299323734
1 I PF1		1.139938567	1.064588890	.9866084896	.9035062845	.8492830293	.8102477664	.7565673768	.7063587578
1 I PQ1		1.408442607	1.329760681	1.295339960	1.309533041	1.356198088	1.418503267	1.574937889	1.892329994
1 I PF2		1.028791797	1.015200571	.9996028100	.9811476105	.9680147409	.9579859765	.9433249157	.9285135711
1 I PQ2		7.051845053	6.505736948	6.208656336	6.168502173	6.333915937	6.593541524	7.285589796	8.721900509
1 I PF3		1.007753664	1.005319106	1.002366353	.9986563302	.9958670904	.9936461638	.9902377932	.9865525477
1 I PQ3		36.60048917	33.20312652	31.09093861	30.19798344	30.49575567	31.32780014	33.90715257	39.65281705
2 I PFC		1.004037411	1.003751264	1.003363179	1.002819156	1.002370959	1.001990127	1.001363137	1.000623656
2 I PQC		338.1494392	284.7172220	246.2264739	218.0272643	205.5284859	199.4831295	196.2911858	202.4711203
1 I ZF1		1.004120419	1.004117042	1.004119914	1.004128944	1.004138113	1.004146508	1.004161101	1.004179371
1 I ZF2		1.006181054	1.006451078	1.006732026	1.007045149	1.007265125	1.007436115	1.007698608	1.007993747
1 I ZF3		1.017649967	1.018564006	1.019594777	1.020827994	1.021739911	1.022470794	1.023624322	1.024958014
1 I ZF4		1.081408730	1.084252626	1.087507407	1.091472224	1.094453366	1.096871434	1.100735269	1.105264804
1 I ZF5		1.486375838	1.496293325	1.507586191	1.521279503	1.531539876	1.539843559	1.553080815	1.568551629

(5) P 9+2	TH = 86	SBE = 1.002441898						MCPER FUNCTIONS	
	RC = 02	RC = 05	RC = 07	RC = 10	RC = 15	RC = 20	RC = 25	RC = 35	RC = 50
AMAX DB		.0108709564	.0213327376	.0436480540	.0988323393	.1772876696	.2802872360	.5675287486	1.249387366
AMIN DB		11.38123820	14.61204407	18.19669378	22.41051065	25.48794702	27.94127370	31.81608699	36.38124753
MU I K		.1009868D 1	.6889445D 0	.4565477D 0	.2835523D 0	.2006916D 0	.1524551D 0	.9881075D-1	.5978397D-1
1 I PFR		1.629014802	1.315148189	1.060228900	.8333211260	.6993895002	.6063296086	.4784744244	.3520713231
1 I PF1		1.168494343	1.097944809	1.021968249	.9387202693	.8835137144	.8434513770	.7830238573	.7359127502
1 I PQ1		1.634889379	1.501199786	1.429684469	1.418333986	1.454129751	1.511244173	1.664859590	1.987440450
1 I PF2		1.029562560	1.018686522	1.005655340	.9896867146	.9780431200	.9690220702	.9556601237	.9419789424
1 I PQ2		9.207352426	8.245868513	7.666931694	7.437739356	7.531519880	7.766729338	8.476603808	10.03431534
1 I PF3		1.006515754	1.004802563	1.002628079	.9997906133	.9975995354	.9958262795	.9930634238	.9900286994
1 I PQ3		53.14788058	46.78838018	42.64500346	40.37844687	40.15570691	40.81462284	43.54381231	50.22849087
2 I PFC		1.002734697	1.002573774	1.002340450	1.001994927	1.001699144	1.001441834	1.001008575	1.000484962
2 I PQC		566.3193481	461.3795886	386.7138209	332.0561425	307.1189649	294.1143497	284.1339766	287.9205153
1 I ZF1		1.002666413	1.002658811	1.002657054	1.002660777	1.002666154	1.002671577	1.002681593	1.002694726
1 I ZF2		1.003981739	1.004187815	1.004406486	1.004653184	1.004827728	1.004963966	1.005174010	1.005411501
1 I ZF3		1.012451299	1.013143965	1.013952172	1.014944940	1.015692444	1.016298186	1.017264426	1.018395045
1 I ZF4		1.064597738	1.066946741	1.069713412	1.073157202	1.075784911	1.077935628	1.081402791	1.085508534
1 I ZF5		1.425148744	1.433995157	1.444331440	1.457094227	1.466768249	1.474650129	1.487294739	1.502176732

$P^{(6)}_{9+4}$

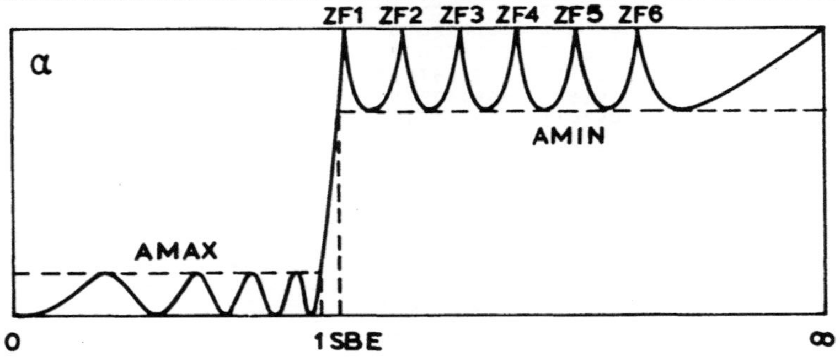

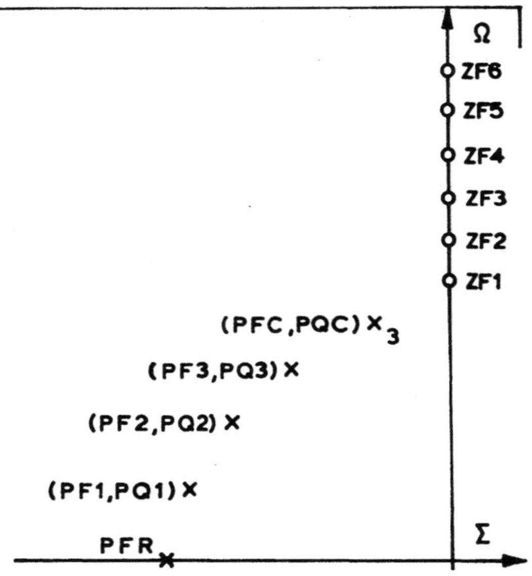

(6) P 9+4	TH = 50	SBE = 1.305407289						MCPER FUNCTIONS	
	RC = 02	RC = 05	RC = 07	RC = 10	RC = 15	RC = 20	RC = 25	RC = 35	RC = 50
AMAX DB	.0017375255	.0108709564	.0213327376	.0436480540	.0988323393	.1772876696	.2802872360	.5675287486	1.249387366
AMIN DB	90.80183310	102.4082578	106.7124746	111.3127457	116.6145143	120.4525892	123.5012311	128.3018870	133.9301879
MU I K	.3527439D-3	.9492451D-4	.5829767D-4	.3460900D-4	.1896649D-4	.1226777D-4	.8677364D-5	.5028633D-5	.2651073D-5
1 I PFR	.6003532389	.4541644946	.4047375152	.3543918505	.2993551608	.2614609667	.2325511152	.1893210013	.1477803778
1 I PF1	.6963577929	.5752950261	.5359694709	.4972110916	.4568069723	.4305071795	.4114594448	.3849331522	.3595124091
1 I PQ1	.6544502996	.7035908790	.7321064778	.7721940671	.8360851877	.8995363249	.9645937599	1.105213465	1.364718625
1 I PF2	.8698376736	.7844468210	.7566729186	.7292858278	.7006401808	.6818392000	.6680607694	.6484428561	.6287151074
1 I PQ2	1.140857745	1.287997552	1.370858908	1.484550128	1.660320517	1.829861832	1.999823524	2.357780154	2.997880518
1 I PF3	1.009704454	.9524480279	.9331965118	.9138461257	.8931191474	.8791566888	.8686807753	.8532718633	.8368965663
1 I PQ3	2.189458502	2.452508059	2.603238752	2.810445004	3.130494819	3.438600071	3.747021641	4.395977063	5.557232694
3 I PFC	1.143851137	1.104126248	1.089783829	1.074835892	1.058223536	1.046693456	1.037881813	1.024731524	1.010674686
3 I PQC	6.213926361	6.257500952	6.355702369	6.516526038	6.785239528	7.046802682	7.303157358	7.814289230	8.630357164
1 I ZF1	1.309991623	1.310270845	1.310366845	1.310465174	1.310573300	1.310648257	1.310705849	1.310792942	1.310888742
1 I ZF2	1.349617982	1.352329640	1.353252088	1.354193046	1.355224033	1.355936860	1.356483657	1.357309310	1.358216065
1 I ZF3	1.448263877	1.456676063	1.459518838	1.462411081	1.465572693	1.467754856	1.469426950	1.471949196	1.474716155
1 I ZF4	1.666771717	1.685555370	1.691925632	1.698416876	1.705523137	1.710433510	1.714198797	1.719882254	1.726121389
1 I ZF5	2.199612930	2.237523906	2.250494252	2.263762427	2.278341325	2.288444357	2.296205690	2.307941583	2.320849218
1 I ZF6	4.022514776	4.111039840	4.141551631	4.172869719	4.207393806	4.231380649	4.249838703	4.277794442	4.308595586

(6) P 9+4		TH = 51 SBE = 1.286759566								MCPER FUNCTIONS
		RC = 02	RC = 05	RC = 07	RC = 10	RC = 15	RC = 20	RC = 25	RC = 35	RC = 50
	AMAX DB	.0017375255	.0108709564	.0213327376	.0436480540	.0988323393	.1772876696	.2802872360	.5675287486	1.249387366
	AMIN DB	88.20509129	99.80269362	104.1044501	108.7024767	114.0020816	117.8388350	120.8865570	125.6859739	131.3131164
MU	K	.4621968D-3	.1246116D-3	.7657524D-4	.4548612D-4	.2494266D-4	.1613982D-4	.1141966D-4	.6620774D-5	.3492085D-5
1	PFR	.6082426885	.4594839809	.4093149251	.3582693402	.3025246428	.2641733941	.2349295346	.1912207441	.1441884004
1	PF1	.7040557995	.5811249447	.5412577421	.5019947201	.4610938800	.4344860987	.4152225561	.3884055775	.3627168574
1	PQ1	.6566749151	.7055397602	.7340047407	.7740694390	.8379795084	.9014815623	.9666079312	1.107405135	1.367279840
1	PF2	.8753895963	.7894680262	.7615157652	.7339533511	.7051274357	.6862114486	.6723512556	.6526220718	.6327910171
1	PQ2	1.154433547	1.300845490	1.383763457	1.497737974	1.674181153	1.844507322	2.015326661	2.375209305	3.018928039
1	PF3	1.011332182	.9546276185	.9355222546	.9163026941	.8957011046	.8818165528	.8713967636	.8560687989	.8397809377
1	PQ3	2.238989012	2.500291252	2.651494646	2.859995358	3.182784572	3.493959131	3.805680164	4.461961975	5.636876648
3	PFC	1.139670768	1.101432791	1.087561347	1.073071571	1.056932136	1.045709333	1.037121561	1.024288521	1.010549489
3	PQC	6.464301043	6.473711750	6.564047430	6.718993218	6.984111839	7.245276668	7.502745510	8.018538594	8.845536099
1	ZF1	1.291116500	1.291394161	1.291489969	1.291588257	1.291696503	1.291771636	1.291829412	1.291916358	1.292013145
1	ZF2	1.328795067	1.331506589	1.332431008	1.333374958	1.334410313	1.335126801	1.335676750	1.336507712	1.337421036
1	ZF3	1.422850202	1.431305356	1.434166616	1.437079726	1.440266532	1.442467539	1.444154877	1.446701445	1.449496921
1	ZF4	1.632359673	1.651279132	1.657706025	1.664260268	1.671441481	1.676407228	1.680216927	1.685970583	1.692291098
1	ZF5	2.146423695	2.184618928	2.197715820	2.211127782	2.225880594	2.236113481	2.243979551	2.255881724	2.268982777
1	ZF6	3.913375931	4.002556468	4.033375132	4.065047987	4.100007316	4.124322343	4.143046609	4.171427086	4.202724592

(6) P 9+4		TH = 52 SBE = 1.269018215								MCPER FUNCTIONS
		RC = 02	RC = 05	RC = 07	RC = 10	RC = 15	RC = 20	RC = 25	RC = 35	RC = 50
	AMAX DB	.0017375255	.0108709564	.0213327376	.0436480540	.0988323393	.1772876696	.2802872360	.5675287486	1.249387366
	AMIN DB	85.64065157	97.22871581	101.5278101	106.1234070	111.4206689	115.2559907	118.3027160	123.1097904	128.7266772
MU	K	.6034347D-3	.1630091D-3	.1002330D-3	.5957517D-4	.3268939D-4	.2116156D-4	.1497752D-4	.8687554D-5	.4584440D-5
1	PFR	.6165424013	.4650584443	.4141063349	.3623238467	.3058353974	.2670049057	.2374112547	.1932017820	.1456558228
1	PF1	.7120970283	.5872044685	.5467693837	.5069777175	.4655569805	.4386270761	.4191379307	.3920172132	.3660486143
1	PQ1	.6590421193	.7076004256	.7360081312	.7760452085	.8399717966	.9035252054	.9687223994	1.109703584	1.369963248
1	PF2	.8810899578	.7946410589	.7665090417	.7387690654	.7097598368	.6907265653	.6767828507	.6569398372	.6370029564
1	PQ2	1.168916169	1.314476165	1.397431236	1.511683478	1.688814438	1.859953314	2.031665111	2.393558669	3.041064503
1	PF3	1.012949721	.9568337108	.9378844948	.9188047391	.8983370674	.8845355944	.8741755402	.8589334717	.8427379380
1	PQ3	2.292113402	2.551287424	2.702915504	2.912716152	3.238333463	3.552708764	3.867884777	4.531860691	5.721151304
3	PFC	1.135439236	1.098696570	1.085300806	1.071274697	1.055614823	1.044704129	1.036344007	1.023333939	1.010418885
3	PQC	6.736779041	6.707476852	6.788848331	6.937017793	7.197823218	7.458270823	7.716721447	8.237197263	9.075555880
1	ZF1	1.273155488	1.273431037	1.273526488	1.273624576	1.273732781	1.273807985	1.273865367	1.273953556	1.274050213
1	ZF2	1.308948004	1.311655400	1.312580542	1.313526271	1.314564732	1.315284048	1.315836531	1.316671907	1.317590865
1	ZF3	1.398544438	1.407033444	1.409910305	1.412841459	1.416050504	1.418268400	1.419969534	1.422538318	1.425360120
1	ZF4	1.599307434	1.618344273	1.624822251	1.631434195	1.638684966	1.643702605	1.647554180	1.653374427	1.659772669
1	ZF5	2.095137823	2.133580550	2.146793388	2.160339300	2.175256357	2.185613078	2.193579627	2.205642733	2.218931068
1	ZF6	3.807857458	3.897610792	3.928712795	3.960718886	3.996092752	4.020723403	4.039705209	4.068498887	4.100282186

(6) P 9+4 — TH = 53 SBE = 1.252135658 — MCPER FUNCTIONS

MU	K	RC = 02	RC = 05	RC = 07	RC = 10	RC = 15	RC = 20	RC = 25	RC = 35	RC = 50
	AMAX DB	.0017375255	.0108709564	.0213327376	.0436480540	.0988323393	.1772876696	.2302872360	.5675287486	1.249387366
	AMIN DB	83.10649992	94.68425761	98.98047244	103.5734404	108.8681660	112.7019377	115.7475835	120.5442036	126.1687299
MU	K	.7851641D-3	.2125355D-3	.1307710D-3	.7777539D-4	.4270451D-4	.2765717D-4	.1958141D-4	.1136350D-4	.5999604D-5
1	PFR	.6252818857	.4709043399	.4191250851	.3665660390	.3092956492	.2699622814	.2400020632	.1952685683	.1471858285
1	PF1	.7205003359	.5935468429	.5525160803	.5121703546	.4702051983	.4429382254	.4232131260	.3957748824	.3695138400
1	PQ1	.6615648933	.7097819220	.7381248341	.7781288987	.8420691058	.9056741573	.9709440681	1.112115971	1.372776829
1	PF2	.8869390645	.7999690074	.7716563779	.7437369963	.7145416986	.6953889949	.6813600697	.6614007359	.6413555476
1	PQ2	1.184390000	1.328955907	1.411924946	1.526446986	1.704279928	1.876260183	2.048900874	2.412894989	3.064367063
1	PF3	1.014552602	.9590640387	.9402816234	.9213512611	.9010266270	.8873137653	.8770173073	.8618664209	.8457684243
1	PQ3	2.349190307	2.605793836	2.757786880	2.968886304	3.297420563	3.615134043	3.933929568	4.605991305	5.810425046
3	PFC	1.131159582	1.095918848	1.083003037	1.069445748	1.054271779	1.043677878	1.035549095	1.023367630	1.010282670
3	PQC	7.034009251	6.960737860	7.031877394	7.172231158	7.427885749	7.687236327	7.946502298	8.471654336	9.321815107
1	ZF1	1.256060729	1.256333598	1.256428519	1.256526245	1.256634239	1.256709405	1.256767313	1.256855126	1.256952033
1	ZF2	1.290026403	1.292725555	1.293650129	1.294596374	1.295636624	1.296357898	1.296912271	1.297751122	1.298674734
1	ZF3	1.375289750	1.383803127	1.386692578	1.389638822	1.392867014	1.395099748	1.396813156	1.399401943	1.402247761
1	ZF4	1.567544436	1.586679299	1.593202523	1.599866586	1.607181224	1.612247076	1.616137846	1.622020868	1.628492963
1	ZF5	2.045652853	2.084304561	2.097622197	2.111291692	2.126362782	2.136836967	2.144899485	2.157116318	2.170586942
1	ZF6	3.705754538	3.795994907	3.827355683	3.859672465	3.895439122	3.920372169	3.939602364	3.968797038	4.001054914

(6) P 9+4 — TH = 54 SBE = 1.236067977 — MCPER FUNCTIONS

MU	K	RC = 02	RC = 05	RC = 07	RC = 10	RC = 15	RC = 20	RC = 25	RC = 35	RC = 50
	AMAX DB	.0017375255	.0108709564	.0213327376	.0436480540	.0988323393	.1772876696	.2802872360	.5675287486	1.249387366
	AMIN DB	80.60068897	92.16731459	96.46041627	101.0505405	106.3425215	110.1746153	113.2190923	118.0141373	123.6371898
MU	K	.1018369D-2	.2762532D-3	.1700911D-3	.1012280D-3	.5562059D-4	.3603887D-4	.2552453D-4	.1481991D-4	.7828656D-5
1	PFR	.6344938540	.4770397663	.4243858145	.3710075992	.3129143843	.2730529188	.2427082687	.1974259516	.1487818794
1	PF1	.7292860675	.6001663919	.5585104704	.5175837414	.4750481849	.4474283198	.4274563123	.3996859596	.3731191905
1	PQ1	.6642578763	.7120943543	.7403639658	.7803288666	.8442792502	.9079360523	.9732805631	1.114650187	1.375729367
1	PF2	.8929367825	.8054549340	.7769614680	.7488613052	.7194775289	.7002034052	.6860876683	.6660096124	.6458536900
1	PQ2	1.200950283	1.344359000	1.427314637	1.542095763	1.720643853	1.893494972	2.067102720	2.433292186	3.088921038
1	PF3	1.016136034	.9613161124	.9427118481	.9239411170	.9037692735	.8901509438	.8799222131	.8648681583	.8488732511
1	PQ3	2.410626911	2.664145636	2.816430011	3.028818927	3.360358461	3.681553814	4.004143109	4.684708654	5.905108147
3	PFC	1.126835050	1.093100997	1.080668953	1.067585257	1.052903224	1.042630634	1.034736783	1.022889449	1.010140632
3	PQC	7.359055704	7.235720626	7.295160715	7.426493201	7.676018390	7.933819801	8.193693816	8.723480643	9.585892071
1	ZF1	1.239788060	1.240057660	1.240151874	1.240249065	1.240356674	1.240431687	1.240489536	1.240577353	1.240674382
1	ZF2	1.271983815	1.274670482	1.275593146	1.276538597	1.277579264	1.278301588	1.278857177	1.279698520	1.280625757
1	ZF3	1.353033839	1.361561728	1.364460628	1.367418875	1.370662969	1.372908390	1.374632471	1.377238934	1.380106334
1	ZF4	1.537005812	1.556218351	1.562780670	1.569490966	1.576863463	1.581973639	1.585900767	1.591842521	1.598384359
1	ZF5	1.997874308	2.036694650	2.050105353	2.063887499	2.079101835	2.089686739	2.097840444	2.110204915	2.123850949
1	ZF6	3.606877234	3.697515726	3.729109638	3.761713513	3.797850119	3.823071629	3.842540552	3.872123308	3.904843853

(6) P 9+4	TH = 55	SBE = 1.220774589						MCPER FUNCTIONS	
	RC = 02	RC = 05	RC = 07	RC = 10	RC = 15	RC = 20	RC = 25	RC = 35	RC = 50
AMAX DB	.0017375255	.0108709564	.0213327376	.0436480540	.0988323393	.1772876696	.2802872360	.5675287486	1.249387366
AMIN DB	78.12132852	89.67593484	93.96567167	98.55272085	103.8417327	107.6720105	110.7152222	115.5085618	121.1300181
MU I K	.1316888D-2	.3580359D-3	.2206016D-3	.1313802D-3	.7224064D-4	.4683041D-4	.3317963D-4	.1927474D-4	.1018759D-4
1 I PFR	.6442146732	.4834846800	.4299046256	.3756613483	.3167014428	.2762849075	.2455367632	.1996792214	.1504477472
1 I PF1	.7384761912	.6070786277	.5647662456	.5232299144	.4800963964	.4521068627	.4318763384	.4037584292	.3768718696
1 I PQ1	.6671376313	.7145490434	.7427357088	.7826544199	.8466109089	.9103193548	.9757403292	1.117314957	1.378830559
1 I PF2	.8990824537	.8111018473	.7824280569	.7541462864	.7245720354	.7051746983	.6909706578	.6707715911	.6505025812
1 I PQ2	1.218704867	1.360768891	1.443678808	1.558705009	1.737980077	1.911732334	2.086347139	2.454832349	3.114821027
1 I PF3	1.017694885	.9635871987	.9451731747	.9265730051	.9065643840	.8930469241	.8828903427	.8679391619	.8520532673
1 I PQ3	2.476887218	2.726721955	2.879207471	3.092866636	3.427498376	3.752325785	4.078893607	4.768409739	6.005658859
3 I PFC	1.122469099	1.090244510	1.078299558	1.065693823	1.051509412	1.041562475	1.033907043	1.022399250	1.009992554
3 I PQC	7.715477915	7.534986935	7.581023384	7.701932223	7.944182353	8.199896262	8.460121840	8.994458594	9.869573787
1 I ZF1	1.224296674	1.224562399	1.224655723	1.224752203	1.224859243	1.224933983	1.224991686	1.225079379	1.225176398
1 I ZF2	1.254777379	1.257447200	1.258366567	1.259309861	1.260349519	1.261071945	1.261628045	1.262470854	1.263400637
1 I ZF3	1.331728559	1.340260704	1.343165772	1.346132793	1.349389393	1.351645242	1.353378315	1.356000001	1.358886417
1 I ZF4	1.507631916	1.526900727	1.533495660	1.540245976	1.547669982	1.552820368	1.556780851	1.562777049	1.569384269
1 I ZF5	1.951714996	1.990661704	2.004153128	2.018036395	2.033382569	2.044071049	2.052310866	2.064815972	2.078630624
1 I ZF6	3.511049072	3.601993567	3.633793866	3.666660129	3.703142694	3.728637987	3.748335439	3.778292607	3.811463174

(6) P 9+4	TH = 56	SBE = 1.206217949						MCPER FUNCTIONS	
	RC = 02	RC = 05	RC = 07	RC = 10	RC = 15	RC = 20	RC = 25	RC = 35	RC = 50
AMAX DB	.0017375255	.0108709564	.0213327376	.0436480540	.0988323393	.1772876696	.2802872360	.5675287486	1.249387366
AMIN DB	75.66657650	87.20820934	91.49431014	96.07803477	101.3638349	105.1921476	108.2339899	113.0254832	118.6452108
MU I K	.1698143D-2	.4627794D-3	.2853504D-3	.1700653D-3	.9358346D-4	.6069679D-4	.4302029D-4	.2500518D-4	.1322404D-4
1 I PFR	.6544848966	.4902611451	.4356992767	.3805413922	.3206676263	.2796671143	.2484950934	.2020341616	.1521875512
1 I PF1	.7480944437	.6143003735	.5712982616	.5291219368	.4853611803	.4569841662	.4364828052	.4080009525	.3807796897
1 I PQ1	.6702229642	.7171587107	.7452514705	.7851159558	.8490737486	.9128334754	.9783327427	1.120119953	1.382091132
1 I PF2	.9053747979	.8169126692	.7880599223	.7595963623	.7298301316	.7103080224	.6960143200	.6756920959	.6553077404
1 I PQ2	1.237776312	1.378279627	1.461105702	1.576359050	1.756371225	1.931055641	2.106719453	2.477606395	3.142172204
1 I PF3	1.019223661	.9658742993	.9476633872	.9292454479	.9094112080	.8960014048	.8859217085	.8710798680	.8553093105
1 I PQ3	2.548502029	2.793953209	2.946529907	3.161427886	3.499236234	3.827852565	4.158595007	4.857540124	6.112590595
3 I PFC	1.118065420	1.087351012	1.075895956	1.063772113	1.050090642	1.040473507	1.033059864	1.021896892	1.009838210
3 I PQC	8.107429896	7.861497531	7.892144456	8.000993292	8.234623824	8.487603829	8.747870158	9.236618008	10.17489077
1 I ZF1	1.209548832	1.209810060	1.209902303	1.209997888	1.210104171	1.210178515	1.210235979	1.210323415	1.210420285
1 I ZF2	1.238367511	1.241016011	1.241930646	1.242870374	1.243907536	1.244629078	1.245184953	1.246028152	1.246959350
1 I ZF3	1.311329568	1.319855297	1.322763111	1.325735527	1.329001072	1.331264976	1.333005274	1.335639602	1.338542327
1 I ZF4	1.479367905	1.498670458	1.505291165	1.512074942	1.519543742	1.524729981	1.528720636	1.534766729	1.541434693
1 I ZF5	1.907094401	1.946123189	1.959682333	1.973654553	1.989120497	1.999904979	2.008225520	2.020863814	2.034839838
1 I ZF6	3.418105765	3.509260870	3.541239654	3.574342447	3.611145771	3.636899379	3.656814585	3.687131689	3.720738839

(6) P 9+4 — TH = 57 — SBE = 1.192363293 — MCPER FUNCTIONS

MU		RC = 02	RC = 05	RC = 07	RC = 10	RC = 15	RC = 20	RC = 25	RC = 35	RC = 50
	AMAX DB	.0017375255	.0108709564	.0213327376	.0436480540	.0988323393	.1772376696	.2802872360	.5675287486	1.249387366
	AMIN DB	73.23462997	84.76226240	89.04443475	93.62456555	98.90689211	102.7330786	105.7734386	110.5629329	116.1807835
	K	.2184044D-2	.5966683D-3	.3681927D-3	.2196056D-3	.1209411D-3	.7848216D-4	.5564807D-4	.3236361D-4	.1712592D-4
1	PFR	.6653498924	.4973936248	.4417894056	.3856632921	.3248248212	.2832092821	.2515915428	.2044971122	.1540058001
1	PF1	.7581664879	.6218499017	.5781226641	.5352740105	.4908548749	.4620714426	.4412861505	.4124229437	.3848511394
1	PQ1	.6735353065	.7199376992	.7479240708	.7877251236	.8516785676	.9154839064	.9810682440	1.123075922	1.385522992
1	PF2	.9118118008	.8228901942	.7938608522	.7652160757	.7352569423	.7156087843	.7012242243	.6807768721	.6602750344
1	PQ2	1.258304424	1.396997561	1.479694849	1.595152760	1.775910008	1.951558289	2.128315132	2.501717938	3.171091851
1	PF3	1.020716493	.9681741275	.9501800264	.9319567720	.9123088516	.8990139752	.8890162387	.8742906637	.8586422022
1	PQ3	2.626081063	2.866329891	3.018864129	3.234954542	3.576019925	3.908588856	4.243714255	4.952601545	6.226480447
3	PFC	1.113627950	1.084422268	1.073459359	1.061820873	1.048647262	1.039363865	1.032195251	1.021382239	1.009677365
3	PQC	8.539782941	8.218689421	8.231624134	8.326497155	8.549925837	8.799416822	9.059326283	9.602279336	10.50415985
1	ZF1	1.195509595	1.195765688	1.195856654	1.195951155	1.196056484	1.196130302	1.196187433	1.196274473	1.196371050
1	ZF2	1.222717627	1.225340221	1.226248645	1.227183343	1.228216468	1.228936096	1.229490978	1.230333443	1.231264871
1	ZF3	1.291796021	1.300304231	1.303211214	1.306185491	1.309456248	1.311725715	1.313471378	1.316115629	1.319031807
1	ZF4	1.452163365	1.471475930	1.478115194	1.484925496	1.492431981	1.497649457	1.501666904	1.507758061	1.514481837
1	ZF5	1.863938119	1.903002590	1.916615763	1.930664087	1.946237027	1.957109474	1.965505014	1.978268565	1.992398231
1	ZF6	3.327894056	3.419161044	3.451289216	3.484601475	3.521699082	3.547694699	3.567816275	3.598477981	3.632507423

(6) P 9+4 — TH = 58 — SBE = 1.179178403 — MCPER FUNCTIONS

MU		RC = 02	RC = 05	RC = 07	RC = 10	RC = 15	RC = 20	RC = 25	RC = 35	RC = 50
	AMAX DB	.0017375255	.0108709564	.0213327376	.0436480540	.0988323393	.1772876696	.2802872360	.5675287486	1.249387366
	AMIN DB	70.82371619	82.33624205	86.61417033	91.19041638	96.46898606	100.2928718	103.3316276	108.1189577	113.7347857
	K	.2802135D-2	.7675116D-3	.4740032D-3	.2829413D-3	.1559518D-3	.1012578D-3	.7182693D-4	.4179803D-4	.2213233D-4
1	PFR	.6768605910	.5049093254	.4481967921	.3910442619	.3291861443	.2869221455	.2548352281	.2070750400	.1559074418
1	PF1	.7687200841	.6297470893	.5852570314	.5417016053	.4965909225	.4673809072	.4462977454	.4170346570	.3890954631
1	PQ1	.6770991792	.7229022364	.7507679657	.7904950181	.8544374661	.9182973823	.9839584932	1.126194847	1.389139383
1	PF2	.9183905830	.8290370412	.7998346174	.7710100799	.7408578084	.7210826618	.7066062452	.6860320099	.6654107064
1	PQ2	1.280449324	1.417043411	1.499558908	1.615193242	1.796700816	1.973345256	2.151241346	2.527279907	3.201711164
1	PF3	1.022167109	.9704830830	.9527203657	.9347050874	.9152562599	.9020841007	.8921737643	.8775718765	.8620527409
1	PQ3	2.710327770	2.944413203	3.096742878	3.313961007	3.658358003	3.995050075	4.334779977	5.054160992	6.347979337
3	PFC	1.109160892	1.081460197	1.070991093	1.059840933	1.047179668	1.038233718	1.031313233	1.020855158	1.009509780
3	PQC	9.018278883	8.610571273	8.603066396	8.681712449	8.893071618	9.138154362	9.397237134	9.944106098	10.86002985
1	ZF1	1.182146591	1.182396898	1.182486385	1.182579606	1.182683779	1.182756937	1.182813635	1.182900135	1.182996267
1	ZF2	1.207793887	1.210385891	1.211286578	1.212214735	1.213242222	1.213958865	1.214511953	1.215352509	1.216282925
1	ZF3	1.273090298	1.281569431	1.284471848	1.287444288	1.290716341	1.292988754	1.294737824	1.297389134	1.300315748
1	ZF4	1.425971990	1.445269557	1.451919748	1.458749239	1.466285879	1.471529693	1.475570344	1.481701428	1.488475764
1	ZF5	1.822177377	1.861228922	1.874881696	1.888992551	1.904658953	1.915610832	1.924075281	1.936955637	1.951230686
1	ZF6	3.240270671	3.331547421	3.363794641	3.397288039	3.434652107	3.460872542	3.481188455	3.512178507	3.546615038

(6) P 9+4			TH = 59	SBE = 1.166633397						MCPER FUNCTIONS
			RC = 02	RC = 05	RC = 07	RC = 10	RC = 15	RC = 20	RC = 25	RC = 35	RC = 50
AMAX DB			.0017375255	.0108709564	.0213327376	.0436480540	.0988323393	.1772876696	.2802872360	.5675287486	1.249387366
AMIN DB			68.43208367	79.92831020	84.20165341	88.77370010	94.04820627	97.86960236	100.9066214	105.6916081	111.3052397
MU		K	.3587024D-2	.9851691D-3	.6089458D-3	.3637956D-3	.2006931D-3	.1303838D-3	.9252733D-4	.5387812D-4	.2854766D-4
1		PFR	.6890743778	.5128386009	.4549456667	.3967034019	.3337661111	.2908175648	.2582362105	.2097756222	.1578979208
1		PF1	.7797852713	.6380135931	.5927205361	.5484216051	.5025839995	.4729258978	.4515300053	.4218472873	.3935227505
1		PQ1	.6809427558	.7260707505	.7537995146	.7934404097	.8573640482	.9212720689	.9870165540	1.129490124	1.392955090
1		PF2	.9251072491	.8353555943	.8059849367	.7769831238	.7466382891	.7267356162	.7121665812	.6914639706	.6707214075
1		PQ2	1.304395185	1.438554731	1.520825888	1.636601861	1.818861601	1.996534946	2.175618818	2.554421464	3.234177388
1		PF3	1.023568826	.9727972243	.9552813846	.9374882635	.9182521968	.9052111058	.8953940045	.8309237632	.8655416949
1		PQ3	2.802057539	3.028848009	3.180776669	3.399035251	3.746830195	4.087822734	4.432392935	5.162861605	6.477824166
3		PFC	1.104668729	1.078466889	1.068492621	1.057833217	1.045688321	1.037083273	1.030413862	1.020315527	1.009335207
3		PQC	9.549722741	9.041841916	9.010681252	9.070444738	9.267522412	9.507102204	9.764777165	10.31516887	11.24555313
1		ZF1	1.169429801	1.169673659	1.169761461	1.169853200	1.169956006	1.170028366	1.170084528	1.170170338	1.170265867
1		ZF2	1.193564977	1.196121616	1.197012997	1.197933051	1.198953240	1.199665784	1.200216242	1.201053663	1.201981767
1		ZF3	1.255177762	1.263615784	1.266509734	1.269476465	1.272745707	1.275018312	1.276768731	1.279424079	1.282357943
1		ZF4	1.400751285	1.420007479	1.426660533	1.433501444	1.441060253	1.446325202	1.450385244	1.456550786	1.463370082
1		ZF5	1.781748591	1.820736284	1.834413451	1.848572487	1.864318006	1.875340248	1.883867120	1.896855266	1.911266964
1		ZF6	3.155101361	3.246282294	3.278616931	3.312261816	3.349863108	3.376290225	3.396787749	3.428088907	3.462916341

(6) P 9+4			TH = 60	SBE = 1.154700538						MCPER FUNCTIONS
			RC = 02	RC = 05	RC = 07	RC = 10	RC = 15	RC = 20	RC = 25	RC = 35	RC = 50
AMAX DB			.0017375255	.0108709564	.0213327376	.0436480540	.0988323393	.1772876696	.2802872360	.5675287486	1.249387366
AMIN DB			66.05799285	77.53663246	81.80502178	86.37252845	91.64263872	95.46134012	98.49647891	103.2789274	103.8901792
MU		K	.4582196D-2	.1262091D-2	.7808156D-3	.4668840D-3	.2578004D-3	.1675868D-3	.1189825D-3	.6932359D-4	.3675959D-4
1		PFR	.7020561675	.5212154345	.4620630758	.4026619729	.3385808353	.2949086838	.2618056271	.2126073430	.1599832450
1		PF1	.7913945605	.6466730481	.6005341293	.5554524757	.5088501653	.4787210111	.4569965180	.4268730857	.3981440414
1		PQ1	.6850985524	.7294642519	.7570373017	.7965780199	.8604736633	.9244277900	.9902571133	1.132976787	1.396986667
1		PF2	.9319567127	.8418479327	.8123154342	.7831400333	.7526041644	.7325739055	.7179117740	.6970796151	.6762142323
1		PQ2	1.330354816	1.461688912	1.543641820	1.659516674	1.842526158	2.021261416	2.201584044	2.583287809	3.268656381
1		PF3	1.024914526	.9751122394	.9578597399	.9403039022	.9212952223	.9083941539	.8986765500	.8843464973	.8691097933
1		PQ3	2.902220234	3.120378690	3.271668266	3.490852255	3.842100150	4.187577014	4.537238660	5.279435837	6.616852463
3		PFC	1.100156248	1.075444613	1.065965547	1.055798754	1.044173744	1.035912781	1.029497220	1.019763230	1.009153390
3		PQC	10.14222668	9.518038556	9.459412155	9.497147051	9.677313734	9.910076337	10.16563224	10.71902388	11.66425111
1		ZF1	1.157331372	1.157568107	1.157654011	1.157744062	1.157845285	1.157916703	1.157972222	1.158057186	1.158151948
1		ZF2	1.180001904	1.182518323	1.183398786	1.184309127	1.185320299	1.186027586	1.186574546	1.187407549	1.188331982
1		ZF3	1.238026546	1.246410923	1.249292327	1.252249295	1.255511415	1.257781319	1.259530918	1.262187119	1.265124867
1		ZF4	1.376462324	1.395649314	1.402296694	1.409140793	1.416713294	1.421993846	1.426069219	1.432263393	1.439121673
1		ZF5	1.742592971	1.781463462	1.795148987	1.809341027	1.825150443	1.836233400	1.844815785	1.857902091	1.872440778
1		ZF6	3.072260024	3.163236038	3.195625117	3.229390451	3.267198226	3.293812888	3.314478558	3.346072525	3.381273616

(6) P 9+4 — TH = 61 SBE = 1.143354068 — MCPER FUNCTIONS

MU	K	RC = 02	RC = 05	RC = 07	RC = 10	RC = 15	RC = 20	RC = 25	RC = 35	RC = 50
	AMAX DB	.0017375255	.0108709564	.0213327376	.0436480540	.0988323393	.1772876696	.2802872360	.5675287486	1.249387366
	AMIN DB	63.69970658	75.15936742	79.42240336	83.98500065	89.25035429	93.06613826	96.09924059	100.8789391	106.4876115
MU	K	.5842287D-2	.1614003D-2	.9994756D-3	.5981825D-3	.3306191D-3	.2150611D-3	.1527608D-3	.8907210D-4	.4726225D-4
1	PFR	.7158797031	.5300780130	.4695793162	.4089437215	.3436482637	.2992101159	.2655558450	.2155796086	.1621700645
1	PF1	.8035831353	.6557512906	.6087207510	.5628144574	.5154070343	.4847822610	.4627121920	.4321254939	.4029714472
1	PQ1	.6896042775	.7331067967	.7605025242	.7999268542	.8637836957	.9277812974	.9936967422	1.136671759	1.401252715
1	PF2	.9389324951	.8485157456	.8188295862	.7894856851	.7587614332	.7386040958	.7238487297	.7028862348	.6818967585
1	PQ2	1.358575321	1.486626845	1.568174020	1.684095393	1.867846866	2.047677050	2.229291963	2.614043433	3.305335667
1	PF3	1.026196644	.9774234130	.9604517338	.9431493084	.9243836673	.9116322265	.9020208436	.8878401541	.8727577156
1	PQ3	3.011928292	3.219868709	3.370230469	3.590190475	3.944931018	4.295082092	4.650102835	5.404721432	6.766020116
3	PFC	1.095628558	1.072395844	1.063411629	1.053738688	1.042636537	1.034722544	1.028563421	1.019198163	1.008964067
3	PQC	10.80552135	10.04572353	9.955095909	9.967058083	10.12717522	10.35153803	10.60410360	11.15981010	12.12021219
1	ZF1	1.145825441	1.146054371	1.146138162	1.146226313	1.146325730	1.146396058	1.146450823	1.146534779	1.146628603
1	ZF2	1.167077811	1.169549090	1.170416985	1.171315955	1.172316332	1.173017161	1.173559718	1.174386968	1.175306304
1	ZF3	1.221607362	1.229925031	1.232789627	1.235732586	1.238983058	1.241247218	1.242993712	1.245647404	1.248585475
1	ZF4	1.353069521	1.372157924	1.378790592	1.385629144	1.393206327	1.398496597	1.402582978	1.408799566	1.415690442
1	ZF5	1.704656175	1.743353582	1.757030548	1.771239531	1.787096682	1.798230081	1.806860605	1.820034775	1.834690408
1	ZF6	2.991627893	3.082286294	3.114695443	3.148548728	3.186530658	3.213312661	3.234132218	3.265999563	3.301555921

(6) P 9+4 — TH = 62 SBE = 1.132570051 — MCPER FUNCTIONS

MU	K	RC = 02	RC = 05	RC = 07	RC = 10	RC = 15	RC = 20	RC = 25	RC = 35	RC = 50
	AMAX DB	.0017375255	.0108709564	.0213327376	.0436480540	.0988323393	.1772876696	.2802872360	.5675287486	1.249387366
	AMIN DB	61.35547991	72.79465511	77.05190436	81.60919124	86.86939609	90.68202028	93.71291611	98.48963383	104.0955091
MU	K	.7435968D-2	.2060784D-2	.1277413D-2	.7652698D-3	.4233988D-3	.2755978D-3	.1958585D-3	.1142846D-3	.6068617D-4
1	PFR	.7306291384	.5394694167	.4775284546	.4155752683	.3489884554	.3037381650	.2695006443	.2187028815	.1644657648
1	PF1	.8163890564	.6652766113	.6173055699	.5705297868	.5222739756	.4911272624	.4686934282	.4376193004	.4080182916
1	PQ1	.6945038863	.7370260529	.7642194660	.8035086044	.8673139136	.9313515966	.9973542115	1.140594169	1.405774217
1	PF2	.9460264911	.8553602293	.8255306562	.7960249748	.7651163094	.7448330736	.7299847406	.7088915857	.6877770912
1	PQ2	1.389345131	1.513577416	1.594615076	1.710518992	1.894998029	2.075955806	2.258919196	2.646875462	3.344428141
1	PF3	1.027407151	.9797255924	.9630532783	.9460214570	.9275156043	.9149240967	.9054261584	.8914046932	.8764860800
1	PQ3	3.132492018	3.328324896	3.477408125	3.697952158	4.056204598	4.411224955	4.771890139	5.539680955	6.926423013
3	PFC	1.091091117	1.069323272	1.060832803	1.051654291	1.041077382	1.033512919	1.027612617	1.018620236	1.008766970
3	PQC	11.55135672	10.63272148	10.50466491	10.48637627	10.62268098	10.83673148	11.08523768	11.64237019	12.61820622
1	ZF1	1.134887981	1.135108422	1.135189881	1.135275914	1.135373297	1.135442382	1.135496280	1.135579060	1.135671768
1	ZF2	1.154767821	1.157188990	1.158042629	1.158928524	1.159916271	1.160609396	1.161146609	1.161966714	1.162879465
1	ZF3	1.205893333	1.214130678	1.216974008	1.219898508	1.223132580	1.225387792	1.227128773	1.229776408	1.232711035
1	ZF4	1.330540443	1.349499224	1.356107597	1.362931329	1.370503613	1.375797330	1.379890103	1.386122467	1.393039104
1	ZF5	1.667887997	1.706353789	1.720004343	1.734213260	1.750100972	1.761273864	1.769944650	1.783195661	1.797957349
1	ZF6	2.913092782	3.003317205	3.035710593	3.069617800	3.107739872	3.134667870	3.155626209	3.187746277	3.223638272

(6) P 9+4	TH = 63	SBE = 1.122326238						MCPER FUNCTIONS	
	RC = 02	RC = 05	RC = 07	RC = 10	RC = 15	RC = 20	RC = 25	RC = 35	RC = 50
AMAX DB	.0017375255	.0108709564	.0213327376	.0436480540	.0988323393	.1772876696	.2802872360	.5675287486	1.249387366
AMIN DB	59.02354929	70.44060453	74.69159631	79.24313657	84.49776563	88.30696587	91.33546975	96.10895481	101.7117950
MU I K	.9449604D-2	.2627580D-2	.1630456D-2	.9777682D-3	.5415453D-3	.3527507D-3	.2508201D-3	.1464666D-3	.7733709D-4
1 I PFR	.7464009795	.5494384532	.4859489536	.4225865732	.3546239158	.3085110881	.2736554354	.2219888415	.1668785767
1 I PF1	.8298534631	.6752800412	.6263162594	.5786229534	.5294723438	.4977754451	.4749583216	.4433708238	.4132992767
1 I PQ1	.6998488990	.7412539978	.7682160768	.8073481398	.8710868946	.9351603422	1.001250871	1.144765724	1.410574929
1 I PF2	.9532286995	.8623819644	.8324216144	.8027627735	.7716752130	.7512680553	.7363275071	.7151039257	.6938639122
1 I PQ2	1.423002819	1.542783073	1.623187761	1.738996135	1.924179969	2.106297189	2.290667991	2.681997716	3.386176548
1 I PF3	1.028537544	.9820131495	.9656598568	.9489169555	.9306888146	.9182683017	.9088915722	.8950399386	.8802954285
1 I PQ3	3.265464292	3.446927842	3.594305585	3.815188584	4.176945057	4.537033633	4.903647487	5.685425828	7.099323589
3 I PFC	1.086549755	1.066229832	1.058231194	1.049546980	1.039497056	1.032284327	1.026645001	1.018029375	1.008561823
3 I PQC	12.39402252	11.28842309	11.11640502	11.06248086	11.17043968	11.37185741	11.61498945	12.17240215	13.16382878
1 I ZF1	1.124496667	1.124707931	1.124786835	1.124870531	1.124965646	1.125033331	1.125086244	1.125167675	1.125259082
1 I ZF2	1.143048883	1.145414937	1.146252600	1.147123674	1.148096898	1.148781029	1.149311923	1.150123431	1.151028041
1 I ZF3	1.190859848	1.199002668	1.201820068	1.204721444	1.207934125	1.210177014	1.211909944	1.214547772	1.217474968
1 I ZF4	1.308845645	1.327642008	1.334215923	1.341014981	1.348572162	1.353862636	1.357956374	1.364197911	1.371132936
1 I ZF5	1.632242084	1.670414968	1.684020263	1.698211090	1.714111095	1.725311795	1.734014419	1.747330456	1.762186486
1 I ZF6	2.836548373	2.926218702	2.958558974	2.992484456	3.030710868	3.057762295	3.078843398	3.111194214	3.147400872

(6) P 9+4	TH = 64	SBE = 1.112601940						MCPER FUNCTIONS	
	RC = 02	RC = 05	RC = 07	RC = 10	RC = 15	RC = 20	RC = 25	RC = 35	RC = 50
AMAX DB	.0017375255	.0108709564	.0213327376	.0436480540	.0988323393	.1772876696	.2802872360	.5675287486	1.249387366
AMIN DB	56.70212069	68.09527985	72.33950167	76.88481999	82.13340742	85.93889523	88.96480449	93.73478139	99.33432630
MU I K	.1199189D-1	.3346243D-2	.2078688D-2	.1247911D-2	.6919442D-3	.4510526D-3	.3208938D-3	.1875366D-3	.9974634D-4
1 I PFR	.7633064846	.5600406717	.4948844293	.4300114964	.3605799976	.3135494115	.2780375205	.2254505774	.1694177084
1 I PF1	.8440207601	.6857956753	.6357833129	.5871209967	.5370257507	.5047483051	.4815268972	.4493981274	.4188306779
1 I PQ1	.7057000618	.7458277825	.7725246871	.8114741110	.8751285444	.9392323187	1.005411116	1.149211167	1.415681870
1 I PF2	.9605269117	.8695807666	.8395050392	.8097038742	.7784447562	.7579165965	.7428851612	.7215320560	.7001665360
1 I PQ2	1.459948230	1.574526760	1.654151140	1.769768659	1.955624061	2.138931125	2.324771070	2.719655687	3.430858972
1 I PF3	1.029578838	.9842799403	.9682664810	.9518320023	.9339007517	.9216631088	.9124159384	.8987455554	.8841862106
1 I PQ3	3.412697681	3.577070251	3.722221196	3.943131687	4.308348528	4.673706134	5.046592924	5.843246127	7.286183667
3 I PFC	1.082010704	1.063118719	1.055609139	1.047418327	1.037896442	1.031037260	1.025660818	1.017425527	1.008348344
3 I PQC	13.35103086	12.02417755	11.80028732	11.70421500	11.77833693	11.96429370	12.20042952	12.75665109	13.76368305
1 I ZF1	1.114630742	1.114832144	1.114908270	1.114989405	1.115082014	1.115148138	1.115199945	1.115279848	1.115369763
1 I ZF2	1.131899638	1.134205564	1.135025501	1.135879967	1.136836720	1.137510525	1.138034086	1.138835488	1.139730332
1 I ZF3	1.176484432	1.184517908	1.187304497	1.190177859	1.193363899	1.195590915	1.197313113	1.199937172	1.202852711
1 I ZF4	1.287958527	1.306557810	1.313086478	1.319850379	1.327381580	1.332661667	1.336752100	1.342994202	1.349939856
1 I ZF5	1.597675692	1.635491494	1.649031624	1.663185246	1.679078096	1.690294122	1.699019565	1.712387940	1.727325699
1 I ZF6	2.761893544	2.850885811	2.883134019	2.917040421	2.955333464	2.982484447	3.003671313	3.036229475	3.072728356

(6) P 9+4 — TH = 65 — SBE = 1.103377915 — MCPER FUNCTIONS

MU		RC = 02	RC = 05	RC = 07	RC = 10	RC = 15	RC = 20	RC = 25	RC = 35	RC = 50
	AMAX DB	.0017375255	.0108709564	.0213327376	.0436480540	.0988323393	.1772876696	.2802872360	.5675287486	1.249387366
	AMIN DB	54.38935667	65.75668490	69.99357775	74.53215511	79.77419162	83.57565129	86.59874393	91.36491075	96.96087473
MU	K	.1519980D-1	.4257174D-2	.2647637D-2	.1591279D-2	.8833826D-3	.5762958D-3	.4102344D-3	.2399514D-3	.1277369D-3
1	PFR	.7814746555	.5713396068	.5043845759	.4378884811	.3668853872	.3188763122	.2826664075	.2291028185	.1720935031
1	PF1	.8589387733	.6968610380	.6457404036	.5960538527	.5449603814	.5120696993	.4884213892	.4557212742	.4246305746
1	PQ1	.7121294596	.7507908114	.7771828977	.8159196970	.8794687368	.9435960324	1.009862950	1.153358835	1.421125910
1	PF2	.9679063519	.8769555071	.8467829951	.8168529218	.7854317214	.7647865968	.7496662894	.7281853655	.7066949730
1	PQ2	1.500656678	1.609140632	1.687808205	1.803118413	1.989598998	2.174124017	2.361497626	2.760132680	3.478795588
1	PF3	1.030521563	.9865192616	.9708676438	.9547623402	.9371484978	.9251064782	.9159978519	.9025210227	.8881587633
1	PQ3	3.576418116	3.720405730	3.862691041	4.083233980	4.451820348	4.822646754	5.202151613	6.014647819	7.488705382
3	PFC	1.077480623	1.059993422	1.052969208	1.045270087	1.036276541	1.029772296	1.024660371	1.016808663	1.008126249
3	PQC	14.44402058	12.85380466	12.56839889	12.42225180	12.45584712	12.62287835	12.85000931	13.40315382	14.42561190
1	ZF1	1.105270905	1.105461769	1.105534891	1.105613241	1.105703101	1.105767500	1.105818077	1.105896267	1.105984493
1	ZF2	1.121300301	1.123541094	1.124341533	1.125177567	1.126115850	1.126777953	1.127293132	1.128082860	1.128966244
1	ZF3	1.162746634	1.170655296	1.173405965	1.176246182	1.179400063	1.181607461	1.183316100	1.185922199	1.188821603
1	ZF4	1.267855221	1.286220781	1.292692741	1.299410324	1.306903939	1.312165999	1.316246983	1.322431993	1.329429779
1	ZF5	1.564149463	1.601541011	1.614994941	1.629091076	1.644956043	1.656174045	1.664912634	1.678319708	1.693325588
1	ZF6	2.689031725	2.777217995	2.809333516	2.843181677	2.881501602	2.908726863	2.930001429	2.962741982	2.999509051

(6) P 9+4 — TH = 66 — SBE = 1.094636279 — MCPER FUNCTIONS

MU		RC = 02	RC = 05	RC = 07	RC = 10	RC = 15	RC = 20	RC = 25	RC = 35	RC = 50
	AMAX DB	.0017375255	.0108709564	.0213327376	.0436480540	.0988323393	.1772876696	.2802872360	.5675287486	1.249387366
	AMIN DB	52.08336184	63.42274563	67.65169828	72.18296657	77.41789411	81.21497930	84.23501164	88.99703662	94.58910556
MU	K	.1924613D-1	.5411714D-2	.3369816D-2	.2027754D-2	.1127098D-2	.7358983D-3	.5241683D-3	.3068657D-3	.1635098D-3
1	PFR	.8010560006	.5834083146	.5145063027	.4462613885	.3735726966	.3245180813	.2875641924	.2329622145	.1749176313
1	PF1	.8746588506	.7085174952	.6562248281	.6054547578	.5533053693	.5197661954	.4956665669	.4623626289	.4307191261
1	PQ1	.7192232253	.7561941043	.7822346940	.8207235364	.8841421088	.9482844442	1.014638693	1.159041341	1.426942493
1	PF2	.9753492625	.8845038957	.8542568835	.8242143250	.7926430303	.7718863011	.7566799553	.7350738801	.7134600004
1	PQ2	1.545697207	1.647017107	1.724515522	1.839375853	2.026418633	2.212186296	2.401160805	2.803757451	3.530357043
1	PF3	1.031355765	.9887238036	.9734572669	.9577032036	.9404287161	.9285960185	.9196356098	.9063656013	.8922132871
1	PQ3	3.759320886	3.878911480	4.017544889	4.237219444	4.609022319	4.985512051	5.372002564	6.201399681	7.708882600
3	PFC	1.072966639	1.056857748	1.050314235	1.043104209	1.034638489	1.028490105	1.023644028	1.016178782	1.007895248
3	PQC	15.69996890	13.79426990	13.43550793	13.22957208	13.21443856	13.35827606	13.57390353	14.12155376	15.15898996
1	ZF1	1.096399211	1.096578870	1.096648765	1.096724102	1.096810968	1.096873476	1.096922695	1.096998983	1.097085316
1	ZF2	1.111232548	1.113403243	1.114182392	1.114998142	1.115915907	1.116564889	1.117070602	1.117847029	1.118717184
1	ZF3	1.149627932	1.157395624	1.160105022	1.162906709	1.166022628	1.168206459	1.169898551	1.172482255	1.175360772
1	ZF4	1.248514497	1.266607594	1.273010664	1.279670037	1.287113665	1.292349517	1.296415001	1.302634154	1.309574980
1	ZF5	1.531627230	1.568524233	1.581869729	1.595886837	1.611701808	1.622907486	1.631648834	1.645079917	1.660139213
1	ZF6	2.617870263	2.705118496	2.737058942	2.770807779	2.809112660	2.836385401	2.857728454	2.890624758	2.927634230

(6) P 9+4	TH = 67 SBE = 1.086360377							MCPER FUNCTIONS	
	RC = 02	RC = 05	RC = 07	RC = 10	RC = 15	RC = 20	RC = 25	RC = 35	RC = 50
AMAX DB	.0017375255	.0108709564	.0213327376	.0436480540	.0988323393	.1772876696	.2802872360	.5675287486	1.249387366
AMIN DB	49.78216659	61.09128987	65.31163217	69.83496800	75.06217343	78.85450322	81.87120701	86.62872469	92.21655221
MU I K	.2434931D-1	.6875247D-2	.4286722D-2	.2582772D-2	.1437500D-2	.9393855D-3	.6695425D-3	.3923411D-3	.2092587D-3
1 I PFR	.8222273127	.5963312870	.5253151435	.4551805282	.3806791914	.3305046930	.2927560265	.2370476773	.1779033230
1 I PF1	.8912358669	.7208107175	.6672779413	.6153607230	.5620932382	.5278674879	.5032901383	.4693472204	.4371189009
1 I PQ1	.7270850540	.7620980267	.7877318558	.8259309009	.8891890547	.9533358830	1.019775853	1.164496432	1.433172534
1 I PF2	.9828344263	.8922222178	.8619272569	.8317921442	.8000856998	.7792242958	.7639357213	.7422083169	.7204732439
1 I PQ2	1.595756351	1.688623010	1.764695519	1.878930946	2.066451887	2.253481948	2.444127105	2.850913801	3.585974938
1 I PF3	1.032071017	.9908856004	.9760286434	.9606492588	.9437375942	.9321289367	.9233271660	.9102782961	.8963498184
1 I PQ3	3.964697111	4.054969640	4.188978480	4.407148992	4.781933272	5.164269583	5.558134966	6.405592975	7.949066132
3 I PFC	1.068476375	1.053715857	1.047647340	1.040922869	1.032983577	1.027191467	1.022612234	1.015535919	1.007655048
3 I PQC	17.15283441	14.86658448	14.41981330	14.14209507	14.06810537	14.18345765	14.38445714	14.92351047	15.97511682
1 I ZF1	1.087998970	1.088166775	1.088233222	1.088305321	1.088388946	1.088449391	1.088497124	1.088571317	1.088655546
1 I ZF2	1.101679420	1.103775116	1.104531176	1.105324760	1.106219915	1.106854320	1.107349448	1.108110885	1.108965972
1 I ZF3	1.137111653	1.144721497	1.147384016	1.150141521	1.153213368	1.155369468	1.157041856	1.159598469	1.162451053
1 I ZF4	1.229917681	1.247697370	1.254018595	1.260607073	1.267987453	1.273188329	1.277231813	1.283425681	1.290349741
1 I ZF5	1.500075843	1.536404776	1.549618325	1.563533513	1.579274869	1.590452888	1.599185821	1.612625067	1.627721852
1 I ZF6	2.548319809	2.634493684	2.666214803	2.699821185	2.738066746	2.765358528	2.786749605	2.819773174	2.856997342

(6) P 9+4	TH = 68 SBE = 1.078534743							MCPER FUNCTIONS	
	RC = 02	RC = 05	RC = 07	RC = 10	RC = 15	RC = 20	RC = 25	RC = 35	RC = 50
AMAX DB	.0017375255	.0108709564	.0213327376	.0436480540	.0988323393	.1772876696	.2802872360	.5675287486	1.249387366
AMIN DB	47.48370860	58.76002369	62.97101868	67.48573592	72.70454368	76.49169790	79.50477701	84.25738356	89.84058653
MU I K	.3078595D-1	.8731254D-2	.5451469D-2	.3288965D-2	.1833119D-2	.1199027D-2	.8551867D-3	.5016239D-3	.2678221D-3
1 I PFR	.8451978018	.6102068594	.5368870194	.4647039415	.3882476929	.3368705068	.2982706932	.2413309020	.1810656553
1 I PF1	.9087280813	.7337911979	.6789457651	.6258130917	.5713604309	.5364068959	.5113232142	.4767031781	.4438552765
1 I PQ1	.7358408071	.7685745139	.7937357577	.8315951838	.8946569832	.9587951983	1.025318223	1.170363058	1.439863544
1 I PF2	.9903366158	.9001050150	.8697935938	.8395899484	.8077667810	.7868094950	.7714436673	.7496001429	.7277472698
1 I PQ2	1.651669374	1.734517861	1.808852300	1.922247149	2.110135396	2.298440631	2.490828334	2.902052706	3.646155079
1 I PF3	1.032656440	.9929959762	.9785743737	.9635945374	.9470707806	.9357019787	.9270700767	.9142578118	.9005681953
1 I PQ3	4.196602348	4.251474017	4.379647935	4.595505612	4.972927472	5.361273751	5.762925318	6.629718093	8.212047232
3 I PFC	1.064017994	1.050572301	1.044971968	1.038728492	1.031313268	1.025877285	1.021565524	1.014380152	1.007405360
3 I PQC	18.84581393	16.09701975	15.54395048	15.17952024	15.03407378	15.11433461	15.29677531	15.82323901	16.38771119
1 I ZF1	1.080054668	1.080209991	1.080272771	1.080341408	1.080421546	1.080479759	1.080525873	1.080597773	1.080679683
1 I ZF2	1.092625229	1.094641123	1.095372293	1.096141814	1.097012227	1.097630560	1.098113949	1.098858653	1.099696756
1 I ZF3	1.125182902	1.132617263	1.135227028	1.137934414	1.140955757	1.143079731	1.144729076	1.147253620	1.150074912
1 I ZF4	1.212048608	1.229471627	1.235697224	1.242201271	1.249504203	1.254660689	1.258675184	1.264333604	1.271730306
1 I ZF5	1.469465021	1.505149008	1.518205734	1.531994651	1.547637135	1.558771021	1.567483503	1.580913781	1.596030774
1 I ZF6	2.480293687	2.565252385	2.596707940	2.630126541	2.668265976	2.695546575	2.716963847	2.750084172	2.787493208

(6) P 9+4 — TH = 69 — SBE = 1.071144994 — MCPER FUNCTIONS

MU		RC = 02	RC = 05	RC = 07	RC = 10	RC = 15	RC = 20	RC = 25	RC = 35	RC = 50
	AMAX DB	.0017375255	.0108709564	.0213327376	.0436480540	.0988323393	.1772876696	.2802872360	.5675287486	1.249387366
	AMIN DB	45.18581173	56.42650351	60.62733797	65.13267896	70.34234225	74.12385590	77.13298234	81.88023012	87.45838334
	K	.3890721D-1	.1108664D-1	.6932255D-2	.4188336D-2	.2337868D-2	.1530682D-2	.1092528D-2	.6415159D-3	.3428865D-3
1	PFR	.8702170552	.6251502685	.5493104647	.4748990125	.3963277055	.3436551440	.3041413253	.2459863397	.1844219064
1	PF1	.9271967598	.7475148255	.6912796103	.6368581992	.5811479404	.5454219633	.5198008851	.4844622612	.4509569238
1	PQ1	.7456446189	.7757099580	.8003196917	.8377798114	.9006019217	.9647152235	1.031317281	1.176707719	1.447071041
1	PF2	.9978259606	.9081446978	.8778540006	.8476106338	.8156932768	.7946511172	.7792144054	.7572616390	.7352956914
1	PQ2	1.714461890	1.785377796	1.857592220	1.969879508	2.157989806	2.347573248	2.541776923	2.957707851	3.711494358
1	PF3	1.033100732	.9950454889	.9810862955	.9665323597	.9504233103	.9393113607	.9308614377	.9183024998	.9048680173
1	PQ3	4.460084346	4.471971794	4.592794501	4.805306336	5.184877262	5.579364781	5.989234289	6.876764069	8.501165644
3	PFC	1.059600238	1.047432066	1.042291923	1.036523787	1.029629225	1.024548607	1.020504537	1.014211609	1.007145892
3	PQC	20.83448685	17.51876778	16.83635906	16.36646559	16.13375159	16.17060886	16.32951131	16.83822116	17.91358197
1	ZF1	1.072551884	1.072694124	1.072753028	1.072817983	1.072894387	1.072950195	1.072994559	1.073063966	1.073143335
1	ZF2	1.084055475	1.085986899	1.086691388	1.087434936	1.088278445	1.088879178	1.089349644	1.090075811	1.090894941
1	ZF3	1.113828510	1.121068963	1.123619815	1.126270846	1.129234910	1.131322117	1.132944887	1.135432085	1.138216382
1	ZF4	1.194893582	1.211914255	1.218029555	1.224434713	1.231644979	1.236746955	1.240724931	1.246836935	1.253694813
1	ZF5	1.439767222	1.474725922	1.487599495	1.501236218	1.516752786	1.527824818	1.536503855	1.549906612	1.565025015
1	ZF6	2.413707263	2.497305183	2.528446817	2.561629948	2.599613704	2.626850947	2.648271088	2.681455431	2.719017158

(6) P 9+4 — TH = 70 — SBE = 1.064177772 — MCPER FUNCTIONS

MU		RC = 02	RC = 05	RC = 07	RC = 10	RC = 15	RC = 20	RC = 25	RC = 35	RC = 50
	AMAX DB	.0017375255	.0108709564	.0213327376	.0436480540	.0988323393	.1772876696	.2802872360	.5675287486	1.249387366
	AMIN DB	42.88616200	54.08810279	58.27787571	62.77300066	67.97269096	71.74804759	74.75285679	79.49424770	85.06687756
	K	.4916015D-1	.1407878D-1	.8816957D-2	.5335155D-2	.2982730D-2	.1954935D-2	.1396419D-2	.8208733D-3	.4392611D-3
1	PFR	.8975854980	.6412975809	.5626894683	.4858445139	.4049768408	.3509045902	.3104063072	.2508931025	.1879920007
1	PF1	.9467054425	.7620435113	.7043368272	.6485481537	.5915020719	.5549551871	.5287629178	.4926605068	.4584564033
1	PQ1	.7566871008	.7836089995	.8075718994	.8445607223	.9070905846	.9711586588	1.037833971	1.183576197	1.454860352
1	PF2	1.005267219	.9163310716	.8861048887	.8558561927	.8238720299	.8027586445	.7872590881	.7652059693	.7431332907
1	PQ2	1.785406020	1.842027221	1.911650982	2.022498369	2.210641022	2.401492198	2.597585806	3.018515706	3.782702538
1	PF3	1.033392220	.9970238701	.9835554059	.9694552484	.9537895187	.9429526884	.9346978095	.9224102950	.9092485963
1	PQ3	4.761495040	4.720854153	4.832411471	5.040251352	5.421289080	5.821999481	6.240535890	7.150349289	8.820451062
3	PFC	1.055232475	1.044300615	1.039611417	1.034311784	1.027933339	1.023206645	1.019430033	1.013530478	1.006876362
3	PQC	23.19126351	19.17424445	18.33316818	17.73402778	17.39402183	17.37692598	17.50593264	17.99013054	19.07352164
1	ZF1	1.065477219	1.065605813	1.065660637	1.065721695	1.065794125	1.065847357	1.065889838	1.065956549	1.066033153
1	ZF2	1.075956778	1.077799236	1.078475270	1.079190938	1.080005356	1.080586933	1.081043262	1.081749033	1.082547130
1	ZF3	1.103036991	1.110064288	1.112549772	1.115137898	1.118037546	1.120083081	1.121675537	1.124119788	1.126861022
1	ZF4	1.178441357	1.195011510	1.201000902	1.207291721	1.214392994	1.219429566	1.223362899	1.229416625	1.236223245
1	ZF5	1.410957531	1.445107034	1.457769573	1.471226478	1.486588136	1.497579213	1.506210750	1.519565847	1.534665166
1	ZF6	2.348477275	2.430563672	2.461340740	2.494238153	2.532013682	2.559173259	2.580571292	2.613784444	2.651464067

(6) P 9+4		TH = 71	SBE = 1.057620681						MCPER FUNCTIONS	
		RC = 02	RC = 05	RC = 07	RC = 10	RC = 15	RC = 20	RC = 25	RC = 35	RC = 50
AMAX DB		.0017375255	.0108709564	.0213327376	.0436480540	.0988323393	.1772876696	.2802872360	.5675287486	1.249387366
AMIN DB		40.58228026	51.74197164	55.91968023	60.40365435	65.59244862	69.36107232	72.36115734	77.09613496	82.66271170
MU	K	.6211571D-1	.1788489D-1	.1121929D-1	.6799850D-2	.3808029D-2	.2498631D-2	.1786253D-2	.1051284D-2	.5632514D-3
1	PFR	.9276683280	.6588107982	.5771471424	.4976332324	.4142626357	.3586725996	.3171104200	.2561342928	.1917990690
1	PF1	.9673186698	.7774458586	.7181816646	.6609417667	.6024753664	.5650549077	.5382546120	.5013390294	.4663909022
1	PQ1	.7692065215	.7923995738	.8155995775	.8520296219	.9142030732	.9782005178	1.044941028	1.191045788	1.463308937
1	PF2	1.012618942	.9246507587	.8945405061	.8643274174	.8323095727	.8111417605	.7955894059	.7734472558	.7512761592
1	PQ2	1.866097333	1.905481291	1.971929768	2.080920844	2.268848264	2.460938012	2.658994513	3.085241797	3.860630654
1	PF3	1.033518922	.9989199640	.9859717756	.9723548301	.9571629416	.9466208618	.9385751286	.9265786396	.9137088968
1	PQ3	5.108924676	5.003616356	5.103470689	5.304925289	5.686486120	6.093426246	6.521089666	7.454895402	9.174810650
3	PFC	1.050924746	1.041183952	1.036935113	1.032095875	1.026227760	1.021852806	1.018342914	1.012837023	1.006596495
3	PQC	26.01178504	21.11833436	20.08083602	19.32195307	18.84903142	18.76446313	18.85538463	19.30639161	20.39352290
1	ZF1	1.058818237	1.058932660	1.058983212	1.059040169	1.059108388	1.059158876	1.059199342	1.059263152	1.059336762
1	ZF2	1.068316803	1.070066013	1.070711853	1.071397745	1.072180875	1.072741714	1.073182665	1.073866133	1.074641061
1	ZF3	1.092798504	1.099592560	1.102005911	1.104524247	1.107351959	1.109350639	1.110908820	1.113304178	1.115995883
1	ZF4	1.162683133	1.178752039	1.184598909	1.190758872	1.197733617	1.202693043	1.206572956	1.212555551	1.219297396
1	ZF5	1.383013574	1.416266297	1.428688267	1.441935891	1.457111502	1.468000999	1.476569806	1.489855326	1.504913151
1	ZF6	2.284521132	2.364939632	2.395299007	2.427857664	2.465369126	2.492414368	2.513763485	2.546967491	2.584727283

(6) P 9+4		TH = 72	SBE = 1.051462224						MCPER FUNCTIONS	
		RC = 02	RC = 05	RC = 07	RC = 10	RC = 15	RC = 20	RC = 25	RC = 35	RC = 50
AMAX DB		.0017375255	.0108709564	.0213327376	.0436480540	.0988323393	.1772876696	.2802872360	.5675287486	1.249387366
AMIN DB		38.27149169	49.38498748	53.54950962	58.02128728	63.19815247	66.95939804	69.95430265	74.68224249	80.24217036
MU	K	.7850581D-1	.2273469D-1	.1428711D-1	.8674277D-2	.4866525D-2	.3196962D-2	.2287493D-2	.1347996D-2	.7231723D-3
1	PFR	.9609143521	.6778845826	.5928305200	.5103753806	.4242649004	.3670225016	.3243063121	.2617490666	.1958701663
1	PF1	.9890998939	.7937978527	.7328862524	.6741056627	.6141277283	.5757764061	.5483278606	.5105450160	.4748031568
1	PQ1	.7835042784	.8022397189	.8245342447	.8602983140	.9220364443	.9859313492	1.052726015	1.199203220	1.472509411
1	PF2	1.019832514	.9330864862	.9031524021	.8730235214	.8410119245	.8198102580	.8042175691	.7820006574	.7597418618
1	PQ2	1.958561962	1.977003729	2.039544138	2.146153138	2.333541594	2.526814877	2.726903875	3.158815547	3.946308551
1	PF3	1.033468646	1.000721663	.9883244542	.9752217230	.9605361977	.9503099622	.9424886014	.9308043912	.9182474632
1	PQ3	5.512816265	5.327218216	5.412234621	5.605073023	5.985857231	6.398922684	6.836173900	7.795861984	9.570281108
3	PFC	1.046687820	1.038088674	1.034268191	1.029879870	1.024514942	1.020488719	1.017244250	1.012131596	1.006306035
3	PQC	29.42430165	23.42304188	22.13990771	21.18170802	20.54270442	20.37315053	20.41533126	20.82151527	21.90645904
1	ZF1	1.052563397	1.052663174	1.052709277	1.052761939	1.052825720	1.052873299	1.052911618	1.052972325	1.053042711
1	ZF2	1.061124207	1.062776145	1.063390101	1.064044346	1.064793988	1.065332493	1.065756805	1.066416013	1.067165570
1	ZF3	1.083104835	1.089644715	1.091978848	1.094420165	1.097168013	1.099114360	1.100634069	1.102974213	1.105609497
1	ZF4	1.147612575	1.163126926	1.168813603	1.174825048	1.181654419	1.186524026	1.190341017	1.196238529	1.202900374
1	ZF5	1.355915450	1.388180043	1.400330142	1.413337026	1.428293103	1.439058706	1.447548231	1.460740264	1.475732022
1	ZF6	2.221756143	2.300344110	2.330229939	2.362393724	2.399581648	2.426473260	2.447744612	2.480898433	2.518697361

(6) P 9+4 — TH = 73 — SBE = 1.045691756 — MCPER FUNCTIONS

MU	K	RC = 02	RC = 05	RC = 07	RC = 10	RC = 15	RC = 20	RC = 25	RC = 35	RC = 50
	AMAX DB	.0017375255	.0108709564	.0213327376	.0436480540	.0988323393	.1772876696	.2802872360	.5675287486	1.249387366
	AMIN DB	35.95089285	47.01369396	51.16376606	55.62217102	60.78594479	64.53908569	67.52829570	72.24849333	77.80110095
MU	K	.9927274D-1	.2892741D-1	.1821372D-1	.1107893D-1	.6227661D-2	.4096350D-2	.2933783D-2	.1731201D-2	.9300600D-3
1	PFR	.9978818614	.6987552451	.6099169176	.5242030911	.4350787928	.3760295580	.3320564139	.2677836619	.2002371995
1	PF1	1.012108161	.8111835253	.7485317162	.6881156074	.6265278123	.5871832679	.5590424766	.5203329795	.4837426174
1	PQ1	.7999666788	.8133269063	.8345390381	.8695045485	.9307094954	.9944615386	1.061295383	1.208153529	1.482573548
1	PF2	1.026851052	.9416162107	.9119287316	.8819416499	.8499843200	.8287739028	.8131562669	.7908824517	.7685496268
1	PQ2	2.065408442	2.058186829	2.115891231	2.219448397	2.405872909	2.600238722	2.802422891	3.240377131	4.040995195
1	PF3	1.033229106	1.002415845	.9906013668	.9780454078	.9639008510	.9540131173	.9464325760	.9350837110	.9228623291
1	PQ3	5.986851848	5.700592397	5.766693634	5.947983869	6.326201601	6.745123447	7.192406461	8.180068120	10.01437270
3	PFC	1.042533245	1.035022049	1.031616407	1.027668049	1.022797686	1.019116276	1.016135311	1.011414656	1.006004746
3	PQC	33.60369700	26.18428552	24.59046602	23.38090294	22.53233677	22.25483413	22.23425072	22.58018241	23.65444409
1	ZF1	1.046702007	1.046786720	1.046828212	1.046876400	1.046935529	1.046980039	1.047016085	1.047073487	1.047140418
1	ZF2	1.054368579	1.055919528	1.056499975	1.057120742	1.057834714	1.058349279	1.058755674	1.059383627	1.060110544
1	ZF3	1.073949378	1.080213308	1.082460809	1.084817517	1.087477150	1.089365371	1.090842158	1.093120371	1.095691375
1	ZF4	1.133225832	1.148129773	1.153637479	1.159481520	1.166145247	1.170911338	1.174655104	1.180452355	1.187019120
1	ZF5	1.329645684	1.360826944	1.372671984	1.385404502	1.400104968	1.410722476	1.419114693	1.432187073	1.447085724
1	ZF6	2.160098671	2.236686351	2.266039749	2.297749135	2.334549993	2.361245745	2.382408185	2.415467302	2.453260582

(6) P 9+4 — TH = 74 — SBE = 1.040299436 — MCPER FUNCTIONS

MU	K	RC = 02	RC = 05	RC = 07	RC = 10	RC = 15	RC = 20	RC = 25	RC = 35	RC = 50
	AMAX DB	.0017375255	.0108709564	.0213327376	.0436480540	.0988323393	.1772876696	.2802872360	.5675287486	1.249387366
	AMIN DB	33.61731858	44.62422447	48.75841274	53.20211313	58.35147969	62.09569295	65.07862553	69.79028163	75.33480465
MU	K	.1256356D 0	.3685530D-1	.2325345D-1	.1417291D-1	.7983466D-2	.5258462D-2	.3769892D-2	.2227832D-2	.1198673D-2
1	PFR	1.039274807	.7217129540	.6286225062	.5392764249	.4468189010	.3857840813	.3404354634	.2742932624	.2049381444
1	PF1	1.036392941	.8296955113	.7652094197	.7030580997	.6397547419	.5993490955	.5704678689	.5307663528	.4932669395
1	PQ1	.8190961826	.8259110504	.8458188035	.8798200447	.9403692819	1.003927142	1.070779941	1.218025271	1.493637641
1	PF2	1.033608141	.9502120309	.9208533792	.8910762468	.8592308425	.8380422390	.8224185920	.8001101180	.7777205694
1	PQ2	2.190047265	2.151064166	2.202743777	2.302387083	2.487286358	2.682603366	2.886933008	3.331341524	4.146247193
1	PF3	1.032788092	1.003988311	.9927891997	.9808140801	.9672472466	.9577223399	.9504003873	.9394119252	.9275509057
1	PQ3	6.549257889	6.135375062	6.177190296	6.343036632	6.716214681	7.140480537	7.598193985	8.616140728	10.51654684
3	PFC	1.038473410	1.031992095	1.028986178	1.025465241	1.021079200	1.017737677	1.015017600	1.010686797	1.005692428
3	PQC	38.79293565	29.53203436	27.54020168	26.00979366	24.89384042	24.47786661	24.37582348	24.64071238	25.69220997
1	ZF1	1.041224171	1.041293462	1.041330206	1.041373758	1.041428037	1.041469327	1.041502976	1.041556878	1.041620126
1	ZF2	1.048040397	1.049486996	1.050032393	1.050617913	1.051294066	1.051783088	1.052170279	1.052774949	1.053466898
1	ZF3	1.065327125	1.071292534	1.073445657	1.075709796	1.078272418	1.080096389	1.081525535	1.083734670	1.086234535
1	ZF4	1.119521582	1.133756810	1.139065618	1.144722070	1.151198338	1.155846093	1.159505447	1.165185894	1.171639471
1	ZF5	1.304189217	1.334188007	1.345692780	1.358114947	1.372520858	1.382963961	1.391239177	1.404163179	1.418938867
1	ZF6	2.099463178	2.173872540	2.202631209	2.233822829	2.270168529	2.296622866	2.317642648	2.350558576	2.388297127

(6) P 9+4		TH = 75 SBE = 1.035276180							MCPER FUNCTIONS	
		RC = 02	RC = 05	RC = 07	RC = 10	RC = 15	RC = 20	RC = 25	RC = 35	RC = 50
AMAX DB		.0017375255	.0108709564	.0213327376	.0436480540	.0988323393	.1772876696	.2802872360	.5675287486	1.249387366
AMIN DB		31.26731449	42.21220537	46.32886776	50.75634366	55.88980300	59.62415015	62.60014021	67.30234187	72.83790667
MU	K	.1591804D 0	.4703601D-1	.2974335D-1	.1816788D-1	.1025686D-1	.6765877D-2	.4855880D-2	.2874120D-2	.1548918D-2
1	PFR	1.085994361	.7471186182	.6492140582	.5557915429	.4596247536	.3963956261	.3495338918	.2813444196	.2100186674
1	PF1	1.061986154	.8494353460	.7830223097	.7190322778	.6539002556	.6123596785	.5826851851	.5419195434	.5034439152
1	PQ1	.8415571612	.8403129964	.8586343077	.8914617018	.9512001518	1.014497931	1.081342396	1.228977661	1.505869815
1	PF2	1.040026384	.9588388350	.9299048425	.9004182276	.8687539271	.8476243086	.8320179137	.8097024158	.7872779542
1	PQ2	2.337015453	2.258272610	2.302384196	2.396990852	2.579617422	2.775673213	2.982177951	3.433487622	4.264013741
1	PF3	1.032133679	1.005423728	.9948732768	.9835144794	.9705643169	.9614283307	.9543841668	.9437833550	.9323098402
1	PQ3	7.224773430	6.646978880	6.657329485	6.802488811	7.167187032	7.595922462	8.064373085	9.115151700	11.08889162
3	PFC	1.034521600	1.029007672	1.026384671	1.023276898	1.019363167	1.016355488	1.013892903	1.009948773	1.005368924
3	PQC	45.33668912	33.64578960	31.13664015	29.19105555	27.72956373	27.13391800	26.92512100	27.08057699	28.09204007
1	ZF1	1.036120743	1.036174329	1.036206209	1.036244989	1.036294238	1.036332169	1.036363305	1.036413517	1.036472856
1	ZF2	1.042130981	1.043470279	1.043979191	1.044527777	1.045164016	1.045625910	1.045992609	1.046566947	1.047226552
1	ZF3	1.057234658	1.062878254	1.064928927	1.067092164	1.069548521	1.071301769	1.072678273	1.074810724	1.077230549
1	ZF4	1.106501080	1.120007046	1.125095853	1.130543162	1.136808493	1.141321857	1.144884629	1.150430197	1.156751256
1	ZF5	1.279533422	1.308246587	1.319373727	1.331446985	1.345516214	1.355756231	1.363892865	1.376636842	1.391256456
1	ZF6	2.039761128	2.111804272	2.139902007	2.170508123	2.206325372	2.232488948	2.253329338	2.286049040	2.323678819

(6) P 9+4		TH = 76 SBE = 1.030613629							MCPER FUNCTIONS	
		RC = 02	RC = 05	RC = 07	RC = 10	RC = 15	RC = 20	RC = 25	RC = 35	RC = 50
AMAX DB		.0017375255	.0108709564	.0213327376	.0436480540	.0988323393	.1772876696	.2802872360	.5675287486	1.249387366
AMIN DB		28.89712682	39.77263229	43.86986631	48.27936685	53.39519482	57.11859731	60.08688012	64.77857673	70.30417662
MU	K	.2019838D 0	.6015817D-1	.3813381D-1	.2334791D-1	.1321354D-1	.8730210D-2	.6273093D-2	.3719278D-2	.2007912D-2
1	PFR	1.139213967	.7754277103	.6720253666	.5739920274	.4736683964	.4079987247	.3594624429	.2890183461	.2155343224
1	PF1	1.088889950	.8705132374	.8020862950	.7361521957	.6690714088	.6263157776	.5957900840	.5538806186	.5143540151
1	PQ1	.8682444787	.8569513681	.8733226806	.9047075822	.9634365176	1.026388715	1.093187904	1.241210532	1.519480187
1	PF2	1.046015732	.9674526021	.9390547981	.9099538910	.8785536831	.8575282510	.8419676760	.8196794505	.7972475057
1	PQ2	2.512468400	2.383290536	2.417800110	2.505887324	2.685235268	2.881715825	3.090391786	3.549084749	4.396770842
1	PF3	1.031254511	1.006705585	.9968374265	.9861316926	.9738393471	.9651202396	.9583746081	.9481911011	.9371348365
1	PQ3	8.047696158	7.256207197	7.225336665	7.342645808	7.694032165	8.125819203	8.605145899	9.691543043	11.74709800
3	PFC	1.030692059	1.026078588	1.023819908	1.021109205	1.017653830	1.014972710	1.012763344	1.009201539	1.005034142
3	PQC	53.73554163	38.77886058	35.58614951	33.09485253	31.18024556	30.34832781	29.99797358	30.00462199	30.95114621
1	ZF1	1.031383287	1.031420963	1.031447895	1.031481792	1.031525861	1.031560307	1.031588823	1.031635167	1.031690380
1	ZF2	1.036632464	1.037861970	1.038333092	1.038843164	1.039437477	1.039870688	1.040215620	1.040757572	1.041382415
1	ZF3	1.049670139	1.054968047	1.056907892	1.058961524	1.061301894	1.062977585	1.064296145	1.066343819	1.068674614
1	ZF4	1.094168203	1.106882458	1.111728987	1.116944170	1.122973294	1.127334863	1.130787793	1.136178694	1.142345952
1	ZF5	1.255668168	1.282988453	1.293698231	1.305381250	1.319068123	1.329073692	1.337048010	1.349576949	1.364003609
1	ZF6	1.980899708	2.050376666	2.077742707	2.107690524	2.142900026	2.168719109	2.189339921	2.221805060	2.259266226

(6) P 9+4 — TH = 77 — SBE = 1.026304108 — MCPER FUNCTIONS

MU	K	RC = 02	RC = 05	RC = 07	RC = 10	RC = 15	RC = 20	RC = 25	RC = 35	RC = 50
	AMAX DB	.0017375255	.0108709564	.0213327376	.0436480540	.0988323393	.1772876696	.2802872360	.5675287486	1.249387366
	AMIN DB	26.50273518	37.29971053	41.37527931	45.76476424	50.86095950	54.57216650	57.53185494	62.21182687	67.72629165
	K	.2567851D 0	.7714637D-1	.4903255D-1	.3009810D-1	.1707914D-1	.1130394D-1	.8132905D-2	.4830893D-2	.2613017D-2
1	PFR	1.200491133	.8072246508	.6974807003	.5941849071	.4891650221	.4207608963	.3703586026	.2974154799	.2215535863
1	PF1	1.117058006	.8930468452	.8225315048	.7545495179	.6853940047	.6413367383	.6098963756	.5667548685	.5260937867
1	PQ1	.9003889482	.8763825107	.8903275186	.9199202257	.9773813172	1.039875604	1.106579189	1.254978520	1.534735276
1	PF2	1.051471564	.9759982552	.9482662442	.9196634730	.8886269619	.8677607267	.8522810814	.8300627138	.8076577771
1	PQ2	2.724946141	2.530797138	2.552976607	2.632554379	2.807252129	3.003695991	3.214485745	3.681076362	4.547715300
1	PF3	1.030140165	1.007816162	.9986638397	.9886489257	.9770576929	.9687853692	.9623606737	.9526267729	.9420204243
1	PQ3	9.066737738	7.991749384	7.906139547	7.985638606	8.316837001	8.749430571	9.239479188	10.36450329	12.51190427
3	PFC	1.027000043	1.023215725	1.021300896	1.018969191	1.015956093	1.013592864	1.011631460	1.008446299	1.004688074
3	PQC	64.73613529	45.29758084	41.18436411	37.96274226	35.44380537	34.29627931	33.75554192	33.55777298	34.40299331
1	ZF1	1.027004038	1.027025690	1.027047624	1.027076562	1.027115328	1.027146184	1.027171985	1.027214297	1.027265177
1	ZF2	1.031537760	1.032655498	1.033087684	1.033557797	1.034108283	1.034511310	1.034833224	1.035340746	1.035928387
1	ZF3	1.042633295	1.047561271	1.049381636	1.051316611	1.053530809	1.055121738	1.056376740	1.058331024	1.060563170
1	ZF4	1.082529505	1.094388225	1.098969064	1.103927675	1.109693418	1.113884310	1.117212932	1.122427455	1.128417399
1	ZF5	1.232585954	1.258401886	1.268652237	1.279900445	1.293155309	1.302892017	1.310677812	1.322952804	1.337145211
1	ZF6	1.922780319	1.989475953	2.016034118	2.045244891	2.079760317	2.105176039	2.125533026	2.157679033	2.194904391

(6) P 9+4 — TH = 78 — SBE = 1.022340595 — MCPER FUNCTIONS

MU	K	RC = 02	RC = 05	RC = 07	RC = 10	RC = 15	RC = 20	RC = 25	RC = 35	RC = 50
	AMAX DB	.0017375255	.0108709564	.0213327376	.0436480540	.0988323393	.1772876696	.2802872360	.5675287486	1.249387366
	AMIN DB	24.07997979	34.78664777	38.83787060	43.20492797	48.27913978	51.97668499	54.92673926	59.59355614	65.09551180
	K	.3272280D 0	.9925528D-1	.6326883D-1	.3894662D-1	.2216482D-1	.1469801D-1	.1058976D-1	.6303013D-2	.3416400D-2
1	PFR	1.271938183	.8432737244	.7261291612	.6167638952	.5063882426	.4348940900	.3823957520	.3066619546	.2281621455
1	PF1	1.146366959	.9171582345	.8445031085	.7743766518	.7030169838	.6575652452	.6251408743	.5806696211	.5387804866
1	PQ1	.9397241810	.8993615787	.9102444198	.9375815576	.9934333344	1.055319989	1.121858734	1.270611774	1.551973924
1	PF2	1.056272481	.9844069203	.9574910678	.9295192038	.8989660606	.8783260844	.8629706021	.8408750738	.8185405889
1	PQ2	2.986605043	2.707230488	2.713345636	2.781692445	2.949839674	3.145567418	3.358328011	3.833353748	4.721052344
1	PF3	1.028781643	1.008736537	1.000332924	.9910472394	.9802024352	.9724088065	.9663292240	.9570801373	.9469596573
1	PQ3	10.35302548	8.894160147	8.734656511	8.762207696	9.063269226	9.493142834	9.993256428	11.16007284	13.41129306
3	PFC	1.023461874	1.020431177	1.018837774	1.016864881	1.014275651	1.012220102	1.010500289	1.007684566	1.004330328
3	PQC	79.48644068	53.74684185	48.36652529	44.14685393	40.80583463	39.22887613	38.42769711	37.94528097	38.63521351
1	ZF1	1.022975362	1.022981477	1.022998404	1.023022345	1.023055730	1.023082911	1.023105918	1.023144054	1.023190411
1	ZF2	1.026840540	1.027845104	1.028237394	1.028666272	1.029171181	1.029542599	1.029840286	1.030311371	1.030859370
1	ZF3	1.036125388	1.040659135	1.042351161	1.044158120	1.046235515	1.047734105	1.048919613	1.050771354	1.052894552
1	ZF4	1.071594246	1.082533025	1.086823719	1.091499857	1.096973036	1.100972764	1.104161276	1.109175544	1.114962123
1	ZF5	1.210282126	1.234477847	1.244223883	1.254989441	1.267758161	1.277188100	1.284756299	1.296733877	1.310645506
1	ZF6	1.865296768	1.928976351	1.954643856	1.983031699	2.016758331	2.041705699	2.061749782	2.093504637	2.130420270

(6) P 9+4	TH = 79	SBE = 1.018716695							MCPER FUNCTIONS
	RC = 02	RC = 05	RC = 07	RC = 10	RC = 15	RC = 20	RC = 25	RC = 35	RC = 50
AMAX DB	.0017375255	.0108709564	.0213327376	.0436480540	.0988323393	.1772876696	.2802872360	.5675287486	1.249387366
AMIN DB	21.62488782	32.22538440	36.24896710	40.59069229	45.64011880	49.32226098	52.26144725	56.91341139	62.40122534
MU K	.4182063D 0	.1282090D 0	.8198922D-1	.5062796D-1	.2890569D-1	.1920851D-1	.1386101D-1	.8268504D-2	.4492042D-2
1 PFR	1.356490209	.8845967099	.7586964671	.6422440360	.5256926254	.4506714724	.3957965436	.3169190080	.2354691058
1 PF1	1.176572824	.9429675111	.8681610584	.7958102554	.7221180831	.6751736695	.6416899838	.5957808768	.5525585298
1 PQ1	.9887590773	.9269389693	.9338920188	.9583467440	1.012179222	1.073204900	1.139482267	1.288547039	1.571663536
1 PF2	1.060277808	.9925923870	.9666668178	.9394826600	.9095568932	.8892251425	.8740472240	.8521406737	.8299315528
1 PQ2	3.315268625	2.921683840	2.906499984	2.959810160	3.118722042	3.312724717	3.527175493	4.011176105	4.922434673
1 PF3	1.027172029	1.009446629	1.001823159	.9933052421	.9832539546	.9759729555	.9702645415	.9615386628	.9519437118
1 PQ3	12.01379939	10.02243652	9.761175750	9.716213803	9.972441483	10.39403762	10.90269713	12.11446371	14.48393902
3 PFC	1.020094980	1.017738409	1.016441991	1.014805468	1.012619149	1.010859345	1.009373489	1.006918251	1.003962673
3 PQC	99.81405355	64.96395496	57.79402273	52.17654671	47.69092348	45.51672982	44.35189839	43.46613510	43.91893991
1 ZF1	1.019292217	1.019281906	1.019293865	1.019312820	1.019340788	1.019364241	1.019384395	1.019418237	1.019459900
1 ZF2	1.022535220	1.023425824	1.023777480	1.024164057	1.024621832	1.024960322	1.025232637	1.025665345	1.026171286
1 ZF3	1.030149153	1.034264786	1.035819510	1.037488859	1.039418425	1.040816740	1.041926499	1.043665979	1.045669206
1 ZF4	1.061374390	1.071329387	1.075304619	1.079670988	1.084820348	1.088606694	1.091637821	1.096425524	1.101979770
1 ZF5	1.188755217	1.211210224	1.220404212	1.230635448	1.242858798	1.251940020	1.259258206	1.270887516	1.284467594
1 ZF6	1.808333055	1.868735892	1.893421706	1.920891956	1.953724869	1.978131465	1.997807691	2.029090316	2.065610777

(6) P 9+4	TH = 80	SBE = 1.015426612							MCPER FUNCTIONS
	RC = 02	RC = 05	RC = 07	RC = 10	RC = 15	RC = 20	RC = 25	RC = 35	RC = 50
AMAX DB	.0017375255	.0108709564	.0213327376	.0436480540	.0988323393	.1772876696	.2802872360	.5675287486	1.249387366
AMIN DB	19.13440877	29.60624534	33.59800306	37.91081231	42.93205161	46.59669006	49.52352230	54.15858966	59.63029464
MU K	.5363679D 0	.1664121D 0	.1068038D 0	.6618020D-1	.3792071D-1	.2525833D-1	.1825805D-1	.1091349D-1	.5946816D-2
1 PFR	1.458337954	.9325952716	.7961656176	.6713150526	.5475480089	.4684528213	.4108530487	.3283970793	.2436162542
1 PF1	1.207244886	.9705804182	.8936774790	.8190548564	.7429111660	.6943726803	.6597488130	.6122826679	.5676086980
1 PQ1	1.051238800	.9606182252	.9624268377	.9831302219	1.034208953	1.094191993	1.160071020	1.309375822	1.594398218
1 PF2	1.063324861	1.000446475	.9757123597	.9495010965	.9203763697	.9004533820	.8855192705	.8638846676	.8418706937
1 PQ2	3.737981426	3.187402499	3.143371475	3.176180372	3.321972319	3.512728950	3.728363902	4.221837855	5.159655843
1 PF3	1.025307409	1.009925335	1.003110967	.9953987349	.9861894019	.9794569399	.9741477111	.9659869102	.9569613433
1 PQ3	14.21689919	11.46535046	11.06051910	10.91224207	11.10135871	11.50580792	12.01999826	13.27953341	15.78482074
3 PFC	1.016917920	1.015152447	1.014126518	1.012801537	1.010994384	1.009516460	1.008255494	1.006149767	1.003584092
3 PQC	128.7481051	80.28563301	70.50992975	62.87687810	56.75282593	53.72594642	52.04058651	50.57050889	50.65880881
1 ZF1	1.015947110	1.015921140	1.015928224	1.015942259	1.015964836	1.015984541	1.016001808	1.016031270	1.016068101
1 ZF2	1.018616948	1.019393481	1.019704027	1.020047489	1.020456819	1.020761206	1.021007098	1.021399592	1.021861125
1 ZF3	1.024708692	1.028383399	1.029791911	1.031313950	1.033084353	1.034374135	1.035401578	1.037018516	1.038889991
1 ZF4	1.051884527	1.060794113	1.064428012	1.068456082	1.073248293	1.076797200	1.079652053	1.084184146	1.089473730
1 ZF5	1.168007452	1.188596207	1.197187264	1.206828273	1.218441202	1.227127045	1.234158847	1.245388593	1.258572776
1 ZF6	1.751760578	1.808590712	1.832193194	1.858640128	1.890461677	1.914245869	1.933492011	1.964210090	2.000237952

(6) P 9+4		TH = 81 SBE = 1.012465126								MCPER FUNCTIONS
		RC = 02	RC = 05	RC = 07	RC = 10	RC = 15	RC = 20	RC = 25	RC = 35	RC = 50
	AMAX DB	.0017375255	.0108709564	.0213327376	.0436480540	.0988323393	.1772876696	.2802872360	.5675287486	1.249387366
	AMIN DB	16.60797338	26.91750341	30.87188169	35.15120769	40.14002869	43.78457808	46.69723346	51.31289964	56.76608457
MU	K	.6908627D 0	.2172776D 0	.1400164D 0	.8709935D-1	.5010830D-1	.3346411D-1	.2423641D-1	.1453392D-1	.7938551D-2
1	PFR	1.583648614	.9892515489	.8399074884	.7049266484	.5725937067	.4887243504	.4279582019	.3413776798	.2527923614
1	PF1	1.237666917	1.000064825	.9212291358	.8443458346	.7656557219	.7154231274	.6795740815	.6304206159	.5841616711
1	PQ1	1.132955561	1.002625908	.9975369452	1.013248874	1.060724675	1.119214457	1.184496434	1.333921946	1.621025767
1	PF2	1.065226322	1.007832867	.9845219137	.9595022657	.9313885600	.9119982133	.8973905482	.8761326634	.8544031835
1	PQ2	4.297453251	3.524401034	3.440264193	3.444470115	3.571354949	3.756521975	3.972457059	4.475772661	5.443789970
1	PF3	1.023188211	1.010150791	1.004170626	.9973003045	.9889820290	.9828358230	.9779557948	.9704057060	.9619981326
1	PQ3	17.23678506	13.36195718	12.74841927	12.44903476	12.53617770	12.90901546	13.42306494	14.73206440	17.39475723
3	PFC	1.013950378	1.012690080	1.011906099	1.010865335	1.009410565	1.008198492	1.007151716	1.005382183	1.003195871
3	PQC	171.5501493	101.9447888	88.23355836	77.59025290	69.04102586	64.75732858	62.30388028	59.96340872	59.48233087
1	ZF1	1.012935033	1.012893892	1.012896260	1.012905509	1.012922787	1.012938771	1.012953148	1.012973186	1.013010084
1	ZF2	1.015081594	1.015744686	1.016013947	1.016313786	1.016673672	1.016942973	1.017161517	1.017512116	1.017927005
1	ZF3	1.019809276	1.023022259	1.024275950	1.025641064	1.027240824	1.028413565	1.029351855	1.030835424	1.032562584
1	ZF4	1.043141637	1.050948789	1.054215420	1.057875740	1.062275501	1.065561000	1.068218940	1.072463316	1.077452031
1	ZF5	1.148045416	1.166636842	1.174570606	1.183560723	1.194491439	1.202729678	1.209433991	1.220199066	1.232919688
1	ZF6	1.695434345	1.748346984	1.770750415	1.796053946	1.826730199	1.849798629	1.868543212	1.898590145	1.934012085

(6) P 9+4		TH = 82 SBE = 1.009827573								MCPER FUNCTIONS
		RC = 02	RC = 05	RC = 07	RC = 10	RC = 15	RC = 20	RC = 25	RC = 35	RC = 50
	AMAX DB	.0017375255	.0108709564	.0213327376	.0436480540	.0988323393	.1772876696	.2802872360	.5675287486	1.249387366
	AMIN DB	14.05066084	24.14487476	28.05407187	32.29383334	37.24480159	40.86599713	43.76218724	48.35531651	53.78696408
MU	K	.8944865D 0	.2857552D 0	.1849984D 0	.1155931D 0	.6680508D-1	.4474806D-1	.3247991D-1	.1953871D-1	.1070668D-1
1	PFR	1.741803536	1.057472124	.8919020942	.7444311642	.6017279597	.5121639083	.4476570104	.3562487759	.2632561921
1	PF1	1.266694770	1.031406435	.9509796938	.8719498779	.7906690538	.7386527336	.7014928676	.6505122036	.6025186152
1	PQ1	1.243237028	1.056396812	1.041785258	1.050672247	1.093227144	1.149635693	1.214023990	1.363372175	1.652751450
1	PF2	1.065768710	1.014578750	.9929566897	.9693869168	.9425389372	.9238347393	.9096573605	.8889096307	.8675801812
1	PQ2	5.064364465	3.964296581	3.822556072	3.785668494	3.884706068	4.060562811	4.275257822	4.788470747	5.791156806
1	PF3	1.020821234	1.010100851	1.004974261	.9989788649	.9916003360	.9860795710	.9816607122	.9747709886	.9670354108
1	PQ3	21.54618830	15.94094888	15.01251603	14.48458717	14.41297556	14.72985887	15.23321185	16.59064692	19.43747015
3	PFC	1.011213098	1.010370101	1.009797556	1.009011109	1.007878653	1.006913979	1.006068823	1.004619435	1.002799215
3	PQC	237.8760176	133.8810005	113.9543231	98.61743460	86.32755914	80.11699500	76.48647893	72.80281228	71.40896009
1	ZF1	1.010250863	1.010195396	1.010193289	1.010197964	1.010210119	1.010222463	1.010233986	1.010254609	1.010281523
1	ZF2	1.011925729	1.012476845	1.012704998	1.012961076	1.013270911	1.013504389	1.013694831	1.014002080	1.014368272
1	ZF3	1.015456975	1.018190819	1.019281742	1.020480719	1.021898469	1.022945542	1.023787645	1.025126541	1.026696052
1	ZF4	1.035164548	1.041820374	1.044694505	1.047957251	1.051927589	1.054921803	1.057360337	1.061281482	1.065928629
1	ZF5	1.128880777	1.145337868	1.152556090	1.160829211	1.170998033	1.178729773	1.185059739	1.195287395	1.207463093
1	ZF6	1.639187078	1.687769154	1.708838460	1.732859196	1.762234680	1.784478602	1.802638049	1.831888546	1.866570403

(6) P 9+4 — TH = 83 — SBE = 1.007509825 — MCPER FUNCTIONS

MU		RC = 02	RC = 05	RC = 07	RC = 10	RC = 15	RC = 20	RC = 25	RC = 35	RC = 50
	AMAX DB	.0017375255	.0108709564	.0213327376	.0436480540	.0988323393	.1772876696	.2802872360	.5675287486	1.249387366
	AMIN DB	11.47931857	21.27107417	25.12333218	29.31493979	34.22076006	37.81433438	40.69109788	45.25765337	50.66388715
MU	K	.1165497D 1	.3792147D 0	.2468237D 0	.1550158D 0	.9006117D-1	.6053302D-1	.4404810D-1	.2659372D-1	.1462671D-1
1	PFR	1.947587942	1.141710187	.9551344623	.7918354308	.6362636104	.5397534457	.4707343630	.3735650091	.2753754321
1	PF1	1.292567353	1.064424743	.9830405450	.9021582438	.8183413713	.7644799548	.7259306643	.6729782464	.6230838344
1	PQ1	1.397810103	1.127501086	1.099256124	1.098486635	1.134106602	1.187536189	1.250571499	1.399508739	1.691370404
1	PF2	1.064714596	1.020463252	1.000832830	.9790176096	.9537454599	.9359189818	.9223035590	.9022378072	.8814597367
1	PQ2	6.162300996	4.559845937	4.331891302	4.233691205	4.290423794	4.450822018	4.661544707	5.184011712	6.226913238
1	PF3	1.018222857	1.009753975	1.005492035	1.000399166	.9940069652	.9891516424	.9852276759	.9790521503	.9720486742
1	PQ3	28.01145728	19.60174946	18.17527109	17.28635204	16.95870172	17.17684626	17.64952612	19.04802951	22.11209187
3	PFC	1.008727710	1.008213564	1.007820156	1.007255540	1.006411807	1.005673374	1.005015123	1.003966620	1.002395933
3	PQC	346.6381920	183.5282105	153.2239120	130.1634423	111.7972684	102.4820906	96.95911805	91.10515246	88.19290329
1	ZF1	1.007889694	1.007821369	1.007815129	1.007815540	1.007822852	1.007831702	1.007840455	1.007856743	1.007878687
1	ZF2	1.009146555	1.009588176	1.009775815	1.009988439	1.010248103	1.010445343	1.010607164	1.010869918	1.011185639
1	ZF3	1.011657942	1.013900692	1.014822104	1.015846612	1.017071528	1.017984423	1.018723244	1.019905838	1.021303668
1	ZF4	1.027972713	1.033441876	1.035900172	1.038736071	1.042238950	1.044912250	1.047106991	1.050665672	1.054925354
1	ZF5	1.110530474	1.124711006	1.131150999	1.138634718	1.147952567	1.155110785	1.161012415	1.170617753	1.182152151
1	ZF6	1.582817978	1.626562005	1.646134468	1.668706019	1.696595668	1.717885390	1.735359697	1.763663130	1.797442463

(6) P 9+4 — TH = 84 — SBE = 1.005508280 — MCPER FUNCTIONS

MU		RC = 02	RC' = 05	RC = 07	RC = 10	RC = 15	RC = 20	RC = 25	RC = 35	RC = 50
	AMAX DB	.0017375255	.0108709564	.0213327376	.0436480540	.0988323393	.1772876696	.2802872360	.5675287486	1.249387366
	AMIN DB	8.933357631	18.27589815	22.05196664	26.18230350	31.03255834	34.59265979	37.44600336	41.98059287	47.35625974
MU	K	.1530692D 1	.5090000D 0	.3334035D 0	.2106573D 0	.1231450D 0	.8310306D-1	.6065029D-1	.3677253D-1	.2031294D-1
1	PFR	2.225177570	1.249164882	1.034351507	.8502760948	.6782174524	.5729860216	.4983755746	.3941568688	.2896967663
1	PF1	1.312702264	1.098611125	1.017385383	.9352611225	.8491525489	.7934485338	.7534548035	.6983940501	.6464192945
1	PQ1	1.624609405	1.225552268	1.176867763	1.161821827	1.187261551	1.236270298	1.297206570	1.445152677	1.739697301
1	PF2	1.061814678	1.025201221	1.007903459	.9882013923	.9648841979	.9481765996	.9352920218	.9161326416	.8961075885
1	PQ2	7.821936472	5.405369044	5.041335353	4.846994822	4.836619805	4.970839834	5.172556623	5.702073109	6.792112432
1	PF3	1.015424138	1.009090849	1.005692706	1.001521331	.9961572540	.9920070188	.9886129368	.9832095607	.9770051477
1	PQ3	38.34417375	25.09709475	22.83332123	21.34064137	20.57896084	20.61859126	21.02121990	22.43857453	25.76016452
3	PFC	1.006516310	1.006244061	1.005996043	1.005618303	1.005025997	1.004489636	1.004001116	1.003130443	1.001983713
3	PQC	537.9017754	266.1523641	217.2482452	180.5660013	151.6463419	136.9958506	123.2339808	118.6613055	113.0812543
1	ZF1	1.005846527	1.005767964	1.005758076	1.005754657	1.005757531	1.005763117	1.005769244	1.005781359	1.005798434
1	ZF2	1.006741714	1.007077744	1.007225955	1.007395974	1.007605964	1.007766974	1.007899962	1.008117511	1.008381396
1	ZF3	1.008416987	1.010165494	1.010912665	1.011755996	1.012778496	1.013549215	1.014177858	1.015192483	1.016404094
1	ZF4	1.021583640	1.025853082	1.027875967	1.030257826	1.033255315	1.035576758	1.037501536	1.040654577	1.044474918
1	ZF5	1.093014601	1.104775958	1.110369894	1.116984343	1.125350723	1.131858261	1.137268507	1.146150906	1.156927784
1	ZF6	1.526066642	1.564341760	1.582213558	1.603129997	1.629306158	1.649482151	1.666148016	1.693317369	1.725992149

(6) P 9+4	TH = 85 SBE = 1.003819838								MCPER FUNCTIONS
	RC = 02	RC = 05	RC = 07	RC = 10	RC = 15	RC = 20	RC = 25	RC = 35	RC = 50
AMAX DB		.0108709564	.0213327376	.0436480540	.0988323393	.1772876696	.2802872360	.5675287486	1.249387366
AMIN DB		15.13835571	18.80373640	22.85067685	27.62913764	31.14716749	33.97138080	38.46642252	43.80434791
MU I K		.6933469D 0	.4576545D 0	.2912734D 0	.1715378D 0	.1163197D 0	.8519323D-1	.5191580D-1	.2882732D-1
1 I PFR		1.392278420	1.137633718	.9249956565	.7308950925	.6142797452	.5324850106	.4193452696	.3070825107
1 I PF1		1.132811279	1.053660873	.9714723994	.8836837827	.8262774720	.7848442578	.7275774618	.6733417983
1 I PQ1		1.368515405	1.287301595	1.249885210	1.259538411	1.301649778	1.359194828	1.505087164	1.802451816
1 I PF2		1.028420540	1.013831003	.9966616846	.9757649283	.9604830799	.9485491821	.9305946613	.9115973372
1 I PQ2		6.685266234	6.090799216	5.735231857	5.611724597	5.699654882	5.882507998	6.413221625	7.558939587
1 I PF3		1.008097519	1.005545050	1.002300598	.9979973257	.9945893660	.9917604012	.9871906900	.9818598392
1 I PQ3		33.99332802	30.20281633	27.61882866	26.06759511	25.76731611	26.01681220	27.39404463	31.01844910
3 I PFC		1.004487984	1.004350761	1.004122786	1.003740851	1.003379086	1.003040305	1.002419894	1.001581573
3 I PQC		416.7362238	331.2470357	268.2213083	219.2523828	194.6018764	179.8081831	163.3154421	152.6897236
1 I ZF1		1.004031691	1.004018855	1.004012210	1.004011212	1.004013866	1.004017588	1.004025803	1.004038221
1 I ZF2		1.004945480	1.005055965	1.005184908	1.005346504	1.005471842	1.005576205	1.005748446	1.005959733
1 I ZF3		1.007000379	1.007571830	1.008230060	1.009042940	1.009664676	1.010176894	1.011012413	1.012023186
1 I ZF4		1.019101198	1.020675840	1.022581102	1.025037513	1.026975820	1.028603112	1.031303467	1.034625833
1 I ZF5		1.085563412	1.090237658	1.095893957	1.103194122	1.108960809	1.113804757	1.121842545	1.131718452
1 I ZF6		1.500585611	1.516489492	1.535483794	1.559653931	1.578511692	1.594210838	1.620006824	1.651313413

(6) P 9+4	TH = 86 SBE = 1.002441898								MCPER FUNCTIONS
	RC = 02	RC = 05	RC = 07	RC = 10	RC = 15	RC = 20	RC = 25	RC = 35	RC = 50
AMAX DB		.0108709564	.0213327376	.0436480540	.0988323393	.1772876696	.2802872360	.5675287486	1.249387366
AMIN DB		11.84555860	15.33271013	19.25418289	23.93230630	27.39425512	30.18042010	34.62441460	39.91392457
MU I K		.9633728D 0	.6420355D 0	.4123468D 0	.2450873D 0	.1671898D 0	.1229885D 0	.7542059D-1	.4214929D-1
1 I PFR		1.594454713	1.280022928	1.025580377	.8002118176	.6679018421	.5763913639	.4514112303	.3290068599
1 I PF1		1.164595843	1.090751874	1.010718375	.9225967089	.8639295386	.8212062717	.7617519591	.7051117032
1 I PQ1		1.593666684	1.456408336	1.381023727	1.364238792	1.394785656	1.446490965	1.588221624	1.888308195
1 I PF2		1.029632411	1.018143072	1.003989851	.9860867605	.9726263776	.9619348501	.9455921363	.9280083899
1 I PQ2		8.806100879	7.782098265	7.129389489	6.797523046	6.797303145	6.940085707	7.456818995	8.667994170
1 I PF3		1.006771807	1.005021272	1.002687816	.9994616119	.9968267939	.9945962947	.9909236570	.9865486915
1 I PQ3		50.01879724	43.11410456	38.32872547	35.18528815	34.17821890	34.07975346	35.25700912	39.21837271
3 I PFC		1.002974696	1.002913858	1.002797058	1.002580880	1.002362698	1.002150460	1.001747347	1.001180630
3 I PQC		728.2373506	561.0696085	440.1317229	347.9498711	302.1049758	274.6450300	243.6360230	222.4028736
1 I ZF1		1.002609258	1.002594553	1.002585535	1.002581446	1.002581637	1.002583273	1.002588003	1.002596123
1 I ZF2		1.003192180	1.003267476	1.003357748	1.003473259	1.003564225	1.003640745	1.003768433	1.003927251
1 I ZF3		1.004420854	1.004820334	1.005294181	1.005894504	1.006362834	1.006753857	1.007400735	1.008196874
1 I ZF4		1.013240971	1.014366343	1.015781466	1.017667297	1.019192910	1.020494955	1.022692495	1.025450969
1 I ZF5		1.067118864	1.070794545	1.075392967	1.081493829	1.086412143	1.090598663	1.097640674	1.106432881
1 I ZF6		1.434531906	1.448100821	1.464805764	1.486571092	1.503833088	1.518350693	1.542445727	1.572024847

$P^{(5)}_{10+0}$

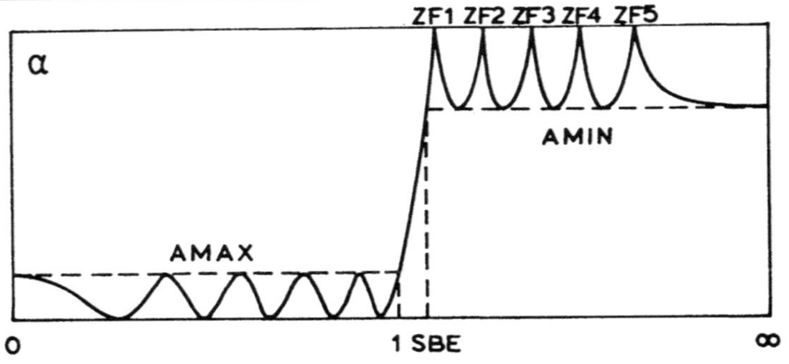

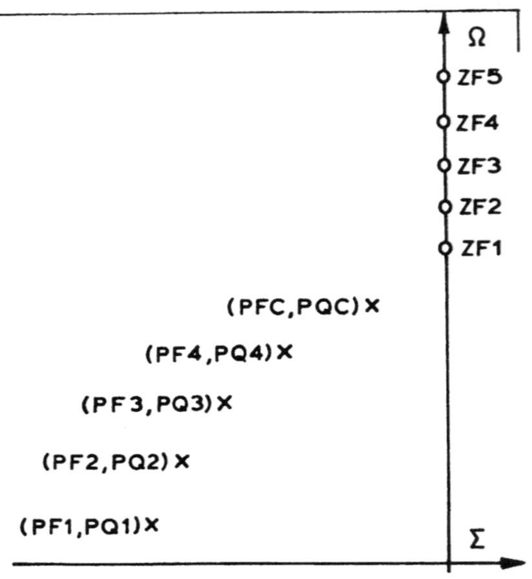

(5) P 10+0	TH = 46	SBE = 1.390163591						CAUER FUNCTIONS	
	RC = 02	RC = 05	RC = 07	RC = 10	RC = 15	RC = 20	RC = 25	RC = 35	RC = 50
AMAX DB	.0017375255	.0108709564	.0213327376	.0436480540	.0988323393	.1772876696	.2807872360	.5675287486	1.249387366
AMIN DB	88.25953319	96.22746678	99.16048928	102.2808438	105.8578533	108.4350833	110.4762831	113.6860854	117.4659832
MU I K	.3863877D-4	.1543927D-4	.1101477D-4	.7690557D-5	.5094568D-5	.3786569D-5	.2993545D-5	.2068692D-5	.1338754D-5
1 I PF1	.6197544237	.4963504035	.4542940972	.4115373990	.3652209623	.3338785248	.3104744850	.2767644958	.2432453094
1 I PQ1	.5395093956	.5542668357	.5629678216	.5754707205	.5960852525	.6173665571	.6399607993	.6911481593	.7922827847
1 I PF2	.7701354224	.6849358623	.6575607513	.6309055157	.6036319163	.5862773491	.5739729261	.5573378139	.5421989645
1 I PQ2	.8592209749	.9644455168	1.022605828	1.102242893	1.225584256	1.345013217	1.465192865	1.719539690	2.177242919
1 I PF3	.9223740220	.8683456867	.8512176473	.8346963505	.8179833426	.8074651553	.8000688461	.7901568719	.7812307699
1 I PQ3	1.612972916	1.877746295	2.020533590	2.213047515	2.506085568	2.785521906	3.063623482	3.645166521	4.677361326
1 I PF4	1.019733241	.9852273344	.9742627460	.9636819377	.9529780715	.9462435752	.9415095751	.9351683870	.9294615993
1 I PQ4	3.402462277	4.003541960	4.325425087	4.757606399	5.412523764	6.034677828	6.652220779	7.939972824	10.21854738
1 I PFC	1.064153937	1.038837066	1.030768448	1.022972506	1.015076890	1.010104940	1.006608052	1.001921695	.9977020760
1 I PQC	11.61493810	13.70507632	14.82232560	16.32082709	18.58905939	20.74177690	22.87713520	27.32690094	35.19443124
1 I ZF1	1.402036353	1.402036353	1.402036353	1.402036353	1.402036353	1.402036353	1.402036353	1.402036353	1.402036353
1 I ZF2	1.509208242	1.509208242	1.509208242	1.509208242	1.509208242	1.509208242	1.509208242	1.509208242	1.509208242
1 I ZF3	1.809701684	1.809701684	1.809701684	1.809701684	1.809701684	1.809701684	1.809701684	1.809701684	1.809701684
1 I ZF4	2.674699978	2.674699978	2.674699978	2.674699978	2.674699978	2.674699978	2.674699978	2.674699978	2.674699978
1 I ZF5	7.503073730	7.503073730	7.503073730	7.503073730	7.503073730	7.503073730	7.503073730	7.503073730	7.503073730

(5) P 10+0	TH = 47 SBE = 1.367327461								CAUER FUNCTIONS
	RC = 02	RC = 05	RC = 07	RC = 10	RC = 15	RC = 20	RC = 25	RC = 35	RC = 50
AMAX DB	.0017375255	.0108709564	.0213327376	.0436480540	.0988323393	.1772876696	.2802872360	.5675287486	1.249387366
AMIN DB	86.13291078	94.10084438	97.03386687	100.1542214	103.7312309	106.3084609	108.3496607	111.5594630	115.3393608
MU I K	.4935765D-4	.1972231D-4	.1407041D-4	.9824013D-5	.6507864D-5	.4837010D-5	.3823992D-5	.2642572D-5	.1710141D-5
1 I PF1	.6261841807	.5010714672	.4584990275	.4152483099	.3684287535	.3367631116	.3131260653	.2790919807	.2452627711
1 I PQ1	.5399073028	.5546215872	.5633127016	.5758088503	.5964211870	.6177055590	.6403058853	.6915119449	.7926908831
1 I PF2	.7757208286	.6896821208	.6620508123	.6351525085	.6076367488	.5901315337	.5777219164	.5609467060	.5456824650
1 I PQ2	.8642946354	.9694751990	1.027736684	1.107572794	1.231295186	1.351137398	1.471757193	1.727079117	2.186606484
1 I PF3	.9254097631	.8713713849	.8542295860	.8376906889	.8209558163	.8104218884	.8030136343	.7930345586	.7841420200
1 I PQ3	1.634655412	1.900542591	2.044292525	2.238285385	2.533798504	2.815736493	3.096408432	3.683476704	4.725729639
1 I PF4	1.019995338	.9859243124	.9750845257	.9646183592	.9540248578	.9473569841	.9426685420	.9363867583	.9307318702
1 I PQ4	3.471861213	4.078556964	4.404336151	4.842203539	5.506309970	6.137562334	6.764350493	8.071787857	10.38588702
1 I PFC	1.062782434	1.038050927	1.030157287	1.022525320	1.014790800	1.009917777	1.006489322	1.001893211	.9977534312
1 I PQC	11.90013670	14.01689216	15.15155785	16.67506743	18.98324300	21.17522631	23.35032766	27.88442717	35.90366817
1 I ZF1	1.378774748	1.378774748	1.378774748	1.378774748	1.378774748	1.378774748	1.378774748	1.378774748	1.378774748
1 I ZF2	1.482254614	1.482254614	1.482254614	1.482254614	1.482254614	1.482254614	1.482254614	1.482254614	1.482254614
1 I ZF3	1.773312683	1.773312683	1.773312683	1.773312683	1.773312683	1.773312683	1.773312683	1.773312683	1.773312683
1 I ZF4	2.614158926	2.614158926	2.614158926	2.614158926	2.614158926	2.614158926	2.614158926	2.614158926	2.614158926
1 I ZF5	7.320320039	7.320320039	7.320320039	7.320320039	7.320320039	7.320320039	7.320320039	7.320320039	7.320320039

(5) P 10+0	TH = 48 SBE = 1.345632730								CAUER FUNCTIONS
	RC = 02	RC = 05	RC = 07	RC = 10	RC = 15	RC = 20	RC = 25	RC = 35	RC = 50
AMAX DB	.0017375255	.0108709564	.0213327376	.0436480540	.0988323393	.1772876696	.2802872360	.5675287486	1.249387366
AMIN DB	84.03690112	92.00483471	94.93785720	98.05821172	101.6352212	104.2124512	106.2536511	109.4634533	113.2433511
MU I K	.6282825D-4	.2510489D-4	.1791048D-4	.1250516D-4	.8283978D-5	.6157117D-5	.4867629D-5	.3363778D-5	.2176870D-5
1 I PF1	.6328936101	.5059889868	.4628765774	.4191096108	.3717648899	.3397621977	.3158823141	.2815106670	.2473587687
1 I PQ1	.5403263000	.5549936441	.5636739724	.5761626525	.5967723037	.6180596411	.6406661526	.6918914984	.7931164356
1 I PF2	.7814919403	.6945887863	.6666931998	.6395440331	.6117782361	.5941174222	.5815991327	.5646791670	.5492853438
1 I PQ2	.8696446961	.9747610876	1.033123458	1.113163233	1.237279673	1.357551229	1.478629236	1.734967792	2.196399195
1 I PF3	.9285035440	.8744668713	.8573141213	.8407599528	.8240052125	.8134566246	.8060371074	.7960917456	.7871334116
1 I PQ3	1.657598390	1.924599731	2.069345437	2.264877709	2.562977004	2.847534351	3.130900364	3.723764586	4.776575884
1 I PF4	1.020249626	.9866281128	.9759174764	.9655700088	.9550908364	.9484920070	.9438507916	.9376306041	.9320295459
1 I PQ4	3.545635382	4.158121956	4.487977425	4.931815429	5.605594315	6.246437026	6.882977488	8.211192436	10.56280720
1 I PFC	1.061387788	1.037250440	1.029534693	1.022069556	1.014499056	1.009726816	1.006368103	1.001863998	.9978056474
1 I PQC	12.20434077	14.34879963	15.50178940	17.05168667	19.40209400	21.63564200	23.85283950	28.47631269	36.65640538
1 I ZF1	1.356667311	1.356667311	1.356667311	1.356667311	1.356667311	1.356667311	1.356667311	1.356667311	1.356667311
1 I ZF2	1.456567377	1.456567377	1.456567377	1.456567377	1.456567377	1.456567377	1.456567377	1.456567377	1.456567377
1 I ZF3	1.738488020	1.738488020	1.738488020	1.738488020	1.738488020	1.738488020	1.738488020	1.738488020	1.738488020
1 I ZF4	2.555995852	2.555995852	2.555995852	2.555995852	2.555995852	2.555995852	2.555995852	2.555995852	2.555995852
1 I ZF5	7.144348217	7.144348217	7.144348217	7.144348217	7.144348217	7.144348217	7.144348217	7.144348217	7.144348217

(5) P 10+0	TH = 49	SBE = 1.325012993						CAUER FUNCTIONS	
	RC = 02	RC = 05	RC = 07	RC = 10	RC = 15	RC = 20	RC = 25	RC = 35	RC = 50
AMAX DB	.0017375255	.0108709564	.0213327376	.0436480540	.0988323393	.1772876696	.2802872360	.5675287486	1.249387366
AMIN DB	81.96958760	89.93752118	92.87054368	95.99089819	99.56790766	102.1451377	104.1863375	107.3961398	111.1760376
MU I K	.7971143D-4	.3185106D-4	.2272337D-4	.1586555D-4	.1051005D-4	.7811656D-5	.6175656D-5	.4267691D-5	.2761837D-5
1 I PF1	.6398973865	.5111125436	.4674349719	.4231283081	.3752352157	.3428809222	.3187478822	.2840245523	.2495367028
1 I PQ1	.5407678791	.5553841071	.5640526405	.5765330566	.5971394718	.6184296472	.6410424368	.6922876686	.7935603630
1 I PF2	.7874537841	.6996606675	.6714925874	.6440846096	.6160607236	.5982392416	.5856087147	.5685392159	.5530115070
1 I PQ2	.8752909366	.9803200743	1.038782583	1.119030509	1.243554383	1.364272060	1.485827209	1.743226056	2.206645682
1 I PF3	.9316539333	.8776319701	.8604714119	.8439045962	.8271322575	.8165702487	.8091402555	.7991795571	.7902061827
1 I PQ3	1.681897781	1.950008265	2.095783575	2.292918340	2.593720835	2.881022066	3.167213298	3.766160895	4.830062325
1 I PF4	1.020495035	.9873380143	.9767609966	.9665364026	.9561756441	.9496483570	.9450560913	.9388997645	.9333545317
1 I PQ4	3.624148563	4.242598931	4.576720191	5.026830670	5.710797450	6.361756575	7.008592202	8.358755247	10.75002109
1 I PFC	1.059970525	1.036435835	1.028900826	1.021605318	1.014201712	1.009532083	1.006244405	1.001834048	.9978587032
1 I PQC	12.52923084	14.70250907	15.87478741	17.45254775	19.84764395	22.12523017	24.38705839	29.10533857	37.45614476
1 I ZF1	1.335646782	1.335646782	1.335646782	1.335646782	1.335646782	1.335646782	1.335646782	1.335646782	1.335646782
1 I ZF2	1.432071632	1.432071632	1.432071632	1.432071632	1.432071632	1.432071632	1.432071632	1.432071632	1.432071632
1 I ZF3	1.705132152	1.705132152	1.705132152	1.705132152	1.705132152	1.705132152	1.705132152	1.705132152	1.705132152
1 I ZF4	2.500059171	2.500059171	2.500059171	2.500059171	2.500059171	2.500059171	2.500059171	2.500059171	2.500059171
1 I ZF5	6.974711915	6.974711915	6.974711915	6.974711915	6.974711915	6.974711915	6.974711915	6.974711915	6.974711915

(5) P 10+0	TH = 50	SBE = 1.305407289						CAUER FUNCTIONS	
	RC = 02	RC = 05	RC' = 07	RC = 10	RC = 15	RC = 20	RC = 25	RC = 35	RC = 50
AMAX DB	.0017375255	.0108709564	.0213327376	.0436480540	.0988323393	.1772876696	.2802872360	.5675287486	1.249387366
AMIN DB	79.92913139	87.89706495	90.83008744	93.95044196	97.52745142	100.1046815	102.1458813	105.3556835	109.1355814
MU I K	.1008192D-3	.4028531D-4	.2874059D-4	.2006679D-4	.1329314D-4	.9880204D-5	.7810987D-5	.5397788D-5	.3493180D-5
1 I PF1	.6472114301	.5164524934	.4721830907	.4273119572	.3788460267	.3461248177	.3217277737	.2866379352	.2518002280
1 I PQ1	.5412336896	.5557941849	.5644498091	.5769210798	.5975236404	.6188164974	.6414356485	.6927013798	.7940236669
1 I PF2	.7936115661	.7049028111	.6764538940	.6487790069	.6204888056	.6025014656	.5897550473	.5725311130	.5568650985
1 I PQ2	.8812552035	.9861707203	1.044732083	1.125192478	1.250137546	1.371318654	1.493371005	1.751376108	2.217372806
1 I PF3	.9348593211	.8808664232	.8637015614	.8471250427	.8303376706	.8197636519	.8123240846	.8023491450	.7933616085
1 I PQ3	1.707659880	1.976868117	2.123707511	2.322510601	2.626139712	2.916316773	3.205472481	3.810809142	4.886366986
1 I PF4	1.020730443	.9880532534	.9776144459	.9675170223	.9572788881	.9508257217	.9462841847	.9401940591	.9347067156
1 I PQ4	3.707806407	4.332390096	4.670976545	5.127680158	5.822385364	6.484024365	7.141737400	8.515105318	10.94831717
1 I PFC	1.058531200	1.035607359	1.028255859	1.021132715	1.013898827	1.009333609	1.006118243	1.001903355	.9979125755
1 I PQC	12.87669034	15.07993090	16.27252390	17.87972768	20.32215581	22.64644699	24.95564157	29.77459837	38.30678000
1 I ZF1	1.315651410	1.315651410	1.315651410	1.315651410	1.315651410	1.315651410	1.315651410	1.315651410	1.315651410
1 I ZF2	1.408698589	1.408698589	1.408698589	1.408698589	1.408698589	1.408698589	1.408698589	1.408698589	1.408698589
1 I ZF3	1.673157185	1.673157185	1.673157185	1.673157185	1.673157185	1.673157185	1.673157185	1.673157185	1.673157185
1 I ZF4	2.446209049	2.446209049	2.446209049	2.446209049	2.446209049	2.446209049	2.446209049	2.446209049	2.446209049
1 I ZF5	6.810998448	6.810998448	6.810998448	6.810998448	6.810998448	6.810998448	6.810998448	6.810998448	6.810998448

(5) P 10+0	TH = 51	SBE = 1.286759566						CAUER FUNCTIONS	
	RC = 02	RC = 05	RC = 07	RC = 10	RC = 15	RC = 20	RC = 25	RC = 35	RC = 50
AMAX DB	.0017375255	.0108709564	.0213327376	.0436480540	.0988323393	.1772876696	.2802872360	.5675287436	1.249387366
AMIN DB	77.91376168	85.88169523	88.81471772	91.93507223	95.51208169	98.08931176	100.1305116	103.3403138	107.1202116
MU K	.1271487D-3	.5080603D-4	.3624634D-4	.2530733D-4	.1676470D-4	.1246047D-4	.9850866D-5	.6807443D-5	.4405441D-5
1 PF1	.6548530440	.5220200478	.4771305356	.4316687193	.3826041152	.3494998491	.3248273813	.2893554452	.2541532779
1 PQ1	.5417255602	.5562252081	.5648666899	.5773278371	.5979258475	.6192211972	.6418467819	.6931336397	.7945074383
1 PF2	.7999706691	.7103205139	.6815822986	.6536322588	.6250673418	.6069088325	.5940427782	.5766593787	.5608505185
1 PQ2	.8875616833	.9923334658	1.050991779	1.131668742	1.257049147	1.378712378	1.501282395	1.760942222	2.228609918
1 PF3	.9381179007	.8841698798	.8670046102	.8504216790	.8336221594	.8230377291	.8155896145	.8056016876	.7966010025
1 PQ3	1.735002754	2.005289807	2.153228345	2.353768497	2.660354552	2.953547478	3.245815781	3.857867200	4.945685601
1 PF4	1.020954678	.9887730216	.9784771429	.9685113128	.9584001442	.9520237598	.9475347900	.9415132852	.9360859669
1 PQ4	3.797062478	4.427943483	4.771205028	5.234842866	5.940875527	6.613799055	7.283015196	8.680940076	11.15856924
1 PFC	1.057070398	1.034765274	1.027599975	1.020651867	1.013590463	1.009131428	1.005989631	1.001771911	.9979672400
1 PQC	13.24883622	15.48320488	16.69720531	18.33554824	20.82815693	23.20203408	25.56155396	30.48754208	39.21265149
1 ZF1	1.296624416	1.296624416	1.296624416	1.296624416	1.296624416	1.296624416	1.296624416	1.296624416	1.296624416
1 ZF2	1.386384966	1.386384966	1.386384966	1.386384966	1.386384966	1.386384966	1.386384966	1.386384966	1.386384966
1 ZF3	1.642482125	1.642482125	1.642482125	1.642482125	1.642482125	1.642482125	1.642482125	1.642482125	1.642482125
1 ZF4	2.394316224	2.394316224	2.394316224	2.394316224	2.394316224	2.394316224	2.394316224	2.394316224	2.394316224
1 ZF5	6.652825318	6.652825318	6.652825318	6.652825318	6.652825318	6.652825318	6.652825318	6.652825318	6.652825318

(5) P 10+0	TH = 52	SBE = 1.269018215						CAUER FUNCTIONS	
	RC = 02	RC = 05	RC = 07	RC = 10	RC = 15	RC = 20	RC = 25	RC = 35	RC = 50
AMAX DB	.0017375255	.0108709564	.0213327376	.0436480540	.0988323393	.1772876696	.2802872360	.5675287486	1.249387366
AMIN DB	75.92176664	83.88970015	86.82272264	89.94307715	93.52008661	96.09731667	98.13851650	101.3483187	105.1282165
MU K	.1599233D-3	.6390208D-4	.4558940D-4	.3183070D-4	.2108607D-4	.1567235D-4	.1239008D-4	.8562174D-5	.5541013D-5
1 PF1	.6628410699	.5278273665	.4822877078	.4362074242	.3865168221	.3530124581	.3280525259	.2921820762	.2566000930
1 PQ1	.5422455232	.5566786449	.5653046172	.5777545535	.5983472314	.6196448475	.6422769247	.6935855494	.7950128681
1 PF2	.8065366489	.7159193350	.6868832557	.6586496816	.6298014774	.6114663653	.5984768385	.5809238140	.5649724444
1 PQ2	.8942372200	.9988308718	1.057583509	1.138480867	1.264311136	1.386475422	1.509585259	1.770450998	2.240389153
1 PF3	.9414276479	.8875418842	.8703805258	.8537948486	.8369864159	.8263933757	.8189378765	.8089383895	.7999257165
1 PQ3	1.764057897	2.035395881	2.184469104	2.386818120	2.696498934	2.992856588	3.288395316	3.907509137	5.008233855
1 PF4	1.021166509	.9894964616	.9793483619	.9695186793	.9595389535	.9532420997	.9488075968	.9428572157	.9374921336
1 PQ4	3.892425367	4.529759514	4.877917219	5.348852603	6.066844060	6.751702234	7.433095238	8.857034734	11.38174804
1 PFC	1.055588737	1.033909856	1.026933370	1.020162900	1.013276689	1.008925575	1.005858585	1.001739710	.9980226709
1 PQC	13.64805509	15.91473455	17.15130751	18.82261211	21.36847774	23.79506007	26.20811279	31.24802766	40.17861067
1 ZF1	1.278513523	1.278513523	1.278513523	1.278513523	1.278513523	1.278513523	1.278513523	1.278513523	1.278513523
1 ZF2	1.365072474	1.365072474	1.365072474	1.365072474	1.365072474	1.365072474	1.365072474	1.365072474	1.365072474
1 ZF3	1.613032213	1.613032213	1.613032213	1.613032213	1.613032213	1.613032213	1.613032213	1.613032213	1.613032213
1 ZF4	2.344260950	2.344260950	2.344260950	2.344260950	2.344260950	2.344260950	2.344260950	2.344260950	2.344260950
1 ZF5	6.499837125	6.499837125	6.499837125	6.499837125	6.499837125	6.499837125	6.499837125	6.499837125	6.499837125

(5) P 10+0	TH = 53 SBE = 1.252135658								CAUER FUNCTIONS
	RC = 02	RC = 05	RC = 07	RC = 10	RC = 15	RC = 20	RC = 25	RC = 35	RC = 50
AMAX DB	.0017375255	.0108709564	.0213327376	.0436480540	.0988323393	.1772876696	.2802872360	.5675287486	1.249387366
AMIN DB	73.95148459	81.91941805	84.85244053	87.97279504	91.54980450	94.12703456	96.16823439	99.37803662	103.1579344
MU I K	.2006438D-3	.8017318D-4	.5719762D-4	.3993560D-4	.2645512D-4	.1966293D-4	.1554491D-4	.1074232D-4	.6951896D-5
1 I PF1	.6711960666	.5338876624	.4876658948	.4409376430	.3905920947	.3566696135	.3314095010	.2951232238	.2591452525
1 I PQ1	.5427958438	.5571561188	.5657650627	.5782025775	.5987890426	.6200886563	.6427272688	.6940583145	.7955412586
1 I PF2	.8133152282	.7217051083	.6923625111	.6638368933	.6346966633	.6161793929	.6030624646	.5853445228	.5692358531
1 I PQ2	.9013116863	1.005687901	1.064531397	1.145652637	1.271947682	1.394633055	1.518305842	1.780431645	2.252745772
1 I PF3	.9447862984	.8909818624	.8738291924	.8572448433	.8404311110	.8298314841	.8223699111	.8123604802	.8033371405
1 I PQ3	1.794972164	2.067322574	2.217566372	2.421799278	2.734720784	3.034401687	3.333379326	3.959927343	5.074249988
1 I PF4	1.021364651	.9902226648	.9802273294	.9705384841	.9606948197	.9544803355	.9501022639	.9442255965	.9389250404
1 I PQ4	3.994467092	4.638398725	4.991685470	5.470305937	6.200934128	6.898427372	7.592724278	9.044253255	11.61893492
1 I PFC	1.054086869	1.033041405	1.026256254	1.019665952	1.012957579	1.008716092	1.005725124	1.001706743	.9980788406
1 I PQC	14.07704619	16.37722800	17.63761698	19.34384554	21.94629741	24.42896958	26.89904022	32.06038165	41.21009561
1 I ZF1	1.261270535	1.261270535	1.261270535	1.261270535	1.261270535	1.261270535	1.261270535	1.261270535	1.261270535
1 I ZF2	1.344707347	1.344707347	1.344707347	1.344707347	1.344707347	1.344707347	1.344707347	1.344707347	1.344707347
1 I ZF3	1.584738339	1.584738339	1.584738339	1.584738339	1.584738339	1.584738339	1.584738339	1.584738339	1.584738339
1 I ZF4	2.295932069	2.295932069	2.295932069	2.295932069	2.295932069	2.295932069	2.295932069	2.295932069	2.295932069
1 I ZF5	6.351702799	6.351702799	6.351702799	6.351702799	6.351702799	6.351702799	6.351702799	6.351702799	6.351702799

(5) P 10+0	TH = 54 SBE = 1.236067977								CAUER FUNCTIONS
	RC = 02	RC = 05	RC = 07	RC = 10	RC = 15	RC = 20	RC = 25	RC = 35	RC = 50
AMAX DB	.0017375255	.0108709564	.0213327376	.0436480540	.0988323393	.1772876696	.2802872360	.5675287486	1.249387366
AMIN DB	72.00129561	79.96922899	82.90225146	86.02260596	89.59961542	92.17684548	94.21804531	97.42784754	101.2077454
MU I K	.2511512D-3	.1003549D-3	.7159578D-4	.4998845D-4	.3311458D-4	.2461261D-4	.1945798D-4	.1344645D-4	.8701873D-5
1 I PF1	.6799405143	.5402153214	.4932773706	.4458697698	.3948385528	.3604788683	.3349051246	.2981847288	.2617937098
1 I PQ1	.5433790535	.5576594300	.5662496548	.5786733975	.5992526582	.6205539518	.6431991242	.6945532577	.7960940372
1 I PF2	.8203122858	.7276839558	.6980261199	.6691998335	.6397586791	.6210535747	.6078052228	.5899119373	.5736460471
1 I PQ2	.9088184189	1.012932245	1.071862154	1.153210340	1.279985459	1.403212912	1.527473058	1.790916314	2.265718555
1 I PF3	.9481913231	.8944891070	.8773503992	.8607718949	.8439568883	.8333529395	.8258867654	.8158692128	.8068367035
1 I PQ3	1.827910054	2.101221757	2.252672195	2.458867405	2.775184347	3.078357609	3.380954363	4.015334995	5.143997815
1 I PF4	1.021547753	.9909506670	.9811132207	.9715700431	.9618672058	.9557380250	.9514184163	.9456181435	.9403844862
1 I PQ4	4.103833071	4.754490883	5.113152031	5.599871523	6.343865812	7.054750345	7.762737419	9.243561222	11.87133776
1 I PFC	1.052565485	1.032160237	1.025568851	1.019161169	1.012633212	1.008503020	1.005589267	1.001673003	.9981357199
1 I PQC	14.53887279	16.87374625	18.15927975	19.90254888	22.56519786	25.10764121	27.63852551	32.92947081	42.31322035
1 I ZF1	1.244850964	1.244850964	1.244850964	1.244850964	1.244850964	1.244850964	1.244850964	1.244850964	1.244850964
1 I ZF2	1.325239935	1.325239935	1.325239935	1.325239935	1.325239935	1.325239935	1.325239935	1.325239935	1.325239935
1 I ZF3	1.557536516	1.557536516	1.557536516	1.557536516	1.557536516	1.557536516	1.557536516	1.557536516	1.557536516
1 I ZF4	2.249226175	2.249226175	2.249226175	2.249226175	2.249226175	2.249226175	2.249226175	2.249226175	2.249226175
1 I ZF5	6.208113116	6.208113116	6.208113116	6.208113116	6.208113116	6.208113116	6.208113116	6.208113116	6.208113116

(5) P 10+0	TH = 55 SBE = 1.220774589								CAUER FUNCTIONS
	RC = 02	RC = 05	RC = 07	RC = 10	RC = 15	RC = 20	RC = 25	RC = 35	RC = 50
AMAX DB	.0017375255	.0108709564	.0213327376	.0436480540	.0988323393	.1772876696	.2802872360	.5675287486	1.249387366
AMIN DB	70.06961323	78.03754647	80.97056894	84.09092343	87.66793289	90.24516295	92.28636278	95.49616500	99.27606282
MU K	.31370350D-3	.1253495D-3	.8942760D-4	.6243870D-4	.4136217D-4	.3074269D-4	.2430423D-4	.1679545D-4	.1086918D-4
1 1 PF1	.6890990500	.5468260405	.4991355090	.4510151160	.3992655652	.3644484246	.3385467961	.3013729258	.2645508339
1 1 PQ1	.5439979910	.5581905805	.5667601995	.5791686606	.5997395982	.6210421983	.6436939331	.6950719337	.7966727718
1 1 PF2	.8275338436	.7338623007	.7038804646	.6747447866	.6449936587	.6260949268	.6127110369	.5946368458	.5782086826
1 1 PQ2	.9167947302	1.020594707	1.079605438	1.161183118	1.288453981	1.412245540	1.537118844	1.801940478	2.279350249
1 1 PF3	.9516399001	.8980627605	.8809438265	.8643761645	.8475643578	.8369586152	.8294894889	.8194658625	.8104258730
1 1 PQ3	1.863056408	2.137263231	2.289956321	2.498195792	2.818072498	3.124918853	3.431327850	4.073968940	5.217770252
1 1 PF4	1.021714403	.9916794447	.9820051560	.9726126219	.9630555302	.9570146854	.9527556417	.9470345402	.9418702411
1 1 PQ4	4.221254013	4.878745802	5.243039847	5.738301131	6.496447763	7.221541849	7.944071362	9.456041011	12.14030976
1 1 PFC	1.051025317	1.031266689	1.024871402	1.018643711	1.012303677	1.008286409	1.005451039	1.001638483	.9981932775
1 1 PQC	15.03702368	17.40776113	18.71985978	20.50245673	23.22922796	25.83545632	28.43129892	33.86078712	43.49488088
1 1 ZF1	1.229213699	1.229213699	1.229213699	1.229213699	1.229213699	1.229213699	1.229213699	1.229213699	1.229213699
1 1 ZF2	1.306624332	1.306624332	1.306624332	1.306624332	1.306624332	1.306624332	1.306624332	1.306624332	1.306624332
1 1 ZF3	1.531367417	1.531367417	1.531367417	1.531367417	1.531367417	1.531367417	1.531367417	1.531367417	1.531367417
1 1 ZF4	2.204046862	2.204046862	2.204046862	2.204046862	2.204046862	2.204046862	2.204046862	2.204046862	2.204046862
1 1 ZF5	6.068778459	6.068778459	6.068778459	6.068778459	6.068778459	6.068778459	6.068778459	6.068778459	6.068778459

(5) P 10+0	TH = 56 SBE = 1.206217949								CAUER FUNCTIONS
	RC = 02	RC = 05	RC = 07	RC = 10	RC = 15	RC = 20	RC = 25	RC = 35	RC = 50
AMAX DB	.0017375255	.0108709564	.0213327376	.0436480540	.0988323393	.1772876696	.2802872360	.5675287486	1.249387366
AMIN DB	68.15487623	76.12280928	79.05583172	82.17618621	85.75319566	88.33042572	90.37162554	93.58142777	97.36132558
MU K	.3910715D-3	.1562642D-3	.1114829D-3	.7783782D-4	.5156324D-4	.3832470D-4	.3029833D-4	.2093768D-4	.1354983D-4
1 1 PF1	.6986987399	.5537369857	.5052549153	.4563860177	.4038833359	.3685872087	.3423425638	.3046946992	.2674224557
1 1 PQ1	.5446558495	.5587518022	.5672987052	.5796901939	.6002515451	.6215550138	.6442132876	.6956156461	.7972791886
1 1 PF2	.8349860467	.7402468803	.7099322749	.6804784054	.6504081176	.6313098513	.6177862181	.5995254242	.5829298013
1 1 PQ2	.9252825152	1.028709651	1.087794272	1.169603355	1.297385988	1.421764791	1.547278564	1.813543380	2.293688106
1 1 PF3	.9551288833	.9017017957	.8846090311	.8680577313	.8512540877	.8406493669	.8331791302	.8231517243	.8141061546
1 1 PQ3	1.900619619	2.175637432	2.329608835	2.539978216	2.863589444	3.174302429	3.484731079	4.136093078	5.295893448
1 1 PF4	1.021863117	.9924079105	.9829021957	.9736654315	.9642591632	.9583097901	.9541134872	.9484744334	.9433820434
1 1 PQ4	4.347560166	5.011966225	5.382165414	5.886442756	6.659591036	7.399782118	8.137780121	9.682909741	12.42737170
1 1 PFC	1.049467139	1.030361125	1.024164166	1.018128747	1.011969066	1.008066310	1.005310463	1.001603175	.9982514803
1 1 PQC	15.57548750	17.98322458	19.32340882	21.14780998	23.94298037	26.61738136	29.28271980	34.86054923	44.76288138
1 1 ZF1	1.214320710	1.214320710	1.214320710	1.214320710	1.214320710	1.214320710	1.214320710	1.214320710	1.214320710
1 1 ZF2	1.288818054	1.288818054	1.288818054	1.288818054	1.288818054	1.288818054	1.288818054	1.288818054	1.288818054
1 1 ZF3	1.506175959	1.506175959	1.506175959	1.506175959	1.506175959	1.506175959	1.506175959	1.506175959	1.506175959
1 1 ZF4	2.160304057	2.160304057	2.160304057	2.160304057	2.160304057	2.160304057	2.160304057	2.160304057	2.160304057
1 1 ZF5	5.933426782	5.933426782	5.933426782	5.933426782	5.933426782	5.933426782	5.933426782	5.933426782	5.933426782

(5) P 10+0	TH = 57 SBE = 1.192363293								CAUER FUNCTIONS
	RC = 02	RC = 05	RC = 07	RC = 10	RC = 15	RC = 20	RC = 25	RC = 35	RC = 50
AMAX DB	.0017375255	.0108709564	.0213327376	.0436480540	.0988323393	.1772876696	.2802872360	.5675287486	1.249387366
AMIN DB	66.25554039	74.22347313	77.15649554	80.27685002	83.85385946	86.43108952	88.47228934	91.68209156	95.46198938
MU I K	.4866570D-3	.1944582D-3	.1387315D-3	.9686291D-4	.6416630D-4	.4769200D-4	.3770384D-4	.2605526D-4	.1686167D-4
1 I PF1	.7087693950	.5609669753	.5116515773	.4619959603	.4087030040	.3729049567	.3463012008	.3081575479	.2704149220
1 I PQ1	.5453562336	.5593455923	.5678674126	.5802400306	.6007903661	.6220941914	.6447589503	.6961864674	.7979151933
1 I PF2	.8426751386	.7468447593	.7161886480	.6864077377	.6560089838	.6367051691	.6230374988	.6045842706	.5878158663
1 I PQ2	.9343289721	1.037315530	1.096465534	1.178507153	1.306817910	1.431808285	1.557991488	1.825768552	2.308784489
1 I PF3	.9586547687	.9054049949	.8883454283	.8718165797	.8550265955	.8444260260	.8369567318	.8269281110	.8178790922
1 I PQ3	1.940835466	2.216558647	2.371843291	2.584432056	2.911963935	3.226751212	3.541422749	4.202002346	5.378731653
1 I PF4	1.021992341	.9931349083	.9838033356	.9747276236	.9654774222	.9596227640	.9554914547	.9499374303	.9449195961
1 I PQ4	4.483698483	5.155063266	5.531454145	6.045256262	6.834325572	7.590578456	8.345053714	9.925540635	12.73423839
1 I PFC	1.047891774	1.029443932	1.023447420	1.017601463	1.011629482	1.007842779	1.005167568	1.001567072	.9983102926
1 I PQC	16.15884280	18.60465238	19.97455063	21.84344238	24.71168367	27.45906644	30.19888236	35.93532394	46.12608554
1 I ZF1	1.200136783	1.200136783	1.200136783	1.200136783	1.200136783	1.200136783	1.200136783	1.200136783	1.200136783
1 I ZF2	1.271781745	1.271781745	1.271781745	1.271781745	1.271781745	1.271781745	1.271781745	1.271781745	1.271781745
1 I ZF3	1.481910928	1.481910928	1.481910928	1.481910928	1.481910928	1.481910928	1.481910928	1.481910928	1.481910928
1 I ZF4	2.117913404	2.117913404	2.117913404	2.117913404	2.117913404	2.117913404	2.117913404	2.117913404	2.117913404
1 I ZF5	5.801801755	5.801801755	5.801801755	5.801801755	5.801801755	5.801801755	5.801801755	5.801801755	5.801801755

(5) P 10+0	TH = 58 SBE = 1.179178403								CAUER FUNCTIONS
	RC = 02	RC = 05	RC = 07	RC = 10	RC = 15	RC = 20	RC = 25	RC = 35	RC = 50
AMAX DB	.0017375255	.0108709564	.0213327376	.0436480540	.0988323393	.1772876696	.2802872360	.5675287486	1.249387366
AMIN DB	64.37006999	72.33800226	75.27102463	78.39137908	81.96838851	84.54561856	86.58681838	89.79662060	93.57651842
MU I K	.6046395D-3	.2416016D-3	.1723649D-3	.1203459D-3	.7972244D-4	.5925419D-4	.4634455D-4	.3237196D-4	.2094952D-4
1 I PF1	.7193439386	.5685366916	.5183430401	.4678597220	.4137367592	.3774123134	.3504322932	.3117606587	.2735351568
1 I PQ1	.5461032274	.5599747527	.5684688291	.5808204395	.6013581395	.6226617235	.6453328772	.6967362622	.7985829950
1 I PF2	.8506074270	.7536633423	.7226570705	.6925402547	.6618036316	.6422881564	.6284720706	.6098204450	.5928739021
1 I PQ2	.9439874632	1.046455516	1.105660543	1.187934879	1.316790405	1.442417958	1.569301353	1.838564424	2.324697600
1 I PF3	.9622136552	.9091709248	.8921522730	.8756525837	.8588823376	.8482893925	.8408233249	.8307963499	.8217462668
1 I PQ3	1.983971721	2.260268851	2.416900430	2.631801989	2.963453060	3.282537912	3.601693159	4.272027429	5.466692968
1 I PF4	1.022109441	.9938592079	.9847075019	.9757962857	.9667095666	.9609529792	.9569889972	.9514230937	.9464825627
1 I PQ4	4.630753421	5.309074983	5.691958840	6.215832156	7.021819961	7.795186218	8.567240534	10.18548855	13.06285020
1 I PFC	1.046300099	1.028515525	1.022721463	1.017067057	1.011285037	1.007615879	1.005022385	1.001530167	.9983696766
1 I PQC	16.79236787	19.27722556	20.67858290	22.59488527	25.54131373	28.36696446	31.18674316	37.09467268	47.59459868
1 I ZF1	1.186629287	1.186629287	1.186629287	1.186629287	1.186629287	1.186629287	1.186629287	1.186629287	1.186629287
1 I ZF2	1.255478919	1.255478919	1.255478919	1.255478919	1.255478919	1.255478919	1.255478919	1.255478919	1.255478919
1 I ZF3	1.458524649	1.458524649	1.458524649	1.458524649	1.458524649	1.458524649	1.458524649	1.458524649	1.458524649
1 I ZF4	2.076795721	2.076795721	2.076795721	2.076795721	2.076795721	2.076795721	2.076795721	2.076795721	2.076795721
1 I ZF5	5.673661054	5.673661054	5.673661054	5.673661054	5.673661054	5.673661054	5.673661054	5.673661054	5.673661054

(5) P 10+0			TH = 59 RC = 02	SBE = 1.166633397 RC = 05	RC = 07	RC = 10	RC = 15	RC = 20	RC = 25	RC = 35	CAUER FUNCTIONS RC = 50
AMAX DB			.0017375255	.0108709564	.0213327376	.0436480540	.0988323393	.1772876696	.2802872360	.5675287486	1.249387366
AMIN DB			62.49692906	70.46486061	73.39788291	76.51823733	80.09524674	82.67247678	84.71367660	87.92347882	91.70337663
MU		K	.7501594D-3	.2997485D-3	.2138483D-3	.1493097D-3	.9890942D-4	.7351503D-4	.5811874D-4	.4016299D-4	.2599149D-4
1		PF1	.7304588365	.5764689283	.5253486102	.4739935402	.4189979754	.3821209470	.3547463420	.3155399924	.2767907330
1		PQ1	.5469014773	.5606424395	.5691057700	.5814339614	.6019571860	.6232598311	.6459372465	.6974172140	.7992846351
1		PF2	.8587892409	.7607103851	.7293454410	.6988838325	.6677999188	.6430665848	.6340976264	.6152415132	.5981110404
1		PQ2	.9543185498	1.056178250	1.115425747	1.197931818	1.327348992	1.453640704	1.581257030	1.852285049	2.341492330
1		PF3	.9658012019	.9129979096	.8960286372	.8795654905	.8628216972	.8522402260	.8447799224	.8347577794	.8257092957
1		PQ3	2.030333714	2.307042304	2.465052642	2.682364410	3.018346779	3.341969817	3.665869197	4.346540370	5.560236180
1		PF4	1.022185703	.9945794993	.9856135453	.9768764351	.9679547929	.9622997494	.9583055128	.9529309374	.9430705632
1		PQ4	4.789972283	5.475188888	5.864882010	6.399414250	7.223405236	8.015034079	8.805874255	10.46452069	13.41541105
1		PFC	1.044693043	1.027576350	1.021986617	1.016525744	1.010935851	1.007385679	1.004874948	1.001492453	.9984295916
1		PQC	17.48217559	20.00691427	21.44160141	23.40849487	26.43872886	29.34847553	32.25427565	38.34632393	49.17998333
1		ZF1	1.173767956	1.173767956	1.173767956	1.173767956	1.173767956	1.173767956	1.173767956	1.173767956	1.173767956
1		ZF2	1.239875723	1.239875723	1.239875723	1.239875723	1.239875723	1.239875723	1.239875723	1.239875723	1.239875723
1		ZF3	1.435972684	1.435972684	1.435972684	1.435972684	1.435972684	1.435972684	1.435972684	1.435972684	1.435972684
1		ZF4	2.036876486	2.036876486	2.036876486	2.036876486	2.036876486	2.036876486	2.036876486	2.036876486	2.036876486
1		ZF5	5.548774783	5.548774783	5.548774783	5.548774783	5.548774783	5.548774783	5.548774783	5.548774783	5.548774783

(5) P 10+0			TH = 60 RC = 02	SBE = 1.154700538 RC = 05	RC = 07	RC = 10	RC = 15	RC = 20	RC = 25	RC = 35	CAUER FUNCTIONS RC = 50
AMAX DB			.0017375255	.0108709564	.0213327376	.0436480540	.0988323393	.1772876696	.2802872360	.5675287486	1.249387366
AMIN DB			60.63457211	68.60250256	71.53552476	74.65587912	78.23288850	80.81011854	82.85131835	86.06112056	89.84101838
MU		K	.9295471D-3	.3714282D-3	.2649865D-3	.1850146D-3	.1225619D-3	.9109490D-4	.7201684D-4	.4976729D-4	.3220691D-4
1		PF1	.7421546027	.5847888794	.5326895930	.4804153066	.4245013664	.3870436828	.3592548821	.3194783832	.2801899551
1		PQ1	.5477562911	.5613522202	.5697814080	.5820834508	.6025901056	.6238909979	.6465744908	.6980817581	.8000230205
1		PF2	.8672268746	.7679940052	.7362620939	.7054470353	.6740062285	.6540487668	.6399224080	.6208555965	.6035355713
1		PQ2	.9653912426	1.066538735	1.125813554	1.208548958	1.338544815	1.465529144	1.593913304	1.866690950	2.359241264
1		PF3	.9694125796	.9168839992	.8999733844	.8835549003	.8668449706	.8562792355	.8488275115	.8388137443	.8297698309
1		PQ3	2.080271098	2.357191102	2.516609329	2.736432743	3.076973360	3.405394464	3.734320314	4.425961275	5.659878937
1		PF4	1.022246325	.9952943864	.9865202339	.9779610123	.9692122281	.9636623243	.9597403398	.9544604209	.9496831687
1		PQ4	4.962796303	5.654769348	6.051603021	6.597427157	7.440603717	8.251754604	9.062706379	10.76465363	13.79443415
1		PFC	1.043071600	1.026626885	1.021243228	1.015977758	1.010582055	1.007152252	1.004725295	1.001453922	.9984899941
1		PQC	18.23537996	20.80062967	22.27065218	24.29160810	27.41183505	30.41212326	33.41065883	39.70141450	50.89555287
1		ZF1	1.161524705	1.161524705	1.161524705	1.161524705	1.161524705	1.161524705	1.161524705	1.161524705	1.161524705
1		ZF2	1.224940734	1.224940734	1.224940734	1.224940734	1.224940734	1.224940734	1.224940734	1.224940734	1.224940734
1		ZF3	1.414213562	1.414213562	1.414213562	1.414213562	1.414213562	1.414213562	1.414213562	1.414213562	1.414213562
1		ZF4	1.998085383	1.998085383	1.998085383	1.998085383	1.998085383	1.998085383	1.998085383	1.998085383	1.998085383
1		ZF5	5.426923983	5.426923983	5.426923983	5.426923983	5.426923983	5.426923983	5.426923983	5.426923983	5.426923983

(5) P 10+0	TH = 61	SBE = 1.143354068							CAUER FUNCTIONS
	RC = 02	RC = 05	RC = 07	RC = 10	RC = 15	RC = 20	RC = 25	RC = 35	RC = 50
AMAX DB	.0017375255	.0108709564	.0213327376	.0436480540	.0988323393	.1772876696	.2802872360	.5675287486	1.249387366
AMIN DB	58.78143420	66.74936297	69.68238502	72.80273929	76.37974863	78.95697865	80.99817846	84.20798066	87.98787847
MU I K	.1150610D-2	.4597607D-3	.3280052D-3	.2290145D-3	.1517094D-3	.1127590D-3	.8914379D-4	.6160287D-4	.3986631D-4
1 I PF1	.7544763956	.5935244788	.5403895725	.4871447952	.4302631687	.3921946594	.3639706206	.3235956549	.2837419559
1 I PQ1	.5486737605	.5621081438	.5704993322	.5827721269	.6032598212	.6245580114	.6472473364	.6987826187	.8008009632
1 I PF2	.8759265161	.7755226892	.7434158229	.7122386520	.6804315152	.6602436072	.6459552599	.6266714291	.6091560023
1 I PQ2	.9772845219	1.077599418	1.136883322	1.219843923	1.350435546	1.478142534	1.607331815	1.881950141	2.378025883
1 I PF3	.9730424159	.9208269335	.9039851406	.8876202435	.8709523511	.8604070671	.8529670448	.8429655911	.8339295576
1 I PQ3	2.134186127	2.411071914	2.571923406	2.794363859	3.139705936	3.473206477	3.807465717	4.510766378	5.766207556
1 I PF4	1.022280412	.9960023795	.9874262466	.9790508744	.9704809228	.9650398830	.9611927499	.9560109435	.9513198961
1 I PQ4	5.150899018	5.849391190	6.253711299	6.811509874	7.675164171	8.507221490	9.339745849	11.08819849	14.20279769
1 I PFC	1.041436829	1.025667648	1.020491671	1.015423352	1.010223794	1.006915682	1.004573468	1.001414568	.9985508375
1 I PQC	19.06030328	21.66641182	23.17391918	25.25273447	28.46978897	31.56777141	34.66650867	41.17220485	52.75665081
1 I ZF1	1.149873453	1.149873453	1.149873453	1.149873453	1.149873453	1.149873453	1.149873453	1.149873453	1.149873453
1 I ZF2	1.210644767	1.210644767	1.210644767	1.210644767	1.210644767	1.210644767	1.210644767	1.210644767	1.210644767
1 I ZF3	1.393208539	1.393208539	1.393208539	1.393208539	1.393208539	1.393208539	1.393208539	1.393208539	1.393208539
1 I ZF4	1.960355866	1.960355866	1.960355866	1.960355866	1.960355866	1.960355866	1.960355866	1.960355866	1.960355866
1 I ZF5	5.307899240	5.307899240	5.307899240	5.307899240	5.307899240	5.307899240	5.307899240	5.307899240	5.307899240

(5) P 10+0	TH = 62	SBE = 1.132570051							CAUER FUNCTIONS
	RC = 02	RC = 05	RC = 07	RC = 10	PC = 15	RC = 20	PC = 25	RC = 35	RC = 50
AMAX DB	.0017375255	.0108709564	.0213327376	.0436480540	.0988323393	.1772876696	.2802872360	.5675287486	1.249387366
AMIN DB	56.93592015	64.90384637	67.83686817	70.95722232	74.53423159	77.11146159	79.15266138	82.36246358	86.14236138
MU I K	.1422997D-2	.5686011D-3	.4056548D-3	.2832298D-3	.1876240D-3	.1394527D-3	.1102470D-3	.7618629D-4	.4930397D-4
1 I PF1	.7674747266	.6027068024	.5484747402	.4942039307	.4363013555	.3975895119	.3689076000	.3279037577	.2874568102
1 I PQ1	.5496609106	.5629148258	.5712636194	.5835036346	.6039696319	.6252640109	.6479588501	.6995228547	.8016217280
1 I PF2	.8848941559	.7833052975	.7508159061	.7192682362	.6870853567	.6666606598	.6522056899	.6326984222	.6149815255
1 I PQ2	.9900891987	1.089431501	1.148702563	1.231882096	1.363086480	1.491547872	1.621582185	1.898139344	2.397937992
1 I PF3	.9766847344	.9248241018	.9080622609	.8917607554	.8751439104	.8646242910	.8571994292	.8472146610	.8381901914
1 I PQ3	2.192543845	2.469094206	2.631399218	2.856565878	3.206970460	3.545855850	3.885783073	4.601497794	5.879888892
1 I PF4	1.022285973	.9967018874	.9883301650	.9801447866	.9717598438	.9664315261	.9626619414	.9575818375	.9529802017
1 I PQ4	5.356233994	6.060881165	6.473047226	7.043557067	7.929104960	8.783595222	9.639307572	11.43781607	14.64331287
1 I PFC	1.039789865	1.024699194	1.019732351	1.014862798	1.009861223	1.006676059	1.004419514	1.001374385	.9986120716
1 I PQC	19.96673585	22.61366343	24.16095748	26.30179401	29.62324952	32.32689174	36.03416436	42.77295666	54.78110720
1 I ZF1	1.138789971	1.138789971	1.138789971	1.138789971	1.138789971	1.138789971	1.138789971	1.138789971	1.138789971
1 I ZF2	1.196960709	1.196960709	1.196960709	1.196960709	1.196960709	1.196960709	1.196960709	1.196960709	1.196960709
1 I ZF3	1.372921373	1.372921373	1.372921373	1.372921373	1.372921373	1.372921373	1.372921373	1.372921373	1.372921373
1 I ZF4	1.923624768	1.923624768	1.923624768	1.923624768	1.923624768	1.923624768	1.923624768	1.923624768	1.923624768
1 I ZF5	5.191499327	5.191499327	5.191499327	5.191499327	5.191499327	5.191499327	5.191499327	5.191499327	5.191499327

(5) P 10+0	TH = 63	SBE = 1.122326238						CAUER FUNCTIONS	
	RC = 02	RC = 05	RC = 07	RC = 10	RC = 15	RC = 20	RC = 25	RC = 35	RC = 50
AMAX DB	.0017375255	.0108709564	.0213327376	.0436480540	.0988323393	.1772876696	.2802872360	.5675287486	1.249387366
AMIN DB	55.09639263	63.06431495	65.99733639	69.11769035	72.69469951	75.27192947	77.31312925	80.52293143	84.30282923
MU I K	.1758654D-2	.7027231D-3	.5013410D-3	.3500382D-3	.2318809D-3	.1723469D-3	.1362522D-3	.9415718D-4	.6093384D-4
1 I PF1	.7812063052	.6123705448	.5569742869	.5016171074	.4426358911	.4032455895	.3740813905	.3324159289	.2913456681
1 I PQ1	.5507258858	.5637775514	.5720789212	.5842821185	.6047232765	.6260125472	.6487124953	.7003059143	.8024889884
1 I PF2	.8941354720	.7913510642	.7584721299	.7265458989	.6939780107	.6733101926	.6586839385	.6389467376	.6210224948
1 I PQ2	1.003910208	1.102116534	1.161348402	1.244737996	1.376571851	1.505821215	1.636743375	1.915345458	2.419081453
1 I PF3	.9803328848	.9288724957	.9122027906	.8959754449	.8794195768	.8689313839	.8615255135	.8515522831	.8425534753
1 I PQ3	2.255884738	2.531730365	2.695502258	2.923507727	3.279255429	3.623858064	3.969819135	4.698775407	6.001684769
1 I PF4	1.022260908	.9973912090	.9892304641	.9812414136	.9730478654	.9678362682	.9641470306	.9591723603	.9546634737
1 I PQ4	5.581094622	6.291369505	6.711752884	7.295770182	8.204767355	9.083379409	9.964072266	11.81658490	15.12130767
1 I PFC	1.038131921	1.023722124	1.018965707	1.014296396	1.009494515	1.006433481	1.004263486	1.001333367	.9986736426
1 I PQC	20.96626394	23.65344297	25.24298510	27.45041454	30.88469132	34.20289778	37.52804448	44.52031537	56.98971859
1 I ZF1	1.128251744	1.128251744	1.128251744	1.128251744	1.128251744	1.128251744	1.128251744	1.128251744	1.128251744
1 I ZF2	1.183863369	1.183863369	1.183863369	1.183863369	1.183863369	1.183863369	1.183863369	1.183863369	1.183863369
1 I ZF3	1.353318131	1.353318131	1.353318131	1.353318131	1.353318131	1.353318131	1.353318131	1.353318131	1.353318131
1 I ZF4	1.887831921	1.887831921	1.887831921	1.887831921	1.887831921	1.887831921	1.887831921	1.887831921	1.887831921
1 I ZF5	5.077529904	5.077529904	5.077529904	5.077529904	5.077529904	5.077529904	5.077529904	5.077529904	5.077529904

(5) P 10+0	TH = 64	SBE = 1.112601940						CAUER FUNCTIONS	
	RC = 02	RC = 05	RC = 07	RC = 10	RC = 15	RC = 20	RC = 25	RC = 35	RC = 50
AMAX DB	.0017375255	.0108709564	.0213327376	.0436480540	.0988323393	.1772876696	.2802872360	.5675287486	1.249387366
AMIN DB	53.26115882	61.22907520	64.16209609	67.28244975	70.85945876	73.43668866	75.47788842	78.68769058	82.46758836
MU I K	.2172411D-2	.8680530D-3	.6192916D-3	.4323919D-3	.2864356D-3	.2128951D-3	.1683083D-3	.1163096D-3	.7526977D-4
1 I PF1	.7957350521	.6225545896	.5659208684	.5094115684	.4492890325	.4091822126	.3795093186	.3371468841	.2954209138
1 I PQ1	.5518781812	.5647024029	.5729505698	.5851123126	.6055250103	.6268076528	.6495121994	.7011356996	.8034068949
1 I PF2	.9036556825	.7996695908	.7663948120	.7340824043	.7011204802	.6802032611	.6654010573	.6454273732	.6272895148
1 I PQ2	1.018869458	1.115748385	1.174909372	1.258496941	1.390976447	1.521049301	1.652905335	1.933667336	2.441574275
1 I PF3	.9839794638	.9329686552	.9164044212	.9002630600	.8837791093	.8733287107	.8659460735	.8560097644	.8470211759
1 I PQ3	2.324840548	2.599528233	2.764771173	2.995730917	3.357123839	3.707806503	4.060202767	4.803311427	6.132469653
1 I PF4	1.022203008	.9980685233	.9901255023	.9823393092	.9743437594	.9692530276	.9656470431	.9607816860	.9563690243
1 I PQ4	5.828189678	6.543354515	6.972335464	7.570721205	8.504881914	9.409490859	10.31716081	12.22808556	15.63973074
1 I PFC	1.036464303	1.022737087	1.018192216	1.013724471	1.009123856	1.006188058	1.004105442	1.001291507	.9987354925
1 I PQC	22.07268779	24.79883538	26.43325118	28.71230600	32.26879894	35.71156391	39.16509405	46.43382600	59.40688558
1 I ZF1	1.118237845	1.118237845	1.118237845	1.118237845	1.118237845	1.118237845	1.118237845	1.118237845	1.118237845
1 I ZF2	1.171329344	1.171329344	1.171329344	1.171329344	1.171329344	1.171329344	1.171329344	1.171329344	1.171329344
1 I ZF3	1.334367004	1.334367004	1.334367004	1.334367004	1.334367004	1.334367004	1.334367004	1.334367004	1.334367004
1 I ZF4	1.852919799	1.852919799	1.852919799	1.852919799	1.852919799	1.852919799	1.852919799	1.852919799	1.852919799
1 I ZF5	4.965802212	4.965802212	4.965802212	4.965802212	4.965802212	4.965802212	4.965802212	4.965802212	4.965802212

(5) P 10+0		TH = 65	SBE = 1.103377919						CAUER FUNCTIONS	
		RC = 02	RC = 05	RC = 07	RC = 10	RC = 15	RC = 20	RC = 25	RC = 35	RC = 50
AMAX DB		.0017375255	.0108709564	.0213327376	.0436480540	.0988323393	.1772876696	.2802872360	.5675287486	1.249387366
AMIN DB		51.42845528	59.39636262	62.32938266	65.44973588	69.02674465	71.60397447	73.64517418	76.85497631	80.63487408
MU	K	.2682732D-2	.1071968D-2	.7647700D-3	.5339655D-3	.3537226D-3	.2629065D-3	.2078458D-3	.1436320D-3	.9295148D-4
1	PF1	.8111333234	.6333026952	.5753511666	.5176178608	.4562856911	.4154209818	.3852107405	.3421130457	.2996963551
1	PQ1	.5531289346	.5656964170	.5738847099	.5859996514	.6063797003	.6276539292	.6503624370	.7020166473	.8043801570
1	PF2	.9134593604	.8082708306	.7745948229	.7418892200	.7085245852	.6873517920	.6723689997	.6521522603	.6337945448
1	PQ2	1.035109395	1.130435668	1.189487618	1.273257108	1.406397579	1.537331519	1.670171027	1.953217967	2.465551162
1	PF3	.9876162249	.9371086065	.9206644381	.9046220466	.8882220687	.8778165011	.8704617941	.8605583794	.8515950779
1	PQ3	2.400154226	2.673126767	2.839832720	3.073864183	3.441228015	3.798387834	4.157661068	4.915928389	6.273252469
1	PF4	1.022109944	.9987318784	.9910135102	.9834369053	.9756461840	.9706806159	.9671609032	.9624088350	.9580960802
1	PQ4	6.100739632	6.819784181	7.257747205	7.871432842	8.832651784	9.765347395	10.70222738	12.67650636	16.20428118
1	PFC	1.034788411	1.021744788	1.017412394	1.013147376	1.008749452	1.005939911	1.003945447	1.001248802	.9987975584
1	PQC	23.30255903	26.06542477	27.74750448	30.10373582	33.79296650	37.37155590	40.96534997	48.53657867	62.06141152
1	ZF1	1.108728821	1.108728821	1.108728821	1.108728821	1.108728821	1.108728821	1.108728821	1.108728821	1.108728821
1	ZF2	1.159336899	1.159336899	1.159336899	1.159336899	1.159336899	1.159336899	1.159336899	1.159336899	1.159336899
1	ZF3	1.316038147	1.316038147	1.316038147	1.316038147	1.316038147	1.316038147	1.316038147	1.316038147	1.316038147
1	ZF4	1.818833178	1.818833178	1.818833178	1.818833178	1.818833178	1.818833178	1.818833178	1.818833178	1.818833178
1	ZF5	4.856131766	4.856131766	4.856131766	4.356131766	4.856131766	4.856131766	4.856131766	4.856131766	4.856131766

(5) P 10+0		TH = 66	SBE = 1.094636279						CAUER FUNCTIONS	
		RC = 02	RC = 05	RC = 07	RC = 10	RC = 15	RC = 20	RC = 25	RC = 35	RC = 50
AMAX DB		.0017375255	.0108709564	.0213327376	.0436480540	.0988323393	.1772876696	.2802872360	.5675287486	1.249387366
AMIN DB		49.59643060	57.56432416	60.49734292	63.61769545	67.19470385	69.77193355	71.81313320	75.02293528	78.80283302
MU	K	.3312672D-2	.1323682D-2	.9443497D-3	.6593488D-3	.4367821D-3	.3246410D-3	.2566512D-3	.1773590D-3	.1147779D-3
1	PF1	.8274833962	.6446643249	.5853065661	.5262703844	.4636538687	.4219861491	.3912073707	.3473328169	.3041874500
1	PQ1	.5544912969	.5667677822	.5748884605	.5869504063	.6072929408	.6285566531	.6512683306	.7029538250	.8054141441
1	PF2	.9235501969	.8171650612	.7830836036	.7499785696	.7162030437	.6947686783	.6796007247	.6591343775	.6405505185
1	PQ2	1.052797512	1.146304777	1.205201652	1.289132087	1.422947525	1.554782357	1.688658910	1.974127151	2.491166657
1	PF3	.9912339751	.9412877891	.9249796604	.9090505003	.8927477814	.8823948230	.8750732482	.8652093563	.8562769783
1	PQ3	2.482705329	2.753275760	2.921420551	3.158641817	3.532328131	3.896401197	4.263039462	5.037581550	6.425203717
1	PF4	1.021979256	.9993791791	.9918925779	.9845324987	.9769536713	.9721177258	.9686874214	.9640529632	.9598437716
1	PQ4	6.402600594	7.124160198	7.571487021	8.201480233	9.191858140	10.15497887	11.12357704	13.16677644	16.82107220
1	PFC	1.033105758	1.020745992	1.016626808	1.012565500	1.008371530	1.005689172	1.003783572	1.001205246	.9988597727
1	PQC	24.67587866	27.47190299	29.20459516	31.64413823	35.47793669	39.20510880	42.95266285	50.65603181	64.98752004
1	ZF1	1.099706589	1.099706589	1.099706589	1.099706589	1.099706589	1.099706589	1.099706589	1.099706589	1.099706589
1	ZF2	1.147865855	1.147865855	1.147865855	1.147865855	1.147865855	1.147865855	1.147865855	1.147865855	1.147865355
1	ZF3	1.298303529	1.298303529	1.298303529	1.298303529	1.298303529	1.298303529	1.298303529	1.298303529	1.298303529
1	ZF4	1.785518792	1.785518792	1.785518792	1.785518792	1.785518792	1.785518792	1.785518792	1.785518792	1.785518792
1	ZF5	4.748337007	4.748337007	4.748337007	4.748337007	4.748337007	4.748337007	4.748337007	4.748337007	4.748337007

(5) P 10+0	TH = 67 SBE = 1.086360377								CAUER FUNCTIONS
	RC = 02	RC = 05	RC = 07	RC = 10	RC = 15	RC = 20	RC = 25	RC = 35	RC = 50
AMAX DB	.0017375255	.0108709564	.0213327376	.0436480540	.0988323393	.1772876696	.2802872360	.5675237486	1.249387366
AMIN DB	47.76312521	55.73099773	58.66401453	61.78436602	65.36137387	67.93860337	69.97980293	73.18960493	76.96950262
MU K	.4091134D-2	.1634745D-2	.1166270D-2	.8142949D-3	.5394253D-3	.4009312D-3	.3169639D-3	.2190381D-3	.1417506D-3
1 PF1	.8448792859	.6566956588	.5958339791	.5354080600	.4714251864	.4289050683	.3975236801	.3528269139	.3089115823
1 PQ1	.5559809079	.5679260870	.5759701204	.5879718567	.6082711985	.6295219081	.6522357756	.7039530518	.8065150088
1 PF2	.9339307024	.8263628407	.7918731772	.7583634890	.7241695637	.7024678894	.6871103172	.6663378334	.6475715863
1 PQ2	1.072132091	1.163503710	1.222189800	1.306254067	1.440756575	1.573534432	1.708506040	1.996544824	2.518599022
1 PF3	.9948224557	.9455009719	.9293463703	.9135461096	.8973552981	.8870635499	.8797808715	.8699638612	.8610686779
1 PQ3	2.573542657	2.840860921	3.010399020	3.250926843	3.631315568	4.002782352	4.377326960	5.169387013	6.589689492
1 PF4	1.021808351	1.000008174	.9927606404	.9856242365	.9782646128	.9735629166	.9702252816	.9657127481	.9616111198
1 PQ4	6.738425464	7.460671884	7.917730970	8.565121200	9.586994894	10.58316881	11.58631567	13.70473544	17.49733917
1 PFC	1.031417977	1.019741532	1.015836072	1.011979268	1.007990338	1.005435990	1.003619898	1.001160837	.9989220621
1 PQC	26.21701325	29.04086113	30.82725542	33.35690283	37.34862623	41.23890123	45.15562740	53.42507200	68.22616378
1 ZF1	1.091154349	1.091154349	1.091154349	1.091154349	1.091154349	1.091154349	1.091154349	1.091154349	1.091154349
1 ZF2	1.136897504	1.136897504	1.136897504	1.136897504	1.136897504	1.136897504	1.136897504	1.136897504	1.136897504
1 ZF3	1.281136796	1.281136796	1.281136796	1.281136796	1.281136796	1.281136796	1.281136796	1.281136796	1.281136796
1 ZF4	1.752925003	1.752925003	1.752925003	1.752925003	1.752925003	1.752925003	1.752925003	1.752925003	1.752925003
1 ZF5	4.642237872	4.642237872	4.642237872	4.642237872	4.642237872	4.642237872	4.642237872	4.642237872	4.642237872

(5) P 10+0	TH = 68 SBE = 1.078534743								CAUER FUNCTIONS
	RC = 02	RC = 05	RC = 07	RC = 10	RC = 15	RC = 20	RC = 25	RC = 35	RC = 50
AMAX DB	.0017375255	.0108709564	.0213327376	.0436480540	.0988323393	.1772876696	.2802872360	.5675287486	1.249387366
AMIN DB	45.92644766	53.89428803	56.82730182	59.94765172	63.52465872	66.10188792	68.14308734	71.35238922	75.13278685
MU K	.5054493D-2	.2019694D-2	.1440904D-2	.1006045D-2	.6664492D-3	.4953425D-3	.3916027D-3	.2706173D-3	.1751300D-3
1 PF1	.8634289831	.6694608371	.6069868572	.5450751468	.4796355313	.4362087453	.4041873829	.3586187714	.3138883961
1 PQ1	.5576165127	.5691826352	.5771394259	.5890725048	.6093219932	.6305567498	.6532715967	.7050210473	.8076898404
1 PF2	.9446018236	.8358749409	.8009761526	.7670578845	.7324389460	.7104645965	.6949131280	.6739282718	.6548732825
1 PQ2	1.093349605	1.182206925	1.240614572	1.324777866	1.459976859	1.593742307	1.729871950	2.020645218	2.548055098
1 PF3	.9983702050	.9497421537	.9337602294	.9181060892	.9020433442	.8918223230	.8845849315	.8748229780	.8659719719
1 PQ3	2.673926674	2.936936058	3.107793652	3.351740593	3.739242633	4.118634368	4.501688229	5.312657380	6.768314549
1 PF4	1.021594487	1.000616437	.9936154609	.9867100995	.9795772427	.9750145982	.9717730241	.9673869742	.9633970229
1 PQ4	7.113875827	7.834370364	8.301501465	8.967464752	10.02344254	11.05563694	12.09654296	14.29735145	18.24170689
1 PFC	1.029726837	1.018732318	1.015040862	1.011389145	1.007606150	1.005180528	1.003454516	1.001115572	.9989843469
1 PQC	27.95591157	30.79983018	32.64312317	35.27040625	39.43520315	43.50519524	47.60879400	56.28339253	71.82672507
1 ZF1	1.083056493	1.083056493	1.083056493	1.083056493	1.083056493	1.083056493	1.083056493	1.083056493	1.083056493
1 ZF2	1.126414515	1.126414515	1.126414515	1.126414515	1.126414515	1.126414515	1.126414515	1.126414515	1.126414515
1 ZF3	1.264513152	1.264513152	1.264513152	1.264513152	1.264513152	1.264513152	1.264513152	1.264513152	1.264513152
1 ZF4	1.721001456	1.721001456	1.721001456	1.721001456	1.721001456	1.721001456	1.721001456	1.721001456	1.721001456
1 ZF5	4.537654265	4.537654265	4.537654265	4.537654265	4.537654265	4.537654265	4.537654265	4.537654265	4.537654265

(5) P 10+0	TH = 69	SBE = 1.071144994						CAUER FUNCTIONS	
	RC = 02	RC = 05	RC = 07	RC = 10	RC = 15	RC = 20	RC = 25	RC = 35	RC = 50
AMAX DB	.0017375255	.0108709564	.0213327376	.0436480540	.0988323393	.1772876696	.2802872360	.5675287486	1.249387366
AMIN DB	44.08414647	52.05193757	54.98494677	58.10529423	61.68229992	64.25952866	66.30072788	69.51052957	73.29042711
MU I K	.6248743D-2	.2496911D-2	.1781364D-2	.1243756D-2	.8239199D-3	.6123836D-3	.4841318D-3	.3345596D-3	.2165103D-3
1 I PF1	.8832572297	.6830335003	.6188264426	.5553222535	.4883258556	.4439325162	.4112300357	.3647350406	.3191402081
1 I PQ1	.5594207709	.5705508529	.5784078817	.5902623492	.6104541269	.6316694133	.6543837436	.7061656183	.8089468563
1 I PF2	.9555624555	.8457122499	.8104057148	.7760765932	.7410272012	.7187753191	.7030259373	.6817725560	.6624727238
1 I PQ2	1.116734362	1.202621582	1.260668252	1.344886063	1.480787214	1.615587326	1.752943559	2.046632114	2.579776434
1 I PF3	1.001864397	.9540044461	.9382161803	.9227271001	.9068102597	.8966705044	.8894854912	.8797876845	.8709886367
1 I PQ3	2.785385265	3.042764870	3.214830552	3.462300844	3.857360776	4.245267015	4.637504701	5.468947379	6.962977348
1 I PF4	1.021334764	1.001201356	.9944546127	.9877878827	.9808896186	.9764710123	.9733290282	.9690742146	.9652002383
1 I PQ4	7.535903962	8.251397746	8.728889226	9.414691478	10.50769495	11.57927690	12.66160353	14.95300436	19.06453641
1 I PFC	1.028034261	1.017719345	1.014241917	1.010795645	1.007219271	1.004922968	1.003287527	1.001069451	.9990465405
1 I PQC	29.92974080	32.78266581	34.68610030	37.41937762	41.77450931	46.04333896	50.35426472	59.47930709	75.84925091
1 I ZF1	1.075398535	1.075398535	1.075398535	1.075398535	1.075398535	1.075398535	1.075398535	1.075398535	1.075398535
1 I ZF2	1.116400867	1.116400867	1.116400867	1.116400867	1.116400867	1.116400867	1.116400867	1.116400867	1.116400867
1 I ZF3	1.248409242	1.248409242	1.248409242	1.248409242	1.248409242	1.248409242	1.248409242	1.248409242	1.248409242
1 I ZF4	1.689698724	1.689698724	1.689698724	1.689698724	1.689698724	1.689698724	1.689698724	1.689698724	1.689698724
1 I ZF5	4.434404353	4.434404353	4.434404353	4.434404353	4.434404353	4.434404353	4.434404353	4.434404353	4.434404353

(5) P 10+0	TH = 70	SBE = 1.064177772						CAUER FUNCTIONS	
	RC = 02	RC = 05	RC = 07	RC = 10	RC = 15	RC = 20	RC = 25	RC = 35	RC = 50
AMAX DB	.0017375255	.0108709564	.0213327376	.0436480540	.0988323393	.1772876696	.2802872360	.5675287486	1.249387366
AMIN DB	42.23377639	50.20149179	53.13449392	56.25483762	59.83184132	62.40906936	64.45026825	67.66006966	71.43996705
MU I K	.7732344D-2	.3089765D-2	.2204323D-2	.1539069D-2	.1019549D-2	.7577859D-3	.5990824D-3	.4139964D-3	.2679178D-3
1 I PF1	.9045089877	.6974987142	.6314233311	.5662075980	.4975431695	.4521168912	.4186877812	.3712062085	.3246925179
1 I PQ1	.5614213318	.5720468164	.5797891885	.5915532365	.6116779770	.6328695778	.6555815408	.7073958964	.8102956449
1 I PF2	.9668088141	.8558856317	.8201755953	.7854354425	.7499516818	.7274180943	.7114671473	.6899394373	.6703888481
1 I PQ2	1.142631223	1.224995646	1.282580130	1.366795616	1.503399429	1.639283800	1.777941460	2.074745558	2.614047091
1 I PF3	1.005290655	.9582799354	.9427083283	.9274051536	.9116539284	.9016071211	.8944823641	.8848588229	.8761204152
1 I PQ3	2.909787969	3.159875820	3.332987975	3.584071556	3.987170282	4.384247884	4.786427860	5.640112932	7.175941206
1 I PF4	1.021026115	1.001760105	.9952754571	.9888551729	.9821995994	.9779302100	.9748914896	.9707728709	.9670193636
1 I PQ4	8.013133190	8.719292416	9.207347969	9.914345324	11.04765880	12.16246966	13.29041756	15.68185835	19.97838049
1 I PFC	1.026342339	1.016703703	1.013440052	1.010199335	1.006830037	1.004663512	1.003119044	1.001022473	.9991085486
1 I PQC	32.18511745	35.03141495	36.99817808	39.84672846	44.41196205	48.90177264	53.44382145	63.07216509	80.36742602
1 I ZF1	1.068167042	1.068167042	1.068167042	1.068167042	1.068167042	1.068167042	1.068167042	1.068167042	1.068167042
1 I ZF2	1.106841787	1.106841787	1.106841787	1.106841787	1.106841787	1.106841787	1.106841787	1.106841787	1.106841787
1 I ZF3	1.232803052	1.232803052	1.232803052	1.232803052	1.232803052	1.232803052	1.232803052	1.232803052	1.232803052
1 I ZF4	1.658967927	1.658967927	1.658967927	1.658967927	1.658967927	1.658967927	1.658967927	1.658967927	1.658967927
1 I ZF5	4.332302650	4.332302650	4.332302650	4.332302650	4.332302650	4.332302650	4.332302650	4.332302650	4.332302650

(5) P 10+0	TH = 71	SBE = 1.057620681						CAUER FUNCTIONS	
	RC = 02	RC = 05	RC = 07	RC = 10	RC = 15	RC = 20	RC = 25	RC = 35	RC = 50
AMAX DB	.0017375255	.0108709564	.0213327376	.0436480540	.0988323393	.1772876696	.2802872360	.5675287486	1.249387366
AMIN DB	40.37265753	48.34025617	51.27324742	54.39358533	57.97058594	60.54781290	62.58901129	65.79381226	69.57870942
MU K	.9580036D-2	.3828135D-2	.2731100D-2	.1906868D-2	.1263196D-2	.9388784D-3	.7422487D-3	.5129315D-3	.3319438D-3
1 1 PF1	.9273538126	.7129553984	.6448594418	.5777985942	.5073417914	.4608086135	.4266022837	.3780673736	.3305746482
1 1 PQ1	.5636522822	.5736899483	.5812998017	.5929593180	.6130058740	.6341687074	.6568760071	.7087226412	.8117474743
1 1 PF2	.9783336265	.8664057286	.8303000154	.7951513062	.7592312318	.7364126751	.7202570099	.6934498174	.6786427025
1 1 PQ2	1.171462600	1.249628550	1.306625965	1.390766482	1.528066361	1.665087041	1.805128046	2.105270518	2.651203706
1 1 PF3	1.008632826	.9625595155	.9472298002	.9321354963	.9165716899	.9066307972	.8995750592	.8900370632	.8813689961
1 1 PQ3	3.049446209	3.290135147	3.464064672	3.718828557	4.130485670	4.537469546	4.950449163	5.828388503	7.409927358
1 1 PF4	1.020665291	1.002289626	.9960751186	.9899093233	.9835048191	.9793900267	.9764583960	.9724311482	.9688528124
1 1 PQ4	8.556378365	9.247401296	9.746090944	10.47572498	11.65305400	12.81550186	13.99392173	16.49635783	20.99858931
1 1 PFC	1.024653358	1.015686596	1.012636167	1.009600843	1.006438822	1.004402388	1.002949195	1.000974642	.9991702680
1 1 PQC	34.78119539	37.59886810	39.63192493	42.60603966	47.40413142	52.14074283	56.94190252	67.13561179	85.47258087
1 1 ZF1	1.061349573	1.061349573	1.061349573	1.061349573	1.061349573	1.061349573	1.061349573	1.061349573	1.061349573
1 1 ZF2	1.097723693	1.097723693	1.097723693	1.097723693	1.097723693	1.097723693	1.097723693	1.097723693	1.097723693
1 1 ZF3	1.217673811	1.217673811	1.217673811	1.217673811	1.217673811	1.217673811	1.217673811	1.217673811	1.217673811
1 1 ZF4	1.628760320	1.628760320	1.628760320	1.628760320	1.628760320	1.628760320	1.628760320	1.628760320	1.628760320
1 1 ZF5	4.231157790	4.231157790	4.231157790	4.231157790	4.231157790	4.231157790	4.231157790	4.231157790	4.231157790

(5) P 10+0	TH = 72	SBE = 1.051462224						CAUER FUNCTIONS	
	RC = 02	RC = 05	RC = 07	RC = 10	RC = 15	RC = 20	RC = 25	RC = 35	RC = 50
AMAX DB	.0017375255	.0108709564	.0213327376	.0436480540	.0988323393	.1772876696	.2802872360	.5675287486	1.249387366
AMIN DB	38.49782552	46.46524333	49.39821770	52.51854665	56.09554247	58.67276775	60.71396537	63.92376567	67.70366246
MU K	.1188800D-1	.4750484D-2	.3389137D-2	.2366316D-2	.1567555D-2	.1165096D-2	.9210893D-3	.6365195D-3	.4119238D-3
1 1 PF1	.9519914111	.7295194242	.6592305272	.5901738708	.5177849352	.4700620047	.4350219135	.3853592293	.3368205550
1 1 PQ1	.5661561279	.5755039468	.5829596733	.5944976524	.6144525976	.6355804957	.6582802733	.7101586336	.8133156929
1 1 PF2	.9901250791	.8772826811	.8407935881	.8052421526	.7688863563	.7457807636	.7294178991	.7073266155	.6872577947
1 1 PQ2	1.203751521	1.276885402	1.333140548	1.417113014	1.555092599	1.693303880	1.834818160	2.138548209	2.691648621
1 1 PF3	1.011872719	.9668326882	.9517725726	.9369124693	.9215602341	.9117396694	.9047627138	.8953228580	.8867359897
1 1 PQ3	3.207250707	3.435845386	3.610271720	3.868747482	4.289522956	4.707238968	5.131992986	6.036483762	7.668238251
1 1 PF4	1.020248850	1.002786601	.9968504566	.9909474229	.9848026563	.9808480511	.9780274965	.9741970261	.9706987868
1 1 PQ4	9.179369661	9.847444929	10.35663241	11.11041648	12.33595698	13.55113235	14.78566549	17.41189715	22.14412706
1 1 PFC	1.022969823	1.014669350	1.011831258	1.009000867	1.006046046	1.004139849	1.002778124	1.000925964	.9992315857
1 1 PQC	37.79401218	40.55210403	42.65392945	45.76499244	50.82228566	55.83602992	60.92904855	71.76205362	91.27917438
1 1 ZF1	1.054934621	1.054934621	1.054934621	1.054934621	1.054934621	1.054934621	1.054934621	1.054934621	1.054934621
1 1 ZF2	1.089034158	1.089034158	1.089034158	1.089034158	1.089034158	1.089034158	1.089034158	1.089034158	1.089034158
1 1 ZF3	1.203001910	1.203001910	1.203001910	1.203001910	1.203001910	1.203001910	1.203001910	1.203001910	1.203001910
1 1 ZF4	1.599026821	1.599026821	1.599026821	1.599026821	1.599026821	1.599026821	1.599026821	1.599026821	1.599026821
1 1 ZF5	4.130769886	4.130769886	4.130769886	4.130769886	4.130769886	4.130769886	4.130769886	4.130769886	4.130769886

(5) P 10+0	TH = 73	SBE = 1.045691756						CAUER FUNCTIONS	
	RC = 02	RC = 05	RC = 07	RC = 10	RC = 15	RC = 20	RC = 25	RC = 35	RC = 50
AMAX DB	.0017375255	.0108709564	.0213327376	.0436480540	.0988323393	.1772876696	.2802872360	.5675287486	1.249387366
AMIN DB	36.60597024	44.57310644	47.50605453	50.62636952	54.20335789	56.78058057	58.82177698	62.03157623	65.81147247
MU I K	.1478092D-1	.5906697D-2	.4214027D-2	.2942263D-2	.1949091D-2	.1448675D-2	.1145279D-2	.7914458D-3	.5121844D-3
1 I PF1	.9786587664	.7473276087	.6746494095	.6034258729	.5289467535	.4799406932	.4440032698	.3931293256	.3434698655
1 I PQ1	.5689865501	.5775180451	.5847932489	.5961890098	.6160360342	.6371214531	.6598101307	.7117191930	.8150162527
1 I PF2	1.002165441	.8885257363	.8516711589	.8157270772	.7789394133	.7555462864	.7389746378	.7165956556	.6962605258
1 I PQ2	1.240153424	1.307216211	1.362534600	1.446219220	1.584848706	1.724306680	1.867393236	2.174991092	2.735867236
1 I PF3	1.014989783	.9710873194	.9563272626	.9417293358	.9266154689	.9169312833	.9100440087	.9007163857	.8922228957
1 I PQ3	3.386863111	3.599880427	3.774357844	4.036523284	4.467017787	4.896398303	5.334042014	6.267751793	7.954923284
1 I PF4	1.019773140	1.003247425	.9975980312	.9919662610	.9860901974	.9823015893	.9795962662	.9759182241	.9725552437
1 I PQ4	9.899777033	10.53430493	11.05353910	11.83302977	13.11154936	14.38537134	15.68263007	18.44773879	23.43869001
1 I PFC	1.021294490	1.013653436	1.011026431	1.008400184	1.005652176	1.003876182	1.002605997	1.000876445	.9992923773
1 I PQC	41.32271966	43.97749844	46.14964708	49.41018213	54.75734999	60.08415207	65.50840442	77.06887456	97.93241623
1 I ZF1	1.048911568	1.048911568	1.048911568	1.048911568	1.048911568	1.048911568	1.048911568	1.048911568	1.048911568
1 I ZF2	1.080761876	1.080761876	1.080761876	1.080761876	1.080761876	1.080761876	1.080761876	1.080761876	1.080761876
1 I ZF3	1.188768818	1.188768818	1.188768818	1.188768818	1.188768818	1.188768818	1.188768818	1.188768818	1.188768818
1 I ZF4	1.569717479	1.569717479	1.569717479	1.569717479	1.569717479	1.569717479	1.569717479	1.569717479	1.569717479
1 I ZF5	4.030927331	4.030927331	4.030927331	4.030927331	4.030927331	4.030927331	4.030927331	4.030927331	4.030927331

(5) P 10+0	TH = 74	SBE = 1.040299436						CAUER FUNCTIONS	
	RC = 02	RC = 05	RC = 07	RC = 10	RC = 15	RC = 20	RC = 25	RC = 35	RC = 50
AMAX DB	.0017375255	.0108709564	.0213327376	.0436480540	.0988323393	.1772876696	.2802872360	.5675287486	1.249387366
AMIN DB	34.69335991	42.66005482	45.59296173	48.71325485	52.29023154	54.86745013	56.90864464	60.11344225	63.89833761
MU I K	.1842180D-1	.7362025D-2	.5252329D-2	.3667222D-2	.2429341D-2	.1805625D-2	.1427473D-2	.9864564D-3	.6383857D-3
1 I PF1	1.007639350	.7665429316	.6912502059	.6176642549	.5409149991	.4905198646	.4536131609	.4014337085	.3505692269
1 I PQ1	.5722123116	.5797687478	.5868308323	.5980589641	.6177780628	.6388116954	.6614847659	.7134228657	.8168684042
1 I PF2	1.014429245	.9001426944	.8629475604	.8266263071	.7894148293	.7657357216	.7489548939	.7262358938	.7056807300
1 I PQ2	1.281500808	1.341182338	1.395317901	1.478559542	1.617790495	1.758552255	1.903320365	2.215103056	2.784451304
1 I PF3	1.017960722	.9753093394	.9608828688	.9465780674	.9317323574	.9222024633	.9154170634	.9062174784	.8978310627
1 I PQ3	3.592988335	3.785873890	3.959783350	4.225535395	4.666388054	5.108490756	5.560309142	6.526318200	8.275005156
1 I PF4	1.019234288	1.003668164	.9983140641	.9929622854	.9873641939	.9837476215	.9811618628	.9776421597	.9744198544
1 I PQ4	10.74068916	11.32714160	11.85549200	12.66223524	13.99916784	15.33857079	16.70637296	19.62929403	24.91226352
1 I PFC	1.019630402	1.012640493	1.010222922	1.007799666	1.005257740	1.003611711	1.002433000	1.000326099	.9993525059
1 I PQC	45.49870081	47.98793818	50.23035251	53.65399835	59.32697031	65.00976346	70.81253082	83.20724708	105.6190545
1 I ZF1	1.043270642	1.043270642	1.043270642	1.043270642	1.043270642	1.043270642	1.043270642	1.043270642	1.043270642
1 I ZF2	1.072896640	1.072896640	1.072896640	1.072896640	1.072896640	1.072896640	1.072896640	1.072896640	1.072896640
1 I ZF3	1.174957012	1.174957012	1.174957012	1.174957012	1.174957012	1.174957012	1.174957012	1.174957012	1.174957012
1 I ZF4	1.540780832	1.540780832	1.540780832	1.540780832	1.540780832	1.540780832	1.540780832	1.540780832	1.540780832
1 I ZF5	3.931402832	3.931402832	3.931402832	3.931402832	3.931402832	3.931402832	3.931402832	3.931402832	3.931402832

(5) P 10+0		TH = 75	SBE = 1.035276180						CAUER FUNCTIONS	
		RC = 02	RC = 05	RC = 07	RC = 10	RC = 15	RC = 20	RC = 25	RC = 35	RC = 50
AMAX DB		.0017375255	.0108709564	.0213327376	.0436480540	.0988323393	.1772876696	.2802872360	.5675287486	1.249387366
AMIN DB		32.75574675	40.72174501	43.65458690	46.77484547	50.35180372	52.92901586	54.97020739	58.18000239	61.95989637
MU	K	.2302569D-1	.9202647D-2	.6565543D-2	.4584138D-2	.3036755D-2	.2257092D-2	.1784389D-2	.1233104D-2	.7980042D-3
1	PF1	1.039275133	.7873614397	.7091939250	.6330203742	.5537945461	.5018892357	.4639312197	.4103390831	.3581740822
1	PQ1	.5759229106	.5823022695	.5891104843	.6001394014	.6197057704	.6406760193	.6633277604	.7152923573	.8188956347
1	PF2	1.026880864	.9121391274	.8746372345	.8379611584	.8003393401	.7763784904	.7593896656	.7364300631	.7155523538
1	PQ2	1.328867204	1.379493558	1.432131502	1.514727639	1.654485618	1.796607877	1.943178446	2.259507012	2.838130725
1	PF3	1.020759012	.9794823706	.9654264488	.9514490762	.9369047109	.9275491484	.9208793019	.9118255306	.9035516358
1	PQ3	3.831769857	3.998487053	4.170966189	4.440080271	4.891961839	5.347992507	5.815477418	6.817404474	8.634794525
1	PF4	1.018628181	1.004044520	.9989943921	.9939315516	.9886210098	.9851827494	.9827210751	.9793658969	.9762899548
1	PQ4	11.73279608	12.25101289	12.78681709	13.62225179	15.02380537	16.43697974	17.88465797	20.98494351	26.60333139
1	PFC	1.017980934	1.011632351	1.009422117	1.007200293	1.004863333	1.003346803	1.002259346	1.000774941	.9994118197
1	PQC	50.49921837	52.73343703	55.04332436	58.64466447	64.68578324	70.77638780	77.01522126	90.37487341	114.5829510
1	ZF1	1.038002876	1.038002876	1.038002876	1.038002876	1.038002876	1.038002876	1.038002876	1.038002876	1.038002876
1	ZF2	1.065429344	1.065429344	1.065429344	1.065429344	1.065429344	1.065429344	1.065429344	1.065429344	1.065429344
1	ZF3	1.161549907	1.161549907	1.161549907	1.161549907	1.161549907	1.161549907	1.161549907	1.161549907	1.161549907
1	ZF4	1.512163132	1.512163132	1.512163132	1.512163132	1.512163132	1.512163132	1.512163132	1.512163132	1.512163132
1	ZF5	3.831948389	3.831948389	3.831948389	3.831948389	3.831948389	3.831948389	3.831948389	3.831948389	3.831948389

(5) P 10+0		TH = 76	SBE = 1.030613629						CAUER FUNCTIONS	
		RC = 02	RC = 05	RC = 07	RC = 10	RC = 15	RC = 20	RC = 25	RC = 35	RC = 50
AMAX DB		.0017375255	.0108709564	.0213327376	.0436480540	.0988323393	.1772876696	.2802872360	.5675287486	1.249387366
AMIN DB		30.78824964	38.75313841	41.68587673	44.80608027	48.38300915	50.96021100	53.00139777	56.21118863	59.99108041
MU	K	.2887936D-1	.1154365D-1	.8235807D-2	.5750373D-2	.3809338D-2	.2831323D-2	.2238361D-2	.1546823D-2	.1001027D-2
1	PF1	1.073982349	.8100215235	.7286759961	.6496533347	.5677121222	.5141570490	.4750534138	.4199257135	.3663510506
1	PQ1	.5802369550	.5851780333	.5916807219	.6024706455	.6218531485	.6427453972	.6653684745	.7173558191	.8211269570
1	PF2	1.039471235	.9245172651	.8867536611	.8497539160	.8117422549	.7875074278	.7703138812	.7470654256	.7259143260
1	PQ2	1.383660951	1.423061100	1.473793514	1.555477065	1.695650919	1.839187701	1.987694739	2.308983357	2.897817738
1	PF3	1.023354308	.9835872579	.9699427105	.9563308738	.9421249261	.9329661809	.9264272800	.9175393801	.9094154877
1	PQ3	4.111377324	4.243798310	4.413637563	4.685705746	5.149303614	5.620643666	6.105541816	7.147672236	9.042336219
1	PF4	1.017950457	1.004371780	.9996344109	.9948696621	.9898565576	.9866031318	.9842702582	.9810960833	.9781624844
1	PQ4	12.91769129	13.33927419	13.87974228	14.74503068	16.21830492	17.71501175	19.25382742	22.55868078	28.56207838
1	PFC	1.016349845	1.010631068	1.008625577	1.006603173	1.004469632	1.003081877	1.002085282	1.000722996	.9994701502
1	PQC	56.56838204	58.41713681	60.78712204	64.58123223	71.04069435	77.60234933	84.34817308	98.83482980	125.1480775
1	ZF1	1.033100079	1.033100079	1.033100079	1.033100079	1.033100079	1.033100079	1.033100079	1.033100079	1.033100079
1	ZF2	1.058351983	1.058351983	1.058351983	1.058351983	1.058351983	1.058351983	1.058351983	1.058351983	1.058351983
1	ZF3	1.148531798	1.148531798	1.148531798	1.148531798	1.148531798	1.148531798	1.148531798	1.148531798	1.148531798
1	ZF4	1.483807372	1.483807372	1.483807372	1.483807372	1.483807372	1.483807372	1.483807372	1.483807372	1.483807372
1	ZF5	3.732288793	3.732288793	3.732288793	3.732288793	3.732288793	3.732288793	3.732288793	3.732288793	3.732288793

(5) P 10+0	TH = 77	SBE = 1.026304108						CAUER FUNCTIONS	
	RC = 02	RC = 05	RC = 07	RC = 10	RC = 15	RC = 20	RC = 25	RC = 35	RC = 50
AMAX DB	.0017375255	.0108709564	.0213327376	.0436480540	.0988323393	.1772876696	.2802872360	.5675287486	1.249387366
AMIN DB	28.78520857	36.74831205	39.68088365	42.80099860	46.37788020	48.95506550	50.99624461	54.20602880	57.98591703
MU I K	.3636969D-1	.1454067D-1	.1037423D-1	.7243527D-2	.4798505D-2	.3566537D-2	.2819602D-2	.1948492D-2	.1260968D-2
1 I PF1	1.112272299	.8348165845	.7499365747	.6677582674	.5828227883	.5274555373	.4870968451	.4302913845	.3751811754
1 I PQ1	.5853148980	.5884738085	.5946044429	.6051045192	.6242635170	.6450591011	.6676440056	.7196486581	.8235987148
1 I PF2	1.052133372	.9372743905	.8993084897	.8620275838	.8236557394	.7991593520	.7817671476	.7582347351	.7368116863
1 I PQ2	1.447766249	1.473075347	1.521365719	1.601780065	1.742206111	1.887204915	2.037797063	2.364524748	2.964669684
1 I PF3	1.025711678	.9876014697	.9744134883	.9612096329	.9473836437	.9384470327	.9320564615	.9233571531	.9153931277
1 I PQ3	4.442903867	4.529883501	4.695368399	4.969703255	5.445691977	5.933931530	6.438307272	7.525773197	9.508056356
1 I PF4	1.017196483	1.004644760	1.000229009	.9957716928	.9910662205	.9880044054	.9858052544	.9827998718	.9800339091
1 I PQ4	14.35301015	14.63722936	15.17781122	16.07354205	17.62664547	19.21863169	20.86233916	24.40404489	30.85514462
1 I PFC	1.014741350	1.009638968	1.007835073	1.006009566	1.004077407	1.002817414	1.001911091	1.000670292	.9995273091
1 I PQC	64.05031434	65.32009211	67.73512148	71.73657430	78.67420483	85.78503470	93.12648547	108.9441549	137.7533392
1 I ZF1	1.028554807	1.028554807	1.028554807	1.028554807	1.028554807	1.028554807	1.028554807	1.028554807	1.028554807
1 I ZF2	1.051657688	1.051657688	1.051657688	1.051657688	1.051657688	1.051657688	1.051657688	1.051657688	1.051657688
1 I ZF3	1.135887800	1.135887800	1.135887800	1.135887800	1.135887800	1.135887800	1.135887800	1.135887800	1.135887800
1 I ZF4	1.455652039	1.455652039	1.455652039	1.455652039	1.455652039	1.455652039	1.455652039	1.455652039	1.455652039
1 I ZF5	3.632113022	3.632113022	3.632113022	3.632113022	3.632113022	3.632113022	3.632113022	3.632113022	3.632113022

(5) P 10+0	TH = 78	SBE = 1.022340595						CAUER FUNCTIONS	
	RC = 02	RC = 05	RC = 07	RC = 10	RC = 15	RC = 20	RC = 25	RC = 35	RC = 50
AMAX DB	.0017375255	.0108709564	.0213327376	.0436480540	.0988323393	.1772876696	.2802872360	.5675287436	1.249387366
AMIN DB	26.74000702	34.70020271	37.63250260	40.75247318	44.32927770	46.90643603	48.94760266	52.15737597	55.93725843
MU I K	.4602562D-1	.1840729D-1	.1313333D-1	.9170148D-2	.6074858D-2	.4515213D-2	.3569603D-2	.2466784D-2	.1596383D-2
1 I PF1	1.154778808	.8621126467	.7732749273	.6875779028	.5993189969	.5419485596	.5002064533	.4415569295	.3847644507
1 I PQ1	.5913789910	.5922934648	.5979647834	.6081088682	.6269930689	.6476677895	.6702020253	.7222161465	.8263571741
1 I PF2	1.064776125	.9504004963	.9123102143	.8748054221	.8361151010	.8113757560	.7937946988	.7699874902	.7482970775
1 I PQ2	1.523761642	1.531122769	1.576253019	1.654914742	1.795352828	1.941848265	2.094690196	2.427415982	3.040180288
1 I PF3	1.027790629	.9914983174	.9788170591	.9660686122	.9526692994	.9439834431	.9377609202	.9292760564	.9214945798
1 I PQ3	4.841779216	4.867705386	5.026370600	5.301852340	5.790838130	6.297814953	6.824133575	7.963160426	10.04572518
1 I PF4	1.016361355	1.004857737	1.000772485	.9966321012	.9922447543	.9893815846	.9873212926	.9844998206	.9819001242
1 I PQ4	16.12068743	16.20785279	16.74119934	17.66686393	19.30900405	21.01055779	22.77618840	26.59513313	33.57287895
1 I PFC	1.013160197	1.008658697	1.007052623	1.005420914	1.003687550	1.002553972	1.001737104	1.000616870	.9995830853
1 I PQC	73.44339765	73.84087273	76.27289838	80.49371578	87.98109578	95.73894285	103.7885742	121.1934964	153.0068477
1 I ZF1	1.024360343	1.024360343	1.024360343	1.024360343	1.024360343	1.024360343	1.024360343	1.024360343	1.024360343
1 I ZF2	1.045340771	1.045340771	1.045340771	1.045340771	1.045340771	1.045340771	1.045340771	1.045340771	1.045340771
1 I ZF3	1.123603797	1.123603797	1.123603797	1.123603797	1.123603797	1.123603797	1.123603797	1.123603797	1.123603797
1 I ZF4	1.427629474	1.427629474	1.427629474	1.427629474	1.427629474	1.427629474	1.427629474	1.427629474	1.427629474
1 I ZF5	3.531062573	3.531062573	3.531062573	3.531062573	3.531062573	3.531062573	3.531062573	3.531062573	3.531062573

(5) P 10+0		TH = 79	SBE = 1.018716695						CAUER FUNCTIONS	
		RC = 02	RC = 05	RC = 07	RC = 10	RC = 15	RC = 20	RC = 25	RC = 35	RC = 50
AMAX DB		.0017375255	.0108709564	.0213327376	.0436480540	.0988323393	.1772876696	.2802872360	.5675287486	1.249387366
AMIN DB		24.64486305	32.60025433	35.53210484	38.65183662	42.22851364	44.80562735	46.84677332	50.05652864	53.83640155
MU	K	.5858101D-1	.2344160D-1	.1672610D-1	.1167907D-1	.7737031D-2	.5750672D-2	.4546334D-2	.3141764D-2	.2033199D-2
1	PF1	1.202294151	.8923733244	.7990699513	.7094191269	.6174435575	.5578425289	.5145646032	.4538741297	.3952262856
1	PQ1	.5987456249	.5967790449	.6018741288	.6115744427	.6301162070	.6506381261	.6731050019	.7251172827	.8294623380
1	PF2	1.077275327	.9638748023	.9257621268	.8881101283	.8491590438	.8242036508	.8064486109	.7823815961	.7604327580
1	PQ2	1.615270065	1.599367433	1.640356472	1.716598166	1.856694486	2.004697566	2.159970879	2.499353805	3.126316129
1	PF3	1.029543807	.9952459218	.9831272323	.9708873873	.9579675209	.9495649309	.9435329378	.9352920967	.9277192155
1	PQ3	5.330075448	5.272522050	5.420755802	5.695582225	6.197999845	6.725843364	7.277083279	8.475334717	10.67393235
1	PF4	1.015439887	1.005004368	1.001258450	.9974446125	.9933861632	.9907289343	.9888128601	.9861837653	.9837563281
1	PQ4	18.34072924	18.14107009	18.65527415	19.60832620	21.34981050	23.17351176	25.08748742	29.23508914	36.84075768
1	PFC	1.011611775	1.007693284	1.006280550	1.004838883	1.003301089	1.002292198	1.001563712	1.000562783	.9996372406
1	PQC	85.49249793	84.56115344	86.95973234	91.40526363	99.52811929	108.0574090	116.9607883	136.3042014	171.7733619
1	ZF1	1.020510683	1.020510683	1.020510683	1.020510683	1.020510683	1.020510683	1.020510683	1.020510683	1.020510683
1	ZF2	1.039396810	1.039396810	1.039396810	1.039396810	1.039396810	1.039396810	1.039396810	1.039396810	1.039396810
1	ZF3	1.111666394	1.111666394	1.111666394	1.111666394	1.111666394	1.111666394	1.111666394	1.111666394	1.111666394
1	ZF4	1.399663646	1.399663646	1.399663646	1.399663646	1.399663646	1.399663646	1.399663646	1.399663646	1.399663646
1	ZF5	3.428715225	3.428715225	3.428715225	3.428715225	3.428715225	3.428715225	3.428715225	3.428715225	3.428715225

(5) P 10+0		TH = 80	SBE = 1.015426612						CAUER FUNCTIONS	
		RC = 02	RC = 05	RC = 07	RC = 10	RC = 15	RC = 20	RC = 25	RC = 35	RC = 50
AMAX DB		.0017375255	.0108709564	.0213327376	.0436480540	.0988323393	.1772876696	.2802872360	.5675287486	1.249387366
AMIN DB		22.49060451	30.43792198	33.36901610	36.48834582	40.06480819	42.64184675	44.68295793	47.89268297	51.67253979
MU	K	.7507058D-1	.3006796D-1	.2145602D-1	.1498245D-1	.9925664D-2	.7377474D-2	.5832464D-2	.4030564D-2	.2608393D-2
1	PF1	1.255814780	.9261959841	.8278101762	.7336772938	.6375087115	.5754024883	.5304051873	.4674373273	.4067269982
1	PQ1	.6078798473	.6021302497	.6064884680	.6156257278	.6337338429	.6540599103	.6764366915	.7284306885	.8329937401
1	PF2	1.089460948	.9776604678	.9396591008	.9019624094	.8628298224	.8376965901	.8197893822	.7954856135	.7732933807
1	PQ2	1.727542036	1.680843090	1.716317203	1.789196700	1.928424158	2.077903941	2.235806683	2.582632522	3.225727325
1	PF3	1.030915269	.9988058103	.9873121105	.9756408010	.9632602993	.9551781220	.9493624482	.9413996896	.9340655199
1	PQ3	5.940427691	5.766203169	5.898587355	6.169850914	6.685770924	7.236947070	7.816753308	9.083826479	11.41847607
1	PF4	1.014426634	1.005077593	1.001679701	.9982020748	.9944835411	.9920398047	.9902735350	.9878446510	.9855968562
1	PQ4	21.19519841	20.56942354	21.04492886	22.01905192	23.87105565	25.84879956	27.92855462	32.47163544	40.83802783
1	PFC	1.010102253	1.006746231	1.005521546	1.004265409	1.002919231	1.002032857	1.001391374	1.000508099	.9996895041
1	PQC	101.3529848	98.35898425	100.6340293	105.2946153	114.1550554	123.6167484	133.5661913	155.2993255	195.3207028
1	ZF1	1.017000528	1.017000528	1.017000528	1.017000528	1.017000528	1.017000528	1.017000528	1.017000528	1.017000528
1	ZF2	1.033822771	1.033822771	1.033822771	1.033822771	1.033822771	1.033822771	1.033822771	1.033822771	1.033822771
1	ZF3	1.100062868	1.100062868	1.100062868	1.100062868	1.100062868	1.100062868	1.100062868	1.100062868	1.100062868
1	ZF4	1.371667035	1.371667035	1.371667035	1.371667035	1.371667035	1.371667035	1.371667035	1.371667035	1.371667035
1	ZF5	3.324561783	3.324561783	3.324561783	3.324561783	3.324561783	3.324561783	3.324561783	3.324561783	3.324561783

(5) P 10+0	TH = 81	SBE = 1.012465126							CAUER FUNCTIONS
	RC = 02	RC = 05	RC = 07	RC = 10	RC = 15	RC = 20	RC = 25	RC = 35	RC = 50
AMAX DB	.0017375255	.0108709564	.0213327376	.0436480540	.0988323393	.1772876696	.2802872360	.5675287486	1.249387366
AMIN DB	20.26648476	28.19995873	31.12975234	34.24839054	37.82448361	40.40139289	42.44244423	45.65211716	49.43194628
MU I K	.9697857D-1	.3890470D-1	.2776594D-1	.1939012D-1	.1284623D-1	.9548395D-2	.7548798D-2	.5216679D-2	.3376002D-2
1 I PF1	1.316595287	.9643653281	.8601388625	.7608730353	.6599251009	.5949765434	.5480350576	.4825010630	.4194762242
1 I PQ1	.6194915224	.6086372570	.6120312214	.6204386806	.6379868138	.6580575149	.6803124913	.7322639505	.8370595671
1 I PF2	1.101098032	.9916963466	.9539824195	.9163784918	.8771731447	.8519159053	.8338880206	.8093818792	.7869699360
1 I PQ2	1.868472842	1.779943625	1.807921555	1.876072034	2.013630930	2.164482174	2.325225145	2.680442662	3.342085272
1 I PF3	1.031838098	1.002130964	.9913323578	.9802974880	.9685248210	.9608057972	.9552362487	.9475911034	.9405307570
1 I PQ3	6.723032550	6.381213676	6.489374591	6.752317430	7.281078408	7.858422314	8.471319677	9.819477732	12.31597416
1 I PF4	1.013315936	1.005069521	1.002028069	.9988962727	.9955288612	.9933064125	.9916957633	.9894753070	.9874149576
1 I PQ4	24.97217780	23.69579217	24.09971530	25.08151411	27.05527678	29.20967145	31.49608230	36.52362474	45.82948238
1 I PFC	1.008638761	1.005821627	1.004778765	1.003702773	1.002543401	1.001776855	1.001220645	1.000452911	.9997395659
1 I PQC	122.8984165	116.6143145	118.6029868	123.4367121	133.1539981	143.7603121	155.0163150	179.7662658	225.5763241
1 I ZF1	1.013825286	1.013825286	1.013825286	1.013825286	1.013825286	1.013825286	1.013825286	1.013825286	1.013825286
1 I ZF2	1.028617187	1.028617187	1.028617187	1.028617187	1.028617187	1.028617187	1.028617187	1.028617187	1.028617187
1 I ZF3	1.088781132	1.088781132	1.088781132	1.088781132	1.088781132	1.088781132	1.088781132	1.088781132	1.088781132
1 I ZF4	1.343536149	1.343536149	1.343536149	1.343536149	1.343536149	1.343536149	1.343536149	1.343536149	1.343536149
1 I ZF5	3.217971632	3.217971632	3.217971632	3.217971632	3.217971632	3.217971632	3.217971632	3.217971632	3.217971632

(5) P 10+0	TH = 82	SBE = 1.009827573							CAUER FUNCTIONS
	RC = 02	RC = 05	RC = 07	RC = 10	RC = 15	RC = 20	RC = 25	RC = 35	RC = 50
AMAX DB	.0017375255	.0108709564	.0213327376	.0436480540	.0988323393	.1772876696	.2802872360	.5675287486	1.249387366
AMIN DB	17.96020183	25.86936589	28.79686474	31.91428195	35.48972277	38.06640367	40.10734929	43.31693015	47.09671036
MU I K	.1264707D 0	.5087820D-1	.3632091D-1	.2536798D-1	.1680792D-1	.1249338D-1	.9877170D-2	.6825799D-2	.4417377D-2
1 I PF1	1.386198834	1.007934686	.8969239678	.7917100017	.6352474518	.6170344435	.5678678805	.4994079361	.4337556632
1 I PQ1	.6347141861	.6167388633	.6188347899	.6262712002	.6430785971	.6628090804	.6848967111	.7367690809	.8418116529
1 I PF2	1.111857580	1.005883706	.9686912011	.9313657041	.8922375097	.8669321492	.8488288443	.8241709615	.8015755341
1 I PQ2	2.050475607	1.903289544	1.920808250	1.982178638	2.116823542	2.268807500	2.432603140	2.797373978	3.480648993
1 I PF3	1.032231090	1.005163008	.9951387001	.9848177373	.9737317650	.9664255012	.9611368438	.9538556390	.9471104731
1 I PQ3	7.758884491	7.167842790	7.238342476	7.484977848	8.024460119	8.631160394	9.282853000	10.72814723	13.42105131
1 I PF4	1.012102051	1.004971290	1.002294216	.9995176817	.9965126950	.9945195430	.9930705536	.9910671361	.9892024842
1 I PQ4	30.15143207	27.84529896	28.12041713	29.08279553	31.18742488	33.55367212	36.09481546	41.72883843	52.22247276
1 I PFC	1.007229647	1.004924302	1.004055949	1.003153689	1.002175308	1.001525288	1.001052197	1.000397341	.9997870683
1 I PQC	153.3358407	141.6047150	143.0042349	147.8998001	158.6057774	170.6423584	183.5683958	212.2264311	265.6020623
1 I ZF1	1.010981082	1.010981082	1.010981082	1.010981082	1.010981082	1.010981082	1.010981082	1.010981082	1.010981082
1 I ZF2	1.023780431	1.023780431	1.023780431	1.023780431	1.023780431	1.023780431	1.023780431	1.023780431	1.023780431
1 I ZF3	1.077809694	1.077809694	1.077809694	1.077809694	1.077809694	1.077809694	1.077809694	1.077809694	1.077809694
1 I ZF4	1.315144782	1.315144782	1.315144782	1.315144782	1.315144782	1.315144782	1.315144782	1.315144782	1.315144782
1 I ZF5	3.108139723	3.108139723	3.108139723	3.108139723	3.108139723	3.108139723	3.108139723	3.108139723	3.108139723

(5) P 10+0 — TH = 83 — SBE = 1.007509825 — CAUER FUNCTIONS

(5) P 10+0	RC = 02	RC = 05	RC = 07	RC = 10	RC = 15	RC = 20	RC = 25	RC = 35	RC = 50
AMAX DB	.0017375255	.0108709564	.0213327376	.0436480540	.0988323393	.1772876696	.2802872360	.5675287486	1.249387366
AMIN DB	15.55856340	23.42381245	26.34713006	29.46232002	33.03657029	35.61283422	37.65358679	40.86299954	44.64269043
MU I K	.1667523D 0	.6742320D-1	.4815523D-1	.3364217D-1	.2229315D-1	.1657137D-1	.1310149D-1	.9054199D-2	.5859566D-2
1 I PF1	1.466500039	1.058352088	.9393706053	.8271693052	.7142499401	.6422313963	.5904801854	.5186347433	.4499567348
1 I PQ1	.6554591606	.6271320188	.6274174643	.6335185950	.6493159906	.6685805515	.6904329921	.7421695074	.8474714526
1 I PF2	1.121270873	1.020063950	.9837076205	.9469144169	.9080722859	.8828266546	.8647132991	.8399782435	.8172542240
1 I PQ2	2.294162450	2.061356773	2.063779559	2.115157894	2.244878374	2.397508956	2.564542973	2.940310795	3.649268382
1 I PF3	1.031994089	1.007828030	.9986681780	.9891502728	.9788427260	.9720074272	.9670406832	.9601733760	.9537977338
1 I PQ3	9.186626634	8.208210908	8.218423009	8.434823030	8.979942043	9.619376731	10.31716972	11.88123175	14.81812059
1 I PF4	1.010779456	1.004772912	1.002467389	1.000055139	.9974238223	.9956681337	.9943870465	.9926096734	.9909494472
1 I PQ4	37.58421143	33.56947836	33.61138939	34.49946546	36.73623655	39.35950758	42.22168647	48.63564294	60.67581890
1 I PFC	1.005884847	1.004060054	1.003357609	1.002621445	1.001817033	1.001279498	1.000886863	1.000341556	.9998315933
1 I PQC	198.5361673	177.3247634	177.5461983	182.2382907	194.0560215	207.9156189	223.0372807	256.9234711	320.5329474
1 I ZF1	1.008464797	1.008464797	1.008464797	1.008464797	1.008464797	1.008464797	1.008464797	1.008464797	1.008464797
1 I ZF2	1.019315113	1.019315113	1.019315113	1.019315113	1.019315113	1.019315113	1.019315113	1.019315113	1.019315113
1 I ZF3	1.067137620	1.067137620	1.067137620	1.067137620	1.067137620	1.067137620	1.067137620	1.067137620	1.067137620
1 I ZF4	1.286333444	1.286333444	1.286333444	1.286333444	1.286333444	1.286333444	1.286333444	1.286333444	1.286333444
1 I ZF5	2.994001069	2.994001069	2.994001069	2.994001069	2.994001069	2.994001069	2.994001069	2.994001069	2.994001069

(5) P 10+0 — TH = 84 — SBE = 1.005508280 — CAUER FUNCTIONS

(5) P 10+0	RC = 02	RC = 05	RC = 07	RC = 10	RC = 15	RC = 20	RC = 25	RC = 35	RC = 50
AMAX DB	.0017375255	.0108709564	.0213327376	.0436480540	.0988323393	.1772876696	.2802872360	.5675287486	1.249387366
AMIN DB	13.04996663	20.83321063	23.74859522	26.85955038	30.43153442	33.00700421	35.04738904	38.25648148	42.03600218
MU I K	.2225880D 0	.9085304D-1	.6494867D-1	.4539651D-1	.3009008D-1	.2236917D-1	.1768604D-1	.1222295D-1	.7910426D-2
1 I PF1	1.559498492	1.117658626	.9892090104	.8686718801	.7480575755	.6715198112	.6167096857	.5408735069	.4686467163
1 I PQ1	.6851706489	.6409944895	.6386367750	.6428221260	.6571872541	.6757903197	.6973016333	.7488104245	.8543780730
1 I PF2	1.128656003	1.033979344	.9988901264	.9629826759	.9247229308	.8996928193	.8816652010	.8569640646	.8341951188
1 I PQ2	2.636176791	2.271784548	2.251422838	2.287494734	2.408883260	2.561193451	2.731554396	3.120139710	3.860287006
1 I PF3	1.031001558	1.010030058	1.001838263	.9932271649	.9838061168	.9775110425	.9729153413	.9665381564	.9605818909
1 I PQ3	11.26193110	9.645727916	9.555097923	9.715627223	10.25498919	10.93012683	11.68350333	13.39655538	16.64590026
1 I PF4	1.009343570	1.004463109	1.002535089	1.000495387	.9982486749	.9967386692	.9956318841	.9940899315	.9926433530
1 I PQ4	48.90718549	41.87427115	41.47981743	42.17849114	44.52575826	47.46352323	50.74130928	58.19317203	72.32479514
1 I PFC	1.004616460	1.003235973	1.002689281	1.002110100	1.001471165	1.001041166	1.000725699	1.000285786	.9998726470
1 I PQC	270.2048024	231.3447560	229.1651970	233.0257254	245.9955119	262.2279681	280.3390650	321.5132467	399.5945168
1 I ZF1	1.006274112	1.006274112	1.006274112	1.006274112	1.006274112	1.006274112	1.006274112	1.006274112	1.006274112
1 I ZF2	1.015226709	1.015226709	1.015226709	1.015226709	1.015226709	1.015226709	1.015226709	1.015226709	1.015226709
1 I ZF3	1.056754503	1.056754503	1.056754503	1.056754503	1.056754503	1.056754503	1.056754503	1.056754503	1.056754503
1 I ZF4	1.256891844	1.256891844	1.256891844	1.256891844	1.256891844	1.256891844	1.256891844	1.256891844	1.256891844
1 I ZF5	2.874084714	2.874084714	2.874084714	2.874084714	2.874084714	2.874084714	2.874084714	2.874084714	2.874084714

(5) P 10+0	TH = 85	SBE = 1.003819838						CAUER FUNCTIONS	
	RC = 02	RC = 05	RC = 07	RC = 10	RC = 15	RC = 20	RC = 25	RC = 35	RC = 50
AMAX DB	.0017375255	.0108709564	.0213327376	.0436480540	.0988323393	.1772876696	.2802872360	.5675287486	1.249387366
AMIN DB	10.43171984	18.05598844	20.95551737	24.05797281	27.62539802	30.19926859	32.23891257	35.44735959	39.22653728
MU I K	.3008943D 0	.1250837D 0	.8958270D-1	.6267601D-1	.4156522D-1	.3090556D-1	.2443736D-1	.1689009D-1	.1093133D-1
1 I PF1	1.666511537	1.188796738	1.049021619	.9183748258	.7883915997	.7063595931	.6478409120	.5671848288	.4906940089
1 I PQ1	.7305850906	.6604834549	.6540257416	.6553020325	.6675263322	.6851433073	.7061374286	.7572610803	.8630848430
1 I PF2	1.132998380	1.047199222	1.013981635	.9794650568	.9422193652	.9176358095	.8998379027	.8753401671	.8526564674
1 I PQ2	3.147585943	2.566794424	2.509886168	2.521190629	2.628070350	2.778089577	2.951587800	3.355308933	4.134482067
1 I PF3	1.029094464	1.011640367	1.004537069	.9969552947	.9885502441	.9828793593	.9787147149	.9729041355	.9674464629
1 I PQ3	14.50393344	11.75395578	11.48345599	11.53720773	12.04496479	12.75643892	13.57774476	15.48397543	19.15004213
1 I PF4	1.007792517	1.004029219	1.002482672	1.000822439	.9989705073	.9977142610	.9967882387	.9954913814	.9942681600
1 I PQ4	67.61440670	54.77831036	53.51546328	53.76641118	56.13685011	59.45805934	63.29163394	72.18818603	89.29522652
1 I PFC	1.003439697	1.002460954	1.002057943	1.001624794	1.001141006	1.000812456	1.000570078	1.000230360	.9999096356
1 I PQC	394.4428423	319.5563379	312.1887538	313.6528117	327.4821880	346.8583781	369.2236553	421.1263161	520.9283568
1 I ZF1	1.004407610	1.004407610	1.004407610	1.004407610	1.004407610	1.004407610	1.004407610	1.004407610	1.004407610
1 I ZF2	1.011524570	1.011524570	1.011524570	1.011524570	1.011524570	1.011524570	1.011524570	1.011524570	1.011524570
1 I ZF3	1.046650430	1.046650430	1.046650430	1.046650430	1.046650430	1.046650430	1.046650430	1.046650430	1.046650430
1 I ZF4	1.226527797	1.226527797	1.226527797	1.226527797	1.226527797	1.226527797	1.226527797	1.226527797	1.226527797
1 I ZF5	2.746245375	2.746245375	2.746245375	2.746245375	2.746245375	2.746245375	2.746245375	2.746245375	2.746245375

(5) P 10+0	TH = 86	SBE = 1.002441898						CAUER FUNCTIONS	
	RC = 02	RC = 05	RC = 07	RC = 10	RC = 15	RC = 20	RC = 25	RC = 35	RC = 50
AMAX DB	.0017375255	.0108709564	.0213327376	.0436480540	.0988323393	.1772876696	.2802872360	.5675287486	1.249387366
AMIN DB	7.729551984	15.03366537	17.89925255	20.98335155	24.54088434	27.11127688	29.14930826	32.35634968	36.13478011
MU I K	.4106978D 0	.1771400D 0	.1273613D 0	.8929609D-1	.5928650D-1	.4409975D-1	.3487664D-1	.2410918D-1	.1560490D-1
1 I PF1	1.785510928	1.276037749	1.122830207	.9797526144	.8380704334	.7491530603	.6859913437	.5993183527	.5175304172
1 I PQ1	.8062928904	.6900037071	.6766294110	.6731236646	.6819020672	.6979438656	.7181017228	.7685476314	.8745764667
1 I PF2	1.132761590	1.058969786	1.028500264	.9961238817	.9605464899	.9367662612	.9194234042	.8953981654	.8730101734
1 I PQ2	3.982885510	3.012187540	2.891370440	2.859260560	2.939310574	3.082746568	3.258406507	3.680183271	4.510239309
1 I PF3	1.026073543	1.012478865	1.006605853	1.000200867	.9929706196	.9880283784	.9843701392	.9792293691	.9743651730
1 I PQ3	20.11147477	15.12215333	14.49875396	14.33294561	14.74620540	15.48590773	16.39058344	18.55849052	22.81287150
1 I PF4	1.006131624	1.003457389	1.002292907	1.001016654	.9995681001	.9985731629	.9978342225	.9967922532	.9958025171
1 I PQ4	102.3001063	76.90357541	73.72987621	72.88572754	74.98965837	78.75536855	83.36068927	94.39571990	116.0488618
1 I PFC	1.002374557	1.001746559	1.001472697	1.001172269	1.000830928	1.000596259	1.000421856	1.000175766	.9999418315
1 I PQC	639.0439297	480.3936640	460.5676789	455.2943061	468.4374972	491.9617334	520.7309225	589.6659493	724.9309487
1 I ZF1	1.002864939	1.002864939	1.002864939	1.002864939	1.002864939	1.002864939	1.002864939	1.002864939	1.002864939
1 I ZF2	1.008223672	1.008223672	1.008223672	1.008223672	1.008223672	1.008223672	1.008223672	1.008223672	1.008223672
1 I ZF3	1.036815955	1.036815955	1.036815955	1.036815955	1.036815955	1.036815955	1.036815955	1.036815955	1.036815955
1 I ZF4	1.194806762	1.194806762	1.194806762	1.194806762	1.194806762	1.194806762	1.194806762	1.194806762	1.194806762
1 I ZF5	2.607120056	2.607120056	2.607120056	2.607120056	2.607120056	2.607120056	2.607120056	2.607120056	2.607120056

$P^{(6)}_{10+2}$

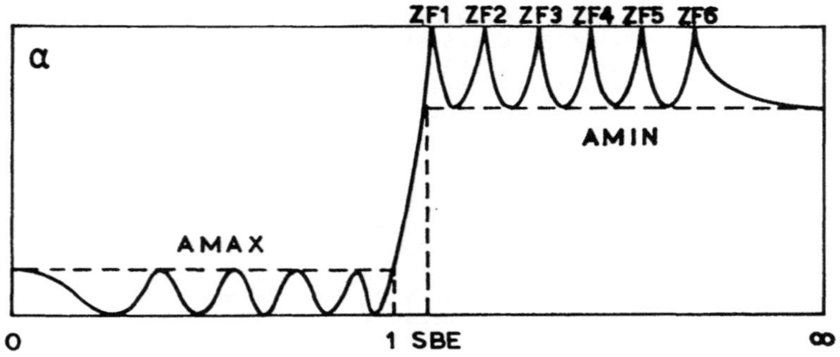

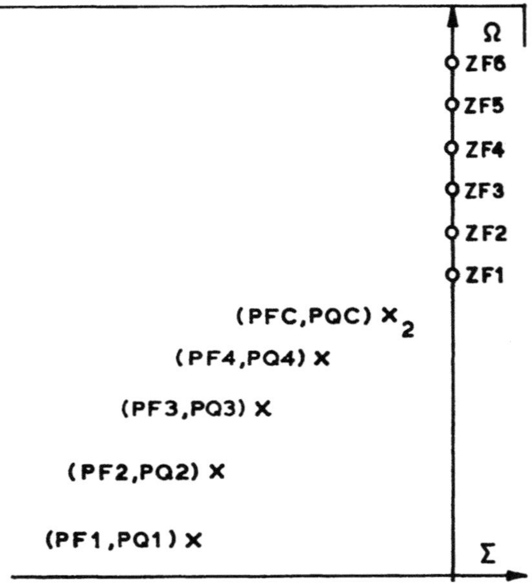

(6) P 10+2		TH = 52	SBE = 1.269018215						MCPER FUNCTIONS	
		RC = 02	RC = 05	RC = 07	RC = 10	RC = 15	RC = 20	RC = 25	RC = 35	RC = 50
AMAX DB		.0017375255	.0108709564	.0213327376	.0436480540	.0988323393	.1772876696	.2802872360	.5675287486	1.249387366
AMIN DB		89.36934549	98.98046277	102.5272729	106.3065116	110.6463079	113.7772836	116.2583140	120.1573039	124.7283208
MU	K	.3400421D-4	.1124545D-4	.7475433D-5	.4838095D-5	.2935517D-5	.2047085D-5	.1538453D-5	.9820527D-6	.5802083D-6
1	PF1	.6038233389	.4771408036	.4345194775	.3914785991	.3451884659	.3140620675	.2909223212	.2577387455	.2248777407
1	PQ1	.5390810108	.5535555506	.5621588888	.5745561572	.5950404913	.6162154678	.6387128449	.6897109266	.7905214822
1	PF2	.7523743195	.6615496035	.6323093607	.6037341730	.5742902405	.5553603901	.5417865009	.5230991191	.5054596421
1	PQ2	.8524834592	.9527552020	1.008807066	1.085860393	1.205564565	1.321692121	1.438668171	1.686439027	2.132604762
1	PF3	.9055642155	.8455750198	.8261197234	.8070216543	.7872023401	.7743306798	.7649974388	.7519159230	.7391278894
1	PQ3	1.567671584	1.802847192	1.932006380	2.107308234	2.375593443	2.632352374	2.888421531	3.424905932	4.378825471
1	PF4	1.009579093	.9708470395	.9580099681	.9452500642	.9318023103	.9229200781	.9163799590	.9070129921	.8974990496
1	PQ4	3.104018299	3.550240304	3.796489400	4.130663693	4.641510292	5.129827277	5.616516333	6.636044839	8.450972608
2	PFC	1.088337529	1.060150835	1.050423031	1.040544951	1.029899270	1.022735706	1.017399134	1.009685799	1.001830615
2	PQC	9.077610323	9.745264558	10.13932220	10.67522185	11.47930215	12.22262789	12.93694609	14.35120505	16.63281264
1	ZF1	1.274752651	1.274926288	1.274984389	1.275043238	1.275107290	1.275151343	1.275185016	1.275235670	1.275291021
1	ZF2	1.325277476	1.326891368	1.327432300	1.327980879	1.328578863	1.328990720	1.329305861	1.329780475	1.330299820
1	ZF3	1.456473267	1.461206310	1.462806468	1.464436287	1.466220827	1.467454584	1.468401092	1.469830400	1.471399317
1	ZF4	1.766277272	1.776565346	1.780088789	1.783699863	1.787678411	1.790443274	1.792571845	1.795797582	1.799352792
1	ZF5	2.626702539	2.648493491	2.656038852	2.663813202	2.672424767	2.678436098	2.683078031	2.690134225	2.697938407
1	ZF6	7.389513308	7.462338767	7.487703625	7.513913798	7.543031563	7.563407114	7.579167260	7.603164776	7.629757577

(6) P 10+2		TH = 53 SBE = 1.252135658								MCPER FUNCTIONS
		RC = 02	RC = 05	RC = 07	RC = 10	RC = 15	RC = 20	RC = 25	RC = 35	RC = 50
AMAX DB		.0017375255	.0108709564	.0213327376	.0436480540	.0988323393	.1772876696	.2802872360	.5675287486	1.249387366
AMIN DB		87.00947343	96.61892272	100.1652607	103.9440650	108.2834389	111.4141549	113.8950037	117.7937481	122.3645349
MU	K	.4461967D-4	.1475890D-4	.9811535D-5	.6350337D-5	.3853258D-5	.2687152D-5	.2019528D-5	.1289177D-5	.7616812D-6
1	PF1	.6114595864	.4826533914	.4394018216	.3957624482	.3488674190	.3173547865	.2939376591	.2603695648	.2271416330
1	PQ1	.5395263793	.5539394503	.5625277961	.5749134629	.5953905457	.6165651276	.6390658666	.6900782127	.7909271056
1	PF2	.7592706946	.6673426431	.6377670632	.6088740983	.5791134237	.5599859674	.5462734807	.5273999257	.5095904943
1	PQ2	.8581975011	.9582246660	1.014318515	1.091512009	1.211530536	1.328020354	1.445391918	1.694053834	2.141905490
1	PF3	.9097443025	.8496928063	.8302030023	.8110662904	.7912035687	.7783030303	.7689491499	.7558402908	.7430294333
1	PQ3	1.592308150	1.827757716	1.957609999	2.134107311	2.404524205	2.663505504	2.921890133	3.463400757	4.426529587
1	PF4	1.010416628	.9722286163	.9595488619	.9469355206	.9336329466	.9248422719	.9183678967	.9090936431	.8996740767
1	PQ4	3.182291859	3.630568421	3.879396193	4.217751941	4.735805984	5.231508761	5.725830331	6.761827064	8.606805283
2	PFC	1.085796151	1.058552139	1.049121323	1.039531127	1.029180957	1.022207984	1.017009211	1.009488893	1.001322972
2	PQC	9.405761460	10.06093016	10.45597247	10.99665368	11.81179027	12.56761548	13.29517574	14.73781018	17.06854344
1	ZF1	1.257613234	1.257788180	1.257846752	1.257906090	1.257970687	1.258015121	1.258049088	1.258100191	1.258156039
1	ZF2	1.305956515	1.307585744	1.308131941	1.308685923	1.309289869	1.309705880	1.310024227	1.310503717	1.311028472
1	ZF3	1.432020506	1.436802585	1.438420177	1.440068194	1.441873169	1.443121368	1.444079132	1.445525729	1.447114036
1	ZF4	1.731338559	1.741730299	1.745293412	1.748947217	1.752975180	1.755775770	1.757932591	1.761202336	1.764307648
1	ZF5	2.566712593	2.588718218	2.596350255	2.604220126	2.612944512	2.619038750	2.623746926	2.630907346	2.638831466
1	ZF6	7.205928087	7.279472477	7.305133987	7.331674233	7.361185385	7.381852065	7.397845890	7.422212699	7.449232490

(6) P 10+2		TH = 54 SBE = 1.236067977								MCPER FUNCTIONS
		RC = 02	RC = 05	RC = 07	RC = 10	RC = 15	RC = 20	RC = 25	RC = 35	RC = 50
AMAX DB		.0017375255	.0108709564	.0213327376	.0436480540	.0988323393	.1772876696	.2802872360	.5675287486	1.249387366
AMIN DB		84.67409019	94.28173068	97.82755650	101.6058898	105.9448058	109.0752401	111.5558919	115.4543700	120.0249073
MU	K	.5838422D-4	.1931583D-4	.1284169D-4	.8312000D-5	.5043822D-5	.3517532D-5	.2643659D-5	.1687647D-5	.9971365D-6
1	PF1	.6194638292	.4884164125	.4445019679	.4002341099	.3527048732	.3207878152	.2970804912	.2631104483	.2294993030
1	PQ1	.5399988008	.5543444563	.5629163600	.5752892421	.5957581554	.6169319886	.6394360198	.6904629956	.7913517224
1	PF2	.7664169083	.6733477040	.6434247558	.6142024372	.5841133381	.5647809250	.5509246587	.5318579371	.5138721317
1	PQ2	.8642670727	.9640068859	1.020136811	1.097470249	1.217911773	1.334677431	1.452460904	1.702053258	2.151668732
1	PF3	.9140042350	.8539085907	.8343883773	.8152163224	.7953129480	.7823850040	.7730114202	.7598763318	.7470437357
1	PQ3	1.618577506	1.854211668	1.984766295	2.162497918	2.435136433	2.696444197	2.957257761	3.504049175	4.476365894
1	PF4	1.011241756	.9736212785	.9611058318	.9486456172	.9354947352	.9267996986	.9203939647	.9112164250	.9018951396
1	PQ4	3.266260730	3.716423613	3.967906455	4.310624118	4.836251181	5.339742718	5.842127249	6.895544353	8.772344907
2	PFC	1.083223981	1.056930177	1.047799587	1.038500790	1.028450154	1.021670586	1.016611754	1.009287608	1.001814217
2	PQC	9.761202665	10.40134372	10.79698314	11.34235911	12.16890511	12.93783248	13.67935960	15.15205426	17.53500906
1	ZF1	1.241295456	1.241471498	1.241530476	1.241590240	1.241655316	1.241700086	1.241734315	1.241785816	1.241842109
1	ZF2	1.287512851	1.289155937	1.289706920	1.290265826	1.290875224	1.291295042	1.291616333	1.292100311	1.292630059
1	ZF3	1.408572774	1.413400205	1.415034054	1.416699105	1.418523283	1.419785097	1.420753495	1.422216467	1.423823195
1	ZF4	1.697671408	1.708159114	1.711759564	1.715453863	1.719528947	1.722363770	1.724547748	1.727859931	1.731513734
1	ZF5	2.508672305	2.530877753	2.538592086	2.546553363	2.555386472	2.561561033	2.566333608	2.573595692	2.581637239
1	ZF6	7.027906635	7.102125032	7.128070093	7.154928259	7.184620903	7.205771371	7.221993738	7.246722890	7.274163143

(6) P 10+2 — TH = 55 SBE = 1.220774589 — MCPER FUNCTIONS

MU		RC = 02	RC = 05	RC = 07	RC = 10	RC = 15	RC = 20	RC = 25	RC = 35	RC = 50
	AMAX DB	.0017375255	.0108709564	.0213327376	.0436480540	.0988323393	.1772876696	.2802872360	.5675287486	1.249387366
	AMIN DB	82.36132339	91.96700256	95.51227308	99.29009563	103.6285150	106.7586438	109.2390820	113.1372714	117.7075382
MU	K	.7619629D-4	.2521447D-4	.1676434D-4	.1085164D-4	.6585280D-5	.4592697D-5	.3451802D-5	.2203619D-5	.1302036D-5
1	PF1	.6278604348	.4944452002	.4498329219	.4049045291	.3567098377	.3243690029	.3003578693	.2659673390	.2319557820
1	PQ1	.5405006612	.5547722393	.5633260830	.5756848590	.5961445717	.6173172491	.6398244786	.6908664478	.7917965786
1	PF2	.7738223596	.6795734613	.6492907605	.6197271316	.5892975083	.5697525095	.5557470797	.5364799174	.5183110531
1	PQ2	.8707237122	.9701272719	1.026286181	1.103758652	1.224431787	1.341687422	1.459900059	1.710464556	2.161926689
1	PF3	.9183419687	.8582227000	.8386767801	.8194732044	.7995324033	.7865787960	.7771866234	.7640267969	.7511735950
1	PQ3	1.646631681	1.882343278	2.013607361	2.192612552	2.467566995	2.731311672	2.994675114	3.547019528	4.530037100
1	PF4	1.012051989	.9750234602	.9626796179	.9503793925	.9373870203	.9287918970	.9224578400	.9133812051	.9041622859
1	PQ4	3.356501956	3.808335416	4.062547482	4.409814979	4.943404043	5.455116704	5.966027676	7.037892940	8.948432137
2	PFC	1.080622524	1.055285647	1.046458303	1.037454259	1.027707031	1.021123601	1.016206803	1.009081920	1.001804279
2	PQC	10.14707243	10.76919133	11.16494963	11.71487757	12.55317435	13.33583612	14.09210416	15.59667654	18.03521510
1	ZF1	1.225758425	1.225935336	1.225994650	1.226054773	1.226120256	1.226165315	1.226199769	1.226251618	1.226308300
1	ZF2	1.269902768	1.271558137	1.272113395	1.272676716	1.273291020	1.273714276	1.274038232	1.274526284	1.275060580
1	ZF3	1.386078893	1.390947754	1.392596607	1.394277454	1.396119521	1.397394069	1.398372440	1.399850813	1.401474925
1	ZF4	1.665207497	1.675783016	1.679418330	1.683150747	1.687270514	1.690137983	1.692347953	1.695700900	1.699401476
1	ZF5	2.452468546	2.474858306	2.482650342	2.490698704	2.499636221	2.505888378	2.510723403	2.518084439	2.526240749
1	ZF6	6.855106607	6.929952622	6.956167660	6.983331097	7.013592840	7.034819432	7.051264966	7.076349177	7.104203030

(6) P 10+2 — TH = 56 SBE = 1.206217949 — MCPER FUNCTIONS

MU		RC = 02	RC = 05	RC = 07	RC = 10	RC = 15	RC = 20	RC = 25	RC = 35	RC = 50
	AMAX DB	.0017375255	.0108709564	.0213327376	.0436480540	.0988323393	.1772876696	.2802872360	.5675287486	1.249387366
	AMIN DB	80.06933397	89.67288649	93.21755488	96.99482359	101.3327045	104.4625021	106.9427086	110.8405850	115.4105584
MU	K	.9920494D-4	.3283641D-4	.2183344D-4	.1413380D-4	.8577580D-5	.5982392D-5	.4496396D-5	.2870587D-5	.1696181D-5
1	PF1	.6366762533	.5007565653	.4554089200	.4097856692	.3608921463	.3281069095	.3037774781	.2689469125	.2345165435
1	PQ1	.5410346492	.5552246654	.5637586384	.5761018293	.5965511813	.6177222353	.6402325411	.6912898637	.7922630480
1	PF2	.7814968893	.6860291333	.6553739471	.6254566664	.5946739893	.5749084872	.5607482982	.5412731262	.5229142381
1	PQ2	.8776029247	.9766142287	1.032793652	1.110403425	1.231416700	1.349077032	1.467737019	1.719317905	2.172714934
1	PF3	.9227550738	.8626353124	.8430690584	.8238383632	.8038638854	.7908866574	.7814772095	.7682942398	.7554219353
1	PQ3	1.676642429	1.912303295	2.044281274	2.224599486	2.501968784	2.768267743	3.034310216	3.592499303	4.586268722
1	PF4	1.012844689	.9764334729	.9642688513	.9521357915	.9393090691	.9308183402	.9245591431	.9155878056	.9064755293
1	PQ4	3.453672325	3.906903215	4.163915120	4.515927816	5.057893220	5.578291775	6.098229285	7.189654683	9.136009970
2	PFC	1.077993386	1.053619300	1.045098009	1.036391885	1.026951776	1.020567129	1.015794400	1.008871809	1.001793087
2	PQC	10.56698950	11.16755009	11.56283925	12.11710679	12.96747668	13.76453506	14.53637142	16.07478584	18.57256828
1	ZF1	1.210964326	1.211141859	1.211201435	1.211261844	1.211327658	1.211372957	1.211407598	1.211459738	1.211516749
1	ZF2	1.253085878	1.254751853	1.255310841	1.255878035	1.256496666	1.256922967	1.257249291	1.257740976	1.258279346
1	ZF3	1.364491578	1.369397695	1.371060219	1.372755542	1.374614100	1.375900442	1.376888080	1.378380814	1.380021205
1	ZF4	1.633883494	1.644538192	1.648205744	1.651973754	1.656135611	1.659034035	1.661268756	1.664660684	1.668406196
1	ZF5	2.397995774	2.420553635	2.428418557	2.436549460	2.445586838	2.451913714	2.456809127	2.464266241	2.472534486
1	ZF6	6.687206943	6.762632671	6.789103591	6.816559166	6.847177088	6.868671796	6.885334874	6.910766506	6.939026746

(6) P 10+2 — TH = 57 SBE = 1.192363293 — MCPER FUNCTIONS

MU	K	RC = 02	RC = 05	RC = 07	RC = 10	RC = 15	RC = 20	RC = 25	RC = 35	RC = 50
	AMAX DB	.0017375255	.0108709564	.0213327376	.0436480540	.0988323393	.1772876696	.2802872360	.5675287486	1.249387366
	AMIN DB	77.79630675	87.39755310	90.94156853	94.71823665	99.05553363	102.1849720	104.6649273	108.5624641	113.1321193
MU	K	.1288797D-3	.4266997D-4	.2837407D-4	.1836911D-4	.1114868D-4	.7775913D-5	.5844584D-5	.3731443D-5	.2204926D-5
1	PF1	.6459409494	.5073689844	.4612455851	.4148906389	.3652625576	.3320108908	.3073477119	.2720561402	.2371875558
1	PQ1	.5416038055	.5557038255	.5642158962	.5765418422	.5969795257	.6181484188	.6406616460	.6917346757	.7927526488
1	PF2	.7894507841	.6927245198	.6616837771	.6314001171	.6002514161	.5802571929	.5659364282	.5462453676	.5276891947
1	PQ2	.8849448183	.9834996083	1.039689461	1.117433827	1.238795613	1.356875975	1.476002503	1.728646808	2.184072969
1	PF3	.9272406854	.8671464318	.8475659564	.8283131856	.8083093639	.7953108927	.7858857052	.7726815202	.7597918129
1	PQ3	1.708804495	1.944261582	2.076954572	2.258625182	2.538513138	2.807491293	3.076350987	3.640697943	4.645812403
1	PF4	1.013617056	.9778494957	.9658720448	.9539136564	.9412600638	.9328784285	.9266974299	.9178359973	.9088348448
1	PQ4	3.558522310	4.012807938	4.272685080	4.629645631	5.180429199	5.710014206	6.239519059	7.351710513	9.336139683
2	PFC	1.075338274	1.051931944	1.043719282	1.035314048	1.026184597	1.020001283	1.015374598	1.008657255	1.001780563
2	PQC	11.02514476	11.59995943	11.99405809	12.55236639	13.41510320	14.22725027	15.01554004	16.58992351	19.15094406
1	ZF1	1.196878149	1.197056042	1.197115800	1.197176417	1.197242481	1.197287965	1.197322754	1.197375125	1.197432401
1	ZF2	1.237024838	1.238699638	1.239261777	1.239832266	1.240454604	1.240883531	1.241211908	1.241706755	1.242248695
1	ZF3	1.343767120	1.348706048	1.350380822	1.352089215	1.353962772	1.355259905	1.356256058	1.357762042	1.359417531
1	ZF4	1.603640637	1.614365358	1.618062361	1.621863280	1.626064465	1.628992043	1.631250191	1.634679195	1.638467680
1	ZF5	2.345155316	2.367864329	2.375797080	2.384005743	2.393138190	2.399536743	2.404490363	2.412040508	2.420417681
1	ZF6	6.523905665	6.599861630	6.626573814	6.654307861	6.685268493	6.707022950	6.723897683	6.749668723	6.778327760

(6) P 10+2 — TH = 58 SBE = 1.179178403 — MCPER FUNCTIONS

MU	K	RC = 02	RC = 05	RC = 07	RC = 10	RC = 15	RC = 20	RC = 25	RC = 35	RC = 50
	AMAX DB	.0017375255	.0108709564	.0213327376	.0436480540	.0988323393	.1772876696	.2802872360	.5675287486	1.249387366
	AMIN DB	75.54044059	85.13918570	88.68249292	92.45850961	96.79517322	99.92422191	102.4039048	106.3010733	110.8703834
MU	K	.1671006D-3	.5534020D-4	.3680233D-4	.2382728D-4	.1446243D-4	.1008762D-4	.7582366D-5	.4841125D-5	.2860756D-5
1	PF1	.6556873894	.5143028194	.4673601054	.4202338379	.3698328712	.3360911982	.3110777618	.2753028623	.2399753424
1	PQ1	.5422115819	.5562120698	.5646999519	.5770067857	.5974313232	.6185974374	.6411133925	.6922024738	.7932670637
1	PF2	.7976947755	.6996700438	.6682303518	.6375672016	.6060390575	.5858075852	.5713201980	.5514050449	.5326440141
1	PQ2	.8927948669	.9908192477	1.047007550	1.124882630	1.246601051	1.365117414	1.484730764	1.738498573	2.196044688
1	PF3	.9317954455	.8717558570	.8521680932	.8328990043	.8128708203	.7998538568	.7904147135	.7771915578	.7642864229
1	PQ3	1.743339539	1.978410227	2.111815214	2.294877158	2.577392705	2.849183212	3.121008282	3.691350171	4.708949835
1	PF4	1.014366122	.9792695641	.9674875823	.9557117168	.9432390927	.9349714809	.9288721853	.9201254930	.9112401625
1	PQ4	3.671913006	4.126826162	4.389626574	4.751744557	5.311817892	5.851129528	6.390787896	7.525056497	9.550019824
2	PFC	1.072659009	1.050224452	1.042322751	1.034221166	1.025405726	1.019426191	1.014947460	1.008438240	1.001766629
2	PQC	11.52641432	12.07050834	12.46253256	13.02447514	13.89983182	14.72778879	15.53347927	17.14613946	19.77476826
1	ZF1	1.183467462	1.183645430	1.183705284	1.183766027	1.183832255	1.183877866	1.183912759	1.183965298	1.184022773
1	ZF2	1.221685105	1.223366835	1.223931509	1.224504677	1.225130065	1.225561172	1.225891263	1.226388773	1.226933745
1	ZF3	1.323865100	1.328832099	1.330517607	1.332237571	1.334124534	1.335431389	1.336435253	1.337953302	1.339622628
1	ZF4	1.574424352	1.585209383	1.588932871	1.592763843	1.597001413	1.599956223	1.602236383	1.605700427	1.609529735
1	ZF5	2.293854719	2.316697158	2.324692425	2.332973815	2.342196271	2.348663285	2.353672800	2.361312743	2.369795645
1	ZF6	6.364917853	6.441352949	6.468291225	6.496289528	6.527578818	6.549584272	6.566664492	6.592766529	6.621816373

(6) P 10+2		TH = 59 SBE = 1.166633397								MCPER FUNCTIONS
		RC = 02	RC = 05	RC = 07	RC = 10	RC = 15	RC = 20	RC = 25	RC = 35	RC = 50
AMAX DB		.0017375255	.0108709564	.0213327376	.0436480540	.0988323393	.1772876696	.2802872360	.5675287486	1.249387366
AMIN DB		73.29993850	82.89597002	86.43850890	90.21381885	94.54979527	97.67842122	100.1578085	104.0545774	108.6235130
MU	K	.2162734D-3	.7164758D-4	.4765128D-4	.3085383D-4	.1872881D-4	.1306408D-4	.9819957D-5	.6270052D-5	.3705308D-5
1	PF1	.6659520961	.5215805720	.4737714435	.4258311268	.3746160632	.3403590934	.3149777179	.2786956228	.2428870510
1	PQ1	.5428619127	.5567520495	.5652131626	.5774987773	.5979084952	.6190711194	.6415895639	.6926950278	.7938081629
1	PF2	.8062400319	.7068767948	.6750244646	.6439683380	.6120468778	.5915693089	.5769090128	.5567612230	.5377874317
1	PQ2	.9012048283	.9986136107	1.054786151	1.132786662	1.254869483	1.373838481	1.493960113	1.748884869	2.208679065
1	PF3	.9364154386	.8764631463	.8568759361	.8375970799	.8175502386	.8045179510	.7950669130	.7818273357	.7689091073
1	PQ3	1.780500900	2.014967279	2.149076124	2.333567427	2.618824864	2.893569896	3.168519515	3.746219950	4.775997410
1	PF4	1.015088739	.9806915579	.9691137066	.9575285782	.9452451399	.9370967264	.9310828138	.9224559402	.9136913617
1	PQ4	3.794836819	4.249847214	4.515618794	4.883110068	5.452977020	6.002599410	6.553048175	7.710823124	9.779008997
2	PFC	1.069957534	1.048497765	1.040909093	1.033113692	1.024615416	1.018841996	1.014513058	1.008214749	1.001751200
2	PQC	12.07649985	12.58394255	12.97280939	13.53784554	14.42601816	15.27053350	16.09463908	17.74308457	20.44911603
1	ZF1	1.170702194	1.170879933	1.170939792	1.171000571	1.171066871	1.171112547	1.171147499	1.171200138	1.171257739
1	ZF2	1.207034712	1.208721359	1.209287911	1.209863104	1.210490839	1.210923652	1.211255098	1.211754737	1.212302168
1	ZF3	1.304748132	1.309738148	1.311432773	1.313162707	1.315061377	1.316376810	1.317387526	1.318916373	1.320598185
1	ZF4	1.546183913	1.557018939	1.560765758	1.564623738	1.568894555	1.571874543	1.574175205	1.577672109	1.581540088
1	ZF5	2.244007163	2.266964476	2.275016672	2.283365487	2.292672609	2.299204678	2.304267637	2.311993940	2.320579162
1	ZF6	6.209973771	6.286835195	6.313983811	6.342231572	6.373834857	6.396082152	6.413361393	6.439785591	6.469217825

(6) P 10+2		TH = 60 SBE = 1.154700538								MCPER FUNCTIONS
		RC = 02	RC = 05	RC = 07	RC = 10	RC = 15	RC = 20	RC = 25	RC = 35	RC = 50
AMAX DB		.0017375255	.0108709564	.0213327376	.0436480540	.0988323393	.1772876696	.2802872360	.5675287486	1.249387366
AMIN DB		71.07299703	80.66608349	84.20778847	87.98233136	92.31756196	95.44572907	97.92479551	101.8211307	106.3896598
MU	K	.2794796D-3	.9261809D-4	.6160424D-4	.3989178D-4	.2421709D-4	.1689326D-4	.1269873D-4	.8103555D-5	.4792002D-5
1	PF1	.6767757848	.5292271842	.4805005818	.4317000262	.3796264438	.3448269835	.3190586879	.2822438178	.2459305355
1	PQ1	.5435593015	.5573267670	.5657581886	.5780201995	.5984131979	.6195715123	.6420921547	.6932143140	.7943780317
1	PF2	.8150981392	.7143565756	.6820776580	.6506147097	.6182856053	.5975527650	.5827130260	.5623236994	.5431288982
1	PQ2	.9102338572	1.006928559	1.063068488	1.141187462	1.263641943	1.383080892	1.503733550	1.759832398	2.222030912
1	PF3	.9410961183	.8812675776	.8616897710	.8424085808	.8223495939	.8093056168	.7998450562	.7865921538	.7736633621
1	PQ3	1.820579387	2.054181263	2.188979466	2.374936623	2.663055838	2.940907448	3.219152998	3.804105205	4.847311782
1	PF4	1.015781569	.9821131881	.9707485067	.9593627104	.9472770746	.9392532932	.9333286305	.9248269126	.9161882634
1	PQ4	3.928442900	4.382894036	4.651670988	5.024756663	5.604955965	6.165522099	6.727454996	7.910298581	10.02465332
2	PFC	1.067235926	1.046752899	1.039479049	1.031992125	1.023813950	1.018248859	1.014071474	1.007986773	1.001734190
2	PQC	12.68210378	13.14579730	13.53017913	14.09760025	14.99870718	15.86055335	16.70416040	18.40112370	21.17983271
1	ZF1	1.158554446	1.158731630	1.158791395	1.158852117	1.158918389	1.158964065	1.158999027	1.159051695	1.159109345
1	ZF2	1.193044064	1.194733491	1.195301223	1.195877743	1.196507079	1.196941091	1.197273510	1.197774709	1.198323987
1	ZF3	1.286381642	1.291389279	1.293091294	1.294829488	1.296738048	1.298060838	1.299077487	1.300615776	1.302308629
1	ZF4	1.518872137	1.529746199	1.533512987	1.537394729	1.541695442	1.544698412	1.547017958	1.550545389	1.554449574
1	ZF5	2.195530922	2.218583678	2.226686928	2.235097575	2.244483712	2.251077226	2.256191026	2.264000034	2.272683940
1	ZF6	6.058817121	6.136050301	6.163392895	6.191874703	6.223776670	6.246256221	6.263727700	6.290464768	6.320270518

(6) TH = 61 SBE = 1.143354068 MCPER FUNCTIONS

MU	P 10+2	RC = 02	RC = 05	RC = 07	RC = 10	RC = 15	RC = 20	RC = 25	RC = 35	RC = 50
	AMAX DB	.0017375255	.0108709564	.0213327376	.0436480540	.0988323393	.1772876696	.2802872360	.5675287486	1.249387366
	AMIN DB	68.85779508	78.44768372	81.98848322	85.76219320	90.09661392	93.22428277	95.70300079	99.59836496	104.1669527
MU	K	.3606702D-3	.1195682D-3	.7953821D-4	.5150986D-4	.3127298D-4	.2181654D-4	.1640023D-4	.1047265D-4	.6189454D-5
1	PF1	.6882039996	.5372703916	.4875708099	.4378599503	.3848798424	.3495085775	.3233329361	.2859578113	.2491144503
1	PQ1	.5443089274	.5579396356	.5663380444	.5785737441	.5989478589	.6201009177	.6426234038	.6937625465	.7949790025
1	PF2	.8242810690	.7221219507	.6894022865	.6575183362	.6247668102	.6037691908	.5887432190	.5681030854	.5486786600
1	PQ2	.9199498623	1.015816283	1.071903618	1.150132059	1.272964774	1.392891690	1.514099509	1.771533686	2.236161794
1	PF3	.9458322235	.8861681006	.8666096666	.8473345583	.8272708391	.8142193295	.8047519666	.7914891317	.7785528466
1	PQ3	1.863910399	2.096336665	2.231801824	2.419258998	2.710365647	2.991486725	3.273213163	3.865343511	4.923296539
1	PF4	1.016441070	.9835319821	.9723899036	.9612124328	.9493336378	.9414401974	.9356088502	.9272379007	.9187306219
1	PQ4	4.074068640	4.527148811	4.798947049	5.177851923	5.768959977	6.341157301	6.915332011	8.124957035	10.28871973
2	PFC	1.064496403	1.044990954	1.038033420	1.030857007	1.023001639	1.017646962	1.013622807	1.007754303	1.001715507
2	PQC	13.35114959	13.76256269	14.14082957	14.70971648	15.62377110	16.50373932	17.36801108	19.11147469	21.97368225
1	ZF1	1.146998323	1.147174602	1.147234169	1.147294731	1.147360871	1.147406478	1.147441396	1.147494016	1.147551635
1	ZF2	1.179685765	1.181375704	1.181943873	1.182520979	1.183151120	1.183585791	1.183918775	1.184420929	1.184971403
1	ZF3	1.268733666	1.273753156	1.275460716	1.277205342	1.279121850	1.280450689	1.281472288	1.283018566	1.284720914
1	ZF4	1.492445116	1.503346558	1.507129732	1.511031770	1.515358793	1.518382395	1.520719092	1.524274545	1.528212341
1	ZF5	2.148348862	2.171476700	2.179624816	2.188091391	2.197550562	2.204201689	2.209363564	2.217251381	2.226030086
1	ZF6	5.911203394	5.988751905	6.016271473	6.044971273	6.077155918	6.099857680	6.117514277	6.144554443	6.174724341

(6) TH = 62 SBE = 1.132570051 MCPER FUNCTIONS

MU	P 10+2	RC = 02	RC = 05	RC = 07	RC = 10	RC = 15	RC = 20	RC = 25	RC = 35	RC = 50
	AMAX DB	.0017375255	.0108709564	.0213327376	.0436480540	.0988323393	.1772876696	.2802872360	.5675287486	1.249387366
	AMIN DB	66.65248179	76.23889607	79.77871178	83.55151680	87.88505756	91.01218503	93.49052446	97.38537684	101.9534849
MU	K	.4649175D-3	.1541896D-3	.1025804D-3	.6643916D-4	.4034104D-4	.2814432D-4	.2115796D-4	.1351158D-4	.7985935D-5
1	PF1	.7002878703	.5457411416	.4950080665	.4443324823	.3903938259	.3544190725	.3278140471	.2898430709	.2524483617
1	PQ1	.5451167758	.5585945532	.5669561603	.5791624634	.5995152229	.6206619319	.6431858333	.6943422155	.7956136940
1	PF2	.8338011284	.7301862969	.6970115845	.6646921539	.6315029924	.6102307507	.5950114944	.5741108988	.5544478521
1	PQ2	.9304311724	1.025336431	1.081347459	1.159673918	1.282890526	1.403324132	1.525112761	1.733898040	2.251141120
1	PF3	.9506176838	.8911632824	.8716354339	.8523759182	.8323158881	.8192615889	.8097905355	.7965217124	.7835813918
1	PQ3	1.910882702	2.141760656	2.277860530	2.466848512	2.761074134	3.045639479	3.331046885	3.931818968	5.004410262
1	PF4	1.017063489	.9849452680	.9740356335	.9630758999	.9514134285	.9436563295	.9379225746	.9296883003	.9213181157
1	PQ4	4.233278944	4.683984650	4.958795836	5.343746119	5.946379893	6.530956687	7.118203052	8.356493198	10.57323664
2	PFC	1.061741346	1.043213121	1.036573080	1.029708936	1.022178829	1.017036509	1.013167166	1.007517337	1.001695057
2	PQC	14.09306130	14.44189135	14.81203782	15.38120614	16.30808122	17.20697337	18.09315463	19.98638066	22.83953083
1	ZF1	1.136009779	1.136184778	1.136244035	1.136304329	1.136370224	1.136415686	1.136450507	1.136502996	1.136560496
1	ZF2	1.166934457	1.168622501	1.169190318	1.169767220	1.170397319	1.170832075	1.171165189	1.171667651	1.172218626
1	ZF3	1.251774672	1.256799848	1.258510979	1.260260081	1.262182457	1.263515945	1.264541441	1.266094153	1.267804334
1	ZF4	1.466861972	1.477778386	1.481574121	1.485492750	1.489842247	1.492883961	1.495235949	1.498816733	1.502785350
1	ZF5	2.102387985	2.125569553	2.133756010	2.142272272	2.151798137	2.158502809	2.163709812	2.171672284	2.180541633
1	ZF6	5.766898298	5.844703770	5.872382629	5.901283677	5.933734265	5.956647708	5.974481940	6.001314910	6.032339064

(6) P 10+2		TH = 63 SBE = 1.122326238								MCPER FUNCTIONS
		RC = 02	RC = 05	RC = 07	RC = 10	RC = 15	RC = 20	RC = 25	RC = 35	RC = 50
AMAX DB		.0017375255	.0108709564	.0213327376	.0436480540	.0988323393	.1772876696	.2802872360	.5675287486	1.249387366
AMIN DB		64.45516319	74.03779987	77.57654595	81.34836704	85.68095101	88.80748981	91.28541759	95.18021346	99.74729997
MU	K	.5987449D-3	.1986598D-3	.1321821D-3	.8562127D-4	.5199391D-4	.3627651D-4	.2727276D-4	.1741764D-4	.1029521D-4
1	PF1	.7130850202	.5546740931	.5028413441	.4511417025	.3961879571	.3595753726	.3325171194	.2939303281	.2559428799
1	PQ1	.5459898010	.5592959912	.5676164566	.5797898341	.6001184052	.6212574960	.6437822961	.6949561318	.7962850569
1	PF2	.8436708865	.7385638545	.7049197400	.6721501050	.6385076808	.6169506397	.6015307810	.5803596715	.5604486055
1	PQ2	.9417685961	1.035557496	1.091464033	1.169874090	1.293479043	1.414438763	1.536835497	1.797042693	2.267047457
1	PF3	.9554455111	.8962512435	.8767665728	.8575333858	.8374865956	.8244349077	.8149637156	.8016934140	.7887530101
1	PQ3	1.961949362	2.190831364	2.327521443	2.518066287	2.815548328	3.103745836	3.393051205	4.002470572	5.091176323
1	PF4	1.017644847	.9863501566	.9756832308	.9649510833	.9535148876	.9459004398	.9402687790	.9321774005	.9239503364
1	PQ4	4.407915592	4.855005052	5.132788857	5.524008905	6.138828872	6.736601535	7.337831120	8.606864367	10.88054389
2	PFC	1.058973304	1.041420694	1.035098981	1.028548565	1.021345902	1.016417726	1.012704681	1.007275379	1.001672738
2	PQC	14.91912073	15.19286133	15.55241282	16.12034268	17.05972345	17.97834007	18.88776131	20.73432502	23.78357555
1	ZF1	1.125566477	1.125739795	1.125798622	1.125858533	1.125924063	1.125969301	1.126003964	1.126056237	1.126113526
1	ZF2	1.154766674	1.156450274	1.157016899	1.157592756	1.158221911	1.158656140	1.158988921	1.159491001	1.160041736
1	ZF3	1.235477408	1.240501665	1.242214254	1.243965737	1.245891753	1.247228389	1.248256653	1.249814128	1.251530362
1	ZF4	1.442084643	1.453002809	1.456807017	1.460738271	1.465106129	1.468163251	1.470528530	1.474131753	1.478128183
1	ZF5	2.057578997	2.080791860	2.089009794	2.097569141	2.107154973	2.113908859	2.119157849	2.127190536	2.136146077
1	ZF6	5.625676233	5.703678243	5.731497991	5.760582814	5.793281828	5.816395895	5.834399894	5.862014317	5.892882763

(6) P 10+2		TH = 64 SBE = 1.112601940								MCPER FUNCTIONS
		RC = 02	RC = 05	RC = 07	RC = 10	RC = 15	RC = 20	RC = 25	RC = 35	RC = 50
AMAX DB		.0017375255	.0108709564	.0213327376	.0436480540	.0988323393	.1772876696	.2802872360	.5675287486	1.249387366
AMIN DB		62.26388730	71.84241315	75.37999528	79.15074564	83.48228841	86.60818661	89.08566638	92.97985659	97.54637543
MU	K	.7705585D-3	.2557875D-3	.1702159D-3	.1102714D-3	.6697081D-4	.4672945D-4	.3513312D-4	.2243919D-4	.1326421D-4
1	PF1	.7266606601	.5641082141	.5111031747	.4583145793	.4022841035	.3649963508	.3374589960	.2982157691	.2596098147
1	PQ1	.5469361290	.5600491054	.5683234359	.5804598344	.6007609577	.6218909557	.6444160327	.6956074816	.7969964299
1	PF2	.8539030688	.7472697796	.7131419749	.6799072373	.6457955471	.6239432028	.6083151558	.5868630730	.5666941718
1	PQ2	.9540679886	1.046558516	1.102326997	1.180802613	1.304798786	1.426304727	1.549338655	1.811044202	2.283970129
1	PF3	.9603076759	.9014295831	.8820022237	.8628074654	.8427847328	.8297417970	.8202745144	.8070078822	.7940719059
1	PQ3	2.017641454	2.243988161	2.381208573	2.573329817	2.874211511	3.166243503	3.459682789	4.078302473	5.184194870
1	PF4	1.018180925	.9877435219	.9773300079	.9668357526	.9556362798	.9481711211	.9426462959	.9347043695	.9266267758
1	PQ4	4.600159812	5.042093462	5.322767439	5.720475128	6.348188152	6.960049537	7.576266810	8.878345681	11.21335456
2	PFC	1.056195019	1.039615076	1.033612165	1.027376614	1.020503285	1.015790873	1.012235497	1.007029938	1.001648447
2	PQC	15.84292852	16.02631162	16.37220371	16.93694871	17.88827092	18.82739372	19.76147406	21.66530196	24.81963162
1	ZF1	1.115647665	1.115818873	1.115877142	1.115936548	1.116001585	1.116046515	1.116080957	1.116132921	1.116189900
1	ZF2	1.143160715	1.144837170	1.145401713	1.145975631	1.146602881	1.147035930	1.147367884	1.147863849	1.148418553
1	ZF3	1.219816762	1.224833023	1.226544804	1.228296422	1.230223688	1.231561862	1.232591681	1.234152128	1.235872498
1	ZF4	1.418077685	1.428983507	1.432791813	1.436731444	1.441113248	1.444182869	1.446559290	1.450181832	1.454202831
1	ZF5	2.013855898	2.037076546	2.045318644	2.053914080	2.063552730	2.070351214	2.075638838	2.083736990	2.092773943
1	ZF6	5.487318790	5.565454741	5.593396210	5.622646557	5.655575642	5.678878718	5.697044196	5.724929618	5.756130277

(6) P 10+2	TH = 65	SBE = 1.103377919						MCPER FUNCTIONS	
	RC = 02	RC = 05	RC = 07	RC = 10	RC = 15	RC = 20	RC = 25	RC = 35	RC = 50
AMAX DB	.0017375255	.0108709564	.0213327376	.0436480540	.0988323393	.1772876696	.2802872360	.5675287486	1.249387366
AMIN DB	60.07662737	69.65067514	73.18698944	76.95657350	81.28698210	84.41218245	86.88917418	90.78270459	95.34860500
MU I K	.9912167D-3	.3292049D-3	.2191041D-3	.1419617D-3	.8622851D-4	.6017151D-4	.4524195D-4	.2889780D-4	.1708322D-4
1 I PF1	.7410889151	.5740875035	.5198302144	.4658814397	.4087068075	.3707031626	.3426585399	.3027212632	.2634623631
1 I PQ1	.5479653148	.5608598721	.5690822959	.5811770392	.6014469495	.6225661350	.6450907408	.6962998931	.7977516070
1 I PF2	.8645104113	.7563201933	.7216946305	.6879798157	.6533825350	.6312240717	.6153799848	.5936360552	.5731990688
1 I PQ2	.9674534751	1.058431189	1.114021540	1.192540251	1.316928468	1.439001374	1.562703534	1.825990158	2.302011157
1 I PF3	.9651949656	.9066952919	.8873410901	.8681983909	.8482119569	.8351847488	.8257259842	.8124688910	.7995424862
1 I PQ3	2.078585410	2.301744527	2.439416091	2.633124430	2.937554467	3.233639161	3.531469657	4.159896645	5.284157613
1 I PF4	1.018667254	.9891219799	.9789730328	.9687274535	.9577756724	.9504667895	.9450537967	.9372682375	.9293468123
1 I PQ4	4.812612300	5.247476013	5.530902281	5.935302517	6.576664493	7.203593470	7.835908922	9.173591022	11.57483211
2 I PFC	1.053409439	1.037797794	1.032113766	1.026193872	1.019651448	1.015156237	1.011759784	1.006779531	1.001622075
2 I PQC	16.88100487	16.95527445	17.28369597	17.84276425	18.80513172	19.76549848	20.72574675	22.69115967	25.95949593
1 I ZF1	1.106234063	1.106402705	1.106460279	1.106519048	1.106583456	1.106627988	1.106662143	1.106713698	1.106770263
1 I ZF2	1.132096535	1.133762986	1.134324499	1.134895528	1.135519851	1.135951024	1.136281625	1.136780691	1.137328518
1 I ZF3	1.204769647	1.209770318	1.211478859	1.213228203	1.215154153	1.216492135	1.217522206	1.219083697	1.220806144
1 I ZF4	1.394808105	1.405686536	1.409494256	1.413437703	1.417828707	1.420907698	1.423292941	1.426931437	1.430973497
1 I ZF5	1.971155598	1.994359242	2.002617828	2.011241929	2.020925787	2.027763943	2.033086614	2.041245135	2.050358363
1 I ZF6	5.351613247	5.429818227	5.457861431	5.487258215	5.520398115	5.543877980	5.562196200	5.590340015	5.621861647

(6) P 10+2	TH = 66	SBE = 1.094636279						MCPER FUNCTIONS	
	RC = 02	RC = 05	RC = 07	RC = 10	RC = 15	RC = 20	RC = 25	RC = 35	RC = 50
AMAX DB	.0017375255	.0108709564	.0213327376	.0436480540	.0988323393	.1772876696	.2802872360	.5675287486	1.249387366
AMIN DB	57.89126257	67.46042634	70.99535806	74.76367026	79.09284200	82.21728130	84.69374078	88.58655158	93.15177746
MU I K	.1274785D-2	.4236222D-3	.2819890D-3	.1827328D-3	.1110089D-3	.7747042D-4	.5825228D-4	.3721109D-4	.2199941D-4
1 I PF1	.7564544456	.5846618677	.5290639511	.4738765385	.4154837331	.3767196238	.3481369677	.3074646377	.2675153332
1 I PQ1	.5490886684	.5617352598	.5698990703	.5819467373	.6021810663	.6232874268	.6458106609	.6970375175	.7985549207
1 I PF2	.8755054594	.7657322280	.7305952603	.6963854474	.6612860032	.6388103237	.6227420871	.6006950205	.5799792515
1 I PQ2	.9820715371	1.071282525	1.126646740	1.205180643	1.329959067	1.452620241	1.577023798	1.841981279	2.321287659
1 I PF3	.9700968220	.9120446489	.8927813629	.8737060671	.8537697750	.8407662123	.8313212085	.8180803425	.8051693720
1 I PQ3	2.145525174	2.364704298	2.502723455	2.698017666	3.006149561	3.306522699	3.609025781	4.247928660	5.391866184
1 I PF4	1.019099099	.9904818639	.9806091042	.9706234829	.9599309125	.9527856622	.9474897710	.9398678782	.9321096931
1 I PQ4	5.048396571	5.473801700	5.759769331	6.171044959	6.826862946	7.469935362	8.119580994	9.495721087	11.96868738
2 I PFC	1.050619741	1.035970516	1.030605032	1.025001211	1.018790916	1.014514143	1.011277733	1.006524684	1.001593506
2 I PQC	18.05358061	17.99553770	18.30172493	18.85192245	19.82399678	20.80626534	21.79427809	23.82603969	27.21841156
1 I ZF1	1.097307761	1.097473348	1.097530083	1.097588075	1.097651711	1.097695748	1.097729544	1.097780587	1.097836527
1 I ZF2	1.121555634	1.123209059	1.123766536	1.124333666	1.124953973	1.125382531	1.125711217	1.126207547	1.126752597
1 I ZF3	1.190314903	1.195291823	1.196994515	1.198738994	1.200660870	1.201996799	1.203025720	1.204536180	1.206308435
1 I ZF4	1.372245209	1.383080174	1.386882281	1.390824645	1.395219738	1.398304722	1.400696284	1.404347096	1.408406418
1 I ZF5	1.929417534	1.952578088	1.960845011	1.969489888	1.979210839	1.986083397	1.991437271	1.999650690	2.008834653
1 I ZF6	5.218351029	5.296557643	5.324681716	5.354204953	5.387535444	5.411179221	5.429640958	5.458030355	5.489860474

(6) P 10+2		TH = 67	SBE = 1.086360377						MCPER FUNCTIONS	
		RC = 02	RC = 05	RC = 07	RC = 10	RC = 15	RC = 20	RC = 25	RC = 35	RC = 50
AMAX DB		.0017375255	.0108709564	.0213327376	.0436480540	.0988323393	.1772876696	.2802872360	.5675287486	1.249387366
AMIN DB		55.70555566	65.26938520	68.80280723	72.56973064	76.89755178	80.02116001	82.49703825	86.38306317	90.95355237
MU	K	.1639541D-2	.5451685D-3	.3629607D-3	.2352412D-3	.1429297D-3	.9975668D-4	.7501500D-4	.4792331D-4	.2833495D-4
1	PF1	.7728544401	.5958881932	.5388515708	.4823387520	.4226462105	.3830726704	.3539182529	.3124660114	.2717854177
1	PQ1	.5503196763	.5626834453	.5707808064	.5827750795	.6029687342	.6240599046	.6465806819	.6978251294	.7994113432
1	PF2	.8869002945	.7755240645	.7398627259	.7051432215	.6695249216	.6467206651	.6304199257	.6080380204	.5870523137
1	PQ2	.9980962411	1.085238179	1.140318538	1.218833004	1.343996352	1.467267534	1.592407960	1.859134031	2.341934814
1	PF3	.9750011534	.9174730996	.8983206228	.8793299986	.8594594992	.8464885664	.8370632857	.8238462656	.8109574111
1	PQ3	2.219350777	2.433582370	2.571814604	2.768677475	3.080668493	3.385585135	3.693069472	4.343187481	5.508255153
1	PF4	1.019471447	.9918191985	.9822347235	.9725208605	.9620995994	.9551257316	.9499525020	.9425019867	.9349145152
1	PQ4	5.311293853	5.724245844	6.012447365	6.430746454	7.101879867	7.762281132	8.430628813	9.848426513	12.39930173
2	PFC	1.047829358	1.034135060	1.029087330	1.023799592	1.017922275	1.013864958	1.010789565	1.006265433	1.001562621
2	PQC	19.38565078	19.16638451	19.44434846	19.98157078	20.96142253	21.96611943	22.98357388	25.08694495	28.61466686
1	ZF1	1.088852126	1.089014140	1.089069881	1.089126948	1.089189656	1.089233098	1.089266459	1.089316878	1.089372274
1	ZF2	1.111520977	1.113158181	1.113710557	1.114272713	1.114887845	1.115312998	1.115639170	1.116131873	1.116673183
1	ZF3	1.176433210	1.181377602	1.183071635	1.184808462	1.186723296	1.188055169	1.189081431	1.190638623	1.192358392
1	ZF4	1.350360467	1.361134774	1.364925871	1.368861877	1.373255549	1.376342878	1.378738050	1.382397235	1.386469697
1	ZF5	1.888583303	1.911673257	1.919939864	1.928597118	1.938346494	1.945247810	1.950628756	1.958891181	1.968139898
1	ZF6	5.087326095	5.165464271	5.193647399	5.223276134	5.256775938	5.280570042	5.299165541	5.327786933	5.359912276

(6) P 10+2		TH = 68	SBE = 1.078534743						MCPER FUNCTIONS	
		RC = 02	RC = 05	RC = 07	RC = 10	RC = 15	RC = 20	RC = 25	RC = 35	RC = 50
AMAX DB		.0017375255	.0108709564	.0213327376	.0436480540	.0988323393	.1772876696	.2802872360	.5675287486	1.249387366
AMIN DB		53.51712675	63.07512072	66.60689174	70.37229639	74.69864052	77.82133986	80.29658241	84.18774780	88.75143123
MU	K	.2109326D-2	.7018495D-3	.4673642D-3	.3029599D-3	.1841060D-3	.1285088D-3	.9664311D-4	.6174654D-4	.3651140D-4
1	PF1	.7904010860	.6078316691	.5492470221	.4913124314	.4302299029	.3897929221	.3600296212	.3177482024	.2762915257
1	PQ1	.5516745476	.5637140895	.5717357892	.5836692635	.6038162746	.6248894633	.6474064732	.6986682517	.8003266129
1	PF2	.8987061647	.7857149569	.7495172984	.7142738646	.6781200169	.6549756470	.6384338330	.6157449900	.5944377284
1	PQ2	1.015735999	1.100446722	1.155173510	1.233625545	1.359164059	1.483067233	1.608982507	1.877583893	2.364109571
1	PF3	.9798941186	.9229751124	.9039557252	.8850692045	.8652821910	.8523540832	.8429553063	.8297708117	.8169116901
1	PQ3	2.301135609	2.509231358	2.647502589	2.845895492	3.161904939	3.471641412	3.784446705	4.446600555	5.634421133
1	PF4	1.019778994	.9931296697	.9838460632	.9744162969	.9642790534	.9574847357	.9524400381	.9451690533	.9377602026
1	PQ4	5.605921147	6.002645044	6.292644838	6.718062859	7.405423052	8.084462558	8.773045908	10.23610404	12.87188484
2	PFC	1.045042004	1.032293418	1.027562162	1.022590079	1.017046178	1.013209094	1.010295532	1.006001826	1.001529295
2	PQC	20.90839522	20.49157709	20.73373835	21.25269088	22.23759710	23.26504424	24.31368350	26.49448199	30.17037649
1	ZF1	1.080851716	1.081009606	1.081064190	1.081120174	1.081181792	1.081224530	1.081257376	1.081307053	1.081361679
1	ZF2	1.101976908	1.103594519	1.104140662	1.104696704	1.105305429	1.105726335	1.106049355	1.106537478	1.107074020
1	ZF3	1.163107022	1.168009432	1.169691782	1.171417956	1.173322548	1.174648205	1.175670179	1.177221689	1.178936337
1	ZF4	1.329127401	1.339822646	1.343596927	1.347520890	1.351907186	1.354992911	1.357388757	1.361052036	1.365133154
1	ZF5	1.848596288	1.871586584	1.879843676	1.888504346	1.898272866	1.905196876	1.910600450	1.918905523	1.928212518
1	ZF6	4.958333220	5.036329972	5.064549312	5.094261533	5.127908225	5.151838295	5.170557218	5.199396177	5.231802601

(6) P 10+2	TH = 69	SBE = 1.071144994						MCPER FUNCTIONS	
	RC = 02	RC = 05	RC = 07	RC = 10	RC = 15	RC = 20	RC = 25	RC = 35	RC = 50
AMAX DB	.0017375255	.0108709564	.0213327376	.0436480540	.0988323393	.1772876696	.2802872360	.5675287486	1.249387366
AMIN DB	51.32342216	60.87501962	64.40498194	68.16872287	72.49344898	75.61515263	78.08969875	81.97992245	86.54272303
MU I K	.2715369D-2	.9041678D-3	.6022141D-3	.3904486D-3	.2373163D-3	.1656694D-3	.1245991D-3	.7961665D-4	.4708297D-4
1 I PF1	.8092246559	.6205674323	.5603123388	.5008484585	.4382756324	.3969153774	.3665021609	.3233372310	.2810551934
1 I PQ1	.5531729352	.5648386930	.5727738294	.5846377711	.6047311012	.6257829960	.6482946512	.6995733125	.8013073927
1 I PF2	.9109329883	.7963252348	.7595807593	.7237999157	.6870940512	.6635979168	.6468062761	.6237780272	.6021571348
1 I PQ2	1.035242412	1.117085122	1.171373701	1.249709827	1.375607937	1.500165046	1.626895856	1.897489495	2.387995334
1 I PF3	.9847598754	.9285440057	.9096826620	.8909221143	.8712385942	.8583648834	.8490003249	.8358582479	.8230375482
1 I PQ3	2.392185600	2.592676287	2.730761507	2.930617099	3.250803679	3.565659676	3.884161051	4.559265919	5.771659956
1 I PF4	1.020016129	.9944085918	.9854389307	.9763061566	.9664662803	.9598601238	.9549501606	.9478673336	.9406454786
1 I PQ4	5.937969243	6.313675249	6.604866651	7.037421394	7.741968601	8.441096202	9.151636807	10.66403262	13.39267913
2 I PFC	1.042261705	1.030447774	1.026031186	1.021373851	1.016163354	1.012547018	1.009795924	1.005733929	1.001493396
2 I PQC	22.66111992	22.00068201	22.19737605	22.69119414	23.67736009	24.72756962	25.80917576	28.07384200	31.91251106
1 I ZF1	1.073292208	1.073445390	1.073498644	1.073553378	1.073613732	1.073655652	1.073687897	1.073736707	1.073790431
1 I ZF2	1.092909083	1.094503545	1.095042256	1.095590972	1.096191978	1.096607743	1.096926930	1.097409457	1.097940133
1 I ZF3	1.150320519	1.155170750	1.156838155	1.158550438	1.160441335	1.161758444	1.162774369	1.164317588	1.166024317
1 I ZF4	1.308521488	1.319117947	1.322869157	1.326774938	1.331147412	1.334227249	1.336620579	1.340283297	1.344368184
1 I ZF5	1.809401273	1.832261160	1.840498938	1.849153445	1.858931150	1.865871326	1.871292731	1.879633570	1.888991812
1 I ZF6	4.831166101	4.908945249	4.937176828	4.966949369	5.000719269	5.024770090	5.043601462	5.072642626	5.105315025

(6) P 10+2	TH = 70	SBE = 1.064177772.						MCPER FUNCTIONS	
	RC = 02	RC = 05	RC = 07	RC = 10	RC = 15	RC = 20	RC = 25	RC = 35	RC = 50
AMAX DB	.0017375255	.0108709564	.0213327376	.0436480540	.0988323393	.1772876696	.2802872360	.5675287486	1.249387366
AMIN DB	49.12167700	58.66624687	62.19422374	65.95613853	70.27908872	73.39969948	75.87348116	79.76267115	84.32450253
MU I K	.3498776D-2	.1165971D-2	.7767635D-3	.5037245D-3	.3062285D-3	.2138036D-3	.1608148D-3	.1027700D-3	.6078198D-4
1 I PF1	.8294773978	.6341826331	.5721192973	.5110055684	.4468304101	.4044302815	.3733715831	.3292629447	.2861010926
1 I PQ1	.5548388956	.5660710594	.5739066359	.5856906723	.6057219718	.6267486204	.6492529908	.7005478447	.8023614706
1 I PF2	.9235886849	.8073762741	.7700764994	.7337459191	.6964720609	.6726125168	.6555621733	.6321817282	.6102346850
1 I PQ2	1.056921976	1.135365902	1.189112903	1.267266403	1.393500949	1.518733480	1.646323430	1.919037905	2.413807927
1 I PF3	.9895802878	.9341717430	.9154963941	.8968864405	.8773290495	.8645228794	.8552013222	.8421129469	.8293405910
1 I PQ3	2.494104975	2.685160293	2.822768383	3.023980711	3.348498539	3.668799308	3.993412490	4.632493758	5.921514681
1 I PF4	1.020176929	.9956508704	.9870087276	.9781864155	.9686579291	.9622490163	.9574803444	.9505948131	.9435688345
1 I PQ4	6.314525184	6.663089724	6.954636141	7.394232184	8.116968211	8.837792935	9.572231933	11.13860468	13.96922763
2 I PFC	1.039492835	1.028600533	1.024496233	1.020152221	1.015274624	1.011879256	1.009291073	1.005461820	1.001454788
2 I PQC	24.69394906	23.73087992	23.86967817	24.32940480	25.31157736	26.38410024	27.50044860	29.85611381	33.87427154
1 I ZF1	1.066160324	1.066308184	1.066359923	1.066413231	1.066472138	1.066513119	1.066544672	1.066592482	1.066645162
1 I ZF2	1.084304413	1.085871979	1.086401989	1.086942093	1.087533984	1.087943652	1.088258282	1.088734127	1.089257763
1 I ZF3	1.138059573	1.142846611	1.144495548	1.146190442	1.148063912	1.149369950	1.150377921	1.151910024	1.153605802
1 I ZF4	1.288520075	1.298996589	1.302717979	1.306598936	1.310950596	1.314019887	1.316407228	1.320064307	1.324147621
1 I ZF5	1.770944049	1.793640915	1.801848912	1.810486989	1.820263165	1.827212458	1.832646502	1.841015639	1.850417471
1 I ZF6	4.705615239	4.783097061	4.811315665	4.841124085	4.874992126	4.899147545	4.919079684	4.947306662	4.980229960

(6) P 10+2 — TH = 71 SBE = 1.057620681 MCPER FUNCTIONS

MU	(6) P 10+2	RC = 02	RC = 05	RC = 07	RC = 10	RC = 15	RC = 20	RC = 25	RC = 35	RC = 50
	AMAX DB	.0017375255	.0108709564	.0213327376	.0436480540	.0988323393	.1772876696	.2802872360	.5675287486	1.249387366
	AMIN DB	46.90886983	56.44569726	59.97148957	63.73139547	68.05239210	71.17180062	73.64474135	77.53279410	82.09355913
MU	K	.4513948D-2	.1505619D-2	.1003288D-2	.6507740D-3	.3957131D-3	.2763185D-3	.2078562D-3	.1328496D-3	.7858181D-4
1	PF1	.8513384823	.6487790540	.5847515140	.5218520183	.4559487349	.4125342180	.3806791791	.3355598050	.2914576710
1	PQ1	.5567021889	.5674279078	.5751483051	.5868400230	.6067993142	.6277959704	.6502906982	.7016007404	.8034980155
1	PF2	.9366782771	.8188904165	.7810296064	.7441386385	.7062816706	.6820472366	.6647292725	.6409835918	.6186974659
1	PQ2	1.081151799	1.155546573	1.208624895	1.286512185	1.413050045	1.538978417	1.667474223	1.942451463	2.441803298
1	PF3	.9943345842	.9398486837	.9213906478	.9029590230	.8835533902	.8708297034	.8615611573	.8485393700	.8358267042
1	PQ3	2.608885426	2.788205628	2.924958864	3.127369773	3.456362398	3.782460946	4.113648366	4.817360822	6.085838327
1	PF4	1.020255144	.9968509612	.9885504030	.9800526122	.9708502434	.9646481568	.9600277126	.9533491643	.9465284910
1	PQ4	6.744515712	7.058041485	7.348794603	7.795172605	8.537126726	9.281437815	10.04197414	11.66763378	14.61072939
2	PFC	1.036740164	1.026754345	1.022959334	1.018926649	1.014380906	1.011206409	1.008781360	1.005185501	1.001413329
2	PQC	27.07161847	25.72947407	25.79423905	26.20809701	27.17902088	28.27272686	29.42551110	31.88006549	36.09694900
1	ZF1	1.059443772	1.059585662	1.059635694	1.059687390	1.059744657	1.059784569	1.059815335	1.059862004	1.059913491
1	ZF2	1.076151012	1.077687738	1.078207701	1.078737829	1.079319123	1.079721678	1.080030976	1.080498983	1.081014322
1	ZF3	1.126311732	1.131023663	1.132650322	1.134324042	1.136176045	1.137468278	1.138466231	1.139984153	1.141665689
1	ZF4	1.269102322	1.279436164	1.283120441	1.286969374	1.291292617	1.294346287	1.296723848	1.300369732	1.304445619
1	ZF5	1.733170978	1.755670150	1.763837159	1.772447772	1.782210856	1.789161633	1.794602679	1.802991985	1.812429038
1	ZF6	4.581465505	4.658566305	4.686745339	4.716563783	4.750503358	4.774746174	4.793766615	4.823161865	4.856316493

(6) P 10+2 — TH = 72 SBE = 1.051462224 MCPER FUNCTIONS

MU	(6) P 10+2	RC = 02	RC = 05	RC = 07	RC = 10	RC = 15	RC = 20	RC = 25	RC = 35	RC = 50
	AMAX DB	.0017375255	.0108709564	.0213327376	.0436480540	.0988323393	.1772876696	.2802872360	.5675287486	1.249387366
	AMIN DB	44.68166712	54.20993552	57.73331773	61.49100794	65.80985014	68.92793282	71.39994610	75.28674457	79.84633341
MU	K	.5833331D-2	.1947616D-2	.1298178D-2	.8422666D-3	.5122801D-3	.3577695D-3	.2691552D-3	.1720532D-3	.1017849D-3
1	PF1	.8750203553	.6644764664	.5983071296	.5334677212	.4656942503	.4211315003	.3884730376	.3422678855	.2971579634
1	PQ1	.5588000633	.5689296919	.5765159702	.5881003386	.6079756542	.6289365774	.6514187664	.7027425340	.8047279091
1	PF2	.9502026796	.8308908158	.7924669315	.7550072903	.7165534545	.6919330364	.6743386089	.6502145140	.6275760171
1	PQ2	1.108401009	1.177942253	1.230194402	1.307710195	1.434505089	1.561147769	1.690599418	1.967996715	2.472287562
1	PF3	.9989989534	.9455632826	.9273576657	.9091356352	.8899108097	.8772866154	.8680825045	.8551420449	.8425020672
1	PQ3	2.739028939	2.903696247	3.039102181	3.242482536	3.576074011	3.908353138	4.246631166	4.967282643	6.266876908
1	PF4	1.020244192	.9980028232	.9900584007	.9818997905	.9730390035	.9670538565	.9625889824	.9561276965	.9495223524
1	PQ4	7.239327138	7.507526884	7.795911531	8.248575090	9.010779144	9.780569251	10.56970620	12.26076951	15.32851767
2	PFC	1.034008898	1.024912141	1.021422744	1.017698769	1.013483240	1.010529157	1.008267226	1.004905400	1.001368869
2	PQC	29.87891608	28.05742517	28.02697481	28.37933717	29.32897933	30.44173430	31.63244563	34.19459826	38.63247844
1	ZF1	1.053131187	1.053266431	1.053314552	1.053364439	1.053419862	1.053458571	1.053488449	1.053533829	1.053583963
1	ZF2	1.068438156	1.069939894	1.070448391	1.070967095	1.071536216	1.071930576	1.072233716	1.072692649	1.073198351
1	ZF3	1.115066226	1.119690145	1.121290397	1.122938842	1.124764995	1.126040456	1.127026149	1.128526559	1.130190271
1	ZF4	1.250249155	1.260415885	1.264055149	1.267864236	1.272150780	1.275183282	1.277546911	1.281175510	1.285237534
1	ZF5	1.696028525	1.718293027	1.726407013	1.734978263	1.744715737	1.751659705	1.757101609	1.765502198	1.774965296
1	ZF6	4.458493277	4.535124840	4.563236157	4.593037184	4.627019964	4.651331805	4.670427198	4.699971891	4.733340237

(6) P 10+2	TH = 73	SBE = 1.045691756						MCPER FUNCTIONS	
	RC = 02	RC = 05	RC = 07	RC = 10	RC = 15	RC = 20	RC = 25	RC = 35	RC = 50
AMAX DB	.0017375255	.0108709564	.0213327376	.0436480540	.0988323393	.1772876696	.2802872360	.5675287486	1.249387366
AMIN DB	42.43635457	51.95512125	55.47583605	59.23107516	63.54753450	66.66415082	69.13513833	73.02054949	77.57883726
MU I K	.7554092D-2	.2524899D-2	.1683481D-2	.1092562D-2	.6646963D-3	.4642933D-3	.3493358D-3	.2233431D-3	.1321473D-3
1 I PF1	.9007769794	.6814169858	.6129022862	.5459470034	.4761418811	.4303359636	.3968096120	.3494341568	.3032406338
1 I PQ1	.5611797513	.5706017146	.5780306775	.5894895461	.6092661845	.6301843753	.6526504438	.7039860872	.8060641809
1 I PF2	.9641570618	.8434011713	.8044171171	.7663837956	.7273213608	.7023045547	.6844250603	.6599093951	.6369049724
1 I PQ2	1.139260420	1.202942811	1.254171902	1.331182672	1.458170788	1.585543003	1.716003859	1.995996276	2.505630297
1 I PF3	1.003546068	.9513017213	.9333878984	.9154107437	.8963996949	.8838943856	.8747677713	.8619255318	.8493731661
1 I PQ3	2.887719957	3.033991450	3.167403890	3.371427111	3.709708649	4.048582511	4.394530032	5.133110759	6.467380957
1 I PF4	1.020137159	.9990998671	.9915265972	.9837224330	.9752194589	.9694619270	.9651603999	.9589272952	.9525479513
1 I PQ4	7.813687674	8.023006216	8.306855801	8.764964711	9.548410914	10.34590086	11.16650304	12.93006591	16.13671319
2 I PFC	1.031304741	1.023077174	1.019888981	1.016470414	1.012582799	1.009848278	1.007749177	1.004621371	1.001321255
2 I PQC	33.22863845	30.79442378	30.64061049	30.91051991	31.82494630	32.95313316	34.18286708	36.86218333	41.54700412
1 I ZF1	1.047212082	1.047339973	1.047385972	1.047433845	1.047487209	1.047524573	1.047553457	1.047597390	1.047646005
1 I ZF2	1.061156256	1.062618647	1.063114177	1.063619926	1.064175199	1.064560213	1.064856316	1.065304855	1.065799484
1 I ZF3	1.104313989	1.108835897	1.110405264	1.112023979	1.113819521	1.115074984	1.116045978	1.117525247	1.119167232
1 I ZF4	1.231943240	1.241916542	1.245502224	1.249262949	1.253503740	1.256508996	1.258854136	1.262458747	1.266499808
1 I ZF5	1.659462722	1.681452982	1.689500973	1.698019986	1.707718245	1.714646354	1.720082395	1.728484517	1.737963549
1 I ZF6	4.336462998	4.412531897	4.440545588	4.470299962	4.504295665	4.528656833	4.547812834	4.577486679	4.611048509

(6) P 10+2	TH = 74	SBE = 1.040299436						MCPER FUNCTIONS	
	RC = 02	RC = 05	RC = 07	RC = 10	RC = 15	RC = 20	RC = 25	RC = 35	RC = 50
AMAX DB	.0017375255	.0108709564	.0213327376	.0436480540	.0988323393	.1772876696	.2802872360	.5675287486	1.249387366
AMIN DB	40.16875159	49.67691344	53.19466491	56.94718301	61.26099798	64.37598717	66.84583644	70.72970810	75.28655194
MU I K	.9807593D-2	.3282119D-2	.2189106D-2	.1421153D-2	.8648685D-3	.6042277D-3	.4546824D-3	.2907466D-3	.1720570D-3
1 I PF1	.9289146360	.6997707949	.6286756907	.5594022187	.4873806293	.4402233062	.4057557656	.3571141598	.3097513329
1 I PQ1	.5639020254	.5724756692	.5797185890	.5910294396	.6106895333	.6315563790	.6540018629	.7053466726	.8075225862
1 I PF2	.9785286228	.8564452963	.8169105576	.7783030436	.7386232116	.7132007242	.6950280310	.6701078956	.6467238659
1 I PQ2	1.174484380	1.231036524	1.280993943	1.357328967	1.484422906	1.612534755	1.744061526	2.026844799	2.542282396
1 I PF3	1.007944513	.9570474551	.9394696224	.9217772045	.9030174129	.8906531420	.8816189904	.8688943769	.8564468047
1 I PQ3	3.059071377	3.182085009	3.312648849	3.516853203	3.859863083	4.205777735	4.560046308	5.318265738	6.690757684
1 I PF4	1.019926805	1.000134896	.9929482308	.9855143828	.9773862461	.9718676000	.9677376618	.9617443473	.9556023803
1 I PQ4	8.486948940	8.619288359	8.895605639	9.357817823	10.16338876	10.99105228	11.84641558	13.69077347	17.05313173
2 I PFC	1.028633962	1.021253069	1.018360859	1.015243649	1.011680917	1.009164665	1.007227803	1.004333709	1.001270329
2 I PQC	37.27348182	34.04631718	33.73121161	33.89021104	34.74993084	35.88772767	37.15687442	39.96376422	44.92594809
1 I ZF1	1.041676794	1.041796604	1.041840259	1.041885904	1.041936987	1.041972856	1.042000634	1.042042956	1.042089875
1 I ZF2	1.054296835	1.055715304	1.056196282	1.056687455	1.057227103	1.057601544	1.057889674	1.058326410	1.058808433
1 I ZF3	1.094047714	1.098452405	1.099986017	1.101570160	1.103329908	1.104561858	1.105515491	1.106969660	1.108585649
1 I ZF4	1.214168986	1.223920482	1.227443263	1.231146330	1.235331449	1.238302776	1.240624406	1.244197627	1.248209863
1 I ZF5	1.623418545	1.645092033	1.653059989	1.661512775	1.671156971	1.678059303	1.683482097	1.691875002	1.701358780
1 I ZF6	4.215122954	4.290529635	4.318413764	4.348090192	4.382066308	4.406455592	4.425656717	4.455437756	4.489171100

(6) P 10+2 — TH = 75 — SBE = 1.035276180 — MCPER FUNCTIONS

MU	K	RC = 02	RC = 05	RC = 07	RC = 10	RC = 15	RC = 20	RC = 25	RC = 35	RC = 50
	AMAX DB	.0017375255	.0108709564	.0213327376	.0436480540	.0988323393	.1772876696	.2802872360	.5675287486	1.249387366
	AMIN DB	37.87410406	47.37034734	50.88479175	54.63427670	58.94514565	62.05832243	64.52690391	68.40906077	72.96429609
MU	K	.1277306D-1	.4280383D-2	.2856015D-2	.1854753D-2	.1129127D-2	.7890125D-3	.5938200D-3	.3797930D-3	.2247942D-3
1	PF1	.9598062364	.7197437735	.6457946939	.5739685548	.4995172842	.4508841928	.4153914808	.3653742201	.3167444903
1	PQ1	.5670463761	.5745918112	.5816126600	.5927475028	.6122688195	.6330736127	.6554928975	.7068432635	.8091223886
1	PF2	.9932935473	.8700464363	.8299792482	.7908031553	.7505012937	.7246655233	.7061922975	.6808553362	.6570781536
1	PQ2	1.215051058	1.262843381	1.311211552	1.386650407	1.513730656	1.642584320	1.775236820	2.061030844	2.582800450
1	PF3	1.012153103	.9627806480	.9455884576	.9282258783	.9097600369	.8975621717	.8886376759	.8760530462	.8637301117
1	PQ3	3.258483722	3.351832457	3.478404268	3.682138129	4.029831177	4.383263121	4.746588957	5.526422471	6.941282311
1	PF4	1.019605579	1.001100042	.9943158201	.9872687501	.9795332916	.9742654298	.9703158194	.9645746509	.9586822088
1	PQ4	9.284989685	9.315818325	9.580420862	10.04465437	10.87300757	11.73359279	12.62753041	14.56246390	18.10056913
2	PFC	1.026003471	1.019443875	1.016841547	1.014020807	1.010779115	1.008479341	1.006703789	1.004042649	1.001215926
2	PQC	42.22424953	37.95623158	37.42790398	37.43679135	38.21426783	39.35254172	40.66028725	43.60589513	48.88136337
1	ZF1	1.036516448	1.036627429	1.036668513	1.036711709	1.036760277	1.036794496	1.036821051	1.036861589	1.036906627
1	ZF2	1.047852522	1.049222267	1.049687023	1.050161905	1.050684043	1.051046609	1.051325767	1.051749198	1.052216973
1	ZF3	1.084261927	1.088532862	1.090025421	1.091569713	1.093288012	1.094492611	1.095425975	1.096350710	1.098436027
1	ZF4	1.196912569	1.206411604	1.209861330	1.213496563	1.217615106	1.220545134	1.222837699	1.226371324	1.230345995
1	ZF5	1.587839187	1.609149947	1.617022602	1.625393886	1.634967729	1.641833362	1.647234749	1.655606523	1.665082595
1	ZF6	4.094199979	4.168837544	4.196557805	4.226122607	4.260044050	4.284438490	4.303667942	4.333532292	4.367413169

(6) P 10+2 — TH = 76 — SBE = 1.030613629 — MCPER FUNCTIONS

MU	K	RC = 02	RC = 05	RC = 07	RC = 10	RC = 15	RC = 20	RC = 25	RC = 35	RC = 50
	AMAX DB	.0017375255	.0108709564	.0213327376	.0436480540	.0988323393	.1772876696	.2802872360	.5675287486	1.249387366
	AMIN DB	35.54694939	45.02967332	48.54040638	52.28649353	56.59406530	59.70521437	62.17237755	66.05261604	70.60605173
MU	K	.1669754D-1	.5604235D-2	.3740931D-2	.2430386D-2	.1480119D-2	.1034521D-2	.7787196D-3	.4991604D-3	.2949154D-3
1	PF1	.9939104787	.7415878335	.6644635194	.5898105318	.5126814309	.4624284388	.4258135071	.3742944253	.3242857225
1	PQ1	.5707187324	.5770020874	.5837550333	.5946785279	.6140331321	.6347624029	.6571483522	.7034993767	.8108874355
1	PF2	1.008412807	.8842262101	.8436564531	.8039257235	.7630030560	.7367489032	.7179690702	.6922041587	.6680205296
1	PQ2	1.262251315	1.299162922	1.345530764	1.419785554	1.546688262	1.676273838	1.810114358	2.099167410	2.627880673
1	PF3	1.016145044	.9684774631	.9517267559	.9347451351	.9166219883	.9046196582	.8958246317	.8834058329	.8712305433
1	PQ3	3.493182451	3.548283024	3.669314309	3.871655136	4.223855728	4.585306746	4.958524693	5.762273497	7.224397469
1	PF4	1.019165654	1.001986690	.9956210673	.9889778015	.9816536924	.9766491731	.9728891606	.9674133021	.9617833777
1	PQ4	10.24311658	10.13859572	10.38557881	10.84864635	11.70002353	12.59656428	13.53351259	15.57066214	19.30865942
2	PFC	1.023420913	1.017654143	1.015334619	1.012804542	1.009879136	1.007793494	1.006177935	1.003748484	1.001157882
2	PQC	48.37850131	42.72165522	41.90770054	41.71155857	42.36738216	43.49195950	44.83547119	47.93137688	53.56284723
1	ZF1	1.031722913	1.031824309	1.031862589	1.031903107	1.031948919	1.031981327	1.032006537	1.032045111	1.032088075
1	ZF2	1.041817056	1.043133037	1.043579812	1.044036592	1.044539225	1.044888531	1.045157654	1.045566177	1.046017949
1	ZF3	1.074953102	1.079072275	1.080518001	1.082016680	1.083687349	1.084860393	1.085770301	1.087160847	1.088710352
1	ZF4	1.180161996	1.189375395	1.192740969	1.196297196	1.200337134	1.203217697	1.205475030	1.208959921	1.212887261
1	ZF5	1.552665156	1.573563210	1.581323867	1.589596873	1.599082394	1.605899226	1.611270137	1.619607495	1.629061917
1	ZF6	3.973392682	4.047145221	4.074664497	4.104081178	4.137909836	4.162284426	4.181523874	4.211445402	4.245447485

(6) P 10+2	TH = 77 SBE = 1.026304108							MCPER FUNCTIONS	
	RC = 02	RC = 05	RC = 07	RC = 10	RC = 15	RC = 20	RC = 25	RC = 35	RC = 50
AMAX DB	.0017375255	.0108709564	.0213327376	.0436480540	.0988323393	.1772876696	.2802872360	.5675287486	1.249387366
AMIN DB	33.18094699	42.64814221	46.15468080	49.89693870	54.20080013	57.30966879	59.77523720	63.65331872	68.20473082
MU I K	.2192566D-1	.7372128D-2	.4923410D-2	.3200023D-2	.1949665D-2	.1363065D-2	.1026214D-2	.6566502D-3	.3888333D-3
1 I PF1	1.031797728	.7656151623	.6849346092	.6071309545	.5270323376	.4749907561	.4371403670	.3839727053	.3324551288
1 I PQ1	.5750632935	.5797747500	.5862005313	.5968673669	.6160196458	.6366562190	.6589996484	.7103446628	.8128476702
1 I PF2	1.023826312	.8990029751	.8579760752	.8177159864	.7761819315	.7495079372	.7304173388	.7042149898	.6796126470
1 I PQ2	1.317823212	1.341044663	1.384871825	1.457561339	1.584060441	1.714349607	1.849441615	2.142035457	2.678407107
1 I PF3	1.019856908	.9741091583	.9578628128	.9413202094	.9235955647	.9118223305	.9031796906	.8909567249	.8789558743
1 I PQ3	3.773044189	3.778176724	3.891538130	4.091169814	4.447498260	4.817483429	5.201542913	6.031911858	7.547147012
1 I PF4	1.018598982	1.002785397	.9968547462	.9906328263	.9837395703	.9790116412	.9754510623	.9702545532	.9649010679
1 I PQ4	11.41061348	11.12311014	11.34401262	11.80104737	12.67495602	13.61075952	14.59590736	16.74927214	20.71662687
2 I PFC	1.020894782	1.015889007	1.013844132	1.011597884	1.008982990	1.007108501	1.005651182	1.003451573	1.001096029
2 I PQC	56.16704654	48.62144038	47.41876248	46.93914000	47.41599186	48.50438973	49.87796501	53.13552291	59.17416043
1 I ZF1	1.027288767	1.027379821	1.027415062	1.027452670	1.027495481	1.027525910	1.027549651	1.027586074	1.027626760
1 I ZF2	1.036185308	1.037442228	1.037869174	1.038305943	1.038786958	1.039121535	1.039379494	1.039771400	1.040205292
1 I ZF3	1.066119812	1.070067602	1.071460183	1.072906947	1.074523213	1.075660089	1.076543038	1.077894164	1.079402189
1 I ZF4	1.163907211	1.172798983	1.176068243	1.179533157	1.183481166	1.186303182	1.188518403	1.191944334	1.195813373
1 I ZF5	1.517833168	1.538263712	1.545893987	1.554050169	1.563427409	1.570181939	1.575512225	1.583800247	1.593217292
1 I ZF6	3.852362599	3.925102846	3.952380607	3.981609295	4.015303453	4.039630761	4.058860049	4.088809958	4.122904140

(6) P 10+2	TH = 78 SBE = 1.022340595							MCPER FUNCTIONS	
	RC = 02	RC = 05	RC = 07	RC = 10	RC = 15	RC = 20	RC = 25	RC = 35	RC = 50
AMAX DB	.0017375255	.0108709564	.0213327376	.0436480540	.0988323393	.1772876696	.2802872360	.5675287486	1.249387366
AMIN DB	30.76866785	40.21771436	43.71946916	47.45737909	51.75703763	54.86332578	57.32709017	61.20273201	65.75195525
MU I K	.2894454D-1	.9752462D-2	.6516682D-2	.4237708D-2	.2583141D-2	.1806482D-2	.1360334D-2	.8706897D-3	.5157120D-3
1 I PF1	1.074185217	.7922182379	.7075245873	.6261835034	.5427686304	.4887388157	.4495193730	.3945305446	.3413518979
1 I PQ1	.5802812248	.5830013565	.5890218904	.5993729348	.6132767273	.6387983569	.6610872739	.7124171323	.8150413030
1 I PF2	1.039444656	.9143892972	.8729715402	.8322228546	.7900983012	.7630082637	.7436056024	.7169591955	.6919274070
1 I PQ2	1.384161756	1.389894580	1.430457956	1.501069168	1.626849599	1.757785808	1.894191480	2.190647511	2.735521800
1 I PF3	1.023237354	.9796409180	.9639698371	.9479323471	.9306703065	.9191649865	.9107013567	.8987092159	.8869141748
1 I PQ3	4.111907795	4.050710501	4.153419742	4.348442022	4.708196755	5.087219929	5.483202294	6.343402286	7.918820871
1 I PF4	1.017897394	1.003485804	.9980065704	.9922239740	.9857818902	.9813445134	.9779938065	.9730216323	.9680295296
1 I PQ4	12.85811486	12.31897086	12.50144157	12.94497058	13.83964368	14.81822725	15.85767100	18.14428334	22.37747437
2 I PFC	1.018434554	1.014154292	1.012374721	1.010404317	1.008093008	1.006425979	1.005124645	1.003152361	1.001030204
2 I PQC	66.23210197	56.05990392	54.31807974	53.44030900	53.65315411	54.67202520	56.06279849	59.49178514	65.99928893
1 I ZF1	1.023207261	1.023287234	1.023319202	1.023353668	1.023393228	1.023421510	1.023443651	1.023477731	1.023515930
1 I ZF2	1.030953320	1.032145606	1.032550780	1.032965529	1.033422695	1.033740985	1.033986579	1.034360048	1.034774057
1 I ZF3	1.057762933	1.061517952	1.062850486	1.064238430	1.065792855	1.066888488	1.067740617	1.069046545	1.070506821
1 I ZF4	1.148140270	1.156671250	1.159830824	1.163190811	1.167032069	1.169785335	1.171950776	1.175306236	1.179104582
1 I ZF5	1.483274732	1.503177039	1.510656516	1.518675209	1.527921820	1.534598867	1.539877078	1.543098370	1.557469649
1 I ZF6	3.730722392	3.802308316	3.829299807	3.858296496	3.891810062	3.916059717	3.935256495	3.965202777	3.999356608

(6) P 10+2			TH = 79	SBE = 1.018716695						MCPER FUNCTIONS
			RC = 02	RC = 05	RC = 07	RC = 10	RC = 15	RC = 20	RC = 25	RC = 35	RC = 50
AMAX DB			.0017375255	.0108709564	.0213327376	.0436480540	.0988323393	.1772876696	.2802872360	.5675287486	1.249387366
AMIN DB			28.30134139	37.72865730	41.22488967	44.95781507	49.25267379	52.35601996	54.81772384	58.69059170	63.23710795
MU		K	.3845324D-1	.1298874D-1	.8684714D-2	.5650791D-2	.3446405D-2	.2411010D-2	.1815990D-2	.1162707D-2	.6898816D-3
1		PF1	1.121984972	.8218985626	.7326372449	.6472908733	.5601422175	.5038848387	.4631367066	.4061211754	.3511009094
1		PQ1	.5866612652	.5868077256	.5923178496	.6022743240	.6208686235	.6412459640	.6634644381	.7147664567	.8175180186
1		PF2	1.055137310	.9303880082	.8886738726	.8474986459	.8048206059	.7773259100	.7576141248	.7305213784	.7050520687
1		PQ2	1.464657446	1.447641295	1.483952613	1.551781609	1.676398045	1.807880977	1.945656642	2.246343203	2.800729828
1		PF3	1.026220505	.9850303176	.9700145811	.9545576585	.9378321323	.9266398341	.9183863031	.9066660315	.8951137594
1		PQ3	4.529731500	4.378758762	4.466545884	4.654169085	5.016133150	5.404639151	5.813774499	6.707659152	8.351942255
1		PF4	1.017052743	1.004076537	.9990650379	.9937400547	.9877702318	.9836380994	.9805083361	.9759165080	.9711618610
1		PQ4	14.69001921	13.79746857	13.92306180	14.34146307	15.25292068	16.27784217	17.37876418	19.81961627	24.36456068
2		PFC	1.016050853	1.012456650	1.010931708	1.009227873	1.007211916	1.005747831	1.004599652	1.002851404	1.000960250
2		PQC	79.56412064	65.64162435	63.13480452	61.68668421	61.50628327	62.40067558	63.78761730	67.39353022	74.44471668
1		ZF1	1.019472286	1.019540477	1.019568946	1.019600042	1.019636103	1.019662066	1.019682478	1.019714018	1.019749515
1		ZF2	1.026118377	1.027240137	1.027621502	1.028012119	1.028443084	1.028743442	1.028975398	1.029328495	1.029720479
1		ZF3	1.049885915	1.053424852	1.054689783	1.056011329	1.057495726	1.058544514	1.059361555	1.060615384	1.062021445
1		ZF4	1.132855607	1.140983000	1.144018125	1.147258062	1.150975995	1.153649194	1.155756044	1.159027096	1.162741544
1		ZF5	1.448914323	1.468220176	1.475525959	1.483383906	1.492474629	1.499056957	1.504270048	1.512406294	1.521692259
1		ZF6	3.608019753	3.678289441	3.704944517	3.733660017	3.766941469	3.791079480	3.810218694	3.840125403	3.874302337

(6) P 10+2			TH = 80	SBE = 1.015426612						MCPER FUNCTIONS
			RC = 02	RC = 05	RC = 07	RC = 10	RC = 15	RC = 20	RC = 25	RC = 35	RC = 50
AMAX DB			.0017375255	.0108709564	.0213327376	.0436480540	.0988323393	.1772876696	.2802872360	.5675287486	1.249387366
AMIN DB			25.76857476	35.16897748	38.65872774	42.38586587	46.67518606	49.77514701	52.23449317	56.10416350	60.64768549
MU		K	.5147203D-1	.1744019D-1	.1166981D-1	.7598130D-2	.4637038D-2	.3245209D-2	.2444980D-2	.1566000D-2	.9281448D-3
1		PF1	1.176368346	.8553089015	.7607975099	.6708716277	.5794788981	.5207027379	.4782323125	.4189416696	.3618624611
1		PQ1	.5946319034	.5913717355	.5962261010	.6056804655	.6238827787	.6440762727	.6662027058	.7174590241	.8203438546
1		PF2	1.070715418	.9469859755	.9051084099	.8635982569	.8204265885	.7925496205	.7725379262	.7450031816	.7190925713
1		PQ2	1.564264914	1.517005071	1.547679352	1.611737592	1.734548814	1.866408994	2.005596790	2.310937575	2.876061582
1		PF3	1.028728864	.9902252666	.9759554832	.9611655413	.9450621335	.9342355623	.9262286563	.9148287234	.9035630915
1		PQ3	5.056294628	4.780894331	4.847480688	5.023519020	5.385631909	5.783914389	6.207592945	7.139843441	8.863831522
1		PF4	1.016057158	1.004545110	1.000017247	.9951682918	.9896924973	.9858810310	.9829839466	.9787195805	.9742897121
1		PQ4	17.06633332	15.66444013	15.70482502	16.07966054	17.00003012	18.07449620	19.24534917	21.86668823	26.78231258
2		PFC	1.013755654	1.010803728	1.009521260	1.008073258	1.006342927	1.005076324	1.004077801	1.002549395	1.000886025
2		PQC	97.75290957	78.30400879	74.68138150	72.39582506	71.61987490	72.30156501	73.64625555	77.42490944	85.11074013
1		ZF1	1.016078342	1.016134118	1.016158875	1.016186382	1.016218703	1.016242180	1.016260732	1.016289534	1.016322112
1		ZF2	1.021679108	1.022724073	1.023079495	1.023443766	1.023846059	1.024126748	1.024343719	1.024674388	1.025042068
1		ZF3	1.042495136	1.045792609	1.046981664	1.048228482	1.049633821	1.050629564	1.051406783	1.052602395	1.053945468
1		ZF4	1.118050441	1.125727225	1.128621524	1.131724522	1.135300484	1.137880644	1.139919052	1.143092612	1.146705184
1		ZF5	1.414666964	1.433298356	1.440404428	1.448075144	1.456981099	1.463448911	1.468581852	1.476610208	1.485796476
1		ZF6	3.483714912	3.552478577	3.578740008	3.607118433	3.640109337	3.664097139	3.683150336	3.712976582	3.747134946

(6) P 10+2		TH = 81	SBE = 1.012465126						MCPER FUNCTIONS	
		RC = 02	RC = 05	RC = 07	RC = 10	RC = 15	RC = 20	RC = 25	RC = 35	RC = 50
AMAX DB		.0017375255	.0108709564	.0213327376	.0436480540	.0988323393	.1772876696	.2802872360	.5675287486	1.249387366
AMIN DB		23.15810884	32.52359869	36.00555923	39.72586040	44.00870228	47.10472213	49.56132253	53.42728654	57.96733428
MU	K	.6951757D-1	.2364940D-1	.1583879D-1	.1032065D-1	.6303255D-2	.4413305D-2	.3326089D-2	.2131256D-2	.1263669D-2
1	PF1	1.238849533	.8933170070	.7927031693	.6974812368	.6012098779	.5395543173	.4951226462	.4332513855	.3738470912
1	PQ1	.6048546098	.5969535171	.6009449148	.6097460277	.6274417081	.6473966325	.6694010113	.7205859679	.8236089047
1	PF2	1.085906267	.9641440680	.9222891617	.8805772549	.8370045953	.8087838470	.7884908220	.7605285457	.7341797093
1	PQ2	1.690503633	1.601952401	1.624989137	1.683848601	1.803908957	1.935864841	2.076477238	2.386960224	2.964333225
1	PF3	1.030670601	.9951611864	.9817400849	.9677164609	.9523348505	.9419359951	.9342189465	.9231970577	.9122706123
1	PQ3	5.737871884	5.284890681	5.320739071	5.478733064	5.837512165	6.245534261	6.685299894	7.661678071	9.479185445
1	PF4	1.014903433	1.004877828	1.000848683	.9964940113	.9915345305	.9880598537	.9854078656	.9814892619	.9774028817
1	PQ4	20.24317046	18.08323536	17.99336606	18.29456170	19.20895283	20.33495391	21.58559379	24.42076191	29.78443663
2	PFC	1.011562544	1.009204390	1.008150579	1.006946018	1.005489865	1.004414184	1.003561037	1.002247213	1.000807415
2	PQC	123.4747903	95.56478409	90.25870952	86.70498553	85.00512847	85.32685685	86.56036769	90.48686154	98.91748126
1	ZF1	1.013020502	1.013063335	1.013084194	1.013107914	1.013136269	1.013157098	1.013173665	1.013199535	1.013228974
1	ZF2	1.017635643	1.018597070	1.018924314	1.019259927	1.019630959	1.019890152	1.020090719	1.020396786	1.020737735
1	ZF3	1.035600366	1.038628808	1.039732930	1.040895861	1.042212160	1.043147977	1.043880102	1.045009046	1.046280918
1	ZF4	1.103725415	1.110899503	1.113634702	1.116581776	1.119994644	1.122467024	1.124425643	1.127483667	1.130976521
1	ZF5	1.380434973	1.398300564	1.405176880	1.412629754	1.421317441	1.427647683	1.432682908	1.440577161	1.449635574
1	ZF6	3.357148383	3.424175336	3.449976207	3.477952740	3.510585586	3.534378669	3.553312999	3.583011522	3.617103056

(6) P 10+2		TH = 82	SBE = 1.009827573						MCPER FUNCTIONS	
		RC = 02	RC = 05	RC = 07	RC = 10	RC = 15	RC = 20	RC = 25	RC = 35	RC = 50
AMAX DB		.0017375255	.0108709564	.0213327376	.0436480540	.0988323393	.1772876696	.2802872360	.5675287486	1.249387366
AMIN DB		20.45581259	29.77314555	33.24542101	36.95744346	41.23256625	44.32392679	46.77729111	50.63889497	55.17486004
MU	K	.9488758D-1	.3245957D-1	.2176351D-1	.1419475D-1	.8677042D-2	.6078601D-2	.4582848D-2	.2938023D-2	.1742838D-2
1	PF1	1.311381496	.9371044763	.8293063497	.7278771648	.6259218680	.5609309183	.5142368179	.4494012465	.3873390743
1	PQ1	.6183998090	.6039494047	.6067711030	.6146988747	.6317231858	.6513614077	.6732007465	.7242764965	.8274400122
1	PF2	1.100314545	.9817805651	.9402089591	.8984878737	.8546546260	.8261526008	.8056109893	.7772512867	.7504782488
1	PQ2	1.855318595	1.708504979	1.720905512	1.772430214	1.888301594	2.019883780	2.161874129	2.478053990	3.069584003
1	PF3	1.031935997	.9997570331	.9873013407	.9741587324	.9596157341	.9497180843	.9423425223	.9317630623	.9212444349
1	PQ3	6.649995627	5.934131956	5.924161737	6.053778342	6.403248994	6.820284585	7.277779813	8.305477086	10.23454110
1	PF4	1.013585688	1.005059724	1.001542968	.9977002470	.9932796110	.9901584741	.9877646796	.9842113953	.9804887509
1	PQ4	24.65265688	21.31815925	21.02325432	21.19990303	22.08030197	23.25669893	24.59829745	27.69037898	33.60651396
2	PFC	1.009487050	1.007669010	1.006828165	1.005852758	1.004657344	1.003764734	1.003051757	1.001945987	1.000724352
2	PQC	161.5022271	120.0137758	112.0587072	106.5081008	103.3268337	103.0329187	104.0287533	108.0340986	117.3402572
1	ZF1	1.010294375	1.010323900	1.010340713	1.010360481	1.010384671	1.010402705	1.010417169	1.010439923	1.010466012
1	ZF2	1.013989833	1.014860358	1.015157085	1.015461630	1.015798703	1.016034493	1.016217166	1.016496340	1.016807986
1	ZF3	1.029215376	1.031944993	1.032954288	1.034023262	1.035239475	1.036107708	1.036788849	1.037842206	1.039033051
1	ZF4	1.089885600	1.096498625	1.099054172	1.101823816	1.105049450	1.107397074	1.109262765	1.112185294	1.115536467
1	ZF5	1.346103507	1.363092941	1.369704058	1.376903027	1.385332896	1.391498244	1.396414867	1.404143728	1.413040505
1	ZF6	3.227493279	3.292489708	3.317749242	3.345246635	3.377441549	3.400987403	3.419764136	3.449279201	3.483246916

(6) P 10+2 — TH = 83 SBE = 1.007509825 — MCPER FUNCTIONS

MU	K	RC = 02	RC = 05	RC = 07	RC = 10	RC = 15	RC = 20	RC = 25	RC = 35	RC = 50
	AMAX DB	.0017375255	.0108709564	.0213327376	.0436480540	.0988323393	.1772876696	.2802872360	.5675287486	1.249387366
	AMIN DB	17.64649166	26.89209960	30.35171919	34.05334736	38.21902680	41.40476191	43.85424009	47.71062523	52.24171402
	K	.1311220D 0	.4522671D-1	.3036785D-1	.1983045D-1	.1213525D-1	.8506715D-2	.6416349D-2	.4115337D-2	.2442948D-2
1	PF1	1.396423461	.9883245228	.8719469813	.7631267415	.6544405110	.5855228287	.5361768252	.4678824968	.4027355040
1	PQ1	.6371010267	.6129946018	.6141708604	.6208903697	.6369965787	.6562020665	.6778124341	.7287212374	.8320228232
1	PF2	1.113362392	.9997427791	.9588217680	.9173708258	.8734887117	.8448043719	.8240688094	.7953664062	.7682009242
1	PQ2	2.078774683	1.846272829	1.843329394	1.884181523	1.993588418	2.123998358	2.267202464	2.589720433	3.197852010
1	PF3	1.032392882	1.003909506	.9925521548	.9804236840	.9668572641	.9575488000	.9505770658	.9405344917	.9304917941
1	PQ3	7.923649476	6.800154642	6.719283663	6.803209419	7.132804518	7.556688131	8.033472125	9.121586921	11.18646867
1	PF4	1.012100522	1.005074553	1.002081584	.9987672337	.9949077647	.9921573888	.9900355237	.9868684198	.9835314617
1	PQ4	31.07295722	25.82266604	25.19160238	25.15320080	25.94578003	27.16393848	28.60818866	32.01382605	38.62811210
2	PFC	1.007547069	1.006209862	1.005564177	1.004801456	1.003851008	1.003132085	1.002552965	1.001647183	1.000636842
2	PQC	220.9701812	156.3675605	144.0153961	135.1588276	129.4941353	128.1170312	128.6336730	132.5525274	142.3800324
1	ZF1	1.007896051	1.007912154	1.007924834	1.007940534	1.007960404	1.007975521	1.007987779	1.008007250	1.008029791
1	ZF2	1.010745574	1.011516989	1.011780749	1.012051727	1.012352051	1.012562457	1.012725685	1.012975570	1.013255205
1	ZF3	1.023358740	1.025757631	1.026661339	1.027625308	1.028729209	1.029521317	1.030144870	1.031112587	1.032211259
1	ZF4	1.076542124	1.082527583	1.084880123	1.087447725	1.090458243	1.092661332	1.094418692	1.097182179	1.100365559
1	ZF5	1.311534310	1.327508685	1.333811594	1.340713107	1.348837379	1.354804729	1.359577367	1.367102663	1.375796378
1	ZF6	3.093682190	3.156251376	3.180867820	3.207790630	3.239450101	3.262684984	3.281257200	3.310521334	3.344296223

(6) P 10+2 — TH = 84 SBE = 1.005508280 — MCPER FUNCTIONS

MU	K	RC = 02	RC = 05	RC = 07	RC = 10	RC = 15	RC = 20	RC = 25	RC = 35	RC = 50
	AMAX DB	.0017375255	.0108709564	.0213327376	.0436480540	.0988323393	.1772876696	.2802872360	.5675237486	1.249387366
	AMIN DB	14.71710277	23.84595490	27.28779683	30.97564220	35.22930488	38.30803918	40.75272655	44.60271664	49.12785405
	K	.1837151D 0	.6422473D-1	.4321258D-1	.2826298D-1	.1731948D-1	.1215961D-1	.9169880D-2	.5886595D-2	.3496289D-2
1	PF1	1.496822178	1.049360003	.9225812594	.8047956305	.6879774574	.6143419177	.5618240306	.4894128696	.4206153036
1	PQ1	.6643232140	.6251744029	.6239223778	.6288945560	.6436930408	.6622846069	.6835662112	.7342159489	.8376426944
1	PF2	1.124195508	1.017756340	.9780096037	.9372383670	.8936286308	.8649181817	.8440781738	.8151277020	.7876310019
1	PQ2	2.396996103	2.031597090	2.005467734	2.030129221	2.129278669	2.257106452	2.401119673	2.730648136	3.358647973
1	PF3	1.031881266	1.007484290	.9973769521	.9864180625	.9739927392	.9653800856	.9588885096	.9494822324	.9400180250
1	PQ3	9.803891087	8.009321521	7.813028719	7.820437463	8.110548724	8.536003893	9.033029826	10.19318833	12.42775137
1	PF4	1.010449116	1.004904979	1.002443597	.9996717518	.9963947982	.9940325715	.9921968938	.9894331253	.9865106720
1	PQ4	41.02537882	32.43499962	31.22098743	30.79576891	31.39231347	32.62554214	34.18169629	37.97640982	45.50078920
2	PFC	1.005763447	1.004841677	1.004370940	1.003801906	1.003077887	1.002521426	1.002068500	1.001352755	1.000545020
2	PQC	320.9885058	213.9580655	193.7806233	179.0777788	168.9908646	165.6146681	165.1628391	168.6074268	180.0869655
1	ZF1	1.005822019	1.005824981	1.005833530	1.005845124	1.005860588	1.005872703	1.005882679	1.005898732	1.005917554
1	ZF2	1.007909244	1.008572196	1.008800416	1.009035254	1.009295970	1.009478964	1.009621153	1.009839269	1.010084066
1	ZF3	1.018054835	1.020089501	1.020876030	1.021722948	1.022701029	1.023407477	1.023966019	1.024836710	1.025830474
1	ZF4	1.063714831	1.068995235	1.071117832	1.073454845	1.076217591	1.078252757	1.079883445	1.082459636	1.085443623
1	ZF5	1.276556585	1.291331835	1.297272308	1.303821996	1.311581187	1.317309262	1.321906180	1.329180012	1.337618410
1	ZF6	2.954289394	3.013857180	3.037694787	3.063919171	3.094919026	3.117762556	3.136071319	3.164999972	3.198495652

(6) P 10+2		TH = 85 SBE = 1.003819838								MCPER FUNCTIONS
		RC = 02	RC = 05	RC = 07	RC = 10	RC = 15	RC = 20	RC = 25	RC = 35	RC = 50
AMAX DB		.0017375255	.0108709564	.0213327376	.0436480540	.0988323393	.1772876696	.2802872360	.5675287486	1.249387366
AMIN DB		11.66662788	20.58684340	24.00103309	27.66901810	31.90641974	34.97602827	37.41457459	41.25645754	45.77410876
MU	K	.2610169D 0	.9346690D-1	.6308823D-1	.4135701D-1	.2539095D-1	.1783194D-1	.1346701D-1	.8653208D-2	.5143924D-2
1	PF1	1.614972748	1.123748347	.9841919056	.8552988934	.7284088750	.6489549439	.5925420541	.5151008836	.4418707442
1	PQ1	.7067820008	.6425040578	.6374322419	.6397230500	.6525540643	.6702293552	.6910161273	.7412506927	.8447672382
1	PF2	1.131538490	1.035329311	.9975170410	.9580378637	.9151968784	.8867101152	.8659131065	.8368766666	.8091609784
1	PQ2	2.880962934	2.294674613	2.231235497	2.229874756	2.311984181	2.434591585	2.578488199	2.915645664	3.568012864
1	PF3	1.030209098	1.010302230	1.001617978	.9920114208	.9809257446	.9731401690	.9672239105	.9585856471	.9498245714
1	PQ3	12.79601295	9.805990505	9.408122736	9.279448670	9.491017578	9.905582705	10.42169609	11.66369362	14.12249960
1	PF4	1.008641319	1.004533172	1.002605479	1.000386285	.9977108995	.9957537981	.9942188211	.9918916949	.9893995672
1	PQ4	57.81755554	42.86484742	40.55730970	39.38890059	39.55529211	40.73089620	42.39586253	46.68029586	55.43993420
2	PFC	1.004160727	1.003582428	1.003263693	1.002866388	1.002346948	1.001939466	1.001603401	1.001065370	1.000449235
2	PQC	506.3689186	313.3768360	277.9008922	251.8889622	233.2389462	225.8930417	223.3925451	225.4153338	238.0451426
1	ZF1	1.004069012	1.004059779	1.004064332	1.004071897	1.004082973	1.004092062	1.004099719	1.004112271	1.004127252
1	ZF2	1.005490219	1.006033923	1.006223902	1.006419975	1.006638214	1.006791760	1.006911303	1.007095135	1.007302193
1	ZF3	1.013334925	1.014971776	1.015628879	1.016345768	1.017183210	1.017793373	1.018278549	1.019039265	1.019913471
1	ZF4	1.051436483	1.055919239	1.057780156	1.059852834	1.062328844	1.064167887	1.065649607	1.068003860	1.070749291
1	ZF5	1.240952241	1.254269646	1.259775592	1.265902394	1.273219347	1.278654641	1.283034632	1.289994597	1.298109249
1	ZF6	2.807326032	2.862996397	2.885859493	2.911213517	2.941386102	2.963731446	2.981698976	3.010181073	3.043284418

(6) P 10+2		TH = 86 SBE = 1.002441898								MCPER FUNCTIONS
		RC = 02	RC = 05	RC = 07	RC = 10	RC = 15	RC = 20	RC = 25	RC = 35	RC = 50
AMAX DB		.0017375255	.0108709564	.0213327376	.0436480540	.0988323393	.1772876696	.2802872360	.5675287486	1.249387366
AMIN DB		8.532949509	17.04731212	20.41220794	24.04776475	28.26086326	31.31764029	33.74780624	37.57335375	42.08665225
MU	K	.3744144D 0	.1404864D 0	.9536513D-1	.6274972D-1	.3863286D-1	.2717177D-1	.2054044D-1	.1321470D-1	.7864432D-2
1	PF1	1.749627325	1.216839573	1.061558269	.9186080895	.7788539647	.6919726335	.6306036209	.5467922793	.4679848520
1	PQ1	.7793081542	.6691706061	.6575341215	.6553531712	.6649866081	.6811925217	.7011824165	.7507138532	.8542327004
1	PF2	1.133493026	1.051557444	1.016811087	.9795677677	.9382876894	.9104366330	.8899322853	.8610932096	.8333645870
1	PQ2	3.686748781	2.698026125	2.568964106	2.522146783	2.573827875	2.685823058	2.827436682	3.172412694	3.855655060
1	PF3	1.027157161	1.012116520	1.005051563	.9970136083	.9875108599	.9807173701	.9754979725	.9677935105	.9599048051
1	PQ3	18.08417779	12.72530841	11.93800186	11.54413218	11.59074980	11.96350202	12.49077449	13.84237128	16.59264209
1	PF4	1.006703889	1.003942518	1.002541330	1.000877996	.9988185325	.9972819852	.9960618926	.9941904187	.9921614498
1	PQ4	89.73808426	61.10595958	56.49975121	53.74924996	52.91840067	53.83323594	55.55685105	60.45720638	70.98703432
2	PFC	1.002768023	1.002454534	1.002261745	1.002010726	1.001669976	1.001395172	1.001164510	1.000738807	1.000350218
2	PQC	899.3960771	507.4297455	437.7955510	386.9171256	349.5439281	333.3341673	326.1295043	324.1772526	337.3658064
1	ZF1	1.002633663	1.002614406	1.002615306	1.002619092	1.002625959	1.002632092	1.002637457	1.002646512	1.002657604
1	ZF2	1.003501111	1.003913635	1.004062553	1.004217247	1.004390226	1.004512360	1.004607703	1.004754794	1.004921234
1	ZF3	1.009237650	1.010447265	1.010962543	1.011535321	1.012216633	1.012718775	1.013121210	1.013757203	1.014494865
1	ZF4	1.039758318	1.043331641	1.044892307	1.046659643	1.048801185	1.050409168	1.051714043	1.053802588	1.056259352
1	ZF5	1.204425468	1.215901616	1.220869834	1.226474640	1.233243254	1.238312526	1.242419165	1.248979600	1.256676112
1	ZF6	2.649832269	2.700109189	2.721684951	2.745906047	2.775005204	2.796699270	2.814215393	2.842094381	2.874647922

$P_{11+0}^{(5)}$

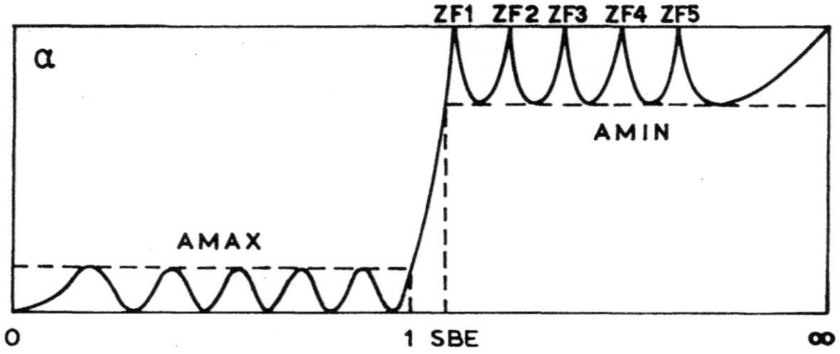

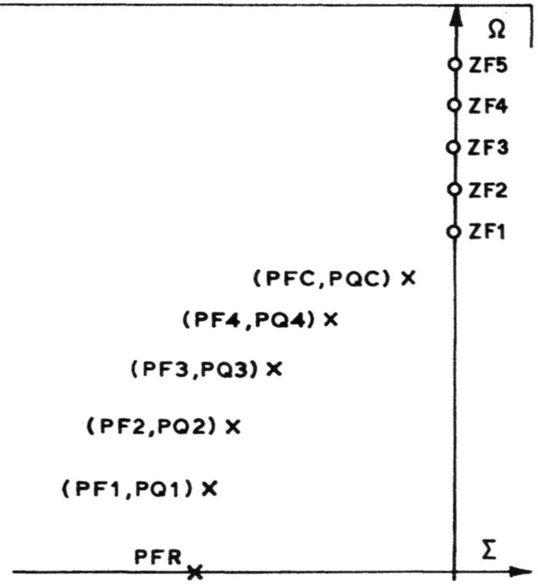

(5) P 11+0	TH = 52	SBE = 1.269018215							CAUER FUNCTIONS
	RC = 02	RC = 05	RC = 07	RC = 10	RC = 15	RC = 20	RC = 25	RC = 35	RC = 50
AMAX DB	.0017375255	.0108709564	.0213327376	.0436480540	.0988323393	.1772376696	.2802872360	.5675287486	1.249387366
AMIN DB	88.11582943	96.08376303	99.01678552	102.1371400	105.7141495	108.2913796	110.3325794	113.5423816	117.3222794
MU I K	.4366041D-3	.1744581D-3	.1244629D-3	.8690049D-4	.5756676D-4	.4278685D-4	.3382597D-4	.2337546D-4	.1512743D-4
1 I PFR	.5687857998	.4424803389	.3984946015	.3529566891	.3022447030	.2666993281	.2391976967	.1973802606	.1512948117
1 I PF1	.6582292955	.5580758960	.5251105220	.4924781034	.4584000802	.4362678680	.4203244655	.3983931005	.3779652127
1 I PQ1	.6567785476	.7091330058	.7388787176	.7803917077	.8461835412	.9112934136	.9779253653	1.121722645	1.386744211
1 I PF2	.8176607098	.7529296688	.7323285195	.7124012159	.6921741040	.6794027912	.6704004330	.6583053736	.6473804824
1 I PQ2	1.154008465	1.326000091	1.419572487	1.546406368	1.740597827	1.926709418	2.112593663	2.502782048	3.198311679
1 I PF3	.9414659212	.9016021948	.8889390652	.8767179543	.8643516661	.8565688180	.8510964466	.8437640164	.8371626864
1 I PQ3	2.211761582	2.592728173	2.797262563	3.072287579	3.489702975	3.886754229	4.281219151	5.104568864	6.562943878
1 I PF4	1.013999321	.9890816795	.9811321968	.9734470734	.9656590515	.9607523008	.9573000295	.9526717973	.9485028211
1 I PQ4	4.689262269	5.528217470	5.976972130	6.579077346	7.490798816	8.356347921	9.215091979	11.00496074	14.17031068
1 I PFC	1.046026146	1.027943856	1.022158410	1.016558139	1.010875906	1.007292396	1.004769554	1.001385359	.9983350717
1 I PQC	16.02979947	18.92553783	20.47297555	22.54805560	25.68833245	28.66809819	31.62342233	37.78095691	48.66614003
1 I ZF1	1.276846163	1.276846163	1.276846163	1.276846163	1.276846163	1.276846163	1.276846163	1.276846163	1.276846163
1 I ZF2	1.346576206	1.346576206	1.346576206	1.346576206	1.346576206	1.346576206	1.346576206	1.346576206	1.346576206
1 I ZF3	1.533538310	1.533538310	1.533538310	1.533538310	1.533538310	1.533538310	1.533538310	1.533538310	1.533538310
1 I ZF4	2.010050474	2.010050474	2.010050474	2.010050474	2.010050474	2.010050474	2.010050474	2.010050474	2.010050474
1 I ZF5	3.660113257	3.660113257	3.660113257	3.660113257	3.660113257	3.660113257	3.660113257	3.660113257	3.660113257

(5) P 11+0		TH = 53 SBE = 1.252135658								CAUER FUNCTIONS
		RC = 02	RC = 05	RC = 07	RC = 10	RC = 15	RC = 20	RC = 25	RC = 35	RC = 50
AMAX DB		.0017375255	.0108709564	.0213327376	.0436480540	.0988323393	.1772876696	.2802872360	.5675287486	1.249387366
AMIN DB		85.94851911	93.91645271	96.84947520	99.96982972	103.5468392	106.1240692	108.1652691	111.3750713	115.1549691
MU	K	.5473973D-3	.2187288D-3	.1560468D-3	.1089525D-3	.7217498D-4	.5364449D-4	.4240970D-4	.2930725D-4	.1896619D-4
1	PFR	.5759757958	.4476280697	.4030121670	.3568601366	.3055052644	.2695320757	.2417108493	.1994241714	.1528412494
1	PF1	.6653350792	.5637256316	.5303200221	.4972697410	.4627735701	.4403793361	.4242516601	.4020626170	.3814201824
1	PQ1	.6587288169	.7109805019	.7407274387	.7822705066	.8481437914	.9133544570	.9801003555	1.124162407	1.389700822
1	PF2	.8231806858	.7579562752	.7371967364	.7171160760	.6967333992	.6838640188	.6747926420	.6626050294	.6515967446
1	PQ2	1.165790967	1.338146804	1.432144431	1.559666275	1.755047181	1.942383352	2.129538143	2.522480572	3.223062219
1	PF3	.9441388420	.9044392773	.8918162886	.8796288374	.8672917502	.8595248695	.8540626439	.8467424427	.8401508157
1	PQ3	2.253141463	2.637200004	2.843955600	3.122251899	3.544987796	3.947327335	4.347177225	5.182013737	6.661152992
1	PF4	1.014144084	.9896675302	.9818481237	.9742840991	.9666142640	.9617796891	.9583771599	.9538142879	.9497029046
1	PQ4	4.811484548	5.662290714	6.118690275	6.731726096	7.660850598	8.543473316	9.419479110	11.24594054	14.47705691
1	PFC	1.044798068	1.027230559	1.021601387	1.016148660	1.010612474	1.007119214	1.004659063	1.001357878	.9983813576
1	PQC	16.51683548	19.46444546	21.04420735	23.16502270	26.37754890	29.42784899	32.45428951	38.76221879	49.91709166
1	ZF1	1.259666167	1.259666167	1.259666167	1.259666167	1.259666167	1.259666167	1.259666167	1.259666167	1.259666167
1	ZF2	1.326856584	1.326856584	1.326856584	1.326856584	1.326856584	1.326856584	1.326856584	1.326856584	1.326856584
1	ZF3	1.507670590	1.507670590	1.507670590	1.507670590	1.507670590	1.507670590	1.507670590	1.507670590	1.507670590
1	ZF4	1.970530632	1.970530632	1.970530632	1.970530632	1.970530632	1.970530632	1.970530632	1.970530632	1.970530632
1	ZF5	3.579193390	3.579193390	3.579193390	3.579193390	3.579193390	3.579193390	3.579193390	3.579193390	3.579193390

(5) P 11+0		TH = 54 SBE = 1.236067977								CAUER FUNCTIONS
		RC = 02	RC = 05	RC = 07	RC = 10	RC = 15	RC = 20	RC = 25	RC = 35	RC = 50
AMAX DB		.0017375255	.0108709564	.0213327376	.0436480540	.0988323393	.1772876696	.2802872360	.5675287486	1.249387366
AMIN DB		83.80331113	91.77124472	94.70426721	97.82462173	101.4016312	103.9788613	106.0200611	109.2298633	113.0097611
MU	K	.6846807D-3	.2735844D-3	.1951822D-3	.1362770D-3	.9027595D-4	.6709815D-4	.5304575D-4	.3665730D-4	.2372278D-4
1	PFR	.5835060562	.4530066977	.4077290629	.3609331535	.3089052870	.2724848241	.2443297517	.2015533231	.1544516697
1	PF1	.6727340771	.5696026377	.5357373852	.5022510719	.4673188282	.4446514763	.4283317875	.4058847492	.3850094999
1	PQ1	.6607910720	.7129262536	.7426721334	.7842445998	.8502011630	.9155161233	.9823804397	1.126718437	1.392796592
1	PF2	.8288539285	.7631318093	.7422116283	.7219750996	.7014340100	.6884646240	.6793278029	.6670406494	.6559471091
1	PQ2	1.178280848	1.350980984	1.445414886	1.573649983	1.770271338	1.958888431	2.147374143	2.543204752	3.249089445
1	PF3	.9468466604	.9073282321	.8947500254	.8826003883	.8702963917	.8625477690	.8570972289	.8497912295	.8432100948
1	PQ3	2.297214575	2.684440427	2.893515859	3.175244426	3.603579746	4.011494504	4.417026655	5.263993154	6.765073536
1	PF4	1.014278848	.9902559035	.9825704436	.9751311966	.9675832099	.9628230824	.9594718769	.9549764630	.9509245404
1	PQ4	4.942391530	5.805555630	6.270018726	6.894619943	7.842198908	8.742950567	9.637297859	11.50266381	14.80373843
1	PFC	1.043553547	1.026506742	1.021035895	1.015732762	1.010344758	1.006943122	1.004546640	1.001329793	.9984282492
1	PQC	17.04049008	20.04259916	21.65663943	23.82608511	27.11558203	30.24111312	33.34345278	39.81197630	51.25495643
1	ZF1	1.243308095	1.243308095	1.243308095	1.243308095	1.243308095	1.243308095	1.243308095	1.243308095	1.243308095
1	ZF2	1.308019247	1.308019247	1.308019247	1.308019247	1.308019247	1.308019247	1.308019247	1.308019247	1.308019247
1	ZF3	1.482834415	1.482834415	1.482834415	1.482834415	1.482834415	1.482834415	1.482834415	1.482834415	1.482834415
1	ZF4	1.932394361	1.932394361	1.932394361	1.932394361	1.932394361	1.932394361	1.932394361	1.932394361	1.932394361
1	ZF5	3.500825919	3.500825919	3.500825919	3.500825919	3.500825919	3.500825919	3.500825919	3.500825919	3.500825919

(5) P 11+0	TH = 55	SBE = 1.220774589						CAUER FUNCTIONS	
	RC = 02	RC = 05	RC = 07	RC = 10	RC = 15	RC = 20	RC = 25	RC = 35	RC = 50
AMAX DB	.0017375255	.0108709564	.0213327376	.0436480540	.0988323393	.1772876696	.2802872360	.5675287486	1.249387366
AMIN DB	81.67846035	89.64639393	92.57941642	95.69977094	99.27678040	101.8540105	103.8952103	107.1050125	110.8849103
MU I K	.8545183D-3	.3414481D-3	.2435979D-3	.1700810D-3	.1126692D-3	.8374210D-4	.6620395D-4	.4575028D-4	.2960730D-4
1 I PFR	.5913988053	.4586302017	.4126570715	.3651855345	.3124526409	.2755642469	.2470602245	.2037723433	.1561295035
1 I PF1	.6804414423	.5757183476	.5413729700	.5074314605	.4720442510	.4490920926	.4325722423	.4098563504	.3887365371
1 I PQ1	.6629744951	.7149776724	.7447198477	.7863208008	.8523624149	.9177352955	.9847727197	1.129398465	1.396040633
1 I PF2	.8346828528	.7684598423	.7473770127	.7269822944	.7062800927	.6932088363	.6839951876	.6716165511	.6604359240
1 I PQ2	1.191537721	1.364557114	1.459438138	1.588412688	1.786328262	1.976286087	2.166167002	2.565028902	3.276484908
1 I PF3	.9495872275	.9102681920	.8977397796	.8856324522	.8733657641	.8656378879	.8602007081	.8529110590	.8463440591
1 I PQ3	2.344225038	2.734689197	2.946188150	3.231520249	3.665753669	4.079551916	4.491086239	5.350375689	6.875166637
1 I PF4	1.014402662	.9908460723	.9832985154	.9759878131	.9685654310	.9638820843	.9605838281	.9561580309	.9521674922
1 I PQ4	5.082840196	5.958886577	6.431861535	7.068712543	8.035884334	8.955910129	9.869771959	11.77655637	15.15215109
1 I PFC	1.042293154	1.025772678	1.020462133	1.015310578	1.010072835	1.006764164	1.004432309	1.001301100	.9984757225
1 I PQC	17.60458338	20.66396225	22.31439192	24.53561190	27.90722482	31.11311517	34.29658144	40.93685625	52.68813532
1 I ZF1	1.227730929	1.227730929	1.227730929	1.227730929	1.227730929	1.227730929	1.227730929	1.227730929	1.227730929
1 I ZF2	1.290019310	1.290019310	1.290019310	1.290019310	1.290019310	1.290019310	1.290019310	1.290019310	1.290019310
1 I ZF3	1.458974702	1.458974702	1.458974702	1.458974702	1.458974702	1.458974702	1.458974702	1.458974702	1.458974702
1 I ZF4	1.895561848	1.895561848	1.895561848	1.895561848	1.895561848	1.895561848	1.895561848	1.895561848	1.895561848
1 I ZF5	3.424851806	3.424851806	3.424851806	3.424851806	3.424851806	3.424851806	3.424851806	3.424851806	3.424851806

(5) P 11+0	TH = 56	SBE = 1.206217949						CAUER FUNCTIONS	
	RC = 02	RC = 05	RC = 07	RC = 10	RC = 15	RC = 20	RC = 25	RC = 35	RC = 50
AMAX DB	.0017375255	.0108709564	.0213327376	.0436480540	.0988323393	.1772876696	.2802872360	.5675287486	1.249387366
AMIN DB	79.57224941	87.54018298	90.47320547	93.59355998	97.17056945	99.74779951	101.7889993	104.9988016	108.7786994
MU I K	.1064335D-2	.4252867D-3	.3034105D-3	.2118424D-3	.1403338D-3	.1043040D-3	.8245954D-4	.5698372D-4	.3687703D-4
1 I PFR	.5996785509	.4645139139	.4178090946	.3696279886	.3161559178	.2787776229	.2499086125	.2060862720	.1578784842
1 I PF1	.6884735568	.5820851258	.5472379787	.5128210326	.4769589161	.4537096217	.4369810175	.4139848278	.3926111799
1 I PQ1	.6652893642	.7171430013	.7468784034	.7885066601	.8546350234	.9201695789	.9872850362	1.132211013	1.399442986
1 I PF2	.8406697926	.7739440455	.7526968501	.7321418464	.7112760133	.6981011109	.6888143041	.6763373006	.6650677964
1 I PQ2	1.205628369	1.378935936	1.474274624	1.604015762	1.803282315	1.994644477	2.185989164	2.588035357	3.305349976
1 I PF3	.9523581763	.9132581602	.9007849532	.8887247984	.8764999909	.8687955636	.8633735646	.8561026047	.8495508481
1 I PQ3	2.394447832	2.788215290	3.002246794	3.291364891	3.731816869	4.151830513	4.569712017	5.443072733	6.991946397
1 I PF4	1.014514526	.9914372649	.9840316553	.9768533556	.9695604319	.9649562630	.9617126281	.9573586693	.9534314956
1 I PQ4	5.233803056	6.123272185	6.605239814	7.255079980	8.243079913	9.183625562	10.11827954	12.06922304	15.52431524
1 I PFC	1.041017496	1.025028661	1.019880315	1.014882252	1.009796786	1.006582386	1.004316094	1.001271798	.9985237516
1 I PQC	18.21347398	21.33304132	23.02214498	25.29856150	28.75791185	32.04977585	35.32009776	42.14436181	54.22607223
1 I ZF1	1.212896724	1.212896724	1.212896724	1.212896724	1.212896724	1.212896724	1.212896724	1.212896724	1.212896724
1 I ZF2	1.272815242	1.272815242	1.272815242	1.272815242	1.272815242	1.272815242	1.272815242	1.272815242	1.272815242
1 I ZF3	1.436040362	1.436040362	1.436040362	1.436040362	1.436040362	1.436040362	1.436040362	1.436040362	1.436040362
1 I ZF4	1.859958682	1.859958682	1.859958682	1.859958682	1.859958682	1.859958682	1.859958682	1.859958682	1.859958682
1 I ZF5	3.351121922	3.351121922	3.351121922	3.351121922	3.351121922	3.351121922	3.351121922	3.351121922	3.351121922

(5) P 11+0	TH = 57	SBE = 1.192363293						CAUER FUNCTIONS	
	RC = 02	RC = 05	RC = 07	RC = 10	RC = 15	RC = 20	RC = 25	RC = 35	RC = 50
AMAX DB	.0017375255	.0108709564	.0213327376	.0436480540	.0988323393	.1772876696	.2802872360	.5675287486	1.249387366
AMIN DB	77.48297961	85.45091315	88.38393564	91.50429016	95.08129962	97.65852968	99.69972951	102.9095317	106.6894296
MU I K	.1323226D-2	.5287342D-3	.3772127D-3	.2633714D-3	.1744689D-3	.1296751D-3	.1025172D-3	.7084456D-4	.4584707D-4
1 I PFR	.6083723965	.4706746947	.4231992961	.3742722535	.3200245203	.2821329097	.2528818472	.2085006103	.1597026840
1 I PF1	.6968481586	.5887163678	.5533445475	.5184307580	.4820726566	.4585132012	.4415667698	.4182782024	.3966398846
1 I PQ1	.6677472202	.7194314362	.7491565088	.7908105681	.8570272812	.9226773994	.9899260694	1.135165509	1.403014746
1 I PF2	.8468169692	.7795881857	.7581752460	.7374581281	.7164263588	.7031461441	.6937849136	.6812077323	.6698476127
1 I PQ2	1.220627845	1.394185379	1.489991830	1.620527637	1.821205170	2.014039432	2.206921187	2.612315604	3.335797203
1 I PF3	.9551569001	.9162969955	.9038848334	.8918771101	.8796991376	.8720210926	.8666162535	.8593665277	.8528322028
1 I PQ3	2.448193775	2.845321462	3.062000184	3.355098970	3.802114028	4.228701230	4.653302873	5.541044886	7.115988310
1 I PF4	1.014613394	.9920286609	.9847691328	.9777271874	.9705676765	.9660451492	.9628578554	.9585780221	.9547162554
1 I PQ4	5.396387861	6.299834288	6.791310975	7.454940709	8.465112550	9.427536651	10.38437787	12.38247627	15.92251150
1 I PFC	1.039727219	1.024275005	1.019290669	1.014447938	1.009516699	1.006397841	1.004198023	1.001241883	.9985723090
1 I PQC	18.87215453	22.05498005	23.78523491	26.12058201	29.67382776	33.05782918	36.42130331	43.44301906	55.87953767
1 I ZF1	1.198770345	1.198770345	1.198770345	1.198770345	1.198770345	1.198770345	1.198770345	1.198770345	1.198770345
1 I ZF2	1.256368577	1.256368577	1.256368577	1.256368577	1.256368577	1.256368577	1.256368577	1.256363577	1.256368577
1 I ZF3	1.413983954	1.413983954	1.413983954	1.413983954	1.413983954	1.413983954	1.413983954	1.413933954	1.413983954
1 I ZF4	1.825515349	1.825515349	1.825515349	1.825515349	1.825515349	1.825515349	1.825515349	1.825515349	1.825515349
1 I ZF5	3.279496026	3.279496026	3.279496026	3.279496026	3.279496026	3.279496026	3.279496026	3.279496026	3.279496026

(5) P 11+0	TH = 58	SBE = 1.179178403						CAUER FUNCTIONS	
	RC = 02	RC = 05	RC = 07	RC = 10	RC = 15	RC = 20	RC = 25	RC = 35	RC = 50
AMAX DB	.0017375255	.0108709564	.0213327376	.0436480540	.0988323393	.1772876696	.2802872360	.5675287486	1.249387366
AMIN DB	75.40896161	83.37689510	86.30991759	89.43027210	93.00728156	95.58451163	97.62571145	100.8355137	104.6154115
MU I K	.1642338D-2	.6562448D-3	.4681820D-3	.3268865D-3	.2165442D-3	.1609478D-3	.1272404D-3	.8792957D-4	.5690364D-4
1 I PFR	.6175104090	.4771311370	.4288432675	.3791312287	.3240687641	.2856398292	.2559875206	.2110213781	.1616065549
1 I PF1	.7055844869	.5956266144	.5597058509	.5242725447	.4873961460	.4635127489	.4463388945	.4227451790	.4008307415
1 I PQ1	.6703610671	.7218532680	.7515638893	.7932418766	.8595484132	.9253181191	.9927054554	1.138272406	1.406768203
1 I PF2	.8531264544	.7853961187	.7638164520	.7429357056	.7217359513	.7083488891	.6989120489	.6862327692	.6747805629
1 I PQ2	1.236620795	1.410381658	1.506665361	1.638024866	1.840176384	2.034555586	2.229052925	2.637971597	3.367951938
1 I PF3	.9579805276	.9193833950	.9070385785	.8950889725	.8829632027	.8753147238	.8699291949	.8627034719	.8561839626
1 I PQ3	2.505815499	2.906349692	3.125796214	3.423083733	3.877033056	4.310581224	4.742307168	5.645303540	7.247938398
1 I PF4	1.014698165	.9926193880	.9855101672	.9786086239	.9715865846	.9671482315	.9640190484	.9598156964	.9560214418
1 I PQ4	5.571861430	6.489850710	6.991391332	7.669679508	8.703483693	9.689276910	10.66983303	12.71837033	16.34932348
1 I PFC	1.038423011	1.023512048	1.018693440	1.014007805	1.009232671	1.006210584	1.004078124	1.001211351	.9986213653
1 I PQC	19.58636793	22.83567297	24.60977027	27.00813237	30.66203818	34.14496381	37.60853132	44.84255350	57.66030006
1 I ZF1	1.185319234	1.185319234	1.185319234	1.185319234	1.185319234	1.185319234	1.185319234	1.185319234	1.185319234
1 I ZF2	1.240643660	1.240643660	1.240643660	1.240643660	1.240643660	1.240643660	1.240643660	1.240643660	1.240643660
1 I ZF3	1.392761375	1.392761375	1.392761375	1.392761375	1.392761375	1.392761375	1.392761375	1.392761375	1.392761375
1 I ZF4	1.792166781	1.792166781	1.792166781	1.792166781	1.792166781	1.792166781	1.792166781	1.792166781	1.792166781
1 I ZF5	3.209841841	3.209841841	3.209841841	3.209841841	3.209841841	3.209841841	3.209841841	3.209841841	3.209841841

(5) P 11+0			TH = 59 RC = 02	SBE = 1.166633397 RC = 05	RC = 07	RC = 10	RC = 15	RC = 20	RC = 25	RC = 35 CAUER FUNCTIONS	RC = 50
AMAX DB			.0017375255	.0108709564	.0213327376	.0436480540	.0988323393	.1772876696	.2802872360	.5675237486	1.249387366
AMIN DB			73.34850572	81.31643915	84.24946163	87.36981614	90.94682560	93.52405566	95.56525549	98.77505771	102.5549555
MU		K	.2035344D-2	.8132818D-3	.5802163D-3	.4051093D-3	.2683624D-3	.1994620D-3	.1576886D-3	.1039708D-3	.7052047D-4
1		PFR	.6271260519	.4839038056	.4347582219	.3842191301	.3282999935	.2893049668	.2592339704	.2136551796	.1635949774
1		PF1	.7147034453	.6028316810	.5663362205	.5303593469	.4929409950	.4687190531	.4513076107	.4273952245	.4051925488
1		PQ1	.6731456115	.7244200515	.7541114411	.7958110422	.8622087135	.9281021716	.9956339243	1.141543330	1.410717014
1		PF2	.8596001257	.7913717802	.7696248660	.7485793462	.7272098621	.7137145749	.7042010350	.6914134473	.6798721654
1		PQ2	1.253703030	1.427610582	1.524380196	1.656593368	1.860287180	2.056287700	2.252484918	2.665117330	3.401954218
1		PF3	.9608258967	.9225158754	.9102452018	.8983598595	.8862921075	.8786766498	.8733127673	.8661140507	.8546219617
1		PQ3	2.567714656	2.971687707	3.194028787	3.495727688	3.957012070	4.397941291	4.837230654	5.756449919	7.398524443
1		PF4	1.014767684	.9932085176	.9862539234	.9794969288	.9726165276	.9682649528	.9651957018	.9610712535	.9573466370
1		PQ4	5.761678638	6.694782840	7.206986537	7.900876378	8.959925283	9.970706868	10.97665561	13.07924246	16.80768307
1		PFC	1.037105603	1.022740151	1.018088895	1.013562033	1.008944807	1.006020676	1.003956432	1.001130202	.9986708887
1		PQC	20.36274954	23.68190392	25.50277324	27.96862975	31.73064862	35.31999447	38.89133141	46.35410390	59.53300368
1		ZF1	1.172513194	1.172513194	1.172513194	1.172513194	1.172513194	1.172513194	1.172513194	1.172513194	1.172513194
1		ZF2	1.225607417	1.225607417	1.225607417	1.225607417	1.225607417	1.225607417	1.225607417	1.225607417	1.225607417
1		ZF3	1.372331583	1.372331583	1.372331583	1.372331583	1.372331583	1.372331583	1.372331583	1.372331583	1.372331583
1		ZF4	1.759851948	1.759851948	1.759851948	1.759851948	1.759851948	1.759851948	1.759851948	1.759351948	1.759851948
1		ZF5	3.142034200	3.142034200	3.142034200	3.142034200	3.142034200	3.142034200	3.142034200	3.142034200	3.142034200

(5) P 11+0			TH = 60 RC = 02	SBE = 1.154700538 RC = 05	RC = 07	RC = 10	RC = 15	RC = 20	RC = 25	RC = 35 CAUER FUNCTIONS	RC = 50
AMAX DB			.0017375255	.0108709564	.0213327376	.0436480540	.0988323393	.1772876696	.2802872360	.5675287486	1.249387366
AMIN DB			71.29991175	79.26784509	82.20086756	85.32122206	88.89823152	91.47546158	93.51666140	96.72646363	100.5063614
MU		K	.2519032D-2	.1006554D-2	.7181017D-3	.5013814D-3	.3321374D-3	.2468631D-3	.1951624D-3	.1348671D-3	.8727930D-4
1		PFR	.6372567000	.4910155195	.4409632219	.3895516725	.3327307465	.2931418875	.2626303790	.2164092814	.1656733160
1		PF1	.7242277870	.6103488075	.5732512811	.5367052894	.4987198635	.4741438765	.4564840603	.4322386602	.4097343980
1		PQ1	.6761175501	.7271448078	.7568114153	.7985297964	.8650197068	.9310412232	.9987234626	1.144991254	1.414876398
1		PF2	.8662396129	.7975191741	.7756050300	.7543940259	.7328534269	.7192487256	.7096575120	.6967699414	.6851282973
1		PQ2	1.271983435	1.445969116	1.543232211	1.676329922	1.881636965	2.079342246	2.277330057	2.693880683	3.437961024
1		PF3	.9636895234	.9256927511	.9135035530	.9016891190	.8896856838	.8821069968	.8767672993	.8695988810	.8631320244
1		PQ3	2.634350660	3.041776857	3.267145646	3.573494575	4.042547787	4.491314777	4.938645966	5.875125904	7.538569368
1		PF4	1.014820738	.9937950602	.9869995071	.9803913090	.9736568238	.9693947055	.9663872621	.9623442291	.9586915812
1		PQ4	5.967517884	6.916309203	7.439820533	8.150341632	9.236387277	10.27395433	11.30714391	13.46776271	17.30096264
1		PFC	1.035775781	1.021959705	1.017477320	1.013110817	1.008653220	1.005828185	1.003832981	1.001148432	.9987208456
1		PQC	21.20900237	24.60151583	26.47235261	29.01062958	32.88899847	36.59307109	40.28069478	47.99048301	61.66499504
1		ZF1	1.160324198	1.160324198	1.160324198	1.160324198	1.160324198	1.160324198	1.160324198	1.160324198	1.160324198
1		ZF2	1.211229153	1.211229153	1.211229153	1.211229153	1.211229153	1.211229153	1.211229153	1.211229153	1.211229153
1		ZF3	1.352656347	1.352656347	1.352656347	1.352656347	1.352656347	1.352656347	1.352656347	1.352656347	1.352656347
1		ZF4	1.728513475	1.728513475	1.728513475	1.728513475	1.728513475	1.728513475	1.728513475	1.728513475	1.728513475
1		ZF5	3.075954242	3.075954242	3.075954242	3.075954242	3.075954242	3.075954242	3.075954242	3.075954242	3.075954242

(5) P 11+0		TH = 61	SBE = 1.143354068						CAUER FUNCTIONS	
		RC = 02	RC = 05	RC = 07	RC = 10	RC = 15	RC = 20	RC = 25	RC = 35	RC = 50
AMAX DB		.0017375255	.0108709564	.0213327376	.0436480540	.0988323393	.1772876696	.2802872360	.5675287486	1.249387366
AMIN DB		69.26145805	77.22939122	80.16241367	83.28276817	86.85977762	89.43700768	91.47820751	94.68800973	98.46790755
MU	K	.3114049D-2	.1244310D-2	.8877232D-3	.6198118D-3	.4105910D-3	.3051743D-3	.2412614D-3	.1667239D-3	.1078954D-3
1	PFR	.6479442544	.4984916850	.4474794482	.3951462831	.3373748698	.2971612721	.2661868906	.2192917020	.1678474854
1	PF1	.7341823267	.6181968291	.5804681083	.5433258104	.5047465900	.4798000766	.4618804220	.4372867672	.4144682717
1	PQ1	.6792959187	.7300422660	.7596776375	.8014113493	.8679943417	.9341483633	1.001987507	1.148630700	1.419263377
1	PF2	.8730462340	.8038423563	.7817616276	.7603849364	.7386722624	.7249571823	.7152874599	.7022935943	.6905552264
1	PQ2	1.291586269	1.465567272	1.563329983	1.697343954	1.904340147	2.103839297	2.303715568	2.724405644	3.476148967
1	PF3	.9665675686	.9289121104	.9168122976	.9050759549	.8931436611	.8856058143	.8802930600	.8731584960	.8667199605
1	PQ3	2.706251364	3.117121668	3.345657874	3.656913004	4.134205647	4.591308328	5.047204070	6.002087086	7.699007356
1	PF4	1.014856050	.9943779602	.9877459594	.9812909097	.9747067331	.9705368264	.9675931223	.9636340792	.9600556673
1	PQ4	6.191324779	7.156366590	7.691882087	8.420158757	9.535133431	10.60146344	11.66393660	13.88699451	17.83299054
1	PFC	1.034434381	1.021171129	1.016859023	1.012654370	1.008358034	1.005633183	1.003707812	1.001116041	.9987711994
1	PQC	22.13411434	25.60362021	27.52791694	30.14404716	34.14789948	37.97593574	41.78933075	49.76649729	63.92272296
1	ZF1	1.148726224	1.148726224	1.148726224	1.148726224	1.148726224	1.148726224	1.148726224	1.148726224	1.148726224
1	ZF2	1.197480368	1.197480368	1.197480368	1.197480368	1.197480368	1.197480368	1.197480368	1.197480368	1.197480368
1	ZF3	1.333700022	1.333700022	1.333700022	1.333700022	1.333700022	1.333700022	1.333700022	1.333700022	1.333700022
1	ZF4	1.698097307	1.698097307	1.698097307	1.698097307	1.698097307	1.698097307	1.698097307	1.698097307	1.698097307
1	ZF5	3.011488671	3.011488671	3.011488671	3.011488671	3.011488671	3.011488671	3.011488671	3.011488671	3.011488671

(5) P 11+0		TH = 62	SBE = 1.132570051						CAUER FUNCTIONS	
		RC = 02	RC = 05	RC = 07	RC = 10	RC = 15	RC = 20	RC = 25	RC = 35	RC = 50
AMAX DB		.0017375255	.0108709564	.0213327376	.0436480540	.0988323393	.1772876696	.2802872360	.5675287486	1.249387366
AMIN DB		67.23138955	75.19932247	78.13234490	81.25269938	84.82970883	87.40693888	89.44813871	92.65794093	96.43733374
MU	K	.3845817D-2	.1536710D-2	.1096329D-2	.7654610D-3	.5070754D-3	.3768869D-3	.2979553D-3	.2059022D-3	.1332497D-3
1	PFR	.6592358804	.5063606943	.4543305194	.4010223561	.3422477639	.3013760781	.2699147484	.2223113192	.1701240270
1	PF1	.7445941799	.6263963740	.5880054101	.5502378284	.5110363413	.4857017459	.4675100442	.4425519097	.4194041587
1	PQ1	.6827025171	.7331291557	.7627257722	.8044706329	.8711472213	.9374383331	1.005441174	1.152477087	1.423897052
1	PF2	.8800209186	.8103454147	.7880994770	.7665574918	.7446722837	.7308461267	.7210972267	.7079959504	.6961596488
1	PQ2	1.312653978	1.486530387	1.584796992	1.719759693	1.928525828	2.129914801	2.331785394	2.756854966	3.516717508
1	PF3	.9694557993	.9321717875	.9201698932	.9085194080	.8966656504	.8891730614	.8838902482	.8767934167	.8703865591
1	PQ3	2.784026182	3.198301521	3.430151467	3.746588196	4.232632132	4.698614940	5.163648156	6.138188587	7.870903382
1	PF4	1.014872275	.9949560907	.9884922513	.9821943081	.9757654513	.9716905914	.9688126165	.9649402238	.9614384362
1	PQ4	6.435366342	7.417200787	7.965473438	8.712737118	9.858772501	10.95605494	12.05007729	14.34046397	18.40820353
1	PFC	1.033082301	1.020374876	1.016234339	1.012192922	1.008059384	1.005435750	1.003580970	1.001083026	.9988219112
1	PQC	23.14862956	26.69885744	28.68043905	31.38043224	35.51993075	39.48224013	43.43200832	51.69934167	66.37873231
1	ZF1	1.137695096	1.137695096	1.137695096	1.137695096	1.137695096	1.137695096	1.137695096	1.137695096	1.137695096
1	ZF2	1.184334591	1.184334591	1.184334591	1.184334591	1.184334591	1.184334591	1.184334591	1.184334591	1.184334591
1	ZF3	1.315429351	1.315429351	1.315429351	1.315429351	1.315429351	1.315429351	1.315429351	1.315429351	1.315429351
1	ZF4	1.668552389	1.668552389	1.668552389	1.668552389	1.668552389	1.668552389	1.668552389	1.668552389	1.668552389
1	ZF5	2.948529044	2.948529044	2.948529044	2.948529044	2.948529044	2.948529044	2.948529044	2.948529044	2.948529044

CAUER FUNCTIONS

(5) P 11+0		TH = 63 RC = 02	SBE = 1.122326238 RC = 05	RC = 07	RC = 10	RC = 15	RC = 20	RC = 25	RC = 35	RC = 50
AMAX DB		.0017375255	.0108709564	.0213327376	.0436480540	.0988323393	.1772876696	.2802872360	.5675287486	1.249387366
AMIN DB		65.20790466	73.17583717	76.10885956	79.22921402	82.80622346	85.38345351	87.42465333	90.63445556	94.41435337
MU	K	.4745707D-2	.1896288D-2	.1352861D-2	.9445726D-3	.6257270D-3	.4650754D-3	.3676744D-3	.2540816D-3	.1644290D-3
1	PFR	.6711849004	.5146544006	.4615428739	.4072015549	.3473665898	.3058007304	.2738264580	.2254779956	.1725102008
1	PF1	.7554930383	.6349700936	.5958837392	.5574599357	.5176057882	.4918643752	.4733876002	.4480476794	.4245551877
1	PQ1	.6863624300	.7364245626	.7659736437	.8077245969	.8744948817	.9409278003	1.009101538	1.156551521	1.428798945
1	PF2	.8871641153	.8170324426	.7946235225	.7729173336	.7508597232	.7369221077	.7270935593	.7138839922	.7019487310
1	PQ2	1.335350654	1.509001908	1.607774292	1.743718793	1.954340961	2.157723335	2.361703081	2.791413389	3.559892854
1	PF3	.9723495460	.9354693308	.9235745620	.9120183332	.9002511268	.8928085925	.8875589800	.8805041024	.8741325814
1	PQ3	2.868382337	3.285985004	3.521301542	3.843216367	4.338569831	4.814029917	5.288830603	6.284410368	8.055477026
1	PF4	1.014867999	.9955282466	.9892372770	.9831020070	.9768321035	.9728552081	.9700450133	.9652620163	.9628393202
1	PQ4	6.702298741	7.701429601	8.263274231	9.030877223	10.21033275	11.34100051	12.46909425	14.83227653	19.03173131
1	PFC	1.031720506	1.019571436	1.015603629	1.011726723	1.007757417	1.005235975	1.003452501	1.001049386	.9988729388
1	PQC	24.26499004	27.89972257	29.94278651	32.73331170	37.01980652	41.12794117	45.22598145	53.80309059	69.05827763
1	ZF1	1.127208345	1.127208345	1.127208345	1.127208345	1.127208345	1.127208345	1.127208345	1.127208345	1.127208345
1	ZF2	1.171767232	1.171767232	1.171767232	1.171767232	1.171767232	1.171767232	1.171767232	1.171767232	1.171767232
1	ZF3	1.297813278	1.297813278	1.297813278	1.297813278	1.297813278	1.297813278	1.297813278	1.297813278	1.297813278
1	ZF4	1.639830372	1.639830372	1.639830372	1.639830372	1.639830372	1.639830372	1.639830372	1.639830372	1.639830372
1	ZF5	2.886971080	2.886971080	2.886971080	2.886971080	2.886971080	2.886971080	2.886971080	2.886971080	2.886971080

CAUER FUNCTIONS

(5) P 11+0		TH = 64 RC = 02	SBE = 1.112601940 RC = 05	RC = 07	RC = 10	RC = 15	RC = 20	RC = 25	RC = 35	RC = 50
AMAX DB		.0017375255	.0108709564	.0213327376	.0436480540	.0988323393	.1772876696	.2802872360	.5675287486	1.249387366
AMIN DB		63.18914047	71.15707233	74.09009466	77.21044909	80.78745851	83.36468855	85.40588837	88.61569059	92.39558841
MU	K	.5852510D-2	.2338544D-2	.1668378D-2	.1164868D-2	.7716604D-3	.5735413D-3	.4534242D-3	.3133391D-3	.2027775D-3
1	PFR	.6838518805	.5234086934	.4691462287	.4137081750	.3527505517	.3104513495	.2779359818	.2288027296	.1750140935
1	PF1	.7669114830	.6439429287	.6041257391	.5650126249	.5244733106	.4983050455	.4795292703	.4537890644	.4299352853
1	PQ1	.6903046712	.7399503622	.7694416267	.8111925667	.8780561293	.9446356911	1.012987967	1.160872151	1.433993400
1	PF2	.8944756799	.8239075049	.8013388209	.7794703354	.7572411508	.7431920701	.7332836388	.7199651827	.7079301577
1	PQ2	1.359866301	1.533146809	1.632423775	1.769383536	1.981953588	2.187441462	2.393655290	2.828291542	3.605932680
1	PF3	.9752436533	.9388019670	.9270242600	.9155713728	.9038994081	.8965121399	.8912992734	.8842909473	.8779587529
1	PQ3	2.960145126	3.380947702	3.619889916	3.947602481	4.452875999	4.938470508	5.423733850	6.441380961	8.254131711
1	PF4	1.014841729	.9960931384	.9899798463	.9840114275	.9779057365	.9740298094	.9712895087	.9675987408	.9642576866
1	PQ4	6.995252631	8.012121764	8.588420793	9.377852153	10.59334852	11.76011538	12.92509950	15.36718075	19.70954313
1	PFC	1.030350031	1.018761339	1.014967288	1.011256044	1.007452293	1.005033953	1.003322461	1.001015123	.9989242368
1	PQC	25.49796925	29.22097659	31.33013820	34.21862061	38.66483808	42.93179732	47.19152239	56.11931345	71.99099179
1	ZF1	1.117245086	1.117245086	1.117245086	1.117245086	1.117245086	1.117245086	1.117245086	1.117245086	1.117245086
1	ZF2	1.159755453	1.159755453	1.159755453	1.159755453	1.159755453	1.159755453	1.159755453	1.159755453	1.159755453
1	ZF3	1.280822788	1.280822788	1.280822788	1.280822788	1.280822788	1.280822788	1.280822788	1.280822788	1.280822788
1	ZF4	1.611885341	1.611885341	1.611885341	1.611885341	1.611885341	1.611885341	1.611885341	1.611885341	1.611885341
1	ZF5	2.826713998	2.826713998	2.826713998	2.826713998	2.826713998	2.826713998	2.826713998	2.826713998	2.826713998

(5) P 11+0		TH = 65	SBE = 1.103377919						CAUER FUNCTIONS	
		RC = 02	RC = 05	RC = 07	RC = 10	RC = 15	RC = 20	RC = 25	RC = 35	RC = 50
AMAX DB		.0017375255	.0108709564	.0213327376	.0436480540	.0988323393	.1772876696	.2802872360	.5675287486	1.249387366
AMIN DB		61.17315597	69.14108679	72.07410903	75.19446340	78.77147280	81.34870283	83.38990265	86.59970486	90.37960268
MU	K	.7214320D-2	.2882695D-2	.2056589D-2	.1435919D-2	.9512165D-3	.7069975D-3	.5589306D-3	.3862494D-3	.2499614D-3
1	PFR	.6973059640	.5326641952	.4771741349	.4205695825	.3584212305	.3153460255	.2822589731	.2322978352	.1776447493
1	PF1	.7788853451	.6533424218	.6127564348	.5729185558	.5316592402	.5050426545	.4859529573	.4597926496	.4355598623
1	PQ1	.6945629861	.7437317564	.7731531263	.8148966827	.8818524531	.9485835963	1.017122529	1.165463596	1.439508078
1	PF2	.9019547407	.8309745953	.8082505223	.7862226048	.7638234944	.7496633866	.7396751189	.7262475131	.7141121871
1	PQ2	1.386422174	1.559155821	1.658932218	1.796940777	2.011556823	2.219271849	2.427856107	2.867730741	3.655131921
1	PF3	.9781324248	.9421665593	.9305166406	.9191769272	.9076096315	.9002832936	.8951110316	.8881542668	.8818657530
1	PQ3	3.060283420	3.484094410	3.726827000	4.060682338	4.576545607	5.073000264	5.569496263	6.611906841	8.468490876
1	PF4	1.014791892	.9966493835	.9907186772	.9849219001	.9789853105	.9752134442	.9725452175	.9689436046	.9656928295
1	PQ4	7.317941606	8.352896537	8.944606024	9.757509989	11.01196898	12.21787445	13.42291315	15.95076103	20.44862607
1	PFC	1.028971994	1.017945160	1.014325744	1.010781184	1.007144190	1.004829791	1.003190908	1.000980235	.9989757561
1	PQC	26.86522760	30.68016978	32.86051341	35.85524917	40.47551888	44.91599680	49.35259623	58.65785231	75.21185654
1	ZF1	1.107785905	1.107785905	1.107785905	1.107785905	1.107785905	1.107785905	1.107785905	1.107795905	1.107785905
1	ZF2	1.148278047	1.148278047	1.148278047	1.148278047	1.148278047	1.148278047	1.148278047	1.148278047	1.148278047
1	ZF3	1.264430763	1.264430763	1.264430763	1.264430763	1.264430763	1.264430763	1.264430763	1.264430763	1.264430763
1	ZF4	1.584673555	1.584673555	1.584673555	1.584673555	1.584673555	1.584673555	1.584673555	1.584673555	1.584673555
1	ZF5	2.767659844	2.767659844	2.767659844	2.767659844	2.767659844	2.767659844	2.767659844	2.767659844	2.767659844

(5) P 11+0		TH = 66	SBE = 1.094636279						CAUER FUNCTIONS	
		RC = 02	RC = 05	RC = 07	RC = 10	RC = 15	RC = 20	RC = 25	RC = 35	RC = 50
AMAX DB		.0017375255	.0108709564	.0213327376	.0436480540	.0988323393	.1772876696	.2802872360	.5675237486	1.249387366
AMIN DB		59.15791274	67.12584191	70.05886400	73.17921829	76.75622764	79.33345766	81.37465747	84.58445968	88.36435749
MU	K	.8890928D-2	.3552633D-2	.2534541D-2	.1769626D-2	.1172279D-2	.8713037D-3	.6888261D-3	.4760137D-3	.3080524D-3
1	PFR	.7116265192	.5424671146	.4856646561	.4278167465	.3644029896	.3205051504	.2868130583	.2359771600	.1804123262
1	PF1	.7914541181	.6631990846	.6218035747	.5812028715	.5391861475	.5120981857	.4926785426	.4660768551	.4414460357
1	PQ1	.6991768593	.7477979376	.7771351731	.8188624423	.3859085313	.9527962684	1.021530497	1.170352976	1.445374561
1	PF2	.9095995346	.8382375814	.8153638448	.7931804830	.7706140628	.7563438932	.7462761700	.7327395574	.7205037142
1	PQ2	1.415277483	1.587250703	1.687516299	1.826606837	2.043373786	2.253448363	2.464552359	2.910008900	3.707829901
1	PF3	.9810095592	.9455595596	.9340490133	.9228331187	.9113807246	.9041214777	.8989940226	.8920942823	.8858542036
1	PQ3	3.169942031	3.596487106	3.843179316	4.183550266	4.710740190	5.218859490	5.727444446	6.796009098	8.700443082
1	PF4	1.014716827	.9971954983	.9914523859	.9858321557	.9800696894	.9764050687	.9738111644	.9703137286	.9671439607
1	PQ4	7.674801324	8.728050569	9.336206361	10.17440378	11.47109582	12.71955932	13.96822052	16.58959272	21.25720826
1	PFC	1.027587601	1.017123524	1.013679461	1.010302467	1.006833299	1.004623605	1.003057909	1.000944725	.9990274437
1	PQC	28.38803076	32.29831420	34.55545048	37.66574281	42.47627250	47.10695977	51.73772173	61.45781350	78.76243013
1	ZF1	1.098812753	1.098812753	1.098812753	1.098812753	1.098812753	1.098812753	1.098812753	1.098812753	1.098812753
1	ZF2	1.137315334	1.137315334	1.137315334	1.137315334	1.137315334	1.137315334	1.137315334	1.137315334	1.137315334
1	ZF3	1.248611840	1.248611840	1.248611840	1.248611840	1.248611840	1.248611840	1.248611840	1.248611840	1.248611840
1	ZF4	1.558153198	1.558153198	1.558153198	1.558153198	1.558153198	1.558153198	1.558153198	1.558153198	1.558153198
1	ZF5	2.709712815	2.709712815	2.709712815	2.709712815	2.709712815	2.709712815	2.709712815	2.709712815	2.709712815

(5) P 11+0	TH = 67	SBE = 1.086360377							CAUER FUNCTIONS
	RC = 02	RC = 05	RC = 07	RC = 10	RC = 15	RC = 20	RC = 25	RC = 35	RC = 50
AMAX DB	.0017375255	.0108709564	.0213327376	.0436480540	.0988323393	.1772876696	.2802872360	.5675237436	1.249387366
AMIN DB	57.14125239	65.10917894	68.04220078	71.16255495	74.73956423	77.31679422	79.35799402	82.56779622	86.34769402
MU I K	.1095692D-1	.4378163D-2	.3123495D-2	.2180836D-2	.1444683D-2	.1073769D-2	.8488893D-3	.5866255D-3	.3796348D-3
1 I PFR	.7269051964	.5528702980	.4946612007	.4354848918	.3707234690	.3259518221	.2916181814	.2398563498	.1833282367
1 I PF1	.8046614316	.6735468304	.6312980361	.5898935736	.5470791851	.5194950301	.4997281921	.4726622209	.4476128948
1 I PQ1	.7041927919	.7521829222	.7814191643	.8231193743	.8902528604	.9573022383	1.026240961	1.175571458	1.451629090
1 I PF2	.9174072079	.8457001348	.8226840393	.8003505399	.7776205688	.7632419280	.7530955289	.7394505349	.7271143444
1 I PQ2	1.446737907	1.617690875	1.718428899	1.858633635	2.077663756	2.290242419	2.504030252	2.955447384	3.764419200
1 I PF3	.9838680776	.9489769529	.9376182947	.9265377515	.9152113731	.9080259228	.9029478552	.8961111022	.8899246556
1 I PQ3	3.290483181	3.719380506	3.970204395	4.317494172	4.856824281	5.377503657	5.899133971	6.995969591	8.957193743
1 I PF4	1.014614778	.9977298877	.9921794763	.9867408145	.9811576300	.9776035350	.9750862726	.9716901366	.9636101996
1 I PQ4	8.071169749	9.142721040	9.768444716	10.63395798	11.97655932	13.27144590	14.56777270	17.29147687	22.14504399
1 I PFC	1.026198158	1.016297112	1.013028950	1.009820247	1.006519835	1.004415525	1.002923537	1.000903595	.9990792420
1 I PQC	30.09218827	34.10075627	36.44088566	39.67720739	44.69641788	49.53637255	54.38108042	64.55834412	82.69243295
1 I ZF1	1.090308858	1.090308858	1.090308858	1.090308858	1.090308858	1.090308858	1.090308858	1.090308858	1.090308858
1 I ZF2	1.126849067	1.126849067	1.126849067	1.126849067	1.126849067	1.126849067	1.126849067	1.126849067	1.126849067
1 I ZF3	1.233342305	1.233342305	1.233342305	1.233342305	1.233342305	1.233342305	1.233342305	1.233342305	1.233342305
1 I ZF4	1.532284142	1.532284142	1.532284142	1.532284142	1.532284142	1.532284142	1.532284142	1.532234142	1.532284142
1 I ZF5	2.652778557	2.652778557	2.652778557	2.652778557	2.652778557	2.652778557	2.652778557	2.652778557	2.652778557

(5) P 11+0	TH = 68	SBE = 1.078534743							CAUER FUNCTIONS
	RC = 02	RC = 05	RC = 07	RC = 10	RC = 15	RC = 20	RC = 25	RC = 35	RC = 50
AMAX DB	.0017375255	.0108709564	.0213327376	.0436480540	.0988323393	.1772876696	.2802872360	.5675237486	1.249387366
AMIN DB	55.12086995	63.08879233	66.02181378	69.14216774	72.71917691	75.29640687	77.33760665	80.54740883	84.32730663
MU I K	.1350569D-1	.5396600D-2	.3850074D-2	.2688137D-2	.1780742D-2	.1323547D-2	.1046356D-2	.7230849D-3	.4679446D-3
1 I PFR	.7432485166	.5639345385	.5042135545	.4436143063	.3774141953	.3317123428	.2966970268	.2439531732	.1864056313
1 I PF1	.8185555963	.6844234884	.6412743097	.5990219729	.5553665000	.5272593732	.5071267251	.4795717515	.4540818225
1 I PQ1	.7096659368	.7569266040	.7860417989	.8277018852	.8949185414	.9621345837	1.031287604	1.181155067	1.458313489
1 I PF2	.9253735716	.8533656421	.8302163421	.8077395648	.7848511533	.7703663752	.7601425551	.7463903828	.7339544789
1 I PQ2	1.481166513	1.650781840	1.751967087	1.893316424	2.114729939	2.329970982	2.546623702	3.004422749	3.825356773
1 I PF3	.9867002394	.9524141920	.9412209521	.9302882623	.9190999815	.9119956334	.9069719507	.9002046997	.8940775717
1 I PQ3	3.423540217	3.854267699	4.109395451	4.464040314	5.016411856	5.550652311	6.086401179	7.213839610	9.226362140
1 I PF4	1.014483889	.9982508343	.9928983288	.9876463734	.9822477689	.9788075784	.9763693516	.9730777431	.9700905610
1 I PQ4	8.513523254	9.603097626	10.24760171	11.14268386	12.53534575	13.88104656	15.22964507	18.06573596	23.12378035
1 I PFC	1.024805082	1.015466667	1.012374767	1.009334916	1.006204030	1.004205693	1.002787872	1.000871848	.9991310884
1 I PQC	32.00929415	36.11832136	38.54830358	41.92249185	47.17142828	52.24253513	57.32396054	68.00879223	87.06197458
1 I ZF1	1.082258636	1.082258636	1.082258636	1.082258636	1.082258636	1.082258636	1.082258636	1.082258636	1.082258636
1 I ZF2	1.116862348	1.116862348	1.116862348	1.116862348	1.116862348	1.116862348	1.116862348	1.116862348	1.116862348
1 I ZF3	1.218599975	1.218599975	1.218599975	1.218599975	1.218599975	1.218599975	1.218599975	1.218599975	1.218599975
1 I ZF4	1.507027707	1.507027707	1.507027707	1.507027707	1.507027707	1.507027707	1.507027707	1.507027707	1.507027707
1 I ZF5	2.596763418	2.596763418	2.596763418	2.596763418	2.596763418	2.596763418	2.596763418	2.596763418	2.596763418

(5) P 11+0 — TH = 69 — SBE = 1.071144994 — CAUER FUNCTIONS

(5) P 11+0	RC = 02	RC = 05	RC = 07	RC = 10	RC = 15	RC = 20	RC = 25	RC = 35	RC = 50
AMAX DB	.0017375255	.0108709564	.0213327376	.0436480540	.0988323393	.1772876696	.2802872360	.5675287486	1.249387366
AMIN DB	53.09428209	61.06219779	63.99521862	67.11557225	70.69258124	73.26981114	75.31101089	78.52081304	82.30071083
NU I K	.1665468D-1	.6654870D-2	.4747757D-2	.3314902D-2	.2195939D-2	.1632145D-2	.1290324D-2	.8916792D-3	.5770504D-3
1 I PFR	.7607811657	.5757302189	.5143791734	.4522513466	.3845113395	.3378168360	.3020755431	.2482879229	.1896591859
1 I PF1	.8331902270	.6958714174	.6517710811	.6086232325	.5640797328	.5354206641	.5149020621	.4868313335	.4608768860
1 I PQ1	.7156622144	.7620760987	.7910462658	.8326503322	.8999442737	.9673318991	1.036709674	1.187145699	1.465476322
1 I PF2	.9334928016	.8612370910	.8379659132	.8153545488	.7923144093	.7777267132	.7674272947	.7535698397	.7410354154
1 I PQ2	1.518997919	1.686885993	1.788482324	1.931003673	2.154929338	2.373006715	2.592724918	3.057373450	3.891178025
1 I PF3	.9894974447	.9558661219	.9448529365	.9340816644	.9230446271	.9160293488	.9110655096	.9043748866	.8983133069
1 I PQ3	3.571087950	4.002939361	4.262539206	4.625011111	5.191426167	5.740351988	6.291429778	7.452265171	9.526023783
1 I PF4	1.014322197	.9987564857	.9936071859	.9885471922	.9833386087	.9800158030	.9776590822	.9744753390	.9715839408
1 I PQ4	9.009789555	10.11670107	10.78129264	11.70846122	13.15589421	14.55742583	15.96357400	18.92359886	24.20743394
1 I PFC	1.023409915	1.014633004	1.011717524	1.008846903	1.005886143	1.003994267	1.002651007	1.000834490	.9991829148
1 I PQC	34.17838657	38.38883303	40.91626091	44.44175143	49.94459218	55.27213680	60.61665831	71.86589085	91.94386054
1 I ZF1	1.074647623	1.074647623	1.074647623	1.074647623	1.074647623	1.074647623	1.074647623	1.074647623	1.074647623
1 I ZF2	1.107339556	1.107339556	1.107339556	1.107339556	1.107339556	1.107339556	1.107339556	1.107339556	1.107339556
1 I ZF3	1.204364112	1.204364112	1.204364112	1.204364112	1.204364112	1.204364112	1.204364112	1.204364112	1.204364112
1 I ZF4	1.482346428	1.482346428	1.482346428	1.482346428	1.482346428	1.482346428	1.482346428	1.482346428	1.482346428
1 I ZF5	2.541573635	2.541573635	2.541573635	2.541573635	2.541573635	2.541573635	2.541573635	2.541573635	2.541573635

(5) P 11+0 — TH = 70 — SBE = 1.064177772 — CAUER FUNCTIONS

(5) P 11+0	RC = 02	RC = 05	RC = 07	RC = 10	RC = 15	RC = 20	RC = 25	RC = 35	RC = 50
AMAX DB	.0017375255	.0108709564	.0213327376	.0436480540	.0988323393	.1772876696	.2802872360	.5675287486	1.249387366
AMIN DB	51.05878861	59.02669362	61.95971345	65.08006654	68.65707525	71.23430505	73.27550476	76.48530638	80.26520464
NU I K	.2055225D-1	.8212259D-2	.5858838D-2	.4090664D-2	.2709838D-2	.2014104D-2	.1592289D-2	.1100352D-2	.7120931D-3
1 I PFR	.7796502319	.5883393975	.5252248160	.4614497050	.3920566679	.3443000197	.3077835979	.2528339169	.1931059616
1 I PF1	.8486249567	.7079382420	.6628319338	.6187370279	.5732546265	.5440121900	.5230857731	.4944702482	.4680253156
1 I PQ1	.7222610825	.7676874800	.7964837724	.8380123964	.9053756218	.9729395291	1.042553213	1.193592409	1.473174350
1 I PF2	.9417570673	.8693169241	.8459377523	.8232026577	.8000194066	.7853330683	.7749605538	.7610005439	.7483694667
1 I PQ2	1.560756861	1.726436637	1.828393672	1.972109800	2.198685461	2.419791028	2.642797968	3.114819828	3.962514310
1 I PF3	.9922501206	.9593268896	.9485096032	.9379144794	.9270430050	.9201254966	.9152274707	.9086212809	.9026320345
1 I PQ3	3.735535938	4.167561483	4.431791580	4.802600628	5.384177685	5.949058060	6.516837373	7.714034757	9.854880156
1 I PF4	1.014127623	.9992448402	.9943041370	.9894414765	.9844285003	.9812266649	.9789540004	.9758915745	.9730890997
1 I PQ4	9.569767874	10.69275377	11.37883495	12.34091150	13.84848883	15.31161692	16.78140093	19.87870772	25.41301797
1 I PFC	1.022014337	1.013797017	1.011057893	1.008356681	1.005566462	1.003781424	1.002513042	1.000796527	.9992346469
1 I PQC	36.64819833	40.95915542	43.59242986	47.28453972	53.06923165	58.68262499	64.32101228	76.20166590	97.42766142
1 I ZF1	1.067462398	1.067462398	1.067462398	1.067462398	1.067462398	1.067462398	1.067462398	1.067462398	1.067462398
1 I ZF2	1.098266287	1.098266287	1.098266287	1.098266287	1.098266287	1.098266287	1.098266287	1.098266287	1.098266287
1 I ZF3	1.190615336	1.190615336	1.190615336	1.190615336	1.190615336	1.190615336	1.190615336	1.190615336	1.190615336
1 I ZF4	1.458203809	1.458203809	1.458203809	1.458203809	1.458203809	1.458203809	1.458203809	1.458203809	1.458203809
1 I ZF5	2.487114429	2.487114429	2.487114429	2.487114429	2.487114429	2.487114429	2.487114429	2.487114429	2.487114429

(5) P 11+0	TH = 71	SBE = 1.057620681							CAUER FUNCTIONS
	RC = 02	RC = 05	RC = 07	RC = 10	RC = 15	RC = 20	RC = 25	RC = 35	RC = 50
AMAX DB	.0017375255	.0108709564	.0213327376	.0436480540	.0988323393	.1772876696	.2802872360	.5675287486	1.249387366
AMIN DB	49.01142551	56.97931329	59.91233151	63.03268375	66.60969201	69.18692165	71.22812128	74.43792333	78.21782106
MU K	.2538687D-1	.1014408D-1	.7237047D-2	.5052935D-2	.3347289D-2	.2487893D-2	.1966852D-2	.1359194D-2	.8796030D-3
1 PFR	.8000307246	.6018584834	.5368286306	.4712720228	.4000987484	.3512021853	.3138558058	.2577681303	.1967656055
1 PF1	.8649262472	.7206777423	.6745062056	.6294083558	.5829317759	.5530717833	.5317137544	.5025218053	.4755580994
1 PQ1	.7295592093	.7738280474	.8024155319	.8438448631	.9112666512	.9790111567	1.048872645	1.200553066	1.481474410
1 PF2	.9501560695	.8776068511	.8541365886	.8312911901	.8079757137	.7931962742	.7827539832	.7686951491	.7559701024
1 PQ2	1.607082886	1.769956400	1.872205070	2.017131796	2.246504845	2.470851017	2.697396389	3.177381044	4.040118732
1 PF3	.9949475872	.9627898377	.9521856179	.9417826564	.9310923621	.9242821367	.9194564627	.9129432688	.9070339668
1 PQ3	3.919853867	4.350778730	4.619778040	4.999474390	5.597466902	6.179742523	6.765789318	8.002957725	10.21739086
1 PF4	1.013897961	.9997137302	.9949871002	.9903272595	.9855156236	.9824384517	.9802524767	.9772949399	.9746046439
1 PQ4	10.20570113	11.34267952	12.05174250	13.05189818	14.62578366	16.15717846	17.69766515	20.94779394	26.76137783
1 PFC	1.020620191	1.012959689	1.010396618	1.007864774	1.005245303	1.003567358	1.002374092	1.000757968	.9992862032
1 PQC	39.48025336	43.88797778	46.63637732	50.51264748	56.61170601	62.54540849	68.51382744	81.10436070	103.6243445
1 ZF1	1.060690530	1.060690530	1.060690530	1.060690530	1.060690530	1.060690530	1.060690530	1.060690530	1.060690530
1 ZF2	1.089629298	1.089629298	1.089629298	1.089629298	1.089629298	1.089629298	1.089629298	1.089629298	1.089629298
1 ZF3	1.177335553	1.177335553	1.177335553	1.177335553	1.177335553	1.177335553	1.177335553	1.177335553	1.177335553
1 ZF4	1.434564069	1.434564069	1.434564069	1.434564069	1.434564069	1.434564069	1.434564069	1.434564069	1.434564069
1 ZF5	2.433288972	2.433288972	2.433288972	2.433288972	2.433288972	2.433288972	2.433288972	2.433288972	2.433288972

(5) P 11+0	TH = 72	SBE = 1.051462224							CAUER FUNCTIONS
	RC = 02	RC = 05	RC = 07	RC = 10	RC = 15	RC = 20	RC = 25	RC = 35	RC = 50
AMAX DB	.0017375255	.0108709564	.0213327376	.0436480540	.0988323393	.1772876696	.2802872360	.5675287486	1.249387366
AMIN DB	46.94890669	54.91676661	57.84978223	60.97013310	64.54714061	67.12436999	69.16556950•	72.37537145	76.15526913
MU K	.3139945D-1	.1254658D-1	.8951055D-2	.6249662D-2	.4140054D-2	.3077121D-2	.2432678D-2	.1681103D-2	.1087926D-2
1 PFR	.8221328585	.6164017146	.5492828557	.4817919686	.4086944988	.3585704503	.3203325863	.2629720013	.2006609770
1 PF1	.8821683016	.7341509350	.6868500396	.6406885337	.5931575590	.5626427035	.5408270726	.5110241353	.4835107253
1 PQ1	.7376754120	.7805793278	.8089153820	.8502159558	.9176820643	.9856108725	1.055732839	1.208096489	1.490455834
1 PF2	.9586764608	.8861076049	.8625667327	.8396275165	.8161934176	.8013279389	.7908201753	.7766674609	.7638521186
1 PQ2	1.658763636	1.818081741	1.920528266	2.066671200	2.298998846	2.526821754	2.757186321	3.245801049	4.124891649
1 PF3	.9975778988	.9662473758	.9558748428	.9456814745	.9351894174	.9284968936	.9237507453	.9173399573	.9115188192
1 PQ3	4.127742617	4.555853818	4.829728498	5.218903171	5.834721755	6.436037849	7.042150493	8.323285252	10.61898754
1 PF4	1.013630876	1.000160804	.9956538018	.9912023787	.9865979642	.9836492598	.9815526930	.9787137417	.9761290018
1 PQ4	10.93306731	12.08078734	12.81440000	13.85620799	15.50351543	17.11094934	18.73040677	22.15159267	28.27831988
1 PFC	1.019229497	1.012122109	1.009734522	1.007371765	1.004923021	1.003352291	1.002234285	1.000718824	.9993374946
1 PQC	42.75319652	47.24966959	50.12340216	54.20401287	60.65553971	66.95025365	73.29157012	86.68681054	110.6729896
1 ZF1	1.054320515	1.054320515	1.054320515	1.054320515	1.054320515	1.054320515	1.054320515	1.054320515	1.054320515
1 ZF2	1.081416465	1.081416465	1.081416465	1.081416465	1.081416465	1.081416465	1.081416465	1.081416465	1.081416465
1 ZF3	1.164507891	1.164507891	1.164507891	1.164507891	1.164507891	1.164507891	1.164507891	1.164507891	1.164507891
1 ZF4	1.411391866	1.411391866	1.411391866	1.411391866	1.411391866	1.411391866	1.411391866	1.411391866	1.411391866
1 ZF5	2.379997178	2.379997178	2.379997178	2.379997178	2.379997178	2.379997178	2.379997178	2.379997178	2.379997178

(5) P 11+0		TH = 73	SBE = 1.045691756						CAUER FUNCTIONS	
		RC = 02	RC = 05	RC = 07	RC = 10	RC = 15	RC = 20	RC = 25	RC = 35	RC = 50
AMAX DB		.0017375255	.0108709564	.0213327376	.0436480540	.0988323393	.1772876696	.2802872360	.5675287486	1.249387366
AMIN DB		44.86755109	52.83536571	55.76837711	58.88872573	62.46573205	65.04296101	67.08416032	70.29396210	74.07385969
MU	K	.3890000D-1	.1554365D-1	.1108924D-1	.7742552D-2	.5129011D-2	.3812169D-2	.3013784D-2	.2082678D-2	.1347805D-2
1	PFR	.8462118137	.6321057382	.5626973591	.4930969508	.4179111990	.3664603831	.3272615338	.2685324724	.2048188885
1	PF1	.9004340739	.7484273954	.6999276850	.6526364470	.6039853051	.5727747467	.5504730297	.5200211899	.4919241211
1	PQ1	.7467574034	.7880411055	.8160732828	.8572084412	.9247000286	.9928159072	1.063211822	1.216305266	1.500213630
1	PF2	.9673011132	.8948186232	.8712318794	.8482189907	.8246831375	.8097405176	.7991727782	.7849326008	.7720318450
1	PQ2	1.716780447	1.871596060	1.974113658	2.121463569	2.356912763	2.588476026	2.822977332	3.320982483	4.217926197
1	PF3	1.000127655	.9696908269	.9595701992	.9496054237	.9393302640	.9327668728	.9281081359	.9218101166	.9160862665
1	PQ3	4.363871543	4.786858420	5.065661441	5.464945053	6.100184088	6.722431844	7.350690656	8.680491274	11.06635581
1	PF4	1.013323889	1.000583503	.9963017513	.9920644504	.9876732861	.9848569658	.9828526143	.9801360749	.9776603962
1	PQ4	11.77169313	12.92522282	13.68499814	14.77249279	16.50148640	18.19408805	19.90227126	23.51609881	29.99616030
1	PFC	1.017844483	1.011285483	1.009072518	1.006878303	1.004600013	1.003136467	1.002093766	1.000679109	.9993884226
1	PQC	46.56895697	51.13970968	54.14992359	58.45819778	65.30718944	72.01141725	78.77691117	93.08892344	118.7500575
1	ZF1	1.048341733	1.048341733	1.048341733	1.048341733	1.048341733	1.048341733	1.048341733	1.048341733	1.048341733
1	ZF2	1.073616752	1.073616752	1.073616752	1.073616752	1.073616752	1.073616752	1.073616752	1.073616752	1.073616752
1	ZF3	1.152116654	1.152116654	1.152116654	1.152116654	1.152116654	1.152116654	1.152116654	1.152116654	1.152116654
1	ZF4	1.388651992	1.388651992	1.388651992	1.388651992	1.388651992	1.388651992	1.388651992	1.388651992	1.388651992
1	ZF5	2.327134251	2.327134251	2.327134251	2.327134251	2.327134251	2.327134251	2.327134251	2.327134251	2.327134251

(5) P 11+0		TH = 74	SBE = 1.040299436.						CAUER FUNCTIONS	
		RC = 02	RC = 05	RC = 07	RC = 10	RC = 15	RC = 20	RC = 25	RC = 35	RC = 50
AMAX DB		.0017375255	.0108709564	.0213327376	.0436480540	.0988323393	.1772876696	.2802872360	.5675287486	1.249387366
AMIN DB		42.76319023	50.73093067	53.66393515	56.78428009	60.36128444	62.93851272	64.97971172	68.13951322	71.96941066
MU	K	.4829069D-1	.1929597D-1	.1376625D-1	.9611648D-2	.6367184D-2	.4732449D-2	.3741329D-2	.2585448D-2	.1673173D-2
1	PFR	.8725810281	.6491357369	.5772043476	.5052917100	.4278291439	.3749381387	.3346992143	.2744933528	.2092719752
1	PF1	.9198163522	.7635868867	.7138131224	.6653201215	.6154767782	.5835256585	.5607065214	.5295640215	.5008458587
1	PQ1	.7569911650	.7963369207	.8240000653	.8649238219	.9324159792	1.000720294	1.071404422	1.225279516	1.510862716
1	PF2	.9760081827	.9037376294	.8801348412	.8570728227	.8334560284	.8184473942	.8078266276	.7935072042	.7805273974
1	PQ2	1.782371933	1.931475169	2.033892495	2.182418645	2.421165375	2.656764624	2.895764120	3.404032442	4.320560427
1	PF3	1.002581772	.9731102383	.9632634991	.9535480592	.9435102489	.9370885585	.9325259205	.9263521079	.9207356378
1	PQ3	4.634213681	5.048939180	5.332639569	5.742697358	6.399166883	7.044535949	7.697366839	9.081339563	11.56782173
1	PF4	1.012974370	1.000979039	.9969282138	.9929108382	.9887390984	.9860591936	.9841499555	.9815597894	.9791968113
1	PQ4	12.74735126	13.89931551	14.68685391	15.82459526	17.64494391	19.43352888	21.24205510	25.07432234	31.95588660
1	PFC	1.016467614	1.010451153	1.008411627	1.006385112	1.004276723	1.002920164	1.001952697	1.000638840	.9994388787
1	PQC	51.06169686	55.69247724	58.34118953	63.40420042	70.70424953	77.87635383	85.12801114	100.4932811	129.0825074
1	ZF1	1.042744401	1.042744401	1.042744401	1.042744401	1.042744401	1.042744401	1.042744401	1.042744401	1.042744401
1	ZF2	1.066220183	1.066220183	1.066220183	1.066220183	1.066220183	1.066220183	1.066220183	1.066220183	1.066220183
1	ZF3	1.140147277	1.140147277	1.140147277	1.140147277	1.140147277	1.140147277	1.140147277	1.140147277	1.140147277
1	ZF4	1.366309022	1.366309022	1.366309022	1.366309022	1.366309022	1.366309022	1.366309022	1.366309022	1.366309022
1	ZF5	2.274588905	2.274588905	2.274588905	2.274588905	2.274588905	2.274588905	2.274588905	2.274538905	2.274538905

(5) P 11+0	TH = 75 SBE = 1.035276180								CAUER FUNCTIONS
	RC = 02	RC = 05	RC = 07	RC = 10	RC = 15	RC = 20	RC = 25	RC = 35	RC = 50
AMAX DB	.0017375255	.0108709564	.0213327376	.0436480540	.0988323393	.1772876696	.2802872360	.5675287486	1.249387366
AMIN DB	40.63104933	48.59866733	51.53166038	54.65199925	58.22900037	60.80622751	62.84742598	66.05722703	69.83712422
MU I K	.6009787D-1	.2401389D-1	.1713213D-1	.1196172D-1	.7923975D-2	.5839543D-2	.4656093D-2	.3217596D-2	.2082268D-2
1 I PFR	.9016306396	.6676937626	.5929647389	.5185031555	.4385451978	.3840833122	.3427135625	.2809071315	.2140554798
1 I PF1	.9404188591	.7797213789	.7285921123	.6788187284	.6277040837	.5949629574	.5715917939	.5397124427	.5103317177
1 I PQ1	.7686142274	.8056217064	.8328339883	.8734880978	.9409478149	1.009439858	1.080427202	1.235141980	1.522543647
1 I PF2	.9847699021	.9128600759	.8892771868	.8661958963	.8425237695	.8274629689	.8167979025	.8024096620	.7893589926
1 I PQ2	1.857124390	1.998950891	2.101035625	2.250676046	2.492903587	2.732871894	2.976783927	3.496325359	4.434452731
1 I PF3	1.004923204	.9764941494	.9669452365	.9575018199	.9477238233	.9414576849	.9370007410	.9309637941	.9254658973
1 I PQ3	4.946529124	5.348695608	5.637131490	6.058651828	6.738415685	7.409461492	8.089718837	9.534376996	12.13380132
1 I PF4	1.012579530	1.001344361	.9975301754	.9937386165	.9897926165	.9872532739	.9854421402	.9829824488	.9807359515
1 I PQ4	13.89410078	15.03352650	15.85031179	17.04345180	18.96655444	20.86406215	22.78691110	26.86378957	34.21023243
1 I PFC	1.015101630	1.009620623	1.007752993	1.005893010	1.003953654	1.002703696	1.001811267	1.000598040	.9994887425
1 I PQC	56.41109531	61.04267063	64.36253584	69.21183183	77.02736670	84.73836986	92.55196464	109.1379660	138.9667392
1 I ZF1	1.037519534	1.037519534	1.037519534	1.037519534	1.037519534	1.037519534	1.037519534	1.037519534	1.037519534
1 I ZF2	1.059217841	1.059217841	1.059217841	1.059217841	1.059217841	1.059217841	1.059217841	1.059217841	1.059217841
1 I ZF3	1.128586300	1.128586300	1.128586300	1.128586300	1.128586300	1.128586300	1.128586300	1.128586300	1.128586300
1 I ZF4	1.344326903	1.344326903	1.344326903	1.344326903	1.344326903	1.344326903	1.344326903	1.344326903	1.344326903
1 I ZF5	2.222241140	2.222241140	2.222241140	2.222241140	2.222241140	2.222241140	2.222241140	2.222241140	2.222241140

(5) P 11+0	TH = 76 SBE = 1.030613629								CAUER FUNCTIONS
	RC = 02	RC = 05	RC = 07	RC = 10	RC = 15	RC = 20	RC = 25	RC = 35	RC = 50
AMAX DB	.0017375255	.0108709564	.0213327376	.0436480540	.0988323393	.1772876696	.2802872360	.5675287486	1.249387366
AMIN DB	38.46559259	46.43300656	49.36598057	52.48630933	56.06330504	58.64053030	60.68172789	63.89152818	67.67142497
MU I K	.7501763D-1	.2997552D-1	.2138531D-1	.1493131D-1	.9891163D-2	.7351667D-2	.5812003D-2	.4016339D-2	.2599207D-2
1 I PFR	.9338536067	.6880302908	.6101769414	.5328869921	.4501776396	.3939928169	.3513871312	.2878374281	.2192179878
1 I PF1	.9623572431	.7969375681	.7443648059	.6932251733	.6407521529	.6071663215	.5832047491	.5505372094	.5204477487
1 I PQ1	.7819348929	.8160926106	.8427499736	.8830598329	.9504431367	1.019119146	1.090425313	1.246045059	1.535430495
1 I PF2	.9935510025	.9221783940	.8986587448	.8755945075	.8518985252	.8368027536	.8261043096	.8116604185	.7985493439
1 I PQ2	1.943103250	2.075601878	2.177036984	2.327684072	2.573579411	2.818293812	3.067597093	3.599591048	4.561686576
1 I PF3	1.007132598	.9798293019	.9706043270	.9614578008	.9519643535	.9458690741	.9415284527	.9356424248	.9302755573
1 I PQ3	5.311078318	5.694728636	5.987534477	6.421205968	7.126627059	7.826358232	8.537434465	10.05056594	12.77801900
1 I PF4	1.012136409	1.001676125	.9981043037	.9945445250	.9908307139	.9884361946	.9867262512	.9844012803	.9822751915
1 I PQ4	15.25780008	16.36832801	17.21554374	18.46988500	20.50929092	22.53137051	24.58556314	28.95517663	36.82814542
1 I PFC	1.013749591	1.008795580	1.007097906	1.005402918	1.003631374	1.002487421	1.001669689	1.000556736	.9995378801
1 I PQC	62.86256558	67.44244364	70.93622044	76.10866935	84.51794048	92.85530130	101.3247117	119.3397241	151.7965620
1 I ZF1	1.032658913	1.032658913	1.032658913	1.032658913	1.032658913	1.032658913	1.032658913	1.032658913	1.032658913
1 I ZF2	1.052601864	1.052601864	1.052601864	1.052601864	1.052601864	1.052601864	1.052601864	1.052601864	1.052601864
1 I ZF3	1.117421357	1.117421357	1.117421357	1.117421357	1.117421357	1.117421357	1.117421357	1.117421357	1.117421357
1 I ZF4	1.322668455	1.322668455	1.322668455	1.322668455	1.322668455	1.322668455	1.322668455	1.322668455	1.322668455
1 I ZF5	2.169959372	2.169959372	2.169959372	2.169959372	2.169959372	2.169959372	2.169959372	2.169959372	2.169959372

(5) P 11+0		TH = 77	SBE = 1.026304108						CAUER FUNCTIONS	
		RC = 02	RC = 05	RC = 07	RC = 10	RC = 15	RC = 20	RC = 25	RC = 35	RC = 50
	AMAX DB	.0017375255	.0108709564	.0213327376	.0436480540	.0988323393	.1772876696	.2802872360	.5675287486	1.249387366
	AMIN DB	36.26031886	44.22738899	47.16033092	50.28064263	53.85762925	56.43485132	58.47604744	61.68584645	65.46574255
MU	K	.9398145D-1	.3755307D-1	.2679133D-1	.1870581D-1	.1239156D-1	.9210106D-2	.7281228D-2	.5031697D-2	.3256264D-2
1	PFR	.9698835775	.7104605872	.6290892119	.5486369932	.4628729049	.4047862613	.3608215875	.2953623739	.2248148130
1	PF1	.9857597097	.8153600376	.7612491092	.7086494859	.6547220323	.6202307637	.5956360218	.5621229455	.5312730420
1	PQ1	.7973607541	.8280046817	.8539729027	.8938417009	.9610895446	1.029941244	1.101582171	1.258180733	1.549741942
1	PF2	1.002306619	.9316809708	.9082769142	.8852739847	.8615928652	.8464834691	.8357653012	.8212823450	.8081241660
1	PQ2	2.043049061	2.163486348	2.263834973	2.415313729	2.665060862	2.914950158	3.170202588	3.716040686	4.704920046
1	PF3	1.009187866	.9831002722	.9742277750	.9654054625	.9562238784	.9503164278	.9461039418	.9403844897	.9351625642
1	PQ3	5.741702548	6.098456622	6.394948375	6.841416853	7.575209848	8.307203287	9.053176304	10.64420727	13.51772723
1	PF4	1.011641877	1.001970654	.9986468995	.9953249152	.9918498630	.9896045403	.9879989679	.9858131119	.9838115129
1	PQ4	16.90152547	17.95857177	18.83677822	20.15880498	22.33076108	24.49657123	26.70310824	31.40774448	39.90143803
1	PFC	1.012414932	1.007977933	1.006447825	1.004915883	1.003310534	1.002271749	1.001528209	1.000514962	.9995861415
1	PQC	70.75890645	75.18782020	78.86726669	84.40598897	93.50510660	102.5780557	111.8212903	131.5283388	167.1055521
1	ZF1	1.028155057	1.028155057	1.028155057	1.028155057	1.028155057	1.028155057	1.028155057	1.028155057	1.028155057
1	ZF2	1.046365465	1.046365465	1.046365465	1.046365465	1.046365465	1.046365465	1.046365465	1.046365465	1.046365465
1	ZF3	1.106641175	1.106641175	1.106641175	1.106641175	1.106641175	1.106641175	1.106641175	1.106641175	1.106641175
1	ZF4	1.301294747	1.301294747	1.301294747	1.301294747	1.301294747	1.301294747	1.301294747	1.301294747	1.301294747
1	ZF5	2.117596671	2.117596671	2.117596671	2.117596671	2.117596671	2.117596671	2.117596671	2.117596671	2.117596671

(5) P 11+0		TH = 78	SBE = 1.022340595						CAUER FUNCTIONS	
		RC = 02	RC = 05	RC = 07	RC = 10	RC = 15	RC = 20	RC = 25	RC = 35	RC = 50
	AMAX DB	.0017375255	.0108709564	.0213327376	.0436480540	.0988323393	.1772876696	.2802872360	.5675287486	1.249387366
	AMIN DB	34.00748796	41.97397092	44.90685805	48.02714065	51.60411174	54.18132836	56.22252197	59.43231878	63.21221372
MU	K	.1182531D 0	.4725152D-1	.3371046D-1	.2353677D-1	.1559181D-1	.1158871D-1	.9161676D-2	.6331183D-2	.4097226D-2
1	PFR	1.010551275	.7353884671	.6500174656	.5659982790	.4768151793	.4166135731	.3711440721	.3035793787	.2309158533
1	PF1	1.010766796	.8351352395	.7793850661	.7252233265	.6697353142	.6342709353	.6089951638	.5745721338	.5429035194
1	PQ1	.8154422007	.8416941984	.8667972513	.9060974199	.9731296464	1.042142039	1.114133478	1.271794375	1.565757218
1	PF2	1.010979461	.9413507299	.9181256908	.8952381346	.8716196158	.8565231411	.8458023364	.8313012132	.8181128320
1	PQ2	2.160678228	2.265341322	2.363993581	2.516027990	2.769795786	3.025349750	3.287207865	3.848551563	4.867604679
1	PF3	1.011063632	.9862889999	.9778002467	.9693322543	.9604927953	.9547920562	.9507208865	.9451355259	.9401241506
1	PQ3	6.257510384	6.575363573	6.874353186	7.334142768	8.099433105	8.867987731	9.653823065	11.33432328	14.37628083
1	PF4	1.011092626	1.002223885	.9991538384	.9960756829	.9928460620	.9907544156	.9892564890	.9872142929	.9853414252
1	PQ4	18.91419386	19.88031344	20.78887499	22.18571328	24.50988120	26.84320312	29.22839163	34.32764368	43.55496346
1	PFC	1.011101534	1.007169856	1.005804418	1.004433103	1.002991877	1.002057151	1.001397112	1.000472760	.9996333582
1	PQC	80.59148285	84.71067547	88.58435122	94.53963056	104.4481009	114.3952218	124.5631590	146.3002971	195.6330765
1	ZF1	1.024001199	1.024001199	1.024001199	1.024001199	1.024001199	1.024001199	1.024001199	1.024001199	1.024001199
1	ZF2	1.040502978	1.040502978	1.040502978	1.040502978	1.040502978	1.040502978	1.040502978	1.040502978	1.040502978
1	ZF3	1.096235607	1.096235607	1.096235607	1.096235607	1.096235607	1.096235607	1.096235607	1.096235607	1.096235607
1	ZF4	1.280164291	1.280164291	1.280164291	1.280164291	1.280164291	1.280164291	1.280164291	1.280164291	1.230164291
1	ZF5	2.064985705	2.064985705	2.064985705	2.064985705	2.064985705	2.064985705	2.064985705	2.064985705	2.064985705

(5) TH = 79 SBE = 1.018716695 CAUER FUNCTIONS

P 11+0			RC = 02	RC = 05	RC = 07	RC = 10	RC = 15	RC = 20	RC = 25	RC = 35	RC = 50
AMAX DB			.0017375255	.0108709564	.0213327376	.0436480540	.0988323393	.1772876696	.2802872360	.5675287486	1.249387366
AMIN DB			31.69774918	39.66321342	42.59600546	45.71623755	49.29318166	51.87038885	53.91157809	57.12137110	60.90126402
MU		K	.1495749D 0	.5976709D-1	.4263939D-1	.2977099D-1	.1972163D-1	.1465822D-1	.1158834D-1	.8008131D-2	.5182464D-2
	1	PFR	1.056971048	.7633417694	.6733716617	.5852868482	.4922404175	.4296660958	.3825164282	.3126120243	.2376095267
	1	PF1	1.037529311	.8564365132	.7989406417	.7431060809	.6859402258	.6494270822	.6234164581	.5880106890	.5554572440
	1	PQ1	.8369409605	.8576144150	.8816159350	.9201773084	.9868835473	1.056031489	1.128388080	1.237208465	1.583839666
	1	PF2	1.019495915	.9511631224	.9281942674	.9054884182	.8819915992	.8669411800	.8562391932	.8417463069	.8285492428
	1	PQ2	2.301160058	2.384891004	2.480980898	2.633140576	2.891060971	3.152840905	3.422085773	4.000945763	5.054314792
	1	PF3	1.012730528	.9893741693	.9813035064	.9732231183	.9647594427	.9592865176	.9553714396	.9500398619	.9451566376
	1	PQ3	6.885602416	7.146970923	7.446458353	7.919826793	8.720210909	9.530556359	10.36239060	12.14679247	15.38526619
	1	PF4	1.010485176	1.002431322	.9996204993	.9967921852	.9938147421	.9918813478	.9904944326	.9886005929	.9868608627
	1	PQ4	21.42478786	22.24183966	23.17789183	24.65710611	27.15750440	29.68832512	32.28570137	37.35607480	47.96269452
	1	PFC	1.009813813	1.006373838	1.005169599	1.003955958	1.002676266	1.001844175	1.001246731	1.000430185	.9996793396
	1	PQC	93.08613529	96.63784948	100.7067804	107.1366412	118.0050631	129.0053917	140.2947427	164.5054178	208.4306753
	1	ZF1	1.020191270	1.020191270	1.020191270	1.020191270	1.020191270	1.020191270	1.020191270	1.020191270	1.020191270
	1	ZF2	1.035009919	1.035009919	1.035009919	1.035009919	1.035009919	1.035009919	1.035009919	1.035009919	1.035009919
	1	ZF3	1.086195678	1.086195678	1.086195678	1.086195678	1.086195678	1.086195678	1.086195678	1.086195678	1.086195678
	1	ZF4	1.259231972	1.259231972	1.259231972	1.259231972	1.259231972	1.259231972	1.259231972	1.259231972	1.259231972
	1	ZF5	2.011931728	2.011931728	2.011931728	2.011931728	2.011931728	2.011931728	2.011931728	2.011931728	2.011931728

(5) TH = 80 SBE = 1.015426612 CAUER FUNCTIONS

P 11+0			RC = 02	RC = 05	RC = 07	RC = 10	RC = 15	RC = 20	RC = 25	RC = 35	RC = 50
AMAX DB			.0017375255	.0108709564	.0213327376	.0436480540	.0988323393	.1772876696	.2802872360	.5675287486	1.249387366
AMIN DB			29.31963056	37.28329358	40.21591743	43.33606016	46.91295657	49.49014707	51.53132858	54.74111486	58.52100420
MU		K	.1903955D 0	.7607817D-1	.5427613D-1	.3789581D-1	.2510387D-1	.1865861D-1	.1475092D-1	.1019364D-1	.6596814D-2
	1	PFR	1.110678528	.7950270857	.6996961595	.6069192188	.5094574585	.4441932265	.3951490003	.3226203313	.2450099517
	1	PF1	1.066202519	.8794703712	.8201194412	.7624932613	.7035201871	.6658734893	.6390672036	.6025959386	.5690820549
	1	PQ1	.8629422000	.8763922555	.8989722321	.9365581435	1.002783659	1.072026409	1.144759999	1.304850039	1.604472635
	1	PF2	1.027760534	.9610832189	.9384649747	.9160227032	.8927211859	.8777584201	.8671023404	.8526512272	.8394730058
	1	PQ2	2.471903353	2.527341432	2.619612623	2.771225392	3.033353962	3.302003993	3.579576274	4.178424605	5.271259527
	1	PF3	1.014154244	.9923303825	.9847156605	.9770598187	.9690095361	.9637881300	.9600457978	.9549402686	.9502551648
	1	PQ3	7.665654730	7.844071238	8.140715637	8.627385835	9.466977995	10.32556259	11.21111085	13.11776210	16.58870475
	1	PF4	1.009815894	1.002587959	1.000041678	.9974691351	.9947506481	.9929801616	.9917077073	.9899670681	.9883650499
	1	PQ4	24.62683119	25.20220199	26.15873763	27.72776236	30.43399596	33.20081647	36.05401441	42.19588393	53.37399731
	1	PFC	1.008556832	1.005592756	1.004545588	1.003486054	1.002364708	1.001633463	1.001107459	1.000387302	.9997238687
	1	PQC	109.3538150	111.9096997	116.1587292	123.1283836	135.1492773	147.4390086	160.1121813	187.3922123	237.0402158
	1	ZF1	1.016719885	1.016719885	1.016719885	1.016719885	1.016719885	1.016719885	1.016719885	1.016719885	1.016719885
	1	ZF2	1.029883090	1.029883090	1.029883090	1.029883090	1.029883090	1.029883090	1.029883090	1.029883090	1.029883090
	1	ZF3	1.076513683	1.076513683	1.076513683	1.076513683	1.076513683	1.076513683	1.076513683	1.076513683	1.076513683
	1	ZF4	1.238447583	1.238447583	1.238447583	1.238447583	1.238447583	1.238447583	1.238447583	1.238447583	1.238447583
	1	ZF5	1.958202636	1.958202636	1.958202636	1.958202636	1.958202636	1.958202636	1.958202636	1.958202636	1.958202636

(5) P 11+0	TH = 81	SBE = 1.012465126						CAUER FUNCTIONS	
	RC = 02	RC = 05	RC = 07	RC = 10	RC = 15	RC = 20	RC = 25	RC = 35	RC = 50
AMAX DB	.0017375255	.0108709564	.0213327376	.0436480540	.0988323393	.1772876696	.2802872360	.5675287486	1.249387366
AMIN DB	26.85883126	34.81923598	37.75155540	40.87153637	44.44834643	47.02550670	49.06667422	52.27644832	56.05633119
MU I K	.2442349D 0	.9759130D-1	.6962416D-1	.4861186D-1	.3220266D-1	.2393483D-1	.1892214D-1	.1307615D-1	.8462239D-2
1 I PFR	1.173858822	.8314175036	.7297338014	.6314590877	.5288809915	.4605282674	.4093220000	.3338165885	.2532679756
1 I PF1	1.096932652	.9044842119	.8431711222	.7836283347	.7227061630	.6838307956	.6561598661	.6185284042	.5839668890
1 I PQ1	.8950461187	.8989219671	.9196265471	.9559077880	1.021430661	1.090702887	1.163819485	1.325307165	1.628316220
1 I PF2	1.035648025	.9710613765	.9489101692	.9268333270	.9038195258	.8889970678	.8784213784	.8640549744	.8509310761
1 I PQ2	2.683914121	2.700208099	2.786788023	2.936791043	3.203034210	3.479291297	3.766339520	4.338272546	5.527110109
1 I PF3	1.015294260	.9951270251	.9880101163	.9808200158	.9732253839	.9682822908	.9647315997	.9598774721	.9554133136
1 I PQ3	8.658005840	8.712271277	9.000445221	9.499094048	10.38252139	11.29741598	12.24657506	14.29233136	18.04988611
1 I PF4	1.009081031	1.002688210	1.000411474	.9981004657	.9956476826	.9940448127	.9928903394	.9913078839	.9898483202
1 I PQ4	28.82265098	29.00384979	29.96604182	31.63079371	34.57961493	37.63290986	40.80008791	47.64859318	60.15876021
1 I PFC	1.007336451	1.004829972	1.003934983	1.003025276	1.002058390	1.001425779	1.000969767	1.000344199	.9997666963
1 I PQC	131.1687165	131.9935679	136.3736248	143.9517641	157.3749076	171.2733781	185.6900389	216.8632054	273.8061185
1 I ZF1	1.013582347	1.013582347	1.013582347	1.013582347	1.013582347	1.013582347	1.013582347	1.013582347	1.013582347
1 I ZF2	1.025120733	1.025120733	1.025120733	1.025120733	1.025120733	1.025120733	1.025120733	1.025120733	1.025120733
1 I ZF3	1.067183335	1.067183335	1.067183335	1.067183335	1.067183335	1.067183335	1.067183335	1.067183335	1.067183335
1 I ZF4	1.217753731	1.217753731	1.217753731	1.217753731	1.217753731	1.217753731	1.217753731	1.217753731	1.217753731
1 I ZF5	1.903514313	1.903514313	1.903514313	1.903514313	1.903514313	1.903514313	1.903514313	1.903514313	1.903514313

(5) P 11+0	TH = 82	SBE = 1.009827573						CAUER FUNCTIONS	
	RC = 02	RC = 05	RC = 07	RC = 10	RC = 15	RC = 20	RC = 25	RC = 35	RC = 50
AMAX DB	.0017375255	.0108709564	.0213327376	.0436480540	.0988323393	.1772876696	.2802872360	.5675287486	1.249387366
AMIN DB	24.29724301	32.25158747	35.18334000	38.30301969	41.87966392	44.45677289	46.49791435	49.70766576	53.48753658
MU I K	.3162960D 0	.1263854D 0	.9016664D-1	.6295470D-1	.4170400D-1	.3099675D-1	.2450508D-1	.1693425D-1	.1095901D-1
1 I PFR	1.249743893	.8738998995	.7645323417	.6596940691	.5510852750	.4791303163	.4254200021	.3464907862	.2625888210
1 I PF1	1.129827668	.9317752726	.8684055352	.8068197714	.7437950429	.7035845600	.6749705288	.6360698271	.6003591972
1 I PQ1	.9357130831	.9265275537	.9446981001	.9791949604	1.043687645	1.112883985	1.186378089	1.349411786	1.656300883
1 I PF2	1.042991190	.9810265632	.9594873683	.9379039906	.9152952087	.9006804539	.8902295453	.8760034424	.8629801127
1 I PQ2	2.954271163	2.914772411	2.992772934	3.139447100	3.409423284	3.694120147	3.992068276	4.641051663	5.834399122
1 I PF3	1.016102058	.9977266626	.9911541028	.9844759448	.9773847654	.9727504997	.9694130620	.9648394517	.9606225638
1 I PQ3	9.958766374	9.822011337	10.09202243	10.59926956	11.53151656	12.51298846	13.53876036	15.76949244	19.86328598
1 I PF4	1.008276803	1.002725808	1.000723152	.9986791526	.9964986964	.9950681622	.9940352392	.9926160707	.9913038656
1 I PQ4	34.50836163	34.03338300	34.97117305	36.73277059	39.96989876	43.37746555	46.93841725	54.68128698	68.88841105
1 I PFC	1.006159534	1.004089457	1.003340868	1.002575868	1.001758731	1.001222036	1.000834221	1.000300988	.9998075336
1 I PQC	161.5147636	159.2911998	163.6813619	171.9280283	187.0820038	203.0336555	219.7031691	255.9486224	322.4531670
1 I ZF1	1.010774649	1.010774649	1.010774649	1.010774649	1.010774649	1.010774649	1.010774649	1.010774649	1.010774649
1 I ZF2	1.020722752	1.020722752	1.020722752	1.020722752	1.020722752	1.020722752	1.020722752	1.020722752	1.020722752
1 I ZF3	1.058200000	1.058200000	1.058200000	1.058200000	1.058200000	1.058200000	1.058200000	1.058200000	1.058200000
1 I ZF4	1.197082759	1.197082759	1.197082759	1.197082759	1.197082759	1.197082759	1.197082759	1.197082759	1.197082759
1 I ZF5	1.847508270	1.847508270	1.847508270	1.847508270	1.847508270	1.847508270	1.847508270	1.847508270	1.847508270

(5) P 11+0	TH = 83	SBE = 1.007509825							CAUER FUNCTIONS
	RC = 02	RC = 05	RC = 07	RC = 10	RC = 15	RC = 20	RC = 25	RC = 35	RC = 50
AMAX DB	.0017375255	.0108709564	.0213327376	.0436480540	.0988323393	.1772876696	.2802872360	.5675287486	1.249387366
AMIN DB	21.61163467	29.55431394	32.48497284	35.60407117	39.18040998	41.75740528	43.79849645	47.00320405	50.78805160
MU I K	.4145371D 0	.1656406D 0	.1181723D 0	.8250835D-1	.5465720D-1	.4062430D-1	.3211633D-1	.2219401D-1	.1436286D-1
1 I PFR	1.343346854	.9245363924	.8056311020	.6927688076	.5768963468	.5006557291	.4439905430	.3610535869	.2732517606
1 I PF1	1.164894705	.9616994658	.8962118570	.8324662549	.7671780029	.7255143895	.6958680980	.6555719595	.6185929541
1 I PQ1	.9889262768	.9612617331	.9758993191	1.007886534	1.070847515	1.139794651	1.213638271	1.378389459	1.689790375
1 I PF2	1.049561995	.9908756286	.9701313239	.9492045785	.9271518656	.9128323681	.9025642553	.8885515416	.8756899951
1 I PQ2	3.310952284	3.188808024	3.253575803	3.394043192	3.666805235	3.960850381	4.271493485	4.952745885	6.212023157
1 I PF3	1.016518559	1.000082692	.9941064943	.9879924612	.9814592624	.9771689042	.9740696888	.9698103905	.9658714799
1 I PQ3	11.73010025	11.28788911	11.52242055	12.03071981	13.01656088	14.07789311	15.19785659	17.65059048	22.17662471
1 I PF4	1.007399547	1.002693700	1.000968964	.9991969758	.9972952003	.9960416630	.9951338716	.9938331780	.9927233800
1 I PQ4	42.54945791	40.94264199	41.79484755	43.64172848	47.22339151	51.07890760	55.14704569	64.05534399	80.49161362
1 I PFC	1.005034223	1.003375976	1.002766952	1.002140540	1.001467457	1.001023353	1.000701520	1.000257915	.9998460425
1 I PQC	205.7444886	197.9738119	202.0951440	211.0267648	228.3477593	246.9929550	266.6662462	309.7459140	389.2292744
1 I ZF1	1.008293498	1.008293498	1.008293498	1.008293498	1.008293498	1.008293498	1.008293498	1.008293498	1.008293498
1 I ZF2	1.016691058	1.016691058	1.016691058	1.016691058	1.016691058	1.016691058	1.016691058	1.016691058	1.016691058
1 I ZF3	1.049561071	1.049561071	1.049561071	1.049561071	1.049561071	1.049561071	1.049561071	1.049561071	1.049561071
1 I ZF4	1.176351973	1.176351973	1.176351973	1.176351973	1.176351973	1.176351973	1.176351973	1.176351973	1.176351973
1 I ZF5	1.789715847	1.789715847	1.789715847	1.789715847	1.789715847	1.789715847	1.789715847	1.789715847	1.789715847

(5) P 11+0	TH = 84	SBE = 1.005508280							CAUER FUNCTIONS
	RC = 02	RC = 05	RC = 07	RC = 10	RC = 15	RC = 20	RC = 25	RC = 35	RC = 50
AMAX DB	.0017375255	.0108709564	.0213327376	.0436480540	.0988323393	.1772876696	.2802872360	.5675287486	1.249387366
AMIN DB	18.77206283	26.69130378	29.61975555	32.73767992	36.31339168	38.89016746	40.93115700	44.14077611	47.92057665
MU I K	.5516738D 0	.2204376D 0	.1572659D 0	.1098037D 0	.7273883D-1	.5406358D-1	.4274102D-1	.2953621D-1	.1911436D-1
1 I PFR	1.462921015	.9865617933	.8554113072	.7324317004	.6075608976	.5260880625	.4658501587	.3781136007	.2857131970
1 I PF1	1.201904624	.9946750418	.9270836801	.8610949502	.7933862158	.7501419194	.7193629589	.6775251595	.6391362602
1 I PQ1	1.061568434	1.006498471	1.016004633	1.044328091	1.104951873	1.173354646	1.247474878	1.414139270	1.730884487
1 I PF2	1.055040368	1.000455131	.9807404133	.9606820613	.9393836867	.9254754777	.9154675425	.9017663260	.8891493223
1 I PQ2	3.803061829	3.552093987	3.595690676	3.724892470	3.998336039	4.302626261	4.628265498	5.348389946	6.690041530
1 I PF3	1.016470323	1.002135745	.9968144574	.9913240150	.9854116629	.9815060270	.9786742535	.9747690308	.9711444358
1 I PQ3	14.26647758	13.30908279	13.47547994	13.96839740	15.01068465	16.16931728	17.40808509	20.14625885	25.23483525
1 I PF4	1.006446045	1.002583932	1.001139921	.9996441922	.9980269545	.9969549049	.9961757744	.9950987638	.9940965290
1 I PQ4	54.57862503	50.91161485	51.54897158	53.43695244	57.42888591	61.86608332	66.60985942	77.09465034	96.57740279
1 I PFC	1.003970362	1.002695347	1.002217784	1.001722632	1.001186702	1.000831123	1.000572541	1.000214879	.9998818207
1 I PQC	274.2839518	255.8539155	259.0572165	268.5460282	288.6090154	310.9097278	334.7511493	387.4456433	485.3614093
1 I ZF1	1.006136357	1.006136357	1.006136357	1.006136357	1.006136357	1.006136357	1.006136357	1.006136357	1.006136357
1 I ZF2	1.013030087	1.013030087	1.013030087	1.013030087	1.013030087	1.013030087	1.013030087	1.013030087	1.013030087
1 I ZF3	1.041266572	1.041266572	1.041266572	1.041266572	1.041266572	1.041266572	1.041266572	1.041266572	1.041266572
1 I ZF4	1.155455864	1.155455864	1.155455864	1.155455864	1.155455864	1.155455864	1.155455864	1.155455864	1.155455864
1 I ZF5	1.729497652	1.729497652	1.729497652	1.729497652	1.729497652	1.729497652	1.729497652	1.729497652	1.729497652

(5) P 11+0	TH = 85 SBE = 1.003819838								CAUER FUNCTIONS
	RC = 02	RC = 05	RC = 07	RC = 10	RC = 15	RC = 20	RC = 25	RC = 35	RC = 50
AMAX DB	.0017375255	.0108709564	.0213327376	.0436480540	.0988323393	.1772376696	.2802872360	.5675287486	1.249387366
AMIN DB	15.74071306	23.61022795	26.53335375	29.64936129	33.22372791	35.80003259	37.84080404	41.05023321	44.82993283
MU I K	.7492147D 0	.2993709D 0	.2135790D 0	.1491217D 0	.9878483D-1	.7342243D-1	.5894553D-1	.4011240D-1	.2595875D-1
1 I PFR	1.623134727	1.065417883	.9178133060	.7815327414	.6450823173	.5569945073	.4922924672	.3986279257	.3006094212
1 I PF1	1.240088930	1.031164088	.9616480542	.8934209050	.8231685217	.7732148204	.7461934610	.7076459158	.6626781329
1 I PQ1	1.166583107	1.068224696	1.069853599	1.092547666	1.149454789	1.216782966	1.291012488	1.459796581	1.783031213
1 I PF2	1.058959784	1.009528406	.9911516416	.9722434149	.9519665238	.9336276203	.9289860013	.9157317938	.9034745409
1 I PQ2	4.524949470	4.058731547	4.066535095	4.174941378	4.444428516	4.759559757	5.103181602	5.873270443	7.319738263
1 I PF3	1.015863771	1.003807873	.9992079871	.9944096862	.9891916758	.9857190035	.9831894493	.9796359388	.9764194930
1 I PQ3	18.15698925	16.26327513	16.29499523	16.73555992	17.83010762	19.10903351	20.50258533	23.62277256	29.47647134
1 I PF4	1.005414221	1.002387550	1.001225494	1.000009073	.9986813581	.9977949263	.9971478224	.9962496039	.9954101190
1 I PQ4	74.01104924	66.28611462	66.41551473	68.21274780	72.67766380	77.89444623	83.57853604	96.30458445	120.1778155
1 I PFC	1.002980135	1.002054850	1.001699081	1.001326361	1.000919183	1.000647128	1.000448420	1.000172456	.9999143824
1 I PQC	389.7829417	349.0972878	349.7788137	359.2444927	382.7602879	410.2359262	440.1726359	507.1975425	632.9311984
1 I ZF1	1.004301517	1.004301517	1.004301517	1.004301517	1.004301517	1.004301517	1.004301517	1.004301517	1.004301517
1 I ZF2	1.009747650	1.009747650	1.009747650	1.009747650	1.009747650	1.009747650	1.009747650	1.009747650	1.009747650
1 I ZF3	1.033320190	1.033320190	1.033320190	1.033320190	1.033320190	1.033320190	1.033320190	1.033320190	1.033320190
1 I ZF4	1.134252523	1.134252523	1.134252523	1.134252523	1.134252523	1.134252523	1.134252523	1.134252523	1.134252523
1 I ZF5	1.665933409	1.665933409	1.665933409	1.665933409	1.665933409	1.665933409	1.665933409	1.665933409	1.665933409

(5) P 11+0	TH = 86 SBE = 1.002441898								CAUER FUNCTIONS
	RC = 02	RC = 05	RC = 07	RC = 10	RC = 15	RC = 20	RC = 25	RC = 35	RC = 50
AMAX DB	.0017375255	.0108709564	.0213327376	.0436480540	.0988323393	.1772376696	.2802872360	.5675287486	1.249387366
AMIN DB	12.47461842	20.23105341	23.14380869	26.25335765	29.82458842	32.39979412	34.44005664	37.64704254	41.42850662
MU I K	.1046493D 1	.4181574D 0	.2983243D 0	.2082912D 0	.1379814D 0	.1025555D 0	.8107725D-1	.5602348D-1	.3625385D-1
1 I PFR	1.852831887	1.171172761	.9999850262	.8451486200	.6929677318	.5960903172	.5255431004	.4242284301	.3190776718
1 I PF1	1.277430151	1.071573469	1.000681477	.9304394104	.8576322467	.8108659382	.7774932241	.7320499429	.6903022113
1 I PQ1	1.331248183	1.158238862	1.146787262	1.160172743	1.210778462	1.275999639	1.349954934	1.521038412	1.852412363
1 I PF2	1.060606433	1.017710377	1.001090396	.9837205253	.9648386451	.9522927125	.9431688796	.9305566230	.9138263315
1 I PQ2	5.681759004	4.818609129	4.760623195	4.328389145	5.083126999	5.408436554	5.773892942	6.608517014	8.197404243
1 I PF3	1.014576536	1.004992462	1.001190318	.9971643943	.9927282855	.9897468413	.9875618738	.9845185793	.9816645750
1 I PQ3	24.74805040	20.95811941	20.70329870	21.00109430	22.12016550	23.54851334	25.15230152	28.31290883	35.77396645
1 I PF4	1.004304799	1.002094624	1.001213267	1.000277231	.9992424952	.9985451070	.9980330258	.9973133308	.9966467067
1 I PQ4	109.0872492	92.37450770	91.25075562	92.56402722	97.49900832	103.7976942	110.8698376	127.0112413	157.7044494
1 I PFC	1.002079158	1.001463895	1.001218270	1.000957237	1.000668475	1.000473734	1.000330677	1.000130944	.9999431269
1 I PQC	609.4838507	516.1056418	509.8269470	517.1645363	544.7375315	579.9298417	619.4435915	709.6292373	881.1187876
1 I ZF1	1.002788240	1.002788240	1.002788240	1.002788240	1.002788240	1.002788240	1.002788240	1.002788240	1.002788240
1 I ZF2	1.006856382	1.006856382	1.006856382	1.006856382	1.006856382	1.006856382	1.006856382	1.006856382	1.006856382
1 I ZF3	1.025731160	1.025731160	1.025731160	1.025731160	1.025731160	1.025731160	1.025731160	1.025731160	1.025731160
1 I ZF4	1.112537730	1.112537730	1.112537730	1.112537730	1.112537730	1.112537730	1.112537730	1.112537730	1.112537730
1 I ZF5	1.597601687	1.597601687	1.597601687	1.597601687	1.597601687	1.597601687	1.597601687	1.597601687	1.597601687

www.ingramcontent.com/pod-product-compliance
Ingram Content Group UK Ltd.
Pitfield, Milton Keynes, MK11 3LW, UK
UKHW051455290926
15471UKWH00008B/1223

9 781580 531269